W9-CBT-498

Enzymatic Reaction Mechanisms

Enzymatic Reaction Mechanisms

CHRISTOPHER WALSH
MASSACHUSETTS INSTITUTE OF TECHNOLOGY

W. H. FREEMAN AND COMPANY
San Francisco

Cover art redrawn after D. Matthews, R. Alden, J. Bolin, S. Freer, R. Hamlin, N. Xuong, J. Kraut, M. Poe, M. Williams, and K. Hoogsteen, 1977, *Science* 197:452. Copyright © 1977 by the American Association for the Advancement of Science.

Sponsoring Editor: Arthur C. Bartlett
Manuscript Editor: Lawrence W. McCombs
Designer: Robert Ishi
Production Coordinator: Linda Jupiter
Illustration Coordinator: Batyah Janowski
Artist: Eric G. Hieber, EH Technical Services
Compositor: The Universities Press Limited
Printer and Binder: The Maple-Vail Book Manufacturing Group

Library of Congress Cataloging in Publication Data

Walsh, Christopher.
Enzymatic reaction mechanisms.

Bibliography: p.
Includes index.
1. Enzymes. I. Title.
QP601.W23 574.1'925 78-18266
ISBN 0-7167-0070-0

Printed in the United States of America

9 8 7 6 5 4 3 2 1

Dedicated to

R. H. Abeles
W. P. Jencks
F. Lipmann
L. B. Spector

CONTENTS

PREFACE

This book has developed from a chemistry course on enzymatic reaction mechanisms that I have taught to chemistry and biology undergraduate and graduate students at M.I.T. The basic approach of analyzing catalysis in biological systems according to the type of chemical reaction involved has in turn derived from the biochemistry course taught primarily by Professors Abeles and Jencks at Brandeis University during my postdoctoral stay there from 1970 through 1972.

The basic premise in this approach is that most of the transformations undergone by specific metabolites in biological systems can be collected into a small number of types of chemical reactions. With a feeling for the underlying patterns of the ways certain functional groups are enzymatically processed, one can see the chemical logic of metabolic sequences and interconversions.

Because enzymes are catalysts, the study of enzymatic reactions must deal both with the kinetics and with the nature and scope of the chemistry that occurs. There are several recent texts, at both beginning and advanced levels, that focus on kinetic mechanisms of enzymatic catalysis. Although kinetic questions are repeatedly considered qualitatively in this book, that subject is not its major focus. Rather, a complementary approach is taken to analyze the kinds of chemical mechanisms that are likely to proceed by low activation-energy barriers (and therefore rapidly) in enzymatic systems. This book includes relatively few kinetic constants and relatively many structures analyzing patterns of electron flow as bonds break and reform during the course of the enzyme-mediated transformation. To this end, much of the book is devoted to coenzyme-dependent enzymatic reactions, analyzing how the particular chemistry open to the specific coenzyme is used to direct flux of reaction molecules to a single set of products and to accelerate the rate of such reactions.

The organization of this book also differs somewhat from many texts on enzymology, which have introductory chapters on structure, several chapters on kinetics, and a concluding chapter or two on specific enzymes, invariably focusing on a few standbys such as chymotrypsin, carboxypeptidase, lysozyme, and alcohol dehydrogenase. Instead, although this book uses hydrolytic enzymes that work on acyl-group–containing substrates as a jumping-off point, the focus here is on a thorough examination of the scope and diversity of enzymatic catalysis. It would be nice to have high-resolution X-ray maps available for each category of enzymes, but that would entail a long wait, especially for some enzymes that carry out the most interesting chemistry. In fact, I began writing this book in 1973 largely because I felt there was no coherent, comprehensible treatment available in any text of the pattern of enzymatic redox reactions, the heart of cellular energetics.

Major sections deal with (1) enzymes that carry out group transfers (of electrophilic substrate fragments to nucleophilic acceptors), (2) enzymes that accelerate oxidation–reduction chemistry, particularly reductive oxygen metabolism, (3) enzymes that promote eliminations, isomerizations, and rearrangements of substrate skeletons, and (4) enzymes that make and break carbon–carbon bonds (the reactions that are the building blocks of biosynthesis).

The pace of new discoveries in biochemistry is encouraging to those engaged in research, but it is a bit dismaying to the author attempting to gauge the state of knowledge in a given area and to provide some interpretations. I have tried to emphasize concepts and ideas about enzymatic chemistry that should have reasonable time constants and that should not rise or fall on the basis of the last (or next) experiment.

The level of the book should make it suitable both as a text and as a reference work (albeit with some selectivity in topics covered) in enzymology. It should be usable by and useful to both advanced undergraduate and graduate students in chemistry, biology, biochemistry, medicinal chemistry, and pharmacology departments, as well as to professional scientists in those areas. Each of the four major sections (and separate chapters within them) can be taught essentially independently, although the group-transfer section is the likely starting place because it contains much basic information. Extensive cross-referencing will aid teacher and student in locating relevant material elsewhere in the text. Most classes should be able to deal comfortably with three of the four sections in a one-semester course. The final chapter in the book can follow any combination of preceding chapters because it attempts to analyze and integrate the chemical logic (and underlying strategies) of metabolic pathways.

I acknowledge substantial intellectual debt to Professors R. H. Abeles and W. P. Jencks of Brandeis University and Professors F. Lipmann and L. B. Spector of The Rockefeller University for both facts and philosophy. I am indebted to the M.I.T. students in Chemistry 5.50 (Enzymatic Reaction Mechanisms) over the past five years for their comments, positive and negative, as they used two stages of course notes (efficiently printed by M.I.T. Graphic Arts Services) that preceded this manuscript. I owe thanks to my colleagues in my research group during this period who carefully read various portions of the manuscript and repeatedly—if somewhat gleefully—pointed out ways of improving the presentation. Michael Johnston deserves particular thanks for reading galley and page proofs. I am delighted to acknowledge the generous support of the Alfred P. Sloan Foundation (1975–1977) and the Camille and Henry Dreyfus Foundation (1976–1978) for fellowships without which the manuscript would not have been completed. Finally, I acknowledge that it would have been no fun at all without Diana and Allison.

November 1978 *Christopher Walsh*

Enzymatic Reaction Mechanisms

Section I

INTRODUCTION

This brief section contains two chapters. The first outlines the scope and organization of the book. The second chapter summarizes some basic concepts about enzymes and enzymatic catalysis—themes that introduce and will underlie later discussions throughout the text.

Chapter 1

Introduction

This introductory chapter is divided in two parts. The first part is a brief discussion of the philosophical orientation and the purpose of this book. It delineates the approach used, explains why the principles are apt for describing enzyme-catalyzed reactions, and sets forth the overall goal in terms of insights into how enzymes work and what reactions they catalyze.

The second part of this chapter summarizes the general outline of the four major sections of the text. It sets out the major premises around which the individual chapters of each section are grouped. It may be useful to review this first chapter again for general orientation as you go through the four major text sections on (1) enzymatic group transfers, (2) oxidation–reduction reactions, (3) eliminations, isomerizations, and rearrangements, and (4) reactions that make or break carbon–carbon bonds.

1.A ORIENTATION AND PURPOSE OF THIS BOOK

The intent of this book is to provide a simple chemical framework for the study and analysis of enzyme-catalyzed reactions. Enzymes are macromolecular protein catalysts, consisting of linear condensed polymers of a small set of α-amino acids joined in amide linkages. Each enzyme has a genetically mandated and unique primary sequence, and it folds in three dimensions into a precise orientation with remarkable catalytic activity. Enzymes are the major agents that effect controlled chemical changes in biological systems. It is fair to say that, if one wishes to understand the chemistry of living systems and the kinds of molecular transformations that occur in them, one must understand both the nature of enzymatic catalysis and the scope of the reactions encountered. This, in turn, entails knowledge of (1) the composition, size, and structure of the subset of proteins that display enzymatic activity and (2) the chemistry of the enzymes' functional groups that actually engage in catalysis. These investigations provide insights into the two most distinguishing characteristics of enzymes as catalysts: the remarkable specificity and rate accelerations of enzyme-mediated transformations.

We shall touch on all of these elements in more or less detail. However, although this book can by no means be considered brief, it has a somewhat narrow focus. The focus is not on protein chemistry per se, nor on three-dimensional protein structure, nor even on the chemistry of catalysis itself. Rather, the purpose of this text is to group the almost bewildering array of enzyme-catalyzed processes into a few major reaction types common in organic chemistry, to develop some simple chemical intuitions about how these reactions occur, and then to see how enzymes modulate and control the chemical paths open to their specific reactants. The focus is primarily on the nature of the chemical changes undergone by substrates as they are converted to products, and on the nature and chemistry of the active-site groups on the enzyme catalyst.

The focus is on the mechanisms by which some stable substrate species—in aqueous solution at neutral pH (for the most part) and room temperature—is converted into some stable product species. Therefore, we shall continually analyze biochemical mechanisms in terms of their two components: (1) the *kinetics* of the process, and (2) the *structural changes in chemical bonds* during the process.

The kinetic question in chemical or biochemical mechanistic study is essentially a question of timing. In what temporal sequence do bonds break and form? Do changes occur sequentially or simultaneously? Because enzymes are such efficient catalysts, and because they always combine with substrates in a specific binding step as a prelude to chemistry, we shall examine how fast both chemical and physical steps occur.

The structural-change question devolves to a question of how chemical bonds break and form in these biological systems. What is the nature of the enzymatic transition state? What kinds of intermediates, with finite lifetimes, form? Do these intermediates involve *covalent bonds* between enzyme and some fragment of a substrate? What catalytic advantage do these mechanisms possess, and how are transition states selectively stabilized? The ultimate goal is to use mechanistic intuition derived from consideration of specific enzyme systems to gain predictive ability—not only to forecast what mechanism will be operant in a given instance, but also to say how a biological system can and will process a substrate with certain structural elements, how it will cope with certain kinds of functional groups.

1.A.1 Modes of Bond Breaking

A major, pervasive didactic device in this book (borrowed from mechanistic organic chemistry) is the analysis of what happens to a shared electron pair in a covalent bond between two atoms as that bond breaks and then reforms with new partner atoms during enzymatic transformations. For example, hundreds of enzymatic reactions involve the fission of carbon–hydrogen bonds at some stage in the reaction. The C—H bond can be cleaved in only two ways. (1) *Homolytic* cleavage, with one electron remaining with carbon and one with hydrogen, produces carbon radicals and hydrogen radicals.

$$-\overset{|}{\underset{|}{C}}\!:\!H \xrightarrow[\text{fission}]{\text{homolytic}} -\overset{|}{\underset{|}{C}}\cdot + H\cdot$$

radicals

In general, radical species in biochemical reactions are unstable and not readily formed. (2) Reactions occur much more often by *heterolytic* cleavage of bonds, one atom retaining the two electrons. When a C—H bond cleaves heterolytically, there are two options: (a) the electrons can remain with carbon, yielding a carbanion intermediate (or a fleeting carbanionic transition state) and a proton, $H^{\oplus}$, or (b) electrons can depart with the hydrogen, producing an electron-deficient *carbonium ion* and a *hydride ion.*

$$\text{(a)}\quad -\overset{|}{\underset{|}{C}}\!:\!H \longrightarrow -\overset{|}{\underset{|}{C}}\!:^{\ominus} + H^{\oplus}$$

carbanion proton

$$\text{(b)}\quad -\overset{|}{\underset{|}{C}}\!:\!H \longrightarrow -\overset{|}{\underset{|}{C}}^{\oplus} + H\!:^{\ominus}$$

carbonium ion hydride ion

Carbon is a more electronegative atom than hydrogen and, in general, pathway (a) is preferred. Indeed, much of enzyme chemistry involves controlled generation of carbanions, as we shall see. But the fragmentation pattern (b), to yield the equivalent of a hydride ion, is a consistent explanation of the facts about dehydrogenases that use nicotinamide coenzymes as redox cofactors, as we note in Chapter 10.

1.A.2 Nucleophiles and Electrophiles

Both of the heterolytic pathways of bond cleavage for C—H (or, more generally, between any two atoms) *involve ionic intermediates* or *transition states*, where full or partial positive or negative charges are developed on reacting atoms. Reactions involving such ionic species are predominant in organic chemistry and enzyme chemistry (we shall come back to unpaired electrons in Section III when we discuss redox catalysis). The predominance of these reactions has led to the broad categorization of various reagents or reactants or enzyme substrates into two broad classes: electron-rich or electron-deficient species.

Electron-rich species have been termed **nucleophiles** (from Greek, meaning "nucleus loving"). Because nuclei of atoms contain the positively charged protons, nucleophilic reagents or atoms are seeking positively charged or electron-deficient species to combine with and to give up electrons to. The recipients are defined as **electrophiles**, those that "love electrons." Nucleophiles, then, are often anionic (negatively charged) or contain a lone or unshared electron pair. They are the *attacking molecules* in chemical (and therefore in enzyme-catalyzed) reactions. Oxygen, sulfur, and nitrogen nucleophiles are important in biological reactions.

$$\mathrm{H\ddot{O}H,\ \ R\ddot{O}H,\ \ R\ddot{O}{:}^{\ominus},\ \ R\ddot{S}H,\ \ R\ddot{S}^{\ominus},\ \ R\ddot{N}H_2}$$

In reactions where carbon–carbon bonds are formed, carbanions act as nucleophiles, generally as enolate ions or eneamines or via the π-electrons of carbon–carbon double bonds (Chapter 26).

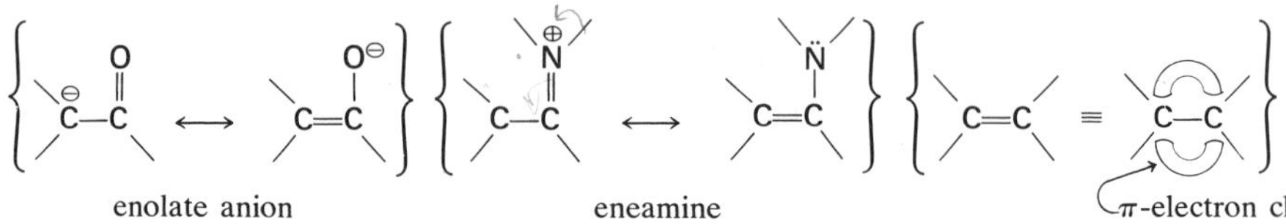

enolate anion

eneamine

π-electron cloud of olefin

In the biological systems under consideration, there are only a few important electrophiles available to the various nucleophilic groups on substrates or enzymes for reaction. The electrophiles are electron-deficient, often cationic (i.e., positively charged) and/or with an unfilled valence electron shell (metal cations). Important biological electrophiles include *protons* and *metal cations* (from the +1 oxidation state of copper to the +6 oxidation state of molybdenum). Additionally, the coenzyme forms of vitamin B_1 and vitamin B_6 (thiamine pyrophosphate and pyridoxal phosphate, respectively) serve as electron sinks (electrophiles) during covalent-adduct formation with various substrates in certain kinds of enzymatic catalysis.

In all discussions of individual enzymes in this book, my goal is to point out the attacking nucleophile of a substrate (or enzyme amino-acid side chain) and to identify the electrophilic center of some cosubstrate that undergoes attack. I shall attempt explicit bookkeeping on the electron pairs shared in covalent bonds as reactants go to products; a minimal goal in every case will be to write a path for continuous electron flow that is chemically reasonable (in accord with simple chemical facts and principles) and that forms intermediates or transition states of reasonable structure. *The major illustrative device of this mechanistic bookkeeping is the use of curved arrows to indicate the flow of electron pairs* in a reacting chemical structure. This allows explicit and accurate account of the progress of an enzymatic reaction. The arrows always originate from nucleophiles (from electron pairs), and the arrowheads point to electrophilic centers as the focus of attack. Thus, for example, we represent hydrolysis of the peptide bond as follows.

$$\underset{\text{amide}}{R{-}C({=}O){-}NHR'} + H_2\ddot{O} \longrightarrow R{-}C(O^{\ominus})(OH){-}\overset{\oplus}{N}HR'(H) \longrightarrow \underset{\text{acid}}{R{-}C({=}O){-}OH} + NH_2R'$$

With such an electron-pushing apparatus we hope to rationalize every enzyme-mediated reaction discussed in this book, as a beginning to more subtle questions and as a starting point, for instance, to inquiries about the timing (the kinetics) of the reaction.

With this underlying framework as flooring for mechanistic intuition, we shall focus on individual enzymes of both physiological significance and chemical interest. We shall use the physical organic chemist's collage of experiments and criteria for chemical mechanism whenever applicable and whenever the delicate biological catalyst permits. Principles of mechanism will be introduced for the most part in the context of a given enzymatic reaction to provide a specific example of the phenomenon. The topics include whether covalent intermediates form, how stable they are, and what their chemical constitution is. We shall use kinetic isotope effects (on both elementary chemical steps and maximal velocity of reaction) to probe which step or steps are rate-determining. We shall use isotopic labeling techniques (both stable and radioactive isotopes) to probe for partial reactions as a clue to complicated sequences. We shall delve into information available both from pre-steady-state (millisecond time scale) and from steady-state kinetics to determine how this information constrains allowable mechanistic formulation. We shall probe the chemistry open to the common organic coenzymes, the metal-ion cofactors, and the enzyme's amino-acid residues to determine their likely roles in catalysis. We shall be on the alert for their actions as acids or bases (reaction with hydrogen) or as nucleophiles or electrophiles (e.g., reaction with carbon). We shall, whenever enzyme structural information is available, attempt the correlation of enzyme active-site structure with catalytic function.

Interwoven through the various chapters and specific enzyme topics is a continuing analysis of two quintessential features of enzymatic catalysis: rate accelerations and specificity. In fact, Chapter 2 is devoted to an introduction to these factors, setting the stage for examination of them at work in individual examples.

1.A.3 Rate Accelerations and Specificity

Because an enzyme often is characterized as much by what it does as by what it is, the subject of rate accelerations is a central one to the chemistry and the biology of an organism. We shall be interested in two major aspects of rate accelerations.

1. How much faster than the uncatalyzed case is the enzyme-mediated reaction? That is, how are enzymatic transition states selectively stabilized or enzyme–substrate initial complexes selectively strained or destabilized so that the activation barrier is lowered? How is $\Delta G^{\ddagger}$, the free-energy difference between the reactants and the highest (rate-determining) transition state, minimized? The lower the energy barrier, the more molecules will have sufficient energy to surmount it, and the faster the rate of reaction. In this connection, we shall seek to determine how enzymes selectively lower one of many possible energy barriers, so that reaction flux is *completely* to the desired product, a situation almost *never* encountered in the chemical laboratory. (This effect is, of course, a component of the enzymatic-specificity question to be discussed next.)
2. From time to time we shall examine the maximal upper limit for the rate of an enzyme-catalyzed reaction. Have any (or many) enzymes evolved to such catalytic efficiency that chemical steps are no longer rate-limiting, but that physical steps (such as diffusion of enzyme and substrate together, or enzyme and product apart, in the aqueous milieu) are rate-limiting? We shall see that a small number of characterized enzymes may in fact be working at the diffusion-controlled limit.

The question of enzyme specificity is multipartite. Enzymes discriminate on the basis of size and shape. Not surprisingly, molecules bigger than the physiological substrate are not acted on but, unexpectedly, many enzymes show almost incredible specificity toward exclusion of molecules smaller than specific substrates. In particular, the key to many biosynthetic reactions is the ability to exclude water or molecular oxygen (two extremely abundant molecules in the biosphere) and to avoid unwanted and adventitious side reactions with them. Specificity can be and is imposed (1) in *initial binding* of substrates to the active site of an enzyme, (2) in *protein conformational isomerizations* thereby induced, and (3) in the *actual chemical steps of catalysis.* The multistep discriminations offer multiplicative factors essential, for instance, to faithful replication of DNA and RNA or to fidelity in protein synthesis. We shall examine the ratio of maximal velocity of reaction to the measured affinity of enzyme for a substrate as an index both of catalytic efficiency and of relative preference of an enzyme for one substrate versus another. Additionally, we shall have a great deal to say about enzyme *stereospecificity* in Sections III, IV, and V of this text. We shall note the ability of enzymes as *chiral* catalysts (composed of only the L-isomers—that is, the *S*-isomers—of α-amino acids) to catalyze reaction with only one of an enantiomeric or diastereomeric pair of chiral substrates. Even more dramatic is the ability of enzymes unerringly to distinguish between chemically like, paired substituents at *meso* or *prochiral* carbon centers (e.g., the methylene groups of citric acid or the C-1 methylene group of ethanol).

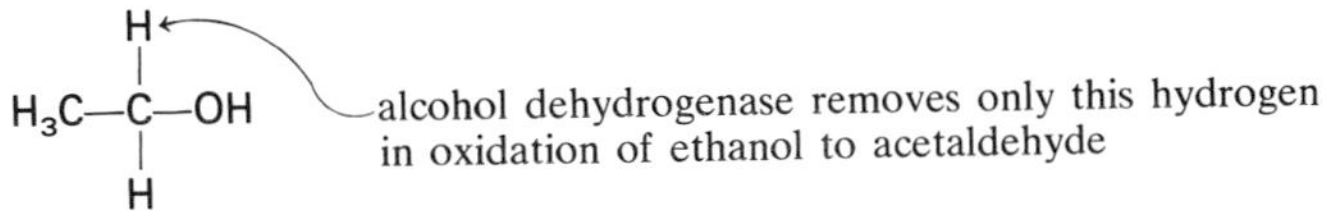

H
|
H_3C—C—OH
|
H

alcohol dehydrogenase removes only this hydrogen in oxidation of ethanol to acetaldehyde

Interspersed among the categorization and mechanistic evaluation of the major classes of enzymatic reactions are comments on the metabolic significance of the processes, on the molecular modes of action of various drugs and antibiotics as they block action of target enzymes, on the development of criteria for patterns of enzyme inhibition and chemical inactivation, and on the chemical logic of linked sequential transformations and metabolic pathways.

1.B GENERAL OUTLINE OF THE FOUR MAJOR SECTIONS OF THIS BOOK

This book is divided into four major sections following this introductory section. Here we summarize the general content of these sections, as a preview and for later use in review and orientation.

1.B.1 Section II: The Enzymology of Group-Transfer Reactions

Chapters 3 through 9 deal with enzymology of functional group transfers. The key concept is that some electrophilic fragment (for instance an acyl, phosphoryl, or glucosyl group) is transferred to some nucleopilic cosubstrate. The cosubstrate's nucleophilic attacking atom may be a specific nitrogen, oxygen, or sulfur atom in a process involved in a *biosynthetic* pathway in the cell. Or alternatively, the nucleophile could be water (at 55.5 M, the most nearly ubiquitous weak nucleophile in cells) in an enzymatic sequence that is *degradative.* In polysaccharide mobilization for energy generation, enzymes will transfer glycosyl groups to an oxygen anion of inorganic phosphate—for instance, producing glucose-1-$PO_3^{2\ominus}$ from glycogen.

Indeed, group transfers are the major categories of enzymatic reactions in both biosynthesis and degradation of the major classes of biological macromolecules. This is especially obvious in considering polymer breakdown. DNA, RNA, proteins, polysaccharides, and complex lipids are all *hydrolyzed* (group transfer to a nucleophilic oxygen of water) for reutilization of the monomeric units by the cell to make new copies of the polymers.

DNA chains
RNA chains
proteins
polysaccharides
lipids

$\xrightarrow{H_2O}$ hydrolysis to monomers for reutilization

(Many complex carbon skeletons also are built up by isoprenyl-group transfers to carbon nucleophiles, but their discussion is delayed until Chapter 26.)

Three major categories are explored in the seven chapters of Section II: acyl-group transfers, phosphoryl-group transfers, and glycosyl transfers.

1.B.1.a Acyl-group transfers

The general stoichiometry for acyl transfers is the following:

$$R{-}C({=}O){-}X + Y^{\ominus} \rightleftharpoons R{-}C(O^{\ominus})(Y){-}X \rightleftharpoons R{-}C({=}O){-}Y + X^{\ominus}$$

When $Y = H_2O$, for instance, the hydrolysis of proteins is involved as water attacks the electron-deficient carbonyl carbon to produce first a tetrahedral adduct, which then eliminates the substituent X (the amine fragment of the polypeptide) and produces the substituted acyl group, in this instance $RCOO^{\ominus}$. Protein hydrolysis is functionally irreversible, but aminoacyl-group transfer is involved in protein biosynthesis as well. The major chemical problem the cell must overcome in protein biosynthesis is chemical activation of the carboxylate oxygen of amino acids for displacement during peptide-bond formation. At physiological pH, carboxylic acids are ionized; for example:

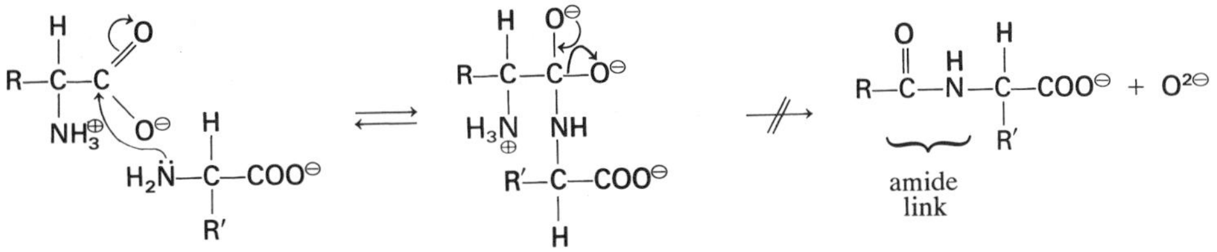

amide link

Attack by the amino group of another amino acid to form a peptide bond would involve formal expulsion of $0^{2\ominus}$, a species so exceedingly unstable that the reaction would not proceed at any reasonable rate or to any reasonable extent. The carboxylate must be chemically modified so that an oxygen can be eliminated with a low energy of activation.

Cells solve the problem by conversion of carboxylates to phosphoric-acid esters. The phosphate group can then be expelled readily from a tetrahedral adduct to give the stable phosphate anion.

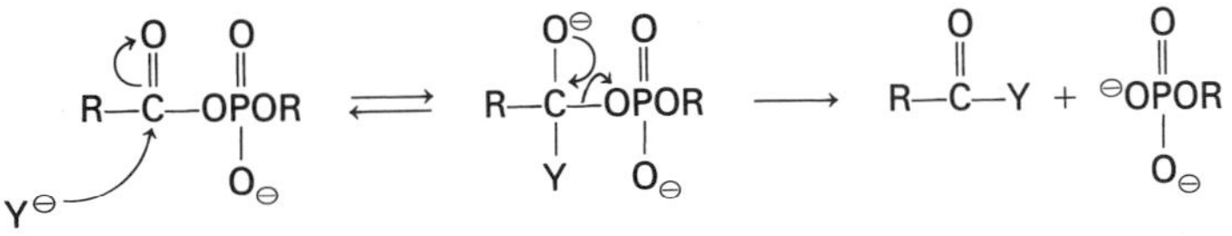

acyl phosphate

stable dianion

1.B.1.b Phosphoryl-group transfers

The formation of the acylphosphoric acid—a *mixed anhydride* between a carboxylic acid and a (substituted) phosphoric acid—introduces the topic of *phosphoryl-group transfers.* The major source of the phosphoryl groups used either for *acyl activation* or for *activation of an alcohol* for C—O cleavage is the nucleoside triphosphate ATP.

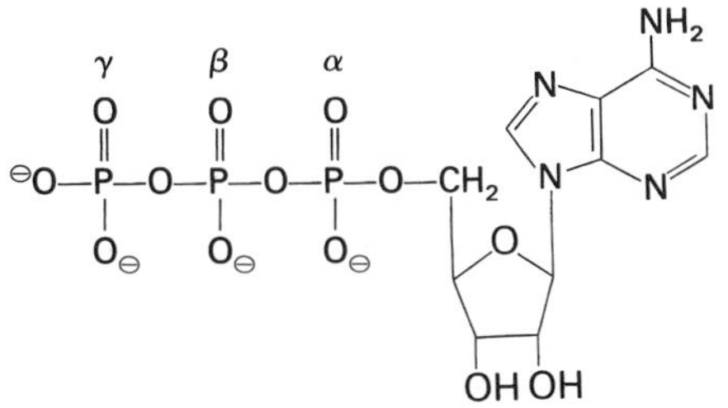

adenosine triphosphate (ATP)

The business end of this molecule consists of two phosphoric-anhydride linkages: that between the α-phosphorus and the β-P, and that between the β-P and the γ-P. ATP, as an anhydride, is thermodynamically destabilized relative to its hydrolysis products. Its *thermodynamic role* in coupled enzymatic reactions (important for biosynthetic pathways) is to provide the driving force for the otherwise unfavorable half-reaction (e.g., amide-bond formation in protein biosynthesis). Its *mechanistic role* is to offer three electrophilic phosphorus atoms for attack by nucleophiles such as carboxylate oxygen anions or alcohol oxygens: *phosphoryl transfer to the nucleophile.* From the point of view of the nucleophile, the nucleophile becomes phosphorylated.

Given three phosphorus atoms in ATP, there exist enzymatic examples of assisted attack by nucleophiles at the α-, the β-, and the γ-phosphorus atoms. Here are examples of these three cases.

L-alanine + $ATP^{4\ominus}$ $\xrightarrow{\text{attack at } \alpha\text{-P}}$ L-alanyl-AMP + pyrophosphate

α-anomer of 5-phosphoribose + $ATP^{4\ominus}$ $\xrightarrow{\text{attack at } \beta\text{-P}}$ 5-phosphoribose-1-pyrophosphate + AMP

either α- or β-anomer of glucose + $ATP^{4\ominus}$ $\xrightarrow{\text{attack at } \gamma\text{-P}}$ glucose-6-phosphate + ADP

Attack by nucleophiles at the β-phosphorus atom of ATP is rather rare, whereas attacks at both the α-P and γ-P are common. The prevalent mode for many biosynthetic enzymes is to specify attack at the α-phosphorous, transferring the AMP group (adenylic acid, hence adenylyl transfer) to the nucleophilic cosubstrate. Although acyl phosphates and acyl adenylates are appropriately activated thermodynamically, they are also kinetically labile to attack by H_2O. The major form of activated acyl groups free in cells is the acylthioester, arising from attack of thiolate anions on the phosphorylated intermediates. For example:

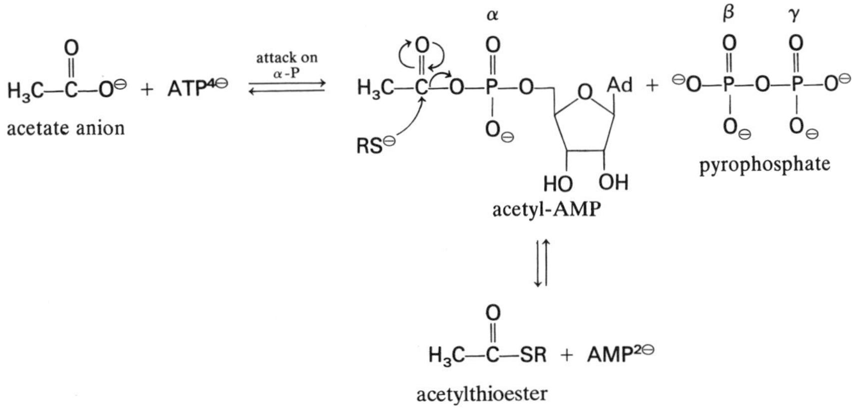

$H_3C{-}C(=O){-}O^{\ominus}$ + $ATP^{4\ominus}$ $\xrightleftharpoons{\text{attack on } \alpha\text{-P}}$ $H_3C{-}C(=O){-}O{-}P(=O)(O^{\ominus}){-}O{-}$[ribose–Ad] + $^{\ominus}O{-}P(=O)(O^{\ominus}){-}O{-}P(=O)(O^{\ominus}){-}O^{\ominus}$

acetate anion; $RS^{\ominus}$; α; β; γ; HO OH; acetyl-AMP; pyrophosphate

$\rightleftharpoons$

$H_3C{-}C(=O){-}SR$ + $AMP^{2\ominus}$

acetylthioester

1.B.1.c Glycosyl transfers

Glycosyl transferases are the enzymes transferring glycosyl (sugar) residues to nucleophiles, both during degradation (cosubstrates = H_2O, $HPO_4^{2\ominus}$) and during biosynthesis of polysaccharides and glycoproteins. In the biosynthetic direction, the activated glycosyl intermediate, set up for (C-1)—oxygen fission, is a nucleoside diphosphosugar.

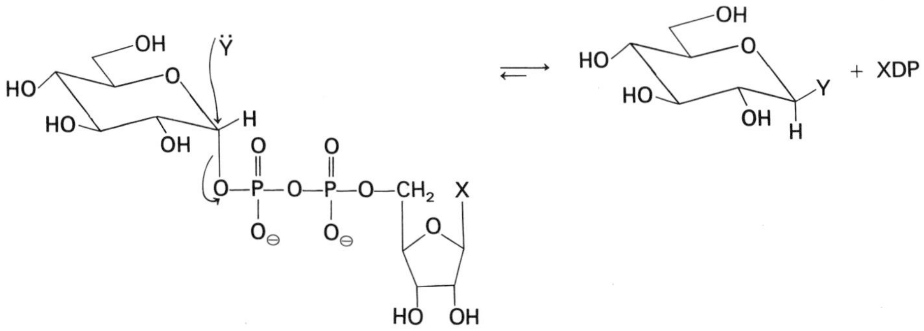

[sugar (OH, HO, HO, OH, H)]$-O{-}P(=O)(O^{\ominus}){-}O{-}P(=O)(O^{\ominus}){-}O{-}CH_2$–[ribose (X, HO, OH)] + $\ddot{Y}$ $\rightleftharpoons$ [sugar (OH, HO, HO, OH, H)]–Y + XDP

1.B.2 Section III: Enzymatic Oxidations and Reductions

Chapters 10 through 16 deal with enzymes categorized according to the nature of the oxidation/reduction processes they catalyze. Most substrates for enzymes catalyzing oxidation or reduction undergo a two-electron change at the reacting center. For instance, an aldehyde may be oxidized by two electrons to an acid, or it may be reduced by two electrons to an alcohol. When two electrons are removed from one substrate (oxidation), they must be transferred to a cosubstrate that concomitantly undergoes two-electron reduction. Thus, the first step in analyzing enzymatic redox transformations is to identify which molecule is oxidized and which reduced. Redox enzymes may often have requirements for transition metals or conjugated organic coenzymes to serve as conduits or intermediate electron carriers between the substrate partners undergoing redox change.

Most organic enzyme substrates are oxidized at carbon centers with C—H bond breakage. The removal of the two electrons in the C—H bond can occur in three distinct ways (see ¶1.A.1): (1) by initial *homolytic* cleavage to radicals, —C—H → —C· + H·, followed by one-electron transfer steps; or (2) by *heterolytic* cleavage with two options: (a) fission of the C—H bond to yield a hydride ion as the species carrying the electrons to the reducible cosubstrate, —C—H → —C$^{\oplus}$ + H:$^{\ominus}$; or (b) initial abstraction of a proton (acid–base chemistry) and subsequent transfer of the two electrons from the transient carbanion, —C—H → —C:$^{\ominus}$ + H$^{\oplus}$.

There are two major categories of redox coenzymes involved in simple enzymatic dehydrogenations. The first group is the nicotinamide coenzymes (NAD, NADP). They appear to be involved in the hydride-ion type of redox process, in which a hydride equivalent from a substrate is attached to C-4 of the pyridine ring of the oxidized coenzyme, generating a dihydropyridine in the reduced form. Direct hydrogen transfer without equilibration with solvent hydrogens is the hallmark of nicotinamide-linked dehydrogenases. Furthermore, the dihydronicotinamide is stable to molecular oxygen, so these coenzymes are not involved in reductive oxygen metabolism.

This behavior is in contrast to the second major category of organic redox

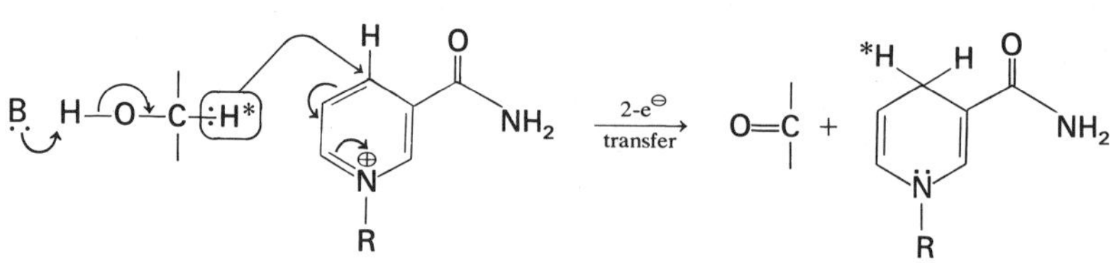

oxidized nicotinamide

2-e$^{\ominus}$ transfer

dihydronicotinamide

coenzymes, the flavin coenzymes. Flavin coenzymes are used for similar types of substrate oxidations, and the half-reaction leading to substrate oxidation and flavin reduction is probably a two-electron process. But the dihydroflavins have chemical versatility in two ways not open to dihydronicotinamides: (1) they react extremely rapidly to reduce molecular oxygen, and (2) they are readily reoxidized in one-electron transfer steps, producing the semiquinone as a stable radical in biological reactions.

2-e⊖ transfer

oxidized flavin

dihydroflavin

1-e⊖ oxidation

1-e⊖ oxidation

flavin semiquinone
(radical)

Just as water is the most prevalent cellular nucleophile, involved in dozens of enzyme-catalyzed hydrolytic reactions, so *oxygen in aerobic organisms is possibly the most pervasive potential acceptor of electrons from substrates undergoing oxidation.* Indeed, reduction of oxygen with oxidation of reduced organic compounds is strongly favored thermodynamically. It is kinetically sluggish because oxygen is a ground-state diradical, whereas organic substrates are spin-paired molecules: reaction to produce nonradical products is spin-forbidden; reaction to produce radical products *is* spin-allowed, but often has a high energy of activation because the substrate radicals have no structural features allowing stabilization of the unpaired spin.

Enzymes that do process oxygen for reductive metabolism have evolved two devices for accelerating reaction: either they employ conjugated organic coenzymes such as flavins or pterin cofactors, or they use transition metals. The conjugated coenzymes in their reduced forms can undergo the radical reactions readily: one-electron transfers yield the delocalized semiquinones and one-electron–reduced oxygen, a radical pair that can then recombine with a second electron transfer to produce oxidized coenzyme and H_2O_2. When transition metals are used, they probably ligand molecular oxygen and form new molecular orbitals with shared electron density between metal and oxygen, sufficient to make triplet O_2 react more like a singlet species.

Thus, oxygen undergoes reduction in enzymatic processes by net one, two, or four electrons (generating superoxide anion, hydrogen peroxide, or water, respectively).

$$O_2 \xrightarrow{1e^{\ominus}} O_2^{\ominus\cdot}$$
$$2\,H^{\oplus} + O_2 \xrightarrow{2e^{\ominus}} H_2O_2$$
$$4\,H^{\oplus} + O_2 \xrightarrow{4e^{\ominus}} 2\,H_2O$$

$$2\,H^{\oplus} + 2\,O_2^{\ominus\cdot} \xrightarrow[\text{dimutase}]{\text{superoxide}} O_2 + H_2O_2$$
$$2\,H_2O_2 \xrightarrow{\text{catalase}} 2\,H_2O + O_2$$

The four-electron reduction product, water, is produced when O_2 serves as terminal electron acceptor for membrane respiratory chains and is the only nontoxic oxygen metabolite. Both superoxide and hydrogen peroxide are deleteriously reactive on the biological scale of reactivity, and aerobic organisms have evolved two enzyme systems as surveillance devices to guard against the dangers of using O_2 as electron acceptor. Superoxide dismutase converts two $O_2^{\ominus\cdot}$ to O_2 and H_2O; catalase reduces H_2O_2 to water, while oxidizing a second H_2O_2 to O_2.

In addition to serving as electron acceptor in enzymatic reactions, molecular oxygen is also employed as oxygen-transfer agent, both for monooxygenations

$$SH_2 + O_2 \longrightarrow SOH + H_2O$$

and for dioxygenations (insertion of both oxygen atoms into product)

$$O_2 + \text{catechol (1,2-dihydroxybenzene)} \longrightarrow \text{HOOC–CH=CH–CH=CH–COOH}$$

We shall examine the variety of mechanisms that enzymes have evolved for oxygen activation and insertion into cosubstrates, including the hemeprotein cytochrome-P_{450} monooxygenases, which (among other things) are the enzymes involved in metabolic activation of certain carcinogens.

At the end of the redox section we shall briefly examine some complex enzymes that carry out apparent simultaneous six-electron transfers during reduction of such inorganic substrates as sulfite ion to sulfide and dinitrogen to two molecules of ammonia.

1.B.3 Section IV: Enzyme-Catalyzed Eliminations, Isomerizations, and Rearrangements

Three topics are taken up in Chapters 17 through 20. **Enzyme-catalyzed eliminations** convert saturated carbon centers in substrates to olefinic centers in the products. Loss of the elements of water, ammonia, or their derivatives HOR and HNR proceeds by loss of a carbon-bound hydrogen at one carbon, and loss of the oxygen or nitrogen leaving group from an adjacent carbon atom. For example:

H, OH $\rightleftarrows$ + H_2O

H, NHR $\rightleftarrows$ + RNH_2

Several highly purified enzymes have been examined for clues to the mechanisms of these eliminations. Among the questions we shall examine is whether chemical steps in the elimination are concerted or stepwise. If stepwise, does the C—H bond break first to generate a carbanion intermediate, or does the C—OH or C—NHR bond break first to produce a carbonium-ion intermediate. We shall see that some enzymes derivatize the —OH and —NH_2 functional groups to make them less basic and thereby lower the activation energy for elimination. We shall also discuss the stereochemistry of enzymatic eliminations—both *trans* (*anti*) and *cis* (*syn*) eliminations are observed—and comment on any mechanistic constraints these stereochemical facts might impose.

The great bulk of **enzyme-catalyzed isomerizations** can be classified, as noted by I. A. Rose (1970), into reactions involving 1,1-, 1,2-, or 1,3-hydrogen shifts (Table 1-1). A formal 1,1-proton shift is simply an inversion at an asymmetric carbon center and describes the behavior of racemases and epimerases. The 1,2-proton shifts involve formal hydrogen transfer between two adjacent carbon atoms, one undergoing oxidation, the other reduction; this describes the stoichiometry of aldose $\rightleftarrows$ ketose isomerases. Some degree of intramolecular proton transfer is, in fact, often observed. The 1,3-proton shifts are formal allylic or azaallylic (when nitrogen is one of the three atoms) isomerizations. The azaallylic isomerization is the chemical reaction involved in enzymes utilizing vitamin B_6 coenzyme (pyridoxal phosphate) to carry out transaminations. These 1,3-shifts can be suprafacial (with some intramolecular proton transfer) or antarafacial (with no intramolecular proton transfer).

Table 1-1
Enzyme-catalyzed isomerizations classified as hydrogen shifts

Type	Examples	Category
1,1-Shifts	$R_2\text{—}C(R_1)(R_3)\text{—}H \rightleftharpoons H\text{—}C(R_1)(R_3)\text{—}R_2$	Epimerases, racemases
1,2-Shifts	$H\text{—}C(=O)\text{—}C(H)(OH)\text{—} \rightleftharpoons H\text{—}C(H)(OH)\text{—}C(=O)\text{—}$ α-hydroxyaldehyde ⇌ β-keto alcohol	Aldose–ketose isomerases
1,3-Shifts	$R_1R_2C(H)\text{—}C{=}C(R_3)(R_4) \rightleftharpoons R_1R_2C{=}C\text{—}C(H)(R_3)(R_4)$	Allylic isomerizations
	$R_1R_2C(H)\text{—}N{=}C(R_3)(R_4) \rightleftharpoons R_1R_2C{=}N\text{—}C(H)(R_3)(R_4)$	Azaallylic isomerizations

Also included in the isomerization-enzyme category are phosphosugar mutases—for instance, interconverting glucose-6-$PO_3^{2\ominus}$ and glucose-1-$PO_3^{2\ominus}$ and thereby connecting fermentative hexose metabolism and aerobic hexose metabolism. Additionally, geometrical isomerizations of *cis*- and *trans*-olefins are briefly examined, an example being the isomerization of 11-*cis*-retinal (vitamin A aldehyde) to the all-*trans* form; this isomerization is the primary photochemical event in mammalian visual processes.

Carbon skeletal rearrangements, common in organic chemistry, are rather rare in enzyme-catalyzed reactions. We shall discuss the three major enzymatic examples: (1) the methyl-group migrations involved in cholesterol biosynthesis; (2) the only example of a Claisen rearrangement in enzymology, the conversion of chorismate to prephenate (also to isochorismate) during microbial aromatic amino-acid biosynthesis; and (3) the coenzyme-B_{12}–dependent 1,2-rearrangements in which a hydrogen substituent on one carbon and either oxygen, nitrogen, or carbon substituents on an adjacent carbon center undergo a formal 1,2-shift. Coenzyme B_{12} (and methyl B_{12}, discussed in Section V) is the only organometallic compound (with a covalent metal–carbon bond) involved in primary biological metabolism.

1.B.4 Section V: Enzymatic Reactions That Make and Break Carbon–Carbon Bonds

Because reactions that make and break carbon–carbon bonds are the key steps in both the biosynthesis and the breakdown of carbon skeletons of molecules in biological systems, the classes of enzymes catalyzing these reactions are in many ways the crucial ones in metabolism. As catalysts, they allow the breakdown of ingested nutrients, the oxidation of reduced carbon compounds for energy metabolism, and the assimilation of carbon fragments in anabolic reactions. They are examined in Chapters 20 through 26.

In any enzyme-catalyzed reaction where a carbon–carbon bond is formed, there must be a *carbanion* (or carbanionic equivalent) to serve as *attacking nucleophile at some other carbon atom that is electrophilic.* The carbon atoms attacked fall into two major categories: either >C=O groups (aldehydes, ketones, esters, CO_2), or sp^3 hybridized, tetrahedral carbon atoms that have a leaving group, X, as a substituent.

(a) $-\text{C}^{\ominus} + \text{C}{=}\text{O} \longrightarrow -\text{C}-\text{C}(\text{OH})-$

(b) $-\text{C}^{\ominus} + -\text{C}-\text{X} \longrightarrow -\text{C}-\text{C}- + \text{X}^{\ominus}$

Thus the major chemical problems these enzymes must solve to achieve rate accelerations are twofold. In both types (a) and (b), methods of forming stabilized, low-energy carbanions must be found. And in category (b), the carbon locus undergoing attack must have a hetero atom (oxygen generally) derivatized so that a low-energy path for its expulsion is available.

Some enzymes, such as decarboxylases, function physiologically in the bond-breakage direction. The reaction mechanism used depends on the structure of the particular carboxylate that serves as substrate. The β-keto acids have a built-in electron sink to stabilize the incipient carbanion formed on loss of CO_2 as an enolate ion.

$$\text{R—}\overset{\text{O}}{\overset{\|}{\text{C}}}\text{—}\overset{\text{H}}{\underset{\text{H}}{\text{C}}}\text{—}\overset{\text{O}}{\overset{\|}{\text{C}}}\text{—O}^{\ominus} \longrightarrow \text{CO}_2 + \left\{ \text{R—}\overset{\text{O}^{\ominus}}{\overset{|}{\text{C}}}\text{=CH}_2 \longleftrightarrow \text{R—}\overset{\text{O}}{\overset{\|}{\text{C}}}\text{—}\overset{\ominus}{\text{C}}\text{H}_2 \right\}$$

β-keto acid — resonance-stabilized enolate anion

When α-keto acids are decarboxylated, the substrate cannot delocalize the transient carbanion effectively, and these enzymes use the coenzyme form of vitamin B_1 (thiamine pyrophosphate) as an *electron sink* that forms an initial covalent adduct with the α-keto acid prior to decarboxylation.

For enzymes that function biosynthetically in carbon–carbon bond formation, there are four discrete types of reaction.

(a) Carboxylases make the enolate anions of keto acids or of acylthioesters as attacking nucleophiles to react with either electrophilic CO_2 or a derivatized form of carbonic acid containing an appropriate leaving group.

(b) There are many enzymes that catalyze aldol condensations or related Claisen condensations, and these are major routes for formation of key metabolites such as citric acid or fructose diphosphate (during gluconeogenesis) or δ-aminolevulinate (for heme biosynthesis). The carbon nucleophile is the stabilized carbanion form of an enolate or an enamine in the aldol condensations, or the α-carbanion of an acylthioester in the Claisen condensation. The electrophilic partner in most cases is a carbonyl carbon of an aldehyde, ketone, or keto acid.

(c) One-carbon–fragment transfers are important alkylation reactions, and two kinds of carrier molecules are used. The pterin coenzyme tetrahydrofolate is a carrier of one-carbon species at varying oxidation levels—e.g., at the methanol, formaldehyde, or formate states. The second carrier is the sulfonium cation *S*-adenosyl methionine, a carrier of methyl groups that can be attacked by cellular oxygen, nitrogen, sulfur, and carbon nucleophiles. The carbon nucleophiles include the enolate anions of pyrimidine bases in RNA and DNA, and the π-electrons of carbon–carbon double bonds in olefinic fatty acids.

(d) The major type of carbon-alkylation reaction involved in carbon-skeleton construction of hundreds of primary metabolites and natural products of secondary metabolism uses C_5-isoprenoid units as the alkylation monomer. We shall examine how the two biological isomeric isoprenyl units (Δ^3-isopentenyl pyrophosphate and 3,3-dimethylallyl pyrophosphate) are biosynthesized, and how the fundamental C_5-unit elongation step occurs enzymatically. The π-electrons of one component apparently attack an allylic cation of the second component in a *head-to-tail condensation.*

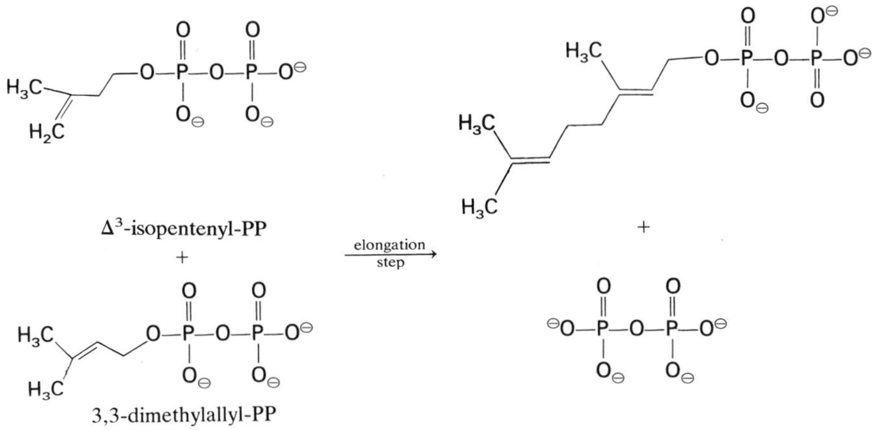

Δ^3-isopentenyl-PP

\+

3,3-dimethylallyl-PP

elongation step →

\+

Then we shall examine an alternate pattern of joining: *tail-to-tail condensation* involved in squalene formation, and subsequent cyclization of the acyclic polyene to the tetracyclic ring system of the sterols.

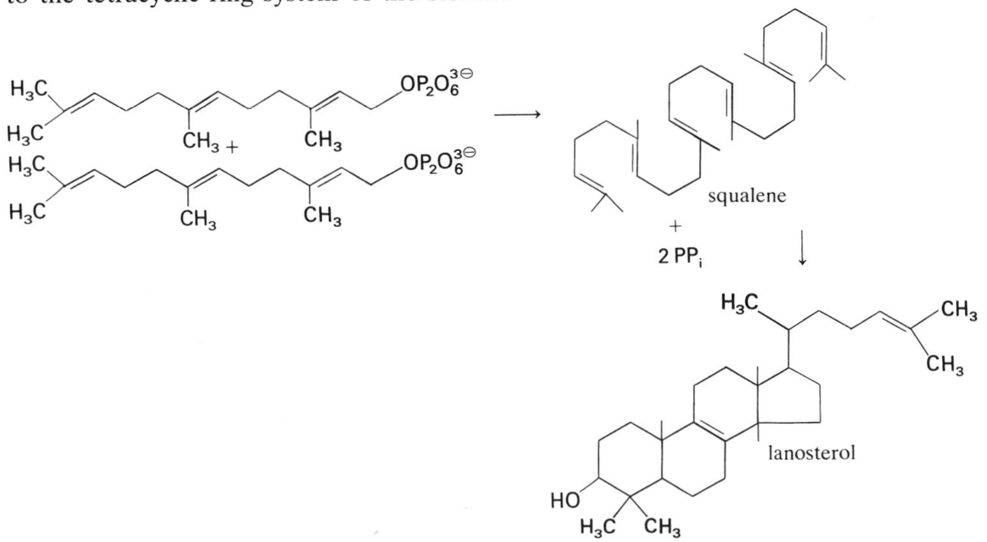

squalene

\+ 2 PP_i

lanosterol

At this juncture, most of the major types of characterized enzymatic reactions will have been analyzed, and the concluding chapter will deal with a brief excursion into a few of the central metabolic pathways to examine the chemical logic for processing of molecules for biosynthesis, degradation, or energy metabolism. We should be able to analyze the structure of a starting substrate and to predict the kinds of products that are likely to arise, given the *few* chemical pathways open to enzymatic catalysts. That predictive and analytic ability is one ultimate goal of the study of enzymatic mechanisms. The mapping-out of the height of every transition state, the estimation of its structure, and the evaluation of rate constants for every elementary step is another such goal—one considerably further removed from realization at present for all but a very few of the best-studied enzymes.

One of the potential virtues of possession of a rational and semiaccurate view of enzyme specificity and mechanism is the design of specific reagents to increase catalytic rates in a genetically impaired enzyme protein or, more easily, to inactivate selectively a given target enzyme for antibiotic effects (e.g., a bacterial target enzyme) or for other therapeutic uses (e.g., selective inhibition of a key enzyme in a rapidly proliferating tumor cell).

1.C ENZYME CATEGORIZATIONS

As the number of known enzymes has climbed into the hundreds in recent years, the International Union of Biochemistry (1964) has suggested a classification scheme and a basis for assigning a unique number to each enzyme. Although we shall not give the numbers for the enzymes discussed in this book, the interested reader can consult the IUB volume to establish these numbers.

The IUB classification scheme includes six main divisions:

1. oxidoreductases;
2. transferases;
3. hydrolases;
4. lyases;
5. isomerases;
6. ligases (synthetases).

Lyases are enzymes that catalyze addition of some group to a double bond in a substrate or, in the reverse direction, generate double bonds. Ligases are enzymes that ligate or join two molecules concomitant with cleavage of a nucleoside triphosphate (often ATP).

We shall not organize our discussion around these categories because the six divisions as such have no obvious didactic value. We shall discuss hydrolases along with other group transferases in Chapters 3 through 9. Oxidoreductases are the subjects of Chapters 10 through 16. Lyases are covered in Chapters 17 and 18, and enzymes carrying out isomerizations and rearrangements are found in Chapters 19 and 20. Ligases are treated in the chapters that deal with the enzymatic chemistry of ATP (Chapters 5, 7, and 8). The IUB categorization does not address systematically the large number of enzymes involved in construction (or breakage) of carbon–carbon bonds; these enzymatic reactions are at the heart of biosynthesis, and they are examined in Chapters 21 through 26.

1.D A WORD ABOUT TOPIC SELECTION AND REFERENCES

It may seem somewhat incongruous at the start of a lengthy book to point out that there is not room to be comprehensive—either in including every known enzyme in a given category, or in summarizing *all* the salient mechanistic information on any given enzyme—but that constraint is quite real. Selectivity has been used on both points, with a hope to provide representative enzymes of clear physiological importance and to discuss experiments that add new information (as the book progresses) to the accumulating mechanistic arsenal.

The decision on referencing was to be parsimonious with references to original literature, favoring listing of recent and comprehensive books and review articles that summarize previous original research results and serve as initial entries into the research literature on a given enzyme or class of enzymatic reactions. Thus, there may be only 30 references in a chapter where 300 references might justifiably be used to document every fact and surmise. However, the references provided will readily lead the reader into the original research literature if absolute documentation is desired. Bibliographic citations for the entire text are grouped in the References section following the text. That section begins with a selective list of texts and review series that are of general utility for the study of enzyme reactions.

Chapter 2

Introductory Remarks About Enzymes and Enzymatic Catalysis

2.A ENZYMES AND CATALYSIS

All enzymes are proteins, but the converse is not true.* Enzymes are the one subset of proteins with catalytic activity—a concept established when Sumner (1926) isolated the enzyme urease (which hydrolyzes urea to CO_2 and NH_3) from jack-bean meal, crystallized it, and validated its protein nature. Proteins are large, linear, condensed biopolymers of α-amino acids as monomeric units. The common structural linkage is the amide or peptide bond between amino group of one monomer and carboxyl group of another.

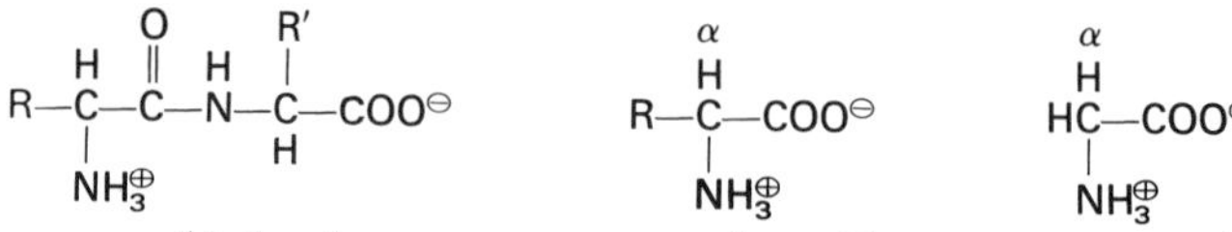

$$R\text{—}CH(NH_3^{\oplus})\text{—}C(=O)\text{—}NH\text{—}CH(R')\text{—}COO^{\ominus}$$

peptide bond

$$R\text{—}C^{\alpha}H(NH_3^{\oplus})\text{—}COO^{\ominus}$$

α-amino acid

$$HC^{\alpha}H(NH_3^{\oplus})\text{—}COO^{\ominus}$$

glycine

There are about 20 α-amino acids that are the common building blocks of proteins. All of them except glycine have at least one asymmetric carbon (the α-carbon) and are thus optically active. There are enantiomeric D- and L-pairs (*R,S*-pairs in the nomenclature proposed by Cahn, Ingold, and Prelog, 1956, 1966). Only the L-amino acids are activated enzymatically and incorporated into proteins (although D-isomers are observed in some low-molecular-weight polypeptide antibiotics and in bacterial cell walls).

---

* It is recommended that readers without a recent perusal of protein chemistry read the relevant chapters in a general biochemistry textbook (such as those listed at the beginning of the References section) as an adjunct to these few comments.

$$\begin{array}{c} COO^{\ominus} \\ | \\ H_3\overset{\oplus}{N}-C-H \\ | \\ CH_3 \end{array}$$

L-isomer of alanine (Fischer projection)

Thus enzymes, composed only of L-amino-acid residues, are asymmetric or chiral reagents—unlike most synthetic chemical reagents, which are achiral. As we shall note, this is the key to several features of enzymatic stereospecificity.

For background information on stereochemistry and structural representations (such as the Fischer projection), see Alworth (1972) and Bentley (1969–1970).

2.B AMINO-ACID STRUCTURES AND FUNCTIONAL GROUPS

Table 2-1 lists the structures of the 20 common amino acids, along with some relevant pK_a values. The pK_a values of the α-amino group range from 8.8 to 10.0; those of the α-COOH group range from 1.8 to 2.3. Thus, at neutral physiological pH values, α-amino acids are dipolar, zwitterionic species and will be represented this way throughout the text (see top of p. 27).

Table 2-1
Amino-acid structures

Acidic	Basic	Free amino acids
$^{\ominus}OOC-\overset{\beta}{C}H_2-CH(\overset{\oplus}{N}H_3)-COO^{\ominus}$ $pK_a = 3.8$ aspartate (Asp)	$H_3\overset{\oplus}{N}-CH_2-CH_2-CH_2-CH_2-CH(\overset{\oplus}{N}H_3)-COO^{\ominus}$ $pK_a = 10.5$ lysine (Lys)	α-COOH: $pK_a = 1.8–2.3$ α-$NH_3^{\oplus}$: $pK_a = 8.8–10.0$
$^{\ominus}OOC-\overset{\gamma}{C}H_2-\overset{\beta}{C}H_2-\overset{\alpha}{C}H(NH_3)-COO^{\ominus}$ $pK_a = 4.2$ glutamate (Glu)	$H_2\overset{\oplus}{N}=C(NH_2)-NH-CH_2-CH_2-CH_2-CH(\overset{\oplus}{N}H_3)-COO^{\ominus}$ $pK_a = 12.5$ arginine (Arg)	
	(imidazolium: $H\overset{\oplus}{N}$, NH ring)$-CH_2-CH(\overset{\oplus}{N}H_3)-COO^{\ominus}$ $pK_a = 6$ histidine (His)	

Table 2-1 *(continued)*

Amides		Secondary
$H_2N-C(=O)-CH_2-CH(NH_3^{\oplus})-COO^{\ominus}$	$H_2N-C(=O)-CH_2-CH_2-CH(NH_3^{\oplus})-COO^{\ominus}$	$\overset{\oplus}{N}H_2$ ring $-CH(COO^{\ominus})$
asparagine (Asn)	glutamine (Glu)	proline (Pro)

Aliphatic (hydrophilic)	Aliphatic (hydrophobic)	
$H-CH(NH_3^{\oplus})-COO^{\ominus}$	$(H_3C)_2CH-CH(NH_3^{\oplus})-COO^{\ominus}$	$(H_3C)_2CH-CH_2-CH(NH_3^{\oplus})-COO^{\ominus}$
glycine (Gly)	valine (Val)	leucine (Leu)
$H_3C-CH(NH_3^{\oplus})-COO^{\ominus}$	$H_3C-CH_2-CH(CH_3)-CH(NH_3^{\oplus})-COO^{\ominus}$	
alanine (Ala)	isoleucine (Ile)	

Aromatic	Polar	
$C_6H_5-CH_2-CH(NH_3^{\oplus})-COO^{\ominus}$	$HO-CH_2-CH(NH_3^{\oplus})-COO^{\ominus}$ $pK_a = 14$	$H_3C-CH(OH)-CH(NH_3^{\oplus})-COO^{\ominus}$
phenylalanine (Phe)	serine (Ser)	threonine (Thr)
$HO-C_6H_4-CH_2-CH(NH_3^{\oplus})-COO^{\ominus}$ $pK_a = 10.1$	$HS-CH_2-CH(NH_3^{\oplus})-COO^{\ominus}$ $pK_a = 8$	$H_3C-S-CH_2-CH_2-CH(NH_3^{\oplus})-COO^{\ominus}$
tyrosine (Tyr)	cysteine (Cys)	methionine (Met)
indole(NH)$-CH_2-CH(NH_3^{\oplus})-COO^{\ominus}$	$S-CH_2-CH(NH_3^{\oplus})-COO^{\ominus}$ $\vert$ $S-CH_2-CH(NH_3^{\oplus})-COO^{\ominus}$	
tryptophan (Trp)	cystine (Cys-S-S-Cys)	

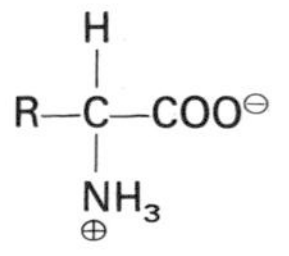

$$R-\underset{\underset{\oplus}{NH_3}}{\overset{H}{C}}-COO^{\ominus}$$

The side chains of the amino acids provide the important functional groups for enzymatic catalysis. The pK_a values of 3.8 for the β-carboxyl of aspartate and 4.2 for the γ-carboxyl of glutamate imply that these are ionized in the neutral pH range inhabited by most proteins and thus can function as either basic or nucleophilic catalytic groups in catalysis (Jencks, 1969, p. 67 ff; Gray, 1971, p. 19 ff). The imidazole group of histidine has a pK_a of 6, so a high concentration of the free base form (a relatively good, unhindered nucleophile despite being a tertiary amine) can exist at neutral pH. Also of importance catalytically will be the thiolate anion of cysteine ($pK_a = 8$), the phenoxide form of tyrosine ($pK_a = 10$), and the free base form of the ε-amino group of lysine ($pK_a = 10.5$).

Proteins with enzymatic activity are macromolecular catalysts, with molecular weights ranging from polypeptide chains of about 9,000 molecular weight for acyl phosphatase from brain to 155, 000 mol wt for the β-subnit of RNA polymerase. Given an average molecular weight for an amino acid of ca. 150, polypeptide chains thus can range from 60 to about 1,000 amino-acid residues long. An enzyme can be composed of several polypeptides (each a subunit) to form oligomeric complexes of up to 5 or 6×10^6 mol wt units (Reed and Cox, 1970). Presumably the enormous size of these biological catalysts is related to the attainment of sufficient local-controlled flexibility on the one hand and precise three-dimensional arrangement of amino-acid side chains on the other to provide the exact spatial array needed to promote efficient and specific catalysis when a substrate molecule (as large as another protein or as small as carbon dioxide) binds to the active site. Some enzymes require metal cations, which will be bound to specific oxygen, nitrogen, or sulfur ligands of the protein. These metals include $K^{\oplus}$, $Na^{\oplus}$, $Cu^{\oplus}$, $Mg^{2\oplus}$, $Zn^{2\oplus}$, $Ca^{2\oplus}$, $Ni^{2\oplus}$, $Fe^{2\oplus}$, $Fe^{3\oplus}$, $Co^{3\oplus}$, $Mo^{6\oplus}$, and inorganic selenium and sulfur (Coleman, 1971; Bender, 1971; Mildvan, 1974; Eichhorn, 1971). Other enzymes depend totally for activity on low-molecular-weight organic coenzymes that may either bind in dynamic equilibrium or be bound so tightly as to form nondissociable, stoichiometric holoenzyme complexes. Some enzymes require lipid bound noncovalently or covalently; others are glycoproteins.

The in vivo milieu of an enzyme is a heterogeneous intracellular (or extracellular) enviroment; but, for the study of catalytic mechanism, enzymes have been purified—often to a state of physical homogeneity to ensure that observed reactions depend only on the single enzyme under study. The advantages of a pure protein for studying catalysis are mitigated by the sometimes herculean labors needed to obtain detectable quantities. A typical purification scheme to isolate a specific enzyme might involve the following steps (all done in aqueous medium near neutral pH, and at low temperatures to minimize loss of catalytic activity): disruption of cells by mechanical or sonic treatment; differential centrifugation to separate "soluble" and particulate fractions; fractionation by differential solubility in solutions of high salt (ammonium sulfate); ion-exchange chromatography to separate species of different charge; adsorption chromatography (i.e., on calcium phosphate crystals); molecular sieve (gel-filtration) chromatography to separate species of different size; and electrophoresis (again separation by charge differences) on various supports.

Analysis for purity can include crystallization, ultracentrifugal analysis, disc-gel electrophoresis, amino-acid composition (and sequence) determination, and various molecular weight determinations. The isolated proteins are often labile with respect to loss of catalytic activity; even minor perturbations in the three-dimensional structure of the protein, as a consequence of the purification sequence or perhaps merely due to removal from its normal milieu, can cause dramatic loss in catalytic function. To try to preserve the native (active) three-dimensional conformation of an enzyme and avoid inactive, denatured conformations, purified proteins are stored at low temperature, anywhere from 4° C to −196° C, and half-lives may range from hours to years in this state. To obtain an enzyme pure of other contaminating proteins, one may have to purify it anywhere from 20-fold to 50,000-fold from the biological starting material. In the latter case, if the overall yield of active enzyme were 10% of that in the crude extract, 500 g of cell protein would yield 1 mg of enzyme. If this enzyme has a molecular weight of 50,000 g/mole, this amount represents 20 nmoles (2×10^{-8} mole), not a large quantity. Indeed some enzymatic purifications result in microgram (10^{-6} g) quantities. This example points out that the general method for analyzing whether a purification procedure is working is by analysis of enzyme activity, not of enzyme mass. The international definition of activity is a unit (U) and corresponds to a rate of conversion of one micromole (10^{-6} mole) of substrate to product per minute.★ The specific activity of an enzyme is measured as U/mg, and this specific activity should rise during purification until a constant value, indicative of pure protein, is obtained. We shall discuss subsequently a number of specific irreversible inactivators for certain enzymes, which can then be used to titrate the number of active enzyme molecules in a test tube and provide a direct measure of how much catalyst is present.

★An alternative unit is the katal (kat), defined as the amount of enzyme that converts one mole of substrate to product per second (1 kat $= 6 \times 10^7$ units).

2.C SPECIFICITY AND RATE ACCELERATIONS

To a student of enzymology, perhaps the most salient features of enzymatic catalysis are the specificity and the rate acceleration when the enzymatic reaction is compared with either the analogous uncatalyzed chemical reaction or the catalyzed nonenzymatic equivalent. Some aspects of these points are noted here and will be examined in the context of specific examples in the next four sections of the text.

2.C.1 Specificity

Enzymatic catalysis always involves prior complex formation between the reactant (**substrate**) and the enzyme, generally in an equilibrium, fast process; this noncovalent, enzyme–substrate complex (ES complex) is also commonly designated as a Michaelis complex, after Lenor Michaelis who enunciated this proposal (Michaelis and Menten, 1913). Subsequent catalysis occurs only from this ES complex and not by direct bimolecular reaction with substrate free in solution.

$$E + S \rightleftarrows E \cdot S \xrightarrow{\text{catalysis}} E \cdot P \rightleftarrows E + P$$

The ES complex does not form in a topologically random manner. Rather, the substrate binds to a specific region on the enzyme (the active-site region) in every catalytic cycle, and catalysis occurs only at the active site.

Specificity can be imposed on the binding step or on any subsequent catalytic steps or on both. Thus an enzyme oxidizing α-hydroxy acids to α-keto acids may bind only one enantiomer (binding specificity) or may bind both but oxidize only one isomer. Compounds that bind to the enzyme active site but do not undergo catalytic conversion are common and function as enzyme inhibitors. They can shed light on reaction paths depending on the type of inhibition they exhibit, as we shall see in Chapter 4.

Specificity can be absolute: there are enzymes for which only a single substrate molecule is acceptable. Or, specificity can be for broad structural types: this is the case, as we shall note explicitly, with the protease chymotrypsin.

As a rule, enzymes demonstrate unerring and complete stereospecificity in catalysis. They can invariably distinguish between optical or geometrical isomers. They (almost) always use one form of an enantiomeric pair, unless their specific function is to interconvert isomers. Even more spectacularly, at first glance, enzymes always distinguish between paired chemically like substituents in cases such as C_{aabc}; e.g., the two hydrogens on a *meso* carbon atom CH_2XY, also known as a prochiral carbon. We shall examine in some detail the chemical basis for this stereospecificity and the mechanistic information it can impart in Section III (Chapter 10).

Why are enzymes such stereospecific catalysts? Because they are asymmetric or chiral reagents, composed uniquely of L-amino-acid centers. Interaction of some normal symmetric chemical laboratory reagent with two enantiomers (e.g., D- or L-lactate) generates transition states that are enantiomeric and of equal energy levels. When a chiral enzyme interacts with each enantiomer during reaction, diastereomeric transition states are formed. These will be of different energies, will have different reactivities, and will partition differently between reactants and products.

2.C.2 Rate Accelerations

As catalysts, enzymes do not participate in the reaction stoichiometry (are not consumed) and cannot affect the equilibrium position of a reaction; they can only hasten the rate of approach to the same equilibrium as that for an uncatalyzed reaction. The equilibrium ratio for a substrate S and a product P, [S]/[P], is of course a function of which compound is more stable—i.e., which has a lower free energy. Most reduced organic molecules are thermodynamically unstable in an oxidizing atmosphere; their oxidations are exergonic processes. For example, consider the oxidation of glucose to CO_2 and H_2O by molecular oxygen.

$$\text{glucose} + 6\,O_2 \longrightarrow 6\,CO_2 + H_2O \qquad \Delta G^0 = -686\ \text{kcal/mole} = -2872\ \text{kJ/mole}$$

This oxidation is favored by an enormous 686 kcal/mole (the negative ΔG^0 reflecting an exergonic process). At chemical equilibrium ($\Delta G^0 = -RT \ln K_{eq}$), there would be essentially no glucose present. But, in fact, glucose is a stable organic compound that can be kept in bottles on the shelf. It is *thermodynamically* labile but *kinetically* stable.

The distinction between effect on thermodynamic stability and effect on kinetic lability of a molecule is clearly indicated in the free-energy diagram (Fig. 2-1). The equilibrium concentrations of substrate and product are determined by their difference in free-energy content: $\Delta G^0 = -RT \ln K_{eq}$. In Figure 2-1 the product is indicated to be more stable than the substrate. Suppose ΔG^0 here were

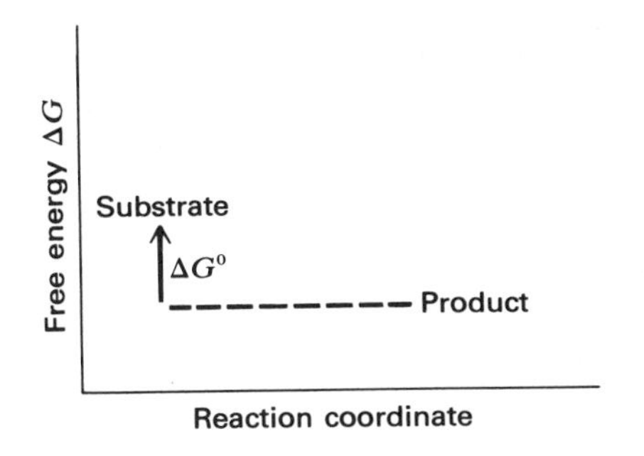

Figure 2-1
Equilibrium position determined by ΔG^0 between substrate and product.

-2.8 kcal/mole $= -11.7$ kJ/mole. Then

$$\ln K_{eq} = (-2.8 \text{ kcal/mole})/(-RT)$$
$$K_{eq} = 100 = [\text{product}]/[\text{substrate}]$$

At chemical equilibrium, there would be 100 product molecules for every substrate molecule. This ratio is what would happen *if* the chemical system *were* at equilibrium, but this *thermodynamic tendency* for product to accumulate provides *no information about how fast* substrate will actually be converted to product. The *rate* at which the conversion occurs, a measure of the kinetic lability of substrate molecules, *is independent of* ΔG^0. Given that the substrate is a stable molecule, say at room temperature, and not instantaneously undergoing complete conversion to product, some energy (e.g., heat) will have to be put into the substrate molecules to convert them to products. That is, there is some activation energy required—some energy barrier that molecules of substrate must surmount before being converted to product. An arbitrary activation energy, $\Delta G^{\ddagger}$, is sketched in Figure 2-2. The highest point on the free-energy curve (surface) is the

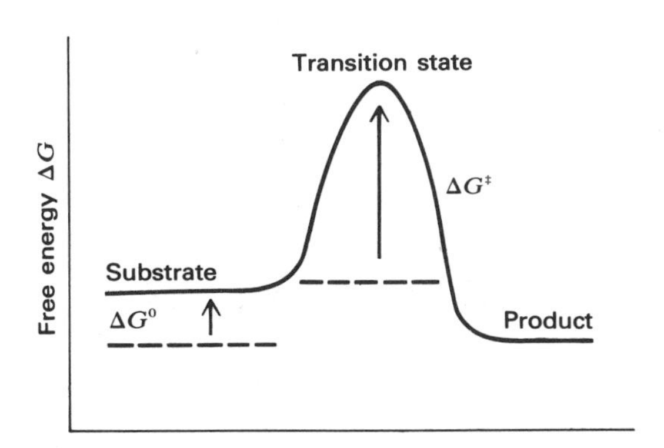

Figure 2-2
Rate of reaction determined by $\Delta G^{\ddagger}$ activation energy barrier between substrate and product.

transition state for the reaction, by definition a fleeting entity, lasting perhaps for one molecular vibration ($\sim 10^{-13}$ sec). A simplified version of transition-state theory for reaction rates relates the rate of a reaction to the height of $\Delta G^{\ddagger}$ through the following simple exponential expression.

$$k_{obs} = (RT/nh)e^{-\Delta G^{\ddagger}/RT}$$

This transition-state formulation, which assumes that reactants and transition state are in equilibrium, is extremely useful in that it allows one to evaluate kinetic behavior on the basis of energetic barriers, potentially predictable from structural data (Fersht, 1977).

Those substrate molecules that have kinetic energy $> \Delta G^{\ddagger}$ can pass over the barrier to products. Any catalyst, chemical or enzymatic, that accelerates a chemical reaction has its effect by lowering the energy barrier between substrate and transition state, reducing $\Delta G^{\ddagger}$, but having no effect on ΔG^0 (no effect on equilibrium position).

A chemical catalyst (e.g., a palladium metal catalyst for hydrogenation of an olefin) probably lowers $\Delta G^{\ddagger}$ almost exclusively by selective stabilization of the transition state (e.g., bringing reactants together at the finely divided metal surface) as suggested in Figure 2-3a. An enzymatic catalyst also will probably

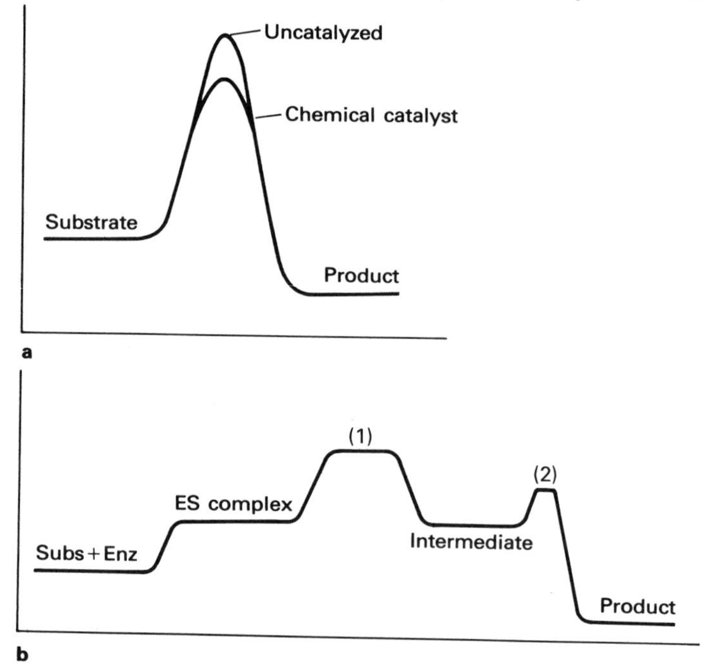

Figure 2-3
(**a**) Action of chemical catalyst to lower $\Delta G^{\ddagger}$ and increase rate of reaction. (**b**) Possible action of enzymatic catalyst to lower $\Delta G^{\ddagger}$ and replace one large activation barrier with multiple lower barriers.

selectively stabilize the transition state relative to reactants (if reactants and transition state were stabilized to the same extent, there would be no lessening of $\Delta G^{\ddagger}$, thus no rate acceleration), but may also selectively destabilize substrates bound at the active site by induction of strain or distortion. Also, enzymatic catalysts probably act to replace a single step having a large $\Delta G^{\ddagger}$ by multiple steps having small $\Delta G^{\ddagger}$ overall. For instance, Figure 2-3b for a hypothetical enzyme catalyst shows destabilization of the ES complex, selective stabilization of a transition state, formation of a finite-lived intermediate (a local energy minimum), and a second shallow $\Delta G^{\ddagger}$ to transition state (2) preceding product formation.

Truly stupendous accelerations can be achieved over nonenzymatic rates. The enzyme urease hydrolyzes urea at a rate estimated to be 10^{14}-fold faster than the nonenzymatic rate of hydrolysis.

$$H_2N{-}C({=}O){-}NH_2 + H_2O \longrightarrow NH_4^{\oplus} + [H_2N{-}COO^{\ominus}] \xrightarrow{H_2O} NH_4^{\oplus} + CO_2$$

This ratio is on the high side, with accelerations of 10^8 to 10^{12} representing an average rate increase brought about by an enzyme. It's worth noting again that these rate accelerations cannot affect the equilibrium constant and thus represent rate acceleration in both directions (Segel, 1976, p. 208). Suppose the equilibrium constant for interconversion of compounds A and B is 10^3 in favor of B.

$$A \underset{k_{-1}}{\overset{k_1}{\rightleftharpoons}} B \qquad K_{eq} = k_{-1}/k_1 = 10^{-3}$$

In the absence of an enzyme, starting with compound B, the observed rate of conversion to A might be $10^{-5}\ \text{min}^{-1}$. Then $k_1(A \rightarrow B)$ must be $10^{-8}\ \text{min}^{-1}$ given the value of K_{eq}. If an enzyme accelerates the rate of B to A by 10^{10}, so that $k_{-1} = 10^5\ \text{min}^{-1}$, then the reverse rate must also be accelerated by 10^{10} up to a value of $10^2\ \text{min}^{-1}$.

An average turnover number for an enzyme is $1{,}000\ \text{min}^{-1}$—i.e., 1,000 moles substrate reacted per minute per mole of enzyme active site. Some enzymes run as high as $10^6\ \text{sec}^{-1}$ (Talalay and Benson, 1972). Such a molecule of purified enzyme in a test tube might maintain full catalytic activity for 24 hours at room temperature and at its optimal pH, so that one might see 1.4×10^8 turnovers per enzyme molecule—emphasizing its behavior as a catalyst.

2.C.3 What Limits the Rate of Enzymatic Reactions?

One can ask what might be the maximal turnover number for an enzyme. Is 10^5 or 10^6 moles substrate reacted per mole enzyme per second an upper limit? As we noted above, enzymatic catalysis involves both physical steps (binding of substrate and debinding of product from the active site of the enzyme) and chemical steps, *as does any chemical reaction.*

$$\text{chemical reaction} \quad X{-}Y + Z \underset{\text{step}}{\overset{\text{physical}}{\rightleftharpoons}} X{-}Y\cdots Z \underset{\text{step}}{\overset{\text{chemical}}{\rightleftharpoons}} X\cdots Y{-}Z \underset{\text{step}}{\overset{\text{physical}}{\rightleftharpoons}} X + Y{-}Z$$

$$\text{enzymatic reaction} \quad E + S \underset{\text{step}}{\overset{\text{physical}}{\rightleftharpoons}} E\cdot S \underset{\text{step}}{\overset{\text{chemical}}{\rightleftharpoons}} E\cdot P \underset{\text{step}}{\overset{\text{physical}}{\rightleftharpoons}} E + P$$

Because few chemical reactions in solution (except some proton transfers) are limited by the rate at which two species diffuse together, the physical steps are rarely important in rate determination. However, because enzymatic rates are so much accelerated, the diffusional steps can reasonably put an upper limit on catalytic rates. The bimolecular rate constant for diffusional approach of small molecules with each other (e.g., $H_3O^{\oplus} + Y^{\ominus}$) is about $10^{10}\ \text{M}^{-1}\ \text{sec}^{-1}$; for small molecules with macromolecules such as enzymes, the diffusion limit may be ca. 10^8 to $10^9\ \text{M}^{-1}\ \text{sec}^{-1}$ (Hammes and Schimmel, 1970; Fersht, 1977). One can then state that the diffusion limit might constrain the rate of enzymatic reactions and examine if this is the fact. First, however, we note that this 10^9 value is $10^9\ \text{M}^{-1}\ \text{sec}^{-1}$, whereas a turnover number of an enzyme equals k_{obs}/[Enz], which is expressed in units of sec^{-1}. One can look at an appropriate bimolecular rate constant for the enzymatic case at low substrate/enzyme ratios in the following simple case.

$$E + S \underset{k_{-1}}{\overset{k_1}{\rightleftharpoons}} E\cdot S \xrightarrow{k_2} E + P$$

Here k_2 is the observed rate of product formation, and $K_S = k_{-1}/k_1$ is the dissociation constant from E · S. The ratio k_2/K_S is essentially an apparent bimolecular rate constant for reaction of enzyme and substrate (in a collisional process corrected for partitioning of E · S forward to product or back to substrate), and the upper limit should be ca. $10^9\ \text{M}^{-1}\ \text{sec}^{-1}$.* Thus, the important factor is the ratio k_2/K_S rather than the absolute size of k_2. Analysis of this (rate constant)/(equilibrium constant) ratio shows that certain enzymes are close to the diffusion-controlled limit. The enzyme triose phosphate isomerase (interconverting glyceraldehyde-3-phosphate and dihydroxyacetone phosphate) has a turnover

* This is so if, in this most simplified kinetic mechanism, $k_2 \gg k_{-1}$. More generally, one can write

$$E{+}S \underset{k_{-1}}{\overset{k_1}{\rightleftharpoons}} E\cdot S \xrightarrow{k_{cat}} E{+}P$$

The k_{cat} can be a complex catalytic rate constant containing several elementary steps. An additional useful constant, as will be discussed explicitly in the next chapter, is K_m (the Michaelis constant); here $K_m = (k_{cat} + k_{-1})/k_1$. Now if $k_{cat} \gg k_{-1}$, then $k_{cat}/K_m = k_1$, the association rate constant, which has the diffusion-controlled upper limit of 10^8 to $10^9\ \text{M}^{-1}\ \text{sec}^{-1}$ (Fersht, 1977). In the prototypic case chosen, $k_2/K_S = k_{cat}/K_m$, although this does not hold in all cases (Fersht, 1977, p. 96).

number (k_{cat}) of ca. 10^3 sec^{-1} and an affinity constant (K_m, which equals K_S only under special kinetic circumstances, as shown in Chapter 3) for glyceraldehyde-3-phosphate of ca. 10^{-5} M (Trentham et al., 1964; Reynolds et al., 1971).

$$\frac{k_{cat}}{K_m} \approx \frac{10^3 \text{ sec}^{-1}}{10^{-5} \text{ M}} = 10^8 \text{ M}^{-1} \text{ sec}^{-1}$$

On the other hand, the bacterial enzyme GTP-cyclohydrolase (Yim, 1975), the first enzyme in formation of the folate coenzymes, has an affinity constant for GTP of 2×10^{-8} M, extremely tight binding. This argues that the turnover rate (k_{cat}) for this enzyme might be in the range of ten catalytic events per second (the actual observed value) but could not be in the range of 10^3 to 10^4 sec^{-1}, because that would exceed the diffusion limit. For example,

$$(10^1 \text{ sec}^{-1})/(2 \times 10^{-8} \text{ M}) = 2 \times 10^9 \text{ M}^{-1} \text{ sec}^{-1}$$

Two points emerge from this brief consideration. First, two enzymes may differ enormously in catalytic turnover numbers even though each may be essentially at the diffusion-controlled limit, because it is k_2/K_S (or k_{cat}/K_m generally), not simply $k_2(k_{cat})$, that is relevant. In these cases, the enzymes are operating at maximal catalytic efficiency: a physical step, not a chemical step, is the slow step in catalysis. If the physical step occurs at the diffusion limit, then the enzyme can do nothing to increase catalytic efficiency. It has been stated that such an enzyme has reached the end of its evolutionary development as a catalyst (Albery and Knowles, 1976). Second, an enzyme that binds substrate loosely can achieve high turnover rates; the tighter the substrate is bound, the more the enzyme must pay in rate of achievable turnover. It should be noted that only a *small fraction* of known enzymes appear to be operating at a diffusion-controlled limit. An additional subset of enzymes have a physical rate-limiting step, but that step is much slower than the diffusion limit. The majority of enzymes probably are limited not by physical steps but by the rate of one or more chemical steps in catalysis.

2.D TYPES OF ENZYMATIC CATALYSIS

Clearly, anyone concerned with how enzymatic catalysis is achieved must give thought to the enabling mechanisms for such effects. We will briefly enumerate the major types of catalysis thought to function. It will be useful to keep in mind the combinations that may be in effect as we discuss individual enzymes in Sections II through V. Jencks (1975*a*) has argued, in a penetrating review article, that specificity and rate acceleration (the two special features of enzymatic catalysis) both derive from the free energy made available on binding of a specific substrate to an enzyme active site. The kernel of his argument is that

> the manifestation of specificity in the maximal velocity of the covalent step of enzymic reactions appears to require the utilization of the free energy that is made available from binding interactions with specific substrates. The observed free energy of binding ordinarily represents what is left over after this utilization.... The principal difference between enzymic and ordinary chemical catalysis is that enzymes can utilize noncovalent binding interactions with substrates to cause catalysis, in addition to the chemical mechanisms utilized by ordinary catalysts. (Jencks, 1975*a*, p. 222)

One must keep in mind then, when actually measuring the free energy of binding of a substrate to an enzyme active site by measuring a binding equilibrium constant K_{eq} ($\Delta G^0 = -RT \ln K_{eq}$), that the observed binding constant is an underestimation of the intrinsic binding energy. Some of the interaction energy will be utilized for rate acceleration, perhaps by inducing strain or distortion into the bound substrate (lowering $\Delta G^{\ddagger}$) by substrate destabilization rather than transition-state stabilization, perhaps by triggering a protein conformational change to produce a more active form of the enzyme or one where productive binding modes of the substrate are maximized, perhaps by freezing out translational and rotational motion of the bound substrate to provide an entropic acceleration to catalysis. How much the observed binding constant underestimates the true interaction energy can be a function of how good the substrate is for a given enzyme. There are substrates for the gastric protease pepsin that display about the same observed binding energy, but their turnover rates are enormously different (Inouye and Fruton, 1967). Presumably, for the substrate with the fast rate of enzymatic reaction, most of the intrinsic binding energy is utilized to bring about an increase in catalysis, whereas for the slow substrate only a small fraction of binding energy is thus used. The result is that specificity shows up, not in tighter binding of a good substrate, but rather as increased reaction rate.

Noting the interrelationship of the physical binding steps and the chemical steps involving covalent change during enzyme-mediated reactions, we can see how these phenomena condition the chemical mechanisms normally invoked in catalysis. Jencks (1969, chaps. 1–3, 5) discusses four categories of chemical catalysis that have been invoked with enzymes:

1. catalysis by approximation (entropic contribution to catalysis);
2. catalysis by covalent intermediates;
3. general acid–base catalysis;
4. catalysis by strain or distortion.

We next discuss these categories in more detail.

2.D.1 Catalysis by Approximation (Entropic Contribution)

Approximation in this context is literally to make reactants *proximal*—adjacent to each other. The term *catalysis by approximation* means the reaction-rate enhancement achieved if two reactants are taken out of dilute solution and held in close proximity to each other, as they will be at the active site of an enzyme. This proximity will raise the effective concentration over that of the reactants free in solution and, intuitively, ought to lead to a rate acceleration.

In an attempt to quantify some propinquity effects, model studies have been performed comparing rates for similar intra- versus intermolecular transformations. For instance, consider the imidazole-catalyzed hydrolysis of *p*-nitrophenyl acetate, which proceeds under a given set of conditions with a rate constant $k_{obs} = 35\ \text{min}^{-1}\ \text{M}^{-1}$, where k_{obs} is a bimolecular rate constant.

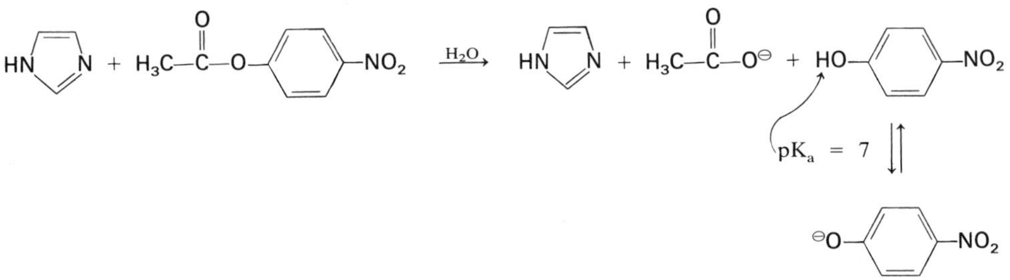

HN N + $H_3C—C(=O)—O—C_6H_4—NO_2$ $\xrightarrow{H_2O}$ HN N + $H_3C—C(=O)—O^{\ominus}$ + $HO—C_6H_4—NO_2$

$pK_a = 7$

$^{\ominus}O—C_6H_4—NO_2$

This second-order rate constant can be compared with the first-order rate constant for a comparable intramolecular case.

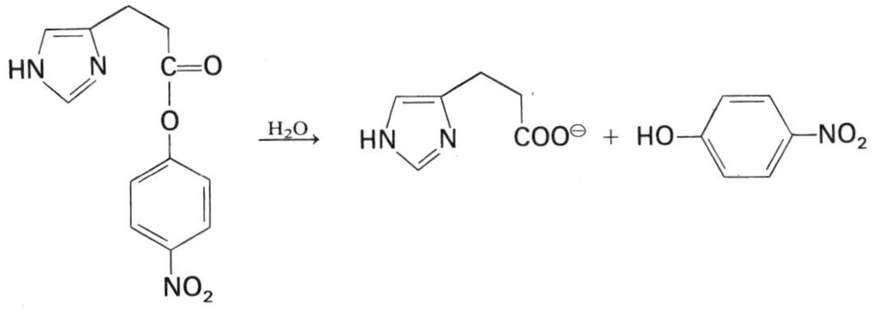

HN N C=O O NO_2 $\xrightarrow{H_2O}$ HN N $COO^{\ominus}$ + $HO—C_6H_4—NO_2$

Here $k_{obs} = 200\ \text{min}^{-1}$, under the same reaction conditions.

To assess the rate acceleration due to the covalently ensured propinquity of the imidazole moiety in the intramolecular case, one can calculate a value called the effective molarity of the imidazole in this intramolecular reaction. Thus,

$$(200\ \text{min}^{-1})/(35\ \text{min}^{-1}\ \text{M}^{-1}) = 5.7\ \text{M}$$

This equation implies that the effective molarity of imidazole is 5.7 M. That is, a concentration of 5.7 M would be required in the intermolecular case to attain the rate observed in the intramolecular reaction (Bender, 1971, p. 304).

Addition of a third methylene group, allowing a six-member rather than five-member transition state (presumably improving the geometry for imidazole participation) causes a jump in the effective molarity to 23.9 M.

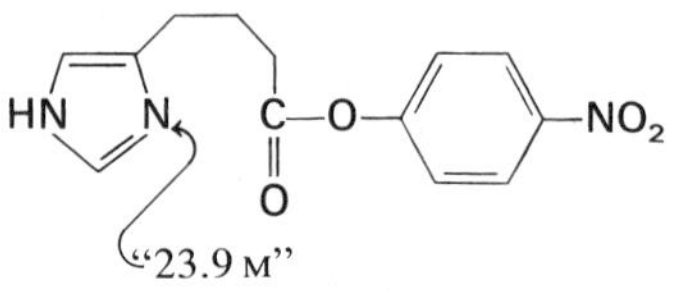

HN N C—O— —NO_2
O
"23.9 M"

Carrying this concept to its logical extreme, Bruice (1970) has compared the bimolecular rate of hydrolysis of phenylacetate by $RCOO^{\ominus}$ with the following bicyclic system:

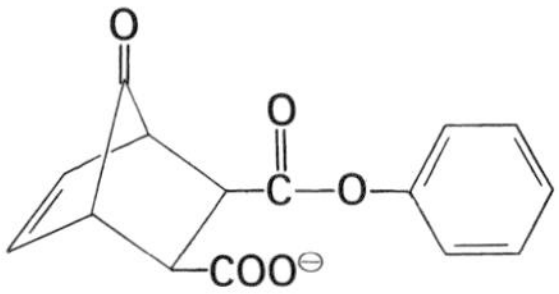

O
O
C—O—
$COO^{\ominus}$

Here the effective molarity calculates to 10^7 M! A concentration of 10^7 M is about six orders of magnitude higher than any realistically attainable concentration and suggests that factors beyond simple concentration effects are involved. Bruice suggests that, in this rigid bicyclic system with *exo* ester and adjacent *exo* carboxylate, the stereochemistry has built-in a favorable geometry for carboxylate assistance (anchimeric assistance, neighbouring-group participation) in the ester hydrolysis. That is, the population of rotamers with unfavorable geometry is much smaller than that in an acyclic system. An *endo,exo* substituent pattern should and does show very low $COO^{\ominus}$ effective molarity.

Reuben (1971) has postulated that catalytic efficiency of enzymes may be enhanced due to the long lifetime of the ES complex relative to the lifetime of a simple bimolecular collisional interaction of chemical reactants. Lifetimes for collisional interactions are thought to be on the order of 10^{-13} second, whereas fast kinetic techniques put the mean lifetimes of ES complexes at 10^{-7} to 10^{-4} second. This means a large probability increase that any ES interaction will generate an activated complex or transition state that will go on to products. Rueben called this "substrate anchoring."

Jencks (1975*a*, p. 268) has argued that these propinquity effects represent the "entropic contribution to enzymic catalysis"—that the probability of reaction, a measure of entropy, between two reactants has been increased enormously when they are bound specifically at the enzyme active site in the proper orientation for reaction. He has compiled a list of the terminology used in the recent enzyme literature in discussing this aspect of enzymatic catalysis. Some of these terms dot the preceding few pages of this book:

entropy loss;
approximation;
orientation;
propinquity;
rotamer distribution (Bruice, 1970);
anchimeric assistance (Winstein et al., 1953);
proximity;
orbital steering (Storm and Koshland, 1972);
stereopopulation control (Milstein and Cohen, 1970);
distance distribution function (DeLisi and Crothers, 1973);
togetherness (Nowak and Mildvan, 1972);
FARCE (freezing at reactive centers of enzymes) (Nowak and Mildvan, 1972).

Experimental evidence for restricted motion of substrates bound at enzyme active sites has been gathered by nuclear magnetic resonance (NMR) relaxation techniques and fluorescence polarization spectroscopy (Haugland and Stryer, 1967), both of which show that the fast rotation or tumbling rates of small molecules in solution are suppressed or slowed to the tumbling rates of the protein macromolecule. The large size of enzymes can thus be used to achieve precise positioning of amino-acid side chains around substrates at the active site to freeze out both translational and rotational motion. The subsequent chemical step from an immobilized substrate in an ES complex will then involve very slight entropy loss, because most of the entropy was lost (and the loss paid for by use of some of the intrinsic binding energy of E to S) in the formation of the ES complex. For a bimolecular reaction of substrates S_1 and S_2 bound immobile and optimally oriented at the active site compared to the bimolecular reaction in solution, it has been estimated that the reaction in solution is less favorable by 35 entropy units (Jencks, 1975*a*, p. 278). At 25° C (298° K) this value is 10.5 kcal/mole (43.9 kJ/mole), which corresponds to a rate decrease for reaction in solution (or a corresponding rate increase for the enzymatic reaction) of 10^8 M.

2.D.2 Covalent Catalysis: Nucleophilic versus Electrophilic Catalysis

The side chains of amino acids found in proteins present a number of nucleophilic (electron-rich) functional groups for catalysis;

$RCOO^{\ominus}$, RNH_2, Ar—OH, (imidazole) N NH, R—OH, $RS^{\ominus}$

These side chains can attack electrophilic (electron-deficient) portions of substrates to form a *covalent* bond between enzyme and substrate as a reaction intermediate. This category of reaction is prevalent in enzyme-catalyzed group transfers discussed in the seven chapters of Section II of this text, where the electrophilic portions of susceptible substrates may be acyl groups, phosphoryl groups, or glycosyl groups.

$$RC(=O)\text{—}Y + \text{Enz—}\ddot{X} \rightleftharpoons R\text{—}C(O^{\ominus})(Y)\text{—}X\text{—Enz} \rightleftharpoons R\text{—}C(=O)\text{—}X\text{—Enz} + Y^{\ominus}$$

acyl enzyme

$$R\text{—}OP(=O)(O^{\ominus})\text{—}OR' + \text{Enz—}\ddot{X} \rightleftharpoons R\text{—}O\text{—}P(O^{\ominus})(O^{\ominus})(OR')\text{—}X\text{—Enz} \rightleftharpoons R\text{—}O\text{—}P(=O)(O^{\ominus})\text{—XEnz} + R'O^{\ominus}$$

phosphoryl enzyme

$$\text{Glucosyl—}Y + \ddot{X}\text{—Enz} \rightleftharpoons \text{Glucosyl—}X\text{—Enz} + Y^{\ominus}$$

glucosyl enzyme

Attack by the enzyme nucleophile (Enz—$\ddot{X}$) can produce acylation, phosphorylation, or glycosylation of the enzyme nucleophile as covalent intermediate. These covalent intermediates can then be attacked in a second step by some low-molecular-weight nucleophile to yield the observed reaction product. When that nucleophile is water, the overall reaction is hydrolysis.

When one examines the ability of the side chains of the amino acids in proteins to behave as electrophiles (i.e., to themselves be attacked by nucleophiles), with the exception of the proton (and cations in metalloenzymes), there are no obvious candidates for nucleophilic groups on a substrate to attack and form a covalent bond. However, this does not mean that covalent electrophilic catalysis is not observed with enzymes. In fact, a number of coenzymes (obligate cofactors for certain enzymes) form covalent adducts with substrates; these adducts generate new electrophilic groups capable of functioning as electron sinks during catalysis. Much of enzyme chemistry is carbanion chemistry, and the adduct-forming coenzymes function to provide routes to low-energy, stabilized, substrate-derived carbanions, thereby providing rate acceleration. This is especially true for pyridoxal phosphate and thiamine pyrophosphate coenzymes.

About 100 enzymes have now been shown to form covalent intermediates during catalysis (Spector, 1973; Bell and Koshland, 1971), suggesting there may be some general advantage to holding on to a portion of the substrate. Jencks has argued that the immobilization achieved in the covalent intermediate should provide significant entropic driving force to account for most of the observed catalysis. Hanson and Rose (1975) have pointed out that covalent catalysis is an efficient process for accommodating multiple steps involving bond breakage and formation in a single active site.

2.D.3 General Acid–Base Catalysis

By definition, general acid–base catalysis is catalysis where a proton is transferred in the transition state. Thus, one may reasonably expect this catalysis in any reaction where proton transfers occur—i.e., whenever proton-transferring agents such as acids or bases are found.

One can make the distinction between specific acid or base catalysis and general acid or base catalysis on an experimental basis (Fig. 2-4), and we note that the latter is very important in enzymatic catalysis because, at pH 7, the concentration of free $H^{\oplus}$ or $OH^{\ominus}$ is a low 10^{-7} M.

For instance, consider specific hydroxide-ion catalysis of a reaction in H_2O. The rate law will have two terms: the uncatalyzed rate and the hydroxide-catalyzed term:

$$k_{obs} = k_0 + k_{OH^{\ominus}}[OH^{\ominus}]$$

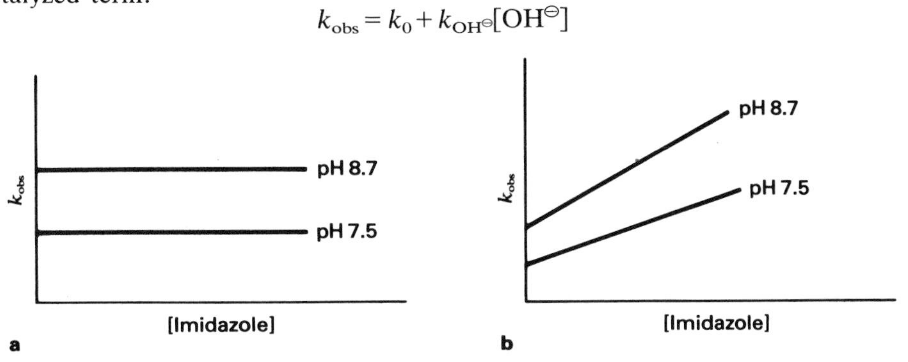

Figure 2-4
(**a**) Plot for specific base catalysts: rate unaffected by increased concentrations of imidazole.
(**b**) Plot for general base catalysis by imidazole: rate increases with increase in imidazole concentration.

More explicitly, consider the case of the hydrolysis of acetyl imidazole

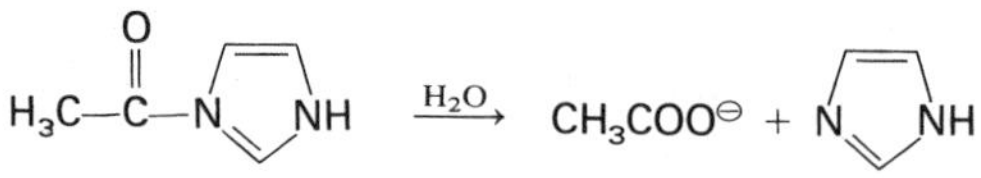

$$H_3C{-}\overset{O}{\overset{\|}{C}}{-}N\langle\text{imidazole}\rangle NH \xrightarrow{H_2O} CH_3COO^{\ominus} + N\langle\text{imidazole}\rangle NH$$

and inquire whether free imidazole can be functioning as a base catalyst in addition to $OH^{\ominus}$. At any given pH, if general base catalysis is occurring, the rate of hydrolysis will increase with increasing concentrations of imidazole.

A possible mechanism for general base catalysis in this instance shows why the process might increase rates.

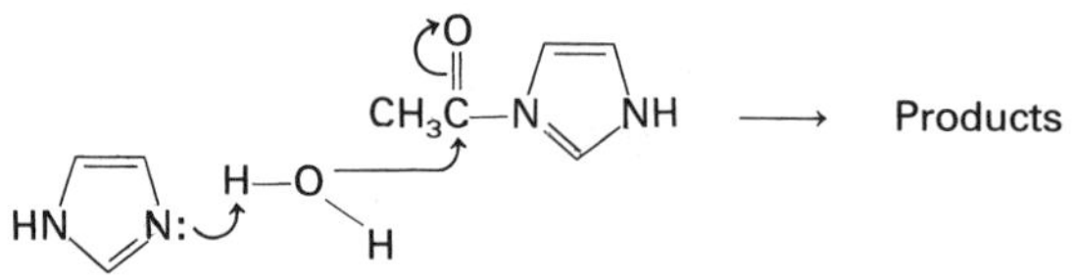

HN N: H—O H CH$_3$C(=O)—N NH ⟶ Products

Catalysis by imidazole as shown implies that the energy barrier for hydrolysis is lowered by proton transfer—that is, proton transfer confers a relative stabilization to the transition state. Why might this transition-state stabilization occur?

For general base catalysis as shown, at any given pH, the nucleophilicity of the water molecule has been greatly enhanced without generation of a high concentration of $OH^{\ominus}$. More generally, hydrogen bonding in the transition state may avoid the formation of unstable high-energy species, thus stabilizing the catalytic transition state. We shall note the importance of this mechanism in chymotrypsin catalysis. It has been estimated that general acid or general base catalysis may provide rate increases in the range of 10- to 100-fold (Jencks, 1975*a*, p. 221).

While on the subject of imidazole and base catalysis, it seems germane to make two comments.

1. The preceding equation implicity notes that imidazole functions both as a general base and as a nucleophile. The nucleophilicity is evident in the existence of the acetyl imidazole itself: it must have arisen by nucleophilic attack of the imidazole nitrogen on the carbonyl carbon of some acetyl derivative.

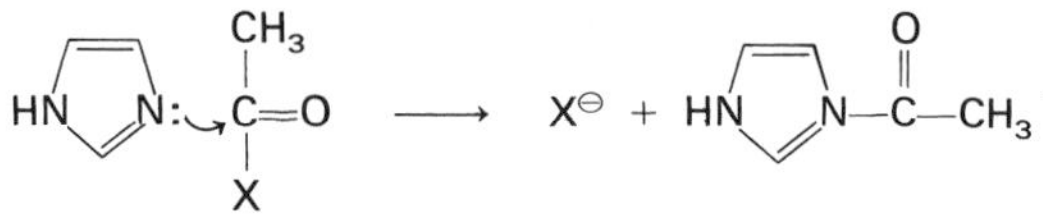

$$HN\langle\text{imidazole}\rangle N: + \underset{X}{\overset{CH_3}{C}}{=}O \longrightarrow X^{\ominus} + HN\langle\text{imidazole}\rangle N{-}\overset{O}{\overset{\|}{C}}{-}CH_3$$

Thus, whenever considering the involvement a priori of a group like imidazole, or for that matter any of the other basic groups in enzyme amino-acid side chains, there is the ambiguous capacity for general base catalysis versus nucleophilic (and therefore covalent) catalysis. However, experimental distinctions can be made, and a good summary of some criteria has been compiled by Bender (1971, p. 101 ff).

2. Because the pK_a of the imidazolium ion (as free imidazole or in histidine) is ca. 6–7 (i.e., 50% protonated at this pH), we have

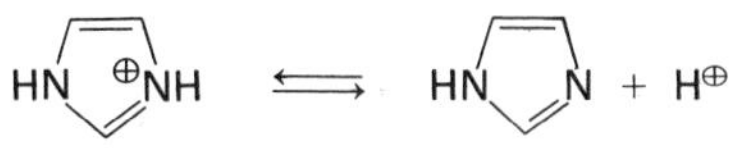

$$\text{HN}\overset{\oplus}{\text{NH}} \rightleftharpoons \text{HN}\;\;\text{N} + \text{H}^{\oplus}$$

This suggests that imidazole may be the most effective base existing at neutral pH in enzymes (pK_a nearest neutrality of any amino-acid side chain). Further, if one correlates basicity with nucleophilicity (a simplification in general permissible in reactions at carbonyl groups), then imidazole is likely to be the most effective nucleophile (i.e., a good base, present at high concentrations). A base stronger than imidazole would clearly be more reactive in the free base form, but at neutral pH will have only a much smaller fraction unprotonated and so, at an equal concentration of total amine species, will react less rapidly than imidazole. In this same connection, the serine alkoxide ion

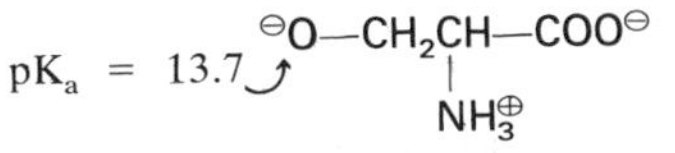

$$pK_a = 13.7 \quad {}^{\ominus}O{-}CH_2CH(NH_3^{\oplus}){-}COO^{\ominus}$$

(as a stronger base) would react much more rapidly than imidazole with unactivated acyl groups. One of the tricks the enzyme chymotrypsin (described in the next chapter) uses in its active site is to generate and stabilize (at macroscopic neutral pH) high concentrations of alkoxide ion or its equivalent as a nucleophile—concentrations that would be vanishingly small in the aqueous bulk phase at pH 7.

In fact, in all the enzymatic acyl transfers catalyzed by proteases so far examined whose active sites have both serine and histidine, the imidazole moiety of histidine functions as a general base, not as a nucleophile, while enough of the stronger alkoxide ion species is generated in the active site milieu to attack the unactivated acyl groups of substrates at fast rates.

Finally, we note that the thiolate anion of cysteine, $pK_a \sim 8$, will exist in appreciable concentrations in enzymes at physiological pH ranges. Due to electronic and polarizability features, this sulfur anion shows 10- to 100-fold higher nucleophilic reactivity than normal oxygen or nitrogen bases of comparable basicity (equal pK_a values).

2.D.4 Strain, Distortion, and Conformational Change

It is clear that strain in starting materials and release of that strain in the transition state to products can provide rate accelerations in chemical reactions. An impressive case is that studied by Westheimer and his colleagues, who examined phosphate-ester hydrolysis rates in the two compounds shown (Covitz and Westheimer, 1963).

HO–CH$_2$CH$_2$–O–P(O)(O$^{\ominus}$)–O$^{\ominus}$ $\xleftarrow{H_2O}$ I (cyclic ethylene phosphate, P with two ring O and two O$^{\ominus}$)

II (dimethyl phosphate: CH$_3$O, CH$_3$O, P=O, O$^{\ominus}$) $\xrightarrow{H_2O}$ CH$_3$OH + CH$_3$O–P(=O)(O$^{\ominus}$)–O$^{\ominus}$

$k_I/k_{II} = 10^8$

The cyclic example (I) relieves considerable ring strain with ring opening on hydrolysis, hydrolyzing at a relative rate 10^8-fold faster than the acyclic example (II). One can imagine that strain and distortion effects are likewise important in enzymatic catalysts where one can consider an effect on the enzyme, such as a change in conformation of the three-dimensional structure of the protein that may convert a low-activity form of the enzyme to a high-activity form. One can also imagine induction of various kinds of strain or destabilization in the substrate as well upon binding to the enzyme, and this point will be discussed first.

As noted above, on binding of a substrate to an enzyme, a portion of this intrinsic interaction energy may be used to accelerate catalysis (Jencks, 1975*a*). Acceleration of catalysis means reduction of the free energy of activation, $\Delta G^{\ddagger}$, and destabilization of the ES complex would have this effect. Destabilization can include geometric distortion of bond angles in the bound substrate or steric compression. It could also involve electrostatic repulsion between a charged group on the substrate and an amino-acid side chain of similar charge at the active site. It may also involve desolvation of a charged substrate molecule in a hydrophobic enzyme active site. All these mechanisms could contribute to a thermodynamic destabilization of bound substrate that lessens the energy barrier to the transition state, *provided* these destabilization forces are released in the transition state—i.e., it is only the ES complex selectively (and not the transition state) that is destabilized.

It has been argued that, should a substrate bind without significant utilization of binding energy for distortion (or induction of a protein conformational change), then the observed binding might be extremely tight. Tight binding of a substrate to an enzyme is not necessarily useful. Suppose a substrate were 50% bound to an enzyme when the substrate was present at 10^{-8} M concentration, but the physiological concentration of the substrate in an organism is 10^{-4} M—then this enzymatic reaction is relatively inefficient. Half of the binding energy could be used instead for destabilization to increase the rate of the catalytic reaction, and the resulting 10^{-4} M binding constant would be in the physiological range. Most enzymes that have been isolated and studied mechanistically do in fact show observed binding affinities at concentrations of substrate in the physiological range, perhaps as a result of evolutionary selection for the most efficient catalysts (Fersht, 1974). Tight binding of substrate would have an additional unfavorable consequence: the resulting tight binding of product would mean that product release could be so slow as to limit the overall rate of catalysis. Such rate-determining product release is observed for several dehydrogenases utilizing nicotinamide coenzymes (Chapter 10). Further, if catalytic efficiency is equated with the ratio k_{cat}/K_m (with a maximal value of 10^9 M^{-1} sec^{-1}), then a smaller value of K_m is correlated with a slower turnover number.

In carbonic anhydrase, which catalyzes an increased rate of equilibration between CO_2 and $HCO_3^{\ominus}$, infrared data indicate no detectable deformation of CO_2 on binding to the enzyme (Riepe and Wang, 1968).

$$CO_2 + H_2O \underset{}{\overset{enz}{\rightleftharpoons}} H_2CO_3 \rightleftharpoons HCO_3^{\ominus} + H^{\oplus}$$

On the other hand, for the pancreatic exopeptidase carboxypeptidase A, X-ray data suggest a twist in the susceptible bond of a bound pseudosubstrate that may mean deformation of that amide linkage out of planarity, a consequent loss of resonance energy, and (by such destabilization) an enhanced susceptibility to enzymatic attack (Blackburn, 1976, p. 169 ff). Experimental evidence in support of distortion in the ES complex has come through the use of enzyme inhibitors that have the common property of tight binding to a particular enzyme. These compounds have been termed *transition-state analogues* (Pauling, 1948; Wolfenden, 1972; Lienhard, 1973; Jencks, 1975*a*, pp. 362–375). To the extent that bound substrate is destabilized in the enzyme–substrate complex, and given that the strain or distortion or charge destabilization is released in the transition state, then the enzyme ought to bind the transition state more tightly than substrates or products. By definition, the transition state for a chemical reaction has such a fleeting existence (i.e., the time for a single vibration is 10^{-13} sec) that one obviously cannot synthesize or isolate it and measure its affinity (the concentration at which 50% is bound) for the enzyme. However, there are stable compounds that are structural analogues for the expected transition state of certain enzymatic reactions. These are the transition-state analogues. The closer the analogue is in actuality to the transition-state structure, the less the intrinsic interaction energy of enzyme with analogue that will be used for destabilization. Then a greater fraction (probably not all in any example) of the intrinsic binding energy should be manifested as observable tight binding. In fact, it has been observed that some transition-state analogues show 50% binding to enzyme at concentrations 10^{-3} that for substrates.

Consider the example of proline racemase, an enzyme purified to homogeneity from the anaerobic bacterium *Clostridium sticklandii.* It interconverts the D- and L-isomers of the amino acid proline.

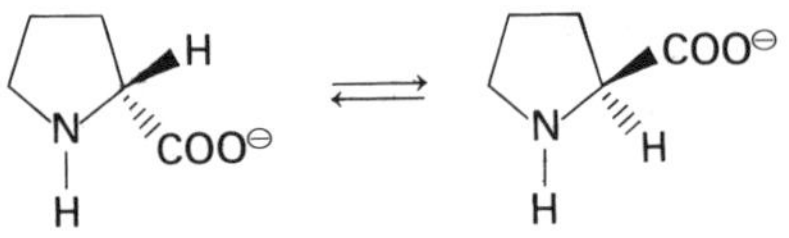

One can imagine that, during this racemization (involving inversion of configuration at the α-carbon), the α-carbon assumes planar geometry in the transition state and that the sp^3 hybridization may partially rehybridize to sp^2 (in fact, good evidence exists for a carbanion mechanism—see Cardinale and Abeles, 1968; Rudnick and Abeles, 1975). A planar analogue of proline might then resemble the transition-state structure of the proline-racemase reaction. The compound pyrrole-2-carboxylic acid is, indeed, an extremely potent inhibitor for proline racemase, binding but not undergoing reaction.

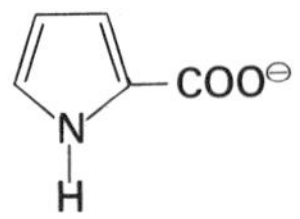

It produces 50% inhibition at a concentration 160-fold lower than the concentration of D-proline or L-proline that produces half-maximal binding (see the same references).

Induced fit vs. nonproductive binding. While on the subject of strain and distortion in enzymatic catalysis, it is appropriate to include the terms *induced fit* and *nonproductive binding* because they are invoked in the literature. Both concepts have been employed to explain how binding of "good" substrates leads to rate accelerations whereas the binding of "poor" substrates or actual inhibitors does not.

In the induced-fit model, it is the enzyme active site that is thought to be flexible; the binding of an appropriate substrate will involve mutual contact of properly aligned substrate groups with correctly oriented enzyme functionalities. This well-developed interaction can induce a conformational change in the enzyme from a form with low intrinsic catalytic activity to one with high catalytic activity. A poor substrate does not align quite properly and induce the correct conformational response from the active site. In this view, the catalytic active site is not actually generated until some isomerization step after the initial E–S interaction:

$$\mathrm{E} + \mathrm{S}_1 \rightleftarrows \mathrm{E}\cdot\mathrm{S}_1 \rightleftarrows \mathrm{E}'\cdot\mathrm{S}_1 \longrightarrow \text{catalysis}$$
$$\mathrm{E} + \mathrm{S}_2 \rightleftarrows \mathrm{E}\cdot\mathrm{S}_2 \not\longrightarrow \text{(little or no catalysis)}$$

Consider the following observation. Hexokinase is an enzyme catalyzing a phosphoryl transfer from ATP (adenosine triphosphate) to C-6 of glucose with a relative maximal rate of 5×10^6.

$$ATP^{4\ominus} + \text{glucose} \longrightarrow ADP^{3\ominus} + \text{glucose-6-}CH_2OPO_3^{2\ominus} \qquad k = 5 \times 10^6$$

This enzyme will also catalyze the transfer of this terminal phosphoryl group of ATP to water at a much slower rate (Colowick, 1973).

$$ATP + HOH \longrightarrow ADP + HO{-}P(=O)(O^{\ominus})_2 \qquad k = 1$$

The basicity and nucleophilicity of water and of the C-6 hydroxyl of glucose are sufficiently similar so that no marked rate difference would be expected on that basis alone.

The argument has been made that binding of glucose induces a conformational change that actually establishes the correct active-site geometry. Water does not induce the enzyme conformational change necessary for efficient catalysis; it is as though, in the reaction with water, the active conformer of the enzyme is at (5×10^{-6})-fold lower concentration than in the reaction with glucose. The good substrate generates an *induced fit* on binding by triggering a movement of the three-dimensional structure of the enzyme, paid for by using some fraction of the binding energy. If the induced fit generates increased contact with both the reacting and nonreacting portions of the substrate, additional interaction energy can be provided for subsequent ES destabilization. This is a positive mechanism for enzyme specificity (Jencks, 1975*a*).

From a different viewpoint, one can explain the above data by assuming that it is not the enzyme but the "poor" substrate that is essentially in a catalytically inactive form when bound. Glucose has the correct number of binding groups in the proper spatial orientation to bind in one catalytically productive mode at the active site. Water (the poor substrate) may, contrarily, bind in many modes, all but one nonproductive. That is, HOH can bind where C-1 through C-5 hydroxyls of glucose bind and not be phosphorylated. Only when at the C-6 locus will it act as phosphoryl acceptor (and it may actually be disfavored from binding in this locus). In the productive-binding view then, one could imagine that binding of a good substrate by the enzyme stabilizes the one proper conformer (rotamer, etc.) that undergoes reaction, thus raising its effective concentration by an entropic contribution. As a poor substrate, water may bind to hexokinase 5×10^6 times nonproductively for each time it binds productively and can act as nucleophile towards the phosphoryl group of ATP.

Section II

ENZYME–CATALYZED GROUP TRANSFERS

In the seven chapters of this section, we discuss classes of enzymes characterized by the common feature that they catalyze reactions in which some electrophilic group of a substrate molecule is transferred to an acceptor that is plentiful in the cellular environment. Mechanistically, these acceptor molecules function as nucleophiles in the catalyses and so turn out to be molecules containing oxygen, nitrogen, or sulfur—three common electronegative elements in the biological environment.

The nearly ubiquitous nucleophile in enzymatic catalysis is water, at 55.5 M nominal concentration, although the effective concentration at a particular enzyme site can appear much higher or, alternatively, much lower; access of solvent water molecules to the enzyme active site can be partially or totally restricted. Much of the enzyme chemistry we discuss in this book involves chemical transformations of carbonyl groups in substrates and products. Accordingly, we start by discussing acyl transfers to water.

$$R{-}\overset{O}{\overset{\|}{C}}{-}X + H_2O \xrightarrow{\text{enzyme}} R{-}\overset{O}{\overset{\|}{C}}{-}OH + HX$$

For reasons that are historical and because of the consequent wealth of classical mechanistic information available, we introduce the subject by considering the action of proteases that cleave peptide (amide) bonds of protein substrates, transferring an aminoacyl moiety of a polypeptide to water.

$$\mathrm{RC(H)(NHR')-C(=O)-X + H_2O \longrightarrow RC(H)(NHR')-C(=O)-OH + HX}$$

We analyze the reactions catalyzed by a specific protease (chymotrypsin) and develop some of the characteristic features of enzymatic catalysis with this example. Then we discuss other examples of acyl transfers to water—hydrolytic enzyme reactions, where the acyl-containing substrates are low-molecular-weight metabolites such as acetylcholine and glutamine—and we use these reactions to discuss how various kinetic data can be used as tools for the study of enzyme mechanisms.

Next we discuss enzyme-catalyzed amino transfers. In these reactions ammonia (NH_3), rather than H_2O, can serve as nucleophile and source of the amino group incorporated into the product, but high concentrations of such a strongly basic nucleophile are deleterious to cells; ammonia is too indiscriminately reactive for proper control in the cellular environment. We shall see that the amide group of glutamine serves as donor of nascent ammonia by undergoing catalyzed hydrolysis at the active site of a specific enzyme, delivering an effective local concentration of the good nucleophile without leakage into bulk solution.

$$\mathrm{H_2N-C(=O)-CH_2CH_2-C(H)(NH_3^{\oplus})-COO^{\ominus} \xrightarrow{H_2O} {}^{\ominus}O-C(=O)-CH_2CH_2-C(H)(NH_3^{\oplus})-COO^{\ominus} + \text{“}NH_3\text{”} \xrightarrow{RX} R-NH_2 + HX}$$

Although the action of hydrolytic enzymes is often for degradative purposes, cells use organic substrates at the acyl oxidation state (RCOX) for various biosynthetic purposes (as we discuss in Section V). Such substrates, either as free acids ($RCOO^{\ominus}$) or as amides ($RCONH_2$), have resonance-stabilized, chemically unreactive acyl groups that require activation before they can be used for biosynthesis. The preferred mode of enzymatic activation is to form phosphate esters from good chemical phosphorylating agents. This introduces the centrally important subject of enzyme-catalyzed phosphoryl-transfer reactions. The premier biological phosphorylating agent is adenosine triphosphate (ATP), with its phosphoric-anhydride linkages acting as reactive donors of phosphoryl groups to acyl anions. Once such acyl-phosphate intermediates have formed, these intermediates show enhanced chemical reactivity over starting materials, transferring the acyl groups to nucleophilic amino and thiolate acceptors, with displacement of the phosphate leaving group and production of the acylated nucleophile. These events, summarized in the two following equations, represent key chemical steps in enzymatic biosynthetic pathways from such important acids as acetate, succinate, formate, and citrate.

$$\mathrm{R{-}C(=O){-}O^{\ominus} + R'O{-}P(=O)(O^{\ominus}){-}OR'' \longrightarrow R{-}C(=O){-}O{-}P(=O)(O^{\ominus}){-}OR' + R''OH}$$

$$\mathrm{R{-}C(=O){-}O{-}P(=O)(O^{\ominus}){-}OR' \xrightarrow{RNH_2} R{-}C(=O){-}NHR + {}^{\ominus}O{-}P(=O)(O^{\ominus}){-}OR'}$$

$$\mathrm{R{-}C(=O){-}O{-}P(=O)(O^{\ominus}){-}OR' \xrightarrow{RSH} R{-}C(=O){-}SR + {}^{\ominus}O{-}P(=O)(O^{\ominus}){-}OR}$$

Molecules containing the phosphate functional group abound in biological systems. In addition to the chemical activating function in the conversion of poor leaving groups to good leaving groups, phosphate linkages provide stabilizing structural elements (e.g., nucleic acids), increase water solubility of hydrophobic molecules (e.g., phospholipids), and offer a reactive handle to biological catalysts. Both phosphate monoesters and phosphate diesters are substrates for hydrolytic enzymes (phosphatases and phosphodiesterases, respectively) with degradative functions physiologically.

While on the subject of phosphoryl transfers as one of the common mechanisms for the activation of other substrate types and for providing the thermodynamic driving force for reactions otherwise endergonic, we note that the common phosphoryl donor ATP undergoes enzymatic attack by nucleophiles at each of the three electrophilic phosphorus atoms of the side-chain triphosphate.

$$\mathrm{{}^{\ominus}O{-}\underset{\gamma}{P}(=O)(O^{\ominus}){-}O{-}\underset{\beta}{P}(=O)(O^{\ominus}){-}O{-}\underset{\alpha}{P}(=O)(O^{\ominus}){-}O{-}CH_2{-}ribose(2'\text{-}OH,\ 3'\text{-}OH){-}adenine\ (NH_2)} \qquad \text{ATP}$$

Attack at the γ-phosphorus (phosphoryl transfer) and at the α-phosphorus (nucleotidyl transfer) are frequent events. Attack at the β-phosphorus atom by oxygen nucleophiles is also known and is categorized as enzyme-catalyzed pyrophosphoryl transfer.

Whereas in the above reactions the chemically versatile phosphate group offers its electrophilic phosphorus to an incoming nucleophile, oxygens of the phosphate moiety can also function as nucleophiles. Group transfer to water is hydrolysis; group transfer to a nucleophilic phosphate oxygen is phosphorolysis. Many of the phosphorylases are enzymes involved in sugar and polysaccharide metabolism. Thus, we discuss glycosyl transfers (transfer of sugar residues) to H_2O or to inorganic phosphate as the last main category of enzyme-catalyzed group-transfer processes.

In discussing the mechanisms of enzymatic group-transfer reactions, we note that frequently (but by no means always) *covalent* derivatives between some nucleophilic side chain of an enzyme amino-acid residue and a portion of a substrate molecule will form as intermediates during the catalytic cycle. In the following equations, X—Z—Enz represents the covalent intermediate.

$$\text{X—Y} + \text{Enz—}\ddot{\text{Z}}\text{H} \rightleftharpoons \text{X—Z—Enz} + \ddot{\text{Y}}\text{H}$$

$$\text{X—Z—Enz} + \text{H}_2\text{O} \rightleftharpoons \text{X—OH} + \text{Enz—}\ddot{\text{Z}}\text{H}$$

Chapter 3

Acyl Transfers to Water: Endopeptidases and Exopeptidases

3.A INTRODUCTION: CLASSIFICATION OF PROTEASES

In this chapter we begin our consideration of enzyme-catalyzed acyl-transfer reactions, a reaction type for which a volume of mechanistic information is available on nonenzymatic behavior. Water is the acceptor molecule to which the substrate acyl moiety is transferred, and the enzymes in question are all hydrolases. Historically, hydrolytic enzymes utilizing protein and oligopeptide substrates were among the first biological catalysts purified and characterized, in part due to the stability of the enzymatic activities to the rather harsh methods of protein purification initially in use during the 1930s and 1940s.

The enzymes cleaving the amide linkages in protein substrates are classified as proteases or (interchangeably) peptidases. Most of the known peptidases fall into one of two categories, depending on the positional specificity exhibited in the cleavage of substrate peptide bonds. When the susceptible peptide bond is some internal bond in a polypeptide, the enzyme is said to be an *endopeptidase.* Pancreatic chymotrypsin and trypsin, and the gastric enzyme pepsin (to be discussed later in this chapter) fall in this category.

$$R^1\text{—}\underset{\underset{NH_3^{\oplus}}{|}}{\overset{H}{C}}\text{—}\overset{\overset{O}{\|}}{C}\text{—}\overset{H}{N}\text{—}\underset{\underset{R^2}{|}}{\overset{H}{C}}\text{—}\overset{\overset{O}{\|}}{C}\text{—}\overset{H}{N}\text{—}\underset{\underset{R^3}{|}}{\overset{H}{C}}\text{—}\overset{\overset{O}{\|}}{C}\text{—}\overset{H}{N}\text{—}\underset{\underset{R^4}{|}}{\overset{H}{C}}\text{—}COO^{\ominus}$$

$$\Big\downarrow \; H_2O,\ \text{endopeptidase}$$

$$R^1\text{—}\underset{\underset{NH_3^{\oplus}}{|}}{\overset{H}{C}}\text{—}\overset{\overset{O}{\|}}{C}\text{—}\overset{H}{N}\text{—}\underset{\underset{R^2}{|}}{\overset{H}{C}}\text{—}COO^{\ominus} + H_3\overset{\oplus}{N}\text{—}\underset{\underset{R^3}{|}}{\overset{H}{C}}\text{—}\overset{\overset{O}{\|}}{C}\text{—}\overset{H}{N}\text{—}\underset{\underset{R^4}{|}}{\overset{H}{C}}\text{—}COO^{\ominus}$$

When the susceptible peptide linkage is at the amino terminus or at the carboxyl terminus of a protein substrate, the enzyme is termed an *exopeptidase.* Depending on the susceptible terminus, an exopeptidase may be further classified as an aminopeptidase or a carboxypeptidase.

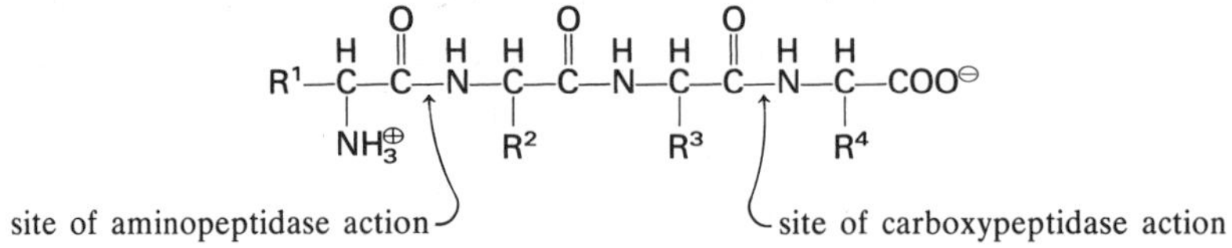

$$R^1-\underset{NH_3^{\oplus}}{\overset{H}{\underset{|}{C}}}-\overset{O}{\overset{\|}{C}}-\overset{H}{N}-\underset{R^2}{\overset{H}{\underset{|}{C}}}-\overset{O}{\overset{\|}{C}}-\overset{H}{N}-\underset{R^3}{\overset{H}{\underset{|}{C}}}-\overset{O}{\overset{\|}{C}}-\overset{H}{N}-\underset{R^4}{\overset{H}{\underset{|}{C}}}-COO^{\ominus}$$

site of aminopeptidase action   site of carboxypeptidase action

Proteases can also be classified on the basis of physiological function. On the one hand, some hydrolytic enzymes function to degrade polypeptides and proteins for digestive and nutritional purposes. Such enzymes function both extracellularly (e.g., in the intestine of animals) and intracellularly (in the hydrolytic subcellular organelles known as lysosomes, prevalent in liver and kidney cells). On the other hand, many proteases act to cause limited proteolysis on protein substrates for control purposes. A single susceptible peptide bond may be cleaved in a protein substrate, with dramatic change in biological activity of the resultant product. Clearly, cleavage of a protein into fragments could switch on a physiological function by converting an inactive precursor into a biologically active protein. Table 3-1 lists some physiological events that are switched on or off by controlled proteolysis, where a distinct and specific protease acts in each case (Neurath, 1975).

Each specific response is initiated by a common chemical step, enzyme-catalyzed hydrolysis of a specific peptide bond in the protein substrate. The hydrolytic reactions catalyzed by proteases are functionally irreversible under physiological conditions. (We examine the chemical activation required for the reverse reaction, peptide-bond formation, in Chapter 8.) Therefore, the limited proteolysis reactions described here are operationally irreversible. Once the single

Table 3.1
Control functions of proteases

Physiological function	Examples
Defense mechanisms	Blood coagulation
Hormone production	Proinsulin → insulin
Macromolecular assembly	Fibrinogen → fibrin Procollagen → collagen
Digestion	Zymogen → active protease
Development	Prococoonase → cocoonase Prochitin synthetase → chitin synthetase

polypeptide chain of the prohormone proinsulin is cleaved to the two-chain active form of the hormone insulin, that hormone molecule cannot be restored to its precursor state.

Because our space is limited, we can analyze here only some of the mechanisms of enzyme-catalyzed acyl transfers to water carried out by the proteases. The digestive enzymes have been most intensely investigated, so we confine our attention to them. The various proteases that have been examined can be placed in five well-characterized mechanistic sets (K. Walsh, 1975):

1. proteases with a nucleophilic serine residue at the active site (endopeptidases);
2. proteases with a nucleophilic cysteine residue at the active site (endopeptidases);
3. metal-containing exopeptidases;
4. metal-containing endopeptidases; and
5. proteases with an acid pH optimum.

Table 3-2 gives examples of specific enzymes, their active-site features, and some inhibitors.

Table 3-2
Examples of proteases, subdivided in mechanistic sets

	Identification		Examples	
Set	Functionality	Inhibitor	Protease	Function
Serine protease	"Active" serine	Fluorophosphates	Trypsin	Digestion
			Thrombin	Blood coagulation
			Plasmin	Lysis of blood clots
			Cocoonase	Mechanical
			Subtilisin	Digestion
			Acrosin	Sperm penetration
Metalloexopeptidase	$Zn^{2\oplus}$	*o*-Phenanthroline	Carboxypeptidase	Digestion
Sulfhydryl protease	CySH	Iodoacetate	Papain	Digestion
			Streptococcal proteinase	Digestion
			Cathepsin B	Intracellular digestion
Acid protease	Acidic pH optimum	*o*-Phenanthroline	Thermolysin	Digestion
			Pepsin	Digestion

SOURCE: K. Walsh, in *Proteases and Biological Control*, ed. E. Reich, D. Rifkin, and E. Shaw (Cold Spring Harbor Conferences on Cell Proliferation, 1975), p. 1.

In this chapter we focus on the mechanism of the serine proteases, because many of the techniques and mechanistic experiments now of general usage in enzymology were developed and used initially on these endopeptidases. We shall make repeated use of these mechanistic criteria in subsequent chapters.

3.B CHYMOTRYPSIN

In the endopeptidase discussions, we shall focus on the enzyme chymotrypsin from mammalian pancreas, but this enzyme (although differing in the specificity of endopeptidase bonds cleaved) is representative mechanistically of the whole class of serine endopeptidases from a variety of diverse biological sources including pancreatic trypsin, subtilisin from the bacterium *Bacillus subtilis*, cocoonase in molting insects, α-lytic protease from bacteria, elastase from smooth muscle, and thrombin involved in blood-clotting processes. Chymotrypsin was isolated by Kunitz in the 1930s (Kunitz and Northrop, 1933, 1935). It has been studied intensively since, and more information is available on its mode of catalysis than for almost any other enzyme.

3.B.1 Zymogen Activation

Chymotrypsin is synthesized in an inactive proenzyme form called a zymogen (in this case, chymotrypsinogen) by cells of the mammalian pancreas and is secreted in that form to the intestine.* Once in the intestine, the single polypeptide chain of chymotrypsinogen undergoes the limited proteolysis indicated in Figure 3-1, catalyzed by a molecule of the endopeptidase trypsin. Thus the chymotrypsin becomes activated only in the extracellular environment of the intestine, where it will act to degrade ingested proteins and polypeptides. As an inactive precursor, while intracellular, chymotrypsinogen is not deleterious to the secreting pancreatic cells.

Trypsin cleaves proteins at internal peptide bonds with specificity such that arginyl or lysyl residues become the carboxyl termini in the peptide fragments (Blackburn, 1976, pp. 105–141). Given an exhaustive incubation of a protein with trypsin in a test tube, cleavage would eventually occur at all the lysyl and arginyl residues. However, under physiological conditions, some Lys—X or Arg—X bonds are cleaved much faster than others. This kinetic preference reflects

* For an excellent recent review of chymotrypsinogen and chymotrypsin, see the book by Blackburn (1976), pp. 11–96.

physical accessibility of the susceptible peptide bond to the attacking protease. There must be complementarity between the topology of the protein-substrate region containing the susceptible bond and that of the attacking trypsin molecule's active site. For an essentially globular protein, loops of polypeptide chain at the exterior surface will be attacked much faster than will peptide bonds in the interior of the protein.

In the proteolytic activation of the serine proteases, the primary cleavage site is a critical peptide bond near the amino terminus of the zymogen—leading to a two-chain product that undergoes a conformational change leading to active enzyme. The inactive chymotrypsinogen is a single polypeptide chain of 245 amino acids, with about 25,000 mol wt. The first product from trypsin cleavage is π-chymotrypsin where the Arg^{15}—Ile^{16} bond has been hydrolyzed, and the two-chain product is catalytically active. This form of the enzyme is not stable in the intestine and suffers two additional cleavages (actually performed by other molecules of chymotrypsin) to yield δ-chymotrypsin and then α-chymotrypsin, a three-chain enzyme where the polypeptide chains are held together by interchain disulfide bonds (Fig. 3-1).

Figure 3-1 also shows the amino-acid sequence of α-chymotrypsin with the three chains labeled A, B, and C from amino to carboxyl end, respectively (Hartley, 1964; Blow et al., 1969; Brown and Hartley, 1966; Meloun et al., 1966). The three-dimensional structure of α-chymotrypsin crystals has been solved to a resolution of 2 Å (Blow, 1969; Matthews et al., 1967; Birktoft and Blow, 1972), and Figure 3-2 shows the position of the polypeptide backbone with parts of three amino-acid residues emphasized: the β-carboxylate of the aspartyl residue at position 102, the imidazole ring of $histidine^{57}$, and the β-hydroxyl of $serine^{195}$ are the three functional groups that are the keys to the chymotrypsin catalytic mechanism. In fact, the catalytic triad of Asp, His, and Ser residues is a characteristic feature of the active site of other serine proteases that have been examined (Stroud et al., 1975). As we proceed through this chapter, we examine chemical evidence for the ways these residues participate in chymotrypsin catalysis, and we suggest a mechanism for peptide-bond hydrolysis.

A detailed X-ray map of chymotrypsinogen also has been accomplished for the purpose of determining the structural bases for difference in catalytic activity between proenzyme and enzyme (Freer et al., 1970). Surprisingly, the active-site geometry is present in chymotrypsinogen, with almost no change in relative positioning of the catalytic triad on zymogen activation. On the other hand, the hydrophobic binding-pocket geometry present in the active enzyme for binding hydrophobic residues (see specificity data in ¶3.B.2) is not present in the zymogen. It is the inability to bind substrate well that renders chrymotrypsinogen inactive. Because the active-site geometry is intact in chymotrypsinogen, one might expect it to possess some weak intrinsic catalytic activity. Indeed, recent measurements show that the zymogen can react with synthetic substrates at 10^{-6} the rate of chymotrypsin (Gertler et al., 1974). For other serine endopeptidases, because the active-site catalytic triad is preserved, it is the geometry of the binding region of each enzyme that determines substrate specificity (Neurath, 1975).

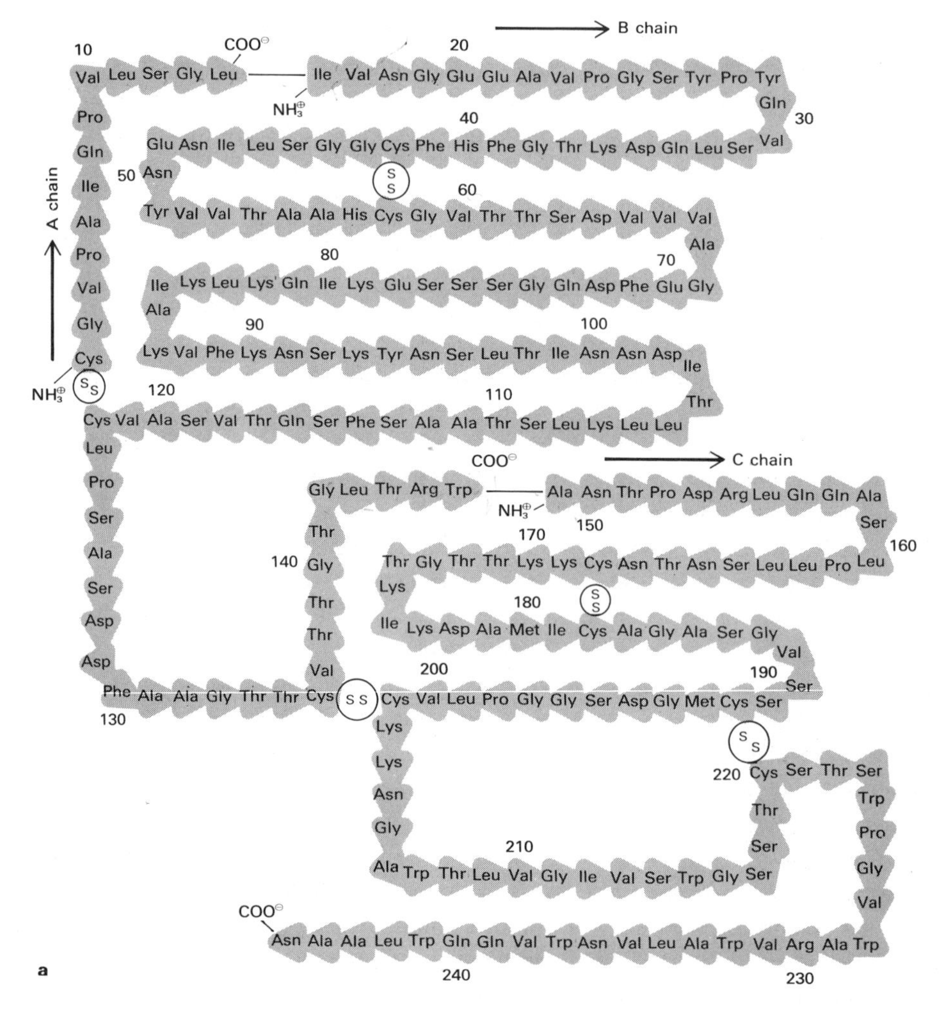

Figure 3-1
(**a**) The amino-acid sequence of α-chymotrypsin. The primary sequences of α- and γ-chymotrypsin are identical, but these forms may exist in different conformations both in solution and when crystalline. (Blow, 1973) (**b**) Scheme for the activation of chymotrypsin. (Blackburn, 1976, p. 14)

Slow activation, chymotrypsinogen/trypsin = 10^4

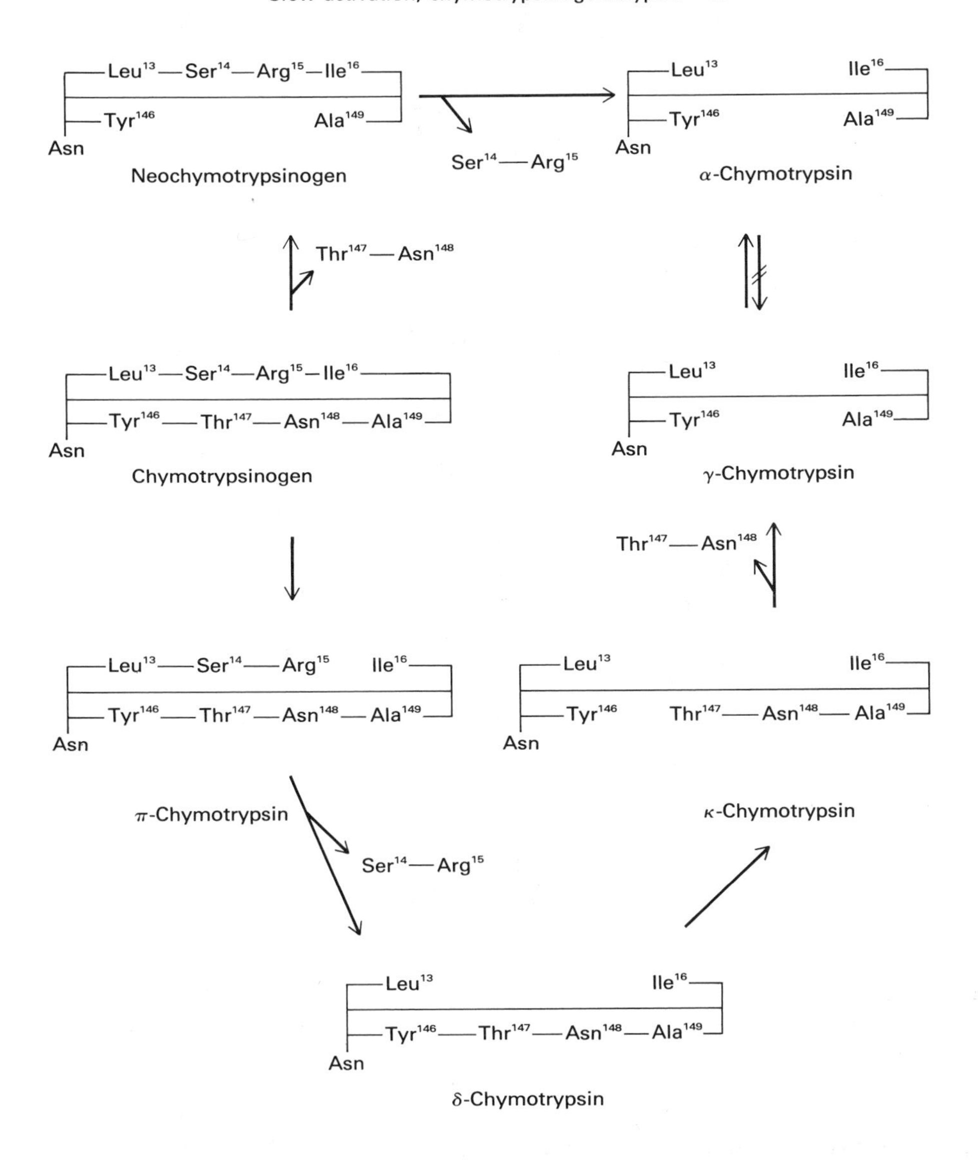

b Rapid activation, chymotrypsinogen/trypsin = 30

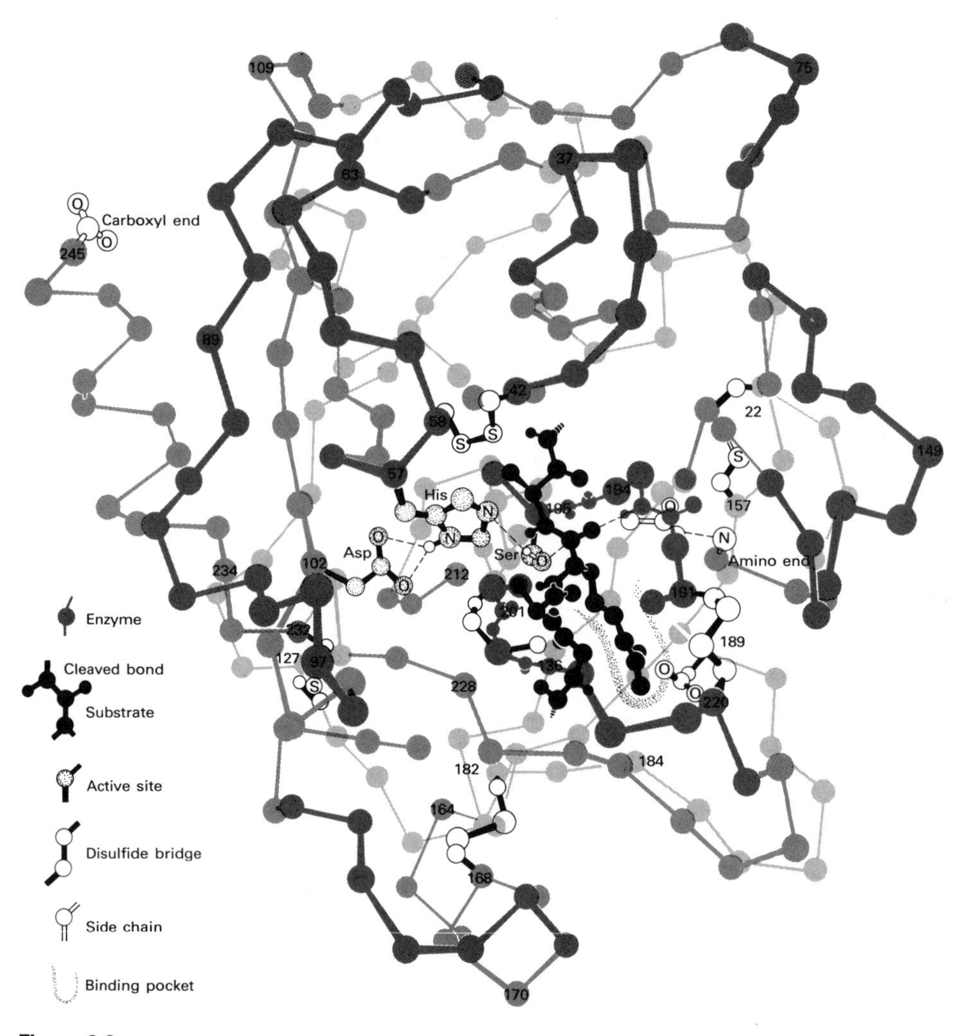

Figure 3-2
Three-dimensional structure of chymotrypsin with Ser^{195}, His^{57}, and Asp^{102} (the catalytic triad of the charge-relay system) clearly indicated. (Courtesy of R. Stroud)

3.B.2 Substrate Specificity of Chymotrypsin

Ingested proteins and polypeptides found in the intestine are the natural, physiological substrates for chymotrypsin, and there it acts as an endopeptidase with broad but clearly definable specificity. This enzyme cleaves at aromatic residues and—less readily, but detectably—at hydrophobic aliphatic residues. It liberates the aromatic residue as a new carboxyl terminus. Thus, it will cleave at Phe, Tyr, and Trp—and, on longer incubation, at Ile, Leu, and Val.

$$\underset{NH_3^{\oplus}}{\text{Ser}}\text{—Ala—Arg—Phe} \wr \text{Glu—Asp—}\underset{COO^{\ominus}}{\text{Gly}} \xrightarrow{\text{chymotrypsin}} \underset{NH_3^{\oplus}}{\text{Ser}}\text{—Ala—Arg—}\underset{COO^{\ominus}}{\text{Phe}} + \underset{NH_3^{\oplus}}{\text{Glu}}\text{—Asp—}\underset{COO^{\ominus}}{\text{Gly}}$$

As noted earlier, the X-ray structure of the enzyme indicates a complementary hydrophobic binding pocket in the active site, with interaction there determining binding recognition and providing most of the binding energy. Not all proteins are equally susceptible substrates for chymotrypsin; again, this is a function of accessibility of the susceptible bonds in the substrate protein to fit into the active-site cleft of the protease.

After the initial enzymatic clips, the product fragments may unfold and lose native structure, thus exposing additional susceptible bonds. If this process continues, limit peptides may eventually result.

For a study of the mechanism of action of chymotrypsin (or any protease) in detail, it is useful to understand what structural features are required in a substrate. Synthetic substrates—small molecules with well-characterized structure and reactivity—are most readily used for such studies. We recognize, of course, that it may not be possible to extrapolate mechanistic details from small synthetic substrates to large polypeptides, but such extrapolation seems relatively safe with chymotrypsin.

It has been established that chymotrypsin also displays esterase activity in addition to its peptidase (amidase) activity. In fact, with model substrates, chymotrypsin is a more efficient esterase than amidase.

Table 3-3 summarizes chymotrypsin's specificity pattern with synthetic substrates. Implicit in the generalized synthetic-substrate structure are two facts:

1. only L-isomers are substrates; and
2. a substrate must have a blocked α-amino group.

In subsequent discussion, we focus only on aminoacyl esters and aminoacyl amides as substrates.

Table 3-3
Specificity pattern for chymotrypsin with synthetic substrates

Generalized substrate structure	X	Substrate type
Y–C_6H_4–CH_2–C(H)(NH–C(=O)–R)–C(=O)–X	NHR, NH_2	Amide
	$NHNH_2$	Hydrazide
	NHOH	Hydroxamate
	OR	Ester
	OCOR	Anhydride
	SR	Thiolester
	Cl	Acid chloride
	CH_2COOEt	β-Acetoacetic ester

3.B.3 Dependence of Reaction Velocity on Substrate Concentration

One model substrate that has been examined with chymotrypsin is the amide *N*-acetyl-L-tryptophanamide; enzymatic hydrolysis produces ammonia (as ammonium ion at physiological pH) and the amino acid *N*-acetyl-L-tryptophan.

$$H_2O + \text{Indolyl–CH}_2\text{–CH(NHCOCH}_3\text{)–C(=O)–NH}_2 \xrightarrow{\text{enz}} \text{Indolyl–CH}_2\text{–CH(NHCOCH}_3\text{)–C(=O)–O}^{\ominus} + NH_4^{\oplus}$$

The reaction velocity for hydrolysis of the amide bond can be measured as a function of varying substrate concentration. The rate of product formation (either product) is measured at fixed enzyme concentration and at a variety of substrate concentrations. At low *N*-acetyltryptophanamide concentrations, a linear dependence is observed, but at high concentrations of substrate a finite limiting velocity is obtained (Fig. 3-3). This limiting rate is the maximal velocity, V_{max}, and is

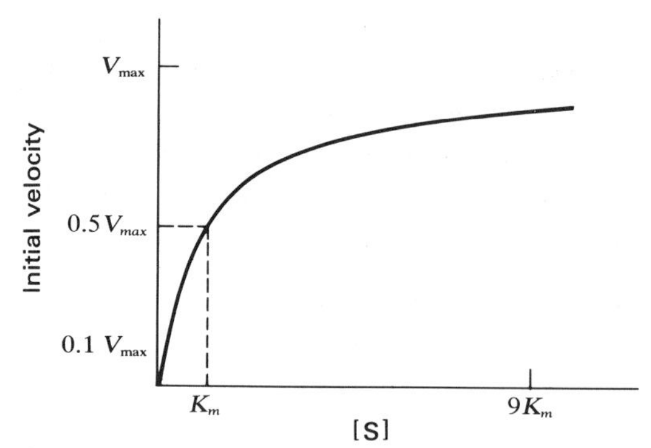

Figure 3-3
Plot of v versus [S] for an enzymatic reaction.

independent of substrate concentration at high [S]. The experimental curve appears to be a rectangular hyperbola. V_{max} is one of the useful experimental kinetic parameters that can be obtained for reaction of an enzyme with a substrate. A second parameter is the concentration of substrate, [S], required to cause half-maximal velocity, $V_{max}/2$. This value of [S] is defined as K_m, the Michaelis constant (after Lenor Michaelis, a pioneering enzyme kineticist). As such, the K_m value is an experimentally defined constant whose mechanistic significance depends on the particular catalytic mechanism employed by the enzyme and on the rates of various elementary chemical and physical steps.

Before looking at the kinetic models consistent with the experimental data, we can comment on the units for V_{max} in an enzymatic reaction. Two expressions are common: specific activity and turnover number.

1. *Specific activity.* In the most common approach, V_{max} is expressed as the specific activity of an enzyme with a given substrate. Catalytic activity is described on a weight basis as μmoles substrate reacted per minute per milligram of enzyme protein at a fixed temperature (usually 30° C):

$$1\ \mu\text{mole min}^{-1}\ (\text{mg enz})^{-1} \equiv 1 \text{ unit of activity.}$$

2. *Turnover number.* When V_{max} is expressed as turnover number, the catalytic activity is related to the number of active sites of enzyme present. This expression can be used *when the enzyme preparation is pure and the molecular weight is known.* The turnover number (T.O.) is expressed in μmoles substrate reacted per minute per μmole of enzyme.

Chymotrypsin hydrolyzes *N*-acetyl-L-trytophanamide with a specific activity of 0.0015 μmole min^{-1} (mg enzyme)$^{-1}$. This is a rather low specific activity, and we shall see that this compound is a very slow substrate for chymotrypsin. (The corresponding *N*-acetyltryptophanyl esters have relative values of V_{max} around 10^3 higher.) The turnover number is readily calculated from the specific activity and the molecular weight of the enzyme. For a molecular weight of 24,000 for chymotrypsin, 1 mg of enzyme protein is about 42 nanomoles (42×10^{-9} mole) of enzyme.

$$\begin{aligned} 0.0015\ \mu\text{mole min}^{-1}\ (\text{mg enz})^{-1} &= 1.5 \text{ nmoles min}^{-1}\ (\text{mg enz})^{-1} \\ &= 1.5 \text{ nmoles min}^{-1}\ (42 \text{ nmoles enz})^{-1} \\ &= 0.036 \text{ mole min}^{-1}\ (\text{mole enz})^{-1} = \text{T.O.} \end{aligned}$$

Again, remember that this is a low turnover number. Several enzymes are known to have turnover numbers greater than 10^5 moles min^{-1} (mole enz)$^{-1}$.

One should note here that the attainment of a maximal velocity is, of course, not unique to chymotrypsin-mediated catalysis but is an absolutely general and characteristic property of enzymatic catalysis. The phenomenon is also known as **saturation kinetics** or **Michaelis–Menten kinetics** (after the two people who proposed the physical model that the kinetic pattern describes—namely, that mentioned earlier, with formation of a noncovalent ES complex as the sole entity from which catalysis proceeds).

3.B.4 Saturation Kinetics: V_{max}, K_m, and the Velocity-Dependence Equation

The central feature of the Michaelis–Menten kinetic model for enzymatic catalysis takes into account the idea that an enzyme interacts with a substrate reversibly to form the noncovalent ES complex, which can then either undergo chemical reaction to products or undergo physical dissociation.

$$\mathrm{E + S} \underset{k_{-1}}{\overset{k_1}{\rightleftharpoons}} \mathrm{ES} \underset{k_{-2}}{\overset{k_2}{\rightleftharpoons}} \mathrm{EP} \underset{k_{-3}}{\overset{k_3}{\rightleftharpoons}} \mathrm{E + P} \quad (1)$$

Under *initial velocity conditions* (the first few percent of the forward reaction), the concentration of P is insignificant, the back reaction can be ignored, and the reaction simplifies (Michaelis and Menten, 1913; see also Segel, 1975*a*, chap. 4).

$$\mathrm{E + S} \underset{k_{-1}}{\overset{k_1}{\rightleftharpoons}} \mathrm{ES} \xrightarrow{k_{cat}} \mathrm{E + P} \quad (2)$$

Originally, Michaelis and Menten made a simplifying assumption that $k_{-1} \gg k_{cat}$ so that the reversible first step is a preequilibrium situation. In fact, for many enzymes k_{cat} may approximate or even be larger than k_{-1}, and the equilibrium assumption does not hold. This difficulty was circumvented by an alternative formulation of Briggs and Haldane (1925) that is independent of the relative rates of k_{-1} and k_{cat} (see also Segel, 1975*a*, 1975*b*). Their simplifying assumption is a steady-state assumption—namely, when enzyme is present in catalytic amounts ([S] $\gg$ total enzyme present) then, at any given instant during the progress of the reaction, the rates of formation and of breakdown of ES are essentially equal. A practically constant steady-state level of [ES] accumulates.

The observed velocity of product formation at any given instant is from conversion of the ES complex:

$$v = k_{cat}[\mathrm{ES}] \quad (3)$$

The total enzyme concentration is the sum of free enzyme and ES:

$$[\mathrm{E_T}] = [\mathrm{E}] + [\mathrm{ES}] \quad (4)$$

The steady-state assumption means $d[\mathrm{ES}]/dt \approx 0$, so the rate of formation of ES can be set equal to its breakdown. According to equation 2,

1. ES forms in the forward reaction $\mathrm{E}+\mathrm{S} \xrightarrow{k_1} \mathrm{ES}$, with

$$\text{rate of ES formation} = k_1[\mathrm{E}][\mathrm{S}] \tag{5}$$

2. ES breaks down by conversion to products or by reversible dissociation, with

$$\text{rate of ES breakdown} = k_{cat}[\mathrm{ES}] + k_{-1}[\mathrm{ES}] \tag{6}$$

For $d[ES]/dt = 0$,

$$k_1[\mathrm{E}][\mathrm{S}] = k_{cat}[\mathrm{ES}] + k_{-1}[\mathrm{ES}] \tag{7}$$

Therefore,

$$[\mathrm{ES}] = k_1[\mathrm{E}][\mathrm{S}]/(k_{cat} + k_{-1}) \tag{8}$$

Although the total amount of enzyme present is known, its partitioning between free E and ES is not. Therefore, we substitute $[\mathrm{E}] = [\mathrm{E_T}] - [\mathrm{ES}]$ (from eq. 4) in equation 8 to obtain

$$[\mathrm{ES}] = k_1[\mathrm{S}][\mathrm{E_T}]/(k_{cat} + k_{-1} + k_1[\mathrm{S}]) \tag{9}$$

Substituting equation 9 into equation 3, we obtain

$$v = k_{cat}[\mathrm{E_T}][\mathrm{S}] \Big/ \left(\frac{k_{-1} + k_{cat}}{k_1} + [\mathrm{S}]\right) \tag{10}$$

This is one form of the rectangular hyperbola, accounting for observed saturation kinetics.

Two comments can be made. First we noted that, at high substrate concentrations, a maximal velocity V_{max} is reached. This V_{max} is attained when all the enzyme $\mathrm{E_T}$ is in the ES form; then,*

$$V_{max} = k_{cat}[\mathrm{E_T}] \tag{11}$$

Substituting equation 11 into equation 10, we obtain

$$v = V_{max}[\mathrm{S}] \Big/ \left(\frac{k_{-1} + k_{cat}}{k_1} + [\mathrm{S}]\right) \tag{12}$$

The second comment is that definition of the ratio of the three rate constants, $(k_{-1} + k_{cat})/k_1 = K_m$, allows recasting of equation 12 in the most familiar form of

* In terms of the two expressions for enzyme activity noted in ¶3.B.3, we see that specific activity $= V_{max} = k_{cat}[\mathrm{E_T}]$, whereas turnover number $= k_{cat} = V_{max}/[\mathrm{E_T}]$.

the velocity-dependence equation:

$$\boxed{v = V_{max}[S]/(K_m + [S])} \tag{13}$$

This equation describes the *initial velocity*, v, relative to V_{max} at a given substrate concentration. In practice, this means 5% to 10% of substrate utilization so that the concentration of S does not change, and product concentration does not become large enough for the reaction in the back direction to become significant.

Although both V_{max} and K_m are experimentally determinable parameters, V_{max} has a clear theoretical and mechanistic significance whereas K_m, as a ratio of rate constants, can have variable mechanistic significance. (For a more complex reaction—with more than one ES complex—K_m can be even more complex.) In the limit (and only at that limit!) where $k_{-1} \gg k_{cat}$ (i.e., where the ES complex is in preequilibrium with E and S), then $K_m = k_{-1}/k_1 = K_S$, the dissociation constant for the ES complex.

Careful examination of the v vs. [S] curve of Figure 3-3 (or that for any other enzyme following Michaelis–Menten kinetics) reveals three distinct regions.

1. At low [S] values, the velocity dependence is directly related to substrate concentration; this is a region of first-order kinetics.
2. At intermediate [S] values, there is extensive curvature.
3. At high [S] values, of course, V_{max} obtains, and the observed velocity is independent of further additions of substrate; this is a region of zero-order kinetics.

In the first case, where $[S] \ll K_m$, equation 13 becomes $v \approx V_{max}[S]/K_m = k[S]$, accounting for the first-order kinetics.

In the third case, where $[S] \gg K_m$, equation 13 becomes $v \approx V_{max}[S]/[S] = V_{max}$, accounting for the observed zero-order kinetics.

A final comment about enzyme kinetic data at this point: because the v vs. [S] plot is a rectangular hyperbola, the V_{max} value is approached asymptotically and is hard to determine accurately. Further, because K_m concentration of substrate is given by $V_{max}/2$, there is uncertainty in this parameter as well. This problem is avoided by plotting kinetic data in one of a variety of linear forms, whose relative merits are discussed at length in texts on enzyme kinetics (Segel, 1975*a*, 1975*b*; Dixon and Webb, 1964; Gutfreund, 1972; Plowman, 1972; Westley, 1969). The most familiar linear form (Fig. 3-4) is the double reciprocal plot of $1/v$ vs. $1/[S]$.

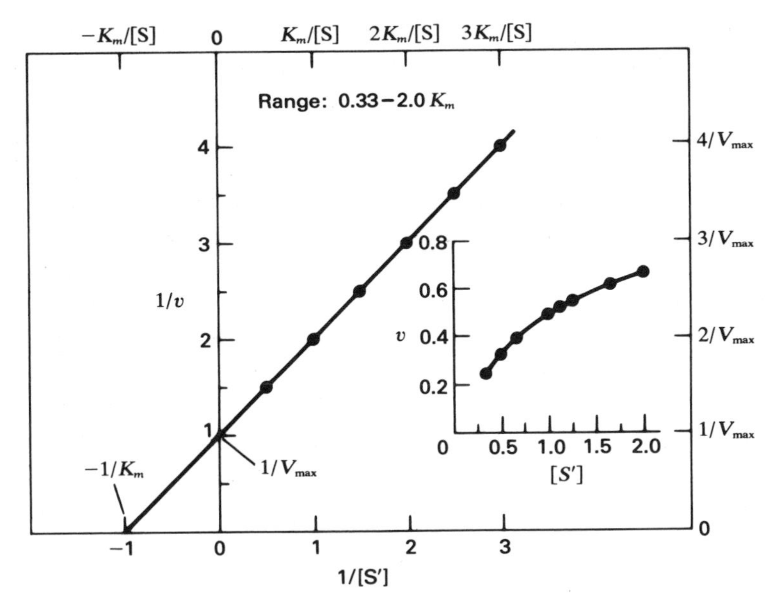

Figure 3-4
Double reciprocal plot (lineweaver–Burk plot), showing $1/v$ versus $1/[S]$. (From I. H. Segel, *Enzyme Kinetics*. Copyright © 1975. Reprinted by permission of John Wiley & Sons, Inc.)

The Michaelis–Menten equation (eq. 13) can be inverted and divided by $1/V_{max}$ to give $1/v = (K_m + [S])/V_{max}[S]$, and separation of terms produces

$$1/v = (1/[S])(K_m/V_{max}) + 1/V_{max} \tag{14}$$

Then, in a plot of $1/v$ vs. $1/[S]$, the slope is K_m/V_{max}, and the vertical intercept is $1/V_{max}$ (see Fig. 3-4). At the horizontal axis, where $1/v = 0$, it is seen that $1/[S] = -(1/K_m)$. For useful hints on how to perform experiments to determine K_m and V_{max}, see the book by Segel (1975*a*).

3.B.5 The Initial-Burst Experiment with *para*-Nitrophenyl Acetate

One of the most telling experiments—of mechanistic significance not only for chymotrypsin catalysis, but also for the general existence of covalent intermediates in enzyme catalysis—was performed by two British scientists (Hartley and Kilby, 1954). These workers tested the reactive ester *p*-nitrophenyl acetate (PNPA) as a substrate for chymotrypsin.

$$H_2O + CH_3{-}\overset{\displaystyle O}{\overset{\|}{C}}{-}O{-}C_6H_4{-}NO_2 \xrightarrow[\text{enzyme}]{} CH_3COO^{\ominus} + {}^{\ominus}O{-}C_6H_4{-}NO_2$$

colorless yellow

At first glance, you might wonder why they attempted this experiment, because PNPA bears no structural resemblance to the aminoacyl-ester substrates known to be specific chymotrypsin substrates. One reason was that the alcohol product produced in hydrolysis, *p*-nitrophenol, is a very acidic phenol with $pK_a = 7$, a low pK_a due to resonance stabilization of the anion via the ring and the nitro group. This stable, delocalized anion is yellow, with an extinction coefficient $\varepsilon = 14{,}000\ L^{-1}\ M^{-1}$ at 405 nm (such that 10 nmoles in 1 ml of solution give an absorbance of 0.140 at 405 nm). This property provides a sensitive assay for monitoring enzyme-catalyzed hydrolysis, because the starting ester is colorless.

As one might have predicted, PNPA is a poor substrate and requires large amounts of enzyme to produce detectable reaction, but that very property revealed a central catalytic capacity of the enzyme. When product formation was monitored with time, a slow rate was detected; curiously, however, extrapolation to zero time gave a nonzero absorbance value. Figure 3-5 shows the dependence of rate of *p*-nitrophenolate production on enzyme concentration and the extrapolated rate at zero time.

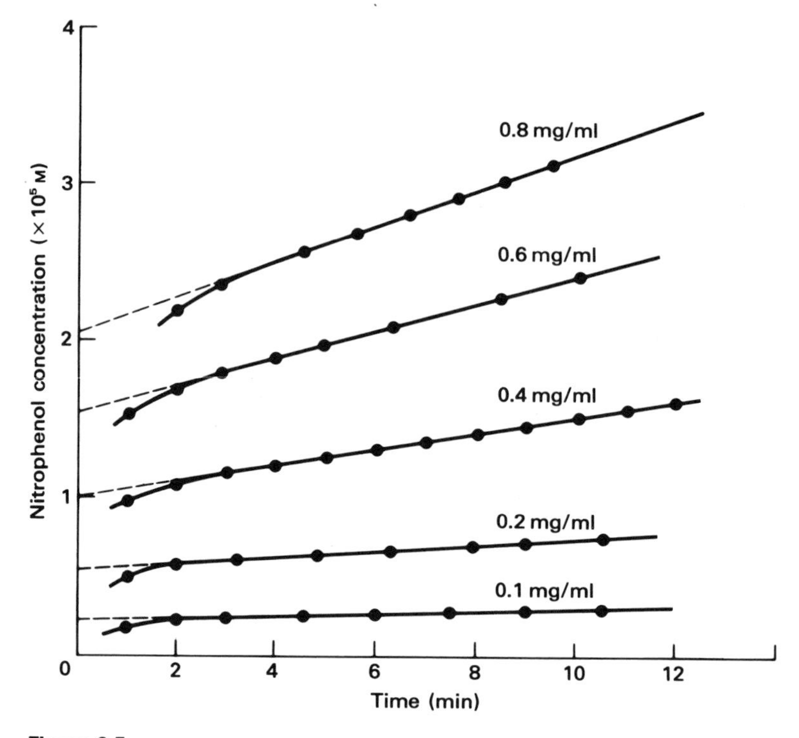

Figure 3-5
Initial-burst experiment of Hartley and Kilby (1954). The liberation of *p*-nitrophenol is shown as a function of time and enzyme concentration (given in mg/ml) in the chymotrypsin-catalyzed hydrolysis of *p*-nitrophenyl acetate.

How can this behavior be interpreted? It is not an experimental artifact. Further study of the experimental data shows that the slope varies with enzyme concentration as expected, but so does the extrapolated nonzero absorbance. In fact, careful examination of the original data produced a clear relationship: when the nonzero absorbance at zero time was converted to nanomoles of nitrophenolate anion, the amount of anion proved to be almost stoichiometric with the amount of enzyme used. At 0.8 mg enzyme per ml, the amount of *p*-nitrophenolate anion apparently liberated at zero time is about 2×10^{-5} M. In a 1 ml volume this is 2×10^{-8} mole, or 20 nanomoles. Given the molecular weight of ~24,000 for chymotrypsin, 0.8 mg enzyme protein corresponds to 33 nmoles of enzyme, *if* the preparation is pure and *fully* active. In fact, chymotrypsin molecules slowly proteolyze one another, so most preparations of crystalline enzyme contain 20% to 30% inactive protein. If the preparation used by Hartley and Kilby in 1953 were 70% active enzyme, the 0.8 mg protein would correspond to 23 nmoles of active enzyme.

Obviously, the nitrophenolate product is not actually being produced instantaneously; presumably it is produced at a fast but finite rate before the second slow phase of product production (observed macroscopically) begins (Fig. 3-6). That is, the kinetics are biphasic: a fast liberation of one mole of *p*-nitrophenolate per mole of enzyme active site is followed by a slow (~10^2-fold slower) steady-state rate of product release. Hartley and Kilby described this behavior as an **initial burst** of *p*-nitrophenolate production. Generally, when this phenomenon is observed in any enzymatic reaction, it is called *initial-burst kinetics* (Gutfreund, 1972).*

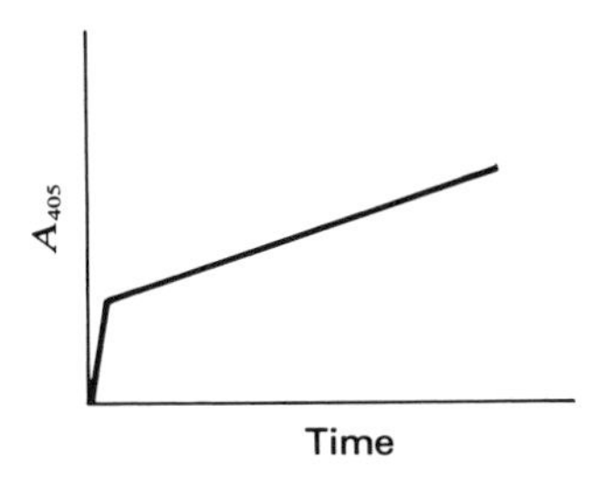

Figure 3-6
Biphasic kinetics in the initial-burst experiment.

*The initial-burst phase is also the pre-steady-state phase, and the usual velocity equation for steady-state kinetics (noted earlier in this chapter) does not hold during this time. However, although one cannot get the two macroscopic constants V_{max} and K_m, one can (from the rate of approach to steady state) obtain rate constants for the elementary steps in the reaction pathway. One can also detect intermediates that form transiently during catalysis. Because *p*-nitrophenolate is such a slow substrate, the approach to the steady state takes several seconds. For physiological substrates with turnover numbers in the region of 10 to 100 turnovers per second, the pre-steady-state interval is in the millisecond range, and rapid kinetic techniques must be used in its examination. Discussions on the measurement and the magnitude of individual rate constants in enzymatic reactions are readily found in the texts by Gutfreund (1972) and by Fersht (1977, chaps. 4, 7).

Note that the experiments of Hartley and Kilby measured only *p*-nitrophenolate anion formation; they provided no evidence about acetate ion, its rate of release, or whether it shows biphasic kinetics. The detection of biphasic kinetics with *p*-nitrophenyl acetate and chymotrypsin cannot be accommodated by the simple assumption of a single ES complex during catalysis. A second kinetically significant complex or intermediate must exist:

$$\mathrm{E + S} \underset{k_{-1}}{\overset{k_1}{\rightleftharpoons}} \mathrm{ES} \xrightarrow{k_2} \underset{\displaystyle \mathrm{P_1}}{\underset{+}{\mathrm{ES'}}} \xrightarrow{k_3} \mathrm{E + P_2} \tag{15}$$

In this scheme, ES′ represents a reaction intermediate that forms from ES *after some chemical change* has occurred and one of the products (P_1) has been released. Subsequently, ES′ breaks down to free enzyme and the second reaction product, P_2. Hartley and Kilby postulated that P_1 is the *para*-nitrophenolate anion and P_2 the acetate anion product, and further that k_2 is fast relative to k_3. Because k_1, the bimolecular rate constant for formation of the Michaelis complex (ES), is generally very fast (e.g., $\sim 10^7\ \mathrm{M}^{-1}\ \mathrm{sec}^{-1}$), these assumptions imply that k_3 is the slowest or *rate-determining* step in *p*-nitrophenyl-acetate hydrolysis catalyzed by chymotrypsin. That is, release of acetate is slow and must control V_{max}. The necessary corollary of k_3 being the rate-determining step is that ES′ must be the predominant enzyme form during steady-state catalysis. It must accumulate, and $[\mathrm{ES'}] \gg [\mathrm{ES}]$.

Hartley and Kilby predicted that ES′ would prove to be a covalent acetyl-enzyme intermediate, and later experiments fully confirmed this prediction. The acetyl group derives from the substrate, and the X group represents some enzyme nucleophile that must have attacked the substrate and displaced the *p*-nitrophenolate (a good leaving group) as first product.

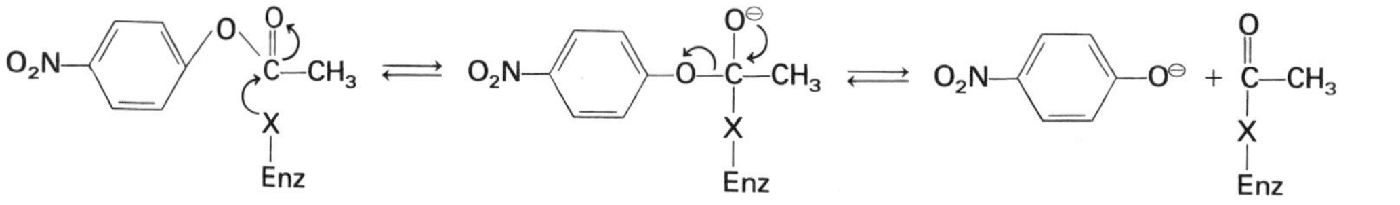

tetrahedral adduct

P_1

acyl-enzyme intermediate

The initial burst of yellow *p*-nitrophenolate anion must represent a rapid acetylation of all the enzyme in solution, providing a stoichiometric amount of chromophoric product. The subsequent slow steady-state rate of *p*-nitrophenolate production reflects the rate-determining hydrolysis of the acetyl enzyme to free enzyme and acetate. The free enzyme molecules are then rapidly acetylated for another catalytic cycle. The V_{max} rate in steady-state turnover is only as fast as the slowest step, acyl-enzyme hydrolysis.

$$\text{Enzyme} + \text{PNPA} \underset{\text{fast}}{\overset{\text{fast}}{\rightleftharpoons}} \text{PNPA} \cdot \text{Enz} \xrightarrow{\text{fast}} \underset{p\text{-Nitrophenolate}}{\text{Acetyl enzyme} + } \xrightarrow{\text{slow}} \text{Acetate} + \text{Enzyme}$$

Note: In all enzymatic reactions involving chemistry where nucleophiles add to carbonyl groups (aldehydes, ketones, acids, esters, amides, etc.), chemical precedent mandates that reaction occurs by intermediate formation of a tetrahedral adduct arising from initial attack of the nucleophile on the carbonyl carbon. The adduct can then partition back to starting material by expulsion of the nucleophile *or*, if there is another potential leaving group, can react by expulsion of the alternate leaving group. *In all subsequent enzymatic examples, tetrahedral adducts are tacit intermediates in such carbonyl-group reactions.*

A velocity equation can be derived for the chymotrypsin acyl-enzyme (ES′) case (eq. 15), either by steady state or by pre-steady-state methods. The equation can be shown to have the same form as the simpler Michaelis–Menten equation (eq. 13) derived earlier. That is (when $k_{-1} > k_2$), the velocity equation has the form

$$v = \left(\frac{k_2 k_3}{k_2 + k_3}[\mathrm{E_T}][\mathrm{S}]\right) \Big/ \left(K_s \frac{k_3}{k_2 + k_3} + [\mathrm{S}]\right),$$

where $K_s = k_{-1}/k_1$, the simple dissociation constant for $\mathrm{ES} \rightleftharpoons \mathrm{E} + \mathrm{S}$. It follows that

$$k_{cat} = k_2 k_3/(k_2 + k_3) \qquad \text{and} \qquad K_m = K_s k_3/(k_2 + k_3)$$

We shall return to these expressions shortly when we discuss the behavior of chymotrypsin with synthetic ester and amide substrates.

3.B.6 Some Criteria for Covalent Intermediates in Chemical and Enzymatic Catalysis

Before proceeding to chemical identification of the acyl-enzyme intermediate in chymotrypsin catalysis, let's consider what criteria can be applied to detect the existence of a covalent intermediate (especially one that accumulates in the steady state) in enzymatic catalysis. The problem is the same in nonenzymatic catalysis, although the relevant methodologies may differ. The excellent chapter on covalent catalysis in the book by Jencks (1969, chap. 2) is recommended for a more complete and comprehensive discussion of these issues.

1. One can isolate the covalent intermediate from reaction mixtures and characterize it.
2. Then the isolated intermediate can be put back into the reaction medium and its *chemical competence* to go to products in either direction can be analyzed.
3. One can see if the isolated intermediate will react to form overall products (either direction) at rates sufficient to account for rates of overall reaction—that is, one can establish its *kinetic competence.*
4. If the specificity of the reaction permits, will the same intermediate be isolable from reactants with different leaving groups?

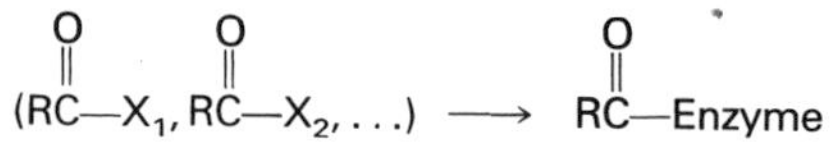

$$(\mathrm{R\overset{O}{\overset{\|}{C}}{-}X_1, R\overset{O}{\overset{\|}{C}}{-}X_2, \ldots}) \longrightarrow \mathrm{R\overset{O}{\overset{\|}{C}}{-}Enzyme}$$

In addition, there are two indirect criteria.

5. One can try to trap a labile intermediate with derivatizing reagents known to scavenge that chemical species under the given reaction conditions. For instance, in the hydrolysis of phthalate half-esters, an anhydride intermediate (arising from intramolecular nucleophilic attack by the adjacent carboxylate) is possible.

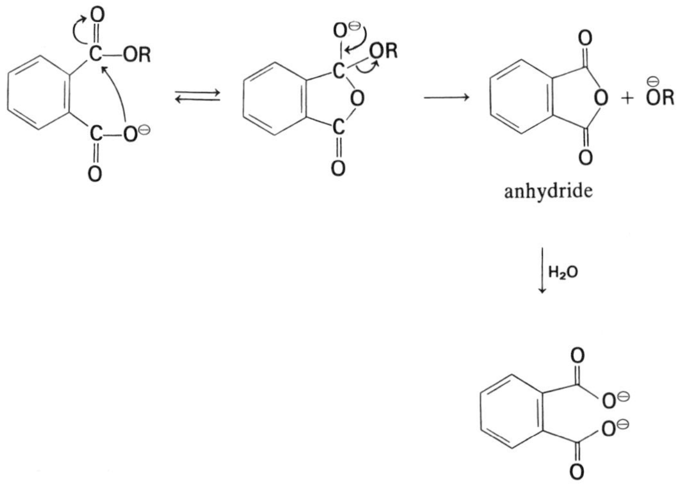

C—OR
C—O⊖
O⊖
OR
O + ⊖OR
anhydride
H_2O
O⊖
O⊖

It is known that aniline reacts rapidly with phthalic anhydride to form the stable monoanilide:

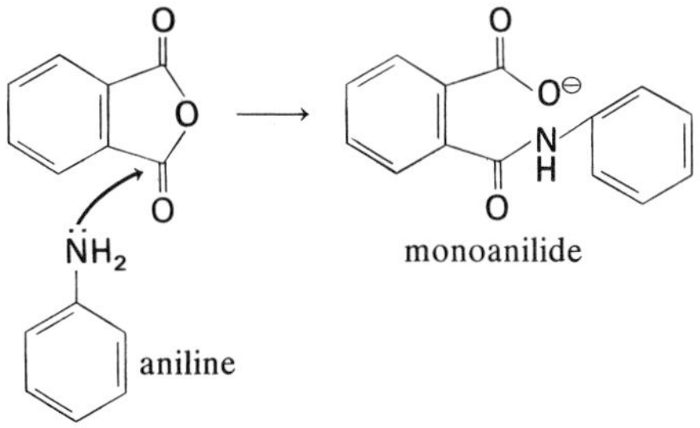

O O O → O $O^{\ominus}$ N H O
$\ddot{N}H_2$ monoanilide
aniline

Then, anilide formation should occur when the phthalate half-ester is hydrolyzed in aqueous aniline, if an anhydride intermediate is involved in the hydrolytic process.

6. Another indirect criterion is to determine whether the enzyme shows a common V_{max} rate for substrates of diverse structure expected to react otherwise at markedly different rates (a kinetic extension of criterion 4 above). However, this criterion simply indicates a common rate-determining step for these substrates, a phenomenon not exclusively explained by formation of a covalent enzyme–substrate intermediate. (For instance, a slow conformational change in the enzyme might control product release: this would also generate a common V_{max}.)

3.B.7 What Criteria Have Been Applied to the Chymotrypsin Acyl-Enzyme Intermediate?

In addition to the initial-burst kinetic experiments of Hartley and Kilby with *p*-nitrophenyl acetate, some of the above criteria have been satisfied for a covalent intermediate in chymotrypsin catalysis.

When *p*-nitrophenyltrimethyl acetate is used as chromophoric substrate, initial-burst kinetics are also observed, but the half-time for hydrolysis of the trimethylacetyl-enzyme intermediate is extremely slow; the half-time at room temperature is around 200 minutes, presumably due to steric hindrance and the inductive effects of the three methyl groups, which slow down the attack of water at the carboryl carbon. This stability has allowed isolation of crystals of the trimethylacetyl-chymotrypsin intermediate (Balls et al., 1958).

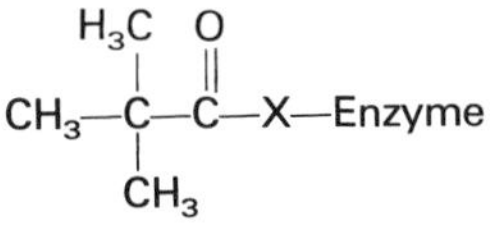

$$CH_3-C(CH_3)_2-C(=O)-X-\text{Enzyme}$$

crystallizable

As isolated, the trimethylacetyl enzyme shows essentially no catalytic activity because the active site is covalently modified by a rather stable acyl group. As the acyl group hydrolyzes slowly, catalytic activity is restored as molecules of enzyme with empty active sites again become available.*

One effective way of regenerating free enzyme from the acyl-enzyme intermediate is to use a stronger nucleophile than water to speed up the rate of deacylation. Hydroxylamine is a small molecule that is powerfully nucleophilic and reacts with a variety of activated acyl groups to form derivatives (acylhydroxamic acids). A 1 M solution of NH_2OH competes for the acyl enzyme very effectively against 55 M water.

$$R{-}\overset{O}{\overset{\|}{C}}{-}X{-}Enz + \ddot{N}H_2OH \rightleftharpoons R{-}\overset{O^{\ominus}}{\overset{|}{\underset{NHOH}{C}}}{-}X{-}Enz \rightleftharpoons R{-}\overset{O}{\overset{\|}{\underset{NHOH}{C}}} + {}^{\ominus}X{-}Enz \rightleftharpoons \underset{\text{acyl hydroxamic acid}}{R{-}\overset{O}{\overset{\|}{C}}{-}NHOH} + \underset{\text{free enzyme}}{HX{-}Enz}$$

Acyl hydroxamates chelate to $Fe^{3\oplus}$ with formation of red-purple complexes, so that their formation is easily detected and quantified.

3.B.8 A Radioactive Method for Detection of Covalent Intermediates

A current method for detection of relatively labile covalent enzyme–substrate intermediates involves use of radioactively labeled substrates. Unstable isotopes of elements decay to more stable forms, often by emission of elementary particles such as α, β, or γ particles. The decay process of these radioactive isotopes can be detected by Geiger counters or by counting photons released when the radioactive material is dissolved in a solution containing material that will phosphoresce as it absorbs energy from collision with an emitted particle (liquid-scintillation counting). Four radioactive isotopes commonly used in enzymology are tritium (^3H), carbon-14 (^{14}C), phosphorus-32 (^{32}P), and sulfur-35 (^{35}S). All are β-emitters, and ejection of a β particle results in formation of a stable isotope of identical atomic weight but one higher atomic number (Segel, 1975*a*, chap. 6). For example,

$$^{3}_{1}H \longrightarrow {}^{3}_{2}He + {}^{0}_{-1}\beta$$
$$^{14}_{6}C \longrightarrow {}^{14}_{7}N + {}^{0}_{-1}\beta$$

---

* It may actually be possible to isolate more reactive and specific acyl-enzyme intermediates by selective stabilization of the intermediate on cooling. For example, a specific acyl-enzyme intermediate produced during action of a related serine protease, elastase, has been crystallized and its X-ray structure determined to 3.5 Å resolution at −55° C (Albert et al., 1976).

Table 3-4
Properties of some radioisotopes commonly used by biochemists

Radioisotope	Half-life ($T_{1/2}$)	Energy of β particle emitted (MeV)
^{3}H	11 years	0.0179
^{14}C	5,200 years	0.154
^{32}P	14 days	1.718
^{35}S	87 days	0.167

The four radioisotopes have different first-order decay rates and emit β particles of different energy (Table 3-4). Because of this energy difference in β particles emitted from ^{3}H and from ^{14}C, for example, it is possible to count ^{3}H and ^{14}C radioactivity independently and simultaneously by monitoring different ends of the energy spectrum, tritium at the low end and carbon-14 at a higher portion.

Finally, before one can perform an experiment, a quantitative assay for radioactive material is necessary. The standard unit of radioactive decay is the curie (Ci), defined as any radioactive substance in which the decay rate is 3.7×10^{10} disintegrations per second (dps), or 2.2×10^{12} disintegrations per minute (dpm).

Because counting efficiency (detection efficiency) experimentally is less than 100%, observed radioactivity is designated in counts per minute (count/min). If 1 mCi (1 millicurie = 2.2×10^{9} dpm) of [^{14}C]-formic acid is counted with 50% efficiency, one will observe 1.1×10^{9} count/min.

In most biochemical experiments with radioactive compounds, only a small fraction (perhaps 1 molecule in 10^{10}) of the compound will actually be radioactive. The label will be present only in trace quantities both for safety reasons and for economic reasons. There need be only enough radioactive molecules in a sample so that accurate counting can be achieved and thus *the radioactivity present can be used as a measure of concentration.* To measure concentration, one must know the *specific activity* of the compound—how much radioactivity is present per unit amount of the substance (Segel, 1975*a*, chap. 6). Specific activity often is given as Ci/mole of compound (μCi/μmole in many enzymatic experiments where μmole quantities of substrates are used). If the specific activity is known, then the count rate can be converted to the amount of compound present in any sample. For the [^{14}C]-HCOOH mentioned above, suppose that the specific radioactivity of a sample is known to be 1 μCi/μmole and that a 1 ml aqueous sample is observed to yield 44,000 count/min. Then the amount of [^{14}C]-HCOOH in the sample is computed as follows:

$$(4.4 \times 10^{4} \text{ count/min})/[2.2 \times 10^{6} \text{ (count/min)}/\mu\text{Ci}] = 2 \times 10^{-2}\ \mu\text{Ci}$$

$$0.02\ \mu\text{Ci}/(1.0\ \mu\text{Ci}/\mu\text{mole}) = 0.02\ \mu\text{mole of } [^{14}\text{C}]\text{-HCOOH present}$$

Two related operating assumptions are normally made in radioactive experiments where the labeled molecules are present only in trace quantities: (1) that the labeled molecules are randomly distributed in the population of unlabeled molecules, and (2) that the labeled molecules show behavior identical to that of the unlabeled molecules. (We shall see in Chapter 4 that the rates at which isotopically labeled compounds undergo chemical and enzymatic reactions *can be slower,* and that these isotope rate effects can provide mechanistic information.)

Hundreds of radioactive organic and inorganic compounds are now available commercially, and others can be synthesized from radioactive precursors by standard methods, although radiochemical syntheses are usually carried out on a microscale to *maximize the specific activity of the product.* High specific activities are often necessary for successful study of enzymatic reactions. Suppose one wishes to examine an enzymatic process where the enzyme availability is limited—for example, such that conversion of 5 μmoles of substrate to products is barely feasible. If a radiochemical synthesis has produced labeled substrate with an activity of 10^{-5} μCi/μmole, the 5 μmoles of labeled product will produce only 55 count/min (at 50% efficiency). In the absence of any sample, the background reading of the liquid scintillation counter may be 30 to 50 count/min, so the signal-to-noise ratio in this case would be unacceptably low.

With these brief comments about radioactive molecules as introduction, we can now consider a possible experiment with chymotrypsin and radioactive *p*-nitrophenyl acetate. In the acetyl-enzyme covalent intermediate, the nitrophenolate part of the substrate will no longer be present. Thus, one might label the two halves of the substrate with different radioisotopes—perhaps ^{14}C at the carbonyl carbon and ^{3}H at one of the aromatic ring positions. Each isotope will be a marker for a different portion of the substrate.

$$CH_3{-}^{14}C(=O){-}O{-}C_6H_3(^3H){-}NO_2 + \text{Enzyme} \longrightarrow CH_3{-}^{14}C(=O){-}X{-}\text{Enz} + {}^{\ominus}O{-}C_6H_3(^3H){-}NO_2$$

On incubation of the labeled substrate with chymotrypsin, if acyl enzyme forms in the steady state, how can it be detected? Separation of macromolecules from small molecules can be achieved by molecular sieve chromatography (gel filtration). Molecular sieve material (Sephadex is a common trade name) is composed of small hollow particles with a meshwork of such dimensions that macromolecules are excluded but small molecules can penetrate. The small molecules therefore can partition into a larger volume and take a "long path" down the column—compared to the macromolecules, which "see" only the "short path" (the excluded volume). As a result, macromolecules emerge from the bottom of the column earlier than the small molecules, which are retarded in their elution.

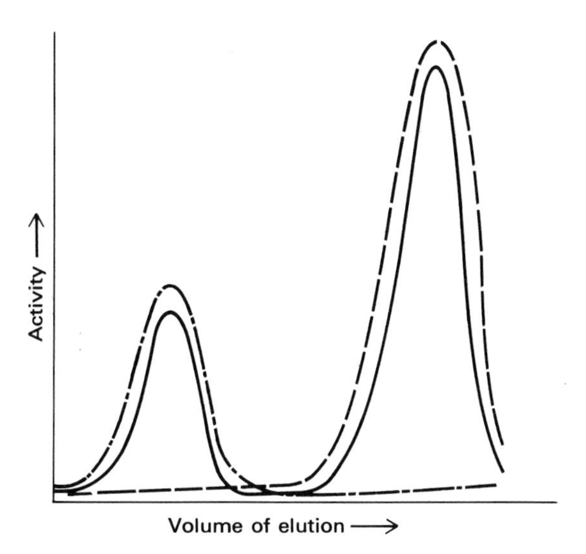

Figure 3-7
Hypothetical gel-filtration column-elution profile for an acetyl-enzyme intermediate on incubation of chymotrypsin with [^{13}C],[^{3}H]-PNPA. The solid line indicates ^{14}C-radioactivity; the dashed line indicates enzyme activity; the "dash/dot" line indicates ^{3}H radioactivity.

If radioactive *p*-nitrophenyl acetate at saturating concentrations ($[S] \gg K_m$) is incubated with chymotrypsin at 30° C for a short interval (~2 minutes—long enough to build up a steady state level of acyl enzyme, but not long enough to hydrolyze all the substrate) and then is placed on a precooled molecular sieve column (2° C) and rapidly chromatographed, the enzyme protein will be collected in fractions eluting before fractions containing small molecules (unreacted [^{14}C], [^{3}H]-PNPA, [^{3}H]-*p*-nitrophenolate, [^{14}C]-acetate). If any acyl enzyme has survived the column procedure, then the enzyme fractions should contain ^{14}C radioactivity. (The reason the column is developed at 2° C is that, at the low temperature, the hydrolysis rate of acetyl enzyme will be slowed even further, thus increasing the chance that some of the intermediate enzyme will survive the column procedure.) A typical column-elution profile might resemble that shown in Figure 3-7.

The fraction of enzyme isolated as the covalent intermediate is calculable. Suppose 2 mg of pure, fully active chymotrypsin (83 nmoles) had been incubated with the [^{14}C],[^{3}H]-PNPA where both isotopes were present at 20 μCi/μmole.

$$(20\ \mu\text{Ci}/\mu\text{mole})(2.2 \times 10^6\ \text{dpm}/\mu\text{Ci}) = 4.4 \times 10^7\ \text{dpm}/\mu\text{mole}$$

$$E_T = 83\ \text{nmoles} = 0.083\ \mu\text{mole}$$

$$(4.4 \times 10^7\ \text{dpm}/\mu\text{mole})(0.083\ \mu\text{mole}) = 3.67 \times 10^6\ \text{dpm}$$

If 3.7×10^6 dpm of [^{14}C] were found in the enzyme-containing fractions, then 100% of the enzyme would have been present as acyl-enzyme intermediate, a stoichiometric isolation. For a labile covalent intermediate subject to hydrolysis during isolation, one might see perhaps only 3.7×10^5 dpm or less. No ^{3}H-radioactivity should be present in the enzyme fractions.* In fact, the first-order rate for breakdown of the acetyl enzyme from PNPA is 0.012 sec^{-1} at room temperature ($T_{1/2} \sim 60$ sec); this nonphysiological acyl enzyme has a half-life (1.6×10^4)-fold longer than the *N*-acetyltyrosyl-enzyme intermediate ($T_{1/2}$ = 3 millisec) formed from specific synthetic substrates (Jencks, 1975*a*).

Finally, before leaving the *p*-nitrophenyl-acetate experiment, we note that (because PNPA is such a poor substrate) very large (substrate) quantities of enzyme had to be used to see a detectable steady-state rate of *p*-nitrophenolate production—amounts of enzyme large enough so that a stoichiometric burst would be detected. With a *good* substrate, a 10^2- to 10^3-fold lower quantity of enzyme would suffice for rate measurements; at zero time the extrapolated nonzero absorbance would be 100-fold less than with the PNPA and probably experimentally indistinguishable from zero, so the key data would be missed in such an experiment.

In general, poor substrates may be quite useful for tricking enzymes into divulging information about component catalytic steps. However, one must then worry whether the poor substrates react by the same mechanisms as do good substrates. In this connection, Bender (1971, p. 500 ff) and his colleagues have shown that the specific substrate *N*-acetyltryptophanyl methyl ester does show initial-burst kinetics when measured at low pH in the millisecond time scale.

3.B.9 Different Rate-Determining Steps for Aminoacyl-Ester and Aminoacyl-Amide Substrates

We have just concluded that hydrolysis of the acyl enzyme (the deacylation step) is rate-determining with the *nonspecific* active ester *p*-nitrophenyl acetate. What about synthetic substrates more structurally related to the natural polypeptide substrates of chymotrypsin, the aromatic aminoacyl esters and amides as *specific* synthetic substrates?

The data of Table 3-5 are available for a series of *N*-acetyl-L-tryptophanyl

---

* If there were equal amounts of ^{3}H and ^{14}C labels with the eluted enzyme, this would suggest that an ES complex had been isolated as the principal enzyme form (i.e., that k_2 is slow). In some enzymatic reactions where ES builds up and $k_{-1}/k_1 = K_D < 10^{-6}$ M, such an ES complex might be isolable by this procedure.

Table 3-5
Kinetic parameters for *N*-acetyltryptophanyl substrates

Generalized structure	X	K_m (M)	k_{cat} (Turnover number) [moles min^{-1} (mole enz)$^{-1}$]
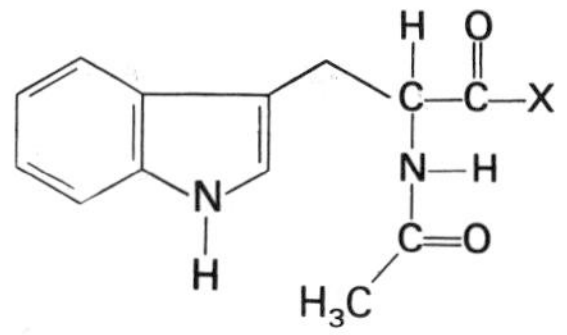 Indolyl–CH_2–C(H)(NH–C(=O)–H_3C)–C(=O)–X	—OEt	9.7×10^{-5}	26.9
	—OMe	9.5×10^{-5}	27.7
	—OPNP	0.2×10^{-5}	30.5
	—NH_2	500×10^{-5}	0.036

derivatives (Bender and Kezdy, 1965). There are three points to be discussed about this table.

1. The three esters are hydrolyzed at a common maximal velocity, despite their possession of leaving groups of differing chemical reactivity. This is a criterion for a common rate-determining step—for instance, the formation and existence of ES′ in Hartley's scheme. We would expect that this intermediate would be the common *N*-acetyl-L-tryptophanyl-enzyme derivative in all three cases:

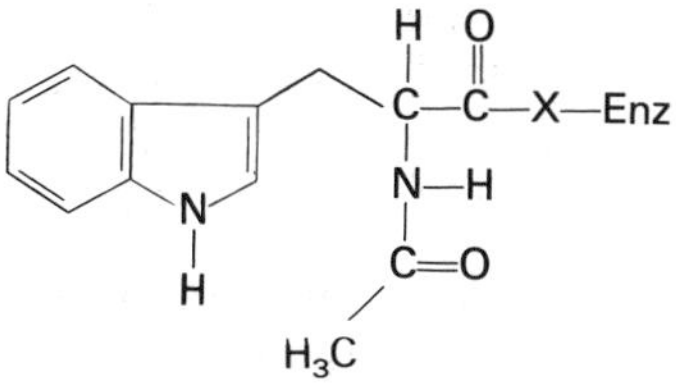

   Indolyl–CH_2–C(H)(NH–C(=O)–H_3C)–C(=O)–X—Enz

   V_{max} obviously is determined by the slowest step in the enzymatic catalysis.

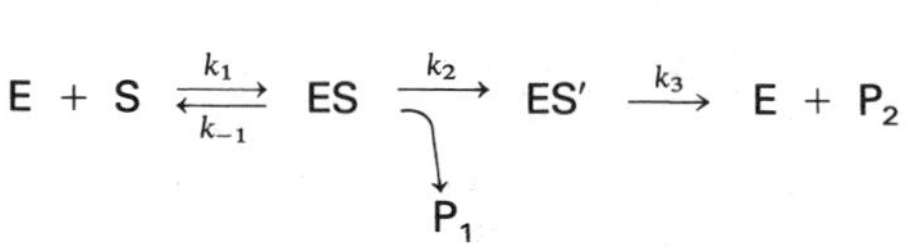

$$E + S \underset{k_{-1}}{\overset{k_1}{\rightleftharpoons}} ES \xrightarrow{k_2} ES' \xrightarrow{k_3} E + P_2$$
$$\qquad\qquad\qquad\quad \downarrow P_1$$

   For the three esters to have a common rate-determining step, it must be postulated that k_3 (the deacylation of the acyl-enzyme intermediate), identical in all three cases, is the rate-determining step.

   In general, for chymotrypsin-mediated amino-acid-ester hydrolysis (and, of course, for *p*-nitrophenyl acetate), k_3 is rate-determining; the acyl enzyme accumulates in the steady state ($k_2 > k_3$), as suggested by the reaction-coordinate diagram (Fig. 3-8).

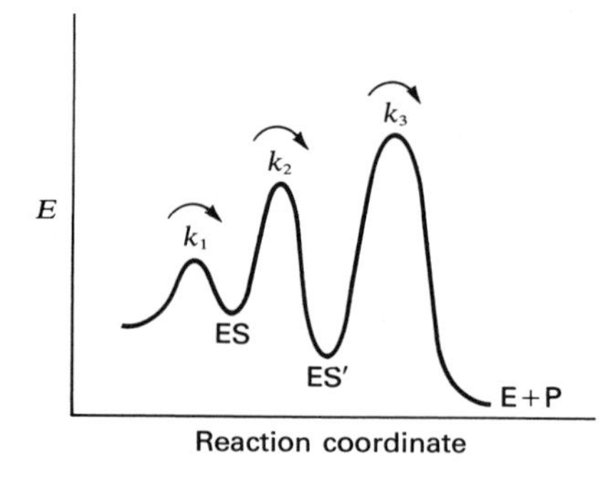

Figure 3-8
Qualitative free-energy profile for reaction of chymotrypsin with synthetic aminoacyl-ester substrates. ES′ accumulates.

As we noted at the end of ¶3.B.5, the Michaelis–Menten equation for the acyl-enzyme case has the form

$$v = \left(\frac{k_2 k_3}{k_2 + k_3}[\mathrm{E_T}][\mathrm{S}]\right) \Big/ \left(K_s \frac{k_3}{k_2 + k_3} + [\mathrm{S}]\right)$$

$$= k_{\mathrm{cat}}[\mathrm{E_T}][\mathrm{S}] / (K_m + [\mathrm{S}]),$$

where $k_{\mathrm{cat}} = k_2 k_3/(k_2 + k_3)$ and $K_m = K_s k_3/(k_2 + k_3)$.

A commonly used measure of an enzyme's relative substrate specificity is the ratio k_{cat}/K_m. For this model, $k_{\mathrm{cat}}/K_m = k_2/K_s$, the ratio of the acylation rate constant to the ES dissociation constant. (Recall ¶2.C.3 for a discussion of this ratio as a measure of an enzyme's catalytic efficiency.)

For the ester substrates, $k_2 > k_3$, so $K_m = K_s(k_3/k_2)$, a complex rate constant in this case.

2. Now let's examine the amide substrate (recall ¶3.B.3) in Table 3-5. The $V_{\max}$ for *N*-acetyl-L-tryptophanamide is 10^3-fold less than the $V_{\max}$ for ester hydrolysis. For reasons discussed above, we attribute the ester $V_{\max}$ of 27 to 30 moles min^{-1} $(\mathrm{mole\ enz})^{-1}$ to k_3, the deacylation of the *N*-acetyltryptophanyl enzyme. If indeed the amide substrate is also hydrolyzed by way of this same *N*-acetyltryptophanyl-acyl enzyme, k_3 for it should be the same. Because the observed $V_{\max}$ is 10^3-fold less, the $V_{\max}$ for amide hydrolysis must be controlled by k_2, the *acylation step.* In general, it does appear that k_2 is rate-determining for *amide*-acyl transfer to water, catalyzed by chymotrypsin. This implies that the acyl enzyme will not accumulate in the steady state, and that ES will be the predominant form in the steady state (Fig. 3-9). (Should an initial burst be expected with amide substrates?)

   Because $K_m = K_s k_3/(k_2 + k_3)$ and for amides $k_3 \gg k_2$, then $K_m \approx K_s$, for amides, and the observed $K_m = 500 \times 10^{-5}\ \mathrm{M} = 5\ \mathrm{mM}$ is probably the true K_s for dissociation of the ES complex and a valid measure of binding activity.

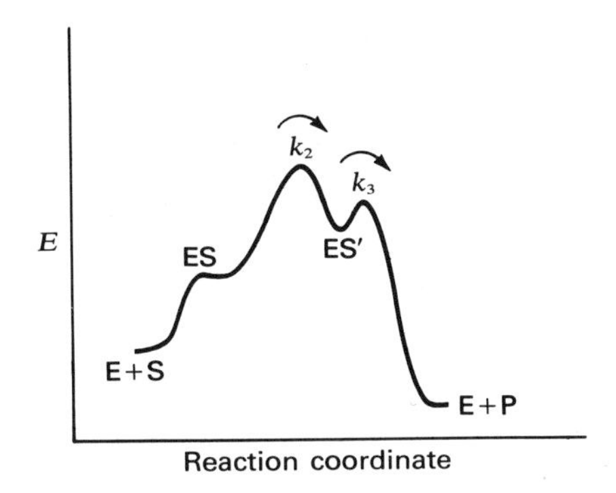

Figure 3-9
Qualitative free-energy profile for reaction of chymotrypsin with synthetic aminoacyl-amide substrates. ES predominates in steady state.

3. We can make some further deductions from the information we have. Assume *first* that the acyl-aromatic moiety is the major binding determinant, and *second* that k_3 is constant for all substrates (as indeed deacylation of a common intermediate should be). Then the principal variable in the measured K_m values is in fact k_2, the acylation rate constant, and differences in observed K_m values reflect differences in k_2 values.

   a. When the substituent X is OMe or OEt, the observed K_m value is about 10×10^{-5} M. Because

   $$K_m = K_s k_3/(k_2 + k_3)$$

   we have

   $$10 \times 10^{-5} = 500 \times 10^{-5}\,(k_3/k_2)$$

   Then $k_3/k_2 = 1/50$, or $k_2 = 50\,k_3$. Clearly $k_3 \ll k_2$ for these ester substrates.

   b. When the substituent X is the *p*-nitrophenoxyl group (—OPNP), we can calculate

   $$0.2 \times 10^{-5} = 500 \times 10^{-5}(k_3/k_2)$$

   and $$k_2 = 2{,}500\,k_3$$

Thus, k_2 for the nitrophenyl ester is much higher than k_2 for the ethyl and methyl esters. This is to be expected, because —OPNP is chemically a much better leaving group (that is, the anion is resonance-stabilized). The tetrahedral adduct should form more readily, and also should partition more toward acyl enzyme.

Also, when comparing esters versus amides as substrates, we note that the amides are lower-energy compounds (more resonance stabilization in the amide functionality; hydroxylamine does not react rapidly with amides), and this is reflected in the relative acylation constants where —OR and —NHR have to act as leaving groups.

One final comment about the scheme we have been using to describe the kinetics of the acyl-enzyme formation and breakdown in chymotrypsin catalysis. Formation of the acyl enzyme and its breakdown (deacylation) undoubtedly proceed via tetrahedral adducts.

$$\text{E—}\ddot{\text{Y}} + \underset{\text{X}}{\overset{\text{R}}{\text{C}}}\text{=O} \rightleftharpoons \text{E—Y—}\underset{\text{X}}{\overset{\text{R}}{\text{C}}}\text{—O}^{\ominus} \rightleftharpoons \text{E—Y—}\overset{\text{R}}{\text{C}}\text{=O} + \text{X}^{\ominus}$$

tetrahedral adduct — acyl enzyme

When we say that k_2 is rate-determining, then it is possible that either the formation or the breakdown of the tetrahedral adduct can be slow in the overall acylation process. Similarly, k_3 (deacylation) involves water addition to the acyl enzyme *and* subsequent decomposition of that tetrahedral adduct.

3.B.10 Trapping an Intermediate Whose Formation or Breakdown Can Be Rate-Determining

We have suggested that a covalent acyl-enzyme intermediate is involved both for aminoacyl-ester and for aminoacyl-amide hydrolysis (and presumably for hydrolysis of protein substrates) by chymotrypsin. We also noted that alternate nucleophiles such as hydroxylamine could compete with water for attack on the intermediate. The intermediate is partitioned between the two nucleophiles, depending on the concentration, the nucleophilicity, and the accessibility of each. Two questions can be asked about catalysis in the presence of an alternate acyl-group acceptor: (1) how much of the intermediate can be trapped by the added nucleophile, and (2) how is the overall rate of the reaction (V_{max} for substrate disappearance) affected? Let's examine these parameters for ester substrates and amide substrates with chymotrypsin.

With ester substrates, $k_2 > k_3$, and the acyl enzyme accumulates because its breakdown is rate-determining. There is sufficient acyl enzyme around to react with every molecule of water that finds its way to the active site, and extra acyl enzyme still is left over at any given instant. Suppose another nucleophile is added to the aqueous solution as potential acyl-group acceptor. A good acceptor (which binds to the enzyme and is thus accessible to attack acyl enzyme at the active site) is L-alanine amide, which undergoes acyl transfer to its free amino group to produce readily detected *N*-acylalanine amide as product (Fastrez and Fersht, 1973; Fersht et al., 1973).

R—C(=O)—X—Enz + H_2N—C(=O)—CH(NH_2)—CH_3 ⟶ R—C(=O)—NH—CH(CH_3)—C(=O)—NH_2 + H—X—Enz

alanine amide   *N*-acylalanine amide

Addition of more alanine amide will raise the total nucleophile concentration and allow partitioning of acyl enzyme between hydrolysis (acyl transfer to water) and aminolysis (acyl transfer to the amine of alanine amide).

$$\text{Acyl enzyme} \xrightarrow[H_2O]{k_3} \text{RCOOH} + \text{Enz}$$

$$\text{Acyl enzyme} \xrightarrow[R'NH_2]{k_4} \text{RCONHR}' + \text{Enz}$$

In the absence of the alanine amide, the rate of breakdown of acyl enzyme (ES) was rate-determining, and $-d[\text{ES}]/dt = k_3[H_2O][\text{ES}]$. In the presence of alanine amide, the breakdown rate of acyl enzyme is increased to the sum of the rate of reaction with H_2O and the rate of reaction with amine:

$$-d[\text{ES}]/dt = (k_3[H_2O] + k_4[R'NH_2])[\text{ES}]$$

The net result is an *increase* in V_{max} of *substrate consumption.* In principle, V_{max} could increase until some other step is slower than the sum of $k_3 + k_4$. For instance, V_{max} could increase until the rate of deacylation exceeds k_2; when $(k_3 + k_4) \gg k_2$, the new V_{max} will be determined by k_2, which has now become the slowest step in the overall reaction. The partitioning ratio depends on competition between the alternate nucleophiles. At 1 M alanine amide, the acyl enzyme is captured by the amino group at least 94 out of every 100 times it breaks down; presumably the efficiency of acyl-enzyme capture is due to the higher nucleophilicity of the amine relative to H_2O *and* to its specific binding at the chymotrypsin active site (Fastrez and Fersht, 1973; Fersht et al., 1973).

With amide substrates, $k_3 > k_2$, and the acyl enzyme does not accumulate detectably in the steady state. Its formation is rate-determining and, once formed, the acyl enzyme hydrolyzes rapidly relative to its formation. Attempts to isolate acyl-enzyme intermediates from amide substrates fail for this reason, and trapping data is all the more important to prove that the hydrolytic mechanism involves the same species as with ester substrates.

For an acyl enzyme formed *after* the rate-determining step, the addition of another nucleophile such as alanine amide cannot increase the V_{max} rate of substrate utilization. V_{max} is controlled totally by the slow k_2 step (acylation), and nucleophiles cannot affect the rate of acyl-enzyme formation On the other hand, added nucleophiles can still compete with water and partition the acyl enzyme. As each molecule of acyl enzyme forms slowly, it can be captured rapidly by a water molecule or by the amino group of an alanine-amide molecule.

If the acyl-enzyme intermediate is the same from an *N*-acetylphenylalanine ester and from an *N*-acetylphenylalanine amide, *the intermediate should be partitioned identically from either substrate.* This expectation has been verified experimentally (Fastrez and Fersht, 1973; Fersht et al., 1973). At 0.84 M alanine amide with an *N*-acetylphenylalanine-amide substrate, more than 94% of the acyl enzyme is captured by the alanine-amide acceptor. This confirms that amide substrates also react by the acyl-enzyme mechanism, and this approach is a *useful method for indirect detection of intermediates that form after the rate-determining step in a catalytic sequence.*

3.B.11 Evidence That Histidine Plays a Catalytic Role at the Active Site

If a covalent acyl-enzyme intermediate exists during chymotrypsin catalysis, this means that some amino-acid side chain in the protein must act as the attacking nucleophile. In Chapter 2 we discussed the groups in proteins that have this potential. At the beginning of this chapter, we commented that the X-ray maps of serine proteases indicate a triad of amino-acid residues that are invariant at the active sites: an aspartyl residue with a β-$COO^{\ominus}$ as potential nucleophile, a histidine with the imidazole group as functional group, and a serine residue with a β-OH as functional group (Stroud et al., 1975). The normal pK_a ranges for the functional groups in the free amino acids in aqueous solution would be around 3–4, 6–7, and 13–15, respectively. Because physiological pH is around 7.8 in the intestine, one might anticipate that the β-carboxylate and the imidazole group are the most likely candidates. This expectation may or may not be incorrect for the "serine" proteases, but we examine here some of the experimental approaches based on this assumption, to evaluate their general applicability and limitations.

First we can examine the pH dependence of the hydrolysis rate (v) for a synthetic substrate (ester or amide). At constant [S], we obtain a bell-shaped curve (Fig. 3-10a). However, it is not easy to interpret this observation, because the catalytic rate can be a complex function, with V_{max} and $1/K_m$ having different pH dependencies. For example, reciprocal pH dependences of V_{max} and $1/K_m$ (Fig. 3-10b) could yield the bell-shaped curve, and in fact this is the case for

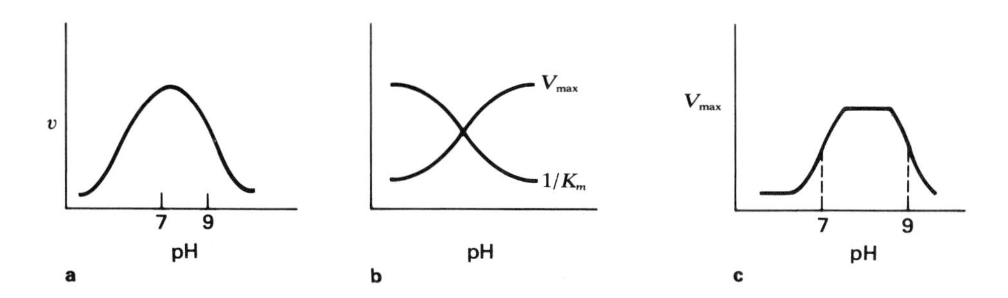

Figure 3-10

(**a**) Rate profile for v vs. pH. (**b**) Rate profiles for V_{max} vs. pH and for $1/K_m$ vs. pH. These curves could yield the profile shown in part a. (**c**) Observed profile for V_{max} vs. pH.

chymotrypsin. Thus any correlations of v with pH must be carried out under V_{max} conditions.

A plot of V_{max} vs. pH (Fig. 3-10c) gives a plateau curve with two inflections (at pH 7 and at pH 9), possibly indicating that some groups on the enzyme with pK_a values of 7 and 9 are important for catalysis. The inflection at pH 9 represents the pK_a of the α-amino group of Ile[16], the amino-terminal residue of the β chain (see ¶3.B.1) of α-chymotrypsin.

$$NH_3^{\oplus}\text{—Ile} \xrightarrow{pK_a = 9} NH_2\text{—Ile} + H^{\oplus}$$

This deprotonation disrupts a salt bridge between Ile[16] and Asp[194]. The loss of this interaction destabilizes the native tertiary structure, and the protein undergoes a conformation change, disrupting the active-site geometry and the catalytic activity. This pK_a, then, is important for structural reasons, not for catalytic function at the active site per se.

The pK_a of 7 strongly implicates a histidine-imidazole moiety, and there are two histidines in chymotrypsin: at positions 40 and 57 from the amino terminus. However, one must always be wary in interpreting apparent pK_a values for identification of enzymatic residues that are functionally important for catalysis (Jencks, 1969, chap. 2). For instance, one might also be titrating a histidine for a major enzyme conformational change. Again, the pK_a may be a perturbed pK_a for a different functional group (as is actually the case in these proteases). Finally, there is the very real possibility that the inflection point in V_{max} does not represent an ionization of some acid or base, but rather a kinetic phenomenon (a change in rate-determining step with the pH change).

As an example, consider the nonenzymatic reaction of a ketone with hydroxylamine to produce an oxime (Jencks, 1969, chap. 2).

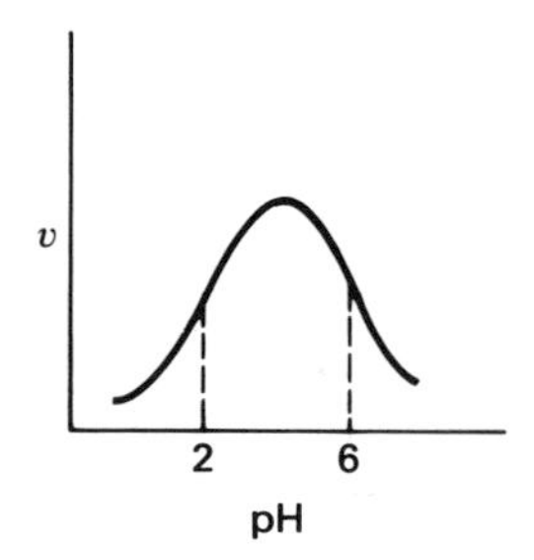

Figure 3-11
Plot of v vs. pH for oxime formation.

$$R_2C{=}O + NH_2OH \underset{k_{-1}}{\overset{k_1}{\rightleftharpoons}} \underset{\text{tetrahedral adduct}}{R_2C(OH){-}NHOH} \xrightarrow[H^{\oplus}]{k_2} \underset{\text{oxime}}{R_2C{=}NHOH} + H_2O$$

The observed rate of oxime formation plotted against pH yields a bell-shaped profile (Fig. 3-11). The inflection at pH 6 is due to ionization of hydroxylamine, but there is *no* possible pK_a of 2 in this system. The inflection at pH 2 is due to a change in rate-determining step. At low pH, k_1 is the rate-determining step; at higher pH, k_2 (which is acid-catalyzed) becomes the rate-determining step. The tetrahedral intermediate accumulates at high pH, but not at low pH.

3.B.12 Active-Site–Directed Inactivating Reagents: Criteria and Specificity

During the past two decades, many different reagents have been used as specific inactivators of target enzymes. The goal has been to cause chemical modification at the actual active site of the enzyme to induce loss of catalytic activity, active-site–directed irreversible inactivation, or affinity labeling (Baker, 1967; Shaw, 1970; Rando, 1974*a*; Maycock and Abeles, 1976; C. Walsh, 1977). The most common approach has been synthesis of molecules that are structurally analogous to a substrate of a target enzyme and that also incorporate a chemically reactive functional group (generally an electrophilic functionality). The affinity-labeling reagent will gain specificity if it binds initially to the enzyme active site rather than reacting by collision in solution. On binding, the electrophilic group of the reagent may be placed in proximity to various amino-acid side chains of the active site that can act as nucleophiles and can participate in displacement reactions to generate *covalent* bonds between the inactivator and the active-site residue. If the nature of this covalent bond and of the amino acid forming it can subsequently be elucidated, this type of experiment can indicate at least some of the nucleophilic groups present at the active site of the target enzyme. In the best of cases, such experiments may identify a nucleophile important in catalysis.

The histidine residue at position 57 of α-chymotrypsin can be covalently modified by a reactive substrate analogue. What are the criteria for proof that the analogue has caused inactivation by reaction at the actual active site, and what mechanistic information does this knowledge convey? Let's look at an example. Shaw (1970) synthesized an *N*-tosylphenylalanine chloromethyl ketone (TPCK) as a structural analogue of the *N*-tosylphenylalanine-ester (TPE) substrates.

$C_6H_5-CH_2-CH(NH-SO_2-C_6H_4-CH_3)-C(=O)-OR$

L-TPE
substrate

$C_6H_5-CH_2-CH(NH-SO_2-C_6H_4-CH_3)-C(=O)-CH_2-Cl$

L-TPCK
affinity alkylation reagent

Shaw used *N*-tosyl as the *N*-acyl group because the simpler *N*-acetyl group produced synthetic problems during the preparation of the chloromethyl-ketone moiety. The electrophilic site in the reagent is the group with a potentially displaceable chloride:

$$-C(=O)-CH_2-Cl \quad \ddot{X}$$

If TPCK is added to solutions of chymotrypsin and the enzymatic activity subsequently measured, a gradual loss of catalytic activity is observed (Fig. 3-12).

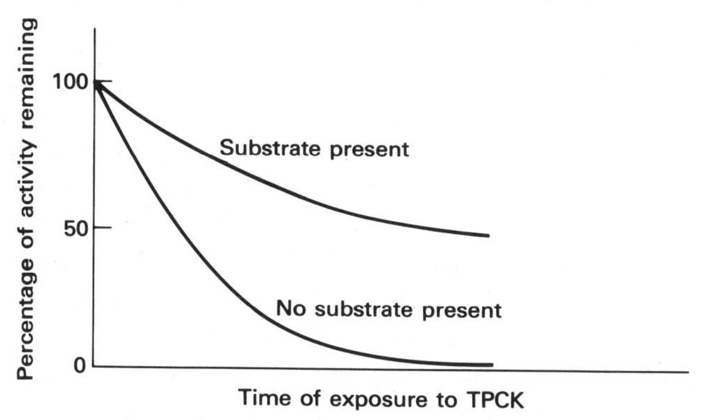

Figure 3-12
Loss of activity of chymotrypsin as a function of time of exposure to a fixed concentration of TPCK.

After a certain length of exposure to TPCK, no detectable catalysis occurs with any substrate subsequently assayed.

A number of criteria can be applied to such an inactivation to determine whether it displays both the specificity and the kinetic course expected for active-site–directed modification. Irreversible loss of chymotrypsin activity caused by [^{14}C]-TPCK proves to be covalent and stoichiometric (one radioactive label is incorporated per active site). Evidence of specificity is provided by the fact that the D-isomer of TPCK binds reversibly but does not inactivate. (Presumably, the chloromethyl-ketone group of the D-isomer aligns improperly at the active site, so that the enzyme nucleophile cannot attack it effectively.) Neither the enzyme precursor chymotrypsinogen nor enzyme denatured in urea is labeled by [^{14}C]-TPCK, indicating that the correct active-site geometry must be present in the enzyme.

If the putative irreversible inhibitor is indeed binding at the active site prior to the chemical step involving covalent linkage to the target enzyme, there are testable *kinetic consequences.* First, the time-dependent loss of catalytic activity should show first-order kinetics (rather than the second-order kinetics expected from a simple bimolecular collision between protein and reagent free in solution). Experimentally, a semilog plot of the percentage of activity remaining versus time (at a fixed concentration of inactivator) should yield data fitting a straight line. The half-life ($T_{1/2}$) for inactivation is related to the rate constant for inactivation by the expression $T_{1/2} = 0.693\ k_{inact}$. When the experiment is performed at several fixed inhibitor concentrations, a family of lines is obtained, producing a series of k_{inact} values. Then a secondary reciprocal plot can be constructed of $1/k_{inact}$ versus 1/[inactivator] (essentially a Lineweaver–Burk plot for the inactivator), yielding the two important kinetic parameters (related to the simple scheme of the following equation) as shown in Figure 3-13.

$$\mathrm{E} + \mathrm{I} \underset{k_{-1}}{\overset{k_1}{\rightleftharpoons}} \mathrm{E \cdot I} \xrightarrow{k_2} \underset{\substack{\text{covalent}\\ \text{inactive}\\ \text{complex}}}{\mathrm{E{-}I}} \qquad K_I = k_{-1}/k_1$$

The vertical intercept of the reciprocal plot is $1/k_2$, the limiting rate constant for inactivation (the observed inactivation rate if *all* the enzyme is in the E·I complex). The *physical* significance of a finite vertical intercept is that the inactivation shows *saturation kinetics* and validates the idea that inactivation occurs from prebound inactivator. The horizontal intercept is $-1/K_I$, corresponding to the dissociation constant of inactivator from the E · I complex, and thus providing some measure of affinity of inactivator for enzyme.

The second kinetic test one can apply to the time-dependent loss of activity is to determine whether addition of substrate retards the rate of inactivation

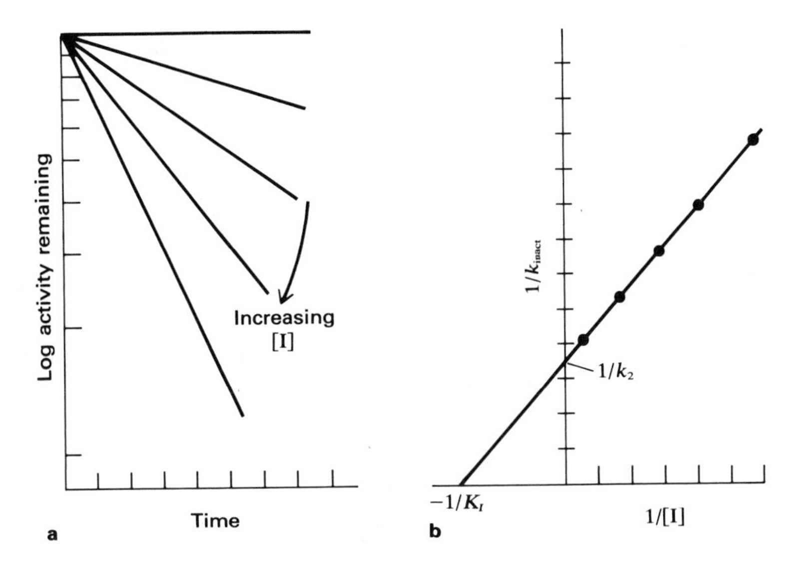

Figure 3–13
Pseudo first-order kinetics of inactivation of an enzyme from covalent reaction with an inactivator prebound at the active site.

produced at a given concentration of the reagent (Fig. 3-12). In the simplest interpretation, this phenomenon of substrate protection arises from a binding competition between substrate and inactivator at the enzyme's active site. To the extent that S draws off enzyme molecules into an ES complex at any given instant, those enzyme molecules cannot react with I. This inaccessibility lowers the concentration of E·I at those instants; a lower [E·I] means a slower observed rate of formation of the inactive covalent EI compound. Note, however, that ES formation is reversible, whereas covalent modification to EI is irreversible; therefore, the presence of substrate will retard the rate of catalytic-activity loss, but cannot completely prevent it.

$$
\begin{array}{ccccc}
E & \overset{I}{\rightleftharpoons} & E\cdot I & \longrightarrow & E\text{—}I \\
{\scriptstyle S}\downarrow\uparrow & & & & \\
ES & & & &
\end{array}
$$

In fact, one sees such protective behavior on the part of substrates for chymotrypsin, suggesting that I and S are competing physically for the active site, and that I (in this case, the TPCK molecule) is an active-site–directed inactivator, implying that identification of the amino acid covalently modified will indeed provide information about the active-site nucleophile.

The covalent nature of the radioactive inactivator can be proved by a gel-filtration experiment of the type described in the text accompanying Figure 3-7, and by the observation that ^{14}C-label is not removed by dialysis before or after unfolding the protein in 8 M urea.

How did this irreversible active-site modification occur? One expects that covalent modification occurred by S_N2 displacement of the chloro substituent of TPCK by an active-site nucleophile. There are two histidine residues in α-chymotrypsin, at positions 40 and 57. By amino-acid analysis, we find that only one remains unmodified after TPCK treatment. Partial hydrolysis (by acid or proteases) indicates only one radioactive peptide that, from knowledge of the primary sequence, contains His^{57} but not His^{40}. Acid hydrolysis (6 N HCl, 110°, 24 hr) to effect complete hydrolysis, followed by performic-acid oxidation, allows isolation of N^3-carboxymethyl histidine (Ong et al., 1964, 1965).

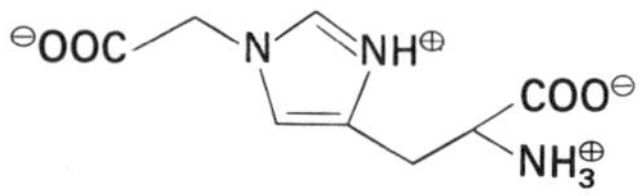

These data all suggest that His^{57} (the active-site histidine by X-ray analysis) acts as nucleophile on L-TPCK bound specifically at the active site, alkylating each enzyme molecule and thereby shutting down each active site. These data strongly implicate His^{57} as an active-site nucleophile.

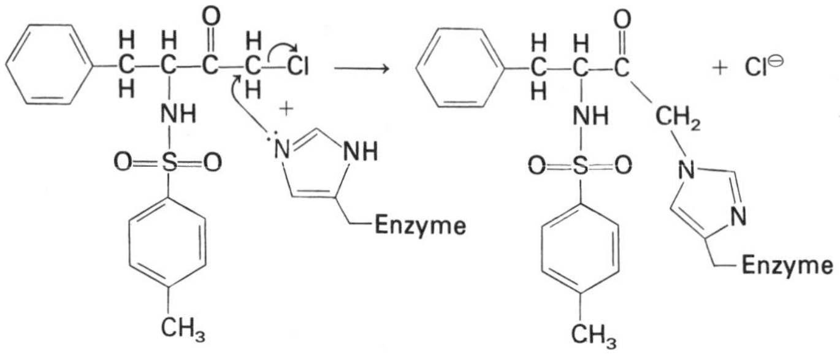

alkylated active site

The inactivation of chymotrypsin by His^{57}-initiated S_N2 displacement of chloride is quite rapid, with a limiting rate constant k_2 of about 0.2 min^{-1}. A similar rate is observed with ZPCK, where the N-tosyl group is replaced by an N-carbobenzoxy (Z) group. When this rate of chloride displacement is compared with the S_N2 displacement by N-acetyl histidine in 80% methanol, the enzymatic process is found to be 10^6-fold faster (Shaw and Ruscica, 1971). Jencks (1975a, p. 341) comments that this dramatic rate increase in the enzymatic alkylation may reflect utilization of the binding energy of ZPCK to decrease $\Delta G^{\ddagger}$ for the S_N2 displacement by 8 kcal/mole (33 kJ/mole). A high local concentration of the chloromethyl ketone held immobilized in proper orientation for reaction with N-3 of His^{57} represents an entropic factor in rate acceleration. It has been argued that α-haloketone functionalities are especially prone to S_N2 displacement because of the possibility for electron overlap (not generally available in S_N2 displacements) between incoming nucleophile and the carbonyl group adjacent to the displacement site, which might stabilize the transition state. Many such haloketone-containing substrate analogues have subsequently been used to label a variety of other enzymes.

In general, we use an affinity label or active-site–directed inactivator such as TPCK to obtain specific inactivation of the target enzyme; the structural resemblance of label or inactivator to substrate takes advantage of the enzyme's binding specificity and discriminates against holding reactive but nonspecific chemical reagents at the active site long enough (or often enough) to inactivate. On this basis alone, we may obtain mechanistically useful information.

However, the TPCK results also offer insight into an unresolved ambiguity that still remains with such reagents—an ambiguity that can lead to incorrect conclusions. This ambiguity arises because the —$COCH_2Cl$ functionality is sufficiently reactive to trap any active-site nucleophile it can. The fact that TPCK reacts essentially exclusively with His^{57} might imply that histidine is the primary active-site nucleophile. In fact (as we have argued already and will again), this is an incorrect conclusion: it is Ser^{195} that plays that role. Apparently His^{57} is favored over Ser^{195} in relative rates of alkylation of TPCK, but Ser^{195} is favored in reactions with substrates. We must fall back on the ad hoc justification that small orientation differences may influence relative rates of alkylation, so that His^{57} predominates as nucleophile with TPCK whereas substrates are attacked by Ser^{195}. A corollary of this statement is the observation that the —$COCH_2Cl$ group is not spatially equivalent to (not isosteric with) the —CONHR or —COOR group of a substrate. To the extent that the attackable electrophile on a pseudosubstrate is aligned differently from the reaction locus of a normal substrate, covalent attachment may select a spurious or secondary nucleophile in the active site.

Consider the mechanistically-related intestinal protease trypsin. Its specificity for cleavage is for basic (Arg, Lys) residues in proteins. With this enzyme, two different affinity labels capture two different active-site nucleophiles. The analogue *N*-tosyllysyl chloromethyl ketone (TLCK) hits His^{46}, a secondary nucleophile at the active site. On the other hand, a guanidinium compound that acts as an arginine analogue exclusively hits Ser^{183}, the residue acylated during normal trypsin catalysis (Shaw, 1970).

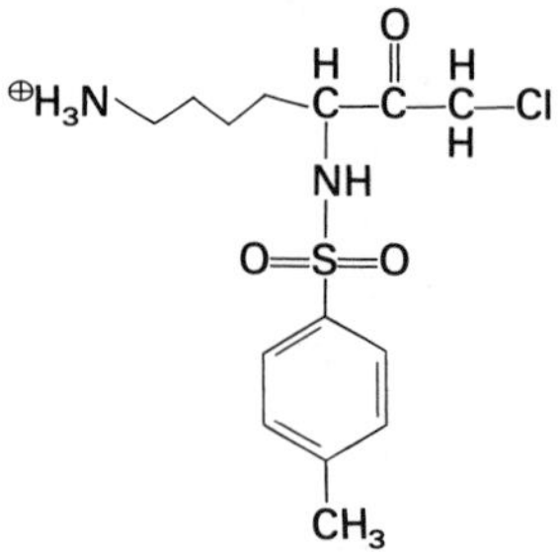

TLCK

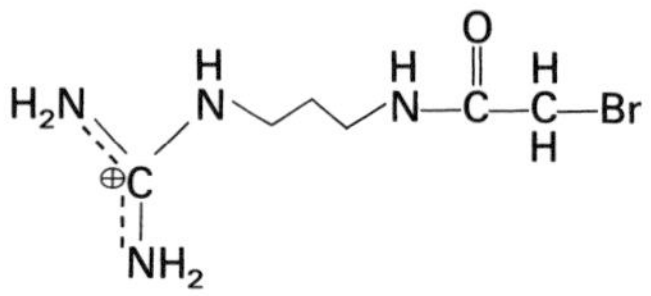

guanidinium compound

The active-site histidine residues are thought to function as general base catalysts rather than as nucleophiles in the general mechanism of action of serine proteases. This uncertainty about the function of the residue trapped by electrophiles is inescapable, and the functional role of the residue must be verified by other lines of experimentation. *Any amino-acid side chain capable of acting as general base can also react as a nucleophile, either in normal catalysis or with a trapping electrophile.*

3.B.13 Identification of the Nucleophilic Serine at the Active Site

In addition to the substrate analogues already discussed, other compounds inactivate chymotrypsin. Reaction with diisopropylphosphofluoridate (DIPF) produces a time-dependent, irreversible inactivation of chymotrypsin, suggesting covalent modification (Jansen, Nutting, Jang, and Balls, 1949).

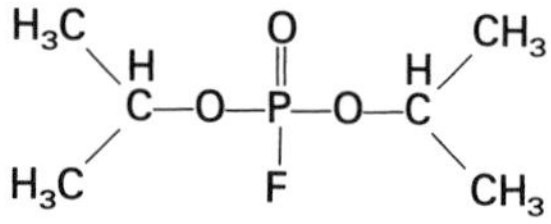

DIPF

Protection is observed with substrates, implying that the process may be active-site directed. Using [^{32}P]-DIPF, Jansen, Nutting, and Balls (1949, 1950) found stoichiometric incorporation of a radioactive phosphate label. The kinetics of labeling follow the kinetics of inactivation, suggesting that the covalent modification is destroying catalytic activity (Fig. 3-14). The covalent adduct is stable to acid hydrolysis in 6 N HCl (110° C, 24 hr), which hydrolyzes all the peptide bonds, liberating the constituent amino acids. One radioactive compound was obtained and characterized as *O*-phosphoryl serine (Schaffer et al., 1953).

$$^{\ominus}O-{}^{32}P(=O)(O^{\ominus})-O-CH_2-CH(NH_3^{\oplus})-COO^{\ominus}$$

It was separately established that authentic serine phosphate is stable to the acid treatment, suggesting that phosphoserine is formed directly on inactivation rather than by some subsequent phosphoryl transfer by internal attack of a second nucleophile at the active site.

Presumably, a serine alkoxide ion of the enzyme displaces $F^{\ominus}$ from DIPF, resulting in phosphorylation of that active-site serine.

$$((H_3C)_2CH-O)_2P(=O)F + Enz-CH_2O^{\ominus} \longrightarrow ((H_3C)_2CH-O)_2P(=O)-O-CH_2-Enz + F^{\ominus}$$

phosphorylated enzyme

Why doesn't this serine derivative break down hydrolytically to liberate free enzyme? Presumably the bulky diisopropylphosphoryl group is attacked only slowly by water.

Because diisopropylphosphofluoridate is hardly a normal substrate for chymotrypsin, other probes for the catalytic significance of this reactive serine were undertaken. Bender (1971, p. 506 ff) summarizes some of this evidence.

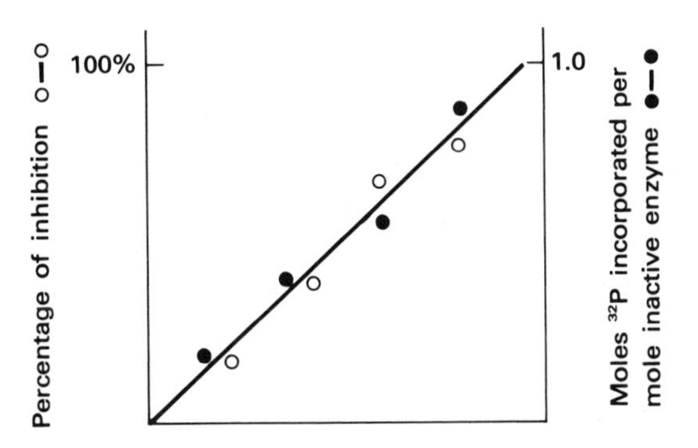

Figure 3-14
Loss of activity of chymotrypsin is proportional to amount of [^{32}P]-DIPF incorporated covalently.

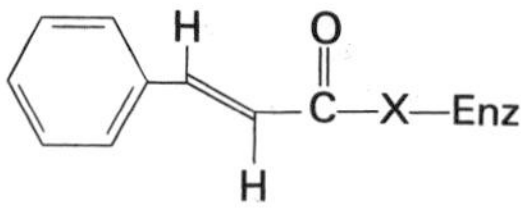

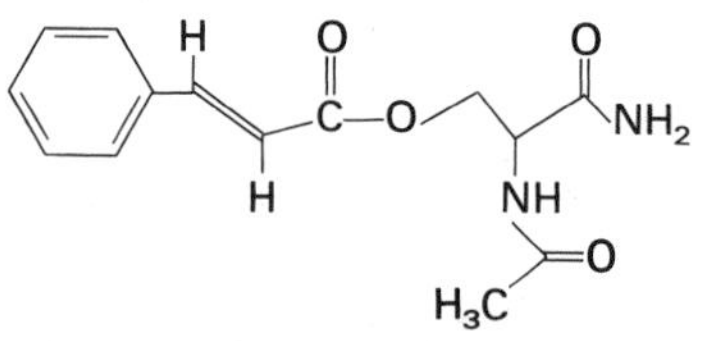

trans-cinnamoyl-acyl enzyme   *trans*-cinnamoyl-*N*-acetyl serinamide

With a *trans*-cinnamoyl-acyl enzyme, the UV-visible spectrum closely parallels that of a model compound, *trans*-cinnamoyl-*N*-acetyl serinamide.
Even more convincingly, the rates of alkaline hydrolysis of both *trans*-cinnamoyl chymotrypsin and acetyl chymotrypsin, after unfolding in 8 M urea to remove steric constraints to $OH^{\ominus}$ accessibility, are similar to rates of hydrolysis of the *O*-cinnamoyl-*N*-acetyl serinamide and *O,N*-diacetyl serinamide as models.

In fact, with the [^{14}C]-DIPF enzyme, enzymatic partial digestion yields one radioactive peptide fragment containing one modified serine—that at position 195 (Hartley, 1964).

3.B.14 A Postulated Mechanism for Hydrolysis Catalyzed by Chymotrypsin and Other Serine Proteases

At this juncture, we can piece together the chemical evidence and the structural evidence and propose a possible mechanism for chymotrypsin catalysis as a prototype for the other "serine" proteases we will not discuss specifically. (For a complete discussion of some other serine proteases, see the book by Blackburn, 1976.)

In the stoichiometric reaction with diisopropylphosphofluoridate, only one serine residue out of 28 has shown nucleophilic properties: the active-site serine. We described the covalent-bond formation as attack by a serine alkoxide ion, but the pK_a for ionization of the β-OH of serine is estimated as 13.7. At neutral pH, a naked alkoxide would be a high-energy species, present in very minute concentrations. Therefore, we must postulate general base catalysis to heighten the nucleophilicity of Ser^{195} without discrete anion formation. A good candidate for the general base is, of course, the imidazolyl group of His^{57}. Turning to the X-ray data, we find that the catalytic triad of Asp^{102}, His^{57}, and Ser^{195} confirms this expectation and suggests an even more extensive charge-relay system in chymotrypsin (and in the other serine proteases). (See Stroud et al., 1975; also Blackburn, 1976, pp. 62–68, and references cited therein.) Figure 3-15 shows a representation of the catalytic triad.

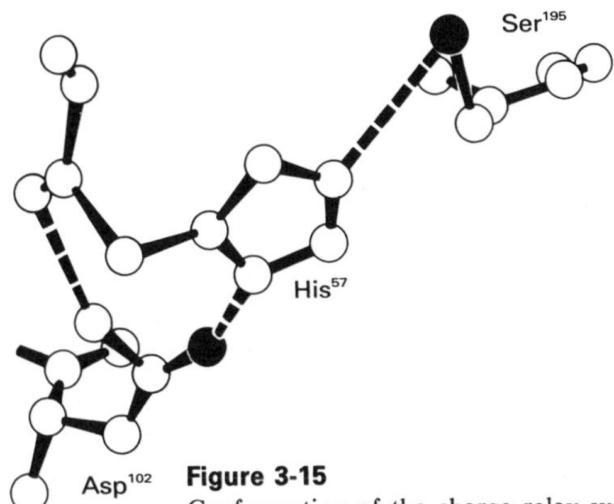

Figure 3-15
Conformation of the charge-relay system in chymotrypsin. (Based on D. M. Blow and T. A. Steitz, "X-ray diffraction studies of enzymes," *Ann. Rev. Biochem.* 39:86. Copyright © 1970 by Annual Reviews Inc. All rights reserved.)

To understand how this catalytic triad participates in catalysis, we must know the microscopic ionization behavior of the system. The V_{max} rates depend on a pK_a of 6.7 to 7.0, and this pK_a was commonly assigned to proton dissociation from the cationic form of His^{57}; however, this pK_a may or may not be assigned correctly. The other candidate is the β-COOH group of Asp^{102}, which normally has a pK_a of around 4.5. However, Asp^{102} is in a hydrophobic domain in the protein interior, shielded from the aqueous medium by His^{57}. This milieu will elevate the β-COOH pK_a, because the charged carboxylate anion will be less stable in a low-dielectric medium. Therefore, Hunkapiller et al. (1973) suggested that the pK_as of the two groups, His^{57} and Asp^{102}, might be reversed and that "the ionization that occurs with $pK_a \sim 6.7$ may actually reflect net loss of a proton from a neutral aspartic acid carboxyl group in the presence of a neutral imidazole ring rather than the more commonly accepted loss of a proton from the imidazolium cation in the presence of an aspartate carboxylate anion." Evidence supporting this postulate has been obtained with a bacterial serine protease, the α-lytic protease from *Myxobacter 495.* There is a single histidine in this enzyme: the one in the charge-relay system at the active site. The microorganism can be grown on [^{13}C]-histidine, and the isolated ^{13}C-labeled protease can be analyzed for ionization behavior of this histidine by carbon magnetic-resonance studies, where changes in chemical shifts and coupling constants can be monitored to estimate pK_a values (Hunkapiller et al., 1973). From these studies, the pK_a for the imidazolium cation was judged to be 4.0, and that for the aspartyl β-carboxylic acid was judged to be 6.7. If this result can be generalized, the active-site

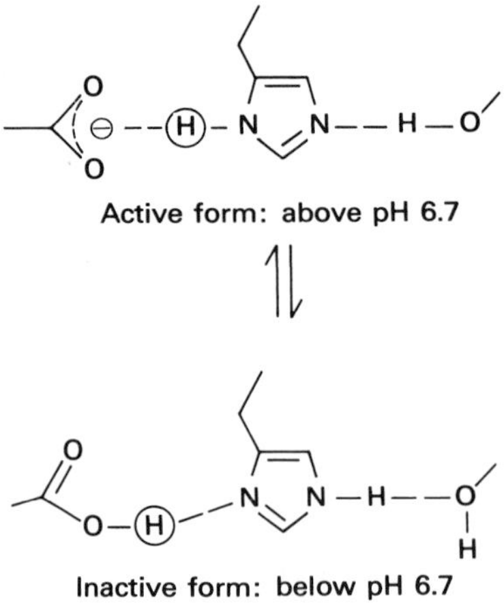

Active form: above pH 6.7

Inactive form: below pH 6.7

Figure 3-16
Ionization of the active center in the pH range 4.0 to 8.0, as discussed in the text. (Stroud et al., 1975, p. 24)

ionization for serine proteases might be as shown in Figure 3-16 (Stroud et al., 1975).*

With these assignments, we can write a scheme for protease hydrolysis as in Figure 3-17, which also is taken from the review by Stroud et al. (1975).

Catalysis is initiated by nucleophilic attack of the Ser^{195} hydroxyl on the susceptible carbonyl carbon of the substrate. This attack is facilitated by general acid–base catalysis as shown in Figure 3-17. The oxyanion of the adduct is held to be stabilized by hydrogen bonding to the backbone NH groups of Gly^{193} and Ser^{195}. Collapse of the tetrahedral adduct in the forward direction involves expulsion of the amine product (protonated from the N-3 of histidine as it leaves, to avoid forming the high-energy amine anion) and generation of the acyl enzyme. Deacylation involves the microscopic reverse of acylation, where the attacking water molecule at the active site similarly acts as a more effective nucleophile due to the proton shuttling of the charge-relay system. Catalytic efficiency can be provided by the charge-relay system through the avoidance of unstable species with charge separation and through the facile transfer of protons from enzymatic residues to and from substrates and departing products. This effect could be achieved by a carboxyl group with elevated pK_a and an imidazole group of lowered pK_a (but see footnote).

* The unusual NMR-based assignment of reversed pK_a values for His and Asp of the catalytic triad has recently been reevaluated using 1-[^{15}N]-His-α-lytic protease and 1,3-[$^{15}N_2$]-His-α-lytic protease (Bachovchin and Roberts, 1978). The ^{15}N NMR studies show unequivocally that the active-site histidine in the catalytic triad of this serine protease has the pK_a value ≈ 7. Additionally, a refined X-ray map for the bacterial protease subtilisin does not support the initial hydrogen-bonding scheme for the catalytic triad (Matthews, Alden, Birktoft, et al., 1977).

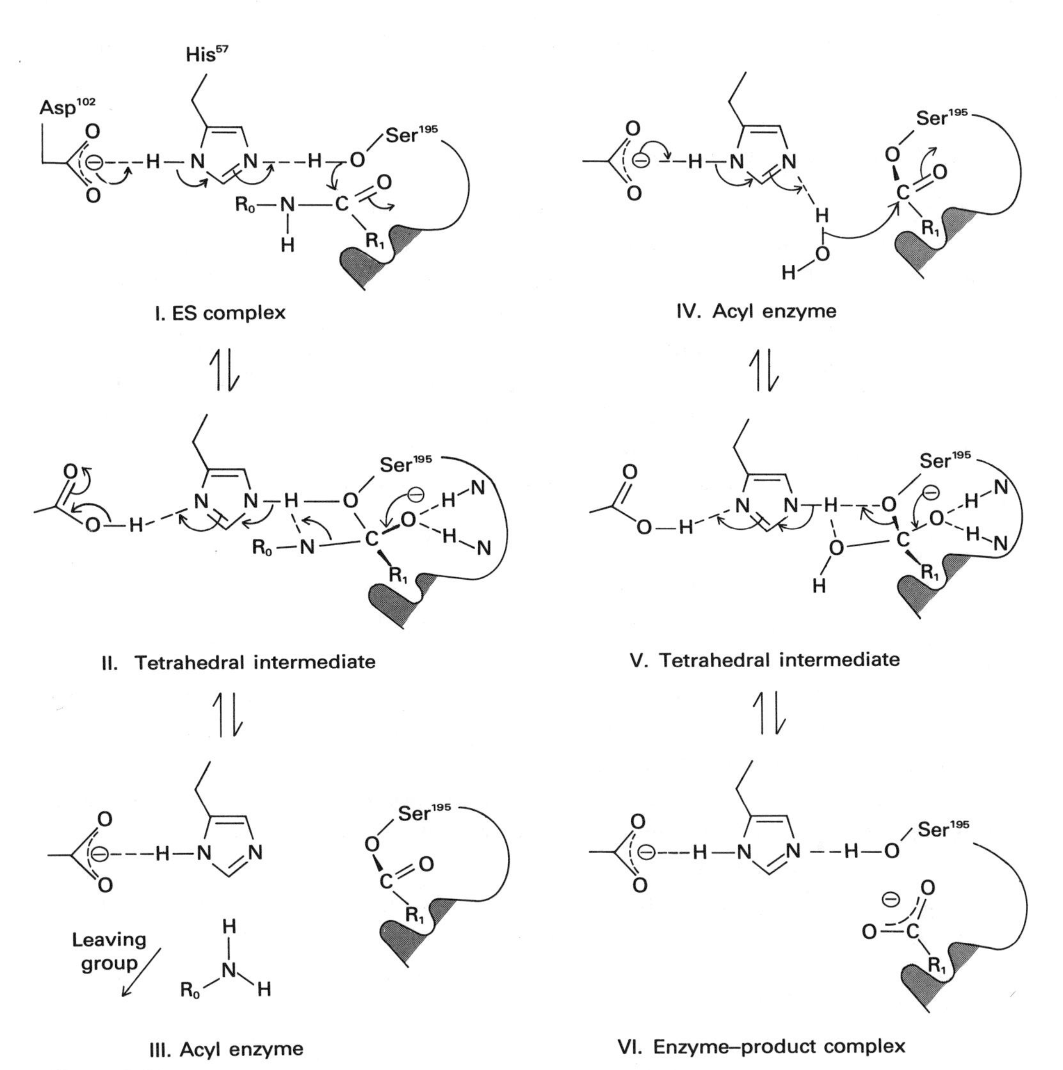

Figure 3-17
A mechanism for serine protease hydrolysis of peptides or amides. In this representation, the proton shuttle is concerted. (Stroud et al., 1975, p. 25)

3.C OTHER TYPES OF PROTEASES

The enzyme chymotrypsin has provided insight into the mechanism of action of the class of endopeptidases that use serine as active-site nucleophile. But, as pointed out in Table 3-2, other proteases do exist that can be placed in distinct mechanistic sets. We shall comment on these only briefly here—because of space limitations, not because of lack of interest or mechanistic evidence (see Blackburn, 1976).

3.C.1 Sulfhydryl Proteases

Related to the serine proteases are the "sulfhydryl" proteases such as papain (from latex or fruit of the papaya), ficin (from figs), bromelain (from pineapple), and cathepsin B (an intracellular protease of animal cells, found in lysosomes, the digestive subcellular organelles).

With these endopeptidases, acyl-enzyme intermediates analogous to those already discussed are generated during hydrolytic catalysis, but the enzyme nucleophile is not a serine hydroxyl group, but rather is the thiolate anion ($pK_a \approx 8$) of an active-site cysteine. The acyl enzyme then is an acylthioester rather than the acyloxygen-ester of the serine proteases.

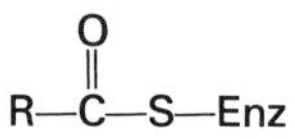

$$\mathrm{R{-}\overset{\overset{\displaystyle O}{\|}}{C}{-}S{-}Enz} \qquad \mathrm{R{-}\overset{\overset{\displaystyle O}{\|}}{C}{-}O{-}Enz}$$

papain acyl enzyme — chymotrypsin acyl enzyme

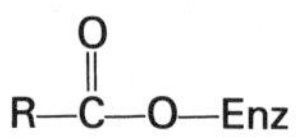

A three-dimensional X-ray map of crystalline papain at 2.8 Å resolution has been obtained, and the active-site cysteine has been identified at position 25 in the 212-residue chain (Drenth et al., 1968, 1970).

Papain shows broad specificity for peptide-bond hydrolysis, and it will also use synthetic *N*-blocked aminoacyl-ester and aminoacyl-thioester substrates. The chloromethyl ketones TPCK (*N*-tosyl-L-phenylalanine chloromethyl ketone) and TLCK (*N*-tosyl-L-lysyl chloromethyl ketone) have been used as affinity labels. In this case, the residue specifically alkylated is the nucleophilic Cys^{25}, and the affinity reagents do in fact identify the correct nucleophile (Bender and Brubacher, 1966; Whitaker and Perez-Villasenor, 1968). The specificity of the enzyme has been probed more carefully using oligopeptides of alanine (Ala_2 to Ala_6) as substrates, and positional specificity and rates of cleavage have been determined (Schechter and Berger, 1967, 1968, 1970). These studies indicate an extended active-site area, up to seven amino-acid residues long, with seven subsites (Fig. 3-18). Hydrolysis occurs at the position between subsites S_1 and S_1' indicated by the arrow in the figure (see Blackburn, 1976, p. 285).

3.C.2 Acid Proteases

The best-characterized representative of the endopeptidases with acid pH optima is pepsin A (one of several related enzymes), which is found in gastric juice (at a

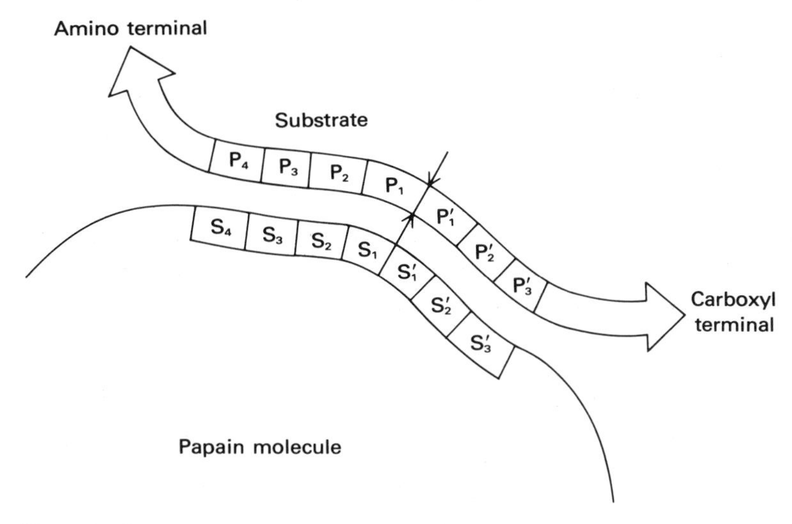

Figure 3-18
The active site of papain, according to Schechter and Berger. (Blackburn, 1976, p. 285)

concentration of 400 mg/liter), where it functions to break down ingested protein (see Clement, 1973; Fruton, 1971, 1974; Tang, 1976). The enzyme was discovered by Schwann in 1834 (see Segel, 1975*b*, p. 1). This enzyme is remarkably acid-stable and shows optimal activity in the pH range 1–5, where the vast majority of proteins would have undergone acid denaturation. The enzyme is secreted by the gastric mucosal cells as an inactive zymogen (pepsinogen), which is then activated in the stomach juice by limited proteolysis. Pepsinogen has a molecular weight of about 38,000 (362 residues), whereas the molecular weight of the active enzyme pepsin is about 35,000 (320 residues). The proenzyme is activated by cleavage of a 42-residue amino-terminal segment, yielding a single polypeptide chain with three cystine disulfides. Pepsin has 29 free carboxyl groups but only four side-chain cationic groups.

In its endopeptidase activity, pepsin shows preference for cleavage at aromatic and hydrophobic loci in proteins. Amino-blocked synthetic aminoacyl amides and aminoacyl esters function as substrates. In addition, organic sulfite esters are hydrolyzed at the active site; bis-*p*-nitrophenyl sulfite, $(p\text{-}NO_2\text{-}\phi\text{-}O)_2SO$,

$$O_2N-C_6H_4-O-\overset{\displaystyle O}{\overset{\|}{S}}-O-C_6H_4-NO_2$$

is the most active pepsin substrate yet found, with a k_{cat} ($V_{max} = k_{cat}[E_T]$) of 60 sec^{-1} at pH 1.91 (Clement, 1973; Fruton, 1971, 1974; Tang, 1976). The pH optimum, in terms of k_{cat} and k_{cat}/K_m for most synthetic substrates, is around 2.

The markedly acid pH optimum was an early indication that side-chain COOH groups might be important in pepsin catalysis, and ionizations near pH 1.0 and pH 4.0 seem to be important in catalysis. A variety of reactive substrate analogues have been used to inactivate and covalently label pepsin. Many of these are diazoketones—for example, *N*-tosyl-L-phenylalanyl diazomethane. This molecule undergoes attack by the β-carboxyl of an active-site aspartate, with expulsion of molecular nitrogen (N_2) and ester formation, leading to inactivation and covalent modification of the active site.

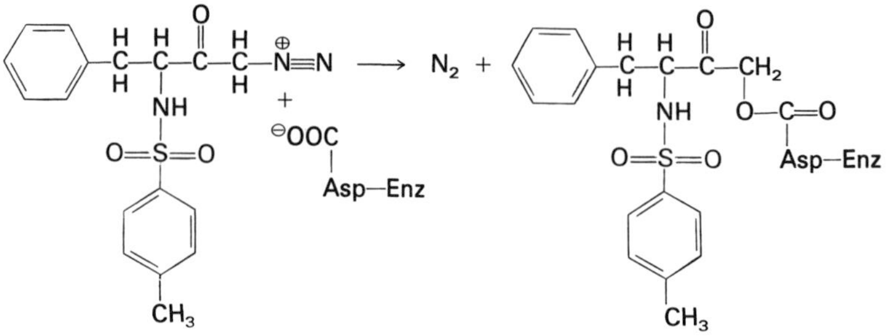

Copper(II) ions accelerate the rate of inactivation, presumably by facilitating loss of N_2 to the carbene-$Cu^{2\oplus}$ complex, which may be the actual reactive electrophile (Delpierre and Fruton, 1966). This same aspartyl residue is labeled by a variety of other *N*-blocked aminoacyl diazoketones. An epoxide, 1,2-epoxy-3-(*p*-nitrophenacyl)-propane, also inactivates the enzyme, modifying a different carboxyl moiety at the active site—consistent with the idea that there are two reactive carboxyl moieties (Tang, 1971).

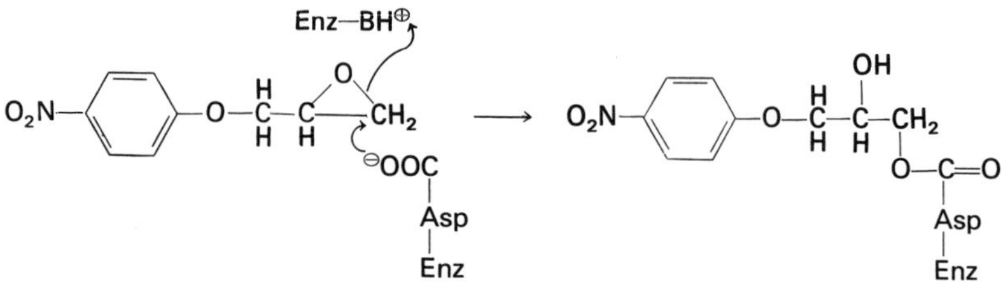

On the basis of experience with chymotrypsin and other active-site serine proteases, one might expect pepsin to catalyze acyl transfer to water by way of covalent acyl-enzyme intermediates. However, unlike chymotrypsin-catalyzed hydrolysis of esters, pepsin-catalyzed hydrolysis yields no detectable initial burst of product formation with chromogenic synthetic substrates. This implies that any covalent intermediates that may form do so after the rate-determining step, thus do not accumulate, and are not easily detectable. This inference is also corroborated by the lack of a common V_{max} for structurally related substrates. Further, whereas chymotrypsin will transfer the acyl moiety of substrates to nucleophiles other than water (such as hydroxylamine or methanol), no such trapping experiments have been successful with pepsin. Either an acyl enzyme does not form, or it is inaccessible to these small simple nucleophiles. Also consistent with a difference in the ways that chymotrypsin and pepsin process substrates is the finding that the ratio of turnover numbers for ester substrates to analogous amide substrates is 1,000 for chymotrypsin but 2 for pepsin.

On the other hand, both pepsin and chymotrypsin catalyze an exchange reaction of $[^{18}O]$-H_2O into the carboxyl group of *N*-blocked L-amino acids but not of the D-enantiomers (Clement, 1973; Fruton, 1971, 1974; Tang, 1976). In the chymotrypsin-mediated exchange, the acyl-enzyme intermediate accounts for this ^{18}O-transfer. The blocked amino acid (an *N*-acyl phenylalanine in this example) can react with free chymotrypsin to form a small amount of acyl enzyme and a molecule of $[^{16}O]$-H_2O.

$$\text{PhCH}_2\text{–CH(NHR)–COO}^{\ominus} + \text{Enz–Ser–OH} \rightleftharpoons H_2O + \text{PhCH}_2\text{–CH(NHR)–C(=O)–O–Ser–Enz}$$

This water molecule diffuses from the active site, while a molecule of the bulk water species $[^{18}O]$-H_2O diffuses in. Now, reversal of the above step in the favored direction means that the acyl enzyme hydrolyzes, with incorporation of ^{18}O into the carboxyl of the *N*-blocked phenylalanine.

$$\text{PhCH}_2\text{–CH(NHR)–C(=O)–O–Ser–Enz} + H_2{}^{18}O \rightleftharpoons \text{PhCH}_2\text{–CH(NHR)–C(=O)–}^{18}O^{\ominus} + \text{Enz–Ser–OH}$$

Repetition of this process many times will lead to macroscopic exchange, as measured by formation of ^{18}O-labeled *N*-acyl phenylalanine. Because ^{18}O is a stable, nonradioactive oxygen isotope, the ^{18}O-incorporation must be measured by mass spectrometry. This mechanism for the $[^{18}O]$-H_2O exchange with chymotrypsin could also be operative in pepsin catalysis, providing an indirect criterion for an acyl-enzyme intermediate.

One of the most interesting facets of pepsin catalysis is the ability to carry out transpeptidation reactions that have provided data in support of, not an acyl enzyme, but rather an unusual *amino-enzyme* intermediate. That is, the moiety normally regarded as the leaving group (the amine fragment of a peptide substrate) is retained on the enzyme while, instead, the acyl fragment departs. The amine moiety can then be transferred to a different acyl acceptor. These conclusions stem from data with model substrates such as *N*-acetyltyrosyl tyrosine (Clement, 1973; Fruton, 1971, 1974; Tang, 1976). Along with the expected hydrolysis products (tyrosine and *N*-acetyl tyrosine), detectable quantities of the dipeptide Tyr–Tyr are formed. The following sequence has been suggested as explanation.

$$\text{Ac—Tyr—Tyr} + \text{Enz—COOH} \rightleftharpoons \text{Ac—Tyr} + \text{Enz—C(=O)—NH—CH(COO}^{\ominus}\text{)—CH}_2\text{—C}_6\text{H}_4\text{—OH} \quad (1)$$

amino enzyme

$$\text{Enz—C(=O)—NH—CH(COO}^{\ominus}\text{)—CH}_2\text{—C}_6\text{H}_4\text{—OH} + \text{Ac—Tyr—Tyr} \longrightarrow \text{Enz—COOH} + \text{Ac—Tyr—Tyr—Tyr} \quad (2)$$

$$\text{Enz—C(=O)—O}^{\ominus} + \text{Ac—Tyr—Tyr—Tyr} \longrightarrow \text{Ac—Tyr} + \text{Enz—C(=O)—NH—CH(CH}_2\text{C}_6\text{H}_4\text{OH)—C(=O)—NH—CH(COO}^{\ominus}\text{)—CH}_2\text{—C}_6\text{H}_4\text{—OH} \quad (3)$$

(enzyme–Tyr–Tyr)

$$\text{Enz—C(=O)—NH—Tyr—Tyr—COO}^{\ominus} \xrightarrow{H_2O} \text{Enz—COOH} + {}^{\oplus}\text{H}_3\text{N—Tyr—Tyr—COO}^{\ominus} \quad (4)$$

In the first turnover, peptide-bond cleavage occurs with release of acetyl tyrosine and retention of the amine fragment (tyrosine) in amide linkage with an active-site carboxyl. Then, in step 2, rather than amine-fragment transfer to H_2O as acceptor, another molecule of acetyl-tyrosine substrate may bind and react to form as initial product the acetylated tripeptide (acetyl-Tyr–Tyr–Tyr). Subsequent enzymatic cleavage of the acetyl-tyrosine trimer (steps 3 and 4) will yield acetyl tyrosine and the observed Tyr–Tyr dipeptide. If the amino enzyme in step 1 is trapped by H_2O rather than by substrate, the expected hydrolysis products (acetyl-Tyr and Tyr) are formed. Examples of "leaving-group transfer" are rare in enzymatic catalysis (only one other case, a reaction catalyzed by an enzyme CoA transferase, is known; see Jencks, 1974), and the generality of this mechanism for pepsin catalysis is not yet established because such transpeptidation occurs only with certain substrates and not with others.

In fact, there is other transpeptidation data available on pepsin catalysis that seems to provide the opposite mechanistic conclusion—evidence for an acyl enzyme (Takahashi and Hofmann, 1972, 1974; Hofmann, 1974). With the tripeptide Leu–Tyr–Leu, one sees, not transfer of the C-terminal amino acid (as above), but rather transfer of the amino-terminal residue—implicating an acyl-enzyme intermediate, as evidenced by the unexpected formation of the dipeptide Leu–Leu as product. This product is explicable if a Leu-enzyme intermediate (acyl enzyme) forms, and then is attacked by the amino terminus of another substrate molecule to yield a molecule of tetrapeptide product. This initial tetrapeptide could then undergo cleavage to two dipeptides: the unique Leu–Leu and the common product Tyr–Leu (Takahashi and Hofmann, 1972, 1974; Hofmann, 1974).

$$\begin{aligned} \text{Leu–Tyr–Leu} + \text{Enz} &\longrightarrow \text{Leu–Enz} + \text{Tyr–Leu} \\ \text{Leu–Enz} + \text{Leu–Tyr–Leu} &\longrightarrow \text{Leu–Leu–Tyr–Leu} + \text{Enz} \\ \text{Leu–Leu–Tyr–Leu} &\xrightarrow{\text{enz}} \boxed{\text{Leu–Leu}} + \text{Tyr–Leu} \end{aligned}$$

Some portion of this apparent contradiction is alleviated by recent experiments (Newmark and Knowles, 1975) using the same Leu–Tyr–Leu tripeptide, but with two radioactive isotopes—for example, [^{14}C]-Leu–Tyr–[^{3}H]-Leu. It can be shown that formation of [^{14}C]-Leu–[^{14}C]-Leu dipeptide must result from acyl transfer, whereas [^{3}H]-Leu–[^{3}H]-Leu must arise from amine transfer. (The reader should verify this.) The experimental results indicate that both processes compete throughout the pH range 2.5–5.2, with acyl transfer accounting for about 80% of the Leu–Leu dipeptide formed, and amine transfer accounting for the rest. Thus, it may be that either half of the cleaved substrate can depart the active site first. One then may be tempted to write mechanisms involving both intermediates, but one should maintain the reservation that the *covalent* nature of either intermediate is unproven to date. It could be that dissociation rates for release of one or the other are slow.

The mechanistic ambiguities may (or may not) be resolved when the three-dimensional structure of pepsin is solved at high resolution. A high-resolution X-ray map (at 2.7 Å) has recently been reported for a related acid protease, penicillopepsin, from a *Penicillium* mold (Hsu et al., 1977). This enzyme and pepsin show similar substrate specificity, similar susceptibility to affinity labels, and related catalytic mechanisms—thus they may have related structures. The active site of resting penicillopepsin may be located inside and at the end of a deep groove that can accommodate perhaps seven to nine amino-acid residues of a polypeptide substrate. There are two carboxylates (of Asp^{32} and Asp^{215}) in intimate contact, probably sharing a hydrogen bond as shown. During catalysis the susceptible substrate bond may lie in between them.

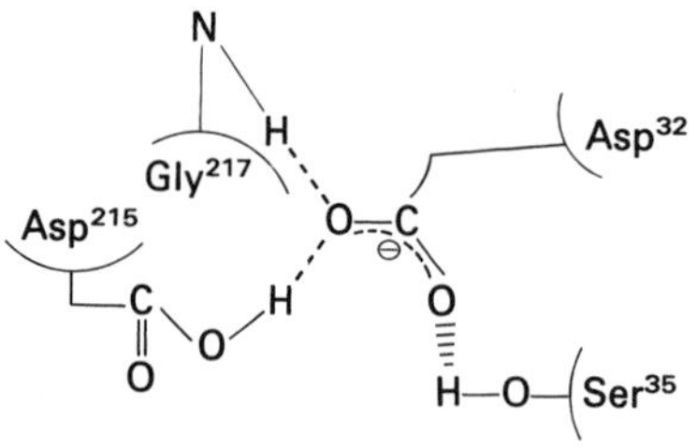

resting form of penicillopepsin

3.C.3 Metallopeptidases

The metallopeptidases require a stoichiometric amount of a divalent cation as cofactor, bound to specific amino-acid residues as ligands, in order to be catalytically active. In this category, more mechanistic information is available on the exopeptidases than on the endopeptidases.

Metallo-exopeptidases are known that cleave either at the amino terminus of a polypeptide substrate or at the carboxyl terminus. Progressive action can lead to the sequential release of subsequent terminal residues. For example,

$$^{\oplus}H_3N\text{—Leu—Ile—Tyr}\sim\sim COO^{\ominus} \xrightarrow[\text{peptidase}]{\text{amino-}} {}^{\oplus}NH_3\text{—Leu—}COO^{\ominus} + {}^{\oplus}H_3N\text{—Ile—Tyr}\sim\sim COO^{\ominus}$$

$$\downarrow \text{amino-peptidase}$$

$$^{\oplus}NH_3\text{—Ile—}COO^{\ominus} + {}^{\oplus}H_3N\text{—Tyr}\sim\sim COO^{\ominus}$$

The best-known *N*-terminal exopeptidase, leucine aminopeptidase (with kinetic preference for hydrophobic terminal residues), has been purified to homogeneity and appears to contain one atom of $Zn^{2\oplus}$ per active site as the essential metal for catalysis. The enzyme is large ($\sim$300,000 mol wt), and its lack of ready availability has slowed both structural and mechanistic investigations, so that little is known about mechanisms of aminopeptidase catalysis (DeLange and Smith, 1971).

The situation is much better with carboxyl-terminus exopeptidases: carboxypeptidase A from bovine pancreas was crystallized in 1937 (Anson, 1937), and an X-ray map at 2.0 Å resolution has been available since 1970 (Lipscomb et al., 1970). The proenzyme (procarboxypeptidase A) is secreted by the pancreas and is activated by removal of a 60–amino-acid *N*-terminal piece to yield the 307-residue active enzyme (Blackburn, 1976, chap. 4). Both proenzyme and enzyme contain one zinc atom per active site, with histidines at positions 69 and 196 and the γ-carboxylate of Glu^{72} acting as ligands. The fourth ligand is H_2O in the free enzyme and is the carbonyl oxygen of the susceptible peptide bond in the ES complex. The apoenzyme is inactive, but the $Zn^{2\oplus}$ can be replaced by cobalt or nickel (and less satisfactorily by other metals) to regenerate activity (Coleman and Vallee, 1960). Chemical modification studies of Tyr^{248} suggest that it is essential for catalysis (Simpson et al., 1963; Riordan et al., 1967), but an irreversible-inactivator substrate analogue, *N*-bromacetyl-*N*-methyl-L-phenylalanine (BMPA), inactivates the enzyme by alkylation of Glu^{270} (Hass and Neurath, 1970). Curiously, although BMPA inactivates both peptidase and esterase activities, some tyrosine modifications lower peptidase activity to 2% but raise esterase activity (with hippuryl-β-phenyl lactate) to 700% of the activity level in native enzyme (Zeffren and Hall, 1973, p. 200).

A current view of carboxypeptidase-A catalysis suggests that binding of the substrate involves charge pairing of the terminal carboxylate group with the guanidinium cation of Arg^{145}, setting up the carbonyl oxygen of the terminal peptide bond to become a ligand of the active-site zinc (Blackburn, 1976, chap. 4). This liganding polarizes the carbonyl group, increasing the electrophilicity of the carbonyl carbon and facilitating nucleophile-mediated hydrolysis.

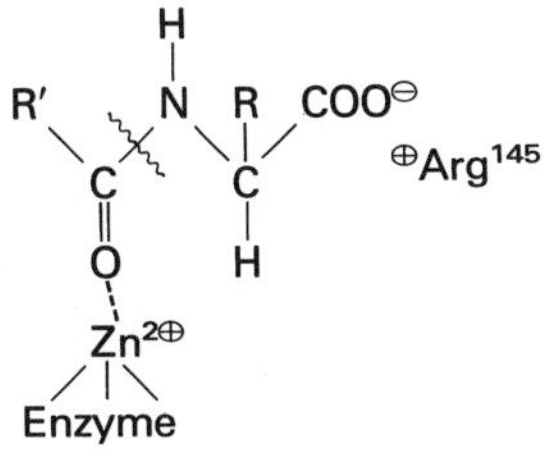

H
R′ N R $COO^{\ominus}$
C C $^{\oplus}Arg^{145}$
O H
$Zn^{2\oplus}$
Enzyme

As noted above, two other residues are implicated in catalysis by chemical-modification and X-ray data. The phenolic hydroxyl of Tyr^{248} is in sufficient proximity to the susceptible amide linkage in the substrate to donate a proton to the amine fragment on cleavage, facilitating its expulsion as a leaving group by general acid catalysis. Additionally, the γ-carboxylate of Glu^{270} is at hand. This carboxylate can act in two ways. It can function as a general base catalyst to heighten the nucleophilicity of a water molecule and thus accelerate hydrolysis:

Or the carboxylate can itself act as nucleophile, generating an anhydride intermediate that would be labile to hydrolysis:

anhydride

$RCOO^{\ominus} + Glu^{270}\text{—}COOH$

Until very recently, attempts to detect anhydride in the steady state or to trap it with alternate nucleophiles had failed (Kaiser and Kaiser, 1972). However, recent experiments with an ester substrate at −25° to −40° C have suggested a covalent anhydride intermediate (Makinen et al., 1976). The data suggest different temperature dependencies for acylation and deacylation rate constants. At −25° C deacylation is rate-determining, and the anhydride accumulates; at room temperature, acylation is rate-determining, and no accumulation of covalent anhydride can occur.

The X-ray map deduced for carboxypeptidase A suggests that the phenolic side chain of Tyr^{248} caps the substrate in the ES complex, sequestering it from solvent (Hartsuck and Lipscomb, 1970, 1973). In binding of substrate or release of product, this tyrosyl residue must move aside. Consequently, the enzyme in crystals is a mixture of the open form and the closed form (where the phenolic-OH is liganded to the active-site zinc). We have noted that chemical modification of Tyr^{248} abolishes all activity toward synthetic amide substrates but does not abolish all activity toward ester substrates (Simpson et al., 1963; Riordan et al., 1967). With amide substrates, Jencks (1975*a*, p. 238) has noted that the phenolic hydroxyl (acting as a general acid catalyst) is essential to avoid high-energy amine anions as leaving groups ($pK_a \approx 35$). With ester substrates, the unprotonated alkoxide ions ($pK_a \approx 15$) are more stable (if still poor) leaving groups,★ and expulsion (an index of ester turnover) may still occur (Vallee et al., 1963).

Breslow and Wernick (1977) have recently proposed a unifying picture for the mechanisms of catalysis by carboxypeptidase A that suggests distinct hydrolytic mechanisms for esters and peptides. They are convinced that peptides are hydrolyzed with the zinc atom coordinated to the peptide carbonyl and Glu^{270} acting as a general base for the attacking water molecule. But they suggest that ester carbonyl groups may be weaker ligands to the zinc and, on binding, may not displace the water molecule coordinated to the zinc in the resting enzyme. Now Glu^{270} could act as nucleophile and form an anhydride intermediate; the aquo-zinc complex would deliver a proton in its role as general acid (a role played by Tyr^{248} when peptides are hydrolyzed). The same groups, a water molecule and a substrate acyl group, would be placed between the Glu^{270} γ-carboxylate and the active-site zinc, but the order of positioning would be reversed in the two substrate classes.

In contrast to the mammalian carboxypeptidases just described, a carboxypeptidase Y purified from bakers' yeast does not have essential metals. Rather, it has both an active-site serine and an active-site histidine, reminiscent of chymotrypsin and other serine active-site endopeptidases (Hayaishi et al., 1975). One might therefore begin to differentiate between "serine carboxypeptidases" and "metal carboxypeptidases."

★The best leaving groups are anions ($X^{\ominus}$) whose conjugate acids have $pK_a < 0$; the anions then are weak bases that can effectively stabilize an electron pair.

Chapter 4

Two Kinetic Tools for the Assay of Enzymatic Mechanisms

Our central purpose in this text is to examine and to categorize the chemistry of the major types of enzyme-catalyzed reactions. As we have seen with chymotrypsin in Chapter 3, this study requires examination of mechanistic experiments designed to identify the most probable reaction path during catalysis. Evaluation of the *kinetic parameters* of a given enzymatic reaction may indicate (among other things) which chemical or physical steps—e.g., binding of a substrate, release of a product, or an enzyme isomerization—are slow (or fast) in the reaction; whether one substrate must bind before another; or whether molecules bind at a single locus (or within some definite region) of the active site. Any proposed chemical mechanism for the transformation must eventually be made consistent with the information from kinetic experiments.

We do not have the space in this book to discuss the complete array of kinetic methods that have been developed for the study of enzyme catalysis; the treatise by Segel (1975*b*) is an excellent and comprehensive recent reference on the subject. In this chapter we limit our attention to two simple kinetic tools that may provide intuitive insights into the chemistry of a given enzymatic reaction.

1. First, we briefly discuss kinetic isotope effects, with attention to the limits of interpretation in enzymatic cases.
2. Then we take up some simple aspects of the kinetic patterns of inhibition caused by *reversible* binding of inhibitors to enzymes. Using the enzyme acetylcholinesterase as a specific example, we examine two limiting cases (competitive inhibition and noncompetitive inhibition) and comment on physical interpretations of these kinetic phenomena.

4.A KINETIC ISOTOPE EFFECTS

Review the mechanistic scheme for chymotrypsin catalysis (Fig. 3-17). The nucleophilicity of the active-site serine is enhanced by the fact that the histidine residue of that charge-relay system acts as a general base catalyst: proton transfers occur in the transition states for formation and breakdown of both tetrahedral adducts and the acyl-enzyme intermediate. Thus we except a proton-transfer step to be involved during the slow step (or steps) of catalysis, both with synthetic (ester and amide) substrates and with natural polypeptide substrates. When the solvent H_2O is replaced by 99.9% deuterium oxide (D_2O), this isotopic substitution has a pronounced effect on the rate of chymotrypsin-mediated hydrolysis. A *kinetic isotope effect* is observed: the V_{max} for substrate hydrolysis is decreased 2- to 4-fold relative to the rate in H_2O (the magnitude of the effect depends on the reaction conditions and the particular substrate).

$$V_{max\,(H_2O)}/V_{max\,(D_2O)} = 2 \text{ to } 4$$

The detection of a solvent-deuterium kinetic isotope effect on V_{max} for chymotrypsin (Bender et al., 1962; Bender and Hamilton, 1962) and for other serine proteases has been interpreted as confirmation that general acid or base catalysis (involving transfer of $H^{\oplus}$—or $D^{\oplus}$ on isotopic substitution) does occur in the rate-determining step. This interpretation is consistent with the catalytic roles assigned to His[57] (general base rather than nucleophile) and Asp[102].

In this section we examine (in a simple way) the physical basis for such a kinetic isotope effect. Our exposition is adapted in part from a discussion of kinetic isotope effects by Jencks (1969, chap. 4); for more comprehensive treatment and for discussion of the limitations of the simplifying assumptions used, the reader is directed to that source and to other reviews (Richards, 1970; Cleland et al., 1976; Eigen, 1964; Bender, 1971, chap. 2).

In many studies of reaction mechanisms, substitution of deuterium for hydrogen has caused a substantial rate reduction in some reaction involving a proton-transfer step. For example, it has been long established that the rate-determining step in the bromination of acetone is the base-catalyzed enolate-ion formation:

$$H_3C{-}C({=}O){-}CH_3 \;(+\;\ddot{B}) \xrightleftharpoons{\text{slow}} \left\{ H_3C{-}C(O^{\ominus}){=}CH_2 \longleftrightarrow H_3C{-}C({=}O){-}CH_2^{\ominus} \right\} \xrightarrow[\text{fast}]{Br_2} H_3C{-}C({=}O){-}CH_2{-}Br + Br^{\ominus}$$

When hexadeuteroacetone (C_3D—CO—CD_3) is used in this reaction, a 6- to 10-fold decrease in bromination rate is observed—i.e., $k_H/k_D = 6$ to 10. The bromination shows a large kinetic isotope effect.

In general, a test for kinetic isotope effect can help determine *whether or not hydrogen transfer occurs in a step that may be partially or fully rate-determining.* Substitute deuterium for the hydrogen, and look for a reduction in rate of reaction. Such a kinetic isotope effect indicates that hydrogen transfer is involved in the rate-determining step or steps. On the other hand, if no kinetic isotope effect is observed ($k_H/k_D = 1$), we cannot jump to the conclusion that no hydrogen transfer is involved in the reaction. Lack of a kinetic isotope effect simply means that hydrogen transfer (if it occurs) is not the slowest step (or one of the slowest steps) in the observed reaction.

4.A.1 Solvent Kinetic Isotope Effects versus Substrate Kinetic Isotope Effects

We make a functional distinction between (1) hydrogen transfers from carbons, and (2) hydrogen transfers from heteroatoms such as oxygen, nitrogen, or sulfur (Jencks, 1969, chap. 4):

(1) C—H
(2) O—H, N—H, or S—H

Hydrogens bonded to carbon are stable to most exchange processes (unless the carbon center is unusually acidic), and these hydrogens do not exchange readily when the compound is dissolved in a protic solvent such as water. On the other hand, hydrogens bonded to oxygen, nitrogen, or sulfur in most cases exchange *very rapidly* with solvent protons. The exchange rate is a function of the pK_a of the functional group.

$$X{-}H + H_2O \underset{k_r}{\overset{k_f}{\rightleftharpoons}} X^{\ominus} + H_3O^{\oplus}$$

For these functionalities, the back-reaction rate constant k_r is essentially that for a diffusion-controlled reaction: ca. $10^{10}\ \text{M}^{-1}\ \text{sec}^{-1}$ (Eigen, 1964; Bender, 1971, chap. 2). The forward-reaction rate constant k_f has been found to be related to pK_a values by the equation

$$k_f = k_r \times 10^{-(15.5-x)}$$

where 15.5 is the pK_a for H_2O, and x is the pK_a for HX. For a lysine-amine N—H bond with pK_a of 10.5, $k_f = 10^{10} \times 10^{-5} = 10^5\ \text{sec}^{-1}$; in 55 M H_2O, the second-order rate constant will be $(10^5\ \text{sec}^{-1})$ (55 M) $= 5.5 \times 10^6\ \text{M}^{-1}\ \text{sec}^{-1}$ for the rate of exchange of N—H with solvent, and the hydrogens bonded to nitrogen thus will come rapidly to equilibrium with solvent hydrogen.

For example, if one dissolves lactic acid in acidic D_2O (below the pK_a of lactic acid), the hydroxyl and carboxylic-acid protons will exchange with solvent deuterons almost immediately after dissolution. Suppose one added 50 μmoles of lactic acid to 1 ml of D_2O, to make a 50 mM solution; the D_2O is 55 M (110 N). Although only a small fraction of D_2O is dissociated into $D^{\oplus}$ and $OD^{\ominus}$ at any instant, the equilibrium, rapidly established, will ensure that essentially all the lactate exists as form 2:

$$\underset{\text{form 1}}{H_3C{-}\overset{\displaystyle H}{\underset{\displaystyle OH}{C}}{-}COOH} \xrightarrow{D_2O} \underset{\text{form 2}}{H_3C{-}\overset{\displaystyle H}{\underset{\displaystyle OD}{C}}{-}COOD}$$

The protons from the exchangeable positions of lactic acid total 100 μeq; the deuterons from 1 ml of D_2O total 110 meq (milliequivalents). The ratio of protons to deuterons is about

$$(10^2\ \mu\text{eq})/(10^2+10^5\ \mu\text{eq}) \approx 1/10^3$$

Thus 99.9% of the lactic acid will exist as form 2.

On the other hand, the carbon-bound hydrogens of the methyl group and the methine carbon (C-2) will show no detectable exchange with solvent deuterons even after months. A proton NMR (nuclear magnetic resonance) spectrum of lactic acid in D_2O confirms that only two kinds of hydrogens (C—H and $C{-}H_3$) are observed. One could record the spectrum of form 1 by dissolving lactic acid in a nonprotic solvent such as hexadeuterodimethyl sulfoxide. Without solvent dissociation, there will be no protons to exchange at the O—H protons, and the NMR spectrum recorded will be that of form 1. In enzymatic catalyses, of course, reactions occur in the *protic solvent water.* Thus, use of D_2O in place of H_2O will equilibrate all rapidly exchangeable hydrogens, and a solvent-deuterium kinetic isotope effect can be used to test a rate-determining role for transfer of some hydrogen bonded to a hetero atom (as with the kinetic isotope effect observed for chymotrypsin-mediated hydrolysis in deuterated solvent).

If one wishes to determine whether C—H bond breakage is the slow step in some chemical or enzymatic transformation, then the compound in question should be synthesized with deuterium in place of that hydrogen. The rate of reaction with the deuterated substrate can be compared to the rate of reaction with protonated substrate. A rate reduction for the deuterated substrate *under V_{max} conditions* signifies a substrate-deuterium kinetic isotope effect (as in the example of the bromination of acetone and hexadeuteroacetone).

4.A.2 The Origin of Kinetic Isotope Effects

A rigorous theoretical explanation of kinetic isotope effects is a complex and difficult venture (Jencks, 1969, chap. 4; O'Leary and Kleutz, 1972; Cleland et al., 1976; Klinman, 1978). Here we limit ourselves to an oversimplification that does seem to have practical and predictive utility, although it may be slightly misleading. This simplification is the argument that (qualitatively and, generally, quantitatively) the most important contribution is the difference in vibrational zero-point energies of bonds to hydrogen vs. bonds to deuterium. Consider, for example, the vibrational energy levels for a C—H bond. They are, of course, quantized, and the lowest level (zero point) has a nonzero energy $E = (\frac{1}{2})h\nu$, where ν is the frequency of the C—H vibration. If we further assume this bond to be represented by a simple harmonic oscillator, then we can envision a Hooke's-law relationship for ν, such that $\nu \approx (\frac{1}{2})\sqrt{K/M}$, where K is a simple force constant, and M is the reduced mass of the system (here, the isolated C—H bond).

The infrared stretching frequencies of C—H and C—D bonds in many compounds cluster at 2,900 cm^{-1} and 2,100 cm^{-1}. These values convert to zero-point energies of 4.15 kcal/mole (17.4 kJ/mole) for C—H and 3.0 kcal/mole (12.6 kJ/mole) for C—D. The zero-point energy for the C—D bond is 1.15 kcal/mole (4.8 kJ/mole) lower, just estimating for one vibrational stretch (Fig. 4-1). Consider a reaction in which the C—H or C—D bond is broken in the transition state—i.e., where this stretching vibrational frequency is frozen out in the transition state (bending-frequency changes are thought to cancel out). In this case, the same transition state is reached for the C—H or for the C—D molecule. However, because the C—D bond is at lower zero-point energy (deeper in the potential well than the C—H bond), more energy will be required to bring C—D to the transition state than for C—H, a difference of 1.15 kcal/mole just on the change in zero-point energy alone (Fig. 4-2). At room temperature, one can

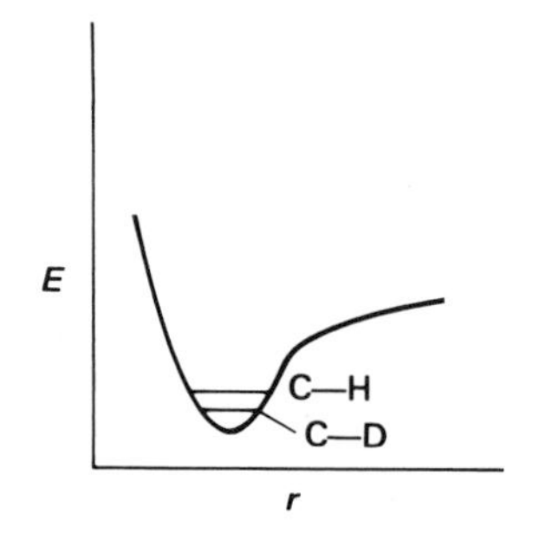

Figure 4-1
Lower zero-point vibrational energy for C—D vibration than for C—H vibration.

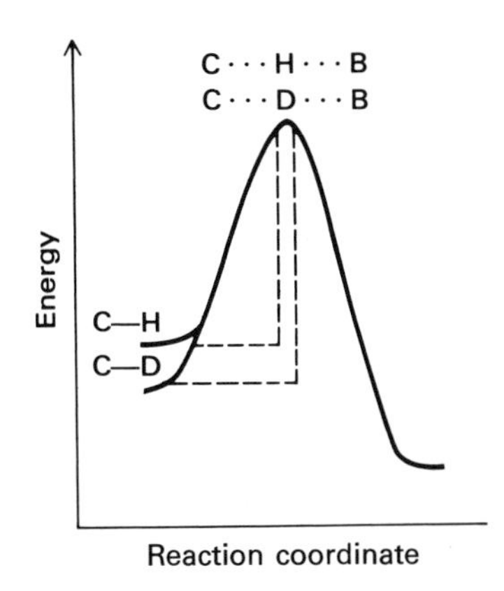

Figure 4-2
Higher activation energy to transition state for C—D bond than for C—H bond.

calculate that this energy difference might show up as a 7-fold rate difference in favor of C—H (Jencks, 1969, chap. 4). In other words, for this effect,

$$k_H/k_D = 7$$

Strictly speaking, one must worry about the smaller mass of C—H than of C—D (or of N—H, O—H, or S—H than N—D, O—D, or S—D) combinations. However, the masses of C, N, O, and S are so much greater than the masses of either H or D that the heavier atoms remain essentially fixed in the vibration, the force constants are about the same, and so the mass difference between H and D is paramount in determining the frequency of such vibrations.

In analogous calculations, one can show that O—H vs. O—D differences in zero-point energies should correspond to a 10-fold rate enhancement for O—H vs. O—D reacting at room temperature (Jencks, 1969, chap. 4). By this analysis, we would expect such a rate difference if, for the reaction in question, such proton (deuteron) transfer were *fully* rate-determining. Again, we stress that this treatment is a crude approximation of the actual bases for isotope effects, in that we have worried only about stretching vibrations and in fact have singled out one stretching vibration from a complex polyatomic molecule.

In fact, k_H/k_D ratios ranging from 2 to 15 are observed in reactions where other evidence suggests that proton transfer occurs in the transition state (i.e., rate-determining proton transfer). Low but real isotope-effect ratios around 2 or less (say, 1.6) may mean that the proton-transfer step is partially rate-determining, one of two or more slow steps of about equivalent rates. For instance, an elementary step involving C—H (C—^{2}H) bond breakage might show an isotope effect of 5 to 6. If some subsequent step in the cycle occurs with a similar rate constant, then expression of the isotope effect will be partially suppressed when overall product formation is examined: the observed isotope effect might be only 2.

$$E + S \rightleftharpoons E\text{—}S \rightleftharpoons E\text{—}S' \rightleftharpoons E\text{—}S'' \xrightarrow{\text{slow}} E\text{—}P \rightleftharpoons E + P$$

C—H cleavage
intrinsic $k_H/k_D = 5$

observed $k_H/k_D = 2$
from monitoring of
product formation

Larger values ($k_H/k_D = 25$ to 30) are seen in some chemical reactions (Jencks, 1969, chap. 4), although almost never in enzymatic reactions (Cleland et al., 1976). These large ratios may reflect quantum-mechanical tunneling through potential barriers, occurring selectively for the light isotope of hydrogen. Isotopic substitution will slow down the rate of any bond-breaking step but, if this rate remains faster than some other step in a reaction, then no detectable isotope effect will show up in the observed reaction rate:

$$k_{obs(H)}/k_{obs(D)} = 1$$

If this ratio of observed rates is greater than one, this observation is diagnostic, indicating that bond breakage is one of the slow steps.

4.A.3 Tritium Kinetic Isotope Effects versus Deuterium Kinetic Isotope Effects

We have discussed two common isotopes of hydrogen, ^1H (protium) and ^2H (deuterium); the third common isotope, of course, is ^3H (tritium). If substitution of deuterium for protium in a chemical reaction produces a detectable kinetic isotope effect, substitution of tritium will produce a larger kinetic isotope effect (k_H/k_T) because of the larger mass of ^3H and the resultant lower zero-point vibrational energy of, for instance, a C—T bond relative to C—D and C—H. In a number of reactions, a simple logarithmic proportionality is observed (Swain et al., 1958):

$$\log(k_H/k_T) = 1.44 \log(k_H/k_D)$$

However, as we shall see, this Swain–Schaad relationship need not hold for some observed kinetic isotope effects at V_{max}, and this inequality probably reflects an overall suppression of the expression of an intrinsic isotope effect on some elementary chemical step (Northrop, 1975).

There is an important experimental distinction between ^2H and ^3H kinetic isotope experiments; this distinction is intuitively apparent, but is worth noting explicitly to avoid problems in interpretation. With deuterium, a stable isotope, all molecules in the population (in most cases) have hydrogen replaced synthetically by deuterium (100% D_2O or 100% C—D molecules). Thus, every molecule in this homogeneous population will experience the same reaction.

In contrast, tritium is an unstable, radioactive isotope (see ¶3.B.8), used only in tracer amounts. In experimental samples of T_2O, only one molecule in 10^{10} (or less) may be ^3H. With synthetic tritiated substrates, where one must work with enough mass to carry out the required synthetic transformations and then purify and characterize the compound adequately, the amount of ^3H may be as low as one molecule in 10^{15}. Clearly, not all the molecules in the population will see the isotope effect. In fact, by measuring the conversion of radioactive molecules to products, we will be monitoring a discrimination process where, for instance, an enzyme will catalyze conversion of 10^{10} protium molecules at one rate and one tritium molecule at a slower rate. Experimentally, one cannot simply measure the reaction rate, because the bulk population is composed of ^1H molecules and thus no detectable macroscopic rate reduction will be observed. One must have a way of following both the ^1H and the ^3H molecules. This can be done by determination of the specific radioactivity of both the substrates and the products (¶3.B.8).

As an example, consider an enzymatic oxidation of lactate to pyruvate, where an isotope effect was suspected from experiments with 2-[^2H]-lactate. One might start with 2-[^3H]-lactate (tritium only at C-2) at 5 μCi/μmole (or $5 \times 2.2 \times 10^6 = 1.1 \times 10^7$ dpm/μmole). During oxidation, the α-[^3H] should be released into the solvent as $^3H^{\oplus}$.

$$H_3C{-}\underset{OH}{\overset{^3H}{C}}{-}COO^{\ominus} \xrightarrow[H_2O]{\text{enzyme}} H_3C{-}\underset{\|\atop O}{C}{-}COO^{\ominus} + H^{\oplus} + 2e^{\ominus} + \boxed{^3H^{\oplus}}$$

One can monitor pyruvate formation by 2,4-dinitrophenylhydrazone formation, and thus can compute a rate of product formation. One can also measure how much ^3H radioactivity has been released into water at any point in time, because [^3H]-H_2O is volatile whereas [^3H]-lactate is not. (To simplify notation, we will write [^3H]-H_2O as T_2O, although no water molecule is likely to have two tritium atoms at once.) If there is no isotope effect, T_2O should be formed at a specific radioactivity equal to that of lactate, or 1.1×10^7 dpm/μmole, because one C—H bond is broken for every molecule of pyruvate formed.

If a kinetic isotope selection occurs against tritiated molecules, ^1H molecules will react faster than ^3H molecules. If $k_H/k_T = 10$, the T_2O formed will have a specific radioactivity one-tenth that of the lactate—i.e., 1.1×10^6 dpm/μmole (see Fig. 4-3).

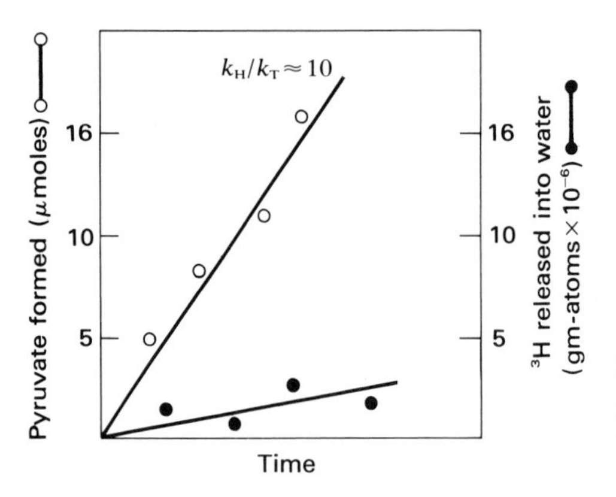

Figure 4-3
Profile for a tenfold tritium isotope selection in lactate oxidation.

Note that (if there is a discrimination against tritiated molecules), as the substrate molecules containing hydrogen react preferentially, the specific radioactivity of remaining substrate molecules will rise and cause errors in analyzing k_H/k_T by T_2O formation. In practice, one determines k_H/k_T at low percentages of substrate conversion (5% to 10%) to minimize this problem. A way of thinking about this isotope selection is to calculate the μmoles of T_2O formed by using the specific radioactivity of the starting [^3H]-lactate. In this example, at 1 μmole of pyruvate produced, one would calculate 0.1 μmole of T_2O produced (for a k_H/k_T of 10). Because the stoichiometry is 1 : 1 in fact, (1.0 μmole)/(0.1 μmole) 10 is the selection exerted against reaction of tritiated substrate molecules.

4.A.4 Heavy-Atom Isotope Effects

To this point, our discussion has concentrated on substitution with isotopes of hydrogen. We can also look at rate effects that occur on substitution with isotopes of heavier atoms, including the stable isotopes ^{13}C for ^{12}C, ^{15}N for ^{14}N, and ^{18}O for ^{16}O. Studies have been performed on heavy-atom isotope effects, both in chemical reactions and in enzyme-catalyzed reactions (Cleland et al., 1976). One immediate problem is that the expected rate differences in heavy-atom isotope effects are much smaller because of the much smaller percentage change in atomic mass on isotopic substitution: ^2H is twice as heavy as ^1H, but ^{13}C is only 8% heavier than ^{12}C, and the zero-point vibrational energy levels for the carbon isotopes are much closer. Maximal observed isotope effects for $^{13}C/^{12}C$ are about 1.06, those for $^{15}N/^{14}N$ about 1.04, and those for $^{18}O/^{16}O$ about 1.06. To obtain credible distinctions, great experimental precision is required in measurements of these small differences in rates. One technique for detection of heavy-atom isotope effects has been to measure isotopic composition of products early in the reaction and again at 100% completion with high-precision ratio mass spectrometers (O'Leary and Kluetz, 1972). Recently, researchers in Cleland's laboratory have developed a sensitive equilibrium-perturbation method for enzymatic experiments where freely reversible reactions are studied (Schimerlik et al., 1975).

Let's look at two examples of studies of heavy-atom isotope effects in enzymatic catalysis.

Nonenzymatic decarboxylation of oxalacetate to CO_2 and pyruvate, catalyzed by $Mn^{2\oplus}$, shows a $^{13}C/^{12}C$ rate isotope effect of 1.06, suggesting that carbon–carbon bond cleavage is rate-determining.

$$^{\ominus}O\text{—}^{13}C(=O)\text{—}CH_2\text{—}C(=O\cdots Mn^{2\oplus})\text{—}C(=O)\text{—}O^{\ominus} \longrightarrow H_2C{=}C(O^{\ominus}\cdots Mn^{2\oplus})\text{—}C(=O)\text{—}O^{\ominus} + {}^{13}CO_2 \overset{H^{\oplus}}{\rightleftharpoons} H_3C\text{—}C(=O)\text{—}C(=O)\text{—}O^{\ominus} + Mn^{2\oplus}$$

On the other hand, enzymatic decarboxylation proceeds without detectable isotope effect, suggesting that some other step (perhaps either a physical binding or release step, or a chemical step such as ketonization of the initial enolate product) is slow in the enzymatic process (Jencks, 1969, chap. 4; Seltzer et al., 1969).

The second study deals with the enzymatic dehydrogenation by glutamate dehydrogenase of the amino acid L-glutamate to 2-iminoglutamate, which then decomposes to 2-ketoglutarate and ammonia, undoubtedly via a carbinolamine tetrahedral adduct, prior to release of products.

$$^{\ominus}OOC\text{—}CH_2CH_2\text{—}CH(^{15}NH_3^{\oplus})\text{—}COO^{\ominus} \xrightarrow{\text{enzyme}} {}^{\ominus}OOC\text{—}CH_2CH_2\text{—}C(={}^{15}NH_2^{\oplus})\text{—}COO^{\ominus} + 2e^{\ominus}$$

$$\xrightarrow{H_2O} {}^{\ominus}OOC\text{—}CH_2CH_2\text{—}C(OH)(^{15}NH_3^{\oplus})\text{—}COO^{\ominus}\ (\text{carbinolamine}) \rightleftharpoons {}^{\ominus}OOC\text{—}CH_2CH_2\text{—}C(=O)\text{—}COO^{\ominus} + {}^{15}NH_3 \overset{H^{\oplus}}{\rightleftharpoons} {}^{15}NH_4^{\oplus}$$

Kinetic studies have suggested that iminoglutamate hydrolysis is the probable rate-limiting step (Cleland, 1975). This supposition has received support from demonstration of a ^{15}N isotope effect of 1.046 with [^{15}N]-glutamate as substrate

(Schimerlik et al., 1975); however, it is not clear whether the slow step is the formation or the decomposition of the carbinolamine (Brown and Drury, 1965).

4.A.5 Kinetic Isotope Effects on V_{max} and V_{max}/K_m in Enzymatic Reactions

As noted in Chapter 3, the two independent kinetic parameters of the Michaelis–Menten equation are V_{max} and K_m. Most enzymologists who have analyzed for isotopic effects have looked for rate reductions in V_{max} because V_{max} (for most enzymes) contains a rate term for the chemical step involving the covalent change being analyzed. In probing with a deuterated substrate, for instance, if $V_{max(H)}/V_{max(D)} > 1$, then V_{max} rates may be controlled totally or in part by a slow catalytic step involving a C—H (C—D) fragmentation. The magnitude of V_{max} difference observed for substrate-deuterium kinetic isotope effects varies widely. The oxidation of L-[2-^{2}H]-lactate by a hydroxyacid oxidase from rat kidney shows a V_{max} isotope effect of 8, suggesting that C—D cleavage is probably fully rate-determining (Cromartie and Walsh, 1975). Such a large isotope effect on V_{max} is relatively rare. In many cases, observed V_{max} isotope effects may be in the range of 1.5 to 2.0, and interpretation can be complex (Walsh, Krodel, et al., 1973). One possible interpretation of the small isotope effects is that the V_{max} of an enzymatic reaction is rarely determined by a *single step* involving chemical change. Rather, the enzyme as catalyst may permit a reaction path where a single high energy barrier is replaced by several smaller barriers of about equivalent heights. That is, multiple steps may occur at similar rates, with each step being partly rate-determining. A low value for a kinetic isotope effect may reflect such behavior. Cleland (1975) argues that the actual chemical steps rarely limit the rates of enzymatic reactions, and that conformational changes in the proteins leading to release of products often place limits on V_{max}; such slow physical steps could, of course, completely suppress expression in V_{max} of the isotope effect from an elementary step.

Interpretation of isotope effects on the other enzyme kinetic constant, K_m, is obscured by the fact that K_m is generally a complex constant composed of several rate constants, which vary depending on mechanism as noted earlier. For the limiting case where $K_m = K_s$, no isotope effect should be detected in the measured Michaelis constants because it is unlikely that a physical binding step would be sensitive to isotopic substitution in the ligand (Fig. 4-4).

Although an isotope effect on K_m per se may not be easily interpretable, Northrop (1975) suggests that one can fruitfully factor an isotope effect in an enzyme-catalyzed reaction into effects on V_{max} and effects on the ratio V_{max}/K_m.

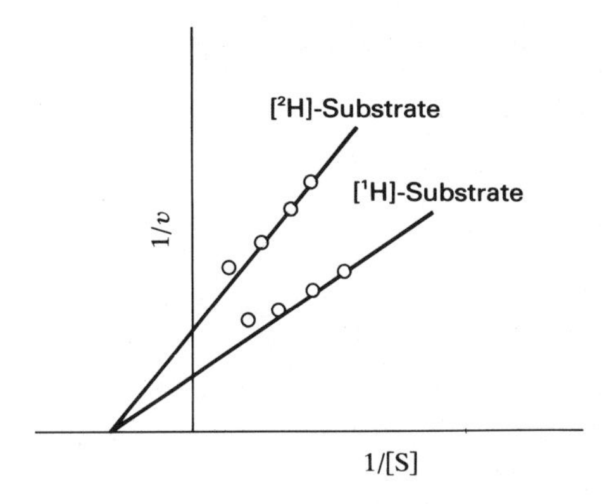

Figure 4-4
Double reciprocal plot indicating a substrate-deuterium kinetic isotope effect on V_{max}, but no effect on K_m.

We recall (¶3.B.4) that V_{max} and V_{max}/K_m represent the zero-order and first-order kinetic regions, respectively, in the velocity vs. [S] dependence curve of enzymes obeying Michaelis–Menten kinetics:

$$\text{at low [S]}, \quad v = (V_{max}/K_m)[S]$$
$$\text{at high [S]}, \quad v = V_{max}$$

Northrop has pointed out that both V_{max} and V_{max}/K_m are complex functions of several rate constants, some in common and some distinct; a reduction in the magnitude of one rate constant, due to an isotope effect on that step, can be expressed differently in V_{max} and V_{max}/K_m, depending on the makeup of these rate constants. The following example was provided by Northrop (1975) to illustrate the approach for a simple scheme,

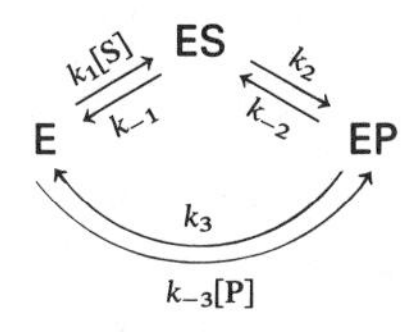

The catalytic step in the front direction is k_2, and this step will experience the isotope effect (e.g., C—H cleavage). Under initial velocity conditions (i.e., k_{-2} and k_{-3}[P] insignificant), from the steady-state assumption, we obtain the following form for the velocity equation:

$$v = \frac{k_1 k_2 k_3}{k_3(k_{-1} + k_2) + k_1(k_2 + k_3)[S]}$$

Expressions for the basic kinetic parameters then have the following forms:

$$V_{max} = \{k_2 k_3/(k_2 + k_3)\}[E_T]$$
$$K_m = \{k_3(k_{-1} + k_2)\}/\{k_1(k_2 + k_3)\}$$
$$V_{max}/K_m = \{k_1 k_2/(k_{-1} + k_2)\}[E_T]$$

How will a kinetic isotope effect on k_2 be expressed in V_{max} or V_{max}/K_m? The extent to which the intrinsic isotope effect is expressed in V_{max} depends on the ratio of k_2 to k_3, the ratio of the rate constant for the chemical step to the rate constant for product release. When k_2/k_3 is small (<1), then the observed isotope effect on V_{max} approaches the true isotope effect; when k_2/k_3 is large (>1), product release is rate-determining, and the isotope effect tends to be suppressed when V_{max} is measured.

$$V_{max(H)}/V_{max(D)} = (k_{2(H)}/k_{2(D)})\{(k_{2(D)} + k_3)/(k_{2(H)} + k_3)\}$$
$$\rightarrow 1 \quad \text{when } k_2 \gg k_3$$
$$\rightarrow k_{2(H)}/k_{2(D)} \quad \text{when } k_3 \gg k_2$$

On the other hand, the expression of the intrinsic isotope effect in V_{max}/K_m depends on a different ratio: the ratio of k_2 to k_{-1} (the off rate from the ES complex). (Let's abbreviate V_{max}/K_m as V/K.)

$$(V/K)_H/(V/K)_D = \{(k_{2(H)}/k_{2(D)}) + (k_2/k_{-1})_H\}/\{(k_2/k_{-1})_H + 1\}$$
$$= (k_{2(H)}/k_{2(D)})\{[1 + (k_{2(D)}/k_{-1})]/[1 + (k_{2(H)}/k_{-1})]\}$$

It can be seen that the apparent isotope effect on V/K varies inversely with the ratio k_2/k_{-1}, between values of 1 and the true intrinsic isotope effect. When $k_2 \gg k_{-1}$, the effect is suppressed; when $k_2 \ll k_{-1}$, the effect is fully expressed.

$$(V/K)_H/(V/K)_D \rightarrow (k_{2(H)}/k_{2(D)})(k_{2(D)}/k_{2(H)}) = 1 \quad \text{when } k_2 \gg k_{-1}$$
$$\rightarrow (k_{2(H)}/k_{2(D)})(1) = k_{2(H)}/k_{2(D)} \quad \text{when } k_{-1} \gg k_2$$

That is, an isotope effect does not show up specifically in V_{max}/K_m when ES breaks down to EP many times for each time it reverts to E+S, a "*high commitment to catalysis.*" Conversely, when dissociation of S from ES occurs much more readily than covalent change to form EP, *commitment to catalysis is low*, and the isotope effect shows up in V_{max}/K_m, but not in V_{max}. Northrop (1975) noted that low values for V_{max}/K_m predominate, indicating "that the commitment to catalysis is generally rather high, and that the rate of dissociation of substrates from enzymes is not as rapid as is often assumed."

Returning to the experimental and theoretical distinctions between deuterium isotope effects and tritium isotope effects, we note that, with a homogeneous population of deuterated substrate molecules, one can clearly measure effects on V_{max} and V_{max}/K_m directly by measuring the rate of bulk species reaction. With tritiated substrates, where the radioactive isotope is in trace amounts, *one can measure the tritium isotope effect only on* V_{max}/K_m, *but not on* V_{max} *directly* (Northrop, 1975).

In many enzymatic reactions where V_{max} isotope effects have been detected, the V_{max}/K_m effect is identical—that is, K_m values are identical for protio and deutero substrates. When K_m is altered for the deuterated substrate, V_{max} and V_{max}/K_m isotope effects will differ. A few enzymes are known where isotope effects are expressed only on V_{max}/K_m and not on V_{max} (Bush et al., 1973; Cheung and Walsh, 1976*a*). One is alcohol dehydrogenase from liver, when acetaldehyde is reduced to 1-[^{2}H]-ethanol with a deuterated redox coenzyme NADH (see Chapter 10) (Bush et al., 1973). This case has been explained by the fact that release of the oxidized coenzyme NAD is obligatorily after ethanol release and is the *slowest* step in catalysis, so that the effect on V_{max} is suppressed (Northrop, 1975).

$$H^{\oplus} + NAD^2H + H_3C\text{—}C(=O)\text{—}H \rightleftharpoons NAD + H_3C\text{—}C(OH)(^2H)\text{—}H$$

On occasion, one may find that the Swain–Schaad relationship is not obeyed in measured tritium and deuterium isotope effects in an enzymatic reaction—i.e., $\log(k_H/k_T) > 1.44 \log(k_H/k_D)$; this observation probably reflects *multiple slow transfers* of the hydrogen isotope during the reaction—e.g., to and from enzymatic basic groups.

4.A.6 Solvent Isotope Effects in Enzymatic Reactions

In experiments with solvent isotope effects, interpretation can be very complex in the case of enzymatic reactions. In D_2O vs. H_2O experiments, there is a change in the equilibrium concentration of acids and bases: $pH = pD + 0.4$ (Jencks, 1969, chap. 4). Thus, in D_2O the ionization behavior of substrate or, more probably, of the multiionic enzyme can be perturbed in unpredictable functional directions. This can change both the numbers and the strengths of hydrogen bonds responsible for active-site geometry. D_2O is 23% more viscous than H_2O, implying a change in solvent structure and polarizability. Also, the O—D bond is shorter than the O—H bond by 0.04 Å, which also can alter enzyme structure (Jencks, 1969, chap. 4).

On the other hand, when the enzymatic reaction in question involves reversible exchange of a carbon-bound hydrogen with solvent at some intermediate stage of catalysis, then solvent-exchange experiments can be quite useful in analysis of how intermediates partition. We shall note in Chapter 19 the elegant studies of Albery and Knowles (1976) and colleagues, in which a systematic use of both substrate deuterium and tritium effects and solvent deuterium and tritium effects was made for evaluation of all the energy barriers in triose-P-isomerase–mediated catalysis.

4.A.7 Secondary Kinetic Isotope Effects

The isotope effects discussed so far are primary kinetic isotope effects: isotopic substitution occurs at the bond broken in the reaction in question, and we have pointed out that, for $k_H/k_D > 1$, the C—H or C—D bond breakage is significantly advanced in the rate-determining transition state. Studies have also been made on the effect of isotopic substitution of deuterium or tritium at a bond *adjacent* to the reacting bond, such that the C—D or C—T bond is not itself broken during the reaction. Isotopic substitution at such α-positions can lead to a rate reduction, a *secondary* kinetic isotope effect. Kirsch (1976) points out that enzymologists have made much less use of 2° kinetic isotope effects than of 1° kinetic isotope effects in mechanistic studies for two reasons. First, the magnitude of 2° effects may range from 1.02 to 1.40 (compared to values of 2 to 10 for 1° effects), requiring much greater experimental precision for detection. Second, the theoretical basis and the interpretation of an observed small 2° effect are considerably less straightforward. Rate changes in 2° isotope effects probably arise from alteration of force constants in a C—H or a C—D bond between ground states and transition states.

The most common experimental studies are those using 2° kinetic isotope effects to probe rehybridization at carbon during a reaction. Chemical studies indicate detectable secondary isotope effects in S_N1 substitution reactions, but not in S_N2 substitutions. For example, the solvolysis of cyclopentyl tosylate shows a 2° kinetic isotope effect of 1.15, whereas the S_N2 displacement of $Br^{\ominus}$ from isopropyl bromide by methoxide gives a value of 1.00 (Streitweiser et al., 1958).

$$\text{C}_5\text{H}_8(\text{H})(\text{OTS}) \xrightarrow{H^{\oplus}, H_2O} \text{C}_5\text{H}_8(\text{H})(\text{OH}) + \text{TSOH}$$

$$k_H/k_D = 1.15$$

$$\text{C}_5\text{H}_8(^2\text{H})(\text{OTS}) \xrightarrow{H^{\oplus}, H_2O} \text{C}_5\text{H}_8(^2\text{H})(\text{OH}) + \text{TSOH}$$

$$\text{(H)C-Br} \xrightarrow{MeO^{\ominus}} \text{MeO-C(H)} + Br^{\ominus}$$

$$k_H/k_D = 1.00$$

$$\text{(}^{2}\text{H)C-Br} \xrightarrow{MeO^{\ominus}} \text{MeO-C(}^{2}\text{H)} + Br^{\ominus}$$

The S_N1 process involves rate-determining formation of the cyclopentyl carbonium ion as an intermediate, whereas the methoxide attack involves concerted arrival of $MeO^{\ominus}$ and departure of $Br^{\ominus}$. The detectable secondary isotope effect in the case of cyclopentyl tosylate is associated with rate-determining conversion of an sp^3 carbon to an sp^2 carbon in the carbonium ion. This is a general association. A variety of sp^3 carbinols have a vibrational frequency at 1,300 cm^{-1} that occurs at 800 cm^{-1} in sp^2 compounds (J. H. Richards, 1970) and, on the basis of this vibrational frequency change, a secondary rate isotope effect of 1.36 has been calculated. A précis of the Streitweiser formulation has been provided by Kirsch (1976). We shall note that most of the enzymatic studies with secondary isotope effects have been carried out when an α-deuterium is placed at a carbon center reversibly converted from sp^3 to sp^2 hybridization (Kirsch, 1976), as in carbonium-ion formation. (Note that, in the reverse direction $sp^2 \rightarrow sp^3$, an inverse secondary isotope effect leads to rate enhancement.) In Chapter 9 we discuss studies of enzyme-catalyzed glycosyl transfers where observed secondary kinetic isotope effects with (C-1)—^{1}H or (C-1)—^{2}H substrates have been interpreted as rate-determining formation of glycosyl carbonium ion species during catalysis.

4.B REVERSIBLE ENZYME–INHIBITION KINETICS WITH ACETYLCHOLINESTERASE AS EXAMPLE

Any compound that acts to diminish the reaction velocity of an enzymatic process can be classified as an inhibitor. The molecular basis of action of antibiotics, of natural toxins (Rando, 1975), and of synthetic poisons is in many instances the inhibition of a specific target enzyme. Further, not all inhibitors are foreign agents: a variety of physiological intracellular inhibition patterns are crucial in regulation of enzyme activity. In Chapter 3, we discussed irreversible inhibitors that produce time-dependent covalent changes in the target enzyme, and we elaborated criteria for physical interpretation. In this section, we discuss some simple patterns of *reversible* enzyme inhibition—inhibition that is essentially instantaneous on mixing enzyme and inhibitor, and that results from noncovalent, reversible binding of the inhibitor substance to the protein. As a specific illustration, we analyze inhibition of acetylcholinesterase, following a few introductory comments about the nature of the enzyme and its physiological role, and a sketch of its mechanism of action.

4.B.1 Some Properties of Acetylcholinesterase

Acetylcholinesterase is another example of an enzyme that has hydrolytic activity, this one as an esterase physiologically rather than as a peptidase, with rather narrow specificity for a small-molecule substrate, acetylcholine.

$$H_3C{-}\overset{\overset{\displaystyle O}{\|}}{C}{-}O{-}CH_2CH_2\overset{\oplus}{N}(CH_3)_3 \xrightarrow[H_2O]{\text{esterase}} H_3C{-}\overset{\overset{\displaystyle O}{\|}}{C}{-}O^{\ominus} + HO{-}CH_2CH_2\overset{\oplus}{N}(CH_3)_3$$

acetylcholine — acetate — choline

Acetylcholine is found as a metabolite in high concentrations in nervous tissue and, in motor nerve tracts, is an active neurotransmitter substance. When a nerve cell enervates a muscle fiber, the nerve-cell ending at the muscle junction forms a morphologically distinct and functionally discrete contact called a synapse. There is a gap (typically of about 500 Å) between the presynaptic nerve-ending membrane and the postsynaptic membrane of the muscle cell. When a nerve impulse travels down the axon to the synapse, acetylcholine is released as a diffusable neurotransmitter in quantized packets from the nerve ending, diffuses across the cleft, and interacts with the postsynaptic receptor for acetylcholine on the muscle-cell membrane. This interaction transmits a depolarization signal to the muscle membrane in a manner imperfectly understood.

The physiological role of acetylcholinesterase is to ensure that the nerve impulse is of finite duration by lowering the acetylcholine concentration in the cleft via hydrolysis. This enzymatic cleavage will lower the concentration of free acetylcholine in the synaptic cleft and, by perturbation of equilibrium, will induce dissociation of the acetylcholine–acetylcholine-receptor complex, turning off the signal to the muscle cell. Active acetylcholinesterase is crucial for normal neuromuscular function, and inhibition of this enzyme produces tetanic shock, with eventual muscle paralysis. In severe cases, asphyxiation can result. Not surprisingly, this enzyme is a target for nerve gases and insecticides such as diisopropylphosphofluoridate (DIPF) and other alkylating agents (irreversible inactivators).

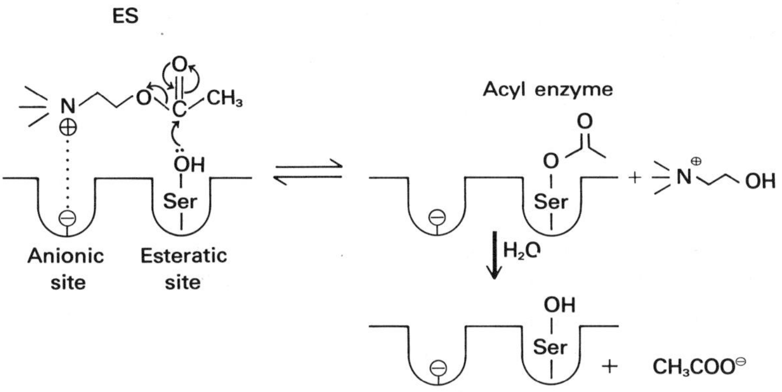

Figure 4-5
Two subsites of the acetylcholinesterase active site.

The enzyme is loosely associated with the postsynaptic membrane. When purified, it has a molecular weight of about 260,000 (tenfold larger than chymotrypsin) and is a functional tetramer of four subunits, each with an active site (Rosenberg, 1975). About 7% of the enzyme by weight consists of carbohydrate, hexose, and hexosamine residues, which may provide specificity for association with the synaptic membranes. As a catalytic entity, acetylcholinesterase is remarkably efficient. With a turnover number (k_{cat}) of 25,000 sec^{-1}, it will cleave a substrate molecule once every 40 μsec. In fact, this high turnover is necessary for postsynaptic membranes to return rapidly to a resting polarized state. Because some nerve fibers can conduct up to 1,000 impulses/sec, this recovery must occur within a fraction of a millisecond. It has been estimated that there are 12,000 enzyme molecules per square μm at some neuromuscular junctions.

Mechanistic experiments with acetylcholinesterase quickly revealed two distinct subregions at the active site. One subsite contains an active-site serine that will undergo acylation to form the acetyl-enzyme intermediate analogous to that of the serine proteases; this is called the esteratic site. The other subsite, called the anionic site, is negatively charged and can provide a salt linkage to enhance binding and recognition of the cationic trimethylammonium portion of acetylcholine in the ES complex (Fig. 4-5). The X-ray structure of acetylcholinesterase has not yet been obtained, and there is no firm evidence for or against the existence of a charge-relay system with this enzyme.

The existence of the anionic binding site can be utilized in the design of therapeutically useful antidotes for treatment of organophosphorus poisoning. In addition to DIPF, other insecticides such as sarin and parathion will covalently block the active-site serine by undergoing nucleophilic attack to produce a serine-phosphoester derivative.

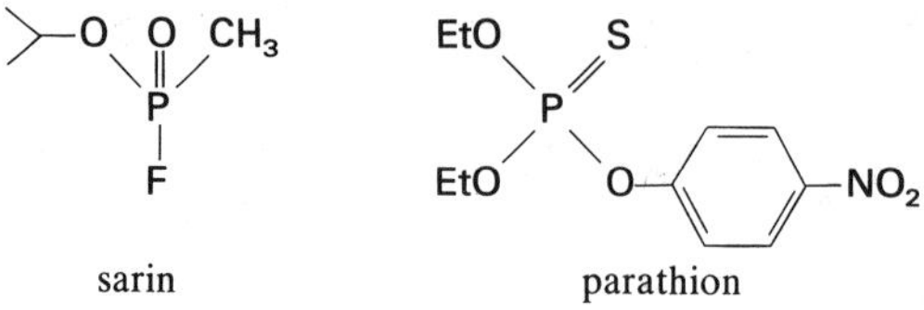

sarin

parathion

These molecules inactivate acetylcholinesterase because the active-site serine-phosphate esters are hydrolyzed only extremely slowly by water (as is true for the proteases). Stronger nucleophiles (hydroxylamine among them) will displace the phosphate and reactivate the enzyme. I. B. Wilson showed that *O*-methylhydroxylamine is ineffective but that *N*-methylhydroxylamine is as good as hydroxylamine itself, suggesting that $NH_2O^{\ominus}$ is the attacking species when NH_2OH is used (Wilson and Ginsburg, 1955; Childs et al., 1955).

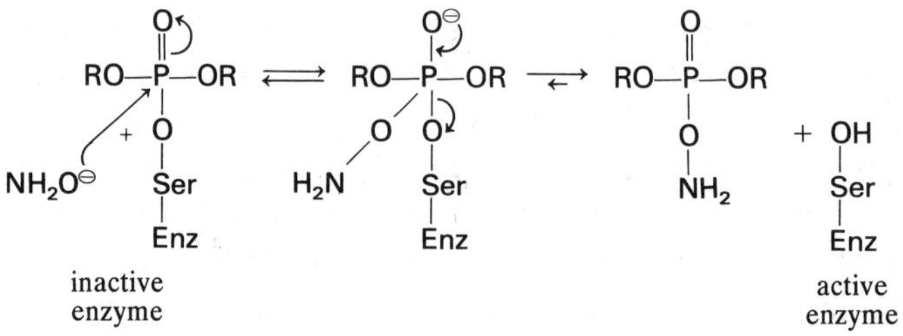

inactive enzyme

active enzyme

This reactivation occurs rapidly only at high concentrations (e.g., 1 M NH_2OH). In the inactive phosphorylated enzyme, as in normal acyl enzyme, the (choline) anionic binding site is unoccupied. If one could synthesize a cationic hydroxylamine derivative that could bind to the anionic site and yet reach the esteratic site, one might see catalysis by approximation—a propinquity effect in enzyme reactivation. One of the most effective reactivation nucleophiles yet found is pyridine aldoximine methiodide or PAM (Frode and Wilson, 1971).

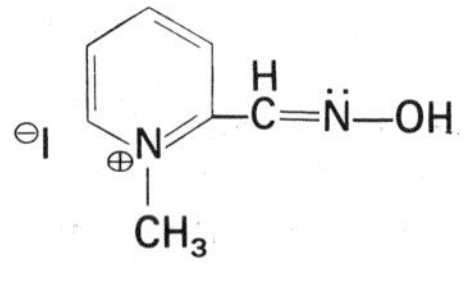

Happily, one gets the same rate of enzyme reactivation at 10^{-6} M PAM as is achieved at 1 M NH_2OH (Fig. 4-6). The propinquity effect is 10^6-fold and makes PAM effective in the treatment of organophosphorus poisonings; it illustrates rational drug design based on a knowledge of enzyme mechanisms.

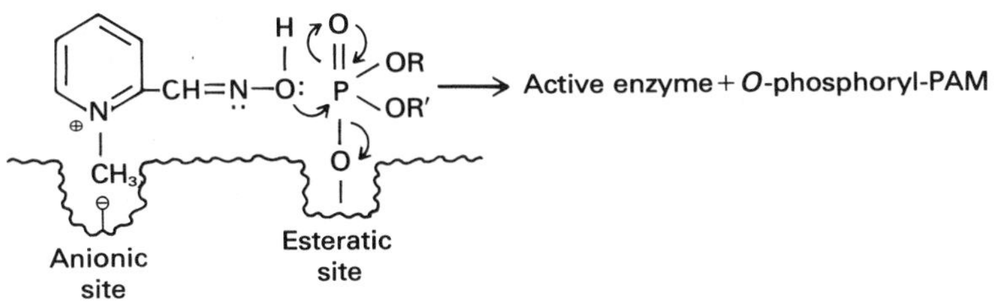

Active enzyme + *O*-phosphoryl-PAM

Anionic site

Esteratic site

Figure 4-6
Enzyme reactivation by PAM.

4.B.2 Reversible Inhibition Patterns

In addition to the irreversible, covalent inhibition of acetylcholinesterase noted above, many cationic ammonium compounds are strong reversible inhibitors, presumably binding at the anionic site. They combine reversibly with the enzyme, do not provide chemical derivatization, and show no time dependence. Because formation of ES and corresponding EI species usually occurs within milliseconds, the inhibition appears instantaneous to the experimenter.

We can distinguish *two* limiting cases for reversible inhibition of acetylcholinesterase by cationic amines, $R_3NH^{\oplus}$, which presumably bind to the anionic (not the esteratic) site.

4.B.2.a Competitive inhibition

Acetylcholinesterase will hydrolyze *N*-methyl-β-amino propylethyl ester

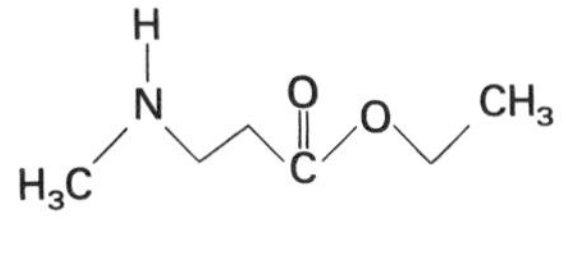

as an alternative substrate *with a low* V_{max}. (What is the acyl enzyme formed from this ester?) When increasing amounts of a tertiary amine cation, $R_3NH^{\oplus}$, are added to enzyme solutions hydrolyzing this substrate, increasing amounts of inhibition of the hydrolytic rate are observed. If a fixed concentration of inhibitor is used and increasing amounts of substrate added, it appears that at very high substrate concentrations the observed velocity approaches the V_{max} rate that would be obtained in the absence of inhibitor. A Lineweaver–Burk plot ($1/v$ vs. $1/[S]$) can be obtained from velocity data in the absence of inhibitor and at some fixed concentration of inhibitor, as shown in Figure 4-7a. If one repeats the

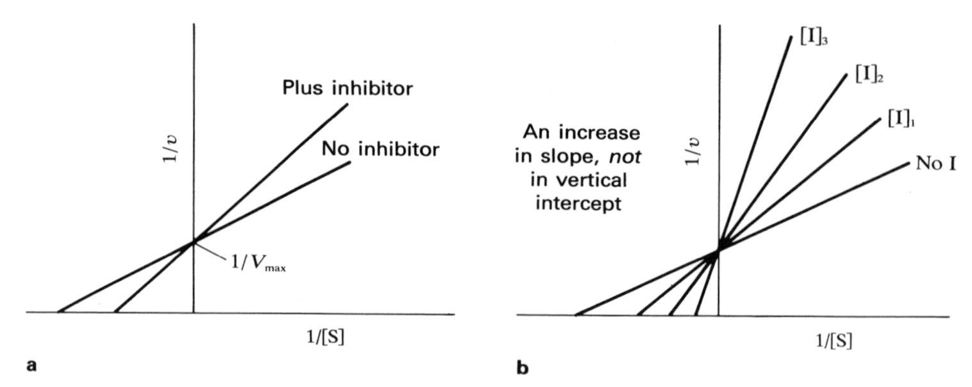

Figure 4-7
(**a**) Double reciprocal plots in absence of inhibitor and at a fixed concentration of inhibitor.
(**b**) Family of plots obtained at various fixed concentrations of inhibitor.

inhibition experiment at several fixed concentrations of $R_3NH^{\oplus}$, one obtains a family of lines in the double reciprocal plot (Fig. 4-7b). Increasing the amount of inhibitor (I) slows down the observed reaction rate at any given [S]. However, at any fixed [I], addition of excess substrate eventually overcomes the inhibition and gives the same V_{max} as is seen in the absence of inhibitor. What has clearly changed is the intercept on the abscissa—corresponding, in absence of I, to $-1/K_m$. Thus, this inhibition affects the observed K_m (increasing it), but does not affect V_{max}.

This kinetic pattern is defined as fully *competitive reversible inhibition.* The physical interpretation of fully competitive inhibition is that I ($R_3NH^{\oplus}$ in this instance) can combine reversibly with E and thus prevent binding of S. Further, it assumes that I binds only to free E—that is, the binding of inhibitor and substrate are mutually exclusive.* This exclusivity may be due to true competition for the same binding region at the enzyme active site, although other physical processes also could produce the competitive pattern.

The relevant scheme is the following:

$$\begin{array}{l} \mathrm{E} + \mathrm{S} \underset{k_{-1}}{\overset{k_1}{\rightleftharpoons}} \mathrm{ES} \xrightarrow{k_2} \mathrm{E} + \mathrm{P} \\ \Big\updownarrow K_i, I \\ \mathrm{EI} \end{array}$$

where $K_i = [E][I]/[EI]$. One can readily show that, taking this new equilibrium for enzyme and inhibitor into account, the Michaelis–Menten equation now can be

* For recent discussions of reversible enzyme-inhibition kinetics, see the books by Segel (1975*a*, 1975*b*) and Zeffren and Hall (1973).

written in the following form:

$$v = V_{max}[S]/(K_m\{1+([I]/K_i)\}+[S])$$

In the reciprocal form for the Lineweaver–Burk plot,

$$1/v = \underbrace{(K_m/V_{max})\{1+([I]/K_i)\}}_{\text{slope}}(1/[S]) + \underbrace{1/V_{max}}_{\text{intercept}}$$

Compare this expression with the reciprocal form in the absence of inhibitor:

$$1/v = (K_m/V_{max})(1/[S]) + 1/V_{max}$$

Note that presence of inhibitor affects the slope term but not the intercept term, consistent with the observation that V_{max} can be attained in the presence of inhibitor. The increase in the slope is the factor $\{1+([I]/K_i)\}$. Clearly, the lower the value of K_i, the more severe the inhibition at any given [S]/[I] ratio (Segel, 1975*a*, 1975*b*).

Thus, for acetylcholinesterase in particular (and for other enzymes in general), one can determine that an inhibitor *is a competitive inhibitor if* V_{max} *is unaffected by the inhibitor while the apparent* K_m *is increased.*

4.B.2.b Noncompetitive inhibition

Now consider the quite different reversible inhibition seen with $R_3NH^{\oplus}$ against the good (high V_{max}) physiological substrate acetylcholine,

$$(CH_3)_3\overset{\oplus}{N}—CH_2CH_2OC(=O)—CH_3$$

In this case, repeated experiments with different fixed amounts of I result in a family of lines with distinct intercept points on the $1/v$ ordinate (Fig. 4-8). Moreover, in this case the lines all intersect at the abscissa. Clearly, there is a change in V_{max} on addition of inhibitor, but no change in K_m. In contrast to the competitive inhibition pattern, this pattern shows that addition of substrate (no matter how much) cannot overcome the inhibition that produces a decreased V_{max}. The inhibitor produces an effect that seems to decrease the amount of active enzyme present.

This double-reciprocal pattern is typical for classical *noncompetitive reversible inhibition,* where the inhibitor and substrate bind "reversibly, randomly, and independently" at different sites. That is, I binds to E and to ES; S binds to E and to EI. However, the resulting ESI complex is catalytically inactive (Segel, 1975*a*, 1975*b*). I and S now are not mutually exclusive.

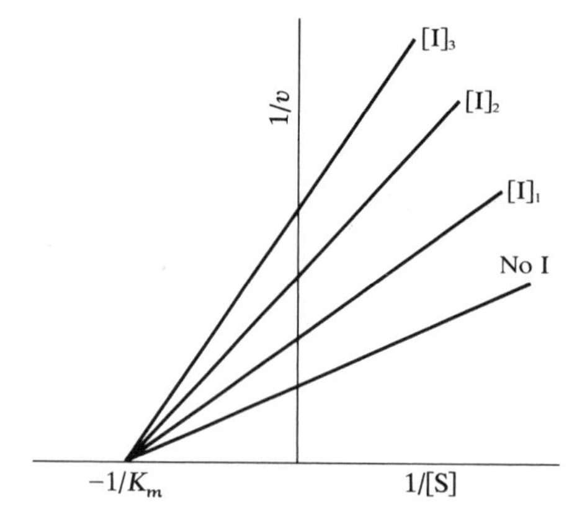

Figure 4-8
Double reciprocal plots for reversible noncompetitive inhibition.

The relevant scheme is the following:

$$\begin{array}{ccccc} E + S & \overset{K_s}{\rightleftharpoons} & ES & \longrightarrow & E + P \\ + & & + & & \\ I & & I & & \\ K_i \downarrow\uparrow & & \uparrow\downarrow K_i & & \\ EI + S & \overset{K_s}{\rightleftharpoons} & ESI & & \end{array}$$

In this simple model, there is an equal affinity of E or ES for I, and also of E or EI for S. The velocity equation now has the following form:*

$$v = V_{max}[S]/(K_m\{1+([I]/K_i)\}+[S]\{1+([I]/K_i)\})$$

Multiplying each side by $\{1+([I]/K_i)\}$, we obtain

$$v = (V_{max}[S]/\{1+([I]/K_i)\})/(K_m+[S])$$

In this form, we see clearly that K_m is unaffected by the inhibitor, and that it is now V_{max} that is decreased by the factor $\{1+([I]/K_i)\}$. The reciprocal equation now has the form

$$1/v = (K_m/V_{max})\{1+([I]/K_i)\}(1/[S])+(1/V_{max})\{1+([I]/K_i)\}$$

---

* Segel points out that noncompetitive inhibition is obtained only under equilibrium conditions (i.e., $K_m = K_s$); a steady-state assumption does not yield an equation of this form.

Note that, because both the slope and the vertical intercept are increased by the same factor, the horizontal intercept ($-1/K_m$) remains unaffected.

4.B.3 A Physical Interpretation of the Kinetic Patterns

One can now ask why the same amine, $R_3NH^{\oplus}$, shows competitive inhibition against a poor substrate and noncompetitive inhibition against a good substrate. The answer stems from the fact that the acetylcholinesterase catalysis involves an acyl-enzyme intermediate. In the Michaelis complex (ES), both the anionic and the esteratic sites are occupied by substrate. In the acyl-enzyme intermediate, only the esteratic site is blocked; because choline has dissociated from the active site, the anionic site is free and could bind $R_3NH^{\oplus}$.

$$
\begin{array}{ccccccc}
E + S & \rightleftharpoons & ES & \xrightarrow{\;\;} & \text{Acyl–Enz} & \longrightarrow & E + P_2 \\
I \downarrow\uparrow & & I \downarrow\uparrow & P_1 & I \downarrow\uparrow & & \\
EI & & ESI & & I \cdot \text{Acyl–Enz} & &
\end{array}
$$

These equilibria a priori predict noncompetitive inhibition. For a good substrate such as acetylcholine, there are numerous data suggesting that deacylation is the rate-determining step; thus the acyl enzyme accumulates as the predominant enzyme species during steady-state catalysis. This acyl enzyme can bind $R_3NH^{\oplus}$ in the free anionic site, rapidly producing the I · Acyl-Enz complex, which apparently does not break down to product. The noncompetitive inhibition pattern is explicable: addition of more S cannot abate the inhibition, because the enzyme is already completely in the acyl-enzyme form.

For a poor substrate, the acylation step is rate-determining, and ES accumulates in reversible fast equilibrium with free E+S. No acyl-enzyme intermediate accumulates during the reaction because its rate of breakdown exceeds its rate of formation. Additional I will compete with S for the active site of the free enzyme and will give fully competitive inhibition. The I cannot combine with acyl enzyme, only because the latter does not accumulate appreciably during catalysis.

If one looks at $R_3NH^{\oplus}$ inhibition vs. a substrate *with an intermediate* V_{max} *value* where neither acylation nor deacylation is fully rate-determining, then appreciable amounts of both E and Acyl-Enz exist during catalysis. In this case, I can combine with both forms and give a noncompetitive inhibition.*

---

* If the affinity for S were different in E and EI, then a mixed pattern of inhibition would result (Segel, 1975*a*, 1975*b*).

Chapter 5

The Versatility of Glutamine: γ-Glutamyl Transfers and Amino Transfers

The amino acid glutamine, in addition to its function as one of the monomeric structural units in proteins, is a central compound in group-transfer metabolism.

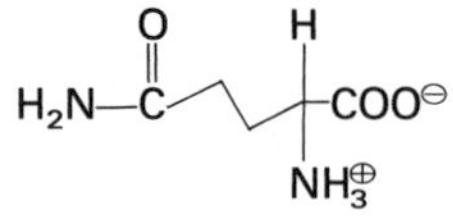

$$H_2N-\overset{\overset{O}{\|}}{C}-CH_2-CH_2-\underset{NH_3^{\oplus}}{\overset{H}{\overset{|}{C}}}-COO^{\ominus}$$

glutamine

Not only does this molecule participate in enzyme-catalyzed acyl transfers, but it is also the major physiological source of ammonia utilized in 13 different biosynthetic reactions (Fig. 5-1; Buchanan, 1973; Prusiner and Stadtman, 1973).

We begin our discussion by examining glutaminases, enzymes that hydrolyze the γ-carboxamide group, carrying out acyl transfer to water (in analogy to the proteases and acetylcholinesterase discussed in preceding chapters).

$$R-\overset{\overset{O}{\|}}{C}-NH_2 + H_2O \rightleftharpoons R-\overset{\overset{O}{\|}}{C}-O^{\ominus} + NH_4^{\oplus}$$

Then we turn to other enzymes that carry out this acyl transfer (γ-glutamyl transfer), not to water, but to a variety of cellular amine acceptors as nucleophiles. These discussions serve as prelude to analysis of the amination reactions that use the nitrogen liberated from the γ-carboxamide group as a source of ammonia. In all these amination reactions, the nonnucleophilic amide nitrogen is rendered nucleophilic by hydrolysis of glutamine to glutamate and ammonia bound at the active site of the specific enzymes; they all possess intrinsic glutaminase activity.

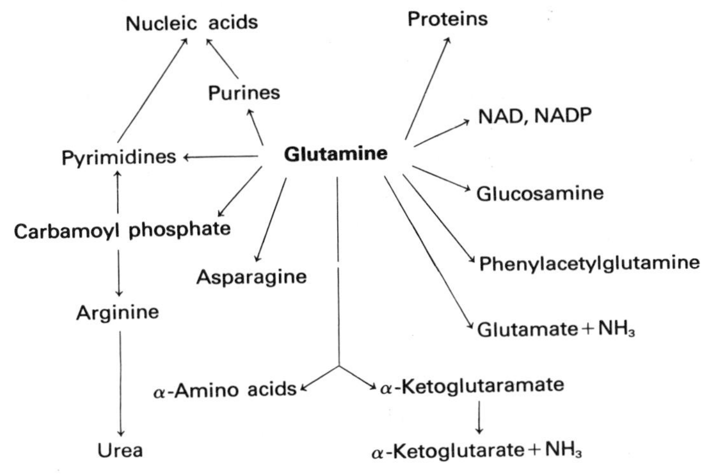

Figure 5-1
Outline of the metabolism of glutamine in mamals.
(Meister, 1974, p. 700)

$$H_3\overset{\oplus}{N}-CH(COO^{\ominus})-CH_2-CH_2-C(=O)-NH_2 \xrightarrow{H_2O} H_3\overset{\oplus}{N}-CH(COO^{\ominus})-CH_2-CH_2-C(=O)-O^{\ominus} + NH_4^{\oplus}$$

L-glutamine → L-glutamate + ammonia

5.A GLUTAMINASE

The *E. coli* glutaminase has been extensively studied. It catalyzes acyl transfer to water, not only from glutamine itself, but also from a variety of glutamyl derivatives:

$$X-C(=O)-CH_2-CH_2-CH(NH_3^{\oplus})-COO^{\ominus}$$

where X may be NHR, $NHNH_2$, NHOH, OR, or SR (Hartman, 1971). The enzyme will also catalyze exchange of glutamate γ-carboxyl oxygens with $[^{18}O]$-H_2O, an experimental result that was used in the chymotrypsin case as evidence for reversible formation of an acyl-enzyme intermediate.

In addition to acyl transfer to water, the enzyme (not surprisingly) catalyzes acyl transfer to hydroxylamine. In the absence of enzyme, glutamine is inert to attack by hydroxylamine. The amide grouping has extensive resonance stabilization and is thermodynamically one of the most stable acyl groups. Thus, the enzyme must at some stage of catalysis convert the γ-carboxamide to a more activated carbonyl derivative that is readily attacked by hydroxylamine—for instance, an oxygen ester or thioester in an acyl-enzyme intermediate.

When a glutamine analogue—the diazoketone 6-diazo-5-oxonorleucine (DON)—is incubated with glutaminase, one sees all the characteristics of an active-site–directed irreversible alkylation process.

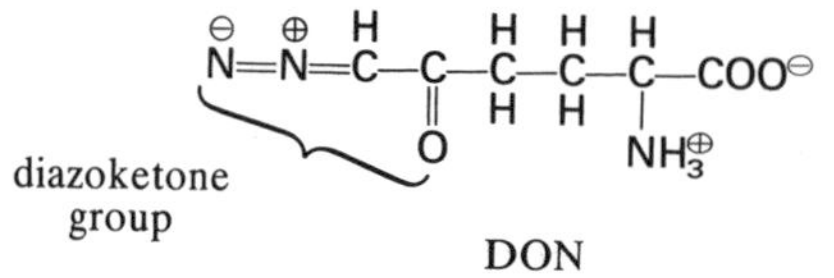

$$\overset{\ominus}{N}{=}\overset{\oplus}{N}{=}CH{-}\underset{\|\atop O}{C}{-}CH_2{-}CH_2{-}\underset{NH_3^{\oplus}}{CH}{-}COO^{\ominus}$$

diazoketone group

DON

However, careful examination of the DON reaction indicates that, prior to undergoing inactivation, each glutaminase molecule can catalyze the hydrolytic cleavage of about 70 molecules of DON (Hartman, 1971). After alkylation of all active sites of the enzyme molecules present, catalytic hydrolysis of DON (or of glutamine) is totally blocked.

When 6-[^{14}C]-DON is used, the hydrolytic reaction yields glutamate and [^{14}C]-methanol. But [^{14}C]-CH_3OH is not the initial product, which instead appears to be [^{14}C]-CH_2N_2 (diazomethane).

$$\left\{ H_2C{=}\overset{\oplus}{N}{=}\overset{\ominus}{N} \longleftrightarrow H_2\overset{\ominus}{C}{-}\overset{\oplus}{N}{\equiv}N \right\}$$

diazomethane

This initial product is indicated by the observation that reaction in benzoate buffer results in production of [^{14}C]-methyl benzoate.

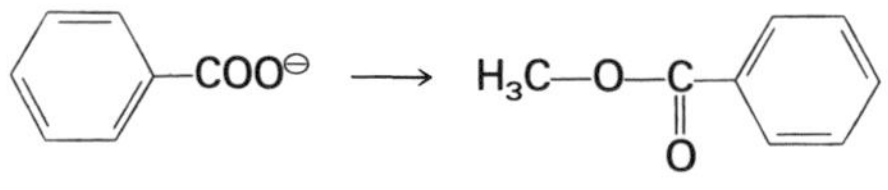

$$C_6H_5{-}COO^{\ominus} \longrightarrow H_3C{-}O{-}\underset{\|\atop O}{C}{-}C_6H_5$$

benzoate   methyl benzoate

The generation of methyl esters is a characteristic reaction of the reactive diazomethane. Presumably, the methanol arises as a secondary product by release of N_2 from diazomethane in aqueous solutions. The observed catalytic hydrolysis of DON could occur by nucleophilic attack at C-5, the carbonyl carbon.

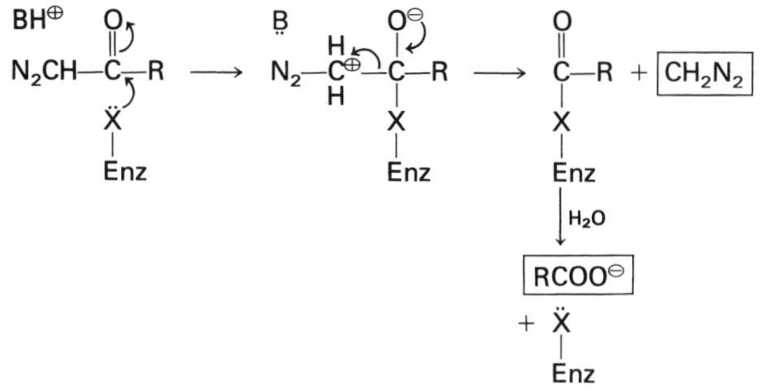

On the other hand, inactivation occurs when a catalytic group of the enzyme attacks C-6 of DON. Attack at this carbon would be facilitated if some active-site $BH^{\oplus}$ acts as a general acid catalyst, converting the diazoketone by protonation to a diazonium species, which can then expel molecular nitrogen easily when attacked nucleophilically at C-6. Attack at C-5, the hydrolytic sequence, occurs 70 times for each attack at C-6, which takes an enzyme molecule out of action. This partitioning may be controlled by the distances between the enzyme nucleophile and C-5 or C-6, and by the frequency with which C-6 is protonated to the diazonium form.

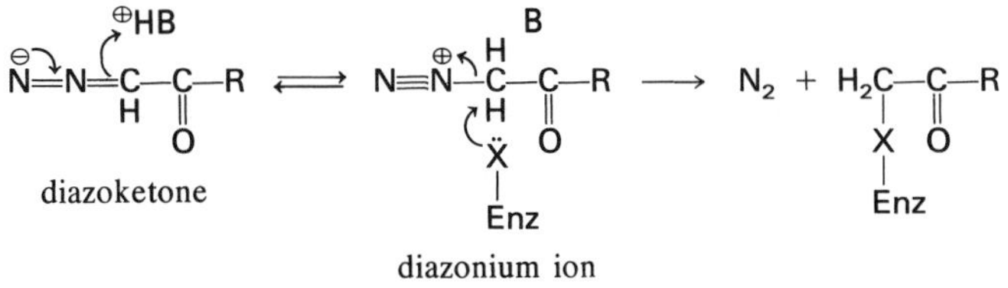

diazoketone

diazonium ion

When glutaminase is inactivated with diazooxonorleucine radioactively labeled at C-1 through C-5, radioactivity is covalently associated with the enzyme, as is label from 6-[^{14}C]-DON. This observation confirms the alkylation as shown and mitigates against the idea that the nascent diazomethane from the hydrolytic sequence is itself the inactivating agent. The nature of the enzyme nucleophile (X) is unknown, although work with other enzymes that perform catalytic reactions on glutamine (see following sections) suggests it could well be a cysteine sulfhydryl.

Although the DON results indicate the existence of a nucleophilic group at the active site, that is insufficient proof that this group does, in fact, add to the amide carbonyl of substrate to form a γ-glutamyl enzyme during hydrolysis. However, the other evidence (^{18}O-exchange and hydroxaminolysis) also suggests a covalent adduct during glutaminase catalysis.

5.B OTHER γ-GLUTAMYL TRANSFERASES: KIDNEY γ-GLUTAMYL TRANSPEPTIDASES

There is a membrane-bound enzyme in mammalian kidney cells that carries out γ-glutamyl transfers to a variety of physiological α-amino acids; it has been termed a γ-glutamyl transpeptidase. Because this activity is localized in the membranes facing the renal lumen, Meister (1973) and his colleagues have suggested that the enzyme has a function in passage of amino acids across the kidney membrane (amino-acid transport) by capturing them as γ-glutamyl derivatives.

The purified enzyme will use a large number of α-amino acids as acceptors (Tate and Meister, 1974, 1975, 1976). As an illustration, consider γ-glutamyl transfer from glutamine to alanine, forming γ-glutamylalanine.

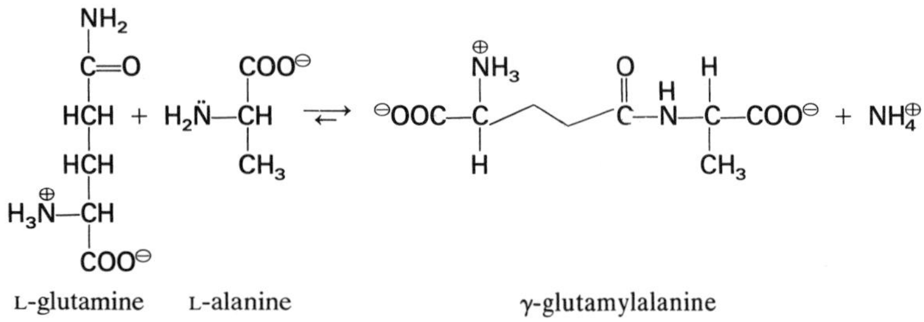

L-glutamine + L-alanine ⇌ γ-glutamylalanine + $NH_4^{\oplus}$

The mechanism is unknown, although it may well be that the alanine amino group attacks a covalent γ-glutamyl-enzyme derivative.

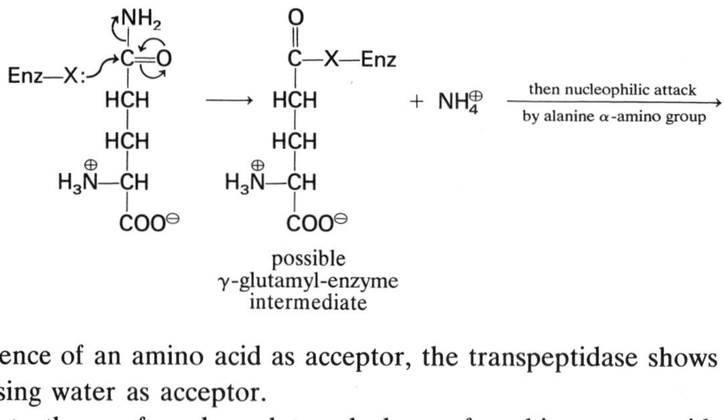

Enz—X: + glutamine ⟶ possible γ-glutamyl-enzyme intermediate + $NH_4^{\oplus}$ ⟶ (then nucleophilic attack by alanine α-amino group)

In the absence of an amino acid as acceptor, the transpeptidase shows hydrolase activity, using water as acceptor.

In fact, the preferred γ-glutamyl donor for this transpeptidase is not glutamine itself, but a ubiquitous γ-glutamyl-containing tripeptide, glutathione (γ-glutamylcysteinylglycine).

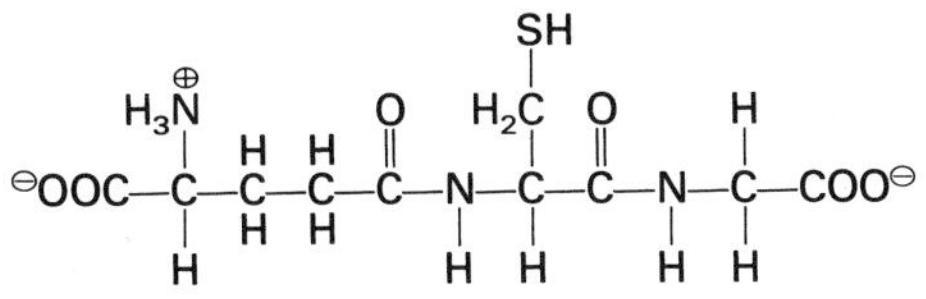

Thus, γ-glutamyl transfer to some α-amino-acid nucleophile will generate, in addition to the new γ-glutamylamino acid, the dipeptide fragment cysteinylglycine.

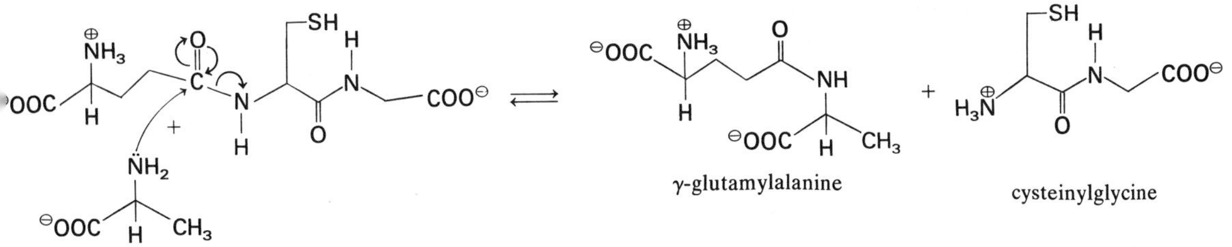

γ-glutamylalanine

cysteinylglycine

The advantage of forming a γ-glutamylamino acid as a putative intermediate in amino-acid transport across kidney membranes would be that the γ-glutamylamino acid, once inside the kidney cell, has a possible ready intramolecular path for breakdown. The α-amino group of the glutamyl moiety can reach, in a five-membered transition state, the γ-glutamyl carbonyl and initiate intramolecular nucleophilic attack. The tetrahedral adduct thus formed can collapse with expulsion of the alanyl residue. There does exist in kidney cells a soluble γ-glutamyl cyclotransferase that accelerates just this reaction with a variety of γ-glutamylamino acids, and that regenerates free alanine while producing a cyclic imino acid, 5-oxoproline, the lactam of glutamic acid.

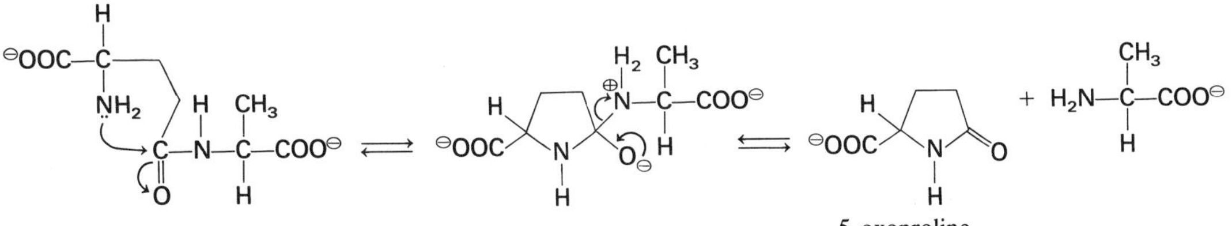

5-oxoproline

The result of the γ-glutamyl transpeptidase and the γ-glutamyl cyclotransferase acting consecutively could be to bring alanine from outside the kidney cell to inside, with concomitant conversion of glutathione to cysteinylglycine and 5-oxoproline. The oxoproline can be opened enzymatically to glutamate, and the tripeptide glutathione can be resynthesized in the kidney cells to permit operation of a γ-glutamyl cycle (Van Der Werf and Meister, 1975). However, the functional significance of a such a cycle in kidney epithelial amino-acid transport has been called into serious question, most recently by the observation that a patient with a genetically defective and nonfunctional transpeptidase nonetheless displayed normal amino-acid transport in that organ (Inoue et al., 1977).

5.C AMINO TRANSFERS FROM THE γ-CARBOXAMIDE OF GLUTAMINE

We have discussed in the preceding sections some enzymes that catalyze γ-glutamyl transfer (acyl transfer) from glutamine to water or to amine acceptors. One of the enzymatic products in every instance is free ammonia in equilibrium with ammonium ion, which predominates at physiological pH ($pK_a = 10$).

$$H\ddot{X} + H_3\overset{\oplus}{N}\text{—}CH(COO^{\ominus})\text{—}CH_2CH_2\text{—}C(=O)NH_2 \longrightarrow H_3\overset{\oplus}{N}\text{—}CH(COO^{\ominus})\text{—}CH_2CH_2\text{—}C(=O)X + NH_3 \underset{}{\overset{H^{\oplus}}{\rightleftharpoons}} NH_4^{\oplus}$$

By such enzymatic hydrolysis of the amide, cells generate controlled amounts of NH_3 required for many essential amino-transfer reactions. Indeed, in biosynthetic pathways for purines, pyrimidines, aromatic amino acids, the cofactors *p*-aminobenzoate and NAD, amino sugars, and the amino acid asparagine, the ultimate source of the cosubstrate ammonia is the glutamine carboxamide nitrogen (Buchanan, 1973; Prusiner and Stadtman, 1973).

It is clear, though, that some activation process must occur in any enzymatic reaction where the glutamine amide nitrogen serves as amino donor—i.e., where it behaves as nucleophile. We have observed earlier that amide functionalities are planar and resonance-stabilized.

$$\left\{ R\text{—}C(=O)\text{—}\ddot{N}H_2 \longleftrightarrow R\text{—}C(\text{—}O^{\ominus})=NH_2^{\oplus} \right\}$$

form 1   form 2

In the canonical form 2, the nitrogen lone pair of electrons is used to form additional bonding interaction with carbon. This tieup of the amide-nitrogen lone pair causes a reduction in the nucleophilicity of that nitrogen.

One mode of increasing the amide-nitrogen nucleophilicity, of course, is to interrupt that resonance interaction. Addition of some nucleophile to the electrophilic carbonyl carbon of the amide generates a tetrahedral adduct, which can then break down to expel the amide nitrogen as nucleophilic ammonia and, in the process, form an acyl derivative of the nucleophile (here shown as an acyl enzyme).

$$\text{Enz—}\ddot{X} + \underset{|}{\overset{NH_2}{\overset{|}{C}}}\text{=O} \rightleftharpoons \text{Enz—X—}\underset{|}{\overset{NH_2}{\overset{|}{C}}}\text{—O}^{\ominus}\ (HB^{\oplus}) \rightleftharpoons \text{Enz—X—}\underset{|}{C}\text{=O} + \ddot{N}H_3$$

In the specific amination examples we discuss in this chapter, it is generally true that both exogenous ammonia and glutamine can serve as amino donors. However, the K_m values for glutamine may be in the physiological range of 10^{-4} to 10^{-3} M, while the value for NH_3 may be 10^{-1} M, suggesting that glutamine is much more efficient in delivering ammonia in effective high local concentrations to the active sites of the biosynthetic enzymes. Furthermore, it is advantageous for cells to avoid high concentrations of free ammonia (ammonium ion) because it is, in the physiological scale, a strong nucleophile and deleteriously reactive. Thus, glutamine is a *carrier* of amino groups in an unreactive, nonnucleophilic form that can be activated by the specific hydrolysis at enzyme active sites to deliver "nascent ammonia" in high local concentrations in a controlled microenvironment, sufficient to aminate the specific second substrate, also held at the active site.

5.C.1 Types of Amination Reactions Using Glutamine as Amino Donor

In general, the amination reactions from glutamine as amino donor occur with three types of substrates: alcohol derivatives, acid or acyl derivatives, and ketones.

$$\text{—}\underset{|}{\overset{|}{C}}\text{—OX} + \text{“}NH_3\text{”} \longrightarrow \text{—}\underset{|}{\overset{|}{C}}\text{—}NH_2 + OX^{\ominus}$$

alcohol derivative

$$\text{—}\underset{\text{—O}}{\overset{}{\underset{|}{C}}}\text{=O} + \text{“}NH_3\text{”} \longrightarrow \text{—}\underset{\text{—O}}{\underset{|}{C}}\text{=NH} \longrightarrow \text{—}\underset{O}{\underset{\|}{C}}\text{—}NH_2$$

acid or acyl derivative

$$\text{>C=O} + \text{“}NH_3\text{”} \longrightarrow \text{>C=}NH_2^{\oplus} + H_2O$$

ketone

The "NH_3" indicates the ammonia released to the active site by the glutaminase function of the enzyme.

A fourth reaction type is a complex reaction involving attack of ammonia on a dihydroaromatic molecule, chorismic acid, to produce *ortho*- or *para*-aminobenzoate. We defer discussion of these processes until Chapter 20.

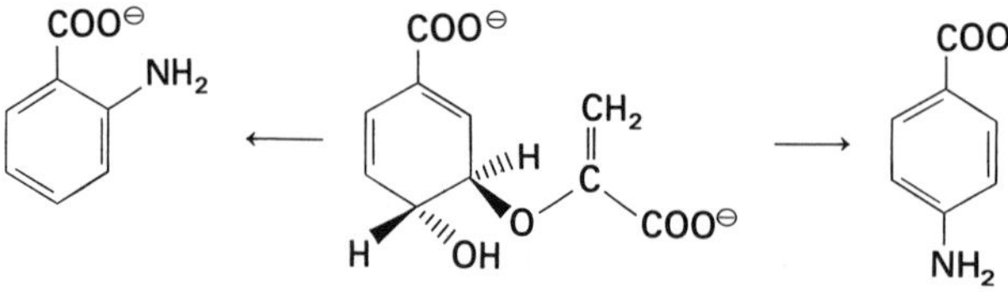

chorismic acid

An alternative classification scheme divides glutamine-dependent aminations into two categories: those that require ATP (adenosine triphosphate) as a cosubstrate and those that do not. Seven of the thirteen enzymatic reactions characterized to date require ATP, which serves as an energy input to force the equilibrium in the amination direction (Buchanan, 1973). We shall use this classification scheme in this chapter, because it offers the first mechanistic insights into the role of ATP in a variety of group-transfer enzymes.

5.C.2 Glutamine-Dependent Aminations with No ATP Requirement

Here we examine two enzymes that have no requirement for ATP: 5-phosphoribosyl-1-pyrophosphate amidotransferase and 2-amino-6-phosphoglucose synthetase.

5.C.2.a 5-Phosphoribosyl-1-pyrophosphate amidotransferase

The enzyme 5-phosphoribosyl-1-pyrophosphate amidotransferase uses the activated D-ribose derivative, 5-phosphoribose-1-pyrophosphate (PRPP), as cosubstrate with glutamine. In terms of the cosubstrate involved, this reaction falls into the first of the four categories described in ¶5.C.1:

$$-\overset{|}{\underset{|}{C}}-OX \xrightarrow{NH_3} -\overset{|}{\underset{|}{C}}-NH_2$$

C-1 of the ribose group of the substrate is activated for displacement of the pyrophosphate substituent by a nucleophile. In general, phosphate and pyrophosphate are effective leaving groups in enzyme-catalyzed displacements, in large part due to the stability of the inorganic phosphate or pyrophosphate products (they are weak bases) and the favorable solvation energies of these products.

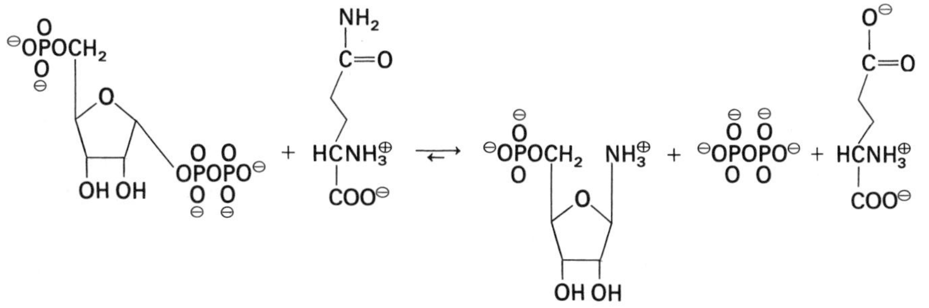

PRPP glutamine 5-phosphoribosylamine PP_i glutamate

PRPP has the pyrophosphate leaving group in the α-configuration (below the plane of the ring). In the product 5-phosphoribosylamine, the amine substituent is in the β-configuration (above the plane of the ring) at C-1, as would be expected for S_N2 attack by NH_3 on PRPP.

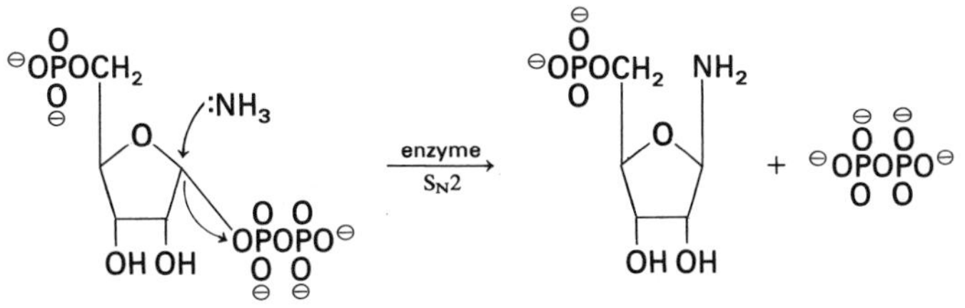

The K_m for PRPP is 2.3×10^{-4} M, and that for glutamine is 1×10^{-3} M. By contrast, NH_3 shows a K_m of 0.4 M, a 400-fold higher concentration to give 50% V_{max} (Buchanan, 1973). In agreement with mechanistic expectations, the enzyme purified from pigeon liver catalyzes the formation of γ-glutamyl hydroxamate in the presence of hydroxylamine, suggesting the trapping of an activated γ-glutamyl-enzyme intermediate. Further, an active-site nucleophile can be alkylated by the diazoketone analogue DON. Curiously, incorporation of one molecule of radioactive diazooxonorleucine into the tetrameric enzyme (four subunits, each of 52,000 mol wt, making up the tetramer of 210,000 mol wt) causes complete inactivation of the glutamine-dependent reaction. This implies that some cooperative conformational change occurs among all the subunits to an inactive state when one is labeled. Even after DON treatment, exogenous ammonia will still produce 5-phosphoribosylamine, suggesting that alkylation with

DON does not block either the NH_3 or the PRPP binding site, a clear indication that the glutaminase activity is at a site distinct from the amination site (Buchanan, 1973).*

5.C.2.b 2-Amino-6-phosphoglucose synthetase

Our second representative of this ATP-independent class is 2-amino-6-phosphoglucose synthetase, the enzyme that converts the 2-ketohexose D-fructose-6-P into the 2-amino-1-aldohexose glucosamine-6-P, which is the predominant amino sugar found in glycoproteins and many polysaccharides. This reaction falls into the third of the four substrate classifications described in ¶5.C.1:

```
   |                  |
H—C—OH              C=O
   |      NH3         |
  C=O     ——→      H—C—NH2
   |                  |
```

In more detail, the overall reaction is the following:

```
   CH2OH              NH2                                         O⊖
     |                 |                                          |
    C=O               C=O                HC=O                    C=O
     |                /                    |                     /
HO—C—H               \                 H—C—NH2                  \
     |                 \                   |                      \
 H—C—OH        +       /       ⇌      HO—C—H           +         /
     |          ⊕     /                    |                     /
 H—C—OH        H3N—C—H                 H—C—OH               H—C—NH3⊕
     |               |                     |                     |
     |     O        COO⊖               H—C—OH                  COO⊖
     |     ||                              |
   CH2OPO⊖                                 |     O
           |                               |     ||
           O⊖                            CH2OPO⊖
                                                 |
                                                 O⊖

D-fructose-6-P      glutamine       D-glucosamine-6-P       glutamate
  (ketose)                               (aldose)
```

The stoichiometry shows that the carbonyl at C-2 of fructose-6-P has been converted to an amino group (a net reduction), while the alcohol at C-1 has been converted to an aldehyde in the glucosamine product (a net oxidation). It is likely that the ammonia generated by glutaminase activity adds reversibly to the C-2

---

* The enzyme from animal cells contains iron required for catalysis (Buchanan, 1973) and the *Bacillus subtilis* enzyme recently has been shown to contain 3 Fe and two sulfur atoms (as $S^{2\ominus}$) required for catalysis (Wong et al., 1977). Dixon et al. (1976) have suggested that most enzymes carrying out animation reactions will be metalloproteins, with the active-site metal chelating NH_3 (and slowing its release before reaction).

keto group to generate the tetrahedral adduct. This adduct could either revert to starting materials or could lose the elements of water to produce an initial product that is the enol of glucosamine. Tautomerization of the enol to the thermodynamically more stable aldehyde form could proceed rapidly with general acid–base catalysis on the part of the enzyme. The net result is a keto-to-aldol isomerization as substrate goes to product.

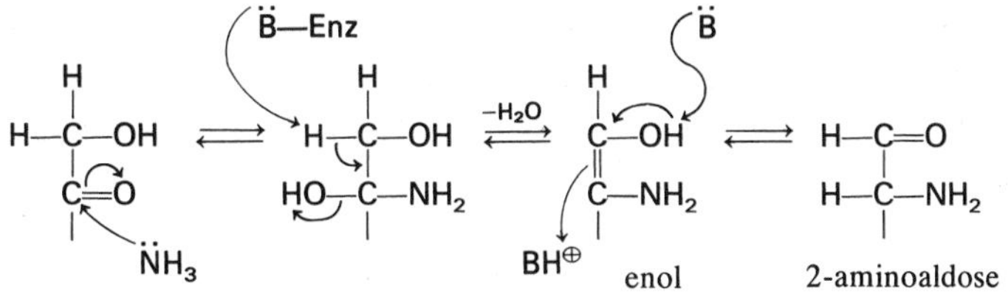

We have written the sugar substrate and product in acyclic form for simplicity in our discussion, although >99% of the molecules in solution exist as the cyclic hemiketal and hemiacetal, respectively.

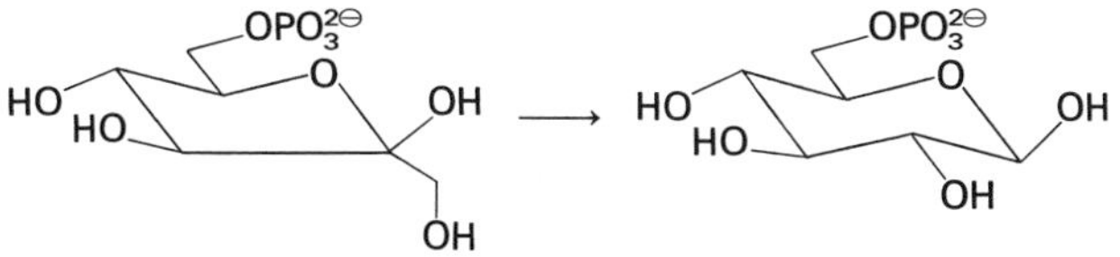

5.C.3 Amino Transfers That Require ATP as Cosubstrate: ATP Hydrolysis as Driving Force

Adenosine triphosphate (ATP) is a purine ribonucleoside triphosphate and is the major energy currency of most living cells. Its structure is the following:

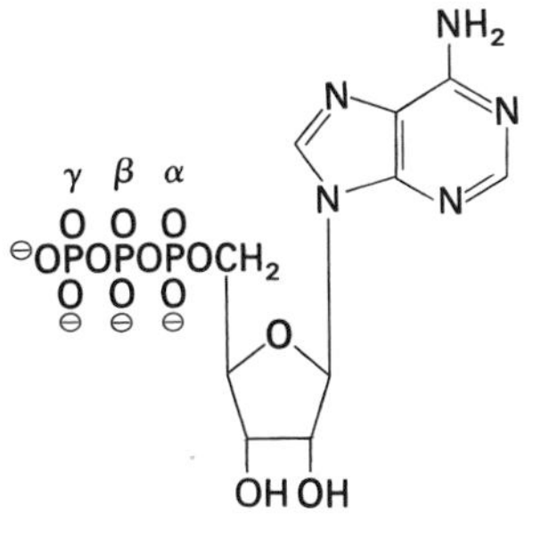

At neutral pH, ATP exists as the tetranion, and all reactions involving ATP require a divalent cation (usually $Mg^{2\oplus}$) complexed with the polyphosphate end of ATP, to reduce the effective negative charge density on the molecule.

The triphosphate side chain is the key structural element in the biological functioning of ATP and other nucleoside triphosphates (purines or pyrimidines). The major fate of ATP in enzymatic reactions is *hydrolysis*, with fragmentation occurring either between the α- and β-phosphorus atoms or between the β- and γ-phosphorus atoms. The hydrolysis of ATP is strongly favored thermodynamically, and this exergonic step is often coupled enzymatically to chemical transformations that may be quite unfavorable and endergonic. The net result (for many biosynthetic reactions discussed in Chapters 7 and 8) is that ATP hydrolysis provides the driving force for displacement of otherwise unfavorable equilibria, often toward biosynthesis of important cellular metabolites.

ATP hydrolysis is exergonic because of the chemistry of the triphosphate structure. There is one oxygen-ester bond, between the 5′-alcoholic carbon of ribose and the α-phosphorus.

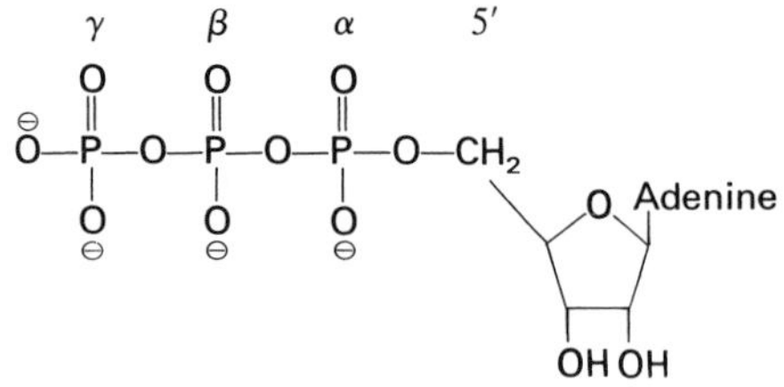

γ β α 5′
$^{\ominus}O-P(=O)(O^{\ominus})-O-P(=O)(O^{\ominus})-O-P(=O)(O^{\ominus})-O-CH_2$ (ribose, Adenine, OH OH)

The other two linkages (from α-P to β-P, and from β-P to γ-P) are phosphoric-acid-anhydride linkages. They represent activated phosphoric-acid derivatives, in the same way that acetic anhydride is an activated acyl compound and, as such, a good chemical acylating agent, undergoing ready acyl transfer to water and other nucleophiles.

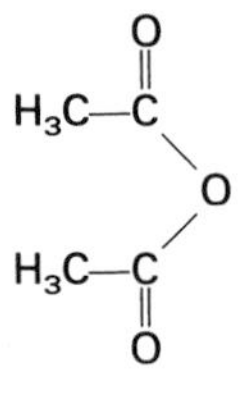

$H_3C-C(=O)-O-C(=O)-CH_3$

Both acetic anhydride and phosphoric anhydrides are thermodynamically destabilized linkages, and their hydrolysis by water is an exergonic process. For hydrolysis of acetic anhydride in water, $\Delta G^{0\prime} \approx -16$ kcal/mole (-67 kJ/mole). Hydrolysis by attack of water at the electrophilic γ-phosphorus of ATP (cleavage between β-P and γ-P), yielding adenosine diphosphate (ADP) and inorganic phosphate (P_i), has $\Delta G^{0\prime} = -7.3$ kcal/mole (-30.6 kJ/mole). The free-energy value corresponds to an equilibrium constant for hydrolysis of about 10^8 in favor of the hydrolysis products, using the standard equation $\Delta G^0 = -RT \ln K_{eq}$ (for instance, see Segel, 1975*a*).

$$\text{ATP} \xrightarrow[\text{H}_2\text{O}]{\gamma\text{-attack}} \text{ADP} + \text{P}_i$$

Similarly, nucleophilic attack by water at the α-phosphorus of ATP (cleavage between α-P and β-P), yielding adenosine monophosphate (AMP) and inorganic pyrophosphate (PP_i), has $\Delta G^{0\prime} = -7.7$ kcal/mole (-32 kJ/mole).

$$\text{ATP} + \text{H}_2\text{O} \xrightarrow[\text{H}_2\text{O}]{\alpha\text{-attack}} \text{AMP} + \text{PP}_i$$

At this point, AMP has no phosphate-anhydride bonds (only a phosphate-ester bond), and AMP is thermodynamically more stable (much less activated) than ATP or ADP. The free energy of hydrolysis of the ester bond between ribose and the α-P is only about -3 kcal/mole (-12.6 kJ/mole).

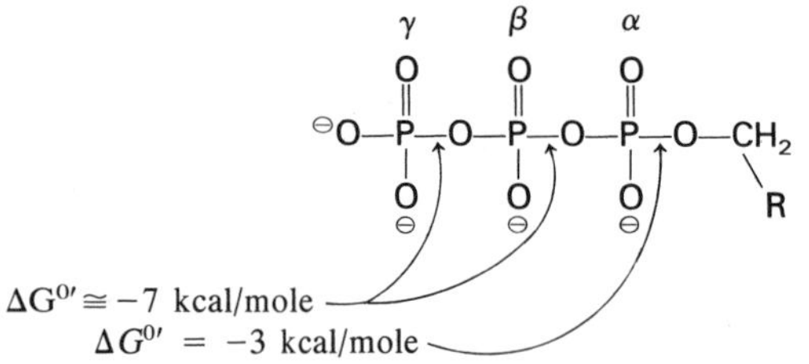

γ β α

$^{\ominus}O-P(=O)(O^{\ominus})-O-P(=O)(O^{\ominus})-O-P(=O)(O^{\ominus})-O-CH_2-R$

$\Delta G^{0\prime} \cong -7$ kcal/mole

$\Delta G^{0\prime} = -3$ kcal/mole

(Attack by nucleophiles at the β-phosphorus of ATP also can occur, but this event is less frequent in enzymatic reactions; see Chapter 8.)

These free-energy values explain the enzymatic cleavage patterns observed with ATP and the molecule's role as an energy currency in biology. But why do cells store potential for driving coupled reactions in phosphoric-anhydride derivatives rather than in activated acyl derivatives (such as acyl anhydrides)? Although both types of compounds are *thermodynamically unstable*, the anionic phosphoric-anhydride derivatives are *kinetically stable* at pH 7 in aqueous solutions, with half-lives of many months, such that they are stable until hydrolyzed under enzymatic control. By contrast, acetic anhydride is kinetically unstable in aqueous solutions, hydrolyzing in water at 25° C with $T_{1/2} = 4.5$ min, to form acetic acid. The tetraanionic character of ATP and PP_i undoubtedly slows down base-catalyzed nonenzymatic cleavage by $OH^{\ominus}$, due to electrostatic repulsion.

In glutamine-dependent amino transfers that require ATP, two major subtypes can be identified: (1) those where ATP is cleaved to ADP and P_i, and (2) those where ATP is cleaved to AMP and PP_i. We shall consider discrete examples of each type to understand how ATP may be accelerating catalysis, and to see what the ATP cleavage pattern may mean mechanistically as well as thermodynamically.

We shall consider three examples where ATP is cleaved to ADP and P_i: FGAR amino transferase (¶5.C.3.a), carbamoyl-P synthetase (¶5.C.3.c), and CTP synthetase (¶5.C.3.e). We shall discuss the ATP $\rightarrow$ AMP $+ PP_i$ pattern of cleavage in ¶5.E.

5.C.3.a FGAR aminotransferase

Formylglycinamide ribonucleotide (FGAR) aminotransferase is our first example of an enzyme catalyzing a glutamine-dependent transfer involving the cleavage pattern ATP $\rightarrow$ ADP $+ P_i$. This enzyme has been extensively characterized by Buchanan (1973) and his associates at MIT. They have purified the enzyme from chicken liver to homogeneity; it has a molecular weight of 133,000, and it is a single polypeptide chain about 1,300 amino-acid residues in length. It catalyzes the following conversion with the indicated stoichiometry.

FGAR + glutamine + Mg · ATP ⟶ formylglycinamidine ribonucleotide (FGAM) + glutamate + Mg · ADP + $HOP^{18}O_3^{2-}$

The sugar substrate bears structural resemblance to the 5-phospho-1-amino ribose discussed in ¶5.C.2.a; an *N*-formylglycyl group has been joined on that 1-amino group in amide linkage.

This enzymatic conversion is that of the substrate-amide carbonyl to the product-amidine imino group:

amide (R–C(=O)–NH–R′) ⟶ amidine (R–C(=NH)–NH–R′)

This clearly is a complex enzyme-catalyzed transformation, and we can dissect it to isolate certain elements for study. Glutamine probably is activated to deliver nascent ammonia at the active site, with formation of a γ-glutamyl thioester derivative of a cysteine residue. Attack by water on this activated thioester yields the observed glutamate product. In support of this hypothesis, we note that the enzyme with glutamine alone has glutaminase activity at 0.5% of the V_{max} for the complete reaction, and that DON is an affinity label for an active-site cysteine nucleophile.

5.C.3.b A mechanistic role for ATP

In the FGAR-aminotransferase reaction, the cosubstrate ATP is cleaved to ADP and inorganic phosphate (P_i). We have discussed the *thermodynamic* role of ATP in such reactions, but what is its *mechanistic* role? In analyzing ATP's mechanistic role, we shall note repeatedly that ATP can function as a *biological dehydrating agent*, facilitating the expulsion of some OH group from a substrate. Elimination of an —OH substituent can occur nonenzymatically in strong acid, where the —OH substituent is first protonated, and OH_2 (not $OH^{\ominus}$) is the actual initial product.

$$\mathrm{ROH} \xrightarrow[\mathrm{H^{\oplus}}]{\text{strong}} \mathrm{R\overset{\oplus}{O}H \atop H} \xrightarrow{\mathrm{HX}} \mathrm{RX} + \mathrm{H_2O}$$

At pH 7 (a reasonable pH for an enzymatic reaction), the expulsion of an alcoholic hydroxyl requires loss of the poor leaving group $OH^{\ominus}$ (a strong base). Although general acid catalysis could assist this process, a common biological solution to this problem is derivatization of the OH group as a phosphate ester.

$$\mathrm{ATP^{4\ominus}} + \mathrm{R{-}OH} \longrightarrow \mathrm{R{-}O\overset{O}{\underset{O^{\ominus}}{P}}O^{\ominus}} + \mathrm{ADP^{3\ominus}} \qquad (1)$$

This esterification represents a chemical activation step, because expulsion of the phosphate moiety is energetically much more favorable than expulsion of $OH^{\ominus}$. The phosphate group is a good leaving group in a displacement reaction because the inorganic phosphate product is a low-energy, resonance-stabilized anion. The net two-step conversion involves cleavage of ATP into ADP+P_i:

$$\mathrm{R{-}O\overset{O}{\underset{O^{\ominus}}{P}}O^{\ominus}} + \mathrm{XH} \longrightarrow \mathrm{R{-}X} + \left\{ \mathrm{O{=}\overset{OH}{\underset{O^{\ominus}}{P}}{-}O^{\ominus}} \longleftrightarrow \mathrm{^{\ominus}O{-}\overset{OH}{\underset{O}{\overset{\|}{P}}}{-}O^{\ominus}} \longleftrightarrow \mathrm{^{\ominus}O{-}\overset{OH}{\underset{O^{\ominus}}{P}}{=}O} \right\}$$

What is the chemical path for such a phosphorylation using ATP as phosphoryl donor? *The oxygen atom of the substrate acts as nucleophile, attacking the electrophilic (electropositive) γ-phosphorus atom of the ATP molecule.* The resultant pentacovalent phosphorane can either expel the added nucleophile and revert to starting materials or expel ADP as leaving group.

This is another example of a group-transfer reaction, a *phosphoryl* (—PO_3) *transfer*, on which we shall elaborate in the following chapters. It bears mechanistic analogies to the acyl-transfer process.

Let us see how the phosphoryl-transfer mechanism from ATP applies to the FGAR-aminotransferase catalysis. Note that, in conversion of the substrate amide to the product amidine, the carbonyl oxygen has been lost, replaced by a nitrogen derived from the ammonia that was produced by glutaminase activity. The carbonyl oxygen of the amide is a terrible leaving group, and phosphorylation by ATP would aid its eventual displacement by ammonia. However, the carbonyl oxygen of the amide, as written, is also a poor nucleophile, and a nucleophilic oxygen is required mechanistically to initiate attack on the γ-phosphorus of ATP. But that oxygen *is* nucleophilic in the isoamide form of the carbonyl group, which is a minor but real contributor to the amide structure.

As long as the isoamide is reactive enough, it can account for the chemical transformation, even though present at small concentrations.

A reasonable mechanism, then, features the isoamide canonical form (2) attacking ATP with formation of a phosphorylated species (3).

The carbonyl carbon of species 3 is electron-deficient and will be the site of nucleophilic addition by the nascent ammonia, to produce the tetrahedral adduct. The lone electron pair on the adjacent nitrogen can then aid in expulsion of inorganic phosphate (and loss of the original carbonyl oxygen atom), with formation of the amidine product. Evidence in support of such a scheme comes from the fact that FGAR labeled in the carbonyl group with ^{18}O undergoes reaction to produce FGAM that (as expected) has lost the ^{18}O. The labeled oxygen does not end up in H_2O, however, but rather in the inorganic phosphate, as indicated in our mechanistic scheme. This indeed suggests a covalent bond between the carbonyl oxygen of FGAR and the γ-phosphorus of ATP at some stage during catalysis.

5.C.3.c Carbamoyl-P synthetase

Carbamoyl phosphate (carbamoyl-P) synthetase is our second example of an enzyme catalyzing a glutamine-dependent transfer involving the ATP → ADP + P_i cleavage pattern.

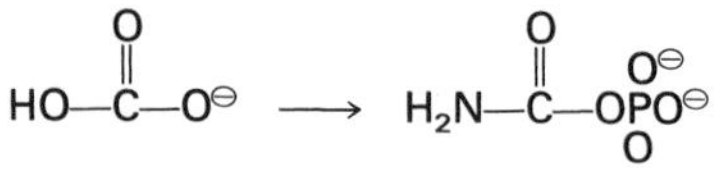

$$HO-C(=O)-O^{\ominus} \longrightarrow H_2N-C(=O)-OPO_2^{\ominus}O^{\ominus}$$

bicarbonate carbamoyl-P

In addition to the amination, this reaction involves a phosphoryl transfer to provide acyl-group activation. Carbamoyl-P is an activated one-carbon compound that is a key intermediate in pyrimidine biosyntheses and in urea synthesis. Carbamoyl-P is a dianion at neutral pH ($pK_1 = 1.1$; $pK_2 = 4.9$) and as such shows a half-life of 40 min at 37° C in neutral aqueous solutions, emphasizing its relative instability to hydrolysis:

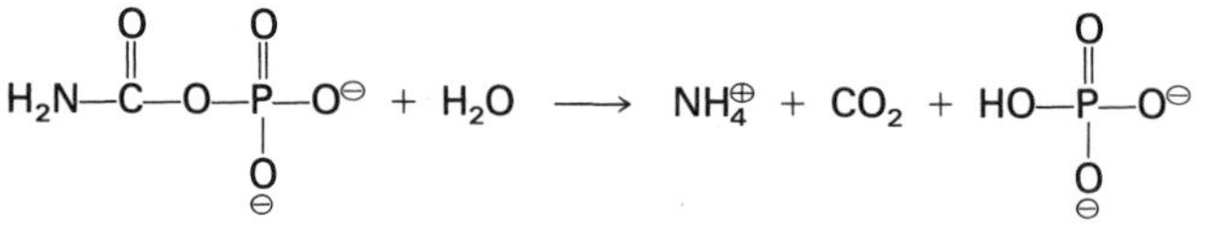

$$H_2N-C(=O)-O-P(=O)(O^{\ominus})-O^{\ominus} + H_2O \longrightarrow NH_4^{\oplus} + CO_2 + HO-P(=O)(O^{\ominus})-O^{\ominus}$$

This chemical lability, both thermodynamic and kinetic, is expected for a molecule that is a *mixed anhydride* of carbamic acid (H_2NCOOH) and phosphoric acid, a structural type referred to as an *acyl phosphate.* Although this is our first example of this structural type, acyl phosphates represent one of the three main structural classes of *activated acyl derivatives* used enzymatically as biosynthetic intermediates (Chapters 7 and 8):

(1) acyl phosphate $R{-}C(=O){-}O{-}P(=O)(O^{\ominus}){-}O^{\ominus}$

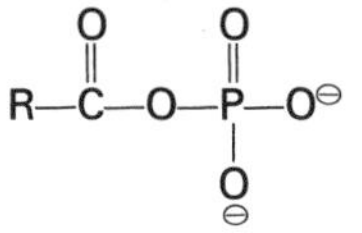

(2) acyl adenylate $R{-}C(=O){-}O{-}AMP$

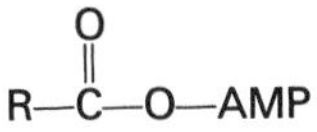

(3) acyl thioester $R{-}C(=O){-}S{-}R'$

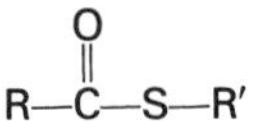

Actually, acyl phosphates are *doubly activated* molecules, in that either half can undergo ready transfer to nucleophiles; the mixed anhydrides are good acylating agents *and* good phosphorylating agents.

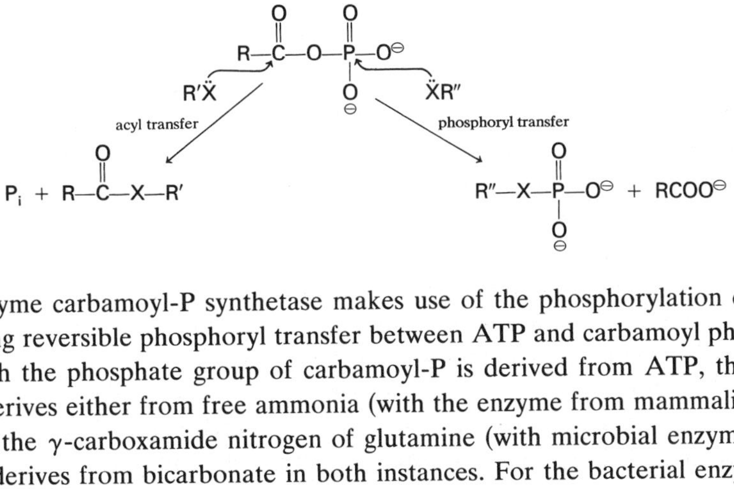

$$R{-}C(=O){-}O{-}P(=O)(O^{\ominus}){-}O^{\ominus}$$

acyl transfer (attack by $R'\ddot{X}$ at C): $P_i + R{-}C(=O){-}X{-}R'$

phosphoryl transfer (attack by $\ddot{X}R''$ at P): $R''{-}X{-}P(=O)(O^{\ominus}){-}O^{\ominus} + RCOO^{\ominus}$

The enzyme carbamoyl-P synthetase makes use of the phosphorylation capacity, catalyzing reversible phosphoryl transfer between ATP and carbamoyl phosphate. Although the phosphate group of carbamoyl-P is derived from ATP, the amino group derives either from free ammonia (with the enzyme from mammalian liver) or from the γ-carboxamide nitrogen of glutamine (with microbial enzymes). The carbon derives from bicarbonate in both instances. For the bacterial enzyme, the stoichiometry of carbamoyl-P synthesis is the following:

$$2\,Mg\cdot ATP + HCO_3^{\ominus} + \text{Glutamine} \xrightarrow{K^{\oplus}} 2\,Mg\cdot ADP + P_i + \text{Carbamoyl-P} + \text{Glutamate}$$

For every molecule of carbamoyl-P produced, two molecules of ATP have been converted to ADP; of the two γ-P released, one appears as P_i and the other in carbamoyl-P.

The *E. coli* enzyme has been studied by Meister and his colleagues (Pinkus and Meister, 1972; Trotta et al., 1974). The protein is a dimer of 340,000 mol wt

Table 5-1
Subunit activities of carbamoyl-P synthetase

Heavy subunit	Light subunit
1. Carbamoyl-P synthesis from NH_3 but not from glutamine.	1. Glutaminase activity.
2. Bicarbonate-dependent ATPase activity; $ATP + HCO_3^{\ominus} \rightarrow ADP + P_i + HCO_3^{\ominus}$. ATP is hydrolyzed catalytically by the heavy subunit. $HCO_3^{\ominus}$ must be bound for the activity to be displayed, but $HCO_3^{\ominus}$ is not consumed.	2. γ-Glutamyl transfer from glutamine to hydroxylamine.
3. Carbamoyl-P–dependent but glutamine-independent ATP synthesis—i.e., $CAP + ADP \rightleftarrows ATP + NH_3 + CO_2$ (phosphoryl transfer from carbamoyl-P to produce ATP).	

in reversible association with monomers of ~170,000 mol wt, with both species active. Some structural insight has been attained with the observation that each monomer in turn is composed of two polypeptide chains: a heavy subunit (130,000 mol wt) and a light subunit (40,000 mol wt). The subunits can be formed reversibly and with retention of catalytic activity by adding 1 M potassium isothiocyanate to enzyme solutions.

Table 5-1 summarizes the activities shown by the separated heavy and light chains. Recombination of the two subunits regenerates the capacity to use glutamine as amino donor in carbamoyl-P syntheses (K_m for glutamine = 0.38 mM; K_m for NH_3 = 80 mM). Addition of the chloroketone analogue of glutamine, L-2-amino-4-oxo-5-chloropentanoate, causes a time-dependent loss of the ability to use glutamine to make carbamoyl-P, but the NH_3-dependent reaction is unaffected.

$$^{14}CH_2(Cl)\text{—}C(=O)\text{—}CH_2\text{—}CH(NH_3^{\oplus})\text{—}COO^{\ominus}$$

chloroketone analogue

The radioactive analogue binds covalently to the light subunit. Curiously, the ATPase activity of chloroketone-treated enzyme is 250% that of intact enzyme; access to adventitious H_2O must have been increased, allowing acceleration of this energetically wasteful partial reaction (Pinkus and Meister, 1972; Trotta et al., 1974).

All these data imply that glutamine is activated on the light subunit, and that the nascent ammonia is used at the active site of the heavy subunit, where it reacts with some form of activated CO_2.

This brings us to another consideration: what is the activated form of CO_2? Comparing the substrate bicarbonate and the product carbamoyl-P, we note substitution of NH_2 for one substrate oxygen and of PO_4 for a second substrate oxygen.

$$HO{-}C({=}O){-}O^{\ominus} \longrightarrow H_2N{-}C({=}O){-}OPO_3^{2\ominus}$$

Convert this oxygen to a leaving group for displacement by NH_3

Phosphorylate this oxygen in product

We have noted that ATP facilitates expulsion of such oxygen atoms during conversion of nonactivated substrates to activated products by phosphoryl transfer. We can understand the requirement for two ATP molecules: there are two activation steps from substrate to product:

$$\underset{\text{bicarbonate}}{HO{-}C({=}O){-}O^{\ominus}} + \underset{\text{ATP}}{ADP{-}O{-}P(O)(O^{\ominus})_2} \rightleftharpoons \underset{\substack{\text{carbonyl-P}\\\text{(carbonic-phosphoric}\\\text{anhydride)}}}{HO{-}C({=}O){-}OPO_3^{2\ominus}} \xrightarrow{\text{"}NH_3\text{"}} \rightleftharpoons \underset{\text{carbamate}}{HO{-}C({=}O){-}NH_2} + ADP + P_i \qquad (1)$$

$$\underset{\text{carbamate ion}}{H_2N{-}C({=}O){-}O^{\ominus}} + \underset{\text{ATP}}{ADP{-}O{-}P(O)(O^{\ominus})_2} \rightleftharpoons \underset{\text{carbamoyl-P}}{H_2N{-}C({=}O){-}OPO_3^{2\ominus}} + ADP \qquad (2)$$

The first step could explain the bicarbonate-dependent ATPase activity of the isolated heavy subunit. The proposed carbonyl phosphate should be hydrolytically labile. In the absence of ammonia, it could be captured by water to give back bicarbonate and inorganic phosphate. A preliminary report has suggested that the enzyme-bound carbonyl-P can be trapped either by reduction with borohydride to yield formate or by esterification with diazomethane to yield the indicated trimethyl ester.

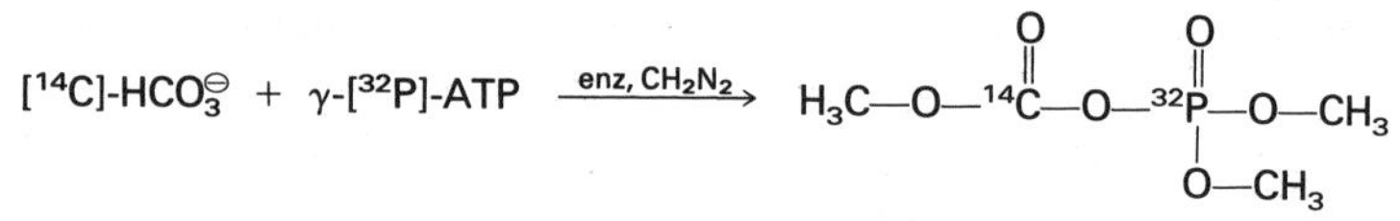

$$[^{14}C]\text{-}HCO_3^{\ominus} + \gamma\text{-}[^{32}P]\text{-ATP} \xrightarrow{\text{enz, } CH_2N_2} H_3C{-}O{-}^{14}C(=O){-}O{-}^{32}P(=O)(O{-}CH_3){-}O{-}CH_3$$

Both derivatizations can also occur in the nonenzymatic controls, however, and the signal-to-noise ratio was only about 3:1 (Powers and Meister, 1976).

5.C.3.d Carbamoyl-P as a carbamoyl-transfer agent

We have noted that carbamoyl-P is a doubly activated intermediate for phosphoryl or acyl transfer. Once formed in the synthetase reaction, it participates in two key enzyme-catalyzed carbamoylations (acyl transfers), each utilizing an amine group of an amino acid as attacking nucleophile.

Aspartate transcarbamoylase is the first enzyme in the pathway for pyrimidine-ring biosynthesis.*

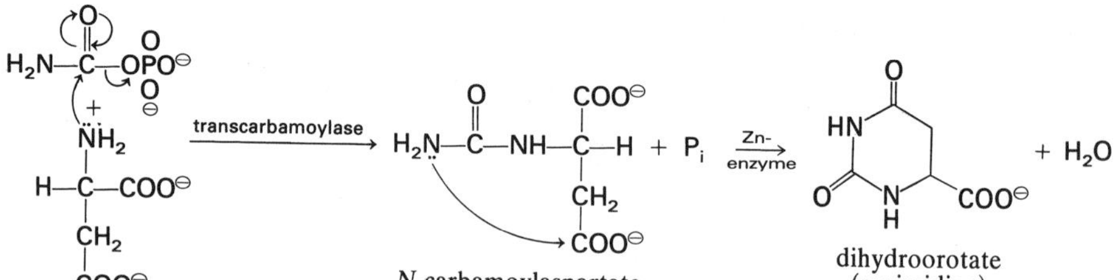

$$H_2N{-}C(=O){-}OPO_3^{\ominus} + \text{aspartate} \xrightarrow{\text{transcarbamoylase}} H_2N{-}C(=O){-}NH{-}CH(COO^{\ominus}){-}CH_2{-}COO^{\ominus} + P_i \xrightarrow[\text{enzyme}]{Zn\text{-}} \text{dihydroorotate} + H_2O$$

aspartate

N-carbamoylaspartate

dihydroorotate (pyrimidine)

Ornithine transcarbamoylase uses carbamoyl-P, not to initiate a biosynthetic pathway, but rather to begin the urea cycle, a pathway involved in generating urea as an excretable metabolite (Ratner, 1973*a*).

---

* The N-carbamoylaspartate is subsequently cyclized and dehydrated by a $Zn^{2\oplus}$-containing dihydroorotase. This enzyme effects net expulsion of an unactivated carboxylate oxygen. This may well occur by zinc chelation and polarization of the —OH group in the tetrahedral adduct.

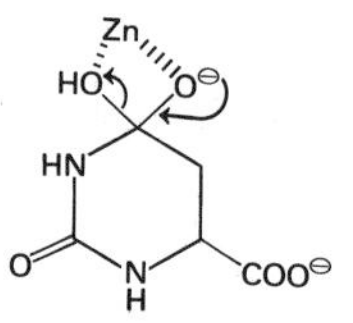

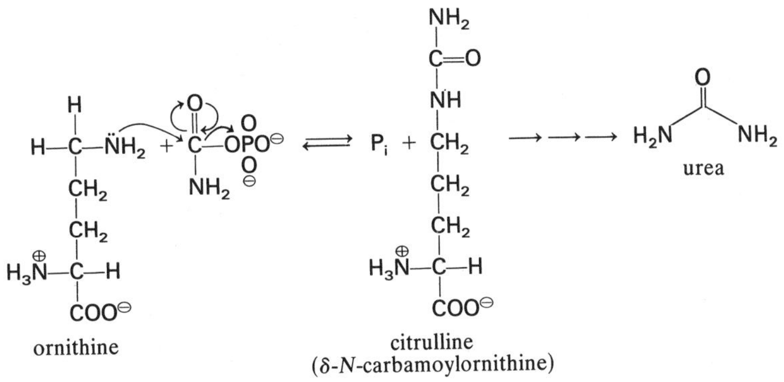

ornithine

citrulline
(δ-*N*-carbamoylornithine)

urea

5.C.3.e CTP synthetase

A third example of an enzyme hydrolyzing ATP to ADP and P_i during a glutamine-dependent amination is the complex enzyme cytidine triphosphate (CTP) synthetase (Levitski and Koshland, 1974). The cosubstrate of ATP is the pyrimidine triphosphate UTP (uridine triphosphate). The carbonyl group at C-4 of UTP is converted to the amidine grouping of CTP in a reaction with strong mechanistic analogy to the FGAR-aminotransferase reaction.

$$\text{UTP} + \text{ATP} + \text{Glutamine} \underset{}{\overset{\text{Mg}}{\rightleftharpoons}} \text{CTP} + \text{ADP} + P_i + \text{Glutamate}$$

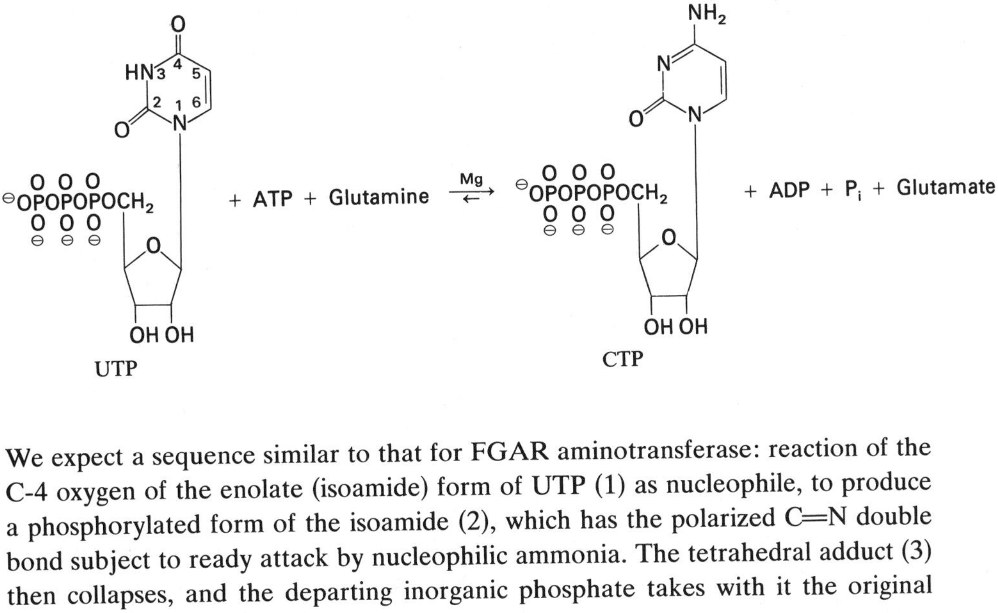

We expect a sequence similar to that for FGAR aminotransferase: reaction of the C-4 oxygen of the enolate (isoamide) form of UTP (1) as nucleophile, to produce a phosphorylated form of the isoamide (2), which has the polarized C═N double bond subject to ready attack by nucleophilic ammonia. The tetrahedral adduct (3) then collapses, and the departing inorganic phosphate takes with it the original C-4 oxygen atom of the UTP.

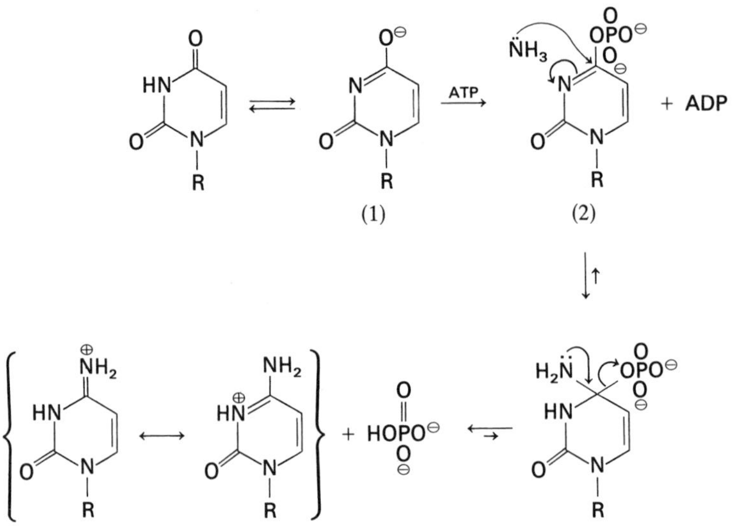

5.C.3.f Some regulatory aspects of CTP synthetase

Like other such enzymes, CTP synthetase is inactivated for glutamine utilization (but not for reaction with ammonia) by the diazoketone substrate analogue diazooxonorleucine (DON). One of the more interesting facets of CTP synthetase is that three of the four common nucleoside triphosphates are substrates or products in the reactions, and the fourth (guanosine triphosphate, or GTP) is a regulatory substance.

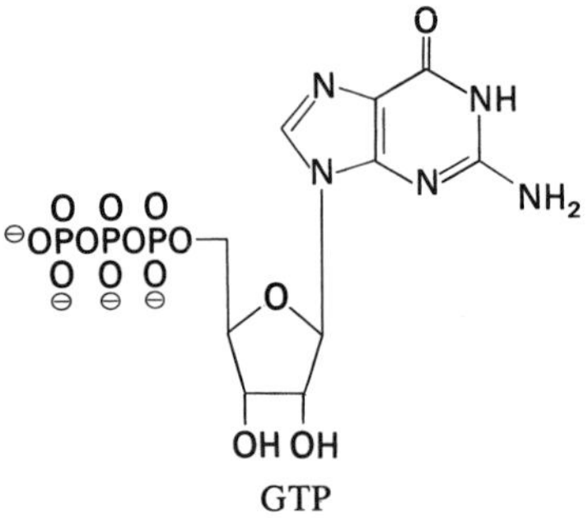

GTP

GTP is an *allosteric effector* (allosteric = "other site"; i.e., at a site distinct from the active site), whose action is focused primarily on the glutamine-activation steps. That is, the presence of GTP has no obvious effect on the synthesis of CTP from ammonia, UTP, and ATP, but it has a pronounced effect when glutamine is the amino donor: the V_{max} increases 10-fold, while the K_m for glutamine is lowered 6-fold. The increase in V_{max} can be dissected into (1) an increase in rate of formation of the covalent glutamyl-S enzyme and (2) more rapid release of "nascent NH_3" (Levitski and Koshland, 1974). It thus would be expected that GTP would accelerate the rate of alkylation of that essential cysteine residue by DON, and this is observed experimentally (Levitski and Koshland, 1974). Thus, the effector molecule GTP seems to have its action focused on the acceleration of one of the several chemical steps in CTP-synthetase catalysis.

How does binding of GTP to an allosteric site produce a change in the activity of CTP synthetase? Presumably this control is exerted by use of some of the binding energy of the enzyme–GTP complex to induce a conformational change in the protein; this change is exerted at the active site to modify the enzyme's kinetic parameters. Indeed, the kinetics of proteins subject to regulation and allosteric control are often complex. Allosteric control and the variety of kinetic equations used to fit observed behavior are described in detail elsewhere (Koshland, 1970; Segel, 1975*b*). In the following section we give only a brief qualitative discussion of these matters.

5.C.3.g Allosteric control and cooperativity in protein regulation

For many enzymes whose activities are modified by allosteric effectors (activators and inhibitors), the velocity vs. substrate curves show marked deviation from rectangular hyperbolas, reflecting the fact that the simple Michaelis–Menten velocity equation no longer describes the kinetic behavior. As we shall note, the whole shape of the v vs. [S] curve is altered in many cases (Fig. 5-2). Many of the enzymes sensitive to allosteric regulation are multisubunit (oligomeric) enzymes, and the kinetic complexity of these enzymes reflects structural interaction or *cooperativity* between individual subunits, each of which generally has an active site. Although monomeric enzymes also can show cooperative kinetic patterns, we shall not discuss them here (see Ainslie et al., 1972).

For an oligomeric enzyme (say, a tetramer of identical subunits such as CTP synthetase) where each subunit has an active site that is functionally and structurally independent, the v vs. [S] curve will be the usual rectangular hyperbola, consistent with Michaelis–Menten kinetics. In other words, one molecule of a tetrameric enzyme with four independent active sites is kinetically indistinguishable from four molecules of a monomeric enzyme, each having an active site. Kinetic evaluation alone cannot determine which of these species is present in the solution. However, kinetic complications arise when the subunits of an oligomeric enzyme interact, with each subunit modulating the responses of other subunits.

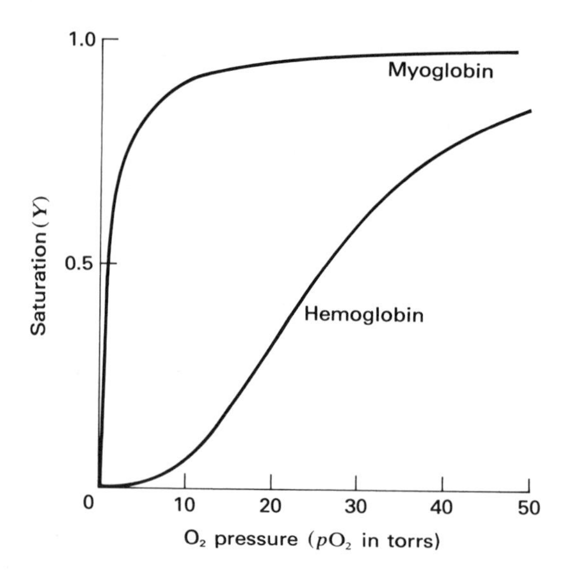

Figure 5-2
Oxygen dissocation curves of myoglobin and hemoglobin. The saturation of the oxygen-binding sites is plotted as a function of the partial pressure of oxygen surrounding the solution. The myoglobin-binding curve follows the hyperbolic curve expected from Michaelis–Menten kinetics; the hemoglobin-binding curve has a distinctively different sigmoid shape, due to cooperative binding. (From L. Stryer, *Biochemistry*, p. 73. W. H. Freeman and Company Copyright © 1975)

In such a case, the presence of a substrate molecule (or some other ligand that activates or inhibits) in one site affects *either the binding* of a subsequent molecule to other vacant sites *or the rate* of product formation from other occupied sites (Ainslie et al., 1972). Then the enzyme no longer will follow Michaelis–Menten kinetics, and the kinetic equations can become quite complex (Segel, 1975*a*, 1975*b*). The structural basis for such interaction in a multimeric enzyme probably is a change in intersubunit contact as a result of ligand binding. If binding of a ligand at one site induces a conformational change in one subunit, the interaction energy between subunits can be altered to induce and stabilize a conformational change in an adjoining subunit. The alteration could result in an increased affinity of a substrate for vacant sites or in a faster rate of product formation from other occupied sites; this effect is called *positive cooperativity.*

Koshland (1970) summarizes some diagnostic tests to determine whether cooperativity is present and whether it is positive or negative cooperativity. We illustrate these considerations here by discussing a simplified velocity equation for allosteric enzymes, the Hill equation. A. V. Hill (1910) originally used this equation to describe the observably sigmoid curves (see Fig. 5-2) obtained for binding of O_2 to the tetrameric protein hemoglobin (a classical allosteric protein), the physiological oxygen carrier in mammalian blood.

Given a high degree of cooperativity in substrate binding for an enzyme with n equivalent binding sites, the velocity equation is approximated by

$$v = V_{max}[S]^n/(K' + [S]^n),$$

where n is the number of binding sites per enzyme molecule (e.g., 4 in tetrameric CTP synthetase), and K' is a complex constant (Segel, 1975*a*, p. 303 ff). K' contains K_s (the intrinsic ES dissociation constant) as well as terms for subunit interactions that alter the intrinsic K_s value. K' differs from K_m in that K' does *not* equal that concentration of substrate producing $V_{max}/2$, except when $n = 1$.

The Hill equation can be rearranged to give the following linear form:

$$\log\{v/(V_{max} - v)\} = n \log[S] - \log K'$$

This form is useful because it allows determination of n. A plot of $\log\{v/V_{max} - v)\}$ vs. $\log[S]$ gives a straight line with a slope equal to n. Actually, the observed n (called n_m, with the m for "measured") obtained experimentally will be less than the theoretical n (the number of binding sites), because an assumption of infinite cooperativity was made in deriving the equation (Segel, 1975*a*, p. 303 ff). The observed value of n_m is known as the Hill coefficient; for hemoglobin (which actually contains four binding sites) $n_m = 2.8$. Generally, the observed Hill coefficient is less than the maximum number of sites, representing a combination of the number of sites and the strength of interaction between them (Koshland, 1970).

When $n_m = 1$, the sites are independent, there is no allosteric interaction, and the enzyme follows Michaelis–Menten kinetics. When $n_m > 1$, the enzyme shows positive cooperativity. When $n_m < 1$, the enzyme shows negative cooperativity. Figure 5-3 shows how these three situations (independent binding sites, positive cooperativity, and negative cooperativity) are evinced in the v vs. [S] plot, the double-reciprocal Lineweaver–Burk plot, and the Hill plot. In the v vs. [S] plot, note that the sigmoidicity (so evident in the hemoglobin-binding data of Fig. 5-2) may be hard to detect; sigmoidicity varies with the size of K_s and the value of n. In the double-reciprocal plot, positive cooperativity gives a line that is concave upward; negative cooperativity gives a line concave downward.

The curves of Figure 5-4 illustrate the physiological utility of positive cooperativity and a sigmoid v vs. [S] curve. A sigmoidal response gives a steeper slope in the intermediate region of [S], precisely the situation that gives a maximal response for a small change in these concentrations of S. Given that $[S]_{0.5}$ values

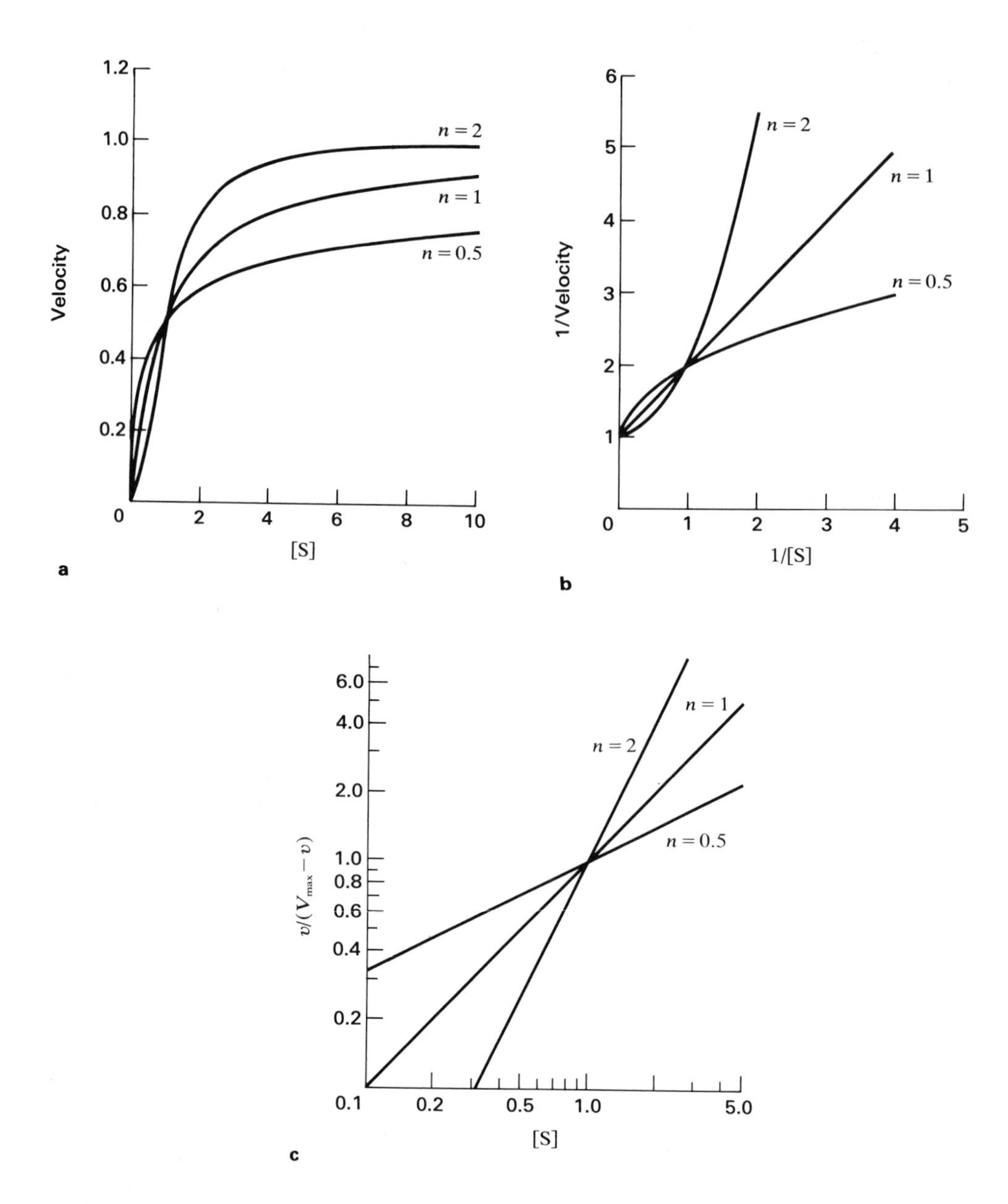

Figure 5-3
Michaelis–Menten, double-reciprocal, and Hill plots for independent binding, positive cooperativity, and negative cooperativity. The data are calculated from the equation $f = V_{max}[S]^{n}H/(K+[S])^{n}H$, in which $V_{max} = 1$, $K = 1$, and ${}^{n}H =$ 0.5, 1, and 2. The same data are then plotted in (**a**) a classic Michaelis–Menten plot, (**b**) a double reciprocal plot, and (**c**) a Hill plot. (Koshland, 1970, p. 352)

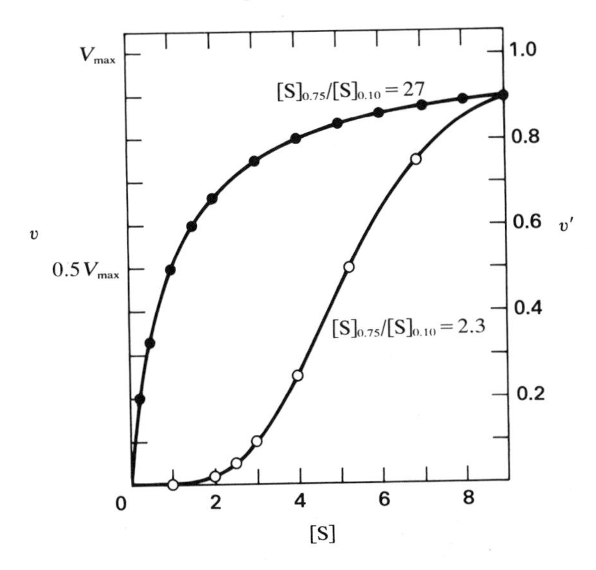

Figure 5-4
Comparison of velocity curves for two different enzymes that coincidentally have the same v at $[S] = 9$. One shows hyperbolic response (*black dots*), the other sigmoidal response (*open dots*). (From I. H. Segel, *Enzyme Kinetics*. Copyright © 1975. Reprinted by permission of John Wiley & Sons, Inc.)

represent substrate concentrations in the physiological range, then this sensitivity represents the optimal possibility for regulation and control of activity.

Let us now consider the behavior of the *E. coli* CTP synthetase and its cooperativity responses. The *E. coli* enzyme is a dimer of 105,000 mol wt in the absence of ATP, UTP, or GTP. Either ATP or UTP alone can induce conversion to the tetramer, and these triphosphates are more effective in combination. Presumably, the tetramer predominates during catalysis. In addition to their effects on the aggregation state of the enzyme, both ATP and UTP exhibit positive cooperativity. For instance, the binding of ATP to one subunit induces a conformational change transmitted to an adjacent subunit, such that the K_D for binding a molecule of ATP to a second subunit of the tetramer is lower; binding is facilitated. Sequentially, the third ATP binds better than the second molecule, and the last ATP more readily than the third. This is, of course, analogous to oxygen binding to hemoglobin. Levitski and Koshland (1974) note that positive cooperativity in oligomeric enzymes is a device for enhancing small effects in the medium—an amplification of a signal.

In contrast, the binding of the allosteric effector GTP displays negative cooperativity: each succeeding molecule of GTP has a harder time binding. The negative cooperativity can also be useful physiologically: it will "dampen the effect of fluctuating ligand concentrations" (Levitski and Koshland, 1974). The existence of both positive cooperativity for a substrate and negative cooperativity for an allosteric ligand indicates the complexity of regulatory behavior.*

The final point we raise about ligand-induced conformational changes in CTP synthetase is the observation that incorporation of two molecules of DON per tetramer causes complete inactivation of the ability of all four subunits to utilize glutamine. Reaction at the cysteine of one subunit induces a conformational change in an adjacent subunit that blocks reaction at that site. This phenomenon has been termed "half-of-the-sites reactivity" (Levitski et al., 1971), and it is not uncommon among oligomeric enzymes—including the PRPP amidotransferase discussed earlier (¶5.C.2a) and glutamine synthetase (discussed in the next section).

5.D GLUTAMINE SYNTHETASE

We have discussed a number of the enzymatic reactions utilizing glutamine as a source of NH_3 for biosynthesis or degradation or as a source of transferable γ-glutamyl groups. In this section we comment on the synthesis of glutamine itself. We shall note (¶5.E.1) that the other common amide-containing α-amino acid, asparagine, is formed from aspartate and glutamine. The carboxamide of glutamine derives instead from free ammonia in reaction with glutamate. In asparagine synthesis, the β-carboxylate of aspartate is activated by reaction with ATP to form β-aspartyladenylate, with resultant net cleavage of ATP to AMP and PP_i (see ¶5.E.1). In the glutamine-synthetase catalysis, the ATP utilized for activation of the γ-carboxylate of glutamate is cleaved instead to ADP and P_i.

$$NH_4^{\oplus} + ATP + {}^{\oplus}H_3N\text{–}CH(COO^{\ominus})\text{–}CH_2CH_2\text{–}COO^{\ominus} \rightleftharpoons {}^{\oplus}H_3N\text{–}CH(COO^{\ominus})\text{–}CH_2CH_2\text{–}C(=O)\text{–}NH_2 + ADP + P_i \qquad K_{eq} = 1{,}200$$

*The discussion of positive and negative cooperativity with CTP synthetase suggests sequential cooperative effects: binding of the first ligand is harder (or easier) than the next, and so on. An alternative model for allosteric effects is the concerted-symmetry model of Monod, Wyman, and Changeux (1965), where only two states exist for a multimeric protein; all subunits switch from one state to the other on binding of a ligand to one subunit. This model does appear valid with some regulatory enzymes, such as the aspartate transcarbamoylase mentioned earlier in this chapter (Jacobson and Stark, 1973).

The equilibrium constant at pH 7 and 39° C is 1,200 in favor of glutamine formation, driven by the thermodynamically favorable splitting of ATP. The cleavage pattern is suggestive of an acyl-phosphate intermediate (i.e., γ-glutamyl phosphate), and much effort has been expended in attempts to verify this possibility (Prusiner and Stadtman, 1973).

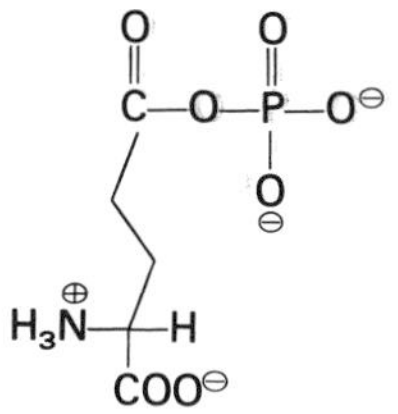

Indeed, an enormous quantity of information is available about glutamine synthetase—concerning both mechanism and regulation. Although the mechanism is probably similar both in mammalian and in microbial cells, the pattern of regulation is different in the two kinds of cells, so we shall consider the two subjects separately—focusing first on the ovine (sheep) brain enzyme with which most mechanistic experiments have been performed (Meister, 1974), and then discussing the complex controls on the activity of the *E. coli* enzyme (Stadtman and Ginsburg, 1974).

5.D.1 Mechanistic Experiments

The ovine-brain enzyme, purified to homogeneity, is an octamer of 350,000 to 400,000 mol wt; it has been studied extensively by Meister (1974) and colleagues, largely in an effort to prove that the acyl phosphate (γ-glutamyl $PO_3^{2\ominus}$) is formed as an enzyme-bound intermediate. Attempts to prepare γ-glutamyl $PO_3^{2\ominus}$ chemically and test its reactivity with the synthetase have failed, due to the lability of the acyl phosphate, which undergoes rapid intramolecular cyclization (with the α-amino group displacing the good leaving group inorganic phosphate, via a tetrahedral adduct, to form the cyclic derivative known as 5-oxoproline—also called pyrrolidone carboxylate, or pyroglutamate).

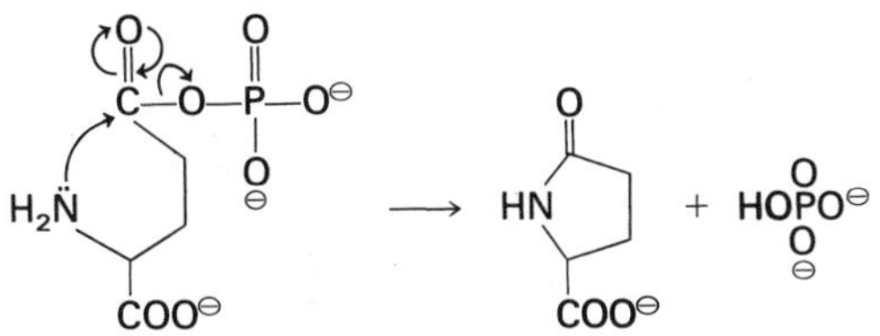

γ-glutamyl-P → 5-oxoproline + P_i

Although these attempts have failed, a number of indirect criteria have been accumulated for the intermediacy of glutamyl phosphate.

First, there is evidence that an activated γ-carboxyl derivative of glutamate is generated. Not only NH_3, but also NH_2OH, can intercept this derivative to form γ-glutamyl hydroxamate (reminiscent of the γ-glutamyl–transfer enzymes). Further, the formation of 5-oxoproline is catalyzed by the enzyme when ATP and glutamate are present in the absence of ammonium ion, at 0.1% of the V_{max} for glutamine formation. To the extent that oxoproline is a fingerprint for the fleeting glutamyl phosphate, this observation supports the hypothesis of glutamate activation prior to amination. Glutamate labeled with ^{18}O in the carboxyl oxygens is converted to glutamine with one atom of ^{18}O ending up in every molecule of inorganic phosphate formed, suggesting the existence of an acyl-phosphate intermediate.

Meister's group determined that β-glutamate (3-aminoglutarate) is a substrate for the synthetase, forming β-glutamine (D-isomer).

$$\mathrm{COO^{\ominus}\text{–}CH_2\text{–}CH(\overset{\oplus}{N}H_3)\text{–}CH_2\text{–}COO^{\ominus}} + \mathrm{ATP} + \mathrm{NH_3} \longrightarrow \mathrm{CONH_2\text{–}CH_2\text{–}CH(\overset{\oplus}{N}H_3)\text{–}CH_2\text{–}COO^{\ominus}} + \mathrm{ADP} + \mathrm{P_i}$$

β-glutamate β-glutamine

An acyl-phosphate intermediate should in this case be slower to cyclize and therefore more amenable to preparation. This is indeed the case, and synthetic β-glutamyl-P can be incubated with ADP and glutamine synthetase with production of ATP (Meister, 1974). Most recently, trapping experiments have been performed with the *E. coli* enzyme, and these experiments support the idea that it is γ-glutamyl-P that is derivatized. It is known that sodium borohydride will reduce activated carboxyl derivatives to aldehydes, and reduce the aldehydes rapidly to alcohols. Borohydride will not reduce the resonance-stabilized γ-carboxyl of glutamate or the γ-amide of glutamine, and Todhunter and Purich (1975) established that 5-oxoproline is also inert to borohydride. Therefore, if borohydride can gain access to an enzyme-bound γ-glutamyl-P and reduce it, δ-hydroxy-α-aminovalerate ought to be the telltale product.

$$\mathrm{{}^{\ominus}O_3PO\text{–}C(=O)\text{–}CH_2\text{–}CH_2\text{–}CH(\overset{\oplus}{N}H_3)\text{–}COO^{\ominus}} \xrightarrow{\mathrm{NaBH_4}} \mathrm{HOCH_2\text{–}CH_2\text{–}CH_2\text{–}CH(\overset{\oplus}{N}H_3)\text{–}COO^{\ominus}} + \mathrm{HOPO_3^{2\ominus}}$$

γ-glutamyl-P δ-hydroxy-α-aminovalerate

With [^{14}C]-glutamate as substrate at saturating levels and 20 μM *E. coli* glutamine synthetase, borohydride addition does allow isolation of 14 μM reduction product as [^{14}C]-γ-hydroxy-α-aminovalerate, suggesting an efficient capture of sequestered acyl phosphate (Todhunter and Purich, 1975). Finally, evidence for reversible cleavage of ATP to ADP (and presumably γ-glutamyl-P) in the presence of glutamate has been provided by elegant ^{18}O-isotope-scrambling experiments (Middlefort and Rose, 1976) discussed in connection with isotope-exchange experiments in ¶5.E.2.

Before discussing the regulation of the *E. coli* enzyme, we shall note how methionine sulfoximine (a convulsion-causing agent that has been isolated from certain batches of bleached flour; see Rando, 1975) inhibits the central-nervous-system glutamine synthetase in animals. Inactivation of the synthetase produces severe perturbation of nervous-system function. The sulfoximine moiety is misrecognized by the enzyme as the γ-carboxylate of glutamate and is phosphorylated enzymatically to form methionine sulfoximine phosphate (Ronzio et al., 1969). Only the 2*S*,5*S*-diastereomer (Manning et al., 1969) is a substrate, and only this diastereomer produces functionally irreversible inactivation. The three other diastereomers bind as reversible inhibitors but are not phosphorylated.

$$\text{ATP} + \text{HN}{=}\underset{|}{\overset{\text{CH}_3}{\overset{|}{\text{S}}}}{=}\text{O}\text{-}\text{CH}_2\text{CH}_2\text{-CH}(\overset{\oplus}{\text{N}}\text{H}_3)\text{COO}^{\ominus} \xrightarrow{\text{enzyme}} \text{ADP} + {}^{2\ominus}\text{O}_3\text{P}{-}\text{N}{=}\overset{\text{CH}_3}{\overset{|}{\text{S}}}{=}\text{O}\text{-}\text{CH}_2\text{CH}_2\text{-CH}(\overset{\oplus}{\text{N}}\text{H}_3)\text{COO}^{\ominus}$$

2*S*,5*S*-methionine sulfoximine — 2*S*,5*S*-methionine sulfoximine phosphate

The inactivation process occurs because the sulfoximine-P *product binds so tightly that it does not dissociate appreciably*, and all the enzyme molecules become tied up as enzyme-product complexes and are nonfunctional for subsequent turnover. This process is mechanistically distinct from any covalent labeling of the enzyme protein that has been discussed for other irreversible inactivators.

5.D.2 Regulation of Glutamine Synthetase

Almost all of the component reactions we have discussed for the sheep brain enzyme are also carried out by the *E. coli* enzyme, and methionine sulfoximine does inactivate the bacterial enzyme in the presence of ATP. The *E. coli* protein is a dodecamer of 12 identical subunits of 50,000 mol wt each, arranged in a double hexagon (one over the other) as shown in the electron micrographs of Figure 5-5. Under some conditions, these double-doughnut dodecamers stack

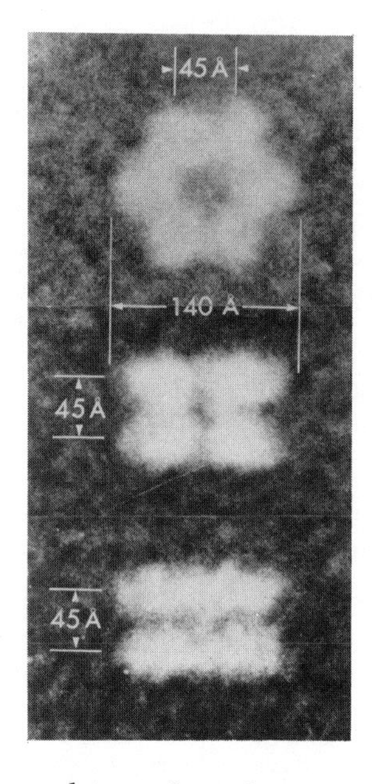

Figure 5-5
A high magnification of a picture of five superimposed images of unfixed glutamine synthetase ($GS_{9.0}$) viewed in three characteristic orientations. The mean dimensions are indicated. When the molecule rests on a face, the subunits appear as a hexagonal ring (*top*). Molecules seen on edge show two layers of subunits, either as four spots (*center*) when viewed exactly down a diameter between subunits or in general as two lines (*bottom*). (Stadtman and Ginsburg, 1974, p. 765)

end-to-end to give cylindrical aggregates (Fig. 5-5). Some physical and chemical properties of the bacterial enzyme are summarized in Table 5-2.

High concentrations of enzyme are induced by bacterial growth on limiting nitrogen; high ammonium-ion concentration in the medium represses enzyme synthesis (an economical control because high levels of ammonium replace

Table 5-2
Physical and chemical characteristics of *E. coli* glutamine synthetase

Property	Observation
Native enzyme	
Molecular weight	600,000
Apparent specific volume	0.707 ml/g
Sedimentation coefficient	20.3 S
α-Helical structures	36%
Appearance in electron microscopy	Dodecamer; double hexagon
Isoionic pH	4.9
Specific absorption at 280 nm ($A^{0.1\%}_{1.0\,\text{cm}}$)	0.385
Subunit	
Molecular weight	50,000
Molecular dimensions	45×45×53 Å
Volume	56,000 Å^3
N-terminal amino acid	Serine
C-terminal amino acid	Valine

Source: E. H. Stadtman and A. Ginsburg, in *The Enzymes*, 3d ed., ed. P. Boyer (New York: Academic Press, 1974), vol. 10, p. 755.

glutamine in the amino transfers we have been considering). The enzyme has been studied from the viewpoint of control by the groups of Stadtman and of Holzer (Stadtman and Ginsburg, 1974; Holzer and Duntze, 1971). The bacterial enzyme is a strategic target for regulation because in the bacterial cell it is involved in the biosynthesis of tryptophan, histidine, CTP, AMP, GMP, *p*-aminobenzoate, NAD, asparagine, carbamoyl phosphate, arginine, pyrimidines, and even glutamate itself. Of the various cellular controls imposed on glutamine synthetase, we shall discuss only two: reversible feedback inhibition by the various metabolic end products on the enzyme, and covalent modification of the protein with consequent change in activity (see Stadtman and Ginsburg, 1974, for discussion of other controls).

It appears that each of the twelve subunits can reversibly bind the following metabolites, experiencing inhibition that can be competitive or noncompetitive: CTP, AMP, histidine, tryptophan, alanine, glycine, serine, glucosamine-6-P, and carbamoyl phosphate. Although alanine, glycine, and serine act at an identical site, the other metabolites in general have separate binding sites for inhibition. This reversible inhibition is termed feedback inhibition, representing inhibition of the first enzyme in a biosynthetic pathway by end products of that pathway. This mechanism avoids overproduction of the final metabolites.

In addition to these freely reversible activity modifications, the *E. coli* glutamine synthetase can be *covalently modified* by reaction with ATP. An adenylyl transferase incorporates an AMP moiety into the enzyme, forming a phosphodiester bond with the phenolic hydroxyl of a specific tyrosyl residue. There is one such site per subunit, and a maximum of 12 AMP molecules per dodecamer can be added to the enzyme.

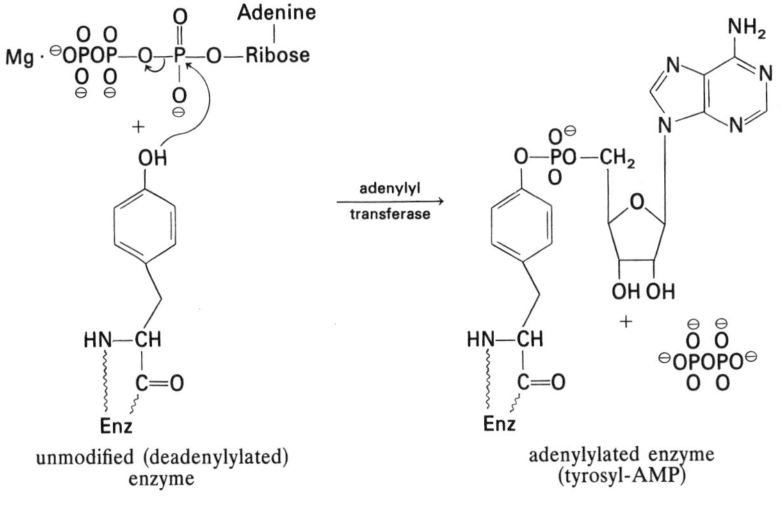

adenylyl transferase

unmodified (deadenylylated) enzyme

adenylylated enzyme (tyrosyl-AMP)

This covalent modification profoundly affects enzymatic activity. Fully adenylylated enzyme is virtually inactive under the normal assay conditions. Enzyme molecules at intermediate stages of adenylylation (e.g., 6 AMP per dodecamer) display intermediate activities. The regulation is of clear physiological importance. Introduction of ammonium ion into cells in medium formerly free of ammonium results in adenylylation of intracellular glutamine synthetase in less than a minute (Stadtman and Ginsburg, 1974). Also, the AMP-enzyme is more sensitive than the unmodified enzyme to feedback inhibition by histidine, tryptophan, CTP, or AMP.

For the adenylylation to be a reasonable control mechanism, the cell must have a reversal mechanism, a deadenylylation pathway. It turns out that this reaction is accomplished by the same specific adenylyl transferase, but the back reaction is not simple hydrolysis; rather, it is a phosphorolysis, requiring inorganic phosphate as the nucleophile and producing ADP rather than free AMP.

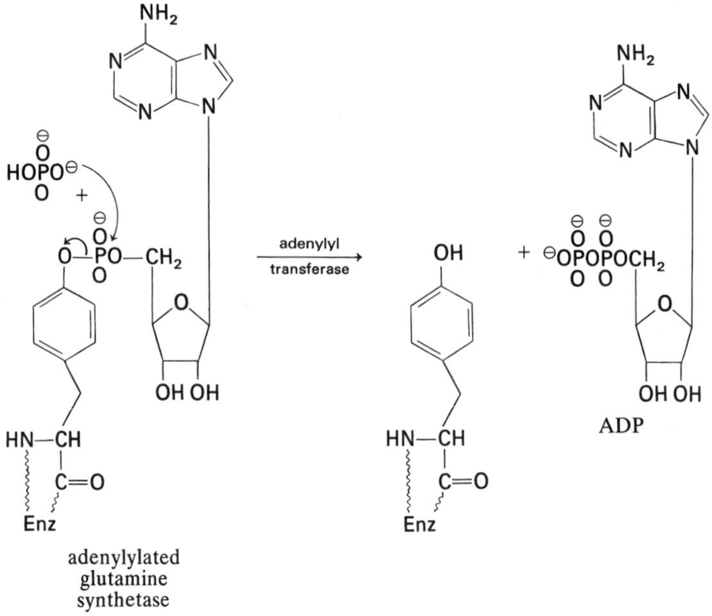

adenylyl transferase

adenylylated glutamine synthetase

ADP

Although the use of the transferase in both adenylylation and the deadenylylation (and reactivation) of glutamine synthetase is economical in one sense, it poses the problem to the cell of how to control in which direction (adding or removing AMP) the adenylyl transferase proceeds in response to a metabolic command. This problem is solved by a regulatory protein (P), which combines with the adenylyl transferase and modulates its directionality (Stadtman and Ginsburg, 1974). The adenylyl transferase is about 130,000 mol wt, and the regulatory protein about 50,000 mol wt. But how is the activity of this two-enzyme complex controlled to add or remove AMP from molecules of glutamine synthetase? In fact, the regulatory protein component itself is covalently modified by yet a third enzyme, a uridylyl (UMP) transferase (160,000 mol wt), which can add the UMP moiety from UTP to the regulatory protein (two UMP molecules introduced per 50,000 mol wt). The UMP–regulatory protein–adenylyl transferase complex performs the adenylylation of glutamine synthetase.

In turn, of course, the UMP residue added to the regulatory component must also be removable, and this time a separate UMP-removing activity has been identified, but not yet well characterized. The overall control of the covalent modification of glutamine synthetase is thus complex and finely tuned by the consecutive action of and interplay of several enzymes.

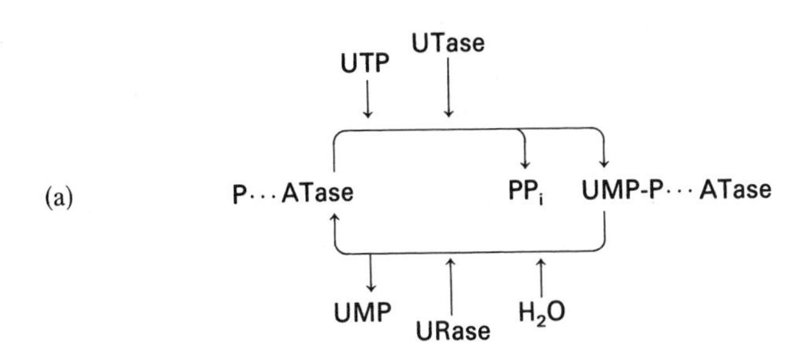

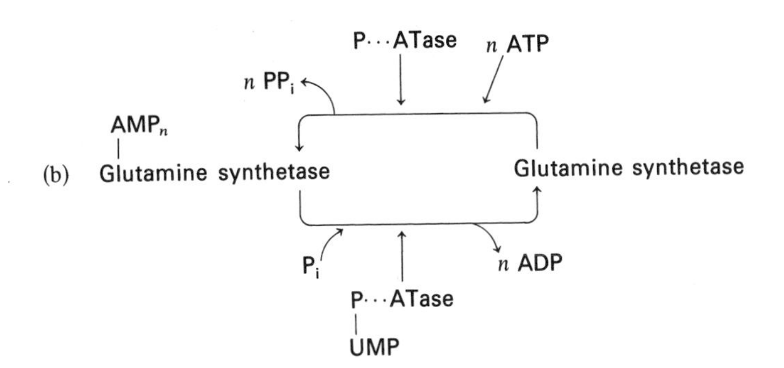

In (a), we indicate the cycle for covalent addition of a UMP group to the P ··· adenylyl-transferase complex. In (b), we show the cyclic action of the UMP—P ··· ATase complex or the unmodified P··· ATase complex on glutamine synthetase itself.

This process is called a *cascade phenomenon* (sequential activation of components), and this is only one example of many such phenomena in enzymatic systems (the process of blood clotting and the mobilization of glycogen being two others) where such complex sequences operate. It has been argued that cascade systems are sensitive ways of amplifying an initial small biological signal. Because each component in the cascade is a catalytic entity, one can have large multiplication factors at each step. The outline presented here is only the bare bones of the cascade scheme involved with glutamine synthetase, because each protein is inhibited or activated by various cations and metabolites such as glutamine, glutamate, and α-ketoglutarate (Stadtman and Ginsburg, 1974).

5.E ANOTHER CLEAVAGE PATTERN FOR ATP IN AMINATION REACTIONS: ATP $\rightarrow$ AMP + PP_i

We have discussed (¶5.C.3) amination reactions in which ATP is cleaved to ADP and P_i. The second mode of enzyme-catalyzed cleavage of ATP in amination reactions is fragmentation between the α- and β-phosphorus atoms to produce adenylic acid (AMP) and inorganic pyrophosphate (PP_i). The thermodynamic driving force is equivalent, but the mechanistic features are distinct. In the example chosen, we shall see that a nucleophilic group of the cosubstrate attacks the α-phosphorus (not the γ-phosphorus) of ATP.

5.E.1 Asparagine Synthetase

In addition to glutamine, there is another common amino acid with an amide group: asparagine. The nitrogen of the β-carboxamide group of asparagine is derived from glutamine. (In contrast to glutamine, asparagine is not metabolically active as an amino donor; there is no obvious chemical reason for this. Perhaps the biological systems simply happened to evolve genes for using glutamine first, and then stuck with them.) The heart of the reaction catalyzed by asparagine synthetase is

$$RCOO^{\ominus} \longrightarrow RCONH_2$$

The stoichiometry of this reaction is the following:

$$\underset{\text{L-aspartate}}{{}^{\ominus}OOC{-}CH_2{-}CH(\overset{\oplus}{N}H_3){-}COO^{\ominus}} + \underset{\text{L-glutamine}}{H_2N{-}C(=O){-}CH_2{-}CH_2{-}CH(\overset{\oplus}{N}H_3){-}COO^{\ominus}} + ATP \underset{}{\overset{Mg}{\rightleftarrows}} \underset{\text{L-asparagine}}{H_2N{-}C(=O){-}CH_2{-}CH(\overset{\oplus}{N}H_3){-}COO^{\ominus}} + \underset{\text{L-glutamate}}{{}^{\ominus}OOC{-}CH_2{-}CH_2{-}CH(\overset{\oplus}{N}H_3){-}COO^{\ominus}} + AMP + PP_i$$

The carboxylate anion of the aspartate is an unactivated, low-energy resonance-stabilized anion that is not easily attacked by nucleophilic NH_3:

$$R{-}C({=}O){-}O^{\ominus} \longleftrightarrow R{-}C(O^{\ominus}){=}O$$

It is extremely difficult to remove one of the carboxylate oxygens in displacement reactions, either enzymatically or nonenzymatically. First the carboxylate group must be activated and a good leaving group introduced. The ATP-cleavage pattern implies that the carboxylate anion, in this instance, attacks the α-phosphorus to generate as covalent intermediate a mixed anhydride of aspartate and AMP: aspartyl AMP. This acyl adenylate (¶5.C.3.c) is destabilized. Attack by nascent ammonia would lead to the stable amide group of asparagine and the observed AMP product. A pentacovalent phosphorane adduct should be involved in formation of the intermediate, and a tetrahedral adduct involved in its aminolytic breakdown.

aspartate + ATP ⇌ aspartyl-AMP + PP_i

aspartyl-AMP ⇌ L-asparagine + AMP + glutamate

5.E.2 Partial Exchange Reactions as Evidence for Reversible Formation of a Covalent Intermediate

Our mechanistic scheme for asparagine synthetase suggests that a covalent intermediate forms during catalysis. One of the ways to test for covalent intermediates is to search for isotopic-exchange reactions that may reflect existence of that intermediate. In fact, the ^{18}O-exchange data from $[^{18}O]$-H_2O into an initially unlabeled amino acid were listed as evidence consistent with reversible formation of a covalent acyl enzyme in chymotrypsin and pepsin catalysis (¶3.C.2).

The question here is how to perform an isotopic exchange that will give evidence for the aspartyl-AMP intermediate. Clearly, if one has *all* the substrates or products present, no information can be obtained, because under these conditions the enzyme will catalyze both the front and back reactions, and the isotopic label will be equilibrated irrespective of mechanism. Information about intermediates could be obtained if a component step proceeds under conditions where the overall reaction cannot proceed. Omission of one of the substrates would ensure that the complete reaction cannot occur; detection of a partial reaction can be facilitated using radioactive isotopes as indicated in the following example.

Suppose aspartate and ATP can react with the enzyme's aid, *in the absence of glutamine*, to form aspartyl-AMP and PP_i *reversibly*. This is a reasonable expectation because (1) ammonia is not postulated to act until the acyladenylate has formed, and (2) other glutamine-utilizing enzymes discussed have separate subsites for hydrolyzing glutamine and for subsequent specific amination reactions. The reversible aspartyl-AMP formation could be deduced if addition of [^{32}P]-pyrophosphate, aspartate, and ATP to enzyme results in formation of radioactive β,γ-[^{32}P]-ATP. One is looking for *flux of radioactive isotope*, a partial exchange reaction, perhaps with little or no detectable mass exchange—e.g., the isotopic exchange could be performed with ATP, [^{32}P]-PP_i, and aspartate at chemical equilibrium (Segel, 1975*b*). This exchange could occur by the following reaction sequence.

$$\text{Mg}\cdot\text{ATP} + \text{Aspartate} + \text{Enz} \rightleftharpoons \text{Enz}\begin{smallmatrix}\cdot\cdot\,\text{Mg}\cdot\text{ATP}\\ \cdot\cdot\,\text{Aspartate}\end{smallmatrix} \quad (1)$$

$$\text{Enz}\begin{smallmatrix}\cdot\cdot\,\text{Mg}\cdot\text{ATP}\\ \cdot\cdot\,\text{Aspartate}\end{smallmatrix} \rightleftharpoons \text{Enz}\begin{smallmatrix}\cdot\cdot\,\text{Aspartyl-AMP}\\ \cdot\cdot\,\text{PP}_i\end{smallmatrix} \quad (2)$$

$$\text{Enz}\begin{smallmatrix}\cdot\cdot\,\text{Aspartyl-AMP}\\ \cdot\cdot\,\text{PP}_i\end{smallmatrix} \rightleftharpoons \text{Enz}\cdot\cdot\,\text{Aspartyl-AMP} + \text{PP}_i \quad (3)$$

$$\text{Enz}\cdot\cdot\,\text{Aspartyl-AMP} + \boxed{[^{32}\text{P}]\text{-PP}_i} \rightleftharpoons \text{Enz}\begin{smallmatrix}\cdot\cdot\,\text{Aspartyl-AMP}\\ \cdot\cdot\,[^{32}\text{P}]\text{-PP}_i\end{smallmatrix} \quad (4)$$

$$\text{Enz}\begin{smallmatrix}\cdot\cdot\,\text{Aspartyl-AMP}\\ \cdot\cdot\,[^{32}\text{P}]\text{-PP}_i\end{smallmatrix} \rightleftharpoons \text{Enz}\begin{smallmatrix}\cdot\cdot\,\beta,\gamma\text{-}[^{32}\text{P}]\text{-ATP}\\ \cdot\cdot\,\text{Aspartate}\end{smallmatrix} \quad (5)$$

$$\text{Enz}\begin{smallmatrix}\cdot\cdot\,\beta,\gamma\text{-}[^{32}\text{P}]\text{-ATP}\\ \cdot\cdot\,\text{Aspartate}\end{smallmatrix} \rightleftharpoons \text{Enz} + \boxed{\beta,\gamma\text{-}[^{32}\text{P}]\text{-ATP}} + \text{Aspartate} \quad (6)$$

The formation of the covalent intermediate cannot be initiated by reaction of [^{32}P]-pyrophosphate with enzyme. Rather, an unlabeled molecule of ATP and aspartate must bind in step 1. The chemical step (2) occurs forming the ternary complex of enzyme, aspartyl-AMP, and unlabeled pyrophosphate. Steps 3 and 4 are *key* steps for detection of isotopic exchange, even though no chemical change is involved in them. Step 3 involves dissociation of the product PP_i molecule to give the enz . . . aspartyl-AMP complex. Once that PP_i molecule is free in solution, it equilibrates with the bulk PP_i molecules, some of which (perhaps 1 in 10^{10}) are labeled with ^{32}P. If a labeled PP_i molecule rebinds in step 4 (the reverse of step 3), then the chemical step can be reversed in step 5, yielding a molecule of bound β,γ-[^{32}P]-ATP at the active site. Release of substrates in step 6 produces the free, radioactive molecule of [^{32}P]-ATP, which can be replaced by an unlabeled molecule of ATP so that the cycle can begin again. As a catalyst, the enzyme will carry out many such cycles, and the specific radioactivity of [^{32}P]-ATP will rise and become detectable.

To measure the progress of the exchange of ^{32}P-isotope from pyrophosphate into ATP, one can follow the exchange in time by removing aliquots from solution, separating ATP and PP_i, and determining their specific radioactivities. The exchange will proceed until the distribution of ^{32}P-radioactivity is proportional to the equilibrium concentrations of ATP and PP_i present.

The initial exchange velocity can be calculated from the following equation, whose derivation is presented by Segel (1975*b*):

$$v = -\{[A][B]/([A]+[B])\}(2.3/t)\log(1-F),$$

where F is the fraction of isotopic equilibrium at time t, and A and B are the pair of molecules undergoing the isotopic exchange. Suppose that [^{32}P]-PP_i and ATP were initially added at their equilibrium concentrations—for instance, 0.4 mM and 1.6 mM, respectively. At isotopic equilibrium, $[[^{32}P]\text{-ATP}]/[[^{32}P]\text{-}PP_i] = \frac{4}{1}$, and one-fifth of the total radioactivity will remain in [^{32}P]-PP_i, while four-fifths will be present as β,γ-[^{32}P]-ATP. Suppose that after five minutes the separated ATP contains 1,000 count/min ^{32}P, and the PP_i has 4,000 count/min ^{32}P. At this point, one-fifth of the total counts per minute are in ATP, whereas four-fifths would be present at equilibrium, so $F = 0.25$, and the initial velocity of exchange can be computed:

$$\begin{aligned} v &= -\{(1.6\times0.4)/(1.6+0.4)\}(2.3/5)\log(1-0.25) \\ &= -1.5\times(-0.125) = 0.19\ \mu\text{mole/min}\,. \end{aligned}$$

This value obviously is a function of the enzyme concentration present.

The positive detection of the partial isotopic exchange expected for a covalent intermediate is a strong indirect criterion for its presence in any enzymatic reaction. However, failure to detect exchange does not necessarily rule out such an intermediate, because a variety of conditions could obscure the meaning of that observation. Two of these conditions are binding phenomena. A multisubstrate enzyme such as asparagine synthetase could follow an *ordered kinetic mechanism*, where glutamine might have to bind first to enable a conformation change so that aspartate and ATP could then bind productively to form the acyl adenylate at detectable rates. Obviously, this was not the case with the asparagine synthetase. Alternatively, at the end of step 3 of the scheme presented earlier in this section, PP_i dissociation might be extremely slow (binding might be very tight), or undetectable. Failure of PP_i to dissociate readily and of $[^{32}P]$-PP_i molecules to reassociate with the enz . . . aspartyl-AMP complex will prevent detection of the intermediate by the isotope-exchange method, even though the intermediate may form reversibly. (It is likely that the acyl adenylate in this and mechanistically related enzymes remains tightly bound; it may resemble the transition state and thus have a strong interaction energy. And, as a mixed anhydride, it would be hydrolytically labile free in solution but could be stabilized in a hydrophobic active site with limited water access.)

Once an exchange reaction has been detected, the velocity-of-exchange equation can be used to establish its kinetic competence. If the acyl adenylate is a normal intermediate in catalysis, the rate of the partial reaction should be as fast as (or faster than) the turnover number for the complete reaction. Some partial exchanges may be slightly slower than overall catalysis because of slow dissociation or association steps that are accelerated when the omitted substrate is present during turnover. However, if a partial exchange is 10^{-3}-fold or 10^{-4}-fold slower than complete turnover rates, the strong possibility exists that one of the substrates or the enzyme may be contaminated with a trace amount of the omitted substrate. Then the apparent partial exchange may be merely the complete reaction proceeding at a very slow rate due to limiting substrate; in such a case, no information on a covalent intermediate is obtained.

As a final caution, we point out that, in the exchange experiment described here (exchange of PP_i and ATP with asparagine synthetase), radioactive $[^{32}P]$-PP_i rather than $[^{32}P]$-ATP was used as the substrate initially labeled. If $[^{32}P]$-ATP had been used, radioactive $[^{32}P]$-PP_i could be formed by way of the aspartyl-AMP species, but it could also form by slow $[^{32}P]$-ATP hydrolysis on the part of the enzyme, an ATPase activity. There is no obvious way for radioactivity to move from PP_i into ATP except for the acyl-adenylate mechanism.

We shall use the partial exchange criterion with other enzymatic reactions where covalent intermediates are possibilities during catalysis. One can generalize the partial exchange reactions expected for any covalent intermediate whose reversible formation is expected. For a substrate X—Y where an enz—X intermediate is suspected, look for a partial exchange of label from Y* into X—Y* with time (the asterisk denotes isotopic label). Conversely, for an E—Y intermediate, an exchange of radioactivity from X* into X*—Y is expected. The reader might verify that, in the carbamoyl-P synthetase reaction (¶5.C.3.c), a bicarbonate-dependent (but glutamine-independent) exchange of radioactivity from [^{14}C]-ADP into ATP is consistent with reversible formation of this intermediate:

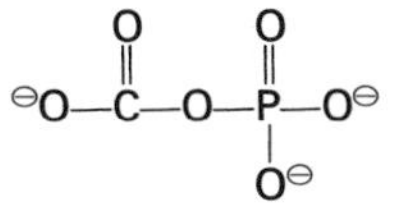

$$^{\ominus}O-C(=O)-O-P(=O)(O^{\ominus})-O^{\ominus}$$

We noted a few paragraphs earlier that a covalent intermediate may go undetected in partial isotope-exchange experiments if the other substrate-derived fragment is not readily released into solution. For this reason there has been a need for an isotope-exchange method that works even when the fragments do not dissociate from the enzyme. Just such a technique for monitoring reversible cleavage of ATP to ADP at the active site of glutamine synthetase has been devised by Middlefort and Rose (1976). They utilized ATP with an ^{18}O-label in the bridge position between the β-P and γ-P atoms. Glutamine-dependent cleavage would yield ADP with ^{18}O as indicated:

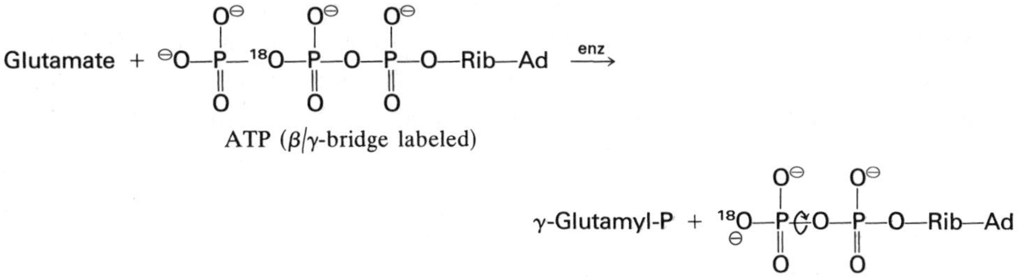

$$\text{Glutamate} + {}^{\ominus}O-P(=O)(O^{\ominus})-{}^{18}O-P(=O)(O^{\ominus})-O-P(=O)(O^{\ominus})-O-\text{Rib}-\text{Ad} \xrightarrow{\text{enz}}$$

ATP (β/γ-bridge labeled)

$$\gamma\text{-Glutamyl-P} + {}^{18}O^{\ominus}-P(=O)(O^{\ominus})-O-P(=O)(O^{\ominus})-O-\text{Rib}-\text{Ad}$$

ADP

Now the β-phosphoryl group of ADP is torsiosymmetric. If it can rotate around the indicated single bond at the active site, then the ^{18}O-atom can be in any of the three rotational positions shown in the following expressions. No such rotational interconversions can occur in ATP. Now if ATP is reformed (prior to release of ADP into solution, such that cleavage has not been detectable by any other technique), then not only will ATP molecules be produced with ^{18}O in the original β,γ-bridge position, but also the two ATP species with ^{18}O in the nonbridge positions on the β-P will be produced.

symmetric β-phosphoryl group of ADP rotates

$^{18}O^{\ominus}$—P—O—P—O—Rib—Ad **1** ⇌ $O^{\ominus}$—P—O—P—O—Rib—Ad **2**

resynthesized ATP

$^{\ominus}O$—P—^{18}O—P—O—P—O—Rib—Ad

$^{\ominus}O$—P—O—P—O—P—O—Rib—Ad

ADP, third rotational isomer

⇌ $^{\ominus}O$—P—O—P—O—Rib—Ad **3**

resynthesized ATP

$^{\ominus}O$—P—O—P—O—P—O—Rib—Ad

All that was needed was a method of detecting this ^{18}O-positional scrambling. The reisolated ATP was converted to inorganic phosphate by a series of reactions allowing distinction between bridge and nonbridge positions, and the ^{18}O-contents were examined by mass spectrometry via a volatile trimethyl-phosphate derivative. Reversible cleavage of ATP, while enzyme bound, was in fact detected.

5.F ASPARTATE AS AN AMINO DONOR

Occasionally the α-amino group of aspartate can be utilized as a transferable amino group, but the mechanistic pathway for such a transfer is markedly distinct from the glutamine aminations noted in preceding sections. We end this chapter with a brief digression on this type of amination reaction.

In ¶5.C.3.d we noted that carbamoyl-P (a doubly activated acyl phosphate) can serve as a carbamoylating agent, and that δ-*N*-carbamoylornithine (citrulline) is formed by acyl transfer to the δ-amino group of ornithine in the first reaction of the urea cycle (Fig. 5-6). The next enzyme in that pathway is argininosuccinate

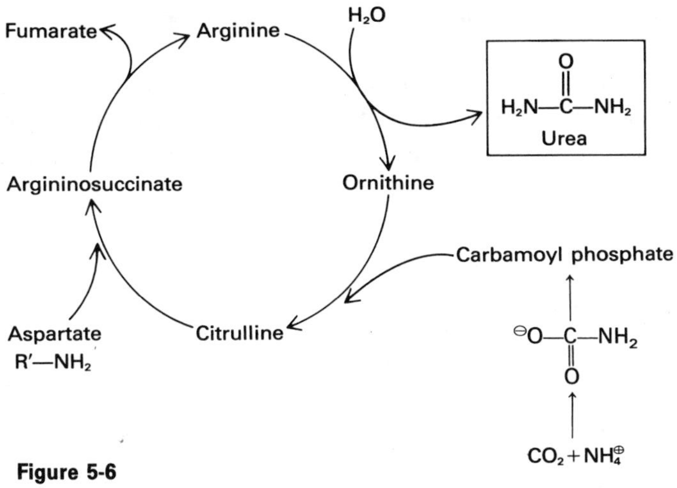

Fumarate ← Arginine → (H_2O) → Urea: $H_2N{-}C(=O){-}NH_2$

Ornithine

Argininosuccinate

Carbamoyl phosphate ← $^{\ominus}O{-}C(=O){-}NH_2$ ← $CO_2 + NH_4^{\oplus}$

Aspartate
$R'{-}NH_2$

Citrulline

Figure 5-6
The urea cycle. (From L. Stryer, *Biochemistry*, p. 437. W. H. Freeman and Company. Copyright © 1975)

synthetase, catalyzing the following interconversion:

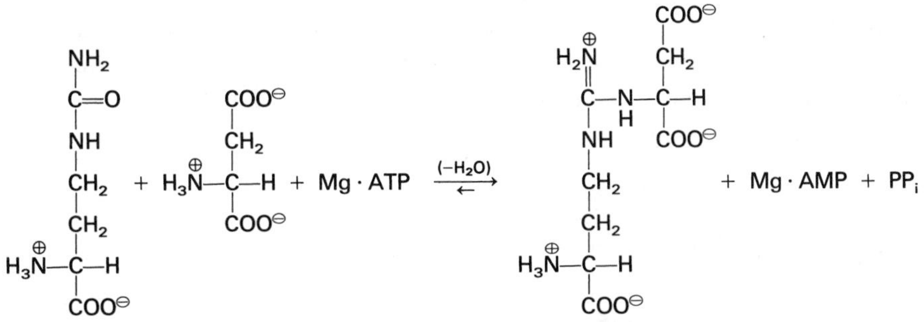

$$\underset{\text{citrulline}}{H_2N{-}C(=O){-}NH{-}CH_2{-}CH_2{-}CH(\overset{\oplus}{N}H_3){-}COO^{\ominus}} + \underset{\text{aspartate}}{\overset{\oplus}{H_3N}{-}CH(COO^{\ominus}){-}CH_2{-}COO^{\ominus}} + Mg\cdot ATP \xrightleftharpoons{(-H_2O)} \underset{\text{argininosuccinate}}{\overset{\oplus}{H_2N}{=}C(NH{-}CH(COO^{\ominus}){-}CH_2{-}COO^{\ominus}){-}NH{-}CH_2{-}CH_2{-}CH(\overset{\oplus}{N}H_3){-}COO^{\ominus}} + Mg\cdot AMP + PP_i$$

Aspartate acts as nucleophile, and ATP is required apparently to enhance the poor leaving-group potential of the carbonyl oxygen—e.g., via citrullyl-AMP formation as intermediate, by analogy to the asparagine-synthetase reaction noted earlier.

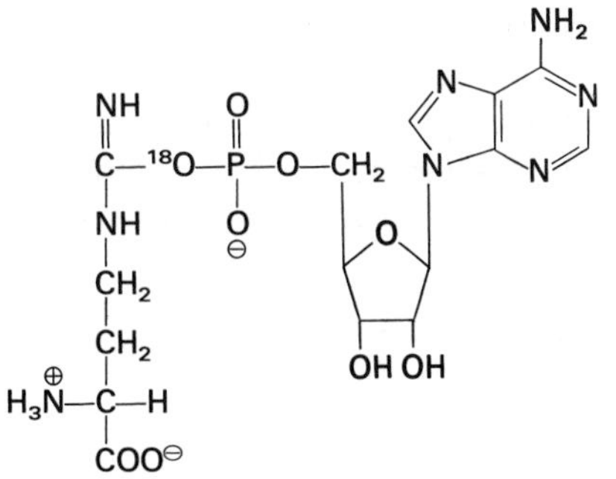

$$HN{=}C({-}^{18}O{-}P(=O)(O^{\ominus}){-}O{-}CH_2{-}\text{ribose(OH, OH)}{-}\text{adenine}(NH_2)){-}NH{-}CH_2{-}CH_2{-}CH(\overset{\oplus}{N}H_3){-}COO^{\ominus}$$

Now the oxygen of the isoureido-phosphoryl group can depart with the AMP product on attack of a nucleophile. This time the nucleophile is not NH_3 but rather, as the stoichiometry indicates, the α-amino group of aspartate.

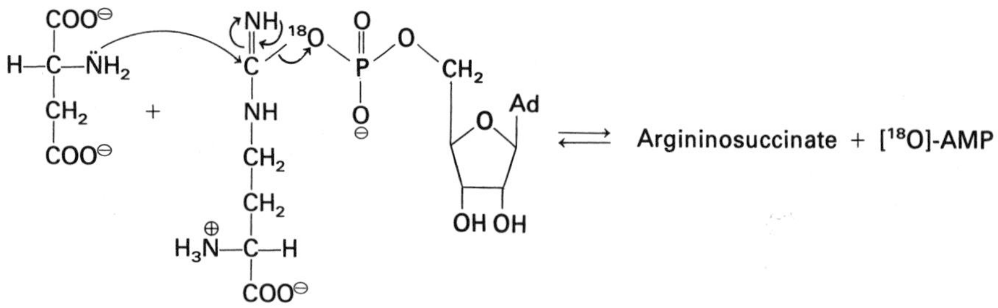

aspartate + citrullyl-AMP ⇌ Argininosuccinate + [^{18}O]-AMP

The adduct thus formed is the observed product, argininosuccinate. The α-amino group has been introduced, but it still has the rest of the aspartate molecule bonded to it. An additional enzyme, argininosuccinase (discussed in Chapter 18), is required to catalyze expulsion of the carbon skeleton of the added aspartate. This expulsion is initiated by abstraction of one of the β-hydrogens of the aspartyl side chain as a proton. Fragmentation then ensues as indicated to produce the olefinic dicarboxylic acid, fumarate (the *trans*-isomer only), and L-arginine.

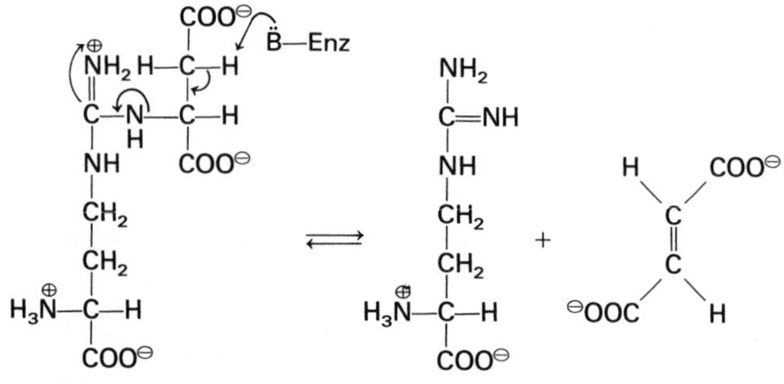

argininosuccinate ⇌ L-arginine + fumarate

In the guanidinium group of arginine then, one nitrogen derives ultimately from the carboxamide nitrogen of glutamine via carbamoyl phosphate, and one other nitrogen was introduced in the two-step process from aspartate.

Chapter 6

Phosphoryl Transfers, I: Phosphatases, ATPases, and Phosphodiesterases

Phosphorus and sulfur are two second-row elements that are common and key components in biochemical structures and in reactions that occur in biological systems. Phosphorus is stable in a variety of oxidation states ranging from −3 to +5, but the stable oxidation state of phosphorus in biological molecules is +5, and no net change in redox state occurs. At this oxidation state, the phosphorus atom does not react readily as a nucleophile, but will act as electrophile, undergoing attack by a variety of other nucleophiles. The overwhelmingly predominant form of phosphorus found in biological systems is the phosphate group, where four electronegative oxygen atoms are attached to the central phosphorus atom. (In contrast, sulfur is found in diverse oxidation states in biological compounds, ranging from the electrophilic +6 oxidation state in sulfate to the nucleophilic −2 oxidation state in H_2S and RSH.)

As a second-row element, phosphorus can form more than four covalent bonds by putting additional electron pairs into unfilled *d* orbitals to form a fifth bond, and stable pentacovalent phosphoranes are known in chemical systems. In the phosphate functional group, the unshared electron pair on one of the oxygen atoms can be shared with a *d* orbital of the phosphorus to form sufficient overlap to produce a d–p_π bond. We then can write the resonance forms of phosphoric acid as

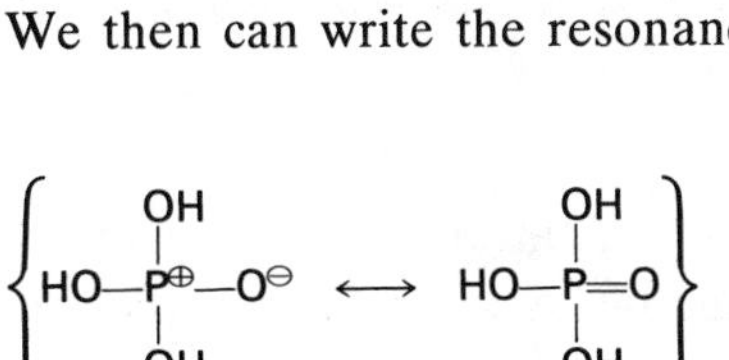

$$\left\{ \mathrm{HO{-}\overset{\displaystyle OH}{\underset{\displaystyle OH}{P^{\oplus}}}{-}O^{\ominus}} \longleftrightarrow \mathrm{HO{-}\overset{\displaystyle OH}{\underset{\displaystyle OH}{P}}{=}O} \right\}$$

form 1 form 2

Resonance form 1 emphasizes the +5 oxidation state of the phosphorus atom, as well as its electrophilicity and sensitivity to attack by nucleophiles. Form 2 indicates the double bond P=O, which comprises a σ-component and a d-p-component and is not a "normal" double bond. The d-p_π double bond does not alter the tetrahedral geometry of the phosphate group. When a nucleophile attacks the electropositive phosphorus, a pentacovalent intermediate is formed; under some conditions, this intermediate is isolable (Westheimer, 1968; Bruice and Benkovic, 1965–1966, vol. 2).

pentacovalent intermediate

The geometry of the pentacovalent intermediate is a trigonal bipyramid with three equatorial and two apical substituents. Normally, the two most electronegative atoms occupy the apical substituents. Expulsion of a leaving group converts the phosphorane species back into a stable four-substituent phosphorus compound. No stable pentacovalent phosphorane species have been found in enzymatic reactions.

The parent phosphate compound is phosphoric acid, which has the following pK_a values:

$$(HO)_3P{=}O \rightleftharpoons (HO)_2P(O^{\ominus}){=}O + H^{\oplus} \qquad pK_{a(1)} \approx 1$$

$$(HO)_2P(O^{\ominus}){=}O \rightleftharpoons (HO)P(O^{\ominus})_2{=}O + H^{\oplus} \qquad pK_{a(2)} = 6.7$$

$$(HO)P(O^{\ominus})_2{=}O \rightleftharpoons P(O^{\ominus})_3{=}O + H^{\oplus} \qquad pK_{a(3)} = 12.7$$

Thus, at pH 7, significant concentrations of inorganic phosphate dianion exist. The pK_a values are not dramatically perturbed in phosphate monoesters and phosphate diesters, which then also are anionic at physiological pH. (Phosphate triesters are not common in biological systems.)

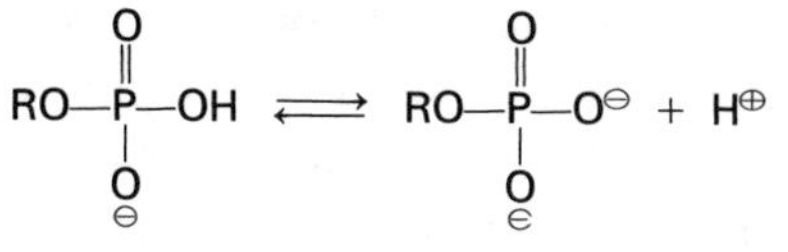

$$RO{-}P(=O)(OH){-}O^{\ominus} \rightleftharpoons RO{-}P(=O)(O^{\ominus}){-}O^{\ominus} + H^{\oplus}$$

monoester

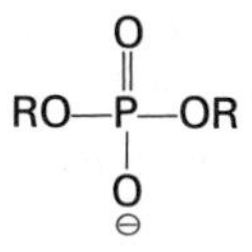

$$RO{-}P(=O)(O^{\ominus}){-}OR$$

diester

The negative charges provide electrostatic repulsion to hydrolysis of phosphate esters, and either protonation or complexation with metal cations to reduce this charge repulsion facilitates attack by water and other nucleophiles. Thus, ATP in enzymatic hydrolyses exists as the $Mg^{2\oplus}$ chelate, with the cation associated to the β- and γ-phosphate groups (Tetas and Lowenstein, 1963; Cohn and Hughes, 1972; Cleland, 1975).

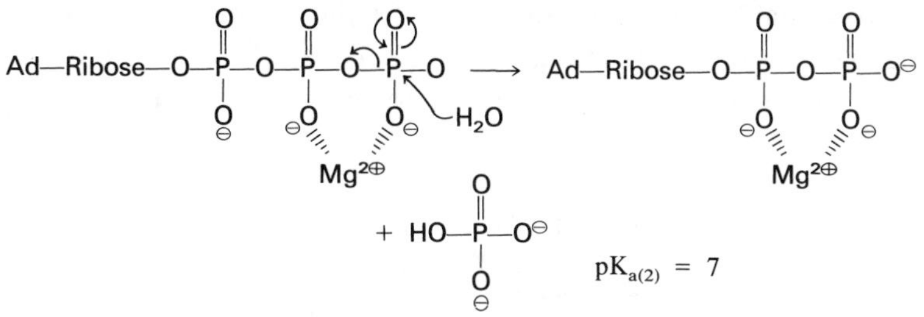

Ad—Ribose—O—P(=O)(O⊖)—O—P(=O)(O⊖)—O—P(=O)(O⊖)—O + H_2O ($Mg^{2\oplus}$ chelated to β- and γ-phosphate O⊖) ⟶ Ad—Ribose—O—P(=O)(O⊖)—O—P(=O)(O⊖)—O⊖ ($Mg^{2\oplus}$) + HO—P(=O)(O⊖)—O⊖

$pK_{a(2)} = 7$

Common phosphate monoesters in biological systems include the metabolites glucose-6-phosphate and glucose-1-phosphate.

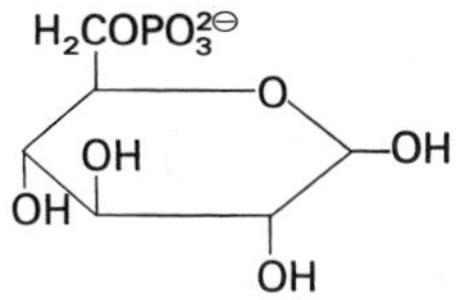

$H_2COPO_3^{2\ominus}$ (ring with OH groups)

glucose-6-P

H_2C—OH (ring with OH groups) —O—P(=O)(OH)—O⊖

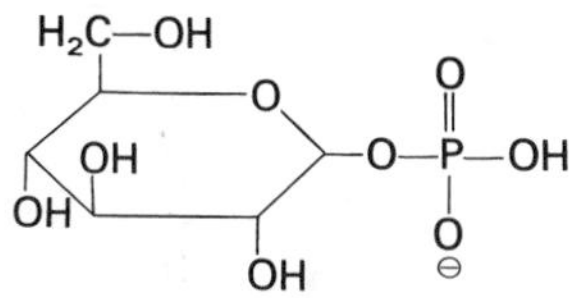

glucose-1-P

Another common biological phosphate monoester is phosphoserine, both as the free *O*-phosphorylated amino acid and as the *O*-phosphoryl-serine residues in phosphoryl-enzyme species such as inactive diisopropyl-fluorophosphate–treated proteases or acetylcholinesterase or the active phosphoryl-enzyme intermediates in enzymes such as alkaline phosphatase (¶6.A.1.b).

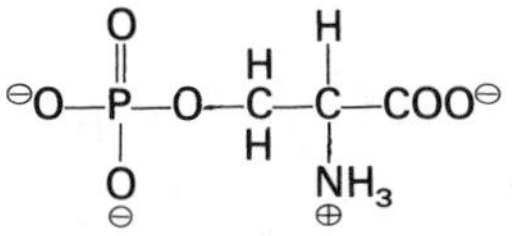

⊖O—P(=O)(O⊖)—O—CH_2—CH($NH_3^{\oplus}$)—$COO^{\ominus}$

O-phosphoserine

⊖O—P(=O)(O⊖)—O—CH(CH_3)—CH(NH—C(=O)—Enz)—C(=O)

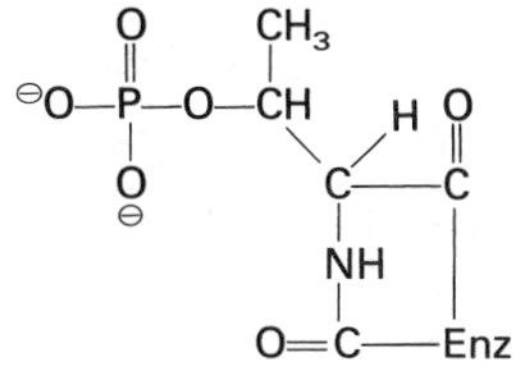

O-phosphothreonyl residue

Phosphorylated serine and threonine residues are found also as apparent storage loci for phosphate in egg proteins such as phosvitin (Rabinowitz and Lipmann, 1960) and as residues phosphorylated in many proteins modified reversibly for control purposes, such as histones (Langan, 1968), ribosomal proteins (Eil and Wool, 1971), and certain protein kinases (Rosen and Erlichman, 1975). These covalent phosphoryl esters were detected initially by virtue of their acid stability. The phosphorylated proteins can be hydrolyzed in 6 N HCl at 110° C for 24 hours, and free phosphoserine or phosphothreonine can be isolated.

By contrast, other phosphoenzyme species are labile to strong acid and resist isolation by these acid degradative procedures. The phosphorylated proteins fall into two basic categories: (1) those in which the phosphoryl linkage to the amino-acid residue is base-stable but acid-labile, and (2) those where the linkage is both acid-labile and base-labile. When the linkage survives alkaline conditions, the E—X—PO_3 compound has X as a nitrogen atom rather than a seryl or threonyl oxygen.

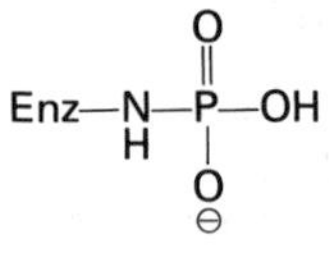

Enz—NH—P(=O)(—OH)—O⊖ pK_a = 6 to 8

Covalent phosphoryl groups attached to the ε-amino group of a lysine residue or to either N-1 or N-3 of the imidazole group of histidinyl residues have been identified (Edlund et al., 1969).

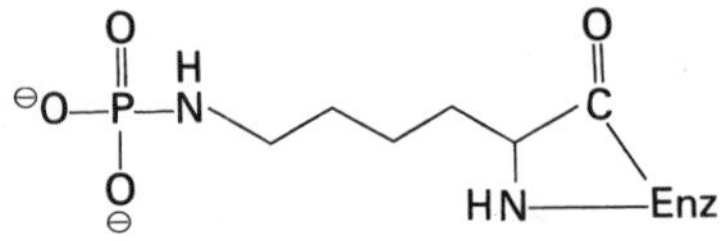

ε-*N*-phosphoryllysyl residue

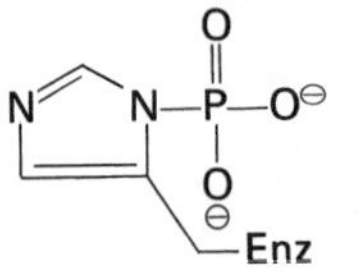

1-*N*-phosphohistidinyl residue

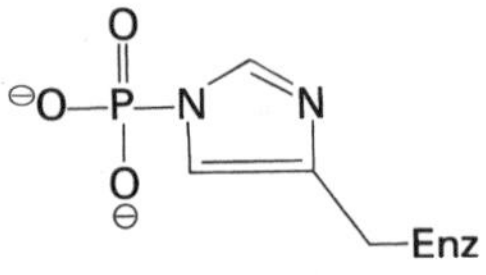

3-*N*-phosphohistidinyl residue

Thus far, it appears that the free *N*-phosphoryl amino acids are not common constituents of metabolism. But low-molecular-weight phosphoramidate-containing compounds are known in biochemistry, including phosphocreatine as a reservoir for activated phosphoryl groups in skeletal muscle and phosphoarginine as the analogous phosphagen in invertebrate muscle.

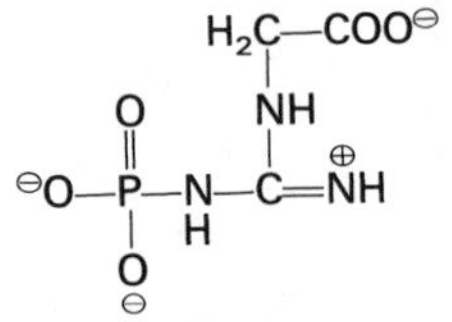

phosphocreatine

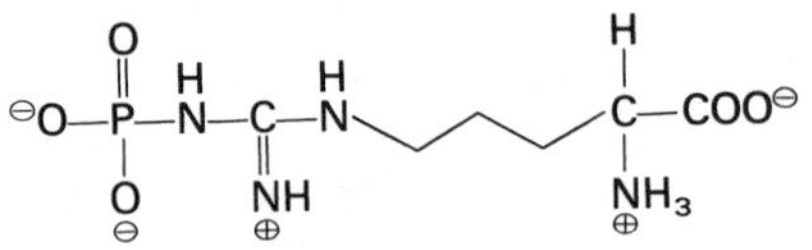

phosphoarginine

These R—N—PO_3 compounds, either as low-molecular-weight species or as enzyme derivatives, are thermodynamically destabilized molecules, and they undergo enzyme-catalyzed phosphoryl transfer to a variety of nucleophiles, *but* generally *not* to H_2O, which would be energetically wasteful (see kinase reactions discussed in Chapter 7).

The third category of covalent E—X—PO_3 (the one that is unstable at either extreme of pH) is the group of acyl phosphates. The enzyme provides the β-carboxylate of an aspartyl residue or the γ-carboxyl of a glutamyl residue as the attacking nucleophile.

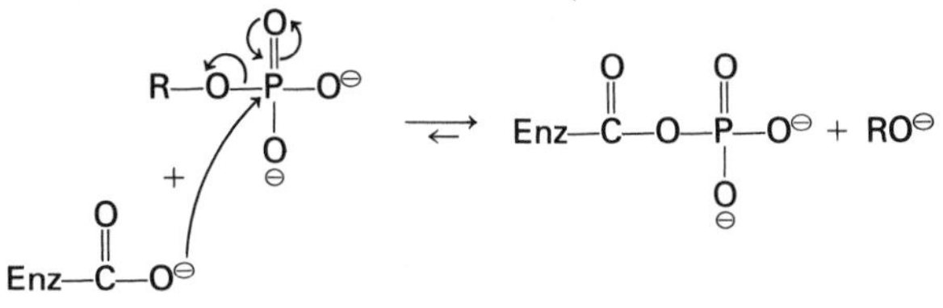

$$\text{R—O—PO}_3^{2\ominus} + \text{Enz—C(=O)—O}^{\ominus} \rightleftharpoons \text{Enz—C(=O)—O—PO}_3^{2\ominus} + \text{RO}^{\ominus}$$

Such an intermediate forms during action of some of the ATPases discussed in ¶6.A.1.c. Low-molecular-weight acyl phosphates, such as carbamoyl phosphate (¶5.C.3.d) and γ-glutamyl phosphate (¶5.D.1), have already been discussed. One report of an —S—$PO_3^{2\ominus}$ (phosphorothiorate) linkage has appeared recently (Pigiet, 1978), as a cysteinyl—S—$PO_3^{2\ominus}$ residue in the *E. coli* protein thioredoxin. (Phosphonates, R—CH_2—$PO_3^{2\ominus}$, have been found in some amoebae and colenterates.)

Many of the phosphodiester linkages found in biochemical systems turn out to be pyrophosphate linkages. The pyrophosphate linkages in molecules such as ATP or ADP serve as energy reservoirs. The pyrophosphate group in molecules such as dolichol pyrophosphate (an 80- to 100-carbon isoprenoid alcoholic pyrophosphate) provides a water-soluble tail to an otherwise hydrophobic and membrane-soluble intermediate. The pyrophosphate group in the coenzyme form of vitamin B_1 (thiamin pyrophosphate) probably provides an electrostatic "handle" for binding interaction, providing specificity to catalysis by supplying interaction energy.

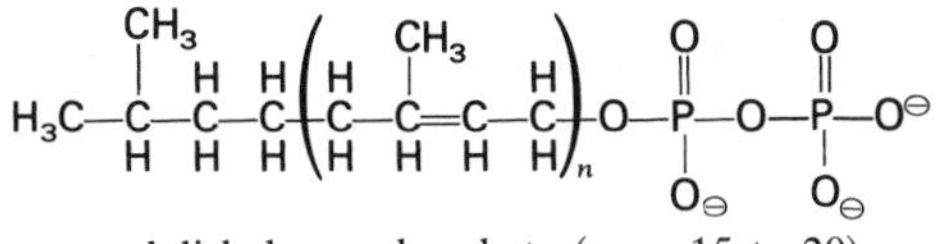

$$\text{H}_3\text{C—C(CH}_3\text{)H—CH}_2\text{—CH}_2\text{—(CH}_2\text{—C(CH}_3\text{)=CH—CH}_2\text{)}_n\text{—O—P(=O)(O}^{\ominus}\text{)—O—P(=O)(O}^{\ominus}\text{)—O}^{\ominus}$$

dolichol pyrophosphate (n = 15 to 20)

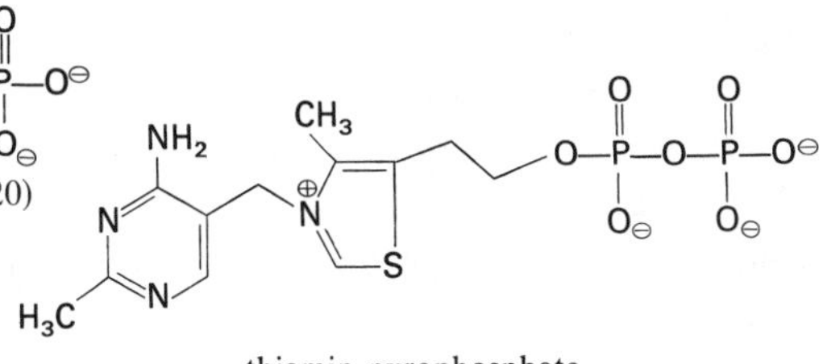

thiamin pyrophosphate

The second linkage is the internucleotide phosphodiester bond that is the structural backbone of the informational biopolymers DNA and RNA. The phosphoryl group is in diester linkage with the 3′-hydroxyl of one nucleoside ribose group and with the 5′-ribose-hydroxyl of an adjacent nucleoside. This internucleotide linkage is the only covalent linkage holding DNA or RNA together.

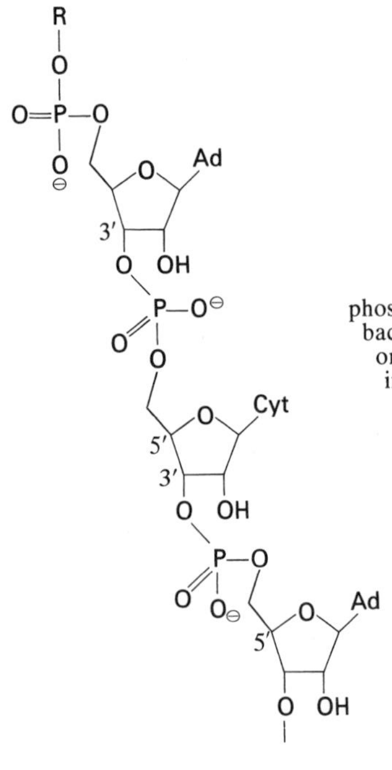

phosphodiester backbone of one chain in RNA

6.A ENZYME–CATALYZED PHOSPHORYL TRANSFERS

Enzyme-catalyzed reactions involving transfer of phosphoryl groups make up a large class of group-transfer reactions that are central to metabolism. We can divide this class of enzymes into four subgroups (based either on the nature of the phosphate-containing substrate or on the nature of the nucleophilic acceptor to which the phosphoryl moiety is transferred): phosphatases, phosphodiesterases, kinases, and phosphorylases.

1. phosphatases

$$R{-}O{-}P(=O)(O^{\ominus}){-}O^{\ominus} \xrightleftharpoons{H_2O} R{-}OH + HO{-}P(=O)(O^{\ominus}){-}O^{\ominus}$$

2. phosphodiesterases

$$R{-}O{-}P(=O)(O^{\ominus}){-}O{-}R' + H_2O \rightleftharpoons R{-}OH + {}^{\ominus}O{-}P(=O)(O^{\ominus}){-}OR$$

3. kinases

$$X{-}P(=O)(O^{\ominus}){-}O^{\ominus} + Y \rightleftharpoons {}^{\ominus}O{-}P(=O)(O^{\ominus}){-}Y + X$$

4. phosphorylases

$$ROR' + HO{-}P(=O)(O^{\ominus}){-}O^{\ominus} \rightleftharpoons R{-}O{-}P(=O)(O^{\ominus}){-}O^{\ominus} + R'OH$$

Categories 1 and 2 use water as acceptor and so involve hydrolysis of phosphate monoesters and phosphate diesters, respectively, as substrates. We examine these reactions in this chapter. The kinases catalyze reactions involving phosphoryl transfer to some nucleophile other than water (Y = oxygen or nitrogen species); we take up these reactions in Chapter 7. The mechanistically common feature of reactions in categories 1, 2, and 3 is transfer of the *phosphoryl group* ($-PO_3^{2\ominus}$), *not a phosphate group* ($-OPO_3^{2\ominus}$). This distinction is crucial and underlines the fact that these reactions represent attack by a nucleophile on the electrophilic phosphorus atom, with subsequent displacement of one of the original oxygen substituents that had been bonded to phosphorus in the starting material.

$$R'\ddot{X} + RO{-}P(=O)(O^{\ominus}){-}O^{\ominus} \longrightarrow {}^{\ominus}O{-}P(=O)(O^{\ominus}){-}X{-}R' + RO^{\ominus}$$

Category 4 is distinct from the other three categories in two ways: (1) the phosphorylated substrate is inorganic phosphate itself; and (2) it is the inorganic phosphate, specifically one of the anionic oxygen atoms, that is the nucleophile in the reaction. The overall reaction then is phosphorolysis. We have seen an example in the deadenylylation of glutamine synthetase, and it illustrates the chemical versatility of the phosphate group to act as electrophile at phosphorus or as nucleophile at oxygen. We discuss some further examples in Chapter 9.

Table 6-1
Comparison of rate constants for hydrolysis and for enzyme-catalyzed transfer

Ester	Monoanion hydrolysis (min^{-1}, ca. 25° C)	Enzyme-catalyzed transfer (min^{-1})	Enzyme
α-Glucose-1-P	3×10^{-8}	6×10^{4}	Phosphoglucomutase
Glucose-6-P	4×10^{-8}	1.9×10^{4}	Phosphoglucomutase
Acetyl-P	2×10^{-3}	1.7×10^{3}	Acylphosphatase
ATP, ADP	3×10^{-6}	3.2×10^{4}	ATPase
Serine-P	4×10^{-8}	1.9×10^{4}	Phosphoglucomutase
Phosphohistidine	6×10^{-3}	1.6×10^{4}	Nucleoside diphosphate kinase
Phosphocreatine	2×10^{-2}	7.2×10^{4}	Creatine kinase
Nucleotide diester analogue (*cis*-tetrahydrofuran-4-ol-3-phenyl phosphate)	6×10^{-6}	1.4×10^{5}	Ribonuclease

SOURCE: Modified from K. Schray and S. Benkovic, in *The Enzymes*, 3d ed., ed. P. Boyer (New York: Academic Press), vol. 8 (1973), p. 201.

Before discussing the phosphatases and phosphodiesterases specifically, we note the rate data in Table 6-1, which compares rate constants for hydrolysis of various phosphorylated substrates (as the monoanions) with rates of enzyme-catalyzed phosphoryl transfers (to water or other nucleophiles) by the indicated enzymes. The rate accelerations are in the range of 10^6 to 10^{12} for the enzymatic reactions.

6.A.1 Phosphatases

We discuss three specific phosphatases in this section: glucose-6-phosphatase, alkaline phosphatase, and ATPase (adenosine triphosphatase) from several sources. All three examples appear to generate covalent phosphoryl-enzyme intermediates during the course of the group transfer to water, in analogy to the covalent intermediates already discussed for acyl-transfer enzymes. It is probable that some other phosphatases carry out direct attack of water on the phosphorylated substrate without covalent intermediate formation.

The chemical hydrolysis of phosphate-monoester dianions is slow but can be accelerated by chelation with metal cations. Both alkaline phosphatase and the ATPase enzymes require divalent cations for catalysis. On the other hand, the chemical hydrolysis of phosphate-monoester monoanions (e.g., at pH 4) can be quite rapid and often proceeds by a dissociative mechanism involving an electrophilic metaphosphate species as a transient intermediate (Schray and Benkovic,

1973; Miller and Westheimer, 1966; Rebeck and Gavina, 1975). The process may involve either synchronous protonation of the leaving group (general acid catalysis) or a fast preequilibrium protonation followed by rate-determining formation of metaphosphate:

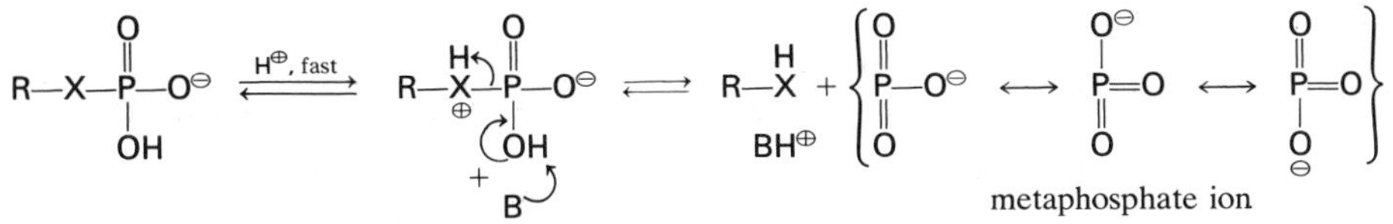

$$\text{R—X—P(=O)(OH)—O}^{\ominus} \xrightleftharpoons{\text{H}^{\oplus},\ \text{fast}} \text{R—X}^{\oplus}\text{(H)—P(=O)(OH)—O}^{\ominus} + \text{B} \rightleftharpoons \text{R—X—H} + \text{BH}^{\oplus} + \left\{ \text{O=P(=O)—O}^{\ominus} \longleftrightarrow {}^{\ominus}\text{O—P(=O)=O} \longleftrightarrow \text{O=P(=O)—O}^{\ominus} \right\}$$

metaphosphate ion

In each instance, protonation provides a more stable leaving group and a lower energy path. The metaphosphate ion then is captured subsequently by water to produce inorganic phosphate. When methanol also is present, the intermediate can be trapped as monomethyl phosphate.

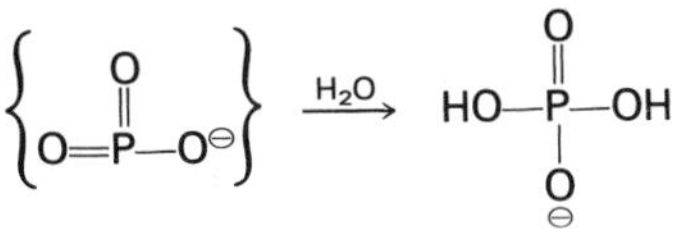

$$\left\{ \text{O=P(=O)—O}^{\ominus} \right\} \xrightarrow{H_2O} \text{HO—P(=O)(O}^{\ominus}\text{)—OH}$$

It is unclear whether this dissociative mechanism occurs frequently during enzymatic catalysis (Schray and Benkovic, 1973). We shall see that alkaline phosphatase will catalyze phosphoryl transfer to nucleophiles other than water, consistent with formation of an intermediate. However, that intermediate is the isolable phosphoryl enzyme. Metaphosphate formation need not be invoked; attack may instead proceed by direct displacement at phosphorus or by way of a discrete pentacovalent intermediate in enzymatic phosphoryl transfers.

6.A.1.a Glucose-6-phosphatase

We typically indicate hexoses and hexose phosphates in the common flat Haworth projection, as we have done in most structural representations thus far. However, we shall also use the conformationally revealing chair form, which indicates that all the hydroxyl substituents in glucose are equatorial (rather than axial):

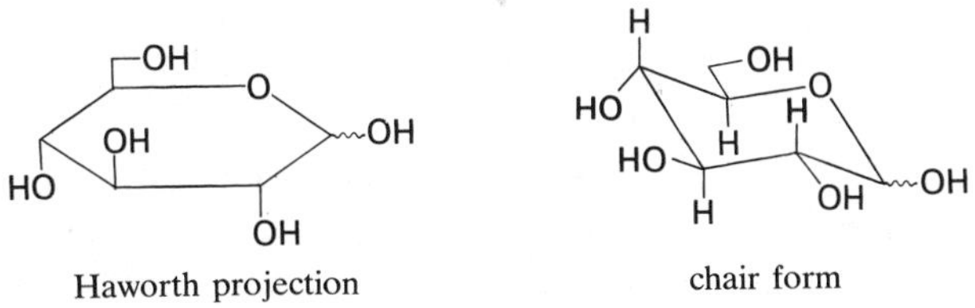

Haworth projection

chair form

Glucose-6-phosphatase converts the hexose phosphate back to free hexose and inorganic phosphate:

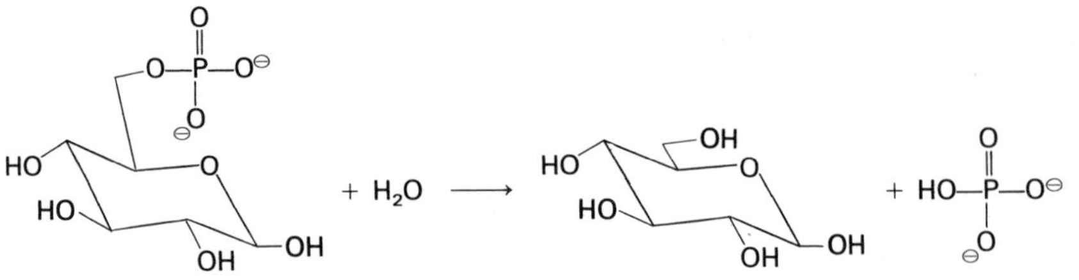

$$\text{glucose-6-phosphate} + H_2O \longrightarrow \text{glucose} + HO{-}PO_3^{2\ominus}$$

In the cyclic form, glucose has an acetal functional group at C-1. The acetal is in equilibrium with the free aldehyde form by a readily reversible ring opening:

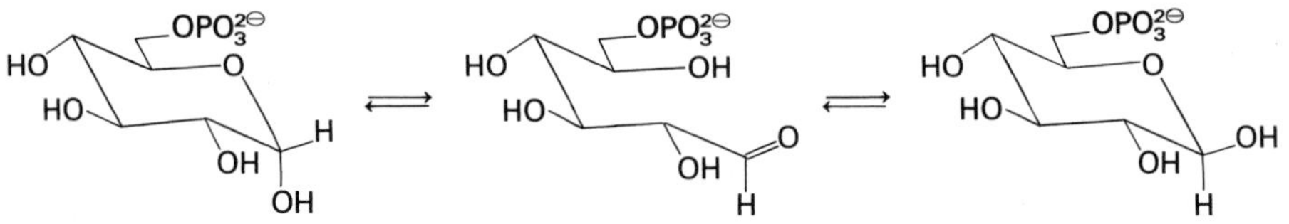

α-anomer $\rightleftarrows$ acyclic aldehyde $\rightleftarrows$ β-anomer

The aldehyde form can be cyclized in two stereochemically distinct ways: the OH group at position 5 can attack the C-1 aldehyde carbon from below the plane of the paper (yielding the β-anomer) or from above the plane of the paper to yield the C-1 hydroxyl down in the acetal (the α-anomer). The two cyclic isomers of glucose-6-$PO_3^{2\ominus}$ are thus distinct compounds, but in aqueous solution are rapidly interconvertible *through* the acyclic aldehyde form. The half-life for $\alpha \rightleftarrows \beta$ interconversion at 25°C and pH 7.6 is estimated to be ~15 sec. The equilibrium concentrations in aqueous solution (shown in Haworth projection in the following formula) indicate that only trace amounts of the acyclic aldehyde form are present at any moment.

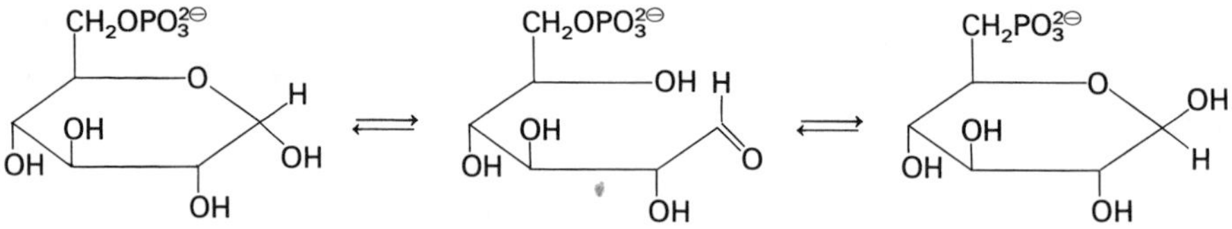

α-anomer (33%) $\rightleftarrows$ aldehyde form (<0.4%) $\rightleftarrows$ β-anomer (66%)

We shall note in subsequent chapters that some enzymes metabolizing hexoses and hexose phosphates use only a single anomer of the substrate. Glucose-6-phosphatase apparently recognizes both anomers of glucose-6-phosphate.

Glucose-6-phosphatase is a membrane-bound enzyme, specifically associated with the endoplasmic reticulum in mammalian cells (Nordlie, 1971). An an integral membrane protein, this enzyme has been particularly resistant to efforts to solubilize and purify it significantly. It has been hypothesized that the physiological role of this enzyme in liver, kidney, and intestine (all tissues that release glucose into the blood) is to hydrolyze intracellular glucose-6-phosphate in a directional manner, cleaving the ester as it passes from cytoplasm through the internal endoplasmic reticular membrane into the cisternae (vacuoles) of the endoplasmic reticulum. The contents of these membrane-enclosed vacuoles eventually are disgorged into the extracellular medium (blood) upon membrane-vesicle fusion with the plasma membrane. The net result is export of free glucose from cells to blood.

The form of the enzyme that has been characterized best (in terms of catalytic properties) is glucose-6-phosphatase from rat-liver membrane fragments (microsomes). One of the earliest studies was by Hass and Byrne (1960), who performed partial-exchange experiments to look for existence of a covalent phosphoryl-enzyme intermediate.

If such a covalent intermediate is formed reversibly, we can predict that a time-dependent exchange of radioactivity should occur from [^{14}C]-glucose into glucose-6-phosphate (recall ¶5.E.2).

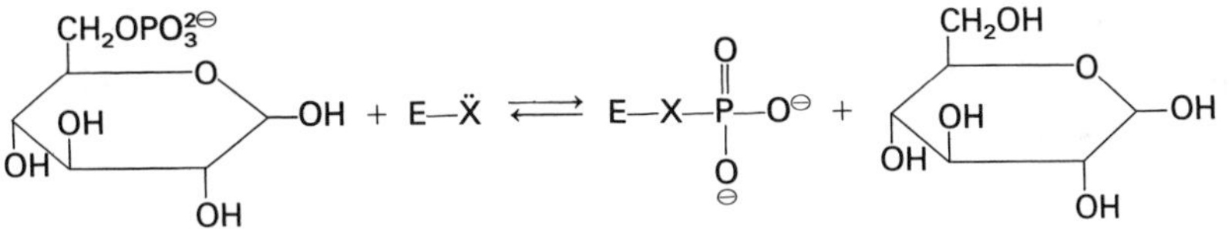

$$\text{Glucose-6-}CH_2OPO_3^{2\ominus} + E\text{—}\ddot{X} \rightleftharpoons E\text{—}X\text{—}P(=O)(O^\ominus)\text{—}O^\ominus + \text{Glucose (}CH_2OH\text{)} \quad (1)$$

$$E\text{—}X\text{—}P(=O)(O^\ominus)\text{—}O^\ominus + HO\text{—}H \rightleftharpoons E\text{—}\ddot{X} + HO\text{—}P(=O)(O^\ominus)\text{—}O^\ominus \quad (2)$$

Glucose is a neutral species; glucose-6-P is anionic. They are easily separable, and Hass and Byrne (1960) were able to detect the isotopic exchange on addition of enzyme.

As we have noted previously, occurrence of a partial-exchange reaction is consistent with the existence of a covalent intermediate. It is a necessary but not a sufficient condition for proof of such a covalent intermediate. It should encourage one to perform a more direct experiment to test for the existence of the covalent intermediate.

There are some alternative explanations (however unlikely) that one should keep in mind for such exchange data. (1) Exchange may occur by direct reaction between glucose and glucose-6-$PO_3^{2\ominus}$ in a ternary complex. This is unlikely, but possible. The enzyme then must be able to bind glucose and glucose-6-$PO_3^{2\ominus}$ at the same time. Because glucose and glucose-6-$PO_3^{2\ominus}$ were found to be strictly competitive inhibitors in glucose-6-phosphatase reactions, we conclude that they are not both bound at the active site, thus ruling out the possibility of direct transfer. (2) *Lack* of partial exchange may not rule out the existence of a covalent intermediate; for example, glucose may be bound extremely tightly and thus may not exchange with free [^{14}C]-glucose in medium. The binding constant for glucose can be determined independently by equilibrium dialysis to test this interpretation.

The putative E—$PO_3^{2\ominus}$ has resisted physical isolation in the native catalytic state, presumably because of extreme lability to hydrolysis by the ubiquitous second substrate, water. Recently, however, researchers have isolated a radioactive ^{32}P-labeled protein from a mixture of [^{32}P]-glucose-6-$PO_3^{2\ominus}$ and crude endoplasmic-reticulum membrane fractions containing enzyme, by rapidly quenching into phenol—a process ensuring rapid denaturation of the active enzyme before hydrolysis of the E—$PO_3^{2\ominus}$ can be catalyzed. With [^{14}C]-glucose-6-$PO_3^{2\ominus}$, no label was associated with protein. Subsequent hydrolysis of the phenol-denatured [^{32}P]-phosphoprotein in alkali yielded 3-*N*-phosphohistidine containing the radioactivity (Feldman and Butler, 1969).

We can speculate that glucose-6-phosphatase undergoes reversible phosphorylation at N-3 of a histidine residue at the active site, forming E—$PO_3^{2\ominus}$ and glucose. In the second step, water regenerates the free histidine residue, forming inorganic phosphate.

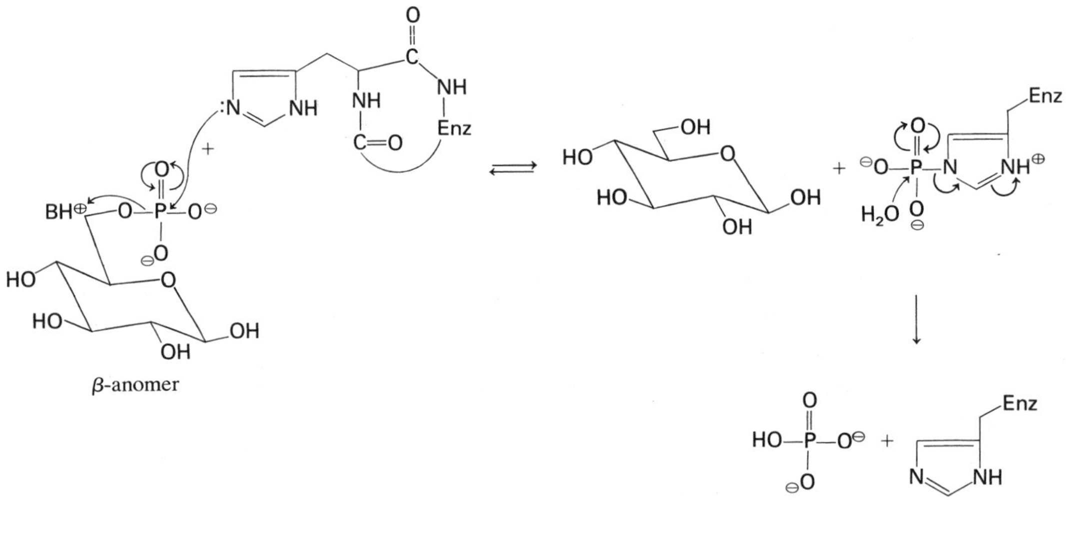

β-anomer

At first, glucose-6-phosphatase was thought to be very specific in recognizing only glucose-6-phosphate and in carrying out phosphoryl transfer only to water as nucleophile; however, the enzyme is exclusively specific in *neither* aspect. In fact, a variety of phosphorylated molecules can be shown to be competitive inhibitors of glucose-6-P hydrolysis. Close examination reveals that these inhibitors are themselves hydrolyzed. They include the following molecules: inorganic pyrophosphate, nucleoside di- and triphosphates, other sugar phosphates (such as mannose-6-$PO_3^{2\ominus}$ and fructose-6-$PO_3^{2\ominus}$), carbamoyl phosphate, and phosphoenol-pyruvate (Nordlie, 1971).

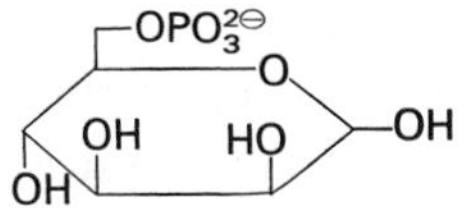

mannose-6-phosphate

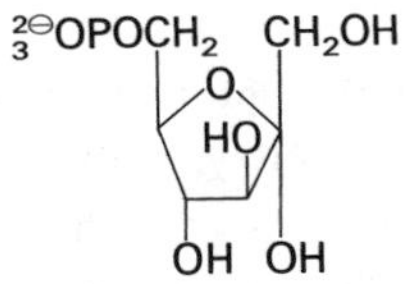

fructose-6-phosphate

Further, when the enzyme catalyzes phosphoryl transfer from these non-specific substrates, glucose can compete successfully with water as nucleophile, resulting in the phosphorylation of the glucose 6-hydroxyl group and synthesis of glucose-6-P (Nordlie, 1971). For example,

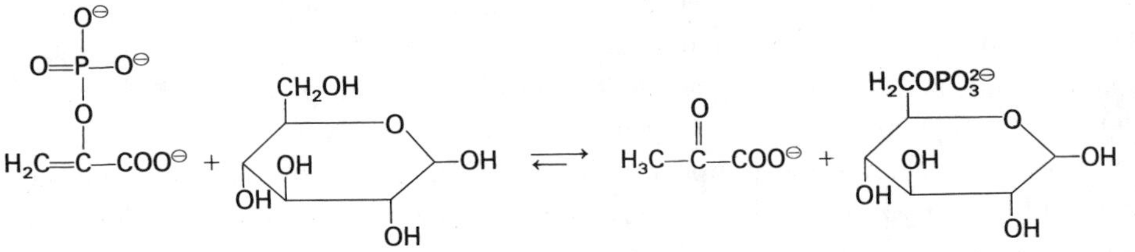

phosphoenolpyruvate + glucose ⇌ pyruvate + glucose-6-P (1)

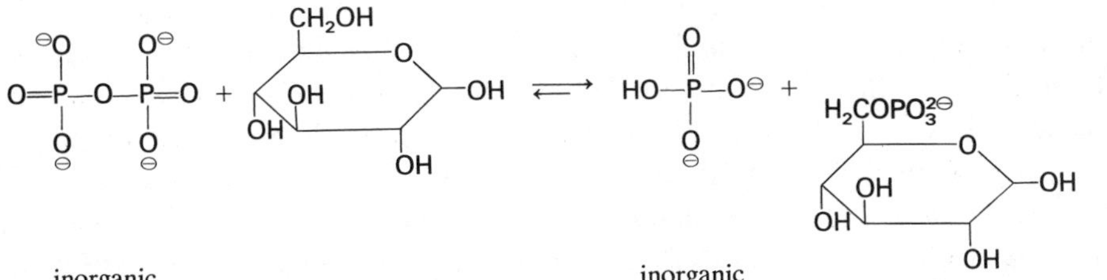

inorganic pyrophosphate + glucose ⇌ inorganic phosphate + glucose-6-P (2)

Presumably, a covalent phosphoryl enzyme forms with the alternative substrates as well.

It may well be that this glucose-6-$PO_3^{2\ominus}$ synthetic activity of the enzyme is an important control element for regulating intracellular glucose-6-P concentration—a balance between hydrolysis of glucose-6-P and its synthesis.

6.A.1.b Alkaline phosphatase

The nonspecific enzyme alkaline phosphatase shows an alkaline pH optimum for V_{max} and is thus distinguished from a phosphomonoesterase activity (less well-studied) that shows an acid pH optimum (and is correspondingly called acid phosphatase).

The alkaline phosphatase from *E. coli* is the best-characterized, although a mechanistically similar enzyme also is found in mammalian systems. The bacterial enzyme can account for up to 6% of the total cellular protein under bacterial growth conditions where inorganic phosphate is limiting (Reid and Wilson, 1971). The enzyme is a zinc-containing dimer of 86,000 mol wt, found in the so-called "periplasmic space" between the inner cell membrane and the cell wall of the bacterium. Only inactive monomers can be found in the cytoplasm; assembly of active dimer occurs in the periplasmic space. The physiological function of the enzyme (with broad structural specificity for diverse phosphate monoesters found in the bacterium's environment) is to provide inorganic phosphate for the cell and to render the remainder of the substrate molecule more lipophilic for its passage across the bacterial cytoplasmic membrane into the cell. The hydrolysis of various $ROPO_3^{2\ominus}$ species to ROH + P_i by the *E. coli* alkaline phosphatase is 10^9 to 10^{11} times faster than the corresponding nonenzymatic hydrolysis.

One can isolate a phosphoprotein intermediate by incubation of the enzyme with either [^{32}P]-P_i or various [^{32}P]-$ROPO_3^{2\ominus}$ substrates at low pH. Acid hydrolysis of the phosphoprotein yields a single phosphopeptide that is further degraded in acid to phosphoserine (Schwartz, 1963; Schwartz and Lipmann, 1961). This serine is thought to be the active-site nucleophile (in contrast to the imidazole nitrogen of a histidine group in glucose-6-phosphate action). It is likely that the $Zn^{2\oplus}$ at the active site chelates to the phosphate-monoester dianion, reducing electrostatic repulsion to attack by the serine hydroxyl. In turn, the serine hydroxyl probably is rendered more nucleophilic by general base catalysis, although the nature of any "charge-relay system" in alkaline phosphatase is unclear. A curious observation is that, although one can isolate the E—$PO_3^{\ominus}$H at pH 4, one finds only 2% of the enzyme as E—$PO_3^{2\ominus}$ at pH 7 (and even less at pH 8) on rapid quench into acid for isolation (Levine et al., 1969). The explanation that has been offered is that the covalent phosphoenzyme and the noncovalent Michaelis complex are in a pH-dependent equilibrium, with the Michaelis

Table 6-2
Fraction of enzyme found as phosphoenzyme or as Michaelis $E \cdot P_i$ complex as a function of pH

pH	Percentage of E as phospho-E	Ratio $[E \cdot PO_3^{2\ominus}]/E \cdot ROPO_3^{2\ominus}]$
8.0	1.7	0.017
7.0	2.7	0.027
6.0	51.0	1.1
5.5	81.0	4.4

complex thermodynamically more stable at high pH and the $E—PO_3^{\ominus}H$ more stable at low pH (Table 6-2).

How does the *E. coli* alkaline phosphatase effect catalysis? We observe that the V_{max} for many substrates of structure $ROPO_3^{2\ominus}$ or $RSPO_3^{2\ominus}$ is identical. The mechanistic interpretation (based by analogy on the interpretation of similar results with chymotrypsin) is that dephosphorylation of a covalent $E—PO_3^{2\ominus}$ is the rate-determining step.

Further, transphosphorylation to an acceptor other than water is detectable; this observation also suggests accumulation of $E—PO_3^{2\ominus}$. For example, the hydroxyl group of ethanolamine reacts with the phosphoryl enzyme to generate *O*-phosphoethanolamine ($^{\ominus}O_3POCH_2CH_2NH_3^{\oplus}$).

Table 6-3 summarizes the data observed for the hydrolysis reaction $ROPO_3^{2\ominus} \rightarrow P_i + ROH$. These data can be interpreted as a partitioning of an intermediate whose breakdown is rate-determining. The difference between formation rates of ROH and inorganic phosphate reflects the transphosphorylation to ethanolamine, forming ethanolamine phosphate.

$$E + ROPO_3^{2\ominus} \rightleftharpoons E \cdot ROPO_3^{2\ominus} \xrightarrow{ROH} E—PO_3^{2\ominus} \xrightarrow{H_2O} E + HO—P(=O)(O)—O^{\ominus}$$

$$E—PO_3^{2\ominus} \xrightarrow{\text{ethanolamine}} E + H_3^{\oplus}N\text{—CH}_2\text{CH}_2\text{—O—P(=O)(O}^{\ominus}\text{)—O}^{\ominus}$$

Table 6-3
Partitioning of phosphoryl-enzyme intermediate

Ethanolamine (mM)	ROH formed (μM min^{-1} mg^{-1})	Inorganic phosphate formed (μM min^{-1} mg^{-1})
0.0	0.86	0.87
0.114	0.96	0.85
0.343	1.19	0.86
0.572	1.35	0.86

With the use of *p*-nitrophenyl phosphate (PNP-phosphate)

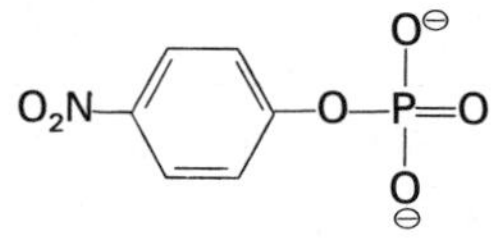

$$O_2N-C_6H_4-O-P(=O)(O^{\ominus})_2$$

as substrate, a burst of *p*-nitrophenolate ion is seen at low pH (5.5), but not at high pH (such as pH 8, which is the pH optimum). This suggests that, at high pH, phosphorylation is the rate-determining step. However, V_{max} data indicate dephosphorylation as rate-determining step, and PNP-phosphate reacts as rapidly as other substrates. Thus, the available data generated contradictory conclusions, and it was then postulated that the slow step is neither phosphorylation nor dephosphorylation of enzyme (chemical steps), but rather is a conformational change or isomerization of the free enzyme from a less active to a more active form. A detectable conformational change has been shown (by rapid kinetic studies) to occur at a rate of $70\ sec^{-1}$, slow enough to limit catalysis. A large number of mechanistic studies, including the initial-burst quantitative data, have been called into question by the observation that crystalline enzyme has inorganic phosphate tightly bound at the active site when isolated (Bloch and Schlessinger, 1974). The existence of the noncovalent product complex requires reevaluation of some earlier kinetic data but probably does not change the overall conclusion that a protein conformational change, a physical step, is rate-determining in catalysis at alkaline pH.

6.A.1.c ATPases

The ATPases, our third representative phosphatase, are actually a family of distinct catalytic entities that carry out hydrolysis of ATP to ADP and P_i (net attack of water at the γ-phosphorus of ATP). Clearly, any enzyme that cleaves ATP to ADP and P_i during catalysis shows ATPase activity, and this is a component reaction during the action of such enzymes as glutamine synthetase, carbamoyl-P synthetase, and the kinases discussed in the next chapter. However, the ATPases discussed in this section hydrolyze ATP without any other obvious accompanying chemical reaction. The hydrolysis of ATP is not coupled to an acyl activation or an alcohol esterification in a cosubstrate. (Strictly speaking, an ATPase is not a phosphomonoesterase, but its mechanism of action is reasonably considered in this context.)

To initial investigators of these enzymes, enzymatic hydrolysis of ATP to no apparent purpose by these ATPases seemed energetically wasteful, and clues were sought for some other, coupled physiological function. It turned out that most of

Table 6-4
ATPases: cation specificity and cellular location

Enzyme	Cation pumped	Cellular location
$Mg^{2\oplus}$-dependent ATPase	$H^{\oplus}$	Inner mitochondrial membrane of animal cells; chloroplast membrane of green-plant cells
$(Mg^{2\oplus},Ca^{2\oplus})$-dependent ATPase	$H^{\oplus}$	Cytoplasmic membrane of bacteria
$(Na^{\oplus},K^{\oplus},Mg^{2\oplus})$-dependent ATPase	$Na^{\oplus}$, $K^{\oplus}$	Plasma membrane of animal cells
$(Mg^{2\oplus},Ca^{2\oplus})$-dependent ATPase	$Ca^{2\oplus}$	Sarcoplasmic membrane of muscle cells

these ATP hydrolases are not soluble proteins but integral hydrophobic protein components of cellular membranes (Boyer, 1970–1976, vol. 10, pp. 375–465). The membrane localization is the key to their physiological function, which is the coupled pumping of cations across the membranes from one intracellullar compartment to another, or from inside the cell to outside the cell. The cations in most cases are pumped against their concentration gradients (active transport of ions); this unfavorable process is driven thermodynamically by coupling to the exergonic hydrolysis of ATP. Four types of ATPases have been extensively characterized on the basis of the ions pumped and the biological source of the enzyme (Table 6-4).

The first two enzymes listed in Table 6-4 are structurally and functionally similar; the solubilized and purified enzyme in each case consists of five nonidentical subunits, and they are thought to be the key enzyme(s) in oxidative phosphorylation or photosynthetic-driven phosphorylation (these topics are beyond the scope of this text; see Boyer, 1970–1976, vol. 10, pp. 375–465; Guidotti, 1976; Oxender, 1972; Boos, 1974; Simoni and Postma, 1975; Pressman, 1976). The role (catalytic or regulatory) of the individual subunits is under investigation; at least one subunit is involved in attachment of the enzyme to membrane.

A distinction between these two related proton-pumping ATPases and the other two ATPases in the table is that the first two give *no detectable* evidence of covalent phosphoryl enzymes during ATP hydrolysis, whereas the last two enzymes in the table do.

Both the $(Na^{\oplus},K^{\oplus})$-pumping ATPase of plasma membranes and the $Ca^{2\oplus}$-pumping ATPase of muscle cells are involved in regulation of ion concentrations. Animal cells maintain high internal $K^{\oplus}$ levels and low intracellular $Na^{\oplus}$ levels by active extrusion of $Na^{\oplus}$. The $K^{\oplus}$ ions come in passively as specific counter ions. The $(Na^{\oplus},K^{\oplus})$-ATPase appears to be the molecular species that is the sodium pump (Dahl and Hokin, 1974; Degani and Boyer, 1973). The enzyme from red cells spans the plasma membrane as an oligomer of unequal subunits; the alkali metal ions may pass through the lipophilic membrane by channels that are the intersubunit spaces of the enzyme (Guidotti, 1976). Similarly, the $Ca^{2\oplus}$-ATPase of muscle pumps $Ca^{2\oplus}$ from the muscle cytoplasm into intracellular vesicles (the sarcoplasmic reticulum), thereby lowering the cytoplasmic $Ca^{2\oplus}$ level. When a nerve impulse reaches the muscle cell, the reticulum membrane is depolarized, $Ca^{2\oplus}$ flows back out rapidly into the cytoplasm, and the high $Ca^{2\oplus}$ level activates the actinomyosin contractile apparatus, inducing contraction of the muscle. After the initial depolarization signal, the reticulum membrane repolarizes rapidly, and the $Ca^{2\oplus}$ ATPase resumes hydrolyzing ATP to provide the driving force for pumping $Ca^{2\oplus}$ from the cytoplasm back into the sequestered vesicles. As $Ca^{2\oplus}$ levels in the cytoplasm fall, muscular contraction (or expansion) ceases. Without the regulatory function of the $Ca^{2\oplus}$ pump, the muscle cell would not be able to relax back to a resting state. As with the other ion-pumping ATPases, the molecular mechanism for ion passage is hypothesized to be through a channel in the oligomeric, membrane-spanning enzyme (Guidotti, 1976).

Covalent phosphoryl enzymes have been detected and characterized as competent reaction intermediates for the $(Na^{\oplus},K^{\oplus})$-ATPase and for the $Ca^{2\oplus}$-ATPase (Dahl and Hokin, 1974; Degani and Boyer, 1973; MacLennan, 1975). In each instance, the enzyme nucleophile that has attacked the γ-P of ATP is a β-carboxylate of an aspartyl residue at the enzyme active site. (This then is the *third structural type of phosphoryl-enzyme intermediate.*) Because acyl-phosphate intermediates are unstable to both acid and base, characterization of the intermediates was extremely difficult and rests on a variety of indirect trapping experiments. Perhaps the most convincing experiments are those involving derivatization of the phosphoryl enzymes by reduction with tritium-labeled sodium borohydride. Acyl phosphates are sufficiently destabilized acyl groups to undergo reduction, first to the aldehyde and then to the alcohol oxidation state, by this water-soluble and water-stable metal-hydride reagent.

$$R-\overset{O}{\overset{\|}{C}}-O-\overset{O}{\overset{\|}{\underset{\underset{\ominus}{O}}{\underset{|}{P}}}}-O^{\ominus} \xrightarrow{[^3H]\text{-}NaBH_4} P_i + R-\overset{O}{\overset{\|}{C}}-{}^3H \xrightarrow{[^3H]\text{-}NaBH_4} R-\overset{OH}{\overset{|}{\underset{^3H}{\underset{|}{C}}}}-{}^3H$$

An enzyme-acyl phosphate would be reduced by this hydride donor to different alcoholic species, depending on whether a γ-glutamyl carboxylate or a β-aspartyl carboxylate was the acyl portion of the phosphoryl-enzyme intermediate. An enzyme-γ-glutamyl phosphate would yield (after $[^3H]$-$NaBH_4$ reduction and then acid hydrolysis to liberate constituent amino acids) tritium-labeled α-amino-δ-hydroxyvalerate.

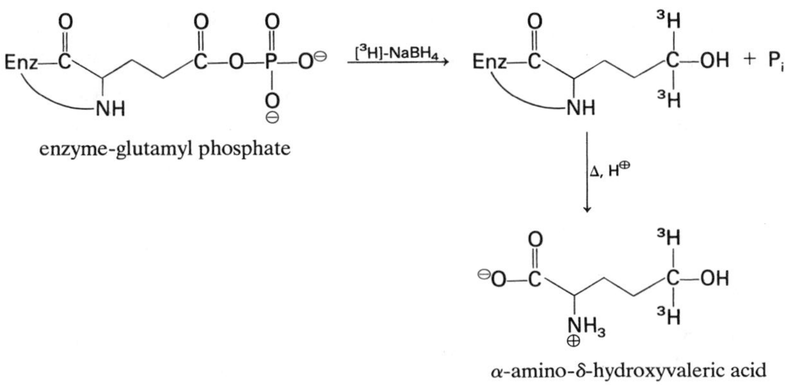

enzyme-glutamyl phosphate

α-amino-δ-hydroxyvaleric acid

An enzyme-β-aspartyl phosphate would produce tritiated homoserine (and homoserine lactone in acid).

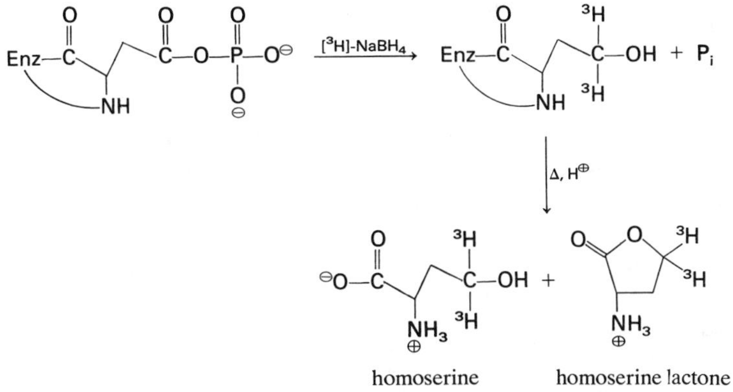

homoserine

homoserine lactone

Radioactive homoserine (plus homoserine lactone) was isolated from both ($Na^{\oplus}$,$K^{\oplus}$)- and $Ca^{2\oplus}$-ATPase phosphoenzymes (Dahl and Hokin, 1974; Degani and Boyer, 1973). It was important to carry out the experiment with dephosphoenzyme controls in each case, because some nonspecific incorporation of carbon-bound tritium results from a small amount of reductive cleavage of peptide bonds in the enzymes by the reducing agent. The borohydride (borotritide) trapping procedure also has been used to derivatize noncovalently-bound γ-glutamyl phosphate as a fleeting intermediate in glutamine-synthetase action (¶5.D.1; Todhunter and Purich, 1975).

It remains to be seen whether there is a true mechanistic dichotomy between the proton-pumping ATPases, on the one hand, and the $Na^{\oplus}$-, $K^{\oplus}$-, and $Ca^{2\oplus}$-pumping ATPases, on the other hand, with respect to covalent phosphoryl intermediates. To date, no evidence for intermediates is available for the proton-pumping ATPases. The proton-pumping ATPases are just as likely to act as ATP synthases under physiological conditions of oxidative phosphorylation (ATP synthesis coupled to electron flow down membrane respiratory chains) in mitochondrial or bacterial cytoplasmic membranes, or under conditions of photophosphorylation in chloroplasts.

6.A.2 Phosphodiesterases

The phosphodiesterases are a class of enzymes specific for the hydrolytic cleavage of a phosphodiester substrate to product phosphomonoester and alcohol moiety. Many phosphodiesterases show broad tolerance for structural diversity in substrates and exercise general degradative functions in cells.

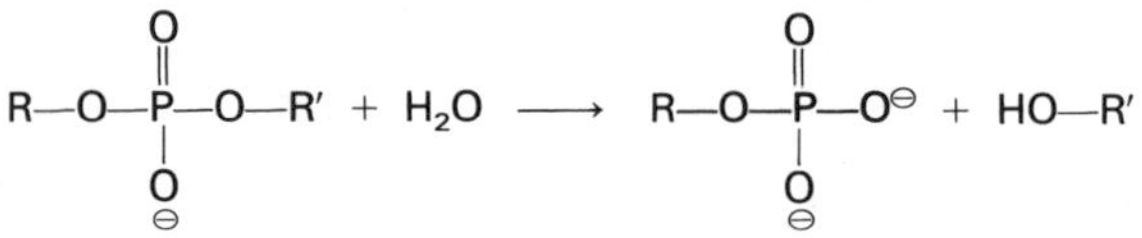

$$R{-}O{-}\overset{\overset{\displaystyle O}{\|}}{\underset{\underset{\displaystyle O^{\ominus}}{|}}{P}}{-}O{-}R' + H_2O \longrightarrow R{-}O{-}\overset{\overset{\displaystyle O}{\|}}{\underset{\underset{\displaystyle O^{\ominus}}{|}}{P}}{-}O^{\ominus} + HO{-}R'$$

Other phosphodiesterases show narrow specificity. For example, 3′,5′-cAMP phosphodiesterase hydrolyzes the 3′,5′-cyclic nucleotide specifically to the 5′-phosphate product AMP.

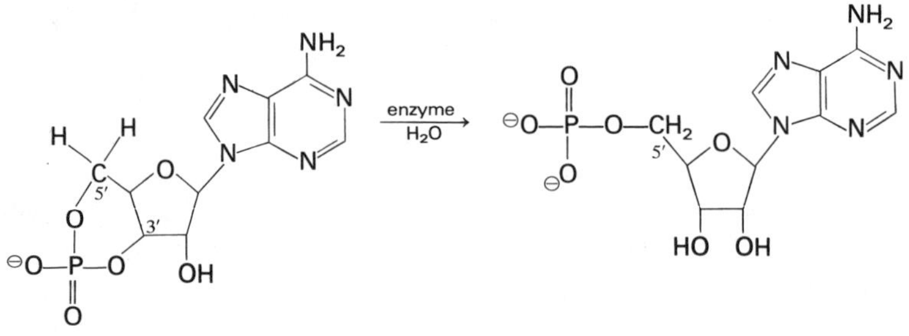

$$\text{3}',\text{5}'\text{-cAMP} \xrightarrow[H_2O]{\text{enzyme}} \text{5}'\text{-AMP}$$

3′,5′-cAMP 5′-AMP

This enzyme is important in regulation of the intracellular concentration of cAMP, a key molecule that may function as an intracellular hormone in both bacterial and animal cells (Pastan et al., 1975). The mechanism of this phosphodiesterase has not been analyzed intensively.

The major class of biomolecules containing phosphodiester linkages consists of the ribonucleic acids and deoxyribonucleic acids, and various phosphodiesterases are involved in the metabolism of these molecules. For instance, deoxyribonucleases (DNases) are enzymes that are specific for hydrolysis of the internucleotide bonds in polydeoxynucleotides. Such phosphodiesterases are classifiable on the basis of their modes of attack. As with the endo- and exopeptidases noted earlier, there are endo- and exonucleases: endonucleases attack at a few or at many phosphodiester loci within a polynucleotide chain, whereas exonucleases generally excise a mononucleotide in stepwise attack from either the 5′-end or the 3′-end of a DNA or RNA substrate. Indeed, this functional categorization may be bridged in that some bacterial nucleases exhibit the ability to cleave either internal or terminal phosphodiester bonds or both, depending on the substrate used (Lehman, 1971).

6.A.2.a Ribonuclease

Probably the best-characterized nuclease is bovine pancreatic ribonuclease A, an endonuclease (Richards and Wyckoff, 1971). It was among the first enzymes to be purified, due to its unusual stability to ranges of pH (e.g., 0.25 N H_2SO_4) and of temperature (10 min at 100° C); it was crystallized by Kunitz in 1939. The enzyme is a small protein composed of a single chain of 124 amino-acid residues, and the small size may contribute to its stability. Because of its ready availability, it was the first enzyme whose primary amino-acid sequence was determined (Fig. 6-1)—a feat that led to development of the methodology and technology for protein sequencing, work for which William H. Stein and Stanford Moore shared

Lys-Glu-Thr-Ala-Ala-Ala-Lys-Phe-Gly-Arg-Gln-His-Met-Asp-Ser-Ser-Thr-Ser-Ala-Ala-
10 20

Ser-Ser-Ser-Asn-Tyr-Cys-Asn-Gln-Met-Met-Lys-Ser-Arg-Asn-Leu-Thr-Lys-Asp-Arg-Cys-
30 40

Lys-Pro-Val-Asn-Thr-Phe-Val-His-Glu-Ser-Leu-Ala-Asp-Val-Gln-Ala-Val-Cys-Ser-Gln-
50 60

Lys-Asn-Val-Ala-Cys-Lys-Asn-Gly-Thr-Asn-Gln-Cys-Tyr-Gln-Ser-Tyr-Ser-Thr-Met-Ser-
70 80

Ile-Thr-Asp-Cys-Arg-Glu-Ser-Thr-Gly-Ser-Lys-Tyr-Pro-Asn-Ala-Cys-Tyr-Lys-Thr-Thr-
90 100

Asn-Ala-Gln-Lys-His-Ile-Ile-Val-Ala-Cys-Glu-Gly-Asn-Pro-Tyr-Val-Pro-Val-His-Phe-
110 120

Asp-Ala-Ser-Val-$COO^{\ominus}$
124

Figure 6-1
The amino-acid sequence of ribonuclease (RNase). Disulfide bonds occurs between residues 26 and 84, 40 and 96, 58 and 110, and 65 and 72. (From S. Blackburn, *Enzyme Structure and Function*, Marcel Dekker, Inc., 1976, p. 328)

the 1972 Nobel Prize in chemistry. This enzyme also has been subjected to intensive X-ray analysis, leading to a map of the three-dimensional structure at 2 Å resolution (Richards and Wyckoff, 1971).

With all of the structural information thus gathered and with its small size, ribonuclease A was a logical target for the first attempt at chemical synthesis of an enzyme—research accomplished simultaneously by Merrifield's solid-phase methods (Gutte and Merrifield, 1969) and by the more classical peptide methods of Hirschmann's group at Merck (Hirschmann et al., 1969).

The classical studies on protein refolding carried out by Christian Anfinsen (1962)—for which he shared the 1972 Nobel Prize with Stein and Moore—were also carried out with ribonuclease (RNase), and they were another spur to the chemical synthesis. Native ribonuclease has four disulfide bonds, which aid in constraining the molecule in its precise, active three-dimensional conformation (Fig. 6-2). The disulfides can be reduced by excess low-molecular-weight thiols (β-mercaptoethanol), and the reduced, denatured, inactive enzyme is produced as an unfolded random coil in concentrated aqueous urea or guanidinium chloride solutions. When the urea or guanidinium chloride is slowly dialyzed away, the cysteinyl groups autoxidize in air back to disulfides. The eight cysteinyl groups in ribonuclease could dimerize in 105 different ways to give 105 different structures; 104 of these are artifactual, and the other is the original native structure. In fact, as dialysis proceeds, most of the ribonuclease activity is regained, and structural analysis indicates that the initial disulfide bonds have specifically reformed; the protein refolds into its native conformation. This observation was highly significant. It proved the hypothesis that the primary sequence of a protein contains the information for three-dimensional folding into the active conformation. In a thermodynamic sense, the native conformation of this particular protein is the most stable one. It is precisely this property that was utilized in the chemical syntheses of ribonuclease: at the end of the abiotic synthetic routes, the synthetic material folds up to an active enzyme.

Ribonuclease also has been subjected to extensive studies involving the directed chemical modification of specific types of amino-acid side chains by chemical reagents, in efforts to determine which functional groups of the enzyme are important in catalysis (Means and Feeney, 1971). Dinitrofluorobenzene rapidly labels a unique lysine (position 41) at its ε-amino group, causing total inactivation.

$$O_2N{-}C_6H_3(NO_2){-}F + H_2N(CH_2)_4CH(NH{\sim}Enz){-}C(=O){\sim} \longrightarrow O_2N{-}C_6H_3(NO_2){-}NH(CH_2)_4CH(NH{\sim}Enz){-}C(=O){\sim} + F^{\ominus}$$

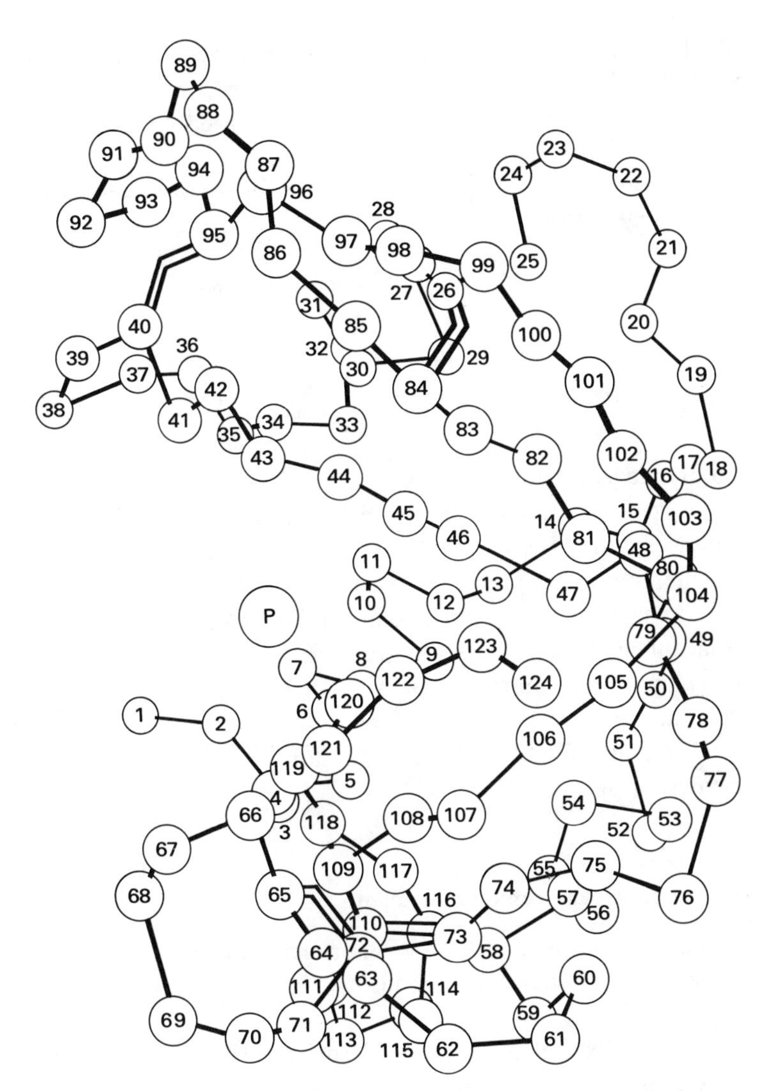

Figure 6-2
Ribonuclease A. The phosphate group in crevice opening P marks the active site. Disulfide bridges are represented only by bent double lines. (Reproduced from *The Structure and Action of Proteins*, by R. E. Dickerson and I. Geis, W. A. Benjamin, Inc., Menlo Park, California, Copyright © 1969 by Dickerson and Geis.)

Also, the alkylating agent iodoacetamide inactivates enzymatic activity maximally at pH 5.5. Under these conditions, histidine residues turn out to be the only amino acids modified. The active halogen is displaced by one of the basic imidozole nitrogens of histidyl residues.

$I{-}CH_2COO^{\ominus}$ + His^{12} (Enz) ⟶ $I^{\ominus}$ + 3-*N*-carboxymethyl histidinyl residue (Enz)

His^{12}

3-*N*-carboxymethyl
histidinyl residue

Structural studies reveal that either His^{12} or His^{119} (positions measured from the amino terminus) can be carboxymethylated in a frequency ratio of 1:8. His^{12} is derivatized at N-3; His^{119} is derivatized at N-1. Formation of one derivatized histidinyl residue apparently precludes reaction of the second, because no dicarboxymethylated enzyme molecules are found. It was argued that His^{12} and His^{119} might be adjacent in the tertiary structure, and that modification of one might interfere with subsequent approach of an iodoacetate molecule to the other. X-ray analysis later confirmed that these two histidine residues and Lys^{41} are in the active site and thus qualify as candidates for both nucleophilic and general acid–base catalysts (Fig. 6-3).

6.A.2.b Ribonuclease catalytic mechanism

Ribonuclease carries out the hydrolysis of internucleotide phosphodiester linkages of RNA in two discrete steps, analogous to the nonenzymatic base-catalyzed hydrolysis.

In the *first step,* RNA chain cleavage occurs to yield an intermediate with a free 5′-hydroxyl group on one side of the cleavage locus, and a 2′,3′-cyclic phosphate on the other side. Although the intimate details of the general acid–base catalysis used and the precise roles of the histidines and lysine are debated (Richards and Wyckoff, 1971), it does seem likely that the nucleophilicity of the ribose 2′-OH (which initiates intramolecular cyclization on the adjacent 3′-position to the 2′,3′-phosphate) is enhanced by one of the active-site histidines acting as a general base. It is also likely that catalytic protonation of the 5′-oxygen of the leaving group occurs, facilitating expulsion (lowered energy of the transition state), by either Lys^{41} or the other active-site histidine acting as general acid catalyst.

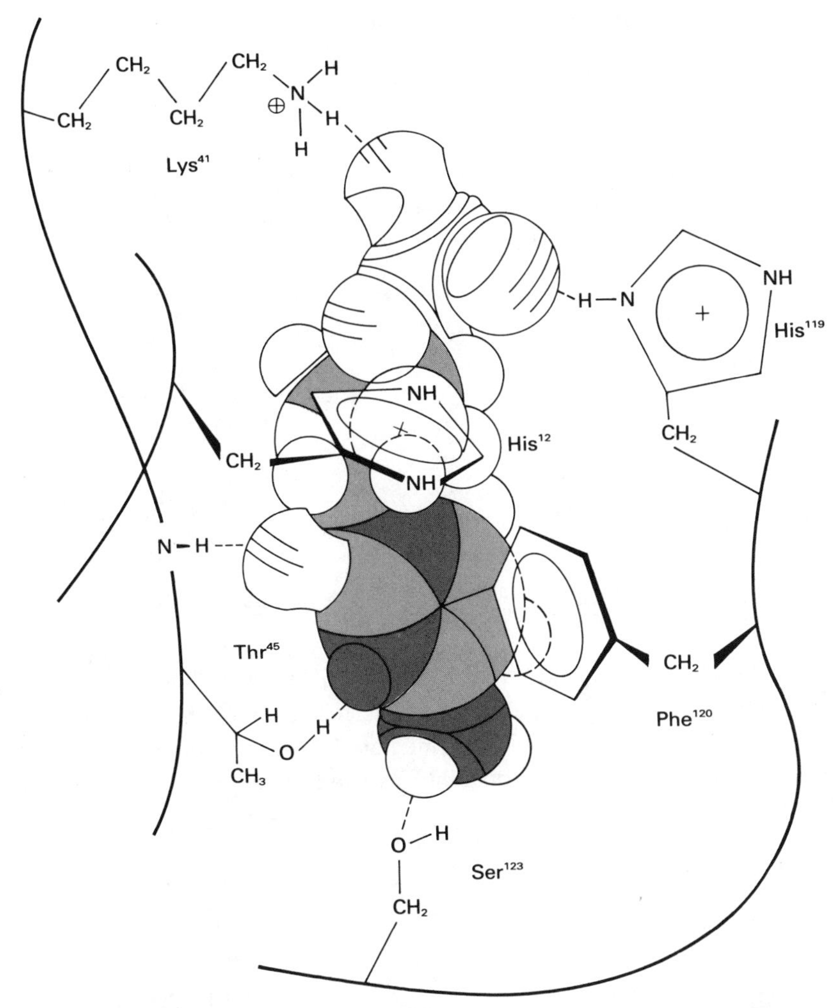

Figure 6-3
Schematic drawing of the cytidine-3′-monophosphate–ribonuclease complex, as seen from the back of the active-site cleft. The relative positions of the various amino-acid residues are taken from the X-ray model of ribonuclease S. (From S. Blackburn, *Enzyme Structure and Function*, Marcel Dekker, Inc., 1976, p. 367)

An imposition of specificity by ribonuclease for cleavage of only certain phosphodiester bonds in an RNA substrate is revealed by the exclusive formation of pyrimidine 2′,3′-cyclic phosphates (cytidine or uridine) but not of purines (adenosine, guanosine). The cleavage initially would generate, for example, a cyclic cytidine 2′,3′-phosphate at the 3′-terminus of the cleavage site and a new 5′-terminus that has a free 5′-hydroxyl group.

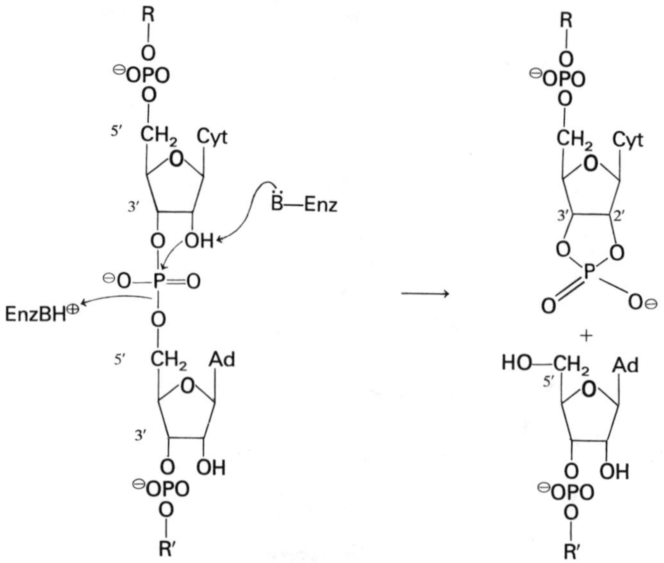

internucleotide region of an RNA chain

The cyclic-phosphodiester intermediate formed in the initial cleavage step can be isolated with appropriate model di- or trinucleotide substrates.

In the *second step*, the cyclic intermediate is ring-opened by reaction with a water molecule, again in all probability assisted by enzymic general acid–base catalysis.

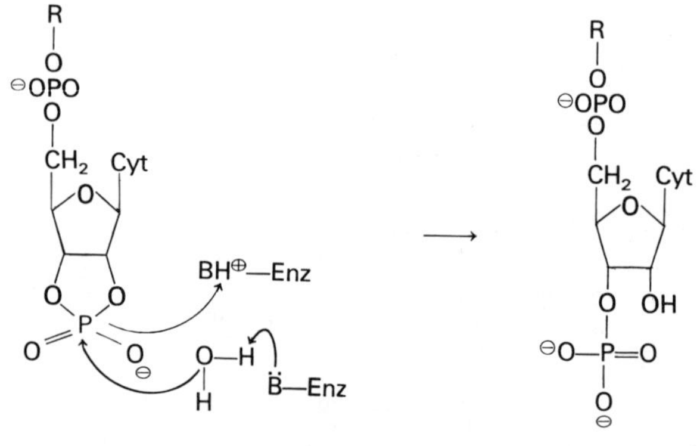

The final product at the 3′-end of the cleavage point is a pyrimidine 3′-phosphate. Although the symmetrical 2′,3′-cyclic phosphate could theoretically open to either 2′-phosphate or 3′-phosphate, the enzyme controls specificity so that only the 3′-phosphate is observed.

Figure 6-3 shows a model for the binding of cytidine 3′-monophosphate to the ribonuclease active site. In both steps of the RNase catalytic sequence—formation and then hydrolysis of the cyclic intermediate—pentacovalent phosphoranes are suggested, either as transition states or as fleeting intermediates.

6.A.2.c Stereochemistry of RNase-catalyzed phosphoryl transfer

It is known that some pentacovalent phosphorane compounds have long enough lifetimes and low enough energies to permit interchange of apical and equatorial ligands (see the opening pages of this chapter). The process is known as pseudorotation and was invoked after detection of stereochemical anomalies in the hydrolysis of optically active phosphodiesters (Boyer, 1970–1976, vol. 10, pp. 375–465). Attack of nucleophiles and departure of leaving ligands is usually from the apical positions in the trigonal-P bipyramidal pentacovalent species, with inversion of configuration expected at phosphorus.

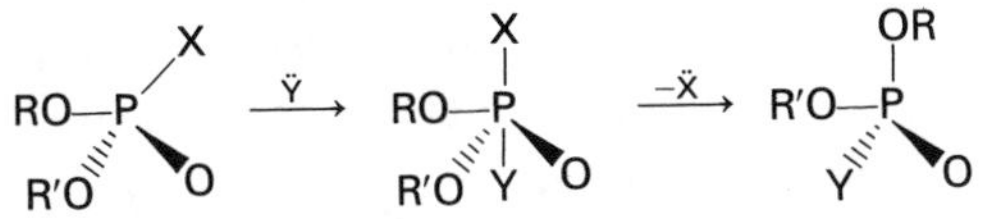

However, retention is sometimes seen, and it has been explained by initial apical attack to form a pentacovalent intermediate where the other apical ligand does not depart readily. Pseudorotation (Westheimer, 1968; Bruice and Benkovic, 1965–1966, vol. 2) can occur to interchange that apical group and an equatorial substituent that can function as a leaving group. Once the ligand reorganization has placed the incipient leaving group in the apical position opposite the incoming nucleophile, departure of that leaving group will produce a net retention.

Efforts have been made to detect this stereochemical feature as an indication of a pentacovalent phosphorus species during ribonuclease action. Usher and colleagues pointed out the stereochemical outcomes for in-line displacements or adjacent displacements (requiring pseudorotation) for the two steps of the RNase-A mechanism, as shown here for a 2′,3′-cyclic uracil (Usher, 1969; Usher et al., 1972; Blackburn, 1976, p. 364).

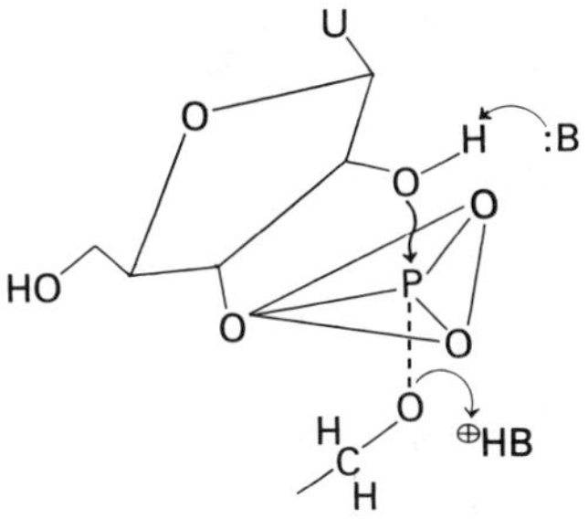

Step 1 (in-line displacement)

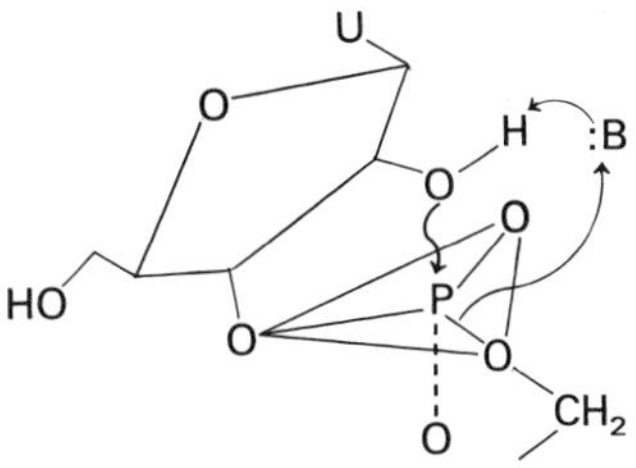

Step 1 (adjacent displacement)

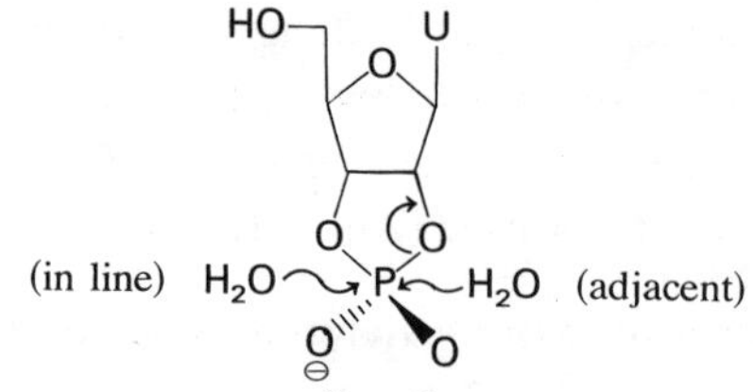

(in line) H_2O → P ← H_2O (adjacent)

Step 2

The adjacent mechanism would use a single base as general base and general acid, whereas the in-line mechanism would have a general base and general acid on opposite faces of the pentacovalent phosphorus transition state. Because the oxygen atoms in the phosphate group are chemically identical, the transformation is stereochemically silent.

The stereochemical outcome of step 2 (and later that of step 1) was shown to be inversion (and so an in-line mechanism) by elegant experiments with a cyclic uracil containing a 2′,3′-thiophosphoryl group instead of the phosphoryl group. The two isomeric forms of this thiophosphate were separated after synthesis, and one diastereomer was used with the enzyme. The strategy was to allow enzymatic ring-opening in [^{18}O]-H_2O to the 3′-[^{18}O]-thiophosphate product. This product was recyclized chemically with diethyl phosphochloridate, to yield equal populations of the two starting cyclic 2′,3′-thiophosphate diastereomers, which were again separated and individually analyzed for ^{18}O content. As the following scheme indicates, if the enzymatic opening is in-line (inversion) and chemical

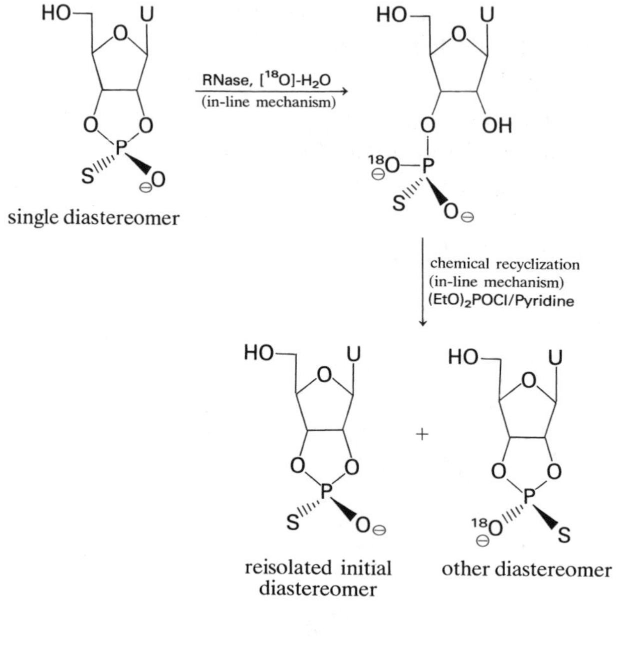

single diastereomer

RNase, [^{18}O]-H_2O (in-line mechanism)

chemical recyclization (in-line mechanism) $(EtO)_2POCl$/Pyridine

reisolated initial diastereomer + other diastereomer

recyclization is known in-line inversion, then the reisolated initial isomer should have no excess ^{18}O. The opposite diastereomer, produced in the symmetric chemical cyclization, should be enriched specifically in ^{18}O. The exactly opposite pattern would result had the RNase-mediated ring opening gone via an adjacent displacement.

When the product samples were analyzed by mass spectrometry for ^{18}O content, the in-line mechanistic pattern was observed. This was a seminal experiment in stereochemical analysis of enzymatic phosphoryl transfers. More recent efforts with ATP-utilizing enzymes are noted in ¶7.E.

6.A.2.d Other RNases

A variety of other nucleases specific for RNA molecules have been purified. Two such are yeast ribonucleases H and bacterial ribonuclease II. There are actually two ribonuclease H enzymes (H_1 and H_2) in eukaryotic cells (Wyers et al., 1976). RNase H_1 is an exonuclease; H_2 is an endonuclease. H_1 binds tightly to single-stranded nucleic acids and may play a role in chromatin structure and function. The *E. coli* ribonuclease II is a potassium-activated exonuclease releasing 5′-phosphorylated mononucleotides sequentially. It shows higher activity toward mRNA molecules, which are less ordered and less base-paired structures than are rRNA or tRNA molecules with their much more extensive base-paired secondary structures (Gupta et al., 1977).

6.B DNases

Dozens of enzymes are known with phosphodiesterase activity toward DNA substrates. Some are exonucleases, excising one nucleotide at a time from the 3′-end or the 5′-end of a DNA molecule. Others are endonucleases, which may cleave DNA at internal phosphodiester bonds repetitively to yield as eventual products small oligonucleotide fragments. There is also a subset of endonucleases that may cleave at only one or a few specific phosphodiester bonds in DNA molecules many thousands of nucleotides in length.

Among the endonucleases are the bacterial restriction endonucleases, which may be used by bacterial cells to cleave foreign DNA molecules (Meselson et al., 1972; Nathans and Smith, 1975). These enzymes are currently used as powerfully specific analytic tools by molecular biologists and biochemists interested in DNA cloning and DNA sequencing.

These restriction enzymes apparently recognize unique (or nearly unique) sites on bacterial DNA molecules; minimal recognition requires a double-strand sequence with twofold symmetry in the antiparallel DNA chains. These symmetrical stretches are known as *palindromes.* The *E. coli* R1 endonuclease (Eco R1) has been purified to homogeneity (Modrich and Zabel, 1976) and is known to recognize the following minimal palindromic sequence:

```
     ↓  *                                              *
5'~pGpApApTpTpCp~3'    Enz—M2⊕     5'~pGOH  +  pApApTpTpCp~3'
3'~pCpTpTpApApGp~5'   ────────→           *
                ↑                  3'~pCpTpTpApAp  +  HOGp~5'
```

Cleavage sites on each chain are indicated by the arrows. The hydrolytic reaction is staggered on the single strands, leading to overlapping or "sticky" ends, useful for subsequent enzymatic rejoining by DNA ligase (see Chapter 8). To avoid cleaving its own DNA molecules, the host cell's DNA is methylated at the adenine residues indicated by the asterisks, and this renders the cleavage sites resistant to endonuclease action (Rubin and Modrich, 1977).

The kinetic course of action of homogeneous Eco R1 endonuclease has been analyzed. At 0° C, an enzyme-bound intermediate with one single-strand break can be isolated, dissociated from the enzyme. At 30° C, both strands are hydrolytically cleaved before the DNA duplex dissociates from the enzyme. The turnover number for the first break is $\sim 40\ \text{min}^{-1}$ at 30° C, whereas the second-strand break has a rate constant of $14\ \text{min}^{-1}$. Both are faster than the overall turnover of $0.72\ \text{min}^{-1}$ (~ 1.3 double-strand breaks per minute), leading to the suggestion that dissociation of doubly cleaved product is rate-determining. Indeed, this hypothesis is consistent with an evaluation of the k_{cat}/K_m ratio. At 37° C, k_{cat} is four scissions per minute, or $0.07\ \text{sec}^{-1}$. The K_m is a low 10^{-9} for susceptible DNA. Thus $k_{cat}/K_m \approx 7 \times 10^7\ \text{M}^{-1}\ \text{sec}^{-1}$. Given that k_{cat}/K_m is essentially the bimolecular rate constant at low substrate ratios, this number of $\sim 10^8\ \text{M}^{-1}\ \text{sec}^{-1}$ approaches the limit for diffusion-controlled associations between macromolecules and small molecules. That is, the observed low K_m mandates a slow k_{cat}, and the k_{cat}/K_m ratio implies that a physical step (product release) does limit the rate of catalysis.

Although there may be many similarities between RNases and DNases, one clear mechanistic distinction is that the cyclic 2′,3′-phosphate cannot form as an intermediate during DNase-mediated hydrolysis. The sugar residues in DNA are, of course, 2′-deoxyribose, and there is no 2′-hydroxyl to initiate attack on the adjacent 3′-phosphate and cause phosphodiester cleavage. Most DNases are activated by divalent cations ($Mn^{2\oplus}$, $Zn^{2\oplus}$), and these metals may coordinate the susceptible phosphodiester bond and increase the nucleophilicity of bound water molecules at the active site for direct internucleotide cleavage by a water molecule.

One interesting DNase activity concerns the *E. coli* enzyme DNA polymerase I, whose basic function is to polymerize deoxynucleotide triphosphates in the presence of a DNA template. The synthetic reaction is a nucleotidyl transfer (Chapter 8) each time a deoxynucleoside monophosphate is incorporated, and the details of mechanism and specificity are beyond the scope of discussion here (see Kornberg, 1974). We simply note that the homogeneous polymerase has two nuclease activities (a 3′→5′-exonuclease and a 5′→3′-exonuclease activity) that are essential components of the polymerase (a single polypeptide chain). As an exonuclease working at the 3′-end, the enzyme acts to excise a base that is mismatched (fails to hydrogen-bond correctly) to the complementary base on the template chain. The mismatched base is removed hydrolytically before polymerization introduces an error in the newly elongating DNA chain. This activity therefore represents a crucial fidelity check on the sequence to keep errors at a very low frequency by allowing two editing steps: one in the initial incorporation, and the second during exonuclease inspection. The second exonuclease activity of the polymerase works from the 5′-end to effect phosphodiester-bond cleavage at the base-paired region. Exonucleolytic degradation here allows excision repair, such as the cutting-out of thymine dimers that may have formed from absorption of ultraviolet light (Kornberg, 1974). The true physiological activity of DNA polymerase I may in fact be hydrolytic excision and then polymerization repair, rather than as the major enzyme for in vivo DNA polymerization in bacteria (Gefter, 1975).

Chapter 7

Phosphoryl Transfers, II: Kinases

Kinases are enzymes that transfer the γ-phosphoryl group of ATP to some nucleophilic acceptor molecule RX (¶6.A). Other purine or pyrimidine nucleoside triphosphates can replace ATP as the phosphoryl donor—for example, guanosine triphosphate (GTP), uridine triphosphate (UTP), or cytidine triphosphate (CTP).

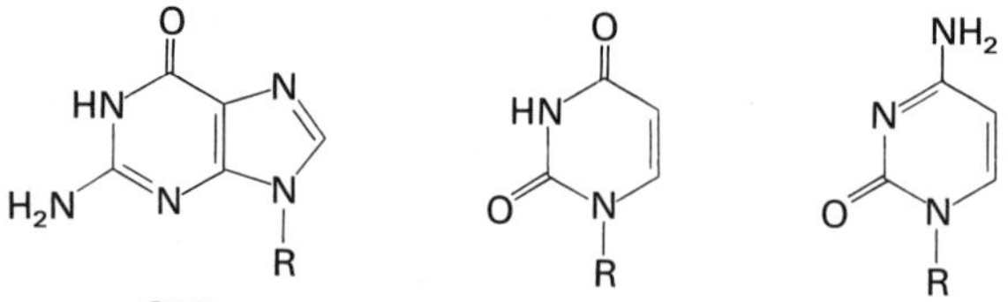

GTP UTP CTP

Note that, by this definition, kinase activity is represented by those reactions using the γ-carboxamide group of glutamine as amino donor, involving the cleavage of ATP to ADP and P_i, where some intermediate $XPO_3^{2\ominus}$ is formed (e.g., glutamine synthetase, carbamoyl-P synthetase).

However, in considering the generic term "kinase," it is worth remembering that names were assigned historically to enzymatic activities before the mechanistic details of the catalyses were determined. We shall see (in Chapter 8) that some enzymes bearing the name kinase do not, in fact, transfer one γ-$PO_3^{2\ominus}$ of a nucleoside triphosphate, but transfer various other portions of the substrate.

Thus, we make a mechanistic division of enzymes that utilize ATP as substrate, based on the particular electrophilic center in the molecule that is attacked by a nucleophile during catalysis. We note four classes in this division.

1. *Attack of nucleophile on* γ-P *of ATP* (*phosphoryl transfer*). These are true kinase reactions.

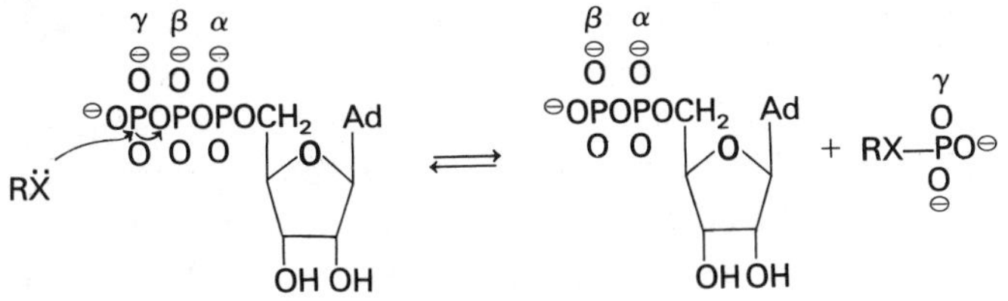

2. *Attack of nucleophile on* β-P *of ATP* (*pyrophosphoryl transfer*).

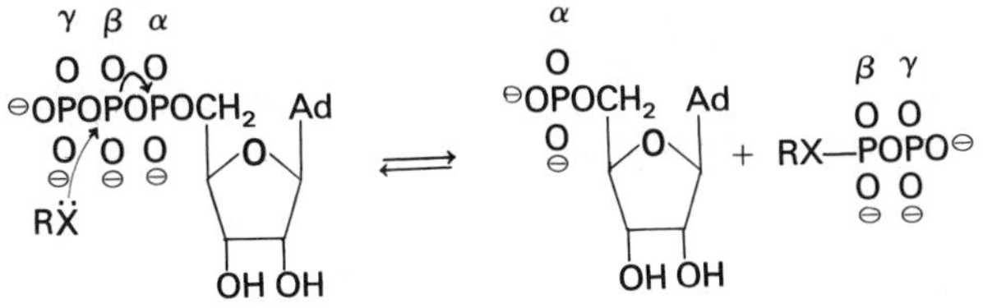

3. *Attack of nucleophile on* α-P *of ATP* (*adenylyl transfer*).

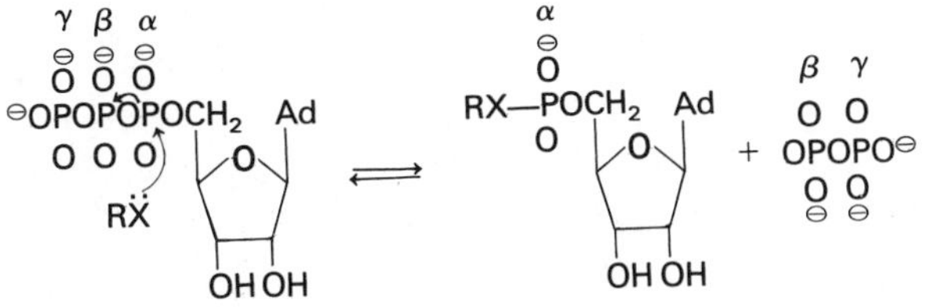

4. *Attack of nucleophile at* C-5′ *of the ribose unit* (*adenosyl transfer*). This type is quite rare, occurring only in methionine biosynthesis and in coenzyme-B_{12} biosynthesis, and we defer discussion until treating those topics in later chapters.

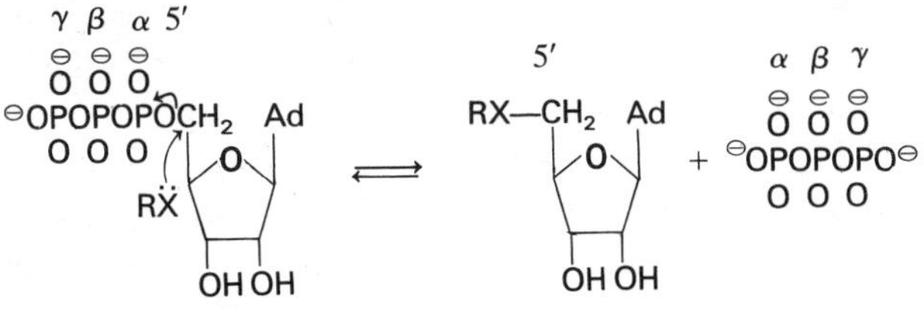

Although we have illustrated these reaction types with ATP as substrate, other nucleoside triphosphates are utilized also (particularly in category 3, which is generalized to the term *nucleotidyl transfers*).

In this chapter, we restrict our attention to reactions of category 1. Reactions of category 2 and category 3 are discussed in Chapter 8.

Looking at the substrates and products of kinase reactions, we note that many of the specific molecules are thermodynamically destabilized (and many are additionally kinetically unstable in aqueous solutions); equilibrium greatly favors the hydrolysis of those compounds. Such molecules have been termed "energy-rich"or "high-energy" compounds, or compounds with "high group-transfer potential." We have already noted the structural basis for destabilization of such molecules as ATP and acyl phosphates in water: they are anhydrides.

Jencks (1962) defined an *energy-rich compound* in biological systems as one that reacts with some molecule X, where X is plentiful in the enviroment, to give products with a large, negative ΔG when the reaction occurs under physiological conditions (see also Klotz, 1967). Such a compound displays *high group-transfer potential.* In most cases X is H_2O, and $\Delta G^{0'}$ for hydrolysis at pH 7 and 25° C is usually −7 to −15 kcal/mole (−29 to −63 kJ/mole) for an energy-rich compound.

For a reaction of the type

$$aA + bB \rightleftarrows cC + dD$$

the free energy (ΔG) is

$$\Delta G = \Delta G^{0'} + RT \ln ([C]^c[D]^d/[A]^a[B]^b)$$

At equilibrium, $\Delta G = 0$, and

$$\Delta G^{0'} = -RT \ln ([C]^c[D]^d/[A]^a[B]^b) = \boxed{-RT \ln K_{eq}}$$

At 25° C and 1 M concentrations, for hydrolysis,

$$\text{when } K_{eq} = 10, \quad \Delta G^{0'} = -1.4 \text{ kcal/mole} = -5.9 \text{ kJ/mole}$$

$$\text{when } K_{eq} = 100, \quad \Delta G^{0'} = -2.72 \text{ kcal/mole} = -11.4 \text{ kJ/mole}$$

$$\text{when } K_{eq} = 1000, \quad \Delta G^{0'} = -4.09 \text{ kcal/mole} = -17.1 \text{ kJ/mole}$$

Clearly, $\Delta G^{0'}$ values of −7 to −15 kcal/mole represent enormous thermodynamic propensities for the reaction to favor buildup of products C and D at equilibrium. That A and B exist in reasonable and detectable quantities at all in the cell reflects their kinetic stability in vivo.

Table 7-1 indicates the free energy of hydrolysis ($\Delta G^{0'}$) for some phosphorylated compounds in biological systems. As indicated, an arbitrary dividing line can be drawn at $\Delta G^{0'} \approx -5$ kcal/mole ≈ -21 kJ/mole between compounds with high phosphoryl-group–transfer potential ($\Delta G^{0'} < -5$ kcal/mole) and compounds with low phosphoryl-group–transfer potential ($\Delta G^{0'} \geq -5$ kcal/mole). Examination of the table indicates that, although ATP may be the prevalent phosphoryl donor (or common cellular energy currency), it is not as destabilized as acyl phosphates or phosphoenolpyruvate. The more negative $\Delta G^{0'}$ values for these structures indicate that the equilibrium for phosphoryl transfer between an acyl phosphate and ATP would favor ATP formation rather than ADP formation.

Table 7-1
Free energy of hydrolysis for some phosphates

Compound ($R—PO_3^{2-}$)		$\Delta G^{0'}$ (kcal/mole)	$\Delta G^{0'}$ (kJ/mole)	Structural type
Phosphoenolpyruvate (PEP) $H_2C{=}C(OPO_3^{2-})COO^{-}$	High group-transfer potential	−14.8	−62.0	Enol-P
1,3-Diphosphoglycerate $^{2-}O_3PO—H_2C—CH(OH)—COOPO_3^{2-}$	High group-transfer potential	−11.8	−49.4	Acyl-P (anhydride)
Phosphocreatine $^{-}OOC—CH(H_3C)—NH—C(=NH_2^{+})—NH—PO_3^{2-}$	High group-transfer potential	−10.3	−43.1	Guanidinium-P
Acetyl phosphate $H_3C—C(=O)—OPO_3^{2-}$	High group-transfer potential	−10.1	−42.3	Acyl-P (anhydride)
Arginine phosphate $^{-}OOC—CH(NH_3^{+})(CH_2)_3CH_2—NH—C(=NH_2^{+})—NHPO_3^{2-}$	High group-transfer potential	−7.7	−32.2	Guanidinium-P
Adenosine triphosphate (ATP^{4-})	High group-transfer potential	−7.3	−30.6	Anhydride
Glucose-1-PO_3^{2-}	Low group-transfer potential	−5.0	−20.9	Hemiacetal
Fructose-6-PO_3^{2-}	Low group-transfer potential	−3.8	−15.9	Alcohol
Glucose-6-PO_3^{2-}	Low group-transfer potential	−3.3	−13.8	Alcohol
Glycerol-1-PO_3^{2-}	Low group-transfer potential	−2.2	−9.2	Alcohol

$$R-\overset{\overset{O}{\|}}{C}-O-\underset{\underset{O^{\ominus}}{|}}{\overset{\overset{O^{\ominus}}{|}}{P}}=O + ADP^{3\ominus} \rightleftharpoons R-\overset{\overset{O}{\|}}{C}-O^{\ominus} + ATP^{4\ominus}$$

Conversely, in phosphoryl transfer from ATP to some sugar substrate, at equilibrium the phosphorylated sugar will predominate.

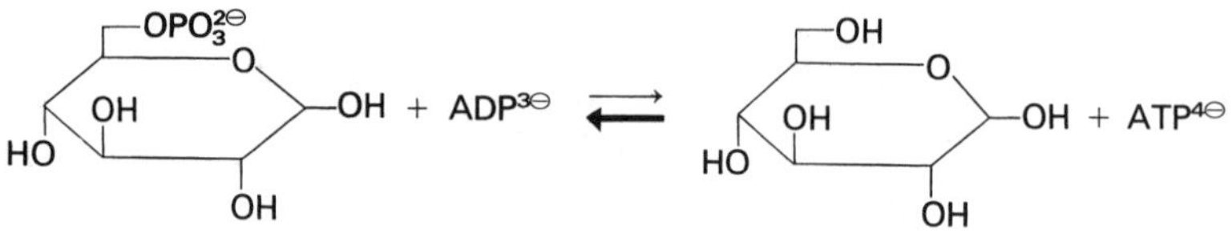

$$\text{sugar-}OPO_3^{2\ominus} + ADP^{3\ominus} \rightleftharpoons \text{sugar-}OH + ATP^{4\ominus}$$

7.A CATION REQUIREMENTS FOR KINASES

A divalent metal cation, usually $Mg^{2\oplus}$ (occasionally $Mn^{2\oplus}$), is required for kinase catalysis and, in this sense, kinases can be included with the metalloenzymes, which represent about a third of known enzymes. However, the $Mg^{2\oplus}$ is used to chelate with the anionic $ATP^{4\ominus}$ substrate, and the Mg–ATP chelate is the actual species undergoing reaction. A subgroup of kinases (including pyruvate kinase, discussed later in this chapter) shows requirements for monovalent cations $K^{\oplus}$ or $NH_4^{\oplus}$; the requirement may be absolute or as an observed stimulation of a basal activity (Suelter, 1970). The $K^{\oplus}$ or $NH_4^{\oplus}$ clearly is binding reversibly at the active site of those enzymes.

Metal ions can function as catalytic agents in enzymatic reactions in several ways (Coleman, 1971; Mildvan, 1970). Bender (1971, pp. 211–280) has suggested the following categorization.

1. *Superacid catalysis.* $M^{n\oplus}$ can function as a proton with a positive charge >1. The metal acts as a superacid in electrophilic catalysis.
2. *Redox catalysis.* Metal ions can act as intermediate electron carriers between one substrate undergoing oxidation and another undergoing reduction.
3. *Metal–carbon bond formation.* A major example of such bond formation currently known is in coenzyme-B_{12} reactions, and there the cobalt also undergoes redox change.
4. *Metals as templates.* Metals can act as templates for binding and positioning of reacting substrates.

There are no detectable redox changes in kinase reactions. Rather, $Mg^{2\oplus}$ clearly acts as a superacid or electrophilic catalyst. Coordination of substrate with metal may well induce electronic distortion via orbital overlap with the metal. As an example, the acidity of a water molecule liganded to Cu^{II} in the hydrated Cu^{II} ion has been measured to be 10^7-fold greater than the acidity of water alone (Bender, 1971).

A metal may facilitate catalysis while acting as superacid by functioning as an electron sink, absorbing electron density generated during reaction and then returning it to products. This action could significantly stabilize the transition state. The metal cation also will neutralize net charge on anions, reducing electrostatic repulsions during reaction, thereby increasing the rate of attack of nucleophiles on the electrophilic phosphorus in phosphate groups. By proper coordination, it can stabilize leaving groups. Metal ions are better than protons at this stabilization because *they are cations that can exist in high concentrations in neutral solutions* and because they can be multidentate coordinators.

7.B TYPES OF KINASE REACTIONS

Kinase reactions can be classified according to the functional group that becomes phosphorylated in the cosubstrate. Table 7-2 provides a selective list of reactions

Table 7-2
Kinases categorized by substrate nucleophile

R—X	Enzyme	Products	Covalent Enz—X—$PO_3^{2\ominus}$
R—OH	Hexokinase	Glucose-6-P; ADP	Possible
	Phosphofructokinase	Fructose-1,6-diP; ADP	No
	Galactokinase	Galactose-1-P; ADP	No
R—$OPO_3^{2\ominus}$	Adenylate kinase (AMP)	ADP; ADP	No
	Nucleoside diphosphokinase (GDP)	GTP; ADP	Yes
R—NH_2	Arginine kinase	Arginine-P: ADP	No
	Creatine kinase	Creatine-P; ADP	No
R—$COO^\ominus$	Acetate kinase	Acetyl-P; ADP	Yes
	Aspartate kinase	β-Aspartyl-P; ADP	No
	Succinate thiokinase	Succinyl coenzyme A; GDP	Yes
R=C(—$O^\ominus$)—$COO^\ominus$	Pyruvate kinase	Phosphoenolpyruvate; ADP	No
Protein—Ser—OH	Protein kinases	Protein-Ser-$OPO_3^{2\ominus}$; ADP	Yes

with specific enzymic examples for phosphorylation of hydroxyl groups, nucleoside phosphates, amino groups, carboxylate groups, an enolate anion, and serine residues in a variety of proteins.* We shall discuss some of these specific enzymes, commenting on their physiological role and their reversibility.

Table 7-2 also contains a column indicating whether evidence has been adduced for covalent phosphoryl-enzyme intermediates. It is clear that kinases fall into two mechanistic categories: those that utilize covalent catalysis, and those where direct phosphoryl transfer occurs between substrate molecules.

A variety of proteins are phosphorylated in cells, often reversibly for regulatory functions (D. A. Walsh and Krebs, 1973). Usually the acceptor proteins are phosphorylated at specific serine residues. A subset of protein kinases includes those enzymes that are activated by 3′,5′-cyclic AMP (which induces dissociation of an inactive dimer of catalytic and regulatory subunits to free the catalytic subunit for catalysis). Recent evidence indicates that cAMP-dependent protein kinase from muscle is itself phosphorylated by ATP during phosphoryl transfer to protein substrates (Rosen and Erlichman, 1975).

7.C BISUBSTRATE ENZYME KINETICS AND THE COVALENT–INTERMEDIATE QUESTION

In Chapter 3 we discussed the simple Michaelis–Menten equation for unireactant enzyme kinetics, and in Chapter 4 the limiting patterns of competitive and noncompetitive reversible inhibition. Many of the enzymes discussed in Chapter 5 and the kinases in this chapter are multisubstrate enzymes, and equations describing their kinetic behavior can be quite complex. However, there are three limiting cases that we can delineate for a bisubstrate → biproduct (BiBi) enzymatic reaction. The three limiting cases represent kinetic mechanisms of two distinct types:

1. two ternary-complex alternatives,
   a. random binding,
   b. ordered binding;
2. one binary-complex alternative.

The *ternary-complex* mechanisms are those where both substrates A and B must be present at the active site before any chemical reaction occurs.

$$\mathrm{A + B \xrightarrow{enzyme} P + Q}$$

*Morrison and Heyde (1972) review information on kinase action. Also see various articles on specific kinases in *The Enzymes* (Boyer, 1970–1976, vols. 8 and 9).

The *binary-complex* mechanism necessarily involves a covalent enzyme intermediate. Substrate A binds to the enzyme to form binary complex EA, which reacts to form a covalent enzyme intermediate and to release a product fragment P. The covalent intermediate then forms a second binary complex with substrate B, which reacts to produce the second observed product Q.

We shall look briefly at the three kinetic alternatives to understand the observed behavior—to see how the kinetic constants $K_{m(A)}$(K_m for A), $K_{m(B)}$, and V_{max} are obtained. We shall determine whether initial-velocity kinetic studies can be diagnostic for a covalent intermediate. The kinetic treatment we use is that described by Segel (1975*a*, 1975*b*), although many others have treated the subject well (see, for example, Dixon and Webb, 1964; Gutfreund, 1972; Plowman, 1972; Westley, 1969). Readers with special interest in determination of kinetic constants and their mechanistic and physical interpretations might consult the excellent recent review by Cleland (1977).

7.C.1 Ternary-Complex Kinetic Mechanisms

We look first at the two alternatives involving formation of a ternary complex by A, B, and the enzyme. In the first alternative, the formation of this complex can proceed randomly; in the second alternative, one substrate *must* bind to the enzyme before the other.

7.C.1.a BiBi reactions with rapid random equilibrium

If both substrates A and B can bind to the enzyme in a random order, and if the two products P and Q can desorb randomly, then the following scheme applies.

$$\begin{array}{ccccccccccc} & \overset{+A}{\nearrow} & E\cdot A & \overset{+B}{\searrow} & & & & \overset{-P}{\nearrow} & E\cdot Q & \overset{-Q}{\searrow} & \\ E & & & & E\cdot A\cdot B & \overset{k_{cat}}{\rightleftharpoons} & E\cdot P\cdot Q & & & & E \\ & \underset{+B}{\searrow} & E\cdot B & \underset{+A}{\nearrow} & & & & \underset{-Q}{\searrow} & E\cdot P & \underset{-P}{\nearrow} & \end{array}$$

This system bears close analogy to the noncompetitive inhibition described in Chapter 4, but E · A · B is active whereas E · S · I in that instance was inactive. Making the assumption that binding steps (on and off steps) are fast relative to the E · A · B ⇄ E · P · Q conversion, one can write the velocity equation, under initial velocity (v_0) conditions:

$$v/[E_T] = k_{cat}[E\cdot A\cdot B]/([E]+[E\cdot A]+[E\cdot B]+[E\cdot A\cdot B])$$

Given our earlier definitions of $K_{m(A)}$ and $K_{m(B)}$, and the dissociation constants K_A (for dissociation of E · A) and K_B (for dissociation of E · B), the equation can be written

$$V_{max}/v_0 = (K_A K_{m(B)} + K_{m(B)}[A] + K_{m(A)}[B] + [A][B])/[A][B]$$
$$= 1 + (K_{m(A)}/[A]) + (K_{m(B)}/[B]) + (K_A K_{m(B)}/[A][B])$$

Note that the random equilibrium mechanism should be symmetric in A vs. B, but the velocity equation has no $K_{m(A)}K_B$ term. Segel (1975*a*, 1975*b*) has pointed out that $K_{m(A)}K_B = K_{m(B)}K_A$ for this mechanism, thus accommodating the symmetry problem.

7.C.1.b BiBi reactions with ordered binding

In some cases, a bireactant enzyme must bind substrate A to induce a conformational change for productive binding of B—i.e., the binding of A must precede binding of B.

$$E + A \rightleftharpoons E\cdot A \overset{+B}{\rightleftharpoons} E\cdot A\cdot B \rightleftharpoons E\cdot P\cdot Q \overset{-P}{\rightleftharpoons} E\cdot Q \overset{-Q}{\rightleftharpoons} E$$

Cleland (1970) developed an alternative shorthand representation:

A ↓		B ↓		P ↑		Q ↑
E	E · A		(E · A · B ⇌ E · P · Q)		E · Q	E

Making the usual steady-state assumption (Chapter 3), one finds that *the kinetic equation for the ordered BiBi system is identical to that derived for rapid random equilibrium* in ¶7.C.1.a:

$$V_{max}/v_0 = 1 + (K_{m(A)}/[A]) + (K_{m(B)}/[B]) + (K_A K_{m(B)}/[A][B])$$

This kinetic equation indicates that, when initial-velocity studies are conducted and the results plotted in the double-reciprocal $1/v$ vs. $1/[S]$ plot, *a family of intersecting lines will be obtained.* Kinetic experiments are conducted by holding fixed the concentration of one substrate (e.g., B) while varying the concentration of the other substrate (e.g., A). At fixed [B], the velocity equation reduces to the Michaelis–Menten form:

$$v/V_{max(app)} = [A]/(K_{m(app)} + [A]),$$

where $V_{max(app)}$ and $K_{m(app)}$ are the observed V_{max} and K_m, respectively, at that concentration of B. A plot of the data in double-reciprocal form gives the usual straight line at one fixed [B] (Fig. 7-1a).

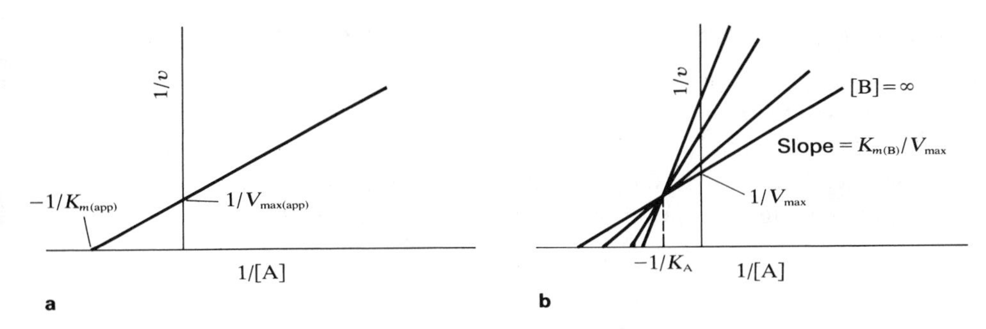

Figure 7-1
Double-reciprocal plot of kinetic data for a sequential BiBi mechanism. (**a**) Data for one fixed concentration of cosubstrate B. (**b**) Data for several fixed concentrations of cosubstrate B.

If one now repeats this experiment at several concentrations of B, a series of lines is generated; the lines intersect to the left of the vertical axis and above the horizontal axis (Fig. 7-1b). The point of intersection represents the value of [A] equal to $-1/K_A$ (the reciprocal of the dissociation constant from $E \cdot A$). If one could go to infinitely high concentrations of B, one could obtain $1/V_{max}$ and $-1/K_{m(A)}$. Because this cannot be achieved experimentally, we can obtain the desired kinetic constants from a replot of the intercept and a replot of the slope data. A replot of the intercept data should yield points fitting a straight line (Segel, 1975*a*, 1975*b*). This intercept replot (Fig. 7-2a) provides two constants, $1/V_{max}$ and $-1/K_{m(B)}$. The slope replot (Fig. 7-2b) provides the third constant, $K_{m(A)}$.

As expected, one can obtain the same kinetic constants by carrying out the complementary velocity experiments—holding [A] as the fixed substrate concentration and using [B] as the variable substrate concentration. The $K_A K_{m(B)}/[A][B]$ term in the velocity equation is the term responsible for the intersecting pattern of lines in the $1/v$ vs. $1/[S]$ plots for ternary-complex mechanisms.

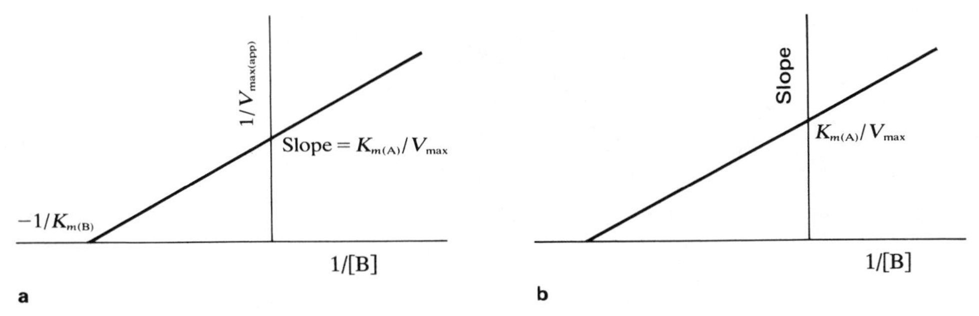

Figure 7-2
Replots of data from Figure 7-1b. (**a**) Intercept replot. (**b**) Slope replot.

7.C.2 Binary-Complex Kinetic Mechanism: Ping-Pong BiBi Reactions

The binary-complex mechanism involves a covalent intermediate as E goes to a modified form of enzyme, F. This was termed a ping-pong mechanism because the enzyme goes back and forth between the two stable forms: free enzyme E and covalently modified enzyme F (see Cleland, 1970, and references cited therein).

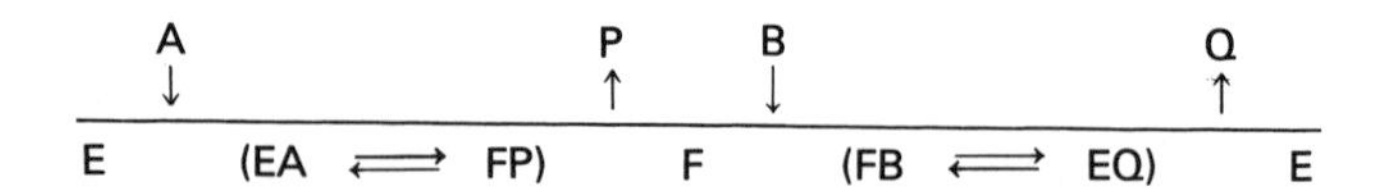

The first product (P) departs the active site before the substrate (B) binds.

From a steady-state assumption, we obtain the following velocity equation:

$$V_{max}/v_0 = 1 + (K_{m(A)}/[A]) + (K_{m(B)}/[B])$$

This is similar to the equation for the ternary-complex mechanisms, but there is one difference: the $K_{m(A)}K_B/[A][B]$ term is lacking, and this absence means that the family of lines obtained in initial-velocity experiments with one substrate (e.g., B) fixed and one (e.g., A) varied *will not intersect.* In fact, a series of *parallel lines* is generated (Fig. 7-3). The vertical intercept at any [B] is $1/V_{max(app)}$, and the horizontal intercept is $-1/K_{m(A)(app)}$. An intercept replot of $1/V_{max(app)}$ vs. 1/[B] yields the true V_{max} and $K_{m(B)}$ (Fig. 7-4a). A replot of $1/K_{m(A)(app)}$ vs. 1/[B] yields the third constant, $1/K_{m(A)}$ as the vertical intercept (Fig. 7-4b).

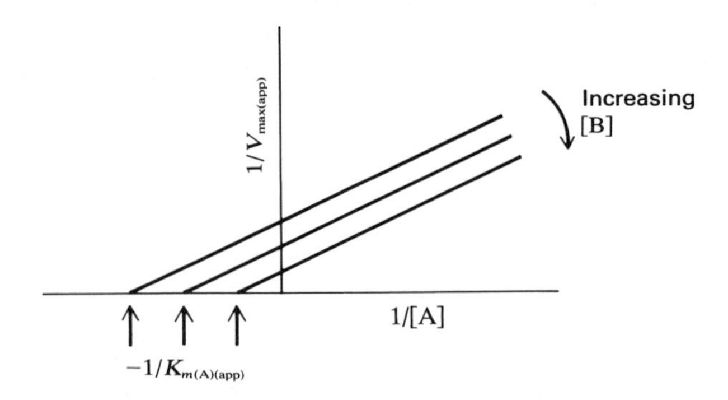

Figure 7-3
Double-reciprocal plot (at several fixed concentrations of cosubstrate B) for a ping-pong BiBi mechanism, illustrating the parallel-line pattern.

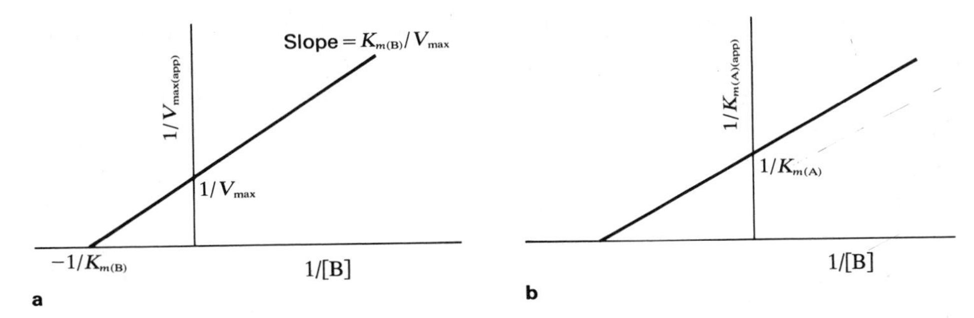

Figure 7-4
Replots of data from Figure 7-3. (**a**) Intercept replot. (**b**) Double-reciprocal plot of $1/K_{m(A)(app)}$ vs. 1/[B].

7.C.3 Conclusions About Bisubstrate Initial-Velocity Kinetic Patterns

Apparently, the most salient mechanistic deduction that can be made from initial-velocity experiments in a BiBi enzymatic reaction is that observation of parallel-line kinetics in double-reciprocal plots is *diagnostic of a covalent enzyme intermediate.* For example, if the parallel-line pattern is observed in studies of the acetate-kinase reaction

$$CH_3COO^{\ominus} + ATP^{4\ominus} \xrightleftharpoons{\text{enz, }Mg^{2\oplus}} H_3C{-}\overset{O}{\overset{\|}{C}}{-}O{-}\overset{O}{\overset{\|}{\underset{O^{\ominus}}{\underset{|}{P}}}}{-}O^{\ominus} + ADP^{3\ominus}$$

we would conclude that a covalent intermediate (in this example, presumably a phosphoryl-enzyme intermediate) is involved in the reaction.

The converse conclusion—that an intersecting family of lines rules out a covalent intermediate—*is not necessarily valid.* An enzyme might generate a covalent phosphoryl-enzyme intermediate on reaction with ATP to produce $E{-}PO_3^{2\ominus}$ but have ADP still bound. The dissociation rate of ADP may be slow enough so that a second substrate binds before ADP departs. In such a case, the kinetics will be that of an ordered ternary complex, even though a covalent intermediate is chemically and kinetically significant in catalysis. Precisely this situation occurs with the *E. coli* succinate-thiokinase reaction (Moffet and Bridger, 1970).

$$ATP^{4\ominus} + \left(\begin{matrix} COO^{\ominus} \\ \\ COO^{\ominus} \end{matrix}\right. + RSH \rightleftharpoons ADP^{3\ominus} + P_i^{2\ominus} + \left(\begin{matrix} \overset{O}{\overset{\|}{C}}{-}SR \\ \\ \underset{O}{\underset{\|}{C}}{-}O^{\ominus} \end{matrix}\right.$$

Nor is the experimental detection of *apparent* parallel-line initial-velocity kinetics ironclad in itself. The enzyme in question may use a sequential mechanism in which the numerical value of the $K_A K_{m(B)}/[A][B]$ term is small but finite. The convergence of the lines may be so slight as to be missed by the experimenter. This event occurred during initial-velocity studies of the flavin-dependent enzyme D-amino-acid oxidase, which we shall discuss in Section III (Koster and Veeger, 1968).

These caveats, then, indicate that initial-velocity experiments should be bolstered by other lines of evidence. In addition to chemical experiments to isolate or trap a putative covalent intermediate (Chapter 3), there are two other complementary lines of kinetic evidence that can be obtained:

1. partial-isotopic-exchange data;
2. product-inhibition data.

The partial-isotopic-exchange technique discussed in ¶5.E.2 monitors *partial chemical reactions that can occur in the ping-pong mechanism but not in either of the ternary mechanisms.* In the following section we briefly discuss the product-inhibition kinetic patterns one can expect.

7.C.4 Product-Inhibition Patterns for Analyzing Kinetic Mechanisms

Let us consider the two sequential mechanisms (ordered and random) versus the ping-pong mechanism, in terms of the product-inhibition patterns (see Cleland, 1970, and references therein).

7.C.4.a Ping-pong BiBi

Recall the ping-pong mechanism:

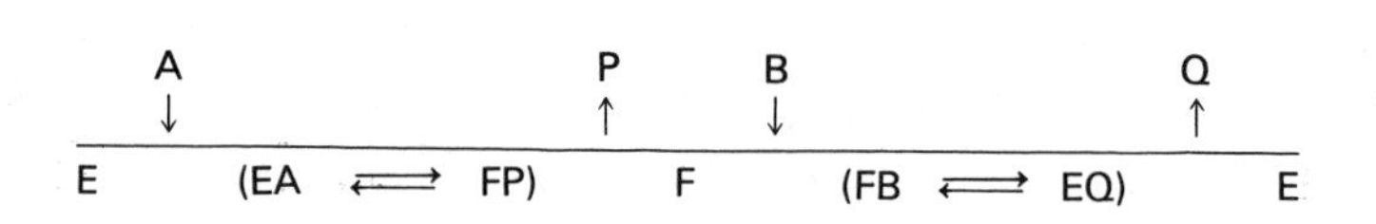

Note that A and Q combine with the same form of enzyme (E) and thus are competitive—i.e., Q is a competitive inhibitor of A (altered K_m but not V_{max}; see Chapter 4). On the other hand, P combines with F, a chemically different form of the enzyme. So P should be a noncompetitive inhibitor of A (altered V_{max} but not K_m). Similarly, we note that P is a competitive inhibitor of B, whereas Q is a noncompetitive inhibitor of B.

Thus, if apparent parallel-line kinetics are seen in initial-velocity experiments, product-inhibition studies should yield two competitive and two noncompetitive patterns. A classical inhibition pattern for ping-pong kinetics is seen with nucleoside diphosphokinase, which forms an $E—PO_3^{2\ominus}$ intermediate (Morrison and Heyde, 1972):

$$Mg \cdot ATP + Mg \cdot GDP \rightleftharpoons Mg \cdot GTP + Mg \cdot ADP$$

In this case, competitive inhibition patterns are observed for GTP vs. ATP and for GDP vs. ADP, whereas noncompetitive patterns are observed for ADP vs. ATP and for GTP vs. GDP.

7.C.4.b Random sequential BiBi

In the random sequential mechanism, all possible binary enzyme–substrate complexes form rapidly and reversibly:

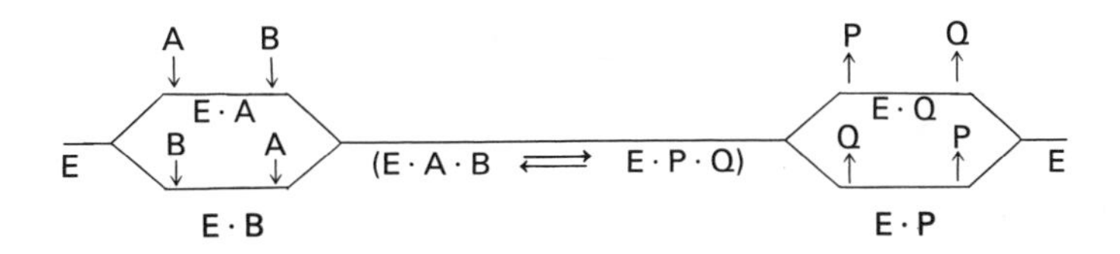

All substrates and products can compete for free E. Therefore, one should see competitive inhibition patterns in all four cases: A vs. P, A vs. Q, B vs. P, and B vs. Q.

7.C.4.c Ordered sequential BiBi

In the ordered sequential mechanism, A must bind before B, and P must leave before Q:

$$\begin{array}{ccccccc} & A\downarrow & & B\downarrow & & P\uparrow & & Q\uparrow & \\ E & & E \cdot A & & (E \cdot A \cdot B \rightleftharpoons E \cdot P \cdot Q) & & E \cdot Q & & E \end{array}$$

A and Q both bind to E and thus are competitive. However, P binds only to E · Q and not to E alone, so P vs. A should show a noncompetitive pattern. Similarly, B combines with E · A only, whereas Q combines with E and P combines with E · Q, so both Q vs. B and P vs. B should be noncompetitive.

A limiting case of the sequential mechanism is that where E · A · B and E · P · Q are at negligibly low concentrations in the steady state; this is known as

Table 7-3
Expected results of initial-velocity and product-inhibition studies for various BiBi mechanisms

Mechanism	Initial-velocity observations	Product-inhibition observations*			
		A vs. P	A vs. Q	B vs. P	B vs. Q
Random sequential BiBi	Intersecting lines	C	C	C	C
Ordered sequential BiBi	Intersecting lines	N	C	N	N
Ping-pong BiBi	Parallel lines	N	C	C	N
Theorell–Chance BiBi	Intersecting lines	N	C	C	N

* C = competitive; N = noncompetitive

the Theorell–Chance mechanism (Theorell and Chance, 1951). In this case, P vs. B becomes competitive:

$$\begin{array}{ccccccc} & A & & B\ P & & Q & \\ & \downarrow & & \searrow\nearrow & & \uparrow & \\ \hline E & & E\cdot A & & E\cdot Q & & E \end{array}$$

Thus, the situation looks ping-pong by product inhibition, but initial-velocity measurements will give intersecting lines.

In analysis of the reaction mechanisms of many kinase enzymes (such as those listed in Table 7-2), the three kinetic techniques of initial-velocity studies, partial-isotopic-exchange studies, and product-inhibition studies have been used to decide on the presence or absence of the indicated $E—PO_3^{2\ominus}$ intermediates. Table 7-3 summarizes the results expected with initial-velocity and product-inhibition studies for each of the BiBi mechanisms.

7.D SPECIFIC KINASES

In the remainder of this chapter, we turn to the reactions of some specific kinases, making use of the general concepts developed thus far.

7.D.1 Phosphorylation of a Substrate Hydroxyl Group

Among the most important sugar substrates phosphorylated enzymatically is glucose, which is converted to glucose-6-phosphate by hexokinase (Colowick,

1973). In animals, hexokinase acts on glucose that has diffused into cells from the blood. The glucose-6-P thus formed is one of the two predominant forms (glucose-1-P is the other) that then undergo intracellular metabolism.

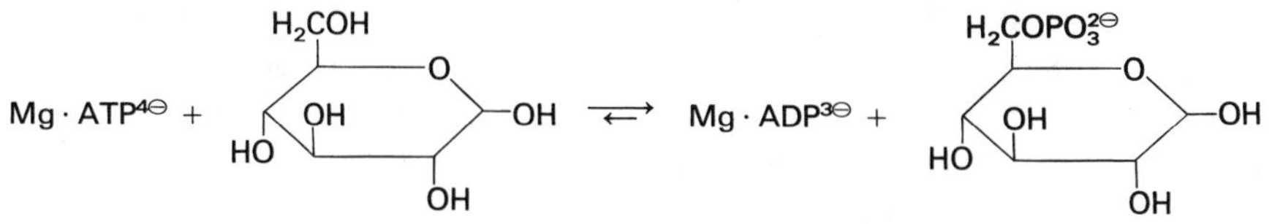

$$Mg \cdot ATP^{4\ominus} + \text{glucose} \rightleftharpoons Mg \cdot ADP^{3\ominus} + \text{glucose-6-}OPO_3^{2\ominus}$$

Hexokinase is of ubiquitous distribution in plants and microorganisms as well, and the yeast enzyme was the first to be purified to physical homogeneity and was used for most mechanistic studies. The pure yeast enzyme is a dimer of 100,000 mol wt. At pH 7 and room temperature, it shows a V_{max} for glucose-6-P formation of 800 μmoles min^{-1} (mg enzyme protein)$^{-1}$. The high group-transfer potential of ATP is utilized to provide an equilibrium constant of 661 in favor of ADP and glucose-6-P formation, ensuring functional unidirectionality in the yeast cell and thus affording maximal concentrations of the phosphate ester of glucose for further metabolic processing.

In addition to the phosphoryl transfer to C-6 of glucose as nucleophile, we have noted (Chapter 2) that the yeast enzyme will catalyze (at much reduced rate) a biologically wasteful transfer to H_2O, resulting in net hydrolysis of Mg · ATP to Mg · ADP and P_i, a useless expenditure of energy. The V_{max} for transfer to H_2O is 0.02 μmole min^{-1} (mg protein)$^{-1}$, 40,000-fold slower than the transfer to glucose. Whereas the K_m for ATP in the normal transfer to glucose is 0.1 mM, the K_m for ATP in the hydrolytic reaction (where glucose is absent) is 4.0 mM, or 40-fold higher. This observation suggests that the binding of glucose may facilitate or tighten ATP binding to the enzyme; it also suggests a cooperative effect of substrates in inducing a protein conformational change—a *substrate synergism.*

The glucose analogue N^2-acetylglucosamine is a competitive inhibitor of glucose in the overall catalysis.

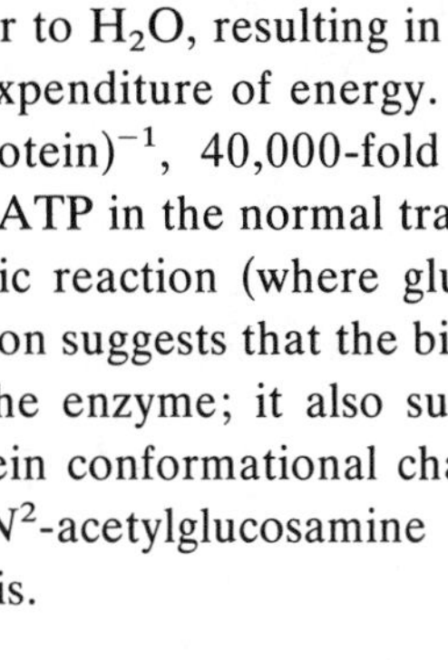

It also inhibits the phosphoryl transfer to water, presumably binding to the active site and either physically excluding water molecules or generating a protein conformation that now restricts solvent access to the active site. In contrast, a 5-carbon aldose, D-lyxose, which is not phosphorylated by the enzyme, actually promotes the hydrolytic activity. At 100 mM D-lyxose concentrations, the V_{max} for ATP hydrolysis (the ATPase activity of hexokinase) is increased by a factor of 20 to about 0.40 μmole min^{-1} mg^{-1}. There is a concomitant reduction of the K_m for ATP from 4.0 mM to 0.1 mM, the same low value seen when glucose is present; again these data imply either induced fit on D-lyxose binding or some decrease in nonproductive binding modes for water.

Colowick (1973) reviewed the literature on the catalytic mechanism for yeast hexokinase and concluded that sugar phosphorylation proceeds via ternary-complex formation—with ATP, glucose, and enzyme interacting prior to occurrence of any chemical process. Somewhat unresolved are the questions of order of addition for the two substrates and of the existence of a phosphorylated enzyme in a ternary complex (i.e., enzyme phosphorylation and dephosphorylation with transfer to glucose before release of ADP).

A variety of kinetic experiments indicate that random binding of substrates can occur; the idea that ATP can bind first is corroborated by the existence of the ATPase reaction. However, data from isotopic exchange experiments at chemical equilibrium suggest that the quantitatively more important pathway is the one where glucose binds first and then the glucose–enzyme binary complex adds ATP (Colowick, 1973). In other words, the flux is greater through the upper pathway in the following reaction scheme:

$$\mathrm{Enz} \begin{array}{c} \nearrow\!\!\swarrow \ \mathrm{Enz\cdot Glucose} \ \searrow\!\!\nwarrow \\ \searrow\!\!\nwarrow \ \mathrm{Enz\cdot ATP} \ \nearrow\!\!\swarrow \end{array} \mathrm{Enz\cdot Glucose\cdot ATP} \rightleftharpoons \text{Product ternary complex}$$

Although kinetic studies have firmly ruled out the existence of a free E—$PO_3^{2\ominus}$ intermediate during hexokinase catalysis, no information is provided on E—$PO_3^{2\ominus}$ formation within substrate and product ternary complexes. The phosphorylated enzyme form, if present, cannot be the predominant species in the steady state, because rapid-quench experiments to isolate [^{32}P]-phosphorylated protein have been negative under a variety of conditions.

An interesting experiment may bear on this problem. Mixing of hexokinase with ATP and the 5-carbon sugars D-xylose or D-lyxose leads, during hydrolysis of ATP, to progressive inactivation of the enzyme for both ATPase and sugar-phosphorylating activities (Dela Fuente, 1970).

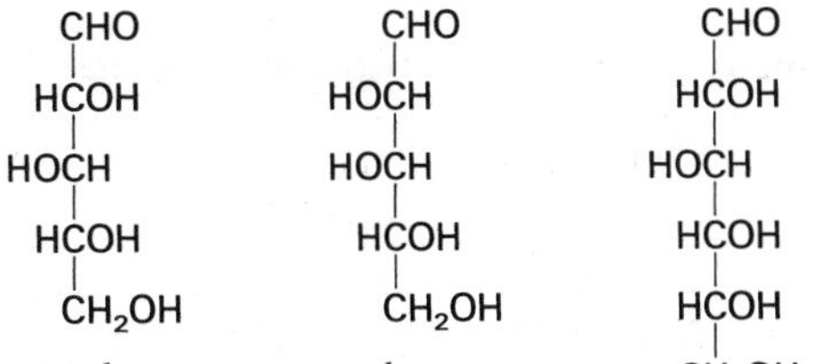

D-xylose	D-lyxose	D-glucose
CHO	CHO	CHO
HCOH	HOCH	HCOH
HOCH	HOCH	HOCH
HCOH	HCOH	HCOH
CH_2OH	CH_2OH	HCOH
		CH_2OH

The inactive enzyme can be shown to have one phosphoryl group (from γ-P of ATP) covalently incorporated as a phosphoserine residue—an *inactive phosphoenzyme.* This inactive E—$PO_3^{2\ominus}$ can be completely revived by addition of ADP and xylose, regenerating active dephosphoenzyme (Menezes and Puckles, 1976). It may be that once every 300 turnovers (the frequency of the inactivation) an active E—$PO_3^{2\ominus}$ rearranges to an inactive form, which then accumulates. Or, it may be that the rare phosphorylation always inactivates the enzyme and reflects an aberrant side reaction not occurring during normal catalysis.

Once glucose-6-P has been produced, it can be isomerized enzymatically to the ketose fructose-6-P, an aldo–keto isomerization sequence discussed in Chapter 19. The keto sugar can then be phosphorylated at another hydroxyl group (at C-1) by the enzyme phosphofructokinase to produce fructose-1,6-diphosphate.

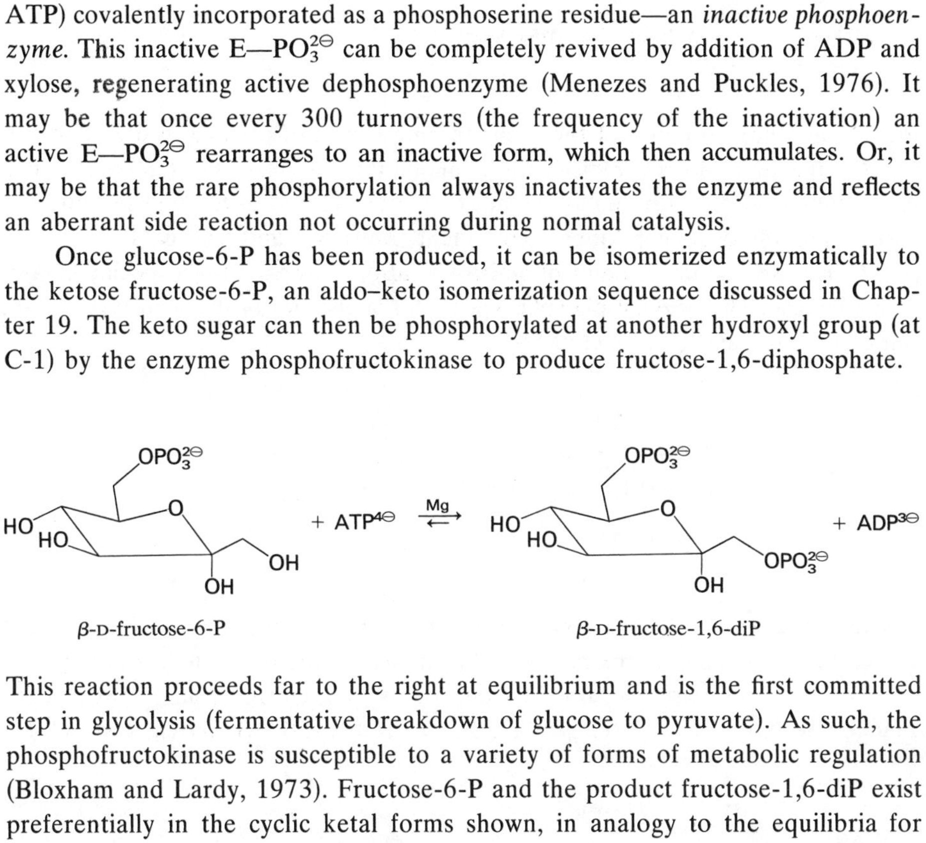

$$\beta\text{-D-fructose-6-P} + ATP^{4\ominus} \underset{}{\overset{Mg}{\rightleftarrows}} \beta\text{-D-fructose-1,6-diP} + ADP^{3\ominus}$$

β-D-fructose-6-P β-D-fructose-1,6-diP

This reaction proceeds far to the right at equilibrium and is the first committed step in glycolysis (fermentative breakdown of glucose to pyruvate). As such, the phosphofructokinase is susceptible to a variety of forms of metabolic regulation (Bloxham and Lardy, 1973). Fructose-6-P and the product fructose-1,6-diP exist preferentially in the cyclic ketal forms shown, in analogy to the equilibria for glucose and glucose-6-P as cyclic hemiacetals. The two isomeric cyclic ketals—the α (CH_2OH down) and the β (CH_2OH up)—are interconvertible via the acylic ketose, with the equilibrium distributions and anomerization half-lives indicated in Table 7-4; the table includes substrate and product, along with other hexoses and hexose phosphates for comparison. Although hexokinase will phosphorylate both the α- and the β-anomers of glucose-6-P, phosphofructokinase appears to

Table 7-4
Distribution of some common hexoses between cyclic and ring-opened forms

Sugar	Cyclic forms α	Cyclic forms β	Ring-opened form (carbonyl)	Anomerization ($\alpha \rightarrow \beta$) half-life
D-Glucose	37%	63%	0.026%	130 min
D-Galactose	31%	69%	——	125 min
D-Glucose-6-P	38%	62%	<0.4%	~15 sec
D-Fructose-6-P	20%	80%	2%	0.4 sec
D-Fructose-1,6-diP	20%	80%	1.7%	1.4 sec

SOURCE: S. Benkovic and K. Schray, in *Advances in Enzymology and Related Areas of Molecular Biology*, vol. 44, ed. A. Meister (New York: Wiley, 1976), p. 139.

utilize only the β-anomer of the cyclic furanose form of D-fructose-6-P. Because the $T_{1/2}$ for interconversion of α- and β-anomers is so short, the stereopreference experiments were performed at icebath temperatures and in the millisecond time scale, using rapid-quench kinetic equipment (Benkovic and Schray, 1976).

Galactose, the C-4 epimer of glucose, is also a substrate for a specific kinase, galactokinase. The enzyme phosphorylates this sugar, not at the hydroxyl of C-6, but at the anomeric hydroxyl at C-1.

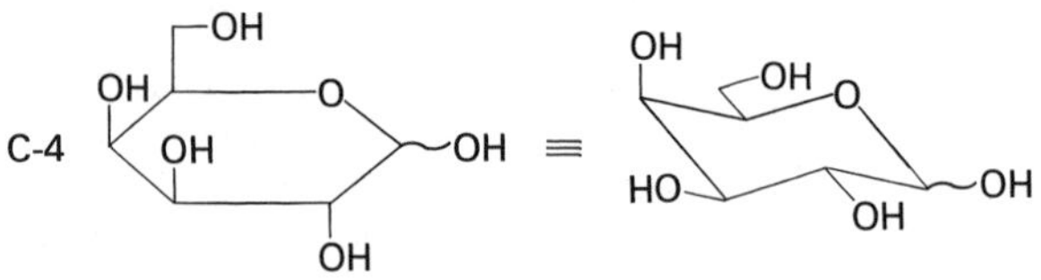

C-4

Because the anomeric hydroxyl must act as nucleophile, the active form of galactose must be one of the hemiacetal isomers, not the open-chain aldose where the C-1 oxygen is a nonnucleophilic aldehyde oxygen. Galactokinase is specific for reaction with D-α-galactose, producing α-galactose-1-P (Benkovic and Schray, 1976).

$+ \ ATP^{4\ominus} \xrightleftharpoons{\text{galactokinase}} \ + \ ADP^{3\ominus}$

α-epimer

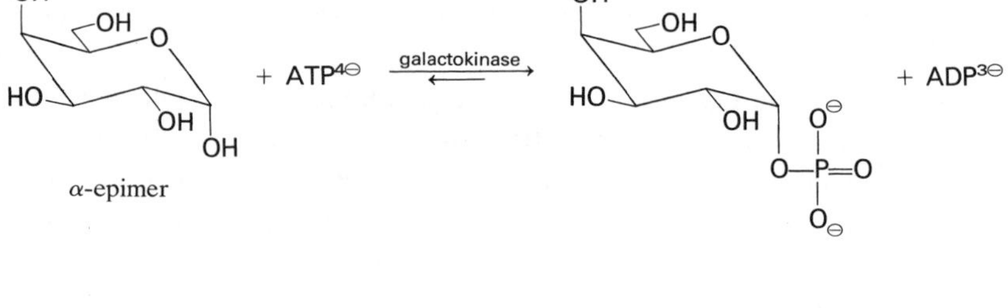

Although the α- and β-anomers of free galactose are interconvertible, the galactose-1-P that is formed is configurationally stable. The free hexose interconverts via the ring-opened aldose form; in the galactose-1-P, no obvious mechanism exists for ring opening. The α-galactose-1-P subsequently participates in a nucleotidyl-transfer process (discussed in Chapter 8), whose ultimate purpose is to set up the galactose ring for enzymatic isomerization at C-4 back into the metabolizable glucose series.

A final example of phosphorylation at a substrate alcoholic carbon is provided by the kinase that converts 5-pyrophosphomevalonate to 3-phospho-5-pyrophosphomevalonate as transient product. This species is rapidly converted to Δ^3-isopentenyl pyrophosphate (the biological isoprene building block) by loss of CO_2 and inorganic phosphate.

$$\text{5-pyrophosphomevalonate} + ATP^{4\ominus} \underset{}{\overset{\text{kinase}}{\rightleftharpoons}} \text{3-phospho-5-pyrophosphomevalonate} + ADP^{3\ominus}$$

$$\xrightarrow{\text{decarboxylase}} \Delta^3\text{-isopentenyl-PP} + CO_2 + O{=}P(OH)(O^{\ominus})O^{\ominus}$$

5-pyrophosphomevalonate

3-phospho-5-pyrophosphomevalonate

Δ^3-isopentenyl-PP

The C-3 tertiary alcoholic group in the 5-pyrophosphomevalonate is a relatively poor leaving group. Phosphoryl transfer from ATP generates a phosphate ester that can depart as phosphate dianion with a much lower activation energy. The introduction of the good leaving group facilitates decarboxylation and olefin formation, probably as indicated above. This is an especially clear indication of how ATP can be used in a phosphoryl-transfer process to activate a cosubstrate for expulsion of an oxygen atom.

A variant of phosphoryl transfer to a substrate hydroxyl group is the reaction catalyzed by pyruvate kinase. ATP and pyruvate interact with phosphoryl transfer, not to a carboxylate oxygen, but rather to the oxygen of the enolate anion of pyruvate, producing phosphoenolpyruvate (PEP).

$$\underset{\underset{BH^{\oplus}}{|}}{Enz}\cdots HC(H)=C(O^{\ominus})-COO^{\ominus} + O=P(O^{\ominus})(O^{\ominus})-O-P(O^{\ominus})(=O)-O-P(=O)(O^{\ominus})-O-Rib-Ad \rightleftharpoons \underset{\underset{BH^{\oplus}}{|}}{Enz}\cdots HC(H)=C(O-PO_3^{2\ominus})-COO^{\ominus} + ADP^{3\ominus}$$

$$\upharpoonleft\downharpoonright$$

$$\underset{\underset{B:}{|}}{Enz}\cdots HC(H)_2-C(=O)-COO^{\ominus}$$

$$\rightleftharpoons \underset{\underset{BH^{\oplus}}{|}}{Enz} + H_2C=C(O-PO_3^{2\ominus})-COO^{\ominus}$$

PEP

The enolate form of pyruvate does not exist free in solution in significant concentrations, but it is generated by general base catalysis from bound pyruvate at the active site. This nucleophile then attacks at the γ-P of ATP. The PEP product represents the trapped form of the thermodynamically destabilized enolate; PEP is a molecule with high phosphoryl-group–transfer potential and with a $\Delta G^{0\prime} = -14.8$ kcal/mole $= 62.0$ kJ/mole, representing a $K_{eq} \approx 10^{10}$ in favor of hydrolysis to pyruvate.

The thermodynamics for phosphoryl-group transfer between ATP and PEP indicate that (at equilibrium) formation of ATP + pyruvate will be enormously favored over PEP + ADP. In fact, this conclusion is borne out by the observation that, under physiological conditions, *the enzyme catalyzes ATP formation irreversibly*; the role of pyruvate kinase in vivo is ATP synthesis during breakdown of glucose to pyruvate during glycolysis (Chapter 27).

$$ADP^{3\ominus} + H_2C=C(OPO_3^{2\ominus})COO^{\ominus} \longrightarrow ATP^{4\ominus} + H_3C-C(=O)-COO^{\ominus} \qquad K_{eq} \approx 3{,}000$$

We have written the pyruvate $\rightarrow$ PEP process as direct transfer to the enolate nucleophile because extensive research failed to reveal any evidence for a covalent E—$PO_3^{2\ominus}$ intermediate. The reaction sequence implies enzyme-mediated enolate formation as an initial step. If this step is reversible, if it can occur in the absence of ATP, and if $BH^{\oplus}$ exchanges the substrate-derived proton rapidly, then a partial isotopic exchange of tritium from [^{3}H]-H_2O into pyruvate should be observable. In fact, no such partial exchange occurs unless an analogue of ATP is present, possibly to evoke the proper conformational change for efficient enolization to occur. Inorganic phosphate serves as an appropriate analogue (no PEP is formed), and the tritium exchange indicative of enolization occurs (Rose, 1960).

$$^{3}H_2O + H_3C{-}\overset{\displaystyle O}{\overset{\|}{C}}{-}COO^{\ominus} + O{=}\underset{\displaystyle O^{\ominus}}{\underset{|}{\overset{\displaystyle O^{\ominus}}{\overset{|}{P}}}}{-}OH \xrightarrow{enz} {^{3}H}{-}\underset{\displaystyle H}{\underset{|}{\overset{\displaystyle H}{\overset{|}{C}}}}{-}\overset{\displaystyle O}{\overset{\|}{C}}{-}COO^{\ominus} + P_i$$

Homogeneous pyruvate kinase shows three other catalytic activities that are probably without clear physiological significance:

1. oxaloacetate-decarboxylase activity (Creighton and Rose, 1976),

$$^{\ominus}OOC{-}\overset{\displaystyle O}{\overset{\|}{C}}{-}\underset{H}{\overset{H}{C}}{-}COO^{\ominus} \longrightarrow {^{\ominus}OOC}{-}\overset{\displaystyle O}{\overset{\|}{C}}{-}CH_3 + CO_2$$

2. "fluorokinase" activity (Kayne, 1973),

$$F^{\ominus} + HCO_3^{\ominus} + ATP^{4\ominus} \longrightarrow F{-}\underset{\displaystyle O^{\ominus}}{\underset{|}{\overset{\displaystyle O}{\overset{\|}{P}}}}{-}O^{\ominus} + HCO_3^{\ominus} + ADP^{3\ominus}$$

fluorophosphate

3. "hydroxylamine-kinase" activity (Kayne, 1973),

$$NH_2OH + HCO_3^{\ominus} + ATP^{4\ominus} \longrightarrow H_2N{-}O{-}\underset{\displaystyle O^{\ominus}}{\underset{|}{\overset{\displaystyle O}{\overset{\|}{P}}}}{-}O^{\ominus} + HCO_3^{\ominus} + ADP^{3\ominus}$$

O-phosphoryl
hydroxylamine

Activities 2 and 3 specifically require $HCO_3^{\ominus}$ in catalytic amounts, presumably to bind at the site normally occupied by the pyruvate carboxylate ion and induce a protein conformation allowing $F^{\ominus}$ or NH_2OH close approach to the γ-P of ATP.

The enzyme pyruvate kinase was one of the first examples where the conformation of bound substrates and the distance between their constituent atoms was analyzed by NMR (nuclear magnetic resonance) studies. Mildvan (1977) in a recent review has summarized the strengths and limitations of the methodology that essentially provides distances from individual atoms to a neighboring paramagnetic reference point (usually a metal or an organic free radical bound in the enzyme active site) by measurement of NMR rates in the presence of that same paramagnet. The NMR method provides distances for ES complexes (or EI complexes) in solution and is complementary to X-ray analyses that provide distances for crystalline ES or EI complexes.

The various distances from a paramagnetic center are evaluated from the effect that an unpaired electron has on the longitudinal relaxation rate of the nearby nuclei (which can be ^{1}H, ^{13}C, or ^{31}P, for example). A variety of experimental conditions must be met if reliable distance calculations are to be made (Mildvan, 1977).

With pyruvate kinase, fifteen separate distances have been measured (Mildvan et al., 1976), allowing construction of the composite picture of intersubstrate distances shown in Figure 7-5. The ATP complex shown is a Cr^{III}–ATP complex—a substitution-inert metal–nucleotide complex that, although inactive in overall catalysis, does facilitate enzymatic enolization of pyruvate. Distances between the paramagnetic Cr^{III} ion and ^{13}C-1, ^{13}C-2, and the methyl protons of bound pyruvate were measured by the nuclear relaxation methods (Gupta et al., 1976). Figure 7-5 indicates that the carbonyl oxygen of pyruvate and the

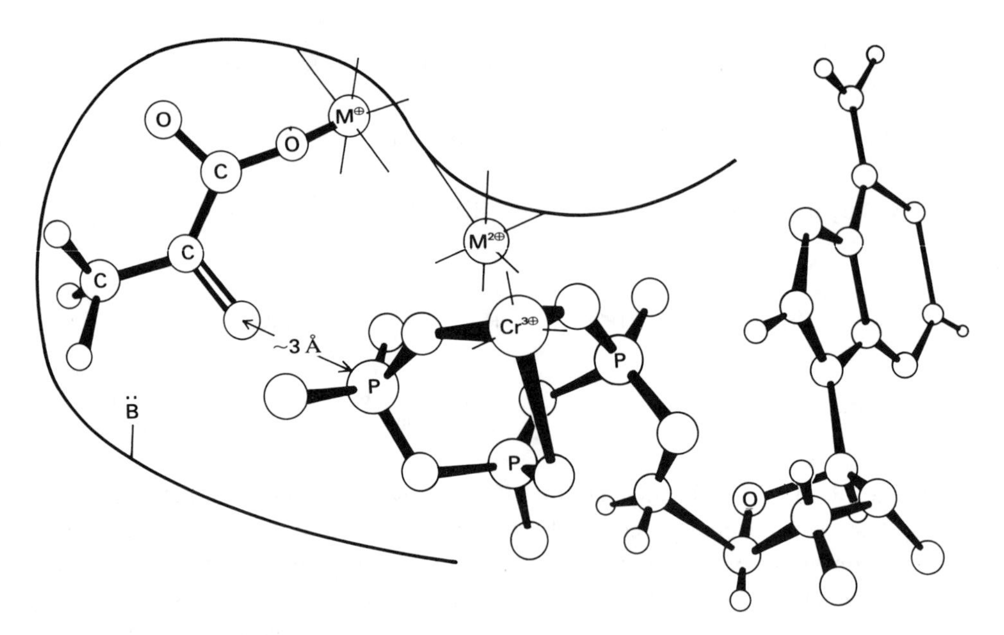

Figure 7-5
Suggested intersubstrate active-site geometry for pyruvate kinase, with chromium-ATP and pyruvate positions indicated. (From Mildvan, 1977, p. 250)

γ-phosphorus of ATP are within 3 Å of each other (essentially within molecular-contact distance), strongly supporting other experiments arguing for direct transfer of the γ-phosphoryl group without phosphoenzyme formation. The placement of the monovalent cation ($K^{\oplus}$, $NH_4^{\oplus}$), also required for full activity of pyruvate kinase, is indicated as chelating the pyruvate carboxylate. We shall discuss other enzyme-bound intersubstrate distances in succeeding chapters—distances also based on such NMR methods.

7.D.2 Nucleoside-Phosphate Kinases

We now consider two different enzymes (adenylate kinase and nucleoside diphosphokinase), both of which catalyze phosphoryl transfer to a nucleoside phosphate acceptor.

7.D.2.a Adenylate kinase

The enzyme adenylate kinase is ubiquitous in free-living organisms; it is commonly called myokinase because of its ready isolation from skeletal muscle. Its physiological role is to take AMP, formed in adenylyl-transfer reactions (cleavage of ATP → AMP), back up to the ADP level (Noda, 1973).

$$\text{Mg}\cdot\text{ATP} + \text{AMP} \xrightleftharpoons{\text{Mg}} \text{Mg}\cdot\text{ADP} + \text{Mg}\cdot\text{ADP}$$

The rabbit-muscle enzyme is a typical adenylate kinase, with ~22,000 mol wt and a specific activity of 2,200 μmoles $\text{min}^{-1}\,\text{mg}^{-1}$, which calculates to a turnover number of 5×10^5 moles $\text{min}^{-1}\,\text{mole}^{-1}$. It is thus a very efficient catalyst and is often a contaminant difficult to remove from other kinases. For instance, for a kinase of specific activity 1 μmole $\text{min}^{-1}\,\text{mg}^{-1}$ and comparable mol wt, one part adenylate kinase in 2,200 parts enzyme gives an equal velocity of reaction by each enzyme (this is a weight contamination of only 0.05%).

All mechanistic studies conducted on adenylate kinases indicate a ternary-complex mechanism, where one of the anionic oxygens of bound AMP attacks the γ-P of an adjacent bound ATP molecule to effect *direct* phosphoryl transfer.

7.D.2.b Nucleoside diphosphokinase

The enzyme nucleoside diphosphokinase probably functions to interconvert any of the four common nucleoside diphosphate–triphosphate pairs:

$$\text{Mg}\cdot\text{ADP} + \text{Mg}\cdot\text{XTP} \rightleftharpoons \text{Mg}\cdot\text{ATP} + \text{Mg}\cdot\text{XDP}$$

where X = adenosine, guanosine, cytidine; or uridine.

One physiological role of the enzyme may be the conversion of ADP back to ATP. The combined consecutive action of adenylate kinase and nucleoside diphosphokinase would be conversion of AMP (a dead end for phosphoryl transfers) back to ATP molecules (utilizable to drive other cellular reactions) at the expense of conversion of an ATP and an XTP molecule to an ADP and an XDP molecule.

In contrast to the adenylate-kinase mechanism, compelling evidence exists that nucleoside diphosphokinase catalyzes phosphoryl transfer by way of an $E—PO_3^{2\ominus}$ intermediate (Parks and Agarwal, 1973). Initial-velocity kinetic experiments give parallel-line kinetics (ping-pong BiBi), and the expected inhibition patterns are observed. An isolable phosphoenzyme derives from incubation of the enzyme with Mg · XTP, and that phosphoenzyme is chemically and kinetically competent in catalysis. Structural studies indicate that $E—PO_3^{2\ominus}$ is an N^3-phosphohistidinyl enzyme.

7.D.3 Acyl Activation: Phosphoryl Transfer to $RCOO^{\ominus}$

A third major mechanistic category of kinases involves carboxylate groups as nucleophiles and represents one of the major biological routes for *acyl-group activation* (¶5.C.3.c). Once the carboxylate group has been chemically activated to an acyl phosphate, the acyl group can be rapidly transferred to a variety of cellular nucleophiles in biosynthetic reactions with expulsion of P_i as leaving group. One example previously discussed (¶5.C.3.c) is glutamine synthetase using a bound γ-glutamyl phosphate attacked by ammonia. Analogously, the bacterial enzyme aspartokinase uses ATP to generate β-aspartyl phosphate as an in vivo product. This acyl phosphate is in turn substrate for a reductase that produces the β-aldehyde by hydride transfer with expulsion of inorganic phosphate. The β-aspartate semialdehyde product is subsequently processed to the amino acids homoserine and threonine.

$$^{\ominus}OOC—CH_2—CH(NH_3^{\oplus})—COO^{\ominus} + ATP^{4\ominus} \xrightarrow[\text{kinase}]{\text{aspartate}} {}^{\ominus}O—P(=O)(O^{\ominus})—O—C(=O)—CH_2—CH(NH_3^{\oplus})—COO^{\ominus} + ADP^{3\ominus}$$

$$\downarrow \text{“}H^{\ominus}\text{,” reductase}$$

$$\underset{\text{threonine}}{H_3C—CH(OH)—CH(NH_3^{\oplus})—COO^{\ominus}} \longleftarrow \underset{\text{homoserine}}{HO—CH_2—CH_2—CH(NH_3^{\oplus})—COO^{\ominus}} \longleftarrow \underset{\text{aspartate semialdehyde}}{HC(=O)—CH_2—CH(NH_3^{\oplus})—COO^{\ominus}} + O{=}P(O^{\ominus})(O^{\ominus})—OH$$

7.D.3.a Acetate activation

The two carbons of acetic acid ($CH_3COO^\ominus$) represent one of the most important *biosynthons* in the cell; the two-carbon fragment is incorporated intact into an enormous number of more complex biological molecules. It is the precursor of all the carbons of fatty acids (C_{12} to C_{30}, typically) and of all the carbons of steroids and other polyisoprenoid compounds. The acetyl group is used as a blocking group in drug detoxification and for a host of biological Claisen condensations (Chapter 23) as well.

However, the initial problem that must be solved by organisms that would utilize acetate is its chemical activation. Again, as a resonance-stabilized carboxylate anion at physiological pH, acetate is not in a usable form as an acylating agent or as a carbanion source. The activation is achieved at the expense of ATP hydrolysis.

We shall see that bacteria use one mechanism, whereas higher organisms use another. We shall examine how "active acetate" is achieved in each case, to see which route is more advantageous for controlled generation of this key biosynthetic intermediate in large amounts.

7.D.3.b Bacterial acetate kinase

Most bacteria have the enzyme acetate kinase, which catalyzes the following reaction:

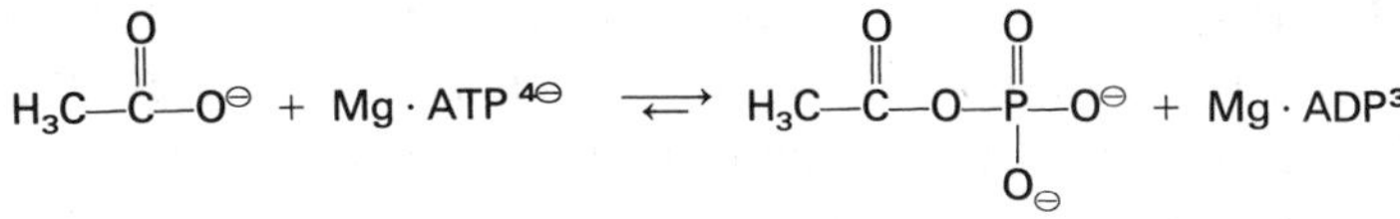

$$H_3C\text{—}\overset{\overset{\Large O}{\|}}{C}\text{—}O^\ominus + Mg\cdot ATP^{4\ominus} \rightleftharpoons H_3C\text{—}\overset{\overset{\Large O}{\|}}{C}\text{—}O\text{—}\underset{\underset{\Large O^\ominus}{|}}{\overset{\overset{\Large O}{\|}}{P}}\text{—}O^\ominus + Mg\cdot ADP^{3\ominus}$$

acetyl phosphate

The product, acetyl phosphate, has a half-life of 10 to 15 hours at pH 7 and 30° C. The equilibrium for the reaction proceeding from left to right is ~0.01, a reflection of the fact that acetyl-P is more destabilized than ATP (Table 7-1).

The enzyme shows ping-pong BiBi initial-velocity kinetics, the expected product-inhibition patterns, and both expected partial exchanges for a phosphoenzyme intermediate that has been identified as an

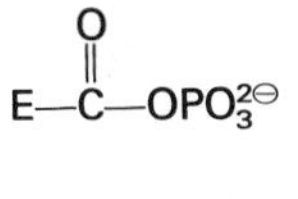

$$E\text{—}\overset{\overset{\Large O}{\|}}{C}\text{—}OPO_3^{2\ominus}$$

species (Anthony and Spector, 1971, 1972; Todhunter and Purich, 1975).

7.D.3.c Phosphotransacetylase

Although action of acetate kinase achieves the primary goal (activation of acetate for subsequent enzyme-catalyzed acetyl transfers), in fact acetyl-P shows too much kinetic instability to be the generally used form of activated acetate in cells, and the equilibrium is hardly favorable in the desired biosynthetic direction. What is needed is to increase the stability of the activated acetate in aqueous solutions kinetically, without giving away the thermodynamic destabilization. This is accomplished by enzymatic transfer of the acetyl group of acetyl-P to a specific thiol acceptor molecule, forming an acylthioester. Acylthioesters retain high group-transfer potentials, but they are relatively resistant to hydrolysis in the pH range 6 to 8. Values of $\Delta G^{0'}$ for acetylthioesters may range from -6 to -10 kcal/mole (-25 to -42 kJ/mole) (Jencks, 1962). *Phosphotransacetylase* is the enzyme responsible for carrying out the acetyl transfer.

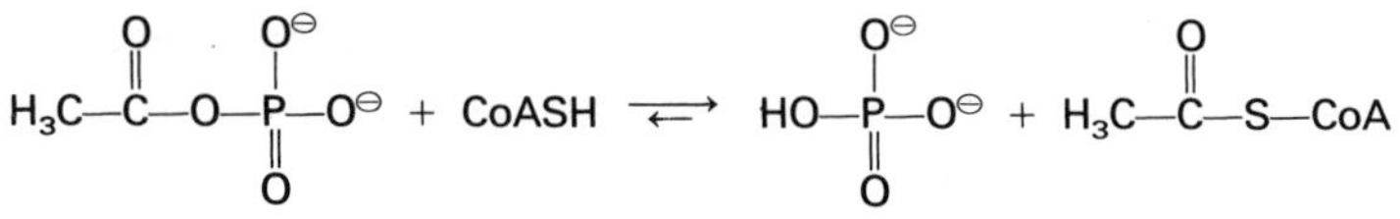

$$H_3C{-}C(=O){-}O{-}P(=O)(O^{\ominus}){-}O^{\ominus} + \text{CoASH} \rightleftharpoons HO{-}P(=O)(O^{\ominus}){-}O^{\ominus} + H_3C{-}C(=O){-}S{-}\text{CoA}$$

coenzyme A acetyl coenzyme A

The thiol acceptor in the transacetylase reaction is coenzyme A (here abbreviated CoASH to emphasize the functional thiol group), a molecule ubiquitously found in living organisms. Its isolation, purification, and structure determination were carried out by F. A. Lipmann and his colleagues in the 1940s—work for which Lipmann shared the 1953 Nobel Prize in medicine. The coenzyme molecule (Fig. 7-6) has two halves in phosphodiester linkage: a 3′,5′-ADP residue, and 4-phosphopantetheine. The phosphopantetheine moiety is itself composed of three structural entities: a branched-chain dihydroxy acid in amide linkage to a β-alanyl residue, which is in turn linked to a cysteamine group. The reactive thiol is in the cysteamine portion.

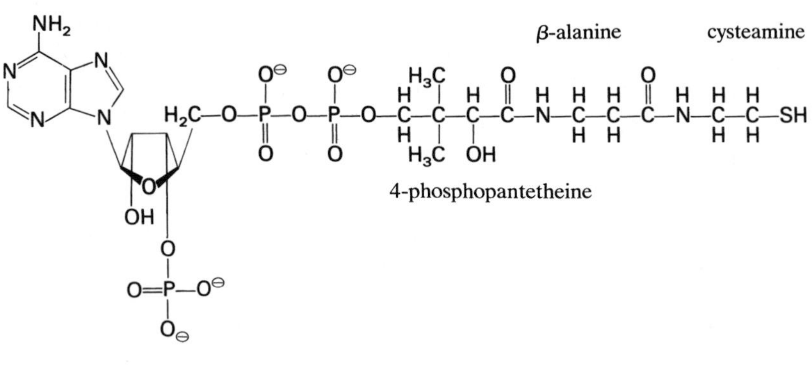

β-alanine cysteamine

4-phosphopantetheine

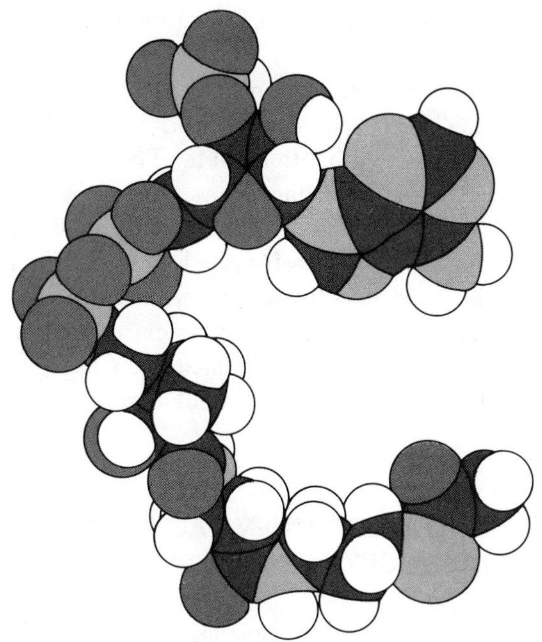

Figure 7-6
Model of acetyl-CoA molecule. (From L. Stryer, *Biochemistry*, p. 367. W. H. Freeman and Company Copyright © 1975.)

We shall write acetyl-CoA as $CH_3CO—S—CoA$, but keep in mind that the CoA moiety is a very large tail to this acylthioester and provides a large surface for binding interaction with the enzymes that use this coenzyme. For $CoASH \rightleftarrows CoAS^{\ominus} + H^{\oplus}$, the $pK_a = 8$, implying a reasonable concentration of thiolate anion (an extremely good nucleophile) at cellular pH values.

Acetyl-CoA is the *common cellular currency for acetyl transfers* (analogous to the role of ATP as the common currency for phosphoryl transfers), and we shall return to acetyl-CoA reactivity, both as an acetylating agent and as a carbanion source, in subsequent chapters. The acetyl transfer catalyzed by phosphotransacetylase appears to be a direct nucleophilic attack of the thiolate anion of $CoAS^{\ominus}$ on acetyl phosphate (Henkin and Abeles, 1976).

The overall stoichiometry of the acetate-kinase–phosphotransacetylase couple is the following:

$$Mg \cdot ATP + H_3CCOO^{\ominus} + CoASH \rightleftarrows Mg \cdot ADP + P_i + H_3C\overset{\overset{\displaystyle O}{\|}}{C}—S—CoA$$

The sequence provides a clearcut example of how ATP can act as a biological dehydrating agent, activating an oxygen anion for elimination via intermediate conversion to a phosphate ester, which is easily displaceable by a thiolate nucleophile.

The *overall equilibrium* for the two-enzyme couple is ~1, because both ATP and acetyl-CoA have $\Delta G^{0'} \approx -7$ kcal/mole (−29 kJ/mole). This fact points up the problem of making acetyl-CoA by this route. For the generation of a key biosynthetic intermediate, this is an inefficient process. A K_{eq} value of 100 to 1,000 would be much more desirable, maximizing the available concentration of acetyl-CoA and lowering the amount of unused and valuable free CoASH.

In this regard, it is noted that this two-enzyme coupled synthesis of acetyl-CoA is restricted to bacteria, and a different system is used in higher organisms that are presumably more advanced and efficient in energy metabolism. This eucaryotic solution to the problem is acetate thiokinase, as discussed in Chapter 8.

7.E AN APPROACH TO STEREOCHEMICAL ANALYSIS OF ENZYMATIC PHOSPHORYL TRANSFERS

The phosphoryl group with its three equivalent oxygens is a torsiosymmetric group (Wimmer and Rose, 1978) and offers no handle to analyze the stereochemical outcome as a tool for mechanistic study. We noted in Chapter 6 that a cyclic phosphothiorate was utilized to probe the stereochemical outcome of RNase action. Substitution of sulfur for one of the phosphate oxygen atoms of ATP has been reported at the γ-P, β-P, and α-P atoms where sulfur is in a nonbridge position to yield ATP-γ-S, ATP-β-S, and ATP-α-S, respectively (Goody et al., 1972; Schlimme et al., 1973; Goody and Eckstein, 1971). Sulfur makes the α-thiophosphate and β-thiophosphate groups asymmetric (but not the γ-thiophosphate), so that ATP-α-S and ATP-β-S are each diastereomeric pairs (e.g., Yount, 1975).

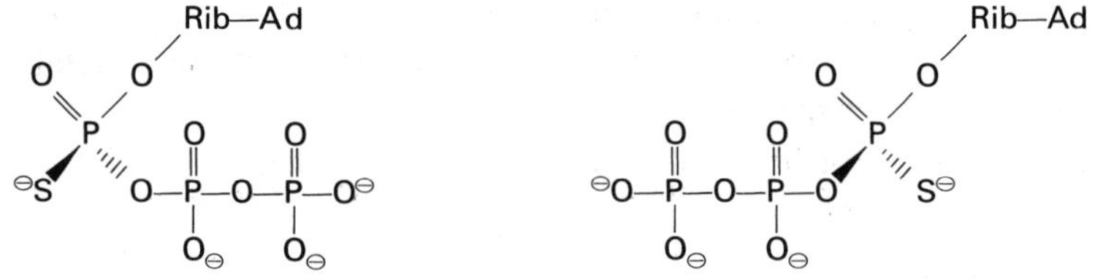

form A (*S*-isomer) form B (*R*-isomer)

ATP-α-S diasteromers

The absolute stereochemistry of ATP-α-S or ATP-β-S isomers has just been determined.* Specific enzymes show kinetic preference for one diastereomeric form. Thus pyruvate kinase will convert 50% of ADP-α-S to one pure diastereomer of ATP-α-S. The opposite diastereomer of ATP-α-S can be obtained with nucleoside diphosphokinase. Similarly, pyruvate kinase and creatine kinase use the different diastereomeric ADP-β-S thiophosphates to make the pure ATP-β-S diastereomers (results reviewed by Yount, 1975).

Two other experiments with clear potential for analyzing the stereochemical outcome of phosphoryl transfer merit mention here. Knowles and colleagues have recently prepared a chiral sample (again, with absolute configuration as yet undetermined) of γ-[^{18}0]-γ-thio-ATP and established that hexokinase, glycerokinase, and pyruvate kinase can transfer the [^{16}O, ^{18}O]-S-P group with chirality intact, and that any pair of transfers leads to overall retention (Orr et al., 1978).

The second line of work, again from Knowles' group, is the asymmetric construction of phosphate monoesters (e.g., glycerol phosphate) with known chirality at phosphorus by virtue of three stable isotopes of oxygen (^{16}O, ^{17}O, and ^{18}O) in the nonbridge position (Abbott et al., 1978). For example,

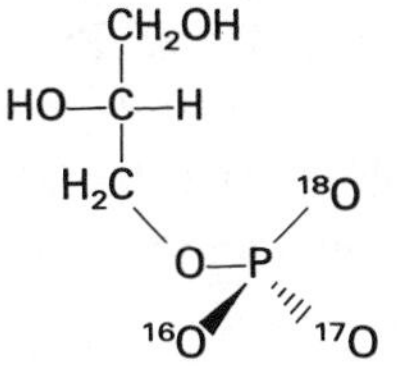

CH_2OH
HO—C—H
H_2C ^{18}O
O—P
^{16}O ^{17}O

This work opens the door for stereochemical analysis of chiral phosphate groups without resorting to thiophosphate groups. It has analogies to chiral [^{1}H,^{2}H,^{3}H]-methyl methodologies described in Chapter 22.

**Note added in proof:* Eckstein and colleagues recently have established the absolute configuration of ATP-α-S diastereomers (Eckstein, personal communication). Pyruvate kinase uses the isomer designated form A on p. 238 (the *S*-isomer). This absolute stereochemical assignment sets the stage for stereochemical analysis in other phosphoryl and nucleotidyl transfers [K. Sheu and P. Frey, *JBC* 253:3378 (1978)].

Chapter 8

Nucleotidyl and Pyrophosphoryl Transfers

In Chapter 7 we discussed examples of enzymes that catalyze attack of various nucleophiles at the electrophilic γ-phosphorus of ATP (and other nucleoside triphosphates)—phosphoryl transfers. In this chapter, we survey the other common modes of attack by nucleophiles on ATP: at the α-phosphorus (constituting nucleotidyl or XMP transfer), and at the β-phosphorus (constituting pyrophosphoryl transfer to the nucleophile). Nucleotidyl transfers are examined first because they represent the *major cleavage pattern for ATP utilized to drive biosynthetic reactions.*

8.A NUCLEOTIDYL TRANSFERS

The generalized stoichiometry for nucleotidyl transfer is the following:

$$\mathrm{XTP} + \mathrm{R\ddot{Y}} \rightleftharpoons \mathrm{R{-}Y{-}\underset{\underset{O^{\ominus}}{|}}{\overset{\overset{O}{\|}}{P}}{-}O{-}CH_2}\text{(ribose, X, OH OH)} + \underbrace{\mathrm{{}^{\ominus}O{-}\underset{\underset{O^{\ominus}}{|}}{\overset{\overset{O}{\|}}{P}}{-}O{-}\underset{\underset{O^{\ominus}}{|}}{\overset{\overset{O}{\|}}{P}}{-}O^{\ominus}}}_{\mathrm{PP_i}}$$

When XTP is ATP, the process is adenylyl transfer (AMP = adenylic acid). The adenylylated nucleophile may be the observed product, but it more often serves as an *activated intermediate* for a subsequent reaction. This type of enzymatic reaction has been cogently reviewed by Stadtman (1973), and the format presented here follows his categorization of the reaction types involved:

1. carboxylate activation;
2. synthesis or metabolism of adenosine phosphodiester derivatives;
3. sugar activation for polysaccharide biosynthesis;
4. phospholipid biosynthesis;
5. sulfate activation, and biosynthesis of imidol adenylate derivatives;
6. adenylylation of enzymes for catalytic or regulatory purposes.

8.A.1 Carboxylate Activation

We have noted (¶5.E.1) that attack of a carboxylate nucleophile can occur at the α-P of ATP in asparagine synthesis. In fact, the adenylyl-transfer mechanism is the principal one for activation of fatty acids and of amino acids to be used for peptide-bond formation.

We have discussed (¶7.D.3) the two-enzyme system (acetate kinase and phosphotransacetylase) used by bacteria to convert acetate to the activated acetyl-CoA. In higher organisms, this conversion is accomplished by one enzyme: *acetate thiokinase.*

$$ATP^{4\ominus} + H_3C{-}\overset{O}{\overset{\|}{C}}{-}O^{\ominus} + CoASH \rightleftarrows H_3C{-}\overset{O}{\overset{\|}{C}}{-}S{-}CoA + AMP^{2\ominus} + PP_i^{4\ominus}$$

The expected enzyme-bound intermediate would be acetyl-AMP, and a variety of experimental evidence has accrued for its existence. When acetate labeled in the carboxylate oxygens with ^{18}O is employed as substrate, AMP containing one atom of ^{18}O is generated as product. This is the expected result of attack by $CoAS^{\ominus}$ on an acetyl-AMP intermediate:

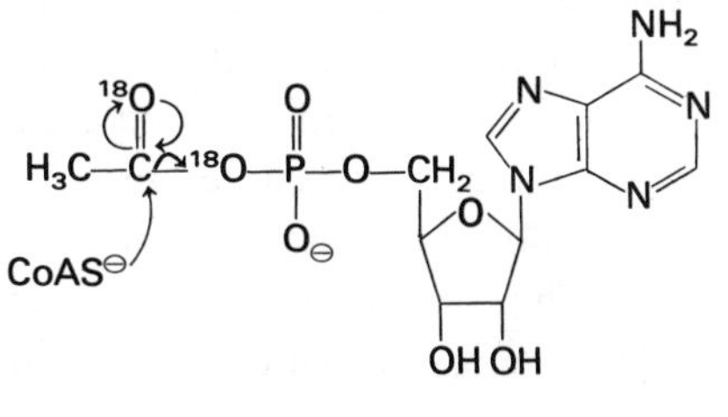

$$CoAS^{\ominus} \rightarrow H_3C{-}\overset{^{18}O}{\overset{\|}{C}}{-}^{18}O{-}\overset{O}{\overset{\|}{P}}(O^{\ominus}){-}O{-}CH_2\text{-ribose(OH, OH)-adenine}(NH_2)$$

Two partial isotopic exchanges suggestive of reversible formation of acetyl adenylate are detectable:

1. $[^{32}P]$-$PP_i \rightleftarrows$ ATP exchange (dependent on acetate, but not on CoASH);
2. $[^{3}H]$-CoA $\rightleftarrows$ acetyl-CoA-$[^{3}H]$ exchange (dependent on AMP).

In particular, the exchange of radioactivity from labeled pyrophosphate into ATP, requiring the presence of the cosubstrate for activation, is the *diagnostic exchange for an adenylate derivative as intermediate.*

A third experiment is the direct detection of enzyme-bound acetyl-AMP by gel-filtration methods. Enzyme, $[^{3}H]$-β,γ-$[^{32}P]$-ATP, and $[^{14}C]$-acetate are mixed for a short time; the enzyme and small molecules then are rapidly separated by molecular-exclusion chromatography (¶3.B.8). The enzyme peak contains stoichiometric amounts of ^{14}C- and ^{3}H-radioactivity, but has no ^{32}P-radioactivity, indicating that the β-P and γ-P of the ATP are no longer present. When acetate thiokinase is incubated singly with either radioactive acetate or the doubly labeled ATP, no radioactivity elutes with the enzyme.

At this juncture, we can ask whether there is an obvious advantage for an organism to activate acetate by phosphoryl transfer from ATP. There is no obvious thermodynamic advantage; the equilibrium favoring hydrolysis is the same for phosphoryl transfer (1) as for nucleotidyl transfer (2):

$$^{\ominus}O{-}\overset{O}{\overset{\|}{P}}(O^{\ominus}){-}O{-}\overset{O}{\overset{\|}{P}}(O^{\ominus}){-}O{-}\overset{O}{\overset{\|}{P}}(O^{\ominus}){-}O{-}Rib{-}Ad \ (+\ R\ddot{X}) \longrightarrow R{-}X{-}\overset{O}{\overset{\|}{P}}(O^{\ominus}){-}O^{\ominus} + ADP^{3\ominus} \qquad \Delta G^{0\prime} \approx -7\ \text{kcal/mole} \qquad (1)$$

$$^{\ominus}O{-}\overset{O}{\overset{\|}{P}}(O^{\ominus}){-}O{-}\overset{O}{\overset{\|}{P}}(O^{\ominus}){-}O{-}\overset{O}{\overset{\|}{P}}(O^{\ominus}){-}O{-}Rib{-}Ad \ (+\ \ddot{X}R) \longrightarrow R{-}X{-}\overset{O}{\overset{\|}{P}}(O^{\ominus}){-}O{-}Rib{-}Ad + PP_i^{4\ominus} \qquad \Delta G^{0\prime} \approx -7\ \text{kcal/mole} \qquad (2)$$

In each case, $\Delta G^{0\prime} \approx -7$ kcal/mole (-29 kJ/mole). What other advantage might there be to the nucleotidyl transfer?

One answer lies in consideration of the methods an organism has for displacing the equilibria of enzyme-catalyzed reactions. The simplest way to displace an equilibrium is to remove one of the products. For instance, in the two-enzyme couple of acetate kinase and phosphotransacetylase, the acetyl-P generated in low equilibrium amounts by acetate kinase is captured by CoASH in the phosphotransacetylase reaction, thus displacing the unfavorable equilibrium of the acetate-kinase reaction. The amount this equilibrium can be displaced depends on the $\Delta G^{0\prime}$ of acetyl-CoA, which happens to be the same as $\Delta G^{0\prime}$ of ATP

(-7 kcal/mole), so that the overall K_{eq} for bacterial formation of acetyl-CoA is approximately 1.

It turns out that all free-living organisms contain a very active enzyme that can displace the equilibria of nucleotidyl-transfer reactions. This enzyme is inorganic pyrophosphatase (PP_iase), which accelerates the hydrolysis of PP_i (to 2 P_i) so effectively that the intracellular concentration of PP_i is generally $<10^{-6}$ M (Josse and Wong, 1971; Butler, 1971).

$$^{\ominus}O-P(=O)(O^{\ominus})-O-P(=O)(O^{\ominus})-O^{\ominus} + H\ddot{O}H \xrightarrow{PP_i ase} 2\ HO-P(=O)(O^{\ominus})-O^{\ominus}$$

For any nucleotidyl transfer generating AMP + PP_i in the presence of PP_iase, (for example, with acetate thiokinase and PP_iase), we can write the extended stoichiometry as follows:

$$ATP^{4\ominus} + H_3C-C(=O)-O^{\ominus} + CoASH \rightleftharpoons H_3C-C(=O)-S-CoA + AMP^{2\ominus} + 2\ P_i^{2\ominus}$$

Now two phosphoric-anhydride bonds of ATP have been hydrolyzed ($\Delta G^{0'} = -14$ kcal/mole), forming one acetyl-CoA product molecule ($\Delta G^{0'} = +7$ kcal/mole). So the net change in $\Delta G^{0'}$ is -7 kcal/mole. Setting -7 kcal/mole $= -RT \ln K_{eq}$, we obtain a value for K_{eq} of more than 10^4, indicating essentially irreversible formation of acetyl-CoA.

Note that there is no corresponding ADPase in the bacterial cells to drive the equilibrium in favor of acetyl-CoA synthesis for the bacterial two-enzyme system.

In general, then, higher organisms use nucleotidyl transfers (rather than phosphoryl transfers) in the biosynthesis of crucial metabolic intermediates because the action of pyrophosphatase (PP_iase) on the PP_i product ensures functional unidirectionality under in vivo conditions.

The activation of amino acids for peptide-bond formation in protein synthesis also involves adenylyl transfer, with the expenditure of one ATP per amino acid activated. There are specific activating enzymes for each of the twenty amino acids incorporated into proteins. These enzymes are called aminoacyl-tRNA synthetases, because the cosubstrate is a specific tRNA (transfer RNA) molecule (Soll and Schimmel, 1974). Each enzyme shows the typical two steps of the nucleotidyl-transfer mechanism: (1) *activation* of the amino acid as the aminoacyl-AMP, and (2) *transfer* of the aminoacyl group to the 2′-hydroxyl or the 3′-hydroxyl of the ribose ring of the terminal adenosine of the tRNA molecule, which is 70 to 80 bases in length (Sprinzl and Cramer, 1975; Hecht and Chinault, 1976; Watson, 1975; Boyer, 1970–1976, vol. 10, chaps. 1–3). The aminoacyl-tRNA product is thus an aminoacyl ester. We illustrate the process with alanyl-tRNA synthetase, where the 2′-hydroxyl is shown as nucleophile in the transfer step (2). Intramolecular migration of the aminoacyl group between the 2′- and 3′-hydroxyl groups occurs very rapidly.

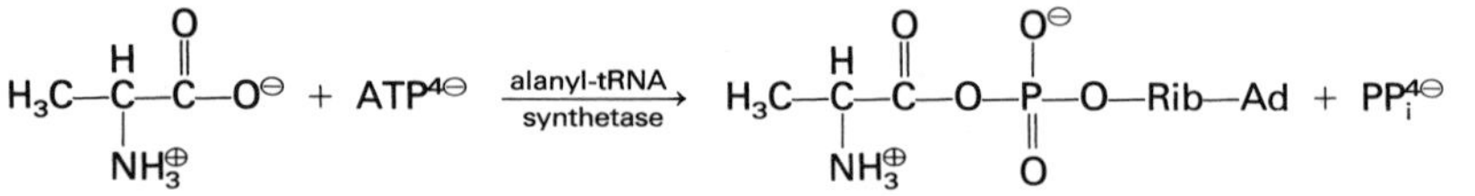

$$H_3C{-}CH(NH_3^{\oplus}){-}C(=O){-}O^{\ominus} + ATP^{4\ominus} \xrightarrow[\text{synthetase}]{\text{alanyl-tRNA}} H_3C{-}CH(NH_3^{\oplus}){-}C(=O){-}O{-}P(=O)(O^{\ominus}){-}O{-}Rib{-}Ad + PP_i^{4\ominus} \quad (1)$$

tRNA^{Ala}

initial formation of 2′-ester

10^3 sec^{-1}

+ AMP

formation of mixture of 2′- and 3′-forms of aminoacyl-tRNA

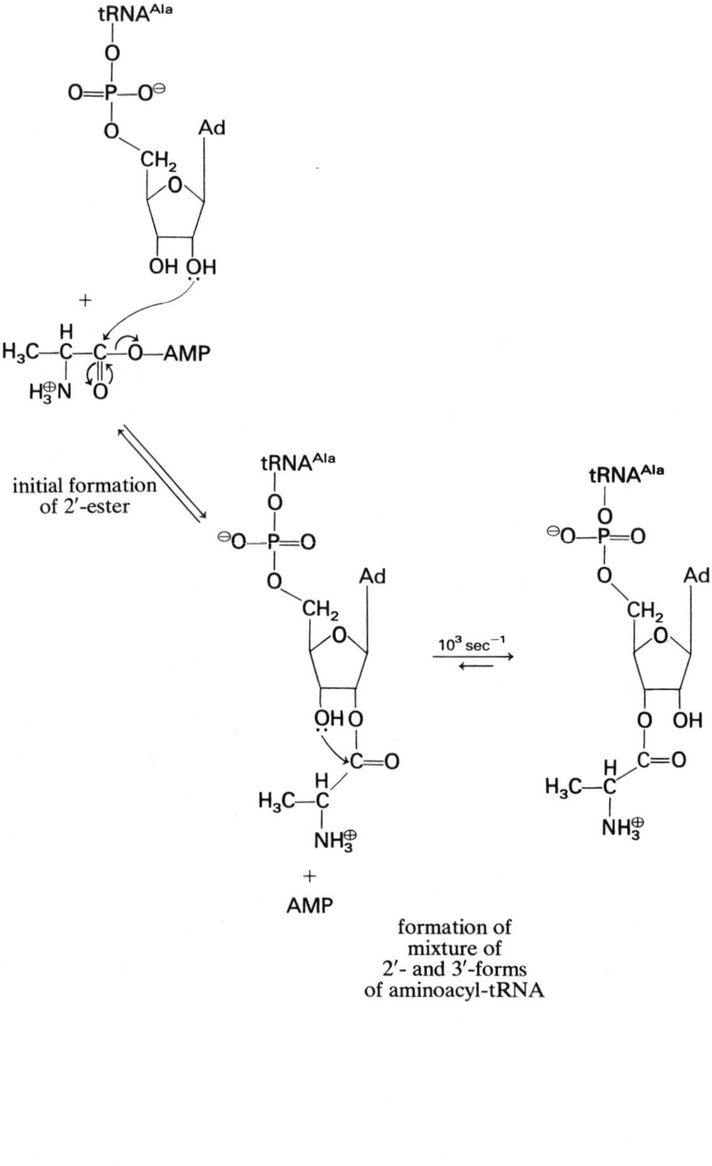

(2)

Two such activated aminoacyl esters can bind next to each other on the ribosomes and undergo peptide-bond formation (Watson, 1975). The peptide-bond–forming step can be illustrated for a typical bacterial protein, where an *N*-formyl methionine is the terminal residue and alanine is the next residue. Details of this spatially and temporally complex process have been elucidated and can be found elsewhere (Watson, 1975). Our present purpose is merely to indicate where the driving force for peptide-bond formation in protein synthesis obtains. The leaving group in this example is the deacylated $tRNA^{fMet}$.

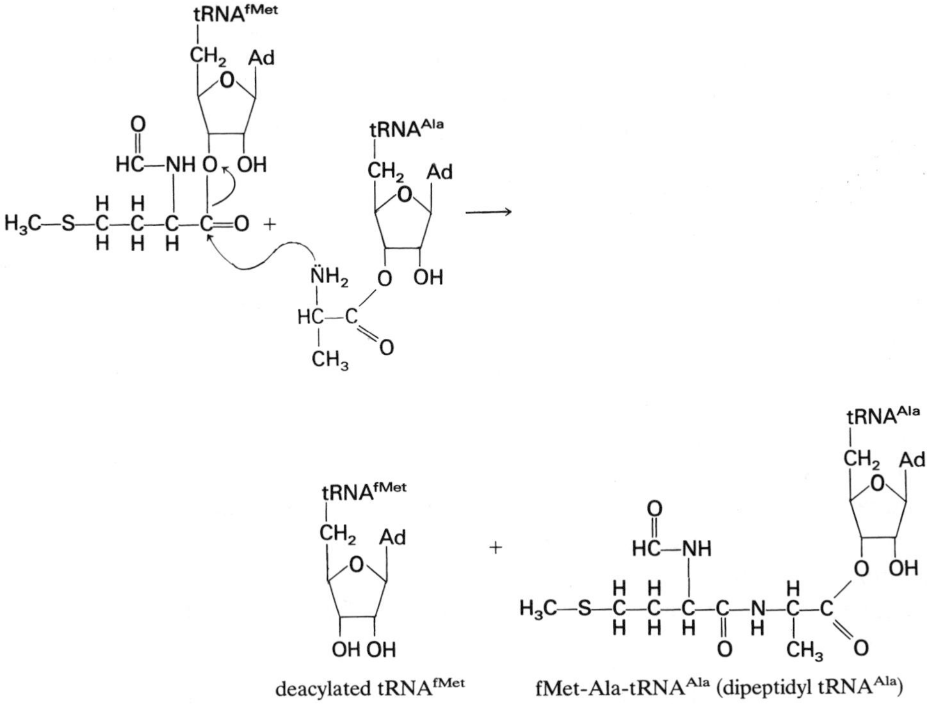

deacylated $tRNA^{fMet}$ + fMet-Ala-$tRNA^{Ala}$ (dipeptidyl $tRNA^{Ala}$)

One thing is particularly worth noting about the specificity of the enzymes involved in amino-acid activation and peptide-bond formation. The linear sequence of amino acids in any given protein is genetically determined, with a given mRNA (messenger RNA) sequence providing template insertion information for each aminoacyl-tRNA species. Insertion of an incorrect amino-acid residue at a given locus of the nascent peptide chain will represent an error and generally will be deleterious to the functioning of that particular protein molecule. For instance, replacement of serine by some other amino acid at position 195 in the chymotrypsinogen chain will have cataclysmic effects for the subsequent activity of that chymotrypsin molecule.

Therefore, fidelity in protein biosynthesis is just as important as fidelity in DNA (or RNA) replication, which we mentioned briefly at the end of Chapter 6. There are two obvious steps where biological editing can occur to ensure accuracy of amino-acid incorporation: (1) during action of aminoacyl-tRNA synthetase; and (2) in the peptide-bond–forming step at the ribosome. If a given tRNA molecule (e.g., $tRNA^{Ala}$) has been misacylated (e.g., with serine to yield seryl-$tRNA^{Ala}$), it is unlikely that the mistake will be correctable in the peptide-bond–forming step, because the $tRNA^{Ala}$–mRNA interaction is the major basis for specificity. Therefore, the chief editing function must be at the level of aminoacyl-tRNA synthetase.

Note that we have written a two-step mechanism for action of aminoacyl-tRNA synthetase. In principle, specificity could be imposed both at the activation step (when the aminoacyl adenylate is formed) and at the transfer step (when the aminoacyl moiety is transferred to the ribose hydroxyl of the terminal adenosine residue of the specific tRNA acceptor molecule).

A third possibility is that, once a misacylated tRNA molecule has been formed, the synthetase might selectively deacylate it in a hydrolytic step. This possibility has been examined with a number of purified aminoacyl-tRNA synthetases, with the following conclusions (Eldred and Schimmel, 1972). Occasional errors do occur in the acyl-activation step—especially for enzymes such as isoleucyl-tRNA synthetase, which will activate valine (differing from isoleucine by only one methylene group) on the average once for each hundred times it activates isoleucine, an estimated error rate of 1% in this step. The expected difference in binding energy for that one CH_2 group is 2 to 3 kcal/mole (8 to 13 kJ/mole). If this increased interaction energy for isoleucine and enzyme were all converted into specificity, a selectivity of around 10^2 is the most one could reasonably expect. Indeed, the isoleucyl enzyme will produce bound valyl-AMP:

$$(H_3C)_2CH{-}CH(NH_3^{\oplus}){-}C(=O){-}O^{\ominus} \longrightarrow (H_3C)_2CH{-}CH(NH_3^{\oplus}){-}C(=O){-}O{-}AMP$$

valine → valyl-AMP

$$H_3C{-}CH_2{-}CH(CH_3){-}CH(NH_3^{\oplus}){-}C(=O){-}O^{\ominus} \longrightarrow H_3C{-}CH_2{-}CH(CH_3){-}CH(NH_3^{\oplus}){-}C(=O){-}O{-}AMP$$

isoleucine → isoleucyl-AMP

One can then ask to what extent this initial error in the activation step is propagated through the transfer step to yield detectable amounts of valyl-$tRNA^{Ile}$. In fact, when $tRNA^{Ileu}$ is added to the valyl-AMP and isoleucyl-tRNA synthetase, there is quantitative hydrolysis to valine and AMP (Baldwin and Berg, 1960). No misacylated tRNA accumulates. This editing result could be due directly to selective hydrolysis of the valyl-AMP before transfer to tRNA, or perhaps the valyl-$tRNA^{Ileu}$ could be formed and then quantitatively hydrolyzed by the enzyme (Eldred and Schimmel, 1972). Direct evidence for hydrolytic editing before release of misacylated tRNA molecules from the synthetase active site has been obtained by elegant rapid-quenching experiments examining the rejection of the isosteric amino acid threonine by valyl-tRNA synthetase from *B. stearothermophilus* (Fersht and Kaethner, 1976).

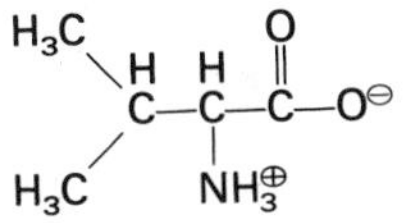

L-valine 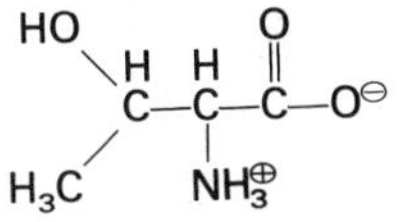 L-threonine

The valyl-tRNA synthetase forms 1:1 complexes of valyl-AMP or threonyl-AMP when mixed with ATP and the respective amino acids. On addition of $tRNA^{Val}$, transfer of the valyl group from valyl-AMP to the tRNA hydroxyl group proceeds with a rate constant of 12 sec^{-1} at 25° C. This valyl-$tRNA^{Val}$ is hydrolyzed at a 1,000-fold slower rate, 0.015 sec^{-1}. So valyl-$tRNA^{Val}$ accumulates.

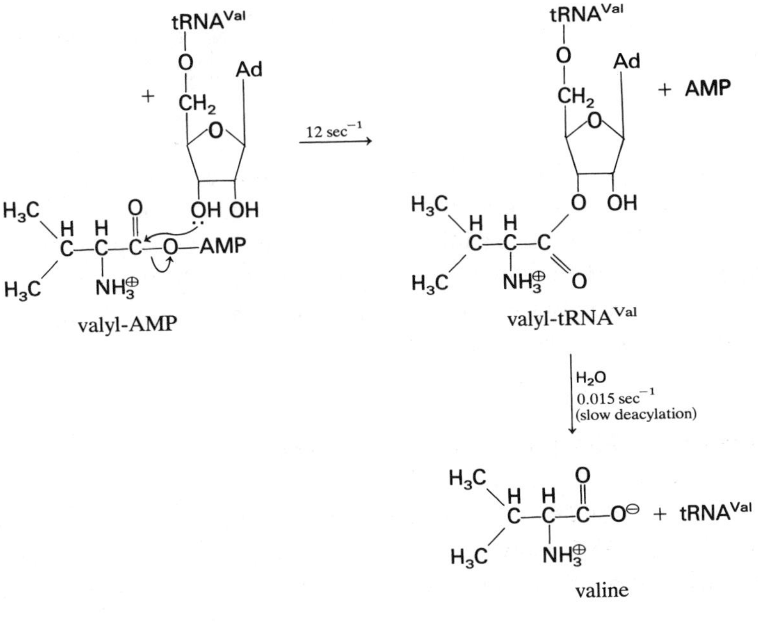

valyl-AMP + $tRNA^{Val}$ $\xrightarrow{12\ sec^{-1}}$ valyl-$tRNA^{Val}$ + AMP

valyl-$tRNA^{Val}$ $\xrightarrow[\text{(slow deacylation)}]{H_2O,\ 0.015\ sec^{-1}}$ valine + $tRNA^{Val}$

valyl-AMP valyl-$tRNA^{Val}$ valine

The fate of threonyl-AMP on addition of $tRNA^{Val}$ is quantitatively distinct. Aminoacyl transfer does occur rapidly, with a rate constant of 36 sec^{-1}. However, hydrolysis is even faster, at 40 sec^{-1}. Thus, threonyl-$tRNA^{Val}$ is actually formed threefold faster than the cognate valyl-$tRNA^{Val}$, but it is then hydrolyzed 3,000-fold faster than hydrolysis of the correctly acylated species. Therefore, there is a transient accumulation of threonyl-$tRNA^{Val}$ as bound product, but it then is hydrolyzed before significant release.

threonyl-AMP + $tRNA^{Val}$ $\xrightarrow{36\ sec^{-1}}$ threonyl-$tRNA^{Val}$ + AMP

threonyl-$tRNA^{Val}$ $\xrightarrow[\text{(fast deacylation)}]{H_2O,\ 40\ sec^{-1}}$ threonine + $tRNA^{Val}$

The misacylated threonyl-$tRNA^{Val}$ does not accumulate. Note that the process of misacylation and its correction is an energy-consuming one in that ATP is hydrolyzed to AMP + PP_i. The editing process occurs in two steps: initial discrimination in activation of an amino acid, and subsequent selective hydrolysis of the misacylated tRNA. The two chances to impose specificity lower the error rates for protein biosynthesis in vivo, multiplicatively.

8.A.2 Adenosine Phosphodiester Derivatives

The enzymes responsible for biosynthesis of the informational macromolecules DNA and RNA (DNA polymerases and RNA polymerases) incorporate 2′-deoxyribonucleotides or ribonucleotides, respectively, by nucleotidyl transfers.

Again, these are exceedingly complex enzymatic catalysts whose mechanisms cannot be discussed adequately here (see Kornberg, 1974; Gefter, 1975). We can give a skeletal representation of the stoichiometry of the RNA-polymerase reaction:

$$m\,\text{ATP} + n\,\text{UTP} + x\,\text{CTP} + y\,\text{GTP} \xrightarrow[\text{Mn}^{2\oplus},\,\text{Mg}^{2\oplus}]{\text{DNA template}} \text{RNA} + (m + n + x + y)\text{PP}_i$$

However, this expression gives no real hint of the complexity of this catalysis in terms of template specificity and movement of the enzyme (or the nascent RNA chain) during formation of the RNA polymer (see Chambon, 1974; Chamberlin, 1974).

Furthermore, there probably are multiple types of DNA and RNA polymerases in cells, each with overlapping specificities and functions. For instance, in *E. coli* cells the DNA polymerase I mentioned in ¶6.B, a 109,000-dalton Zn-enzyme, is present in about 400 copies per cell (Gefter, 1975; Kornberg, 1974). It has associated $5' \rightarrow 3'$ and $3' \rightarrow 5'$ exonuclease activities that may serve editing and error-avoidance functions. (The error rate for incorporating a noncomplementary mononucleotide is less than 1 in 10^5 residues.) DNA polymerase II of *E. coli* is a 90,000-dalton polypeptide, present in 40 copies per cell. It has only $3' \rightarrow 5'$ exonuclease activity and shows about one-tenth the turnover rate of polymerase I. DNA polymerase III is present in only ten copies per cell but may be the major polymerase in DNA replication. It has both $5' \rightarrow 3'$ and $3' \rightarrow 5'$ exonuclease activities. Other proteins and enzymes in addition to polymerases are required for DNA replication; these include polynucleotide ligase, DNA superhelix-relaxation proteins and unwinding proteins, as well as gene products of as-yet-unspecified function (Gefter, 1975).

E. coli DNA polymerase I has been studied with paramagnetic $Mn^{2\oplus}$ in place of $Zn^{2\oplus}$, distances have been mapped to bound ATP and bound TTP, and a computer-generated picture (Fig. 8-1) has been presented to suggest the active-site conformation of bound DNA double helix and adding XMP residue (Sloan et al., 1975). Mildvan (1977) has suggested that, if this model is valid, the stereochemical outcome at the α-P of the nucleoside monophosphate being added to the growing chain via nucleotidyl transfer is inversion—an in-line mechanism akin to that for ribonuclease A (¶6.A.2.c).

A second kind of enzyme involved with adenosine phosphodiester derivatives is the enzyme adenyl cyclase, responsible for converting ATP into the protean intracellular regulatory substance 3′,5′-cyclic AMP. The soluble enzyme has been purified to homogeneity from some bacteria (Umezawa et al., 1974), but the mammalian enzyme is membrane-bound (in keeping with its sensitivity to hormonal activation) and has been more difficult to purify. If the cyclization is direct,

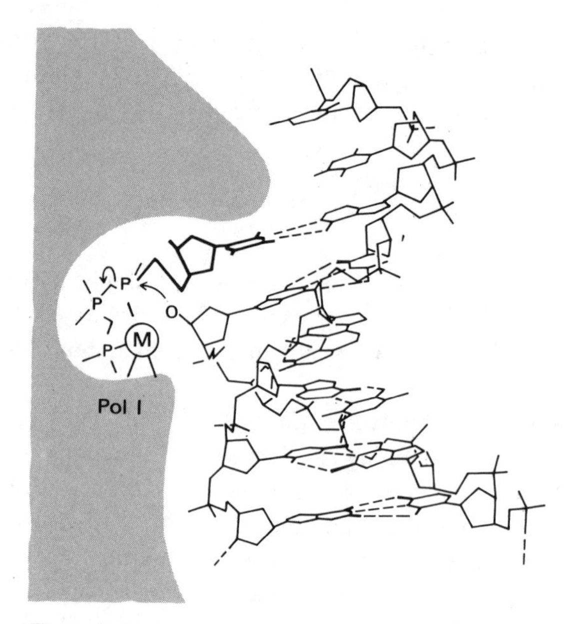

Figure 8-1
Postulated active-site geometry for nucleotidyl-transfer reaction catalyzed by *E. coli* DNA polymerase I. (From Mildvan, 1977, p. 251)

without covalent enzyme intermediates, it must involve specific binding of ATP at the active site in a conformation such that the 3′-OH of the ribose can approach the α-phosphorus.

$$\text{ATP} \xrightarrow[\text{Mg}^{2\oplus}]{\text{cyclase,}} \text{3}',\text{5}'\text{-cAMP} + \text{PP}_i$$

ATP   3′,5′-cAMP PP_i

A third type of enzyme is involved in the biogenesis of adenosine-phosphodiester linkages in various coenzymes that have an AMP residue in phosphodiester bridge to the business end of the cofactor. We illustrate this type with synthesis of dephosphocoenzyme A (deP-CoA), where a phosphate oxygen of 4-phosphopantotheine is nucleophile and becomes adenylylated during catalysis.

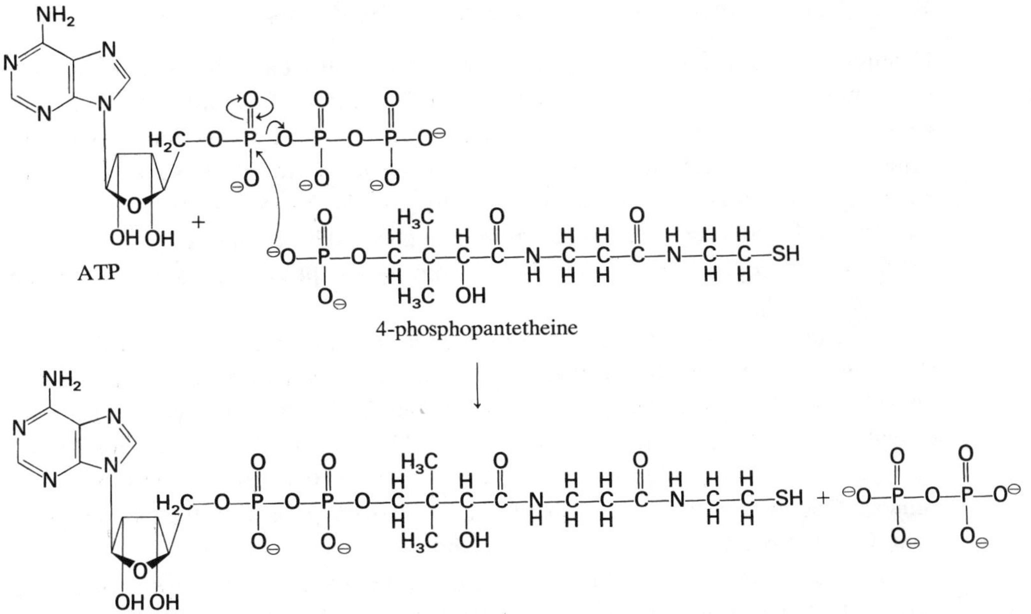

ATP

4-phosphopantetheine

dephosphocoenzyme A

Subsequent phosphorylation (a kinase reaction) at the 2′-OH of the ribose group converts dephospho-CoA into CoA.

Two important redox coenzymes, responsible for the bulk of enzyme-catalyzed dehydrogenation reactions, are NAD (nicotinamide adenine dinucleotide) and FAD (flavin adenine dinucleotide). They are built up by similar adenylyl transfers to the indicated phosphate oxygens of NMN (nicotinamide mononucleotide) and FMN (flavin mononucleotide), respectively. (NAD and FAD are discussed in Chapters 10 and 11.)

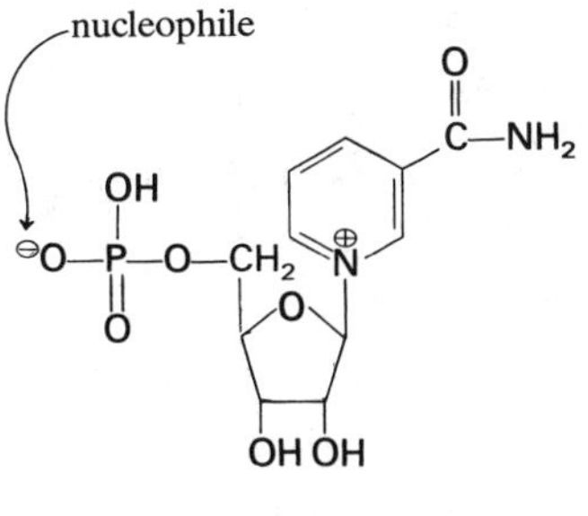

nucleophile

nucleophile

NMN

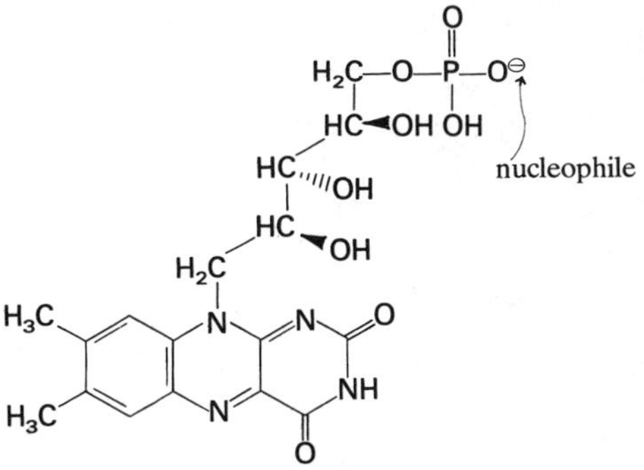

FMN

8.A.3 Sugar Activation for Polysaccharide Synthesis

Chapter 9 deals with some aspects of glycosyl-transfer reactions, and subsequent chapters give examples of the metabolic transformation of activated sugars. Here we simply outline some basic aspects of enzyme-catalyzed *activation of sugars.* In some analogy to the unactivated acyl carboxylates, the free sugars are not chemically activated for condensation into polysaccharides (where linkage of monomers through the C-1 oxygen atom is typical). The anomeric hydroxyl is *not* readily eliminated. It must first be derivatized into a species capable of providing a stable leaving group.

Activation of the common hexoses such as glucose or galactose involves two steps. In the first step, the sugars are enzymatically converted to the hexose-1-phosphates. We have noted (¶7.D.1) that galactokinase phosphorylates galactose directly at C-1 to generate α-galactose-1-P. Hexokinase, however, phosphorylates glucose at C-6. Glucose-1-P can be formed either from glucose-6-P in a mutase reaction (discussed in Chapter 20) or from phosphorolysis of polysaccharides (discussed in Chapter 9). The product is α-glucose-1-P.

In the second and specific step in sugar activation, the α-isomers of the hexose-1-phosphates undergo a *nucleotidyl transfer.* Various nucleoside triphosphates act as cosubstrates, depending on the organism, on the specific sugar-1-P, and on the specific enzyme. In glucose metabolism in animals, UTP (uridine triphosphate) is the usual cosubstrate, whereas ATP may be preferred in bacterial and plant cells. The products from UTP reaction with the specific enzymes that process glucose-1-P and galactose-1-P are UDP-glucose and UDP-galactose, respectively.

UTP + α-glucose-1-P $\longrightarrow$ PP_i + UDP-glucose

The sugar-1-phosphate oxygen is the ultimate nucleophile, attacking the α-P of the nucleotide. Pyrophosphate is the coproduct. Thus, even though UDP-glucose (and other XDP-hexoses) has a nucleoside-diphosphate moiety, this group is formed by nucleotidyl transfer to a phosphate nucleophile, as can be proved by a variety of isotopic labeling experiments.

In any nucleotidyl-transfer reaction catalyzed by an enzyme, two mechanistic possibilities exist: covalent catalysis or direct nucleotidyl transfer (in analogy to the possibilities with phosphoryl-transfer enzymes). For most nucleotidyl-transfer enzymes examined (including the UDP-glucose synthetase just discussed), no evidence has accrued for an AMP-enzyme or a UMP-enzyme formed by attack of an enzyme nucleophile at the α-P of ATP or UTP prior to subsequent attack by the specific cosubstrate. However, recent experiments on the UDP-galactose synthetase provide clear evidence for a uridylyl–enzyme intermediate from ping-pong initial-velocity kinetics, product-inhibition kinetics, partial exchange data, and direct isolation (Wong and Frey, 1974). Thus,

$$\text{Enz—X} + \text{UTP} \rightleftharpoons \text{Enz—X—UMP} + \text{PP}_i$$

$$\text{Enz—X—UMP} + \alpha\text{-galactose-1-P} \rightleftharpoons \text{Enz—X} + \text{UMP—P—galactose}$$

8.A.4 Phospholipid Biosynthesis

Phospholipids are major constituents of biological membranes. Many of them have glycerol backbones, with fatty acids esterified at two alcoholic positions of the glycerol to provide hydrophobicity. The third hydroxyl of glycerol usually is derivatized as a phosphodiester. The other group (R″) on the phosphodiester is a hydrophilic species that may be neutral (e.g., inositol), zwitterionic (serine), cationic (choline, ethanolamine), or anionic (glycerol phosphate). Dioleylphosphatidyl choline is a specific example of a phospholipid.

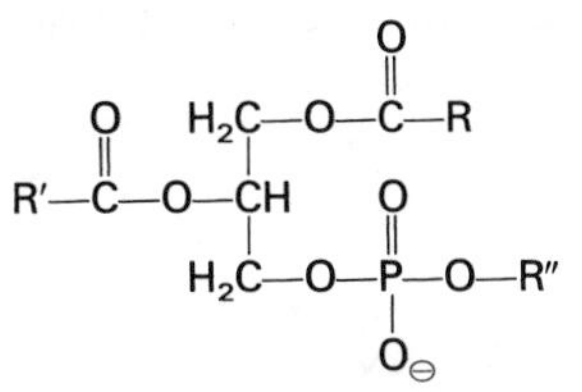

```
                        O
                        ‖
       O    H2C—O—C—R
       ‖      |
R′—C—O—CH     O
              |       ‖
            H2C—O—P—O—R″
                      |
                      O⊖
```

phospholipid

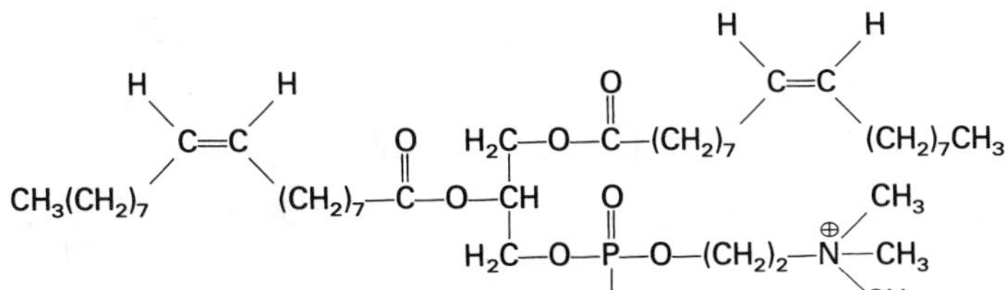

```
                                                   H         H
                                                     \      /
  H          H                         O              C=C
   \        /                          ‖             /    \
    C=C                  O   H2C—O—C—(CH2)7               (CH2)7CH3
   /    \                ‖     |
CH3(CH2)7    (CH2)7—C—O—CH     O                  CH3
                               |       ‖                   ⊕/
                             H2C—O—P—O—(CH2)2—N—CH3
                                       |                      \
                                       O⊖                      CH3
```

dioleylphosphatidyl choline

The enzymatic strategy for phospholipid biosynthesis has analogies to the sugar activation discussed in the preceding section. Choline is first phosphorylated by choline kinase to produce choline phosphate. This phosphate monoester then acts as nucleophile in a nucleotidyl transfer. Whereas UTP is the common nucleoside triphosphate for sugar activation, *CTP (cytidine triphosphate) is used in lipid metabolism.*

$$(H_3C)_3\overset{\oplus}{N}\text{—}CH_2\text{—}CH_2\text{—}OH + ATP^{4\ominus} \rightleftharpoons (H_3C)_3\overset{\oplus}{N}\text{—}CH_2\text{—}CH_2\text{—}O\text{—}PO_3^{2\ominus} + ADP^{3\ominus} \qquad (1)$$

choline — choline phosphate (choline-P)

$$\text{Choline-P} + CTP^{4\ominus} \rightleftharpoons (H_3C)_3\overset{\oplus}{N}\text{—}CH_2\text{—}CH_2\text{—}O\text{—}P(O)(O^{\ominus})\text{—}O\text{—}P(O)(O^{\ominus})\text{—}O\text{—}CH_2\text{-ribose(OH, OH)-cytosine}(NH_2) + PP_i^{4\ominus} \qquad (2)$$

CDP-choline

Note that reactions 1 and 2 represent consecutive phosphoryl and nucleotidyl transfers in a biosynthetic sequence. The CDP-choline product can then be attacked by the hydroxyl group of a diacylglycerol to produce the phosphatidyl choline and CMP as leaving group. The net cleavage of CTP to CMP + PP_i is apparent at this point.

$$(H_3C)_3\overset{\oplus}{N}\text{—}CH_2\text{—}CH_2\text{—}O\text{—}P(O)(O^{\ominus})\text{—}O\text{—}P(O)(O^{\ominus})\text{—}O\text{—Rib—Cyt} + HO\text{—}CH_2\text{—}HC(O\text{—}C(O)\text{—}R')\text{—}CH_2\text{—}O\text{—}C(O)\text{—}R \longrightarrow {}^{\ominus}O\text{—}P(O)(O^{\ominus})\text{—}O\text{—Rib—Cyt} + R\text{—}C(O)\text{—}O\text{—}H_2C\text{—}CH(O\text{—}C(O)\text{—}R')\text{—}H_2C\text{—}O\text{—}P(O)(O^{\ominus})\text{—}O\text{—}CH_2\text{—}CH_2\text{—}\overset{\oplus}{N}(CH_3)_3$$

phosphatidyl choline

In some organisms, a variation on this chemistry is directed by the biosynthetic enzymes. For instance, phosphatidyl serine can be synthesized via a CDP-diglyceride intermediate. This intermediate can arise by attack of a diacylglycerol monophosphate as nucleophile in the cytidylyl-transfer reaction.

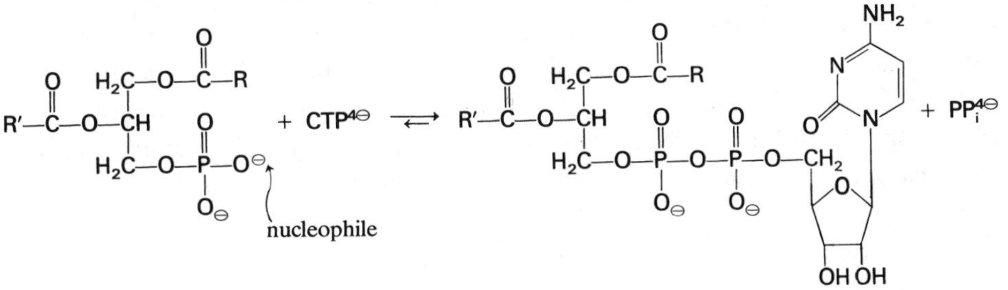

$$\text{diacylglycerol monophosphate} + CTP^{4-} \rightleftharpoons \text{CDP-diglyceride} + PP_i^{4-}$$

nucleophile

CDP-diglyceride

The CDP-diglyceride can be attacked by a serine hydroxyl group (presumably with the assistance of general base catalysis carried out by phosphatidyl-serine synthetase).

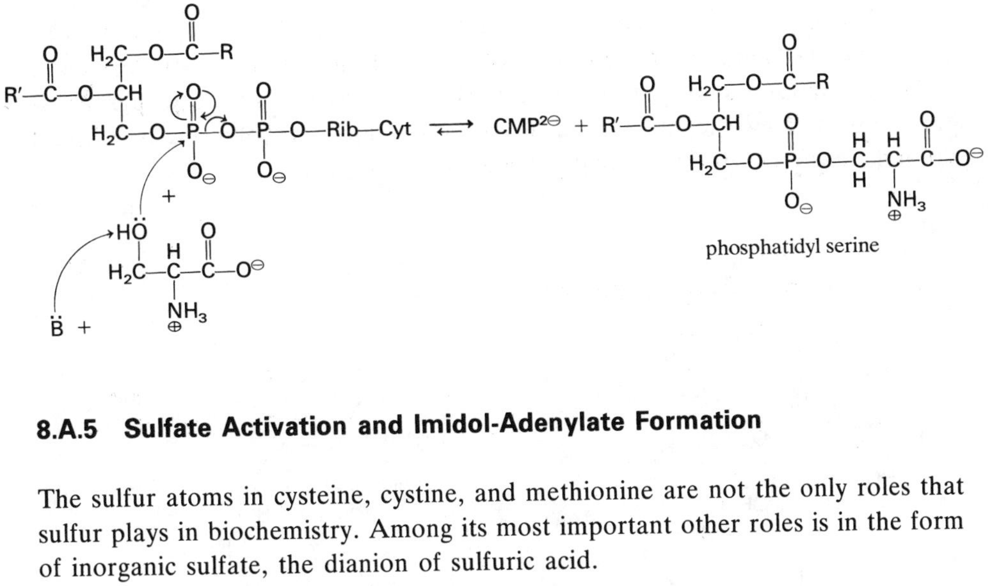

$$\text{CDP-diglyceride (–O–P–O–P–O–Rib–Cyt)} + \text{serine} \rightleftharpoons CMP^{2-} + \text{phosphatidyl serine}$$

phosphatidyl serine

8.A.5 Sulfate Activation and Imidol-Adenylate Formation

The sulfur atoms in cysteine, cystine, and methionine are not the only roles that sulfur plays in biochemistry. Among its most important other roles is in the form of inorganic sulfate, the dianion of sulfuric acid.

$$H_2SO_4 \rightleftharpoons HSO_4^{-} + H^{+} \rightleftharpoons SO_4^{2-} + H^{+}$$

inorganic sulfate (pH 7)

The sulfur atom is in the +6 oxidation state in sulfate ion, and (unlike the phosphate group, which is redox inert in biology) the sulfur can be reduced enzymatically to the +4 oxidation state as sulfite and subsequently to $S^{2\ominus}$ as inorganic sulfide (reactions discussed briefly in Section III). In addition to its role in redox chemistry, sulfate can serve as a donor of $SO_3^{2\ominus}$ (sulfuryl) groups, especially to the hydroxyl and amino groups of sugars and polysaccharides (Stadtman, 1973). However, sulfate itself is not a suitable substrate for $SO_3^{2\ominus}$ transfer (just as inorganic phosphate is not a good phosphoryl donor under physiological conditions). $SO_4^{2\ominus}$ must be activated to a more electrophilic species, and this occurs by adenylyl transfer to one of the oxygens of sulfate.

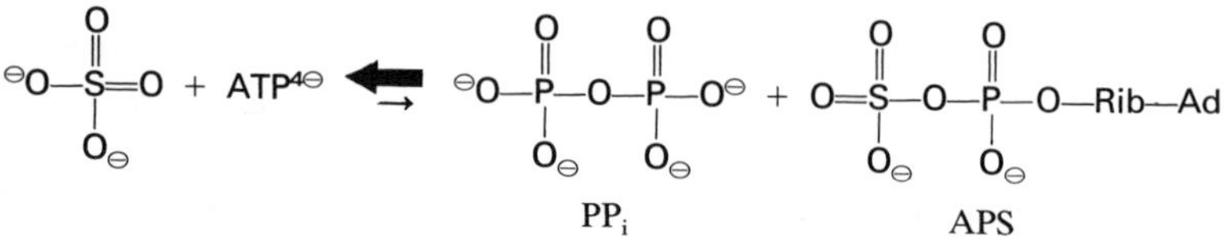

$$SO_4^{2\ominus} + ATP^{4\ominus} \rightleftharpoons PP_i + APS$$

PP$_i$ APS

The product APS (adenosine phosphosulfate, or adenyl sulfate) is a sulfuryl donor in subsequent enzymatic reactions. APS does not accumulate to high concentrations with the isolated enzyme; the equilibrium favors ATP formation by 10^8, allowing an estimate of $\Delta G^{0'}$ for APS of ca. −19 kcal/mole (ca. −80 kJ/mole), a highly destabilized sulfuryl-transfer reagent. The in vivo coupling to pyrophosphatase helps draw the equilibrium to APS formation. The following is a typical sulfuryl transfer, to the 6-hydroxyl of glucose:

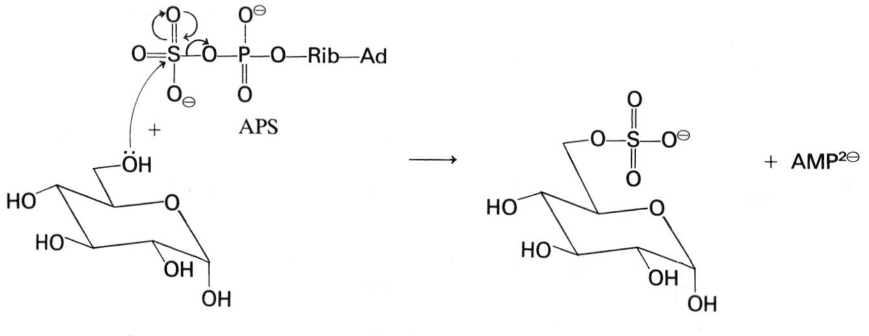

APS + glucose → glucose-6-sulfate + $AMP^{2\ominus}$

glucose glucose-6-sulfate

We have previously noted (¶5.F) the formation of citrullyl-AMP (an imidol adenylate) as an intermediate in argininosuccinate-synthetase catalysis.

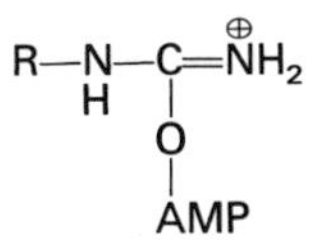

$$R{-}NH{-}C(=\overset{\oplus}{N}H_2){-}O{-}AMP$$

imidol adenylate

The formation of such imidol adenylates is included in this category of nucleotidyl transfers, along with the sulfate-activation reactions.

8.A.6 Adenylylation of Various Enzymes for Catalytic or Regulatory Purposes

There are two major types of enzyme adenylylation. The first type is the use of adenylyl transfers by DNA ligases during DNA repair. These enzymes repair single-strand breaks in DNA molecules, catalyzing the joining of a ribose 3′-OH at one end of the nick and the 5′-phosphate monoester at the other end of the nick to produce a phosphodiester linkage—essentially the reverse stoichiometry of the phosphodiesterase catalysis noted earlier (¶6.B). Repair is an energy-requiring process, and each nick repaired requires a stoichiometric utilization of ATP, with cleavage to AMP + PP_i.

Lehman (1974) and colleagues have purified the DNA ligase induced on infection of *E. coli* with bacteriophage T4. Incubation of $[^3H]$-β,γ-$[^{32}P]$-ATP with the DNA ligase, followed by gel filtration, yields enzyme containing tritium label but no ^{32}P. This species is a covalently adenylylated enzyme, where an active-site ε-amino group of a lysyl residue has acted as nucleophile, forming an N–P (phosphoramidate) linkage.

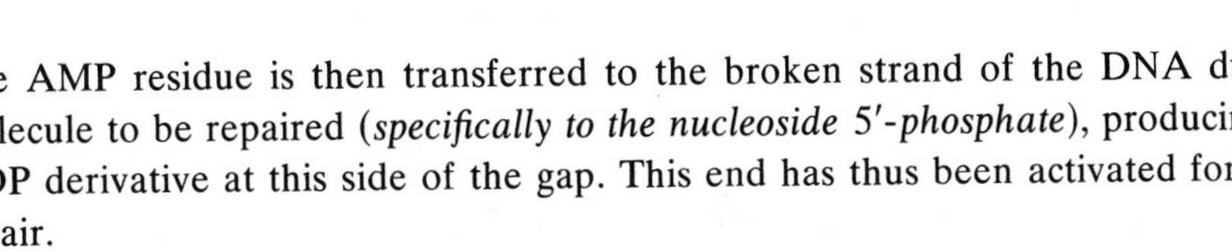

$$Mg \cdot ATP + H_2N{-}(CH_2)_4{-}\underset{\text{Enz}}{CH}{-}C(=O) \rightleftharpoons PP_i + C(=O){-}\underset{\text{Enz}}{CH}{-}(CH_2)_4{-}NH{-}P(=O)(O^{\ominus}){-}O{-}CH_2{-}\text{ribose(Ad)(OH)(OH)}$$

Enz–AMP intermediate

The AMP residue is then transferred to the broken strand of the DNA duplex molecule to be repaired (*specifically to the nucleoside 5′-phosphate*), producing an ADP derivative at this side of the gap. This end has thus been activated for nick repair.

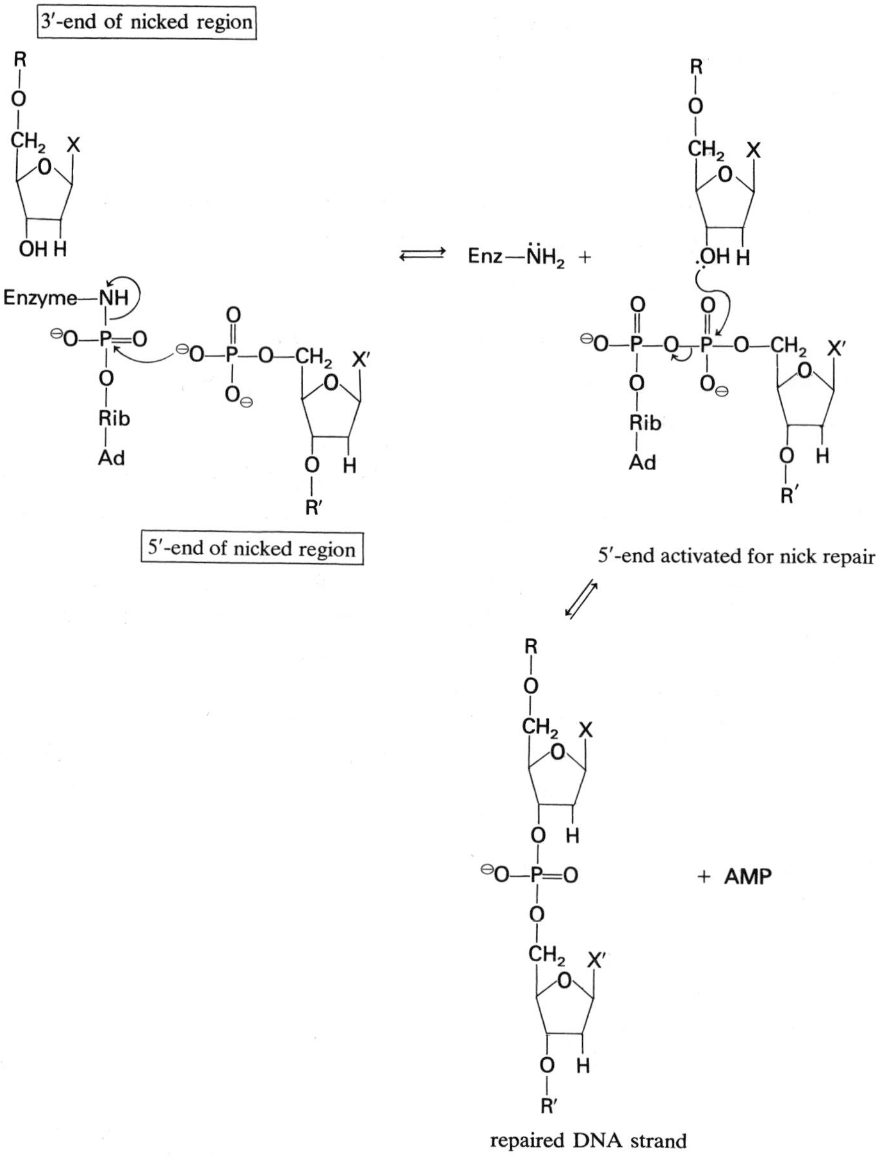

3′-end of nicked region

5′-end of nicked region

⇌ Enz—$\ddot{N}H_2$ +

5′-end activated for nick repair

+ AMP

repaired DNA strand

In the second part of the reaction, attack by the free 3′-OH from the other side of the gap occurs, with expulsion of the introduced AMP residue, regenerating a phosphodiester link in the repaired DNA strand and the good (stable) leaving group AMP.

The second major type of enzyme adenylylation is for regulatory (not catalytic) purposes. We have discussed (¶5.D.2) the adenylylation of the glutamine synthetase from *E. coli*, with catalytic activity being inversely proportional to the number of AMP residues introduced per dodecamer. RNA polymerase of *E. coli* apparently becomes adenylylated and inactivated on infection of the cell by bacteriophage T4 (Stadman, 1973).

8.B PYROPHOSPHORYL TRANSFERS

Having discussed attack by nucleophiles at the α-P of ATP (nucleotidyl transfers) and at the γ-P of ATP (phosphoryl transfers), we now turn to the third class: attack by nucleophiles at the β-P of ATP (pyrophosphoryl transfers).

In fact, although pyrophosphate linkages are common in biological systems, authenticated pyrophosphoryl transfers are rare, with only some four or five reactions known. The generalized stoichiometry involves a substrate with an oxygen nucleophile (hydroxyl, anomeric hydroxyl, or enolate ion) that becomes converted to a pyrophosphate derivative, liberating AMP as coproduct.

$$\text{R—\ddot{O}H} + \text{ATP}^{4\ominus} \rightleftharpoons \text{AMP}^{2\ominus} + \text{R—O—}\overset{\overset{\displaystyle O^{\ominus}}{|}}{\underset{\underset{\displaystyle O}{\|}}{P}}\text{—O—}\overset{\overset{\displaystyle O^{\ominus}}{|}}{\underset{\underset{\displaystyle O}{\|}}{P}}\text{—O}^{\ominus}$$

We might distinguish two classes of such reactions: *overt transfers* (where the pyrophosphate link remains intact in the product) and *crypto-pyrophosphoryl transfers* (where the pyrophosphate link is not intact in the observed product).

8.B.1 Overt Pyrophosphoryl Transfers

We shall briefly discuss here four enzymes known to catalyze overt pyrophosphoryl transfers: PRPP synthetase, TTP synthetase, HPPP synthetase, and PPGPP synthetase.

8.B.1.a PRPP synthetase

PRPP is 5-phosphoribosyl-1-pyrophosphate. We have already encountered this compound in our discussions of glutamine-dependent aminations (¶5.C.2.a), where nascent "NH_3" displaces the α-pyrophosphate group at C-1 with inversion, generating 5-phosphoribosyl-1-(β)-amine. PRPP is synthesized to convert the C-1 of ribose from an anomeric hydroxyl (a poor leaving group) to —OPP_i (an excellent leaving group).

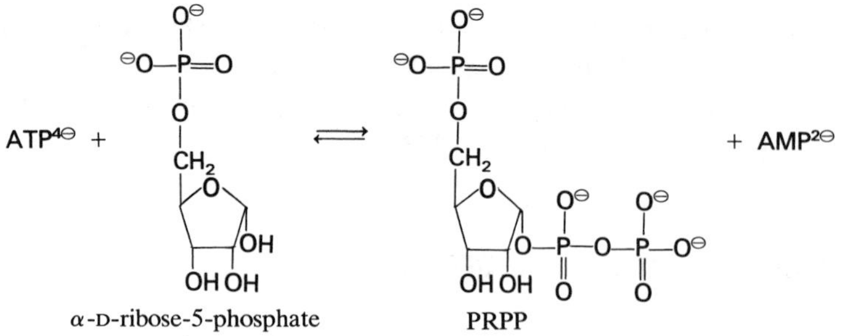

$ATP^{4\ominus}$ + α-D-ribose-5-phosphate ⇌ PRPP + $AMP^{2\ominus}$

Only the α-anomer of D-ribose-5-P is a substrate. The product PRPP also has the pyrophosphate in the the α-position (below the plane of the ring). The derivatization has blocked the ring opening that is possible in the hemiacetal structure.

The first question one might ask is this: how can we determine whether PRPP forms by pyrophosphoryl transfer or by two consecutive phosphoryl transfers (either from one or from two molecules of ATP)? The observation that only one molecule of AMP appears as coproduct rules out the possibility of consecutive phosphoryl transfers from two different ATP molecules, but leaves open the possibility of consecutive phosphoryl transfers from a single ATP molecule.

This uncertainty was resolved by taking advantage of an intramolecular cyclization and displacement undergone nonenzymatically by PRPP in alkali.

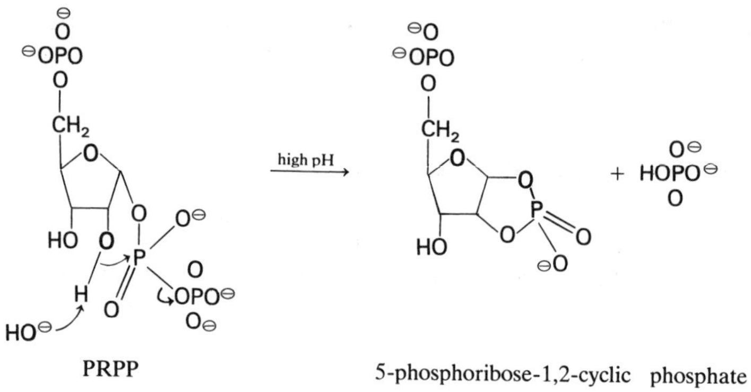

PRPP —high pH→ 5-phosphoribose-1,2-cyclic phosphate + $HOPO_3^{2\ominus}$

(Note that the β-isomer of PRPP would not suffer this displacement for steric reasons: the 2-OH and the 1-PP group would be *trans* in that isomer.)

The following experimental results were obtained. Using γ-[^{32}P]-ATP and ribose-5-P with the enzyme, followed by treatment of the radioactive product in alkali, only inorganic phosphate (note the cyclic ribose derivative) was found to be radioactive. Conversely, from β-[^{32}P]-ATP, only the cyclic phosphate was labeled.

Some thought will indicate that these results establish that PRPP is biosynthesized by attack of the C^1-ribose anomeric hydroxyl group on the β-P of ATP with direct pyrophosphoryl transfer.

8.B.1.b TPP synthetase

TPP (thiamine pyrophosphate) is the active coenzyme form of vitamin B_1 (thiamine). The nutritional roles of vitamin B_1 are mainly satisfied by TPP functioning in vivo as a coenzyme in enzymatic reactions of α-keto acids and ketols (see Chapter 21). TPP arises by pyrophosphoryl transfer to the oxygen of the hydroxyethyl side chain of the thiazolium ring of thiamine. Almost no other mechanistic information is available on this transformation (Switzer, 1974).

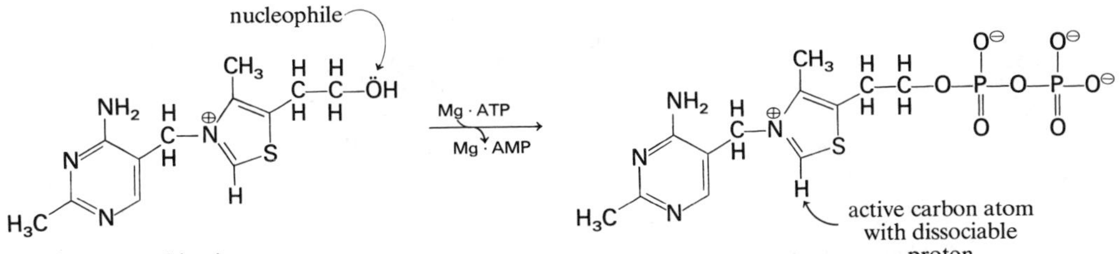

nucleophile

Mg · ATP → Mg · AMP

thiamine

active carbon atom with dissociable proton

TPP (thiamine pyrophosphate)

8.B.1.c HPPP synthetase

Tetrahydrofolic acid is an important coenzyme in one carbon metabolism as we shall discuss in Chapter 26.

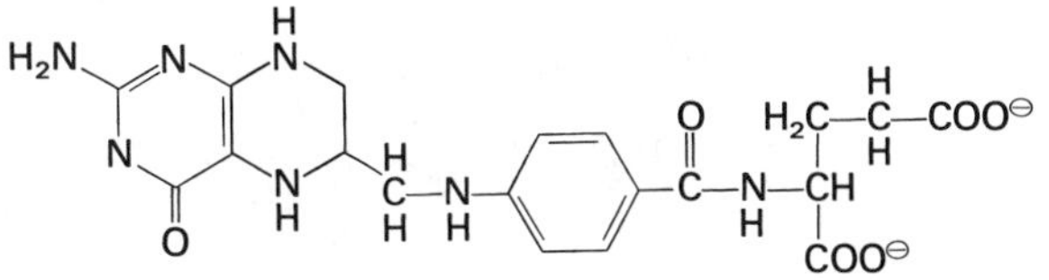

tetrahydrofolate

In the biosynthesis of tetrahydrofolic acid, an alcoholic intermediate (hydroxymethyldihydropterin) is generated. In subsequent enzymatic steps, the alcohol function is displaced by the nucleophilic amino group of p-aminobenzoate, yielding the ring systems of folic acid in the dihydropteroate product. The chemical activation of the alcoholic intermediate occurs in one enzymatic step—a pyrophosphoryl transfer catalyzed by the pyrophosphokinase we have called HPPP synthetase. The intermediate thus formed is hydroxymethylpteridine-PP (HPPP). A second enzyme, dihydropteroate synthetase, then catalyzes displacement of the pyrophosphate by the incoming amino nucleophile of *p*-aminobenzoate. The stoichiometry of the pyrophosphokinase reaction has been established, but no mechanistic information is available (Switzer, 1974).

hydroxymethyldihyropterin + Mg · ATP $\xrightleftharpoons{\text{pterin pyrophosphokinase}}$ HPPP + Mg · AMP (1)

p-aminobenzoate + HPPP $\xrightleftharpoons[\text{synthetase}]{\text{dihydropteroate}}$ dihydropteroate + PP_i (2)

8.B.1.d PPGPP synthetase

During conditions of amino-acid starvation, certain bacterial strains react quickly by slowing down or turning off most macromolecular synthetic machinery. In their study of this cellular response to emergency, Cashel and Gallant (1969) observed that an early event is the appearance of two new compounds on a thin-layer chromatogram of cell extracts—magic spots I and II (see also Cashell, 1975). These spots were identified as 3′-pyrophosphorylguanosine-5′-diphosphate (PPGPP) and 3′-pyrophosphorylguanosine-5′-triphosphate (PPPGPP). These products may then act to shut down cellular biosynthetic machinery by regulation of specific enzymes.

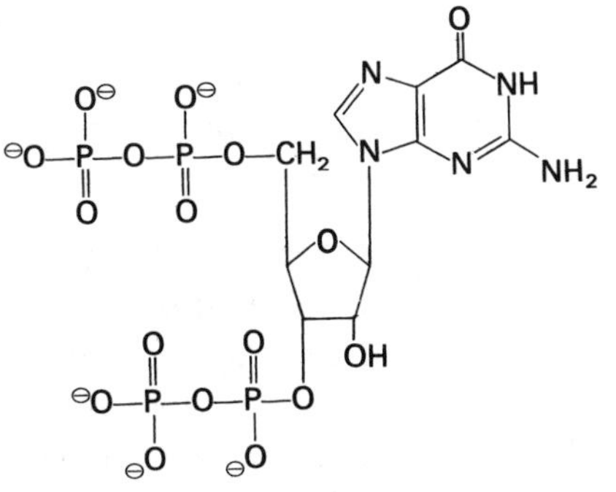

PPGPP

With a partially purified enzymatic activity obtained by washing *E. coli* ribosomes with 0.5 M ammonium chloride, an ATP-dependent synthesis of PPGPP from 5′-GDP was detectable. Use of both β-[^{32}P]-ATP and γ-[^{32}P]-ATP corroborated that the guanosine-tetraphosphate product is formed by a pyrophosphoryl transfer, with the 3′-OH of 5′-GDP attacking the electrophilic β-P of ATP (Sy and Lipmann, 1973).

In none of the four examples of overt pyrophosphoryl transfer discussed here have sufficient quantities of purified enzyme yet been available to determine whether the oxygen of the specific substrate attacks the β-P of ATP directly or whether an enzyme nucleophile reacts first to generate a covalent pyrophosphoryl-enzyme intermediate.

8.B.2 Crypto-Pyrophosphoryl Transfer

Two enzymes are known that synthesize phosphoenolpyruvate (PEP) from pyruvate and ATP where pyrophosphoryl transfers are implicated. The pyrophosphorylated species are thought to be pyrophosphorylated enzyme intermediates that rapidly break down to E—$PO_3^{2\ominus}$ and HO—$PO_3^{2\ominus}$ before the enolate anion of pyruvate acts as nucleophile. These two enzymes that may be examples of crypto-pyrophosphoryl transfers are PEP synthase and pyruvate-P dikinase.

8.B.2.a PEP synthase

PEP synthase is an enzyme that can be induced by growth of *E. coli* (strain B) on lactate as a sole carbon source. PEP formation is favored at equilibrium.

$$H_2O + ATP^{4\ominus} + \underset{\text{pyruvate}}{H_3C{-}\overset{\overset{\large O}{\|}}{C}{-}COO^{\ominus}} \rightleftarrows \underset{\text{PEP}}{H_2C{=}C(OPO_3^{2\ominus}){-}COO^{\ominus}} + AMP^{2\ominus} + P_i^{2\ominus}$$

The PEP formation is achieved at the expense of hydrolysis of *both* phosphoric-anhydride linkages in ATP. Recall (¶7.D.1) that the enol—$PO_3^{2\ominus}$ linkage in PEP is energetically uphill from ATP by 6 to 7 kcal/mole (25 to 30 kJ/mole). Thus, hydrolysis of only one linkage in ATP would not suffice to drive the reaction toward PEP synthesis. This conclusion is corroborated by the observation that the pyruvate-kinase reaction is functionally irreversible in the direction of ATP synthesis, under physiological conditions:

$$PEP + ADP \rightleftarrows ATP + Pyruvate \qquad K_{eq} \approx 3{,}000$$

In the PEP-synthase reaction, it has been reported that a phosphoryl-enzyme intermediate can be formed from PEP, and that mixing the isolated E—$PO_3^{2\ominus}$ with AMP and P_i will produce ATP (Cooper and Kornberg, 1974).

$$H_2C{=}C(O{-}PO(O^{\ominus})_2){-}COO^{\ominus} + Enz \underset{}{\overset{Mg^{2\oplus}}{\rightleftarrows}} Enz{-}PO_3^{2\ominus} + Pyruvate \tag{1}$$

$$Enz{-}PO_3^{2\ominus} + AMP^{2\ominus} + HO{-}\overset{\overset{\large O}{\|}}{\underset{O^{\ominus}}{P}}{-}O^{\ominus} \longrightarrow Enz + ATP^{4\ominus} \tag{2}$$

Similarly, an E—$PO_3^{2\ominus}$ is isolable on incubation of enzyme with ATP. Because γ-[^{32}P]-ATP does not yield radioactive E—$PO_3^{2\ominus}$, whereas β-[^{32}P]-ATP does, it appears that an active-site nucleophile on the enzyme has attacked at the β-P of ATP.

$$\text{Ad—Rib—O—}\overset{\alpha}{\text{P}}(\text{O})(\text{O}^{\ominus})\text{—O—}\overset{\beta}{\text{P}}(\text{O})(\text{O}^{\ominus})\text{—O—}\overset{\gamma}{\text{P}}(\text{O})(\text{O}^{\ominus})\text{—O}^{\ominus} + \text{Enz—}\ddot{\text{X}} \rightleftharpoons \text{Enz—PO}_3^{2\ominus} + \text{P}_i^{2\ominus} + \text{AMP}^{2\ominus}$$

$$\left[\text{Enz—X—}\overset{\beta}{\text{P}}(\text{O})(\text{O}^{\ominus})\text{—O—}\overset{\gamma}{\text{P}}(\text{O})(\text{O}^{\ominus})\text{—O}^{\ominus}\right] + \overset{\alpha}{\text{AMP}^{2\ominus}} \xrightarrow[\text{fast}]{\text{H}_2\text{O}}$$

Intensive efforts have been made to detect an isolable enzyme-PP intermediate, but none has succeeded. It may be that a pyrophosphate derivative of the enzyme forms only transiently as an unstable species, and that it then decays to the isolable E—$PO_3^{2\ominus}$, with expulsion of P_i derived from the γ-P of ATP, as shown in the alternative path in the reaction just given.

8.B.2.b Pyruvate-P dikinase

Pyruvate-phosphate dikinase catalyzes a reaction with stoichiometry similar to that of the bacterial PEP synthase. Pyruvate-P dikinase is found in tropical plants and in certain species of propionibacteria (Cooper and Kornberg, 1974). It uses inorganic phosphate in place of the H_2O of the PEP-synthase reaction. Again the β-P of ATP ends up in PEP.

$$\text{H}_3\text{C—C(=O)—COO}^{\ominus} + \text{ATP}^{4\ominus} + \text{HO—P(=O)(O}^{\ominus}\text{)—O}^{\ominus} \rightleftharpoons \text{H}_2\text{C=C(—O—P(=O)(O}^{\ominus}\text{)—O}^{\ominus}\text{)—COO}^{\ominus} + \text{AMP}^{2\ominus} + \text{PP}_i^{4\ominus}$$

In this case, H. G. Wood and colleagues have succeeded in isolating the first discrete pyrophosphoryl-enzyme intermediate, on mixing purified enzyme (from *Priopionibacterium shermanii*) with Mg·ATP (Milner and Wood, 1972). The pyrophosphoryl group is bound to a histidinyl residue (Wood, 1977). Addition of P_i to the enzyme-PP intermediate yields E—$PO_3^{2\ominus}$ and PP_i in a phosphorolysis reaction. The E—$PO_3^{2\ominus}$, as in the PEP-synthase reaction, can then phosphorylate pyruvate (presumable in its enol form) to produce PEP. The expected partial-isotopic-exchange reactions have been detected.

$$\text{Enz—}\ddot{\text{X}} + {}^{\ominus}\text{O—}\overset{\gamma}{\text{P}}(\text{O})(\text{O}^{\ominus})\text{—O—}\overset{\beta}{\text{P}}(\text{O})(\text{O}^{\ominus})\text{—O—}\overset{\alpha}{\text{P}}(\text{O})(\text{O}^{\ominus})\text{—O—Rib—Ad} \rightleftharpoons \text{Enz—X—}\overset{\beta}{\text{P}}(\text{O})(\text{O}^{\ominus})\text{—O—}\overset{\gamma}{\text{P}}(\text{O})(\text{O}^{\ominus})\text{—O}^{\ominus} + \overset{\alpha}{\text{AMP}^{2\ominus}} \quad (1)$$

$$\mathrm{Enz{-}X{-}\overset{\beta}{P}(=O)(O^{\ominus}){-}O{-}\overset{\gamma}{P}(=O)(O^{\ominus}){-}O^{\ominus} + HO{-}P(=O)(O^{\ominus}){-}O^{\ominus} \rightleftharpoons Enz{-}X{-}\overset{\beta}{P}(=O)(O^{\ominus}){-}O^{\ominus} + {}^{\ominus}O{-}P(=O)(O^{\ominus}){-}O{-}\overset{\gamma}{P}(=O)(O^{\ominus}){-}O^{\ominus}} \tag{2}$$

$$\mathrm{Enz{-}X{-}\overset{\beta}{P}(=O)(O^{\ominus}){-}O^{\ominus} + H_2C{=}C(O^{\ominus}){-}COO^{\ominus} \rightleftharpoons Enz{-}\ddot{X} + {}^{\ominus}O{-}\overset{\beta}{P}(=O)(O^{\ominus}){-}O{-}C(=CH_2){-}COO^{\ominus}} \tag{3}$$

This concludes our discussion of enzymatic reactions in which some cellular nucleophile attacks at the α-P, at the β-P, or at the γ-P of ATP (or of some other nucleoside triphosphate). Chapter 9 deals with glycosyl-transfer reactions, some of which use the phosphate group as nucleophile rather than electrophile. In those reactions, it is usually the oxyanion of an inorganic phosphate that acts as the attacking species. Note that step 2 of the pyruvate-P-dikinase scheme just given represents the combination of these two kinds of reactivity: phosphoryl transfer to P_i yielding PP_i.

Chapter 9

Glycosyl Transfers

Enzyme-catalyzed glycosyl transfers (transfers of sugar moieties) are the key biosynthetic processes in assembly of the rich variety of sugar-containing compounds in biological systems. These compounds include common disaccharides such as lactose and sucrose, and simple polysaccharides (simple in that sugar units are the sole constituents) that serve either structural purposes (e.g., cellulose, chitin) or nutritional purposes (e.g., starch, glycogen). Glycosyl groups are similarly transferred in formation of complex oligosaccharide- and polysaccharide-containing molecules such as sulfated polysaccharides (heparin, chondroitin sulfates), bacterial cell-wall mucopeptides, glycoproteins, and glycolipids. In the complementary breakdown and mobilization of monomeric units from these polymers, glycosyl transfers are similarly utilized, particularly in stepwise phosphorolytic release of glycosyl units from starch or glycogen in responses to cellular energy demands, and in the hydrolysis of various oligosaccharides and polysaccharides.

Most oligo- and polysaccharides, simple or complex, contain sugar units derived from various aldohexoses—although ketohexoses, pentoses, and the 9-carbon *N*-acetylneuraminic acid are also common elements. We shall focus on examples of glycosyl transfers involving hexose residues.

In biosynthetic glycosyl transfers, *the common activated monomeric sugar intermediate is a nucleoside diphosphate sugar.* The intermediate is formed in specific enzymatic nucleotidyl transfers—generally between an α-sugar-1-$PO_3^{2\ominus}$ and a specific nucleoside triphosphate (as discussed in Chapter 8). Although sugar derivatives of UDP are the predominant forms for many oligo- or polysaccharide biosyntheses, other nucleoside diphosphates can be used. ADP-glucose, CDP-glucose, and GDP-glucose are specifically employed as glucosyl donors in starch synthesis in different plants. GDP-mannose is the source of mannosyl groups in many glycoproteins. CMP-*N*-acetylneuraminic acid (CMP-NANA) is the precursor of *N*-acetylneuraminic-acid (NANA) residues (sialic-acid residues) found in cell-surface glycoproteins.

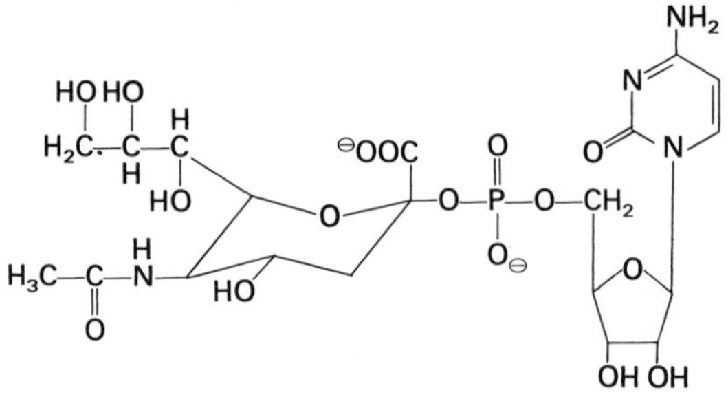

CMP–NANA

Also, dTDP-glucose is a precursor of dTDP-rhamnose in bacterial sugar activation. Such diversity of nucleoside-diphosphoryl grouping may be a reflection of metabolic-pathway specificity. A clearcut case of specificity is found in the synthesis of the O-antigen portion of the lipopolysaccharide part of the outer cell membrane of salmonella bacteria (Robbins and Wright, 1971; Nikkaido, 1973). Tetrasaccharide units are added sequentially to a lipopolysaccharide core; the tetrasaccharide has the sequence abequose–mannose–rhamnose–galactose, where abequose is 3,6-dideoxygalactose, and rhamnose is L-6-methyl-6-deoxyglucose. The activated sugar intermediates are CDP-Abe, GDP-Man, TDP-Rha, and UDP-Gal, respectively.

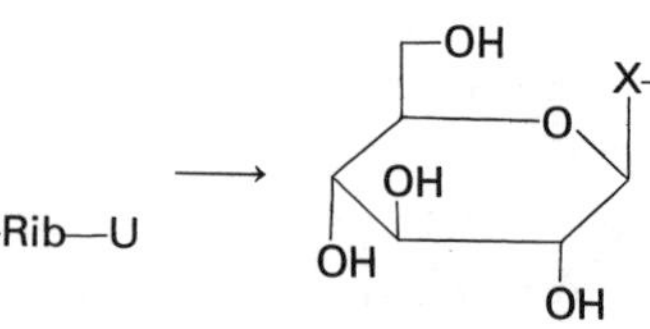

$$\left(\begin{matrix}\text{Abe} & & & & \\ \downarrow & & & & \\ \text{Man} & \rightarrow & \text{Rha} & \rightarrow & \text{Gal}\end{matrix}\right)_n\text{—Core}$$

One can write a typical stoichiometry for a glycosyl transfer from UDP-glucose, in this example, to a nucleophilic biosynthetic acceptor, R—X.

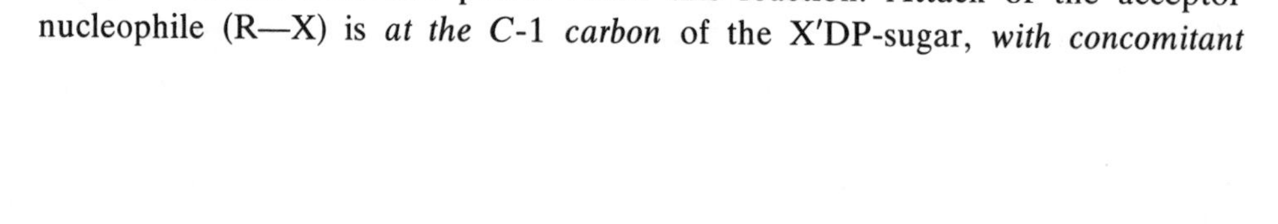

First, we should note two points about this reaction. Attack of the acceptor nucleophile (R—X) is *at the C-1 carbon* of the X′DP-sugar, *with concomitant*

Table 9-1
Categorization of biosynthetic glycosyl transfers by nature of attacking nucleophilic species

Attacking nucleophile (R—X)	Nature of product
Hydroxyl group of sugar (R—OH)	Disaccharide linkage + X′DP
Polyisoprenoid lipid-phosphate monoester 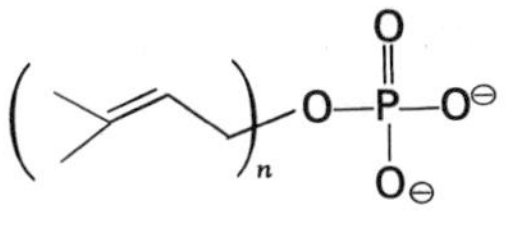	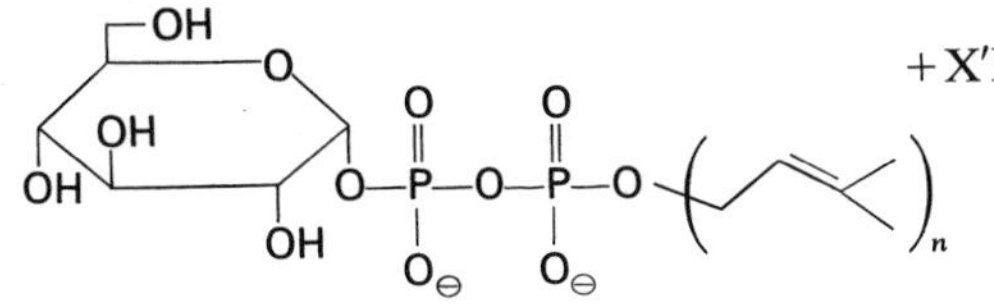 Lipid–PP–sugar intermediate + X′MP
Amine(R—$\ddot{N}H_2$)	*N*-Glycosides + X′DP

fission of the C-1–oxygen bond in all cases. Also, the attack as pictured is a simple S_N2 displacement leading to an inversion of configuration at C-1, from the α-configuration in the X′DP-sugar (UDP-glucose here) to the β-configuration in the glycosylated-nucleophile product. We shall note later in this chapter that stereochemical outcomes can be either inversion or retention, depending on the individual enzyme; we shall examine mechanistic interpretations that have been deduced from this criterion.

In biosynthetic glycosyl transfers, R—X falls into one of three categories, as indicated in Table 9-1. When a hydroxyl group of a sugar is the attacking species, a new disaccharide linkage is produced along with the liberated nucleoside diphosphate. The nucleophilic hydroxyl group is often at C-4 or C-6 (less often at C-3 or C-2) for aldohexose acceptors. For ketohexose acceptors, the hydroxyl group at C-2 of the ketal form is often the nucleophile.

The second category involves a polyisoprenoid phosphate as nucleophile, generally attacking at the phosphorus atom of the X′DP-sugar closest to C-1 of the sugar, producing a glycosylpyrophosphoryl lipid and the X′MP product. This long-chain lipid-glycosyl carrier is involved as substrate for a variety of enzymatic glycosyl transfers that occur in the membrane phases of bacterial, plant, and animal cells. The lipid portion renders the substrate soluble in the hydrophobic lipid milieu of the membrane, where glycosylation of membrane proteins and membrane lipids occurs in animal cells and where polysaccharides in plant cells and lipopolysaccharides in animal cells are biosynthesized (Cook and Stoddart, 1973). We shall elaborate on this point later in this chapter.

The third category involves attack by an amine on the X′DP-sugar to generate an *N*-glycosidic linkage. This linkage is found in several glycoproteins where *N*-acetylglucosamine residues are attached to the carboxamide nitrogen of asparaginyl residues. A mechanistically similar reaction occurs when nascent ammonia displaces the pyrophosphate group from PRPP to form the *N*-glycoside 5-phosphoribosylamine, the precursor of purine nucleosides (¶5.C.2.a).

After discussion of specific examples of biosynthetic glycosyl transfers (¶9.A), this chapter examines degradative glycosyl transfers (¶9.B) where the physiological nucleophile is water or inorganic phosphate attacking a disaccharide linkage in an oligo- or polysaccharide substrate. (Occasionally, an alcoholic molecule can function as alternate acceptor to generate an overall transglycosylation.)

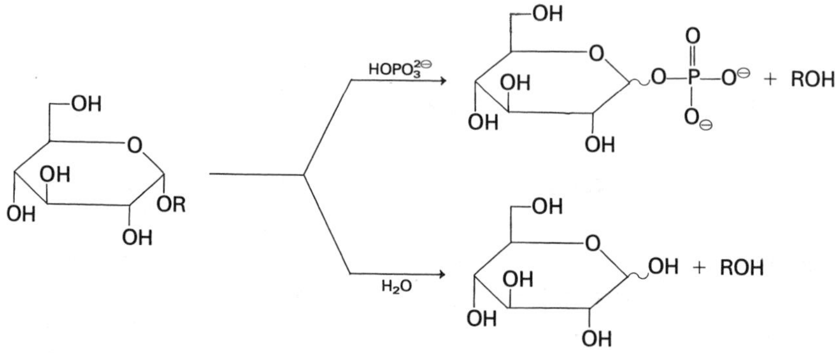

Most of the glycosyl-transfer enzymes that have been subjected to any mechanistic scrutiny fall in this degradative class, in part because these enzymes have been easier to purify.

9.A GLYCOSYL TRANSFERS IN BIOSYNTHETIC PATHWAYS

In this section we examine the three categories of glycosyl transfers (Table 9-1) that occur in biosynthetic pathways.

9.A.1 Disaccharide Formation

Two of the most common disaccharides are lactose (from milk) and sucrose (from green plants, such as sugar cane). Lactose is a galactosylglucose with the C-1 of the galactosyl residue in β-linkage to the C-4 of the glucose residue.

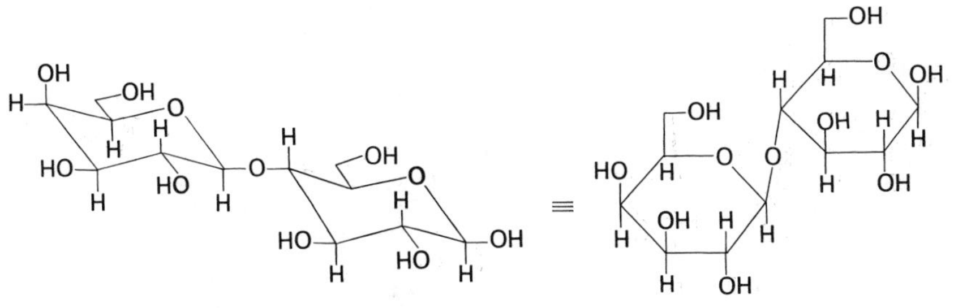

lactose (galactosyl-β-1,4-glucose)

Sucrose is composed of glucose and fructose, with an α-link from C-1 of the glucose unit to C-2 of the fructose unit. The fructose ring is shown in the more stable five-membered furanoside form taken up by ketohexoses (R. Barker, 1971).

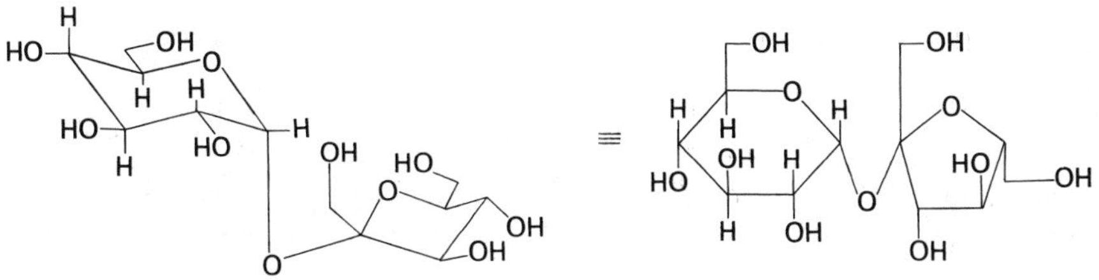

sucrose (glucosyl-α-1,2-fructose)

The lactose and sucrose structures are shown above in both conformational representation (left) and Haworth projections (right).

Sucrose is synthesized in sugar cane by the enzyme sucrose-6-P synthetase by condensation of UDP-glucose with fructose-6-P, producing UDP and sucrose-6-P with $K_{eq} = 3{,}250$ (at 38° C, pH 7.5) for disaccharide synthesis (Hassid and Neufeld, 1962). The sucrose-6-P is subsequently hydrolyzed by a specific phosphatase to sucrose and P_i (see p. 272).

The stereochemical configuration at C-1 of the glucosyl unit in UDP-glucose is α; the configuration is also α at C-1 in the glucosyl units of sucrose-6-P and of sucrose. The most obvious mechanistic conclusion from retention of configuration in a chemical displacement reaction would be that an *even number of displacement steps* must have occurred (for example, two steps) because an odd number (e.g., one step) should lead to inversion of configuration at C-1. A possible two-step

UDP-glucose ($OPOPO$—Rib—U) + fructose-6-$OPO_3^{2\ominus}$

⇅

sucrose-6-$OPO_3^{2\ominus}$ + $UDP^{3\ominus}$

↓ phosphatase

$HO{-}P(=O)(O^{\ominus}){-}O^{\ominus}$ + Sucrose

mechanism (with each step involving inversion of configuration at C-1 of glucose, so that there is overall net retention of configuration) would be the following:

1. formation of a covalent glucosyl enzyme and free UDP;
2. capture of the glucosyl enzyme by the hydroxyl group at carbon-2 of fructose-6-P.

No evidence has been obtained about this mechanism, because neither this enzyme nor an analogous sucrose synthetase (which makes sucrose directly from UDP-glucose and fructose; see Hassid and Neufeld, 1962) has been purified and studied.

On the other hand, the catalytic entity involved in lactose formation (lactose synthetase) has been thoroughly characterized. It represents a unique type of molecular organization and molecular regulation of enzymatic activity (Hill and Brew, 1975). Lactose synthetase activity is present in high levels only in mammalian mammary glands during periods of lactation, and expression of the activity is thus under strict hormonal control (Ebner, 1973). Lactose is the major carbohydrate source for infant mammals whose sole food supply is milk, and the disaccharide represents 2% to 7% by weight of various milks (Ebner, 1973).

Synthesis of lactose results from galactosyl transfer from UDP-galactose to the C-4 hydroxyl group of glucose.

UDP-galactose + glucose ⇌ lactose + UDP^{3-} (1)

The bridging oxygen at C-1 of the galactosyl moiety in lactose is above the plane of the galactose ring. This is a β-1,4-linkage, representing inversion of configuration, and thus consistent with a direct displacement of the UDP leaving group through an S_N2 backside attack by the oxygen at C-4 of the glucose. Reaction 1 is brought about by a 1:1 complex of two structurally and functionally distinct protein subunits: a hydrophobic enzyme (galactosyl transferase) and a soluble protein (α-lactalbumin). During purification of lactose synthetase, the two components dissociate; they must be recombined to detect synthesis of this disaccharide.

The galactosyl transferase is localized in internal membranes of mammary cells. It is found in the Golgi and endoplasmic reticulum membranes, but not in the plasma membranes. This enzyme component is also found in many animal and plant cells (in contrast to the α-lactalbumin, which is specific to milk). As a membrane-associated glycoprotein, the transferase was difficult to purify for mechanistic study until affinity columns were prepared containing lactalbumin covalently attached to a Sephadex resin; in the presence of glucose, the galactosyl transferase selectively associates to the lactalbumin–resin and can thus be separated from other proteins. Alone, the pure galactosyl transferase is essentially inactive with glucose as galactosyl acceptor under physiological conditions. The glucose K_m has been estimated as 2.5 M, indicating extremely weak recognition of this hexose. Instead, the enzyme is highly active for transfer to N^2-acetylglucosamine groups on some glycoproteins. And it is this galactosyl transfer to growing oligosaccharide chains on glycoproteins that is the physiological function of the galactosyl transferase in nonmammary cells.

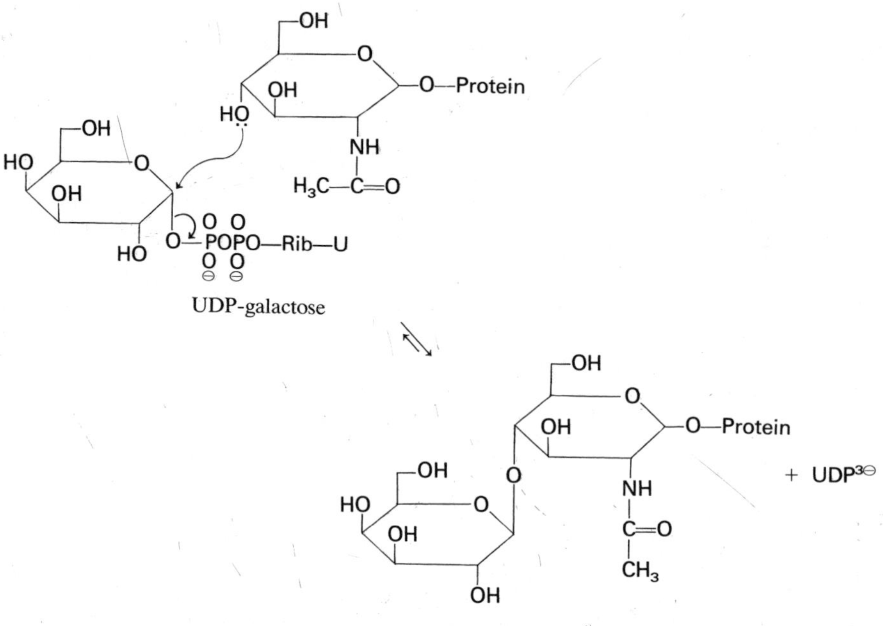

UDP-galactose

galactosyl-*N*-acetylglucosaminyl protein $+ \mathrm{UDP}^{3\ominus}$ (2)

Alternatively, the isolated galactosyl transferase will transfer a galactosyl unit to free *N*-acetylglucosamine to produce *N*-acetyllactosamine.

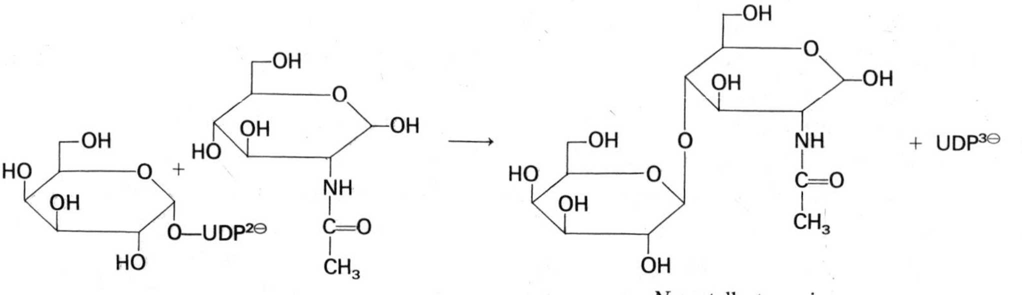

N-acetyllactosamine $+ \mathrm{UDP}^{3\ominus}$ (3)

The companion component of lactose synthetase, α-lactalbumin, by itself has *no detectable catalytic activity.* Apparently, its role is to modulate the catalytic activity of the galactosyl-transferase component by protein–protein interactions whose molecular structural bases are as yet unclear. On addition of lactalbumin to the galactosyl transferase, lactose-synthetase activity is regained at physiological concentrations of glucose; the galactosyl transfer to glycoproteins is correspondingly inhibited (although, curiously, transfer to free *N*-acetylglucosamine is reputedly unaffected). Lactalbumin has been termed a **specifier protein** because it alters the specificity of the galactosyl transferase; it actually *broadens* the enzyme's specificity by lowering the K_m for many substrates that are marginal for the transferase alone. For instance, the K_m for glucose is lowered three orders of magnitude into the mM range when lactalbumin is present.

One of the more interesting features of α-lactalbumin is the size-and-sequence homology to the hydrolytic enzyme lysozyme (¶9.B.3.c). Figure 9-1 shows the sequences for these two proteins. Nonetheless, lactalbumin has no lysozyme activity and is not antigenetically related to lysozyme (Hill and Brew, 1975); a high-resolution X-ray structure will be needed to determine the structural differences from lysozyme that lead to absence of catalytic activity in lactalbumin.

One final comment here concerns regulation of lactose synthetase activity. On sex-hormone stimulation of mammary tissues, it is specifically the lactalbumin component whose biosynthesis is induced. The synthetase complex is active only when the soluble lactalbumin can then interface with the membraneous galactosyl transferase, which otherwise functions in galactosyl transfers to proteins.

9.A.2 Polysaccharide Formation

As an example of a glycosyl transfer in polysaccharide biosynthesis, we can consider the reaction catalyzed by glycogen synthetase. Glycogen is the major storage form of glucose in liver cells, where it serves as a mobilizable reservoir for glucose (both intracellularly and for blood glucose). Glycogen molecules have high molecular weight (averaging several million daltons); they are branched homopolymers of glucose. Two types of glycosidic links are found. The major one is the glucosyl-α-1,4-glucosyl link that forms the backbone of glycogen. About once in every ten glucosyl residues, a 1,6-linkage is found to another α-1,4-linked chain (Figs. 9-2 and 9-3). The large glycogen molecules are often deposited as distinct glycogen granules in liver cells, and glycogen synthetase activity is associated with these particles (Stryer, 1975, chap. 16). The enzyme actually catalyzes an elongation of a glycogen primer or of smaller α-1,4-glucan primers. (Maltose is a very poor acceptor; linear oligosaccharides are intermediate acceptors.)

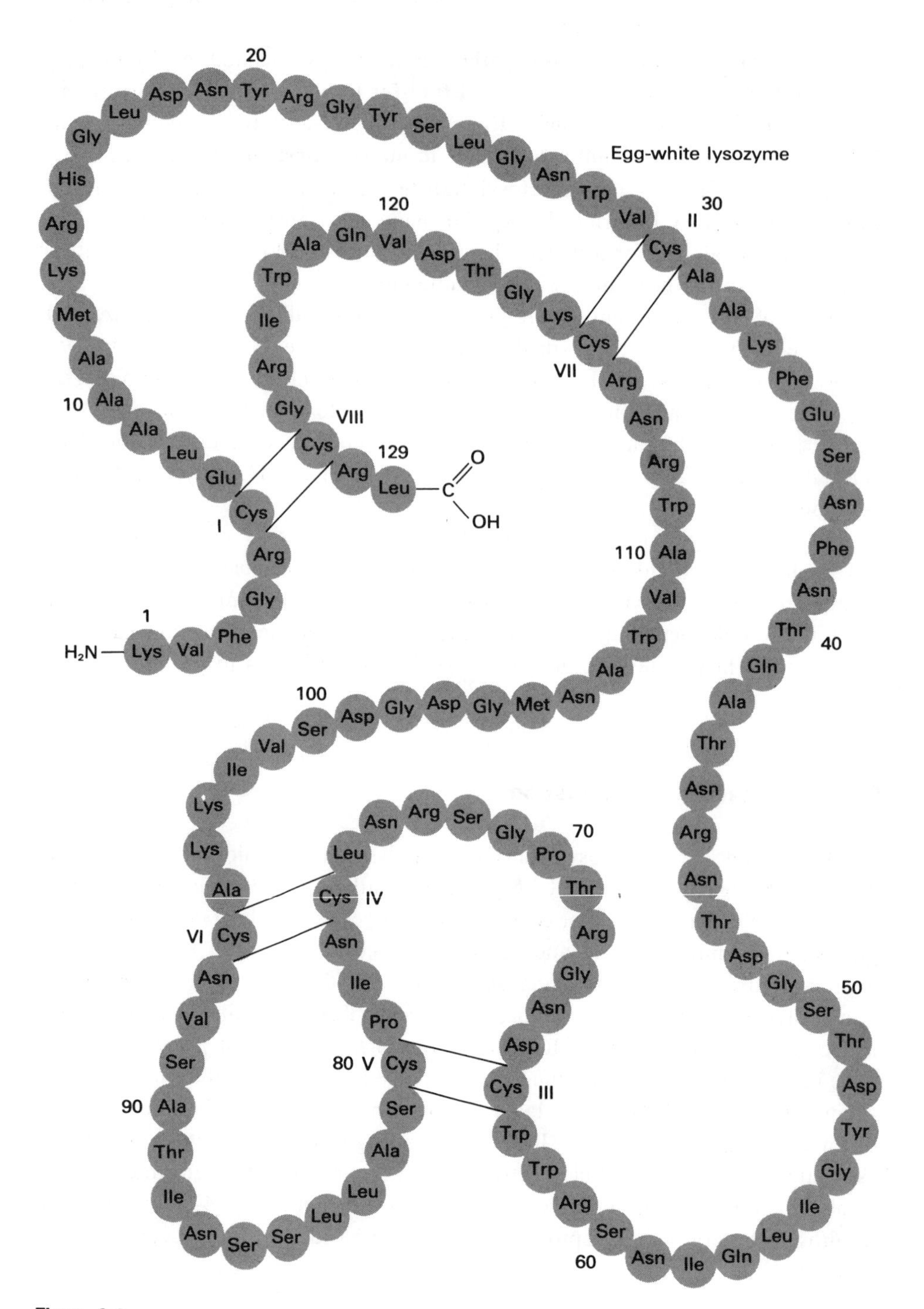

Figure 9-1
Comparison of the complete covalent structures of bovine α-lactalbumin and chicken–egg-white lysozyme. (From T. Vanaman et al., 1970)

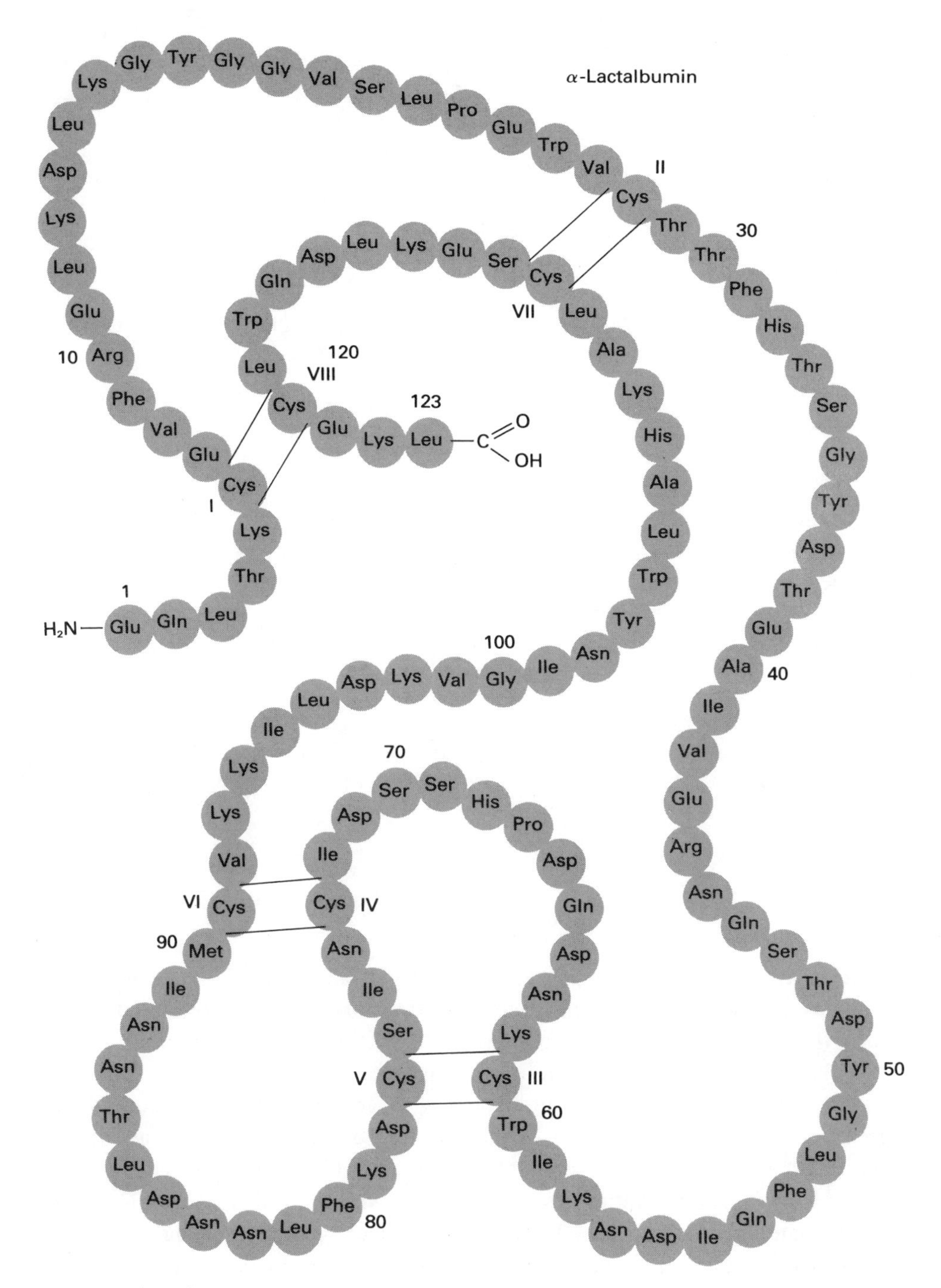

Figure 9-1 *(cont.)*

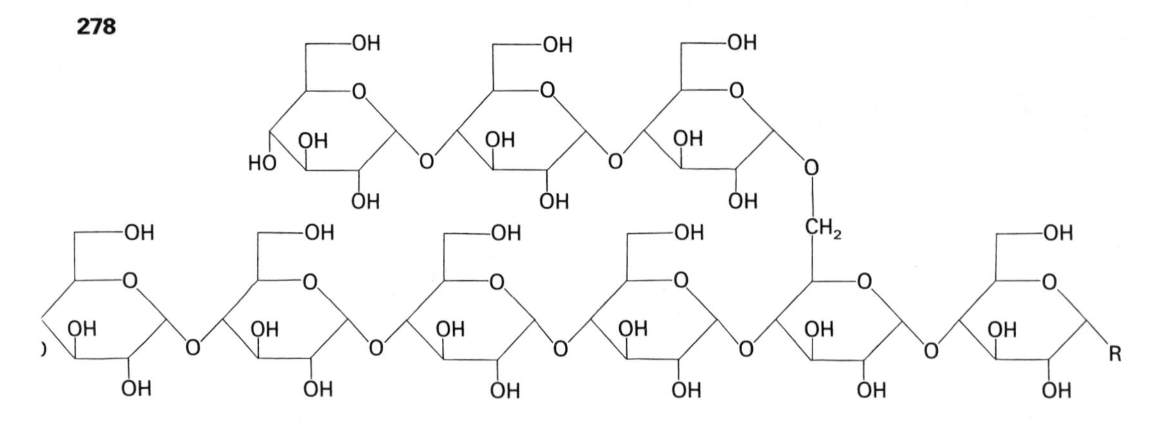

Figure 9-2
Structure of two outer branches of a glycogen particle.

O—UDP$^{2\ominus}$ + R_n ⟶

UDP-glucose

glycogen primer (n + 2 residues)

R_n + UDP$^{3\ominus}$

glycogen product (n + 3 residues)

After an average of seven to ten additions of glucosyl units to the chain by the synthetase, a separate branching enzyme transfers a length of about seven glucosyl residues by breakage of an α-1,4-linkage and formation of a new 1,6-linkage at a more interior site on the glycogen substrate.

The stereochemical outcome of the glycogen-synthetase reaction is net retention of configuration at C-1 of the transferred glucosyl unit. Initial-velocity kinetic experiments indicate parallel-line kinetics (Stalmans and Hers, 1973; Hers, 1976). This observation also is consistent with a covalent glucosyl-enzyme intermediate, but the expected UDP $\rightleftarrows$ UDP-glucose partial exchange could not be detected, suggesting that *reversible* formation of a covalent enzyme intermediate may not occur in the absence of glycogen. It has been claimed that

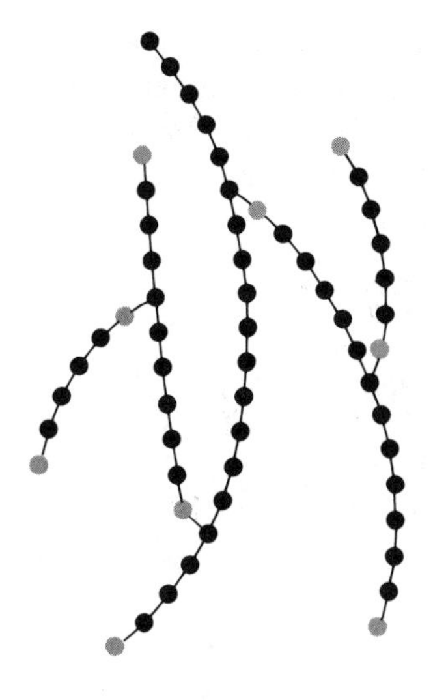

Figure 9-3
Diagram of a cross-section of a glycogen molecule. (From L. Stryer, *Biochemistry*, p. 379. W. H. Freeman and Company. Copyright © 1975)

glycogen synthetase catalyzes multiple glycosyl transfers to one chain of the glycogen (i.e., acts *processively*)★ before diffusing away to another terminus of the branched polymer.

Both the synthesis of glycogen and its phosphorylytic breakdown are subject to precise metabolic regulation in skeletal muscle and in liver. The purified synthetase can be isolated in two forms with markedly different catalytic activity.

1. The D form is almost completely inactive, but it will show full activity when 10 mM glucose-6-P is present as an effector; thus the D form is glucose-6-P–*dependent.* For instance, the D form of the rabbit-muscle enzyme shows a 47-fold increase in V_{max} when glucose-6-P is added (Stalmans and Hers, 1973; Hers, 1976).
2. The I form is completely active with or without added glucose-6-P; this form is glucose-6-P–*independent.*

The structural basis for the changes in catalytic activity has been discovered. The D form has one phosphate residue per subunit; the I form is not phosphorylated. This phosphoryl group is on a unique serine residue and is not involved in catalysis; it serves a regulatory function. The I form is converted to the phosphorylated D form by a glycogen-synthetase kinase, a typical protein kinase

★ Several other enzymes catalyzing formation or breakdown of biopolymers act in this processive fashion. For instance, the DNA polymerase induced on infection of *E. coli* by bacteriophage T5 will move the length of a 400-nucleotide-long DNA primer, adding mononucleotides to the growing complementary strand for 400 cycles before dissociating off the primer strand at the distal end (Das and Fujimura, 1977).

cleaving ATP to ADP + phosphoenzyme. For effective regulation, the phosphorylation must be reversed, and a phosphoprotein phosphatase does readily remove the phosphoryl group as inorganic phosphate (Stalmans and Hers, 1973; Hers, 1976). In the following reaction scheme, "Enz" represents glycogen synthetase.

$$\begin{array}{c} \text{Enz} \\ | \\ \text{Ser} \\ | \\ \text{OH} \end{array} \underset{\text{phosphatase}\;\searrow\;\text{HOPO}_3^{2\ominus}}{\overset{\text{ATP}^{4\ominus}\text{, kinase}\;\nearrow\;\text{ADP}^{3\ominus}}{\rightleftharpoons}} \begin{array}{c} \text{Enz} \\ | \\ \text{Ser} \\ | \\ \text{OPO}_3^{2\ominus} \end{array}$$

The protein kinase is itself a dimer of catalytic and regulatory subunits; the kinase is inactive until the regulatory subunit has been dissociated from the catalytic subunit. This kinase activation is effected by binding of 3′,5′-cAMP to the kinase regulatory subunit; this binding induces dissociation of the two subunits. Because the adenyl cyclase responsible for converting ATP to cAMP is localized in the plasma membrane of liver and muscle cells, and because this cyclase is regulated by binding of hormones to the outer surface of the membranes, the regulation of glycogen-synthetase activity is easily envisaged to be under hormonal control.

The covalent regulation of enzymatic activity by phosphoryl transfer, mediated by a kinase–phosphatase pair, is a major regulatory device not limited to glycogen synthetase. We shall note identical regulation of glycogen phosphorylase and pyruvate dehydrogenase (Chapter 21) in subsequent pages. The reversible covalent modification is also analogous to the adenylylation and deadenylylation sequence with glutamine synthetase (¶5.D.2).

9.A.3 Glycoprotein Formation

A number of enzymes (including some RNases, DNases, and the protease bromelain from pineapple stems) are glycoproteins (Kornfeld and Kornfeld, 1973). So are immunoglobins and a variety of integral membrane proteins of the plasma membranes of animal cells, where the carbohydrate chains are oriented at the outer surface and function as receptors for a variety of hormones and other signals. There are two major types of linkage between proteins and the oligosaccharide chains:

1. an *O*-glycosidic linkage between the β-hydroxyl of a seryl or a threonyl residue and C-1 of an *N*-acetylglucosamine;
2. an *N*-glycosidic linkage to the β-carboxamide nitrogen of an asparaginyl residue (Kornfeld and Kornfeld, 1973).

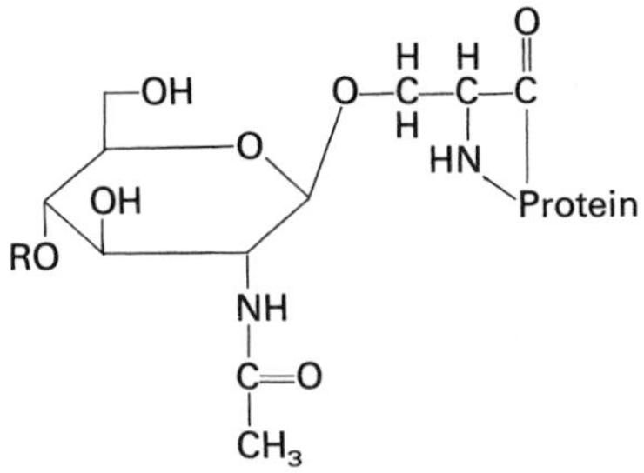

N-acetylglucosamine-*O*-serine linkage

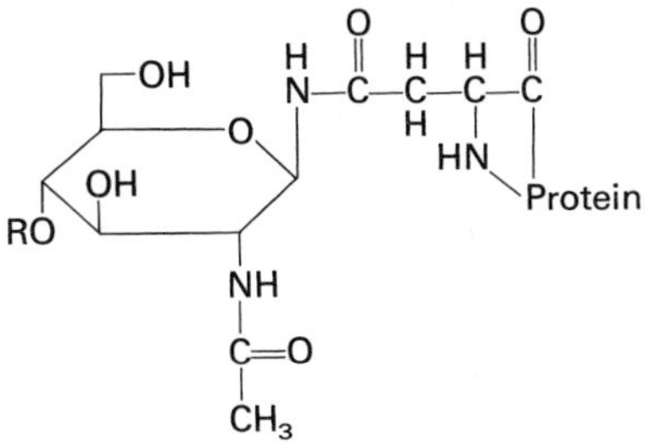

N-acetylglucosamine-*N*-asparagine linkage

Little is known about the biosynthesis of the *O*-glycosidic oligosaccharide chains (which are not the linkage in glycoprotein enzymes); slightly more is clear about the chains linked *N*-glycosidically. Many of the latter have a common core trisaccharide (Kornfeld and Kornfeld, 1973; Waechter and Lennarz, 1973):

$$\text{Man-}\beta\text{-1,4-GlcNAc-}\beta\text{-1,4-GlcNAc—Asn-protein}$$

This core may be added to the asparginyl carboxamide group as a unit. One must invoke the isoamide form of the asparaginyl carboxamide nitrogen as attacking nucleophile. The donor of the trisaccharide appears to be a trisaccharide-pyrophosphoryl-lipid. The lipid is a polyisoprenoid alcohol 80 to 100 carbons in length (i.e., 16 to 20 isoprenyl groups in length), with the first prenyl group saturated.

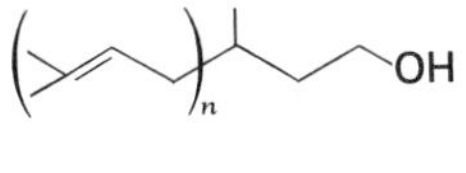

n = 15 to 19

This family of alcohols is termed *dolichols*. The dolichols represent membrane-soluble carriers for the growing saccharide chain.

A reasonable scheme for formation of the intermediate involves initial attack of a dolichol pyrophosphate on UDP-*N*-acetylglucosamine (Waechter and Lennarz, 1973).

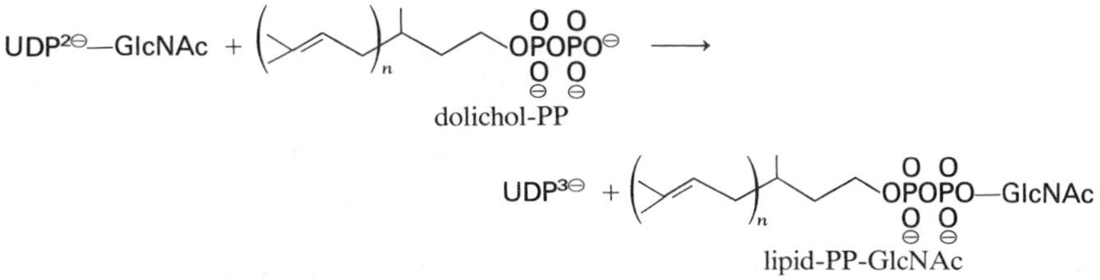

$$\text{UDP}^{2\ominus}\text{—GlcNAc} + \text{dolichol-PP} \longrightarrow \text{UDP}^{3\ominus} + \text{lipid-PP-GlcNAc} \quad (1)$$

The 4-OH of the *N*-acetylglucosaminyl group then can act as nucleophile toward another molecule of UDP-GlcNAc to form the lipid-pyrophosphate-disaccharide.

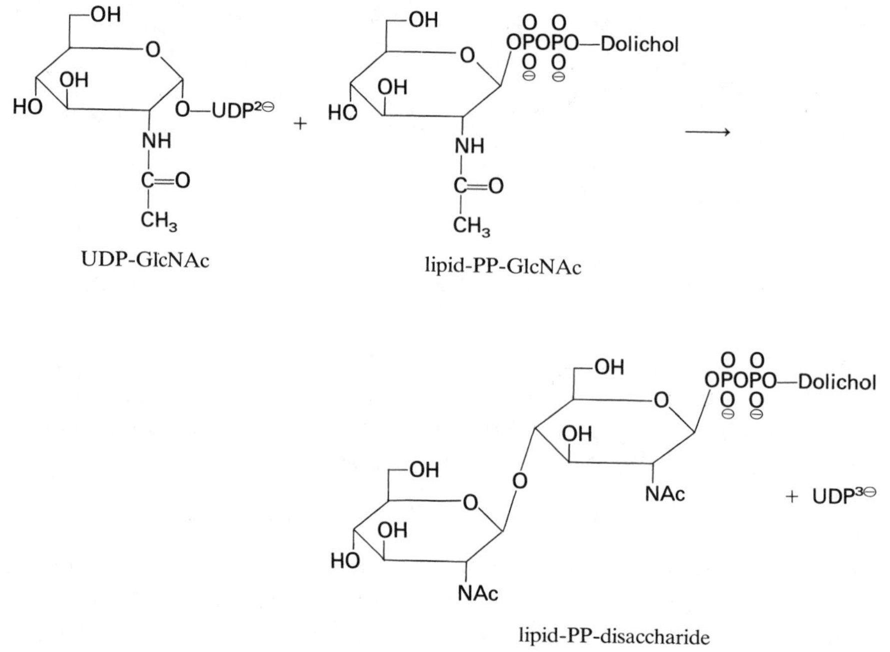

UDP-GlcNAc + lipid-PP-GlcNAc ⟶

lipid-PP-disaccharide + $UDP^{3\ominus}$ (2)

The final step involves mannosyl transfer from GDP-mannose to the 4-OH of the second *N*-acetylglucosaminyl group.

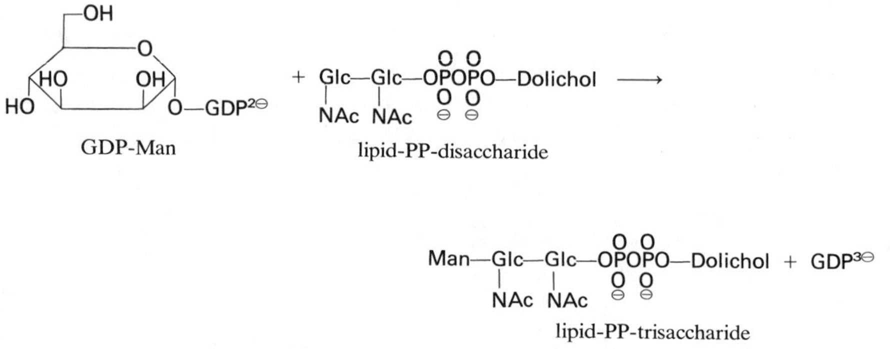

GDP-Man + lipid-PP-disaccharide ⟶

Man—Glc(NAc)—Glc(NAc)—OPOPO—Dolichol + $GDP^{3\ominus}$ (3)

lipid-PP-trisaccharide

The glycosylated lipid intermediate probably then serves as substrate for trisaccharide transfer to the specific asparaginyl group of a protein acceptor—either an endogenous membrane protein or one in the process of being secreted by cells.

Man-β-1,4-GlcNAc-β-1,4-GlcNAc(OH, OH, NAc)–O–POPO–Dolichol ($O^{\ominus}$, $O^{\ominus}$) + H_2N–C(=O)–Asn–Protein ⟶

Man-β-1,4-GlcNAc-β-1,4-GlcNAc—Asn—Protein + Dolichol—OPOPO$^{\ominus}$ ($O^{\ominus}$, $O^{\ominus}$)

Subsequent glycosyl groups can be added to this "core" oligosaccharide chain on the protein from UDP-sugars—as, for instance, in galactosyl transfer from UDP-galactose, carried out by the galactosyl-transferase component of lactose synthetase.

9.B GLYCOSYL TRANSFERS TO INORGANIC PHOSPHATE OR TO WATER

In metabolism of polysaccharides, degradation is as important as biosynthesis to cells that constantly recycle components and alter their disposition according to the needs of the moment. Oligosaccharides and polysaccharides that serve structural purposes, as well as ingested polysaccharides, are hydrolyzed (glycosyl transfer to water) enzymatically to liberate the monomeric sugars. On the other hand, we have noted that (in general) phosphate esters are the predominant intracellular forms of common pentoses and hexoses; glucose is present either as glucose-6-P or as glucose-1-P. Glucose-6-P is the product of hexokinase-mediated phosphorylation of glucose; glucose-1-P is the product when glucosyl groups are transferred to inorganic phosphate (P_i) rather than water as acceptor. Phosphoglucomutase (an enzyme we shall discuss in Section IV) rapidly interconverts the two glucose-phosphate esters, providing a common conduit. Production of glucose-1-P directly from the storage homopolymer glycogen is energetically advantageous to cells. If glucosyl transfer were to H_2O rather than to P_i, then a molecule of ATP would have to be expended to bring the free glucose product back to the phosphorylated form.

We shall briefly examine some disaccharide phosphorylases and glycogen phosphorylase to determine their modes of action; then we shall turn to a consideration of oligosaccharide hydrolases.

9.B.1 Disaccharide Phosphorylases

Three bacterial disaccharide phosphorylases have been studied that use, respectively, cellobiose (glucose-β-1,4-glucose, the basic unit of cellulose), maltose (glucose-α-1,4-glucose), and sucrose (glucose-α-1,2-fructose) as substrate. Table 9-2 lists some data for these three enzymes (Mieyal and Abeles, 1972). All three enzymes give glucose-1-P as product; cellobiose phosphorylase and maltose phosphorylase also yield free glucose as coproduct, while sucrose phosphorylase yields fructose. Let's consider the mechanistic information provided by the data in Table 9-2.

Table 9-2
Some properties of three disaccharide phosphorylases

Enzyme	C-1 configuration in disaccharide substrate	C-1 configuration in glucose-1-P product	Occurrence of $X^* \rightleftarrows$ Glc-X^* exchange	Arsenolysis of glucose-1-P
Cellobiose phosphorylase	β	α	No	No
Maltose phosphorylase	α	β	No	No
Sucrose phosphorylase	α	α	Yes	Yes

We can draw two conclusions from the available stereochemical information.

1. Inversion of C-1 configuration in going from substrate to product in the reactions of cellobiose phosphorylase and maltose phosphorylase implies a direct S_N2 displacement by an oxygen atom of inorganic phosphate at C-1 of the disaccharide moiety being transformed.
2. Retention of C-1 configuration in the reaction of sucrose phosphorylase means either (a) two S_N2 displacements, or (b) an S_N1 (carbonium) mechanism with exclusive frontside attack, perhaps due to shielding of the carbonium ion by the enzyme active site so that random backside and frontside attack cannot occur.

Some additional experimental information (not included in Table 9-2) helps us draw conclusions about glucosyl transfer. If any of the three disaccharides is synthesized with ^{18}O in the bridge oxygen, the glucose-1-P product contains no ^{18}O. This observation is consistent with glucosyl transfer in each case—i.e., cleavage of the oxygen-bridge bond at C-1 of the disaccharide substrate.

The lack of partial exchange with cellobiose and maltose phosphorylases (Table 9-2) is consistent with the ternary-complex mechanism suggested by the stereochemistry. Sucrose phosphorylase shows the following partial exchange:

$$[^{14}C]\text{-fructose} \rightleftarrows \text{glucosyl-}\alpha\text{-1,2-}[^{14}C]\text{-fructose} \equiv (^{14}C]\text{-sucrose})$$

With time, an incubation of [^{14}C]-fructose, sucrose, and enzyme yields radioactivity appearing in sucrose (only in the fructosyl moiety). This exchange occurs in the absence of inorganic phosphate. This observation strongly suggests the double S_N2 displacement mechanism, rather than the shielded-carbonium-ion hypothesis, and further argues specifically for a glucosyl-enzyme intermediate.

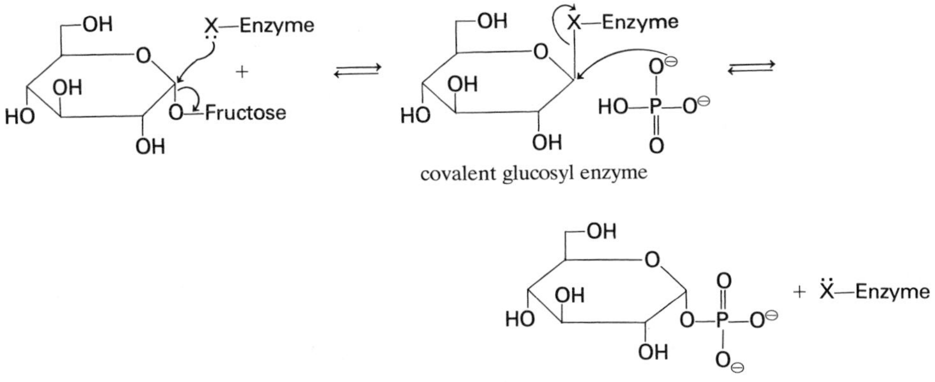

covalent glucosyl enzyme

A notable prediction for such a covalent intermediate is that it should arise by nucleophilic attack of an amino-acid moiety of the enzyme at C-1 of sucrose, giving inversion and thus producing a glucosyl enzyme with the β-configuration at C-1. Subsequent attack by P_i produces a second inversion and gives α-glucose-1-phosphate.

Table 9-2 includes data about arsenolysis, which represents a partitioning experiment. Arsenic esters are kinetically unstable in aqueous solutions. Unlike phosphate esters, they are rapidly decomposed (to alcohol and arsenate) by water attack:

$$R{-}OAsO_3^{2\ominus} \xrightarrow{H_2O} R{-}OH + AsO_4^{2\ominus}$$

The arsenolysis of glucose-1-P implies that the phosphate is converted to free glucose by the enzyme in the presence of arsenate. Note that, in the absence of enzyme, arsenate does not rapidly attack glucose-1-P, probably due to charge repulsion. Further, glucose-1-P does not suffer rapid nonenzymatic hydrolysis under the experimental conditions. These data indicate that the glucose-1-P undergoes the normal back-reaction to glucosyl enzyme, which is then attacked by arsenate dianion ($AsO_4^{2\ominus}$, a structural analogue to $PO_4^{2\ominus}$) to give the unstable arsenic ester, which is broken down to glucose and $AsO_4^{2\ominus}$. The glucosyl enzyme, which is partitioned to glucose-1-$AsO_3^{2\ominus}$, undergoes an irreversible breakdown due to the nonenzymatic hydrolysis reaction.

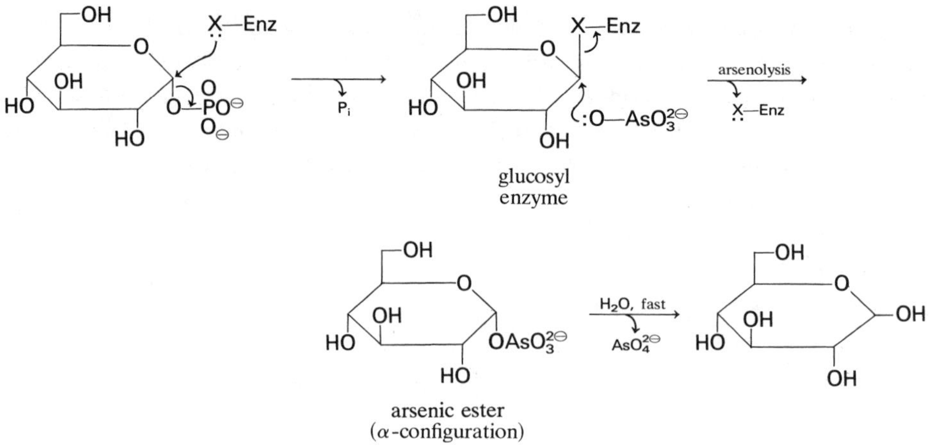

glucosyl
enzyme

arsenic ester
(α-configuration)

The two phosphorylases that use a sequential mechanism show no such arsenolysis.

Two other observations sustain the proposition that catalysis by sucrose phosphorylase involves a covalent glucosyl-enzyme intermediate.

1. The kinetic mechanism is ping-pong BiBi, and the expected product-inhibition patterns are observed (two competitive, two noncompetitive). These data suggest that the fructose is released prior to P_i addition.
2. The glucosyl enzyme is labile to attack by H_2O (55 M) even in the absence of P_i, so it cannot be isolated by gel-filtration methods at neutral pH. However, the enzyme can be quenched during catalysis by rapid addition of acid to pH 3. The protein precipitates and can be analyzed.
   a. Sucrose labeled with ^{14}C in the glucosyl group yields [^{14}C]-protein at 2 moles per 100,000 g.
   b. Sucrose labeled with ^{14}C in the fructosyl group yields no radioactive protein, nor does incubation with [^{14}C]-glucose or [^{14}C]-fructose.

   These data imply that only the glucosyl moiety derivable from sucrose is bound to precipated enzyme (Mieval and Abeles, Abeles, 1972).

The following observations have been made about the glucosyl-enzyme species in catalysis by sucrose phosphorylase.

1. The half-time ($T_{1/2}$) for hydrolysis of the enzyme–glucose complex at room temperature is 80 min at pH 6, and considerably less at alkaline pH. This base lability is consistent with an oxygen ester.
2. Glucose is released rapidly when alkaline (pH 11) hydroxylamine is added. Again, this observation is consistent with an oxygen ester. (The more reactive thioesters react with NH_2OH at acidic and neutral pH.)

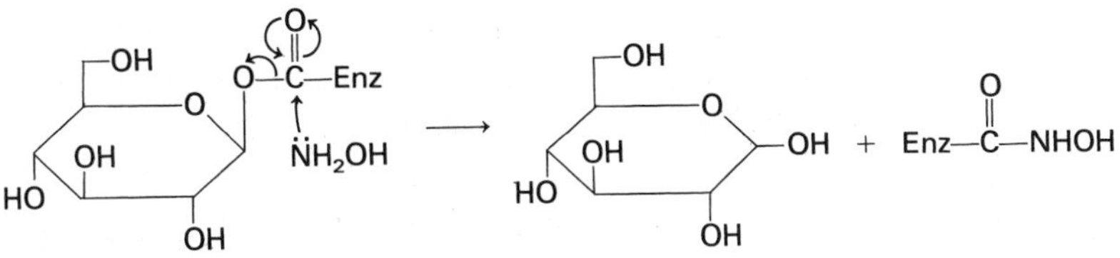

(Methanol also releases glucose when enzyme-glucose is suspended in MeOH, presumably forming an enzyme-methyl ester species.)

3. Differential labeling experiments were carried out with chemical reagents thought to be specific derivatizing agents for enzyme carboxyl groups. One such reagent is a water-soluble carbodiimide that activates carboxyl groups to the *O*-acyl isourea. This intermediate can be attacked by good nucleophiles (such as the amino group of glycine-ethyl ester, a small soluble reactive amine), producing the stable amide derivative.

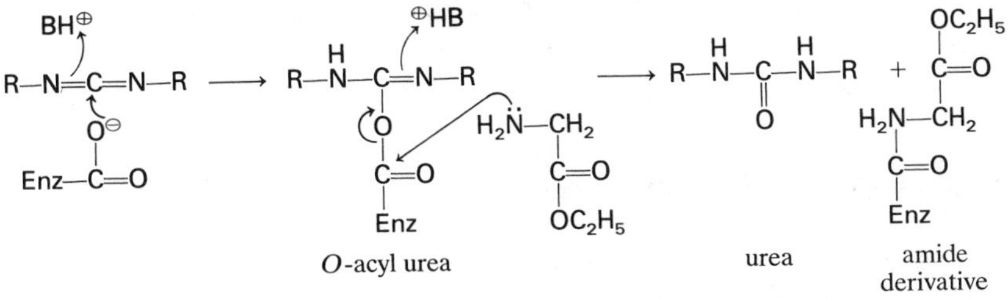

O-acyl urea   urea   amide derivative

The rationale was as follows. Sucrose phosphorylase has about 100 carboxylate groups (β-$COO^{\ominus}$ of Asp, γ-$COO^{\ominus}$ of Glu, and C-terminal $COO^{\ominus}$), Only one group will be blocked in the glucosyl enzyme. Pretreatment of glucosyl enzyme with nonradioactive glycine-methyl ester should derivatize all available $COO^{\ominus}$ groups except the one in ester linkage with glucose. Treatment of free enzyme, similarly, should block all the enzyme $COO^{\ominus}$ groups, including the one at the active site. The glucosyl enzyme can be hydrolyzed by short exposure to pH 8, deblocking the putative active-site $COO^{\ominus}$ group. Addition at this time of [^{14}C]-glycine-ethyl ester and carbodiimide should introduce one radioactive amide linkage. The free enzyme carried through the whole protocol gave no covalent radioactivity at this point, whereas the test sample did in fact yield a radioactive enzyme. Partial digestion yielded a single radioactive peptide whose sequence has not yet been determined.

What is the configuration of the glucosyl enzyme? We predict a β-configuration, because both substrate and product have the α-configuration at C-1. The glucosyl enzyme was trapped at pH 8 in the presence of silylating agents to silylate the liberated glucose before racemization at C-2 could occur nonenzymatically. This product was compared with authentic silyl derivatives of α-D-glucose and β-D-glucose, which have different retention times on gas–liquid chromatographic columns. This analysis confirmed that the product from the enzyme is a β-derivative. It is noted that the release of glucose from glucosyl enzyme that had first been denatured is by a hydroxide attack on the carbonyl oxygen.

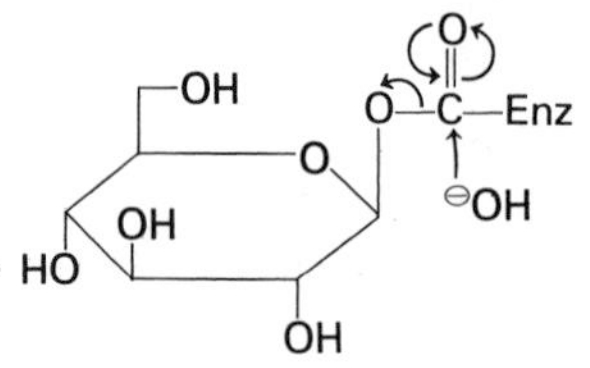

The nature of this attack is inferred from the fact that addition of denatured glucosyl enzyme to methanolic solutions produces only glucose and no methyl) glucoside. In contrast, when active enzyme is glycosylated from sucrose in methanol–water mixtures, the glucosyl-enzyme intermediate can be transferred both to water and to methanol, yielding the α-anomer of methylglucoside.

9.B.2 Glycogen Phosphorylase

The three phosphorylases described in ¶9.B.1 use disaccharides as substrates. Various polysaccharide phosphorylases are known as well. Of these, the enzyme from rabbit skeletal muscle (utilizing glycogen as substrate) is the best-studied. The rabbit-muscle enzyme was crystallized in 1942 and has since been studied exhaustively (Stalmans and Hers, 1973; Hers, 1976).

Its physiological function is degradative—the generation of monomer glucose-1-P units from the glycogen polymer in animal tissues.

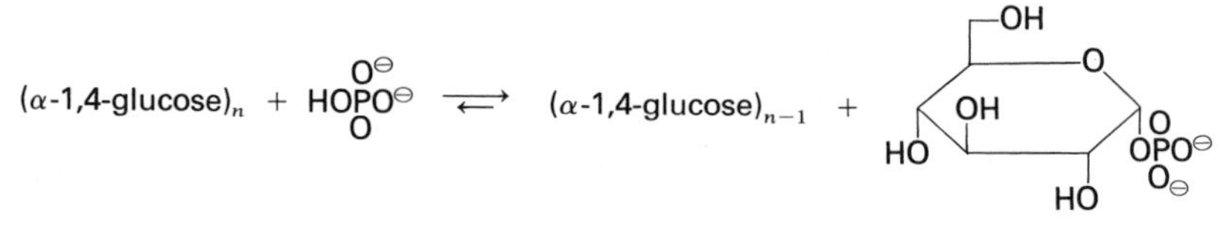

$$(\alpha\text{-1,4-glucose})_n + HOPO_3^{2-} \rightleftharpoons (\alpha\text{-1,4-glucose})_{n-1} + \alpha\text{-glucose-1-P}$$

α-glucose-1-P

In comparing the possible catalytic mechanism of glycogen phosphorylase with those presented for the various disaccharide phosphorylases, the following data can be considered (Graves and Warn, 1972).

1. Initial-velocity measurements reveal a random kinetic mechanism, at odds with the idea of an obligate glucosyl-enzyme formation.

2. The α-configuration of the glucose residue in the glycogen polymer is retained in the glucose-1-P product, consistent with either a double-displacement mechanism or the existence of an enzyme-glucosyl C-1 carbonium ion that is shielded from backside attack and is accessible only to frontside attack by an oxygen of inorganic phosphate.
3. No partial exchanges (i.e., incorporation of [^{32}P]-P_i into glucose-1-P) are detectable. No covalent intermediates are isolable.

The first and third data are consistent with a shielded C-1 carbonium ion rather than a double-displacement mechanism.

oxocarbonium form

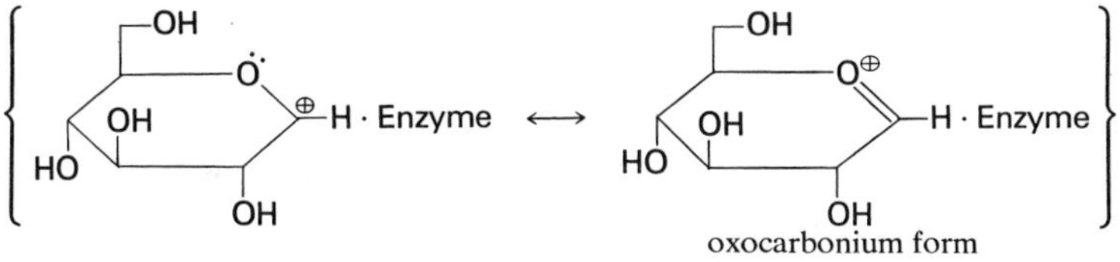

The carbonium-ion species would be a resonance hybrid with charge distribution on the oxygen, and the principal contributor to the structure is probably the oxocarbonium form.

A neighbouring $COO^{\ominus}$ from the enzyme could provide electrostatic stabilization for such a carbonium ion. This ionic interaction could assist in maintenance of the initial geometry at C-1 while the glycogen$_{n-1}$ leaving group departs and a molecule of inorganic phosphate diffuses into the active site and captures the oxocarbonium ion specifically from the accessible frontside, with resultant overall retention of configuration at C-1.

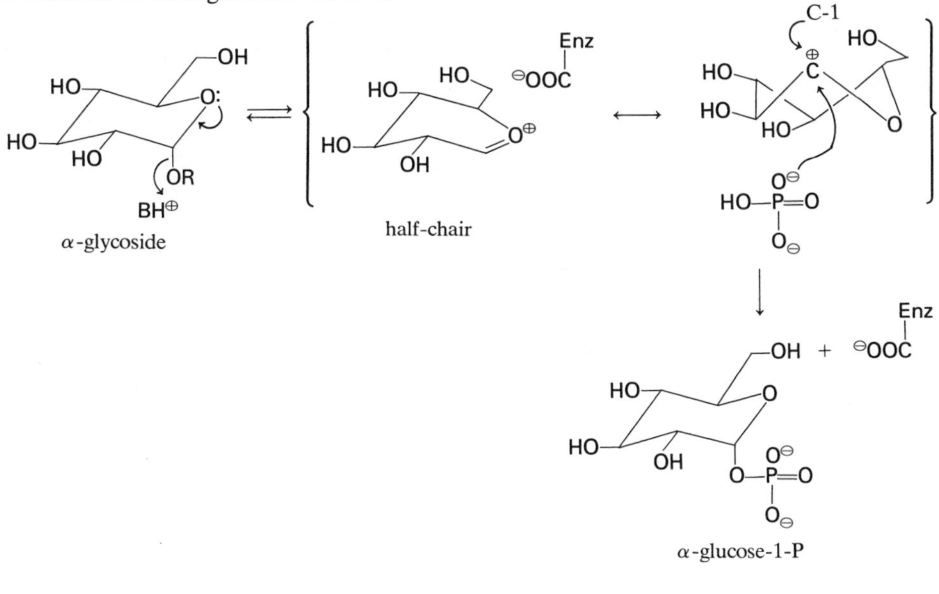

α-glycoside

half-chair

α-glucose-1-P

The geometry of the stabilized sp^2 C-1 carbonium ion could be a puckered ring or a "half-chair" form. In this connection, it has been observed that 1,5-gluconolactone is a very powerful reversible inhibitor of glycogen phosphorylase (Graves and Warn, 1972). One can argue that this compound may be a transition-state analogue, because X-ray analysis of gluconolactone crystals indicates a half-chair conformation.

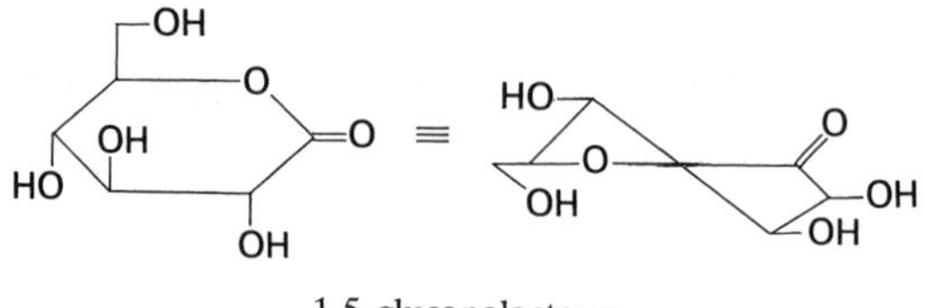

1,5-gluconolactone

We noted earlier in this chapter that the enzyme glycogen synthetase (responsible for sequential addition of glucosyl units to termini of glycogen molecules) is under tight metabolic regulation, including reversible stoichiometric phosphorylation and dephosphorylation by protein kinase or phosphoprotein phosphatase action. An equivalent regulatory situation exists with glycogen phosphorylase, and it was the Nobel-prize-winning investigations by Cori and Cori on this enzyme that initially unearthed covalent regulation of enzyme activity and cascade phenomena. Phosphorylase exists in two forms in liver or muscle cells. Phosphorylase *a* is active; phosphorylase *b* is relatively inactive. The muscle enzyme has three subunits (A, B, and C) with provisional stoichiometry $A_4B_4C_8$ and about 1.3×10^6 mol wt (Stalmans and Hers, 1973; Hers, 1976). Phosphorylase *a* has one phosphorylated serine residue per B subunit, and one per C subunit. The active enzyme is converted to inactive phosphorylase *b* by phosphatase action. The activation from phosphorylase *b* back to *a* is effected by the now-familiar cAMP-activated protein kinase, and this activation accounts for hormonal sensitivity of glycogen phosphorolysis by the mechanism outlined in ¶9.A.2.

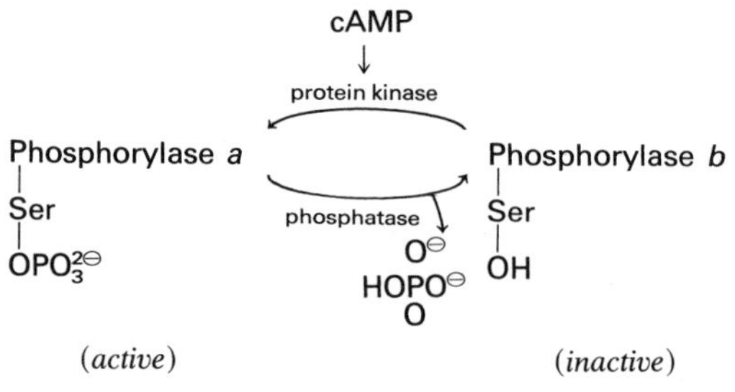

cAMP → protein kinase

Phosphorylase *a* — Ser — $OPO_3^{2\ominus}$ (*active*)

Phosphorylase *b* — Ser — OH (*inactive*)

phosphatase → $HOPO_3^{2\ominus}$

As a final comment about glycogen phosphorylase, we note that the enzyme contains a tightly bound cofactor (pyridoxal phosphate) bound in imine linkage to an ε-amino group of a lysyl residue (Graves and Warn, 1972). The enzyme resolved of this cofactor is inactive. In Chapter 25, we shall discuss the role of this coenzyme as an electron sink for stabilizing substrate carbanions in enzymatic transformation of α-amino acids. It does not fulfill such a mechanistic role in glycogen phosphorylase. Rather, it appears to have a structural role in induction of active enzyme conformation.

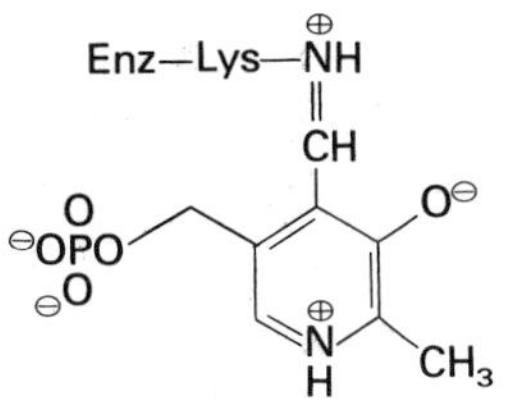

Enz—Lys—$\overset{\oplus}{N}H$ = CH ... $^{\ominus}OPO(O)(O^{\ominus})$ ... $O^{\ominus}$ ... $\overset{\oplus}{N}H$ ... CH_3

9.B.3 Oligosaccharide Hydrolases

We can make certain comparisons between glycoside hydrolases and other hydrolytic enzymes discussed previously, such as peptidases and nucleases. Just as we categorized endo- and exopeptidases, and endo- and exonucleases, so we can divide glycosidases into endoglycosidases and exoglycosidases. For instance, the highly branched α-1,4-glucose polymer amylose is cleaved at random internal positions by the endoglycosidase α-amylase, producing oligosaccharide fragments. On the other hand, the exoglycosidase β-amylase successively removes glucosyl-α-1,4-glucosyl (maltosyl) units from the nonreducing end of the polymer to yield the disaccharide maltose with an OH group in β-configuration at C-1. As implied in these two examples, an additional specification with glycosidases is their preference for α- or β-configuration at the susceptible C-1 linkage of the polysaccharide (or of the product released).

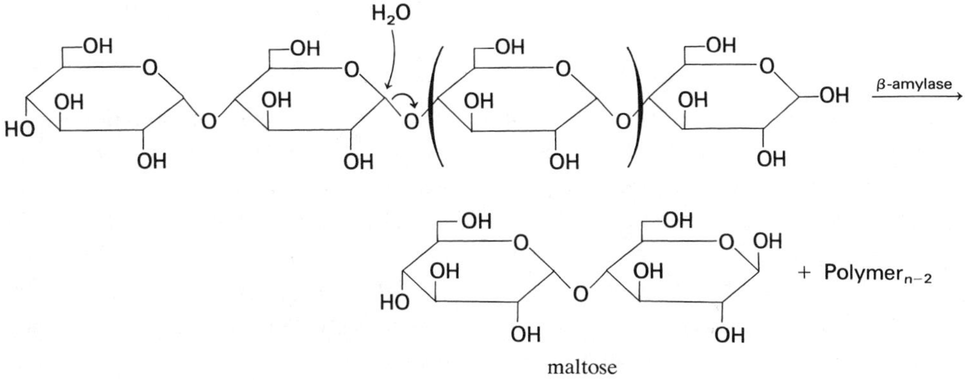

H_2O β-amylase → + $Polymer_{n-2}$

maltose

The functional group at C-1 of glycosides is an acetal, and hydrolysis produces the alcohol and the hemiacetal at C-1 of the monomeric sugar (Gray, 1971).

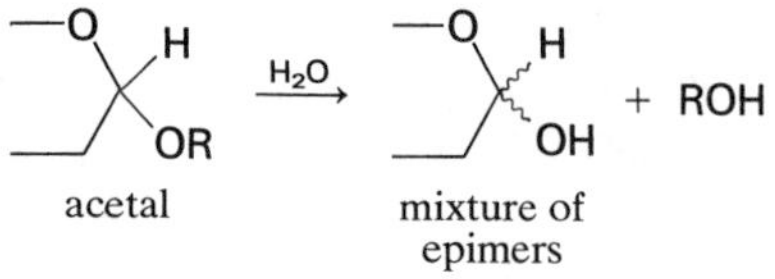

acetal $\xrightarrow{H_2O}$ mixture of epimers + ROH

The chemical (and enzymatic) hydrolysis of glycosides proceeds via cleavage of the (C-1)—OR bond, and this rate is facilitated by acid-catalyzed protonation of the departing alcohol.

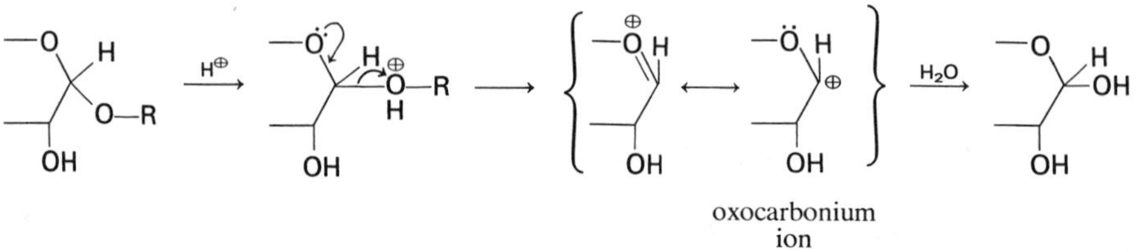

oxocarbonium ion

Loss of ROH produces a glycosyl C-1 carbonium ion, which is stabilized by the lone pair on the ring oxygen, and the major form contributing to the resonance hybrid is probably the oxocarbonium ion. The stabilized carbonium ion can then be attacked by water to yield the hemiacetal form of the product. The presence of partial —C=O$^{\oplus}$— double-bond character in the oxocarbonium ring system will distort this species from a chair form into a twisted or half-chair form to maximize the planarity required for resonance stabilization.

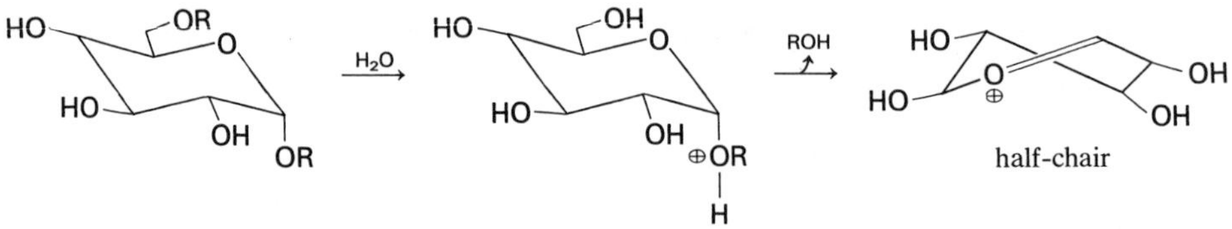

half-chair

Some major questions in enzymatic hydrolysis of glycosides have developed from these model data. Does the oxocarbonium mechanism apply? Does an oxocarbonium intermediate collapse to a glycosyl-enzyme covalent intermediate on attack by an anionic enzyme nucleophile? What is the rate-determining step in these enzymes? If covalent adducts form, do they accumulate, and are they then directly detectable? We shall try to provide information on these questions by examining data for three of the better-studied glycosidases: β-galactosidase, sucrase-isomaltase (an α-glucosidase), and lysozyme.

9.B.3.a β-Galactosidase

The β-galactosidase from *E. coli* is the most readily available and best-characterized representative of the widespread class of β-galactosidases. These enzymes are responsible for degradation of galactosyl β-linkages in glycolipids, glycoproteins, polysaccharides, and such disaccharides as lactose. Some *E. coli*, induced for the lactose-utilization operon (see Beckwith and Zipser, 1970), have amounts of β-galactosidase up to 5% of the total cellular protein content. The *E. coli* enzyme is a tetramer of 520,000 mol wt, so each subunit is about 1,100 amino acids long. The enzyme is specific for β-galactosides; α-galactosides are competitive inhibitors. Most thiogalactosides

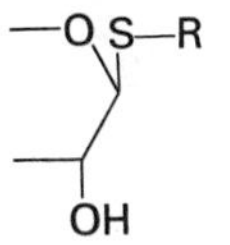

—O S—R

OH

are very poor substrates, whereas β-D-galactosylazide is an excellent substrate:

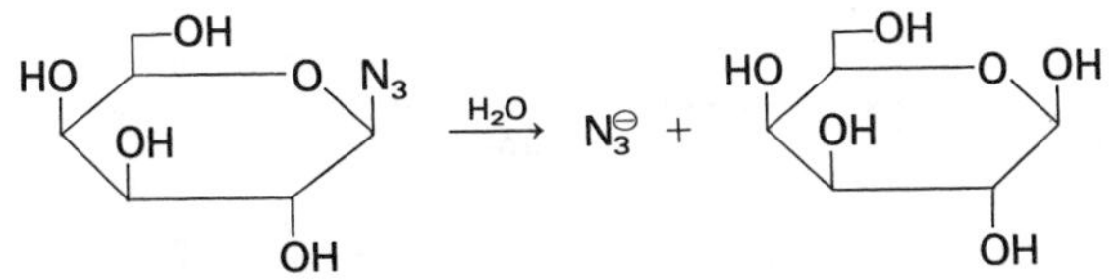

$$\text{HO, OH, OH, OH (galactosyl)}-N_3 \xrightarrow{H_2O} N_3^{\ominus} + \text{galactose (OH)}$$

As with some other enzymes (proteases, alkaline phosphatase) that use water as acceptor, alternative nucleophiles such as alcohols can function as cosubstrates for β-galactosidase, resulting in transglycosylation sequences.

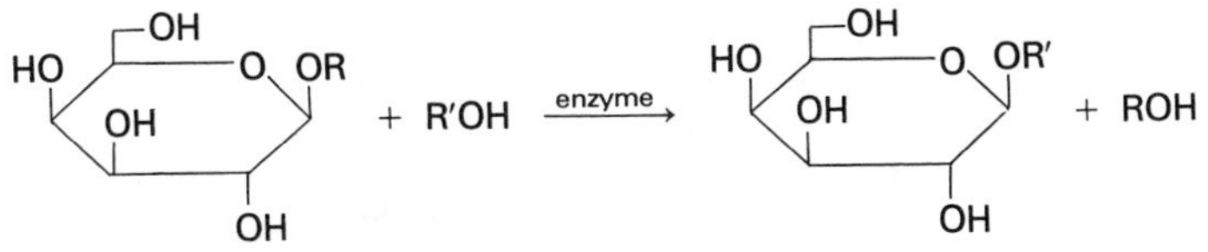

$$\text{Gal}-OR + R'OH \xrightarrow{\text{enzyme}} \text{Gal}-OR' + ROH$$

In both the hydrolysis and the alcoholysis processes, overall retention of configuration at the galactosyl C-1 is observed, consistent with either of two possible mechanisms: a double displacement (two S_N2 steps), or a shielded oxocarbonium ion with frontside attack by the incoming nucleophile.

Many of the kinetic experiments on β-galactosidase have been carried out with chromogenic β-galactosides such as *ortho*- or *para*-nitrophenyl-β-D-galactoside. Reaction is easily monitored by appearance of yellow color due to liberation of the nitrophenolate anions (recall the *p*-nitrophenyl-acetate experiments with chymotrypsin, ¶3.B.5).

HO OH O HO OH O NO_2 ONPG —enzyme→ HO OH O HO OH OH + $^{\ominus}O$ NO_2 yellow

Remember that this assay yields information on the rate of appearance of the phenolate species, but not on the rate of free β-D-galactose appearance. With *ortho*-nitrophenyl-β-D-galactoside (ONPG) as substrate and β-mercaptoethanol as acceptor competing with H_2O, the data of Figure 9-4 were obtained (Wallenfels and Weil, 1972). The data indicate that the rate of free galactose production declines with increasing thiol, suggesting partitioning of an intermediate between water (galactose formation) and mercaptoethanol (β-thiogalactoside formation), consistent with the scheme written for proteases.

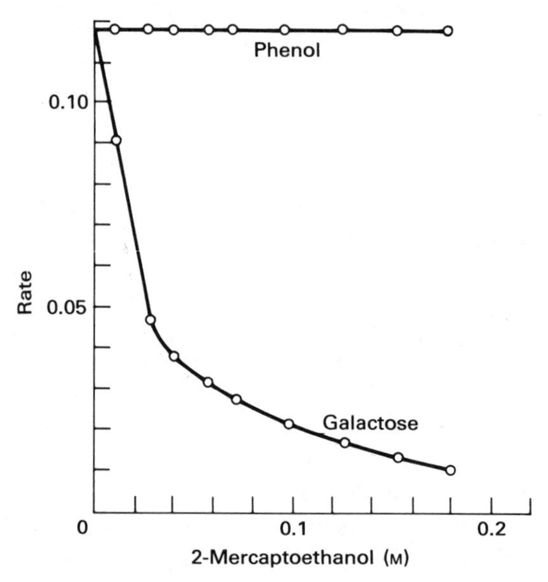

Figure 9-4
Effect of 2-mercaptoethanol on the total reaction rate (liberation of phenol) and on the hydrolysis rate (liberation of galactose of phenyl β-D-galactoside cleavage). (From Wallenfels and Weil, 1972, p. 649)

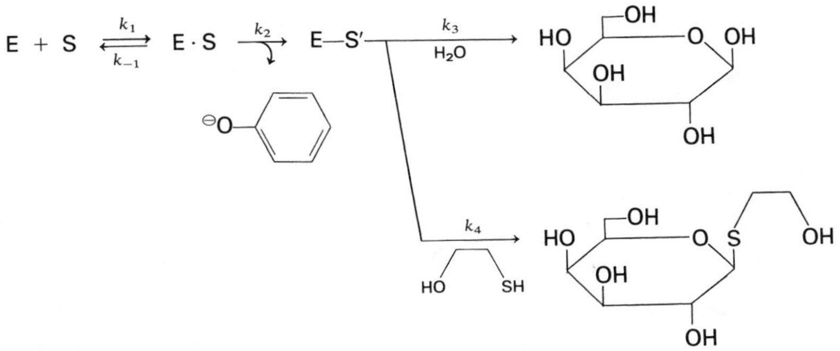

$$E + S \underset{k_{-1}}{\overset{k_1}{\rightleftharpoons}} E\cdot S \xrightarrow{k_2} E{-}S'$$

Further, the β-mercaptoethanol increases nitrophenolate production by only about 1.3-fold, so $k_2 \approx k_3$ for *ortho*-nitrophenyl-β-galactoside. At lower temperatures (such as $-30°$ C), k_2 is again somewhat larger than k_3, because an initial burst of *o*-nitrophenolate (corresponding to 61% of the enzyme active sites) is observed (Angelides and Fink, 1976; Fink, 1977). These data suggest that ES′ is a galactosyl-enzyme complex. This complex could be a galactosyl C-1 carbonium ion stabilized by electrostatic interaction, or a covalent galactosyl-enzyme compound (which presumably would have an α-linkage at C-1). Sinnott and colleagues have examined a variety of related aryl-β-galactosides

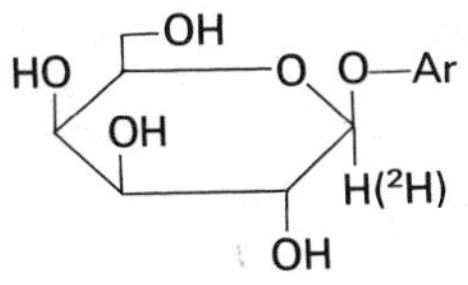

for secondary deuterium effects (see ¶4.A.7) and have found a correlation of increasing k_H/k_D to a high of 1.34 for 2,4-dinitrophenyl-β-D-[^{1}H or ^{2}H]-galactoside in the presence of methanol as coacceptor (Sinnott and Souchard, 1973). This is the highest deuterium kinetic secondary isotope effect seen in an enzymatic reaction (Kirsch, 1976). It suggests that rehybridization from sp^3 (tetrahedral) to sp^2 (trigonal) at C-1 occurs in the rate-determining transition state; i.e., oxocarbonium-ion formation (subsumed in k_2) is slow.

Wallenfels has raised another possible structure for the kinetically detectable ES′ in hydrolysis of the *model* β-galactosides, based on the fact that the 2-OH in the α-configuration is requisite for good substrate ability (Wallenfels and Weil, 1972). The suggestion is that the two-step sequence for overall retention involves a galactose-1,2-epoxide, and that the C-2 hydroxyl acts as intramolecular nucleophile to assist departure of the leaving group. Wallenfels argues that such an epoxide would be in a twisted conformation resembling a half-chair.

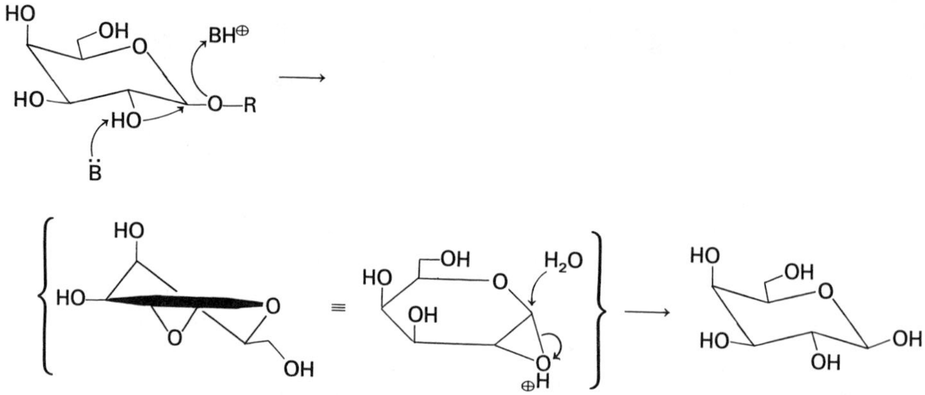

Subsequent opening of the epoxide by water would be from the top at C-1 to give β-D-galactose. This possibility cannot yet be ruled out.

One additional activity of β-galactosidase deserves mention. An olefinic analogue of galactose, β-D-galactal, is a powerful reversible inhibitor of β-galactosidase (Wentworth and Wolfenden, 1974). X-ray analysis indicates that galactal exists as a half-chair form, and it was argued that galactal may thus inhibit the enzyme as a transition-state analogue (Wallenfels and Weil, 1972).

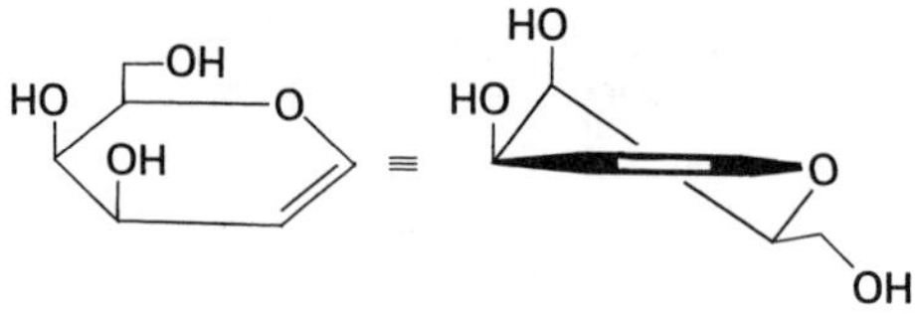

D-galactal

The picture is somewhat more complicated, though, because β-galactosidase actually recognizes D-galactal as a substrate, converting it (at 10^{-4} the rate for a normal β-galactoside) to β-2-deoxy-D-galactose. This is a hydration reaction on a double bond.

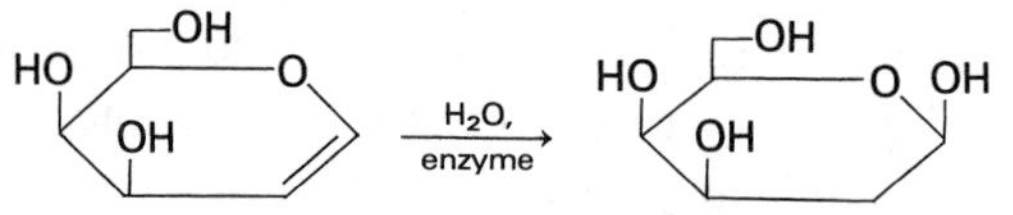

β-D-galactal $\xrightarrow{H_2O,\ \text{enzyme}}$ β-2-deoxy-D-galactose

On close examination, Wentworth and Wolfenden (1974) noted that, in the inhibition behavior, D-galactal produces a slow time-dependent inhibition that is reversible on dilution into galactal-free solutions. The rate constant for onset of inhibition is $2.7 \times 10^2\ \text{M}^{-1}\ \text{sec}^{-1}$, a bimolecular rate constant 10^4-fold slower than most bimolecular rate constants for association of enzyme and β-galactoside. This time dependence could mean that inhibition arises from covalent addition of an active-site enzyme nucleophile to galactal. Dilution favors elimination of the enzyme and release from inhibition. The even slower hydration to 2-deoxygalactose could occur by occasional displacement of Enz—X by H_2O bound at the active site.

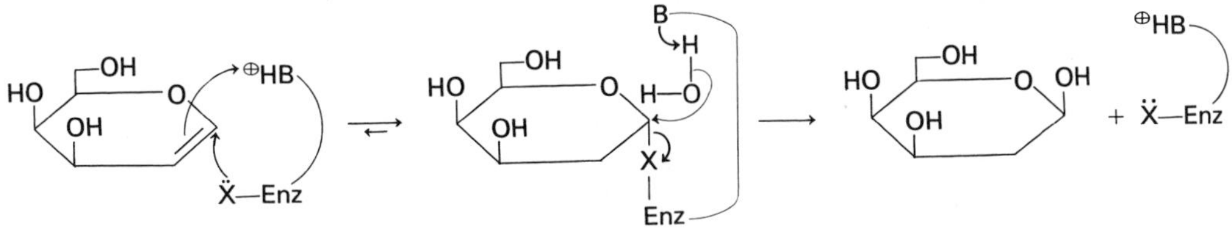

inhibited enzyme 2-deoxy-D-galactose

9.B.3.b Sucrase-isomaltase

The bifunctional enzyme sucrase-isomaltase is an α-glucosidase that can be solubilized from the brush-border membranes of intestinal epithelial cells by papain treatment. The resultant enzyme is a lipid-free glycoprotein of 220,000 mol wt, containing two subunits. The subunits are dissociable in mild alkali into isolated A and B chains that are active (Quaroni et al., 1974). Table 9-3 indicates the specificities for the A and B subunits. Isomaltose is a glucosyl-α-1,6-glucose disaccharide.

Sucrase-isomaltase has been investigated because of its possible role in transport of the disaccharides from the intestinal lumen across the epithelial membrane and in deposition of the glucose (and fructose) monomers inside the epithelial cells.

Both sucrase and isomaltase activities lead to retention of configuration at the transferring glucosyl C-1. Initial-velocity kinetics are ping-pong BiBi for each

Table 9-3
Substrate specificities of A and B subunits of sucrase-isomaltase

A subunit	B subunit
Maltose	Maltose
Aryl-α-glycosides	Aryl-α-glycosides
Isomaltose	Sucrose

specific disaccharide. These two data suggest a covalent intermediate in a double-displacement sequence.

Aryl-α-glucosides, labeled with ^1H or ^2H at C-1, were examined for secondary kinetic deuterium-isotope effects during hydrolysis. With the substrate

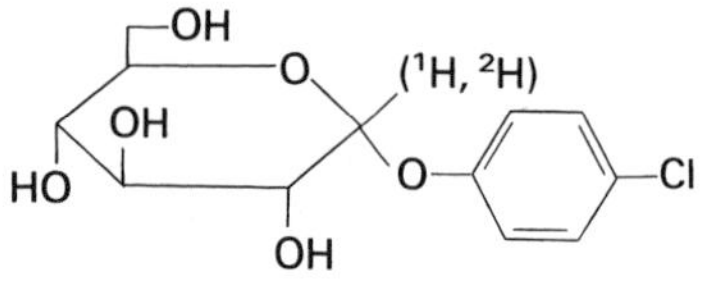

k_H/k_D ratios of 1.16 to 1.21 were obtained for the intact enzyme, and a ratio of 1.14 was obtained for the isomaltase subunit (subunit A). This formation suggests that the transition state leading to the glucosyl-oxocarbonium ion is the rate-determining step for this substrate. (The ping-pong kinetic data could hold if a covalent glucosyl enzyme formed *after* the rate-determining step.)

In an effort to determine whether an enzyme carboxylate group might be at the active site as a potential partner in ion-pair formation with the glucosyl-oxocarbonium ion to stabilize it (and/or as a potential nucleophile), Semenza and colleagues added a sugar epoxide (conduritol-β-epoxide) to the bifunctional enzyme. The epoxide behaved as an active-site–directed irreversible inactivator, labeling both the sucrose and the isomaltose subunits stoichiometrically (Cogoli and Semenza, 1975; Stefani et al., 1975). Lability of tritiated inactivator–enzyme compound to NH_2OH suggests an oxygen-ester functionality, and suggests also that a carboxyl oxygen from the enzyme glucosyl-binding site acts as nucleophile. Sugar epoxides have similarly been used to trap active-site carboxyl groups in β-glucosidases and lysozyme (Legler, 1966, 1970; Thomas et al., 1969).

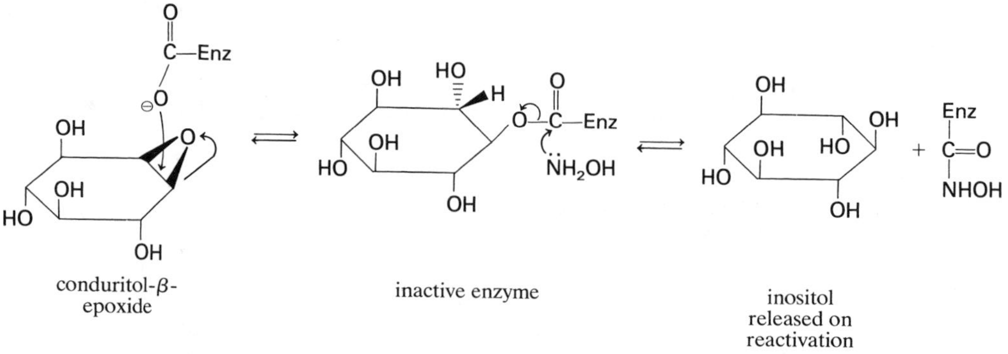

conduritol-β-epoxide

inactive enzyme

inositol released on reactivation

The carboxylate is pictured as attacking at C-1 from above the plane of the epoxycyclohexane ring, although neither of these points has been established.

Given an Enz—COO$^{\ominus}$, we can sketch a scheme for sucrase action (Quaroni et al., 1974).

Enz HO OH O C=O O$^{\ominus}$ HO HO O Fructose BH$^{\oplus}$ ⇌ Enz HO OH O C=O O$^{\ominus}$ HO HO OH$^{\oplus}$ Fructose B

slow

Products ⟵ Enz C=O HO OH O O HO OH ⟵ fast Enz HO OH O$^{\oplus}$ C=O O$^{\ominus}$ HO HO B + Fructose

Whether the ion-pair–stabilized carbonium ion collapses to a covalent glucosyl-enzyme intermediate after the rate-determining step is open to speculation and, even then, it may be a nonproductive side path. Any compound forming after the rate-determining step will not accumulate and cannot be detected easily.

9.B.3.c Lysozyme

The enzyme lysozyme was discovered by Alexander Fleming (1922), who noted that a drop of his nasal mucus (he had a cold that day) would dissolve, or *lyse*, bacteria on a culture plate. The enzyme activity proved not to have clinical value, although Fleming did find penicillin seven years later. Lysozyme is present in quantity in egg white, and it has been crystallized from this source, its primary sequence of 129 amino acids elucidated (Canfield, 1963, 1965), and an X-ray map determined at 2 Å resolution—the first X-ray map of an enzyme (Blake et al., 1967; Blackburn, 1976, chap. 10). The physiological activity of bacterial-cell lysis is secondary to the glycosidase action of lysozyme; the enzyme hydrolyzes bacterial cell-wall polysaccharide at specific loci, and the disintegration of this rigid layer in the cell wall leaves the bacteria osmotically fragile and sensitive to lysis.

The basic repeating unit of bacterial cell walls is a disaccharide of two types of modified glucose residues in β-1,4 linkage. Both glucosyl units have *N*-acetylamine groups at C-2. In addition, one of the glucosyl units has an ether linkage to lactic acid at C-3 and is known as *N*-acetylmuramic acid. The *N*-acetylglucosamine (NAG) and *N*-acetylmuramic acid (NAM) alternate in the polymer. In the cell wall, the lactic-ether carboxyl group is in amide linkage to peptides, indicated here by NHR.

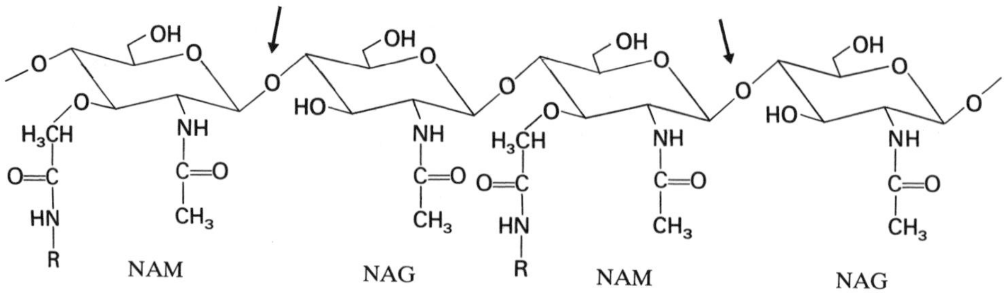

NAM NAG NAM NAG

When lysozyme acts on this physiological substrate, it acts specifically as an *N*-acetylmuramidase, transferring the *N*-acetylmuramyl group to H_2O and so cleaving at the points indicated by the bold arrows. The enzyme *does not cleave* the polysaccharide at C-1 of the *N*-acetylglucosaminyl residues. On the other hand, the enzyme will cleave chitin (β-1,4-poly-*N*-acetylglucosamine, the exoskeletal polysaccharide of many insects) into oligosaccharide fragments. Many kinetic experiments have been performed with model substrates of well-characterized structure, such as *N*-acetylglucosamine oligosaccharides and β-aryl-*N*-acetylglucosides. The X-ray analysis indicates a cleft in the lysozyme tertiary structure, and analysis of the crystal structure of an enzyme–$(NAG)_3$ complex shows this to be the substrate-binding region (Blackburn, 1976, chap. 10). There is a considerable degree of nonproductive binding of small oligosaccharides at the lysozyme active site, indicating that these small substrates can bind in several modes, many of them nonproductive for hydrolysis. Larger oligosaccharides apparently bind predominantly in productive modes for hydrolysis, as indicated in Table 9-4 by the rate data for hydrolysis.

Apparently, the NAG hexamer is long enough to fill the active region with proper positioning for effective hydrolysis. The initial products from lysozyme action on the $(NAG)_6$ substrate are the two fragments $(NAG)_4$ and $(NAG)_2$. The susceptible glycosidic linkage is between the fourth and fifth residues of the bound NAG hexamer.

$$(NAG)_6 \rightarrow (NAG)_4 + (NAG)_2$$

Earlier crystallographic model building had suggested that the hexamer could fit into the known crystal structure, with the glycosyl moieties A through F (Fig. 9-5) positioned near the indicated protein residues, and that sugar residue D would be

Table 9-4

Relative rates of enzymatic hydrolysis for NAG polymers

Substrate	Relative rate
$(NAG)_2$	0
$(NAG)_3$	1
$(NAG)_4$	4,000
$(NAG)_6$	30,000
$(NAG)_8$	30,000

SOURCE: Data from W. Stalmans and H. Hers, in *The Enzymes*, 3d ed., ed. P. Boyer (New York: Academic Press), vol. 9 (1973), p. 310; H. Hers, *Ann. Rev. Biochem.* 45 (1976): 167; S. Blackburn, *Enzyme Structure and Function* (New York: Marcel Dekker, 1976), chap. 10.

distorted from its normal chair configuration due to steric crowding at that subsite (Blake et al., 1967). This strain induction would make the D—O—E glycosidic bond the kinetically susceptible link—in agreement with the experimental results cited earlier. We shall return to this distortion shortly.

When the *N*-acetylglucosamine tetramer is incubated with lysozyme in the presence of alternate acceptors (ROH), transglycosylation is detectable. Methanol will replace water, producing β-methylglycosides. The $(NAG)_2$ molecule, itself

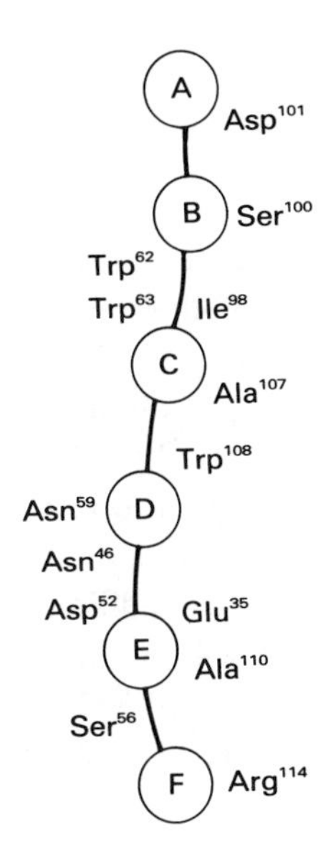

Figure 9-5
A schematic illustration of the active site in the cleft region of lysozyme. **A** through **F** represent the glycosyl moieties of a hexasaccharide. Some of the amino acids in the cleft region near these subsites of the active site are shown. (From S. Blackburn, 1976, p. 455).

inactive for cleavage, will act as acceptor for transglycosylation, leading to hexamer production (Blackburn, 1976, chap. 10).

$$(NAG)_4 + (NAG)_2 \rightarrow (NAG)_6$$

The kinetics can quickly become complicated because hexamer thus formed is also a substrate for cleavage. The existence of the transglycosylation sequence suggests a partitionable intermediate during lysozyme catalysis.

The stereochemical outcome of transglycosylation is overall retention of the transferred *N*-acetylglucosaminyl residue because β-glycosidic bonds are formed in the products. Although the retention result might suggest a double-displacement mechanism, secondary kinetic isotope data indicate that the rate-determining transition state has much sp^2 character at C-1 of the glycosyl group undergoing transfer; i.e., the data suggest that the formation of the C-1 oxocarbonium-ion species is the slow step in catalysis. The initial experiments by Raftery and colleagues were, in fact, the first use of secondary isotopic effects to probe an enzymatic mechanism, and they involved the use of β-arylglycosides in a clever double-label experiment (Dahlquist et al., 1969). The substrates were disaccharides with a β-phenoxy leaving group.

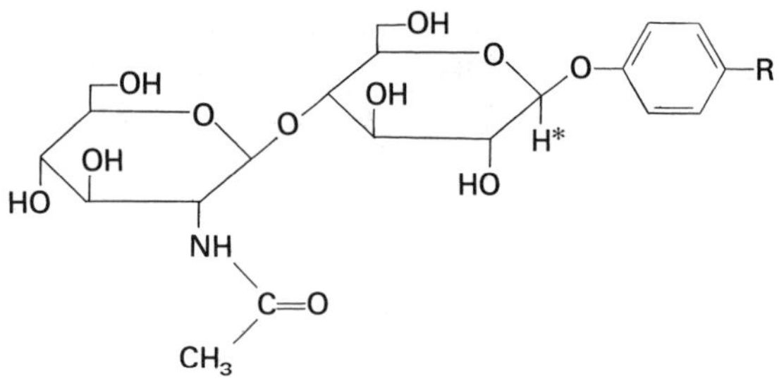

OH
O
O
OH
R
OH
O
O
OH
H*
HO
HO
NH
C=O
CH3

The hydrogen represented by H* was hydrogen or deuterium for measurement of the rate difference due to kinetic isotopic effects on substitution of the heavier hydrogen isotope in this secondary position. The C-1–protio-containing molecules also had tritium in the phenolic leaving group; the C-1–deutero-containing molecules had ^{14}C label in the phenolic group. Any isotopic effect could then be measured by isolating the phenol products from incubations containing protio and deutero materials and determining the $^3H/^{14}C$ ratios in the products compared to starting materials. Experiments were performed under initial-velocity conditions. If k_1 is greater than k_2, the $^3H/^{14}C$ ratio in the substrate mixture will be larger than the ratio in the product mixture. In fact, the observed ratio change is 1.11 (in the product phenol mixture); thus 1.11 is the magnitude of the secondary kinetic isotopic effect.

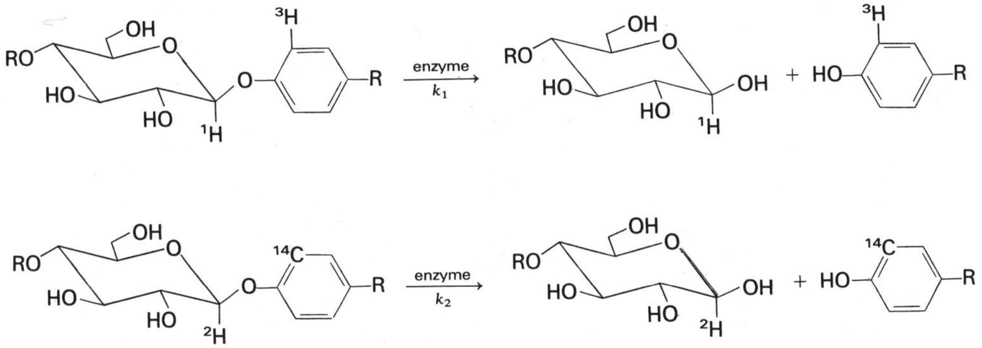

Chemical experiments were performed on the hydrolysis of these substrates in acid (which is known to proceed by the S_N1-type oxocarbonium mechanism) and in base (where an S_N2-type displacement by $OH^{\ominus}$ occurs); Table 9-5 summarizes the results. The value for lysozyme approaches that for a known oxocarbonium mechanism: the β-glucosidase from almonds shows essentially no secondary kinetic isotopic effect, indicating that the rate-determining step is not an $sp^3 \rightarrow sp^2$ change at C-1 of the glycosyl group in that enzyme. Although these initial results were with model aryl-β-glycoside substrates and thus open to some debate whether the same rate-determining step would operate in oligosaccharide hydrolysis, subsequent work from the same research group with similarly labeled $(NAG)_3$ molecules confirmed a secondary deuterium kinetic isotopic effect of 1.14 with the chitotrioses (Smith et al., 1973).

The X-ray extrapolations indicated in Figure 9-6 suggest that two enzymatic acidic residues, Asp^{52} and Glu^{35}, are on either face of the susceptible glycosyl group in the hexameric NAG substrate. Selective chemical modification with radioactive triethyloxonium fluoroborate produced complete inactivation with stoichiometric labeling; structural analysis indicated that a β-ethylaspartyl enzyme had formed, with unique derivatization of the β-carboxyl group of Asp^{52}.

Table 9-5
Secondary isotope effects in hydrolysis of disaccharides with β-phenoxy leaving group

Reagent	Secondary isotope effect (k_H/k_D)
Acid	1.13
Base (methoxide)	1.03
Lysozyme	1.11
β-Glucosidase	1.01

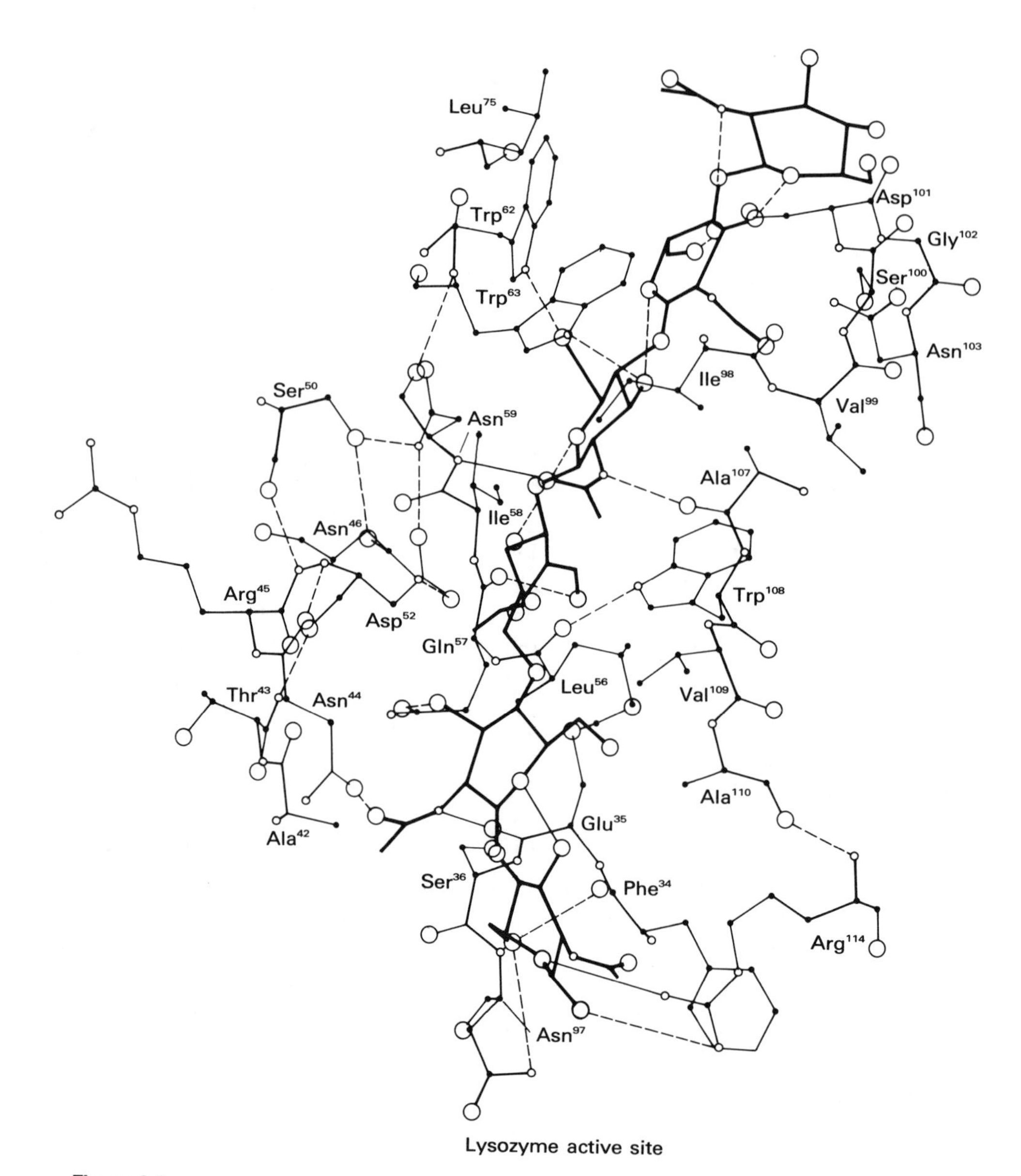

Figure 9-6
Binding to lysozyme of hexa-*N*-acetylglucosamine. The three sugar residues in the upper half of the diagram are bound in the way observed for tri-*N*-acetylglucosamine. The mode of binding to the three in the lower half was derived by model building. The positions of atoms in the enzyme are those observed in the native structure. (From S. Blackburn, 1976, p. 456).

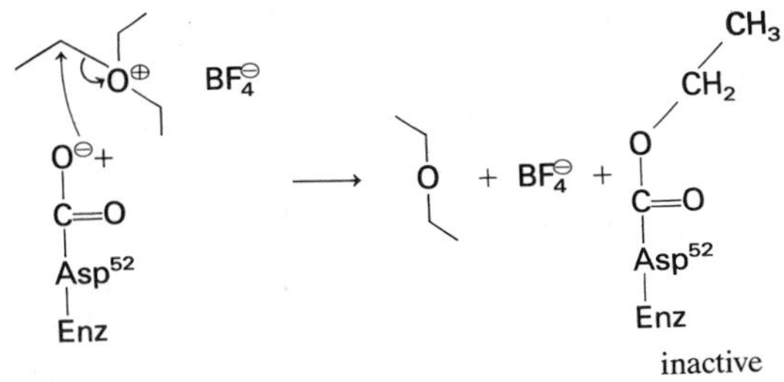

It has been argued that Asp^{52} is in a hydrophilic microenvironment, whereas Glu^{35} is located more internally in a hydrophobic pocket. Two pK_a values of 4.7 and 6.1 have been titrated (Blackburn, 1976, chap. 10) that are essential to enzyme activity, and these may represent ionizations of Asp^{52} and Glu^{35}; the pK_a of 6.1 is strongly perturbed for a carboxylic acid, but the hydrophobic milieu of Glu^{35} would have this influence. If the pK_a assignments are correct, at pH 5 (where lysozyme is optimally active) Asp^{52} would have a β-$COO^{\ominus}$ whereas Glu^{35} would have a γ-COOH. The former group is a good candidate for electrostatic stabilization of the developing glycosyl-oxocarbonium ion; the latter group is a good candidate for a general acid catalyst to protonate the leaving group.

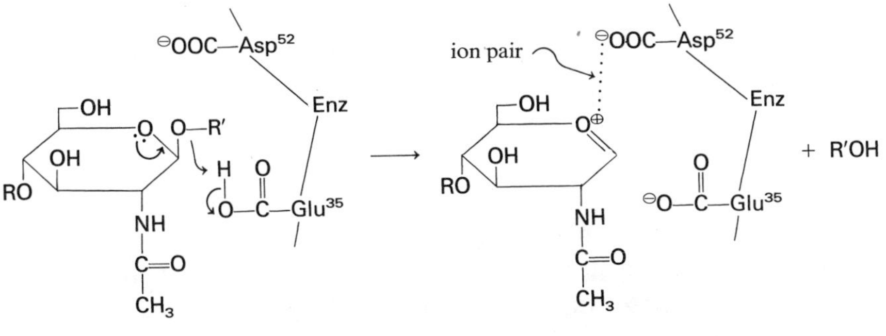

Estimated distances of closest approach are 3 Å for the β-aspartyl carboxylate oxygen anion to the C-1–O-5 carbonium ion, and also about 3 Å for the γ-COOH proton to the leaving group oxygen.

As indicated earlier, several pieces of evidence have suggested that the glycosyl residue that will be transferred to H_2O (or to an ROH) is strained on binding to the enzyme. The X-ray model-building indicated that the C-6 hydroxymethyl group of this saccharide residue is bulky enough to interfere with binding to the protein in the normal chair conformation that is the low-energy form in solution. The hydroxymethyl group may serve as a lever used by the enzyme to force this glycosyl group into a distorted, half-chair conformation—a destabilization that puts the ES complex at higher energy than unbound E and S, thus effectively lowering the activation energy ($\Delta G^{0\ddagger}$). This effect will be expressed as a rate acceleration.

Equilibrium-binding studies with oligosaccharides indicate that, in the hexosaccharide $(NAG)_6$, binding energies from rings A, B, and C are favorable at −1.8, −3.7, and −5.7 kcal/mole, respectively. Ring D, however, gives a net positive ΔG of binding of +2.9 kcal/mole; this apparent unfavorable interaction reflects the fact that the intrinsic binding energy of interaction between enzyme and the D-ring has been used for destabilization toward a distorted, strained ring. This distortion is paid for; the *observed* binding energy is unfavorable. If a xylosyl group is placed into a substrate at this position, the observed binding energy due to this saccharide is −2.2 kcal/mole (Van Eikeren and Chipman, 1972)—confirmation that the hydroxymethyl group at C-6 is the structural lever to induce ring strain (Jencks, 1975). Finally, Lienhard and colleagues have prepared a tetrasaccharide substrate analogue that competitively inhibits lysozyme with an estimated greater affinity than $(NAG)_4$ by 3,600-fold (Secemski and Lienhard, 1971; Secemski et al., 1972). The analogue is $(NAG)_3$-*N*-acetylglucono-δ-lactone, and this tight binding may qualify it as a transition-state analogue.

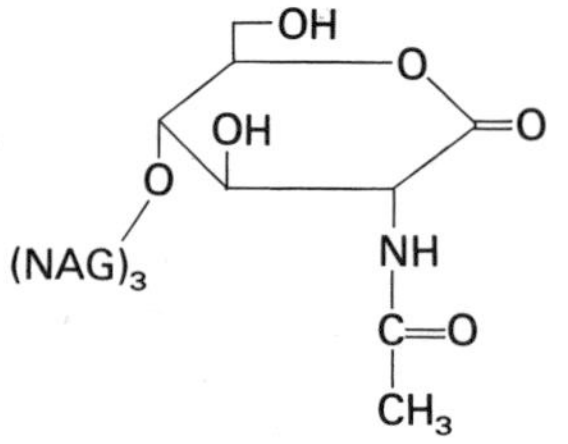

OH
O
OH
=O
O
$(NAG)_3$
NH
C=O
CH_3

X-ray analysis of the lysozyme–analogue complex indicates that the lactone ring is in a boat form. The tighter observed binding may reflect a more complete expression of the intrinsic energy of interaction between enzyme and substrate, because distortion does not have to be induced and paid-for from binding energy; it preexists in the lactone analogue in solution.

Thus, any attempt at describing the lysozyme mechanism of catalysis should feature induction of ring strain on binding, slow formation of the half-chair oxocarbonium, and attack by water or other alchols specifically at the face from which the leaving group departed (overall retention). It is not clear that one can rule out a post–rate-determining collapse to a covalent glycosyl enzyme that will be rapidly hydrolyzed and so will not accumulate. The transglycosylation reaction requires an intermediate: that intermediate could be such a covalent species *or* a long-lived oxocarbonium ion.

Throughout the last part of this chapter, we have discussed evidence that in aggregate suggests that several oligosaccharide hydrolases form carbonium-ion species as intermediates. In fact, carbonium-ion intermediates may be otherwise quite rare in enzymatic catalysis. Thus, these carbonium ions are specially stabilized by delocalization via the oxocarbonium-ion contributing forms. A second carbonium-ion case appears to be in enzymatic alkylations carried out by prenyl transferases (discussed in Chapter 26), where stabilized allylic carbonium-ion species are formulated. (The hypothesis that fumarase, discussed in Chapter 17, eliminates the elements of H_2O from *S*-malate via a carbonium ion is not overwhelming.) We shall note in Section V of this book that stabilized carbanion species are much more common entities in enzymatic catalysis.

This look at lysozyme action completes both our examination of enzyme-catalyzed glycosyl transfers and our survey of the more general topic of Section II, enzyme-catalyzed group transfers.

Section III

ENZYMATIC OXIDATIONS AND REDUCTIONS

The focus in this section of the text (Chapters 10 through 16) shifts from enzymes that catalyze group-transfer processes to enzymes that catalyze the oxidation of some substrate molecule and the concomitant reduction of some cosubstrate or cofactor. Most organic substrates undergo two-electron oxidations; we shall note a few that undergo four-electron oxidations. When one molecule is oxidized with the enzyme as catalyst, another must be reduced: the electrons removed from the molecule undergoing oxidation must be transferred to some acceptor. Molecular oxygen is a common electron acceptor in biological systems, undergoing 2-$e^{\ominus}$ reduction to H_2O_2 and also 4-$e^{\ominus}$ reduction to H_2O. Among the most complex of enzyme-mediated redox processes are the reductions of inorganic compounds—e.g., the 6-$e^{\ominus}$ reduction of sulfite to sulfide by sulfite reductase.

A substrate is oxidized when electrons are removed from it, but where do the electrons go? Some electron sink is required, and we have noted before that there are no obvious electrophilic groups in the side chains of amino acids found in proteins. Thus, enzymes carrying out redox transformations show (virtually without exception) obligate requirements for either an organic coenzyme molecule to act as electron acceptor (or donor in the reverse direction) or a transition metal, as a conduit for passage of electrons out of substrate to some acceptor molecule and to facilitate the rate of such transfer.

We shall discuss oxidations including nicotinamide, flavin, and pterin coenzymes, and then discuss metalloenzymes that use either copper or iron as the redox component in catalysis. We shall also note how molecular oxygen may be reduced and, in some enzymatic cases, activated to insert either one (monooxygenation) or both oxygen atoms (dioxygenation) into the product molecules that are formed.

Chapter 10

Enzymatic Oxidations and Reductions via Apparent Hydride Transfers: Nicotinamide Coenzymes

The bulk of this chapter is devoted to discussion of our first category of redox enzymes: those that use nicotinamide coenzymes. Before we begin that discussion, however, we comment on some general properties of redox processes.

10.A OXIDATION–REDUCTION PROCESSES: SOME INTRODUCTORY COMMENTS

For most organic molecules in biological systems, there are four stable oxidation states at carbon atoms; oxidation consists of the removal of electrons from the molecule in question. In each of the two-electron transfers indicated, there must also be an acceptor molecule present that concomitantly undergoes reduction.

$$\underset{\text{alkyl}}{R{-}CH_3} \underset{+2e^{\ominus}}{\overset{-2e^{\ominus}}{\rightleftarrows}} \underset{\text{alcohol}}{R{-}CH_2{-}OH} \underset{+2e^{\ominus}}{\overset{-2e^{\ominus}}{\rightleftarrows}} \underset{\text{aldehyde}}{R{-}CH{=}O} \underset{+2e^{\ominus}}{\overset{-2e^{\ominus}}{\rightleftarrows}} \underset{\text{acid}}{R{-}C(OH){=}O}$$

reduced ⟵——————————⟶ oxidized

In biological systems, oxidation is often coupled to energy production; that is, the energy released during oxidation is not lost to the organism, but is trapped in a chemically useful form. We shall discuss an example of this in enzymatic oxidation of aldehydes to the oxidation level of acids, where the energy released is coupled to the formation of molecules that are "energy-rich," and that have high group-transfer potential. In fact, it is in the harnessing of oxidative energy that cells make a living.

Most stable organic compounds have singlet ground states—that is, all their electrons are spin-paired. Unless substrates have some structural feature that can stabilize radical species, one-electron transfers have high energy barriers. In fact, the stoichiometry of most oxidations of organic substrates to products in biological systems is in two-electron steps. Some four-electron and six-electron oxidations occur in enzyme-catalyzed redox processes.

10.A.1 Oxidation–Reduction (Redox) Potentials

As we have emphasized, no compound can become more reduced and gain electrons unless some other substance gives up electrons and becomes more oxidized. Oxidation and reduction sequences are coupled, and we can regard them as occurring in paired half-reactions:

$$\begin{array}{rcl} X_{ox} + ne^{\ominus} & \rightleftharpoons & X_{red} \\ Y_{red} & \rightleftharpoons & Y_{ox} + ne^{\ominus} \\ \hline X_{ox} + Y_{red} & \rightleftharpoons & X_{red} + Y_{ox} \end{array}$$

One can measure the tendency for a molecule (e.g., X_{ox}) to acquire electrons and become reduced. This reduction potential is measured by determining the electrical potential difference that is detected when an electrochemical half-cell containing oxidized and reduced forms of X is connected to some reference half-cell (e.g., the hydrogen electrode) (Glasstone, 1962; Mahler and Cordes, 1966). In this way, a table of reduction potentials (Table 10-1) can be constructed for redox processes of enzymatic importance. Values listed for the reduction potential (E_0') are relative to the half-reaction $2\,H^{\oplus} + 2\,e^{\ominus} \rightarrow H_2$, for which $E_0' = -0.42$ volt at pH 7 (Segel, 1975*a*). (This half-reaction is defined to have $E_0' = 0.00$ volt at standard conditions of 1 M $H^{\oplus}$ and 1 atmosphere of H_2, but that is not a physiological set of conditions.)

To interpret the values of the reduction potentials, we must look at what happens for the coupling of any two of the half-reactions. The half-reaction with the more positive potential (greater tendency to accept electrons) proceeds as a reduction—that is, in the direction shown in the table. Suppose we want to know whether pyruvate will be reduced to lactate by the coenzyme NADH (which we

Table 10-1
Standard reduction potentials of some oxidation–reduction half-reactions

Reaction	Half-reaction (written as a reduction)	E_0' at pH 7.0* (volts)
1	$\frac{1}{2}O_2 + 2\,H^{\oplus} + 2\,e^{\ominus} \rightarrow H_2O$	0.816
2	$Fe^{3\oplus} + 1\,e^{\ominus} \rightarrow Fe^{2\oplus}$	0.771
3	$2\,I^{\ominus} + 2\,e^{\ominus} \rightarrow I_2$	0.536
4	Cytochrome-a_3-$Fe^{3\oplus} + 1\,e^{\ominus} \rightarrow$ cytochrome-a_3-$Fe^{2\oplus}$	0.55
5	$SO_4^{2\ominus} + 2\,H^{\oplus} + 2\,e^{\ominus} \rightarrow SO_3^{2\oplus} + H_2O$	0.48
6	$NO_3^{\ominus} + 2\,H^{\oplus} + 2\,e^{\ominus} \rightarrow NO_2^{\ominus} + H_2O$	0.42
7	$\frac{1}{2}O_2 + H_2O + 2\,e^{\ominus} \rightarrow H_2O_2$	0.30
8	Cytochrome-a-$Fe^{3\oplus} + 1\,e^{\ominus} \rightarrow$ cytochrome-a-$Fe^{2\oplus}$	0.29
9	Cytochrome-c-$Fe^{3\oplus} + 1\,e^{\ominus} \rightarrow$ cytochrome-c-$Fe^{2\oplus}$	0.25
10	2,6-Dichlorophenolindophenol$_{(ox)} + 2\,H^{\oplus} + 2\,e^{\ominus} \rightarrow$ 2,6-DCPP$_{(red)}$	0.22
11	Crotonyl-S-CoA $+ 2\,H^{\oplus} + 2\,e^{\ominus} \rightarrow$ butyryl-S-CoA	0.19
12	$Cu^{2\oplus} + 1\,e^{\ominus} \rightarrow Cu^{\oplus}$	0.15
13	Methemoglobin-$Fe^{3\oplus} + 1\,e^{\ominus} \rightarrow$ hemoglobin-$Fe^{2\oplus}$	0.139
14	Ubiquinone $+ 2\,H^{\oplus} + 2\,e^{\ominus} \rightarrow$ ubiquinone-H_2	0.10
15	Dehydroascorbate $+ 2\,H^{\oplus} + 2\,e^{\ominus} \rightarrow$ ascorbate	0.06
16	Metmyoglobin-$Fe^{3\oplus} + 1\,e^{\ominus} \rightarrow$ myoglobin-$Fe^{2\oplus}$	0.046
17	Fumarate $+ 2\,H^{\oplus} + 2\,e^{\ominus} \rightarrow$ succinate	0.030
18	Methylene blue$_{(ox)} + 2\,H^{\oplus} + 2\,e^{\oplus} \rightarrow$ methylene blue$_{(red)}$	0.011
19	Pyruvate $+ NH_3 + 2\,H^{\oplus} + 2\,e^{\ominus} \rightarrow$ alanine	−0.13
20	α-Ketoglutarate $+ NH_3 + 2\,H^{\oplus} + 2\,e^{\ominus} \rightarrow$ glutamate $+ H_2O$	−0.14
21	Acetaldehyde $+ 2\,H^{\oplus} + 2\,e^{\ominus} \rightarrow$ ethanol	−0.163
22	Oxalacetate $+ 2\,H^{\oplus} + 2\,e^{\ominus} \rightarrow$ malate	−0.175
23	FAD $+ 2\,H^{\oplus} + 2\,e^{\ominus} \rightarrow FADH_2$	−0.18†
24	Pyruvate $+ 2\,H^{\oplus} + 2\,e^{\ominus} \rightarrow$ lactate	−0.190
25	Riboflavin $+ 2\,H^{\oplus} + 2\,e^{\ominus} \rightarrow$ riboflavin-H_2	−0.200
26	Cystine $+ 2\,H^{\oplus} + 2\,e^{\ominus} \rightarrow$ 2 cysteine	−0.22
27	GSSG $+ 2\,H^{\oplus} + 2\,e^{\ominus} \rightarrow$ 2 GSH	−0.23
28	$S^0 + 2\,H^{\oplus} + 2\,e^{\ominus} \rightarrow H_2S$	−0.23
29	1,3-diphosphoglyceric acid $+ 2\,H^{\oplus} + 2\,e^{\ominus} \rightarrow$ GAP $+ P_i$	−0.29
30	Acetoacetate $+ 2\,H^{\oplus} + 2\,e^{\ominus} \rightarrow \beta$-hydroxybutyrate	−0.290
31	Lipoate$_{(ox)} + 2\,H^{\oplus} + 2\,e^{\ominus} \rightarrow$ lipoate$_{(red)}$	−0.29
32a	$NAD^{\oplus} + 2\,H^{\oplus} + 2\,e^{\ominus} \rightarrow NADH + H^{\oplus}$	−0.320
32b	$NADP^{\oplus} + 2\,H^{\oplus} + 2\,e^{\ominus} \rightarrow NADPH + H^{\oplus}$	−0.320
33	Pyruvate $+ CO_2 + 2\,H^{\oplus} + 2\,e^{\ominus} \rightarrow$ malate	−0.33
34	Uric acid $+ 2\,H^{\oplus} + 2\,e^{\ominus} \rightarrow$ xanthine	−0.36
35	Acetyl-S-CoA $+ 2\,H^{\oplus} + 2\,e^{\ominus} \rightarrow$ acetaldehyde $+$ CoA	−0.41
36	$CO_2 + 2\,H^{\oplus} + 2\,e^{\ominus} \rightarrow$ formate	−0.420
37	$H^{\oplus} + 1\,e^{\ominus} \rightarrow \frac{1}{2}H_2$	−0.420
38	Ferredoxin-$Fe^{3\oplus} + 1\,e^{\ominus} \rightarrow$ ferredoxin-$Fe^{2\oplus}$	−0.432
39	Gluconate $+ 2\,H^{\oplus} + 2\,e^{\ominus} \rightarrow$ glucose $+ H_2O$	−0.45
40	3-Phosphoglycerate $+ 2\,H^{\oplus} + 2\,e^{\ominus} \rightarrow$ glyceraldehyde-3-phosphate $+ H_2O$	−0.55
41	Methylviologen$_{(ox)} + 2\,H^{\oplus} + 2\,e^{\ominus} \rightarrow$ methylviologen$_{(red)}$	−0.55
42	Acetate $+ 2\,H^{\oplus} + 2\,e^{\ominus} \rightarrow$ acetaldehyde	−0.60
43	Succinate $+ CO_2 + 2\,H^{\oplus} + 2\,e^{\ominus} \rightarrow \alpha$-ketoglutarate $+ H_2O$	−0.67
44	Acetate $+ CO_2 + 2\,H^{\oplus} + 2\,e^{\ominus} \rightarrow$ pyruvate	−0.70

SOURCE: I. H. Segel, *Biochemical Calculations*, 2d ed. (New York: Wiley, 1975), pp. 414–415.

* Standard conditions: Unit activity of all components except $H^{\oplus}$, which is maintained at 10^{-7} M. Gases are at 1 atm pressure.

† The value given is for free FAD/$FADH_2$. The E_0' of the protein-bound coenzyme varies.

shall discuss later in this chapter). The two half-reactions are the following:

reaction 24 $$H_3C{-}\overset{O}{\overset{\|}{C}}{-}COO^{\ominus} + 2\,H^{\oplus} + 2\,e^{\ominus} \longrightarrow H_3C{-}\underset{OH}{\underset{|}{\overset{H}{\overset{|}{C}}}}{-}COO^{\ominus} \qquad E_0' = -0.190\text{ V}$$

reaction 32a $$NAD^{\oplus} + 2\,H^{\oplus} + 2\,e^{\ominus} \longrightarrow NADH + H^{\oplus} \qquad E_0' = -0.320\text{ V}$$

The pyruvate → lactate half-reaction has the more positive (less negative) E_0' and so proceeds from left to right as a reduction. The other half-reaction must then proceed as an *oxidative* half-reaction: NADH → $NAD^{\oplus}$: The sum of the two half-reactions gives the overall reaction:

$$H_3C{-}\overset{O}{\overset{\|}{C}}{-}COO^{\ominus} + NADH + H^{\oplus} \longrightarrow H_3C{-}\underset{OH}{\underset{|}{\overset{H}{\overset{|}{C}}}}{-}COO^{\ominus} + NAD^{\oplus}$$

Of equal interest is the extent to which the reaction will proceed from left to right; what is the equilibrium position? This relative tendency can be calculated from the simple difference of the E_0' values for the two half-reactions:

$$\Delta E_0' = E_{0(\text{red})}' - E_{0(\text{ox})}'$$

where $E_{0(\text{red})}'$ is the value of E_0' for the half-reaction that proceeds as a reduction, and $E_{0(\text{ox})}'$ is the value of E_0' for the half-reaction that proceeds as an oxidation. In this example,

$$\Delta E_0' = (-0.190\text{ V}) - (-0.320\text{ V}) = +0.130\text{ V}$$

If the reaction *does proceed from left to right,* $\Delta E_0'$ will always have a *positive value.* The $\Delta E_0'$ value can be directly correlated with K_{eq} because

$$\Delta G' = -nF\,\Delta E_0'$$

where n = number of electrons transferred per mole, and F = Faraday's constant = 23.063 kcal volt^{-1} equivalent^{-1} = 96.542 kJ volt^{-1} equivalent^{-1} (Segel, 1975*a*).

A spontaneous reaction (proceeding as written) must have a negative $\Delta G'$ and so a positive $\Delta E_0'$. Because

$$\Delta G' = -RT \ln K_{eq}$$

we have

$$\Delta G_0' = -(2.3\,RT/nF) \log K_{eq}$$

At 25° C,

$$\Delta E_0' = (0.059/n) \log K_{eq}$$

For the reaction pyruvate + NADH ⇆ lactate + $NAD^{\oplus}$ at pH 7, we have

$$0.130 \text{ V} = (0.059/2) \log K_{eq}$$

$$\log K_{eq} = 4.3$$

$$K_{eq} \approx 20{,}000 \text{ in favor of pyruvate reduction}$$

Thus, knowledge of redox potentials for oxidation–reduction half-reactions allows one to calculate the thermodynamic tendency for a coupled redox reaction to proceed in one direction or the other, and provides a quantification of what the equilibrium position will be *if a mechanism for the redox process exists.* In the remainder of Section III, we shall focus on what those mechanisms might be and on how the enzymes (as catalysts) can accelerate kinetically the approach to the equilibrium positions.

An additional point about the reduction potentials of Table 10-1 is that the $NAD^{\oplus}$/NADH couple has the most negative redox potential of the common cofactors or redox active metals used in biological oxidations—i.e., NADH will tend to pass electrons and become reoxidized to the other species. The following rank order is evident in the table and describes a thermodynamically favored cascade for transfer of electrons out of NADH (2 $e^{\ominus}$ at a time) to flavins, then coenzyme Q (ubiquinone) to cyt-*c* and then cyt-a_3 (1 $e^{\ominus}$ at a time), and finally to O_2 (4 $e^{\ominus}$ at a time).

$NAD^{\oplus}$	⇌	NADH	−0.32 V
Riboflavin	⇌	Dihydroflavin	−0.20 V
Ubiquinone	⇌	Dihydroubiquinone	+0.10 V
Cyt-*c*-$Fe^{3\oplus}$	⇌	Cyt-*c*-$Fe^{2\oplus}$	+0.30 V
Cyt-a_3-$Fe^{3\oplus}$	⇌	Cyt-a_3-$Fe^{2\oplus}$	+0.55 V
O_2	⇌	H_2O	+0.81 V

This is precisely the path of electron flow in animal, plant, and procaryotic membrane respiratory chains, underscoring the chemical logic of this sequence. It will also be no surprise then that many soluble enzymes use NADH to reduce enzyme-bound flavins (Chapter 11) and that additional iron chromophores or quinones may be the bound dihydroflavin reoxidants (Chapter 13). We shall note in Chapter 13 that many copper redox potentials are in the range (+0.2 to +0.5 V) where only molecular oxygen is a likely electron acceptor.

10.A.2 Two-Electron Oxidations

The stoichiometry of most enzyme-catalyzed redox processes indicates that two electrons are transferred. One can therefore envision two mechanistic alternatives:

1. two 1-$e^{\ominus}$ steps;
2. one 2-$e^{\ominus}$ step.

If two 1-$e^{\ominus}$ steps occur, then a free-radical intermediate must form. Free radicals, including the hydrogen atom, are highly reactive and very energetic species, and discrete long-lived radical intermediates are infrequently seen in biological systems, because most substrates and enzymes lack suitable structural elements to provide substantial stabilization and thus lower the free energy of formation of such radicals. We shall see radical intermediates when they can be stabilized, as in the following two main categories:

1. flavin-coenzyme–mediated reactions, and also reactions using other quinoid-coenzyme structures (vitamins C, E, and K, and coenzyme Q);

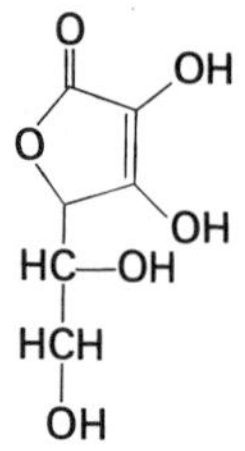

vitamin C

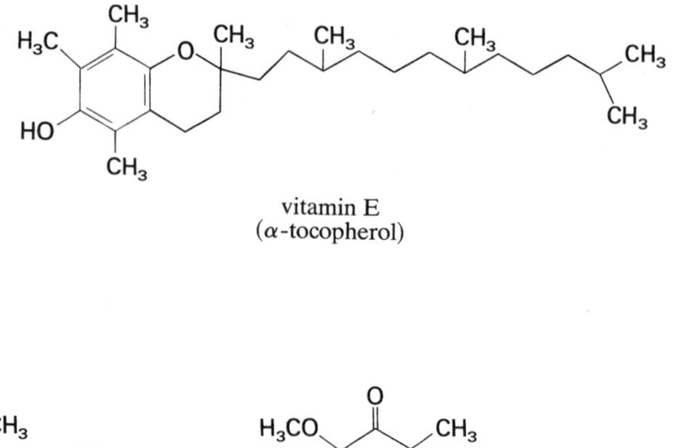

vitamin E
(α-tocopherol)

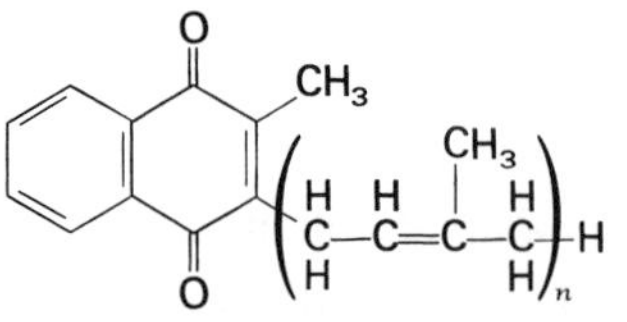

vitamin K

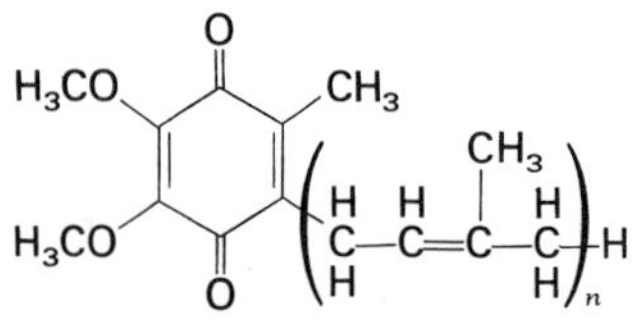

coenzyme Q
(ubiquinone)

2. enzymes containing transition metals (Fe, Mo, Cu) that participate in the redox reaction.

Given that two 1-$e^{\ominus}$ transfer steps look (a priori) somewhat unfavorable except in the special situations just noted, one can analyze the options for the more common single 2-$e^{\ominus}$ step. These options are either

1. hydride ($H:^{\ominus}$) transfer, or
2. proton ($H^{\oplus}$) abstraction, followed by 2-$e^{\ominus}$ transfer.

At least in a formal sense, the hydride mechanism implies that a substrate carbonium ion forms (if only transiently), whereas the proton-abstraction mechanism implies that a substrate carbanion (or carbanionic transition state) is generated:

$$\text{hydride transfer:}\quad X{-}\overset{|}{\underset{|}{C}}{-}H \longrightarrow \left[H:^{\ominus} + X{-}\overset{|}{\underset{|}{C}}^{\oplus}\right] \xrightarrow{Y} HY + X{=}\overset{|}{\underset{|}{C}}$$

$$\text{proton transfer:}\quad X{-}\overset{|}{\underset{|}{C}}{-}H \longrightarrow \left[H^{\oplus} + X{-}\overset{|}{\underset{|}{C}}:^{\ominus}\right] \xrightarrow{Y} HY + X{=}\overset{|}{\underset{|}{C}}$$

In this chapter we look at enzymatic dehydrogenations where hydride transfer may be the oxidation mechanism. In subsequent chapters we proceed to those reactions where proton abstraction and 2-$e^{\ominus}$ transfer appear to be the rule. In an attempt to determine in what kinds of enzyme-catalyzed transformations oxidation–reduction may proceed by hydride transfer, we first examine briefly some chemical precedents.

10.A.3 Some Nonenzymatic Hydride Transfers

Reduction of carbonyl compounds by metal hydrides ($NaBH_4$, $LiAlH_4$) proceeds by hydride transfer (House, 1972, chap. 2).

$$R{-}\underset{\underset{O}{\|}}{C}{-}X \xrightarrow{NaBH_4} R{-}\overset{H}{\underset{OH}{\overset{|}{\underset{|}{C}}}}{-}X$$

In these molecules, the polarity is such that hydrogen is the most electronegative element, and thus a likely $e^{\ominus}$ carrier during reduction. The reaction can be written as a transfer of an $H^{\ominus}$ equivalent, although *free* hydride ion as a distinct entity is probably not involved. Free hydride is such a strong base ($pK_a \sim 40$) that it would not exist in significant concentrations in water.

$$H_3B^{\ominus}{:}H\ Na^{\oplus} + {>}C{=}O \longrightarrow H{-}C{-}O^{\ominus}Na^{\oplus} + BH_3$$

A second example of the oxidative interconversion of carbonyl compounds and alcohols is provided by the Meerwein–Pondorff–Oppenhauer reaction with Al^{III} salts as catalysts (Gray, 1971, pp. 98–100). The aluminate salt of the alcohol can transfer a hydride equivalent to an aldehyde or ketone.

$$Al(OCHR_2)_2(OCR_2H^*) + R'R''C{=}O \rightleftharpoons [\text{complex}] \rightleftharpoons R'R''CH^*{-}O{-}Al{-}(OCHR_2)_2 + R_2C{=}O$$

The aluminum may coordinate to and polarize the carbonyl group in the reactive complex to favor the transfer of a hydride equivalent, simultaneous with the indicated exchange of metal ligands. The metal-ion–mediated hydride transfer has been suggested as a model system for the action of the $Zn^{2\oplus}$-dependent alcohol dehydrogenase discussed later in this chapter.

Also relevant to enzymatic catalysis is the *Cannizzaro reaction*: aldehydes without β-hydrogens (so that no aldol reaction is possible) react in base to disproportionate to one acid and one alcohol product molecule (Cannizzaro, 1912; Fieser and Fieser, 1961). One aldehyde molecule undergoes oxidation, the other reduction.

$$2\ C_6H_5{-}CHO \xrightarrow{OH^{\ominus}} C_6H_5{-}COOH + C_6H_5{-}CH_2OH$$

benzaldehyde → benzoic acid + benzyl alcohol

The observed rate law is first-order in benzaldehyde and second-order in base:

$$-d[\phi\text{CHO}]/dt = k[\phi\text{CHO}]^1[\text{OH}^\ominus]^2$$

A likely scheme to account for participation of two moles of hydroxide involves the dianion of the tetrahedral adduct as a hydride donor.

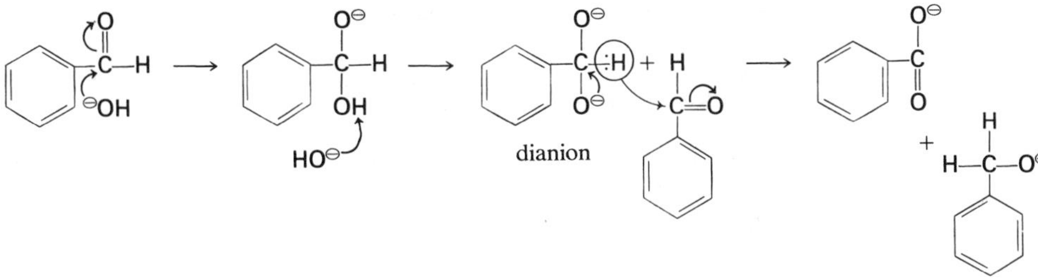

dianion

Thus, in the Cannizzaro reaction, a direct hydride transfer (constituting the actual redox step) occurs from one aldehyde molecule (which gets oxidized) to the second aldehyde molecule (which gets reduced). This conclusion is supported experimentally by the observation that there is no equilibration of the transferred hydrogen with solvent hydrogen.

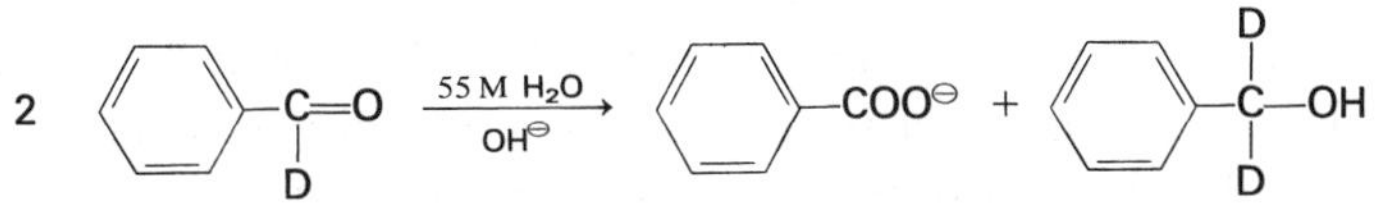

$$2\ \phi\text{–CDO} \xrightarrow[\text{OH}^\ominus]{55\text{ M H}_2\text{O}} \phi\text{–COO}^\ominus + \phi\text{–CD}_2\text{–OH}$$

In chemical reactions involving transfer of hydrogen in aqueous solution, the failure to equilibrate transferable hydrogen with solvent protons usually is taken as a representation of a hydride-ion mechanism. We shall note that, when this criterion is applied to enzymatic catalysis, straightforward mechanistic interpretation as hydride transfer can be somewhat obscured by the knowledge that some enzymes can exclude solvent water from the active site. Shielded proton transfers can then proceed without equilibrium with solvent hydrogen, so such a mechanism could also explain a direct hydrogen-transfer result (I. A. Rose, 1970).

10.B GLYOXYLASE: AN ENZYMATIC HYDRIDE TRANSFER?

It has been suggested that an enzymatic analogy to the Cannizzaro reaction may occur in the transformation carried out by the consecutive actions of the enzymes glyoxylase I and glyoxylase II. Enzyme I converts methylglyoxal (pyruvaldehyde) in the presence of the tripeptide glutathione to lactylglutathione, an acylthioester product. Enzyme II is a hydrolytic activity (a thioesterase), converting lactylglutathione to lactate and free glutathione (the latter tripeptide is thus required only in catalytic amounts for the overall reaction). In the following equations, GSH represents the reduced (SH) form of glutathione.

$$\underset{\text{methylglyoxal}}{H_3C{-}\overset{O}{\overset{\|}{C}}{-}\overset{O}{\overset{\|}{C}}H} + GSH \xrightarrow{\text{enzyme I}} \underset{\text{lactylglutathione}}{H_3C{-}\underset{H}{\overset{HO}{\overset{|}{C}}}{-}\overset{O}{\overset{\|}{C}}{-}S{-}G}$$

$$\underset{\text{lactylglutathione}}{H_3C{-}\underset{H}{\overset{HO}{\overset{|}{C}}}{-}\overset{O}{\overset{\|}{C}}{-}S{-}G} + H_2O \xrightarrow{\text{enzyme II}} \underset{\text{lactate}}{H_3C{-}\underset{H}{\overset{HO}{\overset{|}{C}}}{-}\overset{O}{\overset{\|}{C}}{-}O^{\ominus}} + GSH$$

The net reaction can be written as follows:

$$H_3C{-}\overset{O}{\overset{\|}{C}}{-}\overset{O}{\overset{\|}{C}}H + H_2O \xrightarrow{\text{glyoxylase}} \underset{\text{lactate}}{H_3C{-}\underset{H}{\overset{HO}{\overset{|}{C}}}{-}\overset{O}{\overset{\|}{C}}{-}O^{\ominus}}$$

Clearly, the step catalyzed by enzyme I is the oxidative step; enzyme II is simply a hydrolase, a group-transfer catalyst of the type we've discussed in Section II. Note that the conversion of methylglyoxal to lactate is an intramolecular redox reaction, analogous to an intramolecular Cannizzaro reaction. C-1 of the substrate has undergone a two-electron oxidation (aldehyde to acid) while C-2 has concomitantly experienced a two-electron reduction (ketone to alcohol).

This transformation then can be formulated as a hydride transfer. Note that the actual substrate for such a hydride transfer from C-1 to C-2 is not free aldehyde of methylglyoxal, but in fact is the thiohemiacetal that arises from prior nucleophilic attack by the thiolate anion of glutathione on the aldehyde carbonyl. No oxidation has occurred in this adduct formation. The oxidation step then occurs, *producing the acylthioester directly.*

$$H_3C{-}\overset{O}{\overset{\|}{C}}{-}\overset{O}{\overset{\|}{C}}H + {}^{\ominus}S{-}\text{Glutathione} \longrightarrow H_3C{-}\overset{O}{\overset{\|}{C}}{-}\underset{O^{\ominus}}{\underset{|}{\overset{H}{\overset{|}{C}}}}{-}S{-}\text{Glutathione} \xrightarrow{\text{oxidation}} H_3C{-}\underset{H}{\overset{O^{\ominus}}{\overset{|}{C}}}{-}\underset{O}{\underset{\|}{C}}{-}S{-}\text{Glutathione}$$

$$\xrightarrow[\text{protonate C-2 alcohol anion}]{H^{\oplus},} H_3C{-}\underset{H}{\overset{OH}{\overset{|}{C}}}{-}\underset{O}{\underset{\|}{C}}{-}S{-}\text{Glutathione}$$

Note that, because the oxidation occurs from the thiohemiacetal and not the aldehyde, the product is the activated acyl derivative (the acylthioester) rather than the low-energy, stabilized carboxylate anion. We shall return to this point later.

It is also possible to formulate the oxidation in an alternative manner. Rather than hydride transfer, one can postulate *proton abstraction* from C-1 of the thiohemiacetal by an enzymatic base, to form the relatively stable enediolate species, with subsequent reprotonation at C-2 rather than C-1 of the enolate.

$$H_3C{-}C({=}O){-}CH(O^{\ominus})(S{-}\text{Glutathione}) + \ddot{B}{-}\text{Enz} \longrightarrow \left\{ H_3C{-}C(O){-}\overset{\ominus}{C}(O^{\ominus})(S{-}\text{Glutathione}) \longleftrightarrow H_3C{-}C(O^{\ominus}){=}C(OH)(S{-}\text{Glutathione}) \right\}$$

enediolate ion

$+\ {}^{\oplus}HB{-}\text{Enz}$

Now the enolate ion can be protonated, either at C-1 to give back the starting material, or at C-2 to give the thioester. Protonation at C-2 would proceed as follows:

$$H_3C{-}C(O^{\ominus}){=}C(O{-}H\cdots{:}B){-}S{-}\text{Glutathione} + {}^{\oplus}H{-}B{-}\text{Enz} \longrightarrow H_3C{-}CH(HO){-}C({=}O){-}S{-}\text{Glutathione} + \ddot{B}{-}\text{Enz}$$

An experimental distinction is sometimes possible between (1) hydride transfer and (2) proton transfer with enolization. Simplistically, in the proton-transfer mechanism, there is the conjugate acid $BH^{\oplus}$ at the active site. In general, because enzymes are in 55 M H_2O, one expects rapid equilibration of $BH^{\oplus}$ (where $B=O$, N, S) with solvent protons (10^3 to $10^7\ sec^{-1}$) (see ¶4.A.1; Eigen, 1964). Thus, in D_2O and with a proton-transfer mechanism, one would expect incorporation of deuterium at C-2 of lactate. Conversely, for a hydride-transfer mechanism, one would expect ^1H at C-2 of lactate, even in D_2O. The latter has been the actual experimental result. Also, [^3H]-(C-1)-methylglyoxal is reported to form [^3H]-(C-2)-lactate without loss of tritium—a direct hydrogen transfer. Because one could imagine shielded proton transfer in the enzyme active site, this is not definitive evidence for hydride transfer. In fact, very recent preliminary evidence indicates that, in D_2O at elevated temperature (e.g., 40° C), glyoxylase action does incorporate solvent deuterium at C-2 of the lactate product (Hall, 1976). This result, if correct, mandates the shielded-proton explanation. At higher temperatures, the shielding of Enz—$BH^{\oplus}$ from water is reduced.

Note that direct oxidation of the aldehyde would require 2-e$^{\ominus}$ removal from an already electrophilic, electron-deficient center. Prior addition of the thiol group of the glutathione cofactor to the aldehyde carbonyl generates a species (the thiohemiacetal) not electrophilic, from which 2-e$^{\ominus}$ removal is more favorable, and gives as product the acylated nucleophile, a product with high group-transfer potential.

10.C ANOTHER ALDEHYDE OXIDATION INVOLVING APPARENT HYDRIDE SHIFT; NICOTINAMIDE COENZYMES

A metabolically central oxidation of an aldehyde, coupled to ATP production, occurs during the fermentation of glucose by consecutive action of a dehydrogenase (glyceraldehyde-3-P dehydrogenase) and a kinase (3-phosphoglycerate kinase).

$$H_2C(O{-}PO_3^{2\ominus}){-}CH(OH){-}CHO + HO{-}P(O)(O^{\ominus})_2 \underset{\text{dehydrogenase}}{\overset{\text{oxidation step}}{\rightleftharpoons}} H_2C(O{-}PO_3^{2\ominus}){-}CH(OH){-}C(O){-}O{-}P(O)(O^{\ominus})_2 \overset{ADP^{3\ominus}\ \text{kinase}}{\rightleftharpoons}$$

3-phospho-glyceraldehyde; 1,3-diphosphoglycerate (acyl phosphate)

$$H_2C(O{-}PO_3^{2\ominus}){-}CH(OH){-}C(O){-}O^{\ominus} + ATP^{4\ominus}$$

3-phosphoglycerate

The oxidation from the aldehyde to the acid oxidation state occurs in the dehydrogenase reaction; the stoichiometry of the kinase reaction is similar to that of acetate kinase (¶7.D.3.b), and it represents a conversion of the high group-transfer potential in the 1,3-diphosphoglycerate into the more common currency of ATP.

Perhaps the first question to ask about the reaction catalyzed by glyceraldehyde-3-P dehydrogenase is: where do the electrons go when the aldehyde carbon of the substrate is oxidized to the acyl-phosphate level in the product (a 2-$e^{\ominus}$ oxidation)? What is the cosubstrate or coenzyme that must be reduced by two electrons? The answer to these questions was provided when the enzyme was purified to homogeneity, a relatively easy task because this dehydrogenase represents 10% of the soluble protein in skeletal muscle cells. The homogeneous muscle enzyme is active as a tetramer of identical subunits, each about 36,000 mol wt and each composed of 334 amino acids. The active enzyme was isolated as a 1:1 complex with an organic coenzyme, tightly but not covalently bound. The cofactor is *nicotinamide adenine dinucleotide* (NAD). NAD makes up one of the two main classes of redox coenzymes (flavins being the other class) that function in simple enzyme-catalyzed dehydrogenations.* The cofactor is composed of an AMP moiety in phosphodiester linkage to a 5-phosphoribosyl-1-nicotinamide portion. (Both sugars are D-riboses.)

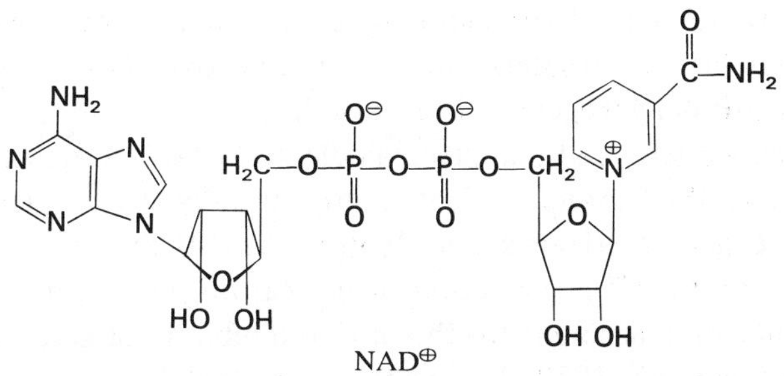

$NAD^{\oplus}$

It is the nicotinamide ring system that is redox active; the oxidized pyridinium system of NAD can accept two electrons reversibly to produce a dihydropyridine ring. Two-electron reduction involves addition of one hydrogen to form the dihydropyridine system; chemical reduction gives a mixture of all three species: the 1,2-, the 1,6-, and the 1,4-dihydronicotinamide.

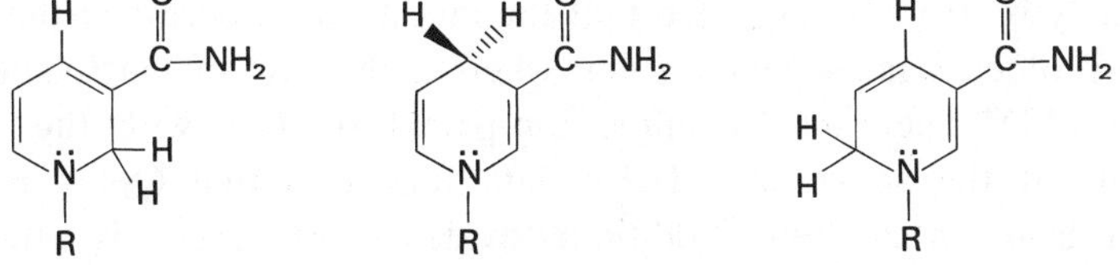

1,2-dihydro-NAD

1,4-dihydro-NAD (NADH)

1,6-dihydro-NAD

* In the older biochemical literature, NAD is called DPN (for diphosphopyridine nucleotide). For a review of nicotinamide bioorganic chemistry, see Bruice and Benkovic (1965–1966, vol. 2, p. 301). Also Boyer (1970–1976, vols. 11 and 13) includes articles on dehydrogenases requiring nicotinamide coenzymes.

However, in the 250 or so known enzymatic redox reactions involving NAD, it appears that the 1,4-dihydronicotinamide is the exclusive isomer formed biologically. This isomer is known as NADH, or reduced NAD.

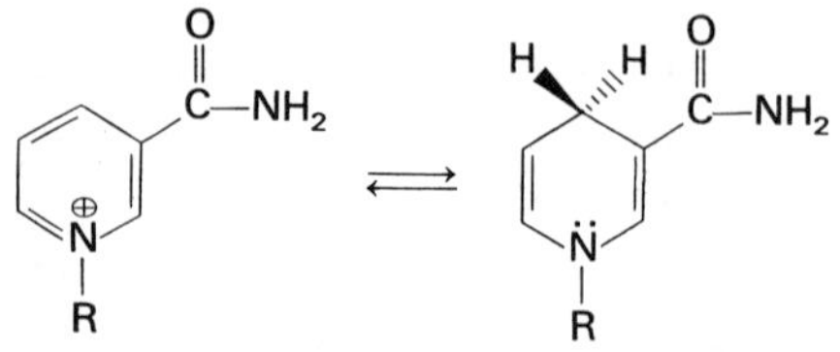

NADH
($2\text{-}e^{\ominus}$ reduced)

We shall have some general comments to make about nicotinamide coenzymes shortly, but first let us consider how the specific oxidation brought about by glyceraldehyde-3-P dehydrogenase is achieved.

It was quickly discovered that glyceraldehyde-3-P dehydrogenase contains an unusually reactive thiol group, stoichiometrically alkylated by iodoacetate with loss of catalytic activity (Harris and Waters, 1976). Structural determination indicated that only Cys^{149} is converted to the carboxymethyl derivative. In fact, this single modification accounts for the in vivo inhibition of glucose fermentation by iodoacetate (Rapkine, 1938; J. I. Harris et al., 1963).

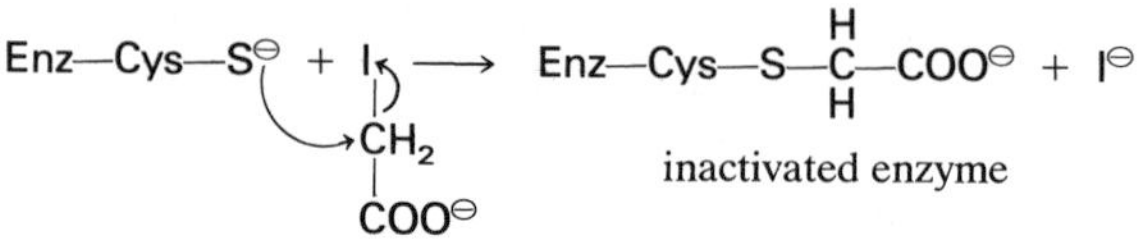

$$\text{Enz—Cys—S}^{\ominus} + \text{I—CH}_2\text{—COO}^{\ominus} \longrightarrow \text{Enz—Cys—S—CH}_2\text{—COO}^{\ominus} + \text{I}^{\ominus}$$

inactivated enzyme

Kinetic analysis strongly suggested that the modification occurs at the active site. Substrate protects against inactivation, whereas the rate of inactivation is facilitated by $NAD^{\oplus}$, perhaps by initial ion-pair formation with the iodoacetate carboxylate at the active site. These data suggested that Cys^{149} might be an important group in catalysis. Additional evidence that Cys^{149} is adjacent to the bound NAD at the active site came from the observation that the NAD–enzyme complex has a yellow color due to a broad absorption band centered around 365 nm. The carboxymethylated enzyme is no longer yellow. Racker suggested that the color arises from a charge-transfer complex between the Cys^{149} thiolate anion as electron-rich donor and the electron-deficient oxidized pyridinium ring of $NAD^{\oplus}$ as acceptor (Racker and Krimsky, 1953; Kosower, 1956).

In the back direction (reduction of acyl phosphate to aldehyde), 1,3-diphosphoglycerate is not the only molecule recognized. The enzyme will reduce acetyl phosphate to acetaldehyde.

$$H^{\oplus} + H_3C{-}C(=O){-}O{-}P(=O)(O^{\ominus}){-}O^{\ominus} + NADH \xrightleftharpoons{\text{enzyme}} H_3C{-}CH(=O) + {}^{\ominus}O{-}P(=O)(O^{\ominus}){-}OH + NAD^{\oplus}$$

acetyl phosphate acetaldehyde

A partial isotope exchange, revealing of mechanism, can be detected with acetyl phosphate, which is more stable than 1,3-diphosphoglycerate ($T_{1/2}$ at pH 7 and 25° C is ~20 hr for acetyl phosphate vs. 30 min for the diphosphoglycerate) and thus easier to work with. The enzyme catalyzes an exchange of ^{32}P radioactivity from $[^{32}P]$-P_i into acetyl phosphate.

$$^{\ominus}O{-}{}^{32}P(=O)(O^{\ominus}){-}OH + H_3C{-}C(=O){-}O{-}P(=O)(O^{\ominus}){-}O^{\ominus} \xrightarrow{\text{enzyme}} H_3C{-}C(=O){-}O{-}{}^{32}P(=O)(O^{\ominus}){-}O^{\ominus} + {}^{\ominus}O{-}P(=O)(O^{\ominus}){-}OH$$

This observation indicates reversible formation of an *acetyl-enzyme* intermediate and, by extension, of a 3-phosphoglyceryl-enzyme intermediate during physiological catalysis.

$$H_3C{-}C(=O){-}X{-}Enz \qquad {}^{\ominus}O{-}P(=O)(O^{\ominus}){-}O{-}H_2C{-}CH(OH){-}C(=O){-}X{-}Enz$$

acetyl enzyme 3-phosphoglyceryl enzyme

The most likely candidate for the required enzyme nucleophile is the thiolate anion of Cys^{149}. In fact, Segal and Boyer (1953) had suggested that the initial chemical step in catalysis (in the aldehyde-to-acyl-phosphate direction) is nucleophilic addition of the Cys^{149} thiolate anion to the aldehyde carbonyl of the bound glyceraldehyde-3-P substrate, producing a thiohemiacetal (the sulfur analogue of a hemiacetal), which is the actual species undergoing oxidation to an acylthioester enzyme intermediate.

$$^{\ominus}O\text{-}P(=O)(O^{\ominus})\text{-}O\text{-}CH_2\text{-}CH(OH)\text{-}CH{=}O \;\; + \;\; ^{\ominus}S\text{-}Cys^{149}\text{-}Enz \rightleftharpoons {}^{\ominus}O\text{-}P(=O)(O^{\ominus})\text{-}O\text{-}CH_2\text{-}CH(OH)\text{-}CH(O^{\ominus})\text{-}S\text{-}Cys^{149}\text{-}Enz \xrightarrow{\text{“}-2e^{\ominus}, H\text{”}} {}^{\ominus}O\text{-}P(=O)(O^{\ominus})\text{-}O\text{-}CH_2\text{-}CH(OH)\text{-}C(=O)\text{-}S\text{-}Cys^{149}\text{-}Enz$$

thiohemiacetal acylthioester intermediate

How does the oxidation step proceed? As the thiohemiacetal form of the ES complex is oxidized by two electrons, the bound NAD coenzyme is reduced by two electrons to the dihydronicotinamide form, NADH. This reduction is easily monitored because the reduced form (NADH) has substantial absorbance in the visible region of the spectrum whereas the oxidized form ($NAD^{\oplus}$) does not (Fig. 10-1). The extinction coefficient for NADH at 340 nm is 6,220 M^{-1} cm^{-1}, whereas the coefficient for $NAD^{\oplus}$ is essentially zero. Thus 0.1 μmole of NADH per 1 ml of solution has 0.662 absorbance units and represents a very sensitive and convenient assay for redox change (Fig. 10-2). As a practical matter, many enzymes are assayed by this spectrophotometric assay by coupling the product of a given enzymatic reaction to a dehydrogenase that utilizes $NAD^{\oplus}$/NADH. As glyceraldehyde-3-P oxidation proceeds, the absorbance of the solution at 340 nm increases with time, because one mole of NADH is generated per mole of aldehyde oxidized.

The most pertinent observation about the mechanism of the enzymatic oxidation is that the hydrogen at the oxidizable locus of glyceraldehyde-3-P is transferred *directly* and *quantitatively* to C-4 of the dihydropyridine ring of the

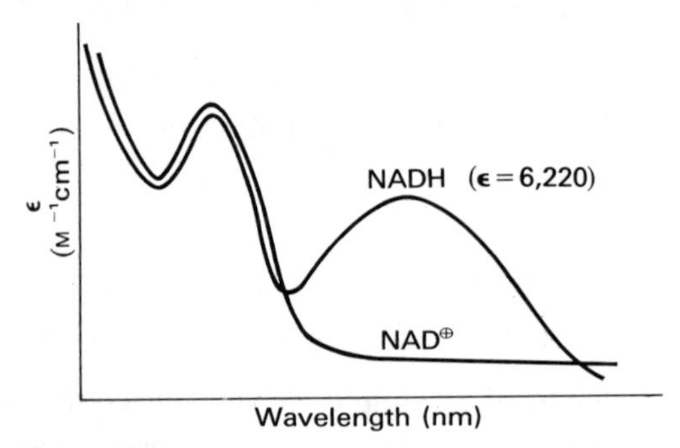

Figure 10-1
Electronic absorption spectra for $NAD^{\oplus}$ and NADH.

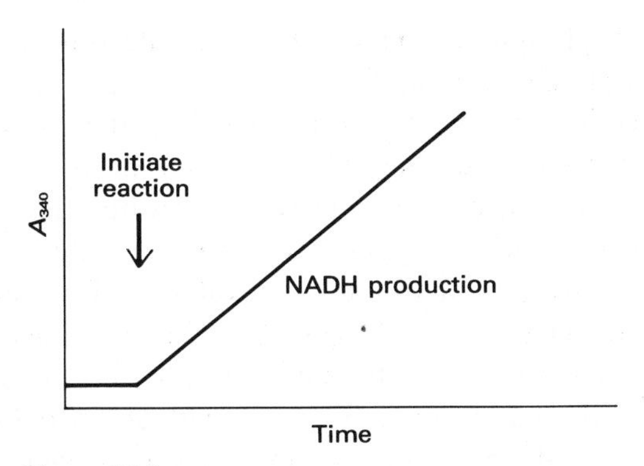

Figure 10-2
Assay for NADH formation by monitoring increase in absorbance at 340 nanometers.

NADH formed. With [1-^{3}H]-glyceraldehyde-3-P, all the tritium ends up in NADH and none in water.

$$^{\ominus}O{-}P(=O)(O^{\ominus}){-}O{-}CH_2{-}CH(OH){-}C(=O){-}{}^3H + {}^{\ominus}O{-}P(=O)(O^{\ominus}){-}OH + \text{Enz—NAD} \rightleftharpoons {}^{\ominus}O{-}P(=O)(O^{\ominus}){-}O{-}CH_2{-}CH(OH){-}C(=O){-}O{-}P(=O)(O^{\ominus}){-}O^{\ominus} + \text{4-[}^3\text{H]-NADH} + \text{Enz}$$

4-[^{3}H]-NADH

Conversely, when the experiment is repeated with nonradioactive substrate in [^{3}H]-H_2O, no tritium ends up in the NADH product. These observations are *generally valid* for essentially all dehydrogenases utilizing nicotinamide coenzymes. These results prompt interpretation of the redox step as *hydride-ion transfer* rather than as proton abstraction with subsequent electron transfer. Model oxidations with *n*-alkyl nicotinamides also show direct hydrogen transfers in redox reactions (Abeles et al., 1957; Kurz and Frieden, 1975). The results with model oxidations argue against interpretation of the enzymatic data as shielded proton transfers at inaccessible active sites.

Before writing a minimal schematic mechanism, we must note the last phase of catalysis; the involvement of the inorganic phosphate cosubstrate. The 3-phosphoglycerylenzyme formed after the oxidation step is a destabilized acylthioester. It undergoes group transfer (3-phosphoglyceryl transfer) to one of the oxygens of inorganic phosphate bound at the active site, a phosphorolysis generating the 1,3-diphosphoglycerate-acyl-P product with preservation of the thermodynamic acyl activation. At this point, one sees the strategy of having the enzyme catalyze formation of the thiohemiacetal of the aldehyde substrate. Oxidation of the thiohemiacetal produces an activated acylthioester as initial product and activated acyl phosphate as released product. In the very next enzymatic step of glycolysis, the doubly activated acyl phosphate transfers its phosphoryl group to ADP during phosphoglycerate-kinase action to make ATP. The chemical energy stored in the terminal phosphoric-anhydride bond of ATP was initially *released and captured* in the acylthioester-dehydrogenase intermediate during the oxidation step. This two-enzyme sequence, then, affords a particularly clear example of how organisms use controlled enzymatic oxidations to capture energy in a chemically useful form, as activated acyl or phosphoryl groups. If the glyceraldehyde-3-P dehydrogenase oxidized the aldehyde instead of the thiohemiacetal, the oxidation product would be the low-energy, resonance-stabilized 3-phosphoglycerate, and none of the energy released during oxidation would have been trapped. One might compare the energetic cost to the organism for (1) acyl-phosphate production by action of glyceraldehyde-3-P dehydrogenase with (2) the formation of acetyl phosphate by the acetate-kinase–phosphotransacetylase couple of bacteria (¶7.D.3).

We are now ready to propose a reasonable catalytic scheme:

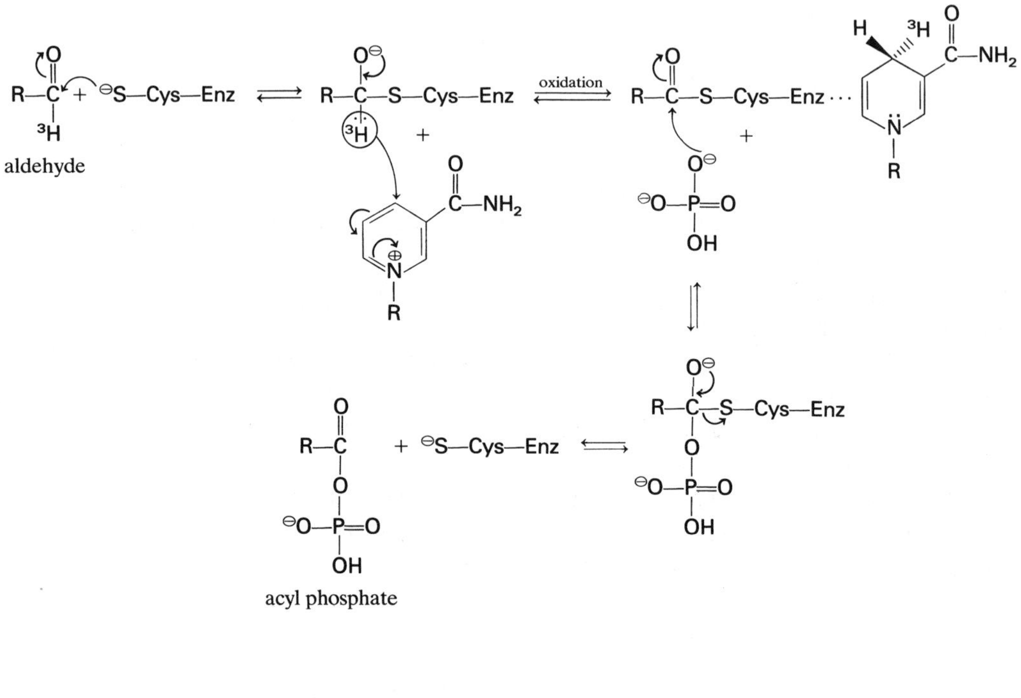

aldehyde

oxidation

acyl phosphate

Finally, let us note two points of a kinetic nature. When [^{3}H]-C_1-aldehyde is used, the 4-[^{3}H]-NADH formed has specific radioactivity identical to that of the substrate. There is no kinetic isotope effect observed; the hydride transfer (the oxidative step) cannot be rate-determining. Kinetic studies suggest that, in every turnover, NADH dissociates and NAD$^{\oplus}$ replaces it at the active site *before* the phosphorolysis step can proceed at an effective rate, suggesting that a conformational change occurs. At pH > 7.5, NADH dissociation is the slow step in catalysis; however, at pH < 7, the phosphorolysis step appears to be rate-limiting (Harris and Waters, 1976).

The preceding paragraph suggests that the nicotinamide coenzyme affects the conformation of the protein as it binds. This is only one indication of such an effect. NAD binding somewhat alters the viscosity of enzyme solutions, the optical rotatory dispersion spectrum, and the sedimentation pattern (Harris and Waters, 1976). Careful analysis of NAD binding to the muscle enzyme indicates that the affinity for NAD is a function of how much NAD is already bound to the enzyme tetramer (Conway and Koshland, 1968). The binding constants for the first through fourth molecules added to apoenzyme show a progressive decrease in affinity amounting to a millionfold drop between the first and the last NAD added:

$$K_1 = 10^{-11}\ \text{M}; \quad K_2 = 10^{-9}\ \text{M}; \quad K_3 = 10^{-7}\ \text{M}; \quad K_4 = 2.6 \times 10^{-5}\ \text{M}$$

This is a startling example of stepwise negative cooperativity; each successive coenzyme molecule is bound less tightly. (In contrast, the yeast enzyme appears to show concerted cooperative behavior; see Patel, 1969; Sarma et al., 1968.) X-ray data of the NAD-binding site on the enzyme indicate that the binding domain may include a region of intersubunit contact, and these data suggest that the cooperative-binding data reflect transmission of a protein conformational change, on binding of a NAD molecule, to an adjacent vacant subunit that alters the binding site of that adjacent subunit to a lower affinity for coenzyme.

A final point about glyceraldehyde-3-P dehydrogenase concerns recent experiments suggesting that some reagents modifying Cys^{149} do not completely inactivate the enzyme, but alter the specificity, converting it from a dehydrogenase to an acyl phosphatase. Allison (1976) has recently summarized evidence that mild oxidants such as *O*-iodosobenzoate convert the Cys^{149}—SH, not to a disulfide with some other cysteine in the enzyme, but rather initially to a sulfenic acid.

$$\text{Enz—Cys}^{149}\text{—SH} \xrightarrow{\text{oxidant}} \text{Enz—Cys}^{149}\text{—S—OH}$$

The sulfenyl derivative no longer reacts productively with bound aldehyde substrate, but is apparently nucleophilic enough to displace phosphate from acyl phosphates and produce an analogue of the normal acyl enzyme. This covalent intermediate is not an acylthioester, but an acylsulfenate that is hypothesized to be hydrolytically unstable at the active site. Water attack regenerates the Enz—Cys—SOH linkage for catalytic turnover.

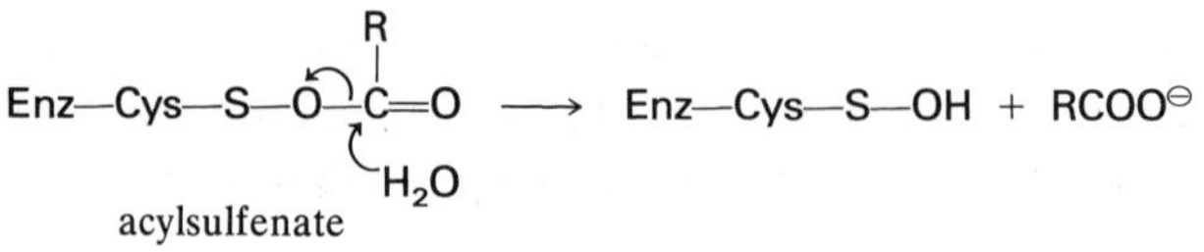

$$\text{Enz—Cys—S—O—C(R)=O} \xrightarrow{H_2O} \text{Enz—Cys—S—OH} + RCOO^{\ominus}$$

acylsulfenate

Allison notes that both H_2O_2 and O_2, components of aerobic metabolism, could effect oxidation of enzyme sulfhydryls to sulfenates, which may function as active species in catalysis. (In particular, they could behave as electrophiles in certain oxidative catalyses; see Allison, 1976.)

10.C.1 Some General Properties of Nicotinamide Coenzymes and Their Enzymatic Functions

The NADH molecule was the first enzymatic cofactor identified—detected in 1904 by Hardin and Young, who were then studying the fermentation of glucose to ethanol. A similar molecule, detected by Warburg and Christian in 1934, was found to be required for glucose-6-P oxidation by erythrocytes. This molecule contains an additional phosphate group beyond the two found in $NAD^{\oplus}$. Structural studies indicated that this phosphate group is at the 2′-OH of the ribose of the AMP moiety. This molecule is $NADP^{\oplus}$ (nicotinamide adenine dinucleotide phosphate; called TPN, for triphosphopyridine nucleotide, in the older literature). It is functionally indistinguishable from $NAD^{\oplus}$ as a redox cofactor, undergoing 2-$e^{\ominus}$ reduction to give the 1,4-dihydropyridine form, NADPH. Specific enzymes, though, usually show a preference (which may be absolute) for either $NAD^{\oplus}$ or $NADP^{\oplus}$.

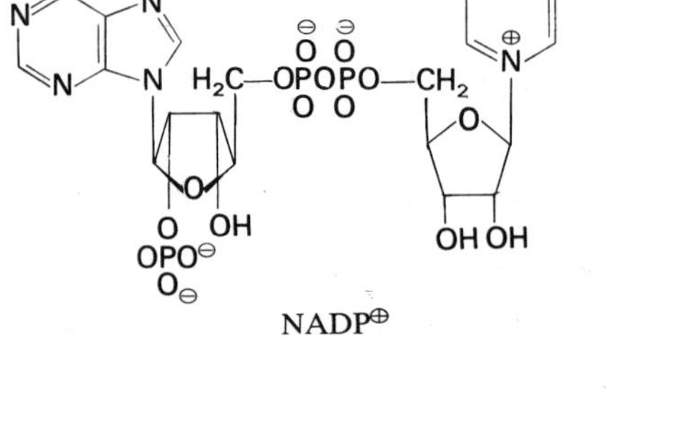

NH₂ N N N N H₂C—OPOPO—CH₂ O O O O O C—NH₂ N⊕ O O OH OPO⊖ O⊖ OH OH

$NADP^{\oplus}$

The synthesis of NAD$^{\oplus}$ was completed in 1957 by Todd's group in England. The nicotinamide coenzymes have two planar heterocyclic ring systems (the purine ring and the pyridine ring system) separated by a more-or-less flexible arm containing the D-ribose units and the pyrophosphate group. Nuclear magnetic resonance (NMR) studies suggest that the coenzyme molecules in aqueous solutions exist predominantly as folded forms with the heterocyclic rings stacked intramolecularly in parallel planes, a form presumably in ready equilibrium with an open or extended form of the molecule (Patel, 1969; Sarma et al., 1968).

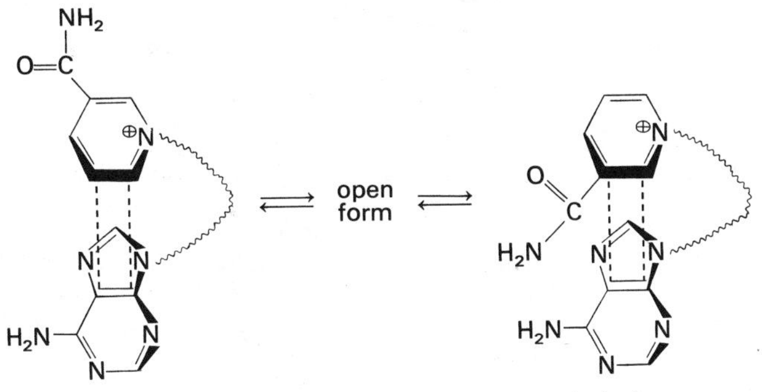

$$\text{stacked} \rightleftharpoons \text{open form} \rightleftharpoons \text{stacked}$$

In contrast, the coenzyme is bound in an extended form at the active site of several NAD-requiring dehydrogenases examined by X-ray analysis (Rossman et al., 1975). Figure 10-3 shows representations of the NAD$^{\oplus}$ at the active site of malate dehydrogenase (MDH) from pig heart-muscle cytoplasm; Figure 10-4 shows the schematic view for NAD$^{\oplus}$ binding to glyceraldehyde-3-P dehydrogenase (GAPDH) (notice the proximity to Cys^{149}). In the MDH picture, both the AMP and the nicotinamide groups are *anti* ("away from") about the respective *N*-glycosidic bonds. In contrast, the GAPDH-bound coenzyme has the nicotinamide ring rotated 180° to a *syn* position about the glycosidic bond. We shall note later in this chapter that hydride is in fact transferred to opposite faces of the nicotinamide ring of NAD in these two enzyme cases.

At the outset of this chapter, we noted that the half-reaction for NAD$^{\oplus}$ reduction has a reduction potential of −320 mV, a strongly negative value.

$$\text{NAD}^{\oplus} + 2\,\text{e}^{\ominus} + \text{H}^{\oplus} \longrightarrow \text{NADH} \qquad E_0' = -320\,\text{mV}$$

Most reduction potentials for redox-active dehydrogenase substrates are more positive, suggesting that the equilibrium position will be oxidation of NADH to NAD$^{\oplus}$ and reduction of the specific substrate (e.g., pyruvate → lactate, and acetaldehyde → ethanol, at neutral pH). That is, the enzymes may function as

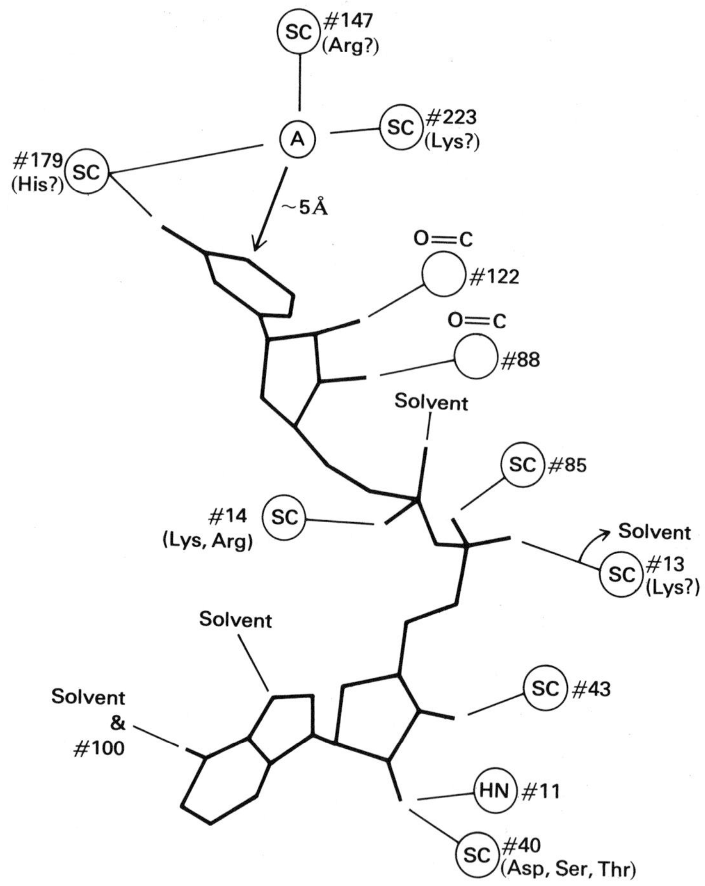

Figure 10-3
Schematic drawing of the conformation of NAD$^{\oplus}$ when bound to pig-heart *S*-MDH. The adenine end of NAD$^{\oplus}$ is at the bottom of the drawing, and the pyridine end of the dinucleotide is near the top. Also shown on the drawing are contacts made with regions of the enzyme molecule. The letters "SC" indicate that atoms from an amino-acid side chain are involved in the interaction, as opposed to groups such as C=O that arise from the polypeptide chain. (From Banaszak and Bradshaw, in *The Enzymes*, 3d ed., ed. P. Boyer, New York: Academic Press, 1975, vol. 11.)

pyruvate and acetaldehyde reductases, even though they are named lactate and alcohol dehydrogenases.

The roles of nicotinamide coenzymes in biological redox processes are conditioned by various chemical features; one of the key features is that the two-electron-reduced dihydropyridine ring system of NADH *is stable in air.* It is not rapidly autoxidized to the radical or the oxidized form. Thus, NAD and NADP

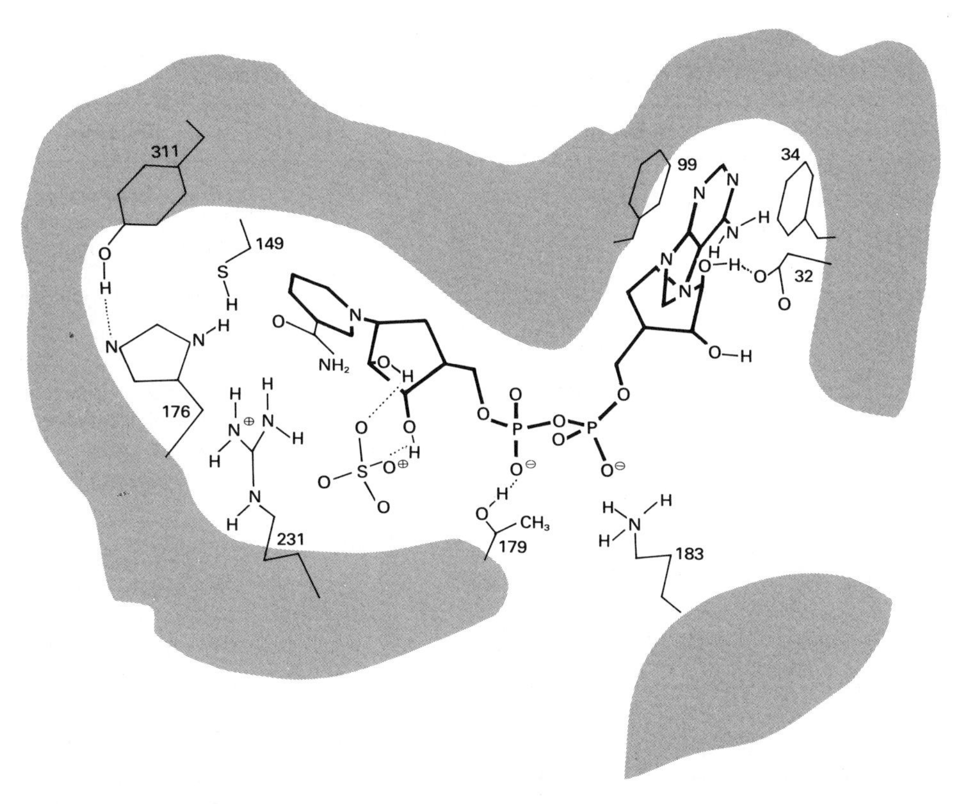

Figure 10-4
Schematic view of the NAD$^{\oplus}$ binding to GAPDH, showing functional amino acids. (From M. Rossman et al., in *The Enzymes*, 3d ed., ed. P. Boyer, New York: Academic Press, 1975, vol. 11, p. 87.)

coenzymes *are not cofactors for redox enzymes utilizing molecular oxygen.* This represents a fundamental distinction from the dihydro-forms of flavin coenzymes (Chapters 11 and 12), which are catalytically reoxidized by O_2. Pyridinyl radicals are known compounds in chemistry and, although most people believe that the redox processes involving nicotinamides are hydride transfers, Bruice has recently revived Kosower's arguments that fast H· and 1-$e^{\ominus}$ transfer steps are at least thermodynamically possible alternatives (Kosower, 1975; R. F. Williams et al., 1975).

One method of classification of simple NAD- and NADP-dependent dehydrogenases is by the chemical reaction type. Table 10-2 lists some categories with examples for each type. Category 3 involves a decarboxylation after an oxidation of a β-hydroxy acid to a β-keto-acid intermediate; this reaction type is discussed in Chapter 21. Space does not permit specific discussion here of many particular

Table 10-2
Classification of NAD- and NADP-dependent dehydrogenases by reaction type

Category	Reaction type	Examples
1	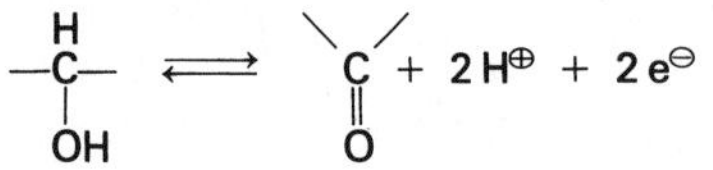 $\text{H—C(—)—OH} \rightleftharpoons \text{>C=O} + 2\,H^{\oplus} + 2\,e^{\ominus}$	Malate dehydrogenase Lactate dehydrogenase Alcohol dehydrogenase
2	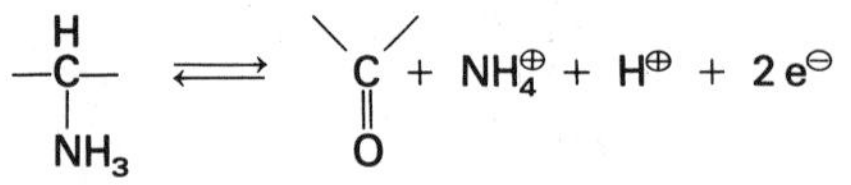 $\text{H—C(—)—NH}_3 \rightleftharpoons \text{>C=O} + NH_4^{\oplus} + H^{\oplus} + 2\,e^{\ominus}$	Glutamate dehydrogenase
3	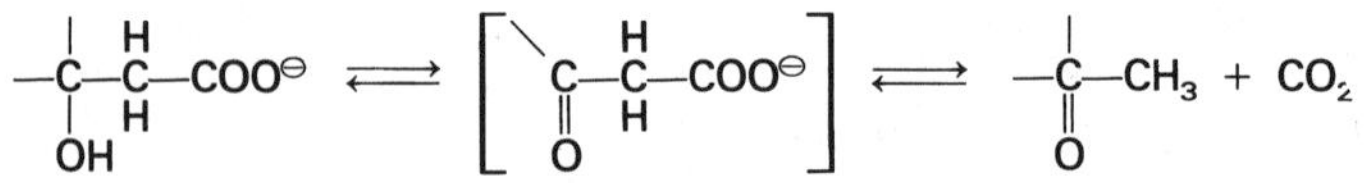 $\text{—C(OH)—CH}_2\text{—COO}^{\ominus} \rightleftharpoons \left[\text{>C(=O)—CH}_2\text{—COO}^{\ominus}\right] \rightleftharpoons \text{—C(=O)—CH}_3 + CO_2$	Isocitrate dehydrogenase 6-Phosphogluconate dehydrogenase
4	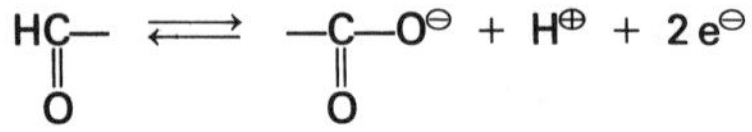 $\text{HC(=O)—} \rightleftharpoons \text{—C(=O)—O}^{\ominus} + H^{\oplus} + 2\,e^{\ominus}$	Aldehyde dehydrogenase
	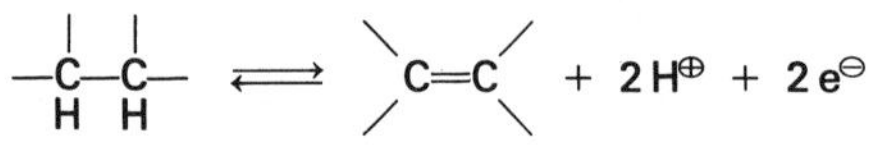 $\text{—CH—CH—} \rightleftharpoons \text{>C=C<} + 2\,H^{\oplus} + 2\,e^{\ominus}$	Dihydrosteroid dehydrogenase (Steroid reductase)
6	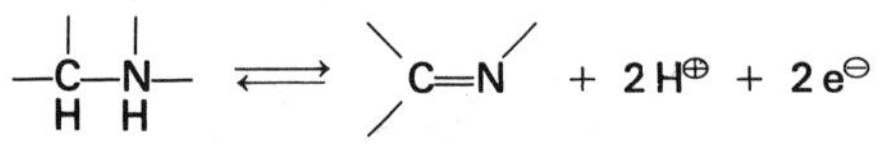 $\text{—CH—NH—} \rightleftharpoons \text{>C=N—} + 2\,H^{\oplus} + 2\,e^{\ominus}$	Dihydrofolate reductase

dehydrogenases, but recent reviews on several of the better-characterized enzymes are available (Boyer, 1970–1976, vols. 11 and 13). For all the enzymes listed in Table 10-2, the nicotinamide coenzyme functions as a readily dissociable cofactor, entering into the overall reaction stoichiometry as a substrate. This point also will be contrasted subsequently with the mechanism of flavoenzyme catalysis.

10.C.2 Alcohol Dehydrogenase

One further simple NAD-dependent dehydrogenase, alcohol dehydrogenase, deserves a closer look for three reasons: (1) it is a $Zn^{2\oplus}$-requiring enzyme; (2) it shows compulsory ordered binding, consistent with the kinetic behavior of many (but not all) of these enzymes; and (3) most importantly, classical experiments on enzyme stereochemistry and the recognition of prochiral centers were performed with this enzyme. Crystalline alcohol dehydrogenases have been isolated both from liver cells and from yeast, and an X-ray map at 2.4 Å has been compiled of the horse-liver enzyme (Branden et al., 1975). There are two zinc atoms per subunit of this dimeric enzyme. One of the zinc atoms functions in catalysis, the other in a structural role. The catalytically active zinc atom is liganded to the thiol groups of Cys^{46} and Cys^{174}, the imidazole nitrogen of His^{67}, and a water molecule in the resting enzyme; it is 4.5 Å from C-4 of the bound NAD coenzyme. The water molecule is probably displaced by the alcohol group of ethanol in the ES complex, or by the carbonyl oxygen of acetaldehyde in the ES complex from the opposite direction. The second (structural) zinc atom is substitutionally inert, in distorted tetrahedral geometry with four other cysteine groups as ligands (Branden et al., 1975).

Cys[174], Cys[46], Zn, H_2O or ROH, His[67] — catalytic zinc

Cys[111], Cys[103], Zn, Cys[97], Cys[100] — structural zinc

The catalytic zinc may polarize the oxygen of the substrate, facilitating direct hydride transfer (in analogy to the aluminum catalyst in Meerwein–Pondorff–Oppenauer catalysis).

(No stereochemistry is indicated in this scheme)

Theorell and Chance (1951) proposed that, in the alcohol-oxidation direction, $NAD^{\oplus}$ must bind first before ethanol can be bound productively (an ordered BiBi mechanism; ¶7.C.1.b). Detailed kinetic studies have validated this suggestion (Dalziel, 1975). The ternary complex then formed does not accumulate in the steady state but is converted rapidly to the product ternary complex, then to the E·NADH binary complex. *The rate-determining step is dissociation of* E·NADH; the physical step of reduced-coenzyme desorption limits V_{max} when primary alcohols are the substrates.

$$E + NAD^{\oplus} \rightleftharpoons E \cdot NAD^{\oplus} \rightleftharpoons (E \cdot NAD^{\oplus} \cdot \text{ethanol}) \rightleftharpoons (E \cdot NADH \cdot \text{acetaldehyde}) \rightleftharpoons \text{acetaldehyde} + E \cdot NADH \overset{\text{slow}}{\rightleftharpoons} E + NADH$$

Table 10-3
Dissociation constants for the binary complex E·NAD or E·NADH

Enzyme	K_D for E·NAD$^{\oplus}$	K_D for E·NADH
Liver alcohol dehydrogenase	2.66 μM	0.3 μM
Malate dehydrogenase	280 μM	1.0 μM
Lactate dehydrogenase	200 μM	0.8 μM

SOURCE: K. Dalziel, in *The Enzymes*, 3d ed., ed. P. Boyer (New York: Academic Press, 1975), vol. 11, p. 40.

(Secondary alcohols such as isopropanol are slow substrates. In these cases, the hydride-transfer step is rate-determining, as evinced by a detectable kinetic isotope effect on V_{max} of acetone formation using 2-[^{2}H]-isopropanol.) Consistent with the fact that release of NADH from the enzyme is a slow step for various NAD-dependent dehydrogenases is the observation that several of these enzymes bind NADH two orders of magnitude more tightly than they bind NAD$^{\oplus}$ (Table 10-3). This observation is not valid for glyceraldehyde-3-P dehydrogenase; recall that GAPDH can be isolated as a 1:1 E·NAD$^{\oplus}$ complex.

10.C.3 Stereochemistry of Hydrogen Transfer Catalyzed by NAD-Dependent Dehydrogenases

We noted briefly in Chapter 2 that enzymes (and other proteins) are biosynthesized only from the L-isomers of α-amino acids and thus are asymmetric or chiral macromolecular catalysts. (If an enzyme contains 200 residues that are not glycyl residues, then it has 200 chiral centers, all of them L.) We commented that this asymmetry constitutes the structural basis for stereospecific action of enzyme catalysts; we now confront our first example in detail of this aspect of enzyme stereospecificity. To explain how an enzyme can catalyze the reaction with only one enantiomer of a racemic mixture (such as D,L-lactate), one must examine any potential for difference in the activation energy for reaction—both in binding steps and in subsequent catalytic steps. Clearly, D- and L-lactate have mirror-image positioning of substituents in three dimensions and will on binding (for example, to L-lactate dehydrogenase) have different energies of interaction with specific amino-acid side chains at the active site.

HO COO$^{\ominus}$ C H CH$_3$

D-lactate

$^{\ominus}$OOC OH C H$_3$C H

L-lactate

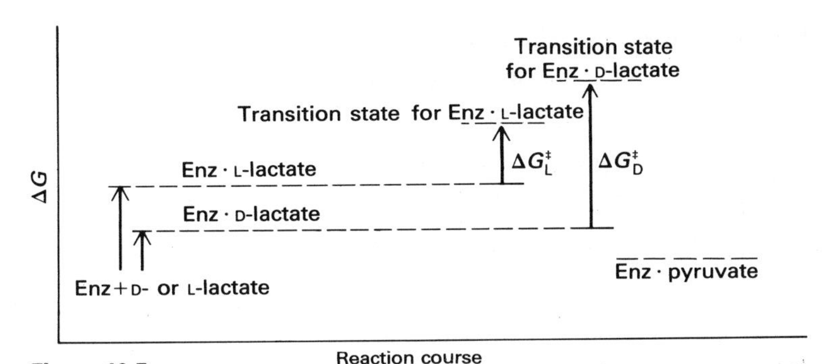

Figure 10-5
Hypothetical reaction coordinate for specific catalysis by an enzyme of L-lactate oxidation.

Thus the two different complexes (L-LDH•L-lactate and L-LDH•D-lactate) are *diastereomeric, not enantiomeric* (as they would be if D- and L-lactate interacted with a *symmetrical* chemical reagent). Not only the Enz•D-lactate and Enz•L-lactate ES complexes, but also the respective transition states for each complex will be diastereomeric and thus will differ in activation energy. If $\Delta G^{\ddagger}$ from Enz•L-lactate to a transition state is considerably lower than the $\Delta G^{\ddagger}$ from Enz•D-lactate to a corresponding transition state, then specific catalysis of the L-lactate oxidation will occur. Figure 10-5 illustrates a hypothetical reaction coordinate. The converse specificity could occur; an enzyme active site could obtain maximal interaction energy on binding D-lactate and thus selectively lower $\Delta G^{\ddagger}$ for its oxidation. Indeed, NAD-dependent dehydrogenases are known that are absolutely specific for L-lactate oxidation, and others exist that are specific for D-lactate oxidation.

Stereospecific action of a chiral enzyme molecule on a substrate chiral center, such as C-2 of lactate, is thus expected. Somewhat less obvious at the outset is the observation that enzymes also distinguish unerringly between two chemically like paired substituents at a C_{aacd} center (a *meso* carbon, for example). Examples of these centers include C-1 of ethanol and C-4 of NADH.

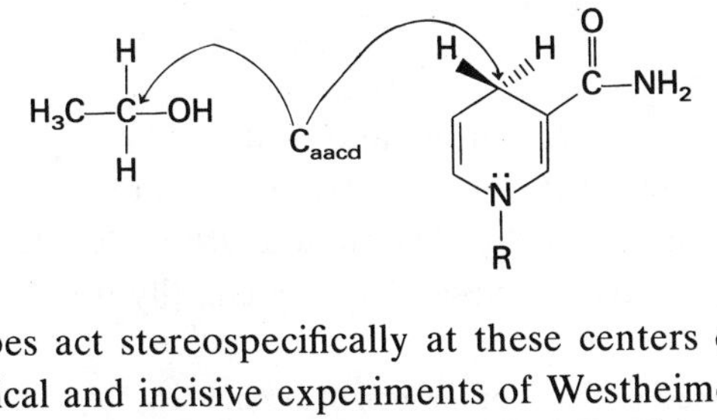

That an enzyme does act stereospecifically at these centers during catalysis was proved by the classical and incisive experiments of Westheimer, Vennesland, and colleagues (Loewus et al., 1953; Levy et al., 1957; H. Fisher et al., 1953).

First, they incubated synthetic 1-dideuteroethanol with alcohol dehydrogenase and NAD and isolated the monodeutero-NADH thereby generated.

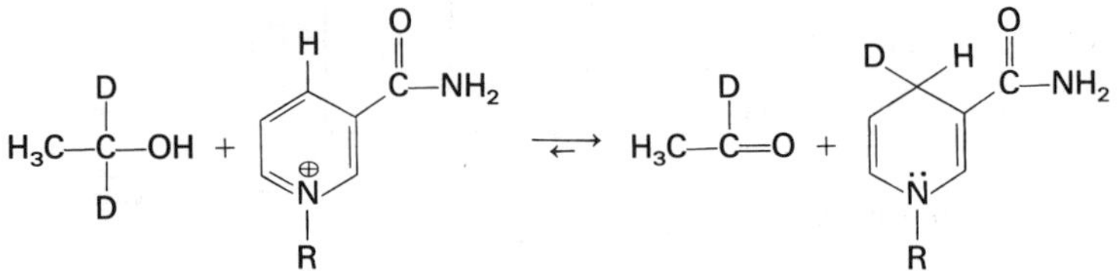

This monodeutero-NADH was then mixed with acetaldehyde and alcohol dehydrogenase, the ethanol isolated, and its deuterium content analyzed by mass spectrometry. A symmetric chemical reagent would not have distinguished between the chemically like D and H at C-4 of NADH; it would have transferred each isotope 50% of the time (given the absence, experimentally verified, of a kinetic isotope effect).* In fact, the ethanol contained all the deuterium; none remained in the NAD formed as coproduct.

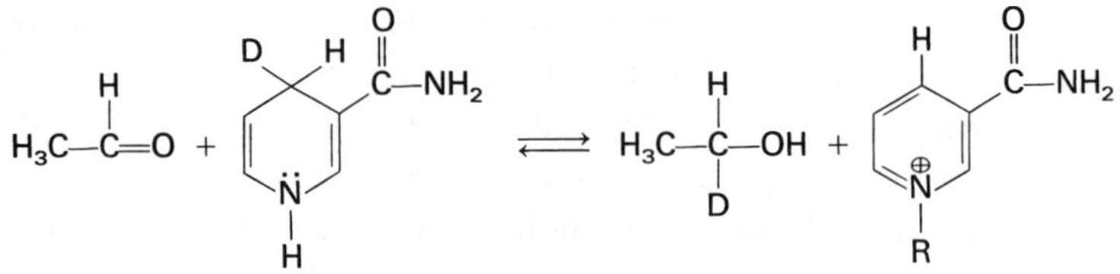

Obviously, alcohol dehydrogenase has recognized and removed only one of the two hydrogen positions at C-4 of NADH. The enzymatically generated monodeuteroethanol has four different substituents and might then be a chiral sample (only one isomer) or a mixture of the two possible mirror-image isomers:

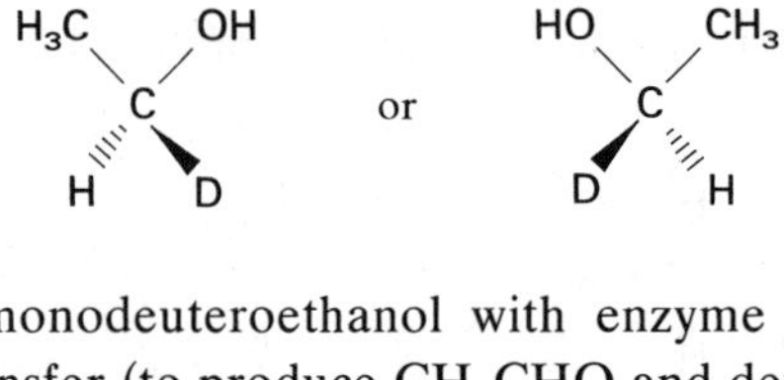

Reincubation of this monodeuteroethanol with enzyme and NAD again gave complete deuterium transfer (to produce CH_3CHO and deutero-NADH), proving three points: (1) the enzymatically generated deuteroethanol was only one isomer and therefore (2) the dehydrogenase also acts chirally toward the C-1 methylene

* There is no isotope effect on V_{max} for reaction of alcohol dehydrogenase with NADH deuterated at C-4 of the nicotinamide ring during acetaldehyde reduction.

hydrogens of ethanol, and (3) the enzyme selects back the same hydrogen it originally transferred from deutero-NADH. These and companion experiments were the first demonstration of chiral enzyme action at certain *meso* carbon atoms, and they spurred a spectacular development of knowledge about enzyme stereochemistry in the ensuing 25 years (see Alworth, 1972, and references therein).

10.C.4 Chirality and Prochirality Assignments

Before we indicate *which* hydrogen at C-1 of ethanol or at C-4 of NADH is recognized by alcohol dehydrogenase, we make a brief digression to introduce the Cahn–Ingold–Prelog system for absolute stereochemical designation and the rules that apply for designation in the simple organic substrates under consideration.

For a typical chiral molecule such as lactate, we may be accustomed to writing two-dimensional Fischer projections for L-lactate and D-lactate.

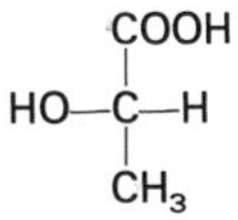

```
   COOH            COOH
    |               |
HO—C—H          H—C—OH
    |               |
   CH3             CH3
 L-lactate       D-lactate
```

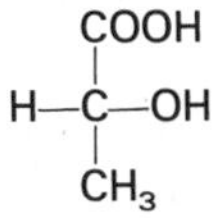

A more general (and, in complex cases, less ambiguous) system for stereochemical designation is the *R*,*S*-system suggested by Cahn, Ingold, and Prelog (1956, 1966; also see Alworth, 1972, p. 69 ff). For simple sp^3 carbon centers, we can consider three basic rules.

1. Assign priority to functional groups in order of decreasing atomic number. For example,

   $O > N > C > H$

   $^3H > ^2H > ^1H$

   $C—OH > C—CH_2Cl > C—CH_2OH > C—CH_3 > C—H$

   Unsaturated centers should be treated as though carbon atoms were attached. For example,

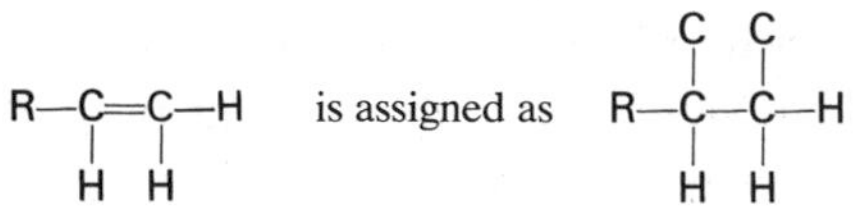

```
                                  C   C
                                  |   |
R—C═C—H   is assigned as    R—C—C—H
  |  |                            |   |
  H  H                            H   H
```

2. Orient the center under examination so that either (a) the viewer is farthest away from the lowest-priority substituent in a tetrahedral projection or (b) the lowest-priority substituent occupies the bottom position in a Fischer projection. For example,

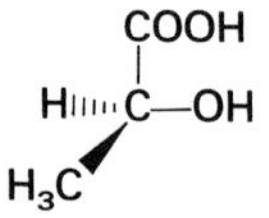

COOH
H····C—OH
H_3C

tetrahedral projection
(H behind plane of paper)

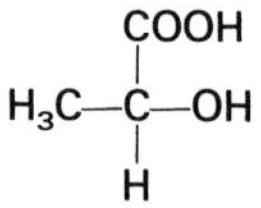

COOH
H_3C—C—OH
H

Fischer projection
(H at bottom)

3. With the center thus oriented, count around the remaining three substituents in order of decreasing priority.
   a. If these three substituents thereby describe a *clockwise turn*, the center is designated *R* (for *rectus*, Latin "right[handed]").
   b. If these three substituents thereby describe a *counterclockwise turn*, the center is designated *S* (for *sinister*, Latin "left[handed]").

In the example cited for rule 2, the three remaining substituents in the tetrahedral projection (and in the Fischer projection) describe a counterclockwise turn. Let's represent priorities of groups as $1 > 2 > 3 > 4$.

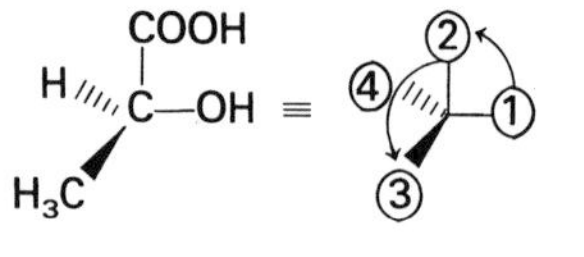

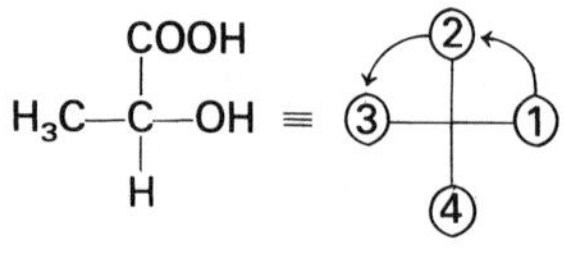

COOH
H····C—OH ≡ ④ ② ① ③
H_3C

COOH
H_3C—C—OH ≡ ③ ② ① ④
H

The chiral center therefore is *S* in either structural representation, and the two structures are different projections of the same chiral isomer, *S*-lactate. Examination of the Fischer projection indicates that this structure also is that of L-lactate in the Fischer system, so L-lactate ≡ *S*-lactate.

A comment is in order about using Fischer projections and rotating the molecule to put the lowest-priority group at the bottom to assign the center as *R* or *S*. Because Fischer projections are two-dimensional projections (in the plane of the paper) of three-dimensional structures, one must remember the rules that govern use of such projections. Because the flat projection represents a structure that does not lie entirely in the plane of the paper, one *may not* simply rotate the entire structure as written. Instead, one *may* rotate any three attached groups of the Fischer projection stepwise, while keeping the position of the central atom and the fourth group fixed (see Alworth, 1972).

For example, consider the representation of *S*-lactate just given. We can hold the COOH group fixed and rotate the other three groups to obtain the following equivalent representations of *S*-lactate:

$$\begin{array}{c}\text{COOH}\\ |\\ \text{H}_3\text{C—C—OH}\\ |\\ \text{H}\end{array} \equiv \begin{array}{c}\text{COOH}\\ |\\ \text{HO—C—H}\\ |\\ \text{CH}_3\end{array} \equiv \begin{array}{c}\text{COOH}\\ |\\ \text{H—C—CH}_3\\ |\\ \text{OH}\end{array}$$

By holding other groups fixed, still other equivalent Fischer representations can be obtained. As an exercise to test your understanding of these rules, verify that all of the following are equivalent Fischer representations of an *S*-center:

$$\begin{array}{c}2\\3{+}1\\4\end{array}\;\begin{array}{c}1\\2{+}3\\4\end{array}\;\begin{array}{c}3\\1{+}2\\4\end{array}\;\begin{array}{c}4\\2{+}1\\3\end{array}\;\begin{array}{c}1\\4{+}2\\3\end{array}\;\begin{array}{c}2\\1{+}4\\3\end{array}\;\begin{array}{c}3\\4{+}1\\2\end{array}\;\begin{array}{c}1\\3{+}4\\2\end{array}\;\begin{array}{c}4\\1{+}3\\2\end{array}\;\begin{array}{c}4\\3{+}2\\1\end{array}\;\begin{array}{c}2\\4{+}3\\1\end{array}\;\begin{array}{c}3\\2{+}4\\1\end{array}$$

You should also be able to write the twelve equivalent representations of an *R*-center.

Let us determine the chirality of another example.

$$\begin{array}{c}\text{COO}^{\ominus}\\ |\\ \text{H—C—NH}_3^{\oplus}\\ |\\ \text{CH}_3\end{array} \equiv \begin{array}{c}\text{COO}^{\ominus}\\ |\\ \text{H}_3\overset{\oplus}{\text{N}}\text{—C—CH}_3\\ |\\ \text{H}\end{array} \equiv \begin{array}{c}②\\ |\\ ①\text{—+—}③\\ |\\ ④\end{array}$$

D-alanine ≡ *R*-alanine

One can assign several chiral centers in the same molecule. For example, consider threonine, which has two chiral centers (C-2 and C-3).

$$\begin{array}{c}\text{COOH}\\ |\\ \text{H}_2\text{N—C—H}\\ |\\ \text{H—C—OH}\\ |\\ \text{CH}_3\end{array} \equiv \begin{array}{c}\text{COOH}\\ |\\ \text{H}_2\text{N—C—H}\\ |\\ \text{CHOHCH}_3\end{array} \equiv \begin{array}{c}②\\ |\\ ①\text{—(C-2)—}④\\ |\\ ③\end{array} \equiv \begin{array}{c}②\\ |\\ ③\text{—(C-2)—}①\\ |\\ ④\end{array} \equiv S$$

$$\equiv \begin{array}{c}\text{CHNH}_2\text{COOH}\\ |\\ \text{H—C—OH}\\ |\\ \text{CH}_3\end{array} \equiv \begin{array}{c}②\\ |\\ ④\text{—(C-3)—}①\\ |\\ ③\end{array} \equiv \begin{array}{c}②\\ |\\ ①\text{—(C-3)—}③\\ |\\ ④\end{array} \equiv R$$

This molecule is (2*S*,3*R*)-threonine. For common biochemically important molecules such as L-lactate, D-alanine, and D-glucose, the D,L-nomenclature of the Fischer projections is still in common use. However, for the only slightly more complex threonine, it is clear that explicit designation of the stereochemistry at each chiral center is both more informative and more useful. The virtues of the *R*,*S*-system are even more evident in a consideration of configuration at prochiral centers.

Carbon centers of the type C_{aacd} (such as C-1 of ethanol and C-4 of NADH) are designated *prochiral centers* because they are precursors to chiral centers in the following sense. If one of the chemically identical a-groups at the center is replaced by a new group b ($b \neq c,d$), then C_{abcd} (a *chiral center*) is produced. Because chiral centers can be either *R* or *S*, replacement of either a-group will generate one of the two mirror-image isomers in the C_{abcd} product.

A rule for assignment of configuration at a prochiral center is mentally to replace one of the a-groups with some group b, so that the priorities of the substituents remain in the same order; that is,

$$\text{if } d > c > a,a \qquad \text{then} \qquad d > c > b > a.$$

Now, if the chiral center generated in such a mental manipulation is *S*, then the a-group replaced mentally by b is pro*S*. If the chiral center generated is *R*, then the a-group replaced mentally occupies the pro*R* position.

As an example, consider assignment of configuration to the two prochiral hydrogens at C-1 of ethanol. Try replacing one of the H groups with a ^2H group, for example, and then assign the configuration of the deuteroethanol structure. (When two groups have equal atomic numbers, the group with higher atomic mass has priority.)

H_3C–C(H_R)(H)–OH ⟶ H_3C–C(^2H)(H)–OH ≡ (b) top, (c) left, (d) right, (a) bottom, or (3) top, (2) left, (1) right, (4) bottom

So the substituted molecule is *R*-[^2H]-ethanol. Because the enantiomer of the chiral 1-[^2H]-ethanol generated in such a substitution is *R*, the hydrogen replaced by deuterium must be pro*R*, designated H_R. Similarly, replacement of the other hydrogen in ethanol would yield *S*-1-[^2H]-ethanol (verify the assignments).

H_3C–C(H)(H_S)–OH ⟶ H_3C–C(H)(^2H)–OH ≡ H, CH_3, C–OH, ^2H

The following is the proper assignment of absolute configuration of the prochiral center in ethanol:

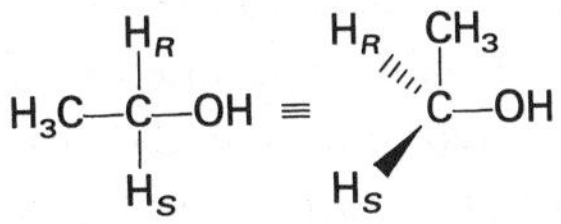

$$H_3C-C(H_R)(H_S)-OH \equiv$$

Ethanol has one prochiral center and no chiral centers. Thus, the prochiral hydrogens at C-1 can be considered *enantiotopic* prochiral hydrogens, because their specific replacement (e.g., with ^2H) leads to *enantiomers* of the chiral compound (e.g., 1-[^2H]-ethanol). If a molecule has at least one existing chiral center in addition to the prochiral center of interest, then the two paired a-groups are *diastereotopic* prochiral groups, because generation of a chiral center by replacement with some group b would produce *diastereomers*, not enantiomers (the molecule would then have at least *two* chiral centers).

Glyceraldehyde presents such an example. In *R*-glyceraldehyde (*R* at C-2), the C-3 hydrogens are a diastereotopic pair.

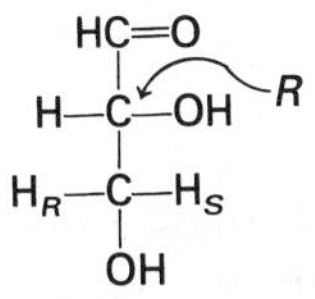

R-glyceraldehyde

We bring up this stereochemical subtlety because the prochiral methylene hydrogens at C-4 of the dihydronicontinamide ring of NADH also are diastereotopic. (There are several asymmetric centers in NADH, and in NAD$^{\oplus}$—most obviously in the two D-ribose rings.)

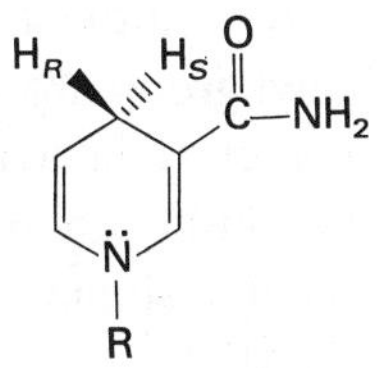

Note that it is the carboxamide group at C-3 that allows C-4 to be a C_{aacd} center and permits ready designation of absolute configuration as indicated.

10.C.5 Chirality of Alcohol-Dehydrogenase Catalysis

Now that we have learned how to assign prochiral configurations to ethanol and NADH, we can return to the consideration of the stereospecificity of alcohol-dehydrogenase catalysis. The elegant experiments of Westheimer, Vennesland, and collaborators clearly indicated that the dehydrogenase removes only one prochiral hydrogen from alcohol substrate and reduced coenzyme, but these experiments per se could not establish absolute configuration. Later experiments involved chemical synthesis of authentic *S*- and *R*-1-[^{2}H]-ethanols by an unambiguous route, and subsequent testing of these ethanols for deuterium transfer. The results clearly indicated that alcohol dehydrogenase removes the pro*R* hydrogen at C-1 of ethanol. The NADH determination was more complicated, because unambiguous synthesis of 4*R*-[^{2}H]-NADH or of 4*S*-[^{2}H]-NADH is not easily envisaged. In fact, the enzymatically deuterated 4-[^{2}H]-NADH eventually was degraded chemically, with conversion of the dihydronicotinamide ring to 2-[$^2H_1$]-succinate (Cornforth et al., 1966). Comparison with authentic chiral monodeuterosuccinates established the sample's configuration as *R*-deuterosuccinate, thus indicating the configuration of the original enzymatically deuterated NADH as 4*R*-[^{2}H]-NADH.

10.C.6 *sp*^{2} Chiral Centers

We have learned how to assign absolute stereochemistry of alcohol dehydrogenase action at the *sp*^{3} centers. But, of course, the cosubstrates acetaldehyde and NAD$^{\oplus}$ have *sp*^{2} hybridization at C-1 and C-4, respectively, to which the itinerant hydride ion comes. If only *R*-[^{2}H]-ethanol is generated enzymatically when deuteride is transferred from 4*R*-[^{2}H]-NADH to acetaldehyde, it should be intuitively obvious that *the deuteride must attack the acetaldehyde carbonyl from only one specific face*: the corollary is that the carbonyl group of acetaldehyde must be *an sp*^{2} *prochiral center.* Similar logic mandates that C-4 of NAD$^{\oplus}$ is also an *sp*^{2} prochiral center.

In general, if a trigonal carbon atom has three distinct substituents (C_{abc}), then the two faces of the carbonyl group are sterically nonequivalent and must be prochiral. If the planar *sp*^{2} center is in a molecule without an existing chiral center, the two faces of that prochiral center are enantiotopic faces (e.g., acetaldehyde); if at least one chiral center already exists in the molecule, the prochiral faces are diastereotopic (e.g., NAD$^{\oplus}$).

To assign configuration for *sp*^{2} centers by the Cahn–Ingold–Prelog system, we use a simple sequence of rules, somewhat analogous to those for *sp*^{3} centers (see Alworth, 1972).

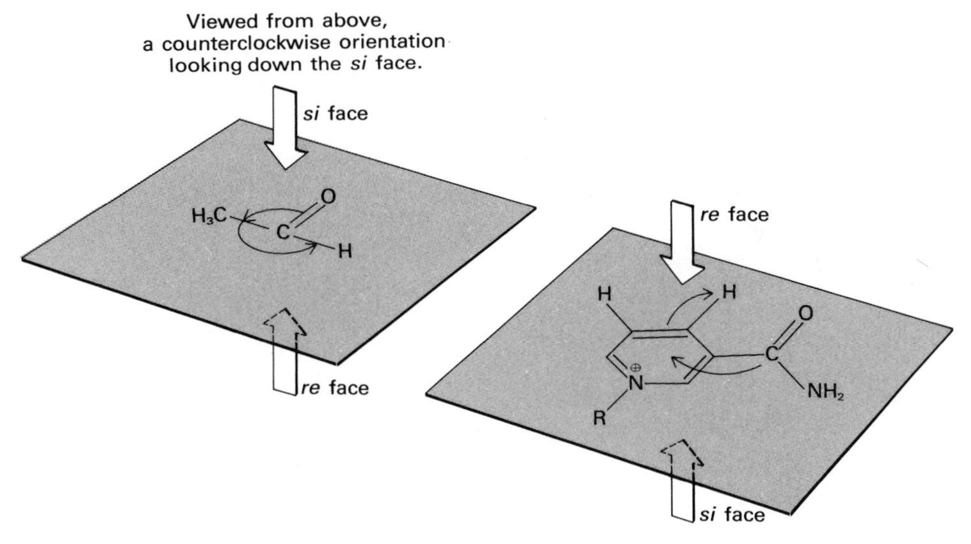

Figure 10-6
Assigning conformation of sp^2 chiral centers.

1. Determine the three substituents of C_{abc} in order of decreasing priority.
2. Orient the trigonal center so the viewer is either above or below the plane of the sp^2 center.
3. If the priority sequence is *clockwise*, the viewer is closest to the *re* (rectus) face. If the sequence in decreasing order is *counterclockwise*, the viewer is closest to the *si* (sinister) face (Fig. 10–6).

A complete stereochemical description of the reaction would then be as follows: the pro*R*-hydrogen is transferred from C-4 of NADH to the *re* face of acetaldehyde, thus introducing the pro*R*-hydrogen in the ethanol product. In the direction of ethanol oxidation, $H_R^{\ominus}$ attack must also proceed at the *re* face of the bound $NAD^{\oplus}$. From the X-ray map's placement of the catalytic zinc atom, one guesses that the $Zn^{2\oplus}$ must coordinate the acetaldehyde from the *si* face during the redox process (Fig. 10-7).

Once it had been established that alcohol dehydrogenase transfers a hydride equivalent to and from one face of the bound nicotinamide coenzyme, several other nicotinamide-requiring dehydrogenases were tested to see if the same prochiral hydrogen (H_R) at C-4 of NADH is invariably transferred. Actually, many dehydrogenases were examined in the ten-year period that ensued between the initial alcohol-dehydrogenase experiments and the elucidation of the absolute prochiral assignment in NADH by the chemical-degradation route outlined in

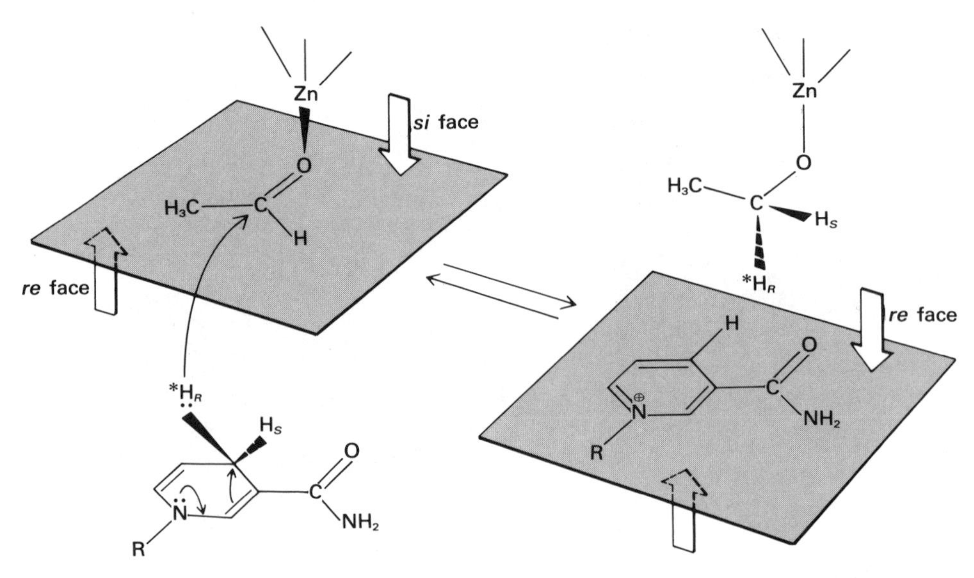

Figure 10-7
Proposed stereochemical mechanism for redox process at active site of alcohol dehydrogenase.

¶10.C.5. During that period, relative stereochemical designations had to be used, and these designations still surface today. Alcohol dehydrogenase was defined arbitrarily as catalyzing transfer to and from the "A" face of $NAD^{\oplus}$ and NADH (now known, as we have seen, to be the *re* face). It was soon discovered that some other enzymes use the opposite ("B") face (now known to be the *si* face), among them the glyceraldehyde-3-P dehydrogenase discussed earlier in this chapter. Table 10-4 gives an annotated list of the stereospecificity of some of the more common NAD-dependent dehydrogenases of oxidative metabolism. A much

Table 10-4
Stereospecificity at C-4 for some nicotinamide-dependent dehydrogenases

Enzyme	Nucleotide	C-4 prochiral hydrogen transferred	Specific oxidized product
Alcohol dehydrogenase	NAD	H_R	Acetaldehyde
UDP-glucose dehydrogenase	NAD	H_S	UDP-glucuronate
L-Lactate dehydrogenase	NAD	H_R	Pyruvate
Malate dehydrogenase	NAD	H_R	Oxalacetate
Isocitrate dehydrogenase			
cytoplasmic	NADP	H_R	α-Ketoglutarate
mitochondrial	NAD	H_R	α-Ketoglutarate
Glyceraldehyde-3-P dehydrogenase	NAD	H_S	1,3-Phosphoglycerate
Glutamate dehydrogenase	NADP, NAD	H_S	α-Ketoglutarate$+NH_4^{\oplus}$

more extensive list of chirality of nicotinamide-dependent enzymes (ca. 50 examples) has recently been compiled by Simon and Kraus (1976). There is currently no known mechanistic imperative for this stereochemical dichotomy toward both faces of the NAD or NADP coenzymes. There is at least one dehydrogenase—an $NAD^{\oplus} + NADPH \rightleftarrows NADP^{\oplus} + NADH$ transhydrogenase from animal-cell mitochondria (Rydstrom et al., 1976)—that uses pro*R* for one molecule and pro*S* for the other.

10.C.7 Complex NAD-Dependent Transformations

Before concluding this survey of nicotinamide-coenzyme function in simple dehydrogenations, let's consider three examples* of nicotinamide-dependent enzymatic transformations where the redox function is not so apparent in the stoichiometry of the transformation. In these examples, the oxidized coenzyme is tightly bound as a 1:1 complex at the outset, remains tightly bound, and functions *catalytically*. The oxidized form is regenerated at the end of each catalytic turnover cycle, so this role of $NAD^{\oplus}$/NADH is distinct from the cosubstrate role of $NAD^{\oplus}$/NADH discussed in preceding sections of this chapter.

10.C.7.a UDP-glucose epimerase

UDP-glucose epimerase interconverts UDP-glucose and its 4′-epimer UDP-galactose, necessary for normal galactose metabolism in animals.

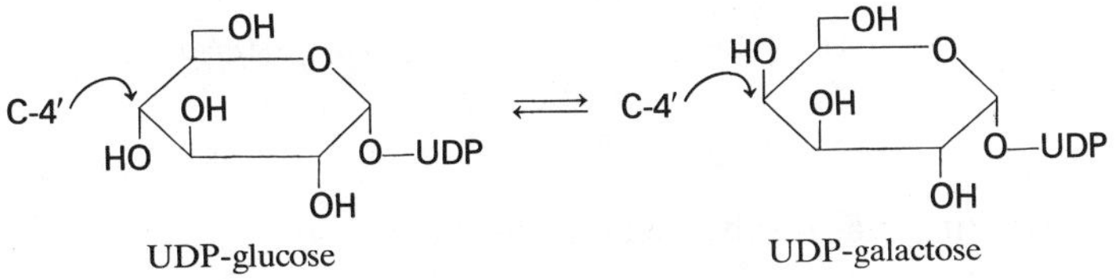

C-4′ OH HO OH O O—UDP OH ⇌ C-4′ HO OH OH O O—UDP OH

UDP-glucose UDP-galactose

At first glance, this appears to be an epimerization of C-4′, categorizable as a 1,1-hydrogen shift, and not an oxidation or reduction reaction at all. One would not a priori predict any requirement for $NAD^{\oplus}$ or $NADP^{\oplus}$, and none need be added exogenously with purified enzyme.

However, in $[^3H]$-H_2O, starting with either substrate, there is no observed tritium incorporation into either UDP-glucose or UDP-galactose—suggesting no

*A probable fourth example is provided by dehydroquinate synthase (¶27.C.3; Simon and Kraus, 1976).

Table 10-5
Properties of UDP-glucose epimerases

Property	Yeast enzyme	*E. coli* enzyme
Subunits	2	2
Molecular weight	125,000	79,000
NAD content (moles per subunit)	0.5	0.5

SOURCE: L. Glaser in *The Enzymes*, 3d ed., ed. P. Boyer (New York: Academic Press, 1972), vol. 6, p. 355.

exchange of C-4 with solvent. Such lack of exchange, of course, is a hallmark of nicotinamide-coenzyme–dependent catalysis.

Enzymes from both *E. coli* and yeast have been purified to homogeneity. Each was examined for tightly bound NAD$^{\oplus}$, with the results shown in Table 10-5. The coenzyme content suggests one active site per dimer of apparently identical subunits.

Given an NAD–enzyme functional unit, one expects that the enzyme achieves epimerization at the 4′-carbon by an oxidation–reduction process. For instance, a 4′-keto intermediate could participate:

H* / —C— / OH ⇌ NADH* + —C(=O)— ⇌ OH / —C— / H*

A number of experiments have been performed with results that are consistent with the formation of just such an intermediate species:

O= (4-keto sugar ring with –OH, O, OH, O—UDP, OH) · Enzyme · NADH

These experimental results include the following:

1. no exchange with solvent protons;
2. no ^{18}O in either of the UDP-sugars in [^{18}O]-H_2O, ruling against a dehydration–rehydration sequence;
3. addition of either substrate to the *E. coli* enzyme causes transient increase and then decay of absorbance at 340 nm, indicative of bound NADH forming and then reoxidizing;
4. the E·NAD$^{\oplus}$ can be treated with [^{3}H]-$NaBH_4$ to yield Enz·[^{3}H]-NADH with a high degree of stereoselectivity (only one face accessible to

solvent?). This E·[^{3}H]-NADH can be incubated with an analogue of the predicted intermediate; the analogue is dTDP-4′-keto-6′-deoxy-D-glucose.

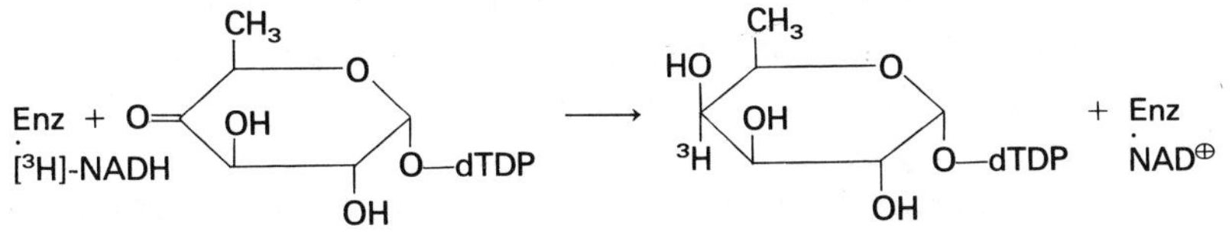

dTDP-4′-keto-6′-deoxy-D-glucose

dTDP-4′-[^{3}H]-6′-deoxy-D-galactose

This tritium transfer shows the E·NADH species to be catalytically competent, although why only the galactose and not the glucose product forms is not clear. Secondly, the tritium transfer suggests that the enzyme can recognize a 4′-keto-sugar nucleotide as a species fit for reduction. However, a 4′-keto species is not the only reasonable intermediate one can imagine. For instance, reversible oxidation could occur at C-3′. The 3′-keto species will have an acidic H at C-4′, which might be abstracted by a fully sequestered base, with stabilization of charge density at C-4′ by the enolate delocalization.

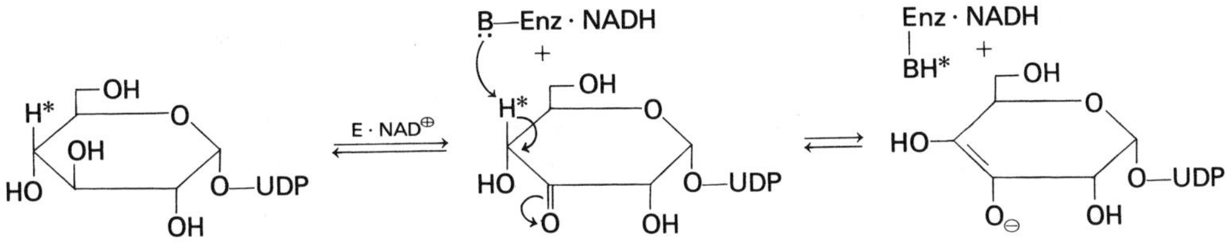

UDP-glucose

(3′-keto-glucose species)

enolate

Reprotonation of the planar enolate from the opposite face by the fully shielded base would give the 3′-keto-galactose species, which then can be reduced to yield UDP-galactose and to regenerate NAD$^{\oplus}$.

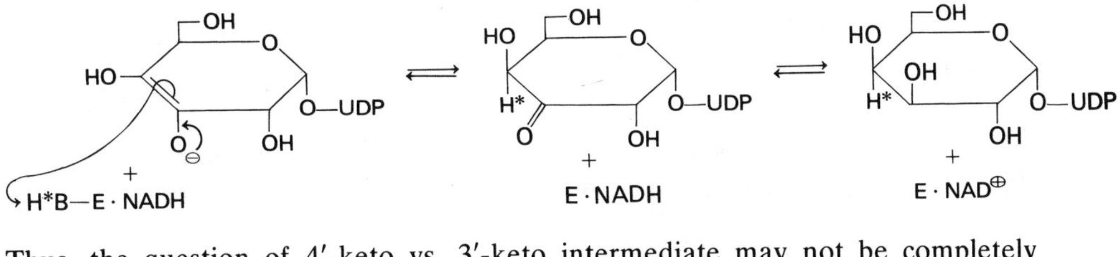

Thus, the question of 4′-keto vs. 3′-keto intermediate may not be completely settled, although most workers in the area favor the 4′-keto intermediate. A distinction between these hypotheses was attempted using TDP-D-glucose, also an epimerizable substrate for UDP-galactose epimerase. Gabriel and colleagues prepared 3′-[^{3}H]- and 4′-[^{3}H]-TDP-D-glucose and found that only the 4′-tritio substrate experienced a tritium kinetic isotope selection ($k_H/k_T = 2.6$), a positive index that the (C-4′)—H bond is broken (Adair et al., 1971). However, that bond would be broken in either mechanism, so this result per se does not rule against the 3′-keto species. It merely indicates that, if the (C-3′)—H bond *is* broken, that is not a rate-limiting step.

One of the more intriguing features of this NAD-dependent epimerase is precisely the mechanistic requirement that the enzyme must act *nonstereospecifically* at C-4′ of the UDP-keto sugar, in contrast to the uniformly stereospecific hydrogen transfers noted earlier in this chapter for nicotinamide-dependent dehydrogenases.* The suggestion has been made that the 4′-keto sugar moiety moves relative to the NADH at the active site. Kang et al. (1975) argue that rotation around the saccharide C-1 oxygen atom and the β-P atom of the pyrophosphoryl linkage should be relatively unhindered. (The uridine portion of the intermediate would remain tightly anchored.)

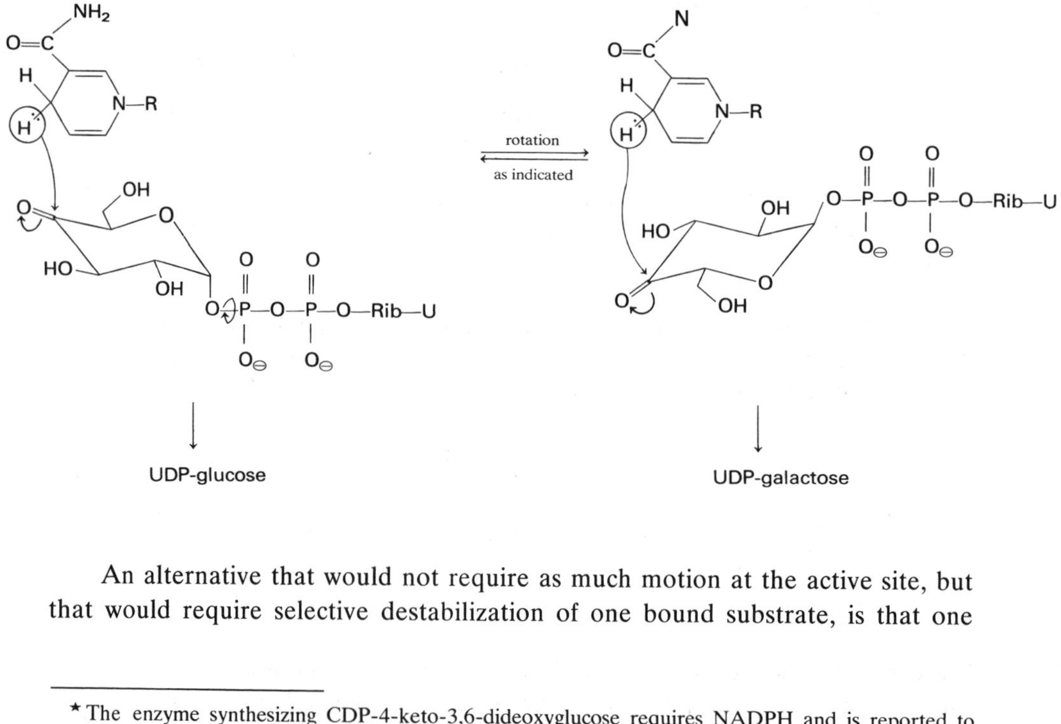

NH2 O=C H H N—R OH O HO OH O O O—P—O—P—O—Rib—U O⊖ O⊖

rotation ⇄ as indicated

N O=C H H N—R O O O—P—O—P—O—Rib—U O⊖ O⊖ OH HO O O OH

↓ UDP-glucose

↓ UDP-galactose

An alternative that would not require as much motion at the active site, but that would require selective destabilization of one bound substrate, is that one

* The enzyme synthesizing CDP-4-keto-3,6-dideoxyglucose requires NADPH and is reported to transfer either the 4*R* or the 4*S* hydrogen (see ¶17.B.4.a).

epimer reacts from a chair conformation and the other from a boat conformation (Glaser and Ward, 1970).

10.C.7.b Ornithine cyclase

Ornithine cyclase (deaminating) has been obtained from *Clostridium sporonges.* The homogeneous enzyme has 81,000 mol wt, is a dimer of identical subunits, and again has only one NAD$^{\oplus}$ tightly bound per dimer ($K_D = 6\ \mu$M). The two subunits may contribute complementary halves to an active-site structure.

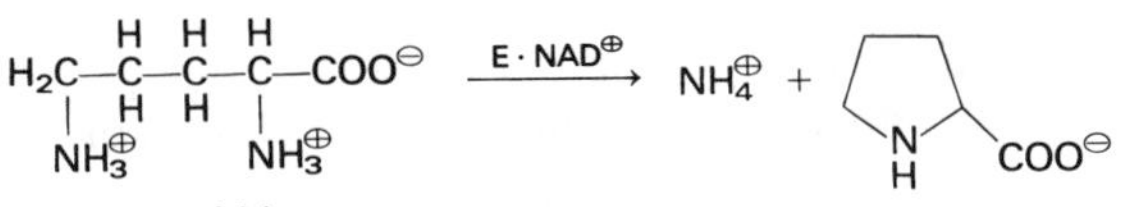

L-ornithine (2,5-diaminopentanoate) —E · NAD$^{\oplus}$→ $NH_4^{\oplus}$ + L-proline

Mass-spectrometric data suggest that 5-[^{15}N]-ornithine goes to [^{15}N]-proline, implying that the 2-amino group of ornithine is lost. A direct displacement of the 2-amino group (a poor leaving group) by attack of the 5-amino group on C-2 is unlikely and without enzymatic or chemical precedent. Recall that imino acids in aqueous solutions dissociate to keto acids and ammonia. Thus, we expect an intermediate NAD-dependent oxidation at C-2 of ornithine, followed by attack of the C-5 amino group on the now-electrophilic C-2 carbonyl carbon and re-reduction.

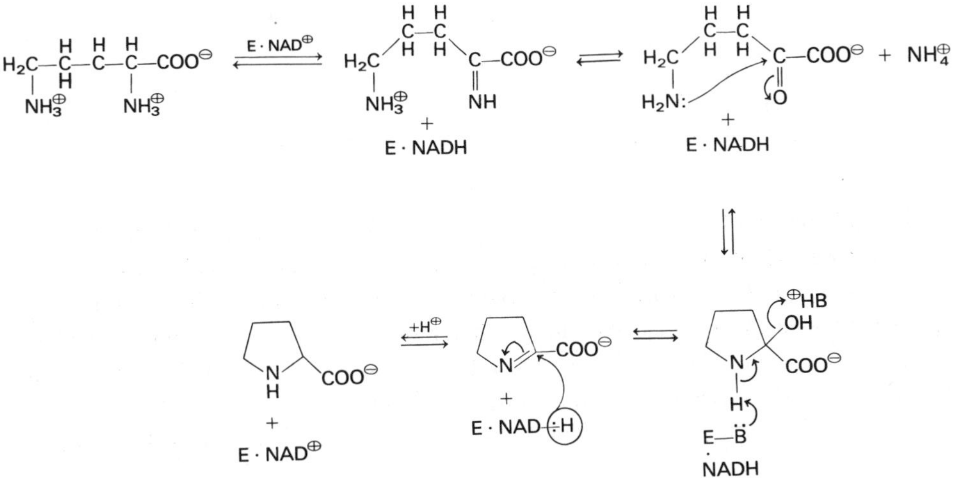

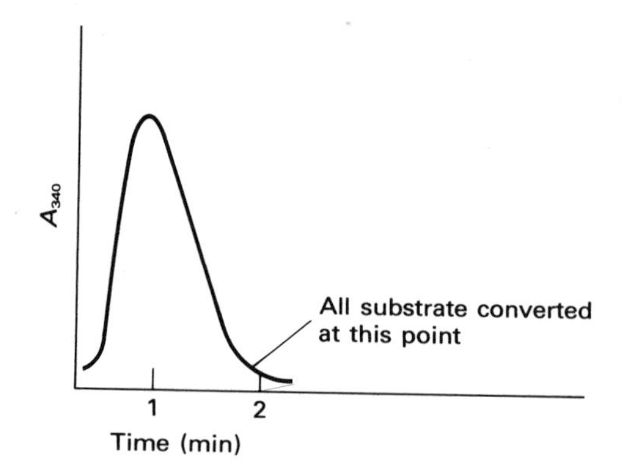

Figure 10-8
Absorbance at 340 nm for ornithine-cyclase reaction with limiting substrate.

One would predict 2-[^{3}H]-ornithine → 2-[^{3}H]-proline, but this has not been reported. However, one does see a transient absorbing peak at 340 nm due to E·NADH when a limiting quantity of substrate is used (Fig. 10-8).

10.C.7.c Inositol-1-P cyclase

Our third example from this class of complex NAD-dependent enzymes is inositol-1-P cyclase, which is responsible for the conversion of glucose-6-P to myoinositol-1-P (a carbocyclic pentahydroxy-1-phosphocyclohexane).

$OPO_3^{2\omines}$ O OH OH HO OH $\xrightarrow{E \cdot NAD^{\oplus}}$ OH HO OH $OPO_3^{2\ominus}$ OH OH

All available data suggest the oxidation of C-5 of glucose-6-P to the 5-ketoglucose-6-P, which then undergoes intramolecular aldol condensation, with the now-acidic C-6 hydrogen undergoing enzyme-mediated abstraction as a proton to give the C-6 carbanion, which attacks the aldehyde carbonyl at C-1. The resulting aldol product is a carbocyclic system. Reduction of the keto group by the E·NADH regenerates the initial E·NAD$^{\oplus}$ complex, prevents the aldol from reopening, and produces the observed inositol product (Muth and Costillow, 1974).

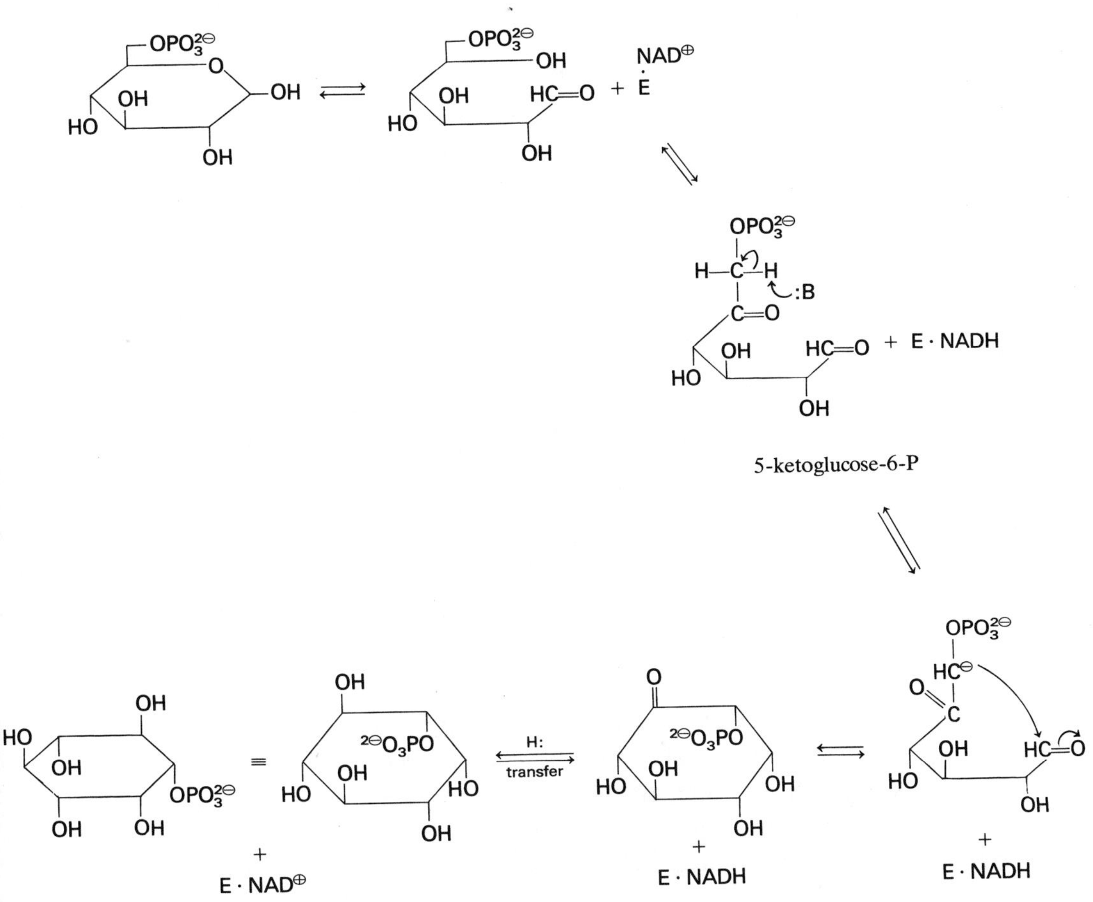

5-ketoglucose-6-P

10.C.8 A Nonredox Role for NAD: Group Transfer of the ADP-Ribosyl Portion

In addition to the ubiquitous participation of nicotinamide coenzymes in simple or complex enzymatic oxidoreduction processes, oxidized NAD$^{\oplus}$ undergoes fragmentation at the glycosidic bond between N-1 of the nicotinamide and C-1 of the adjacent ribose in several enzymatic reactions. These are transfers of the ADP ribosyl group to some acceptor nucleophile, generating free nicotinamide and the ADP-ribosylated nucleophile, entirely analogous to the enzyme-mediated group-transfer chemistry noted in preceding chapters. (This glycosidic linkage is activated for such group transfer; $\Delta G^{0'}$ of hydrolysis = −8.2 kcal/mole = −34 kJ/mole, pH 7, 25° C. See Zatman et al., 1953.)

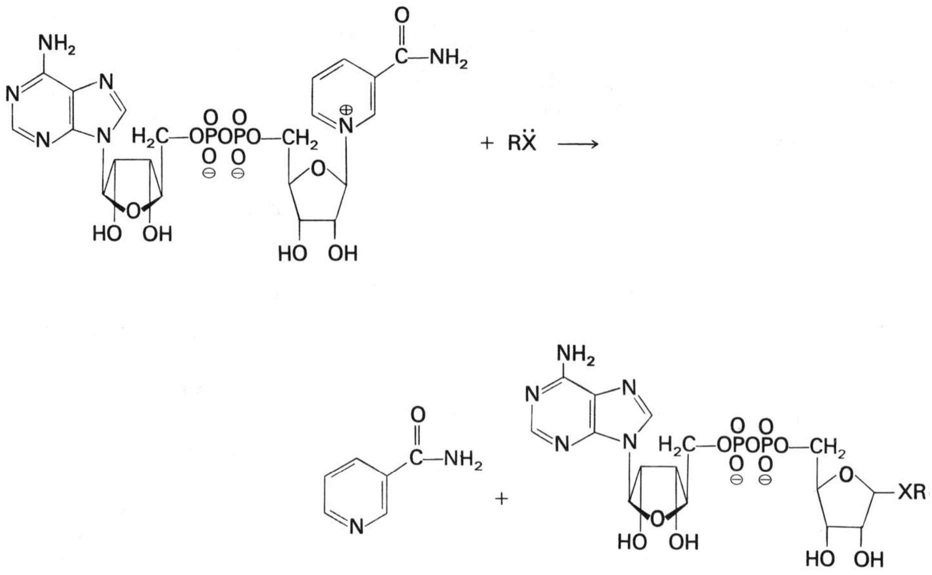

ADP-ribosylated product

Two functional categories have been noted: NAD glycohydrolases (NADases) on the one hand, and protein-ADP-ribosylating enzymes on the other. However, the underlying chemistry is probably the same with H_2O as nucleophilic acceptor in the first case ($RX = H_2O$) and with some nucleophilic amino-acid side chain of the target protein as acceptor in the second case. A variant of this latter type involves biosynthesis of ADP ribose polymers in animal-cell nuclei, as we shall note.

For years it has been known that NADases carry out ADP-ribosyl transfer not only to water but also to pyridine derivatives, a capacity that has often been utilized to synthesize NAD analogues on a microscale (Zatman et al., 1953, 1954; Kaplan and Ciotti, 1956). Because the resultant NAD analogues also have the β-configuration at C-1 of the ribose participating in the glycoside link, the retention of configuration may signal a double-displacement mechanism and possibly a covalent ADP-ribosyl-enzyme intermediate (see the discussion of sucrose phosphorylase in Chapter 9) that can be captured either by H_2O or by the nitrogen of the alternative pyridine cosubstrate.

Several bacterial toxins—among them the potentially lethal diphtheria toxin and cholera toxin—appear to catalyze ADP-ribosyl transfers as key molecular events in their inhibition of animal cells. The diphtheria and cholera toxins each have two kinds of subunits, designated as A and B proteins. The B component appears to bind to surface receptors on membranes of susceptible cells (the cholera-toxin receptor is the ganglioside GM 1; see Kohn, 1977), and then the A component may either pass through the membrane (diphtheria toxin) or move in the plane of the membrane until it interacts with the membrane adenyl cyclase (cholera toxin). In both cases the A component can then use NAD to ribosylate a specific protein. The diphtheria-toxin A protein specifically transfers the ADPR group of NAD to elongation factor II, a key component required for translocation of the nascent peptide chains on the ribosome during protein synthesis. The modification inactivates the elongation factor and blocks protein biosynthesis (Collier, 1967; Honjo et al., 1968). Similarly, it is hypothesized that cholera toxin may ADP-ribosylate plasma-membrane adenyl cyclase, this time activating it and setting off a series of reactions that may change membrane permeability and generate the water fluxes that are the pathophysiological consequences of cholera-toxin action (Moss et al., 1977; Moss and Vaughn, 1977; Kohn, 1977).

A third example of an NAD-derived ADP-ribosylation of a protein leaves enzyme activity unaffected, and its purpose in unclear. Within minutes of infection of *E. coli* cells by bacteriophage T4, the α subunits of the multicomponent *E. coli* RNA polymerase are specifically ADP-ribosylated at a single arginine residue. One of the guanidino nitrogens must have acted as nucleophile (the stereochemistry at the ribosyl carbon has not been reported) (Goff, 1974; Zillig et al., 1975).

NAD$^{\oplus}$ + H_2N—C(=NH)—NH—(CH$_2$)$_3$—CH(NH—)—C(=O)—Enz ⟶

nicotinamide (pyridine—C(=O)—NH_2) + ADP—CH_2—ribose(OH, OH)—$\overset{\oplus}{N}$=C(NH)—NH—Arg—Enz

In addition to these enzymatic transfers of one ADP-ribose group onto specific proteins, there are enzymes located in the nuclei of eukaryotic organisms that can polymerize the ADP-ribose functionality of NAD into a repeating (ADP-ribose)$_n$ homopolymer attached to chromatin-associated proteins, probably histones (Hayaishi, 1976).

$n\text{NAD}^{\oplus}$ + Nuclear protein $\xrightarrow[\text{synthetase}]{\text{poly-ADP-ribose}}$ n (nicotinamide) + poly-ADP-ribosylated protein

The value of n may be as high as 50, and the connecting linkage is a ribose $(1' \rightarrow 2')$ bond (Hayaishi, 1976). The physiological consequences of such enzymatic action are unclear, but it would conceivably play some regulatory role in eukaryotic DNA replication.

A final nonoxidative role for NAD is found in its enzymatic cleavage, not to ADPR and nicotinamide, but to AMP and nicotinate mononucleotide during action of *E. coli* DNA ligase. We commented on the adenylyl-transfer mechanism used by DNA ligase of coliphage T4 during closure of single-strand breaks in DNA (¶8.A.6). In that instance, ATP was the source of the AMP residue undergoing transfer, first to yield an AMP-enzyme, then an ADP derivative on the 5′ side of the DNA gap. In the case of *E. coli* DNA ligase, it is $\text{NAD}^{\oplus}$ rather than ATP that is the source of the transferring AMP residue (Lehman, 1974). The subsequent steps in *E. coli* and T4 DNA ligase action are probably identical. This is an unusual nucleotidyl-transfer role for NAD.

adenylyl transfer

AMP-enzyme

NMN

This completes our coverage of NAD-dependent dehydrogenases in isolation. In the next chapter, we take up flavin coenzymes. However, we shall note that several flavin-dependent enzymes use NADH as a specific substrate, catalyzing its oxidation to $NAD^{\oplus}$ during reduction of the flavin coenzyme. These enzymes are dehydrogenases (Chapter 11) in some cases and hydroxylases (Chapter 12) in others.

Chapter 11

Flavin-Dependent Dehydrogenases and Oxidases

In Chapter 10 we noted repeatedly that direct hydrogen transfer between substrate and coenzyme is a hallmark of simple enzymatic catalyses utilizing nicotinamide coenzymes. We shall now discuss some enzymatic transformations involving nicotinamides where such transfer does not occur; the explanation for the observed solvent exchange in these reactions introduces the second major type of redox coenzymes: the flavin coenzymes.

11.A NICOTINAMIDE–REQUIRING ENZYMES THAT DO NOT SHOW DIRECT HYDROGEN TRANSFER

In this section we shall briefly consider two examples of enzymatic reactions that do not show direct hydrogen transfer: glutathione reductase and dihydroorotate dehydrogenase.

11.A.1 Glutathione Reductase

We have previously encountered the thiol-containing tripeptide glutathione (γ-Glu–Cys–Gly), and we have examined both its acyl-transfer function in the γ-glutamyl cycle proposed for amino-acid transport in the kidney and its role as sulfur nucleophile in glyoxylase catalysis. As with other thiol-containing molecules, the polarizable sulfur in glutathione is susceptible to autoxidation (two-electron oxidation by O_2, probably by catalysis from trace amounts of metals) in air-saturated aqueous solutions, forming the intermolecular disulfide called oxidized glutathione (GSSG).

$$2\,\text{GSH} \longrightarrow \text{GSSG} + 2\,\text{H}^{\oplus} + 2\,\text{e}^{\ominus}$$

In the oxidized form, the glutathione dimer is inactive in those biological functions where its thiol nucleophile is required. The enzyme glutathione reductase recycles oxidized glutathione (GSSG) to reduced glutathione (GSH) at the expense of a molecule of reduced NADPH. The reduction potential for $\text{GSSG} \rightarrow 2\text{GSH} + 2\text{e}^{\ominus}$ is -0.23 V, compared to -0.32 V for the $\text{NADP}^{\oplus}$/NADPH potential, confirming that the equilibrium position will favor production of reduced glutathione.

$$\text{H}^{\oplus} + \text{NADPH} + \text{GSSG} \rightleftharpoons 2\,\text{GSH} + \text{NADP}^{\oplus}$$

When 4-[^{3}H]-NADPH of the appropriate chirality (such that it is the tritium that gets transferred—in this case, 4*S*; see Simon and Kraus, 1976) is mixed with enzyme and oxidized glutathione, tritium ends up in [^{3}H]-H_2O. However, the fact that transferred tritium ends up as solvent hydrogen does not rule out hydride transfer here. In fact, a direct-hydrogen-transfer process would be obscured, because the hydrogen transferred to the thiol of one of the reduced glutathione molecules that form as product is now attached to a sulfur; because of this attachment to a heteroatom, it exchanges rapidly with solvent protons ($\sim 10^2\ \text{sec}^{-1}$). Note that, with all nicotinamide enzymes dicussed thus far, hydrogen from C-4 of NADH (or NADPH) is transferred to a carbon atom of the substrate (and not to a heteroatom), from which an acid-base dissociation reaction is *exceedingly slow*. Here is a *minimal scheme* to explain tritium washout into solvent in the GSSG → GSH reaction. (The pK_a of the thiol of glutathione is ~8.)

4-[^{3}H]-NADH + GSSG ⇌ $\text{NAD}^{\oplus}$ + [^{3}H]-GSH $\overset{H_2O}{\rightleftharpoons}$ GSH + [^{3}H]-H_2O

+ $\text{GS}^{\ominus}$ $\overset{H_2O}{\rightleftharpoons}$ GSH

This proposal would be a consistent explanation for tritium washout *if* the hydride equivalent were transferred *directly* to one of the sulfur atoms of oxidized glutathione. However, the hydride equivalent could also be transferred first to some group on the enzyme to form some reduced intermediate. A second step could involve reoxidation of the enzyme intermediate and reduction of oxidized glutathione. For example,

$$\text{NADH} + \text{Enz—Y} \rightleftharpoons \text{NAD}^{\oplus} + \text{Enz—YH} \quad (1)$$

$$\text{Enz—YH} + \text{H}^{\oplus} + \text{GSSG} \rightleftharpoons \text{Enz—Y} + 2\,\text{GSH} \quad (2)$$

Each step is a two-electron transfer. If the Enz—Y group were oxygen, nitrogen, or sulfur, then exchange with solvent could occur at the Enz—YH stage as well. In fact, this two-step mechanism for reduction of oxidized glutathione to two molecules of thiol form is the actual mechanism for action of glutathione reductase (Williams, 1976). A priori, the putative group Y in the Enz—Y complex must be distinct from the normal amino-acid–side-chain functional groups whose roles we have discussed in previous chapters. If NADH transfers H: (a nucleophilic hydride) to Enz—Y, then the Y, group must be electrophilic, and none of the side chains of amino acids in proteins are obviously electrophilic (unless the group Y were an enzymatic disulfide, which could qualify for such chemistry).

Indeed, on purification, glutathione reductase from all sources examined is a yellow 1:1 complex of protein and a tightly bound *flavin coenzyme*; the coenzyme undergoes two-electron reduction in the first step of glutathione-reductase catalysis and reoxidation in the second step, as oxidized glutathione undergoes net two-electron reduction.

We shall spend the rest of this chapter examining the structure, function, and mechanism of flavin coenzymes in enzymatic redox reactions, but first we shall note a second example where a flavin coenzyme is an intermediate redox carrier between a nicotinamide and the specific substrate (in this case, orotate/dihydroorotate).

11.A.2 Dihydroorotate Dehydrogenase

Dihydroorotate is a cyclic biosynthetic precursor of the pyrimidine bases uracil, thymine, and cytosine—metabolic end products functioning as elements of DNA, RNA, and nucleotide coenzymes. In the biosynthetic pathway, dihydroorotate is dehydrogenated (oxidized by two electrons) to orotate.

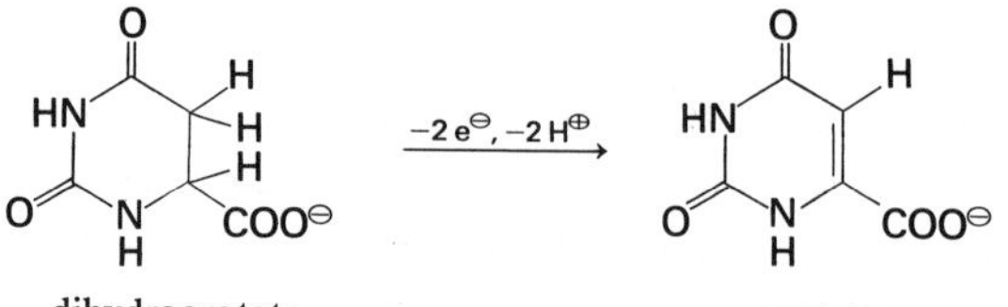

$$\xrightarrow{-2\,e^{\ominus},\,-2\,\text{H}^{\oplus}}$$

dihydroorotate — orotate

The N-1 of orotate subsequently acts as nucleophile toward the activated C-1 of 5-phosphoribosyl-1-pyrophosphate (PRPP) in an enzymatic phosphoribosylation to yield the nucleotide orotidine-5′-monophosphate (OMP). Subsequent enzymatic decarboxylation generates the familiar pyrimidine UMP.

PRPP + orotate (N-1) ⟶ PP_i + OMP ⟶ (CO_2) UMP

The biosynthetic dihydroorotate dehydrogenase (Taylor et al., 1974) may be a simple flavoprotein dehydrogenase of the type discussed in ¶11.C. However, the enzyme most carefully studied is one isolated from *Clostridium oroticum* grown on orotate as a sole carbon source (Singer et al., 1973). The function of this induced bacterial enzyme is probably degradative—orotate to dihydroorotate, which is then opened in a hydrolytic reaction to *N*-carbamoylaspartate, which can be used as a normal metabolite. Thus this degradative dihydroorotate dehydrogenase actually functions as an orotate reductase and uses NADH as reductant.

$H^{\oplus}$ + NADH + orotate ⟶ dihydroorotate + $NAD^{\oplus}$

dihydroorotate ⟶ (H_2O, dihydroorotase) *N*-carbamoylaspartate

$NH_4^{\oplus}$ + CO_2 + Aspartate ⟵ $H_2N{-}C(=O){-}NH{-}CH(COO^{\ominus}){-}H_2C{-}COO^{\ominus}$

N-carbamoylaspartate

Starting either with dihydroorotate tritiated at an oxidizable carbon or (in the other direction) with tritiated 4*R*-NADH (Simon and Kraus, 1976), all of the tritium ends up in H_2O; no direct hydrogen transfer is demonstrable. One suspects (as discussed earlier with glutathione reductase) that the direct transfer is obscured by washout to solvent from some rapidly exchangeable species. Here, this washout cannot be due to exchange from NADH or from the dihydroorotate product, because the transferred H is bonded to carbon (either C-5 or C-6, depending on mechanism).

Again, the orotate reductase (degradative dihydroorotate dehydrogenase) contains stoichiometric amounts of a flavin coenzyme as the reversibly reducible intermediate between nicotinamide and the orotate substrate. In this case, it *must* be the reduced flavin coenzyme that exchanges the transferable hydrogen with solvent protons.

11.B FLAVIN COENZYMES

In order to understand how the tightly bound cofactor functions in glutathione reductase and dihydroorotate dehydrogenase, we must delve briefly into the structure and chemistry of flavin coenzymes. The generic term "flavin" derives from the Latin word *flavius* ("yellow"), which was bestowed on these molecules because of the brilliant yellow color they exhibit as solids and in neutral aqueous solutions.

The coenzyme forms are enzymatically modified versions of vitamin B_2 (riboflavin). The structural assignment for riboflavin was completed in 1935.

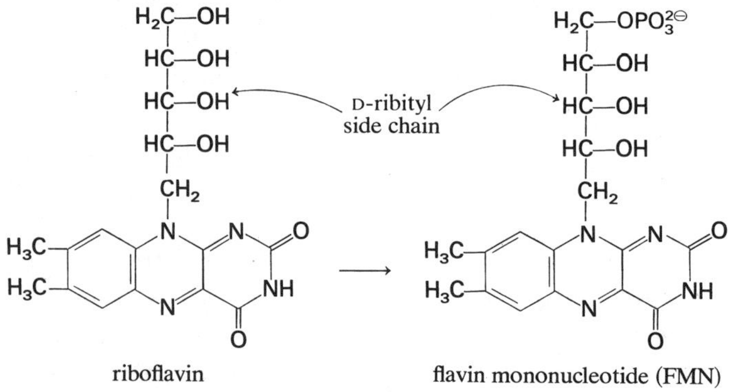

D-ribityl side chain

riboflavin → flavin mononucleotide (FMN)

The vitamin is converted enzymatically to two active coenzyme forms. Phosphorylation at the ribityl 5′-OH yields FMN (flavin mononucleotide), and adenylylation of FMN yields FAD (flavin adenine dinucleotide).

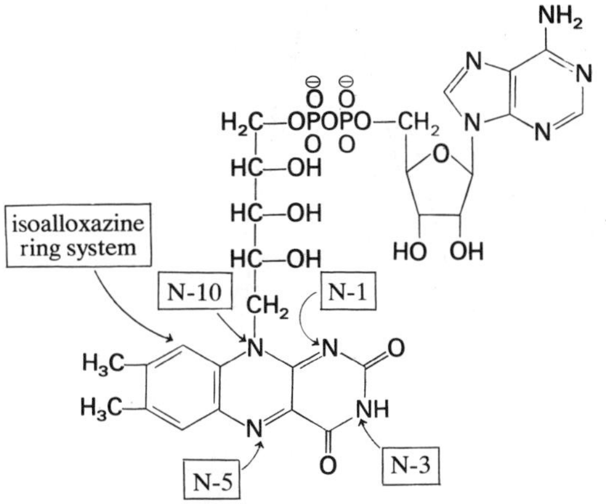

flavin adenine dinucleotide (FAD)

FMN and FAD appear to be functionally equivalent as redox coenzymes. FAD exists predominantly in a stacked conformation in solution (adenine ring over isoalloxazine ring), which results in quenching of the intrinsic fluorescence of the isoalloxazine chromophore. The conformation of FAD bound to apoenzymes is unknown except in glutathione reductase (as noted below).

The electron sink is the same in both FMN and FAD. It is the highly conjugated, tricyclic isoalloxazine ring system. This extensive conjugation results in the observed *yellow* chromophore.

Oxidized coenzyme can undergo ready, reversible two-electron reduction chemically or enzymatically to give specifically the 1,5-dihydroflavin isomer, with interrupted conjugation; this is a leuco form, and reduction can be monitored by loss of absorbance at 450 nm (Fig. 11-1).

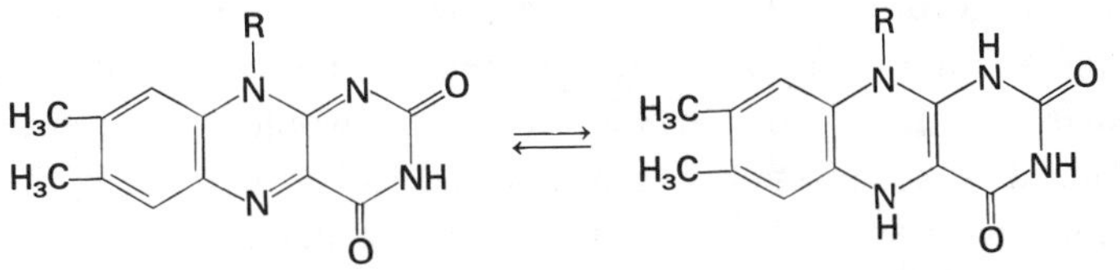

FAD or FMN

$FADH_2$ or $FMNH_2$
(*N*-1,5-dihydroflavin coenzyme)

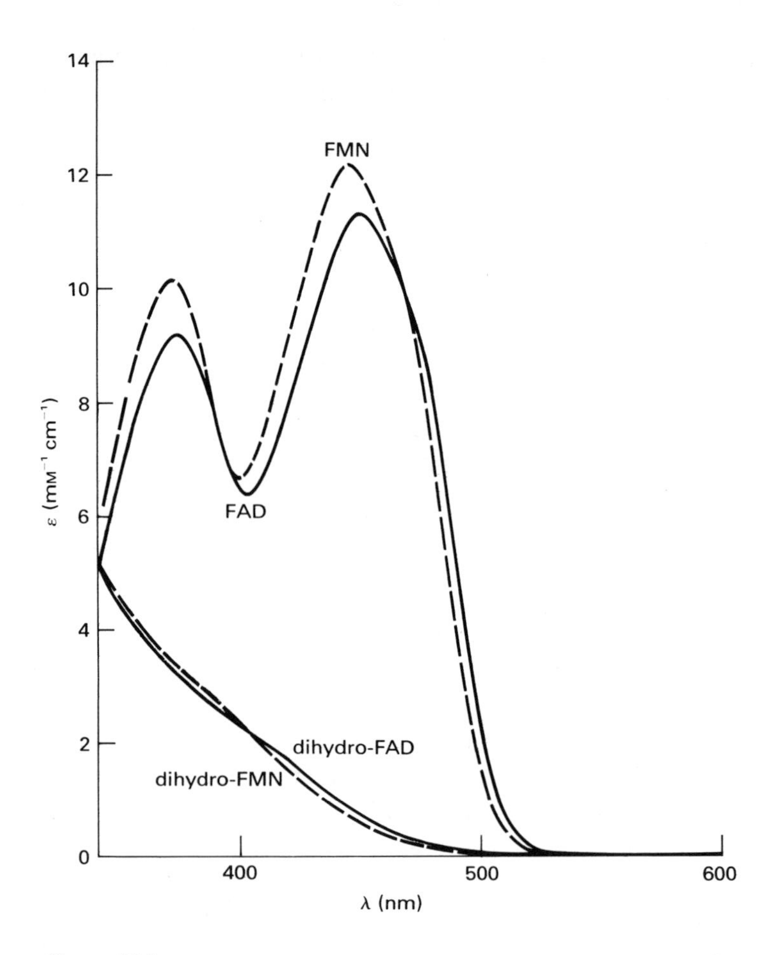

Figure 11-1
Visible spectra of FMN, and of FAD, and of the 2-e$^{\ominus}$–reduced forms.

The oxidized form of flavins is planar, but in the reduced (dihydro) form the central ring is a dihydropyrazine ring system, containing eight electrons. If the dihydroflavin system were planar, the central ring would thus be antiaromatic and be destabilized. A lower-energy form appears to be available by bending into a butterfly form with an angle of 10° to 30° out of plane (Fig. 11-2). As we shall note later in this chapter, flavin cofactors can also undergo facile one-electron transfers to produce radical species such as the flavin semiquinone.

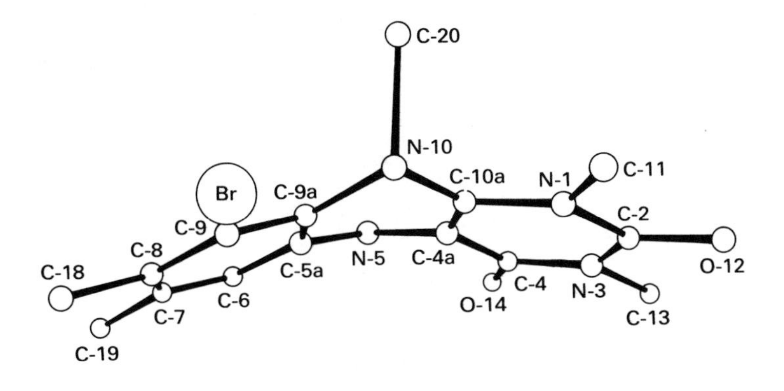

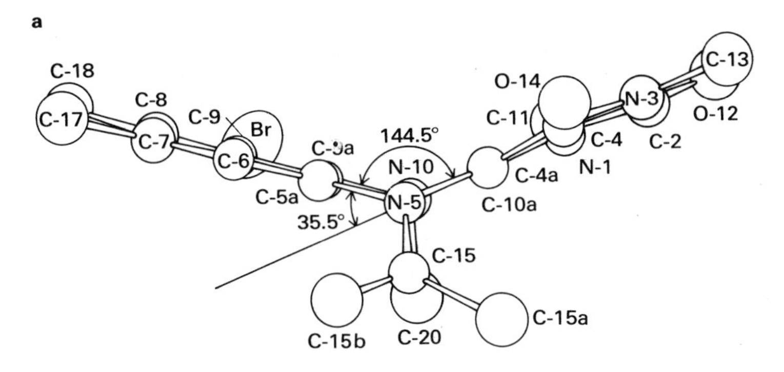

Figure 11-2
(**a**) The structure of 9-bromo-1,3,7,8,10-pentamethyl-1,5-dihydroisoalloxazine. (**b**) The structure of 5-acetyl-9-bromo-1,3,7,8,10-pentamethyl-1,5-dihydroisoalloxazine. (From P. Kierkegaard, in *Flavins and Flavoproteins*, ed. H. Kamin, vol. 3, p. 13. Copyright © 1971 by University Park Press, Baltimore)

11.B.1 Flavin Coenzymes and Dihydroorotate Dehydrogenase

The purified dihydroorotate dehydrogenase (¶11.A.2) is a tetramer of 115,000 mol wt, with a total of two active sites. Each dimeric unit possesses one FMN, one FAD, and two redox-active irons that shuttle between $Fe^{3\oplus}$ and $Fe^{2\oplus}$ during catalysis. Thus, the dehydrogenase is actually a metalloflavoprotein using the nicotinamide as a substrate (in stoichiometric quantities). We shall defer comment on the functions of the iron atoms until a later chapter devoted to metalloflavoprotein catalysis. For now, we isolate the flavin coenzyme as the major redox element available during electron transfer to the holoenzyme.

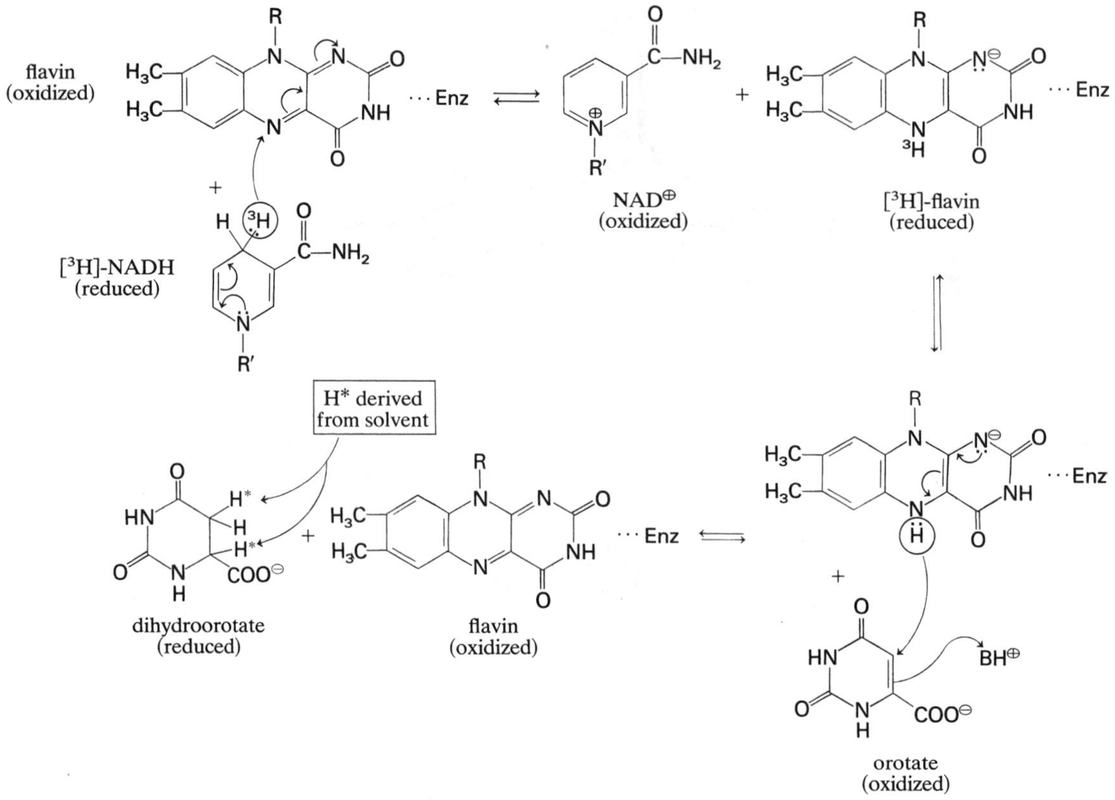

The suggested path of electrons is indicated for the reaction catalyzed by dihydroorotate dehydrogenase, and oxidized and reduced species are identified. The scheme suggests that transfer of the hydrogen and the two electrons between nicotinamide, flavin, and orotate molecules may be a hydride transfer. The pK_a at N-1 in dihydroflavins is about 6 to 7, and the N-1 anion is the redox-active form.

This scheme then places the transferred hydrogen from C-4 of NADH at N-5 of the Enz · $FADH_2$ complex formed on two-electron oxidation of the NADH to $NAD^{\oplus}$. With $4R$-[^{3}H]-NADH, this reaction produces N^5-[^{3}H]-$FADH_2$ · Enz. Because protonic-exchange processes from basic nitrogens (N-5 is a secondary amine in the $FADH_2$ state) with water protons are very fast (10^3 to 10^7 sec^{-1}), it is likely that the tritium will equilibrate with solvent hydrogens at this stage. Because H_2O is 55 M and the Enz · $FADH_2$ may be 10^{-8} to 10^{-10} M in these experiments, essentially no tritium will remain at exchange equilibrium on any Enz · $FADH_2$ complexes. Solvent exchange may or may not be mediated by some active-site basic group.

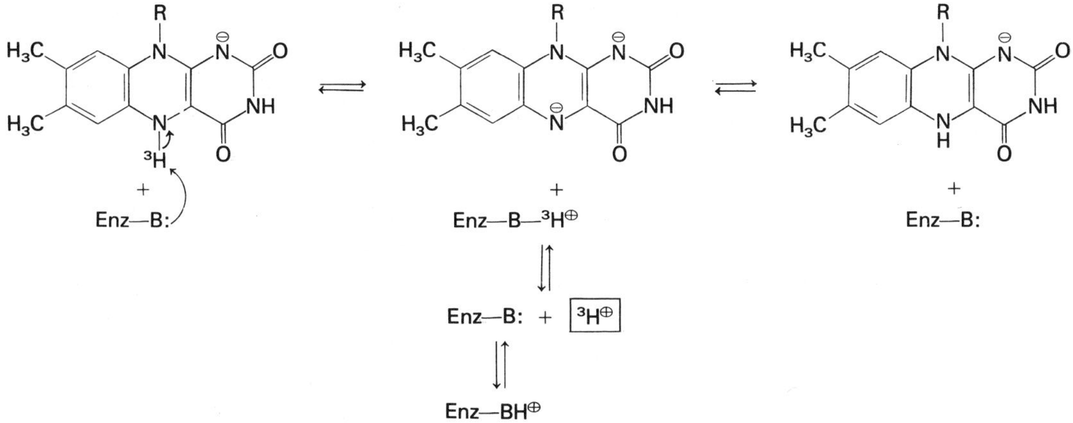

Enz—B: ⇌ Enz—B—$^3H^{\oplus}$ ⇌ Enz—B:

Enz—B: + $^3H^{\oplus}$

Enz—$BH^{\oplus}$

The second half-reaction then proceeds with two-electron reoxidation of Enz · $FADH_2$ back to Enz · FAD ready for another catalytic cycle, and with concomitant two-electron reduction of orotate to dihydroorotate, resulting in net incorporation of solvent hydrogen at both the methine and the methylene carbons of the dihydroorotate product. (In the scheme shown at the beginning of this subsection, a hydride transfer is hypothesized from N-5 of the dihydroflavin to C-5 of bound orotate, but this is by no means proved. The hydrogen species could be transferred to C-6 of orotate instead. The fact that solvent exchange occurs at the dihydroflavin level means that both C-5 and C-6 hydrogens arise from solvent and obscure the distinction.) The catalytic turnover number is 650 moles min^{-1} (mole enzyme-bound flavin)$^{-1}$ at V_{max} conditions (Singer et al., 1973). This turnover is certainly slow enough for tritium washout to occur at the E · $FADH_2$ stage.

Unlike most enzymes using nicotinamides as redox coenzymes, flavoenzymes generally exist as functional holoenzymes (and dihydroorotate dehydrogenase is no exception). The flavin coenzymes are not in dynamic equilibrium with free flavin coenzymes in solution (whose concentration in cells is quite low). The K_D for flavin dissociation ranges from 10^{-7} M to 10^{-11} M. Thus, *the flavin coenzyme must be catalytically reoxidized at the end of each cycle*, whereas the nicotinamide coenzyme can be released in an altered oxidation state to be replaced by another molecule in the correct initial oxidation state (in the simple dehydrogenases, but not in the complex ones noted at the end of Chapter 10).

One advantage of tight binding between flavin coenzymes and apoproteins may be the possibility for *enzyme control over the nature of the physiological oxidant of the reduced flavin.* If $FADH_2$ were released into solution, it would reoxidize extremely rapidly from reaction with dissolved O_2 ($T_{1/2} < 1$ sec) or would disproportionate with a molecule of oxidized flavin to form radical species (bimolecular rate constant $\sim 10^8\ \text{M}^{-1}\ \text{sec}^{-1}$).

Another advantage of tight binding is the possibility for *shifting the redox potential* if the oxidized, the dihydro (two-electron reduced), and the semiquinone (one-electron reduced) forms of flavin coenzymes are bound differentially to apoproteins. From Table 10-1, note that the redox potential for $FAD/FADH_2$ in aqueous solutions (at pH 7 and 25° C) is −208 mV. If an enzyme selectively binds the dihydro form more tightly, the redox potential for coenzyme reduction in that flavoenzyme will be more negative than −208 mV. If the flavin semiquinone is selectively stabilized at the active site of an apoprotein, the two one-electron reduction potentials can become widely separated; then the enzyme may operate physiologically as a one-electron transfer agent either between oxidized FMN and FMNH · (semiquinone) or between FMNH · and $FMNH_2$. The redox potential of a given flavoenzyme helps determine which molecules the flavoenzyme can use as electron donors or acceptors in redox catalysis; therefore, variable redox potentials allow different functional niches for flavoproteins. The observed range of flavoenzyme redox potentials is from −465 mV to +149 mV. The E_0' of FAD bound to D-amino-acid oxidase (11.D.1) is ~0.0 V, a 200 mV change from free FAD, arising from $\sim 10^7$-fold tighter binding of dihydroflavin ($FADH_2$) ($K_D \approx 10^{-14}$ M) compared to oxidized FAD ($K_D \approx 10^{-7}$ M).

11.B.2 Flavin Coenzymes and Glutathione Reductase

The catalytic mechanism for glutathione reductase (¶11.A.1) is probably similar to that for dihydroorotate dehydrogenase, with a hydride transfer from NADH to C-5 of the bound FAD coenzyme in the first half-reaction. However, reoxidation of the Enz · $FADH_2$ and reduction of the glutathione-disulfide linkage may proceed through a covalent intermediate in this case.

Model studies for the oxidation of dithiols to disulfides by oxidized flavins suggest that thiolate anions can add to C-4a of oxidized flavins, producing an adduct that is now set up for disulfide formation with an obvious path for transfer of electrons into the flavin (Yokoe and Bruice, 1975; Loechler and Hollocher, 1975).

adduct

In the reverse direction, attack presumably would be initiated from the eneamine moiety of the dihydroflavin on the electrophilic disulfide.

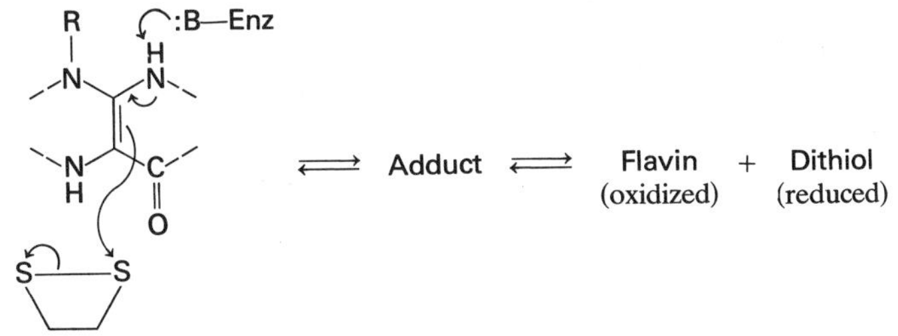

$\rightleftharpoons$ Adduct $\rightleftharpoons$ Flavin (oxidized) + Dithiol (reduced)

One might speculate that this is the pathway for reaction of E · $FADH_2$ of glutathione reductase with the oxidized glutathione. However, the active site of glutathione reductase is somewhat more complex in terms of redox capacity, adding variation to the possible mechanisms (C. H. Williams, 1976). In addition to FAD, there is a disulfide grouping (potentially reducible) at the active site of the enzyme. Initial hydride transfer from NADH generates a dihydroflavin at first, but apparently the enzyme cystinyl disulfide can undergo subsequent electron transfer to set up an equilibrium between forms 1 and 2 (and perhaps 3). Form 2 has oxidized FAD and the dithiol form of the active-site cysteinyl residues.

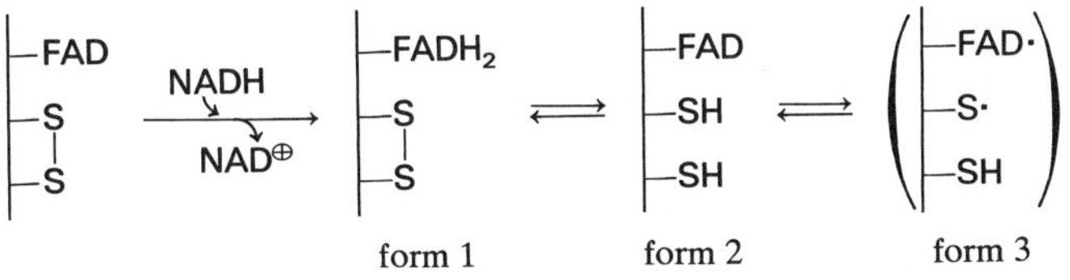

form 1 form 2 form 3

Figure 11-3a shows the UV–visible spectrum of 2-$e^{\ominus}$–reduced glutathione reductase from yeast. Clearly, the spectrum is not that of a simple $FADH_2$ chromophore. If form 2 exists with a thiolate anion in close proximity to oxidized FAD, charge-transfer absorption bands could arise and account for the spectrum (Massey and Ghisla, 1974). Presumably the dihydroenzyme structures (forms 1 and 2) would interconvert by the mechanism of addition at C-4a, as outlined above for the model compounds. Obviously, this interconversion of forms 1 and 2 is a reversible two-electron transfer between reduced flavin and disulfide; it is the same chemical reaction as reaction of $FADH_2$ with GSSG. If form 2 is the predominant structure for the dihydro form of glutathione reductase, then the passage of electrons into oxidized glutathione might go by disulfide interchange.

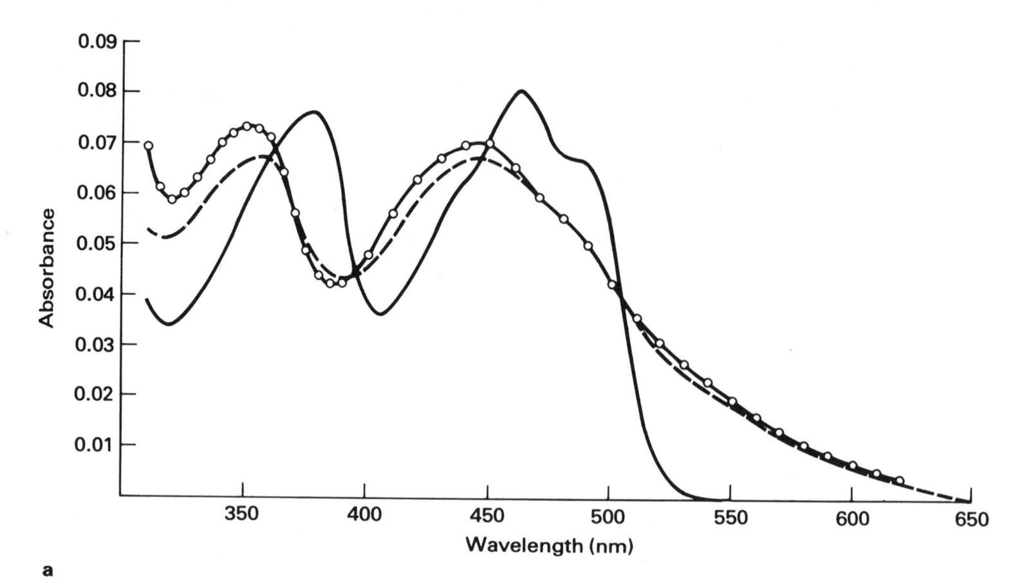

a

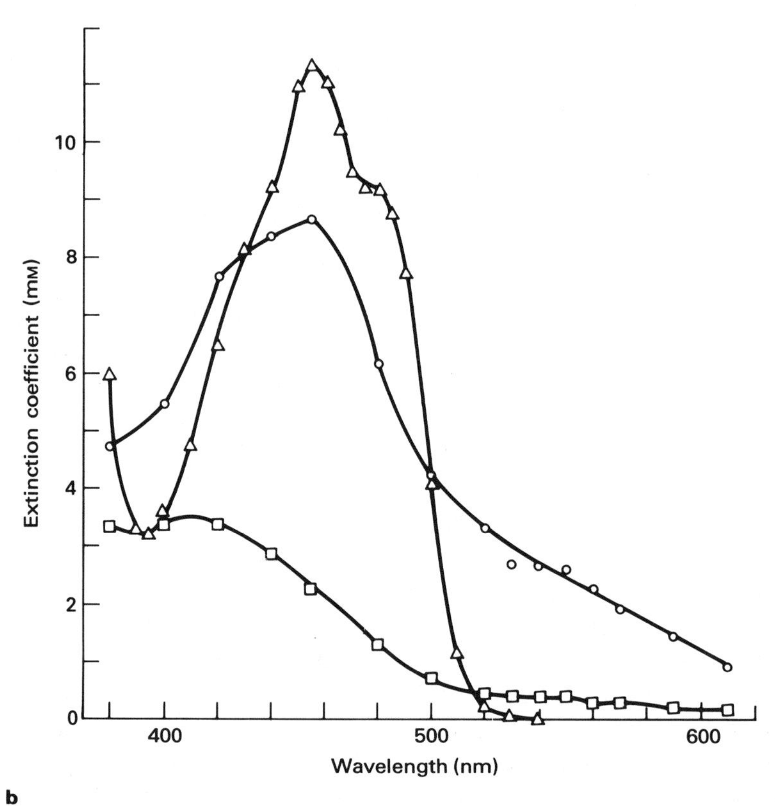

b

Figure 11-3
(**a**) Yeast glutathione reductase. Spectra of the oxidized and 2-e$^{\ominus}$–reduced forms of the enzyme recorded anaerobically. The solid line represents oxidized enzyme, the dashed line represents 526 moles GSH, and the line marked by circles represents 1 mole TPNH in the presence of DPNase. (**b**) *Escherichia coli* lipoamide dehydrogenase. The line marked by triangles represents the spectrum of the oxidized enzyme, the line marked by circles represents that of the 2-e$^{\ominus}$–reduced enzyme as generated from rapid-reaction spectrophotometry, and the line marked by squares represents the spectrum of the 4-e$^{\ominus}$–reduced enzyme as produced by anaerobic reduction by 12 moles dihydrolipoamide per mole FAD. (From C. H. Williams, 1976, pp. 95, 115)

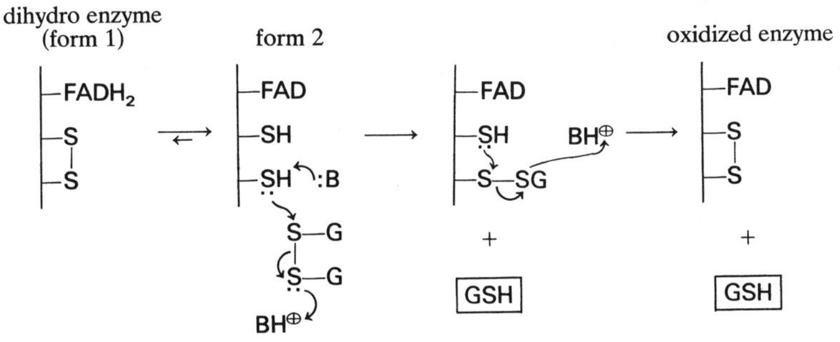

dihydro enzyme (form 1) — form 2 — oxidized enzyme

-FADH2, -S-S ⇌ -FAD, -SH, -SH :B + S—G, S—G, BH⊕ ⟶ -FAD, -SH, -S—SG, BH⊕ + GSH ⟶ -FAD, -S-S + GSH

Form 3 is an alternative electronic formulation for the dihydro state of glutathione reductase, featuring a 1-e$^{\ominus}$–reduced flavin (the semiquinone form) and a thiolate radical. This structure takes explicit account of the fact that flavin semiquinones are *stable radicals,* because the unpaired electron is highly delocalized through the conjugated isoalloxazine structure. It is not clear whether form 3 is the principal contributing structural form to the dihydro enzyme, but we shall note in subsequent pages that the semiquinone's stability is the key to flavin action in many flavoenzymes. The semiquinone is blue at neutrality and red above pH 8.4.

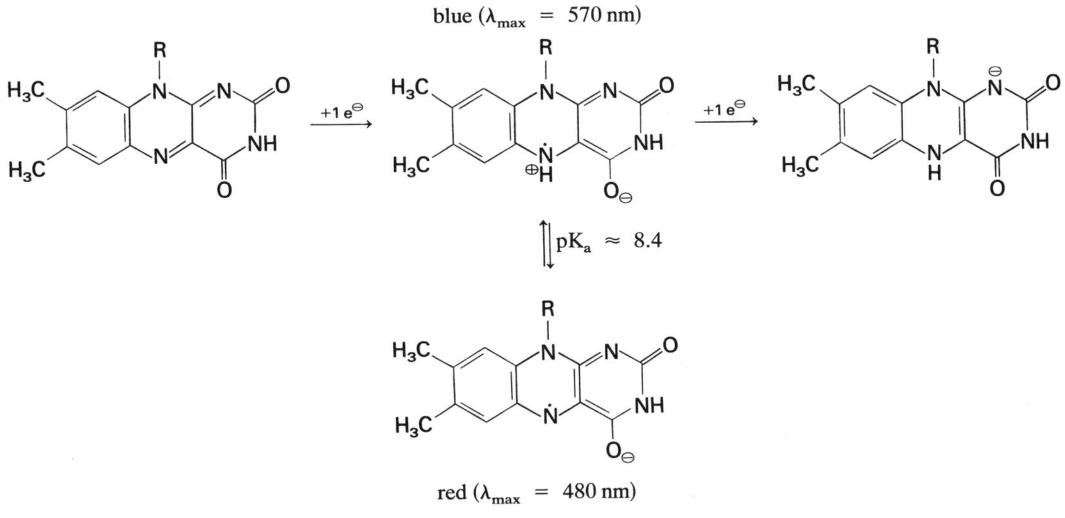

blue (λ_{max} = 570 nm)

$\xrightarrow{+1\,e^{\ominus}}$ $\xrightarrow{+1\,e^{\ominus}}$

$pK_a \approx 8.4$

red (λ_{max} = 480 nm)

11.B.3 Flavoenzymes Related to Glutathione Reductase

There are two other flavoenzymes related mechanistically to glutathione reductase, with the physiological function of transferring electrons from NADH to a disulfide-containing substrate (C. H. Williams, 1976). These are dihydrolipoamide dehydrogenase and thioredoxin reductase.

We shall discuss *dihydrolipoamide dehydrogenase* in Chapter 21, when we take up decarboxylations of α-keto acids mediated by thiamine pyrophosphate. For now, we simply note that dihydrolipoamide is the dithiol form of 1,5-dithiooctanoamide. Oxidation is, in this instance, intramolecular to a five-membered dithiane ring (the cyclic disulfide found in oxidized lipoamide).

NAD⊕ + HS HS ... CNHR (dihydrolipoamide) $\xrightleftharpoons{E \cdot FAD}$ S–S ... CNHR (lipoamide) + NADH + H⊕

dihydrolipoamide lipoamide

The active site of this enzyme appears entirely analogous to that of glutathione reductase (Fig. 11-3b; C. H. Williams, 1976). Thorpe and Williams have recently accrued spectroscopic evidence for a 4a-thiol-FAD adduct in dihydrolipoamide dehydrogenase action by selective alkylation of one of the two active-site cysteinyl thiol groups with iodoacetamide. Now the 4a-thiol adduct once formed cannot undergo closure to the disulfide because the second cysteine has its potentially nucleophilic sulfur atom carboxymethylated.

Thioredoxin is small bacterial protein (~12,000 mol wt) that, in its oxidized form, has a single cystine disulfide (C. H. Williams, 1976). *Thioredoxin reductase* uses NADPH to generate the dithiol form of the enzyme. In the dithio oxidation state, thioredoxin is a specific reductant in the enzymatic conversion of ribonucleotides to the 2′-deoxyribonucleotides, which are precursors of DNA (C. H. Williams, 1976, p. 143). (The reduced thioredoxin is also a specific protein cofactor for catalytic action of the DNA polymerase elaborated by the bacterial virus T7; see Mark and Richardson, 1976.)

11.B.3.a Absolute stereochemistry of glutathione reductase catalysis

While on the subject of NADH-dependent disulfide-reducing flavoprotein dehydrogenases, we should note the recent X-ray structural map obtained by Schirmer et al. (1978) for glutathione reductase from bovine erythrocytes. It is signal for two reasons. First, it allows determination of the absolute stereochemistry of electron transfer to and from a flavin at the active site of an enzyme and, second, it proves that both faces of the flavin are used in glutathione reductase action.

We noted in Chapter 10 that the methylene hydrogens at C-4 of NADH (NADPH) are prochiral, and that nicotinamide-dependent dehydrogenases can be characterized by stereochemical selectivity for transfer of the 4*R* or 4*S* hydrogen. We have alluded to the fact that, when a hydride equivalent is transferred to oxidized flavin, that hydrogen rapidly equilibrates with solvent, presumably from transfer between N-5 and bulk water. This kinetic exchange prevents stereochemical categorization of flavoproteins, despite the fact that the two faces of flavin cofactors are in fact not stereochemically equivalent and are in principle differentiable as *si* and *re* faces.

One attempt at stereochemical elucidation has been the use of 5-deazaflavins, with carbon replacing nitrogen at position 5 (Hersh and Jorns, 1975; Jorns and Hersh, 1974; Fisher and Walsh, 1974; Spencer et al., 1976; J. Fisher et al., 1976). Now on enzymatic reduction (observed in seven different enzymes), C-5 is prochiral if hydrogen is transferred (chiral if deuterium or tritium is transferred) from substrate undergoing oxidation, and the hydrogen incorporated *will be kinetically stable because it is bound to carbon.* In such experiments, direct hydrogen transfer (e.g., from NAD^{3}H; see Fisher and Walsh, 1974) is observed at C-5 (incidentally proving that N-5 and not N-1 is the acceptor locus in the natural flavin), and C-5 is chiral upon reoxidation (Jorns and Hersh, 1974). However, a rapid disproportionation between oxidized and reduced 5-deazaflavins, once they are liberated into solution, scrambles chiral integrity at C-5 and so frustrates chirality assignment at C-5 of the deazaflavin (Spencer et al., 1976).

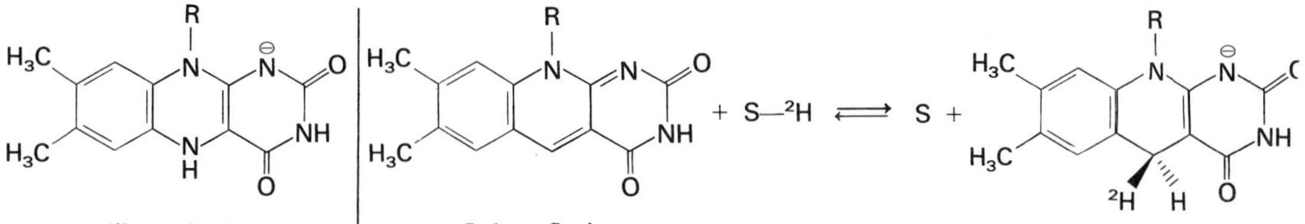

dihydroflavin

5-deazaflavin (oxidized) + S—^{2}H ⇌ S + chiral 5-[^{2}H]-dihydro-5-deazaflavin (one enantiomer

Thus, only an X-ray map of a flavoenzyme with bound substrate is likely to give stereochemical information (relative and absolute chirality) on flavin involvement. The Schirmer–Schultz X-ray map (Fig. 11-4) suggests that NADH approaches the *si* face of the flavin as hydride transfer is initiated (the 4*S*-hydrogen is transferred; see Simon and Kraus, 1976). Interestingly, the reducible active-site Cys–Cys disulfide is on the opposite (*re*) face of the bound FAD, pointing out that the flavin cofactor is sandwiched geometrically between the substrate undergoing oxidation and that undergoing reduction. Knowledge of the disulfide/dithiol disposition relative to bound flavin allows one to intuit that the 4a-flavin adduct in the dithiol-oxidation direction occurs with attack of thiolate anion on the *re* face.

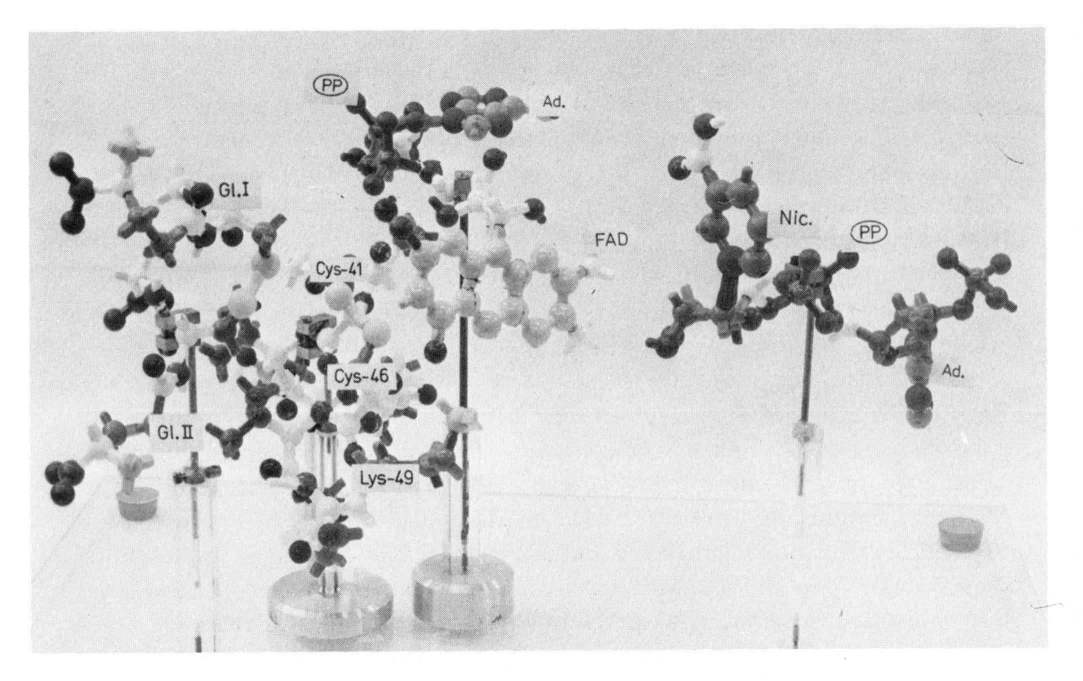

Figure 11-4
Active-site model for glutathione reductase from X-ray diffraction studies. (From Schultz et al., 1978)

All three NADH-dependent disulfide-reducing flavoprotein dehydrogenases (glutathione reductase, dihydrolipoamide dehydrogenase, and thioredoxin reductase) use the 4*S*-hydrogen of NADH, suggesting common geometry in the nicotinamide-binding domain (and common evolution?). On the other hand, *C. oroticum* dihydroorotate dehydrogenase is 4*R*-specific vis a vis NADH. Also, the several NADH-dependent (NADPH-dependent) flavoprotein monooxygenases (discussed in Chapter 12) are reported to be 4*R*-specific, suggesting a distinct active-site evolution there.

11.B.3.b Other NADH-oxidizing (NADPH-oxidizing) flavoenzymes

In addition to the disulfide-reducing flavoenzymes that use NADH or NADPH as source of the two electrons as reducing equivalent, there are two other physiologically important categories of flavoenzymes where the two-electron reduction of bound flavin cofactor by dihydronicotinamide (acting as oxidizable cosubstrate) is worth mention here. One type is represented by flavoprotein monooxygenases (Chapter 13), both the aromatic hydroxylases and those (such as luciferase) that work on aldehyde substrates. The other category includes liver cytochrome-P_{450} reductase, bacterial NADPH-putidaredoxin reductase, adrenal NADPH-adrenodoxin reductase (all discussed in Chapter 15), and xanthine oxidase (Chapter 13); dihydroorotate dehydrogenase also fits in this second group. The reoxidant is either a nonheme iron–sulfur cluster on the same or on a different protein, or a heme group on the same or on a second enzyme. We shall discuss their physiological and mechanistic properties in the subsequent chapters cited.

11.B.4 Classification of Flavoenzymes

Thus far we have noted that nicotinamide and flavin redox interactions may involve transfer of hydride equivalents to and from N-5, whereas redox interactions between flavins and thiols/disulfides could involve covalent addition compounds to the isoalloxazine nucleus at C-4a of the coenzyme. In fact, the nature of the electron acceptors and donors provides a possible categorization for flavoenzymes. If we concentrate on the nature of the acceptor molecule reoxidizing the enzyme-bound dihydroflavin, we can distinguish between enzymes that do use molecular oxygen (O_2) as physiological acceptor and those that do not.

If O_2 is the acceptor, it can suffer four fates, as indicated in categories 2 through 6 of Table 11-1. The reduction potential for riboflavin free in solution is

Table 11-1
Classification of flavoenzymes by electron acceptors

Category	Comments
1 Dehydrogenases	Do not use O_2 as acceptor. Acceptors often are 1-$e^{\ominus}$ acceptors such as quinones, cytochromes, or nonheme-iron–sulfur clusters.
2. Oxidases	The acceptor is O_2, which is reduced by 2 $e^{\ominus}$ to H_2O_2.
3 Oxidase-decarboxylases	The acceptor is O_2, which is reduced by 4 $e^{\ominus}$ to H_2O.
4 Monooxygenases	The acceptor is O_2; one oxygen atom ends up in H_2O and the other in the hydroxylated product.
5 Dioxygenases	The acceptor is O_2; both oxygen atoms end up in the product.
6 Metalloflavoenzymes	The acceptor may be O_2. Bound transition-metal ions (Fe^{II} or Fe^{III}, and Mo^{VI}) are required for catalysis. May be dehydrogenases or oxidases.
7 Flavodoxins	These are 1-$e^{\ominus}$–transfer proteins; the semiquinone form clearly is involved.

-200 mV; the reduction potential for O_2 to H_2O_2 is $+300$ mV; and the reduction potential for O_2 to H_2O is $+810$ mV. Thus, the thermodynamics strongly favor oxygen reduction. The quinone reduction potential is -100 mV, whereas reduction potentials for cytochromes are in the range of $+200$ to $+300$ mV, so dihydroflavin reoxidation by these compounds is also exergonic. (The one-electron reduction potential for O_2 to $O_2^{\ominus\cdot}$, superoxide ion, is -300 mV, and electron transfer will be endergonic by that route.)

Category 7 represents a small number of low-molecular-weight proteins (flavodoxins) of bacterial origin that contain FMN as redox cofactor (Mayhew and Ludwig, 1975). The FMN bound to these proteins has perturbed reduction potentials as noted earlier (in the range of -400 to -500 mV) that place these redox proteins among the strongest biological reductants. The flavodoxins probably function in photosynthetic electron-transport chains. We shall note only two additional features of them here. They are one of the two kinds of flavoproteins (glutathione reductase is the other) whose structures have yet been determined at atomic resolution, and it would appear that the accessible site for potential redox reaction is the dimethyl-benzenoid end of the isoalloxazine ring (Fig. 11-5). The

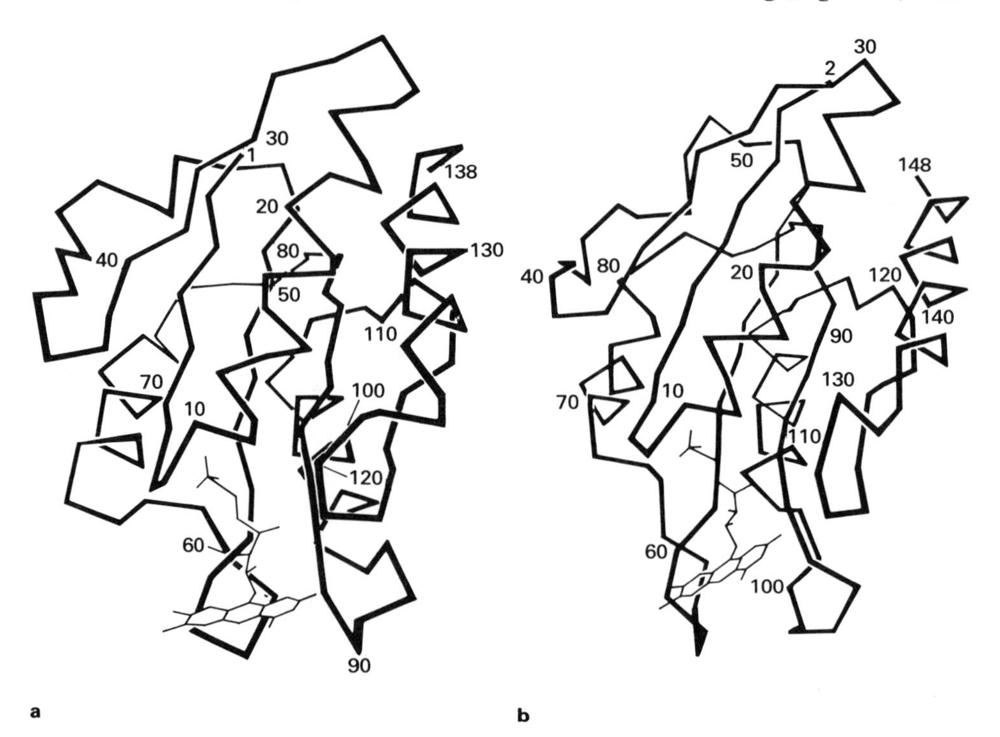

Figure 11-5
Drawings of the α-carbon skeleton and FMN atoms of (**a**) oxidized *Clostridium* MP flavodoxin and (**b**) oxidized *D. vulgaris* flavodoxin. (From Mayhew and Ludwig, 1975, p. 72)

second feature is that the apoprotein must selectively stabilize the 2-e$^{\ominus}$–reduced and the semiquinone forms of the coenzyme, because the functional redox cycle involves shuttling between $FMNH_2$ (dihydroflavin) and FMNH · (semiquinone).

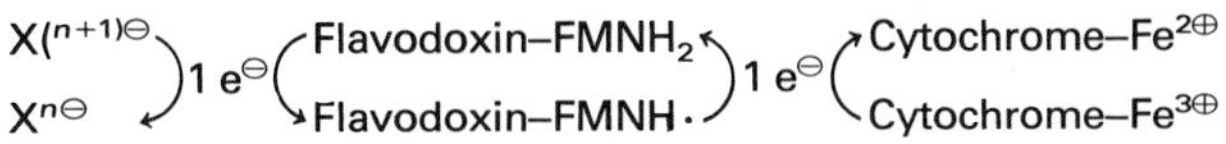

$$X^{(n+1)\ominus} \rightleftarrows X^{n\ominus} \quad 1\,e^{\ominus} \quad \text{Flavodoxin–FMNH}_2 \rightleftarrows \text{Flavodoxin–FMNH}\cdot \quad 1\,e^{\ominus} \quad \text{Cytochrome–Fe}^{2\oplus} \rightleftarrows \text{Cytochrome–Fe}^{3\oplus}$$

An alternative way to categorize diverse flavoenzymes is by the structural types of organic functional groups oxidized. Table 11-2 lists seven such classes, with specific examples of flavoenzymes in each category. If one makes a comparison

Table 11-2
Classification of flavoenzymes by structure of substrates oxidized

Category	Reduced substrate form	Oxidized substrate form	Examples
1	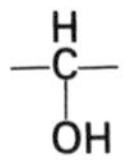 H–C(–)–OH	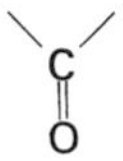 >C=O	D-Lactate dehydrogenase from *E. coli* membrane Glucose oxidase, thiamine dehydrogenase (alcohol → acid)
2	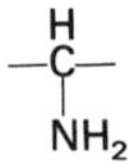 H–C(–)–NH_2	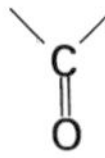 >C=O + $NH_4^{\oplus}$	D- and L-amino-acid oxidases Amine oxidases
3	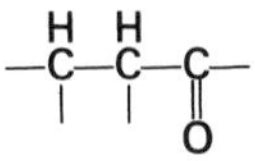 –CH–CH–C(=O)–	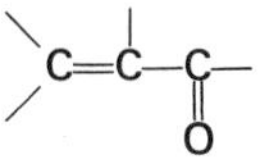 >C=C–C(=O)–	Succinate dehydrogenase Acyl-CoA dehydrogenases Dihydroorotate dehydrogenases
4	NADH	$NAD^{\oplus}$	NADH dehydrogenases Transhydrogenases Dihydroorotate dehydrogenases NADH-dependent monooxygenases
5	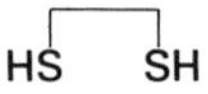 HS SH (dithiol)	 S—S (disulfide)	Lipoamide dehydrogenase Glutathione reductase
6	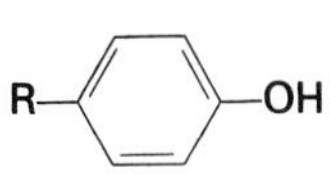 R–C$_6$H$_4$–OH	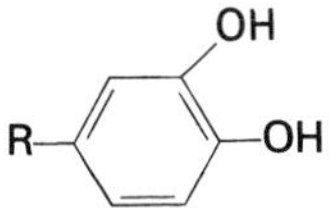 R–C$_6$H$_3$(OH)–OH	*p*-Hydroxybenzoate hydroxylase Phenol hydroxylase
7a	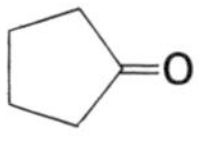 cyclopentanone (=O)	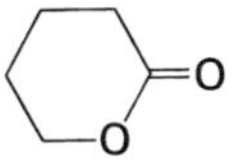 six-membered lactone (–O–, =O)	Cyclopentanone monooxygenase
7b	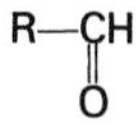 R–CH(=O)	R–$COO^{\ominus}$	Luciferase

with the functional groups directly susceptible to nicotinamide oxidation (Table 10-2), one notes close analogy in the types of oxidative transformations that can be accomplished by enzymes using either a nicotinamide or a flavin coenzyme. The major distinction between nicotinamides and flavins is that free dihydroflavins are not stable in air; they are reoxidized within a fraction of a second.

We shall use the first categorization scheme (Table 11-1) in our attempt to analyze certain general features about the initially bewildering array of oxidative transformations carried out by flavoenzymes. In the rest of this chapter, we shall examine dehydrogenases and oxidases. Chapters 12 and 13 take up hydroxylases (monooxygenases) and metalloflavoenzymes, respectively.

11.C FLAVOPROTEIN DEHYDROGENASES

The distinction between flavoprotein dehydrogenases and flavoprotein oxidases is that the oxidases react rapidly with molecular oxygen, whereas the dehydrogenases do not. Instead, the physiological acceptors of many dehydrogenases are obligate $1\text{-e}^{\ominus}$–transfer agents; flavin reoxidation ($FlH_2 \rightarrow Fl_{(ox)}$) must necessarily use two $1\text{-e}^{\ominus}$ steps.

We shall defer comment on the mechanism of reoxidation of $E \cdot FlH_2$ by O_2 until our discussion of oxidases in ¶11.D, but we point out here that the pseudo-first-order nonenzymatic rate of O_2-dependent reoxidation of dihydroflavins is about $2\,\text{sec}^{-1}$ in air. The unreactivity of flavoprotein dehydrogenases with O_2 clearly represents a suppression of the normal chemical autoxidation on the part of the apoenzyme, conferring a new pattern of reactivity and allowing electron transfer to other physiological acceptors. Either O_2 could be excluded from binding to the active site, or favorable geometry for oxidation could be avoided.

The ability of reduced flavoprotein dehydrogenases to react easily with $1\text{-e}^{\ominus}$ acceptors reflects the fact that the $1\text{-e}^{\ominus}$–reduced ($\equiv 1\text{-e}^{\ominus}$–oxidized) flavin is a stable free radical. The flavin free radical can exist near physiological pH either as a red anionic semiquinone or as a protonated blue zwitterionic form with a pK_a of 8.4. The active-site environment of the flavoprotein dehydrogenases thus far examined apparently stabilizes the blue form of the semiquinone.

In addition to characteristic visible spectra, flavin semiquinones are also detectable by the electron paramagnetic resonance signal due to the unpaired electron. Because of the thermodynamic stability of flavin semiquinone, it is a significant intermediate in enzymatic catalysis. This reveals a critically important role of flavin coenzymes in biological oxidations: they are *mediators* (transformers) *between* $2\text{-}e^{\ominus}$ *oxidations* (essentially all normal organic-substrate enzymatic oxidations) *and the* $1\text{-}e^{\ominus}$ *oxidations carried out by cytochrome components of*

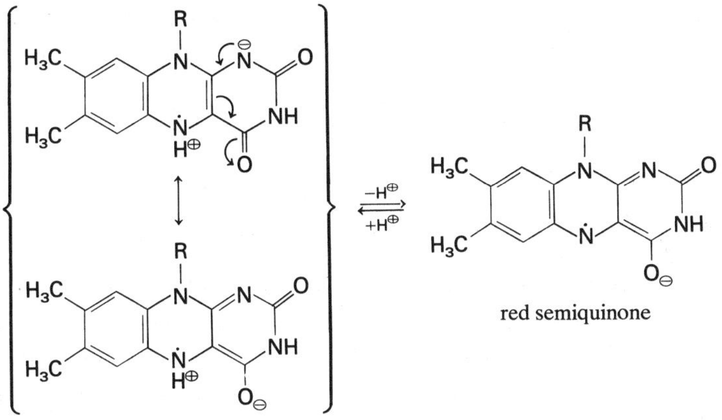

blue semiquinone

red semiquinone

membrane respiratory chains. In accord with this physiological function, many of the flavoprotein dehydrogenases are particulate, membrane-bound enzymes.

We next examine two typical flavoprotein dehydrogenases. (Note that the four enzymes discussed in ¶11.B are all representatives of flavoprotein dehydrogenases where the flavin-reoxidation step could actually occur in a two-electron step.) Other flavin-linked dehydrogenases that are complex metalloflavoenzymes in multienzyme complexes are noted in subsequent chapters.

11.C.1 D-Lactate Dehydrogenase from *E. coli* Membranes

E. coli and other bacteria contain a flavoprotein, D-lactate dehydrogenase, that is an integral component of the cytoplasmic (inner) membrane and is involved in the membrane respiratory chain. The enzyme recently has been solubilized with nonionic detergents, purified to homogeneity, and shown to contain one FAD molecule per 75,000 mol wt (Kohn and Kaback, 1973). The purified enzyme oxidizes D-lactate to pyruvate, while forming enzyme-bound $FADH_2$. The $FADH_2$ bound to the enzyme is not autoxidizable; it must be sequestered from O_2 access. Artificial dyes such as dichlorophenol indophenol (DCIP) can be used to reoxidize the $E \cdot FADH_2$; this reaction provides a convenient assay, because oxidized DCIP is blue ($\varepsilon_{600} = 22{,}000$), whereas reduced DCIP is colorless.

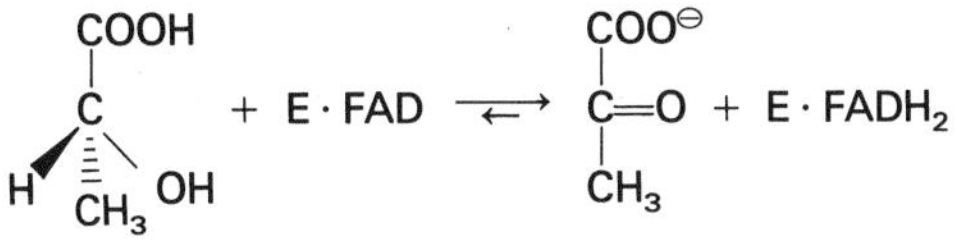

$$\text{D-lactate (COOH, H, OH, CH}_3\text{)} + E \cdot FAD \rightleftharpoons \text{COO}^{\ominus}\text{—C=O—CH}_3 + E \cdot FADH_2$$

When the enzyme is in its membrane milieu in vivo, the physiological reoxidant is thought to be a 1-e$^{\ominus}$ acceptor such as coenzyme Q or the heme-iron–containing cytochrome b_1.

$$\text{Enz} \cdot \text{FADH}_2 \xrightarrow[\text{2 cyt-}b\text{-(Fe}^{3\oplus}) \rightarrow \text{2 cyt-}b\text{(Fe}^{2\oplus})]{\text{DCIP}_{(ox)} \rightarrow \text{DCIP}_{(red)}} \text{Enz} \cdot \text{FAD}$$

Electrons subsequently pass down the membrane electron-transport chain to the copper- and iron-dependent cytochrome oxidase that catalyzes the complex 4-e$^{\ominus}$ reduction of O_2 to H_2O (Caughey et al., 1976). Experimentally, one can add D-lactate to *E. coli* membrane fragments and see O_2 consumption as an assay for the overall process. The reduction potential for pyruvate to lactate is -190 mV; that for $O_2 \rightarrow H_2O$ is $+810$ mV. The electron-transport chain thus represents a potential drop to pass electrons from D-lactate to H_2O in a strongly exergonic process. The energy released can be converted into an electrochemical potential across the membrane (Mitchell, 1966), which can be used to drive active transport of solutes into the cell (Harold, 1974) or converted into stored chemical potential during ATP synthesis.

Almost no mechanistic studies have been carried out in terms of how electrons are removed from α-hydroxy acid and passed into the oxidized FAD (this type of conversion *has* been studied with other enzymes discussed later in this chapter). An inactivating substrate analogue, the acetylenic 2-hydroxy-3-butynoate, partitions between catalytic turnover and irreversible inactivation after about 35 turnovers (C. Walsh et al., 1972).

$$HC{\equiv}C{-}CH(OH){-}COO^{\ominus} \begin{cases} \xrightarrow{35} HC{\equiv}C{-}C({=}O){-}COO^{\ominus} \\ \xrightarrow{1} \text{Inactivation} \end{cases}$$

Inactivation modifies the tightly bound coenzyme, not the apoprotein, of this and other lactate-oxidizing flavoenzymes, apparently by combination at positions C-4a and N-5 of the coenzyme (C. Walsh, 1977; Schonbrunn et al., 1976).

H$_3$C, H$_3$C, R, N, N$^{\oplus}$, N, O, NH, O, $^{\ominus}$OOC

It is not yet clear whether the inactivation occurs (1) by rearrangement of an acetylenic carbanion and attack on oxidized flavin,

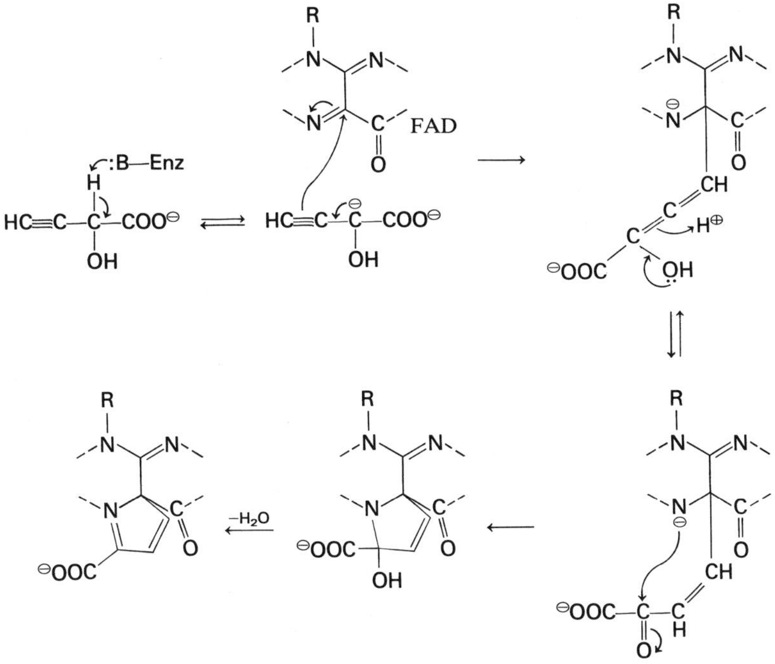

or (2) by attack of dihydroflavin on the oxidized keto acid through Michael attack before the keto acid can depart from the active site.

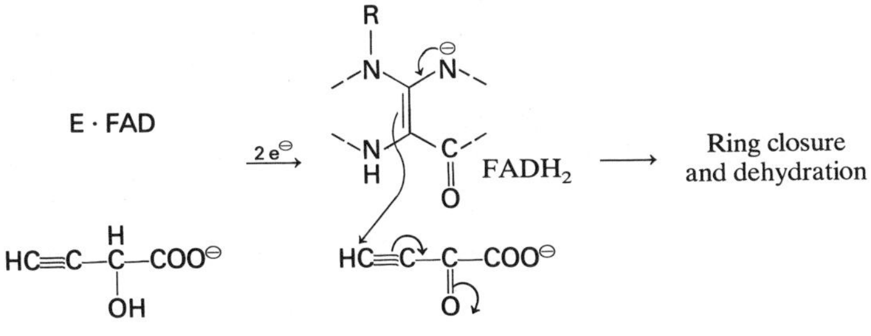

Because the acetylenic substrate *partitions*, it may be that the mechanism for inactivation of catalytic activity is similar to the normal mode of substrate oxidation. Independent of inactivation mechanism, the structure of the cyclic adduct bolsters the idea that both C-4a and N-5 are loci proximal to and reactive with substrates and substrate analogues.

11.C.2 Succinate Dehydrogenase

Perhaps the best-studied flavoenzyme dehydrogenase historically is succinate dehydrogenase, an enzyme of the Krebs cycle. In higher organisms, it is found in the mitochondrial inner membrane; in bacteria, it is located in the cytoplasmic membrane (Singer et al., 1973; Hatefi and Stigall, 1976; Boyer 1970–1976, vol. 13, p. 22 ff and references therein). The enzyme from beef-heart mitochondria has been solubilized and purified to homogeneity. It is a dimer of unequal subunits, the larger containing one FAD and four nonheme-iron atoms, and the smaller having a similar nonheme-iron chromophore but no flavin. The purified enzyme is assayed by electron transfer from succinate to artificial dyes.

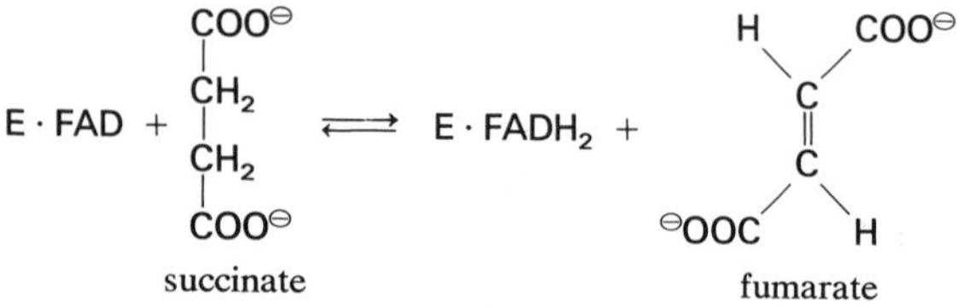

$$\text{E}\cdot\text{FAD} + \underset{\text{succinate}}{{}^{\ominus}\text{OOC}-\text{CH}_2-\text{CH}_2-\text{COO}^{\ominus}} \rightleftharpoons \text{E}\cdot\text{FADH}_2 + \underset{\text{fumarate}}{{}^{\ominus}\text{OOC}(\text{H})\text{C}=\text{C}(\text{H})\text{COO}^{\ominus}} \tag{1}$$

$$\text{E}\cdot\text{FADH}_2 + \left(\begin{matrix}\text{Coenzyme Q}_{(ox)},\\ 2\text{ Cytochrome } b\ (\text{Fe}^{3\oplus})\end{matrix}\right) \longrightarrow \text{E}\cdot\text{FAD} + \left(\begin{matrix}\text{Coenzyme Q}_{(red)},\\ 2\text{ Cytochrome } b\ (\text{Fe}^{2\oplus})\end{matrix}\right) \tag{2}$$

Physiologically, the acceptor of electrons from reduced succinate dehydrogenase is some component (quinone or cytochrome) of the mitochondrial inner-membrane respiratory chain.

When the binding of the FAD coenzyme to succinate dehydrogenase was examined, it was observed that the coenzyme does not dissociate even when the protein is denatured to a random coil by heat or acid treatment (conditions sufficient to release most flavin coenzymes from their apoproteins). Subsequently, Singer and colleagues established that FAD is covalently linked to the enzyme, the linkage occurring via a methylene at C-8a of FAD to N-3 of a histidinyl residue of the protein (Kearny and Kenney, 1974; Singer and Edmondson, 1974).

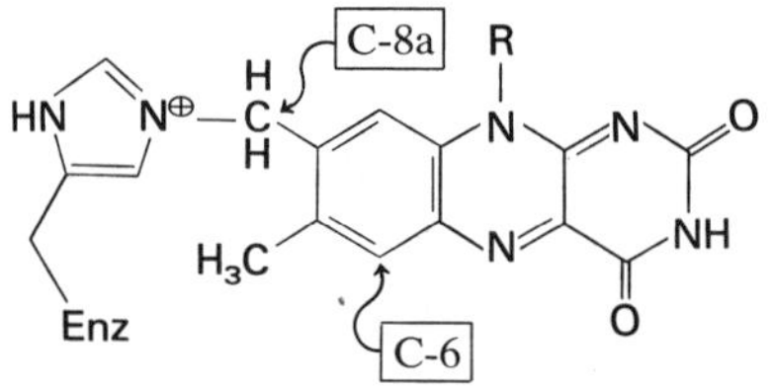

C-8a
R
HN N⊕ — CH₂
H₃C
N
NH
O
Enz
C-6

Covalent linkages have subsequently been found for about a dozen other flavoenzymes. In most instances, the methyl group at C-8a of the flavin coenzyme has been converted to a methylene and is the atom linked to some residue on the protein. The other main structural types are a thioether linkage to a cysteine sulfur in monoamine oxidase (Kearny and Kenney, 1974; Singer and Edmondson, 1974) or to N-1 of the imidazole ring of a histidine residue (Kenney et al., 1976). A novel linkage between C-6 of one flavin and an enzymatic cysteinyl sulfur has been reported for a bacterial trimethylamine dehydrogenase (Kenney et al., 1978). The advantage gained by covalent attachment of the flavin cofactor to an apoenzyme is unclear as yet.

Mechanistically, the reaction catalyzed by succinate dehydrogenase is dehydrogenation α,β to a carbonyl (in this case, a carboxylate on both the α- and the β-carbons) to yield an α,β-unsaturated carbonyl system, analogous to the dihydroorotate $\rightleftarrows$ orotate interconversion. One of the immediate questions raised concerned the extent to which this dehydrogenation is stereospecific: is the stereochemistry of H_2 removal *cis, trans,* or nonstereospecific? Succinate is a symmetrical molecule with a plane of symmetry between C-2 and C-3, each of which has an *R,S*-pair of methylene hydrogens.

$$\begin{array}{ccc} & COO^{\ominus} & \qquad \text{C-1} \\ & | & \\ H_S— & C & —H_R \qquad \text{C-2} \\ \text{-----} & | & \text{-----} \\ H_R— & C & —H_S \qquad \text{C-3} \\ & | & \\ & COO^{\ominus} & \qquad \text{C-4} \end{array}$$

Because of this plane of symmetry, the H_S at C-2 is indistinguishable (even to a chiral reagent) from the H_S at C-3, and a similar situation prevails for the two H_R atoms. However, either of the *R,S*-pairs at C-2 or at C-3 is enantiotopic and prochiral, so the two hydrogens at a single carbon can be distinguished.

Thus, the stereochemical problem can be phrased by asking whether—

1. the enzyme removes an adjacent *R,S*-pair (H_R at C-2 and H_S at C-3, or H_S at C-2 and H_R at C-3), which would constitute a *trans* elimination of H_2;
2. the enzyme removes an adjacent *R,R*-pair or *S,S*-pair of hydrogens, which would constitute a *cis* elimination of H_2; or
3. the elimination is a random, nonstereospecific mixture of the two alternatives.

Experiments with 2-*R*-monochlorosuccinate and 2-*S*-monochlorosuccinate gave a preliminary indication that succinate dehydrogenase acts with clear stereospecificity in its catalysis. The 2-*R*-isomer is a competitive inhibitor of succinate dehydrogenation, but it does not undergo reaction. The 2-*S*-isomer is a substrate, generating monochlorofumarate. Tchen and Van Milligan (1960) examined the stereochemical question by preparing various dideuterosuccinates via D_2/Pd reduction (a chemical *cis* addition) of fumarate or maleate and then submitting the dideuterosuccinates to enzyme action.

First they carried out the *cis* addition of D_2 to fumarate. Hydrogenation with D_2 gas will give equal probabilities of addition to the *si–si* face or to the *re–re* face of succinate, yielding equal populations of (2*S*,3*S*)- and (2*R*,3*R*)-D_2-succinate.

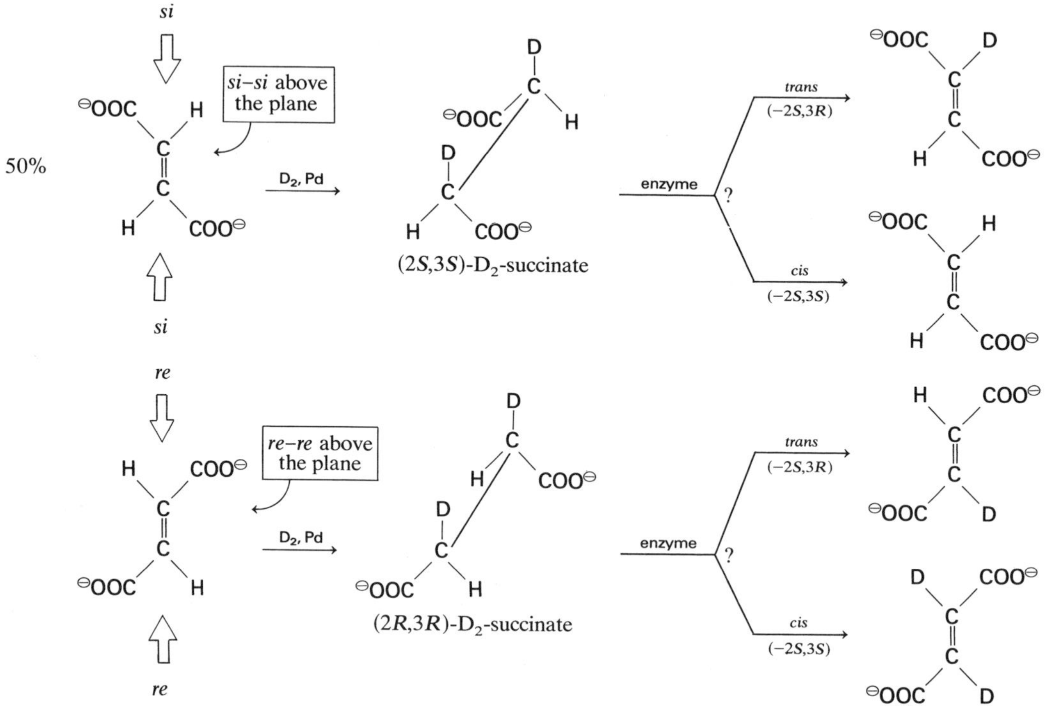

50%

si–si above the plane

D_2, Pd

(2*S*,3*S*)-D_2-succinate

enzyme ?

trans (−2*S*,3*R*)

cis (−2*S*,3*S*)

re–re above the plane

D_2, Pd

(2*R*,3*R*)-D_2-succinate

enzyme ?

trans (−2*S*,3*R*)

cis (−2*S*,3*S*)

From the reactions shown, we can make the following predictions:

1. if the enzyme catalyzes *trans* elimination (−2*S*,3*R*), the product fumarate should be entirely monodeuterofumarate;
2. if the enzyme catalyzes *cis* elimination (−2*S*,3*S*), 50% of the product fumarate should be dideuterofumarate and 50% nondeuterated fumarate, with no monodeuterofumarate;
3. if the enzyme catalyzes nonstereospecific elimination, 50% of the product fumarate should be monodeuterated, 25% dideuterated, and 25% nondeuterated.

The experiment was carried out with molecules 2.4% enriched in D_2 (50% as 2*S*,3*S* and 50% as 2*R*,3*R*). The product fumarate contained only about 0.1% dideuterated molecules, leading to the conclusion that the enzyme-catalyzed elimination of H_2 is a *trans* elimination.

Tchen and Van Milligan then carried out the complementary experiment: the production of the *meso* isomer, (2*R*,3*S*)-D_2-succinate, by maleate reduction. In this case, *trans* removal by the enzyme should yield product molecules with 50% of the molecules retaining both atoms of deuterium.

D_2, Pd → (2*R*,3*S*)-D_2-succinate (*meso*) → enzyme ?

trans → 50% dideuterofumarate, 50% nondeuterated fumarate

cis → 100% monodeuterofumarate

The experimental results confirmed that the enzyme elimination is *trans.*

Gawron and coworkers carried out elegant independent experiments leading to the same conclusion. They prepared (2*S*,3*S*)-2-chloro-3-monodeutero-succinate. *Trans* elimination would yield monodeuterochlorofumarate. *Cis* elimination (H,D removal) would yield protiochlorofumarate.

(2*S*,3*S*)-2-chloro-3-monodeuterosuccinate → *trans*

Note that *trans* elimination of H,D or *cis* elimination of H,H would lead to the deuterochloromaleates or protiochloromaleates (carboxyls *cis* to each other), but the enzyme does *not* form chloromaleate products. As predicted, the sole product is monodeuterochlorofumarate. Using various deuterated *S*-chlorosuccinate molecules as substrated, it was also possible to show that removal of the H_R at C-3 of the chlorosuccinate is the rate-determining step in catalysis. The following compound shows a kinetic isotope effect:

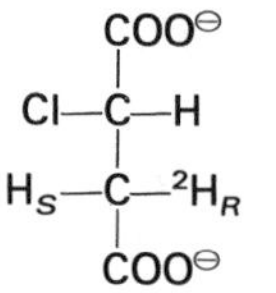

$$\begin{array}{c} COO^{\ominus} \\ | \\ Cl-C-H \\ | \\ H_S-C-{}^2H_R \\ | \\ COO^{\ominus} \end{array}$$

From the experiments described, it remains unclear how the two hydrogens and two electrons are removed during dehydrogenation (i.e., abstraction of hydride ion or of proton). In this connection, it has been possible to study the partial exchange of succinate hydrogens with D_2O under anaerobic conditions in the absence of electron acceptors (Retey, Seibl, et al., 1970). The experiments confirmed the movement of deuterons from D_2O into the protiosuccinate added initially—under conditions where no [^{14}C]-fumarate radioactivity appeared in succinate, ruling out the idea that the exchange data simply represent back reaction. With soluble preparations of enzyme after one hour in D_2O, Retey and coworkers found that 5% of the molecules were dideuterosuccinate and 1% were trideuterosuccinate. Analysis indicated that the monodeuterated molecules are *R*-monodeuterosuccinate, consistent with initial pro*R*-hydrogen removal, and that the dideuterated molecules are the *meso* (2*R*,3*S*)-species predicted if exchange of two hydrogens occurs by the same *trans* process as that operating during normal catalysis.

We have noted that succinate dehydrogenase is representative of a class of flavoprotein dehydrogenases that form α,β-unsaturated carbonyl systems. Presumably, if a proton-abstraction mechanism were involved, charge pairing by ionic interaction might produce a rise in the acidity of the hydrogen at the α-carbon to the carbonyl (by electrostatic interaction of the carboxylates with cationic groups at the active site), such that initial abstraction might occur there. This consideration would not be relevant for hydride-transfer processes, where carbonium-ion stability could instead be important.

Two interesting additional substrates have been tested with succinate dehydrogenase. An illuminating experiment has been reported using 2,2-difluorosuccinate as substrate for the enzyme (Tober et al., 1970). It appears to be a very slow substrate, losing HF to generate fluorofumarate.

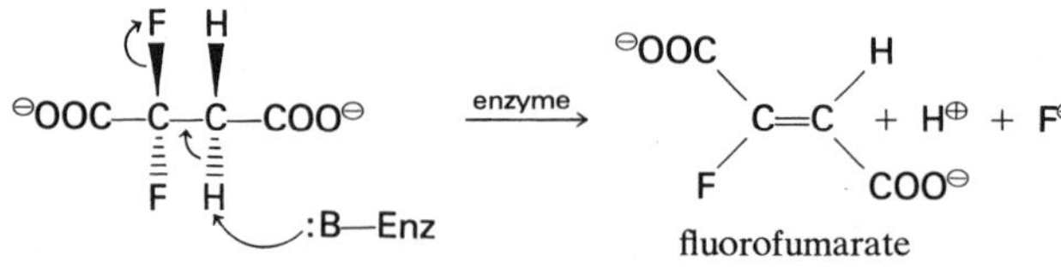

$$^{\ominus}OOC-CF_2-CH_2-COO^{\ominus} \;\; (:B-Enz) \xrightarrow{\text{enzyme}} {}^{\ominus}OOC(F)C{=}C(H)COO^{\ominus} + H^{\oplus} + F^{\ominus}$$

fluorofumarate

HF elimination strongly suggests that catalysis is initiated by a proton abstraction at C-3 and loss of a fluoro substituent as $F^{\ominus}$ from C-2. Whether this means that catalysis normally is initiated by proton abstraction and hydride transfer to the FAD coenzyme remains to be proved, because the HF elimination should occur in the absence of electron acceptor. No reducing equivalents end up in the E · FAD in the HF elimination; rather, the two electrons depart with the fluoride ion.

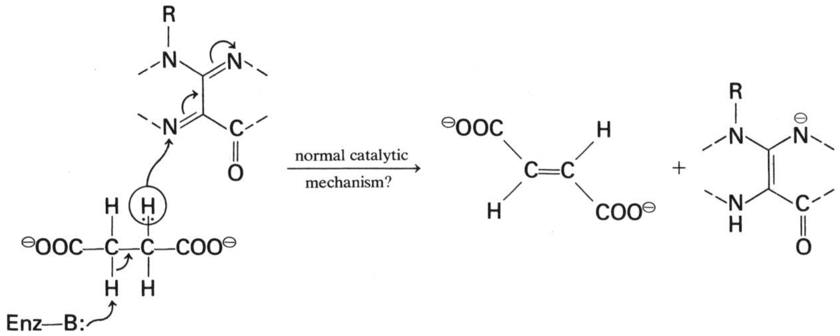

normal catalytic mechanism?

The second set of experiments dealt with the C-3 carbanion of 3-nitropropionate, an analogue of succinate.

3-nitropropionate succinate

Nitropropionate is a natural product in certain plants that are toxic to grazing livestock. The molecular basis of toxicity has now been suggested to be irreversible inactivation of the bound flavin in succinate dehydrogenase, leading to eventual impairment of cellular respiration and oxidative-phosphorylation capacity. Bright and colleagues used the kinetically stable C-3 carbanion to inactivate the enzyme still in mitochondrial membranes, and they hypothesized an initial reversible adduct between the carbanion and N-5 of the FAD (Alston et al., 1977).

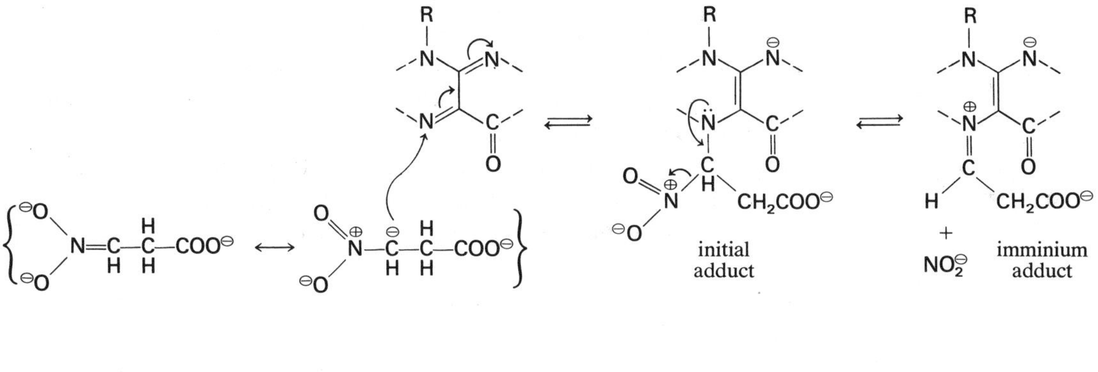

initial adduct

imminium adduct

N-5 in the initial adduct is now basic, its electron lone pair available to assist intramolecularly in expulsion of the nitro group as nitrite anion, simultaneously generating an imminium adduct. The imminium adduct could be in prototropic equilibrium with the eneamine tautomer by proton loss from C-2 of the propionate side chain to yield a stable eneamine adduct, accounting for functionally irreversible loss of activity. (Recall that the modified coenzyme is covalently bound at the active site.)

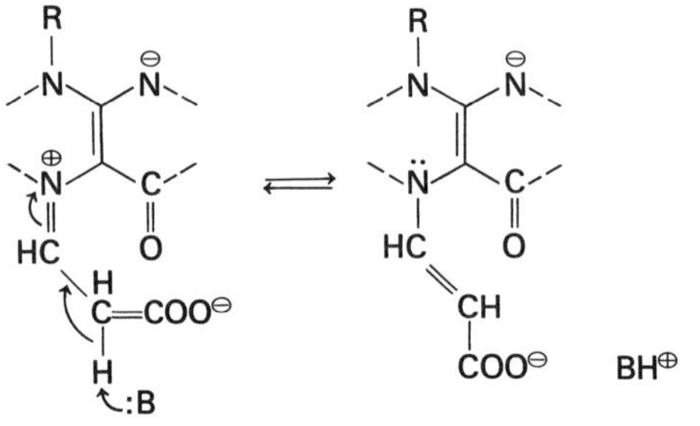

imminium adduct ⇌ eneamine adduct $BH^{\oplus}$

Precedent for this speculation derives from related experiments using nitroalkane anions and D-amino-acid oxidase noted later in this chapter (Porter et al., 1973). However, very recent studies suggest that the actual group modified is not the flavin but a cysteinyl sulfhydryl at the active site (T. Singer, personal communication).

11.D FLAVOPROTEIN OXIDASES

The distinguishing feature in the reactions of flavoprotein oxidases is the catalytic reoxidation of the reduced flavoenzyme by molecular oxygen. The reactions in which O_2 oxidizes most reduced organic molecules are thermodynamically favorable but kinetically sluggish. The predominant kinetic barrier is the fact that ground-state O_2 is a triplet (diradical) molecule with two unpaired spins. Most stable organic compounds (as well as H_2O_2 and H_2O) are ground-state, spin-paired singlets. Reaction of triplet O_2 with these organic molecules is spin-forbidden and so kinetically slow (G. Hamilton, 1971). The product would be a spin-allowed triplet (radical product), which is likely to have a high activation energy (spin inversion to the singlet state may take between 10^{-9} sec and 1 sec). Alternatively, triplet O_2 could absorb energy and go to the first excited singlet state ($^1O_2$, now spin-paired) and then react rapidly—for instance, with conjugated olefins to form cyclic peroxides, or with alkenes to form allylic hydroperoxides.

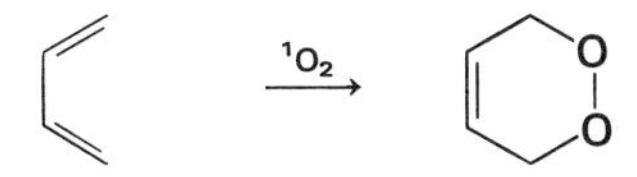

$^1O_2$ →

conjugated diene   cyclic peroxide

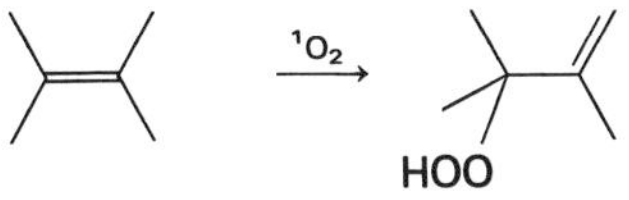

$^1O_2$ →

HOO

alkene   allylic hydroperoxide

However, it takes 22 kcal/mole (92 kJ/mole) to effect the state interconversion, and oxidases and oxygenases probably don't generally have this much energy available to make this kind of transition state accessible.

Catalysis of oxygen reactions by enzymes has evolved into basic types.

1. *Metalloenzymes* use a transition metal. The O_2 can complex to the transition metal, which itself has unpaired electrons, and generate new molecular orbitals via overlap of the metal *d* orbital and the oxygen *p* orbital, making the O_2 more singletlike in its reactivity. This avoids the high energy requirement for free singlet $^1O_2$ and thus promotes reaction. These enzymes include various copper- and iron-containing enzymes, as we shall note in Chapters 14, 15, and 16.
2. Alternatively, triplet O_2 may react by its *radical mechanism* (triplet O_2 + singlet cosubstrate → 2 free radicals). The two radicals, if stable enough to form, can recombine to give singlet products in a spin-allowed reaction. The first step (radical production) usually is highly endothermic, due to the unstable character of most substrate radicals in enzymatic systems.

This energy barrier to radical production will be lowered if the radical can be stabilized by structural features such as delocalization in the π-cloud of an extensive conjugated system—just the features available to the flavin semiquinone (and, as we shall see in Chapter 12, to the semiquinone of the pteridine coenzyme as well). Indeed, one observes the kinetically rapid chemical reaction of reduced flavins with O_2 using a radical mechanism. The 1-$e^{\ominus}$–reduced species ($O_2^{\ominus}$) is known as the *superoxide anion* ($pK_a = 4.5$). It is easily identifiable by its characteristic EPR (electron paramagnetic resonance) signal and can be detected in flavin reoxidations, as can the other odd-electron species in reoxidation, the flavin semiquinone.

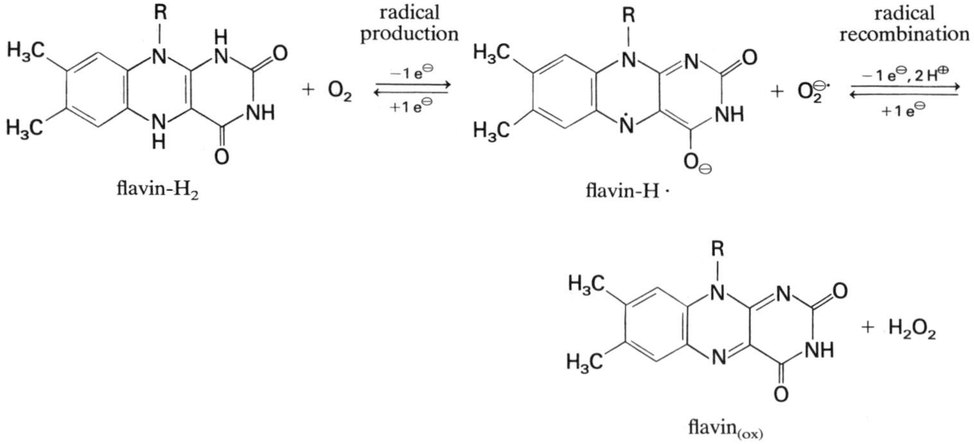

radical production

radical recombination

flavin-H_2 $+ O_2$ $\underset{+1e^{\ominus}}{\overset{-1e^{\ominus}}{\rightleftharpoons}}$ flavin-H · $+ O_2^{\ominus\cdot}$ $\underset{+1e^{\ominus}}{\overset{-1e^{\ominus}, 2H^{\oplus}}{\rightleftharpoons}}$ flavin$_{(ox)}$ $+ H_2O_2$

For riboflavin-H_2 reoxidation in air-saturated solution, the pseudo-first-order rate for appearance of oxidized flavin is about 200 min^{-1}. The turnover number for dihydroflavin reoxidation by O_2 in flavoprotein oxidases is 10^4 to 10^5 min^{-1}, suggesting some acceleration on the part of the enzyme. Conversely, for flavoprotein dehydrogenases (where O_2 reoxidation is not catalytically significant), the slow reoxidation by O_2 (e.g., that for yeast NADPH dehydrogenase with $FMNH_2$ is about 50 min^{-1}) yields detectable amounts of superoxide anion. Presumably, the second 1-$e^{\ominus}$ transfer (radical recombination) is slow.

When the O_2-dependent reoxidation is catalytically significant (as in the flavoprotein oxidases), no superoxide is detectable, suggesting that the radical-recombination step in these enzymes is very efficient (fast); alternatively (and less likely), oxygen reduction by the reduced flavoprotein oxidases might not proceed in two discrete single-electron steps. The radical recombination between FlH · and $O_2^{\ominus\cdot}$ may yield a covalent adduct (a flavin-4a-hydroperoxide) as initial product; this adduct then fragments heterolytically to $Fl_{(ox)}$ and H_2O_2. We shall discuss this adduct species further in Chapter 12 when we focus on the oxidative half-reaction and examine alternate fates of this 4a-hydroperoxide.

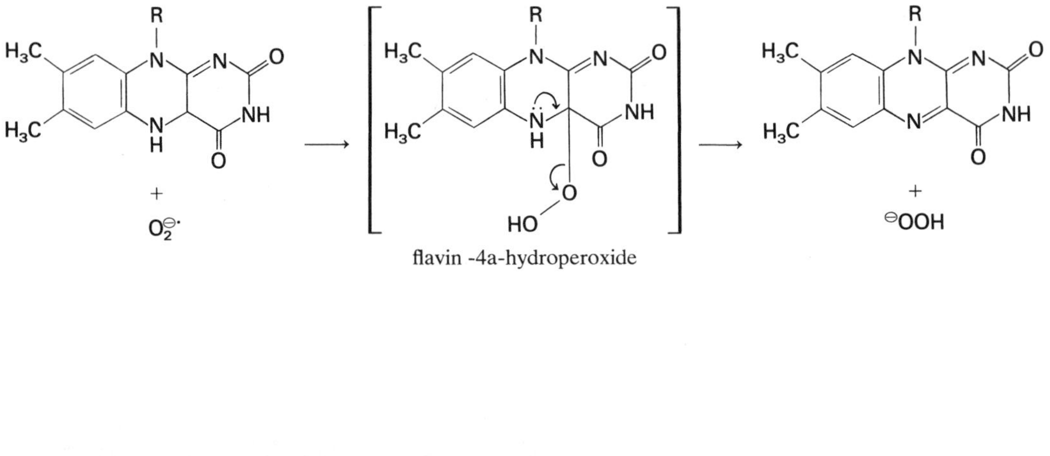

$+ O_2^{\ominus\cdot}$ → [flavin -4a-hydroperoxide] → $+ {}^{\ominus}OOH$

flavin -4a-hydroperoxide

11.D.1 Flavoenzyme Reductive Half-Reactions

The stoichiometry of reactions catalyzed by simple flavoprotein oxidases is

$$SH_2 + O_2 \xrightarrow{\text{enzyme–flavin}} S + O_2$$

This reaction occurs in two discrete steps: the reductive half-reaction

$$SH_2 + E \cdot Fl_{(ox)} \longrightarrow S + E \cdot FlH_2 \quad (1)$$

and the oxidative half-reaction

$$E \cdot FlH_2 + O_2 \longrightarrow E \cdot Fl_{(ox)} + H_2O_2 \quad (2)$$

Oxygen is involved only in the second half-reaction. In dehydrogenases, the specific reoxidants are likewise confined to the second step.

So far in this chapter, we have looked at three structural types of oxidizable substrate (SH_2): categories 3, 4, and 5 of Table 11-2 (dihydronicotinamides, dithiols, and two adjacent methylene groups oxidizable to an olefin). We have suggested a possible hydride transfer to N-5 for NADH and a covalent adduct at C-4a for thiol substrates. Oxidation of succinate we tentatively postulated as a proton/hydride process from the HF elimination data of 2,2-difluorosuccinate. Categories 1 and 2 of Table 11-2 include for SH_2 polarized carbon centers such as alcohols, amines, and aldehydes, which are generally oxidized by two electrons to ketones or aldehydes, imines, and acids, respectively.*

Given that 1-$e^{\ominus}$ steps are often involved in the oxidative half-reaction, what about the flavin-reduction step (≡substrate-oxidation step) for these groups? Are the oxidations occurring by way of hydride ions, proton abstractions (and substrate carbanions), or radical pairs? Clearly, carbanions and radicals are not mutually exclusive; there is much chemical precedent for oxidation of an electron-rich carbanion, once formed, in 1-$e^{\ominus}$ steps via radical anion species. In comprehensive model studies, Bruice and colleagues have accrued ample evidence for carbanionic intermediates in isoalloxazine (model flavin) oxidations of α-ketols and of *trans*-methyldihydrophthalate. Further, they have shown that radical anions clearly are formed as subsequent intermediates in a variety of flavin model reactions and in glucose oxidase catalysis with artificial substrates (Bruice, 1976; Chan and Bruice, 1977). A parallel question, raised by the evidence of 4a-flavin

---

* Thiamine dehydrogenase (Neal, 1970) from a soil bacterium carries out the net 4-$e^{\ominus}$ oxidation of the hydroxyethyl side chain of thiamine to the acetic side chain of thiamine acetate, while O_2 is reduced by 4$e^{\ominus}$ to H_2O. Presumably this process goes in two 2-$e^{\ominus}$ steps, with a thiamine-aldehyde equivalent and H_2O_2 as bound intermediates that react further before dissociation.

covalent adducts during dithiol oxidation and by the speculative proposal for 3-nitropropionate-carbanion inactivation of succinate dehydrogenase by a flavin-N^5-adduct, is whether flavin–substrate adducts form subsequent to carbanionic species and, if so, whether they are chemically and kinetically competent intermediates in these reaction types.

Much of the enzymatic evidence for reductive half-reaction mechanism has been obtained with D-amino-acid oxidase as catalyst.

11.D.2 D-Amino-Acid Oxidase

The flavoenzyme D-amino-acid oxidase (purified from hog kidneys) is an unusually well-studied representative because of its ease of isolation as a homogeneous protein in large amounts. Its normal physiological function is obscure because D-amino acids are not normal metabolites in mammalian cells, but it may process D-amino acids liberated from bacterial cells that have been lysed by macrophages. We can illustrate the typical stoichiometry with D-alanine, although a wide variety of D-isomers of α-amino and imino acids (D-proline) are oxidized. The amino acids give imino acids as primary products, but the favored equilibrium position in aqueous solutions is toward the α-keto acid and ammonia. Again we emphasize that O_2 is involved only in the flavin-reoxidation step, not in substrate oxidation or flavin reduction.

$$H_3C\text{—}CH(NH_3^{\oplus})\text{—}COO^{\ominus} + \text{Enz}\cdot\text{FAD} \rightleftharpoons \text{Enz—FAD}\cdots H_3C\text{—}CH(NH_3^{\oplus})\text{—}COO^{\ominus} \xrightleftharpoons[\text{step}]{\text{redox}} \text{Enz—FADH}_2\cdots H_3C\text{—}C(=NH_2^{\oplus})\text{—}COO^{\ominus}$$

$$\text{Enz(FADH}_2)\cdots H_3C\text{—}C(=NH_2^{\oplus})\text{—}COO^{\ominus} + O_2 \rightleftharpoons \text{Enz(FAD)}\cdots H_2O_2 \cdots H_3C\text{—}C(=NH_2^{\oplus})\text{—}COO^{\ominus} \rightleftharpoons \text{Enz}\cdot\text{FAD} + H_2O_2 + H_3C\text{—}C(=NH_2^{\oplus})\text{—}COO^{\ominus}$$

$$H_3C\text{—}C(=NH_2^{\oplus})\text{—}COO^{\ominus} \xrightleftharpoons{\text{nonenzymatic}} H_3C\text{—}C(=O)\text{—}COO^{\ominus} + NH_4^{\oplus}$$

For many years it was believed that the imino acid is released from E · $FADH_2$ prior to binding of oxygen, on the basis of parallel-line initial-velocity kinetic patterns (Palmer and Massey, 1968). However, this evidence was misleading, as product-inhibition studies later showed; pyruvate is a competitive inhibitor of D-alanine oxidation, indicating that they both bind to the same form of the holoenzyme. The parallel lines in initial-velocity analyses derive from a small numerical value in the $K_{AB}/[A][B]$ term of the steady-state kinetic equations:

$$V_{max}/v_0 = 1 + (K_A/[A]) + (K_B/[B]) + (K_{AB}/[A][B])$$

Intersecting lines will result only when the value of the last term is significant. These results stress the point (made in Chapter 7) that initial-velocity kinetic studies alone are insufficient to determine a kinetic mechanism.

The reaction with D-alanine has been monitored by a variety of pre-steady-state and steady-state techniques (Porter and Bright, 1975; Porter et al., 1977). Flavin reduction is very rapid, and turnover is controlled by rate-determining dissociation ($10\ sec^{-1}$) of the product pyruvate imine from the enzyme—a clear case where a physical step limits the V_{max} of an enzymatic reaction (Cleland, 1975). There are three lines of experimental evidence that, taken together, suggest that amino-acid oxidation may begin by enzyme-mediated abstraction of the hydrogen at C-2 (the oxidizable locus) as a proton and formation of a transient carbanion.

Initial results indicative of carbanionic pathways in D-amino-acid oxidase catalysis were obtained in Hellerman's laboratory at Johns Hopkins (Neims et al., 1966). A number of ring-substituted phenylglycines (with —X either *meta* or *para*) were examined with the enzyme.

$$X{-}C_6H_4{-}CH(NH_3^{\oplus}){-}COO^{\ominus} \longrightarrow [X{-}C_6H_4{-}C^{\ominus}(NH_3^{\oplus}){-}COO^{\ominus}] \longrightarrow X{-}C_6H_4{-}C({=}NH_2^{\oplus}){-}COO^{\ominus}$$

Electron-withdrawing substituents (which would stabilize carbanionic species) increased V_{max} values, whereas electron-releasing substituents (which would destabilize an α-carbanion because the electron-rich ring could not effectively delocalize the electron density at the α-carbon) decreased V_{max} values (Neims et al., 1966). However, some of the substrates were at the limit of their low solubilities in water, and this may have weakened the validity of the results and conclusions.

11.D.2.a Reaction of D-β-chloroalanine with D-amino-acid oxidase

One can further probe the possible pathway with a carbanion/proton-abstraction mechanism by using a nonphysiological derivative of alanine, β-chloroalanine.

$$H_2C(Cl){-}CH(NH_3^{\oplus}){-}COO^{\ominus}$$

Unlike alanine, β-chloroalanine has a good leaving group ($Cl^{\ominus}$) at the β-carbon; under appropriate conditions (i.e., if next to an adjacent carbanion), this group might depart.

Incubation of homogeneous hog-kidney D-amino-acid oxidase with D-β-chloroalanine produces two detectable keto-acid products, chloropyruvate and pyruvate (C. Walsh et al., 1971).

$$H_2C(Cl)\text{—}CH(NH_3^{\oplus})\text{—}COO^{\ominus} \xrightarrow{\text{enzyme}} \begin{cases} \xrightarrow{O_2} H_2C(Cl)\text{—}C(=O)\text{—}COO^{\ominus} & \text{chloropyruvate} \\ \xrightarrow{N_2} H_3C\text{—}C(=O)\text{—}COO^{\ominus} & \text{pyruvate} \end{cases}$$

The ratio of the two products is a function of molecular oxygen concentration, but the total amount of keto-acid product formed is not (Table 11-3). From these data, we can draw the following points.

1. The constant rate of total product formation (although product composition is changing) strongly suggests a rate-determining formation of a common precursor to the two keto-acid products (as we noted in earlier partitioning experiments with chymotrypsin). Or the slow step may be rate-determining release of the product imino acids from the active site in each case.
2. Only pyruvate is formed anaerobically, and essentially only chloropyruvate is formed at 100% O_2; the presence of oxygen diverts the intermediate to chloropyruvate formation.
3. The ability of chloroalanine to form a product catalytically under anaerobic conditions is unique; all other substrates yield stoichiometric product and reduced enzyme, indicating that this anaerobic reaction does not produce reduced coenzyme (does not put $e^{\ominus}$ into the coenzyme).

Table 11-3
Products of reaction between D-amino-acid oxidase and D-β-chloroalanine at varying concentrations of molecular oxygen

Percentage of O_2	V_{max} (Total keto acid formed) ($\mu M\ min^{-1}\ mg^{-1}$)	Product Distribution	
		Chloropyruvate	Pyruvate
0%	5.0	0%	100%
20%	4.9	60%	40%
100%	5.1	98%	2%

SOURCE: Data from C. Walsh et al. (1971).
NOTE: "Percentage of O_2" refers to the gas phase in equilibrium with the solutions in which the reactions are carried out.

Look at the anaerobic reaction (exclusive pyruvate formation). It must go through a carbanion or its equivalent (C. Walsh et al., 1971).

$$\begin{array}{c} :B\text{—Enz—Fl}_{(ox)} \\ + \\ H_2C(Cl)\text{—}C(H)(NH_3^{\oplus})\text{—}COO^{\ominus} \end{array} \rightleftharpoons \begin{array}{c} ^{\oplus}HB\text{—Enz—Fl}_{(ox)} \\ + \\ H_2C(Cl)\text{—}\overset{\ominus}{C}(NH_3^{\oplus})\text{—}COO^{\ominus} \end{array} \rightleftharpoons \begin{array}{c} \text{Enz—Fl}_{(ox)} \\ + \\ \text{eneamine} \\ H_2C{=}C(NH_2)\text{—}COO^{\ominus} \\ + \\ Cl^{\ominus} \end{array} \rightleftharpoons \begin{array}{c} \text{Enz—Fl}_{(ox)} \\ + \\ \text{pyruvate imine} \\ H_3C\text{—}C({=}NH_2^{\oplus})\text{—}COO^{\ominus} \end{array}$$

The substrate does undergo a 2-$e^{\ominus}$ oxidation at C-2, but the electrons leave with the β-substituent, so the flavin coenzyme remains oxidized and functional for another catalytic cycle. The bound FAD is required for this anaerobic elimination process even though it undergoes no *net* reduction; the apoenzyme is inactive for either oxidation or elimination of HCl.

The α-carbanion is one likely candidate for a common precursor involved also in chloropyruvate production.

$$H_2C(Cl)\text{—}C(H)(NH_3^{\oplus})\text{—}COO^{\ominus} \rightleftharpoons H_2C(Cl)\text{—}\overset{\ominus}{C}(NH_3^{\oplus})\text{—}COO^{\ominus} \underset{\text{oxidation}}{\overset{\text{normal}}{\rightleftharpoons}} H_2C(Cl)\text{—}C({=}O)\text{—}COO^{\ominus} + NH_4^{\oplus} + E\cdot FADH_2$$

$$\downarrow \text{elimination, irreversible}$$

$$H_3C\text{—}C({=}O)\text{—}COO^{\ominus} + NH_4^{\oplus} + HCl + E\cdot FAD$$

Such a scheme would explain both the anaerobic and the aerobic reaction products:

1. anaerobically, all product molecules are pyruvate if the normal oxidation pathway is freely reversible;
2. aerobically, O_2 draws off intermediate on the oxidation pathway by reoxidizing $E \cdot FADH_2$ and displacing that equilibrium.

These data suggest that α-carbanionic species may be the intermediates in normal amino-acid oxidations carried out by D-amino-acid oxidase. Also, this mechanism predicts, in the elimination reaction, obligate formation of an eneamine, whose presence should be detectable in D_2O; the pyruvate produced should have 1.0

atom of deuterium. In fact, one finds 0.5 atom. Using α-[^{3}H]-β-chloroalanine, one sees some ^{3}H in the β-carbon of the pyruvate formed. This observation indicates—

1. competition between (a) release of eneamine into solvent where it is protonated achirally and (b) specific ketonization on the enzyme specifically using the substrate-derived α-H; and
2. restriction of access to the active site for solvent protons; the initial ^{3}H abstracted by an active-site base exchanges slowly or not at all with the solvent before recapture by eneamine intermediate (Porter and Bright, 1975; Porter et al., 1977; C. Walsh, Krodel, Massey, and Abeles, 1973; Cheung and Walsh, 1976*a*).

$$\text{B—Enz} + \alpha\text{-}[^3\text{H}]\text{-Chloroalanine} \rightleftharpoons {}^{\oplus 3}\text{HB—Enz} + \text{H}_2\text{C}{=}\text{C}(\ddot{\text{N}}\text{H}_2)\text{—COO}^{\ominus} \rightleftharpoons \text{H}_2\text{C}(^3\text{H})\text{—C}(=\text{NH}^{\oplus})\text{—COO}^{\ominus}$$

Because chloroalanine undergoes not only elimination but also normal O_2-dependent oxidation as well at rates comparable to enzymatic D-alanine oxidation, it is reasonable to hypothesize that identical mechanisms (i.e., via carbanionic species) apply for oxidations of physiological amino acids catalyzed by D-amino-acid oxidase.

When β-fluoro- and β-bromo-D-alanine were also examined as substrates, the partitioning rates between oxidation to β-halo-α-imino acid and α,β-elimination of HX to aminoacrylate (the pyruvate precursor) was controlled by the leaving-group tendency of the halogen (Dang et al., 1976; C. Walsh, 1978). β-Bromoalanine experienced HBr elimination exclusively, whereas β-fluoroalanine yielded only β-fluoropyruvate, supporting the idea that the partitioning intermediate is the α-carbanion. With α-amino-β-chlorobutyrates (D-erythro and D-threo), the stereochemistry of the internal tritium transfer during the HX-elimination pathway could be determined. Surprisingly, both D-erythro- and D-threo-2-[^{3}H]-2-amino-3-chlorobutyrate diastereomers generate the 2-imino-3*R*-[^{3}H]-ketobutyrate as enzymatic product (Cheung and Walsh, 1976*a*), indicating a net *syn* and a net *anti* elimination geometry, respectively, to a common transoid-bound enamine (aminocrotonate) as initial product at the D-amino-acid oxidase active site.

11.D.2.b Experiments with nitroalkane anions

Preformed nitroalkane anions, such as nitroethane anion, are also substrates for D-amino-acid oxidase. The nitroalkane is converted to the carbanion with a slight excess of base and then is incubated with the flavoenzyme. The substrate carbanion is stabilized by charge delocalization and so is slow to protonate.

$$\left\{ H_3C{-}\overset{\ominus}{\underset{H}{C}}{-}\overset{\oplus}{N}{\begin{smallmatrix}\nearrow O \\ \searrow O^{\ominus}\end{smallmatrix}} \longleftrightarrow H_3C{-}\underset{H}{C}{=}N{\begin{smallmatrix}\nearrow O^{\ominus} \\ \searrow O^{\ominus}\end{smallmatrix}} \right\} + O_2 \xrightarrow{E \cdot FAD} CH_3CHO + NO_2^{\ominus} + H_2O_2$$

preformed carbanion

catalytic turnover

The reaction is an O_2-dependent catalytic turnover, producing one mole of acetaldehyde and one mole of nitrite per mole of $FADH_2$ formed (Porter et al., 1973). O_2 is used for normal reoxidation of the $E \cdot FADH_2$. These data clearly show that the enzyme will oxidize the carbanionic form of substrates when presented with them. So, all three lines of evidence and the chemical precedents from model data suggest that substrate α-carbanions form early in flavoenzyme oxidative catalysis at susceptible carbon centers. But carbanion formation does not constitute an oxidation. How are electrons transferred from this electron-rich locus of the substrate to the oxidized flavoenzyme? In addition to the model work of Bruice (1976), which suggests sequential 1-$e^{\ominus}$ transfers by way of radical anions, two other (possibly competing) paths are likely candidates.

11.D.2c Charge-transfer complexes and covalent intermediates

Electron transfer could occur from a charge-transfer complex. We have already raised the possibility of charge-transfer complexes to account for the long-wavelength species in the 2-$e^{\ominus}$–reduced glutathione reductase (Fig. 11-3). There is a good deal of evidence for complex formation of electron-rich donor molecules and the electron-poor oxidized flavin with other flavoenzymes (Massey and Ghisla, 1974), including D-amino-acid oxidase. For example, addition of anthranilate to the D-amino-acid oxidase causes the yellow enzyme solution to turn green; the 450-nm peak shifts 20 nm to the blue, and a broad transition develops between 600 and 700 nm, the classical long-wavelength absorbance of π–π charge-transfer complexes. These complexes require a sandwich-type geometry with orbital overlap to produce new, low-lying antibonding orbitals (Fig. 11-6). The long-wavelength transitions represent partial electron transfer to this new, low-energy excited state.

It may be that electron transfer from normal substrates (anthranilate is not oxidized by the enzyme) to flavin may proceed from such a close molecular complex of oxidized coenzyme and electron-rich substrate carbanion. From such

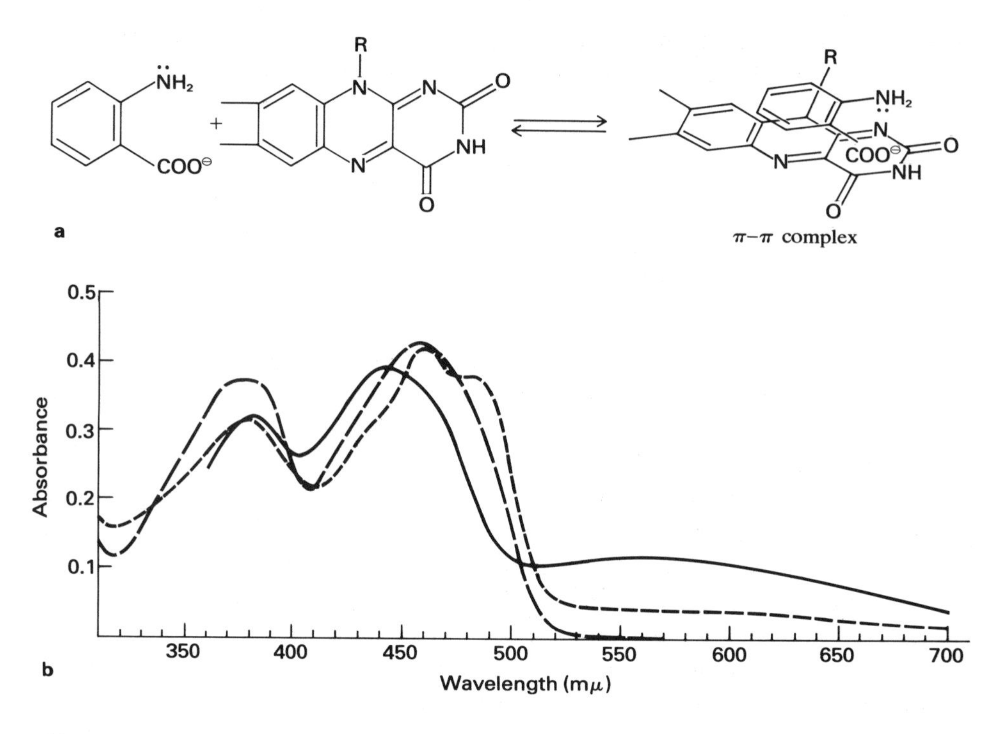

Figure 11-6
(**a**) Formation of charge-transfer complex. (**b**) Comparison of the spectra of the complexes of D-amino-acid oxidase with 2-aminobenzoate and 3-aminobenzoate. The spectra shown are (———) D-amino-acid oxidase, no additions; (——) enzyme plus 300 moles 2-aminobenzoate; and (-----) enzyme plus 300 moles 3-aminobenzoate. (Reprinted with permission from Massey and Ganther, *Biochemistry* 4(1965): 1168. Copyright © 1968 by the American Chemical Society.)

an intimate complex, the electrons could pass two at a time (or perhaps one at a time and yet not be detectable by EPR assay for stable radical species).

Given that charge-transfer complexes may form during catalysis, do they collapse to covalent adducts at any time during catalysis—and in particular *before* redox processes, before electron transfer between a substrate carbanion and the coenzyme? Strong evidence that covalent adducts can form during catalysis by D-amino-acid oxidase comes from the nitroethane-anion experiments mentioned earlier. If a nucleophile such as cyanide ion is added while the nitroethane anion is being catalytically oxidized to acetaldehyde and nitrite, irreversible inactivation of the enzyme ensues. Inactivation is accompanied by modification of the coenzyme's spectrum; use of [^{14}C]-cyanide indicates that loss of activity coincides with stoichiometric modification of the bound FAD. By correlation with known derivatives, the cyano derivative of FAD was characterized to be an N^5-substituted flavin, specifically the following aminoacetonitrile derivative (Porter et al., 1973):

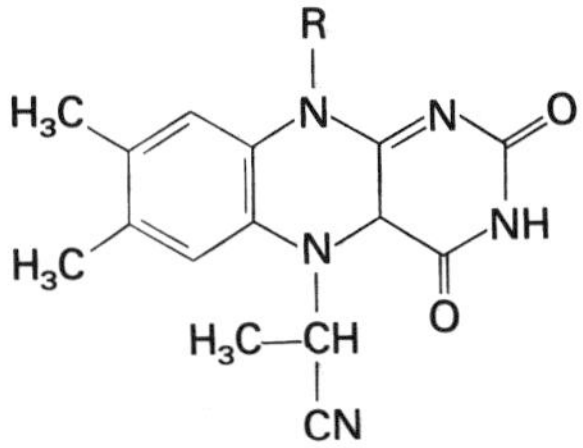

This structure suggested a common mechanism for turnover of the nitroalkane ion and for inactivation, a mechanism in which the nitroalkane anion (perhaps in initial charge-transfer complex) adds covalently to N-5 of the oxidized FAD. The lone electron pair at N-5 in this dihydro derivative can now be used to help expel the product nitrite ion and generate a product imine.

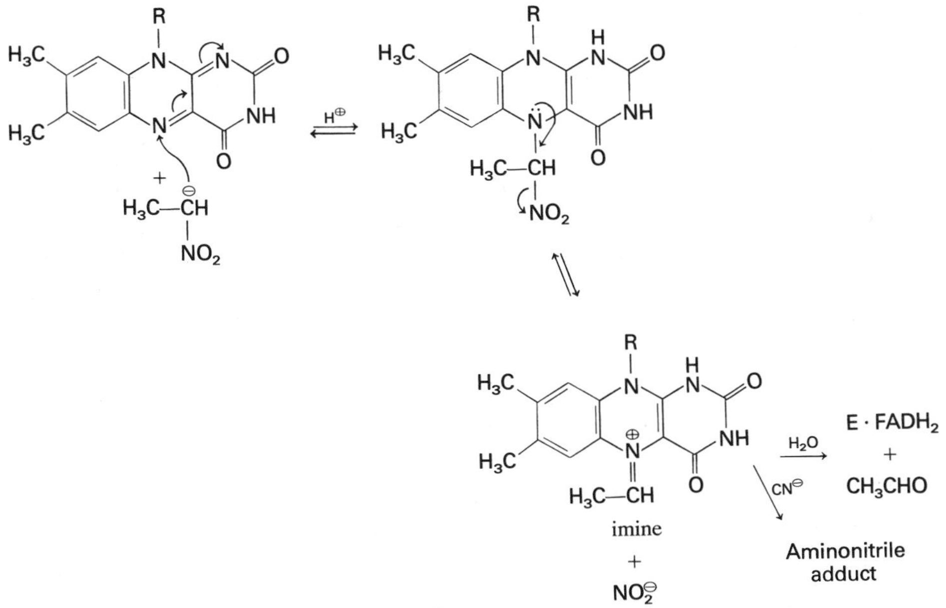

imine

+

$NO_2^{\ominus}$

$\xrightarrow{H_2O}$ E · $FADH_2$ + CH_3CHO

$CN^{\ominus}$ → Aminonitrile adduct

Nucleophiles, such as water and cyanide, could compete for the covalent product imine, one route producing released acetaldehyde and the other producing the covalent flavin–aminoacetonitrile adduct. For a normal amino-acid substrate involving a carbanion intermediate, one could write a similar N-5 adduct on the reaction pathway.

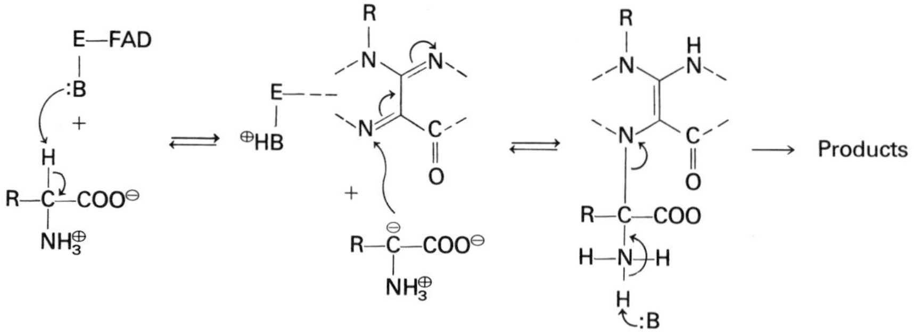

Bruice (1976) has criticized these proposals, suggesting that—in an analogous model reaction, the oxidation of methanol to formaldehyde—the N-5 adduct may be a nonproductive adduct that accumulates as a side product off the main reaction pathway.

Also, in a recent study with N-5 of a model flavin blocked by alkylation, carbanion oxidation (and concomitant N^5-alkylflavin reduction) occurred, mitigating in that instance against a covalent path and in favor of radical-anion–1-$e^{\ominus}$ electron-transfer pathways (T. Bruice, personal communication). Probably the best evidence for collapse of a substrate carbanion to an N-5 flavin covalent adduct that may be kinetically and chemically competent is during oxidation of glycollate by the flavin monooxygenase L-lactate oxidase (noted in the next chapter). Metastable species are detectable during catalysis. Massey and Ghisla argue that an initial glycollate carbanion, presumably formed by chiral abstraction of the pro*S*-hydrogen at C-2, produces an N^5-glycollyl adduct, which then rapidly collapses to glyoxylate and reduced flavin (Massey and Ghisla, 1975; Ghisla and Massey, 1978). They suggest that an observed progressive inhibition of enzyme develops from occasional enzymatic abstraction of the other (pro *R*?) C-2 hydrogen, to yield a carbanion that reacts to give an N^5-glycollyl adduct of opposite, diastereomeric chirality, and that this isomeric unnatural adduct is slow to undergo oxidation (see top of p. 401).

Even if substrate carbanions do collapse to such covalent adducts in at least some flavoenzymes, it has not yet been determined whether the transition state for two 1-$e^{\ominus}$ steps is lower than that for one 2-$e^{\ominus}$ step, or whether both routes to the covalent adduct could be in kinetic competition.

At this juncture, we have implicated two positions, C-4a (thiol addition) and N-5, as sites in oxidized flavin molecules that may participate in adduct formation during substrate oxidations. Calculations indicate that both sites are electrophilic (Sun and Song, 1973). Mono-substituted derivatives at both C-4a and N-5 have been prepared synthetically. Massey and colleagues also noted that sulfite ion attacks at N-5 in an equilibrium process, both with oxidized flavins free in

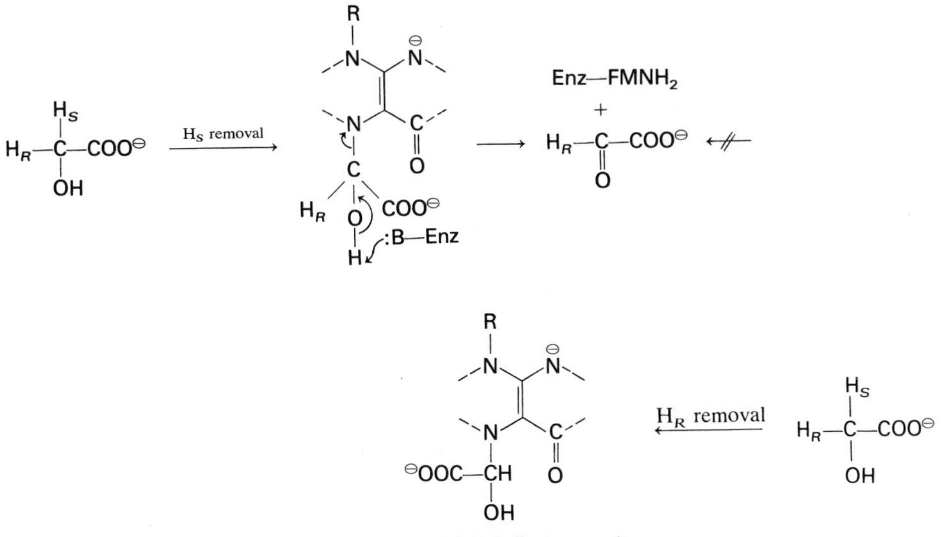

postulated diasteromeric adduct

solution and also with a variety of flavoprotein oxidases (but, curiously, not with dehydrogenases) (Massey, Muller, et al., 1969).

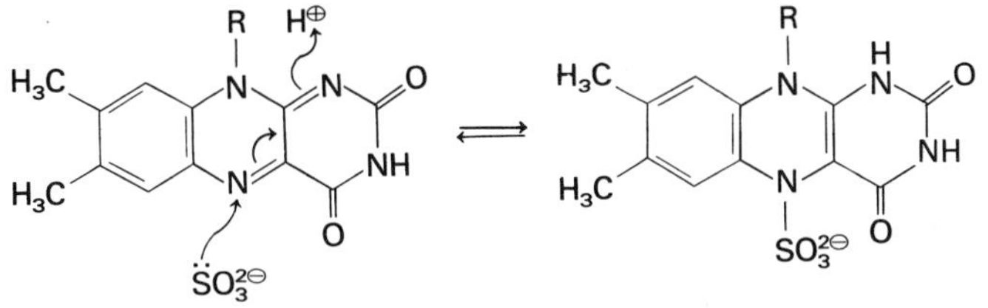

These adducts clearly substantiate that N-5 of the oxidized coenzyme is electrophilic. It may be that this N^5-sulfite adduct plays a catalytic role in one flavoenzyme's catalysis, indicated below.

11.D.3 AMP-Sulfate Reductase

The enzyme AMP-sulfate reductase is an FAD-dependent enzyme from *Desulfovibrio*, a bacterial genus that uses sulfide as terminal electron acceptor (in place of O_2) for membraneous electron-transport chains. This enzyme catalyzes the first step in reduction of sulfur at the sulfate oxidation state to inorganic sulfide. It converts an activated ester of sulfate, 5′-AMP-$SO_4^{\ominus}$ or adenosine phosphosulfate (APS; see ¶8.A.5), to AMP and sulfite (a two-electron reduction) in the presence of an oxidized dye that becomes reduced as it accepts electrons from the reduced FAD (Bramlett and Peck, 1975). There are also 12 moles each of nonheme iron and of acid-labile sulfide, reflecting the probable presence of a ferredoxin-type electron sink (see Chapter 12). In the following equation, left to right is the nonphysiological direction.

Enz—FAD + $SO_3^{2\ominus}$ ⇌ N^5-sulfite adduct + AMP—$O^\ominus$ ⇌ E · $FADH_2$ + APS

E · $FADH_2$ → (oxidized dye → reduced dye) → E · FAD

In the nonphysiological direction, incubation of E · FAD with $SO_3^{2\ominus}$ produces bleaching of the oxidized flavin, and an N^5-sulfite adduct can be isolated. Addition of AMP results in formation of APS, presumably via nucleophilic attack of a phosphate oxygen on the sulfur atom, displacing E · $FADH_2$ as a good leaving group. The reduced flavin would then be reoxidized by the dye (acceptor) molecule. In the physiological direction, conversion of APS to $SO_3^{2\ominus}$ represents a 2-$e^\ominus$ reduction of sulfur. We shall discuss (in Chapter 15) a sulfite-reductase complex that adds an additional $6e^\ominus$ to sulfite, completing the formal 8-$e^\ominus$ biological reduction of sulfate to sulfide.

The last part of this chapter notes some experiments that have been carried out to inactive another flavoprotein oxidase (monoamine oxidase) with a suicide substrate, again showing that N-5 of the flavin is close to the catalytic action.

11.D.4 Monoamine Oxidase

Monoamine oxidase is a flavoprotein oxidase of purported central metabolic importance, converting neuroactive amines into inactive aldehydes. The aldimine is the initial oxidation product and suffers hydrolysis to the observed aldehyde and ammonia.

$$H_2O + R{-}CH_2{-}NH_2 + {}^{18}O_2 \xrightarrow{E \cdot FAD} R{-}CH{=}O + H_2{}^{18}O_2 + NH_3$$

(via [RCH=NH], hydrolyzed by H_2O)

The flavin-linked monoamine oxidase is localized in the outer mitochondrial membrane of animal cells.

This mitochondrial flavoenzyme is a representative of those enzymes with the coenzyme bound in covalent linkage; in this case, the methylene at C-8a is in thioether linkage to an active-site cysteine residue (Kearny and Kenney, 1974; Singer and Edmondson, 1974).

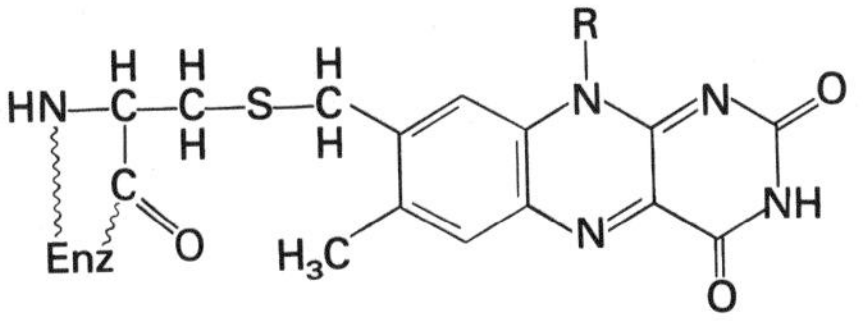

The enzyme is irreversibly blocked by the drug pargyline (*N*-benzyl-*N*-methylpropynylamine), which appears to modify the flavin cofactor, although the structure of the adduct has not been identified (Chuang et al., 1974).

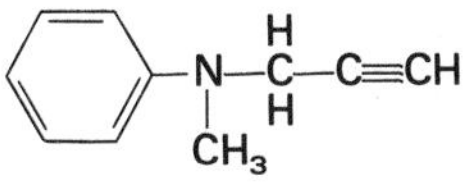

pargyline

The propynyl moiety is likely to be the relevant functionality, which is activated at the amine-oxidase active site. Recent studies with [^{14}C]-dimethylpropynylamine as substrate for the enzyme indicate that a radioactive, covalently modified coenzyme is formed on loss of catalytic activity.

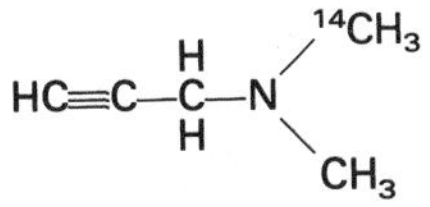

dimethylpropynylamine

Isolation of the flavin adduct allowed determination of an N^5-substituted acyclic structure (Maycock et al., 1976).

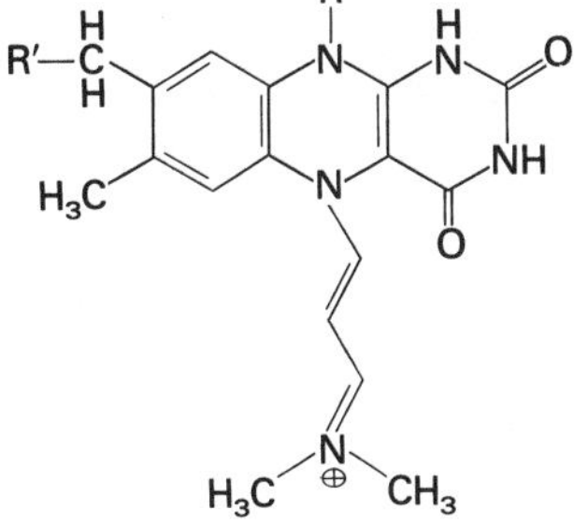

Such an adduct could arise from enzymatic oxidation of the acetylenic tertiary amine to a product imine, which is attacked in a nucleophilic conjugate addition reaction by the now-nucleophilic N-5 of the reduced $FADH_2$ (bound to enzyme).

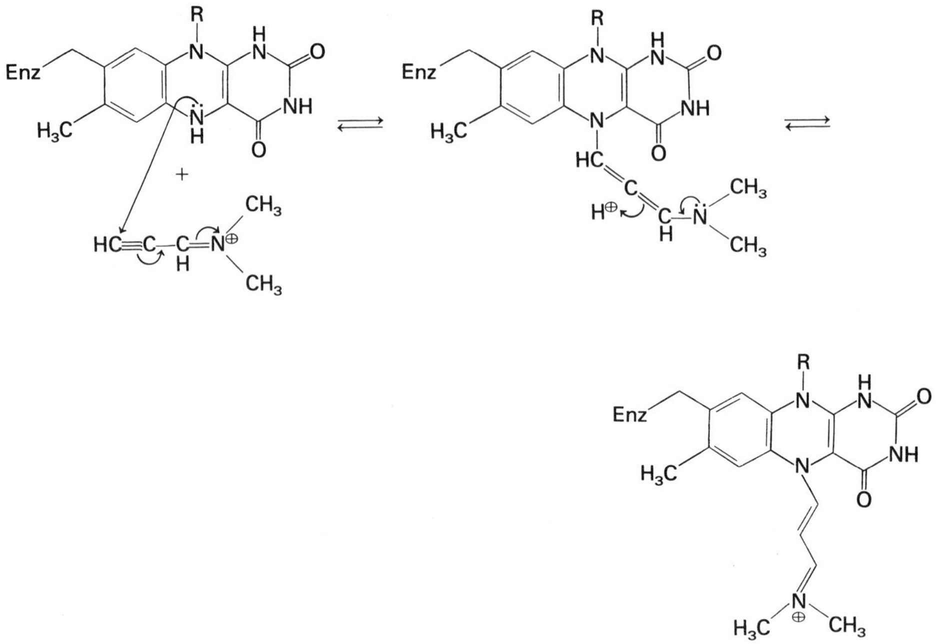

Alternatively, if monoamine-oxidase catalysis proceeds by a carbanion mechanism, formation of a carbanion in α-position to the acetylene generates, in essence, an allenic anion that could attack the still-oxidized, electrophilic coenzyme at N-5. It is currently unclear which route is responsible for inactivation (see top of p. 405).

Recall that earlier in this chapter we noted another acetylenic substrate (2-hydroxy-3-butynoate) that partitions between normal oxidation and inactivation with flavin modification in a number of cases. In the one enzyme (L-lactate oxidase from *Mycobacterium smegmatis*) where the adduct structure was determined (see C. Walsh, 1977), the sites of attachment are to the two positions of the flavin on which we have focused: C-4a and N-5 (¶11.C.1).

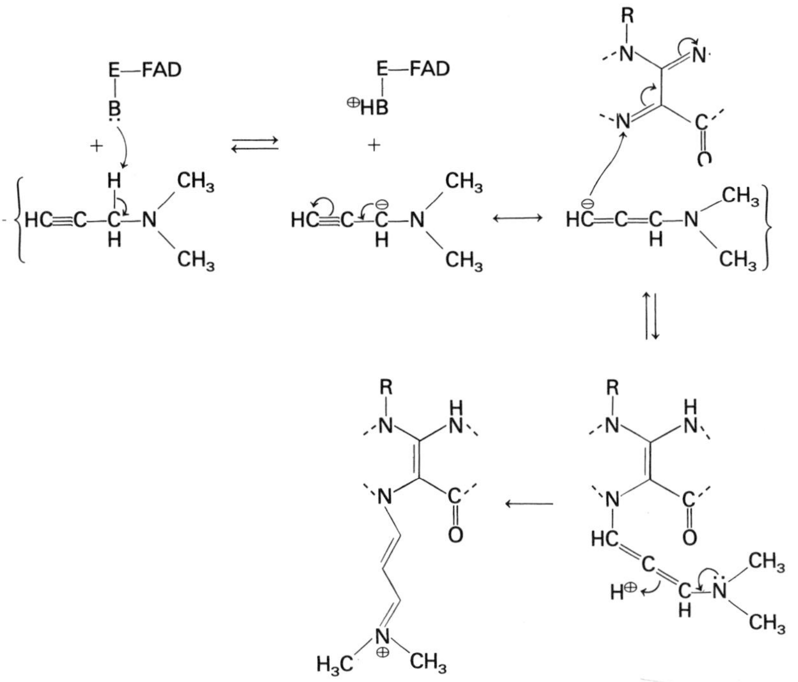

In the next chapter, we shall use the mechanistic intuition we have now accumulated on flavin enzymology to look at a class of enzymes where the oxygen cosubstrate is not merely reduced by two electrons to H_2O_2, but is in some way activated for transfer of one of its oxygen atoms to hydroxylate an activated aromatic cosubstrate, or to oxygenate an aldehyde or ketone to a carboxylic acid.

Chapter 12

Flavin- and Pterin-Dependent Monooxygenases

In four of the next five chapters we shall be dealing with some aspect of reductive oxygen metabolism, where O_2 serves not only as electron acceptor but also as oxygen-transfer agent. Therefore, we begin this chapter with a general discussion of reductive oxygen metabolism.

12.A REDUCTIVE OXYGEN METABOLISM

During biological reduction, dioxygen is activated for transfer of one or both oxygen atoms into a cosubstrate undergoing concomitant oxidation. In quantitative terms, the major reduction product of O_2 in aerobic organisms is H_2O, the 4-$e^{\ominus}$–reduced form. For consideration of reaction intermediates in oxygen chemistry, we can indicate the species that lie between O_2 and H_2O by successive 1-$e^{\ominus}$ reductive steps.

$$\underset{\text{dioxygen}}{\tfrac{1}{2}O_2} \xrightarrow{1\text{-}e^{\ominus}} \underset{\text{superoxide}}{\tfrac{1}{2}O_2^{\ominus\cdot}} \xrightarrow[H^{\oplus}]{1\text{-}e^{\ominus}} \underset{\text{hydrogen peroxide}}{\tfrac{1}{2}H_2O_2} \xrightarrow{1\text{-}e^{\ominus}} \underset{\text{hydroxyl radical}}{OH\cdot} \xrightarrow[H^{\oplus}]{1\text{-}e^{\ominus}} \underset{\text{water}}{H_2O}$$

We encountered superoxide and hydrogen peroxide in Chapter 11. Addition of one electron per oxygen to H_2O_2 generates two hydroxyl radicals, and the final electron input produces water. All three intermediates ($O_2^{\ominus\cdot}$, H_2O_2, and $OH\cdot$) are markedly reactive with the probable order $OH\cdot > O_2^{\ominus\cdot} > H_2O_2$ in terms of potential toxicity and instability. Free hydroxyl radicals are so deleterious that they are probably always avoided in normal metabolism, the second electron input (to yield nontoxic water) occurring rapidly and to an enzyme-bound species. The following are the redox potentials (per electron transferred) for the steps in oxygen reduction.

$$\tfrac{1}{2}O_2 + 1\,e^{\ominus} \rightleftharpoons \tfrac{1}{2}O_2^{\ominus\cdot} \qquad E^{0\prime} = -330\,\text{mV}$$
$$H^{\oplus} + \tfrac{1}{2}O_2^{\ominus\cdot} + 1\,e^{\ominus} \rightleftharpoons \tfrac{1}{2}H_2O_2 \qquad E^{0\prime} = +\ 94\,\text{mV}$$
$$\tfrac{1}{2}H_2O_2 + 1\,e^{\ominus} \rightleftharpoons OH\cdot \qquad E^{0\prime} = +136\,\text{mV}$$
$$H^{\oplus} + OH\cdot + 1\,e^{\ominus} \rightleftharpoons H_2O \qquad E^{0\prime} = +233\,\text{mV}$$

As noted earlier (Table 10-1), the overall reduction potential for $O_2 \rightarrow 2H_2O$ is +820 mV.

Table 12-1 gives a brief summary of oxygen-insertion reactions to be discussed in this or succeeding chapters, according to reaction type. We shall note

Table 12-1
Types of oxygen-insertion reactions (in monooxygenase catalyses)

Type	Generalized reaction
1	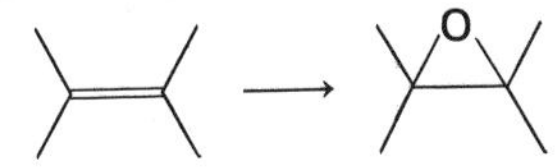 $>C{=}C< \longrightarrow$ epoxide (O bridging the two carbons)
2	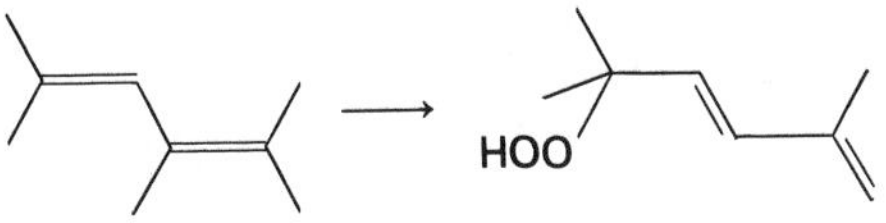 conjugated diene $\longrightarrow$ HOO–C (allylic hydroperoxide, with shifted double bond)
3	$-\overset{\vert}{\underset{\vert}{C}}-H \longrightarrow -\overset{\vert}{\underset{\vert}{C}}-OH$
4	$\overset{\vert}{\underset{\vert}{N}}-H \longrightarrow \overset{\vert}{\underset{\vert}{N}}-OH$
5	$-S-H \longrightarrow -S-OH$
6	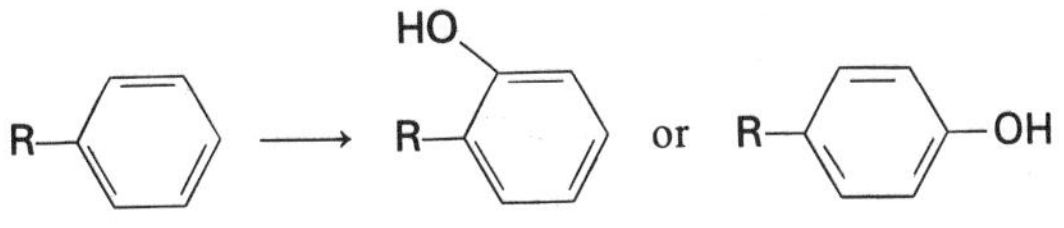 R–C₆H₅ $\longrightarrow$ R–C₆H₄–OH (ortho, HO) or R–C₆H₄–OH (para)
7	$R-\overset{O}{\overset{\Vert}{C}}-R' \longrightarrow R-\overset{O}{\overset{\Vert}{C}}-OR' \quad \left(R' = H, \overset{\vert}{\underset{\vert}{C}}-\right)$

examples of types 1, 4, 5, 6, and 7 in this chapter. G. Hamilton (personal communication, 1978) has suggested that all of these oxygenation reactions are likely to arise from an oxygen species at the oxidation state of peroxide. In model chemistry, peroxides can be utilized to effect all these kinds of transformations (types 1 through 7). Superoxide is in general insufficiently reactive, whereas hydroxyl radicals are much too reactive to be controlled. In this view, O_2 might be considered as a latent source of enzymatically generated peroxide equivalents (usually bound in some organic or metal-complexed form) that carry out exothermic oxygen-transfer chemistry on (for the most part) chemically unactivated substrates.

This chapter will focus on flavin-dependent and pterin-dependent monooxygenases. First, we discuss a transitional class of flavin oxidases/decarboxylases that emphasize the mechanistic similarities between oxidation and monooxygenation flavoenzymes.

12.B FLAVOENZYME OXIDASES/DECARBOXYLASES

There are variants of the normal flavoenzyme oxidase pattern in which the substrate undergoes, not a 2-$e^{\ominus}$ oxidation, but rather a net 4-$e^{\ominus}$ oxidation. O_2 is still the oxidant, but rather than a 2-$e^{\ominus}$ reduction to H_2O_2, it now undergoes 4-$e^{\ominus}$ reduction to H_2O. Important to note is the fact that the other oxygen atom from O_2 ends up in one of the products formed; in effect, an oxygen-transfer reaction has been achieved, although mechanistically it is quite primitive compared to most hydroxylases (Flashner and Massey, 1974; Massey and Hemmerich, 1975).

One well-studied example is the L-lactate oxidase from *Mycobacterium smegmatis.*

$$\underset{\substack{\text{L-lactate} \\ (S\text{-isomer})}}{\text{HO—CH(COO}^{\ominus}\text{)—CH}_3} + O_2 \xrightarrow{\text{E—FAD}} \underset{\text{acetate}}{CH_3COO^{\ominus}} + CO_2 + H_2O$$

The C-2 carbon of L-lactate is oxidized from an alcohol (through the ketone level) to an acid. One would expect that this net 4-$e^{\ominus}$ removal occurs through the normal 2-$e^{\ominus}$–removal sequence (repeated a second time), involving an intermediate enzyme-bound pyruvate and 2-$e^{\ominus}$–reduced O_2 (i.e., hydrogen peroxide). This supposition is supported by the following observations pertinent to the mechanism of catalysis (Flashner and Massey, 1974; Massey and Hemmerich, 1975).

1. Under anaerobic conditions, one observes stoichiometric formation of $E \cdot FMNH_2$ along with one equivalent of pyruvate—a standard one-turnover sequence: 2-$e^{\ominus}$ oxidation of substrate, 2-$e^{\ominus}$ reduction of bound FMN.

2. Mixing $E \cdot FMNH_2$ and radioactively labeled pyruvate, *followed* by addition of O_2, leads to the following reaction:

$$E \cdot FMNH_2 + H_3C{-}^{14}C(=O){-}COO^{\ominus} \xrightarrow{O_2} E \cdot FMN + H_3C{-}^{14}COO^{\ominus} + H_2O + CO_2$$

If O_2 is added to $E \cdot FMNH_2$ *before* adding radioactive pyruvate, there is no decarboxylation of pyruvate, but now a stoichiometric H_2O_2 formation from O_2.

3. With $^{18}O_2$, one ^{18}O ends up in the acetate carboxyl group (the oxygen transfer), and the other in [^{18}O]-H_2O (Hayaishi and Sutton, 1957).

4. A model reaction for this oxygen transfer is known in the decarboxylation of α-keto acids by hydrogen peroxide (Hamilton, 1971):

$$H_3C{-}C(=O){-}COO^{\ominus} + H_2O_2 \longrightarrow H_3C{-}COO^{\ominus} + H_2O + CO_2$$

These data support the following oxidation scheme for catalysis by lactate oxidase. Whereas the H_2O_2 and pyruvate would be released by a normal flavoprotein oxidase, they are retained at the active site of L-lactate oxidase in high local concentration long enough for the α-keto-acid decarboxylation process to occur.

$$H_3C{-}CH(OH){-}COO^{\ominus} + E \cdot FMN \rightleftharpoons H_3C{-}C(=O){-}COO^{\ominus} + E \cdot FMNH_2 \quad (1)$$

$$E \cdot FMNH_2 \cdots H_3C{-}C(=O){-}COO^{\ominus} + {}^{18}O_2 \rightleftharpoons (E \text{-----} FMN)\; H_2{}^{18}O_2 \; H_3C{-}C(=O){-}COO^{\ominus} \quad (2)$$

$$\text{FMN}\cdots E\cdots H{-}{}^{18}O{-}{}^{18}O{-}H \;\; H_3C{-}C(=O){-}COO^{\ominus} \rightleftharpoons \text{FMN}\cdots E \;\; H_3C{-}C(O^{\ominus})(COO^{\ominus}){-}{}^{18}O{-}{}^{18}O{-}H \;\; (BH^{\oplus}) \rightleftharpoons \text{FMN}{-}E \;\; H_3C{-}C(=O^{\ominus}){=}{}^{18}O + H_2{}^{18}O + CO_2 \quad (3)$$

An enzyme apparently analogous to lactate oxidase is the L-lysine oxygenase crystallized from *Pseudomonas fluorescens* (Takeda and Hayaishi, 1966). An FAD-enzyme, it converts the dibasic amino acid lysine, via oxidative decarboxylation, to δ-aminovaleramide.

$$H_2C(NH_3^{\oplus})\text{—}CH_2\text{—}CH_2\text{—}CH_2\text{—}CH(NH_3^{\oplus})\text{—}COO^{\ominus} + {}^{18}O_2 \xrightarrow{\text{E—FAD}} H_2C(NH_3^{\oplus})\text{—}CH_2\text{—}CH_2\text{—}CH_2\text{—}C(={}^{18}O)\text{—}NH_2 + CO_2 + H_2{}^{18}O$$

δ-aminovaleramide

The following observation supports the idea that this enzyme also carries out two 2-e$^{\ominus}$ oxidation steps: with L-ornithine (the five-carbon analogue of L-lysine), the enzyme carries out, not an oxygen-transfer reaction, but a typical oxidase reaction, L-ornithine → 2-keto-5-aminopentanoate (Nakazawa et al., 1972). Indeed, modification of the native enzyme's sulfhydryl groups with *p*-chloromercuribenzoate converts the enzyme from a monooxygenase for lysine to a simple oxidase for lysine. When the mercurial is removed from the enzyme SH groups, oxygenase activity is restored. In the oxidase mode,

$$H_2C(NH_3^{\oplus})\text{—}CH_2\text{—}CH_2\text{—}CH(NH_3^{\oplus})\text{—}COO^{\ominus} + O_2 \xrightarrow[\text{enzyme}]{\text{modified}} H_2C(NH_3^{\oplus})\text{—}CH_2\text{—}CH_2\text{—}C(=O)\text{—}COO^{\ominus} + NH_4^{\oplus} + H_2O_2$$

L-ornithine

nonenzymatic cyclization

$$\rightleftharpoons \text{(pyrroline ring)—}COO^{\ominus} + H_2O$$

Δ¹-pyrroline carboxylate
(stable cyclic imine)

Apparently, when sulfhydryl groups are blocked, the active-site geometry is perturbed sufficiently that the "active oxygenating intermediate" (possibly a flavin hydroperoxide) breaks down to H_2O_2 instead of decomposing with oxygen transfer.

When the *M. Smegmatis* L-lactate oxidase is incubated with β-chloro-L-lactate to test for a carbanion mechanism, both elimination and normal oxidation processes occur in competition (C. Walsh, Lockridge, Massey, and Abeles, 1973), similar to the reaction of D-amino-acid oxidase with β-chloroalanine. The product from O_2-independent elimination is pyruvate, whereas O_2-driven oxidation yields the 4-e$^{\ominus}$–reduced, decarboxylated chloroacetate. The elimination of HCl yields the enol of pyruvate as initial bound product. No electrons are transferred into the FMN coenzyme so that anaerobic turnover can proceed. In the O_2-dependent arm, enzyme-bound chloropyruvate and H_2O_2 are the likely intermediates.

$$H_2C(Cl)-CH(OH)-COO^{\ominus} \xrightarrow{E-FMN} \begin{cases} \xrightarrow{N_2} H_3C-C(=O)-COO^{\ominus} + Cl^{\ominus} + H^{\oplus} \\ \xrightarrow{O_2} H_2C(Cl)-COO^{\ominus} + H_2O + CO_2 \end{cases}$$

Flashner and Massey (1974) have referred to representatives of the lactate-oxidase and lysine-oxygenase class as internal flavoprotein monooxygenases. The species that inserts oxygen into the product may be hydrogen peroxide. Hydrogen peroxide is an effective oxygenating agent only by virtue of the fact that the molecule with which it reacts is an α-keto acid; during its decarboxylation, fission of the weak peroxide bond results in leaving one oxygen behind in the product acid. Because H_2O_2 and keto (or imino) acid are the initial products, Massey has suggested that the monooxygenase function in these enzymes may be somewhat adventitious.

In the case of the hydroxylases (external monooxygenases) to be discussed next, the oxygen-insertion process may be quite different from that in this transitional class of flavoenzymes.

12.C FLAVOPROTEIN HYDROXYLASES

The flavoprotein hydroxylases are all monooxygenases; one atom from O_2 is introduced into substrate as a hydroxyl moiety, and the other ends up as H_2O (Flashner and Massey, 1974; Massey and Hemmerich, 1975). Most of these enzymes so far studied are bacterial in origin, often induced to high levels by growth of the organism on the aromatic substrate as major or sole carbon source, resulting in high yields of enzyme. These enzymes are simple flavoproteins in the sense that no other detectable organic coenzymes or transition metal elements are required for catalysis. The generalized stoichiometry is

$$S-H + O_2 + XH_2 \xrightarrow{E-FAD} S-OH + H_2O + X$$

XH_2 is an obligatory reducing agent, either NADH or NADPH. The substrates are aromatic or heteroaromatic and are usually activated by some function such as a preexisting phenolic hydroxyl group; where possible structurally, *ortho hydroxylation* is essentially the exclusive reaction. The role of the reduced nicotinamide coenzyme is to generate the reduced holoenzyme E—$FADH_2$, which is the reactive species combining with O_2 to form the active hydroxylating agent.

$$\text{NADH} + \text{E—FAD} \rightleftharpoons \text{NAD}^{\oplus} + \text{E—FADH}_2 \xrightarrow{^{18}O_2} \text{"Hydroxylating species"}$$

"Hydroxylating species" + (4-X-phenol) ⟶ (2-^{18}OH-4-X-phenol) + $H_2{}^{18}O$

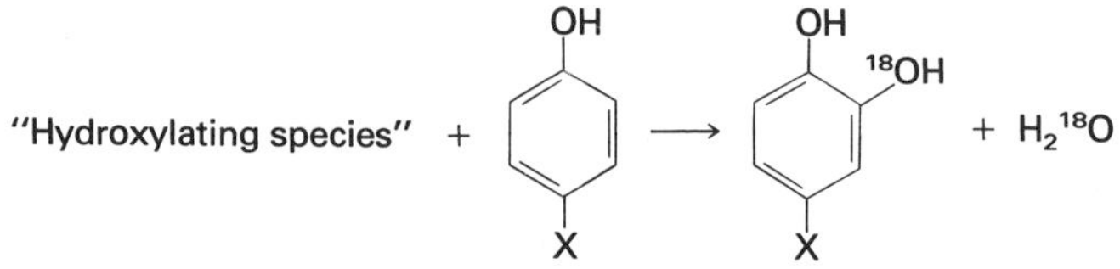

12.C.1 *p*-Hydroxybenzoate Hydroxylase

Obtainable from various pseudomonads, *para*-hydroxybenzoate hydroxylase carries out the following reaction:

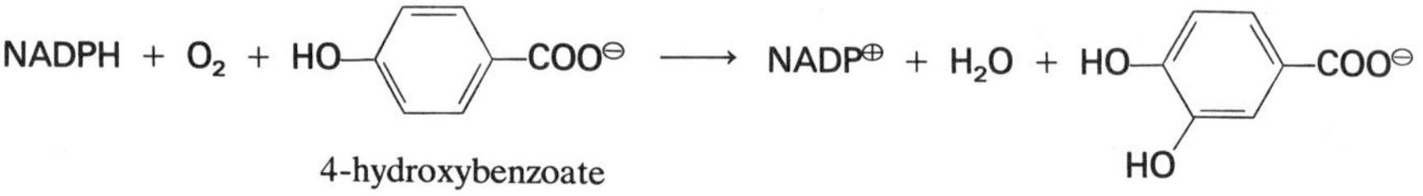

NADPH + O_2 + HO–C$_6$H$_4$–COO$^{\ominus}$ (4-hydroxybenzoate) ⟶ NADP$^{\oplus}$ + H_2O + (HO)$_2$C$_6$H$_3$–COO$^{\ominus}$ (3,4-dihydroxybenzoate)

Howell and Massey (1971; Howell et al., 1972) found that, although NADPH will reduce the bound FAD in the absence of substrate ($k = 0.41\ \text{min}^{-1}$), the addition of saturating *p*-hydroxybenzoate enhances anaerobic flavin reduction by a factor of about 40,000 ($k = 1.52 \times 10^4\ \text{min}^{-1}$). NADPH is also bound 13-fold more tightly in the presence of hydroxylatable substrate. This is a good example of substrate synergism in enzymatic catalysis.

This effector role can be fulfilled even by molecules that do not undergo hydroxylation. For instance, the 3,4-dihydroxybenzoate, added to solutions of NADPH and E—FAD, accelerates the rate of flavin reduction although the dihydroxybenzoate is recoverable unchanged. Aerobically, this results in increased rate of O_2 consumption, with all of the O_2 being reduced to H_2O_2. That is, *hydroxylation* (oxygen insertion) *has been uncoupled from coenzyme reoxidation*: the hydroxylase has been converted back to a flavoprotein oxidase.

The effector molecules then activate the hydroxylase, but *only as a catalyst for oxidase activity*. It is reasonable to speculate that, in both oxidases and hydroxylases, a common oxygen–flavin species forms. In the absence of hydroxylatable substrate, this active species breaks down to oxidized holoenzyme (E—FAD) and H_2O_2. A key question, then, is the structure of the common oxygen–flavin intermediate that can function as a hydroxylating agent. Rapid kinetic studies with *p*-hydroxybenzoate hydroxylase suggest transient formation of three closely related, spectroscopically detectable complexes (Entsch, Ballou, and Massey, 1976). The first complex was postulated to be the 4a-hydroperoxyflavin mentioned in the last chapter as a potential intermediate in oxidase catalysis. Consistent with this tentative assignment is the recent report that the N^5-methyl-4a-hydroperoxy adduct can be prepared as light green crystals (Kemal and Bruice, 1976).

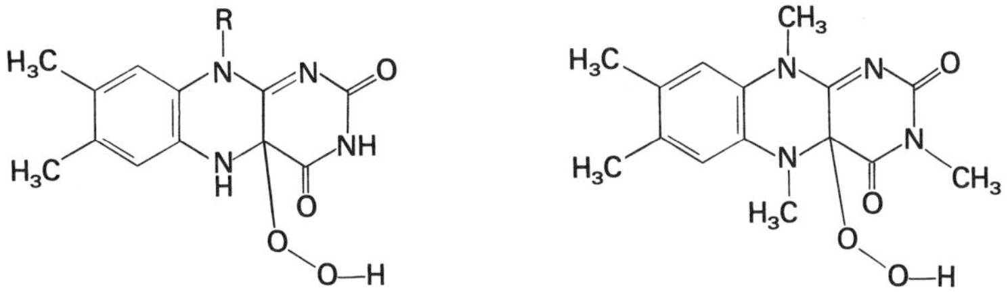

4a-hydroperoxyflavin

The adduct shows a spectrum almost identical to that of the first intermediate in the hydroxybenzoate hydroxylase catalysis (Fig. 12-1). The 4a-hydroperoxide could arise from the two-step radical mechanism when O_2 and dihydroflavin interact. The question of whether such a peroxide could function as an oxygen-transfer agent has been debated; little mechanistic evidence is available (Hamilton, 1974; Orf and Dolphin, 1974), although Bruice and colleagues have shown oxygen transfer to thiols in model systems (Chan and Bruice, 1977). Massey and colleagues have suggested the following hypothesis (Entsch, Ballou, and Massey, 1976).

Noting that only activated aromatic rings are hydroxylated, specifically in the *ortho* position, they speculate that fission of the weak peroxy O—O bond is initiated by nucleophilic attack of the π-electrons of the aromatic ring adjacent to

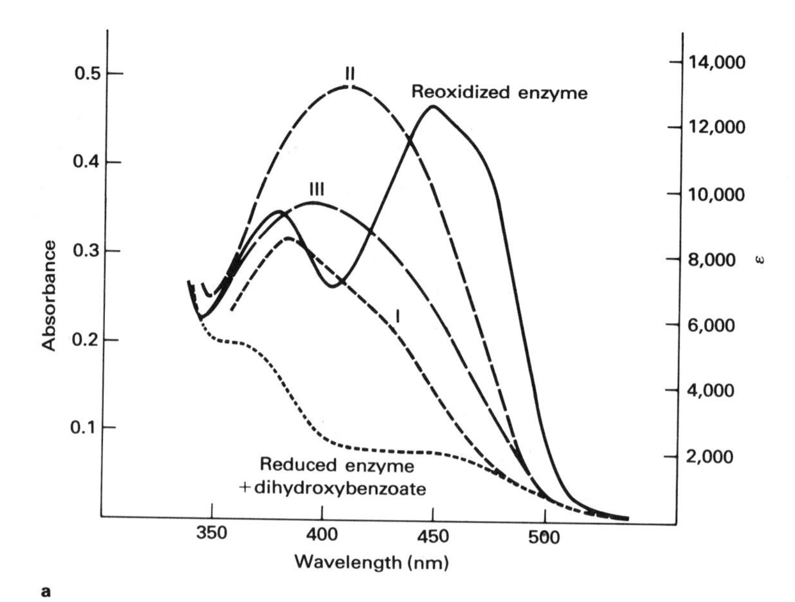

a

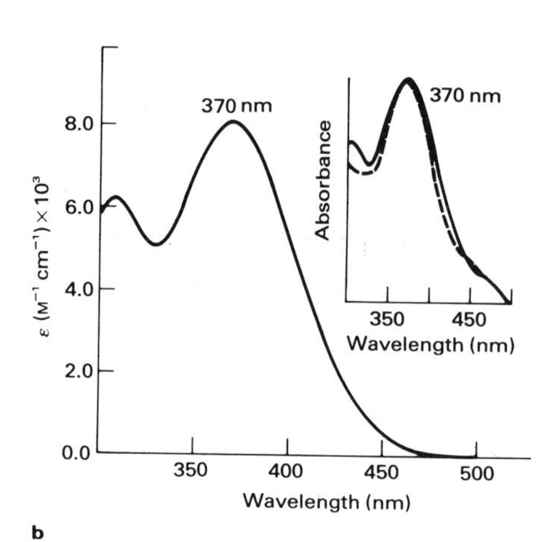

b

Figure 12-1

(**a**) Intermediates in the reaction of oxygen with reduced *p*-hydroxybenzoate hydroxylase complexed with 2,4-dihydroxybenzoate. Species I could be a 4a-peroxyflavin adduct, species II a ring-opened form, and species III a 4a-hydroxyflavin pseudo base. See Entsch, Ballou, and Massey (1976). (From Massey and Hemmerich, 1975, p. 215) (**b**) Spectrum of FlC_2H_5—OOH in MeOH. The inset shows a comparison of the spectrum of $FlCH_3$—OOH in dioxane (*solid line*) with that of the bacterial-luciferase-bound oxygenated flavin (E—FMN—OOH) intermediate (*dashed line*). The two curves were normalized at the absorption peak at 370 nm. The shoulder at about 460 nm in the $FlCH_3$—OOH spectrum is due to about 5% $Fl_{(ox)}$ impurity. It is likely that E—FMN—OOH is contaminated with a few percent of FMN. (From Kemal and Bruice, 1976, p. 996)

the phenolate. Cleavage of the O—O bond is coordinate with flavin ring opening and generation of the cyclohexadienone tautomer of the product. This oxygen-transfer step is the crucial postulate.

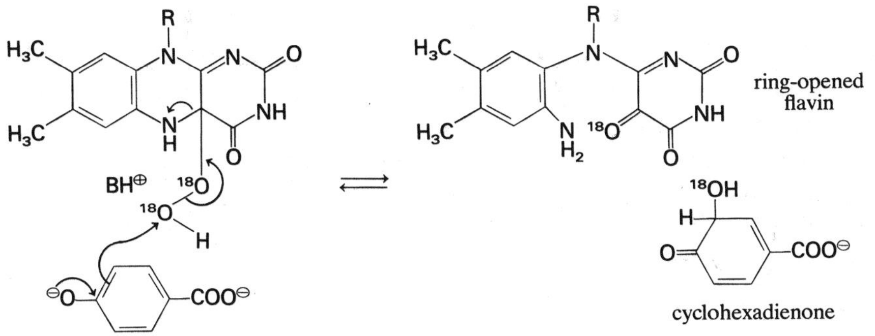

ring-opened flavin

cyclohexadienone

The ring-opened form of the flavin is in the oxidized form; ring closure will proceed via the carbinolamine to normal oxidized flavin.

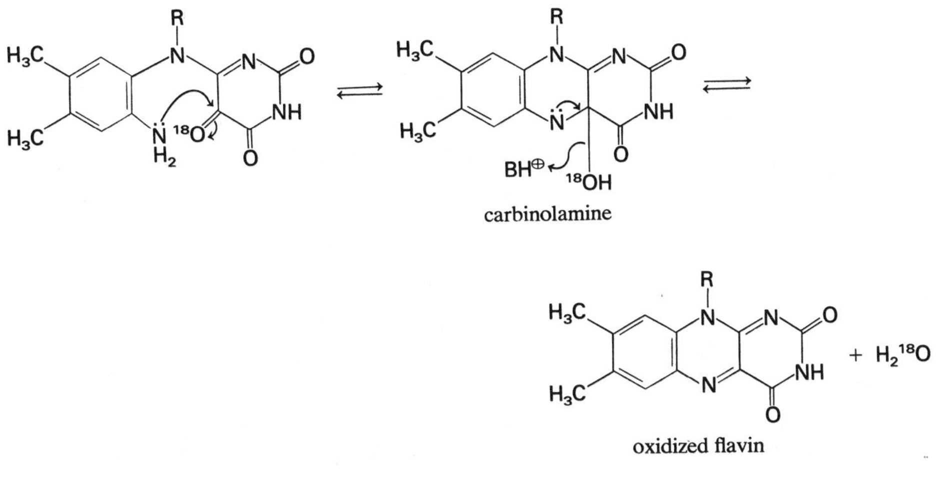

carbinolamine

$+ H_2{}^{18}O$

oxidized flavin

The cyclohexadienone form of the product will aromatize readily.

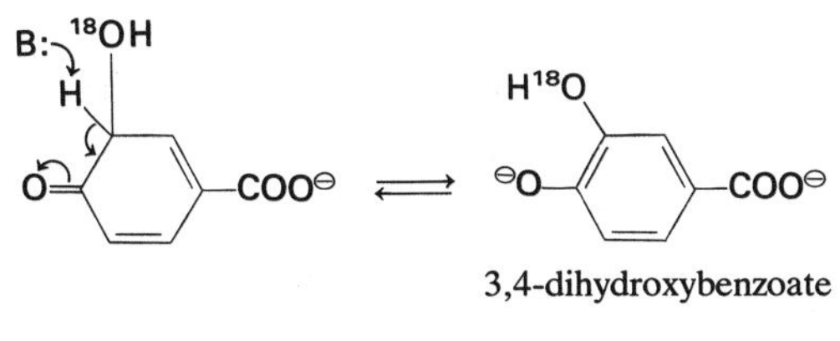

3,4-dihydroxybenzoate

Massey and colleagues tentatively suggested that the ring-opened flavin and the carbinolamine might be the second and third spectroscopically detectable species, respectively, that form transiently.

Hamilton (1974) has argued that alkyl hydroperoxides, such as the flavin hydroperoxide, in the absence of transition metal catalysts, are chemically poor hydroxylating agents. Rather, he speculates rearrangement to some other form of adduct that can deliver the needed electrophilic species of oxygen for hydroxylation. One such speculation involves a ring-opened form Hamilton terms a carbonyl oxide that, he argues, is essentially a vinylogous ozone and thus a chemically feasible oxygen-insertion agent.

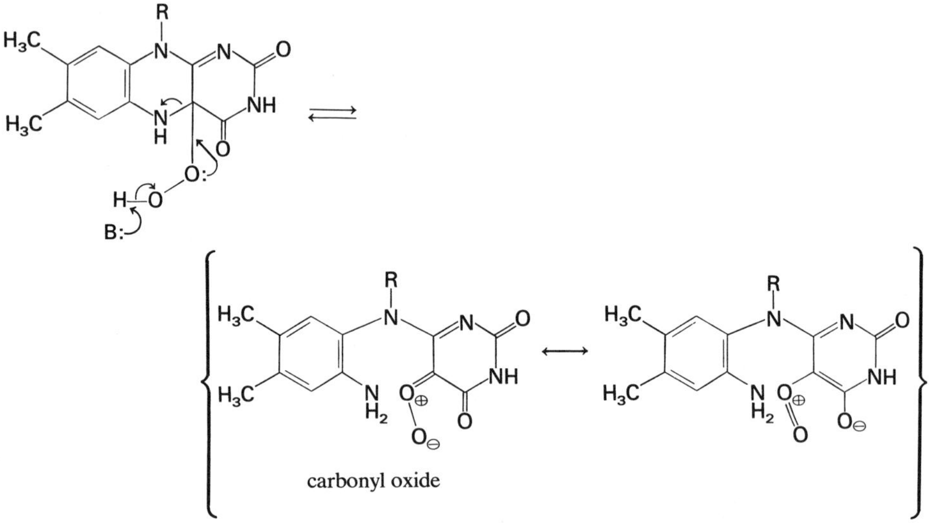

carbonyl oxide

12.C.2 Salicylate Hydroxylase

We can illustrate how such a species might operate in the reaction carried out by salicylate hydroxylase, where a *net* displacement of the salicylate $COO^{\ominus}$ occurs by the OH introduced into the product catechol (Flashner and Massey, 1974; Massey and Hemmerich, 1975; Katigiri et al., 1965; White-Stevens and Kamin, 1972).

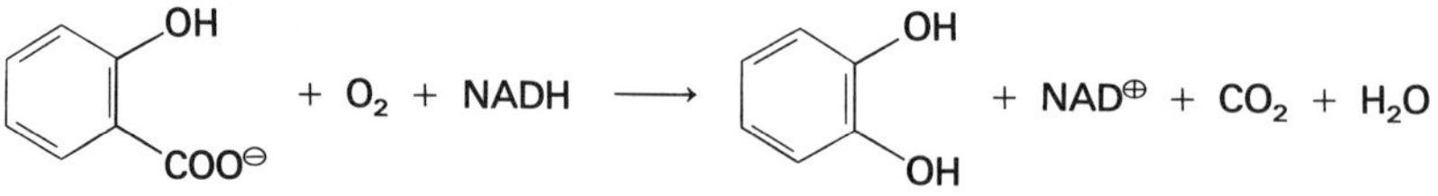

$$\text{salicylate} + O_2 + \text{NADH} \longrightarrow \text{catechol} + \text{NAD}^{\oplus} + CO_2 + H_2O$$

salicylate catechol

The proposed mechanism involving a carbonyl oxide is the following:

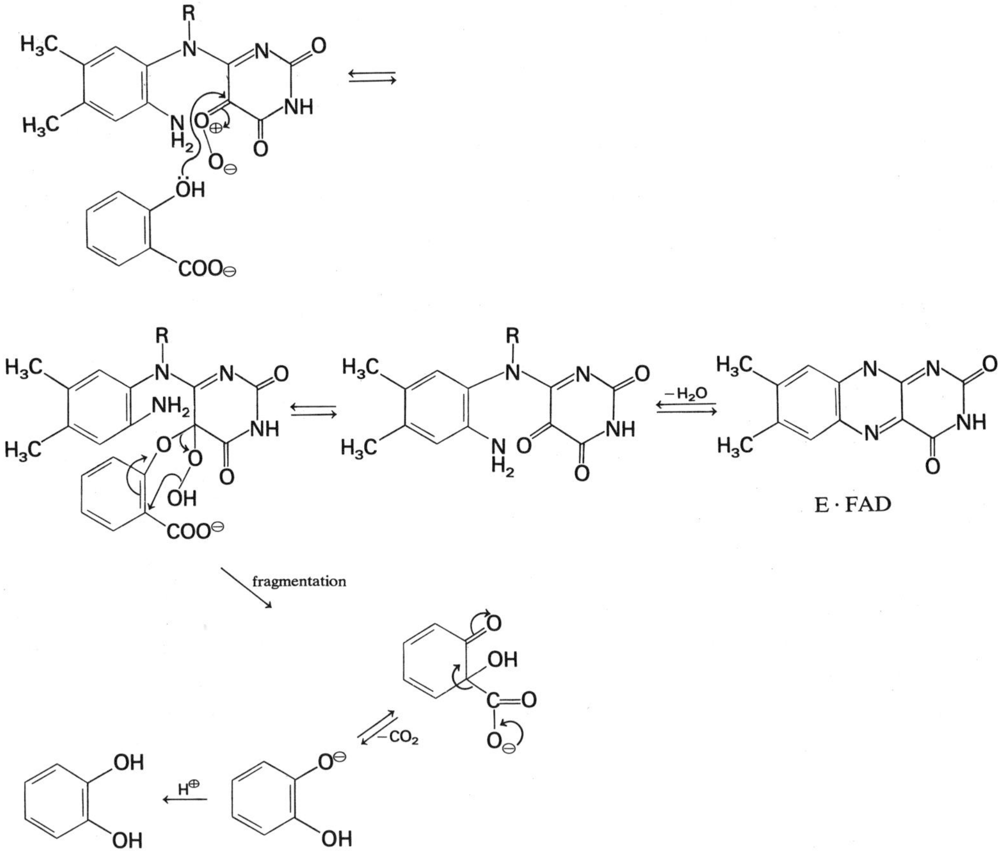

fragmentation

$-H_2O$

$-CO_2$

$H^{\oplus}$

E · FAD

This "paper chemistry" has some analogy to the scheme we have just written for *p*-hydroxybenzoate hydroxylase, in that the fragmentation shown would again yield the ring-opened form of the oxidized flavin coenzyme, which would cyclize with ring closure and loss of H_2O derived initially from O_2. The other product is likewise pictured as the cyclohexadienone tautomer of a 1,2-dihydroxybenzoate species that, as a β-keto acid, would have a low-energy path for decarboxylation because the resulting carbanion can be delocalized via enolate formation as indicated. This enolate is the phenolate anion of the observed enzymatic hydroxylated product, catechol.

On the other hand, the scheme pictures attack on the carbonyl oxide by the phenolate oxygen of the substrate, in this instance. A major problem in this formulation is that the carbonyl oxide is postulated to be captured by this oxygen nucleophile, yet the intramolecular concentration of amine nucleophile should be so high that reversion to the flavin hydroperoxide should be much more favorable.

Clearly, further experimentation is needed to determine the nature of the oxygen-transfer agent.

12.C.3 Flavoenzyme-Mediated Oxygen Transfer to Sulfur or Nitrogen

Recent studies with *p*-mercaptobenzoate and *p*-hydroxybenzoate hydroxylase have suggested that hydroxylation can occur at sulfur in that substrate. The observed product is the dimeric disulfide, with an overall stoichiometry suggesting that only one mercaptobenzoate was hydroxylated.

$$NADPH + O_2 + 2\,HS\text{-}C_6H_4\text{-}COO^{\ominus} \xrightarrow{\text{enzyme}} 2\,H_2O + NADP^{\oplus} + {}^{\ominus}OOC\text{-}C_6H_4\text{-}S\text{-}S\text{-}C_6H_4\text{-}COO^{\ominus}$$

The same transient spectroscopic intermediates are seen with both the thiol substrate and with the normal *p*-hydroxybenzoate, leading Entsch, Ballou, Hussain, and Massey (1976) to suggest that the flavin-4a-hydroperoxide is in this case intercepted by the nucleophilic sulfur anion to give a sulfenate rather than hydroxylation at C-3. Sulfenates are unstable in aqueous solutions and could react rapidly in a displacement with a molecule of excess mercaptobenzoate to produce the disulfide.

R N N N H BH$^{\oplus}$ C O O H—O + $^{\ominus}$S—C$_6$H$_4$—COO$^{\ominus}$ + BH$^{\oplus}$ $^{\ominus}$S—C$_6$H$_4$—COO$^{\ominus}$ ⟶ R N N NH$_2$ C O O + HO—S—C$_6$H$_4$—COO$^{\ominus}$ + BH$^{\oplus}$ $^{\ominus}$S—C$_6$H$_4$—COO$^{\ominus}$ ⟶ H$_2$O B: S—C$_6$H$_4$—COO$^{\ominus}$ | S—C$_6$H$_4$—COO$^{\ominus}$

The only known flavin-dependent monooxygenase found in animal cells is a membraneous enzyme in liver microsomes that oxidizes secondary and tertiary amines to hydroxylamines and *N*-oxides, respectively.

$$>NH \longrightarrow >N{-}OH \qquad -N< \longrightarrow -\overset{\oplus}{N}{-}O^{\ominus}$$

It also oxidizes thiol-containing compounds to disulfides (again presumably via sulfenic-acid intermediates)—for instance, cysteamine to the corresponding disulfide (Prough and Ziegler, 1977). Such an oxygen-dependent enzymatic formation of an unstable sulfenic-acid intermediate could be involved in directed disulfide-bond formation in specific folding of newly biosynthesized proteins (Paulson and Ziegler, 1976).

12.C.4 Ketone Monooxygenases

A further variant of flavin-dependent monooxygenases is exemplified by FAD-dependent ketone monooxygenases from various bacteria (Griffin and Trudgill, 1972; Britton and Markovetz, 1977). Simple ketones such as cyclohexanone and cyclopentanone are converted to the corresponding cyclic lactones shown (a formal insertion of an oxygen atom into a carbon–carbon bond during oxidation of the ketone substrate).

$$NADPH + {}^{18}O_2 + \text{(cyclopentanone)} \longrightarrow NADP^{\oplus} + H_2{}^{18}O + \text{(}{}^{18}O\text{-lactone)} \underset{}{\overset{H_2O}{\rightleftarrows}} \text{(}{-}COO^{\ominus}\text{, }{}^{18}OH\text{)}$$

$$NADPH + O_2 + \text{(cyclohexanone)} \longrightarrow NADP^{\oplus} + H_2O + \text{(lactone)} \overset{H_2O}{\rightleftarrows} \text{(}{-}COO^{\ominus}\text{, }OH\text{)}$$

The cognate model reaction is the Baeyer–Villiger oxidation of cyclic ketones to the same ring-expanded cyclic lactones by hydrogen peroxide. This analogy suggests a 4a-peroxyflavin intermediate acting as oxygen-transfer agent:

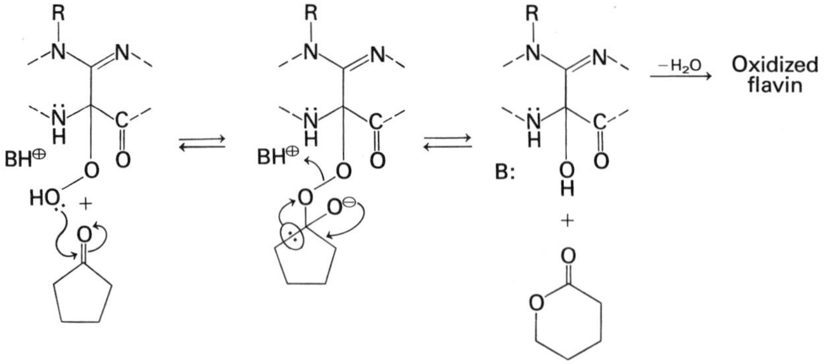

$\xrightarrow{-H_2O}$ Oxidized flavin

This scheme pictures initial *nucleophilic* attack by the distal oxygen of the peroxyflavin adduct on the ketone carbonyl carbon. The resultant tetrahedral adduct collapses with migration of the electrons in the adjacent C—C bond to oxygen and with expulsion of the 4a-pseudo-base of oxidized flavin. It is worth explicit notice that this ketone monooxygenation is distinct from phenolic ring hydroxylations (e.g., salicylate, *p*-hydroxybenzoate, phenol) or oxygen transfer to the thiol function of mercaptobenzoate because in those cases the distal oxygen of the peroxyflavin adduct is acting formally as an *electrophilic* oxygen. A clear functional reversal of polarity in the two variants suggests the flexibility of the peroxyflavins as monooxygenating agents.

12.D BIOLUMINESCENCE

A variety of organisms, from fireflies to coelenterates to marine bacteria, carry out enzyme-catalyzed production of light. The enzyme-mediated light-forming reaction is an oxidative process and represents an interesting variant of monooxygenases. Light is generated from an enzymatic product in its first excited singlet electronic state. Advances have come in recent years through identification of the structure of the luciferins (the light-generating molecules) and analysis of their chemistry. The bacterial luciferase, a luciferase from limpets (Shinomura and Johnson, 1978), and one from earthworms (Ohtushka et al., 1976) are the only flavin-dependent monooxygenases. We shall comment on the bacterial luciferase after a short digression on the light-emitting systems in coelenterates and fireflies (Cormier et al., 1973; McCapra, 1976).

12.D.1 Bioluminescence in Coelenterates and Fireflies

Several of the luciferins of coelenterates have an imidazolopyrazine ring system, such as the one from the sea pen *Renilla.*

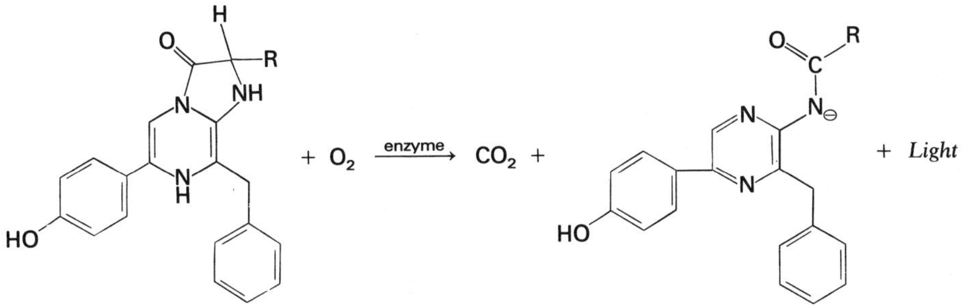

Renilla luciferin + O_2 $\xrightarrow{\text{enzyme}}$ CO_2 + oxyluciferin anion + *Light*

Reaction with O_2 and the *Renilla* luciferase yields blue light ($\lambda_{max} = 490$ nm) with a 5% quantum yield. The products are CO_2 and oxyluciferin. Analogous nonenzymatic oxidative reaction in DMF yields a similar chemiluminescence due to the same oxyanion. Cormier (1975) and colleagues have summarized evidence for the following scheme for enzymatic and nonenzymatic light emission, in which initial proton abstraction generates the carbanion that attacks O_2. They claim that ^{18}O-labeling studies in $[^{18}O]$-H_2O and $^{18}O_2$ favor either pathway (1) or the dioxetanone pathway (2), depending on the organism used.

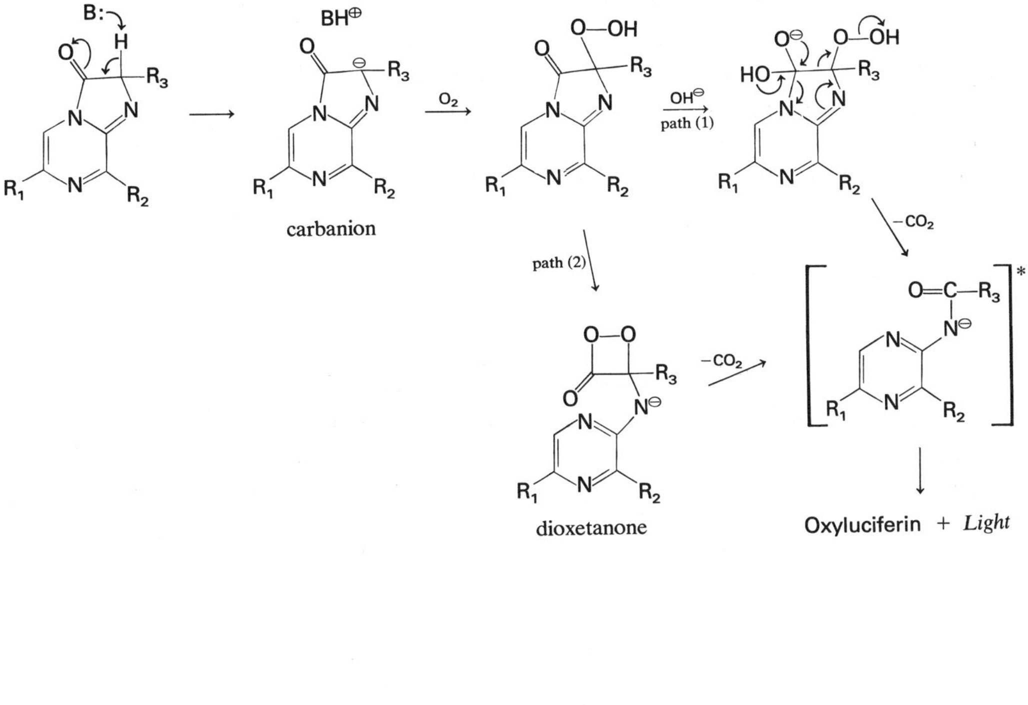

carbanion $\xrightarrow{O_2}$ ... $\xrightarrow{OH^{\ominus}}$ path (1) ... $-CO_2$

path (2) → dioxetanone $\xrightarrow{-CO_2}$ [ ]*

Oxyluciferin + *Light*

In the firefly system, the luciferin is first activated by adenylyl transfer to the substrate carboxylate group, producing a bound luciferyl-AMP intermediate. Molecular oxygen is then used to generate an oxyluciferin in excited state, which gives off light and returns to ground state.

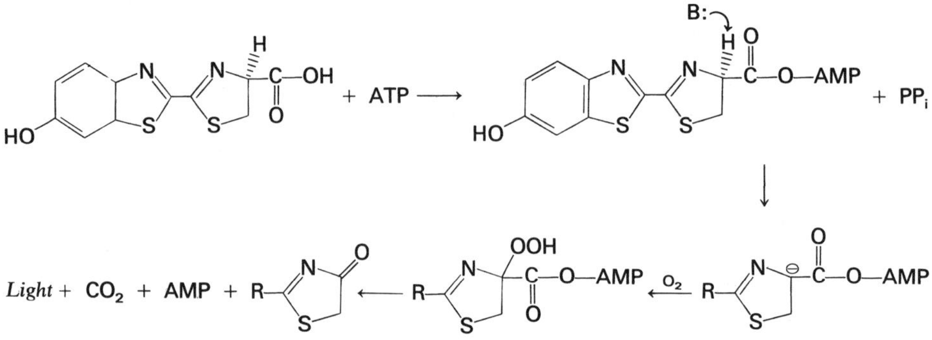

A mechanism similar to that written for the *Renilla* luciferin is favored, involving initial carbanion formation at the carbon α to the adenylate group, attack by the carbanion on O_2, and subsequent rearrangement and peroxidative fission to the excited-state ketone, CO_2, and AMP. (This could be initiated by addition of the peroxy group to the acyl carbonyl, followed by collapse of the tetrahedral adduct to the ketone excited state and transient mixed anhydride of CO_2 and AMP, which would hydrolyze rapidly.) These enzymatic oxygenation/decarboxylation sequences occur without redox-active metal-ion or conjugated-cofactor assistance and so are *quite unusual* (see also the monooxygenase activity of ribulose-diphosphate carboxylase in Chapter 22). McCapra (1976) has noted that these processes look like directed autoxidation (by O_2) of the substrate carbanions (probably by one electron to the substrate radical anion and $O_2^{\ominus\cdot}$, which could recombine to the indicated peroxy intermediates).

12.D.2 Bacterial Luciferase

The enzyme responsible for bioluminescence in marine bacteria has been better characterized than have the coelenterate or firefly enzymes. The bacterial luciferase has 80,000 mol wt and is a metal-free dimer of unequal subunits. The enzyme binds $FMNH_2$ at one subunit and, on addition of O_2 and a long-chain aldehyde, catalyzes light formation with λ_{max} from 480 to 505 nm (Hastings, Eberhard, et al., 1973).

$$O_2 + RCHO + FMNH_2 \xrightarrow{\text{luciferase}} \textit{Light} + FMN + H_2O + RCOO^{\ominus}$$

Luciferase is thus an unusual flavoenzyme because it binds oxidized FMN poorly (10^3-fold less well than $FMNH_2$) and cannot be isolated as a holoenzyme. In the intact luminous bacterium, a feeder enzyme supplies the $FMNH_2$ at the expense of NADH oxidation. This NADH:FMN oxidoreductase is also unusual in using flavins as freely dissociating substrates rather than as nondissociable bound coenzymes. In view of the rapid autoxidation rate of free $FMNH_2$ to FMN ($T_{1/2} \approx 1$ sec), it is likely that oxidoreductase and luciferase form a functional complex in vivo.

The complete reaction stoichiometry of luciferase catalysis appears to be established as a flavin-linked monooxygenase. The aldehyde undergoes 2-$e^{\ominus}$ oxidation to the carboxylic acid. But no H_2O_2 is detected; rather O_2 appears to yield H_2O. This would be a 4-$e^{\ominus}$ reduction, and the second electron pair comes from $FMNH_2$.

$$RCHO + O_2 + FMNH_2 \longrightarrow RCOOH + H_2O + FMN + H^{\oplus} + h\nu$$

Recent low-temperature experiments (−30° C, in mixed aqueous-organic solvents) by Hastings and Douzou have allowed isolation of enzyme-bound 4a-peroxy-FMN. On warming the solution in the presence of aldehyde, oxidation to the acid occurs with production of light at the usual efficiency of 10% quantum yield (Fig. 12-1b); Hastings, Balny, et al., 1973; Hastings and Balny, 1975, 1976). The oxygen-transfer chemistry is reminiscent of the ketone-monooxygenase type (¶12.C.3) rather than the aromatic ring hydroxylations and suggests that the distal oxygen of the flavin peroxide must behave formally as a nucleophile. However, a simple Baeyer–Villiger process does not seem to explain the observed facts. For example, stopped-flow kinetic studies with E·FMN—OOH and 1-[^{2}H]-aldehyde (decanal) show a large deuterium kinetic isotope effect on formation of a transient, spectroscopically detectable intermediate ($k_H/k_D = 5.4$) but a much smaller effect ($k_H/k_D = 1.6$) on light production (Shannon et al., 1978). This suppression of isotope effect could be interpreted to mean that cleavage of the C—H (C—^{2}H) bond at C-1 in aldehyde oxidation occurs in a step separate from and prior to formation of the bound light-emitting species, ruling out concerted mechanisms. A mechanistic speculation proposing a nonconcerted oxidative process has been proposed to involve binding of aldehyde in imine linkage at N-5 of the 4a-peroxyflavin, followed by a series of steps to produce a cyclic four-membered oxazetidine, which could fragment in a 2+2 cycloreversion, producing FMN in an excited state (as the emitter) and the carboxylate product (Lowe et al., 1976). The following is a variant of this hypothetical scheme.

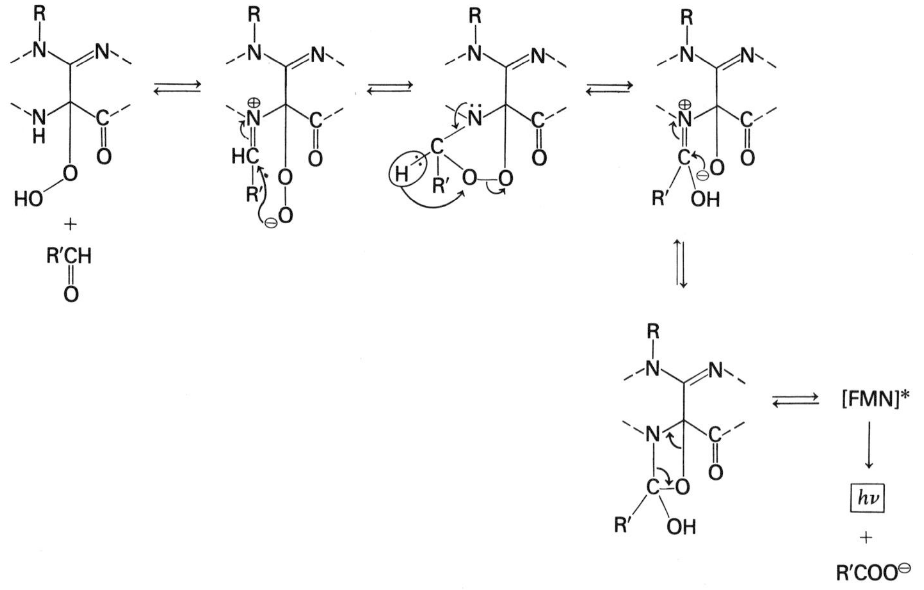

Model studies by Bruice and colleagues, however, show that light emission (in quite low yield) can occur with N^5-methyl-4a-peroxyflavins and aldehydes, arguing against necessary involvement of the nitrogen at position 5 in luminescence (Kemal and Bruice, 1976).

12.E MONOOXYGENASES REQUIRING PTERIN COENZYMES

The flavin-dependent hydroxylases discussed so far (with one exception noted in ¶12.C.3) are restricted to bacteria and eukaryotic microorganisms. There are monooxygenase reactions in animal cell metabolism, but they do not in general involve flavin enzymes in the actual hydroxylation step. The three main types of hydroxylases that do function in mammalian cells are (1) pterin-requiring monooxygenases, (2) copper-requiring monooxygenases, and (3) iron-requiring monooxygenases. Categories 2 and 3 are the topics of Chapters 14 and 15; we shall take up the first category here.

The three known pterin-dependent hydrolyases in animal cells* are—

1. phenylalanine hydroxylase,
2. tyrosine hydroxylase,
3. tryptophan hydroxylase.

* Other pterin- and Fe^{II}-requiring hydroxylases have recently been purified from microorganisms (C. Walsh, 1978).

Note that 1 and 3 represent hydroxylations of *unactivated* aromatic substrates and are, in this sense, distinct from the flavin-dependent hydroxylations. The phenylalanine hydroxylase has been the most extensively characterized, and essentially all mechanistic information about this category derives from studies with that enzyme (Flashner and Massey, 1974; Massey and Hemmerich, 1975; Kaufman and Fisher, 1974; Kaufman, 1976). All three enzymes are important in metabolism of neuroactive amines. Phenylalanine is converted to tyrosine and then to 3,4-dihydroxyphenylalanine (dopa) by consecutive action of enzymes 1 and 2. Dopa is the precursor for the neurotransmitters epinephrine (adrenalin) and norepinephrine (noradrenalin).

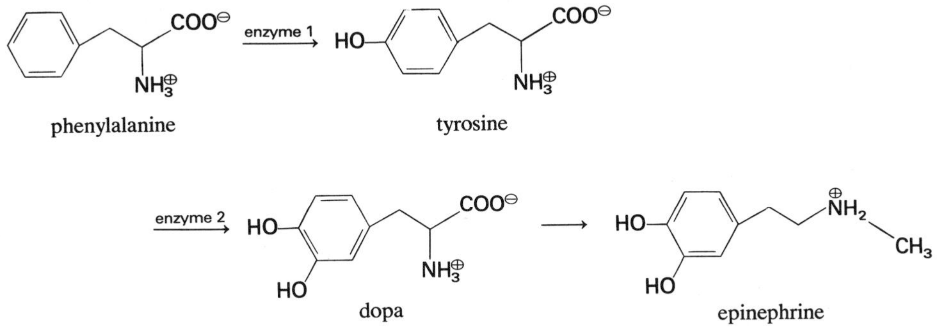

phenylalanine —enzyme 1→ tyrosine

—enzyme 2→ dopa ⟶ epinephrine

Tryptophan hydroxylase (enzyme 3) hydroxylates its substrate at C-5 to produce 5-hydroxytryptophan, which is precursor to the central-nervous-system transmitter, serotonin (5-hydroxytryptamine).

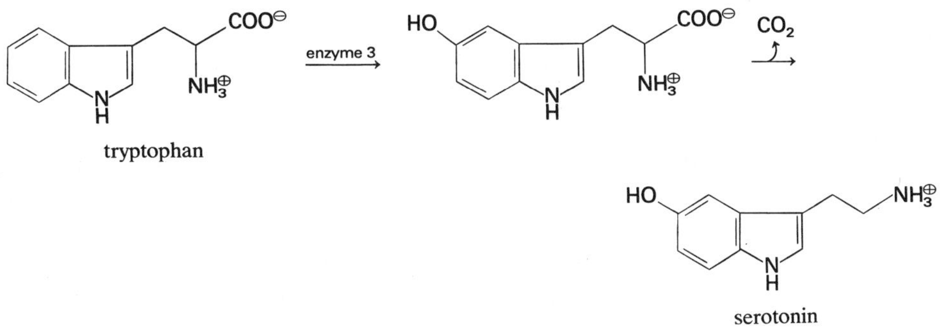

tryptophan —enzyme 3→ 5-hydroxytryptophan —(CO_2)→ serotonin

There is a great amount of intrinsic interest in how phenylalanine hydroxylase produces tyrosine because the enzyme is inactive in the genetically determined disease phenylketonuria, a disease producing severe mental retardation and characterized by elevated blood and urinary levels of phenylalanine. Immunoassays for cross-reacting protein in phenylketonuric individuals revealed, in one case, 0.25% of the normal catalytic activity, apparently reflecting an altered enzyme arising from a mutation in the structural gene for the enzyme (Stanbury et al., 1972). In liver, the activity of phenylalanine hydroxylase is regulated in at least two ways: (1) by covalent phosphorylation/dephosphorylation, and (2) by tight noncovalent interaction with a protein inhibitor. The enzyme/protein inhibitor can be purified together and then dissociated by detergent treatment or by protease action (Kaufman and Fisher, 1974).

The structure of the pteridine cofactor involved in these hydroxylations has similarities to the isoalloxazine ring system of flavins, possessing a structure similar to the B+C rings of that chromophore (the dispositions of the four ring nitrogens are identical).

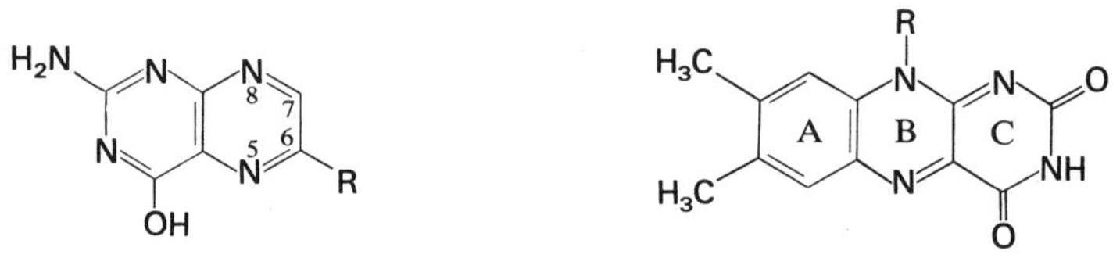

pteridine ring

flavin ring

The pteridine ring can be reduced chemically or in vivo to the 7,8-dihydro isomer (shown below for biopterin, the natural coenzyme in native phenylalanine hydroxylase). Also shown is an isomeric dihydro form, the quinoid dihydropterin, which isomerizes to the other form in a few minutes under physiological conditions.

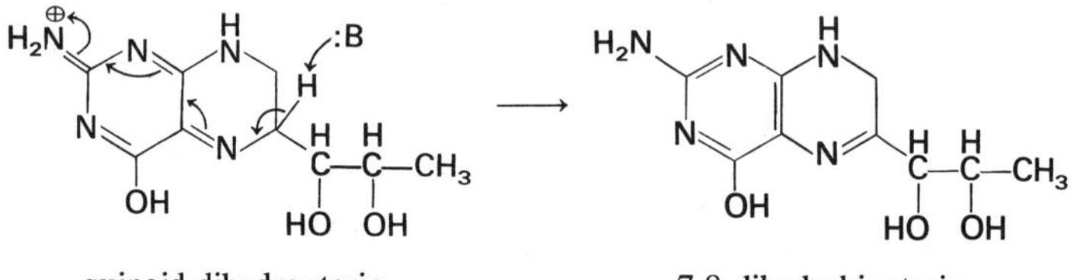

quinoid dihydropterin

7,8-dihydrobiopterin

However, in either dihydro oxidation state, the coenzyme is inactive for hydroxylation. It is functionally equivalent to oxidized flavin. Rather, the coenzyme must be converted to the tetrahydro form before it will combine with O_2 to form an active hydroxylating species. The tetrahydro form is generated enzymatically by an NADH-dependent dehydrogenase, a dihydropteridine reductase. The tetrahydrobiopterin is functionally equivalent to dihydroflavin in that this is the oxidation state used in the hydroxylation step, although it is much less labile to autoxidation than is 1,5-dihydroflavin. The dihydropteridine reductase is specific for the quinoid dihydropterin as substrate; it will not reduce 7,8-dihydrobiopterin.

$NADP—^3H$ + quinoid dihydropterin $\xrightarrow[\text{reductase}]{\text{dihydropteridine}}$ $NADP^{\oplus}$ + 5,6,7,8-tetrahydrobiopterin

The stoichiometry for phenylalanine hydroxylase is described by the following equation. (Note that the product of hydroxylase catalysis is the *unstable* dihydroquinoid.)

Tetrahydropterin + $^{18}O_2$ + phenylalanine $\xrightarrow{\text{enzyme}}$ $H^{18}O$–tyrosine + $H_2{}^{18}O$ + Quinoid dihydropterin

Unlike the tightly bound flavin coenzymes (and more analogous to nicotinamides), the pterin cofactors dissociate from the enzyme at the end of each catalytic cycle and are regenerated by action of the dihydropteridine reductase. If we consider the consecutive action of the hydroxylase/reductase couple, the overall stoichiometry is now analogous to that of the flavin monooxygenase.

NADPH + O_2 + phenylalanine $\xrightarrow[H_4\text{-pterin}]{\text{reductase, hydroxylase}}$ $NADP^{\oplus}$ + H_2O + HO–tyrosine

Augmenting the similarities is the observation that uncoupling of substrate hydroxylation occurs if either substrate or coenzyme structure is modified (Table 12-2). The ratio of NADPH oxidized to tyrosine formed increases to a ratio greater than one, the excess O_2 undergoing reduction by 2 $e^{\ominus}$ to H_2O_2 (Kaufman and Fisher, 1974). This is the same kind of partitioning of an "oxygenating intermediate" as noted for *p*-hydroxybenzoate hydroxylase. We might speculate that the hydroxylating species in the phenylalanine-hydroxylase case might also be

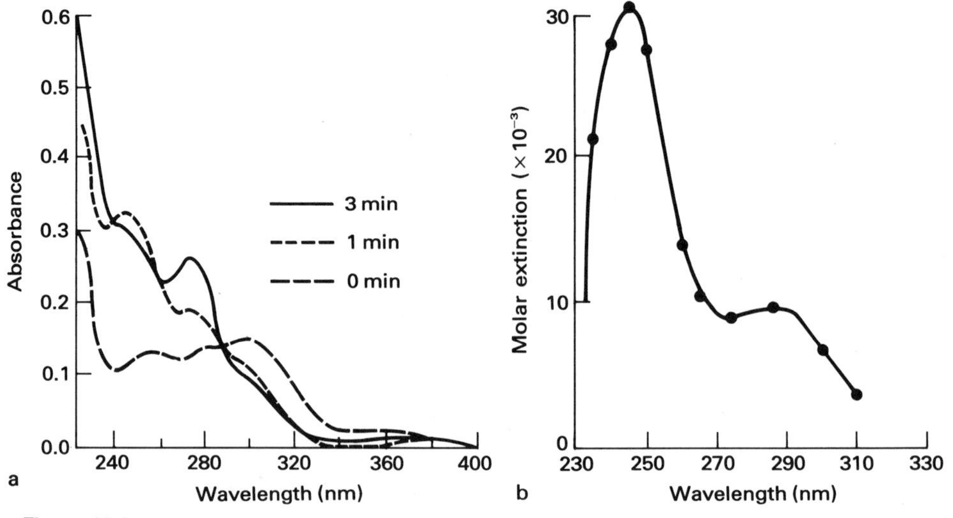

Figure 12-2
(**a**) Spectral changes during the enzymatic conversion of phenylalanine to tyrosine. The complete reaction mixture contained the following components (in μmoles) in a final volume of 1.0 ml: potassium phosphate (pH 8.2), 30; TPNH, 0.08; phenylalanine, 1.2; tetrahydrobiopterin, 0.015; glucose-6-P, 6.0; phenylalanine hydroxylase (90% pure), 80 μg; dihydropteridine reductase, catalase, and glucose-6-P dehydrogenase, in excess. Temperature 22° to 23° C. The control cuvette contained the same components without tetrahydrobiopterin. (From Kaufman, 1976, p. 94) (**b**) The absorption spectrum of the intermediate. The reaction mixture was the same as that described for part a. At each wavelength, the observed changes in absorption have been corrected (using known extinction coefficients) for the contribution due to the conversion of phenylalanine to tyrosine and for the disappearance of tetrahydrobiopterin. (From Kaufman, 1976, p. 96)

a peroxy-coenzyme adduct (Fig. 12-2). The intermediate detected also could be a hydroxypterin adduct accumulating after the hydroxylation step. One can write a pterin tautomer that could lead to an analogous pterin hydroperoxide (Kaufman, 1976).

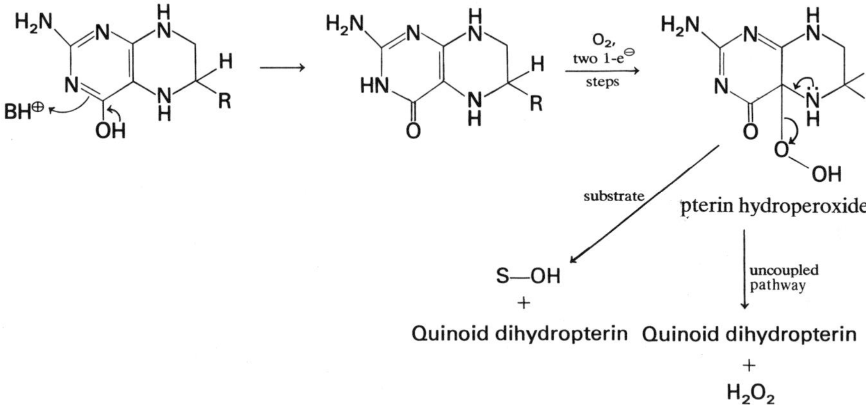

O_2, two 1-$e^{\ominus}$ steps

substrate

pterin hydroperoxide

S—OH + Quinoid dihydropterin

uncoupled pathway

Quinoid dihydropterin + H_2O_2

Table 12-2
Uncoupling of phenylalanine hydroxylase by substrate or coenzyme analogues

Substrate	Pterin coenzyme	(NADPH oxidized) / (tyrosine formed)
Phenylalanine	6-Methyltetrahydropterin	1.1
Phenylalanine	7-Methyltetrahydropterin	5.31
4-Fluorophenylalanine	6-Methyltetrahydropterin	5.11

This hydroperoxide might behave as an oxygen-transfer agent by analogy to one of the schemes listed for *p*-hydroxybenzoate hydroxylase, or it could decompose to H_2O_2 and quinoid dihydropterin. If such a species is involved in oxygen transfer, one must note that (in contrast to the flavin mechanism) phenylalanine has no hydroxyl group on the aromatic ring to initiate attack leading to fission of the O—O bond in the peroxide or to a ring-opened "pterin carbonyl oxide." We shall return to this point in a moment.

Three issues cloud the assumption that hydroxylations mediated by pterin coenzymes proceed in exact analogy with the FAD-dependent processes. First, tetrahydropterins are (compared to dihydroflavins) quite sluggish to O_2-induced oxidation ($T_{1/2}$ in minutes rather than milliseconds) (Flashner and Massey, 1974; Massey and Hemmerich, 1975); the apoproteins would have to accelerate this conversion dramatically. Second, homogeneous phenylalanine hydroxylase contains two atoms of $Fe^{3\oplus}$, as yet structurally uncharacterized as to ligands (it is not heme iron), per active site (Kaufman and Fisher, 1974). The iron is redox active, shuttling from $Fe^{3\oplus}$ to $Fe^{2\oplus}$ oxidation state during catalysis. It may well be that some *oxoiron* complex is the actual oxygen-transfer species at the active site. The role of the tetrahydropterin cofactor could be ancillary, as a reductant to generate $Fe^{2\oplus}$ at the active site. As we shall note in Chapter 15, an $Fe^{3\oplus} \cdots O_2^{\ominus\cdot}$ complex could possibly serve as an oxygen-insertion reagent. The third issue is the subject of the following section.

12.F THE N.I.H. SHIFT

An additional set of mechanistic data is available that distinguishes phenylalanine hydroxylase from flavin-mediated hydroxylases. When 4-[^{3}H]-phenylalanine is used as a substrate, the product tyrosine is tritiated, with the ^{3}H having undergone migration to the 3-position, rather than simply being lost to solvent.

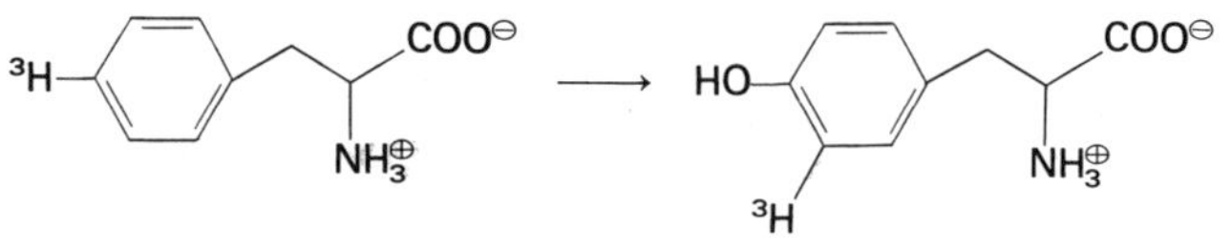

This type of substituent migration is also detected if, rather than ^{3}H in the *para* position of the substrate, the substituent is ^{2}H, Cl, Br, or an alkyl group. This intramolecular migration of a substituent during hydroxylation was first detected by Witkop, Daly, and Jerina at the National Institutes of Health (N.I.H.) in hydroxylations carried out by poorly characterized, membranous microsomal hydroxylases (Guroff et al., 1967; Jerina, 1973); it has been dubbed the "N.I.H. shift." An exhaustive study has linked this type of experimental result with the intermediacy of an epoxide species (an arene oxide). In particular, peracids (e.g., *m*-chloroperbenzoic acid) are chemical hydroxylating agents that form intermediate arene oxides with aromatic substrates. The N.I.H shift occurs in these chemical reactions. The model catalysis involves electrophilic substitution and then epoxide formation.

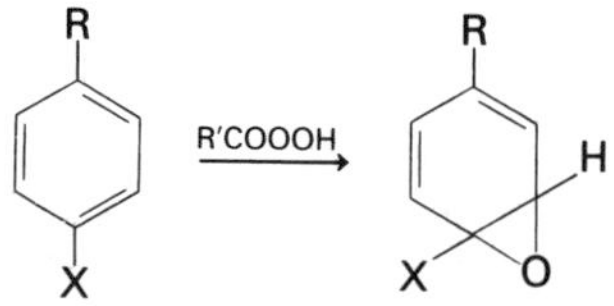

Acid-catalyzed ring opening of the epoxide would generate an initial C^{3}-carbonium ion, which can undergo a hydride shift to the more stable oxonium ion.

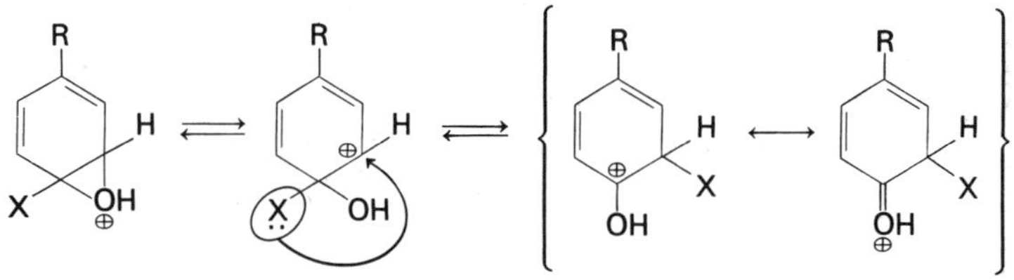

oxonium ion

The oxonium ion then deprotonates easily to the cyclohexadienone, which can now aromatize to the phenolic product. Such aromatization requires abstraction of a hydrogen at C-3. When X = ^{2}H (or ^{3}H), the ^{1}H proton will be abstracted because of the lower energy barrier for C—^{1}H bond breakage; this will cause a kinetic isotope effect in this step and a selective retention of the heavier isotope of hydrogen (^{2}H or ^{3}H), consistent with the observed results.

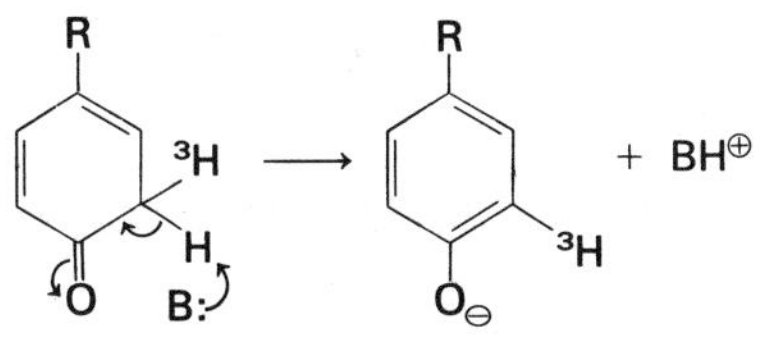

Given that an arene oxide exists in the nonenzymatic reaction (and in the microsomal hydroxylase systems responsible for aromatic hydrocarbon hydroxylation and detoxification that are discussed in Chapter 15), it is as likely that a discrete epoxide indeed forms during enzymatic phenylalanine hydroxylation, but is rapidly processed to the phenolic product *before* dissociation from the active site can occur.

In the phenylalanine-hydroxylase case, an epoxide intermediate would presumably have the following structure:

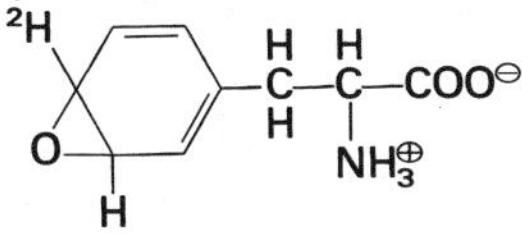

(It is known that oxoiron-containing enzymes will act as epoxidizing agents.) The postulation of a possible epoxide intermediate, in this case, raises the question of whether discrete epoxides might form in flavin-linked hydroxylases and why there is no N.I.H. shift observed. One cannot rule out an α-hydroxy-epoxide species as a catalytic entity in the flavin-dependent hydroxylases. It would presumably always open as indicated, so that no N.I.H. shift would be expected.

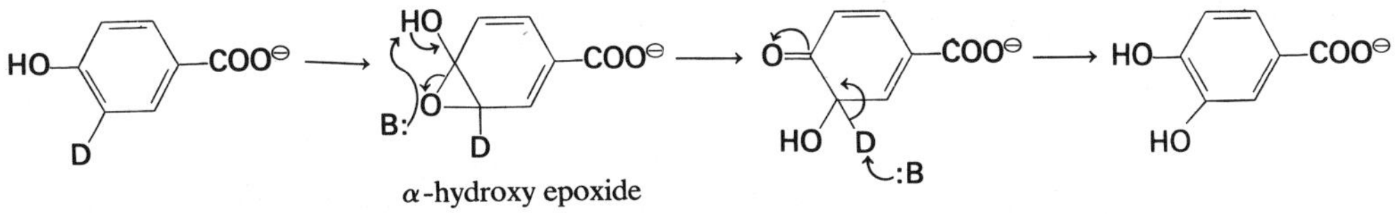

α-hydroxy epoxide

One might expect a deuterium kinetic isotope effect on 3-[^{2}H]-4-hydroxybenzoate hydroxylation, and a small one ($k_H/k_D = 1.45$) has been detected (Entsch, Ballou, and Massey, 1976).

Chapter 13

Metalloflavoprotein Oxidases and Superoxide Dismutase

13.A METALLOFLAVOPROTEIN OXIDASES

Xanthine oxidase and aldehyde oxidase are two well-characterized flavoenzymes that also contain transition metals required for catalysis: both molybdenum and iron are present (Massey, 1973; Bray, 1975). The stoichiometry of the reaction catalyzed by either enzyme is the following, where X = N or O.

$$\underset{\displaystyle \overset{\Vert}{X}}{R{-}C{-}H} + H_2O + O_2 \longrightarrow \underset{\displaystyle \overset{\Vert}{X}}{R{-}C{-}OH} + H_2O_2$$

For xanthine oxidase, it is the C-8 position that undergoes oxidation in production of uric acid, which is then excreted.

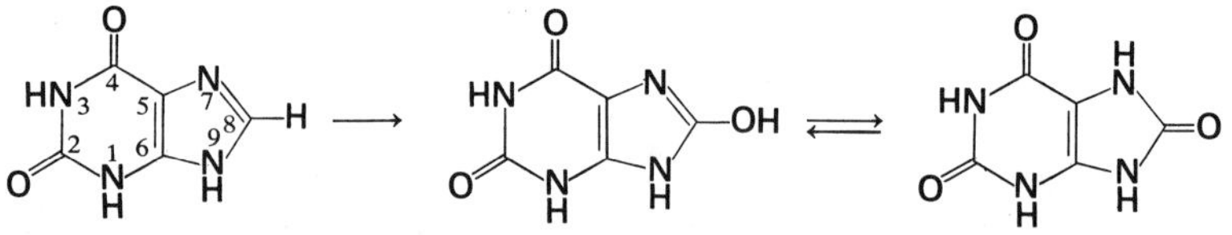

xanthine → uric acid ⇌ (more stable keto tautomer)

High levels of urate are characteristic in gout, where deposition of calcium urate crystals in synovial fluid of joints can contribute to some of the acute pain of the disease (Stanbury et al., 1972). One clinically effective drug, allopurinol, is a powerful inhibitor of xanthine oxidase.

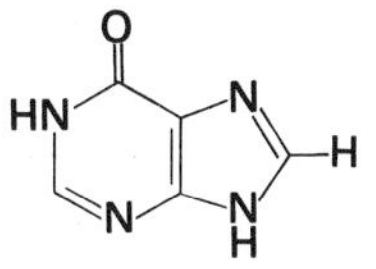

hypoxanthine

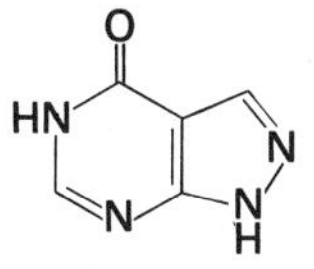

allopurinol
(a pyrazolo pyrimidine)

Hypoxanthine is a physiological substrate of the enzyme, being oxidized to xanthine and then to uric acid, proving that xanthine oxidase can introduce H_2O-derived hydroxyl (keto) groups at C-2 as well. This feature is the key to the mechanism of allopurinol, which will be addressed later in this chapter.

The stoichiometry of the xanthine-oxidase reaction suggests oxidation of the aldehyde substrate by 2 $e^{\ominus}$ to the level of an acid, with introduction of a hydroxyl group into the product at the locus of oxidation; in this sense, the reaction might qualify a priori as a monooxygenation, but experiments with $^{18}O_2$ dispelled this idea. With $^{18}O_2$, all the ^{18}O-label ends up in $H_2{}^{18}O_2$ and none in the acid product, suggesting a straightforward 2-$e^{\ominus}$ reduction of O_2, analogous to a simple flavoprotein oxidase. Complementary experiments in $H_2{}^{18}O$ show that the reaction water is the source of the —OH introduced into the product, so the mechanism clearly must be distinct from the flavin- and pterin-requiring hydroxylases we have just discussed.

Xanthine oxidase has been purified to homogeneity from buttermilk (and other sources). The milk enzyme is a dimer of 260,000 mol wt. Each subunit has one molecule of FAD and one molecule of molybdenum ion at the Mo^{VI} oxidation state; there are also four atoms of iron and an equivalent amount of inorganic sulfide releasable with acid and so termed acid-labile sulfur. To understand the possible redox role or the iron and sulfur components, we must digress a moment to introduce (and categorize as biological redox elements) the nonheme iron–sulfur clusters.

13.A.1 Nonheme Iron–Sulfur Chromophores

The presence of iron (not associated with the heme porphyrin macrocycles) and equimolar amounts of sulfide (accompanied by long-wavelength, visible absorbance) characterize iron–sulfur chromophores that act as *one-electron redox shuttles.* Iron–sulfur proteins can be categorized on the basis of iron content (Lovenberg, 1973; Palmer, 1975), as we shall briefly note here.

13.A.1.a One iron per center

The anaerobic bacterium *Clostridium pasteurianum* elaborates a low-molecular-weight (~6,000 mol wt) reddish protein, rubredoxin, which has one Fe atom liganded to the sulfur atoms of four cysteine residues. There is no inorganic sulfide in this unique representative. The X-ray structure has been solved by Jensen (1974) and colleagues and is shown in Figure 13-1, indicating a distorted tetrahedron for this iron–sulfur cluster. The reduction potential is an amazing −0.57 V for $Fe^{3\oplus} \rightleftarrows Fe^{2\oplus}$ (the normal $Fe^{3\oplus} \rightleftarrows Fe^{2\oplus}$ potential is +0.77 V). Although the redox function is unclear in *C. pasteurianum*, an analogous rubredoxin from *Pseudomonas oleovorans* is an electron carrier in alkane hydroxylations (Chapter 15; Loge and Coon, 1973).

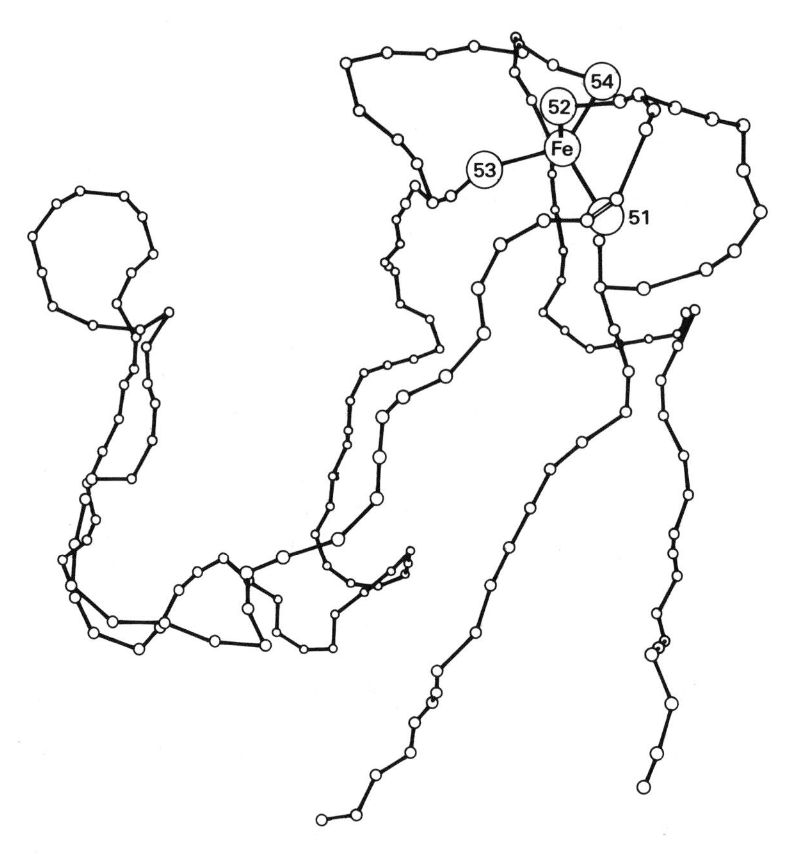

Figure 13-1
Chain model of the three-dimensional structure of rubredoxin. (From Palmer, 1975, p. 5)

13.A.1.b Two irons per center

The archetypal example of the two-iron category is spinach ferredoxin (ferredoxin = "iron-containing redox material") of 10,000 mol wt, containing two iron atoms and two atoms of inorganic sulfur. Figure 13-2 shows the spectra of oxidized and 1-e$^{\ominus}$–reduced spinach ferredoxin.

Included in this category are structurally analogous proteins purified from mitochondria of adrenal cells (adrenodoxin) and from the bacterium *Pseudomonas putida* (putidaredoxin), both of these function in multicomponent monooxygenase systems described in Chapter 15.

Spinach ferredoxin is thought to accept electrons from a chlorophyll reaction center and pass electrons on to NADP$^{\oplus}$, generating NADPH. Because the reduction potential for NADPH/NADP$^{\oplus}$ is −0.32 V (Table 10-1), the ferredoxin potential would have to be more negative to function as suggested—and it is (−0.42 V).

Chlorophyll $\xrightarrow{h\nu}$ Excited chlorophyll → (1 e$^{\ominus}$) → Ferredoxin-Fe$^{3\oplus}$ → Ferredoxin-Fe$^{2\oplus}$; Excited chlorophyll → Oxidized chlorophyll (1)

$$2\ \text{Ferredoxin-Fe}^{2\oplus} + \text{NADP}^{\oplus} \xrightarrow{2\text{-e}^{\ominus}\ \text{transfer}} 2\ \text{Ferredoxin-Fe}^{3\oplus} + \text{NADPH} \quad (2)$$

On 1-e$^{\ominus}$ reduction, there is one Fe$^{3\oplus}$ and one Fe$^{2\oplus}$ in the cluster. Exact synthetic analogues of these (2 Fe)/(2 S) clusters have been synthesized and characterized by Holm's group (Holm, 1977).

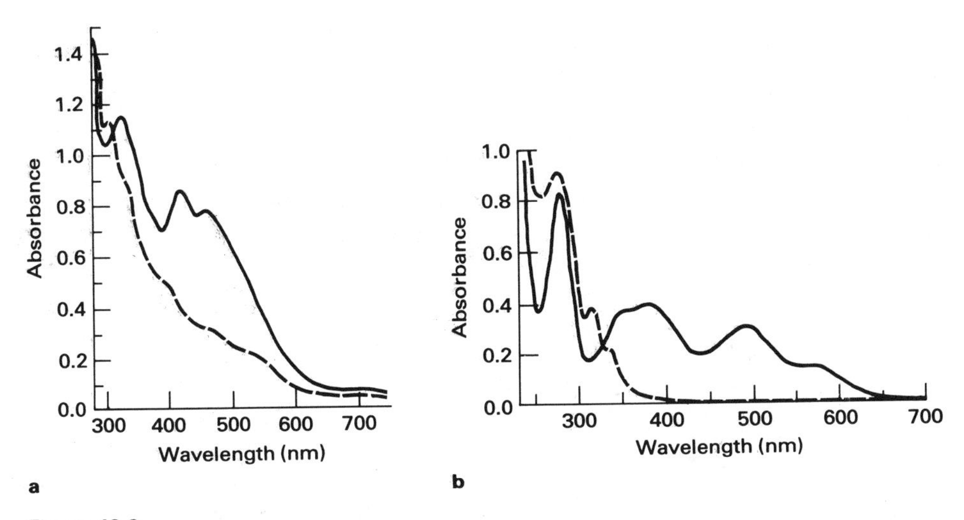

Figure 13-2
(**a**) The optical spectra of ferredoxin. (**b**) the optical spectra of rubredoxin. Solid lines represent the oxidized forms; dashed lines the reduced forms. (From Palmer, 1975, p. 7)

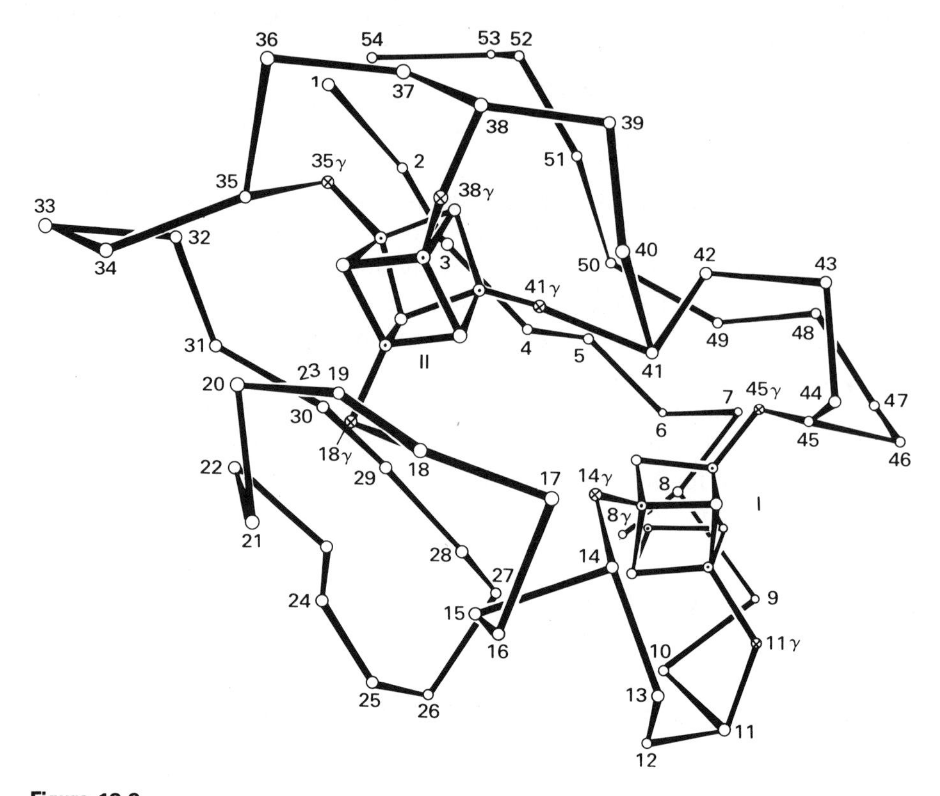

Figure 13-3
X-ray structure of the ferredoxin from *P. aerogenes*, showing the positions of carbon, iron, and sulfur. (From Palmer, 1975, p. 38)

13.A.1.c Four irons per center

The bacterial ferredoxins fall in the (4 Fe)/(4 S) category, and Figure 13-3 shows an X-ray structure of the ferredoxin from *Peptococcus aerogenes* (Adman et al., 1973). The protein actually has two (4 Fe)/(4 S) clusters as distorted cubic arrays about 12 Å away from each other. The relative positioning of cysteinyl sulfur and inorganic sulfur in the distorted cube is the following:

Schematic of tetrameric nonheme-iron (4 Fe)/(4 S) cluster

Schematic of nonheme-iron (2 Fe)/(2 S) cluster

Figure 13-4 shows X-ray structures of synthetic model (4 Fe)/(4 S) and (2 Fe)/(2 S) clusters prepared and characterized by Holm and colleagues (Holm, 1977). Chemical models for the (4 Fe)/(4 S) clusters have recently been synthesized, and these models have allowed excellent structural and electronic characterization of the chromophore, free of the protein (Holm, 1975, 1977). Each of the two (4 Fe)/(4 S) clusters in bacterial ferredoxin can accept one electron in reversible reduction. The exact electronic nature of the iron atoms in the reduced cluster is unclear; significant electronic delocalization probably exists.

The specific (4 Fe)/(4 S)-containing proteins noted above function as electron-transfer intermediates, but do not themselves have demonstrable catalytic activity. On the other hand, there is a large class of redox enzymes that contain iron–sulfur

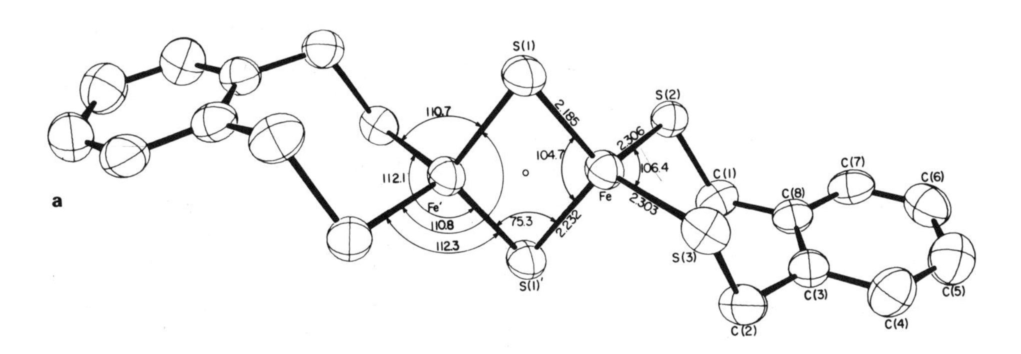

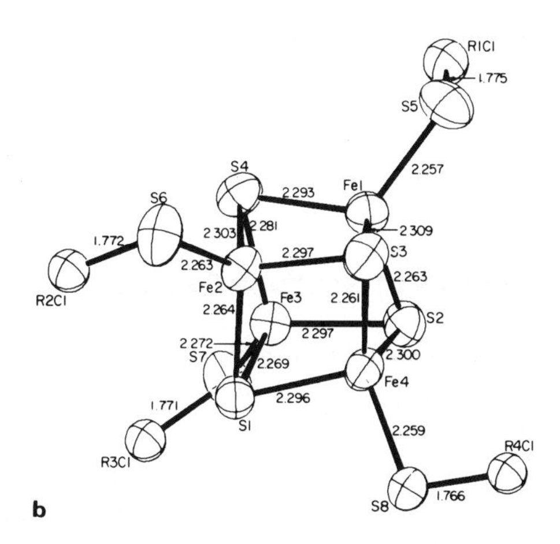

Figure 13-4
(**a**) X-ray structure for synthetic Fe_2S_2 cluster with $(S_2\text{–O–xylene})_2$ ligands. (**b**) X-ray structure for synthetic Fe_4S_4 cluster with $(S\text{–phenyl})_4$ ligands. (From Holm, 1977)

chromophores as constituent redox elements. Table 13-1 is a partial list of such enzymes. The symbol Fd represents a separate bacterial-type ferredoxin component that interacts with the enzyme protein. Table 13-1 recalls that a number of the flavin-dependent enzymes we have previously discussed (such as succinate dehydrogenase, dihydroorotate dehydrogenase, and AMP-sulfate reductase) contain the (4 Fe)/(4 S) cluster (or multiples of such clusters). Electron transfer from the reduced flavin chromophores of these dehydrogenases may occur one at a time through the Fe/S cluster. Note that, if this happens in the case of dihydroorotate dehydrogenase, it reopens the question of 1-$e^{\ominus}$–transfer steps in reactions with nicotinamide coenzymes. Sulfite reductase and nitrogenase catalyze 6-$e^{\ominus}$ reductions of inorganic compounds and are discussed in Chapter 15. Hydrogenase is anomalous in lacking any other redox cofactor; it serves as a terminal electron acceptor in respiratory chains of anaerobes, liberating H_2 gas while reoxidizing the ferredoxins. The pyruvate dehydrogenase of *C. urici* represents an α-keto-acid decarboxylase, which will be mentioned in Chapter 21. We note it here as a thiamine-PP-dependent representative using ferredoxins as a functional electron sink. The xanthine oxidase and aldehyde dehydrogenase modes of action are examined in later sections of this chapter.

Clearly, both the (2 Fe)/(2 S) and the (4 Fe)/(4 S) ferredoxin-type clusters represent extremely important *one-electron redox systems* in a host of enzyme-catalyzed redox reactions, among them the most complex electron transfers known in biological systems.

The exact mode of electron transfer to and from iron–sulfur clusters is not yet known but is under intensive scrutiny (Holm, 1977). Especially intriguing is the clostridial hydrogenase with three (4 Fe)/(4 S) clusters, only one of which is redox-active for reversible conversion of 2 $H^{\oplus}$ to H_2 (Mortensen, 1978). Whether (and in what mode) H_2 binds to the face or edge of the iron–sulfur cube is under study.

13.A.2 Mechanistic Studies of Xanthine Oxidase

One of the most useful probes of the functioning of xanthine oxidase has been rapid-reaction EPR spectroscopy. (EPR = electron paramagnetic resonance; also known as ESR = electron spin resonance.) In rapid-reaction spectroscopy, substrates and enzyme are mixed for short intervals (milliseconds) and then rapidly frozen in liquid isopentane to freeze the enzyme at various stages in catalysis; the frozen samples are then analyzed for the nature and amount of radical signals (Massey, 1973; Bray, 1975). In this manner, three distinct radical species are detectable:

Table 13-1
Properties of some iron–sulfur enzymes

Enzyme	Mol wt	Number of Fe,S	Other cofactors	Catalytic reaction
Pyruvate dehydrogenase (*Clostridium urici*)	2.4×10^5	6	Thiamine-*PP*	Pyruvate + CoA + 2 Fd-$Fe^{3\oplus}$ $\rightarrow$ Acetyl-CoA + CO_2 + 2 Fd-$Fe^{2\oplus}$
Nitrate reductase	1.4×10^5	8	Mo^{VI}	$NO_3^{\ominus} + NADPH + H^{\oplus} \rightleftarrows NO_2^{\ominus} + H_2O + NADP^{\oplus}$
Sulfite reductase	6.7×10^5	16	4 FAD, 4 FMN, 4 siroheme	$3\ NADPH + SO_3^{2\ominus} + 3\ H^{\oplus} \rightleftarrows$ $3\ NADP^{\oplus} + 3\ H_2O$ + Sulfide
AMP-$SO_4^{2\ominus}$ reductase	2×10^5	8	FMN	AMP-$SO_4^{2\ominus} \rightarrow$ AMP + $SO_3^{2\ominus} + 2\ e^{\ominus}$
Hydrogenase	6×10^4	12	None	$2\ H^{\oplus}$ + 2 Fd-$Fe^{2\oplus} \rightleftarrows H_2$ + 2 Fd-$Fe^{3\oplus}$
Nitrogenase (*C. pasteurianum*)				
a. Fe–protein	5.5×10^5	4	None	N_2 + 3 Fd-$Fe^{2\oplus}$ + 12 ATP $\rightleftarrows$
b. Fe,Mo–protein	2.2×10^4	24–32	Mo^{VI}	$2\ NH_3$ + 12 ADP + 12 P_i + 3 Fd-$Fe^{3\oplus}$
Xanthine oxidase	3.6×10^5	8	2 FAD, 2 Mo^{VI}	Xanthine + O_2 + $H_2O \rightleftarrows$ Uric acid + H_2O_2
Aldehyde oxidase	2.8×10^5	8	2 FAD, 2 Mo^{VI}	$RCHO + O_2 + H_2O \rightleftarrows RCOO^{\ominus} + H_2O_2$
Dihydroorotate dehydrogenase	1.1×10^5	4	2 FAD, 2 FMN	Dihydroorotate + $NAD^{\oplus} \rightleftarrows$ NADH + Orotate
Mitochondrial NADH dehydrogenase	—	16	FMN	NADH + Coenzyme Q $\rightleftarrows NAD^{\oplus}$ + Coenzyme QH_2
Mitochondrial succinate dehydrogenase	—	8	8a-Histidinyl-FAD	Succinate + Coenzyme Q $\rightleftarrows$ Fumarate + Coenzyme QH_2

SOURCE: Data from G. Palmer (1975) and W. H. Orme-Johnson (1973).

1. conversion of initial Mo^{VI} to Mo^{V}, with $g = 1.97$;
2. appearance of flavin semiquinone, $g = 2.00$;
3. formation of reduced iron–sulfur cluster, $g = 1.94$ (telltale EPR signal for the presence of an iron–sulfur cluster).★

The kinetic studies indicate a path of electron transfer:

$$\text{Xanthine} \rightarrow \text{Mo} \rightarrow \text{Flavin} \rightarrow \text{Fe–S cluster}$$

However, it is not clear that electron transfer to and from the three electron sinks in xanthine oxidase need necessarily be linear. There appear to be branch paths in rapid equilibrium during steady-state catalysis.

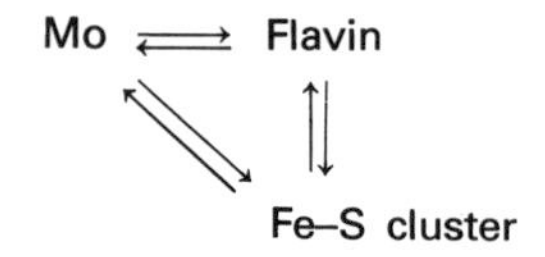

In this connection, there has been some controversy about how electrons are transferred from the 2-$e^{\ominus}$–reduced enzyme to oxygen. On the basis of initial experiments indicating that xanthine oxidase reduces oxygen by two single-electron transfers, as we shall note below, Fridovich and colleagues favored the Fe–S cluster as the oxygen-reactive species (Misra and Fridovich, 1972). However, it was then confirmed that reduced flavins can also form superoxide ion on reoxidation by O_2, as we have remarked earlier (Massey, Strickland, et al., 1969). This question seems to have been resolved by Massey's group; they prepared FAD-free xanthine oxidase by dialysis against 2 M $CaCl_2$, $Ca^{2\oplus}$ catalyzes hydrolysis of the small amounts of dissociated FAD to FMN and so displaces the equilibrium until apoenzyme is obtained. The deflavo enzyme retained full activity to dehydrogenate xanthine in the presence of artificial electron acceptors (such as the dye methylene blue), but showed no catalytic activity with O_2 as reoxidant. Thus, electrons exit to O_2 from the $FADH_2$ chromophore in xanthine oxidase.

Before asking how oxidation of substrate might proceed, we must introduce an additional functional component that is at the active site of xanthine oxidase: a group suggested to be a persulfide. It had been known for some time that xanthine oxidase slowly becomes inactive on storage; this turned out to be due to formation of desulfo enzyme as the persulfide decomposes.

★ Recent NMR studies reveal that the 4 Fe and 4 $S^{2\ominus}$ atoms are actually present as two (2 Fe)/(2 S) clusters (R. Holm, private communication, 1978).

$$E{-}S{-}S^{\ominus} \longrightarrow E{-}S^{\ominus} + {}^{\ominus}SH$$

Inactivation proceeds very rapidly in the presence of cyanide ion. With [^{14}C]-CN$^{\ominus}$, stoichiometric formation of radioactive thiocyanate ($^{\ominus}$S—^{14}C≡N) accompanies destruction of catalytic activity.

$$E{-}S{-}SH + {}^{14}C^{\ominus}{\equiv}N \longrightarrow E{-}S^{\ominus} + {}^{\ominus}S{-}{}^{14}C{\equiv}N$$

Reincubation with [^{35}S]-sulfide regenerates some active enzyme; cyanide addition releases the label as [^{35}S]-thiocyanate. Although this is consonant with a persulfide linkage, the structure is by no means determined unambiguously.

Nonetheless, Massey and colleagues attribute a catalytic role to the supposed persulfide linkage in substrate oxidation, by its initiating nucleophilic attack on the C-8 imine group of substrate, with resultant transfer of the C-8 hydrogen as *hydride ion* to the complex electron sink of the enzyme (Komai et al., 1969).

Enz—S—S$^{\ominus}$ + [imidazole ring, C-8—H] + Mo^{VI} ⇌ [Enz—S—S adduct at C-8, H transferred] + Mo^{VI} ⇌ [H₂O + ring—S—S—Enz] + Mo^{IV}—H ⇌ [ring]—OH + $^{\ominus}$S—S—Enz

Mo^{IV}—H → (1 e$^{\ominus}$) Mo^{V} → (1 e$^{\ominus}$) Mo^{VI}

Rapid transfer of e$^{\ominus}$ to Fe-S cluster and FAD components of holoenzyme electron sink

They have shown that $NaBH_4$, a classical hydride-transfer agent, is a specific 2-e$^{\ominus}$ reductant of xanthine oxidase. When 8-[^{2}H]-xanthine is used, the initial molybdenum EPR signal due to Mo^{V} shows different hyperfine coupling, which has tentatively been regarded as indicating the initial formation of a metal-hydride bond between Mo^{VI} and either the C-8 hydride or deuteride (yielding Mo^{IV}—H species). The total capacity of the Mo,(Fe/S),FAD electron sink in xanthine oxidase is six electrons (2 e$^{\ominus}$ for Mo^{VI} → Mo^{IV}, 1 e$^{\ominus}$ for each Fe/S cluster, and

2 $e^{\ominus}$ into FAD). With purified enzyme and excess xanthine, one can show that the 6-$e^{\ominus}$–reduced enzyme can indeed form, although this is not likely to be the predominant reduced species during turnover.

Once the Mo^{IV} has been generated by initial 2-$e^{\ominus}$ removal from xanthine, electrons can be funneled rapidly out, one at a time, to both Fe/S clusters and to FAD. Massey considers that the Fe/S centers serve as reserve electron sinks, to reoxidize Mo^{IV} and Mo^{V} back to Mo^{VI}, so it can function again to allow subsequent reaction with another substrate molecule or to reduce FAD fully to $FADH_2$ for reaction with molecular oxygen. The aldehyde oxidase has not been subjected to as intense a set of mechanistic scrutinies, but it would appear to be an analogous catalyst to xanthine oxidase (Massey, 1973; Bray, 1975). This mechanistic postulate suggests that the enzyme-bound molybdenum is the key locus for redox entry of substrate-derived electrons; one might suspect that the other electron components could be substituted, and Table 13-1 seems to confirm that idea.

Another variant occurs with formate dehydrogenase, found in large amounts in the cytoplasmic membranes of anaerobically grown *E. coli.* This enzyme feeds electrons into the membrane respiratory chain while converting formate to CO_2.

$$HCOO^{\ominus} \longrightarrow CO_2 + (2\,e^{\ominus} + H)$$

The enzyme is extremely labile to molecular oxygen when purified, due to apparent structural decomposition of a nonheme-iron–sulfur cluster (Enoch and Lester, 1975). Additionally, the enzyme contains molybdenum and (interestingly) selenium ion, whose role in catalysis and actual structure in the enzyme (inorganic selenium as selenide, selenite, or a seleno amino acid?) remain unknown. No mechanistic studies have been reported, but it is quite likely that the oxidation of $HCOO^{\ominus}$ proceeds by hydride transfer to the molybdenum, as featured for xanthine oxidase.

$$H{:}\!-\!C(=O)\!-\!O^{\ominus} + Mo^{VI}\!-\!Enz \longrightarrow CO_2 + H\!-\!Mo^{IV}\!-\!Enz$$

Future study should reveal whether the selenium in this enzyme (and in other selenoenzymes; see T. Stadtman, 1974) is redox active. The nature of the ligands to molybdenum in these enzymes is unknown, but heat denaturation yields a low-molecular-weight molybdenum-containing complex whose purification is in process. Molybdenum is also present in nitrogenase (Chapter 15), but there its ligand environment may be distinct from that in these molybdoflavoproteins.

Mechanism of allopurinol inhibition of xanthine oxidase. As noted at the outset of this chapter, the pyrazolo pyrimidine allopurinol is a clinically important xanthine-oxidase inhibitor. Actually it functions first as a substrate, undergoing hydroxylation at C-2 of the pyrimidine ring in analogy to hypoxanthine→ xanthine conversion, to yield alloxanthine; this product remains tightly bound to the enzyme's active site, with molybdenum still at the Mo^{IV} oxidation state (Massey et al., 1970).

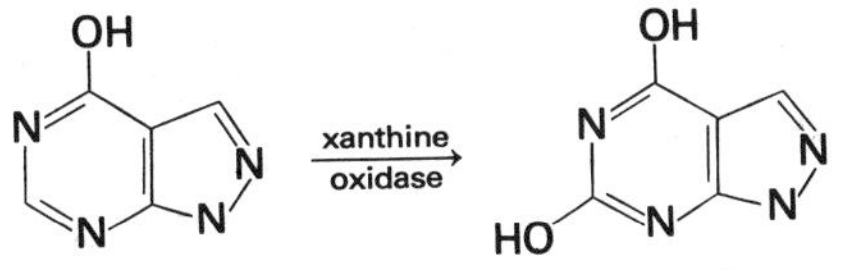

OH N N N N —xanthine oxidase→ OH N N N N HO

allopurinol alloxanthine

That the tight noncovalent binding of alloxanthine is specific for this oxidation state can be demonstrated with $[^{14}C]$-alloxanthine: it binds stoichiometrically to the Mo^{IV}–enzyme but not to native Mo^{VI}–enzyme. Thus, allopurinol acts as a "suicide" substrate (C. Walsh, 1977) in a fashion analogous to that of methionine sulfoximine (¶5.D.1). The bound product alloxanthine probably chelates (via the pyrazolo portion?) to the Mo^{IV}. The half-time for dissociation and recovery of catalytic activity is 300 min at 25° C (Massey et al., 1970).

13.B SUPEROXIDE DISMUTASE

We have commented that xanthine oxidase can reduce oxygen in single-electron steps. Indeed, xanthine oxidase was the first enzyme found that is capable of superoxide-anion formation (by McCord and Fridovich, 1969). The superoxide-ion formation was tracked down because xanthine oxidase was found to display some properties during its catalysis that were initially mystifying. Xanthine oxidase, in the presence of xanthine, was found to initiate the aerobic oxidation of sulfite—a known free-radical chemical chain reaction, readily initiated by several free-radical species. It also causes chemiluminescence of luminol (a radical process); and, in perhaps the easiest reaction to assay, xanthine oxidase during turnover causes catalytic reduction of the heme cytochrome *c*, reducing the heme iron from $Fe^{3\oplus}$ to $Fe^{2\oplus}$, a 1-$e^{\ominus}$ reduction (easily monitored because the $Fe^{2\oplus}$-form has absorbance at 550 nm, and the oxidized form does not). These processes do not occur when artificial electron acceptors replace O_2 in the xanthine-oxidase reaction. Although it was hypothesized that univalent reduction of O_2 to $O_2^{\ominus\cdot}$ by xanthine oxidase generates the culprit, no specific assay for detecting trace amounts of short-lived superoxide was available for a positive detection system.

The problem became amenable to further observation when Fridovich (1972*a*, 1975) chanced to find that a preparation of bovine carbonic anhydrase (which accelerates the reversible hydration of CO_2 to bicarbonate) could suppress this xanthine-oxidase–dependent cytochrome-*c* reduction, competitively and with a K_i value of 3×10^{-9} M. It turned out that the suppressing agent was a trace impurity (<0.8%) in the preparation of carbonic anhydrase. Assiduous purification of the impurity resulted in a homogeneous protein, assayable by this suppression activity; because it removes superoxide anion, it was designated superoxide dismutase. The purified protein could be used to show what percentage of O_2 when reduced by xanthine oxidase proceeded through $O_2^{\ominus\cdot}$ molecules that could diffuse away and be scavenged by dismutase before the second $e^{\ominus}$ could be transferred to give H_2O_2. It appears that xanthine oxidase reduces about 20% of the O_2 at pH 7 via univalent reduction to $O_2^{\ominus\cdot}$ that can get away, whereas at pH 8 as much as 80% of the O_2 reduced goes via detectable superoxide anion (Fridovich, 1972*a*, 1975).

We have noted previously that molecular oxygen is a stable triplet diradical in the ground state, with a kinetic preference for undergoing radical reactions such as initial univalent reduction in enzymatic reactions where H_2O_2 is formed. Superoxide dismutase acts to scavenge the molecules of superoxide anion that form during such biological reductions but escape from a specific enzyme's active site, thus generating an adventitious potential oxidizing agent of cellular constituents.

$$O_2 \xrightarrow{1\,e^{\ominus}} HO_2^{\cdot} \quad \begin{cases} \xrightarrow{1\,e^{\ominus}} H_2O_2 \\ \underset{pK_a\,=\,4.9}{\rightleftharpoons} O_2^{\ominus\cdot} + H^{\oplus} \end{cases}$$

hydroperoxy radical ($HO_2^{\cdot}$); superoxide anion ($O_2^{\ominus\cdot}$) (The predominant form at physiological pH)

The dismutase removes the free superoxide-anion radical by accelerating its bimolecular recombination. The reaction is indeed a dismutation, one molecule of superoxide undergoing 1-$e^{\ominus}$ reduction to H_2O_2, the other undergoing 1-$e^{\ominus}$ oxidation back to molecular oxygen.

$$O_2^{\ominus\cdot} + O_2^{\ominus\cdot} \xrightarrow{2\,H^{\oplus}} O_2 + H_2O_2$$

The enzyme has a very high turnover number, estimated at 3×10^6 moles min^{-1}

(mole enzyme)$^{-1}$, with a K_m for superoxide of 5×10^{-4} M (Fridovich, 1972*a*, 1975). Expressing the catalytic coefficiency as k_{cat}/K_m, we obtain a bimolecular rate of 10^8 M^{-1} sec^{-1}, a value that is slightly less than the diffusion limit. As a catalyst, of course, superoxide dismutase simply accelerates the nonenzymatic rate of chemical dismutation. These nonenzymatic rates have been measured for the various protonated and anionic recombination paths.

$$HO_2^{\cdot} + HO_2^{\cdot} \longrightarrow O_2 + H_2O_2 \qquad 7.6 \times 10^5\ \text{M}^{-1}\ \text{sec}^{-1} \qquad (1)$$
$$HO_2^{\cdot} + O_2^{\ominus\cdot} \xrightarrow{H^{\oplus}} O_2 + H_2O_2 \qquad 8.5 \times 10^7\ \text{M}^{-1}\ \text{sec}^{-1} \qquad (2)$$
$$O_2^{\ominus\cdot} + O_2^{\ominus\cdot} \xrightarrow{2H^{\oplus}} O_2 + H_2O_2 \qquad < 10^2\ \text{M}^{-1}\ \text{sec}^{-1} \qquad (3)$$

The nonenzymatic disproportionation of two molecules of anion is relatively slow, probably due to electrostatic repulsions. The enzyme presumably overcomes these by binding anion and protonating it at the active site before dismutation.

Superoxide dismutases have now been purified to homogeneity from both eucaryotic and procaryotic sources. The bovine-erythrocyte enzyme has 32,000 mol wt, has two $Cu^{2\oplus}$ and two $Zn^{2\oplus}$, and is blue-green from the copper content. Figure 13-5a shows its visible spectrum. The Zn and Cu atoms are in close proximity, sharing a common imidazole group of a histidine residue as ligand. Undoubtedly, the metals may be involved in both binding and electron transfer between oxygen species. A reasonable scheme for redox function of the copper (the zinc is not essential for activity) can be written (Fridovich, 1972*a*, 1975):

$$\text{Enz-Cu}^{II} + O_2^{\ominus\cdot} \rightleftharpoons \text{Enz-Cu}^{I} + O_2 \qquad (1)$$
$$\text{Enz-Cu}^{I} + O_2^{\ominus\cdot} \rightleftharpoons \text{Enz-Cu}^{II} + H_2O_2 \qquad (2)$$

A low-resolution X-ray map of the Cu,Zn–superoxide dismutase has recently been published; Figure 13-6 represents the three-dimensional structure.

When the enzyme from *E. coli* was isolated, the solutions of enzyme were found to be wine-red rather than blue-green in color (Fig. 13-5b). This enzyme is a dimer of 40,000 mol wt with one manganese ion (apparently as Mn^{III}; Fee et al., 1976) per subunit, but with no copper or zinc ions. It has the same spectrum of activity as the enzyme from eucaryotic cytoplasm.* Subsequent studies established that eucaryotic mitochondria contain a manganoenzyme analogous to the major *E. coli* superoxide dismutase. And, as a further variant, *E. coli* also has an iron-containing superoxide dismutase (Fig. 13-5c), apparently located in the periplas-

* Model studies have shown that copper ions in aqueous acid will actually cause $O_2^{\ominus\cdot}$ dismutation at a rate (embarrassingly) faster than that at which the dismutase functions at pH 7 (Rabans et al., 1973; Fielden et al., 1974). The model may be more efficient than the enzyme in this one instance.

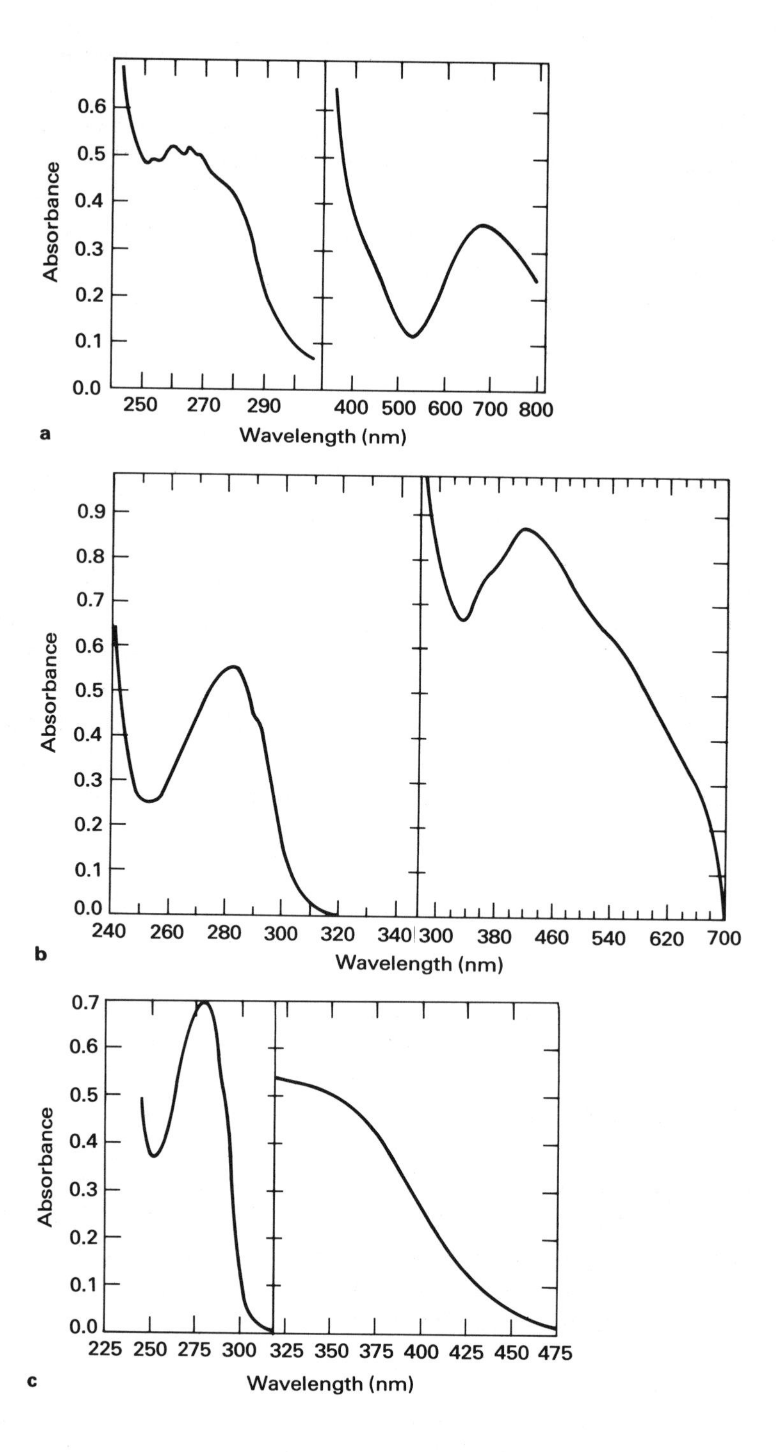
0.6
0.5
0.4
0.3
0.2
0.1
0.0
Absorbance
250
270
290
400
500
600
700
800
a
Wavelength (nm)
0.9
0.8
0.7
0.6
0.5
0.4
0.3
0.2
0.1
0.0
Absorbance
240
260
280
300
320
340
300
380
460
540
620
700
b
Wavelength (nm)
0.7
0.6
0.5
0.4
0.3
0.2
0.1
0.0
Absorbance
225
250
275
300
325
350
375
400
425
450
475
c
Wavelength (nm)

Figure 13-5
(**a**) Absorption spectrum of bovine superoxide dismutase. The concentration of the enzyme was 1.5 mg ml^{-1} for the ultraviolet spectrum, 47 mg ml^{-1} for the visible spectrum. A path length of 1 cm was used. This is the Cu,Zn-enzyme. (From McCord and Fridovich, 1969, p. 6052) (**b**) Absorption spectrum of *E. coli* superoxide dismutase. The ultraviolet spectrum was obtained with a solution containing 0.7 mg ml^{-1} of the enzyme in 0.05 M potassium phosphate at pH 7.8; the visible spectrum was obtained from a solution containing 7.0 mg ml^{-1} of enzyme in the same buffer. This is the manganoenzyme. (From Keele et al., 1970, p. 6178) (**c**) Absorption spectrum of *E. coli* iron-containing superoxide dismutase. The ultraviolet spectrum was obtained with a solution containing 0.5 mg ml^{-1} in 0.05 M potassium phosphate at pH 7.8 and 25° C; the visible spectrum was obtained from a solution containing 11.6 mg ml^{-1} of the enzyme in the same buffer. This is the iron enzyme. (From Yost and Fridovich, 1975, p. 1351)

mic space between the inner and outer cell walls (Fridovich, 1972*a*, 1975). The iron-enzymes and manganoenzymes are now receiving close examination. Clearly, copper, iron, and manganese must play similar electron-transfer roles.

Because the enzyme suppresses reactions caused specifically by $O_2^{\ominus\cdot}$ and is effective in the ng ml^{-1} concentration range, it is an extremely sensitive and specific assay for detecting whether $O_2^{\ominus\cdot}$ is formed during some biological process. Among the systems thus found to produce superoxide are—

1. aldehyde oxidase, sulfite oxidase;
2. flavoenzyme dehydrogenases during their slow reoxidation by O_2;
3. reduced iron–sulfur chromophores being autoxidized in ferredoxins;

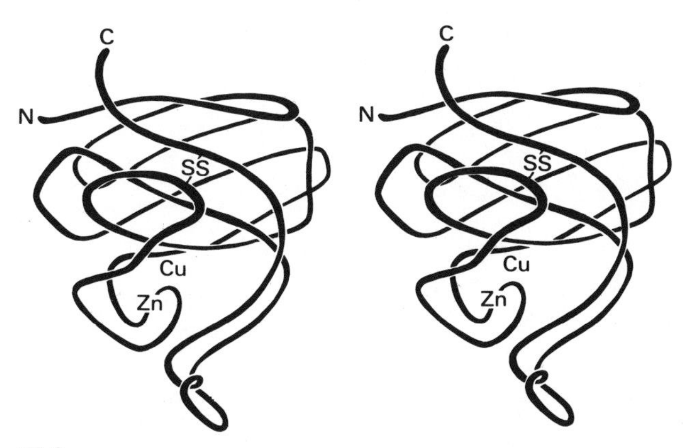

Figure 13-6
Simplified schematic for Cu–Zn superoxide dismutase main-chain folding of one subunit; this is a stereoscopic pair of images. (From Richardson et al., 1975, p. 1351)

4. a dioxygenase (tryptophan 2,3-dioxygenase; Chapter 16) is inhibited by superoxide dismutase;
5. in fruit, ethylene acts as a ripening hormone; its production is attended by formation of $O_2^{\ominus\cdot}$ (Fridovich, 1972*a*, 1975).

Fridovich (1972*a*, 1975) has proposed the general theory that superoxide dismutase was evolved as a surveillance device against the toxic molecule oxygen (toxic as the precursor to $O_2^{\ominus\cdot}$ and H_2O_2). This enzyme protects oxygen-metabolizing enzymes against the deleterious effects of free $O_2^{\ominus\cdot}$ such as sulfhydryl oxidation or unsaturated-lipid oxidations. It has long been known that obligately anaerobic microorganisms are killed on exposure to oxygen. Fridovich felt that this oxygen toxicity might arise from the absence of superoxide dismutase in the anaerobes; in their normal metabolism they don't generate $O_2^{\ominus\cdot}$, and thus they have no need for a scavenging device. Indeed, in a species survey, all aerobic bacteria examined had detectable dismutase, but the anaerobes had no superoxide dismutase activity. Fridovich has suggested that aerobic organisms evolved this enzyme in response to the challenge of O_2 production generated when green plants began to photosynthesize and to convert H_2O to O_2.

Chapter 14

Copper-Containing Oxidases and Monooxygenases

There are about two dozen well-characterized cuproproteins that function in redox systems (Peisach et al., 1966; Malkin, 1973; Vaneste and Zuberbühler, 1974; Malmstrom et al., 1975). All of these (except plastocyanin) use molecular oxygen as electron acceptor; O_2 is produced by superoxide dismutase. The use of O_2 may, in part, be due to the highly positive reduction potentials for cuproproteins (+0.3 to +0.6 V), which rule out almost all known biological electron acceptors except O_2 ($E_0' = +0.8$ V). The identity and the ligand structures of copper at the active sites of many of these proteins remain undetermined★ (in contrast to the nonheme-iron cases noted in Chapter 13).

One can distinguish some enzymes that have ESR signals indicative of the normal parameters of paramagnetic $Cu^{2\oplus}$ and other copper enzymes that are ESR silent. These ESR-silent systems may contain $Cu^{\oplus}$, or perhaps dimagnetic $Cu^{2\oplus}$·$Cu^{2\oplus}$ pairs. ESR measurements have shown also that some copper enzymes have shifted spectroscopic parameters. These proteins are intensely blue, with visible absorption maxima around 600 nm and extinction coefficients from 700 to 10,000 $\text{M}^{-1}\,\text{cm}^{-1}$ (Fig. 14-1; Malmstrom et al., 1975). Although the precise reasons for the blue color of this class of $Cu^{2\oplus}$ proteins are unknown,★ it has been proposed that distortion of the $Cu^{2\oplus}$ coordination complex from its square coplanar geometry to an asymmetric geometry, and the possibility of charge-transfer interaction with cysteinyl sulfur as ligand, may be responsible

★ The ligands in the blue copper-protein plastocyanin have recently been determined to be two histidine-imidazolyl nitrogens, a cysteinyl sulfur, and a methionyl sulfur (H. Freeman, private communication, 1978).

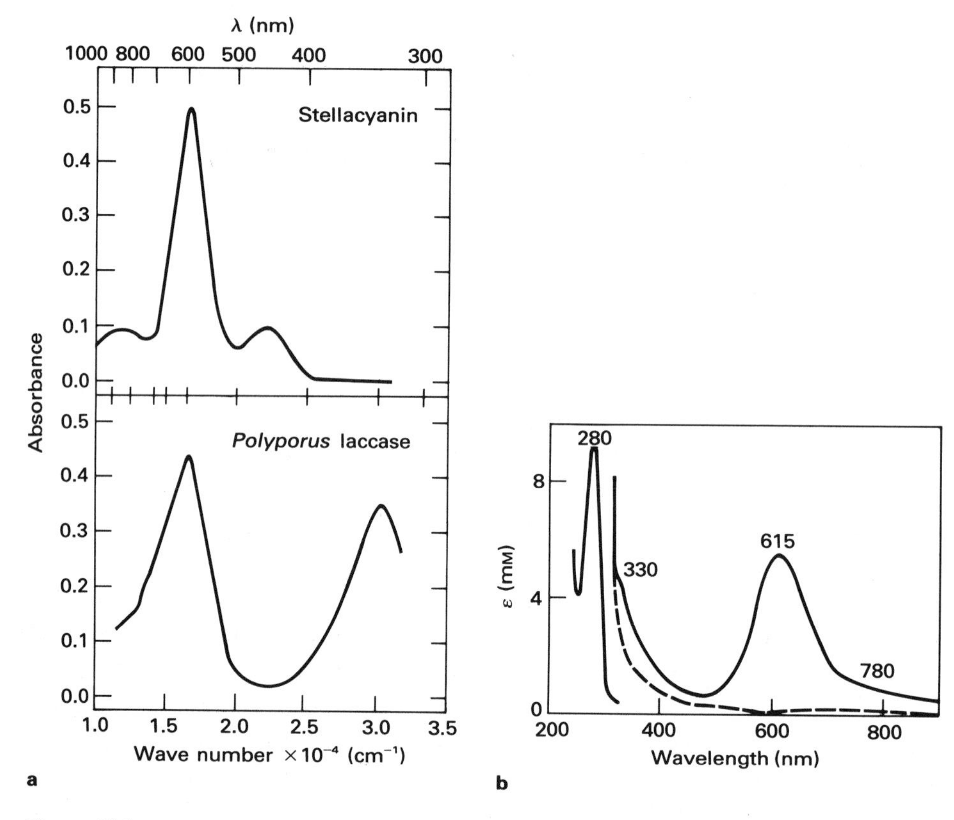

Figure 14-1
(**a**)Visible and near-ultraviolet absorption spectra of stellacyanin and laccase. The spectra are presented as the difference between the oxidized and reduced proteins, with the latter obtained by addition of ascorbate. (From Eichhorn, 1971, p. 691) (**b**) Absorption spectrum of *Rhus verniciferca* laccase in phosphate buffer (pH 7.0), showing oxidized form (*solid line*) and reduced form (by ascorbate, *dashed line*). The extinction coefficient at 615 nm peak used was 5.4 (mM protein)$^{-1}$. (From G. Hamilton et al., 1976, p. 409)

(Malmstrom et al., 1975). The copper oxidases that reduce O_2 to H_2O_2 are not colored; ascorbate oxidase, which reduces O_2 by four electrons to two H_2O molecules, is brilliantly blue. The two monooxygenases we shall discuss (dopamine-β-hydroxylase and tyrosinase) again are nonblue.

One can, with some reservation, divide cuproenzymes into two redox categories: those in which no redox charge is detectable at the metal center, and those where the copper clearly is redox active, shuttling between $Cu^{2\oplus}$ and $Cu^{\oplus}$. On the other hand, the failure to detect redox change in bound copper at the active site of some enzymes may be a kinetic problem. Presumably the dioxygen molecule binds near the copper atom or atoms at the active sites of the enzymes, and the metal serves as a conduit for electrons between triplet O_2 and singlet substrate.

An alternative categorization, which we shall use here, can be made on the basis of how much of the oxidizing power of the O_2 substrate is used by the copper enzymes.

1. Some copper enzymes reduce O_2 to H_2O_2 (with 2-$e^{\ominus}$ oxidation of the specific substrate).
2. Other copper enzymes reduce O_2 all the way to H_2O but do not carry out substrate monooxygenation; this is a 4-$e^{\ominus}$ reduction of O_2.
3. A third group of copper enzymes shows monooxygenase activity.
4. The fourth category comprises the single known copper-requiring dioxygenase, quercitinase.

The first three categories are discussed in this chapter. Quercitinase is mentioned in Chapter 16.

14.A COPPER ENZYMES REDUCING O_2 TO H_2O_2

The copper enzymes that reduce O_2 to H_2O_2 carry out 2-$e^{\ominus}$ oxidations of the specific substrate (S); these reactions have a stoichiometry analogous to that of the flavoenzyme oxidases:

$$SH_2 + O_2 \longrightarrow \longrightarrow S + H_2O_2$$

We shall comment on two examples in this class: amine oxidase and galactose oxidase.

14.A.1 Amine Oxidase

We noted in Chapter 11 that animal cells contain a monoamine oxidase as an integral protein of mitochondrial membranes and that this amine oxidase contains covalently bound FAD. There exists a different group of amine oxidases (some oxidizing monoamines, some oxidizing diamine substrates) that do not use FAD as redox cofactor, but instead have $Cu^{2\oplus}$ at the active site (Malmstrom et al., 1975; Mondovi et al., 1971). The generalized stoichiometry might be written as follows, with the imine form of the aldehyde released from the enzyme as initial product.

$$R-CH_2-NH_2 + O_2 \xrightarrow{Enz\text{-}Cu^{2\oplus}} R-CH=NH_2^{\oplus} \xrightarrow[H_2O]{\text{nonenzymatic}} R-CH=O + NH_4^{\oplus}$$

Numerous reports in the literature have shown that these copper-dependent amine oxidases have an additional prosthetic group tentatively identified as pyridoxal phosphate (Malmstrom et al., 1975). The pyridoxal-P would provide an aldehyde group to engage in imine formation with the substrate and facilitate generation of a stabilized carbanion as an intermediate in substrate oxidation. (See Chapter 24 for discussion of enzymatic catalysis mediated by pyridoxal-P.)

resonance-delocalized carbanion

Protonation at the methine carbon of the coenzyme would yield a product imine that could rehydrate to the aldehyde product and the amine form of the coenzyme.

Such a scheme would leave the pyridoxamine coenzyme in a 2-$e^{\ominus}$–reduced form, incompetent for subsequent catalytic cycles. G. Hamilton (1971) has speculated that O_2 might reoxidize the pyridoxamine back to pyridoxal while undergoing 2-$e^{\ominus}$ reduction to H_2O_2. The electron-transfer process could be mediated by the $Cu^{2\oplus}$ ions. This would be a unique example of an O_2-mediated reoxidation of pyridoxamine to pyridoxal, with no precedents for it. This picture is further complicated by claims that the other prosthetic group besides the $Cu^{2\oplus}$ is *not* pyridoxal-P but some as-yet-unidentified functionality; Allison (1976) has suggested a sulfenic-acid grouping. Trapping experiments suggest that the imine, not the aldehyde, is the product released from the active site; if these experiments are correct, this observation conflicts with the scheme using pyridoxal-P, because that scheme produces aldehyde and pyridoxamine-P products at the end of the substrate-oxidation phase of the catalytic cycle. Additional experiments are needed to determine whether pyridoxal-P is involved at all and, if so, in what fashion.

Despite this mechanistic uncertainty, it does appear that a carbanion mechanism may be involved in the $Cu^{2\oplus}$-dependent amine oxidase purified from bovine plasma. It converts β-chlorophenyethylamine to phenylacetaldehye and chloride ion in an O_2-independent elimination of HCl (Neumann et al., 1975), which is most easily construed as an $H^{\oplus}$,$Cl^{\ominus}$ elimination, initially generating an eneamine that ketonizes and hydrolyzes.

$$\text{C}_6\text{H}_5\text{—CH(Cl)—CH}_2\text{—NH}_2 \;(\text{:B—Enz}) \xrightarrow[\text{Cl}^{\ominus}]{\text{Enz}} \text{C}_6\text{H}_5\text{—HC=CH—}\ddot{\text{N}}\text{H}_2 + \text{BH}^{\oplus} \rightleftharpoons \text{C}_6\text{H}_5\text{—H}_2\text{C—CH=}\overset{\oplus}{\text{N}}\text{H}_2 \xrightleftharpoons{\text{H}_2\text{O}} \text{C}_6\text{H}_5\text{—H}_2\text{C—CH=O} + \text{NH}_4^{\oplus}$$

phenylacetaldehyde

Because the two electrons removed from substrate depart with the $Cl^{\ominus}$, the enzyme–$Cu^{2\oplus}$ prosthetic-group complex is not reduced, and O_2 is not required for reoxidation. On the other hand, when β-bromoethylamine is used as substrate, the enzyme is rapidly inactivated with all the kinetic signs of an active-site–directed covalent modification. It has been suggested that the bromoethylamine does not eliminate HBr, but is instead oxidized to β-bromoacetaldehyde (with the reactive α-haloketone group; see ¶3.B.12), which is then attacked by a nucleophilic amino-acid side chain before release from the active site (Neumann et al., 1975; Maycock and Abeles, 1976).

$$\text{BrH}_2\text{C—CH}_2\text{—NH}_2 \xrightarrow{\text{O}_2,\ \text{Enz}} \text{BrH}_2\text{C—CH=O} \;(\text{:X—Enz}) + \text{H}_2\text{O}_2 + \text{NH}_4^{\oplus} \longrightarrow \text{Enz—X—H}_2\text{C—CH=O} + \text{Br}^{\ominus}$$

A last comment about the plasma $Cu^{2\oplus}$-dependent enzyme is that Abeles and colleagues also found that aminoacetonitrile-1-[^{14}C] inactivates the enzyme with covalent incorporation of radioactive label (Maycock et al., 1975; Maycock and Abeles, 1976). The enzyme could catalyze its own inactivation by recognition of aminoacetonitrile as a substrate, with abstraction of the α-H to give a carbanion analogous to normal processing of substrates. This carbanion could, however, rearrange to a highly electrophilic ketenimine that could alkylate a residue at the active site.

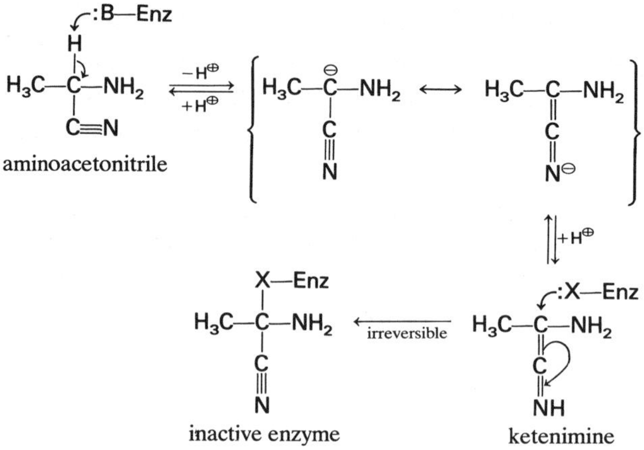

aminoacetonitrile

ketenimine

irreversible

inactive enzyme

Such a process may be the molecular mechanism by which β-aminopropionitrile produces lathrytic defects in collagen (Rando, 1975). Grazing animals ingest the nitrile in their food; it inactivates a $Cu^{2\oplus}$-dependent lysine ε-amine oxidase responsible for generating ε-aldehyde groups as precursors to cross-links in collagen. With the lysyl oxidase inactivated, cross-linking does not occur, and the collagen does not acquire its normal tensile strength.

The FAD-linked amine oxidase (Chapter 11) is not susceptible to aminoacetonitrile-induced inactivation. Conversely, the plasma amine oxidase is not affected by *N*-methyl-*N*-benzylpropynylamine (pargyline), which modifies the bound FAD of the flavoenzyme. This exclusive and complementary specificity has allowed evaluation in vivo of the role of these two amine oxidases in neuroactive amine metabolism, including the processing of hallucinogenic amines (Riceberg et al., 1973).

14.A.2 Galactose Oxidase

Fungal galactose oxidase has been purified and shown to be a colorless protein requiring copper ion for enzymatic activity (Kelly-Falcoz et al., 1965). Carbon-6 of galactose (and various galactosides) is oxidized to the aldehyde, while O_2 is reduced to H_2O_2.

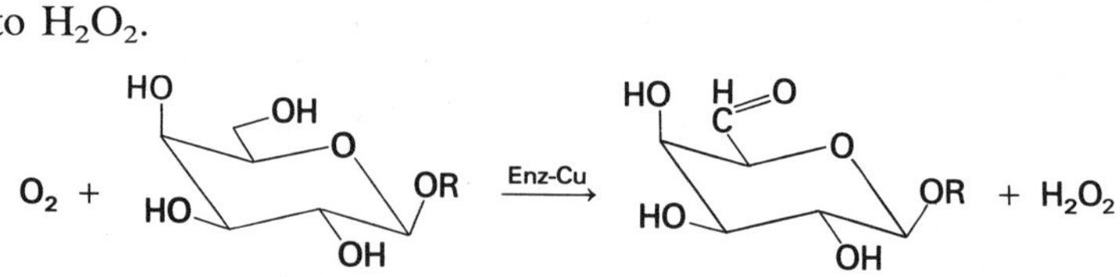

Enz-Cu

The enzyme has been used to label externally accessible galactosyl groups in glycoproteins of various biological membranes. The aldehyde group so generated can be reduced with NaB^3H_4 to regenerate the galactosyl group now tritiated at C-6. The role of the copper and its redox state during catalysis have been matters of recent controversy, and little is known with certainty about the mechanism (Malmstrom et al., 1975).

The predominant form of Cu in the isolated enzyme is Cu^{II} (70% by paramagnetic measurements). Hamilton and colleagues have observed that addition of ferricyanide activates the enzyme and causes dissipation of the EPR signal (G. Hamilton et al., 1976). The disappearance of the EPR signal could be a line-broadening phenomenon from apposition of $Fe^{III}(CN)_6$ and the Cu^{II}. Alternatively, the ferricyanide could be reduced by one electron to ferrocyanide, while oxidizing the enzyme-Cu^{II} to enzyme-Cu^{III}. This interpretation is favored by Hamilton, because it explains why ferricyanide addition *activates* the enzyme as well as silencing the EPR signal; it is nonetheless disputed by others (Ettinger, 1974; Ettinger and Kosman, 1974; Giordano et al., 1974). This would be the first example of Cu^{III} involvement in enzymatic catalysis, although Cu^{III} compounds are known (Margerum et al., 1975; Burce et al., 1975).

A possible scheme for electron transfer has been advanced by Hamilton, involving a shuttling of the enzymatic copper between Cu^{III} and Cu^{I} oxidation states. Given that activated enzyme might be in the Cu^{III} oxidation state, a 2-$e^{\ominus}$ oxidation of the C-6 alcoholic group of galactose (with loss of the pro*S* hydrogen; see Maradufu et al., 1971) would produce aldehyde directly and Cu^{I}. The Cu^{I} may then pass electrons into O_2 one at a time, yielding a transient $Cu^{II}—O_2^{\ominus\cdot}$ intermediate. A molecule of $O_2^{\ominus\cdot}$ occasionally leaks out of the active site, leaving the enzyme in an inactive Cu^{II} form. This happens about once out of 2,000 to 5,000 times; ferricyanide then reactivates the enzyme again.

Enz—B: HO H_S Cu^{III} C HO H_R O HO OR OH

postulated hydride transfer to copper → O=C(H)R′ + Enz—Cu^{I}···H

$\downarrow O_2$

Enz—Cu^{I}···O_2

$\downarrow$ 1-$e^{\ominus}$ transfer

Enz—Cu^{III} + H_2O_2 ← (1-$e^{\ominus}$ transfer, 2 $H^{\oplus}$) — $Cu^{II}—O_2^{\ominus\cdot}$

The H_S-abstraction step is fully rate-determining, with a V_{max} isotope effect of 7.7 when the 6*S*-[^{2}H]-D-galactose is used as substrate (Maradufu et al., 1971).

14.B COPPER ENZYMES REDUCING O_2 TO H_2O WITHOUT OXYGENATION

Our second category of cuproenzymes includes those that reduce O_2 all the way to H_2O, but do not carry out substrate monooxygenation. We shall briefly examine two examples of enzymes that carry out these double oxidations (4-$e^{\ominus}$ reduction of O_2): ascorbate oxidase and cytochrome oxidase.

14.B.1 Ascorbate Oxidase

Ascorbate oxidase is an enzyme found in plants; the enzyme from squash is a glycoprotein (ca. 10% carbohydrate) and has been purified to homogeneity. It contains six copper atoms per 150,000 mol wt and is a typical blue copper protein. The λ_{max} for the blue color is just above 600 nm, and the extinction coefficient has been estimated to be $\varepsilon = 3{,}300\ \text{M}^{-1}\ \text{cm}^{-1}$. There appear to be three types of copper from EPR measurements: normal $Cu^{2\oplus}$, an abnormal $Cu^{2\oplus}$, and EPR-silent copper (Malmstrom et al., 1975). The abnormal $Cu^{2\oplus}$ is responsible for the blue color, and it has been suggested that the color represents transitions due both to distorted ligand geometry and to charge-transfer interactions with a cysteinyl-thiolate anion of the apoenzyme. The EPR-silent copper does not appear to be $Cu^{\oplus}$, as might have been predicted, but perhaps magnetically coupled $Cu^{2\oplus} \cdots Cu^{2\oplus}$ dimers, which may represent the functional redox group in catalysis. The substrate, ascorbate, is the γ-lactone of a hexonic acid, with an enediol functionality at C-2 and C-3. Primates cannot synthesize ascorbate, and for them it is an essential vitamin (vitamin C). One of its predominant physiological roles may be as a redox cofactor in the dopamine-β-hydroxylase system we shall discuss in ¶14.C.1.

During catalysis, the oxidase removes two electrons from the ascorbate-enediol system, generating a 2,3-α-diketone moiety in the dehydroascorbate product. Yet, the external electron acceptor (O_2) undergoes a 4-$e^{\ominus}$ reduction to water; thus, the overall stoichiometry should be (and is) two molecules of ascorbate dehydrogenated per one O_2 undergoing reduction to two water molecules.

$$2\ \text{ascorbate} + O_2 \xrightarrow{\text{Enz-Cu}^{2\oplus}} 2\ \text{dehydroascorbate} + 2\,H_2O$$

ascorbate

dehydroascorbate

During catalysis, the free-radical signal due to 1-e$^{\ominus}$–reduced ascorbate-enolate radical (a delocalized free radical, and thus a relatively low-energy intermediate) is detectable. Concomitantly, the copper becomes EPR-silent, consistent with reduction of cupric to cuprous ion (Malmstrom et al., 1975). These data suggest that the transfer of electrons from substrate to enzyme-bound $Cu^{2\oplus}$ occurs one at a time. Anaerobically, one can titrate the enzyme, reducing 4 $Cu^{2\oplus}$ to 4 $Cu^{\oplus}$ while oxidizing two molecules of ascorbate.

$$\text{Enz}\cdot(4\,Cu^{2\oplus}) + 2\ \text{Ascorbate} \xrightarrow[\text{transfers}]{\text{two 1-e}^{\ominus}} \text{Enz}\begin{cases}(2\,Cu^{\oplus})\\(2\,Cu^{2\oplus})\\(2\ \text{Ascorbate radicals})\end{cases}$$

$$\xrightarrow[\text{transfers}]{\text{two 1-e}^{\ominus}} \text{Enz}\begin{cases}(4\,Cu^{\oplus})\\2\ (2\ \text{Dehydroascorbate})\end{cases}$$

Electron transfer from bound ascorbate to Enzyme·$Cu^{2\oplus}$ undoubtedly is facilitated by chelation of an ascorbate molecule to one metal, which may be in electronic communication with another metal center—the postulated EPR-silent $Cu^{2\oplus}\cdots Cu^{2\oplus}$ couple.

$$\text{Ascorbate}\cdots Cu^{2\oplus}\cdots Cu^{2\oplus} \xrightarrow{1\,e^{\ominus}} \left\{\text{Ascorbate radical}\cdots Cu^{\oplus}\cdots Cu^{2\oplus} \longleftrightarrow \text{Ascorbate radical}\cdots Cu^{\oplus}\cdots Cu^{2\oplus}\right\} \xrightarrow{1\,e^{\ominus}} \text{Dehydroascorbate} + Cu^{\oplus}\cdots Cu^{\oplus}$$

In the reoxidation phase, one does not detect any EPR signal from superoxide anion, nor is any H_2O_2 released, suggesting a rapid and efficient 4-e$^{\ominus}$ transfer from 4 $Cu^{\oplus}$ to O_2. This is a poorly understood process for which Hamilton (1969) has proposed the following scheme. (Reoxidation by two 2-e$^{\ominus}$ steps would avoid both $O_2^{\ominus\cdot}$ and, probably more importantly, discrete formation of hydroxy radicals.)

$Cu^{\oplus}$ $Cu^{\oplus}$ $Cu^{\oplus}$ $Cu^{\oplus}$ $\rightleftharpoons$ $Cu^{\oplus}$ $Cu^{\oplus}$ $Cu^{\oplus}$ $Cu^{\oplus}$–O=O $\xrightarrow[\text{transfer}]{2\text{-}e^{\ominus}}$ $Cu^{\oplus}$ $Cu^{\oplus}$ $Cu^{2\oplus}$ $Cu^{2\oplus}$–O–OH

$\rightleftharpoons$

$Cu^{2\oplus}$ $Cu^{2\oplus}$ $Cu^{2\oplus}$ $Cu^{2\oplus}$ + H_2O $\underset{}{\overset{H^{\oplus}}{\rightleftharpoons}}$ $Cu^{2\oplus}$ $Cu^{2\oplus}$ $Cu^{2\oplus}$ $Cu^{2\oplus}$–OH $\xleftarrow[\text{transfer}]{2\text{-}e^{\ominus}}$ $Cu^{\oplus}$ $Cu^{\oplus}$ $Cu^{2\oplus}$ $Cu^{2\oplus}$=O + H_2O

4-$e^{\ominus}$-oxidized enzyme

14.B.2 Cytochrome Oxidase

The electronic description of the fast 4-$e^{\ominus}$ transfer from 4 $Cu^{\oplus}$ to O_2 in ascorbate oxidase remains inadequate. A parallel 4-$e^{\ominus}$ transfer with reduction of O_2 to H_2O is the last redox step in both mitochondrial and bacterial respiratory chains catalyzed by cytochrome oxidase (Caughey et al., 1976) and is quantitatively the major route for O_2 reduction in aerobic organisms. This membrane enzyme has two coppers and two iron atoms (in slightly distinct heme groups) for a 4-$e^{\ominus}$ storage capacity. Again, almost nothing is known about how this redox transformation, essential to aerobic life, occurs. One of the copper atoms is Cu^{II}, the other possibly Cu^{I} with a cysteine-thiolate radical as ligand. The cytochrome-oxidase problem has been complicated by the fact that the enzyme is a hydrophobic membrane species. Recent purification to homogeneity suggests at least seven polypeptides in the yeast enzyme, some coded for by mitochondrial DNA, others by nuclear DNA (Poynton and Schatz, 1975). This may be one of the most structurally complex enzymes known.

14.C COPPER ENZYMES REDUCING O_2 TO H_2O: MONOOXYGENASES

The third category of cuproenzymes are those that show monooxygenase activity while reducing O_2 to H_2O. Here we consider dopamine-β-hydroxylase and the tyrosinases.

14.C.1 Dopamine-β-Hydroxylase

We noted earlier (Chapter 12) that tyrosine hydroxylase is a pterin-dependent monooxygenase, producing 3,4-dihydroxyphenylalanine (dopa) as product. The dopa is subsequently decarboxylated by a pyridoxal-phosphate–dependent enzyme to dopamine (the role of pyridoxal phosphate is discussed in Chapter 24).

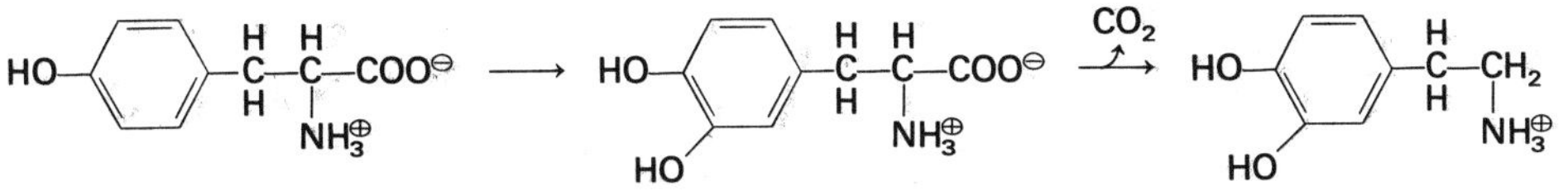

L-tyrosine $\longrightarrow$ L-dopa $\xrightarrow{CO_2}$ dopamine

Continuing in this biosynthetic pathway for neurotransmitter biosynthesis, the next enzyme hydroxylates dopamine at C-2 (the benzylic position) to produce norepinephrine, one of the adrenergic transmitters.

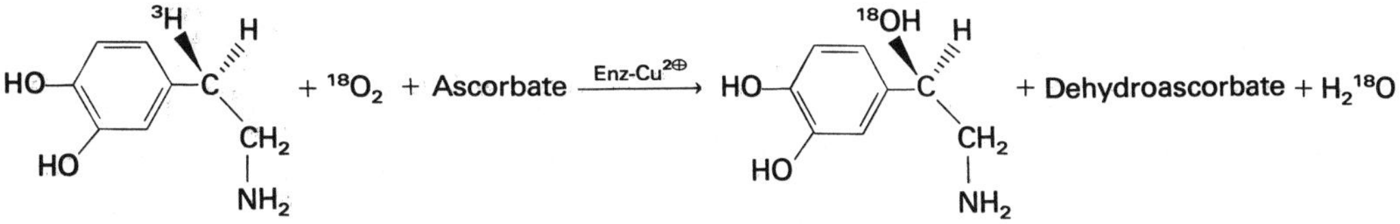

dopamine + $^{18}O_2$ + Ascorbate $\xrightarrow{\text{Enz-Cu}^{2\oplus}}$ norepinephrine (noradrenalin) + Dehydroascorbate + $H_2{}^{18}O$

This enzyme is a copper-dependent monooxygenase, dopamine-β-hydroxylase (Vaneste and Zuberbühler, 1974). The mammalian enzyme is found localized in storage vesicles in nerve endings of the sympathetic nervous system and in the adrenal glands, in the chromaffin cells of the adrenal medulla (Kirshner, 1957). The enzyme from bovine adrenal glands is a multimer of about 290,000 mol wt (subunits ~36,000 mol wt); assays for copper range from 4 to 7 per multimer. In the isolated enzyme, the copper is nonblue cupric ion.

In the reactions catalyzed by flavin- or pterin-dependent hydroxylases, there is a requirement for an external reducing agent (NADH or NADPH), required to produce reduced coenzyme, which is the active species combining with O_2 to form the hydroxylating agent. Analogously, in the dopamine-β-hydroxylase reaction, ascorbate is specifically required as reducing agent to generate two atoms of $Cu^{\oplus}$ as the electronic sink presumably interacting with O_2 to activate it for substrate hydroxylation. Kinetic studies confirm that the initial phase of catalysis is ping-pong BiBi (Goldstein et al., 1968). The ascorbate binds to the $Cu^{2\oplus}$–enzyme and reduces it to $Cu^{\oplus}$–enzyme; the dehydroascorbate departs prior to binding of O_2 or dopamine. The EPR signal disappears in the ascorbate-treated enzyme, as expected. In the absence of any excess ascorbate, the reduced enzyme can hydroxylate one molecule of substrate, undergoing reoxidation back to catalytically incompetent $Cu^{2\oplus}$–enzyme. One might then propose the following cycle, with two copper atoms in electronic contact.

$$\text{Enz} \cdot (2\,Cu^{2\oplus}) + \text{Ascorbate} \rightleftharpoons \text{Enz} \cdot (2\,Cu^{\oplus}) + \text{Dehydroascorbate}$$

$$\updownarrow \; O_2, \text{ dopamine}$$

$$\text{Enz—}(2\,Cu^{2\oplus}) \cdots H_2O + \text{Norepinephrine} \leftrightarrows \text{Enz—}(2\,Cu^{\oplus}) \cdot O_2 \cdots \text{Substrate}$$

The chemical nature of the hydroxylating species remains obscure—although, by analogy to the other coenzyme-dependent hydroxylases discussed, one suspects that an electrophilic oxygen species is required (in Hamilton's terminology, an "oxenoid" reagent; see Hamilton, 1969, 1971), and that it could be at the formal oxidation state of peroxide if two electrons are transferred from the copper atoms before substrate oxygenation.

Hamilton has speculated, on the other hand, that the O_2 bound to a cuprous ion at the active site will have an electronic structure to which the $Cu^{2\oplus}$-superoxide anion is a contributing resonance form:

$$\{Cu^{\oplus} \cdot O_2 \longleftrightarrow Cu^{2\oplus} \cdot O_2^{\ominus\cdot}\}$$

The superoxide anion, however, is nucleophilic, not electrophilic as required. So this anion per se is unlikely to be the hydroxylating species we are seeking. Rather, a peracid or analogous electronic structure would be appropriate. Hamilton (1974) then proposes that an adjacent enzyme carboxylate or amide functionality might be attacked nucleophilically by the $O_2^{\ominus\cdot}$, generating either a peracid or peramide, respectively, at the enzyme active site that would qualify as an oxygen-inserting agent, perhaps by an epoxide intermediate. Further study will be required to validate these speculations.

$$\text{Enz}(\cdots Cu^{2\oplus} \cdot O_2^{\ominus\cdot})\text{—C(=O)OH} \rightleftharpoons \text{Enz}(\cdots Cu^{2\oplus} \cdot O\text{—O—})\text{C(OH)(OH)} \xrightleftharpoons{-H_2O} \text{Enz}(\cdots Cu^{2\oplus} \cdots O\text{—O—})\text{C=O} \text{ (peracid)} \xrightarrow{\text{substrate}}$$

$$\text{Enz}(\cdots Cu^{2\oplus} \cdot O_2^{\ominus\cdot})\text{—C(=O)HNR} \rightleftharpoons \text{Enz}(\cdots Cu^{2\oplus} \cdot O\text{—O—})\text{C(HO)(NHR)} \xrightleftharpoons{-H_2O} \text{Enz}(\cdots Cu^{2\oplus} \cdots O\text{—O—})\text{C=NR} \text{ (peramide)} \xrightarrow{\text{substrate}}$$

$$\text{3,4-(HO)}_2C_6H_3\text{—CH(OH)—CH}_2\text{—NH}_2$$

14.C.2 Tyrosinases

The tyrosinases are cuproenzymes of ubiquitous distribution. They function in the biosynthesis of the skin pigment melanin in vertebrates, in sclerotization of insect cuticles, and in the biosynthesis of polyphenolic compounds (many of which possess antibiotic activity) in plants and microbes (Vaneste and Zuberbühler, 1974; Malmstrom et al., 1975).

Tyrosinases will oxidize *ortho*-diphenols (e.g., catechol), by two electrons, to the *ortho*-quinones, which are often unstable primary products and susceptible to decomposition and polymerization. The observed stoichiometry with catechol is 2 moles of diphenol dehydrogenated per mole of O_2 consumed, with the O_2 undergoing net 4-$e^{\ominus}$ reduction to water (two molecules).

$$2\ \text{catechol } (o\text{-diphenol}) + O_2 \longrightarrow 2\ o\text{-quinone} + 2\,H_2O$$

catechol (*o*-diphenol) — *o*-quinone

In addition to this oxidase activity, tyrosinases will demonstrate monooxygenase activity when presented with an appropriate *monophenol* substrate, yielding again the *ortho*-quinone product, but in this instance using one substrate molecule per O_2 and catalyzing oxygen insertion at the *ortho* position of the phenol.

$$\text{phenol} + O_2 \longrightarrow o\text{-quinone} + H_2O$$

o-quinone

The enzyme has been purified from a diverse variety of sources including mushroom, silkworm, and a mouse melanoma. Subunit sizes are similar, suggesting one Cu per subunit of 32,000 mol wt, and it seems that the functional catalytic unit may be dimeric, involving two coppers per active site. It is likely that the copper (EPR-silent) in the native enzyme is $Cu^{\oplus}$. It was known for some time that tyrosinase is irreversibly inactivated during catalysis by some substrates. When [^{14}C]-phenol was used, labeled enzyme was obtained on inactivation (B. Wood and Ingraham, 1965). This observation is consistent with the idea that, on an occasional turnover, some nucleophilic group adds to the electrophilic *o*-quinone product before it is released from the active site. This inactivation blocks both oxidase and oxygenase activities.*

Despite much effort, the catalytic mechanisms for both activities remain unclear. It is known that, for oxygenase action on a monophenol substrate, there is a continuous requirement (both in initiating the first turnover and during steady-state catalysis) for a diphenol (e.g., catechol) molecule (Vaneste and Zuberbühler, 1974). Thus, a diphenol is required for both activities of the enzyme. Mason (1965) has suggested that the diphenol is a cosubstrate for the oxygenation of monophenols, being required in each catalytic cycle to produce the active hydroxylating agent. Hamilton (1974) has been willing to suggest a scheme for the monooxygenase activity in line with his proposals for the flavin hydroxylases that we have noted earlier, and consistent with the observations that $Cu^{\oplus}$ does not undergo detectable valence change and that no substrate radical signals are seen.

In the initial step, O_2 would complex with the active-site cuprous ion (a spectroscopically detectable complex) and from this complex undergo attack by the bound catechol. Loss of water from such an initial adduct could yield a molecule with the vinylogous ozone moiety that Hamilton champions.

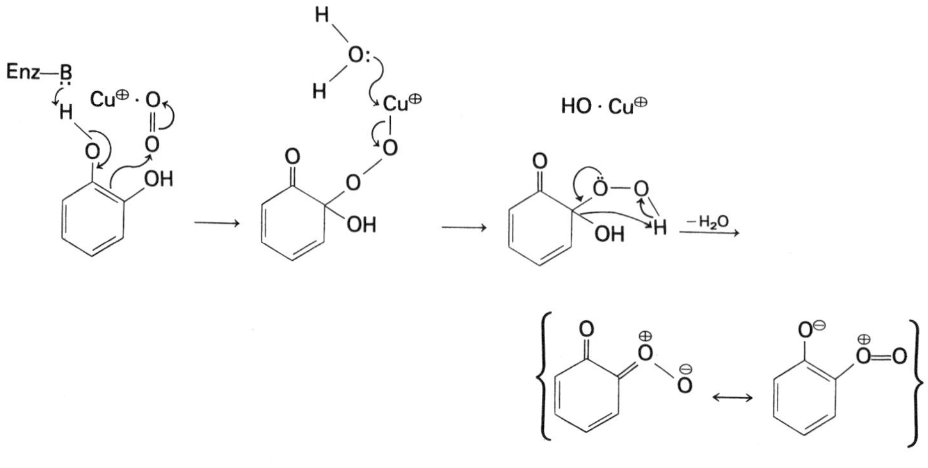

This electrophilic species, in analogy to his postulate for flavin hydroxylase (¶12.B.1), could be attacked by the anion of the bound monophenol substrate to generate an intermolecular adduct whose breakdown is analogous to that in the flavin case, yielding one molecule of *ortho*-quinone (derived from the initial

* However, the stoichiometry of such an inactivating process may be variable (K. Lerch, personal communication, 1978).

o-diphenol) and the cyclohexadienone tautomer of the diphenol product, containing one of the oxygen atoms from O_2.

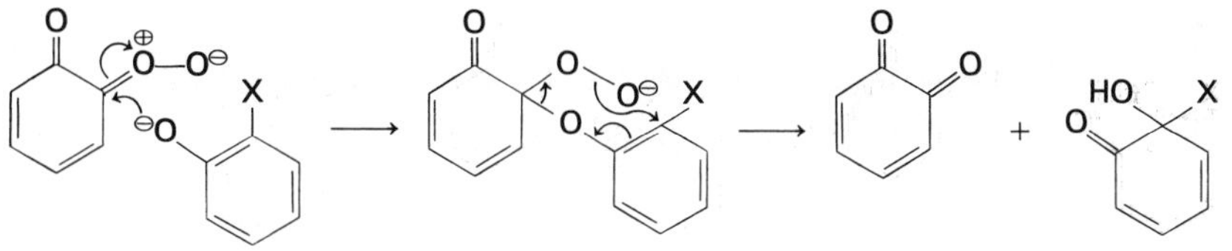

When X is OH (i.e., a second molecule of catechol), loss of H_2O from the gem diol would yield another molecule of *o*-quinone. Hamilton (1974) has pointed out that, given freely dissociating species (which may not be the case), this mechanism predicts, during each turnover, loss of one catechol oxygen to water and its replacement by a molecule of O_2 during the functioning of the enzyme as oxidase.

Chapter 15

Hemoprotein Oxidases, Monooxygenases, and Reductases

Proteins containing heme prosthetic groups are present in all free-living organisms. They play such diverse roles (not all in each organism) as reversibly binding dioxygen for transport (hemoglobin and myoglobin), transferring electrons one at a time in membraneous respiratory chains (cytochromes), reducing peroxides (catalases and peroxidases), and acting as terminal component in multienzyme systems involved in hydroxylations. The latter systems hydroxylate steroid hormones, steroidal bile acids, fatty acids, xenobiotic hydrocarbons for detoxification, and (in a well-studied bacterial system) camphor (Gunsalus et al., 1974).

In almost all instances, the prosthetic group (tightly and sometimes covalently bound to apoenzyme) involves an iron atom ($Fe^{3\oplus}$ or $Fe^{2\oplus}$) coordinated to a macrocyclic tetrapyrrole ring (protoporphyrin IX). The pyrrole nitrogens make up four ligands to the iron in equatorial positions, leaving two axial positions to be filled by other ligands.

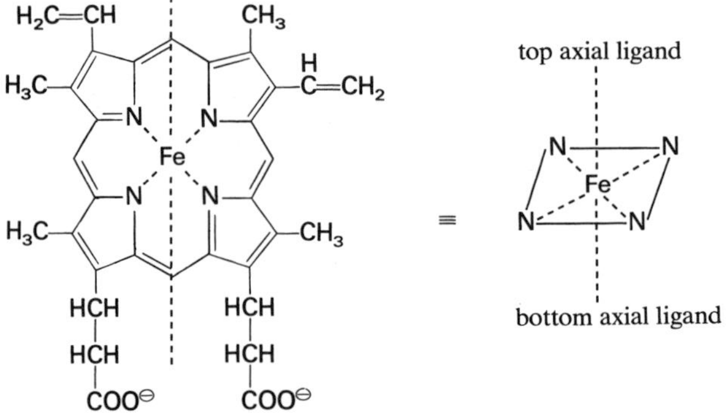

protoporphyrin IX

top axial ligand

bottom axial ligand

Generally, at least the bottom axial position is filled by a ligand from apoprotein. The top axial position may be filled by a ligand from the protein, H_2O, or O_2 or, adventitiously, from CO or CO_2 (inactivators). We shall not discuss here the structure or function of the oxygen-binding proteins hemoglobin or myoglobin where the functional forms are $Fe^{2\oplus}$ (and redox inactive), except to note that the ability to bind O_2 reversibly as an $Fe^{2\oplus}$–O_2 complex and to release the O_2 in parts of the body where oxygen tension is low (without oxidation of the iron from $Fe^{2\oplus}$ to $Fe^{3\oplus}$) is a crucial capability for multicellular aerobic life forms. Before discussing catalytic functions of heme–protein monooxygenases, though, we must make a brief digression to understand the nature of the iron in the heme group and the way that the tetrapyrrole macrocycle conditions its size, spin states, and reactivity.

Two oxidation states of iron—$Fe^{2\oplus}$ (Fe^{II}) and $Fe^{3\oplus}$ (Fe^{III})—are stable in aqueous solutions, and these are the major redox forms in iron proteins. (Both higher and lower oxidation states of iron—Fe^{I}, Fe^{IV}, Fe^{V}—are known in special inorganic complexes.) However, iron can have more than one spin state (Spiro and Saltman, 1974). Fe^{II} has six valence electrons in the $3d$ shell; Fe^{III} has five $3d$ electrons. Because there are five $3d$ orbitals, in the absence of perturbing influences (for instance, in the gas phase), we would expect single occupancy of all five $3d$ orbitals in Fe^{III} and a single electron pairing for Fe^{II}. (Recall that the lowest energy state is for electrons to fill orbitals singly; when electrons pair up, they must take opposite spins, requiring a certain amount of spin-pairing energy, *P*.)

Fe^{III}	↑	↑	↑	↑	↑	$3d$ orbitals (gas phase)
Fe^{II}	↑↓	↑	↑	↑	↑	

The situation is different when the iron has ligands coordinated to it, as it does in any of the biological situations we are discussing. In this instance, the orbitals are no longer degenerate; there is ligand field-splitting (Spiro and Saltman, 1974). In an octahedral ligand field (of the type found in heme), the orbitals are split into two discrete groups of three low-lying and two higher-lying orbitals, separated by a field-splitting energy, Δ.

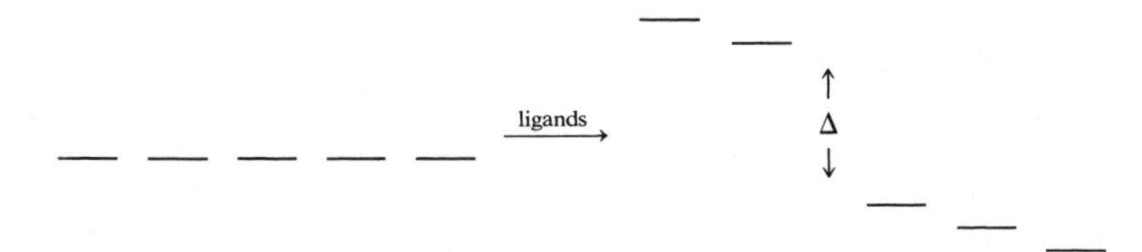

When Δ is small relative to the energy P required for spin-pairing of electrons, then all the five $3d$ orbitals will be populated; this gives *high-spin iron.* High-spin Fe^{III} has five unpaired electrons; high-spin Fe^{II} has four unpaired electrons.

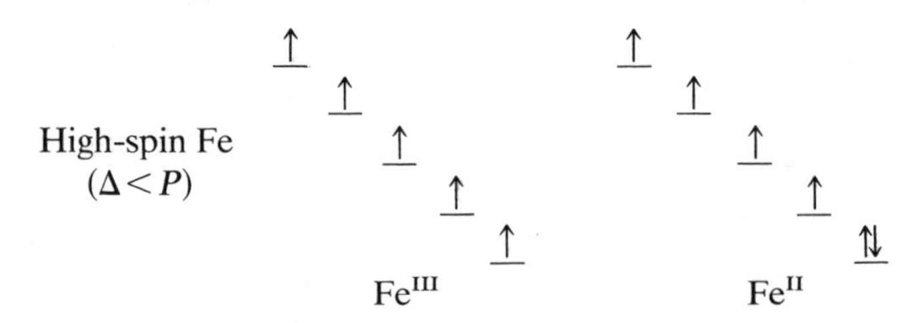

When the magnitude of the field-splitting energy is greater than the spin-pairing energy ($\Delta > P$), the two higher orbitals are not populated, and the five $3d$ electrons distribute themselves in the three low-lying orbitals, giving *low-spin iron* (Spiro and Saltman, 1974).

Low-spin Fe ($\Delta > P$) Fe^{III} Fe^{II}

In the low-spin state, Fe^{III} has only one unpaired electron, and Fe^{II} has none; low-spin Fe^{II} is EPR silent. In the low-spin states, the $3d$ electrons are not in the upper, antibonding orbitals, and low-spin iron (Fe^{II} or Fe^{III}) is more compact than high-spin iron. This change in size with change in spin state is exceedingly important for heme-iron chemistry. The cavity in the tetrapyrrole macrocycle is 2.02 Å in diameter. Low-spin Fe^{II} and Fe^{III} (~1.91 Å) can just fit in this cavity and so be in the plane of the macrocycle. But high-spin iron is about 2.06 Å in diameter (Eichhorn, 1971), and experimental results show that it does not sit within the heme cavity (R. J. P. Williams, 1974).

High-spin Fe^{III} is about 0.3 Å out of the plane; high-spin Fe^{II} (which has one more electron) is about 0.7 Å out of the plane. The size of the ligand field-splitting in heme-iron systems appears to be balanced at the transitions between high-spin and low-spin states, and the axial ligands can determine the equilibrium position. For example, in the resting form of the oxygen-carrying hemoproteins myoglobin and hemoglobin, the iron is high-spin $Fe^{2\oplus}$. Myoglobin is monomeric, but hemoglobin is a functional tetramer and shows positive cooperativity in oxygen binding (¶5.C.3.g). When hemoglobin binds oxygen, the oxygen coordination induces a shift from high-spin $Fe^{2\oplus}$ to low-spin $Fe^{2\oplus} \cdot O_2$; the iron atom can sink 0.7 Å back into the equatorial plane of the macrocycle. Perutz (1970) has argued that this movement on oxygenation may trigger the protein conformational changes that engineer the binding cooperativity as subsequent subunits of the tetramer bind oxygen. We shall suggest that the transitions between high-spin and low-spin states also are important in hemoprotein monooxygenases and oxidases, to be discussed in this chapter.

The heme-dependent monooxygenase component has been termed generically a cytochrome ("pigment in the cytoplasm") as a representative of the family of red, heme-containing proteins first observed by the British pathologist Charles McMunn in 1886. For many years, the cytochromes were thought to be localized to mitochondrial membranes in animal cells, where they function as one-electron carriers between various iron-sulfur flavoprotein dehydrogenases (e.g., succinate dehydrogenase, NADH dehydrogenase) and molecular oxygen. Figure 15-1 is a schematic view of electron transfer from succinate to O_2 in the mitochondrial membrane, with reduction potentials indicated.

The protoporphyrin macrocycle has been modified differently in the side chains of cytochromes *b*, *c*, and *a*, and the distinct environments in the apoproteins account for the different reduction potentials of the three cytochromes. We shall have no more to say about these cytochromes except to note that, although enormous effort has revealed chemistry and X-ray structures of the various cytochromes (Nicholls and Elliot, 1974; Dickerson and Timkovich, 1975; Hagihara et al., 1975), the exact routes whereby electrons pass in and out of the heme iron from donor molecule to acceptor molecule are not understood. Among the debates are whether electrons are passed in by way of aromatic or thiol-containing amino-acid residues reversibly reduced and oxidized, and/or whether electrons enter through the tetrapyrrole π-system of the heme macrocycle.

More recently, two different kinds of cytochromes were discovered in animal cells (initially in liver cells). These cytochromes are not localized in the mitochondria (the subcellular organelles concerned with energy production and transduction), but rather in the endoplasmic reticulum (also known as microsomal membranes). These microsomal cytochromes have functions different from those of the classical mitochondrial cytochromes and distinct from those of one another. Cytochrome b_5 is a specific component of a multienzyme complex in microsomal membranes involved in desaturation of stearyl-CoA (a C_{18} fatty acyl-CoA) to the

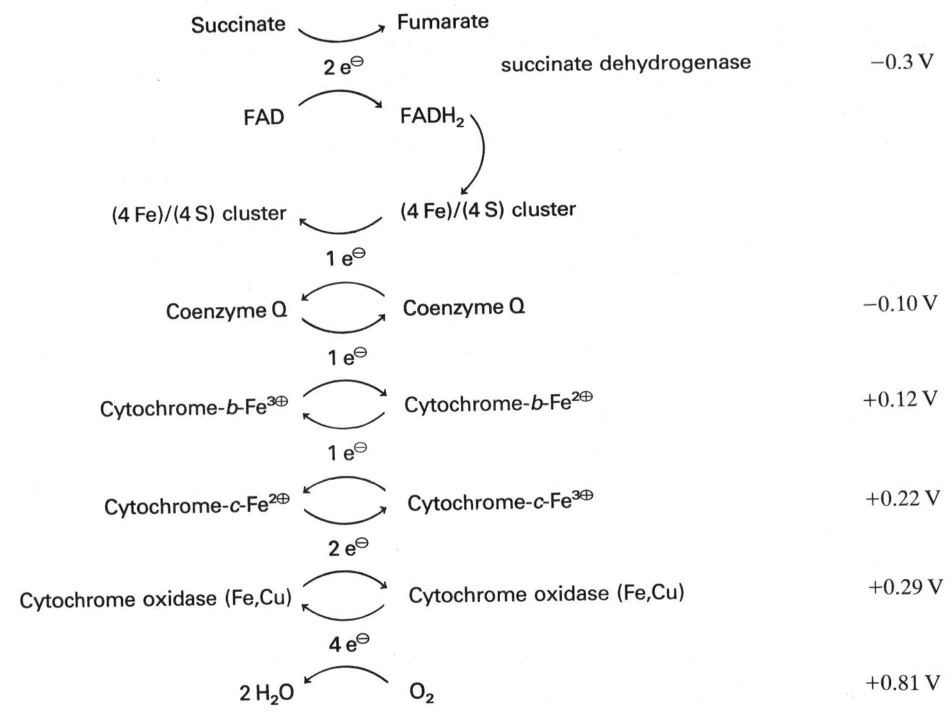

Figure 15-1
Schematic representation of electron transfer from succinate to O_2 in the mitochondrial membrane.

cis-Δ^9-monounsaturated oleyl-CoA. We shall examine this sequence later in this chapter. The other microsomal cytochrome is cytochrome P_{450}, the heme-iron–dependent oxygenase, and we shall discuss this enzymatic activity first.

15.A HEME-DEPENDENT HYDROXYLATION SYSTEMS

In this section we shall discuss the activities of the two microsomal cytochromes just mentioned, the bacterial hydroxylation of camphor, and the action of aryl-hydrocarbon hydroxylase and carcinogen activation.

15.A.1 Cytochrome P_{450}

In 1958 a novel hemoprotein was observed in liver preparations. It shows a cytochrome-*b*-type spectrum, with an absorption maximum at 420 nm (Soret band); in the $Fe^{2\oplus}$ form, it complexes with carbon monoxide to absorb at 450 nm, an anomalous λ_{max}. The new hemoprotein was termed cytochrome P_{450} (Omura and Sato, 1964). Investigations on this new chromophoric entity revealed it to have monooxygenase activity and, in fact, to be the terminal component in discrete multienzyme systems funneling electrons initially from NADH or NADPH, eventually to molecular oxygen, reducing and activating it such that one oxygen atom ends up in H_2O and the other undergoes oxygen transfer to substrate. In the sense that the ferrous porphyrin cofactor of cytochrome P_{450} binds to O_2, it resembles hemoglobin and myoglobin, the reversible-O_2-binding heme proteins. Unlike these oxygen carriers, though, the P_{450} hydroxylases also "activate" O_2 for hydroxylation. Chang and Dolphin (1975) have noted that the "axial ligands of the heme iron in cytochrome P_{450} are of great interest because they hold the key to our understanding of the enzymatic function and the underlying principles that enable the simple complex protoheme to perform various functions ranging from oxygen transport, oxidation catalysis, to electron transport." The bottom axial ligands contributed by hemoglobin and myoglobin are imidazole nitrogens of histidine residues; the bottom ligand in cytochrome P_{450} is thought to be a cysteinyl sulfur. How $Fe^{2\oplus}$-heme reactivity (either to bind O_2 reversibly or to activate it for insertion) is conditioned by the nature of the bottom axial ligand is a problem under scrutiny in both enzymatic and model studies. Presumably, interaction of the sulfur axial ligand on the bottom in P_{450} must affect the coordinated O_2 at the top axial position to polarize the O—O bond effectively and set it up for fission.

There is a family of P_{450}-type cytochromes that are found in animal cells, but initially they were refractory to extensive solubilization and purification. At that point a bacterial P_{450} system, the *Pseudomonas* camphor hydroxylase, appeared to be the system of biochemical choice because all three components of that multienzyme complex are soluble proteins, and each has been purified to homogeneity. We shall examine the basic features of P_{450} action in the bacterial system and then return to the animal enzymes.

15.A.2 Camphor Hydroxylation

The stoichiometry for the three-enzyme complex catalyzing bacterial hydroxylation of camphor is the following (Gunsalus et al., 1974).

$$NADPH + O_2 + \text{camphor} \longrightarrow \text{5-}exo\text{-alcohol} + NADP^{\oplus} + H_2O$$

camphor 5-*exo*-alcohol

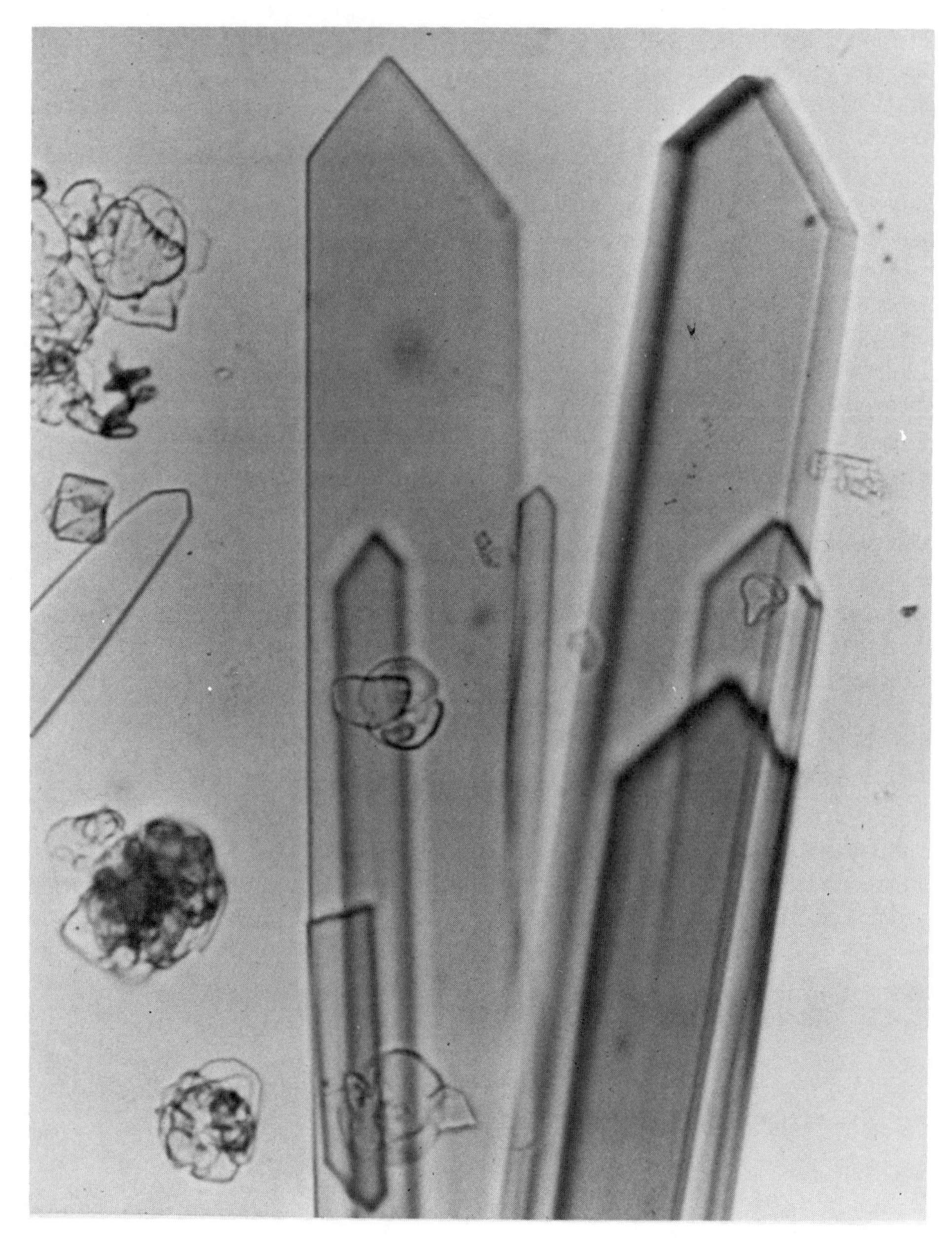

Figure 15-2
Crystalline cytochrome P_{450}–D-camphor complex (From Yu and Gunsalus, 1974, p. 97)

The three enzymes are (1) an NADH-oxidizing flavoprotein dehydrogenase; (2) a (2 Fe)/(2 S)-cluster protein (putidaredoxin; see Table 13-1); and (3) the heme–protein P_{450} hydroxylase (Fig. 15-2). The camphor substrate is hydroxylated at an unactivated methylene position (a rare biochemical reaction type occurring at unactivated carbons) to produce the 5-*exo*-alcohol specifically.

The roles of the first two enzymes are reasonably clear. The flavoenzyme oxidizes NADPH and then is reoxidized by passing one electron at a time into the (2 Fe)/(2 S) cluster (one-electron capacity) of the putidaredoxin. The putidaredoxin is the feeder system for passing electrons into the hemoprotein hydroxylase one at a time. The O_2 undergoes a formal 4-$e^{\ominus}$ reduction in the P_{450} step; two electrons are ultimately provided by the NADPH and two by C-5 of camphor, which is oxidized by 2 $e^{\ominus}$ to the alcoholic product.

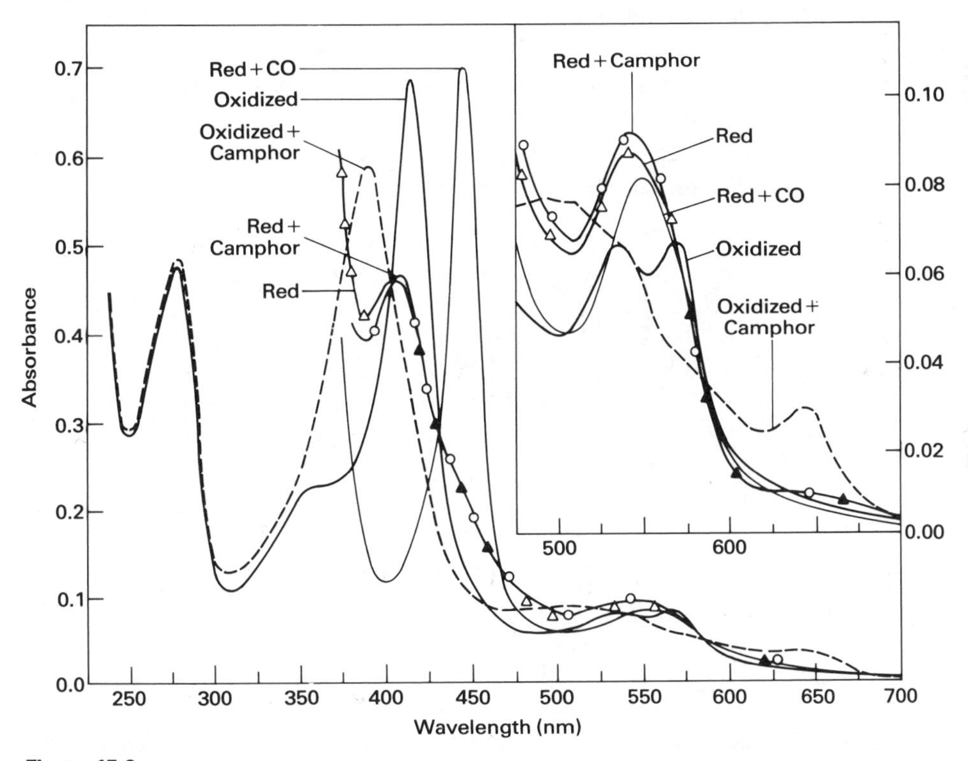

Figure 15-3
Absorption spectra of P_{450} from the camphor hydroxylase system. A Cary model-14 spectrophotometer with a light path of 1 cm was used; the *inset* shows results with a light path of 5 cm at 25° C. The protein concentration was 340 μg per ml in 50 mM potassium phosphate buffer, pH 7.0. In the sample labeled Oxidized + Camphor, the substrate was 100 μM D-camphor. In the sample labeled Red + Camphor, solid $Na_2S_2O_4$ was added under argon, then gassed with CO for 20 sec and sealed under CO at 1 atm, thus converting the P_{450} to its *reduced* (ferrous) state. (From Yu and Gunsalus, 1974, p. 98)

In the camphor system, a barrage of sophisticated and physical techniques have been unleashed to study the P_{450} involvement in substrate hydroxylation (Gunsalus et al., 1974). They have yielded the kinetic mechanism we shall now describe. The $Fe^{3\oplus}$-P_{450} resting state is low-spin $Fe^{3\oplus}$ by optical and EPR spectra (Fig. 15-3). The first redox step is passage of 1 $e^{\oplus}$ from putidaredoxin's iron–sulfur chromophore (by unknown means) into the Michaelis complex of camphor and hemoprotein to generate substrate–$Fe^{2\oplus}$ complex where the Fe^{II} is now high-spin by optical and ESR analysis. These spin interconversions represent movement of iron out of the heme equatorial plane and may induce protein conformational changes. The rate for the first electron in is 35 sec^{-1}. At this stage, O_2 can bind reversibly as the top axial ligand. There are large optical changes, and the $Fe^{2\oplus} \cdots O_2$–S complex is diamagnetic. The spin state of the iron is unclear but, in analogy to hemoglobin oxygenation, the O_2 may induce the low-spin condition again. The hydroxylase system is now poised to receive the second electron from the feeder enzymes, and this input is the rate-determining step at 17 sec^{-1}. After input of the second electron, then, species do not accumulate, and the picture is murky. Hydroxylation occurs rapidly, with regeneration of starting low-spin heme-Fe^{III} and alcohol product.

E---S, Fe^{III} (N, N, N, N) $\xrightarrow[35\ sec^{-1}]{1\ e^{\ominus}}$ E---S, Fe^{II} (N, N, N, N) $\xrightarrow{O_2}$ E---S, O=O---Fe^{II} (N, N, N, N)

$\downarrow$ 1 $e^{\ominus}$, 17 sec^{-1}

E---S, $O_2^{\ominus\cdot}$---Fe^{II} (N, N, N, N) $\xrightarrow{rapid}$ S—OH + Fe^{III} (N, N, N, N)

A major question is: what is the electronic nature of the heme–O_2 complex after the second electron is introduced? We have written one possible structure as Fe^{II}–$O_2^{\ominus\cdot}$, but coordinated superoxide—as Hamilton (1974) has noted—is more likely to be nucleophilic than electrophilic as required for the hydroxylating agent. The coordinated oxygen species could be at the level of peroxide, a good chemical candidate for an oxygen-transfer agent. This view is supported by recent studies showing that organic peroxides (and H_2O_2 at higher concentrations) can react with animal P_{450} monooxygenases and substrates to elicit specific substrate hydroxylation—essentially a peroxide shunt around the flavoprotein–iron/sulfur-protein electron-input system. It is also possible that, after input of the second electron, an Fe^{III}-OOH species undergoes O—O cleavage of the peroxide to yield a transient ferryl-ion derivative of P_{450}, a high iron-oxidation state analogous to those proposed for catalase and peroxidases later in this chapter. We can summarize these possibilities as follows:

$$Fe^{II}—O_2 \xrightarrow{1\,e^{\ominus}\ ?} \begin{cases} \xrightarrow{a} \{Fe^{I}—O_2 \longleftrightarrow Fe^{II}—O_2^{\ominus\cdot}\} \\ \xrightarrow{b} Fe^{III}—(OOH)^{\ominus} \\ \xrightarrow{c} Fe^{III}—O—OH\ \ {}^{\oplus}H \longrightarrow \{Fe^{V}—O^{2\ominus} \longleftrightarrow Fe^{IV}{=}O\} \end{cases}$$

Thus, either possibility b or c could be regarded as an oxygenating intermediate. Recent model studies, where cyclohexanol has been converted in moderate yield to the *cis*-1,3-cyclohexanediol, have been interpreted as proceeding via such a coordinated $Fe^{IV}{=}O$ as oxygen-transfer reagent (Groves and Van der Puy, 1976). This fascinating enzymatic problem may be somewhat clarified when the X-ray structure of the *Pseudomonas* P_{450} monooxygenase is completed. Still unclear, also, is the way in which the multienzyme hydroxylation complex controls the temporally discrete delivery of one electron at a time to the axial oxygen coordinated to the heme iron. Perhaps the interconversion of iron spin states may induce conformational changes that control the access of the second electron from the (2 Fe)/(2 S) reservoir.

15.A.3 Functions of P_{450} Monooxygenases

Now that we have some ideas about the composition and mechanism of P_{450}-mediated hydroxylation in the bacterial system, we can categorize the involvement of P_{450} monooxygenase according to three broad functional classes (Table 15-1), and these bring us back to animal cells.

In the *Pseudomonas* reaction, the ultimate fate of camphor following its conversion to the 5-*exo*-alcohol is degradation and energy provision to the cell. An analogous hydroxylation of *n*-octane occurs in other pseudomonads to provide a functional group that activates the molecule for eventual enzymatic degradation and release of energy to the cell. Thus the function of the P_{450} monooxygenase in bacterial cells is that of dissimilation or energy provision.

In animal cells, the P_{450} systems can be used in biosynthetic conversions or in detoxification of xenobiotic compounds. In general, the P_{450} systems in animal

Table 15-1
Functional categorization of P_{450} monooxygenase activities

I. Dissimilatory or energy-yielding functions (bacterial cells)
   - A. Camphor → *exo*-alcohol → → isobutyrate
   - B. *n*-Octane → 1-octanol → degradation

II. Synthetic or assimilatory functions (adrenal cells)
   - A. Processing of cholesterol to corticosteroids
     1. 17-α-Hydroxylase
     2. 21-Hydroxylase
     3. 11-β-Hydroxylase
     4. Side-chain dismutase (cleavage activity)

III Detoxifying functions (liver cells)
   - A. Hydroxylation of xenobiotic compounds

cells may be two-component or three-component systems, correlatable with the particular intracellular membrane systems in which they are localized. The three-enzyme sequence of flavoprotein dehydrogenase, iron–sulfur protein, and P_{450} hemoprotein occurs both in bacteria and in animal mitochondrial membranes. A two-enzyme sequence, flavoprotein dehydrogenase followed by P_{450} hemoprotein hydroxylase, exists in the endoplasmic reticulum of adrenal cells and liver cells.

15.A.3.a Assimilatory function

In adrenal cells, the initial product in sterol biosynthesis, cholesterol, is converted into the adrenal steroid hormones (aldosterone, hydrocortisone, and corticosterone) by a series of P_{450}-mediated hydroxylation steps.

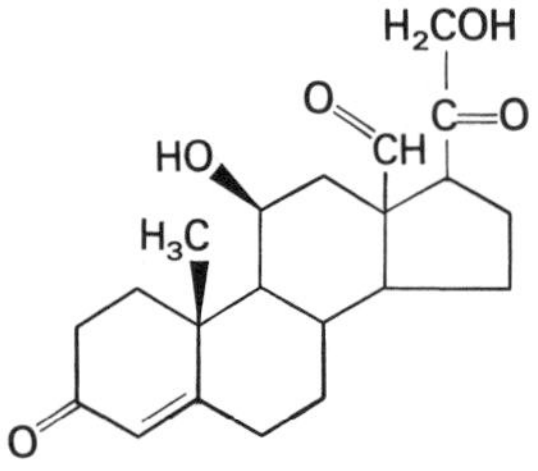

aldosterone

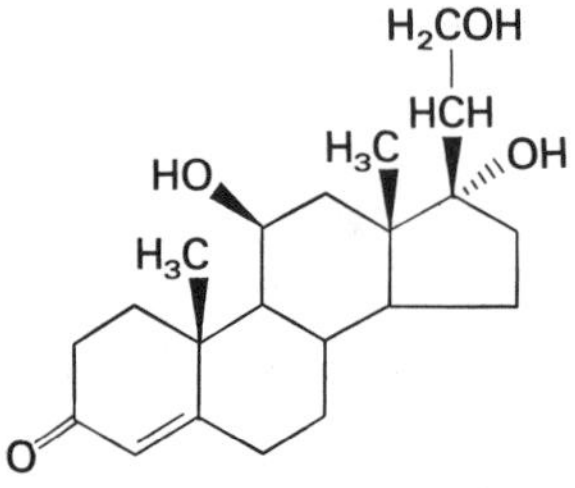

hydrocortisone (cortisol)

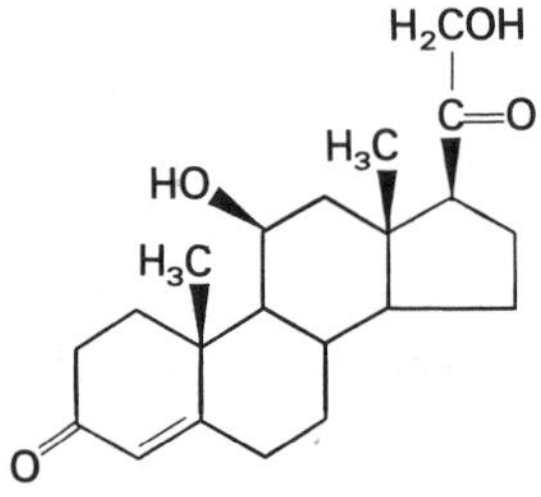

corticosterone

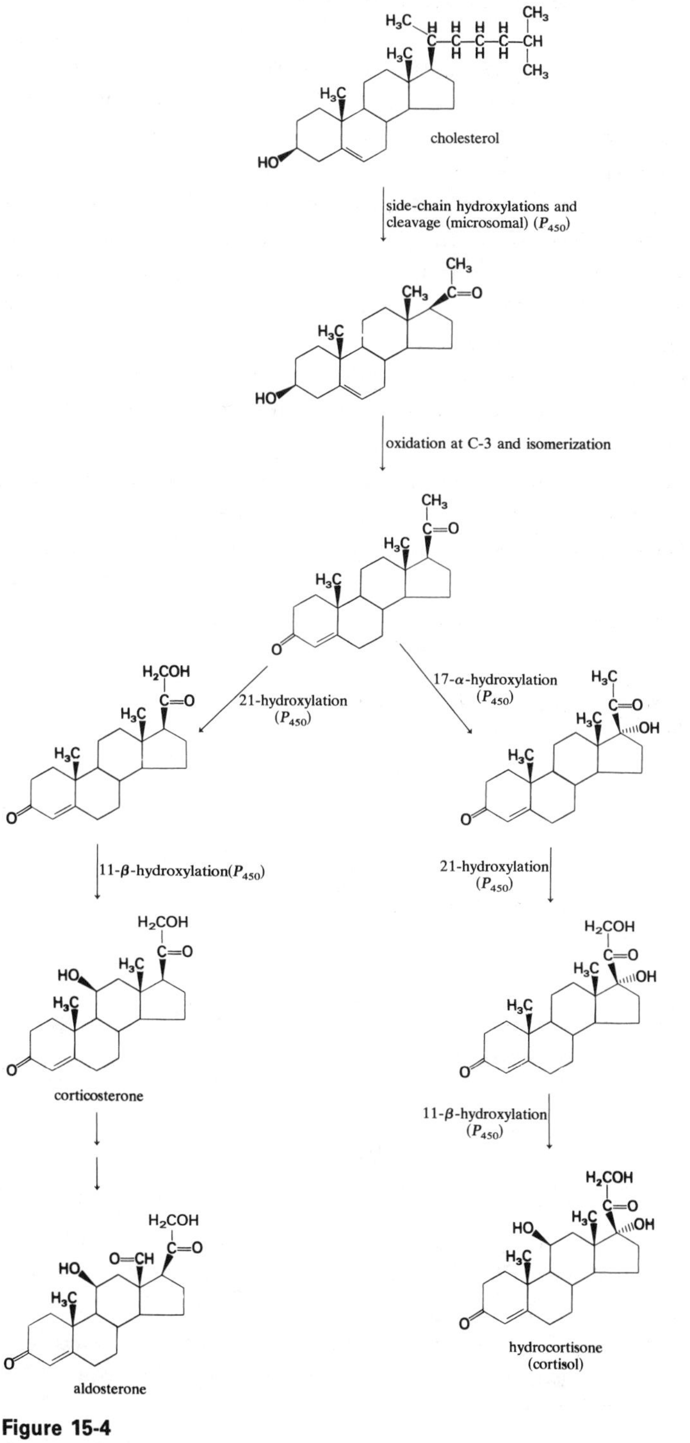

cholesterol

side-chain hydroxylations and cleavage (microsomal) (P_{450})

oxidation at C-3 and isomerization

21-hydroxylation (P_{450})

17-α-hydroxylation (P_{450})

11-β-hydroxylation(P_{450})

21-hydroxylation (P_{450})

corticosterone

11-β-hydroxylation (P_{450})

aldosterone

hydrocortisone (cortisol)

Figure 15-4
One path for conversion of cholesterol into the steroid hormones.

Figure 15-4 shows one path for conversion of cholesterol into these steroid hormones (Hamberg, Samuelsson, et al., 1974). Note the massive involvement of P_{450}-mediated hydroxylations in production of adrenal hormones. Both the 17-α-hydroxylase and the 21-hydroxylase activities are in the endoplasmic reticulum of the adrenal cortex and do *not* involve an Fe/S-cluster protein. On the other hand, the side-chain–cleavage activity and the 11-β-hydroxylase activity are in the mitochondria of the adrenal cortical cell and *do* involve an Fe/S protein, specifically the (2 Fe)/(2 S) protein *adrenodoxin*, as an intermediate between the flavoprotein (NADPH–adrenodoxin reductase) and the P_{450} hydroxylase. This three-component system is functionally analogous to the camphor hydroxylase. Again, most of the hydroxylations on these steroid molecules represent hydroxylation at unactivated aliphatic positions.

The cholesterol side-chain–cleavage activity that represents the first step in steroid hormone biosynthesis is a remarkable 6-$e^{\ominus}$ oxidation and cleavage of a pair of adjacent methylene groups to two carbonyl fragments, pregnenelone and isocaprylaldehyde (C. Walsh, 1978). Three moles of NADH and three moles of O_2 are consumed in the process.

3 NADH + 3 O_2 + [cholesterol: HO, H_3C (18), H_3C (19), H_3C (21), (20), (22), CH_3, CH_3] $\xrightarrow{E \cdot FAD}$ $\xrightarrow{E \cdot (2\,Fe)/(2\,S)}$ $\xrightarrow{P_{450(SCC)}}$

3 $NAD^{\oplus}$ + 4 H_2O + [pregnenolone: HO, H_3C, H_3C, H_3C, O] + [isocaprylaldehyde: H, O, CH_3, CH_3]

Three discrete steps are proposed, the first two involving hydroxylations at C-22 and C-20 to produce a 20,22*R*-diol species, which then cleaves across the bond between C-20 and C-22 in the last oxidative step (SCC = side-chain cleavage). As noted in ¶15.A.2, a ferryl-oxoiron species may be involved, and its possible role in the oxidative cleavage of the diol intermediate is pictured as follows (VanLier and Rousseau, 1976).

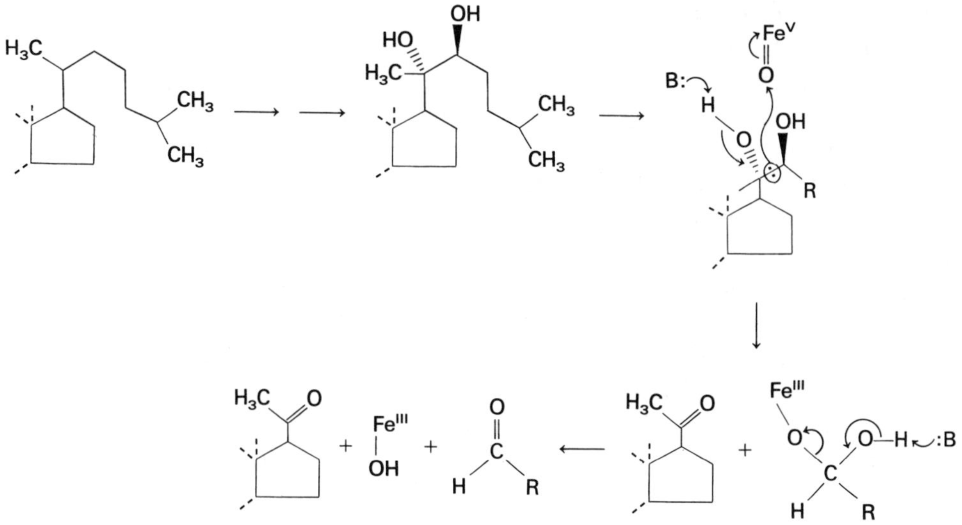

15.A.3.b Detoxifying function

The P_{450} hydroxylases in liver, which act as detoxification agents for various foreign compounds circulating in the blood, are inducible hepatic monooxygenases. The presence of the xenobiotic compound induces de novo synthesis of hydroxylase, which acts to detoxify the compound by hydroxylation to render the compound more polar and water soluble, so that its excretion via the kidney will be more favorable. One can use different foreign compounds as inducers and observe genetically distinct P_{450} hydroxylase components being produced, a state of microheterogeneity. Two compounds that typically have been used as inducers are phenobarbital (5-ethyl-5-phenylbarbituric acid) and 3-methylcholanthrene.

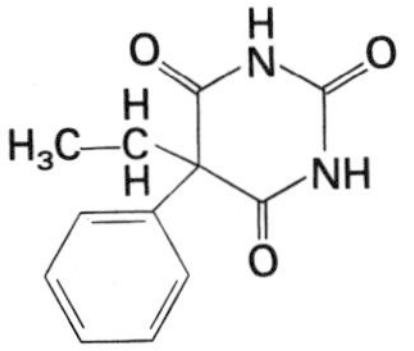

phenobarbital

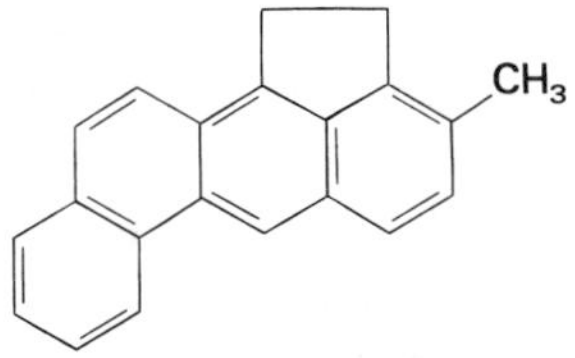

3-methylcholanthrene

Phenobarbital induces a two-component system that hydroxylates various amines and thiol zenobiotics *but works less well on polycyclic hydrocarbons.* Complementarily, 3-methylcholanthrene induces a P_{448} system specialized for hydroxylation of polycyclic hydrocarbons; this system has been termed *aryl-hydrocarbon hydroxylase* (AHH) (Heidelberger, 1975). Under maximal induction, up to 10% of the microsomal membrane proteins of liver can be of the P_{450} family, which is thus a major set of membrane proteins.

In fact, purified AHH shows four protein bands on gel electrophoresis; the heterogeneity may stem from different amounts of carbohydrate (mannose, glucosamine) in this membrane glycoprotein (Ernster et al., 1976). The flavoprotein dehydrogenase (NADPH–P_{450} reductase) is the same in both systems; it is an enzyme containing FAD and FMN (C. H. Williams, 1976).

In recent years, the two subfamilies of liver cytochrome P_{450} have been purified to homogeneity. Coon and associates term the P_{450} induced by phenobarbital "cytochrome P_{450} LM2" (for liver-membrane band 2 on acrylamide gels) and that induced by polycyclic aromatics "P_{450} LM4." The LM2 enzyme has a subunit mol wt of 48,700, LM4 one of 55,300, and they show no immunological cross-reactivity. Both require phospholipid (as a mimic of the membrane phase?) for activity. The LM2 enzyme will hydroxylate alkanes; will dealkylate nitrogen, sulfur, or oxygen functionalities;

$$R'\text{—}X\text{—}CH_2\text{—}R \longrightarrow \left[R'\text{—}X\text{—}CH(OH)\text{—}R \;\; \ddot{B}\right] \longrightarrow R'\text{—}X^{\ominus} + HC(=O)\text{—}R + BH^{\oplus} \qquad X = O, N, S$$

and will catalyze deaminations, desulfurizations, sulfenylations, sulfoxidations, *N*-hydroxylations, dehalogenations, alkene peroxidations, and epoxidations. The enzyme will also catalyze azogroup and nitrogroup reductions, where these functionalities must actually act as replacements, not for the oxidizable substrate, but for O_2. That is, they act as alternate electron acceptors.

$$R'\text{—}N{=}N\text{—}R \xrightarrow[2\,H^{\oplus}]{+2\,e^{\ominus}} R'\text{—}NH\text{—}NH\text{—}R$$

$$R\text{—}NO_2 \xrightarrow[6\,H^{\oplus}]{+2\,e^{\ominus}} R\text{—}NH_2 + 2\,H_2O$$

In keeping with its metabolic role of all-purpose detoxifying agent, cytochrome P_{450} LM2 may catalyze the widest variety of reactions of any known enzyme.

Recent elegant studies with norbornane as substrate for the pure complex of NADPH-P_{450} reductase and P_{450} LM2 suggest that the enzyme is not completely stereospecific and that the substrate oxidation may proceed by radical intermediates. Norborane itself is hydroxylated to a 3.4:1 mixture of *exo* and *endo* alcohols, indicating an apparent incomplete stereospecificity for *exo* hydrogen abstraction, possibly suggesting that the oxoiron-oxidizing species is positioned in some angle between the *exo* and *endo* hydrogens.

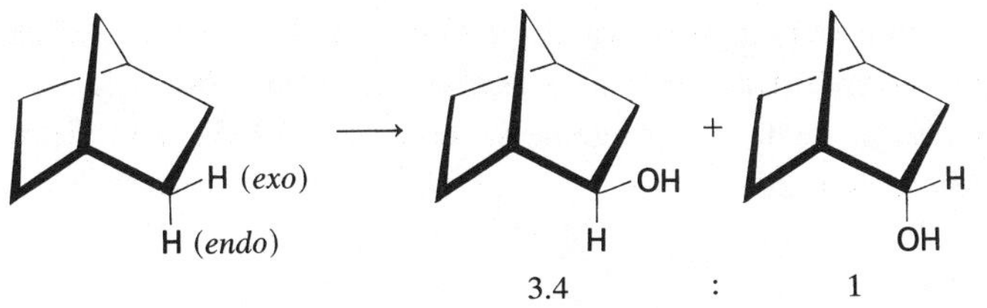

H (*exo*) H (*endo*) ⟶ OH H + H OH

3.4 : 1

With all-*exo*-D_4-norbornane, the *exo*-alcohol:*endo*-alcohol product ratio actually inverts to 0.76 : 1, reflecting the intrinsic kinetic isotope effect, although the step is not rate-determining in catalysis because there is no V_{max} isotope effect (Groves et al., 1978).

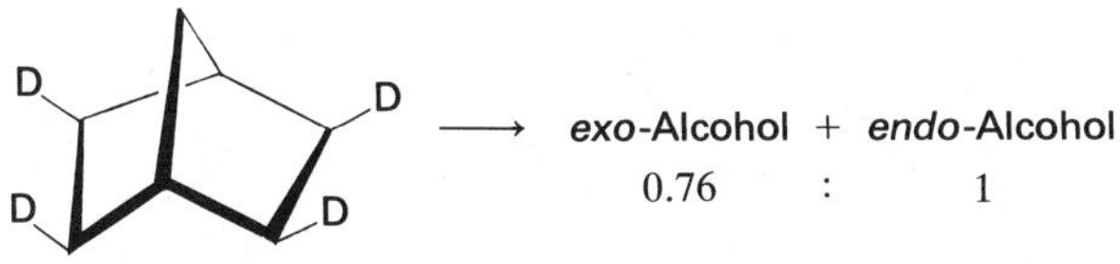

⟶ *exo*-Alcohol + *endo*-Alcohol

0.76 : 1

When the separated *exo*- and *endo*-alcohols were examined for deuterium content, both D_3- and D_4-species were observed. In particular, 25% of the *exo*-alcohol molecules surprisingly retained all four deuteriums, suggesting that in those instances (at a minimum) an *endo* hydrogen is abstracted as a hydrogen atom to yield a norbornyl radical that can epimerize and then be trapped by the oxygenating agent to yield an *exo*-alcohol rather than the *endo*-alcohol anticipated if *complete* retention of stereochemistry proceeded (and if the intermediate remained chirally pure).

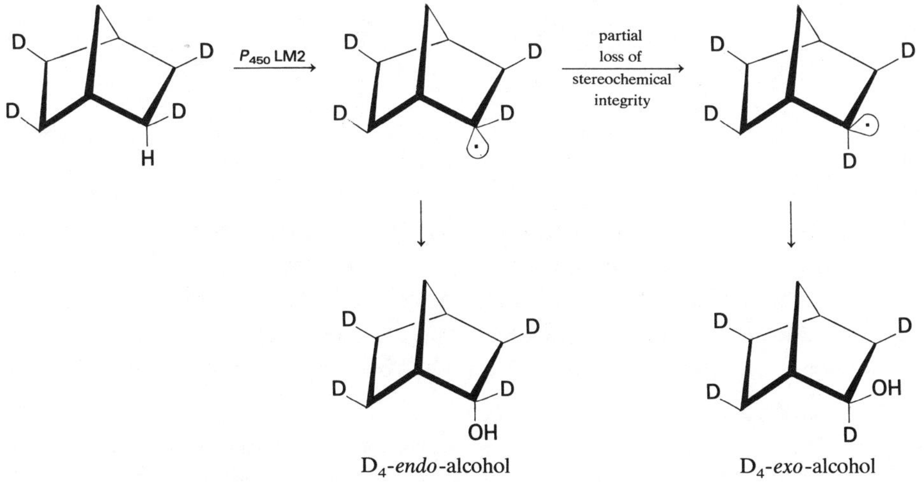

P_{450} LM2

partial loss of stereochemical integrity

D_4-*endo*-alcohol

D_4-*exo*-alcohol

The results obtained by Groves et al. (1978) argue against an oxidation mechanism invoking hydride transfer and a norbornyl cation, because such cations are known to undergo facile rearrangements (Bartlett, 1965), and no rearrangement products were observed.

15.A.4 Aryl-Hydrocarbon Hydroxylase and Carcinogen Activation

If one examines simple aromatic hydrocarbon substrates such as naphthalene, one finds that the naphthalene-1-ol produced by AHH action has been hydroxylated with a concomitant N.I.H. shift.

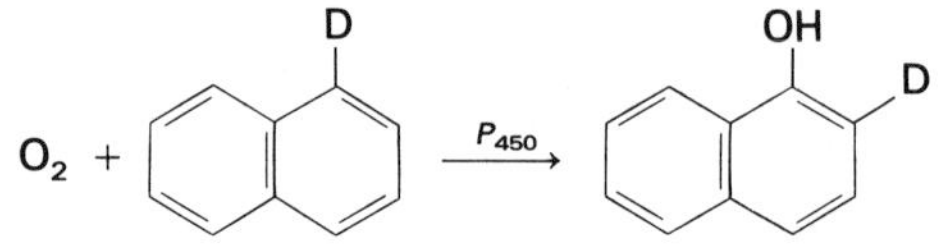

D
O_2 +
P_{450}
OH
D

The 1,2-shift, we have argued in Chapter 13, is diagnostic of an epoxide intermediate, an arene oxide.

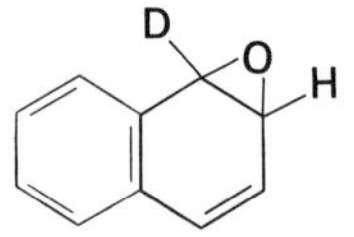

D
O
H

In fact, careful examination reveals that the initial enzymatic product is the arene oxide itself (Heidelberger, 1975). For example,

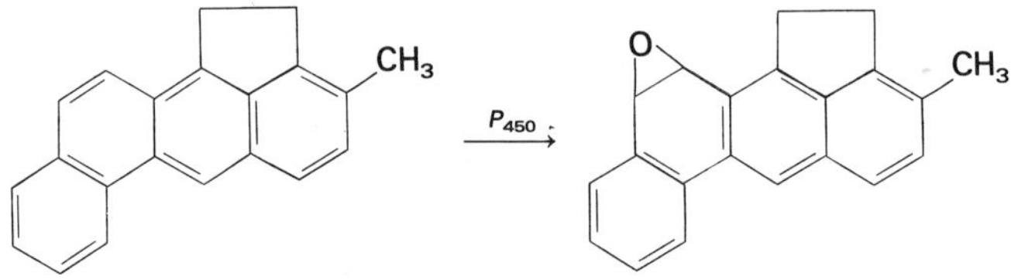

CH_3
P_{450}
O
CH_3

Thus, aryl-hydrocarbon hydroxylase is actually an aromatic epoxidase; the arene oxides are the products. The enzymatic product is actually a relatively reactive epoxide.* This enzymatic process reveals the dilemma that aromatic hydrocarbons pose to cells. *Enzymatic detoxification involves metabolic activation to reactive intermediates* (the arene oxides). The arene-oxide products can suffer various

* Epoxidation may occur at several different sites on polycyclic hydrocarbons such as benzpyrene to yield a mixture of isomeric monoepoxides that turn out to display distinct carcinogenic potential.

metabolic fates. Some are enzymatic, some are nonenzymatic. Some are harmless, some are pernicious. We can delineate two enzymatic routes and three nonenzymatic routes.

15.A.4.a Enzymatic routes for epoxide products

In the first enzymatic route, epoxide hydrase (a microsomal membrane activity) catalyzes *trans* addition of water to the epoxide (Lu et al., 1975).

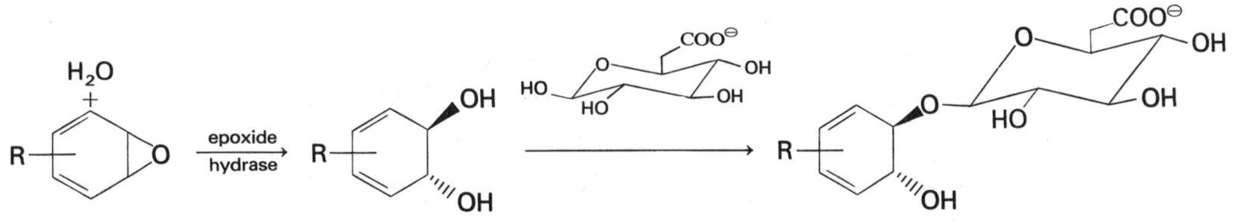

trans-dihydrodiol

glucuronide

The resultant *trans*-dihydrodiol can undergo further metabolism, including conjugation to a glucuronic acid for facilitated excretion.

The second enzymatic route involves glutathione-*S*-epoxide transferases; these are soluble enzymes that catalyze addition of the thiolate anion of glutathione to the epoxide (Habig et al., 1974).

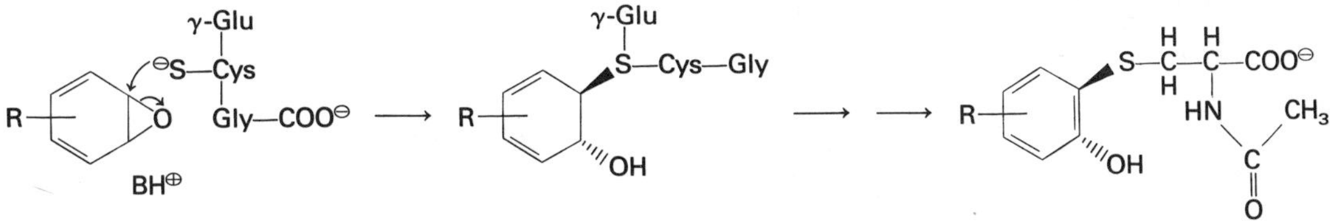

mercapturic acid

The *trans*-thiodiol adduct can be sequentially degraded with hydrolysis of the γ-Glu and Gly components, and *N*-alkylation of the cysteinyl α-amino group, to yield a mercapturic acid (an *N*-acetyl-*S*-alkylated cysteine), which is excretable. Both the epoxide hydrase and the glutathione-*S*-epoxide transferases are enzymes induced along with the P_{450} system. These two routes of epoxide opening represent controlled enzymatic deactivation of the arene oxides and successful detoxification pathways.

15.A.4.b Nonenzymatic routes for epoxide products

The three nonenzymatic routes are reviewed by Bruice and Bruice (1976).

In the first route, acid-catalyzed ring opening (S_N1) to the phenol can occur after initial protonation to generate a feasible leaving group in the ring-opening step.

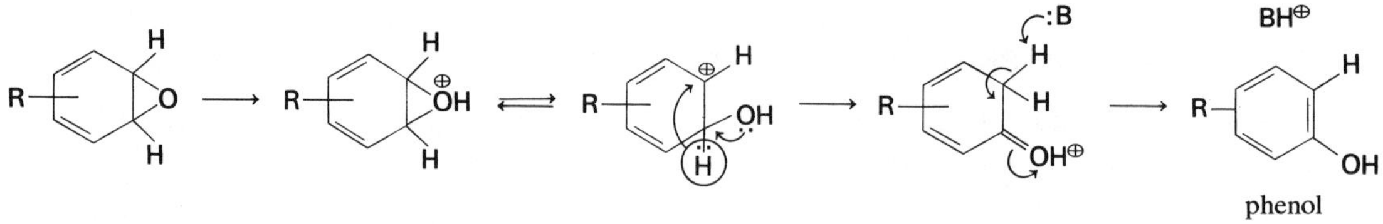

phenol

The carbonium ion can undergo a 1,2-hydride shift to the cyclohexadiene tautomer of the phenolic product. This sequence accounts for observed phenol production and the N.I.H. shift attending its formation.

In the second route, water can add nonenzymatically to give the *trans*-dihydrodiol, the same product formed from epoxide hydrase.

Neither of the first two nonenzymatic routes is deleterious to the organism. The third nonenzymatic route involves alkylative capture of the arene oxide or the carbonium ion by cellular nucleophiles. When these nucleophiles are groups in DNA, RNA, or essential proteins, this represents an extremely deleterious fate and may be the initiation step in chemical carcinogenesis. Both strong nucleophiles and weak nucleophiles can be alkylated—the good nucleophiles by S_N2 opening of the arene oxides, and the weak nucleophiles by S_N1 opening of the arene oxide to the carbonium ion.

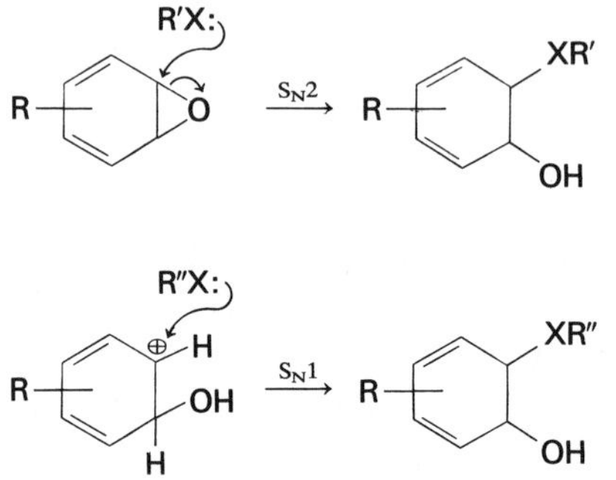

The nature of the sites in DNA and RNA that become modified in cells that activate polycyclic hydrocarbons is a subject of intensive scrutiny. The structures of two adducts between activated polycyclic arene oxides and a base in DNA have recently been determined. Both the 7,8-dihydroxy-9,10-epoxide of benzo[*a*]pyrene (Weinstein et al., 1976; see also ¶15.A.4.c) and the 4,5-epoxide of 7,12-dimethylbenz[*a*]anthracene (Jeffrey et al., 1977) have been attacked by the N-2 amino group of a guanyl residue. A third adduct structure results after metabolic activation of aflatoxin B_1, a metabolite (that is a potent liver carcinogen) from the mold *Aspergillus flavus.* The aflatoxin is epoxidized (and rendered electrophilic) at the 2,3 double bond; then the epoxide ring is opened by attack of N-7 of a guanine residue in DNA (Essigman et al., 1977).

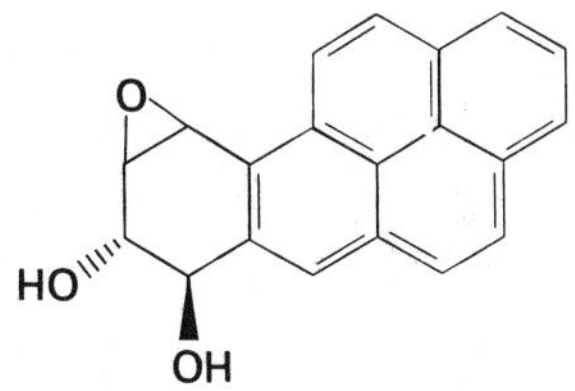

7,8-dihydroxy-9,10-epoxide of benzo[*a*]pyrene

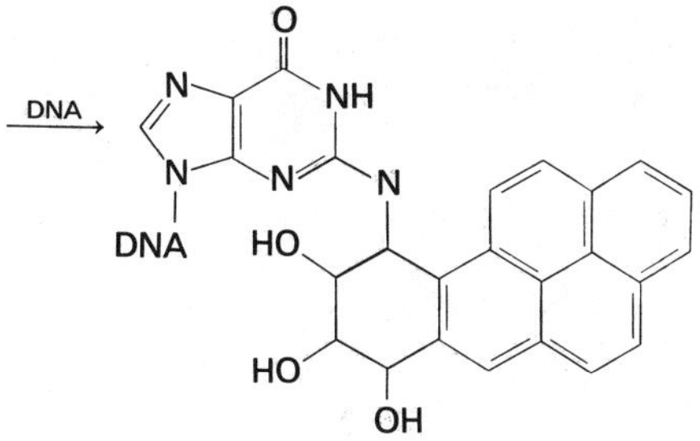

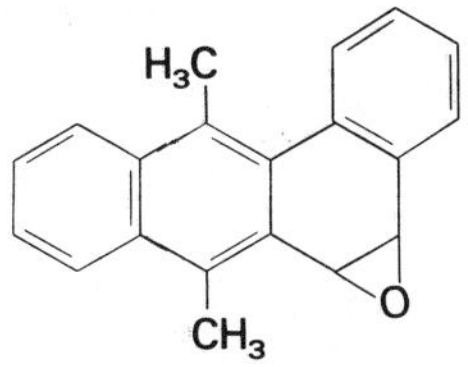

4,5-epoxide of 7,12-dimethylbenz[*a*]anthracene

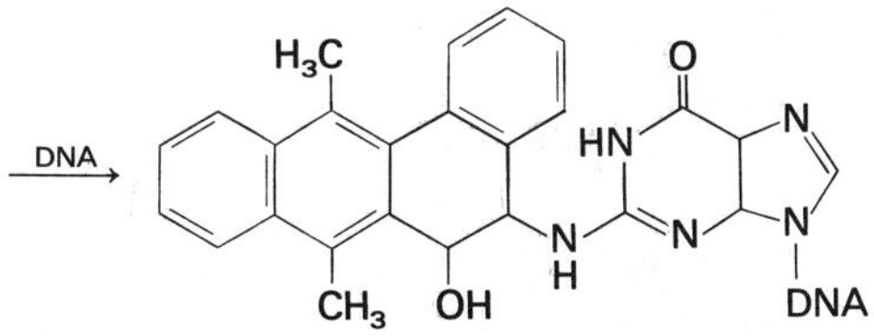

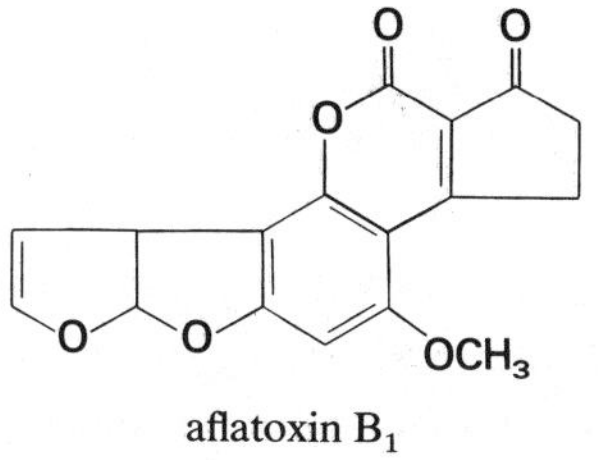

aflatoxin B_1

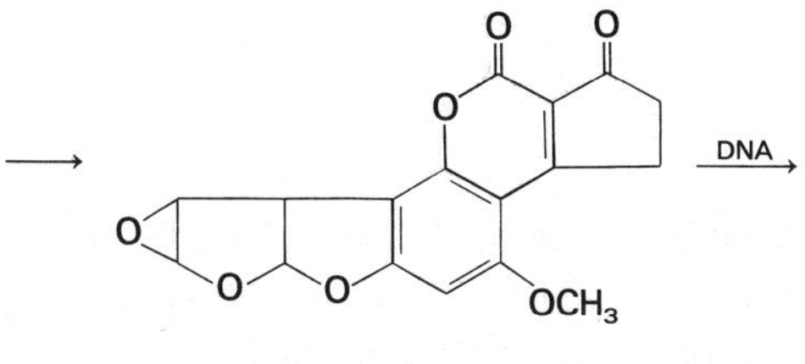

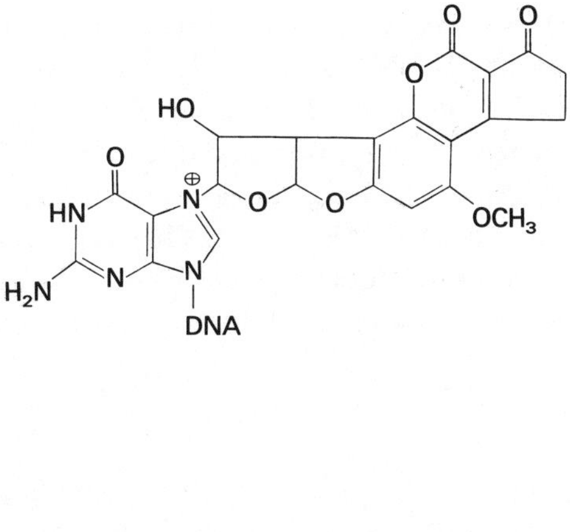

15.A.4.c Benzo[*a*]pyrene activation and epoxide-hydrase complicity

Benzo[*a*]pyrene is probably the principal active precarcinogen in coal tar. Although it can be converted to several epoxides, phenols, or quinones (secondarily) by microsomal activation, a major primary oxygenated product is the 7,8-epoxide. The 7,8-epoxide is opened as expected to a *trans*-7,8-dihydrodiol by epoxide hydrase. However, this supposed detoxification step actually appears to put the organism *at greater risk* in this instance, for the 7,8-dihydrodiol is again a substrate for the liver microsomal P_{450} monooxygenase, undergoing stereospecific epoxidation to the 7,8-dihydroxy-9,10-epoxide (Yang et al., 1976).

P_{450}, O_2 → epoxide hydrase →

7,8-*trans*-dihydrodiol

↓ P_{450}, O_2

Covalent adduct ← DNA (fast)

Tetrols ← H_2O

7,8-dihydroxy-9,10-epoxide

Although this dihydroxyepoxide can again be processed by epoxide hydrase to the tetrahydrotetrol products (2 diastereomers), it appears that—before this can occur—the dihydroxyepoxide can covalently label nucleic acids and protein even faster than the initial 7,8-epoxide. In the Ames test for bacterial mutagenesis, the dihydroxyepoxide is a more potent mutagen than the initial epoxide, suggesting that it may be the actual carcinogen.

It has been noted for many years that some polycyclic aromatics are more carcinogenic than others, and electronic correlations of reactivity with carcinogenicity have been sought more or less unsuccessfully. One recent suggestion (Jerina et al., 1977) is that the "bay region" epoxides of aromatic polycyclics may be the most deleterious. The "bay region" refers to the region of the molecules where an angular phenyl group makes a hindered or bay region with the main polycyclic surface. The dotted lines indicate the bay regions of the following structures.

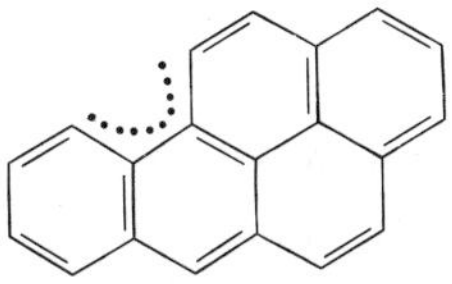

benzo[*a*]pyrene

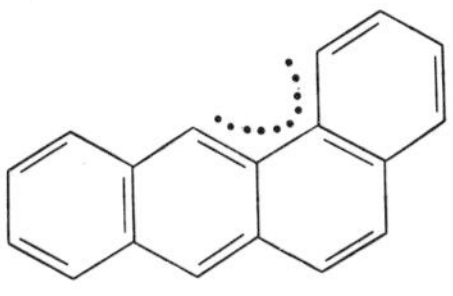

benz[*a*]anthracene

Jerina and colleagues have argued that molecular orbital calculations may suggest that carbonium ions can form most easily in these bay regions. This might imply that epoxides formed in these regions will open more often to carbonium ions and enter into S_N1 alkylation processes with DNA. It remains to be seen how predictive this concept will be. The 9,10-epoxide of benzo[*a*]pyrene might be construed loosely to be in the bay region.

15.A.5 Cytochrome b_5

The functional three-enzyme complex for oxygen-dependent hydroxylations involving a hemoprotein has its variants in other oxidative processes, such as the conversion of saturated acyl chains to monounsaturated chains. In the O_2-dependent stearyl-CoA-desaturase system discussed in this section, we note that the functional positions of the hemoprotein and nonheme-iron components are reversed from those in the P_{450} systems. Whereas cytochrome P_{450} is the entity that interacts with O_2 in the reactions just discussed, in the oxidative dehydrogenation of fatty acylthioesters we find that the cytochrome (cytochrome b_5) is the middle component and does not react directly with O_2. An X-ray structure is available for cytochrome b_5; the heme is buried in the interior of the protein, presumably inaccessible to O_2.

We look now at the NADH-dependent stearyl-CoA desaturase activity that involves cytochrome b_5. The microsomal fraction of liver is the subcellular membrane site in tissue extracts that is responsible for the oxidative desaturation of long-chain acyl-CoA-thioesters to monounsaturated acylthioesters. In rat-liver microsomes, stearyl-CoA is a preferred substrate, undergoing O_2-dependent dehydrogenation to the C_{18} *cis*-Δ^9-monolefinic product, oleyl-CoA. The *cis*-olefinic acylthioester is then incorporated into phospholipids, often at the C-2 position of the glycerol backbone. Unsaturated phospholipids have lower melting transitions and increase the fluidity of biological membranes. This dehydrogenation also requires NADH, which is stoichiometrically oxidized to $NAD^{\oplus}$.

$$\mathrm{NADH} + \mathrm{O_2} + \mathrm{CH_3(CH_2)_7CH_2CH_2(CH_2)_7\overset{O}{\overset{\|}{C}}SCoA} \longrightarrow \mathrm{CH_3(CH_2)_7CH{=}CH(CH_2)_7\overset{O}{\overset{\|}{C}}SCoA} + \mathrm{NAD^{\oplus}} + 2\,\mathrm{H_2O}$$

stearyl-CoA → oleyl-CoA (cis, H and H on the same side of C=C)

Careful investigations ruled out the existence of the likely intermediates, free 9- or 10-hydroxystearates, in the microsomal or in similar bacterial desaturase systems. Thus, in the 4-$e^{\ominus}$ reduction of O_2, two $e^{\ominus}$ apparently come from NADH and two from the acyl-CoA undergoing desaturation. The liver microsomal system requires three specific enzymes to accomplish this task, analogous to some of the P_{450} systems discussed earlier. Although this desaturase system requires a cytochrome, it is not the P_{450} hemoprotein, but one with distinct spectroscopic parameters—cytochrome b_5, which more closely resembles mitochondrial b cytochromes.

In recent years, thanks largely to the efforts of Strittmatter and colleagues, remarkable progress has been made—not only the purification of the three enzymes of the desaturase system, but also an understanding of the structural elements of the proteins that render them membrane soluble, as well as insights into the interactions of the purified enzymes in the membrane phase (Strittmatter, 1974; Rogers and Strittmatter, 1973, 1974; C. H. Williams, 1976). In addition to the hemoprotein cytochrome-b_5 component, there is a flavoprotein dehydrogenase (NADH–cytochrome-b_5 reductase) and a third enzyme with the actual stearyl-CoA desaturase activity.

NADH / $NAD^{\oplus}$ —(2 $e^{\ominus}$)→ FP / FPH_2 —(1 $e^{\ominus}$)→ Cyt-b_5-$Fe^{2\oplus}$ / Cyt-b_5-$Fe^{3\oplus}$ → Desaturase-$Fe^{3\oplus}$ / Desaturase-$Fe^{2\oplus}$ —(S, O_2)→ Active complex → H_2O + Oleyl-CoA

Initial isolation of cytochrome b_5 involved solubilization of this membrane protein by limited trypsin treatment. The cytochrome could then be purified as a hydrophilic protein by conventional techniques to yield a protein of 11,000 mol wt. This molecule would not rebind to microsomal membranes but was catalytically reduced by the flavoprotein, cytochrome-b_5 reductase. The flavoenzyme, in turn, was originally extracted from microsomes following a snake-venom treatment, and was purified as a hydrophilic protein containing one FAD per polypeptide chain of 33,000 mol wt. Again, the enzyme would not rebind to microsomal membranes.

More recent isolations of both the cytochrome and the flavoprotein dehydrogenase have employed nonproteolytic solubilizations with nonionic detergents such as deoxycholate and Triton-X-100, which solubilize the hydrophobic proteins as components of detergent micelles. This extraction procedure yields a cytochrome b_5 of 16,700 mol wt, containing 40 amino acids more than the material isolated by earlier procedures. As expected, trypsin cleaves this preparation to yield the 11,000-mol-wt hydrophilic fragment and the 40-amino-acid residue, which is quite hydrophobic and prone to self-association. This suggests two domains in the intact hemoprotein, one hydrophilic and one hydrophobic, producing an amphipathic protein. The hydrophobic tail enables the 16,700-mol-wt form of cytochrome b_5 to rebind to the microsomal membranes. The two forms of cytochrome b_5 isolated (intact and proteolyzed) appear indistinguishable with respect to catalytic criteria involved in interaction with the flavoprotein and reoxidation by various electron acceptors.

Encouraged by the results with cytochrome b_5, Strittmatter and Spatz found they could also purify the cytochrome-b_5 reductase by detergent extraction rather than proteolytic solubilization. This treatment led to isolation of reductase of 43,000 mol wt rather than the 33,000 mol wt obtained earlier. They noted that this mass increase is due to an additional 99-amino-acid segment of high hydrophobicity, which again permits the enzyme to rebind to microsomal membranes. Again, proteolytic cleavage (by chymotrypsin) can produce the hydrophilic fragment (33,000 mol wt). The catalytic properties, assayed by K_m values for substrates NADH and cytochrome b_5, are similar for either form of enzyme. The V_{max} for NADH remains at 29,000 min^{-1}, but the turnover number with cytochrome b_5 as reoxidant falls from 29,000 to 14,100 min^{-1}, which may be due to artifactual polymerization of the detergent-solubilized reductase during assay. Elegant studies have shown that the intact molecules of cytochrome b_5 and of b_5 reductase, held to the membrane surfaces by the hydrophobic tails of each of the proteins, can diffuse rapidly in the plane of the membrane and do not exist as tight functional complexes.

Purification of the third enzyme, the stearyl-CoA desaturase, has recently been completed by detergent extraction, producing an enzyme of 33,000 mol wt, a single polypeptide chain of 456 residues. Unlike the other two amphipathic components, the desaturase appears hydrophobic throughout the chain. The protein contains one atom of tightly bound nonheme iron, which is essential for catalysis (judged by the fact that activity is destroyed by chelators of $Fe^{2\oplus}$). Until more information is accumulated on the nature of this single atom of nonheme iron at the desaturase active site and its structure (is it like rubredoxin?), it is difficult to comment on how the O_2-utilizing oxidative desaturation process is effected. It could still involve initial formation of hydroxylated intermediates that are never released before subsequent loss of H_2O occurs. This would provide some mechanistic continuity with other Fe-dependent systems where O_2 is split.

$$R{-}\overset{H}{\underset{H}{C}}{-}\overset{H}{\underset{H}{C}}{-}R' + Fe^{2\oplus}\cdot O_2 \xrightarrow[H_2O]{} \left[R{-}\overset{OH}{\underset{H}{C}}{-}\overset{H}{\underset{H}{C}}{-}R'\right] \xrightarrow[H_2O]{} R{-}\underset{H}{C}{=}\underset{H}{C}{-}R'$$

Similar fatty acylthioester desaturase systems have been observed in microbial metabolism, and Bloch and colleagues have partially purified the three enzymatic components of a soluble system from *Euglena gracilis.* Again, a sequence of electron transfer from NADPH to iron-protein to desaturase is indicated, although in this case the middle component does not have heme iron as cytochrome b_5 does, but has a nonheme-iron ferredoxin with the Fe/S cluster as electron sink.

15.B HEME PROTEINS WITH HYDROPEROXIDASE ACTIVITY

There is a class of ferri-(Fe^{III})-hemoproteins that have been termed collectively "hydroperoxidases"; the preferred substrates of these enzymes are either H_2O_2 or alkyl peroxides. Enzymes preferring H_2O_2 as substrate are called catalases (Schonbaum and Chance, 1976), and those preferring alkyl peroxides are called peroxidases (Yamazaki, 1974), although each type *can* use either H_2O_2 or alkyl peroxides.

The generalized stoichiometry is the following:

$$AH_2 + ROOH \longrightarrow ROH + H_2O + A$$

where AH_2 = phenols, aryl and alkyl amines, hydroquinones, ascorbate, cytochrome *c*, or glutathione (i.e., easily oxidized functionalities).

Catalases use H_2O_2 preferentially (i.e., $AH_2 = H_2O_2$ and $ROOH = H_2O_2$) in a bimolecular dismutation where the second molecule serves as electron donor to reduce the first molecule.

$$2\,H_2O_2 \longrightarrow O_2 + 2\,H_2O$$

This dismutation is analogous to the dismutation of $O_2^{\ominus\cdot}$ catalyzed by superoxide dismutase. Indeed, just as the dismutase can be regarded as a surveillance enzyme to scavenge one of the reactive molecules that is a consequence of O_2 activation, so catalase functions to detoxify another reactive product of oxygen metabolism. In liver and kidney cells there are high concentrations of catalase in discrete subcellular organelles termed *peroxisomes*, a functional name. These organelles also contain D- and L-amino-acid oxidases and α-hydroxy-acid oxidases, the flavin-linked enzymes responsible for the bulk production of H_2O_2. The segregation of H_2O_2 generation and enzymatic decomposition from the general cytoplasm lessens the chance of toxicity before H_2O_2 can be dismuted to harmless O_2 and H_2O. These enzyme activities are ubiquitous in animals, plants, and aerobic microorganisms, where they undoubtedly function physiologically to remove toxic H_2O_2 generated by the action of various flavoprotein oxidases indigenous to the organism.

In contrast, peroxidases are relatively more rare in the animal world, but are common in plants. There, they may function in polyaromatic biosyntheses: during cooxidation of phenols and amines (as the peroxide is reduced), phenolic and aromatic amine radicals are generated; these radicals can couple and/or polymerize to observed polyphenolic products. For example,

ROOH + 2 (OH) ⟶ ROH + 2 { (O·) ⟷ (O) } ⟶ (OH OH)

radical coupling

Beef-liver catalase is a tetramer of 284,000 mol wt. With a turnover number of 2.8×10^6 moles min^{-1} (mole enzyme)$^{-1}$, it qualifies as one of the most efficient enzymatic catalysts known ($k_{cat}/K_m \approx 4 \times 10^7\ \text{M}^{-1}\ sec^{-1}$). Peroxidases from milk (lactoperoxidase), leucocytes (myeloperoperoxidase), bacteria (NADH peroxidase, an FAD enzyme), and horseradish represent peroxidative enzymes that have been characterized. Additionally, a chloroperoxidase that displays peroxidative halogenating activity has been isolated from the mold *Caldanomyces fumage* (Hollenberg et al., 1974). All peroxidases and catalases are Fe^{III} hemoproteins in the resting state. These enzymes are brown in color and show characteristic optical spectra with α, β, and Soret bands around 600, 500, and 400 nm, respectively—characteristic of high-spin, out-of-plane Fe^{III} heme (Fig. 15-5).

Much effort has been expended in elucidating the pathway of hydroperoxidase breakdown of peroxide substrates, and discrete intermediates have been characterized by many physical methods (and even crystallized, in the horseradish-peroxidase case).

From a variety of experiments with different hydroperoxidases, one can write a skeletal framework, including two spectroscopically discrete intermediates, called compounds I and II. The resting high-spin Fe^{III}–enzyme is oxidized by two electrons as the first molecule of ROOH is split and a molecule of H_2O thereby generated. The RO fragment is retained in compound I, which is green. The

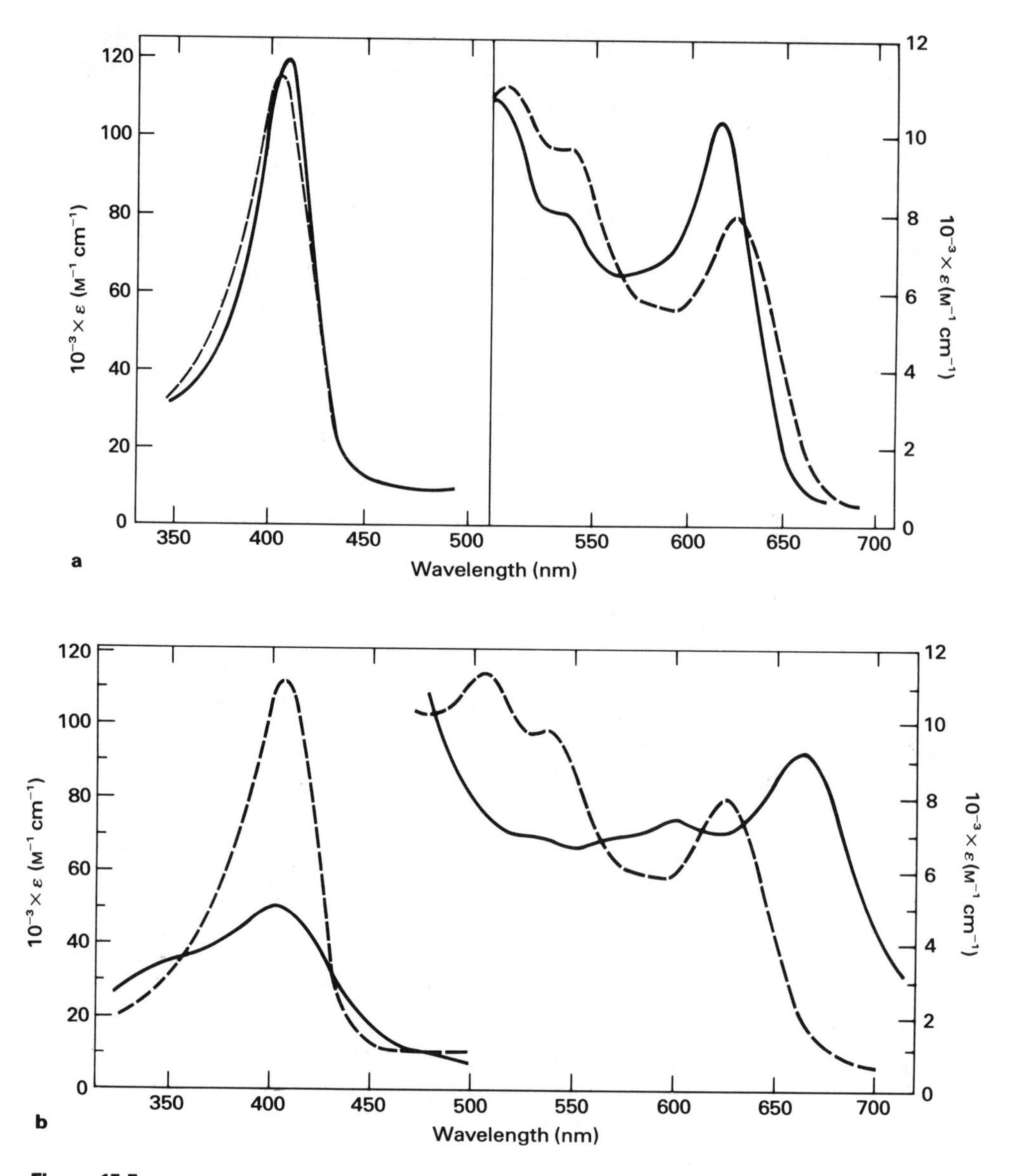

Figure 15-5
(**a**) Spectra of horse-erythrocyte catalase (*dashed line*) and its formate derivative (*solid line*); pH 4.6, 25° C. (From Schonbaum and Chance, 1976, p. 374) (**b**) Spectra of horse-erythrocyte catalase (*dashed line*) and its peracetic-acid derivative (compound I) (*solid line*); 0.08 M phosphate, pH 7.23, 25° C. (From Schonbaum and Chance, 1976, p. 397)

formal oxidation state is Fe^{V} if the electron deficiency is all localized at the iron (which is unlikely).

$$\text{ROOH} + \text{Enz—N}_4\text{Fe}^{III} \longrightarrow \underset{\text{Compound I}}{\text{RO—}} + H_2O$$

compound I (green)
(2-e$^{\ominus}$ oxidized)

Given that the RO group is at the alcohol oxidation state, this phase of catalysis has entailed reduction of the peroxide and 2-e$^{\ominus}$ oxidation of the heme. The second substrate molecule can then enter the sequence. For catalase, this second molecule is ROOH = HOOH; for peroxidase, this is the reductant cosubstrate AH_2. We shall examine the second half of the reaction separately for the peroxidases and then for the catalases.

15.B.1 Peroxidase Cycle

Binding of AH_2 displaces the coordinated ROH group from complex I. We have noted that the AH_2 substrates are donors that give stable radicals and so can donate one electron at a time, reducing compound I by one electron and generating a red compound II.

$$AH_2 + \underset{\text{Compound I}}{\text{RO—}} \longrightarrow [AH_2 \cdot \text{Compound I}] + \text{ROH} \xrightarrow[\text{transfer}]{1\text{-}e^{\ominus}} \underset{\substack{\text{(red)} \\ (1\text{-}e^{\ominus}\text{ oxidized})}}{\text{Compound II}} + AH\cdot \quad (1)$$

The formal oxidation state in compound II is Fe^{IV}, one e$^{\ominus}$ above resting enzyme. Now, a second molecule of AH_2 can bind and donate one e$^{\ominus}$ to compound II. This electron transfer will regenerate Enz-Fe^{III} (ready for another cycle) and produce another molecule of AH·.

$$\text{Compound II} + AH_2 \longrightarrow AH\cdot + \text{Enz—N}_4\text{Fe}^{III} \quad (2)$$

The sum of equations 1 and 2 yields

$$2\,AH_2 + \text{Compound I} \longrightarrow \text{ROH} + \text{Native enzyme} + 2\,AH\cdot$$

The stoichiometry of peroxidase action is completed, in a formal sense, by writing the disproportionation (generally fast) between the two AH· molecules,

$$AH\cdot + AH\cdot \longrightarrow AH_2 + A$$

thus reducing the overall ratio of AH_2 used to one per catalytic cycle. Although both compounds I and II are detectable during action of horseradish peroxidase, only compound I accumulates detectably during catalase action. It is suggested that breakdown of compound I is rate-determining in the catalase case—i.e., that compound II may form, but not accumulate.

A reasonable electronic description of the two high-oxidation states of the iron-porphyrin system has been proposed recently by Dolphin, based on model experiments utilizing electrochemical 2-$e^{\ominus}$ oxidation of Co^{II}-porphyrins (Dolphin et al., 1971). The products were characterized as Co^{III}-porphyrin cation species (1 $e^{\ominus}$ removed from the cobalt, 1 $e^{\ominus}$ removed from the porphyrin π-electron system). The diperchlorate salt of the Co^{III}-octaethyl-porphyrin cation radical is essentially identical in optical absorption spectrum to the compound I in catalysis by horseradish peroxidase, whereas the dibromide salt is spectroscopically identical to catalase compound I (Fig. 15-6). This is evidence that compound I with

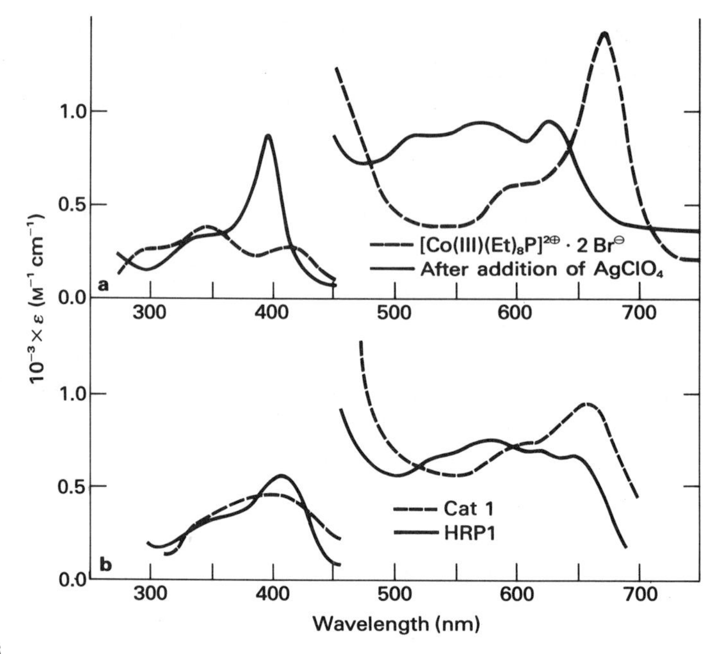

Figure 15-6
A comparison of the optical absorption spectra (**a**) of [Co(III) Et)$_3$P]$^{2\oplus}$·2Br$^{\ominus}$ and [Co(III)(Et)$_3$P]$^{2\oplus}$·2ClO$_4$ in CHCl$_3$, and (**b**) of catalase 1 and horseradish peroxidase compound 1 (HRP 1). Catalase is tetrameric, and the ε shown is per hematin. (From Dolphin et al., 1971, p. 616)

formal Fe^{V} oxidation state is likely to be an *Fe^{IV}-porphyrin cation.* The unpaired electron in the tetrapyrrole macrocycle is highly delocalized, and this contributes to overall stability of the complex.

compound I — compound II ⇌ ferryl

Conversion of compound I to compound II (by 1-$e^{\ominus}$ input) generates a formal Fe^{IV} oxidation state that may indeed be the ferryl iron species ($FeO_2^{2\oplus}$) initially postulated by George (1956) on the basis of model studies. Subsequent 1-$e^{\ominus}$ reduction of compound II by a second molecule of reductant would regenerate native enzyme and H_2O, while producing a second molecule of 1-$e^{\ominus}$–reduced substrate.

15.B.2 Catalase Cycle

Before suggesting a similar scheme for the second half of the catalase reaction, we note that catalase also shows efficient alcohol oxidase activity toward ethanol and other simple alcohols *when in the compound-I oxidation state.* Ethanol is oxidized to acetaldehyde, while the 2-$e^{\ominus}$–oxidized compound I reverts to resting enzyme. The alcohol oxidation is stereospecific, proceeding with removal of the pro*R*-hydrogen at C-1 of ethanol, the same chirality as the NAD-dependent alcohol dehydrogenase.

$$\text{Enz-Fe}^{3\oplus} + H_2O_2 \rightleftharpoons H_2O + \text{Compound-I–Fe}^{IV} \text{ cation radical}$$

$$H_3C\text{–}CH_2\text{–}OH \rightarrow CH_3CHO$$

$$\text{Enz-Fe}^{3\oplus} + H_2O$$

The intimate details of how the electrons are removed in the catalytic alcohol oxidation are unclear, although it can be written as a proton abstraction and 2-$e^{\ominus}$ flow to the iron.

Thus, we have a working model for the catalase version of the hydroperoxide mechanism (Schonbaum and Chance, 1976). We begin with the formation of compound I with the first H_2O_2.

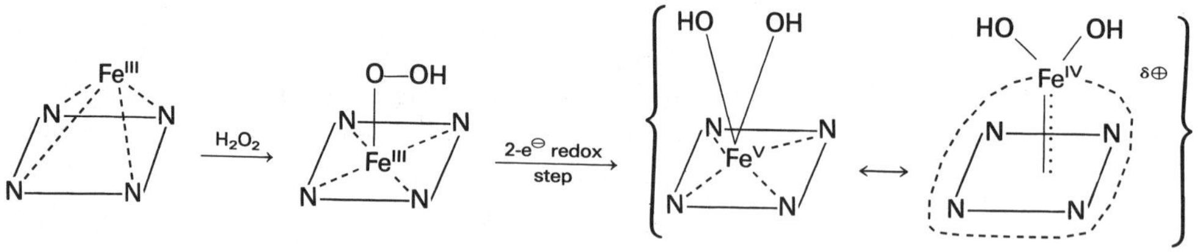

compound I

Compound I now reacts with the second H_2O_2 and breaks down to products.

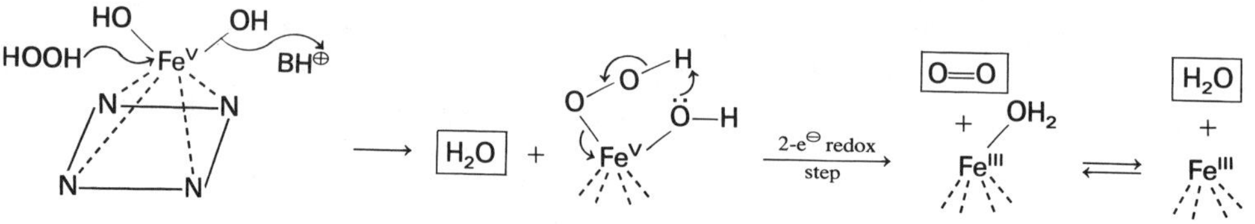

(1)

In the alcohol oxidation, compound I reacts instead with the alcohol.

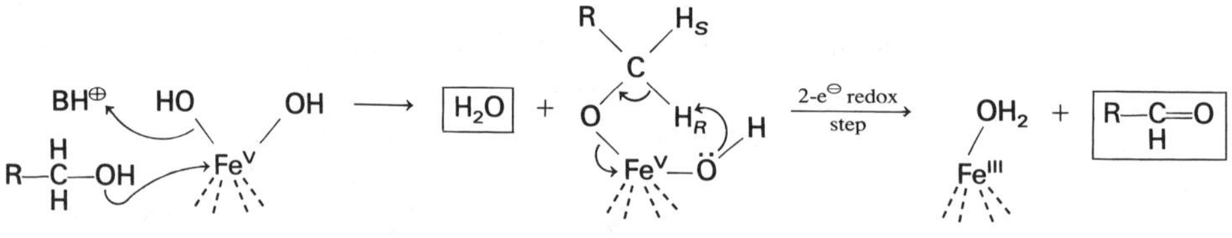

(2)

In the alternate product-forming sequences (equations 1 and 2), the second H_2O_2 (equation 1) or the RCH_2OH (equation 2) displaces a coordinated H_2O from compound I, and reduction then occurs. Reduction is shown here as a 2-e$^{\ominus}$ step, but it could proceed by rapid 1-e$^{\ominus}$ transfers. Also, no attempt is made to indicate all structures that contribute to the likely electronic distributions of the various species.

15.B.3 Chloroperoxidase

An interesting variant in peroxidative catalysis is the ability shown by fungal chloroperoxidase to catalyze formation of carbon–halogen bonds. This enzyme was discovered by Hager and coworkers, who were in fact searching for an enzymatic chlorination process as a source for the halogenated molecules found in various microorganisms (especially marine organisms). The enzyme will use $Cl^{\ominus}$, $Br^{\ominus}$, or $I^{\ominus}$ (but not $F^{\ominus}$) as $X^{\ominus}$ in the following reaction:

$$AH + X^{\ominus} + H_2O_2 \longrightarrow AX + 2\,H_2O$$

The substrate molecule to be halogenated must possess some activated site capable of acting as nucleophile with an electrophilic halogenating reagent. Suitable substrates include β-keto acids, and the general assay substrate is a cyclic β-diketone.

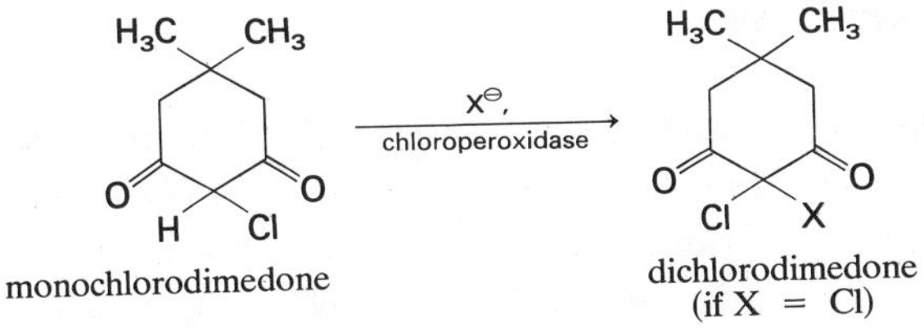

H_3C CH_3 O O H Cl — $X^{\ominus}$, chloroperoxidase → H_3C CH_3 O O Cl X

monochlorodimedone

dichlorodimedone (if X = Cl)

Recently Hager's group has demonstrated that sodium chlorite ($NaClO_2$) will function as a dual source of peroxidative and halogen capacity (Hollenberg et al., 1974). With [^{36}Cl]-chlorite, the dichlorodimedone derivative is produced containing ^{36}Cl label. Even in the presence of 200 M excess of chloride ion, there is no dilution of specific radioactivity in the product. Hager has suggested that the active chlorinating species delivers a chloronium ion or its equivalent. Chlorite generates an intermediate with spectral properties extremely similar to those of compound I generated from H_2O_2. These workers speculate that it may represent

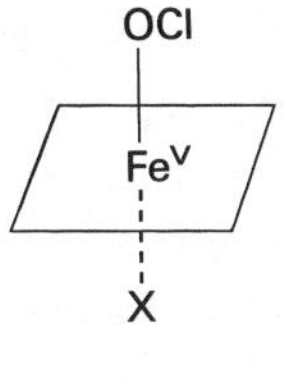

OCl

Fe^{V}

X

which will retain the polarization of a positive halogenium ion.

15.C SIX-ELECTRON REDUCTIONS OF INORGANIC COMPOUNDS

Most oxidations of organic substrates by enzymes are 2-$e^{\ominus}$ processes, with some 4-$e^{\ominus}$ oxidations we have noted; oxygen undergoes 2-$e^{\ominus}$ reduction and 4-$e^{\ominus}$ reduction (to H_2O) in a variety of enzymatic systems. There are a few authenticated examples of enzymatic redox processes involving transfer of six electrons to or from a substrate without release of free intermediates. Those 6-e$^{\ominus}$ redox systems so far discovered involve the reduction of inorganic substrates, such as nitrogenase catalyzing the reduction of dinitrogen to two molecules of NH_3, or sulfite reductase involved in the reduction of sulfite to sulfide.

$$6e^{\ominus} + 6H^{\oplus} + N_2 \xrightarrow{\text{nitrogenase}} 2\,NH_3$$

$$6e^{\ominus} + 6H^{\oplus} + SO_3^{2\ominus} \xrightarrow[\text{reductase}]{\text{sulfite}} S^{2\ominus} + 3\,H_2O$$

15.C.1 Sulfite Reductase

In *E. coli*, the sulfite reductase functions physiologically to provide sulfide utilized in cysteine biosynthesis. The donor of six electrons is NADPH (three separate molecules consecutively). Evidence for a discrete enzyme complex responsible for the complete 6-e$^{\ominus}$ reduction of sulfite had been obtained by several laboratories, but recently Siegel purified the *E. coli* NADPH–sulfite reductase to homogeneity and demonstrated it to be a multimeric complex with a functional molecular weight of 670,000 (Fig. 15-7; Siegel et al., 1974; Siegel and Davis, 1974; Faeder et al., 1974). This unit contains four FAD molecules, four FMN molecules, 21 atoms of nonheme iron, 15 to 16 molecules of acid-labile sulfide, and about four atoms of heme iron. The heme cofactor is modified from the usual protoporphyrin-IX structure and is, rather, an octacarboxylic-tetrahydroporphyrin designated "siroheme." Encouragingly for the biochemist, this multimeric unit is composed of only two distinct kinds of polypeptide chains, designated α and β, and has the stoichiometry $\alpha_8\beta_4$. Each monomer has a molecular weight of around 60,000, with the α-chain slightly larger. In 5 M urea, the multimer dissociates into the component polypeptides, allowing separation by ion-exchange chromatography. The β-chains contain the heme cofactor and all the nonheme iron and labile sulfide, whereas the α-chains have the FMN and FAD coenzymes and exist as the α_8 octamer. Remixing of chains reconstitutes the capacity for 6-e$^{\ominus}$ transfer from NADPH to sulfite.

With the purified protein in hand, Siegel and colleagues proceeded to map the path of electrons in transit from NADPH to sulfite. It turns out the FMN is bound less tightly than FAD, allowing FMN-depleted enzyme to be prepared and its remaining catalytic activity determined. NADPH still reduces the bound FAD at rates equal to those with native enzyme, but no sulfite reduction occurs, suggesting that FMN serves as the conduit of electrons between FAD and the β-subunit containing nonheme and heme iron. Addition of FMN restored the ability of depleted enzyme to reduce sulfite. This is the first case where a functional distinction has been found between FAD and FMN when both are

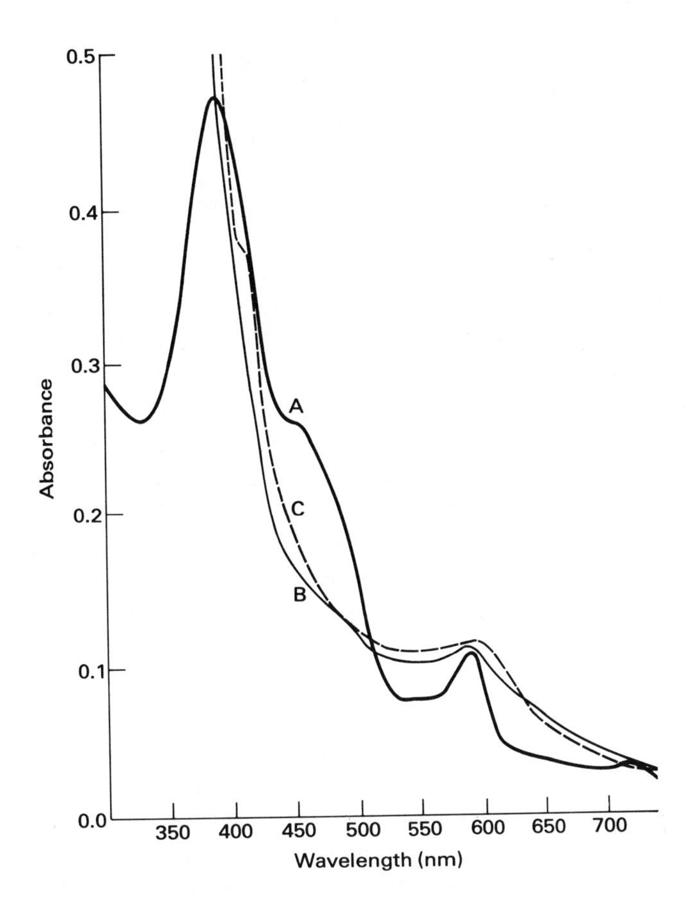

Figure 15-7
Absorption spectra of *E. coli* sulfite reductase in the presence of reducing agents. All experiments contained enzyme at a final concentration of 1.54 μM in the sample cell. Spectra were recorded versus a buffer blank as soon as possible after addition of components. Sample A was enzyme in buffer; sample B was enzyme plus 0.3 mM NADPH (0.1 ml of 23.1 M enzyme was added to 1.4 ml of a solution of NADPH that had been bubbled with N_2 for 30 min); and sample C was enzyme plus sodium dithionite. (From Hatefi and Stigall, 1976, p. 289)

bound to an apoenzyme. Only FAD accepts electrons from NADPH in this instance, and the $FADH_2$ is specifically reoxidized by FMN. The FAD is reduced in a 2-$e^{\ominus}$ process, whereas the subsequently produced $FMNH_2$ clearly is reoxidized in 1-$e^{\ominus}$ transfers.

Agents such as carbon monoxide or cyanide, which specifically complex with the $Fe^{2\oplus}$ form of iron bound to the heme cofactor, inhibit sulfite reduction but not the reoxidation of bound reduced flavins by artificial electron acceptors, confirming that the port of exit of electrons to sulfite is at the heme site in the β-subunit. Siegel proposes a linear sequence of electron transport.

$$\text{NADPH} \longrightarrow \text{FAD} \longrightarrow \text{FMN} \longrightarrow \text{Heme} \longrightarrow \text{Sulfite}$$

Unplaced in this scheme are the nonheme-iron/sulfur clusters. Given their physical location in the β-subunit, and if they are functional in a linear sequence, they would seem likely to be interposed between FMN and heme. The $\alpha_8\beta_4$ system might then be a functional tetramer (four $\alpha_2\beta$), with one α-subunit containing FMN and one α-subunit containing FAD interacting with one β-subunit. It has been noted that such a hypothetical $\alpha_2\beta$ unit—containing FAD, FMN, a (4 Fe)/(4 S) cluster, and heme—has a 6-$e^{\ominus}$ storage capacity and might then be constructed to ensure that electrons from three NADPH molecules could be stored up and then transferred very rapidly to reduce one molecule of bound sulfite to sulfide without any compounds of intermediate oxidation state having sufficient time to diffuse away from the active site.

15.C.2 Nitrogenase

Nitrogen-fixing bacteria can carry out the 6-$e^{\ominus}$ reduction of molecular nitrogen (N_2) to ammonia and so convert an inert form of nitrogen into one readily assimilated by biological systems (Eady and Postgate, 1974; Palmer, 1975). Among other reasons for interest, this reaction is one of enormous economic consequence.

$$6\,H^{\oplus} + N_2 \xrightarrow{6\,e^{\ominus}} 2\,NH_3$$

Nitrogenase activity is displayed by a two-enzyme complex that has been difficult to purify and study because of its extreme lability to oxygen-mediated inactivation (substantial to complete and irreversible inactivation on exposure to air for less than a minute). Recently, though, intact and functional complex has been purified from various nitrogen-fixing bacteria by careful isolation under inert gas (e.g., argon) and in the presence of chemical reducing agents (1 mM $Na_2S_2O_4$, dithionite). The stoichiometry of the nitrogenase from *Clostridium pasteurianum* and *Klebsiella pneumoniae* is the following:

$$N_2 + 3\,\text{Ferredoxin-Fe}^{II} + 12\,ATP^{4\ominus} \xrightarrow{\text{nitrogenase}} 2\,NH_3 + 3\,\text{Ferredoxin-Fe}^{III} + 12\,ADP^{3\ominus} + 12\,P_i^{2\ominus}$$

Two features of the stoichiometry warrant comment. The electron-donating cosubstrate is reduced bacterial ferredoxin (Chapter 13), which contains two (4 Fe)/(4 S) clusters, each capable of 1-$e^{\ominus}$ transfer. Thus 3 ferredoxin-Fe^{II} represent six electrons. ATP also is a required cosubstrate, undergoing enzyme-catalyzed hydrolysis to ADP and P_i. A ratio of 12 ATP molecules per N_2 molecule reduced has been obtained consistently (Eady and Postgate, 1974). The mechanistic role of ATP in nitrogenase action is unclear at present; the thermodynamic driving force is $12 \times (-7) = -84$ kcal (-352 kJ) per mole of N_2 reduced—a costly process on that basis alone. It could be used in some way to "ratchet down" the potential of redox sinks in nitrogenase—i.e., in some energy-transduction apparatus.

Purified nitrogenase contains two protein components, termed molybdoferredoxin (MoFd) and azoferredoxin (azoFd) (see Table 13-1), each containing Fe/S clusters as the names indicate. The *K. pneumoniae* molybdoferredoxin has one molybdenum atom per (16 Fe)/(16 S), suggesting four (4 Fe)/(4 S) clusters. This component has 220,000 mol wt, and it appears to be an $\alpha_2\beta_2$ tetramer. Each subunit could have an Fe/S cluster, but the placement of the single Mo atom is unclear.* In analysis of how the molybdoferredoxin component may function in electron transfer, the simpler metalloenzymes discussed in Chapter 13 may suggest some precedents. The azoferredoxin component of the *K. pneumoniae* enzyme has 69,000 mol wt with *no* molybdenum and one (4 Fe)/(4 S) cluster. The half-times for oxygen inactivation are 10 min for MoFd and 45 sec for azoFd.

In addition to nitrogen, a variety of other substrates can undergo reduction. Acetylene can be reduced by 2 $e^{\ominus}$ to ethylene and also (less well) by 4 $e^{\ominus}$ to ethane.

$$HC{\equiv}CH \xrightarrow{2\,e^{\ominus}} H_2C{=}CH_2 \xrightarrow[\text{slow}]{2\,e^{\ominus}} H_3C{-}CH_3$$

The nitrile function of hydrogen cyanide can be reduced to methyl amine.

$$HC{\equiv}N \xrightarrow{2\,e^{\ominus}} [H_2C{=}NH] \xrightarrow{2\,e^{\ominus}} H_3C{-}NH_2$$

---

*Recent X-ray–absorption fine structural studies on a molybdenum cofactor isolated from nitrogenase indicate that the molybdenum atom has at least four sulfur ligands and no oxygen ligands (Cramer et al., 1978). These researchers consequently propose that Mo could be present as substitute for an iron atom in a modified (2 Fe)/(2 S) or (4 Fe)/(4 S) cluster such as the following:

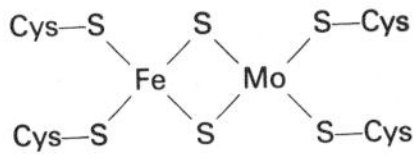

Cys—S S S—Cys
Fe Mo
Cys—S S S—Cys

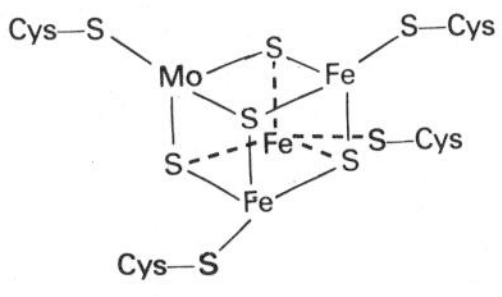

Cys—S S—Cys
S
Mo Fe
S
Fe S—Cys
S S
Fe
Cys—S

Additionally, the two-enzyme complex shows hydrogenase activity—2-$e^{\ominus}$ reduction of 2 $H^{\oplus} \rightarrow H_2$, using one molecule of reduced ferredoxin as electron donor.

$$\text{Ferredoxin-Fe}^{II} + 2\,H^{\oplus} \xrightarrow{\text{nitrogenase}} H_2 + \text{Ferredoxin-Fe}^{III}$$

Palmer (1975) has suggested a sequence for the nitrogenase components during action in the hydrogenase mode. The azoFd component is the electron-input component; the MoFd contains the N_2 substrate-binding site.

$$2\,\text{AzoFd} + 2\,e^{\ominus} \rightleftharpoons 2\,\text{AzoFd}^{\ominus} \quad (1)$$

$$2\,\text{AzoFd}^{\ominus} + 4\,\text{ATP} \rightleftharpoons 2\,(\text{ATP})_2\text{—AzoFd}^{\ominus} \quad (2)$$

$$(\text{ATP})_2\text{—AzoFd}^{\ominus} + \text{MoFd} \rightleftharpoons \text{MoFd}^{\ominus} + \text{AzoFd} + 2\,\text{ADP} + 2\,P_i \quad (3a)$$

$$\text{MoFd}^{\ominus} + (\text{ATP})_2\text{—AzoFd}^{\ominus} \rightleftharpoons \text{MoFd}^{2\ominus} + \text{AzoFd} + 2\,\text{ADP} + 2\,P_i \quad (3b)$$

$$\text{MoFd}^{2\ominus} + 2\,H^{\oplus} \rightleftharpoons \text{MoFd} + H_2 \quad (4)$$

Note that electron transfer to the MoFd component and ATP hydrolysis are indicated as two 1-$e^{\ominus}$ steps (3a and 3b); the 2-$e^{\ominus}$–reduced form of MoFd is written as $\text{MoFd}^{2\ominus}$. (It is likely that in the back direction of step 4 the molybdenum is reduced by transfer of one H from H_2 as a hydride ion to yield an Mo^{IV}H metal hydride, from which the H subsequently leaves as a proton.) The ratio of ATP hydrolyzed to H_2 produced is 4:1 for this 2-$e^{\ominus}$ reduction. For the 6-$e^{\ominus}$ reduction of N_2 it is 12:1.

One problem in formulation of the N_2-reduction mechanism is the question of how all six electrons are passed into N_2 before release of any partially reduced intermediates (e.g., hydrazines or diimides), a problem also posed for the sulfite reductase discussed earlier. The MoFd protein could store four electrons in the four (4 Fe)/(4 S) clusters and two electrons in molybdenum ($\text{Mo}^{VI} \rightarrow \text{Mo}^{IV}$), thus preparing for rapid 6-$e^{\ominus}$ transfer.

Chapter 16

Dioxygenases

Many of the enzymes discussed in Chapters 13 through 15 are monooxygenases; the activated oxygen is split, with incorporation of one atom into product and with the other coming out as H_2O. The enzymes discussed in this chapter are dioxygenases; both atoms of some activated form of the substrate O_2 are incorporated into product (Hayaishi, 1974). When both oxygen atoms end up in the same product molecule, the enzyme might be termed an *intramolecular* dioxygenase (Hayaishi, Nozaki, and Abbott, 1975). *Intermolecular* dioxygenases, on the other hand, incorporate one atom of oxygen into each of *two* separate products; almost invariably, these dioxygenases use α-ketoglutarate as a cosubstrate, converting it to succinate with oxygen-atom incorporation.

We have seen that monooxygenase enzymes require various cofactors—flavins, pteridines, copper, nonheme iron, and/or heme iron. Most of the 35 known dioxygenase enzymes use nonheme iron whose structural nature has not yet been characterized (the iron is not in Fe/S clusters). A few enzymes use heme iron as prosthetic group. And, as we note at the end of the chapter, there is one known copper-dependent dioxygenase and two known flavin-dependent dioxygenases.

Hayaishi has also suggested classification of dioxygenases by the category of chemical transformation achieved (Hayaishi, Nozaki, and Abbott, 1975). We enumerate such a classification here, and then we shall look at some specific examples in the classes for which useful mechanistic information is available (classes 1, 2, 6, and 7).

1. Hydroxylation of unactivated alkyl carbons in α-ketoglutarate-requiring reactions.

2. Cleavage of aromatic rings or heteroatomic rings in substrates. Dioxygenases represent the only general way that organisms open aromatic ring systems to catabolize them.
3. Production of catechol or catecholic derivatives. The first three examples noted here involve benzenoid substrates, and each enzymatic conversion is reported to require added $Fe^{2\oplus}$ and NADPH (Hayaishi, Nozaki, and Abbott, 1975):

   a. anthranilate hydroxylase (decarboxylating)

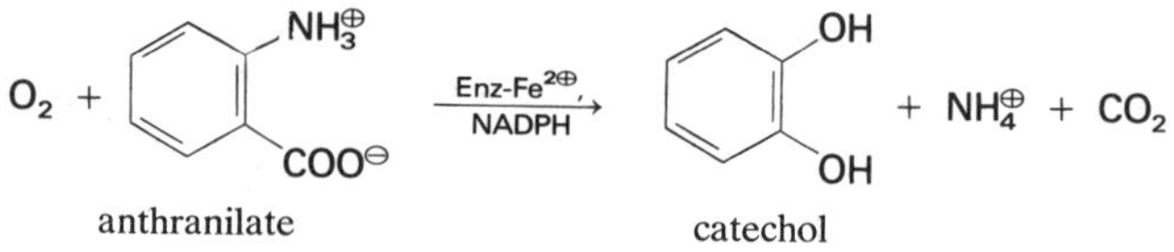

$$O_2 + \text{anthranilate } (NH_3^{\oplus}, COO^{\ominus}) \xrightarrow[\text{NADPH}]{\text{Enz-Fe}^{2\oplus},} \text{catechol } (OH, OH) + NH_4^{\oplus} + CO_2$$

anthranilate catechol

   b. benzoate oxygenase

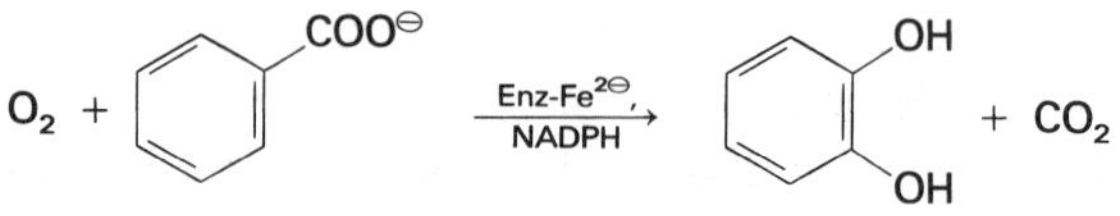

$$O_2 + \text{benzoate } (COO^{\ominus}) \xrightarrow[\text{NADPH}]{\text{Enz-Fe}^{2\ominus},} \text{catechol } (OH, OH) + CO_2$$

   c. benzene oxygenase

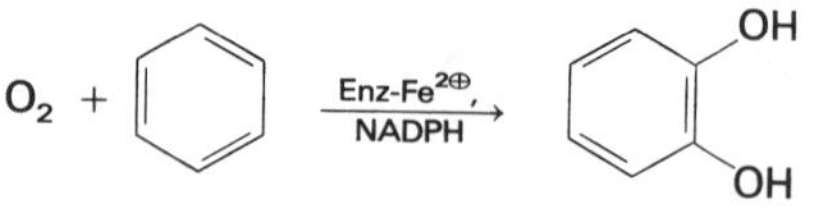

$$O_2 + \text{benzene} \xrightarrow[\text{NADPH}]{\text{Enz-Fe}^{2\oplus},} \text{catechol } (OH, OH)$$

A fourth example studied by Jeffrey et al. (1975) is the conversion of naphthalene to its 1,2-diol by an enzyme from *Pseudomonas putida*:

   d. naphthalene dioxygenase and dihydrodiol dehydrogenase sequence

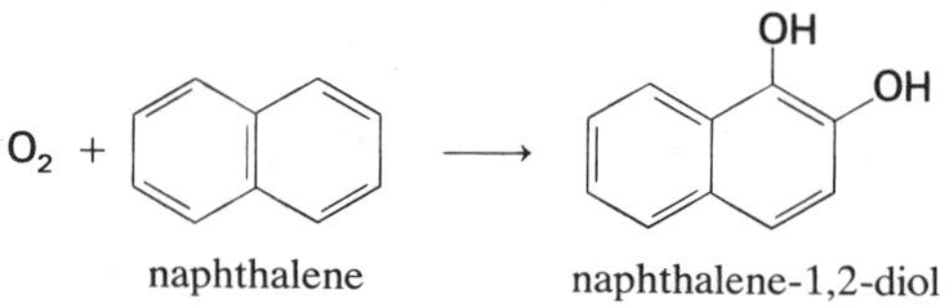

$$O_2 + \text{naphthalene} \longrightarrow \text{naphthalene-1,2-diol } (OH, OH)$$

naphthalene naphthalene-1,2-diol

Jeffrey and colleagues demonstrated that the product of the NADPH-requiring dioxygenase is actually the *cis*-1,2-dihydrodiol and established its absolute stereochemistry as *cis*-1*R*,2*S*-dihydrodiol. They hypothesized the following scheme, where O_2 reacts with the aromatic ring to produce a transient dioxetane that is reductively opened (hence the NADPH requirement).

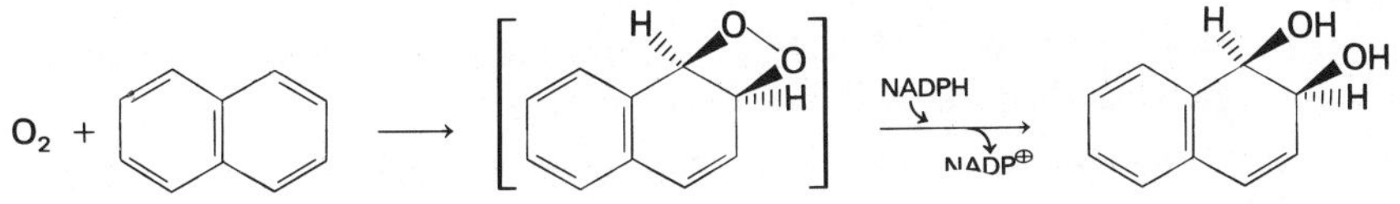

O_2 + naphthalene ⟶ [dioxetane intermediate] $\xrightarrow[NADP^{\oplus}]{NADPH}$ *cis*-1,2-dihydrodiol

cis-1,2-dihydrodiol

A second enzyme, an NAD$^{\oplus}$-requiring dihydrodiol dehydrogenase, is required to oxidize the dihydroaromatic species by 2 e$^{\ominus}$ up to naphthalene-1,2-diol with concurrent production of NADH. For example,

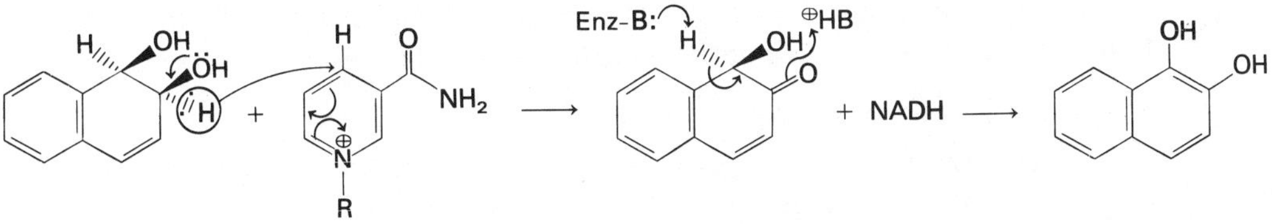

Enz-B: ⟶ $^{\oplus}$HB + NADH ⟶

It is likely that similar two-enzyme sequences are involved in the three cases noted above for dioxygenation of benzene rings to catechols, and that dihydroaromatic diols are the primary products of dioxygenase action. We shall shortly comment on a possible mode of formation of the putative dioxetane intermediates.

4. Conversion of cysteine or cysteamine to the sulfinate products.

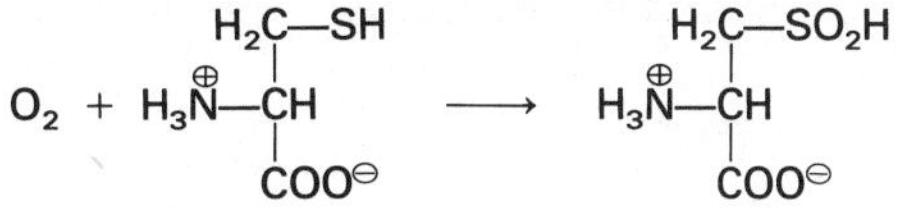

$$O_2 + H_3\overset{\oplus}{N}\text{—CH(CH}_2\text{SH)COO}^{\ominus} \longrightarrow H_3\overset{\oplus}{N}\text{—CH(CH}_2\text{SO}_2\text{H)COO}^{\ominus}$$

cysteine ⟶ cysteine sulfinate

5. β-Carotene 15,15′-dioxygenase converts a molecule of the C_{40} polyconjugated terpene into two molecules of all-*trans*-retinal, the aldehyde form of vitamin A.

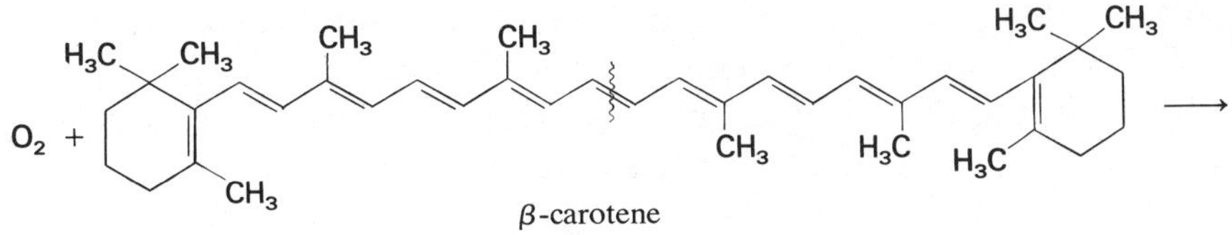

O_2 + β-carotene ⟶ 2 retinal

β-carotene

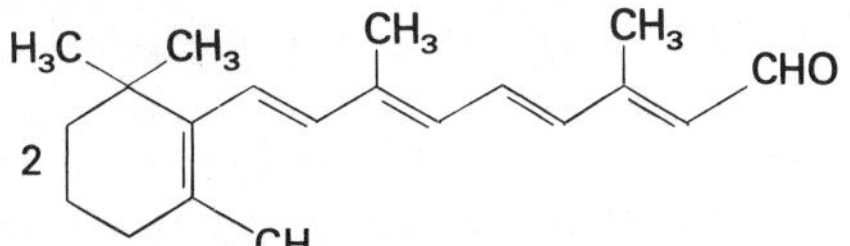

retinal

6. Prostaglandin synthetase generates the oxycyclopentane ring of prostaglandins from acyclic polyenoic precursors.
7. Lipoxygenase catalyzes the hydroperoxidation of polyunsaturated acids and esters containing a *cis,cis*-1,4-pentadiene system.

16.A DIOXYGENASES REQUIRING AN α-KETO-ACID COSUBSTRATE

In this section we discuss enzymes from class 1 of the classification scheme just given. Recent review articles summarize knowledge about these enzymes (Hayaishi, Nozaki, and Abbott, 1975; Abbott and Udenfriend, 1974; Hamilton, 1974). We shall discuss examples of three variants of the catalytic process:

1. prolyl and lysyl hydroxylases;
2. thymine-5-methyl oxygenase, 5-hydroxymethyluracil oxygenase, and 5-formyluracil oxygenase;
3. *p*-hydroxyphenylpyruvate hydroxylase.

Type 1 involves the hydroxylation of an unactivated methylene group in the prolyl substrates. The three reactions of type 2 may all be catalyzed by the same enzyme; the sequential oxidation of a methyl moiety to a carboxyl function utilizes an oxygen atom from O_2 at each step. In type 3, the reaction is an intramolecular version, with hydroxylation of an aromatic locus.

The generalized stoichiometry of these α-ketoglutarate-requiring enzymes is the following:

$$S + {}^{18}O_2 + \begin{array}{c} COO^{\ominus} \\ | \\ HCH \\ | \\ HCH \\ | \\ C{=}O \\ | \\ COO^{\ominus} \end{array} \xrightarrow[\text{ascorbate}]{E\text{—}Fe^{2\oplus}} S^{18}OH + \begin{array}{c} COO^{\ominus} \\ | \\ HCH \\ | \\ HCH \\ | \\ C{=}O \\ | \\ {}^{18}O^{\ominus} \end{array} + CO_2$$

where S = Substrate. In reactions of types 1 and 2, the α-keto-acid cosubstrate is obligatorily α-ketoglutarate; in type 3, it is the pyruvyl side chain of the substrate. The requirement for ascorbate (or other reducing agents) most likely reflects the need to maintain the iron in an active, ferrous state.

16.A.1 Prolyl and Lysyl Hydroxylases

Although the prolyl and lysyl hydroxylases were named before their modes of catalysis were appreciated, they obviously represent the intermolecular variant of dioxygenation.

Prolyl hydroxylase is the enzymatic activity responsible for formation of the 4-hydroxyprolyl residues in animal collagen and in plant cell-wall proteins. The animal enzymes have been better characterized; they act on certain prolyl residues of nascent chains of collagen still bound to ribosomes. The rat-skin enzyme has been purified; a dimer of 130,000 mol wt, it appears to recognize the sequence X–Pro–Gly–Y and prefers high-molecular-weight polypeptides. The product from prolylhydroxylase action is the *trans*-4-hydroxyproline moiety.

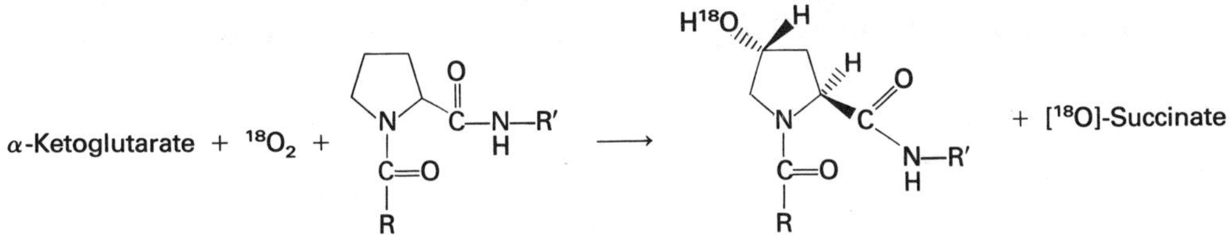

Lysyl hydroxylase is a similar enzyme that acts exclusively on collagen or elastin as substrate to transform some of the lysine residues to 5-hydroxylysine groups.

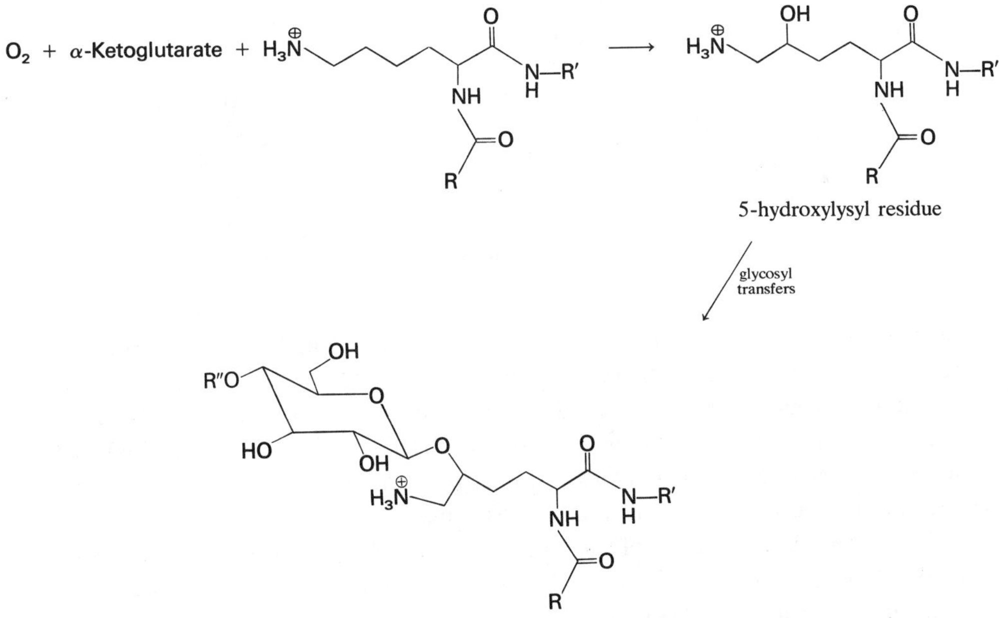

This activity is not as well characterized as the prolyl hydroxylase, but these 5-hydroxyl groups in the product collagen appear to be the nucleophilic group in subsequent glycosylations that result in the formation of galactosylglucosyllysyl collagen. It has been argued that this dioxygenase-mediated hydroxylation and the subsequently-enabled glycosylations may be prerequisites for the secretion of collagen monomers from fibroblasts into the extracellular matrix, where they undergo polymerization and perform their structural functions.

16.A.2 Thymidine→Uracil-5-Carboxylate

Extracts of the mold *Neurospora crassa* convert the 2′-deoxyribonucleotide thymidine to uracil-5-carboxylate in the following sequence of enzymatic steps.

thymidine (2′-deoxy) —1→ thymidine ribonucleoside —2 (P_i → Ribose-P)→ thymine —3→ 5-hydroxymethyluracil —4→ 5-formyluracil —5→ uracil-5-carboxylate ($COO^{\ominus}$)

The enzymes responsible for step 1 and for steps 3, 4, and 5 have been purified and characterized. Each of these four steps requires O_2 and α-ketoglutarate as cosubstrate, producing succinate, CO_2, and the indicated specific products; these then are intermolecular dioxygenase reactions. The protein thus far purified that carries out reaction 3 also performs reactions 4 and 5, and it may be that the identical enzyme carries out all three transformations.

It is likely that oxidation of the alcohol to the aldehyde in step 4 proceeds by initial conversion of the hydroxymethyl moiety to the hydrated aldehyde, from which loss of water will proceed.

$$R{-}CH_2{-}OH \xrightarrow[\alpha\text{-ketoglutarate}]{^{18}O_2,} R{-}CH(OH)(^{18}OH) \xrightarrow{-H_2O} RCHO \text{ or } RCH^{18}O$$

One might expect a similar scheme for production of the carboxylic acid from the aldehyde.

16.A.3 *p*-Hydroxyphenylpyruvate Hydroxylase

Our third example of this class of dioxygenase is *p*-hydroxyphenylpyruvate hydroxylase, an enzyme involved in the degradation of tyrosine.

HO— —COO⊖ (NH$_3^{\oplus}$) ⟶ HO— —COO⊖ (O) $\xrightarrow[\text{Enz-Fe}^{2\oplus}]{O_2,}$ HO— —OH (—COO⊖) + CO_2

tyrosine — *p*-hydroxyphenylpyruvate — homogentisate

The homogentisate product from the hydroxylase (dioxygenase) reaction is subject to ring cleavage by a second dioxygenase of the variety we shall discuss in ¶16.B.

The hydroxylase-mediated production of homogentisate is a complex reaction. It has been established that the carboxymethyl side chain of homogentisate derives from the pyruvyl side chain of substrate, which undergoes an *ortho* migration during catalysis and concomitant decarboxylation. The new hydroxyl group derives from O_2 and is the one *para* to the original hydroxyl and *ortho* to the carboxymethyl side chain. With $^{18}O_2$, the other oxygen atom ends up in the homogentisate carboxylate group. In this reaction, the pyruvyl group fulfills the function that external α-ketoglutarate does in the earlier intermolecular examples.

16.A.4 Mechanistic Schemes

We now ask whether these various enzymatic reactions described in the preceding three subsections can be described by some common and satisfactory mechanistic proposal. We shall discuss the alkyl hydroxylations first.

Linstedt has proposed initial formation of a substrate hydroperoxide at the methylene carbon to be hydroxylated, followed by attack of the hydroperoxy anion on the electrophilic carbonyl of α-ketoglutarate to form a bridged peroxy species (Holme et al., 1968). Decarboxylation would thus proceed with fission of the O—O peroxide linkage to yield observed products.

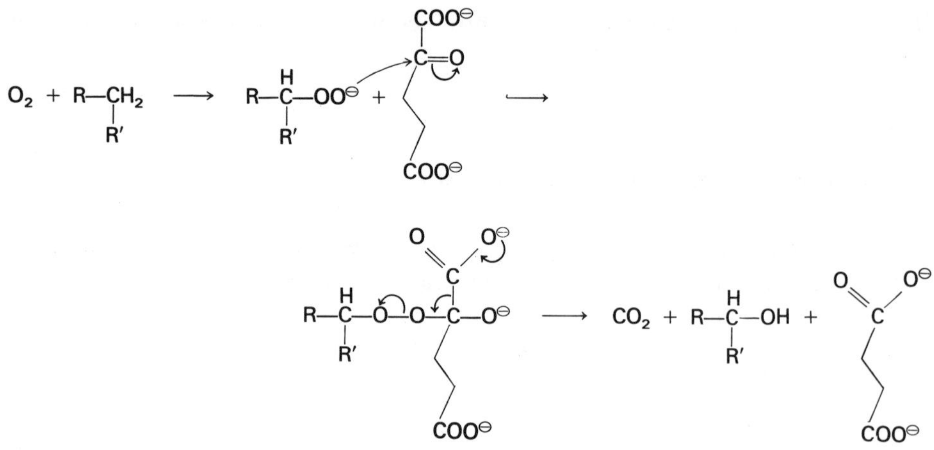

$O_2 + R{-}CH_2R' \longrightarrow R{-}CH(R'){-}OO^{\ominus} + {}^{\ominus}OOC{-}CO{-}CH_2CH_2{-}COO^{\ominus} \longrightarrow$

$R{-}CH(R'){-}O{-}O{-}C(O^{\ominus})(COO^{\ominus}){-}CH_2CH_2{-}COO^{\ominus} \longrightarrow CO_2 + R{-}CH(R'){-}OH + {}^{\ominus}OOC{-}CH_2CH_2{-}COO^{\ominus}$

This proposal suffers from the drawback that the initial attack of the enzyme-bound form of O_2 at the unactivated methylene carbon of prolyl residues is hard to visualize.

A somewhat more attractive proposal (on paper at least) has been put forth by Hamilton (1974); it quite nicely encompasses the alkyl hydroxylations and the *p*-hydroxylphenylpyruvate reaction. Hamilton argues for the initial formation of a persuccinic-acid species at the active site, resulting from reaction of enzyme-bound $Fe^{2\oplus} \ldots O_2$ and the α-ketoglutarate, which might be coordinated to the active-site iron. The coordinated tetrahedral adduct is postulated to break down as indicated. The coordination to the iron is presumed to direct electron flow as outlined to obtain the persuccinic acid.

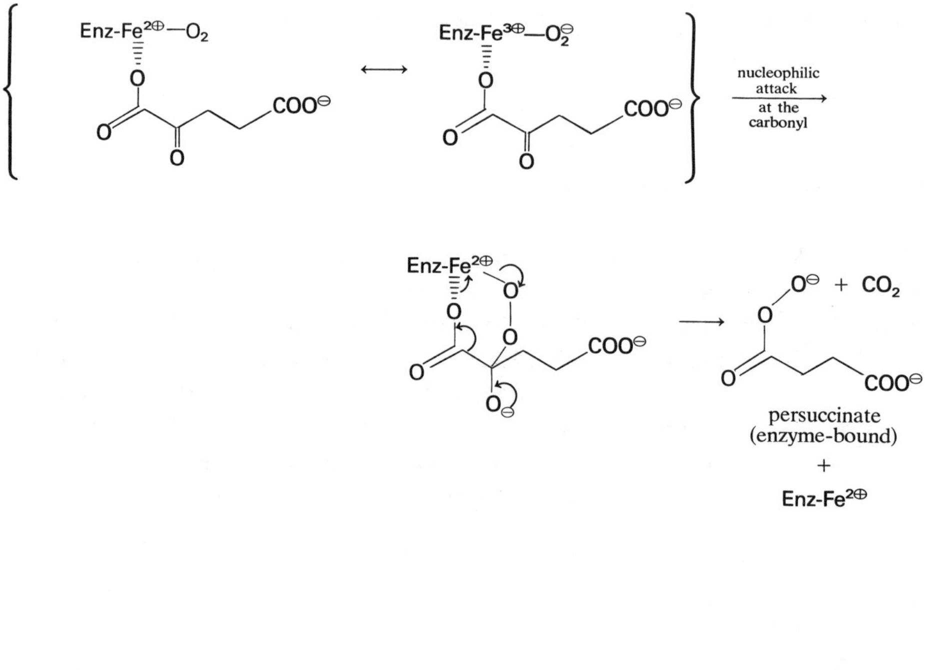

Enz-$Fe^{2\oplus}$—O_2 $\longleftrightarrow$ Enz-$Fe^{3\oplus}$—$O_2^{\ominus}$

nucleophilic attack at the carbonyl

Enz-$Fe^{2\oplus}$ $\longrightarrow$ $O^{\ominus}$ + CO_2

persuccinate (enzyme-bound)

+

Enz-$Fe^{2\oplus}$

The possibility exists that such a peracid could act to insert an oxenoid oxygen at an unactivated C—H bond in the way that nitrenes and carbenes insert. No more detailed description of this hypothetical oxygen transfer has been postulated. Some experimental evidence for it and against it is gained with prolyl hydroxylase.

In favor of the mechanism is the uncoupling behavior of prolyl hydroxylase. Under certain conditions, with [^{14}C]-carboxy-α-ketoglutarate, nine molecules of $^{14}CO_2$ are produced per prolyl residue hydrolylated (Hayaishi, Nozaki, and Abbott, 1975). This result suggests excess decarboxylation of α-ketoglutarate prior to substrate hydroxylation (analogous to uncoupling of flavin-dependent hydroxylases) and could be explained by spurious breakdown of persuccinate molecules prior to hydroxylation.

On the other hand, the proposal of persuccinate as oxygen-inserting agent was tested by Udenfriend and colleagues directly. Synthetic monoperoxysuccinate was added to prolyl hydroxylase and substrate, but no hydroxylation ensued (Hayaishi, Nozaki, and Abbott, 1975). To keep the proposal viable, one would have to postulate that persuccinate must be delivered to the active site by *in situ* generation before it is utilized.★

For the *p*-hydroxyphenylpyruvate conversion to homogentisate, the idea of an enzyme-bound peracid species leading to an arene-oxide species from which side-chain migration occurs on ring opening is consonant with this reaction being cast as an example of the N.I.H.-shift phenomenon we have notea previously for hydroxylases. Thus, Hamilton (1974) suggests the intermediate generation of the *p*-hydroxyphenyl peracetic acid (on loss of CO_2 from the initial O_2 adduct as indicated above for the α-ketoglutarate case), which can epoxidize the aromatic ring (or yield the analogous electrophilic aromatic substitution sigma complex).

O_2 + HO–C$_6$H$_4$–CH$_2$–C(=O)–COO$^\ominus$ $\xrightarrow{-CO_2}$ HO–C$_6$H$_4$–CH$_2$–C(=O)–O–O–H (peracid) ⟶ HO–(epoxide)–CH$_2$–COO$^\ominus$ (epoxide)

peracid

epoxide

Ring opening of the epoxide with 1,2-migration of the acetic-acid side chain would yield the cyclohexadienone tautomer of homogentisate, a ketonization away from the observed product.

★The hydroxylation outcome at an unactivated methylene carbon is reminiscent of alkyl-group hydroxylation by heme-P_{450} monooxygenases (Chapter 15). In that instance, we noted evidence interpreted in favor of substrate radicals. No evidence yet rules against such a path in these intermolecular dioxygenations.

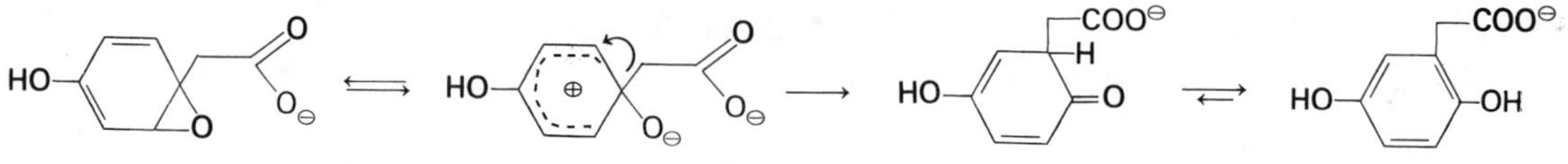

homogentisate

16.B INTRAMOLECULAR DIOXYGENASES CLEAVING AROMATIC OR HETEROAROMATIC RINGS

The intramolecular dioxygenases that act to cleave aromatic and heteroaromatic rings (indoles, pyridines) are extremely important routes (if not the only routes) by which these molecules are degraded biologically (Hayaishi, Nozaki, and Abbott, 1975; Nozaki, 1974; Feigelson and Brady, 1974). The bulk of these enzymes are bacterial in origin, although homogentisate oxidase and tryptophan 2,3-dioxygenase are found in animal cells. Various drugs, insecticides, and carcinogens as well as natural alkaloids and lignin are ring-opened and subsequently metabolized by these systems. The following is a generalized stoichiometry at an aromatic "double" bond:

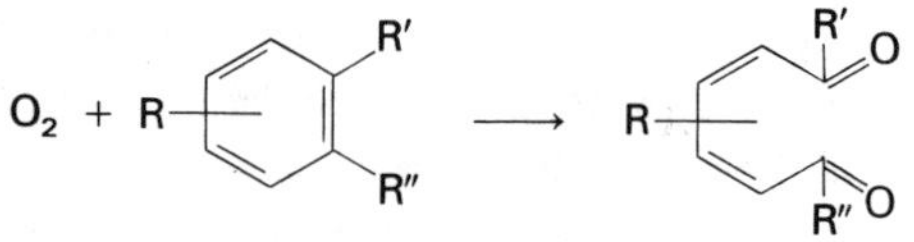

There are a variety of purified dioxygenases known that carry out oxidative ring-cleavage reactions on *ortho*-diphenolic (catecholic) substrates. Such ring cleavages are probably exclusively catalyzed by dioxygenases and these enzymes are thus exceedingly important in removing aromatics from the biosphere. (The formation of catecholic systems from simple aromatics was noted in the introduction to this chapter.)

Nozaki has recently categorized these enzymes by the positional specificity of cleavage on the catecholic substrate (Hayaishi, Nozaki, and Abbot, 1975; Nozaki, 1974).

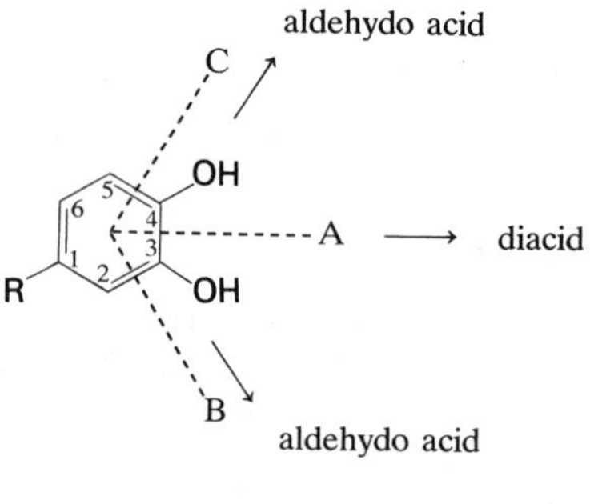

aldehydo acid

A ⟶ diacid

aldehydo acid

A. Intradiol cleavage (between C-3 and C-4) yields a ring-opened diacid.
B. Proximal extradiol cleavage (between C-2 and C-3) yields an aldehydo acid product.
C. Distal extradiol cleavage (between C-4 and C-5) also yields a ring-opened acid-semialdehyde product.

In all cases, both oxygens of O_2 end up in the product, one on either side of the cleavage site.

Here are four examples of reactions carried out by crystalline enzymes. Protocatechuate 3,4-dioxygenase is an intradiol enzyme (category A):

O_2 + protocatechuate → (protocatechuate 3,4-dioxygenase) → *cis,cis*-β-carboxymuconate

Protocatechuate 4,5-dioxygenase is a distal extradiol enzyme (category C):

O_2 + protocatechuate → (protocatechuate 4,5-dioxygenase) → α-hydroxy-γ-carboxymuconic ε-semialdehyde

Metapyrocatechase is a proximal extradiol enzyme (category B):

O_2 + catechol → (metapyrocatechase) → α-hydroxymuconic ε-semialdehyde

Another intradiol enzyme (category A) is pyrocatechase, which works on catechol itself:

O_2 + catechol → (pyrocatechase) → *cis,cis*-muconate

Interestingly, the *intradiol-cleaving* dioxygenases are *bright red* with broad absorption between 390 and 650 nm and a peak at about 440–450 nm. EPR studies indicate that the structurally uncharacterized nonheme iron (no acid-labile sulfide) is ferric iron. In contrast, the *extradiol-cleaving* dioxygenases are *colorless*, have no peaks in the visible spectrum, are EPR silent, and contain iron in the ferrous oxidation state. How this dichotomy relates to positional specificity in ring opening is unclear. Initial-velocity kinetic studies indicate prior combination of enzyme and catecholic substrate, followed by binding of O_2 to form the catalytically competent ternary complexes.

A variant of the phenolic dioxygenases is homogentisate oxygenase, utilizing for ring cleavage the product of *p*-hydroxyphenylpyruvate hydroxylase action noted in ¶16.A.3.

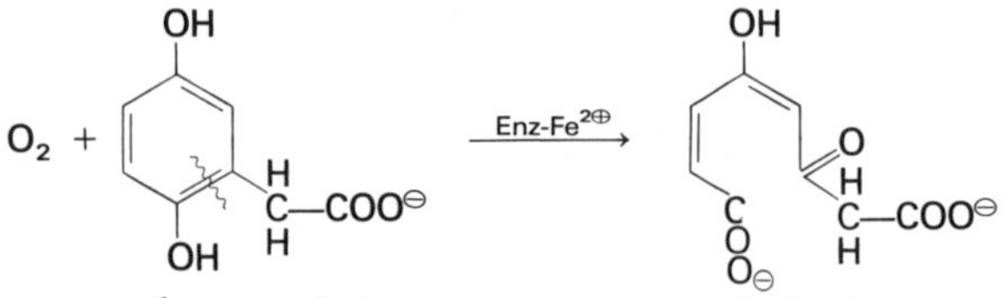

O_2 + homogentisate $\xrightarrow{\text{Enz-Fe}^{2\oplus}}$ maleylacetoacetate

homogentisate maleylacetoacetate

Only the most rudimentary mechanistic data are available for these purified nonheme-iron-requiring dioxygenases. It may be that the $Fe^{2\oplus}$ enzymes combine with O_2 to yield $Fe^{2\oplus} \ldots O_2 \leftrightarrow Fe^{3\oplus} \ldots O_2^{\ominus\cdot}$ as active species, whereas the $Fe^{3\oplus}$ enzymes undergo discrete electron transfer from phenolic substrate to produce $Fe^{2\oplus}S^{\cdot}$, which rapidly undergoes reaction with O_2. A number of people have written cyclic endoperoxide structures to explain the patterns of ^{18}O incorporation from $^{18}O_2$ during ring cleavage, one atom on each side of the cleavage point. Thus, for pyrocatechase, one could speculate about the following scheme:

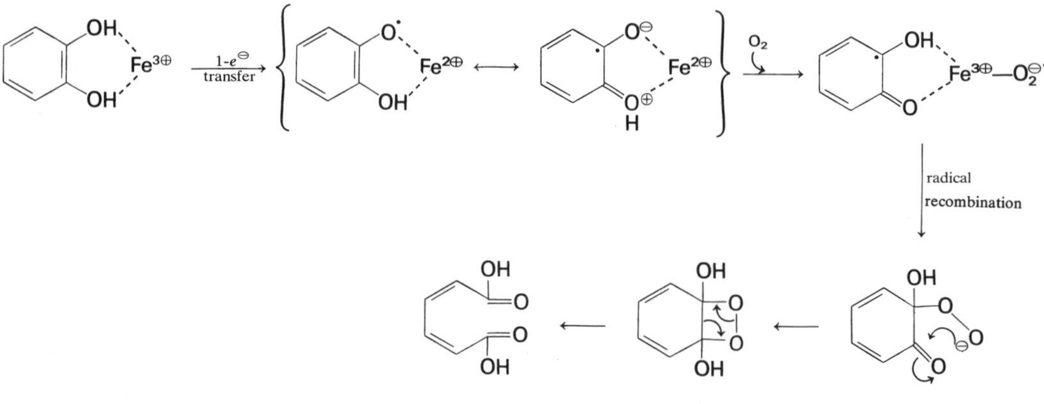

$Fe^{3\oplus}$ $\xrightarrow{1\text{-}e^{\ominus}\ \text{transfer}}$ $Fe^{2\oplus}$ $\longleftrightarrow$ $Fe^{2\oplus}$ $\xrightarrow{O_2}$ $Fe^{3\oplus}—O_2^{\ominus\cdot}$

radical recombination

One can write similar schemes for the extradiol cleavage patterns. The idea of cyclic endoperoxide intermediates in dioxygenase catalysis is strengthened by data from Samuelson's laboratory on prostaglandin formation (¶16.C), but no direct evidence has actually been accumulated for such dioxetane species during catecholic ring opening.*

Before we proceed to prostaglandin synthetase, however, there is one *heteroaromatic* ring-cleaving dioxygenase that is worth mentioning (Feigelson and Brady, 1974). This is tryptophan 2,3-dioxygenase. It is one of the few heme-iron–containing dioxygenases so far discovered. It functions physiologically in the catabolism of tryptophan, converting this indole derivative into a substituted anthranilate, known as *N*-formylkyneurinine.

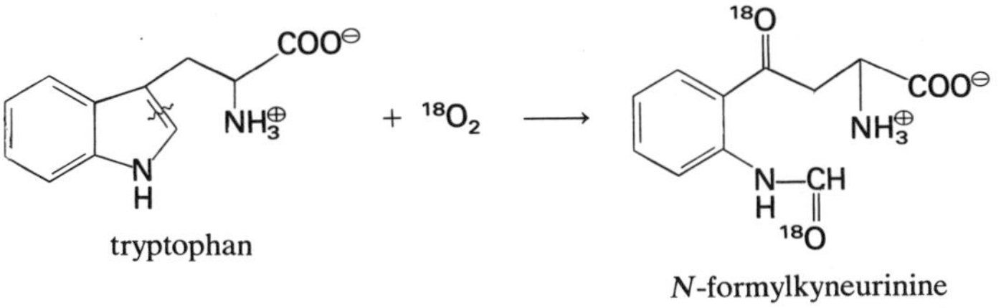

tryptophan + $^{18}O_2$ ⟶ *N*-formylkyneurinine

* An alternative formulation has been argued by Que et al. (1977) for reaction of the initial substrate-peroxy adduct (after radical recombination), purportedly avoiding the highly strained dioxetane structure (Hamilton, 1974) and yielding instead a tetrahedral adduct of a cyclic anhydride, coordinated to the active-site iron. Que and colleagues point out that such rearrangements have precedent in rearrangements of *trans*-decalin peresters (Gould, 1959). An interpretation of their scheme for protocatechuate-3,4-dioxygenase catalysis is the following, where the Fe^{III} atom is suggested as chelating the 4-hydroxyl specifically.

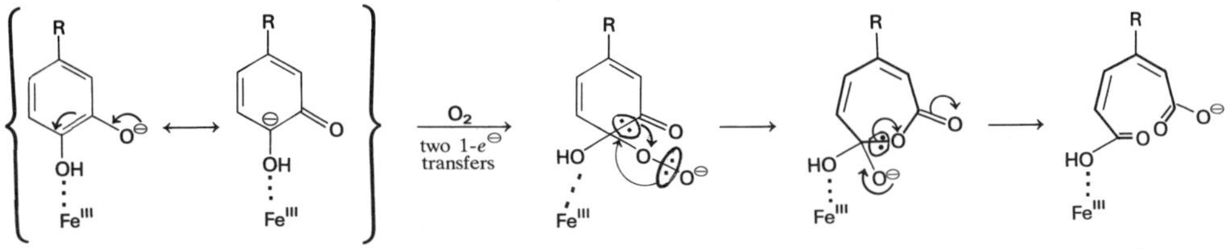

O_2, two 1-$e^{\ominus}$ transfers

product

However, it is worth noting that the peroxy adduct and the somewhat suspect dioxetane are in fact simply resonance forms, and one must argue that the four-ring contributor is a very minor contributor to the overall structure.

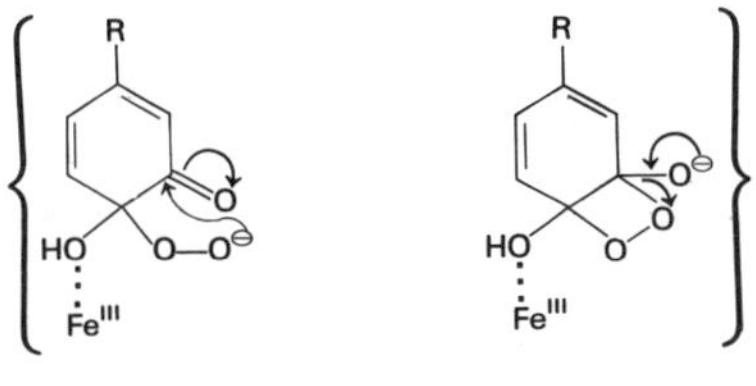

The enzymes from *Pseudomonas acidovorans* and from rat liver have been purified extensively. Each contains two heme groups with the iron in the $Fe^{2\oplus}$ oxidation state. Each also contains two gram-atoms (per mole) of copper as cupric iron. It has been assumed that the ferrous iron of the porphyrin is the coordinating site for O_2 binding and for reaction, on the basis of iron function in other dioxygenases.*

Recent studies indicate that tryptophan 2,3-dioxygenase leaks $O_2^{\ominus\cdot}$ into solution (detectable with superoxide dismutase) during catalysis, consistent with the idea ventured above for the catecholic enzymes—that coordinated superoxide might be an initial oxygen nucleophile attacking a substrate radical.

16.C PROSTAGLANDIN CYCLOOXYGENASE

Prostaglandins are fatty acids composed of 20 carbons with a substituted cyclopentane ring. They are active in a diversity of pharmacological roles and appear to modulate the action of many hormones on target tissues. Among the physiological effects of prostaglandins are effects on nerve transmission, on peripheral blood circulation, on ion transport across cell membranes, and on inflammatory responses (Stryer, 1975, p. 821). There are four major classes of prostaglandins, differing in position of the double bonds and/or the oxygen substituents on the

* This precedent cannot be adopted without question, though, in view of the recent isolation of an inducible enzyme, quercetinase, from the mold *Aspergillus flavus*; a glycoprotein of 111,000 mol wt, this enzyme contains two atoms of nonblue cupric ion per molecule as the only prosthetic group (Vaneste and Zuberbühler, 1974).

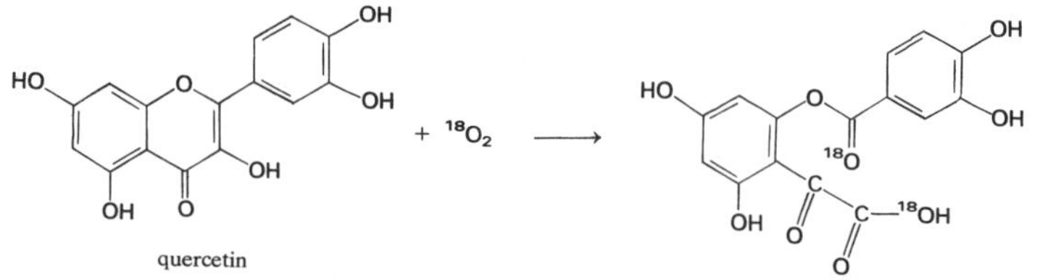

HO OH OH OH OH O + $^{18}O_2$ ⟶ HO OH OH O ^{18}O OH O ^{18}OH O

quercetin

It is likely in each case (tryptophan 2,3-dioxygenase and quercetinase) that an endoperoxide adduct (either I or II) forms and then collapses with ring fragmentation.

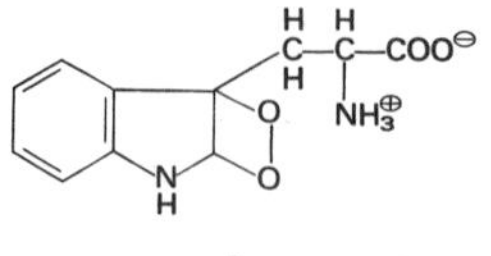

H H C—C—COO$^{\ominus}$ H NH$_3^{\oplus}$ O O N H

I

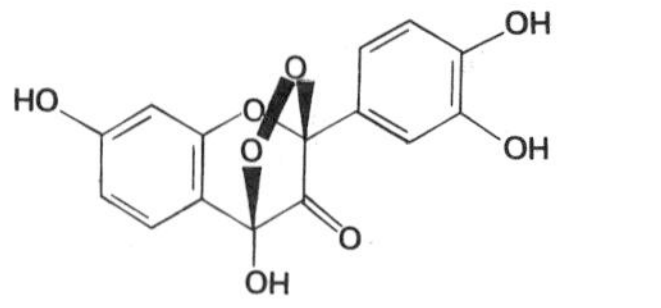

HO O O OH OH O OH

cyclopentane ring: PGA, PGB, PGE, and PGF (Samuelsson and Paoletti, 1975). Subscripts (e.g., PGE_1, PGE_2) refer to the number of exocyclic double bonds in the particular prostaglandin. The following are representative structures from the four classes.

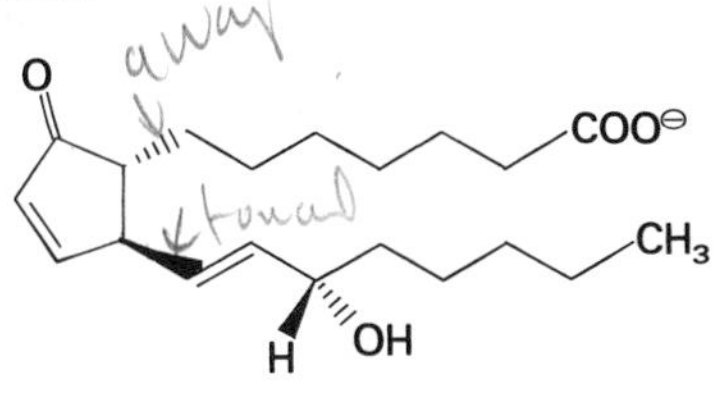

PGA_1 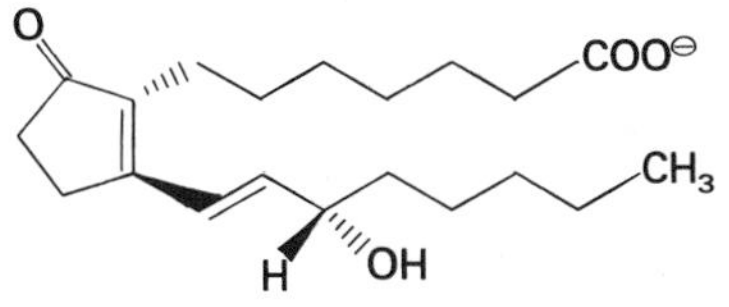 PGB_1

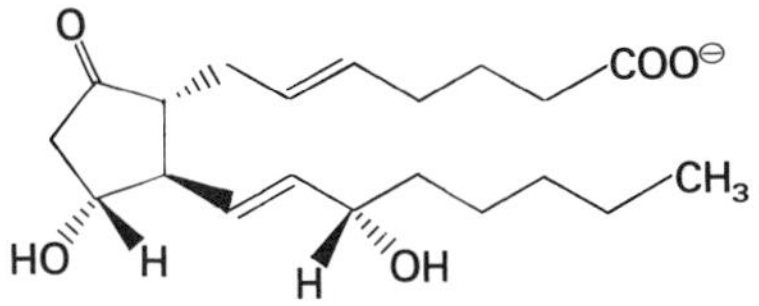

PGE_2 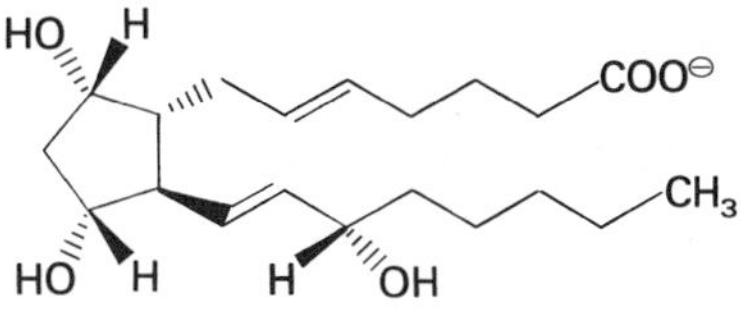 $PGF_{2\alpha}$

The biosynthetic precursors of the prostaglandins are acyclic, polyolefinic C_{20} fatty acids. The prostaglandins with one exocyclic double bond derive from *cis*-$\Delta^8,\Delta^{11},\Delta^{14}$-eicosatrienoate, whereas those with two exocyclic double bonds derive from the tetraenoate $\Delta^5,\Delta^8,\Delta^{11},\Delta^{14}$-eicosatetraenoate (trivial name, arachidonate).

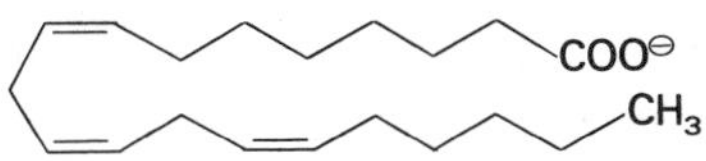

$\Delta^8,\Delta^{11},\Delta^{14}$-eicosatrienoate 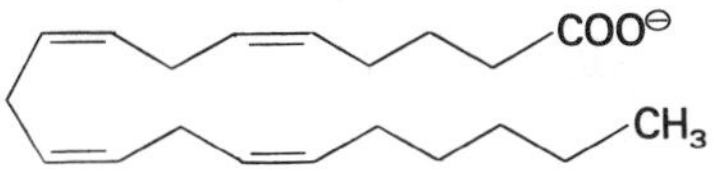 archidonate

All the oxygen atoms (except the $COO^\ominus$ group) in the various prostaglandins derive from O_2. There are three such oxygen atoms—the two in the cyclopentane ring derive from the same molecule of O_2. The 15-α-hydroxyl group also is introduced by dioxygenase action: an initial 15-α-hydroperoxy adduct is formed and the O—O bond then is fragmented to form the alcohol and H_2O.

Initial mechanistic speculations had suggested that the cyclopentane-ring oxygens are introduced via a cyclic endoperoxide adduct, whose fragmentation and rearrangement then lead to the observed products (Hamberg, Samuelsson, et al., 1974). This speculation was strengthened by subsequent isolation of two unstable intermediates ($T_{1/2} \approx$ minutes) from extracts making prostaglandins. When the $\Delta^8,\Delta^{11},\Delta^{14}$-eicosatrienoate is substrate, the two intermediates have the structures indicated in prostaglandin G_1 and prostaglandin H_1 (Hamberg and

Samuelsson, 1973; Hamberg, Svensson, et al., 1974). The key structural element is the unstable five-membered cyclic endoperoxide.

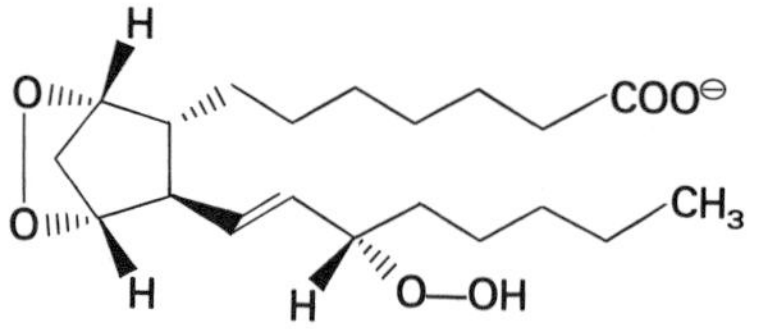

PGG$_1$

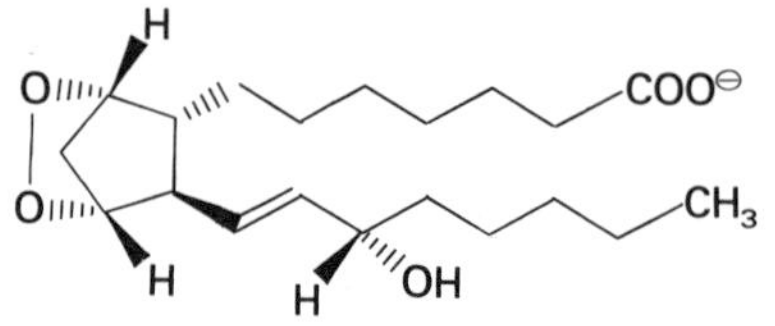

PGH$_1$

The enzyme system responsible for conversion of the trienoate and tetraenoate precursors to prostaglandins has been termed prostaglandin cyclooxygenase (or prostaglandin synthetase). Progress in mechanistic study has been slow because the activity is integrally bound to microsomal membranes and is resistant to solubilization and purification. However, Hayaishi and colleagues have now succeeded in purifying the complex from bovine vesicular gland membranes and have separated it into two enzyme fractions (I and II) responsible for converting $\Delta^8,\Delta^{11},\Delta^{14}$-eicosatrienoate to PGE$_1$ (Miyamoto et al., 1976).

Enzyme I has 3.5×10^5 mol wt and requires added heme for activity (perhaps for reconstituting a heme dioxygenase activity?) to convert eicosatrienoate to PGG$_1$. This homogeneous protein carries out *two types* of dioxygenase reaction: (1) cyclic endoperoxide formation, and (2) allylic hydroperoxide formation at C_{15}. One could write several schemes analogous to the following.

Addition of first molecule of $\mathbf{O_2}$**.** Catalysis could be initiated by proton abstraction and double-bond attack on the coordinated O_2 to yield an allylic peroxide.

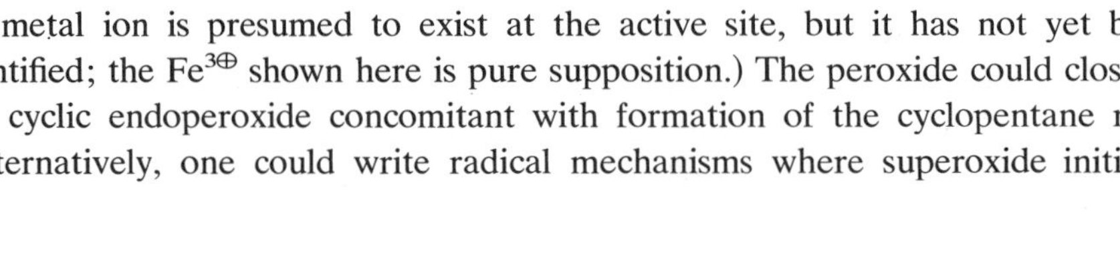

(A metal ion is presumed to exist at the active site, but it has not yet been identified; the $Fe^{3\oplus}$ shown here is pure supposition.) The peroxide could close to the cyclic endoperoxide concomitant with formation of the cyclopentane ring. (Alternatively, one could write radical mechanisms where superoxide initiates attack, and a hydrogen atom is expelled.*) Consistent with the hypothesis of the acyclic hydroperoxy species as intermediate is detection in tissue extracts of side products that could arise by fission of the O—O bond.

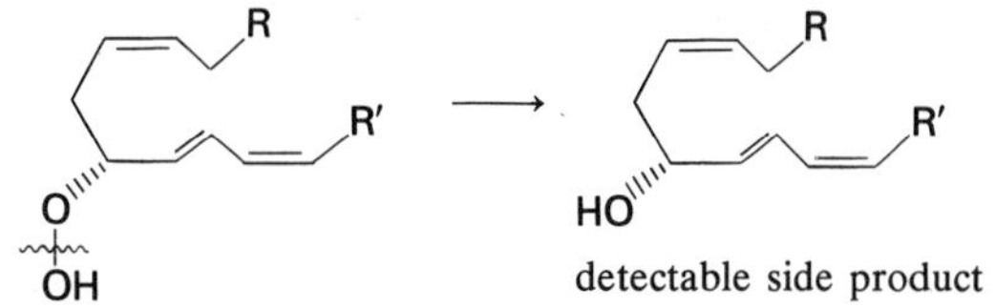

R R' O OH → R R' HO
detectable side product

Addition of second molecule of O_2. Economically, one can write an identical mechanism for the second dioxygenative step, again an allylic hydroperoxylation to the detected product, prostaglandin G_1. This hydroperoxide does not cyclize as the first one did. The turnover number for conversion of $\Delta^{8,11,14}$-eicosatrienoate to PGG_1 is 740 min^{-1}.

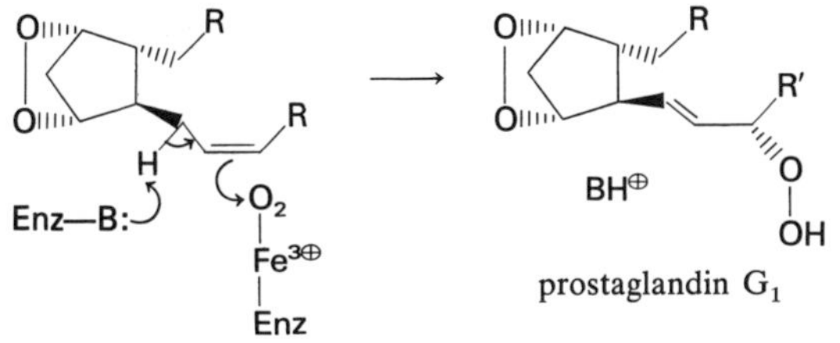

R R H Enz—B: O_2 $Fe^{3\oplus}$ Enz → R R' $BH^{\oplus}$ O OH
prostaglandin G_1

Enzyme I will show an additional catalytic activity if tryptophan is added (in addition to heme). Now the enzyme will convert prostaglandin G_1 (15-α-hydroperoxy) to prostaglandin H_1 (15-α-hydroxy), a peroxidase-type cleavage at a rate of 900 min^{-1} (see Chapter 15). The fate of the added tryptophan has not been determined (the indole ring could act as reductant). (Recall that the stoichiometry for heme-dependent peroxidases is $AH_2 + ROOH \rightarrow ROH + H_2O + A$.)

---

* In support of radical intermediates are consistent observations that cyclooxygenases undergo self-destruction during turnover within 30 to 60 sec. (Eagon et al., 1976) have found that reductive processing of the PGG_2 intermediate generates radicals that inactivate the enzyme. Phenol and methional, good radical scavengers (of OH˙, for example) increase both initial rates and extents of catalytic reaction before inactivation, and they promote formation of PGH_2 at the expense of PGG_2 (starting initially from the tetraene arachidonate). Preliminary EPR studies show radicals forming during turnover in the absence of radical scavengers; about 5,000 turnovers occur before an inactivating event. The PGG_2-derived oxygen radical (adventitious breakdown of a highly reactive chemical intermediate) may be reacting with a disulfide at the enzyme active site.

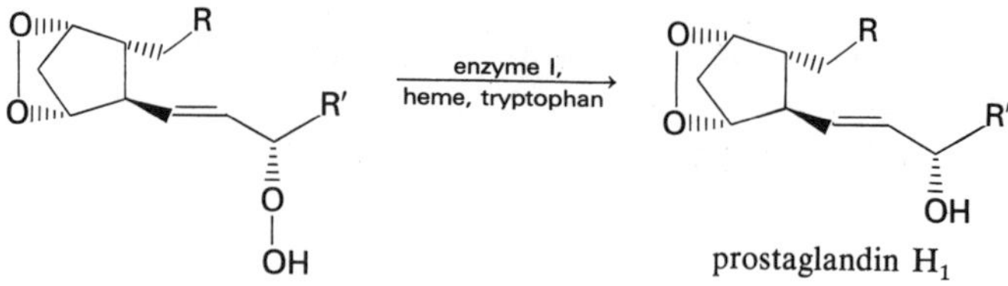

enzyme I, heme, tryptophan

prostaglandin H_1

The prostaglandin H_1 still has the endoperoxide linkage intact. It is the job of enzyme II to cleave this endoperoxide (Miyamoto et al., 1976). Enzyme II requires reduced glutathione for activity. Whether gluthathione is acting as a base, as a nucleophile, or as a reductant for the enzyme is not yet known. There may be analogy to the known enzyme glutathione peroxidase where alkyl hydroperoxides are reduced at the expense of oxidation of reduced glutathione (Arias and Jakoby, 1976). One simple paper scheme for generation of the overall product, prostaglandin E_1, is the following:

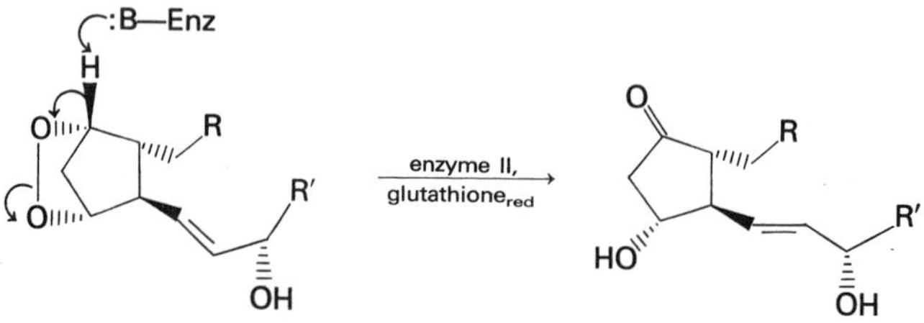

enzyme II, glutathione$_{red}$

prostaglandin H_1 prostaglandin E_1

The isolated two-enzyme prostaglandin-cyclooxygenase complex thus shows at least three activities: (1) two dioxygenase steps that may each be allylic-hydroperoxide formations; (2) peroxidase activity toward the 15-α-OOH; (3) endoperoxide isomerase activity.

A final comment about the cyclooxygenase: both aspirin and indomethacin, classical antiinflammatory drugs, may exert their pharmacological effects primarily through inhibition of the prostaglandin cyclooxygenase (Roth et al., 1975).

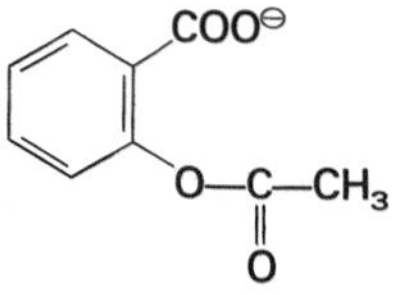

aspirin

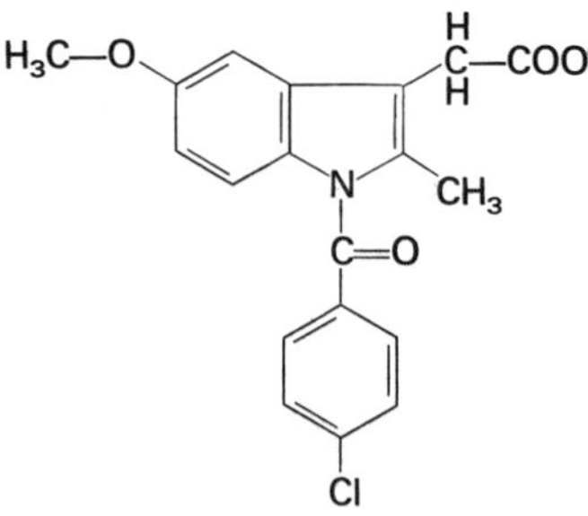

indomethacin

Hayaishi has shown that both drugs inhibit only at the first step, somewhere in the sequence leading to PGG_1, the hydroperoxy-endoperoxide. Indomethacin produces 50% inhibition at 1 μM, aspirin 50% inhibition at 15 mM. Previous studies have suggested that each drug produces time-dependent, irreversible inactivation of cyclooxygenase (Vane, 1971; Roth et al., 1975). Aspirin appears to acetylate the complex by virtue of its chemical reactivity as an acetylating agent.

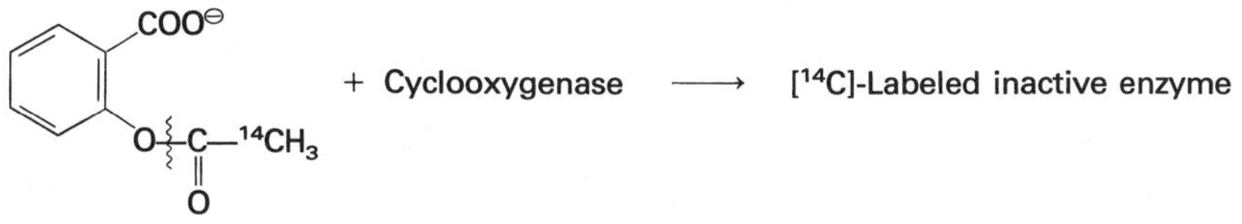

$COO^{\ominus}$, $O-C(=O)-{}^{14}CH_3$ + Cyclooxygenase ⟶ [^{14}C]-Labeled inactive enzyme

With indomethacin, the inactivation process is unclear; covalent labeling may not be involved.

An alternative enzymatic fate for the cyclic endoperoxide PGH_2 (in platelets, at least) is its enzymatic conversion—by thromboxane-A_2 synthetase (Hammarstrom and Falardeau, 1977)—into a rearranged compound, thromboxane A_2, a molecule containing an unstable oxane oxetane structure that reacts rapidly ($T_{1/2} = 32$ sec at 37° C; Hamberg et al., 1975) in water to form the stable but pharmacologically inactive thromboxane B_2.

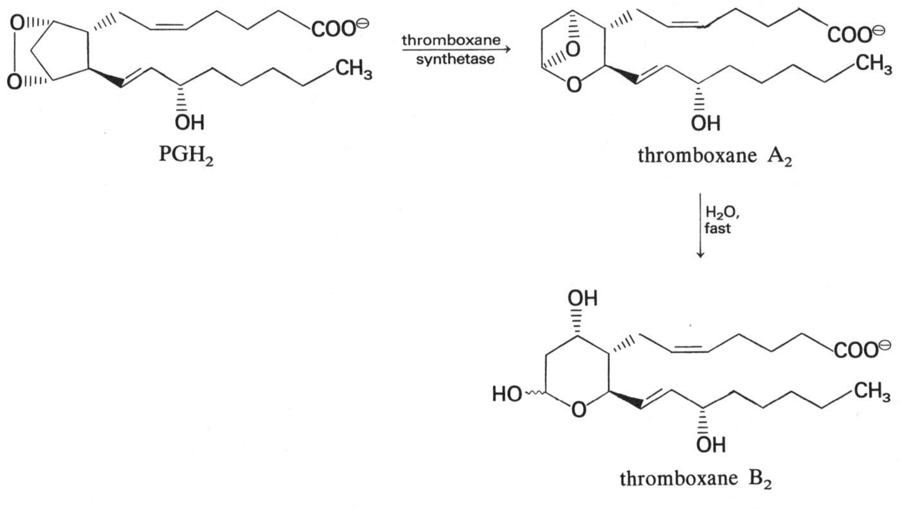

PGH_2 —thromboxane synthetase→ thromboxane A_2 —H_2O, fast→ thromboxane B_2

The mechanism of enzymatic conversion of the cyclopentane endoperoxide to the oxane oxetane of the initial thromboxane is not yet understood, but the following speculation has been offered (Dicfalusy et al., 1977).

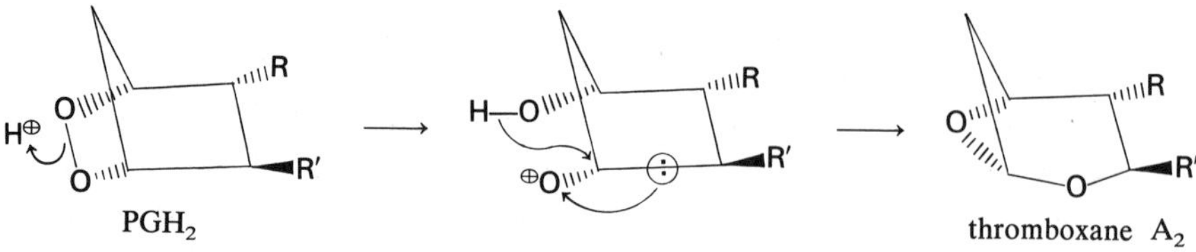

PGH$_2$ → → thromboxane A$_2$

16.D LIPOXYGENASE

Enzymes catalyzing the formation of allylic hydroperoxides of polyunsaturated fatty acids are found in many plants; the enzyme from soybeans has been purified and found to require iron for catalysis (Hamberg, Samuelsson, et al., 1974). The soybean enzyme introduces both atoms of O_2 as a peroxy group at one terminus of a *cis*-1,4-diene system. The natural substrates are hydroperoxidized six carbons in from the methyl terminus of the fatty acid (i.e., at C-13 for C_{18} fatty acids, at C-15 for C_{20} fatty acids). The reaction is illustrated for linoleate (Δ^9,Δ^{12}-octadecadienoate).

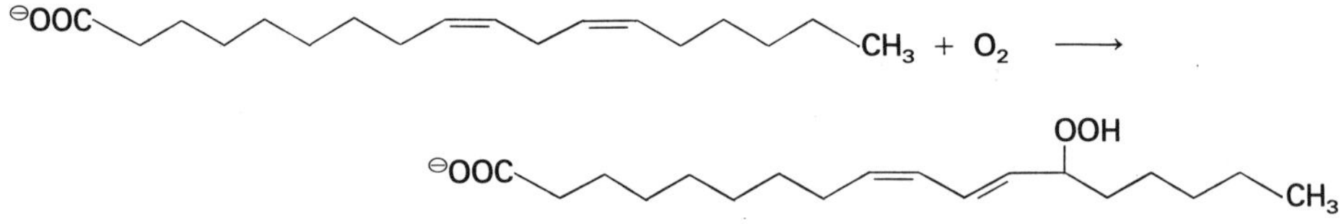

$^{\ominus}$OOC–…–CH$_3$ + O$_2$ ⟶ $^{\ominus}$OOC–…(OOH)…–CH$_3$

The methylene-interrupted 9,12-diene system of the substrate is isomerized to the 9,11-conjugated *cis,trans*-diene attendant with —OOH introduction. This conversion is chemically identical to the addition of the second O_2 molecule in the prostaglandin-cyclooxygenase system noted earlier. There we suggested that either ionic or radical mechanisms might apply. Here we show a possible radical mechanism for lipoxygenase, where H-atom abstraction would yield a delocalized dienyl radical that could react with coordinated $O_2^{\ominus}\cdot$.

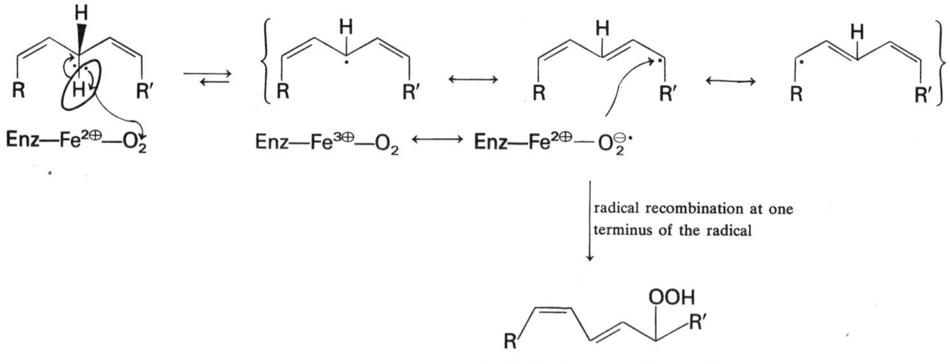

Enz—Fe$^{2\oplus}$—O$_2$ ⇌ Enz—Fe$^{3\oplus}$—O$_2$ ⟷ Enz—Fe$^{2\oplus}$—O$_2^{\ominus}\cdot$

radical recombination at one terminus of the radical

allylic hydroperoxide product

With soybean lipoxygenase, the pro*S* hydrogen is removed from C-13, and the migrating double bond ends up *trans*. In a recent review, Gibian and Galway (1977) note that it may be that O_2 reacts from the side of the molecule opposite that from which H_S is abstracted.

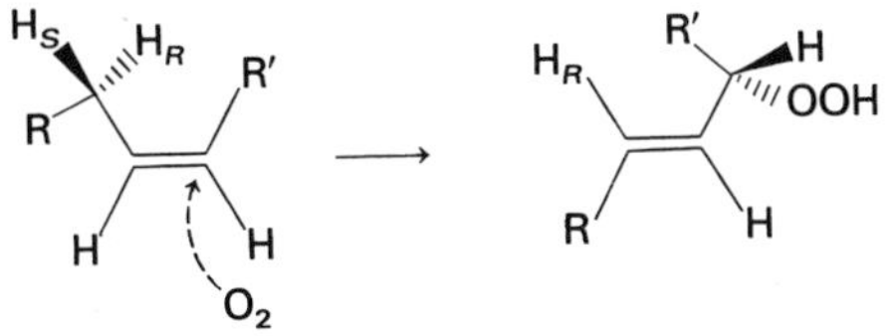

16.E FLAVOPROTEIN DIOXYGENASES

Because much of Section III has dealt with flavin-cofactor chemistry, we note here the two dioxygenases that are known to use O_2 to effect addition of both oxygen atoms to product. One enzyme catalyzes ring opening of the pyridine ring of 2-methyl-3-hydroxypyridine-5-carboxylate during bacterial degradation of pyridoxal (Sparrow et al., 1969). The overall transformation resembles the aromatic fissions carried out by the nonheme dioxygenases.

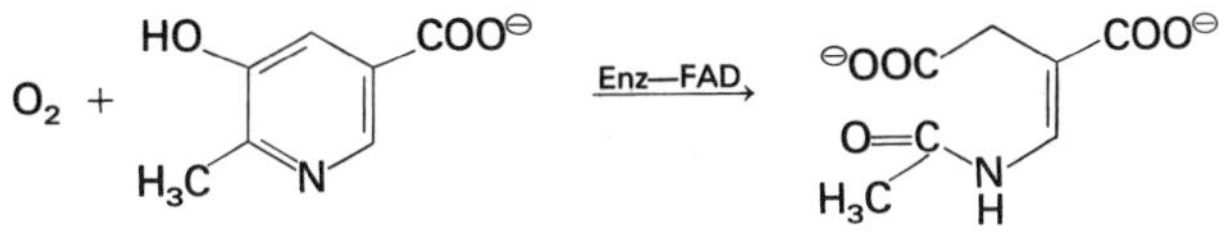

The second example is the recently reported 2-nitropropane dioxygenase from yeast (Kido et al., 1976). Two molecules of nitropropane are converted to two molecules of acetone and nitrite per O_2 consumed.

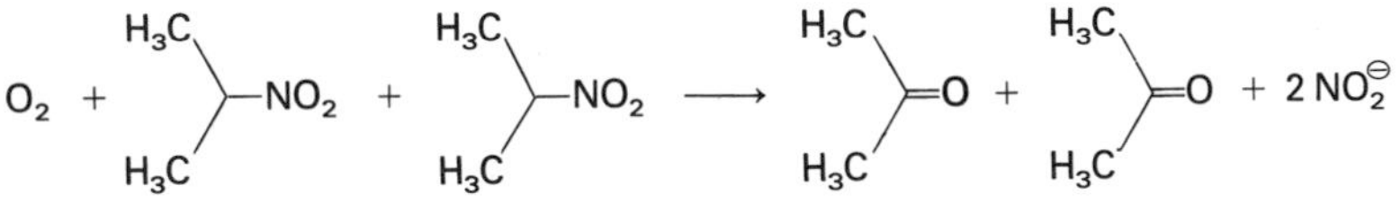

$$O_2 + (H_3C)_2CH{-}NO_2 + (H_3C)_2CH{-}NO_2 \longrightarrow (H_3C)_2C{=}O + (H_3C)_2C{=}O + 2\,NO_2^{\ominus}$$

The pure enzyme contains FAD and one Fe^{III} atom but no labile sulfur. How the dioxygen molecule is fragmented and transferred to the acetone products has not yet been uncovered.

This concludes our discussion of enzymatic redox reactions. However, we shall encounter a variety of redox steps in the enzymatic catalyses discussed in the remaining two sections—in eliminations, isomerizations and rearrangements, and in reactions that make and break carbon–carbon bonds.

Section IV

ENZYME–CATALYZED ELIMINATIONS, ISOMERIZATIONS, AND REARRANGEMENTS

To this point we have discussed two major topics: (1) we examined group-transfer reactions, focusing on the chemical problems that enzymes face and solve while activating substrate groups to facilitate reaction; and (2) we have just concluded a survey of biological oxidations, examining the devices that enzymes use to remove electrons from or add electrons to molecules undergoing redox catalysis. In some senses, the material we shall cover in this section differs in being more of a potpourri of apparently isolated and diverse transformations, without such obvious connecting threads.

Certainly, this grouping of eliminations, isomerizations, and rearrangements is more arbitrary, but we can argue that the enzymatic reactions discussed here are primarily acid–base chemistry. For the most part, they constitute the enzymology of hydrogen-transfer processes, where the hydrogens are abstracted as protons. There are exceptions, such as the NAD-dependent isomerization sequences and most notably the coenzyme-B_{12}–dependent rearrangements, which are surely not proton-transfer processes.

Enzymatic eliminations of HX generally are characterized by initial abstraction of a carbon-bound hydrogen as a proton. Subsequent loss of a good leaving group X (= OH, OR, NH_2, NHR) from a carbanionic species generates an olefinic product.

$$\mathrm{-\underset{H}{\overset{|}{C}}-\overset{X}{C}-} \;\xrightarrow[\ HX\ ]{}\; \mathrm{>C{=}C<}$$

Occasionally, data have been accumulated suggesting S_N1 dissociation of substrate to $X^{\ominus}$ and a carbonium ion, which decomposes to product with subsequent loss of $H^{\oplus}$. It is not clear if those data are entirely compelling, as we shall note in a discussion of fumarase catalysis. The X moiety can be an oxygen- or nitrogen-containing group. When X = OH, the enzymes catalyze loss of the elements of water from adjacent carbons. We shall see that, physiologically, a number of these enzymes function in the reverse direction—adding water to double bonds (hydratase activity). The X group can also be OR—often the phosphate group, which (providing a more stable anion than $OH^{\ominus}$) can lower the energy barrier for elimination. When $X = NH_2$, loss of ammonia is effected. Some of these ammonia lyases resort to covalent intermediates; to lower the activation energy, the departing NH_2 (a strong base, difficult to displace) is converted into a better leaving group.

In enzymatic isomerizations, again, the initiation step often is proton abstraction by an active-site base of the enzyme, followed by acid-base chemistry (proton transfers) and controlled reprotonation of the desired product tautomer. The stereochemical outcomes in these catalyses have been examined in some detail. Other epimerases utilize nicotinamide coenzymes and proceed via redox mechanisms.

Rearrangements are relatively common occurrences in organic chemistry, but notably infrequent in biochemical systems. The nonenzymatic reactions can occur in strong base (anionic processes), or in strong acid (hydride transfers and carbonium-ion intermediates), or thermally (electrocyclic reactions or radical rearrangements). However, the optimal pH of the bulk solution for the vast majority of enzymes is between 5 and 9, mitigating against the acid- or base-dependent rearrangements as frequent mechanisms. Furthermore, organisms are for the most part nearly isothermal, and enzymes don't have the energy available to overcome the large thermal barriers to most concerted rearrangements.

We shall look at the kinds of rearrangements that do occur in the biosynthesis of branched-chain amino acids and in biosynthesis of aromatic amino acids such as phenylalanine. Then we shall examine the dozen or so known 1,2-rearrangements in which a substrate hydrogen migrates from one carbon to an adjacent one while an X group concomitantly migrates in the opposite direction.

$$\begin{array}{cc} | \ \ | & | \ \ | \\ -C^1-C^2- & -C^1-C^2- \\ | \ \ | & | \ \ | \\ H \ \ X & X \ \ H \end{array} \quad \text{(left)} \longrightarrow \text{(right)}$$

X can represent oxygen, nitrogen, or carbon moieties. We shall see that all these enzymes require the cobalt-containing corrinoid cofactor, coenzyme B_{12}, and that these rearrangements are mechanistically quite complex.

Chapter 17

Enzymatic Eliminations and Additions of Water

We begin this chapter with a discussion of four enzymes that catalyze freely reversible dehydration/rehydration reactions on substrates that are important cellular metabolites. Then we examine dehydration/hydration sequences in polyhydroxy compounds and compare the mechanistic pathways, noting the redox sequences in some of the dehydrosugar biosyntheses.

17.A ENZYMATIC ELIMINATION AND ADDITION OF WATER TO MONOHYDROXY COMPOUNDS

The four enzymes discussed in this section are aconitase, fumarase, enolase, and crotonase. The generalized reaction is the following:

$$\mathrm{R{-}C(H)(R'){-}C(H)(OH){-}R''} \underset{+H_2O}{\overset{-H_2O}{\rightleftharpoons}} \mathrm{R{-}C(R'){=}C(H){-}R''}$$

Characterization of these reactions as *cis* or *trans* eliminations, and of the olefinic product as *cis* or *trans*, provides information relevant to the mechanism of the eliminations.

17.A.1 Aconitase

The first enzyme we shall discuss is aconitase, found in all organisms that show aerobic metabolism (Glusker, 1968, 1971). It is one of the component enzymes of the citric-acid cycle. The catalytic activity is the reversible interconversion of the isomeric monohydroxy tricarboxylates, citrate and isocitrate, via the intermediacy of the *cis*-olefinic tricarboxylate *cis*-aconitate. In this sense the enzyme is more complex than many of the other dehydratases discussed in this chapter, because it is also an isomerase (one that proceeds by a dehydration–rehydration sequence). The enzyme has also been designated aconitate hydratase and citrate (isocitrate) hydrolase.

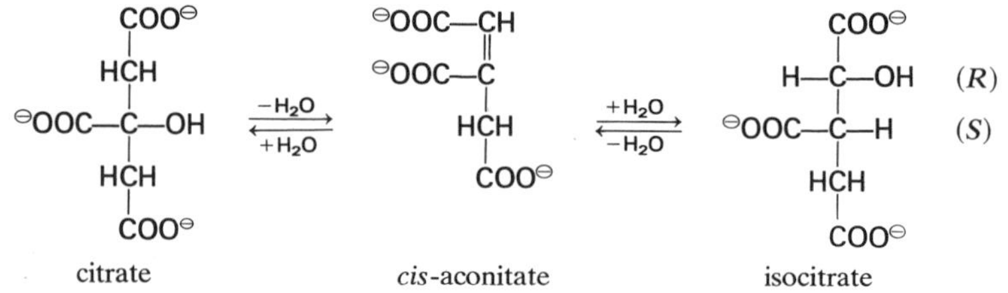

COO⊖ | HCH | ⊖OOC—C—OH | HCH | COO⊖ $\underset{+H_2O}{\overset{-H_2O}{\rightleftharpoons}}$ ⊖OOC—CH ‖ ⊖OOC—C | HCH | COO⊖ $\underset{-H_2O}{\overset{+H_2O}{\rightleftharpoons}}$ COO⊖ | H—C—OH (*R*) | ⊖OOC—C—H (*S*) | HCH | COO⊖

citrate — *cis*-aconitate — isocitrate

At equilibrium, under physiological conditions, the components are citrate (88), *cis*-aconitate (4), isocitrate (8). However, in cells the enzyme primarily converts citrate to isocitrate, which is removed from the equilibrium by isocitrate dehydrogenase (discussed in Section V). The pig-heart aconitase enzyme has been purified and is a brown protein of 89,000 mol wt, the color stemming from $Fe^{2\oplus}$, which apparently represents an Fe/S cluster (Gawzon et al., 1974). There is an additional requirement for $Fe^{2\oplus}$, which must be added if catalysis is to occur, possibly to reduce the Fe/S cluster (Ruzicka and Beinert, 1978).

An examination of the substrates reveals that citrate has a plane of symmetry through C-3 and no asymmetric centers, whereas isocitrate has asymmetric carbons at C-2 and C-3 and thus can exist as $2^2 = 4$ diastereomers. The only one recognized by aconitase is the 2*R*,3*S*-isomer shown above. On the other hand, although citrate is not a chiral species, it clearly is a *prochiral* molecule; each methylene carbon has a pro*R* and a pro*S* hydrogen. Replacement of the pro*R* hydrogen with deuterium, for instance, would generate a chiral *R* species. Additionally, the carboxymethyl arms of citrate ($—CH_2COO$) are prochiral moieties: a pro*R* and a pro*S* arm.

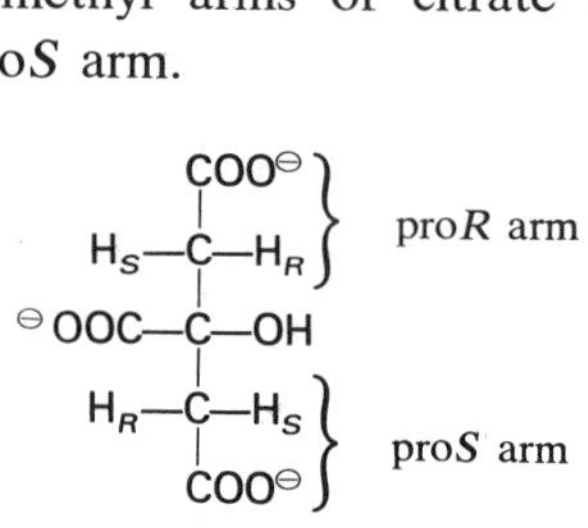

COO⊖ | H_S—C—H_R } pro*R* arm | ⊖OOC—C—OH | H_R—C—H_S | COO⊖ } pro*S* arm

The five well-characterized enzymes of citrate metabolism (the other four are discussed in Section V) all show the expected ability to distinguish between the pro*R* and the pro*S* arms. Stereochemical experiments have established that, when aconitase dehydrates citrate to *cis*-aconitate, it specifically removes the pro*R* hydrogen at C-2 of the pro*R* arm of citrate, and then the OH at C-3 is lost (see Alworth, 1972, and references therein).

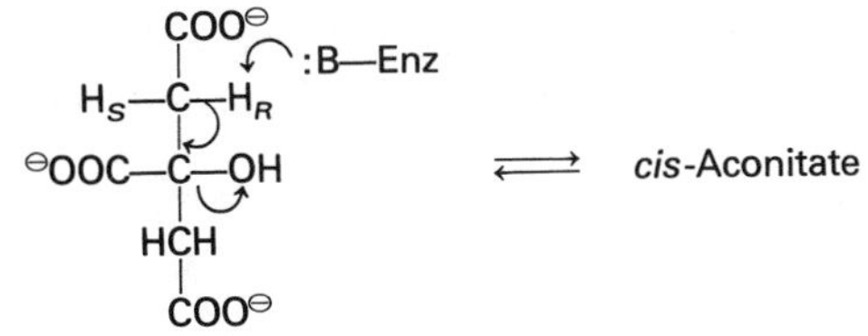

COO⊖ :B—Enz
H_S—C—H_R
⊖OOC—C—OH ⇌ *cis*-Aconitate
HCH
COO⊖

This knowledge, plus the fact that it is the *cis* isomer of aconitate that is the dehydration product, characterizes the reversible elimination of H_2O as a *trans* elimination of H and OH. In the other half-reaction (*cis*-aconitate⇌isocitrate), the addition or elimination of H_2O is again a *trans* process. *cis*-Aconitate has two sp^2 carbons at C-2 and C-3 and, using the rules for stereochemical assignments at trigonal centers (¶10.C.6), one can show that one face of the planar double-bond system is *re–si* (C-2–C-3) and the other *si–re* (C-2–C-3). Thus, attack of hydroxide on C-3 of *cis*-aconitate at the *re–si* face produces citrate ($H^{\oplus}$ adding to the *si–re* face at C-2).

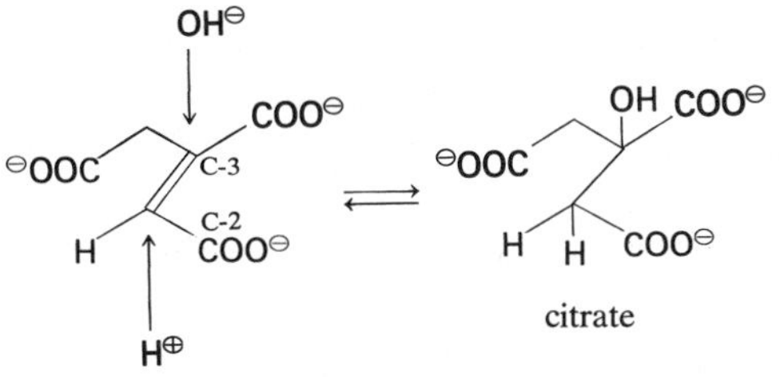

OH⊖
⊖OOC COO⊖ C-3 C-2 COO⊖ H H⊕ ⇌ ⊖OOC OH COO⊖ H H COO⊖
citrate

Conversely, to produce 2*R*,3*S*-isocitrate from *cis*-aconitate, hydroxide addition must occur at C-2 of the *si–re* face of the double bond (with $H^{\oplus}$ adding, *trans*, at the other face).

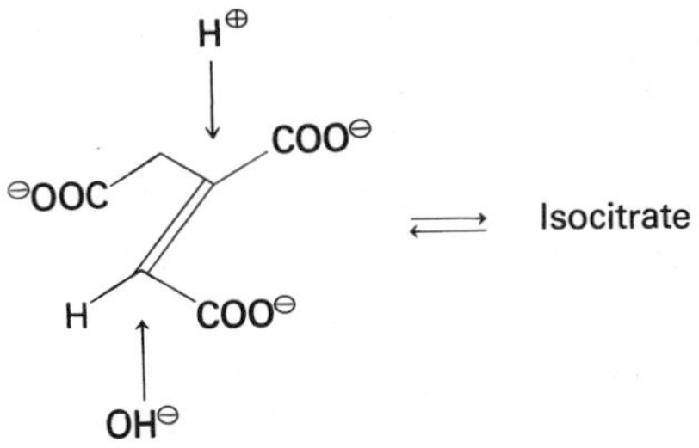

H⊕
⊖OOC COO⊖ H COO⊖ OH⊖ ⇌ Isocitrate

One consequence of this is that C-2 *is always attacked from one side of the plane* ($OH^{\ominus}$ *or* $H^{\oplus}$ *addition*), *whereas* C-3 *is always attacked in trans manner from the other side of the plane, as indicated in the structures above.* The data to be presented next show that the pro*R* hydrogen at C-2 of the pro*R* arm of citrate can be added back to C-3 of isocitrate, suggesting that this proton does not move. If this is so, then the *cis* aconitate molecule *must flip over* to be able to go between citrate⇄isocitrate, and still to have $H^{\oplus}$ added from a stationary position:

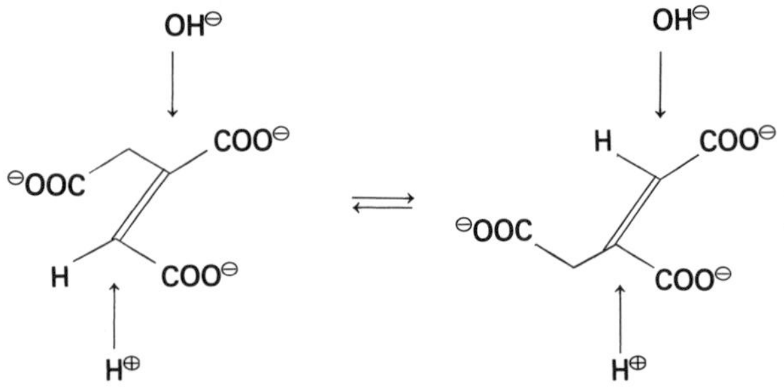

bound *cis*-aconitate flips end over end

Initial isotopic experiments with the aconitase reaction in D_2O turned up some unexpected data. Speyer and Dickman (1956) converted isocitrate to citrate in D_2O, or *cis*-aconitate to citrate in D_2O. In each case, the deuterium content of the citrate was analyzed. As expected, the citrate formed from *cis*-aconitate contained 1.0 atom of carbon-bound deuterium per molecule (at C-2). On the other hand, the citrate formed from isocitrate had only 0.1 atom of deuterium incorporated per molecule.

$$\text{Isocitrate} \xrightarrow[D_2O]{\text{enzyme,}} \text{Citrate} \quad (0.1 \text{ atom D})$$

$$cis\text{-Aconitate} \xrightarrow[D_2O]{\text{enzyme,}} \text{Citrate} \quad (1.0 \text{ atom D})$$

The incorporation of less than a full atom of deuterium must mean that the hydrogen abstracted from C-3 of isocitrate is retained at the active site, not equilibrated with the 55 M D_2O bulk solvent, and is specifically reutilized to protonate C-2 of *cis*-aconitate to form citrate. That is, the enzyme shows *shielded proton transfer,*★ sequestering the proton from exchange with solvent protons. Further, these data show that the *cis*-aconitate formed (from citrate or isocitrate) must remain largely enzyme-bound during normal turnover, because exogenous *cis*-aconitate, when added, allows full deuterium incorporation from solvent.

★We have noted enzyme shielding of substrate hydrogens previously, as in the reaction of D-amino-acid oxidase with β-halo-α-amino-acid substrates in Chapter 11.

In contrast to these data on intramolecular hydrogen transfer, other results show that the hydroxyl group is lost to the medium in every turnover. Isocitrate labeled with ^{18}O in the C-2 hydroxyl produces no [^{18}O]-citrate.

An additional experiment complementary to the D_2O experiments involves labeling one of the hydroxyacid substrates with tritium in the transferable position and checking to see if the isomerized product retains the tritium. Indeed, Rose and Hanson (1975) demonstrated that, starting with 3-[^3H]-isocitrate, citrate formed in the initial-velocity stage of the enzymatic reaction contains tritium at C-2 with no dilution of specific radioactivity and no loss of tritium to the solvent, implying 100% internal transfer.

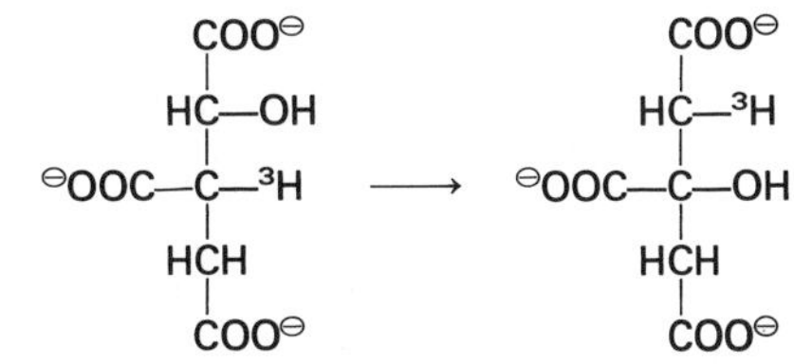

```
        COO⊖                     COO⊖
         |                        |
        HC—OH                    HC—³H
         |                        |
  ⊖OOC—C—³H      ⟶        ⊖OOC—C—OH
         |                        |
        HCH                      HCH
         |                        |
        COO⊖                     COO⊖
```

3-[^3H]-isocitrate   2-[^3H]-citrate

A further elegant experiment confirmed the existence of a slowly dissociating $EB^3H^{\oplus}$ at the active site of aconitase by proof of *intermolecular* transfer of tritium (Rose and O'Connell, 1967). Incubation of 2*R*-[^3H]-citrate with aconitase in the presence of 2-methyl-*cis*-aconitate produced *cis*-aconitate and some 2-methyl-3-[^3H]-isocitrate.

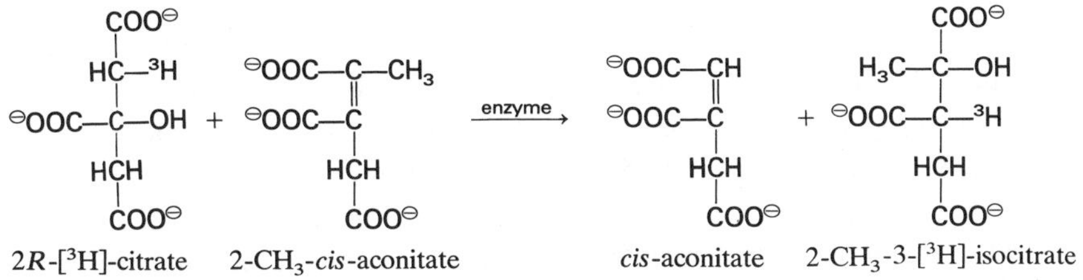

```
       COO⊖                                                              COO⊖
        |                                                                 |
       HC—³H        ⊖OOC—C—CH₃                ⊖OOC—CH              H₃C—C—OH
        |                 ||                         ||                   |
 ⊖OOC—C—OH  +  ⊖OOC—C        —enzyme→       ⊖OOC—C        +  ⊖OOC—C—³H
        |                 |                          |                    |
       HCH               HCH                        HCH                  HCH
        |                 |                          |                    |
       COO⊖              COO⊖                       COO⊖                 COO⊖
```

2*R*-[^3H]-citrate   2-CH_3-*cis*-aconitate   *cis*-aconitate   2-CH_3-3-[^3H]-isocitrate

Formation of the radioactive methylisocitrate must proceed as follows. The tritium at C-2 of citrate is abstracted by an enzymatic base during the dehydration that produces *cis*-aconitate and $EB^3H^{\oplus}$. The *cis*-aconitate dissociates from the active site, and the methyl-*cis*-aconitate binds *in its place before* $EB^3H^{\oplus}$ *can deprotonate.* Rehydration now yields the radioactive methyl analogue of isocitrate.

Rose and O'Connell thus wrote a mechanistic scheme with the aconitate flipping over during a normal catalytic cycle. Occasionally enzyme-bound *cis*-aconitate can dissociate, and only then will the $EBT^{\oplus}$ exchange with solvent.

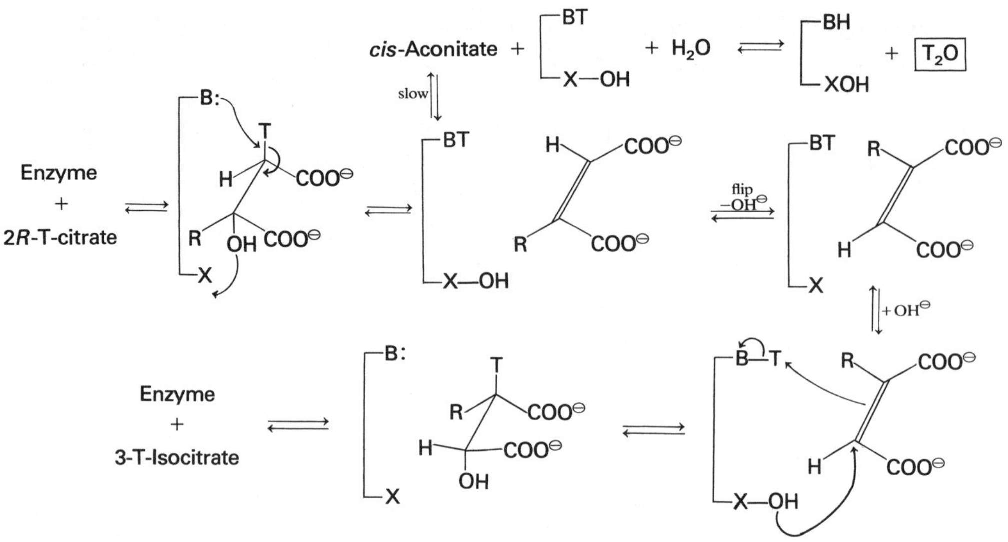

Rose (1970) has estimated the turnover number of aconitase to be about 15 sec^{-1}; this means that in a single turnover, consuming 70 msec, $EBH^{\oplus}$ does not dissociate and reprotonate with solvent hydrogens. Yet the OH group removed from substrate does exchange with water oxygens, and so H_2O (or $OH^{\ominus}$) must be accessible to the active site. Rose has speculated that $EBH^{\oplus}$ could be recalcitrant to exchange if it is hydrogen-bonded to some other basic species at the active site.

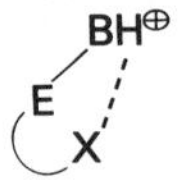

A slow conformational change (e.g., brought on by the rare dissociation of *cis*-aconitate) disrupting such a hydrogen bond may be the factor that would be required before $EBH^{\oplus}$ could exchange rapidly. All of the mechanistic information gathered thus far about this dehydratase says nothing about whether the actual elimination step proceeds by a carbanion mechanism, proceeds by a carbonium-ion mechanism, or is a concerted loss of HOH. (We shall see that this kind of information is available for fumarase and enolase.)

Some conjectures have been made about the way in which bound *cis*-aconitate might flip to change the face of the double bond so that the sequestered proton can be added to the face opposite that from which it was abstracted. We noted above that aconitase preparations are inactive without the addition of ferrous ion (VillaFranca and Mildvan, 1971, 1972). Glusker (1968, 1971) has proposed that the aconitate intermediate, and both citrate and isocitrate substrates also, are chelated to the bound $Fe^{2\oplus}$ through the tertiary carboxyl, the tertiary hydroxyl, and one of the two primary carboxyl groups. Evidence for such chelation has been obtained by VillaFranca and Mildvan from NMR studies on relaxation rates of water in the analogous $Mn^{II} \cdot$ Enzyme.

citrate–Fe^{II}–enzyme ternary complex

Chelation of the tertiary —OH to the Fe^{II} should facilitate its expulsion during *cis*-aconitate formation; the *cis*-aconitate will originally be in a "citrate-like" conformation, still bound to the iron by the two carboxylate functionalities. Now, to effect the rehydration to isocitrate, the plane of the olefinic *cis*-aconitate must rotate from a "citrate-like" to an "isocitrate-like" conformation bound to the active-site Fe^{II}. Three possible mechanisms have been described for this rotation.

In the "ferrous wheel" scheme, the aconitate rotates while both $COO^{\ominus}$ groups stay liganded to the Fe^{II} at the active site (Glusker, 1968, 1971).

citrate–Fe^{II}–enzyme

"citrate-like" *cis*-aconitate

ferrous wheel rotation

"isocitrate-like" *cis*-aconitate

exchange of H_2O ligand

2*R*,3*S*-isocitrate–Fe^{II}–enzyme

One of the Fe^{II} ligand bonds must break, and the H_2O (derived from C-3 of citrate) can be displaced by an incoming H_2O molecule, which is positioned to attack the "isocitrate-like" conformer of the coordinated *cis*-aconitate. The ligand-exchange processes at Fe^{II} centers may be up to $10^6\ sec^{-1}$ in inorganic models; they are at least $10^4\ sec^{-1}$ for metal-substrate–complex dissociation in aconitase and so qualify kinetically in turnover ($13.5\ sec^{-1}$) (VillaFranca and Mildvan, 1971, 1972).

In the "reverse ferrous wheel" variant, it is the tertiary carboxylate ligand in the Fe^{II}–*cis*-aconitate complex that is thought to dissociate, rotate, and then religand to the Fe^{II}, setting up the geometry for *anti* addition to form isocitrate.

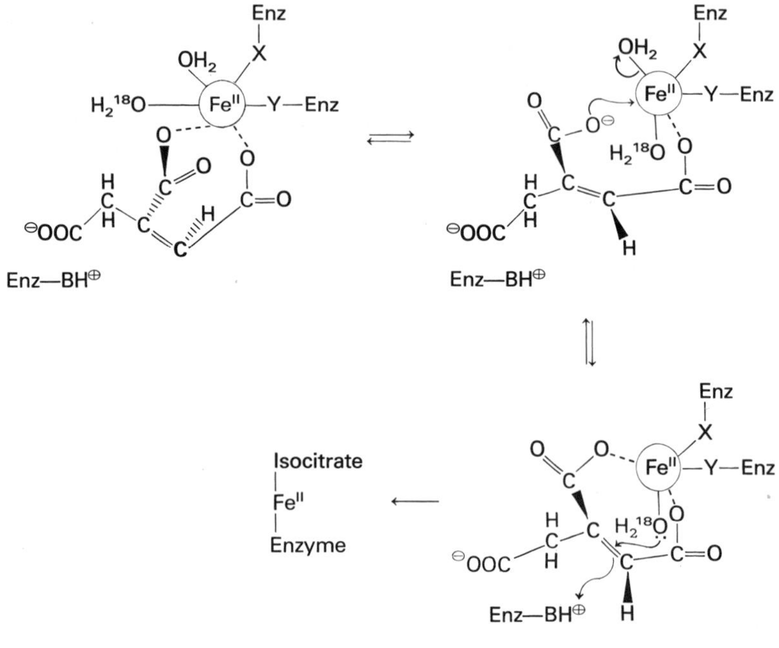

A third variant, the "Bailar twist" mechanism, has also been suggested by VillaFranca and Mildvan; it represents a similar "turnstyle" process, although without discrete breakage of any ligands to iron (Bailar, 1958). No experimental distinctions for these three possibilities (or for still others) have been reported.

17.A.2 Fumarase

The enzyme fumarase (or fumarate hydratase) catalyzes the equilibration of fumarate and *S*-malate (L-malate).

$$^{\ominus}OOC\text{–}CH\text{=}CH\text{–}COO^{\ominus} \underset{-H_2O}{\overset{+H_2O}{\rightleftharpoons}} {}^{\ominus}OOC\text{–}CH_2\text{–}CH(OH)\text{–}COO^{\ominus}$$

Like aconitase, fumarase is a component of the tricarboxylate cycle, and thus is widely distributed (Hill and Tiepel, 1971). Unlike aconitase and enolase (¶17.A.3), fumarase shows no requirements for metals during catalysis. The enzyme appears to be tetrameric (194,000 mol wt) in its active form. The addition of water to fumarate is a *trans* process, as was the addition to aconitate. In D_2O, a monodeuteromalate is formed. *Cis* addition would yield 2*S*,3*S*-malate-3-D, whereas *trans* addition would produce 2*S*,3*R*-malate-3-D. Synthetic threo-D,L-malate (a racemic mixture of *S*,*S*- and *R*,*R*-3-D-malates) was synthesized, and NMR analysis showed it to be diastereomeric to the enzymatic product, which thus was the erythro diastereomer (Gawron and Fondy, 1954). Because only 2*S*-malate is produced, the product must be 2*S*,3*R*-malate-3-D in D_2O, a *trans* addition of water.

Alberty and colleagues demonstrated that dehydration of 3-*R*-D-malate produces no isotope effect, nor is there any rapid exchange of tritium from 3-*R*-[^{3}H]-malate into solvent faster than the overall dehydration (Alberty et al., 1957).

Boyer's group has elegantly examined the mechanisms by which both fumarase and enolase catalyze dehydrations; they used isotopic exchanges measured as initial rates and at chemical equilibrium (Hansen et al., 1969). For the fumarase-catalyzed reaction, they measured the rates of exchange (1) of the carbon skeleton of malate with fumarate, (2) of the *R*-hydrogen at C-2 of malate with solvent, and (3) of the hydroxyl group at C-3 with solvent in an effort to deduce whether evidence for a nonconcerted elimination could be obtained. The rate of exchange of radioactive [^{14}C]-malate to [^{14}C]-fumarate can be calculated from the following equation:★

$$R = -\frac{[\text{fumarate}][\text{malate}]}{[\text{fumarate}]+[\text{malate}]} \times \frac{\ln(1-F)}{t}$$

where R = rate of exchange, and F = fraction of isotopic equilibrium attained at time t. When exchange with either solvent hydrogen or oxygen species is involved, Boyer pointed out that the high concentration of water allows simplification to

$$R = -[\text{malate}][\ln(1-F)]/t$$

★ Noted earlier in its generic form in ¶5.E.2.

Table 17.1
Isotopic exchange rates at chemical equilibrium (fumarase)

Isotopic exchange	Relative rate
[^{18}O]-Malate ⇄ [^{18}O]-H_2O	4.0
[^{14}C]-Malate ⇄ [^{14}C]-Fumarate	2.5
2*R*-[^{3}H]-Malate ⇄ [^{3}H]-H_2O	1.0

SOURCE: Data from Hansen et al. (1969).

Initial-velocity exchanges showed that the rate of conversion of [^{14}C]-malate to [^{14}C]-fumarate is equal to the rate at which either 2*R*-[^{2}H]- or 2*R*-[^{3}H]-3*S*-malate loses deuterium or tritium to solvent, consistent with the earlier work from Alberty's laboratory, indicating the absence of a primary deuterium isotope effect.

Exchange experiments were also performed at chemical equilibrium. With chemical equilibrium established, a trace amount of labeled component (e.g., [^{14}C]-malate, [^{3}H]-malate) can be added without adding enough mass to perturb chemical equilibrium. One can then monitor the rate of appearance of isotope in the other component (e.g., [^{14}C]-fumarate, [^{3}H]-H_2O) as an isotope-exchange process proceeding at chemical equilibrium (¶5.E.2). Comparison of the rates of the various possible exchanges is revealing (Table 17.1).

The fastest reaction is the loss of the malate hydroxyl group, faster than can be explained by back reaction (indicated by ^{14}C-exchange rate). Further, the rate of C-3 hydrogen exchange is slower than the carbon-skeleton exchange rate. Two conclusions were reached from these data. First, Hansen, Dinovo, and Boyer (1969) assumed that the data support a carbonium-ion intermediate whose formation is fast, to account for the rapid ^{18}O-exchange. Second, the slow proton exchange suggests that, analogous to the aconitase reaction, the $EB^3H^{\oplus}$ can remain undissociated long enough for [^{12}C]-fumarate to be liberated to solvent and to be replaced at the active site by a [^{14}C]-fumarate molecule. The turnover number for fumarase is $2 \times 10^3\ \text{sec}^{-1}$, which is similar to rates at which a protonated imidazolium ion dissociates and reprotonates. If the enzymatic base were histidine, the turnover number for catalysis is faster than $BH^{\oplus} \rightarrow B + H^{\oplus}$, and this inequity would limit the loss of proton to solvent.

Enz···[$^{-}$OOC–CH(OH)–CH(^{3}H)–COO^{-}] ⇄ (−OH$^{\ominus}$, fast) [$^{-}$OOC–CH$^{\oplus}$–CH(^{3}H)–COO^{-} + :B–Enz] (carbonium ion) ⇄ [H(COO$^{\ominus}$)C=C($^{\ominus}$OOC)···Enz–H B^{3}H$^{\oplus}$] ⇄ H(COO$^{\ominus}$)C=C(H)($^{\ominus}$OOC) + Enz—B^{3}H$^{\oplus}$

Enz—B^{3}H$^{\oplus}$ ⇄ (slow) Enz—B + ^{3}H$^{\oplus}$

Rose (1970) has commented, however, that these data do not support carbonium-ion formation unequivocally, nor do they rule against carbanion or concerted mechanisms. It could actually be that carbon–hydrogen–bond cleavage at C-3 of malate precedes C—OH cleavage (and so generates a carbanionic intermediate), but that proton release is so slow (c.f., aconitase) that it *looks* as though the C—OH bond has broken first by isotopic-exchange velocities. (The C—H cleavage, of course, could not be slow enough to be rate-determining in any case, because there is no observed primary kinetic isotope effect.) Thus, the sequence of chemistry at the active site could, in principle, be obscured by physical processes.

A number of other substances related to fumarate are hydrated by fumarase. Interestingly, acetylene dicarboxylate is converted to oxaloacetate, presumably via the enol as initial product of hydration, which then tautomerizes to the more stable observed product, oxaloacetate, in a nonenzymatic step (Tiepel et al., 1968).

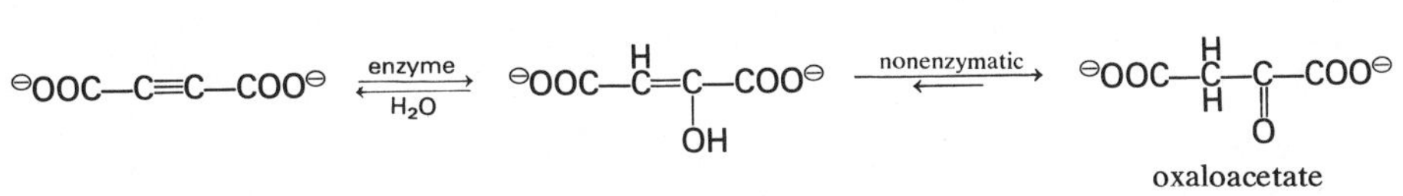

$$^{\ominus}OOC—C{\equiv}C—COO^{\ominus} \underset{H_2O}{\overset{enzyme}{\rightleftharpoons}} {}^{\ominus}OOC—\underset{}{\overset{H}{C}}{=}\underset{OH}{C}—COO^{\ominus} \xrightleftharpoons{nonenzymatic} {}^{\ominus}OOC—\underset{H}{\overset{H}{C}}—\underset{O}{\overset{\|}{C}}—COO^{\ominus}$$

oxaloacetate

17.A.3 Enolase

The enzyme known as enolase (systematic name, 2-phospho-D-glycerate hydroxylase) interconverts the D-isomer of 2-phosphoglycerate (2-PGA) and phosphoenolpyruvate (PEP) with an equilibrium constant $K_{eq} = 6.7$ for PEP formation at pH 7.8. The enzyme is ubiquitous, as expected for a key enzyme in glucose fermentation (Wold, 1971). Catalytic activity is absolutely dependent on a divalent cation such as $Mg^{2\oplus}$, which may chelate to the phosphate group of substrate and product at the active site.

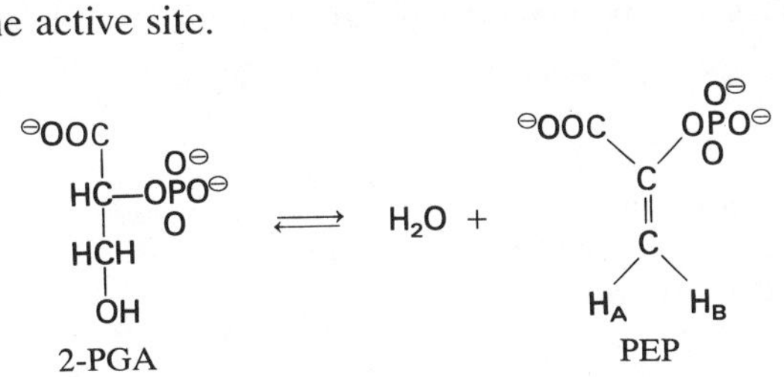

$$^{\ominus}OOC—\underset{\substack{|\\HCH\\|\\OH}}{HC}—OPO_3^{2\ominus} \rightleftharpoons H_2O + (^{\ominus}OOC)(^{2\ominus}O_3PO)C{=}CH_AH_B$$

2-PGA PEP

The yeast enzyme, a well-studied representative, is a dimer of 88,000 mol wt. The stereochemistry of elimination of H_2O from 2-PGA was determined to be *trans* (*anti*) by NMR spectroscopy. In the PEP structure above, the vinyl hydrogen *trans* to the phosphate is H_A and that *cis* to the phosphate is H_B. The proton NMR spectrum of PEP in D_2O consists of two multiplets, one at 5.15 ppm (δ), the other at 5.33 ppm (δ). The low-field multiplet was tentatively assigned as the H_A signal on a correlation with known vinyl compounds; this assignment was confirmed using ^{13}C–H coupling constants with 1-[^{13}C]-PEP (Cohn et al., 1970). The stereochemistry of enolase catalysis could then be established if one started with a chiral sample of 3-[^{2}H]-2-PGA. In fact, 3*R*-3-[^{2}H]-2-PGA was prepared by action of various glycolytic enzymes and then submitted to enolase action to yield a 3-[^{2}H]-PEP product. The proton NMR spectrum showed only one peak (split to a doublet by the ^{31}P) centered at 5.14 ppm. Therefore, the deuterium at C-3 is *trans* to the phosphate, and thus elimination of the OH at C-3 must be *anti* to the C-2 proton lost (Tiepel et al., 1968). In the back reaction, the product 2-PGA is D- or 2*R*-2-PGA, and $H^{\oplus}$ must add to the *si* face of the vinyl double bond of PEP at C-2. (It follows that OH addition at C-3 must be at the *re* face.)

3*R*-[^{2}H]-2-PGA ⇌ 3-[^{2}H]-PEP

In isotopic exchange experiments of the type described earlier with fumarase, Dinovo and Boyer (1971) established the relative rates of isotopic exchanges at chemical equilibrium shown in Table 17-2. With [^{3}H]-PGA, the relative rate of exchange into H_2O was 0.9, suggesting that a kinetic isotope selection is occurring, and allowing one to calculate the anticipated [^{1}H]-PGA⇄H_2O exchange

Table 17-2
Isotopic exchange rates at chemical equilibrium (enolase)

Isotopic exchange	Relative rate
[^{2}H]-PGA⇄[^{2}H]-H_2O	1.9
[^{18}O]-PGA⇄[^{18}O]-H_2O	1.3
[^{14}C]-PGA⇄[^{14}C]-PEP	1.0

SOURCE: Data from Dinovo and Boyer (1971).

from the following two equations (¶4.A.3; Swain et al., 1958):

$$k_H/k_T = (k_D/k_T)^{3.26}$$

$$k_H/k_T = (k_H/k_D)^{1.44}$$

On this relative scale, the protonic exchange rate from C-2 of PGA to H_2O is 12, establishing this as the fastest partial exchange. This evidence is consonant with a carbanionic species forming during enzyme-catalyzed dehydration. On the other hand, with 2-[^{2}H]-2-PGA there is no deuterium kinetic isotope effect on V_{max} of PEP formation (the overall reaction), in contrast to the large isotope discrimination in the exchange reaction. This suggests that proton abstraction at C-2 of 2-PGA, involved in overall PEP formation, is a fast step in catalysis. Boyer suggested the following sequence, in which the carbanion is formed in a fast, quasi-equilibrium step, to account for the experimental data (Dinovo and Boyer, 1971).

$E\ddot{B} + H_2O$

fast

EB· · ·H—C(COO⊖)(OPO$_3$⊖)—HCH—OH $\overset{\text{fast}}{\rightleftharpoons}$ EBH⊕· · ·⊖C(COO⊖)(OPO$_3$⊖)—HCH—OH (BH⊕) $\overset{\text{slow}}{\rightleftharpoons}$ H_2O + C(COO⊖)(OPO$_3$⊖)=HCH

first step is at equilibrium

carbanionic intermediate

This scheme would predict that, if 2-PGA and enolase were added to ^3H_2O and the reaction monitored under conditions where the back reaction does not occur (PEP, once formed, does not return to 2-PGA), then ^3H should be incorporated into the "unreacted" 2-PGA as a consequence of its going to the carbanion and back many times (picking up solvent ^3H) before going on to PEP. Indeed, when PEP was trapped with ADP and pyruvate kinase to eliminate any possibility of ^3H incorporation by overall back reaction, reisolated 2-PGA acquired ^3H from solvent rapidly (Dinovo and Boyer, 1971). Thus, it is likely that enolase dehydrates by initial abstraction of a proton from C-2 to form the C-2 carbanion, whose breakdown (with expulsion of OH from C-3) is rate-determining. Figure 17-1 shows a three-dimensional projection of the enolase active site, interpreted from distance data acquired by NMR measurements (Mildvan, 1977).

Two other facts about enolase catalysis bear comment. One is the recent observation that the C_4 molecule 2-phospho-3-butenoate is isomerized catalytically by enolase to the phosphoenolketobutyrate (Stubbe, 1975). (The back reaction could not be demonstrated.)

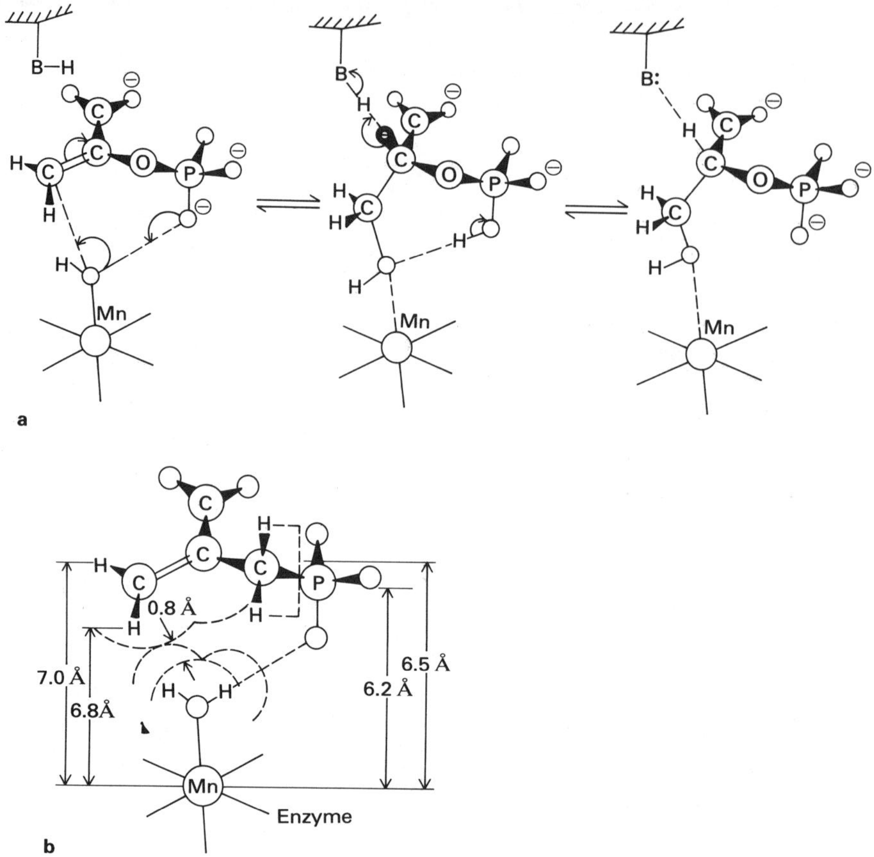

Figure 17-1
(**a**) Mechanism of enolase, consistent with NMR data, isotope exchange studies, and stereochemistry. PEP is in the second coordination sphere of the Mn and reacts with the H_2O molecule in the first coordination sphere. (**b**) Composite NMR distance map for PEP bound to $Mn^{2\oplus}$-enolase. (From Mildvan, 1974. Reproduced with permission from *Annual Review of Biochemistry*, vol. 43. Copyright © 1974 by Annual Reviews Inc. All rights reserved.)

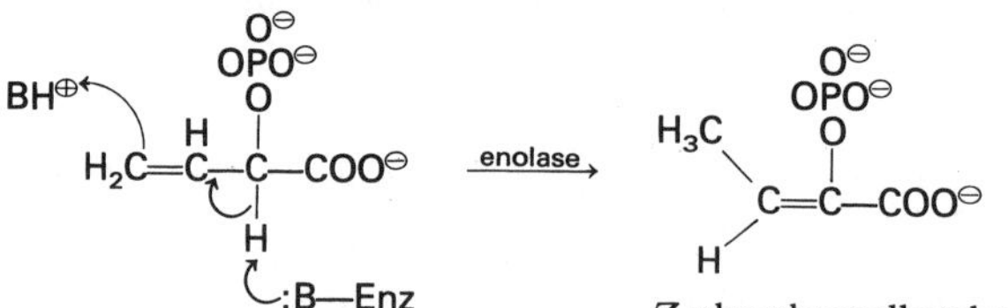

$$BH^{\oplus} \quad H_2C{=}C(H){-}CH(H)(OPO_3^{2-}){-}COO^{\ominus} \;\xrightarrow{\text{enolase}}\; H_3C(H)C{=}C(OPO_3^{2-}){-}COO^{\ominus}$$

:B—Enz

Z-phosphoenolketobutyrate

This reaction presumably occurs by abstraction of the C-2 proton and subsequent isomerization with protonation at C-4; only the *Z*-isomer of the product was generated. This activity points out the similarity between H_2O eliminations and the enzymatic isomerizations we shall be discussing subsequently.

The second fact is the observation made by Rose and O'Connell (1973) that enolase is irreversibly inactivated by glycidol phosphate (2,3-epoxypropanol-1-$PO_3^{2\ominus}$).

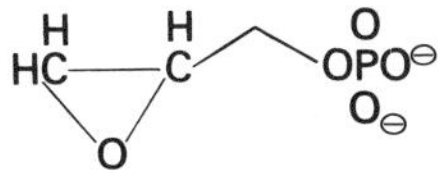

The epoxide was prepared as an affinity label for triose-P isomerase (as we shall note later); unexpectedly, it inactivated enolase as well. C-2 of glycidol-P is asymmetric, and Rose and O'Connell have found that both D- and L-enantiomers inactivate, the D-isomer with a K_m of 9 mM and maximal inactivation rate of 2.35% min^{-1}, and the L-isomer with K_m of 15 mM and maximal inactivation rate of 0.67% min^{-1}. The inactivation is covalent, but the radioactive inactivator is released by hydroxylamine, suggesting an ester linkage (with either enantiomer) and implicating an enzymatic $COO^{\ominus}$ group as the attacking nucleophile in epoxide ring opening (shown here as the likely acid-catalyzed ring opening).

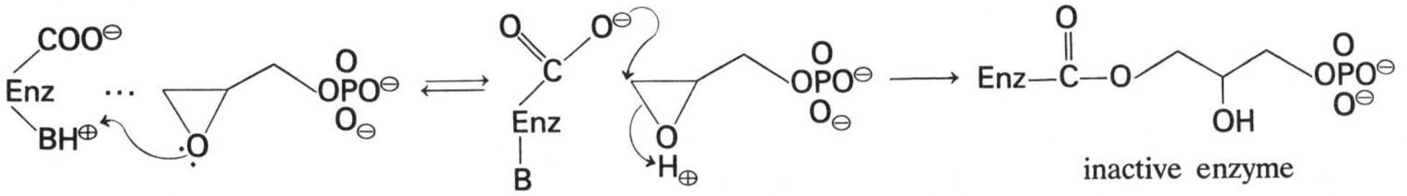

inactive enzyme

It is not known whether the $COO^{\ominus}$ opens the ring by attack at C-2 or at C-3; here it is arbitrarily shown as C-3 for diagrammatic purposes only. Rose and O'Connell suggest that this enzyme $COO^{\ominus}$ may be the active-site base in normal isomerization catalysis.

17.A.4 Crotonase

The enzyme enoyl-CoA hydrase was termed crotonase for historical reasons (Hill and Tiepel, 1971). Crotonyl-CoA (Δ^2-butenyl-CoA) was the first substrate recognized. Physiologically, this enzymatic activity functions as a hydratase in fatty-acid oxidations, utilizing as substrate the α,β-unsaturated acyl-CoA produced by the action of flavin-dependent acyl-CoA dehydrogenases noted in Section III. The β-hydroxyacyl-CoA product from crotonase action undergoes nicotinamide-dependent oxidation to the β-keto acyl-CoA by a subsequent enzyme, and finally thiolytic cleavage by CoASH, catalyzed by thiolase (discussed in Section V).

$$\mathrm{R{-}CH_2{-}CH_2{-}\overset{O}{\overset{\|}{C}}{-}S{-}CoA} \xrightarrow[\text{(FAD)}]{\text{acyl-CoA dehydrogenase}} \mathrm{RCH{=}CH{-}\overset{O}{\overset{\|}{C}}SCoA} \xrightarrow{\text{enoyl-CoA hydrase}} \mathrm{RCH(OH){-}CH_2{-}\overset{O}{\overset{\|}{C}}SCoA}$$

$$\downarrow \beta\text{-hydroxyacyl-CoA dehydrogenase (NAD)}$$

$$\mathrm{R\underset{\|}{\underset{O}{C}}SCoA + H_3C\overset{O}{\overset{\|}{C}}SCoA} \xleftarrow[\text{thiolase}]{\text{CoASH,}} \mathrm{R\underset{\|}{\underset{O}{C}}{-}CH_2{-}\overset{O}{\overset{\|}{C}}SCoA}$$

The ox-liver enoyl-CoA hydrase is a hexamer of 164,000 mol wt. Curiously, it will hydrate both *cis* and *trans* isomers of Δ^2-enoyl-CoA to the corresponding L- or D-β-hydroxyacyl-CoA products. It shows a turnover number of 340,000 moles min^{-1} (mole enzyme)$^{-1}$ for the C_4 enoyl-CoA; the value decreases progressively to 2,300 for the Δ^2-hexadecenoyl-CoA substrate (C_{16}), with little change in K_m values (2×10^{-4} M); for crotonyl-CoA, $V_{max}/K_m \approx 6\times10^7\ \text{M}^{-1}\text{sec}^{-1}$, close to the diffusion limit. Apparently, one enzyme is used for all the species formed during complete oxidation of long-chain fatty acyl-CoAs to CO_2. Rather little is known about the mechanism of this hydration reaction. The enzyme does not require a metal ion for catalytic activity, and in this sense it resembles fumarase more closely than it does aconitase or enolase.

17.B ENZYMATIC DEHYDRATIONS OF DIHYDROXY OR POLYHYDROXY SUBSTRATES

In our discussion of enzymatic eliminations of H_2O, we now switch from monohydroxy substrates to substrates where two or more hydroxyl functions are present. We begin with the loss of water from vicinal glycol groups of dihydroxy or polyhydroxy molecules, where a ketone functionality is generated in the product.

$$\mathrm{{-}\underset{HO}{\overset{H}{C}}{-}\underset{OH}{\overset{H}{C}}{-}} \longrightarrow \mathrm{H_2O + {-}\underset{H}{\overset{H}{C}}{-}\underset{\underset{O}{\|}}{C}{-}}$$

Then we shall look at dehydrations in deoxysugar formation; these reactions combine elimination and redox chemistry.

17.B.1 Dihydroxy Substrates

For dihydroxy substrate molecules, we can make a fundamental distinction between two classes on the basis of mechanism. First, there are discrete dehydrases known for glycerol and for 1,2-propanediol that have been purified from *Klebsiella* strains of bacteria (see Abeles and Dolphin, 1976, and references therein). The products are β-hydroxypropionaldehyde and propionaldehyde, respectively.

$$\underset{\text{glycerol}}{\begin{matrix} H_2COH \\ | \\ HCOH \\ | \\ H_2COH \end{matrix}} \longrightarrow \begin{matrix} HC{=}O \\ | \\ HCH \\ | \\ H_2COH \end{matrix} + H_2O \qquad \underset{\text{1,2-propanediol}}{\begin{matrix} H_2COH \\ | \\ HCOH \\ | \\ CH_2 \end{matrix}} \longrightarrow \begin{matrix} HC{=}O \\ | \\ HCH \\ | \\ CH_3 \end{matrix} + H_2O$$

These enzymatic eliminations of water require coenzyme B_{12} and are accompanied by a 1,2-hydrogen-shift process; we defer discussion of them until Chapter 20, where we discuss the generalized role of coenzyme B_{12} in intramolecular shift reactions.

The other basic category of enzymatic eliminations from dihydroxy substrates (and this is true also for most polyhydroxy substrates) is generalized as an elimination of the elements of H_2O to *yield an enol as primary product*; the enol then tautomerizes to the more stable methylene-keto moiety observed in the product.

In a sense, this type of transformation is represented by the two-enzyme sequence of enolase and pyruvate kinase. Enolase dehydrates 2-phosphoglycerate to an enol, albeit a phosphorylated one, PEP. The kinase then deblocks this trapped enol with phosphoryl transfer, yielding ATP and the keto product, pyruvate.

$$\begin{matrix} & H \\ & | \\ H_2C & -C- & COO^{\ominus} \\ | & | \\ OH & OX \end{matrix} \longrightarrow H_2C{=}\underset{OX}{\underset{|}{C}}{-}COO^{\ominus} \xrightarrow[\text{ATP}]{\text{ADP}} H_3C{-}\underset{O}{\underset{\|}{C}}{-}COO^{\ominus} \qquad X = PO_3^{2\ominus}$$

The —CH_2OH moiety at C-3 of 2-PGA is converted to the —CH_3 of pyruvate. It is likely that this two-step mechanism occurs at a single enzyme active site in the following two examples.

17.B.1.a Dihydroxyacid dehydrase

Dihydroxyacid dehydrase is involved in valine and isoleucine biosynthesis. An enol intermediate is formed on initial dehydration and then ketonized at the active site *before* release to the medium (Hill et al., 1973). This can be determined by the fact that addition of a proton to the double bond of the enol is stereospecific, giving only one of the two possible chiral arrangements at C-3 in the product formed.

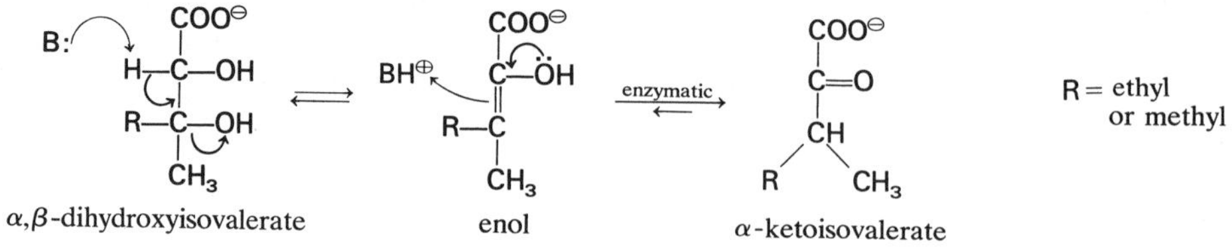

α,β-dihydroxyisovalerate ⇌ enol ⇌ (enzymatic) α-ketoisovalerate

R = ethyl or methyl

17.B.1.b Imidazole-glycerol-P dehydrase

An analogous dehydration occurs during the microbial biosynthetic pathway to histidine, where a glycerol-$PO_3^{2\ominus}$ moiety is converted to an acetol-$PO_3^{2\ominus}$ grouping. Little is known about this enzyme except that a solvent proton is incorporated at C-3 of the imidazole-acetol phosphate.

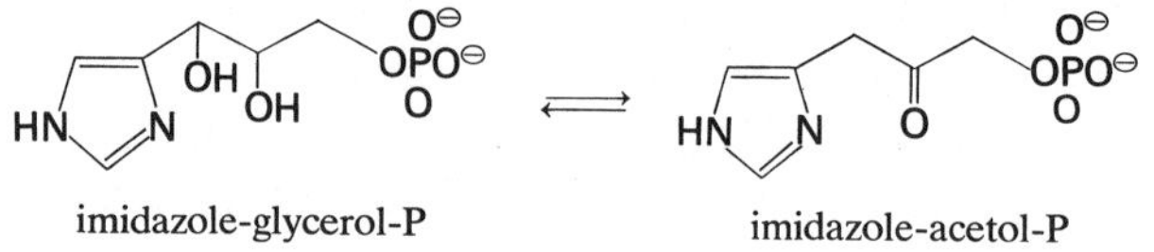

imidazole-glycerol-P ⇌ imidazole-acetol-P

17.B.2 Dehydrases for Sugar Substrates

When the substrates are polyhydroxy compounds, a large number of hexose and pentose dehydrases are known, mostly for aldonic acids (C^1-acids) or their phosphate esters (Glaser and Zarkowsky, 1971; W. A. Wood, 1971). The majority of these enzymes carry out the following transformation, specifically yielding the 2-keto-3-deoxy product.

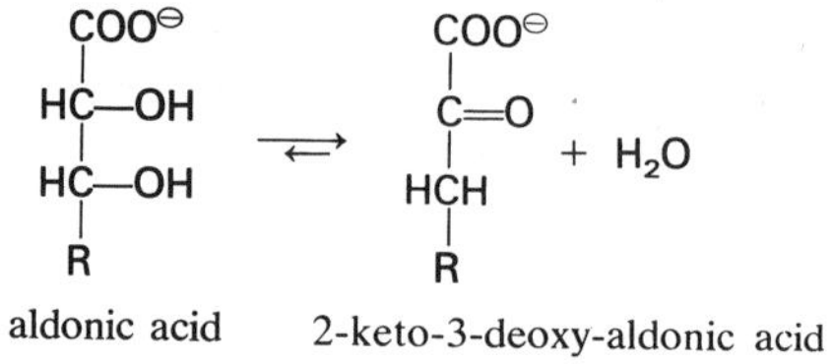

aldonic acid ⇌ 2-keto-3-deoxy-aldonic acid + H_2O

It is likely that all of these enzymes utilize mechanisms where carbanionic species are generated at C-2, followed by loss of the OH moiety from C-3 to yield the enol product, which then ketonizes. We shall look at two examples.

17.B.2.a 6-Phosphogluconate dehydrase

Probably the most carefully characterized example of a sugar dehydrase is 6-phosphogluconate dehydrase, purified from extracts of *Pseudomonas saccharophilia* (Meloche and Wood, 1964).

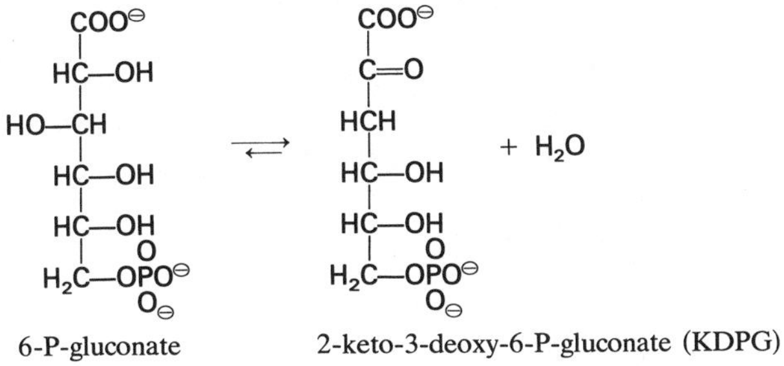

6-P-gluconate ⇄ 2-keto-3-deoxy-6-P-gluconate (KDPG) + H_2O

When the dehydration is carried out in ^3H_2O, one atom of ^3H from the solvent is incorporated at C-3 of the product (ruling out a hydride mechanism), starting from 6-P-gluconate. Incubation of enzyme, ^3H_2O, and the ketodeoxy product produced no tritium in the KDPG, nor any 6-P-gluconate, showing that the enzyme does not enolize C-3 of KDPG nor use it for back reaction. Further, when "unreacted" 6-P-gluconate is reisolated from incubations with enzyme and ^3H_2O after dehydration has proceeded on some of the substrate molecules, it contains carbon-bound tritium. This exchange is similar to that described earlier for enolase and is consistent with a rapid, preequilibrium formation (i.e., reversible) of a carbanion (or its enolic equivalent) at C-2 of 6-P-gluconate many times before loss of the OH from C-3 proceeds. Finally, monitoring the formation of KDPG by specific absorption of the α-keto-acid group, its rate of formation is slower than substrate disappearance, suggesting that ketonization of the initial enolic product is rate-determining. This observation further suggests that turnover produces released enol as product, which then ketonizes nonenzymatically—a slow step off the enzyme, not affecting V_{max}. This supposition was corroborated by conducting the enzymatic dehydration in D_2O. The 3-[^2H]-KDPG formed is an equal mixture of both diastereomers, the result expected for nonenzymatic, achiral addition of a proton to either face of the enol on ketonization.

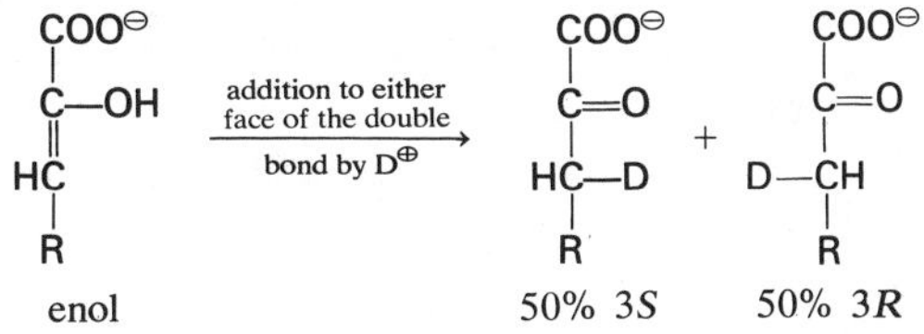

enol —(addition to either face of the double bond by $D^{\oplus}$)→ 50% 3*S* + 50% 3*R*

Thus, the scheme for dehydration can be represented as follows:

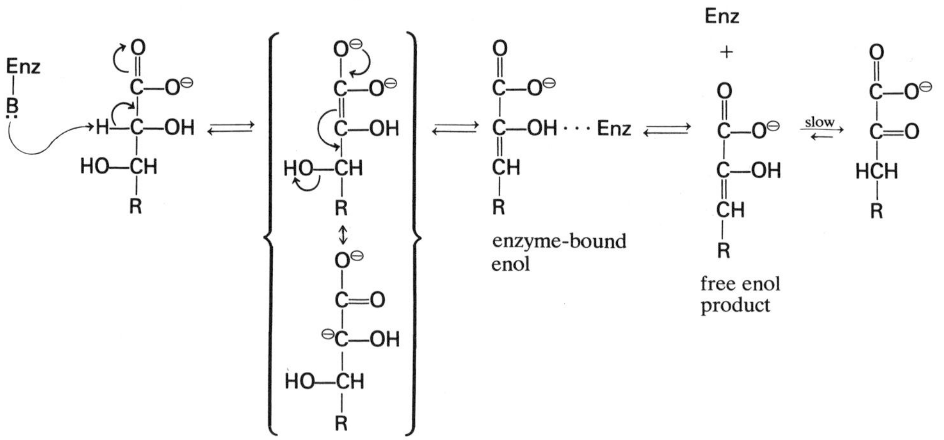

Enz—B: → H—C(OH)(COO$^\ominus$)—HO—CH—R ⇌ [$^\ominus$O—C(—O$^\ominus$)=C(OH)—HO—CH—R ↕ $^\ominus$O—C(=O)—$^\ominus$C(OH)—HO—CH—R] ⇌ O=C(—O$^\ominus$)—C(—OH···Enz)=CH—R ⇌ Enz + O=C(—O$^\ominus$)—C(—OH)=CH—R $\underset{\longleftarrow}{\xrightarrow{\text{slow}}}$ O=C(—O$^\ominus$)—C=O—HCH—R

enzyme-bound enol

free enol product

17.B.2.b L-2-Keto-3-deoxyarabonate dehydrase

A second sugar dehydrase of interest is L-2-keto-3-deoxyarabonate dehydrase. Purified to homogeneity, again from *Pseudomonas saccharophilia*, this enzyme has 85,000 mol wt and requires no metal cations for conversion of the L-2-keto-3-deoxyarabonate to α-ketoglutaric semialdehyde. The reaction is unusual in that the elimination of H_2O appears to occur β,γ to a carbonyl group rather than α,β as in all the examples we have discussed so far.

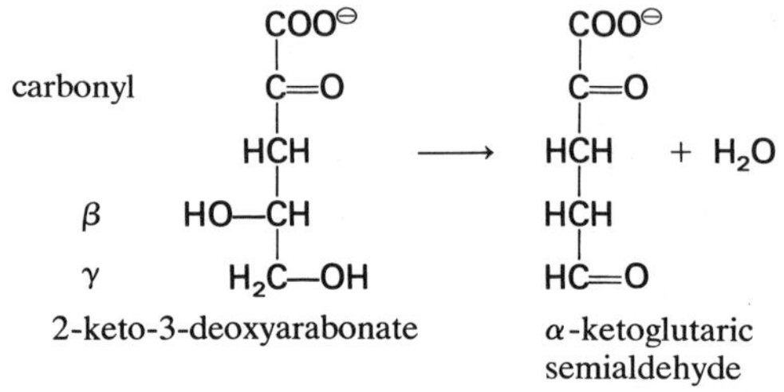

	2-keto-3-deoxyarabonate		α-ketoglutaric semialdehyde	
	COO$^\ominus$		COO$^\ominus$	
carbonyl	C=O		C=O	
	HCH	⟶	HCH	+ H_2O
β	HO—CH		HCH	
γ	H_2C—OH		HC=O	

In the matter of the β,γ-elimination, this reaction is related to the kind of reactions carried out by the dehydrases requiring coenzyme B_{12} (¶20.C; Abeles and Dolphin, 1976), but the B_{12} coenzyme is not used here. The chemical problem the enzyme faces in accelerating the dehydration is that, in the loss of water, the hydroxyl is removed from C-4 and the proton from C-5, and the C-5 hydrogen is chemically nonactivated. It is nonacidic, in marked contrast to all the hydrogens removed by previously described dehydrases, hydrogens that were adjacent to carbonyl groups, where such proton loss would lead to a carbanion stabilized by the adjacent carbonyl. Removal of a C-5 proton from 2-keto-3-deoxyarabonate would generate a carbanion without such possibility for stabilization and would seem to be prohibitively high in energy.

Abeles and colleagues thus searched for a chemical path by which the enzyme could increase the acidity of C-5 hydrogens in the substrate (Stoolmiller and Abeles, 1966; Portsmouth et al., 1967). If the enzyme could generate a stable carbanion at C-3, then this electron pair might be used for expulsion of the C-4 hydroxyl to yield a 3,4-dehydro species, and now the C-5 hydrogens might be sufficiently acidic. *One favored route for generating the equivalent of low-energy carbanions is via eneamine formation,* and this dehydrase uses that mechanism for stabilizing carbanions. (So do many other enzymes utilizing carbanion chemistry in catalysis, and we shall explore this topic further in Section V.) The enzyme has a lysine residue at the active site, a lysine whose nucleophilic ε-amino group can attack C-2 of the substrate to form an *imine* where the nitrogen is sufficiently basic to undergo ready protonation at physiological pH.

$$\underset{\text{2-keto-3-deoxyarabonate}}{COO^{\ominus}\text{-}C(=O)\text{-}HCH\text{-}HO\text{-}CH\text{-}HCH\text{-}OH} + H_2\ddot{N}\text{-Enz} \underset{}{\overset{-H_2O}{\rightleftharpoons}} \underset{\text{imine}}{COO^{\ominus}\text{-}C(=\ddot{N}\text{-Enz})\text{-}HCH\text{-}HO\text{-}CH\text{-}HCH\text{-}OH} + H_2O \xrightarrow{H^{\oplus}} \underset{\text{iminium cation}}{COO^{\ominus}\text{-}C(=\overset{\oplus}{N}H\text{-Enz})\text{-}HCH\text{-}HO\text{-}CH\text{-}HCH\text{-}OH} \quad (1)$$

This iminium cation is an excellent electron sink; it increases the acidity of hydrogens at C-3 because the resulting C-3 carbanion, on proton loss, is stabilized by delocalization with charge quenching as indicated.

$$\text{Enz—B:} + \underset{\text{iminium cation}}{COO^{\ominus}\text{-}C(=\overset{\oplus}{N}H\text{—Enz})\text{-}H\text{—}CH\text{-}HO\text{—}CH\text{-}HCH\text{-}OH} \rightleftharpoons \text{Enz—BH}^{\oplus} + \underset{\text{eneamine}}{COO^{\ominus}\text{-}C(\text{—}\ddot{N}H\text{—Enz})(=CH)\text{-}HO\text{—}CH\text{-}HCH\text{-}OH} \quad (2)$$

The resulting eneamine has a lone pair of electrons available at the nitrogen to be used in facile expulsion of the C-4 hydroxyl group. This elimination of H_2O occurs α,β to the carbonyl, in analogy to the previous dehydration examples.

eneamine ⇌ dehydrated iminium cation (3)

In this dehydrated cation, the C-5 hydrogens have been rendered acidic, because abstraction of a C-5 proton yields an anion in extended conjugation all the way to the iminium nitrogen.

dehydrated iminium cation ⇌ conjugated anion (4)

The enolic form of this conjugated anion could ketonize with protonation at C-4 to yield a dehydrated eneamine.

enolic form of anion ⇌ dehydrated eneamine (5)

Now this eneamine can reprotonate at C-3 to generate a product imine, which hydrolyzes to yield free enzyme and the observed semialdehyde product.

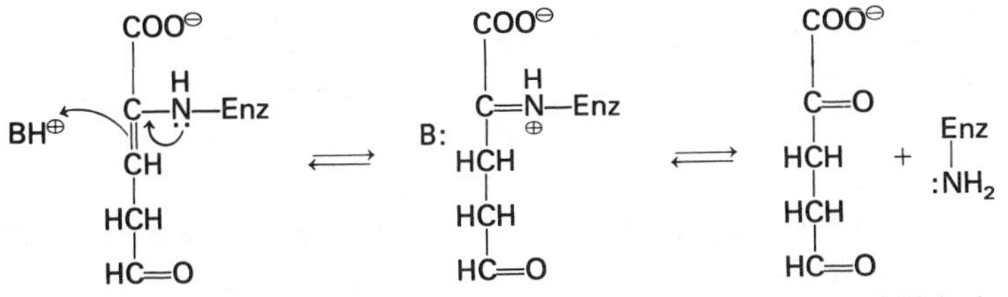

$$\text{BH}^{\oplus}\ \ \text{COO}^{\ominus}\text{—C(—NH—Enz)=CH—HCH—HC=O} \rightleftharpoons \text{B:}\ \ \text{COO}^{\ominus}\text{—C(=}\overset{\oplus}{\text{N}}\text{H—Enz)—HCH—HCH—HC=O} \rightleftharpoons \text{COO}^{\ominus}\text{—C(=O)—HCH—HCH—HC=O} + \text{Enz—:NH}_2 \qquad (6)$$

dehydrated eneamine — product imine — free semialdehyde product

This ingenious scheme is supported by the observation that, in D_2O, deuterium ends up at C-3 and C-4 of the product, as predicted; presumably both C-3 and C-4 of the dideutero product are chiral, but this has not been determined. This sequence is a clear demonstration of how eneamines can be used to stabilize carbanions and facilitate activation of a position that is not activated in the starting substrate.

17.B.3 Dehydrations in Nucleotide-Linked Sugar Biosynthesis

Dehydration reactions in deoxysugar formation are of two fundamentally distinct mechanistic types. One variant is typified by the reaction of 6-phosphogluconate dehydrase (¶17.B.2.a). The other variant is exemplified in the biosynthesis of 6-deoxyhexoses, which are prominent components in complex carbohydrates of bacterial and plant cell walls and membranes. The enzymatic formation of these molecules proceeds, not at the free sugar level, but rather at the level of a nucleoside diphosphate hexose (Chapter 9). These dehydrations proceed by intramolecular oxidation–reduction sequences, involving 4-keto sugar species as intermediates between the C-6 hydroxymethyl functionality of the substrate and the C-6 methyl group of the product.

17.B.3.a dTDP-glucose oxidoreductase

A well-studied example is the conversion of D-glucose to L-rhamnose as the dTDP (2′-deoxythymidine diphosphate) derivatives (Glaser and Zarkowsky, 1971).

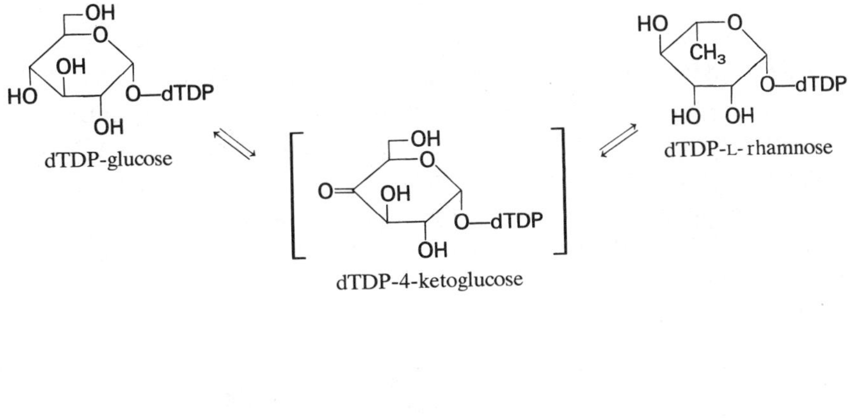

dTDP-glucose ⇌ [dTDP-4-ketoglucose] ⇌ dTDP-L-rhamnose

The knowledge that 4-keto intermediates appear to form in this overall reductive dehydration immediately brings to mind the UDP-glucose epimerase reaction discussed earlier (¶10.C.7.a) where tightly bound NAD⊕ was reversibly reduced and reoxidized reciprocally with formation of keto-sugar intermediate. And, expectedly, enzyme-bound NAD⊕ is required for the formation of intermediates on the way to dTDP-L-rhamnose and other 6-deoxy sugars, functioning in a catalytic rather than a stoichiometric role.

The dTDP-glucose oxidoreductase responsible for the first step of dTDP-rhamnose synthesis has been purified from *E. coli* (Glaser and Zarkowsky, 1971). A dimer of 78,000 mol wt contains one molecule of bound NAD⊕, and removal of NAD⊕ causes dissociation of the active enzyme into monomeric subunits. The enzyme converts dTDP-D-glucose into the dTDP-4-keto-6-deoxy-D-glucose, with loss of H_2O; in a formal sense, C-4 has been oxidized and C-6 reduced.

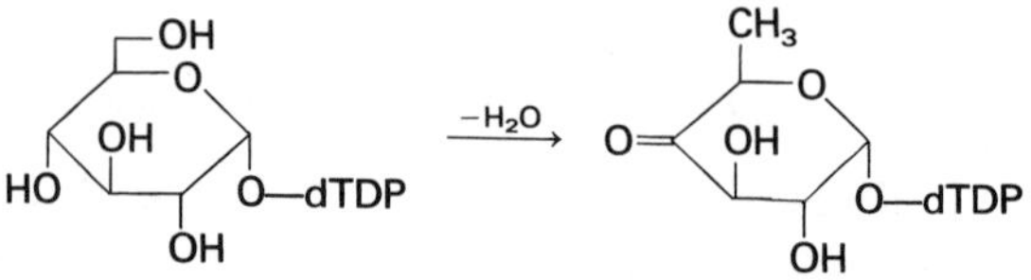

OH O OH HO O—dTDP OH $\xrightarrow{-H_2O}$ CH_3 O O= OH O—dTDP OH

dTDP-D-glucose dTDP-4-keto-6-deoxy-D-glucose

Subsequent enzymes, less well characterized, must catalyze the epimerizations that invert configuration at C-5 and C-3, facilitated by the presence of the 4-keto group that increases the acidity of both the C-3 and the C-5 hydrogens needed for epimerization via enolate intermediates. The final step in constructing the L-rhamnose moiety would be stereospecific reduction of the 4-keto group.

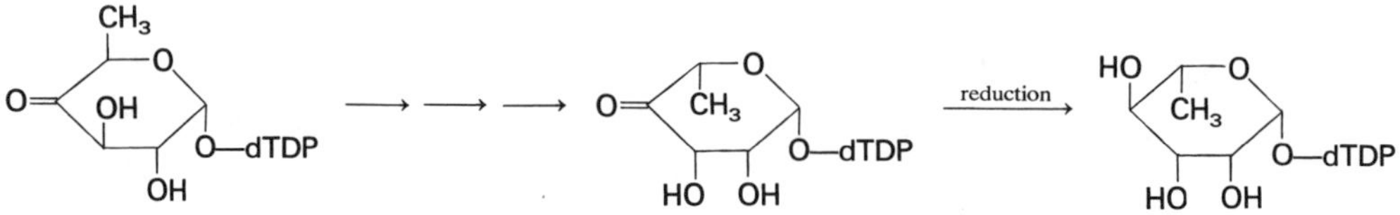

CH_3 O O= OH O—dTDP OH ⟶ ⟶ ⟶ O O= CH_3 O—dTDP HO OH $\xrightarrow{\text{reduction}}$ HO O CH_3 O—dTDP HO OH

dTDP-L-rhamnose

Glaser and colleagues have proposed the following scheme, which features a 4-keto-5,6-glucoseen intermediate. Initial oxidation at C-4 of the glucose moiety with hydride transfer to bound NAD⊕ generates enzyme-bound NADH and the 4-ketoglucose derivative. Now the adjacent C-5 hydrogen is acidic and is easily abstracted by an active-site base. Subsequent or concomitant loss of the C-6 hydroxyl completes the elimination of H_2O and forms the 5,6 (exomethylene) double bond in the glucoseen intermediate.

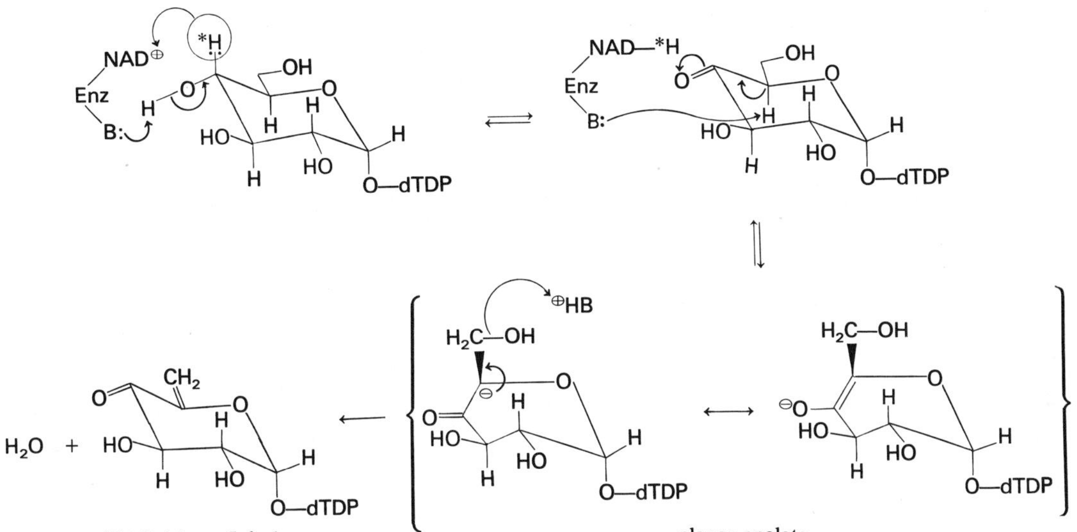

dTDP-4-keto-5,6-glucoseen

planar enolate

Glaser then postulates reduction of the 5,6 double bond of this conjugated intermediate by delivery of hydride ion from the bound NADH to C-6 of the glucoseen, producing the enolate form of the product and regenerating E·NAD$^{\oplus}$ (Glaser and Zarkowsky, 1971).

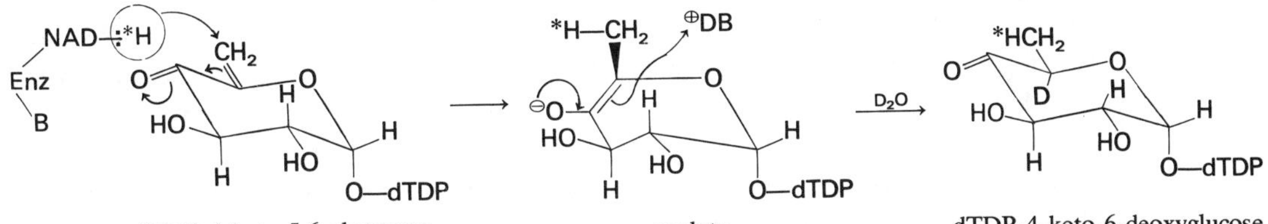

dTDP-4-keto-5,6-glucoseen

enolate

dTDP-4-keto-6-deoxyglucose

This scheme explains the various experimental observations made by Glaser's group and others. Substrate labeled at C-4 of the glucose moiety with ^{2}H or ^{3}H yields product with ^{2}H or ^{3}H at C-6, with no loss of isotope to the solvent. In D_2O, the enzymatic incubation produces dTDP-4-keto-6-deoxyglucose containing one deuterium at C-5, as predicted by this mechanism. When apoenzyme is reconstituted with 4-[^{3}H]-NAD$^{\oplus}$, no tritium ends up in the product, showing the expected stereospecificity of transfer (i.e., the NAD^{3}H generated as intermediate will have tritium in the nontransferable position). In the following, S = substrate and P = product.

4-[^{3}H]-NAD$^\oplus$ + Apoenzyme $\rightleftharpoons$ [4-^{3}H-pyridinium, C(=O)—NH$_2$, N$^\oplus$—R] $\xrightleftharpoons{S—H}$ [4-H,4-^{3}H-dihydropyridine, C(=O)—NH$_2$, N—R] $\xrightleftharpoons{P}$ [4-^{3}H-pyridinium, C(=O)—NH$_2$, N$^\oplus$—R] + P—H

When a substrate analogue, dTDP-6-deoxyglucose-4-[^{3}H], is used, tritium is transferred to enzyme-bound NAD$^\oplus$, but the NAD^{3}H is not reoxidized because C-6 of this analogue is already a methyl group. The bound 4-[^{3}H]-NADH can be isolated from the oxidoreductase and shown to be 4*S*-[^{3}H]-NADH.

The mechanistic scheme predicts that, during steady-state catalysis, some fraction of the enzyme might be present in a form or forms containing reduced nicotinamide. Indeed, analysis at 340 nm shows that about 5% of the Enz · NAD molecules are as Enz · NADH during turnover. This information can be utilized in determining which step might be slow in catalysis. When the V_{max} for conversion of 4-deutero-dTDP-D-glucose is compared with V_{max} of the protio form, there is a k_H/k_D of 3.3, suggesting that transfer of this hydrogen is at least partly rate-determining. This slow step could be formation of the 4-ketoglucose and NADH, or the step involving reoxidation of NADH and reduction of the glucoseen species, because this hydrogen is transferred in each of these steps. Analysis of the amount of absorbance at 340 nm due to E · NADH during V_{max} turnover with the 4-deutero substrate shows that now 20% to 25% of the enzyme is present as E · NADH, compared to 4% to 5% with the 4-protio substrate. That is, the reduced nicotinamide–enzyme accumulates preferentially when it is 4*S*-[^{2}H]-NADH rather than 4*S*-[^{1}H]-NADH that is formed. Therefore, the NADH-reoxidation step (reduction of the glucoseen) is rate-limiting. In agreement with this deduction, when substrate is converted enzymatically in H_2O or D_2O, there is no solvent deuterium isotope effect, showing that transfer of D$^\oplus$ to C-5 of the product is not slow. In ^3H_2O, there *is* a tenfold selection against tritium incorporation at C-5, consistent with ketonization of the enolate form of the product occurring *after* the rate-determining step (Glaser and Zarkowsky, 1971).

17.B.3.b Biosynthesis of 3,6-dideoxyhexoses

Of the various possible 3,6-dideoxyhexose isomers, five have been found in the lipopolysaccharides of gram-negative bacteria. CDP-D-glucose is the precursor for paratose, abequose, tyvelose, and ascarylose, whereas GDP-D-mannose is precursor for colitose. Rubenstein and Strominger (1974) have elucidated the enzymatic steps involved.

$$\text{CDP-D-glucose} \xrightarrow{E\cdot NAD^\oplus} \text{CDP-4-keto-6-deoxy-D-glucose} \quad (1)$$

$$\text{CDP-4-keto-6-deoxy-D-glucose} \xrightarrow{NADH} \text{CDP-4-keto-3,6-dideoxy-D-glucose} \quad (2)$$

$$\text{CDP-4-keto-3,6-dideoxy-D-glucose} \xrightarrow{NADPH} \text{CDP-3,6-dideoxyglucose} \quad (3)$$

Reaction 1 is presumed to be analogous to that of the enzyme just discussed. Reaction 3 is a straightforward reduction consuming NADPH in stoichiometric quantities. Reaction 2 is obviously more complex, and Strominger and colleagues have shown that a two-enzyme complex is required for this transformation. NADPH is required, as is another cofactor, a derivative of vitamin B_6.

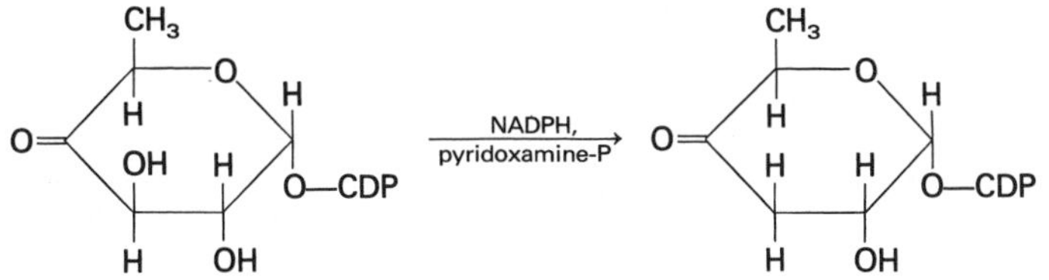

NADPH, pyridoxamine-P

CDP-4-keto-6-deoxy-D-glucose → CDP-4-keto-3,6-dideoxy-D-glucose

Vitamin B_6 is a substituted pyridine, pyridoxine. The active coenzyme form for most enzymatic catalysis using the coenzyme is the aldehyde form, pyridoxal phosphate, with the primary alcohol phosphorylated; we shall discuss pyridoxal-dependent enzymes in Chapter 24.

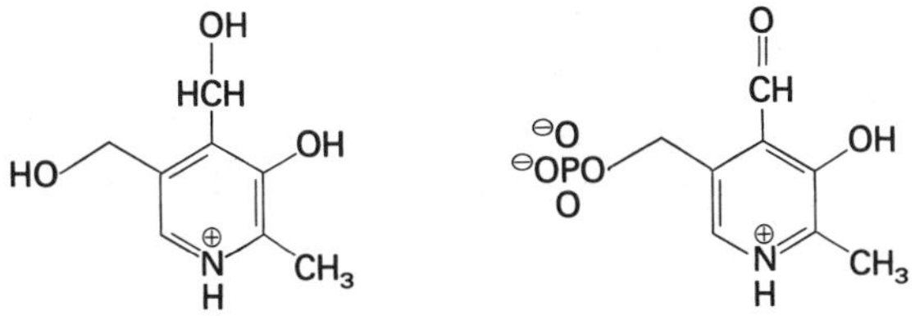

pyridoxine (vitamin B_6)

pyridoxal-P

The other common oxidation state of the cofactor is the pyridoxamine form, a two-electron–reduced form of the pyridoxal-P imine. It is pyridoxamine phosphate that is specifically required in this sugar-dehydration sequence.

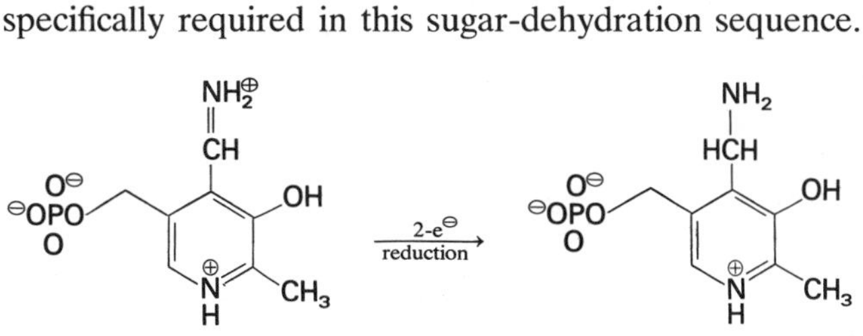

$2\text{-}e^{\ominus}$ reduction

imine of pyridoxal-P → pyridoxamine-P

This vitamin-B_6 derivative is bound to one of the two enzymes, whereas the second enzyme shows the NADPH oxidase activity. The pyridoxamine-P coenzyme probably is utilized to form an imine with the 4-keto group of the substrate, increasing the acidity of the C-3 hydrogen of the sugar and also fixing the substrate firmly in place. Consistent with the imine as an intermediate is the finding that, in $H_2{}^{18}O$, the ^{18}O exchanges into the C-4 carbonyl of the substrate, as it would in reversible imine formation with the pyridoxamine coenzyme at the active site of the first enzyme.

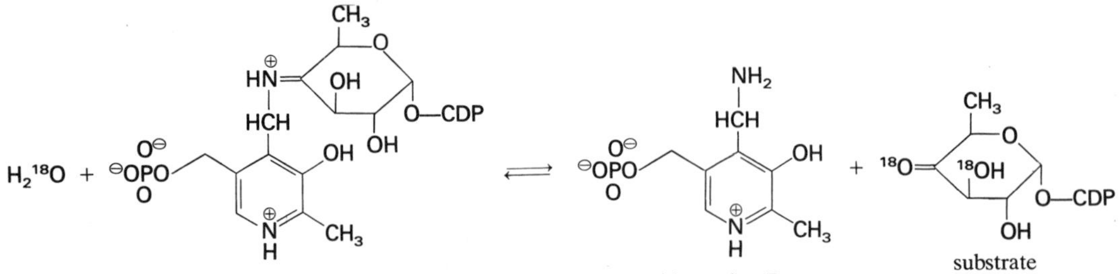

imine

pyridoxamine-P

substrate

Oxygen-18 also is incorporated into the C-3 hydroxyl of the substrate, suggesting its reversible expulsion and readdition during catalysis. Studies with substrate labeled with ^3H at C-3 of the glucose ring indicate no loss of this hydrogen during product formation, indicating that it is not labilized. Studies with methylene-labeled [^3H]-pyridoxamine-P show exchange with the solvent, indicating that isomerization of the initial Schiff-base (imine) linkage can occur and so labilize (render acidic) that proton. Thus, Strominger postulated that (in the tautomeric, isomerized imine form) one can readily envisage loss of the C-3 hydroxyl reversibly, accounting for the ^{18}O-exchange data. As shown, this would lead to a 3,4-glucoseen species on such dehydration.

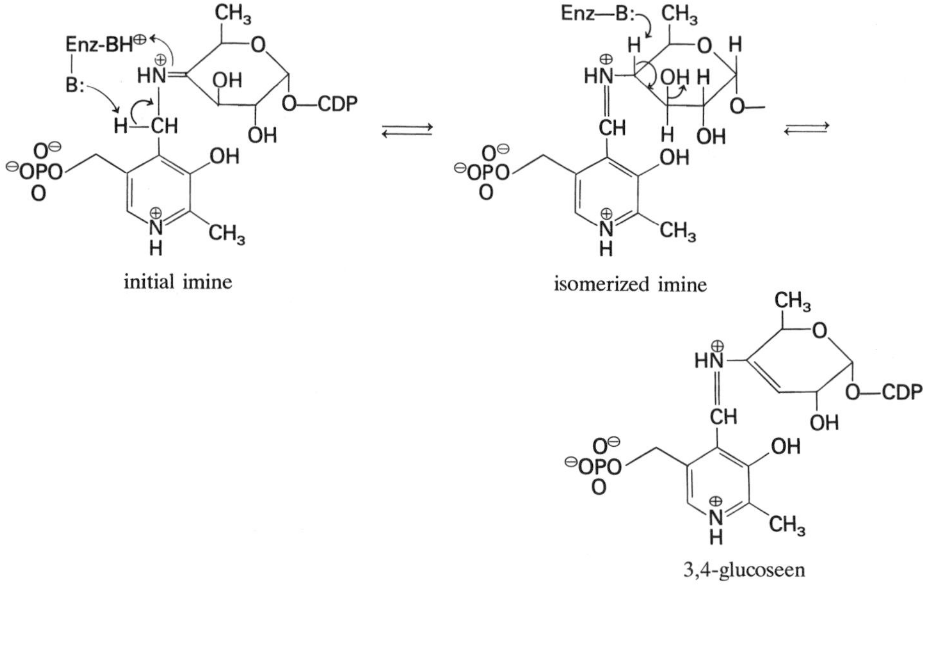

initial imine

isomerized imine

3,4-glucoseen

The expectation would be that the second enzyme of this couple would transfer hydride to C-3 of the glucoseen in a conjugate addition, analogous to reduction of the 5,6-glucoseen in the catalysis discussed earlier with dTDP-glucose oxidoreductase. This transfer should result in tritium incorporation from one enantiomer of 4-[^{3}H]-NADPH. However, incubation of the enzymes with either 4*R*-[^{3}H]-NADPH or 4*S*-[^{3}H]-NADPH resulted in no tritium incorporation, neither at C-3 of the dideoxy product nor in the pyridoxamine-P methylene group (which might occur from reduction of a tautomer of the 3,4-glucoseen). Rather, all the tritium ended up in solvent. This result is not readily explicable, although one could hypothesize that hydride transfer does go to C-3, generating the product still in imine linkage. Now one would have to postulate that this imine undergoes stereospecific exchange of the just-introduced C-3 ^{3}H with solvent protons many times before imine hydrolysis (and product release from the enzyme) to result in washout of label.

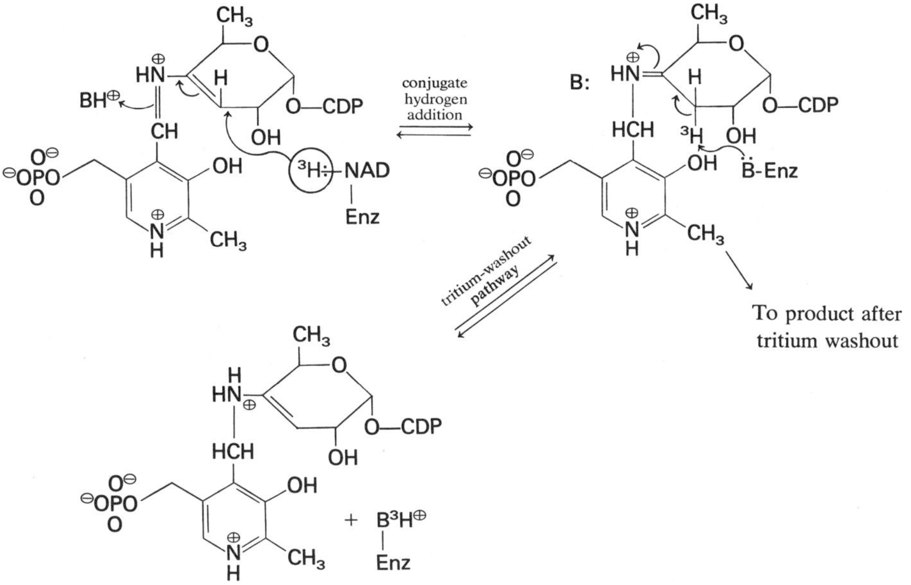

conjugate hydrogen addition

tritium-washout pathway

To product after tritium washout

A curious aspect of this enzymatic conversion is that the enzyme uses both the 4*R* and the 4*S* hydrogens of NADPH in the ultimate transfer to water, *a rare lack of stereospecificity in nicotinamide-dependent catalysis.*

17.B.4 *syn* Eliminations of Water

In all the enzymatic eliminations of water we have discussed thus far where the stereochemistry of elimination is known, it is *anti* or "*trans*." This is true as well for the overwhelming majority of other enzymatic eliminations we have not explicitly discussed and for the carbon–nitrogen lyases we shall discuss in Chapter 18 (Hanson and Rose, 1975). It has been argued that steric considerations favor the *anti* transition state, with much less eclipsing of groups than in the *syn* arrangement.

B: H X B: X H

transition states for HX elimination

Estimates of 10^3 rate advantage in favor of *anti* eliminations have been made for a concerted process. In less synchronous eliminations (i.e., carbanion or carbonium-ion processes) where maximal orbital overlap is not as necessary, the advantage for the *trans* process could be less. Electronic considerations have dissected the concerted elimination into components of (1) nucleophilic substitution of electrons on one carbon, C_α, from the adjacent C_β, and (2) the complementary electrophilic substitution on C_β by C_α. Because S_N2 processes produce inversion and S_E2 retention, an overall *anti* process is indicated. However, Ingold noted that early C—H bond fission would allow enough leakage of electron density for the S_E2-like component to go with inversion and lead to overall *syn* elimination generally. Recent experimental evidence suggests that *syn* eliminations may be considerably more common than was generally realized a few years ago, especially in cycloalkane systems and bicycloalkane systems (see Saunders and Cockerell, 1972, and references therein). Additionally, in acyclic systems, although *cis*-olefins form almost exclusively by *anti* eliminations, *trans*-alkenes may form via *syn* pathways, especially in producing 1,2-disubstituted ethylenes. Still debated is the question of whether *syn* eliminations ever occur by a concerted mechanism or always involve a two-step (e.g., carbanion) mechanism; this question is difficult to settle, because kinetically invisible steps could be occurring.

Four or five enzymatic eliminations of water are known that proceed with overall *syn* geometry. We shall discuss two examples.

The first example is 3-methylglutaconyl-CoA hydratase, which is involved in leucine degradation, hydrating the *E*-isomer of the α,β-enoyl-CoA to 3*S*-hydroxymethylglutaryl-CoA.

H_2O + $^{\ominus}OOC$ H_3C O CSCoA H ⇌ $^{\ominus}OOC$ H_3C OH O CSCoA

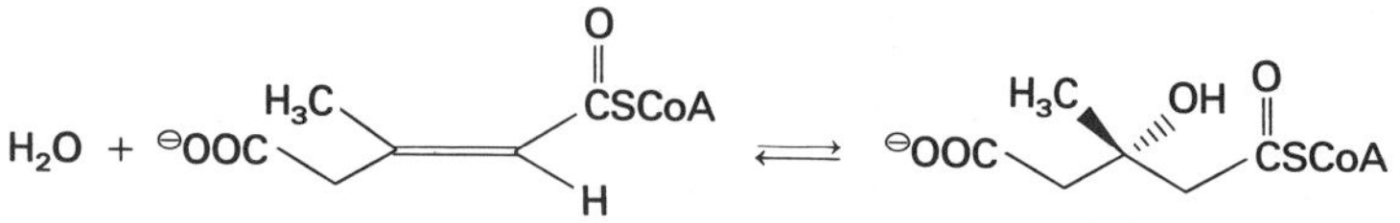

E-3-methylglutaconyl-CoA 3*S*-hydroxymethylglutaryl-CoA

The second example comes from the microbial pathway for biosynthesis of aromatic amino acids, where the enzyme 5-dehydroquinate dehydrase interconverts 5-dehydroshikimate and 5-dehydroquinate, operating physiologically in the dehydration direction.

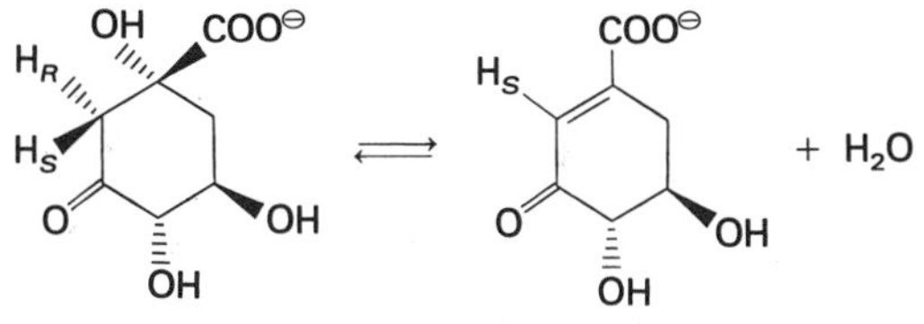

5-dehydroshikimate

Hanson and Rose (1963) determined that loss of water occurs with removal of the prochiral pro*R* proton, constituting a *syn* elimination. In ^3H_2O, hydration yields the 5*R*-tritiodehydroquinate.

Rose (1970) had suggested that the unusual enzymatic *syn* eliminations may all be two-step eliminations via carbanionic intermediates. In a two-step mechanism, the assignment of overall *syn* dehydration poses no special problems. Recently, experimental data have surfaced supporting a carbanionic intermediate in the dehydroquinase-catalyzed reaction, specifically via initial imine formation between the keto group of the substrate and an amino group (probably an ε-amino of a lysine residue) and subsequent formation of a stable carbanion prior to loss of OH (Butler et al., 1974). Once the initial imine forms, the enzyme can abstract the H_R hydrogen specifically to yield a carbanion stabilized by the iminium electron sink.

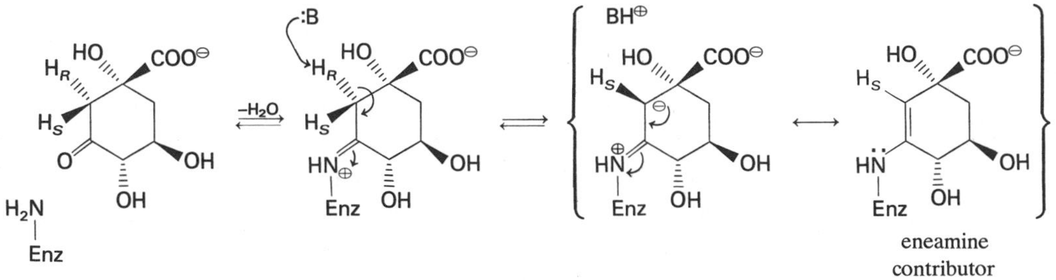

eneamine contributor

carbanion intermediate

From the eneamine form of the intermediate, there is a ready path for facilitated expulsion of the OH group, completing the loss of HOH and producing the product imine, which can then hydrolyze to 5-dehydroshikimate.

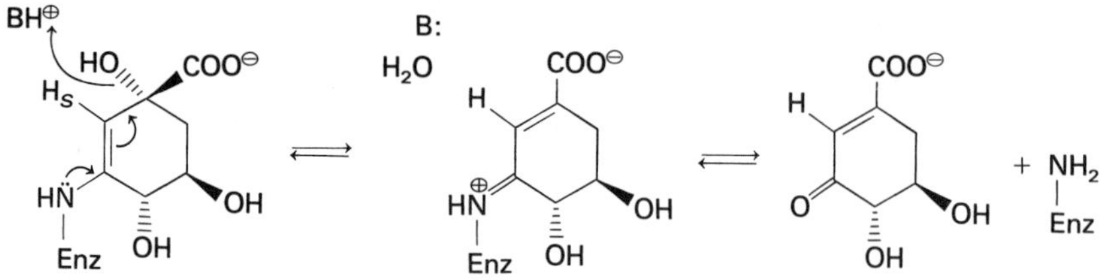

Supporting this mechanism is the fact that $NaBH_4$ inactivates the enzyme in the presence of substrate but not in its absence (Butler et al., 1974). $NaBH_4$ is an avid hydride donor to imines, reducing them to secondary amines, which (unlike imines) are hydrolytically stable linkages. Reduction of either substrate or product imine would lead to a stable, covalently modified active site, thus producing inactivation. We shall note in Section V that *such* $NaBH_4$ *inactivation is diagnostic evidence for an imine intermediate.*

Analogously, in the enoyl-CoA *syn* hydration–dehydration sequence, a two-step carbanionic mechanism could occur and possibly lower the energy barrier for reaction compared to an *anti* synchronous pathway (Hanson and Rose, 1975; Messner et al., 1975). In this connection, the α-H of acylthioesters is reasonably acidic, with the carbanion stabilized by interaction with the COSR moiety:

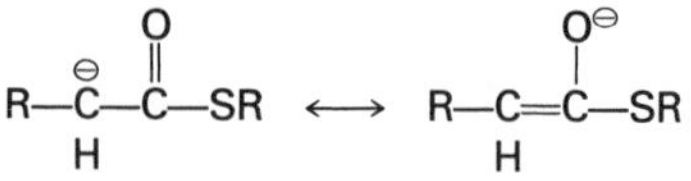

$$R-\underset{H}{\overset{\ominus}{C}}-\overset{O}{\overset{\|}{C}}-SR \longleftrightarrow R-\underset{H}{C}=\overset{O^{\ominus}}{\overset{|}{C}}-SR$$

17.B.5 Elimination of the Elements of Inorganic Phosphate

We noted earlier (Section II) that ATP is often used to activate low energy, stabilized species for reaction, as in the conversion of a carboxylate anion to an acyl phosphate, which undergoes facile displacement of inorganic phosphate by incoming nucleophiles. We also noted that $—OPO_3^{2\ominus}$ is a better leaving group than —OH because the activation energy for $OH^{\ominus}$ expulsion is likely to be larger than that for loss of the resonance-stabilized inorganic-phosphate anion. Thus, some molecules are activated by one enzyme converting a hydroxyl substituent to a phosphate monoester; a second enzyme then catalyzes elimination of the elements of phosphoric acid. We note two examples here.

The first is an enzyme further along (from dehydroshikimate) in the biosynthetic pathway to phenylalanine, namely chorismate synthetase.

3-enoylpyruvylshikimate-5-P → chorismate + $HOPO_3^{2-}$ + $H_R^{\oplus}$

The overall geometry of the 1,4-elimination of phosphoric acid is *anti*, a result claimed to violate orbital symmetry expectations in a 1,4-diolefinic system (Rose, 1970); therefore hypothetical two-step mechanisms have been advanced. Curiously, the enzyme requires exogenous $FMNH_2$, although the stoichiometry includes no redox function (Welch et al., 1974). Perhaps the $FMNH_2$ reduces an active-site disulfide to a thiol that acts as initial nucleophile in an addition–elimination mechanism.

A second example involves a decarboxylation concerted with 1,4-loss of —$OPO_3^{2\ominus}$ to form an olefinic product. The olefin, Δ^3-isopentenyl pyrophosphate, is a central intermediate in biosynthesis of steroids and other polyisoprenoid molecules (Chapter 26). It is generated enzymatically from a six-carbon precursor, 3-phosphomevalonate-5-pyrophosphate.

3-phosphomevalonate-5-PP → Δ^3-isopentenyl-PP $+ CO_2 +$ $HOPO_3^{2-}$

The electron density at C-2 generated on loss of CO_2 is used to form a π–π bond between C-2 and C-3 as —$OPO_3^{2\ominus}$ is expelled from C-3. The overall geometry of decarboxylation and loss of P_i is *anti*.

There are two other enzymes classified as dehydrases that deserve consideration: β-hydroxydecanoyl-thioester dehydrase and serine dehydrase. The first enzyme also carries out an allylic isomerization, and we shall discuss it in Chapter 19 with other allylases. The serine dehydrase requires pyridoxal-P for catalysis, and is an example of a typical β-elimination enzyme of that variety; we shall defer this topic to Chapter 24.

In the next chapter, we look at carbon–nitrogen lyases, where H and NHR are eliminated from adjacent carbons.

Chapter 18

Carbon–Nitrogen Lyases

The enzymes that catalyze the breakage of substrate carbon–nitrogen bonds can be divided into two classes on the basis of their substrates:

$$\begin{array}{c} R' \\ | \\ R\text{—}CH \\ | \\ NH_2 \end{array} \qquad\qquad \begin{array}{c} R' \\ | \\ R\text{—}CH \\ | \\ HN\text{—}R'' \end{array}$$

generalized substrate of first category of enzymes

generalized substrate of second category of enzymes

Enzymes in the first category eliminate ammonia. Enzymes in the second category catalyze the second and the terminal steps in the donation of the α-amino group of aspartate to some acceptor molecule, a process we noted in Section II (¶5.F). The first enzyme of such a couple catalyzes adduct formation between aspartate and some acceptor molecule; the second enzyme catalyzes breakdown with C—N bond cleavage. We shall discuss the two enzyme categories separately in this chapter.

18.A ENZYMATIC ELIMINATIONS OF AMMONIA

Havir and Hanson (1973) have cogently reviewed the topic of enzymatic eliminations of ammonia. They summarize experimental evidence on the four ammonia lyases we shall discuss here: aspartate ammonia lyase (aspartase), β-methylaspartate ammonia lyase (β-methylaspartase), histidine ammonia lyase (histidase), and phenylalanine ammonia lyase.

The β-methylaspartase prefers the *threo* diastereomer of the substrate, but will accept the *erythro* form at 100-fold lower V_{max}. The phenylalanine ammonia lyase from various sources will also convert tyrosine to the olefinic product *trans*-*p*-coumarate. Hanson has noted that the olefinic products all are conjugated systems, and he has speculated that such stabilizing groups may be needed to lower the activation energy into a range attainable in enzymatic catalysis.

L-aspartate $\underset{\text{aspartase}}{\rightleftharpoons}$ fumarate $+ NH_4^{\oplus}$

L-*threo*-β-methylaspartate $\underset{\beta\text{-methylaspartase}}{\rightleftharpoons}$ mesaconate (methylfumarate) $+ NH_4^{\oplus}$

L-histidine $\underset{\text{histidase}}{\rightleftharpoons}$ urocanate $+ NH_4^{\oplus}$

L-phenylalanine $\underset{\text{phenylalanine ammonia lyase}}{\rightleftharpoons}$ *trans*-cinnamate $+ NH_4^{\oplus}$

L-tyrosine $\underset{\text{phenylalanine ammonia lyase}}{\rightleftharpoons}$ *trans*-*p*-coumarate $+ NH_4^{\oplus}$

Each of these four lyases removes from the β-position of the substrate the hydrogen that is *erythro* (*syn*) to the α-amino group, generating in each instance the olefinic product with a *trans* double bond. In no case is the α-hydrogen labilized significantly, thus *ruling out* ylid mechanisms and β-hydrogen shifts.

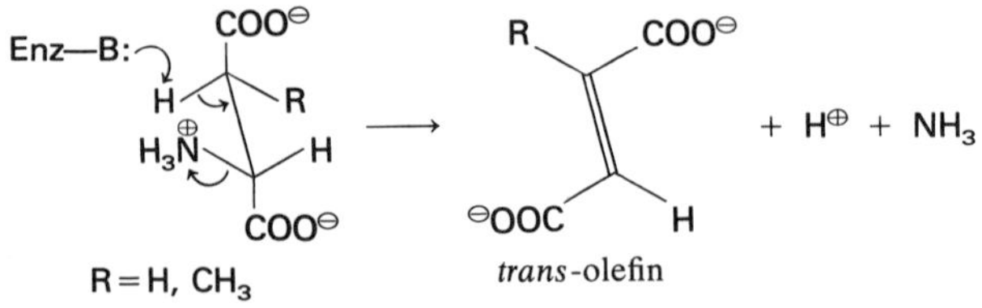

Enz—B: COO⊖ H R $H_3\overset{\oplus}{N}$ H COO⊖ → R COO⊖ ⊖OOC H + H⊕ + NH_3

R = H, CH_3 *trans*-olefin

In these enzymatic eliminations (as in those discussed in Chapter 17), a major mechanistic problem is whether the eliminations are concerted or stepwise. For aspartase, histidase, and phenylalanine ammonia lyase there is no compelling evidence available on which to base a decision. Given that concerted eliminations may be exclusively antiperiplanar, the stereochemical results are compatible with this expectation. (In the reverse direction for such *anti* eliminations of $NH_4^{\oplus}$ or HOH, addition of $H^{\oplus}$ and $X^{\ominus}$ to the olefin is antarafacial, to the opposite faces of the planar system; *syn* additions, of course, would be suprafacial.)

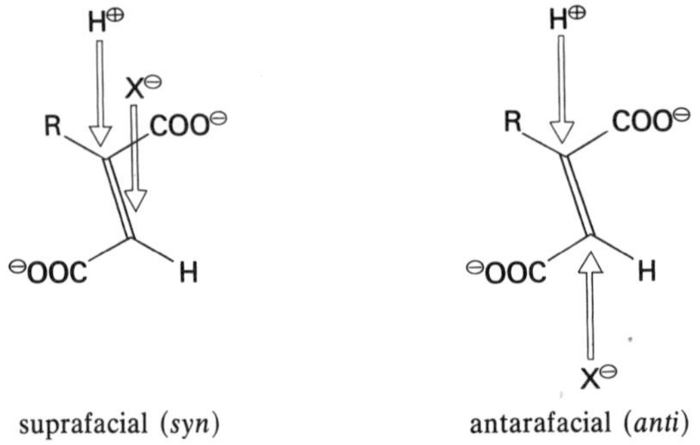

H⊕ X⊖ R COO⊖ ⊖OOC H

suprafacial (*syn*)

H⊕ R COO⊖ ⊖OOC H X⊖

antarafacial (*anti*)

For β-methylaspartase, however, the fact that both *threo* and *erythro* diastereomers of β-methylaspartate give the same *trans* olefin (monomethylfumarate ≡ mesaconate) suggests that a two-step mechanism may be involved in this instance (and quite possibly for the other three enzymes); we shall note data consistent with a carbanionic mechanism.

Havir and Hanson (1973) have noted that —$NH_3^{\oplus}$ of the α-amino acids is not a good leaving group ($\ddot{N}H_3$ is a strong base) and that reactions involving poor leaving groups often result in carbanionic transition states (i.e., the proton dissociation precedes departure of the leaving group). They point out that the enzymatic elimination of —$N^{\oplus}(CH_3)_3$ from ergothionine to yield mercaptourocanate and trimethylammonium ion resembles a straightforward base-catalyzed Hofmann elimination with the cationic nitrogen drawing off electrons from the C—N bond inductively.

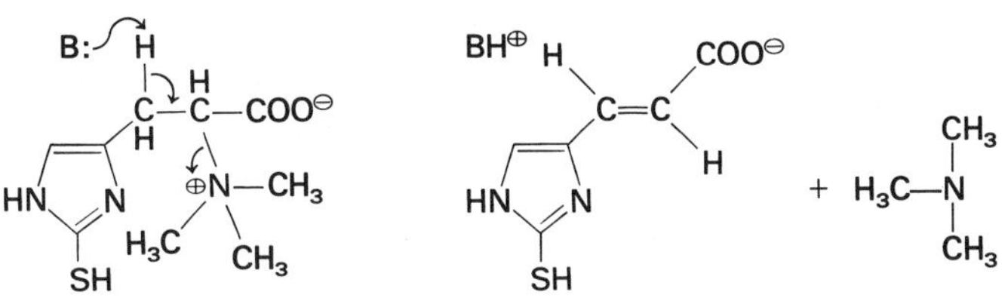

ergothionine mercaptourocanate

This example introduced their remark that catalytic efficiencies of the ammonia lyases for $NH_4^{\oplus}$ elimination might be improved by modifying the amino group. This is in the nature of a post facto comment because histidase and phenylalanine ammonia lyase are in fact known to contain prosthetic groups that react with the α-amino group of the specific substrate prior to (and thus setting up) the elimination step. This does not appear to be the case for the first two enzymes we shall discuss in detail, aspartase and β-methylaspartase.

18.A.1 Aspartase and β-Methylaspartase

The aspartase from various sources has been examined, but problems in stabilizing its activity in vitro have rendered it difficult to study. The enzyme from *Bacillus cadaveris* has been purified and shown to be a tetrameric species of 180,000 mol wt, absolutely specific for aspartase. More is known about the β-methylaspartase catalysis, although the enzyme is of more restricted biological distribution (it is found predominantly in anaerobic bacteria that ferment glutamate). The enzyme from *Clostridium tetanomorphum* has been crystallized and is a dimer of 100,000 mol wt.

As noted earlier, the enzymatic conversion of the *erythro* diastereomer of β-methylaspartate occurs at 1% of the rate at which the *threo* diastereomer is converted, although this slower rate still produces a turnover of about 2 sec^{-1}. Formation of the common *trans*-olefinic product may be explicable by formation of a *common carbanionic species from either diastereomer.* (A concerted *anti* elimination would be expected to yield different product geometries from the two diastereomers.) Havir and Hanson comment that the initial tetrahedral *erythro* carbanion might become planar to achieve maximal orbital overlap (and so maximal stabilization) with the π-system of the adjacent carboxylate group. An identical planar carbanion could be produced from the initial tetrahedral carbanion derived from the *threo* substrate (Cram, 1965). Subsequent expulsion of —$NH_3^{\oplus}$ from such a common carbanion would account for the stereoconvergent results: a single product olefin. The following scheme for stereoconvergence by way of a common planar β-carbanion is modified from Havir and Hanson (1973).

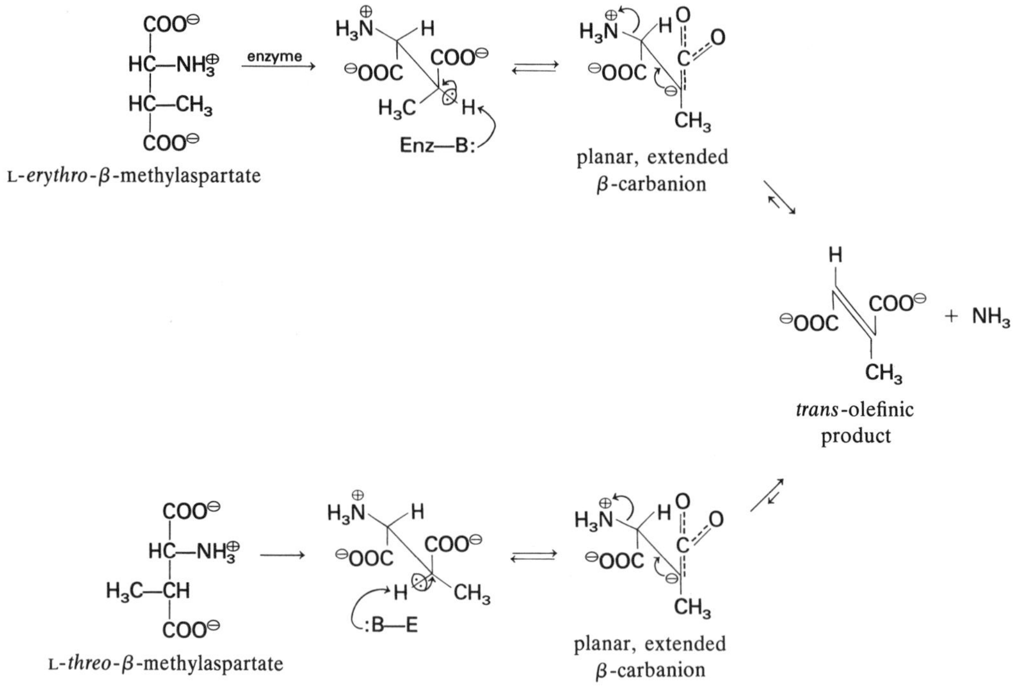

L-*erythro*-β-methylaspartate

enzyme

Enz—B:

planar, extended β-carbanion

trans-olefinic product

+ NH_3

L-*threo*-β-methylaspartate

:B—E

planar, extended β-carbanion

When 3-[^{2}H]-3-methylaspartate is tested, no primary kinetic isotope effect is seen, so C—H cleavage is not rate-determining. We have already discussed evidence with aconitase (¶17.A.1) that C—H *bond breakage and release of* $H^{\oplus}$ *into solvent* during enzymatic catalysis *are distinct and experimentally differentiable events.* (This, of course, also obscures direct assignment of observed preequilibrium exchange as a direct proof of a carbanion mechanism; olefin release may be slower than $BH^{\oplus}$ deprotonation.) But, examination of the ability of β-methylaspartase to catalyze β-proton exchange produced a positive result. Using 3-[^{2}H]-3-methylaspartate and enzyme, aliquots were removed at intervals and the amount of ^{2}H remaining in *recovered substrate* was analyzed by infrared spectroscopy (Bright, 1964, 1967; Bright et al., 1964; Wu and Williams, 1968). The rate of hydrogen exchange exceeded the rate of mesaconate formation by as much as 3.3-fold, evidence for a preequilibrium exchange. Under identical conditions, no

$^{15}NH_4^{\oplus}$ equilibrated into the methylaspartate, showing that the results cannot be trivially explained by occurrence of the back reaction. Supported by the stereochemical argument noted earlier, we conclude that a carbanion is indicated and that C—N bond fission is probably the slow step.

$$^{\ominus}OOC\text{—}\underset{\underset{H}{|}}{\overset{\overset{H_3C}{|}}{C}}\text{—}\underset{\underset{NH_3^{\oplus}}{|}}{\overset{H}{C}}\text{—}COO^{\ominus} \underset{}{\overset{fast}{\rightleftharpoons}} {}^{\ominus}OOC\text{—}\overset{\overset{H_3C}{|}}{\underset{\ominus}{C}}\text{—}\underset{\underset{NH_3^{\oplus}}{|}}{C}\text{—}COO^{\ominus} \overset{slow}{\rightleftharpoons} \overset{^{\ominus}OOC}{\underset{H_3C}{}}\!\!>C{=}C<\!\!\overset{H}{\underset{COO^{\ominus}}{}} + NH_4^{\oplus}$$

Enz—B:

For the aspartase-mediated production of fumarate, almost no information is available on the route of elimination. Deamination of 3-[^{2}H]-aspartate occurs without isotope effect, but there is no detectable exchange of C-3 hydrogen with solvent; thus, if a carbanion forms, the proton is not rapidly, reversibly, and detectably released before completion of elimination.

18.A.2 Histidine Ammonia Lyase and Phenylalanine Ammonia Lyase

Histidase (histidine ammonia lyase) activity is common in animals and microbes as the first enzyme in histidine catabolism. Most mechanistic studies have used the *Pseudomonas* enzyme, a tetramer of about 220,000 mol wt, with a turnover per active site of about 130 sec^{-1}. The phenylalanine-deaminating enzymes have been studied from both maize and potatoes, the latter source producing a tetramer of 320,000 mol wt with an active-site rate of 3 molecules sec^{-1} (molecule enzyme subunit)$^{-1}$. The product *trans*-cinnamate is a precursor in plants for phenylpropanoid constituents of secondary metabolism, such as alkaloids, lignin, anthocyanins, and flavanoid compounds such as the quercetin noted earlier (¶16.C). Hanson has postulated that this enzyme represents a switch for rerouting phenylalanine from protein synthesis and primary metabolism to those secondary plant products.

Each enzyme shows UniBi initial-velocity kinetic patterns, where the olefinic acid is released before the ammonia. Early investigations by Peterkofsky (1962) on histidase detected a partial exchange reaction of [^{14}C]-urocanate label into histidine in the absence of NH_3. Similarly, ^3H_2O exchanged into histidine, both partial exchanges occurring in the absence of significant back reaction. These data led Peterkofsky to propose the kinetic existence of an "amino-enzyme" intermediate.

$$E\cdot S \rightleftharpoons \begin{matrix}E\cdot\text{Olefin}\\ \text{\\}\\ NH_3\end{matrix} \rightleftharpoons \begin{matrix}\text{“}E\cdot NH_3\text{”}\\ +\\ \text{Olefin}\end{matrix} \rightleftharpoons E + NH_3$$

He postulated the existence of a prosthetic group on the enzyme to combine with the amino group and release it only after urocanate release. The idea of a covalent derivative of NH_3 was later borne out, and it differentiates these two enzymes from those that generate noncovalently-bound "nascent ammonia" during use of the γ-carboxamide of glutamine as an amino donor (Chapter 5). Similar partial exchanges were noted by Hanson and Havir (1970) with potato phenylalanine amino lyase. The nature of the prosthetic group was probed by Givot, Smith, and Abeles (1969) on histidase and by Hanson and Havir (1970) on the potato phenylalanine-deaminating enzyme. They used radioactive nucleophiles to label the electrophilic prosthetic group of the enzyme, causing irreversible inactivation. The labeled protein could then be digested and degraded, and radioactive fragments were identified to piece together structural information on the intact electrophile.

Addition of NaB^3H_4 to the potato phenylalanine ammonia lyase followed by acid hydrolysis allowed isolation of tritio-D,L-alanine, radioactive predominantly at C-3, although some label was at C-2. This result supported the proposal that a dehydroalanyl residue is present that has been reduced predominantly by $H^{\ominus}$ attack at the β-carbon. The D,L-product may result from racemization at C-2 during acid hydrolysis, suggesting that C-2 was originally activated.

$$^3H_3\overset{\ominus}{B}\text{—}(:{}^3H) + H_2C{=}C(NHR')\text{—}C(=O)\text{—}R \;(H^{\oplus}) \longrightarrow {}^3H\text{—}H_2C\text{—}CH(NHR')\text{—}C(=O)\text{—}R \xrightarrow{\Delta,\, H^{\oplus}} {}^3H\text{—}H_2C\text{—}CH(NH_3^{\oplus})\text{—}COO^{\ominus}$$

dehydroalanyl residue

D,L-product

Subsequent data suggested that, in the histidine case, such activation involves the nitrogen of the dehydroalanyl group in imine linkage to some other group. Use of [^{14}C]-$H_2C^{\ominus}$—NO_2 to label and inactivate the enzyme, followed by denaturation and acid hydrolysis, produced radioactive 4-[^{14}C]-aspartate. The same product was obtainable from maize phenylalanine ammonia lyase. Addition of either $CN^{\ominus}$ or $^{\ominus}CH_2NO_2$ to the β-carbon of the dehydroalanyl group, followed by acid hydrolysis, converts both the cyano (—CN) and the nitro groups (—C—NO_2) to —$COO^{\ominus}$ of the aspartate isolated. If the nitromethane-inactivated enzyme is enzymatically digested and the C—NO_2 moiety reduced with H_2 to an amino group, then 2,4-diaminobutyrate, 4-amino-2-hydroxybutyrate, and β-alanine are all obtained. A reasonable route can be envisaged from a dehydroalanyl-imine partial structure.

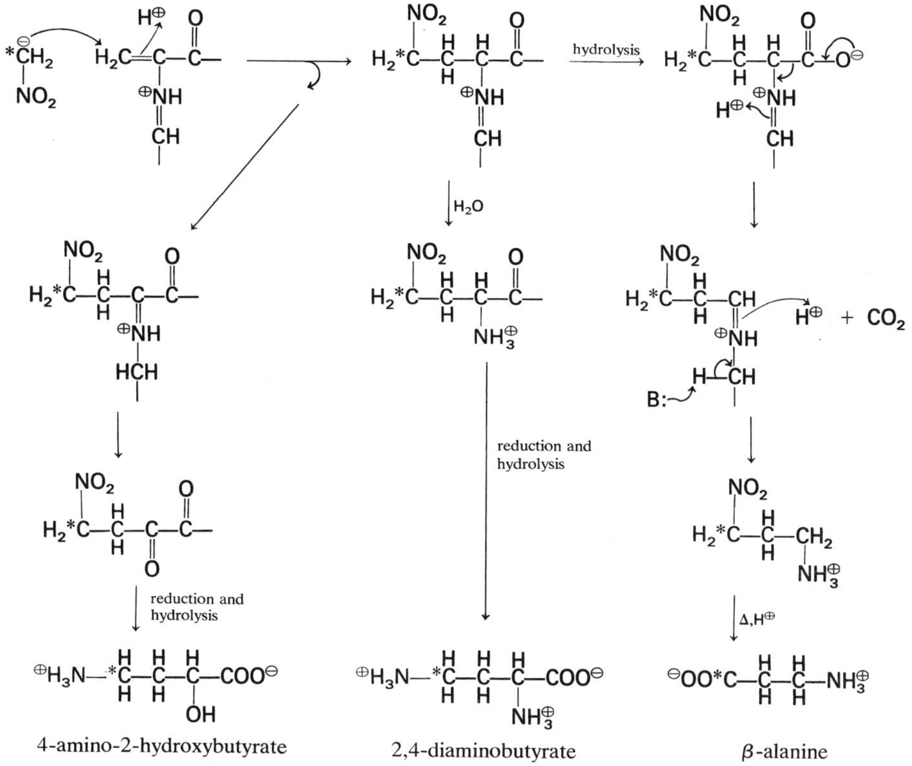

hydrolysis

H_2O

reduction and hydrolysis

reduction and hydrolysis

$\Delta, H^{\oplus}$

4-amino-2-hydroxybutyrate

2,4-diaminobutyrate

β-alanine

It may be that two enzyme subunits of histidase combine to form one active site, perhaps by one *N*-terminal dehydroalanyl residue on one chain interacting with a serine-aldehyde *N*-terminus on another subunit (each arising from a post-ribosomally modified *N*-terminal serine) to form an extensively conjugated electrophilic center.

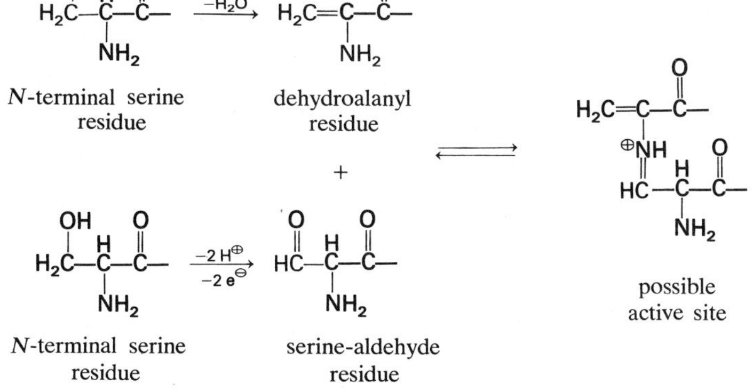

$-H_2O$

N-terminal serine residue

dehydroalanyl residue

+

$-2\,H^{\oplus}$, $-2\,e^{\ominus}$

N-terminal serine residue

serine-aldehyde residue

possible active site

In their review article, Havir and Hanson (1973) speculate on how such an electrophilic prosthetic group might function in catalytic deamination. Attack of the substrate amino group at the β-carbon of the dehydroalanyl residue could occur in a net 1,4-addition step (**1**$\rightarrow$**2**).

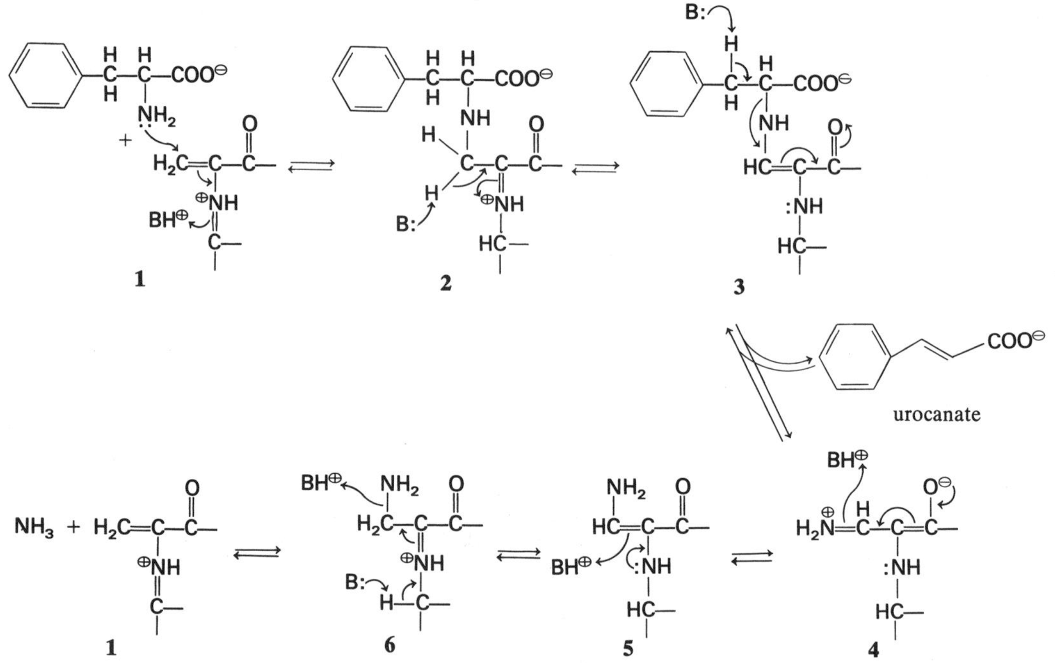

To activate the initial complex (**2**) for loss of the amino moiety, a 1,3-prototopic shift could occur facilely, producing an eneamine (**3**) with the conjugated α,β-unsaturated carbonyl system. Formation of the eneamine then has established the requisite extended electron sink for ready C—N bond cleavage, and formation of *trans*-cinnamate (proton abstraction and olefin formation may or may not be concerted), which can leave reversibly (**3**$\rightarrow$**4**) before free ammonia is released. Expulsion of ammonia from **4** could proceed along with regeneration of the initial dehydroalanyl residue at the active site by a 1,3-prototropic shift (**5**$\rightarrow$**6**) and finally 1,4-elimination of NH_3 (**6**$\rightarrow$**1**). If this is indeed the catalytic mechanism, histidase and phenylalanine ammonia lyase have gone to great lengths to assure a low-energy path for ready expulsion of NH_3.

18.A.3 Oxidative Deaminations of Amino Acids

There are a variety of enzymes known that catalyze the oxidative deamination of α-amino acids to keto acids and ammonia. These are generally half-reactions, either in 2-e$^{\ominus}$ oxidations to the imino acids (e.g., the flavoenzymes D- and L-amino-acid oxidases noted in Chapter 11), or as half-reactions in overall transaminations—that is, NH_3 is not lost to the medium but is transferred to some other keto-acid acceptor.

$$R{-}CH(NH_3^{\oplus}){-}COO^{\ominus} \longrightarrow R{-}C(=O){-}COO^{\ominus} + \text{“}NH_4^{\oplus}\text{”}$$

These are generally enzymes requiring pyridoxal phosphate as a cofactor, and they will be discussed in Chapter 24.

18.B ENZYMATIC ELIMINATION OF —NHR

Although the γ-carboxamide nitrogen of glutamine is the common cellular source of amino groups introduced in enzymatic biosynthesis, we noted that the β-carboxamide of asparagine is not used and that in only a few cases is the α-amino moiety of L-aspartate used (Chapter 5). The three known examples of use of this α-amino moiety (reviewed by Ratner, 1973*b*) are (1) in the formation of the guanidino group in arginine, (2) in formation of an amide nitrogen during buildup of the purine ring system, and (3) in the conversion of inosinic acid (IMP) to adenylic acid (AMP). In each of these three transformations, two enzymatic steps are required. The first is a condensation where ATP is cleaved, with intermediate adenylylation or phosphorylation of a substrate oxygen to activate it for subsequent nucleophilic displacement by the aspartate amino group. We have discussed this type of reaction in the case of argininosuccinate synthetase (¶5.F).

$$R{-}C(=X){-}OH + ATP \rightleftharpoons R{-}C(=X){-}O{-}AMP + PP_i \;\;(+\; {}^{\ominus}OOC{-}CH_2{-}CH(\ddot{N}H_2){-}COO^{\ominus}) \rightleftharpoons R{-}C(=X){-}NH{-}CH(CH_2COO^{\ominus}){-}COO^{\ominus} + AMP \qquad (1)$$

After this step of adduct formation, the second enzyme in the sequence is (in each case) the carbon–nitrogen lyase. One of the β-hydrogens of the aspartate moiety is abstracted, followed by C—N bond cleavage.

$$BH^{\oplus}\;\; R{-}C(=X){-}NH{-}CH(COO^{\ominus}){-}CH_2{-}COO^{\ominus} \;\; (:B{-}Enz) \rightleftharpoons R{-}C(XH){=}\overset{\oplus}{N}H_2 + {}^{\ominus}OOC{-}CH{=}CH{-}COO^{\ominus} \qquad (2)$$

fumarate

This leaves the nitrogen attached to the acceptor group and liberates the carbon chain of aspartate as the olefinic diacid—specifically, the *trans* isomer, fumarate. Argininosuccinase catalyzes the following reaction:

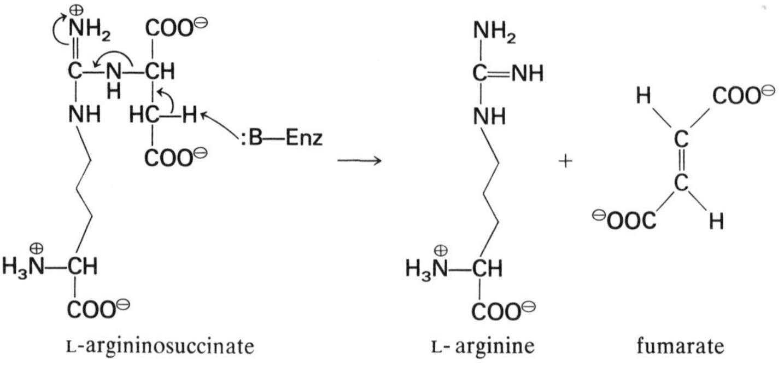

L-argininosuccinate → L-arginine + fumarate

Adenylosuccinase carries out carbon–nitrogen bond cleavage on both adenylosuccinate and 5-aminoimidazole-4-*N*-succinocarboxamide ribonucleotide (AICAR succinate).

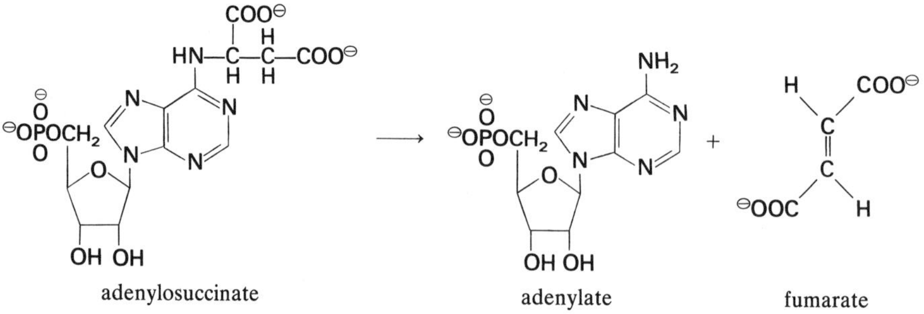

adenylosuccinate → adenylate + fumarate

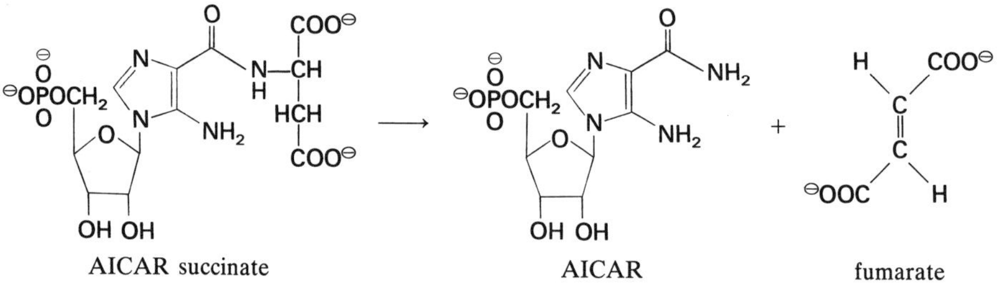

AICAR succinate → AICAR + fumarate

AICAR is 5-aminoimidazole-4-carboxamide ribonucleotide.

Unlike the carbon–nitrogen lyases of ¶18.A, which eliminate NH_3 in *degradative* reactions, arginosuccinase and adenylosuccinase are involved in *biosynthesis* of key components of intermediary metabolism. The stereochemistry of H removal and NHR loss from the α,β positions of the aspartyl moieties of the various substrates is *trans* (antiperiplanar; see Havir and Hanson, 1973), consistent with (but not establishing) elimination occurring by a concerted path. There is no evidence for prosthetic electrophilic centers such as those found in histidase and phenylalanine ammonia lyase. This is probably a reflection of the fact that the leaving groups in the reactions discussed here are better leaving groups than ammonia; the acceptor moieties possess electron sinks (amide groups) that lower the activation energy for (and so facilitate) C—N breakage. Havir and Hanson (1973) postulate concerted reactions for adenylosuccinase and argininosuccinase catalysis.

This concludes our two-chapter survey of enzyme-catalyzed reactions of HX (where X = OR or NR) during olefin formation.

Chapter 19

Enzyme-Catalyzed Isomerizations

When one examines individually the many enzymatic reactions that result in inversion of configuration at one center in a substrate to form some isomer as product, the transformations seem bewildering in their complexity. However, one can separate them into a small number of classes according to the type of chemical isomerization achieved (I. A. Rose, 1970, 1972). We shall sequentially take up enzymatic reactions involving 1,1-hydrogen shifts, 1,2-hydrogen shifts, and 1,3-hydrogen shifts. (We continue to postpone until Chapter 20 discussion of 1,2-hydrogen shifts mediated by coenzyme B_{12} that accompany rearrangements.) Then we shall examine the reactions of the phosphosugar mutases and finally discuss briefly the topic of enzyme-mediated *cis-trans* isomerization.

19.A 1,1–HYDROGEN SHIFTS

The generalized expression for a 1,1-hydrogen shift is the following:

$$R^1\text{—}\overset{\displaystyle R^2}{\underset{\displaystyle R^3}{C}}\text{—}H \rightleftharpoons H\text{—}\overset{\displaystyle R^2}{\underset{\displaystyle R^3}{C}}\text{—}R^1$$

Enzymes interconverting configuration at a single center of a substrate can be classified as racemases or epimerases (Table 19-1). A *racemase* acts on substrate with a single asymmetric center; the racemized product is the enantiomer of the substrate. An *epimerase* acts on substrate with multiple asymmetric centers; the

Table 19-1
Classification of enzymes catalyzing 1,1-hydrogen shifts

Number of chiral centers in substrate	Enzyme category	Product isomer
One	Racemase	Enantiomer
Two or more	Epimerase	Diastereomer

epimerized product is a diastereomer of the substrate. An alternative mode of classification involves assessment of whether the isomerization proceeds by proton abstraction or by hydride transfer. In the 1,1-hydrogen shifts, certain enzymes isomerize substrates with exchange of the abstracted hydrogen with solvent protons—clearly a proton-transfer mechanism. A few enzymes require nicotinamide coenzymes (we have discussed UDP-glucose 4′-epimerase in Chapter 10) and carry out isomerization by redox catalysis and reversible hydride transfer between substrates and tightly bound coenzyme. In addition two enzymes discussed in this section are enigmatic as yet in that they isomerize without solvent hydrogen exchange, but do not require NAD or NADP.

19.A.1 Proton Transfers

Racemases are known for many of the common α-amino acids formed in proteins and for some not found in proteins (Adams, 1970, 1976). Those that have been fully and carefully purified and characterized contain pyridoxal phosphate, and we shall defer their discussion until Chapter 24. On the other hand, racemases for the two common α-imino acids, proline and hydroxyproline, do not proceed via pyridoxal-phosphate–mediated reactions (the enzyme for hydroxyproline is actually an epimerase because hydroxyproline has two asymmetric centers). Nor does the ATP-dependent racemization of L-phenylalanine to D-phenylalanine, associated with synthesis of the cyclic decapeptide antibiotic gramicidin S (Lipmann, 1971). Racemases for an α-hydroxy acid, mandelate (Hegeman et al., 1970; Kenyon and Hegeman, 1970), and an acylthioester, methylmalonyl-CoA (Mazumder et al., 1962; Overath et al., 1962), have been purified and show exchange with solvent.

Of the various epimerases in this category, we shall mention the ribulose-5-P 3-epimerase (McDonough and Wood, 1961; Hurwitz and Horecker, 1956)—in part to contrast its action subsequently with ribulose-5-P 4-epimerase (Wolin et al., 1957; Burma and Horecker, 1958)—and also the enzyme responsible for converting UDP-*N*-acetylglucosamine to its 2′-epimer, UDP-*N*-acetylmannosamine (Ghosh and Roseman, 1965).

19.A.1.a Proline racemase and hydroxyproline epimerase

Proline racemase will convert either D- or L-proline to a racemic mixture of isomers with an equilibrium constant of 1. This enzyme has been purified almost to homogeneity by Cardinale and Abeles (1968) from the anaerobic bacterium *Clostridium sticklandii* (where the D-isomer is reductively cleaved subsequently to δ-aminovalerate in a reaction catalyzed by D-proline reductase, an enzyme that may be membrane-bound and energy-generating) (Seto and Stadtman, 1976).

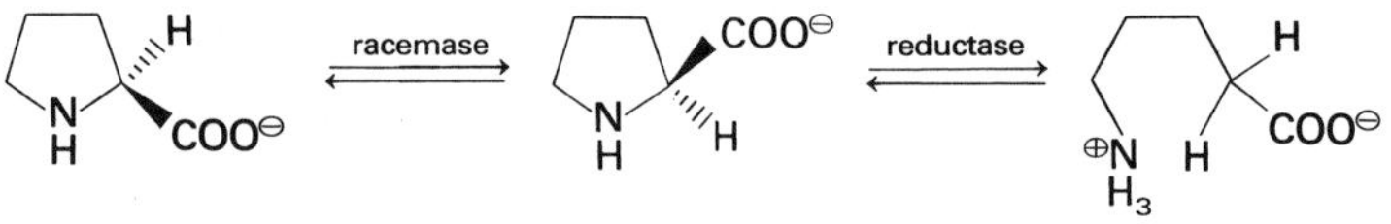

racemase reductase

L-proline D-proline δ-aminovalerate

No detectable prosthetic groups have been found bound to the enzyme, although a divalent cation (probably $Zn^{2\oplus}$) is required (Rudnick and Abeles, 1975).

Hydroxyproline epimerase is an inducible enzyme purified by Findlay and Adams (1970) from *Pseudomonas putida.* It equilibrates the C-2 configuration of its substrates, L-4-hydroxyproline and D-4-allohydroxyproline (the C2 diastereomer). The enzyme will work on both the *cis* and *trans* isomers of 4-hydroxyprolines (the *trans*-L-isomer is the one formed by prolyl hydroxylase; see ¶16.A.1).

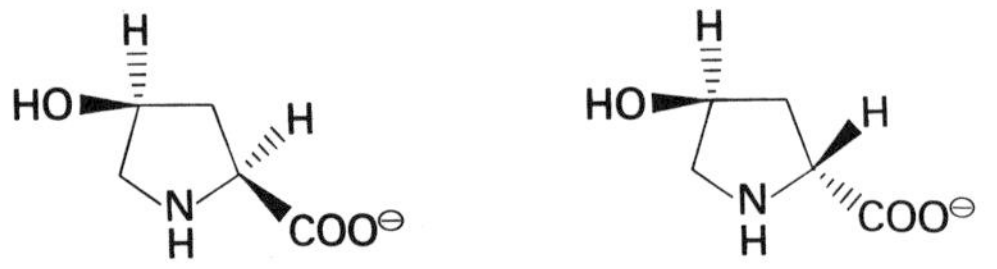

L-*cis*-4-hydroxyproline D-*trans*-4-hydroxyproline

D-*cis*-4-hydroxyproline L-*trans*-4-hydroxyproline

The enzyme also acts on the various diastereomers of 3-hydroxyproline. The molecular weight of the epimerase is 64,000; it contains no detectable coenzymes or prosthetic groups.

These enzymes invert configuration at C-2 of the substrates; the isotope-exchange data indicate that they probably do so via proton abstraction. Therefore, an early mechanistic question broached was whether the active site contains a single base for such abstraction or two bases (one set up to work on either epimer of the substrate). In a single-base mechanism, one can envisage the basic group on a flexible arm, or one can imagine an intermediate that is able to flip over (as in aconitase, for example) to accomplish abstraction of a C-2 hydrogen from one face of the molecule and its readdition to the opposite face. On the other hand, no such active-site motion is required in a two-base mechanism where the two groups can be set up antarafacially (one at each face of the cyclic substrate molecule). The following scheme (adapted from Adams, 1970, 1976) indicates how a two-base mechanism might proceed, either through a discrete carbanionic intermediate or by concerted proton transfer and a carbanion-like transition state.

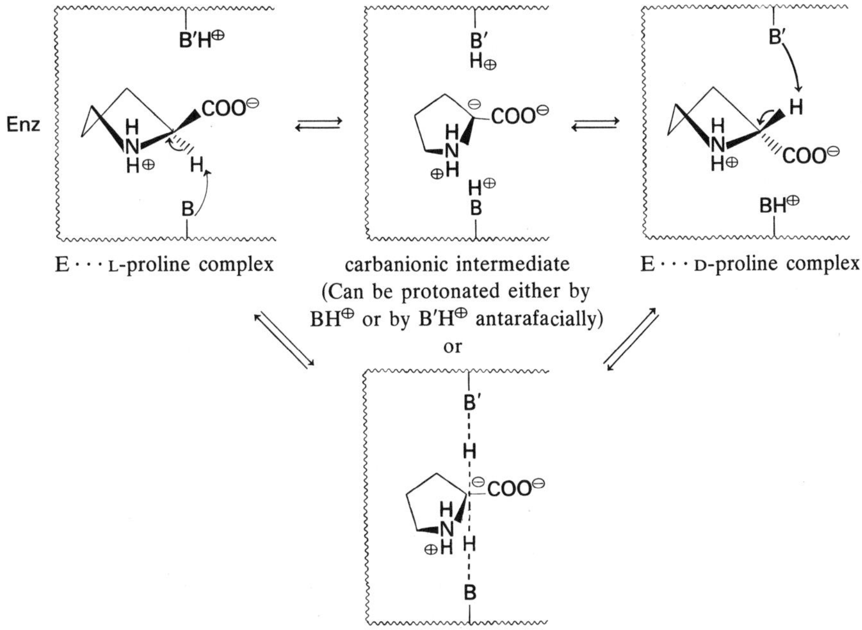

E · · · L-proline complex

carbanionic intermediate (Can be protonated either by $BH^{\oplus}$ or by $B'H^{\oplus}$ antarafacially) or

E · · · D-proline complex

carbanionic transition state (with concerted transfer)

The base written as $B'H^{\oplus}$ may be Cys—SH; the base written as B may be Cys—$S^{\ominus}$.

Cardinale and Abeles (1968) initially adduced results supporting the two-base mechanism. Starting with L-proline and the racemase in D_2O, they allowed the reaction to proceed to a few percent of complete conversion and then isolated the isomeric prolines and analyzed them for deuterium content. They found that the rate of deuterium incorporation exactly parallels the rate of racemization (D-proline formation, measured with D-amino-acid oxidase). There is no dilution of the ^2H by substrate proton, as might be expected in a single-base mechanism (provided $BH^{\oplus}$ exchanges slowly relative to racemization rates).

In these experiments, measurements were made during the initial-velocity phase (up to 10% conversion), so the back-reaction rate (D-proline→L-proline) should be negligible. Interestingly, all the of ^2H incorporated from solvent appeared in the D-proline product and none in the L-proline starting material. This result is inconsistent with the hypothesis of a single $BH^{\oplus}$ that exchanges with solvent deuterons (because there is no reason then to posit asymmetric partitioning to one isomer and not the other), but it *is* consistent with the two-base mechanism. This result further suggests that the substrate-derived proton does not exchange with solvent until *after* the racemized product is released.

Findlay and Adams (1970) obtained identical results from a D_2O experiment with hydroxyproline epimerase, suggesting the existence of a two-base mechanism in that case as well. Alkylation with iodocetamide inactivates the enzyme; one mole of $[^{14}C]$-ICH_2CONH_2 produces two moles of carboxymethyl-$[^{14}C]$-cysteine per mole enzyme on acid hydrolysis.

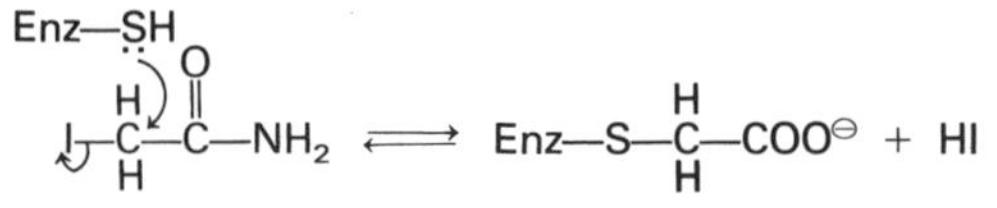

$$\text{Enz—SH} + \text{I—CH}_2\text{—C(=O)—NH}_2 \rightleftharpoons \text{Enz—S—CH}_2\text{—COO}^{\ominus} + \text{HI}$$

The inactivation appears active-site–directed in that either diastereomer of hydroxyproline protects enzymatic activity. When the $[^{14}C]$-alkylated enzyme is digested proteolytically, two radioactive peptides (each containing a unique alkylated cysteine in the primary sequence) are detected. It is possible that the sulfur atoms of these two cysteines are the two active-site bases, lined up on either face of hydroxyproline when it is bound. Similarly, recent studies on proline racemase suggest that there is one active site per two subunits; each of the subunits may contribute a cysteinyl sulfur for a total of two active-site bases (Rudnick and Abeles, 1975). Presumably, one base would be as the conjugate acid RSH and the other as the basic $RS^{\ominus}$ at the opposite face of the bound substrate.

We have noted (¶2.D.4) that the planar molecule of pyrrole-2-carboxylate

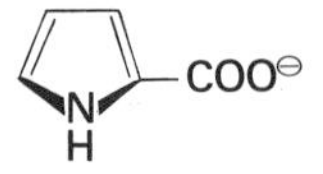

$$\text{pyrrole-2-COO}^{\ominus}\ (\text{N—H})$$

is a potent inhibitor of proline racemase, presumably acting as an analogue of the transition state for proline racemization, where the C-2 undergoing inversion must approach the plane of the five-membered ring. Recently, it has been found that aziridine-2-carboxylate is an active-site–directed reagent for proline racemase, possibly undergoing attack by the free-base form of one base and protonation-assisted ring opening by the conjugate acid of the second base (D. Steffens, G. Rudnick, C. Walsh, and R. Abeles, unpublished observations). For example,

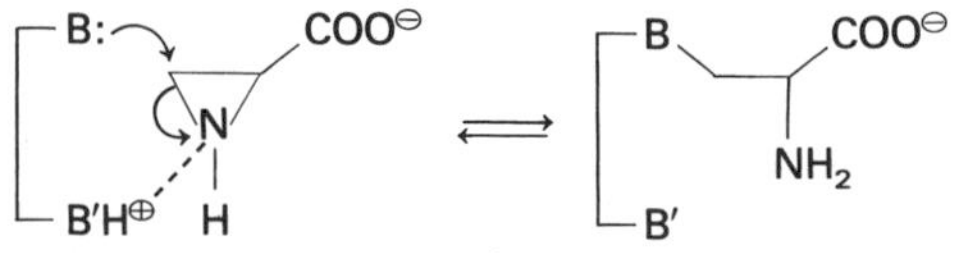

$$[\text{B:},\ \text{B'H}^{\oplus}] + \text{aziridine-2-COO}^{\ominus}\ (\text{N—H}) \rightleftharpoons [\text{B—CH}_2\text{—CH(NH}_2\text{)—COO}^{\ominus},\ \text{B'}]$$

Again we point out that this two-base mechanism, used by the proline racemase and by the hydroxyproline epimerase to equilibrate the configuration at C-2 of the imino acids, has not yet been tested with α-amino-acid racemases, which may all be pyridoxal enzymes.

19.A.1.b ATP-dependent racemization of phenylalanine

All of the amino-acid racemases thus far characterized are found only in bacteria, where their physiological role may be the supply of D-amino acids. D-Amino acids such as D-alanine and D-glutamate are common bacterial cell-wall constituents. Other D-amino-acid isomers are found in various microbial polypeptides that display antibiotic activity. One such antibiotic is gramicidin S, a cyclic decapeptide from *Bacillus brevis* that contains two repeating pentameric units, each with a D-phenylananine residue (Lipmann, 1971). (In the following representation, arrows point from the amino end of one amino-acid residue to the carboxyl end of the next.)

L-Leu → D-Phe → L-Pro → L-Val → L-Orn → L-Leu → D-Phe → L-Pro → L-Val → L-Orn → (back to first L-Leu)

Initial incorporation studies on biosynthesis of gramicidin S indicated that both L- and D-phenylalanine can serve as precursors of the bound D-residue. Subsequent studies indicated that this decamer is biosynthesized nonribosomally, and the mode of peptide-bond formation has been scrutinized. For each peptide bond formed, a molecule of ATP is cleaved to AMP and PP_i (reminiscent of amino-acid activation as aminoacyl adenylates in ribosome-dependent protein synthesis). Purification of the synthesizing enzyme yielded a light enzyme component (100,000 mol wt) and a heavy component (400,000 mol wt) (Lipmann, 1971). The light enzyme shows phenylalanine-dependent exchange between PP_i and ATP, but no exchange stimulated by the other four amino acids. Conversely, the heavy subunit activates the other four amino acids and displays the specific exchanges between PP_i and ATP. Thus, the light enzyme apparently activates phenylalanine only. Alone, neither fraction stimulates any peptide-bond formation, although addition of light enzyme to heavy enzyme leads to formation of gramicidin S. Apparently, the D-Phe is the amino-terminal residue in each of two pentamers that are somehow cyclized head-to-tail before release (as free linear pentamers).

The homogeneous light enzyme will convert L- to D-phenylalanine (in a reaction cleaving ATP to AMP and PP_i) at a rate that is slow compared to the rate of formation of gramicidin S by light and heavy fractions. Further, the enzyme can use either L-phenylalanine or the D-isomer in the PP_i–ATP exchange assay, suggesting that both L-phenylalanine-AMP and D-phenylalanine-AMP are prepared by the enzyme. When [^{3}H]-ATP and L-[^{14}C]-phenylalanine are mixed with the light enzyme and subjected to gel filtration, the isolated enzyme contains twice as many moles of ^{14}C label as it does of ^{3}H label. Lowering the pH to an acid value removes the tritium and half of the ^{14}C label; these acid conditions hydrolyze aminoacyl-AMP molecules, and this reaction presumably explains the stoichiometry of label loss from the enzyme. What about the remaining [^{14}C]-phenylalanyl moiety? Lipmann (1971), Gevers, and Kleinkauf indicated that it is present as an aminoacyl thioester bound to an enzyme cysteinyl sulfur atom:

$$[^{14}\text{C}]\text{—Phe—}\overset{\overset{\displaystyle \text{O}}{\|}}{\text{C}}\text{—S—Enzyme}$$

It is possible, on mixing this complex with charged heavy fraction, to make [^{14}C]-gramicidin S, showing that the labeled species is a catalytically competent intermediate.

$$\text{C}_6\text{H}_5\text{—CH}_2\text{—}{}^{14}\text{CH}(\text{NH}_3^{\oplus})\text{—COO}^{\ominus} + \text{ATP} + \text{Enz—SH} \rightleftharpoons \text{C}_6\text{H}_5\text{—CH}_2\text{—}{}^{14}\text{CH}(\text{NH}_3^{\oplus})\text{—}\overset{\overset{\displaystyle \text{O}}{\|}}{\text{C}}\text{—O—AMP} \;(\text{Enz—SH}) + \text{PP}_i \rightleftharpoons$$

[^{14}C]-L-phenylalanine

$$\text{C}_6\text{H}_5\text{—CH}_2\text{—}{}^{14}\text{CH}(\text{NH}_3^{\oplus})\text{—}\overset{\overset{\displaystyle \text{O}}{\|}}{\text{C}}\text{—S—Enz} + \text{AMP}$$

Hydrolysis of the Phe—S—Enz and characterization of the phenylalanine chirality with D-amino-acid oxidase indicated a mixture of D- and L-isomers. Racemization occurs, presumably after the free L-phenylalanyl-carboxylate anion has been activated (either as the aminoacyl-AMP or as the aminoacyl thioester). We have noted the increased acidity of hydrogen in α-relation to thioester carbonyls in our discussion of the dimethylglutaconyl-CoA hydratase reaction (¶17.B.4), and we shall discuss this topic again in Section V when we review enzymatic Claisen reactions. Thus, the bound L-phenylalanine–enzyme has a C-2 hydrogen of requisite acidity for extraction by a base and formation of a carbanion stabilized as the *aci* tautomer.

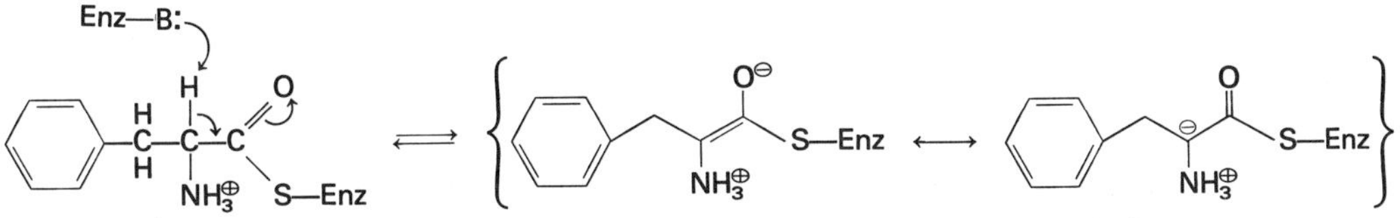

aci form of carbanion

stabilized carbanion

Reprotonation in a sterically random manner could regenerate the L-isomer or form the D-isomer of the Phe—S—Enz. The D-isomer may be more stable in this active-site environment or may be drawn off from the equilibrium by preferential or sole utilization in the subsequent initial peptide-bond-forming step of synthesis of gramicidin S.

This mode of acyl activation and racemization may explain the formation of both D-α-amino and D-α-hydroxy acid residues in other peptide antibiotics and depsipeptide antibiotics (e.g., valinomycin). The review by Lipmann (1971) offers additional insight into the workings of gramicidin synthetase, including the role of a phosphopantetheine moiety in the heavy subunit as a flexible swinging arm bearing the growing peptide chain prior to cyclization.

19.A.1.c Methylmalonyl-CoA racemase

A counterpart to the phenylalanine-racemizing activity of the gramicidin-synthesizing enzyme is found in methylmalonyl-CoA racemase (Mazumder et al., 1962; Overath et al., 1962). The enzyme has been studied from sheep liver and from *Propionibacterium shermanii*; in both cases, it is a heat- and acid-stable protein of 29,000 mol wt. Methylmalonyl-CoA is a key intermediate connecting the metabolism of odd chain fatty acids (which yield propionate on β-oxidation) and the tricarboxylate cycle (Chapter 27). We shall note in Section V that carboxylation of propionyl-CoA yields only the *S*-isomer of methylmalonyl-CoA. On the other hand, the coenyzme-B_{12}–dependent methylmalonyl-CoA mutase (Chapter 20) takes only the *R*-isomer to succinyl-CoA, a readily usable metabolite.

"CO_2" + $H_3C{-}CH_2{-}C(=O){-}SCoA$ $\xrightarrow{\text{carboxylase}}$ $H{-}C(CH_3)(COO^{\ominus}){-}C(=O){-}SCoA$

propionyl-CoA

S-isomer of methylmalonyl-CoA

$^{\ominus}OOC{-}C(CH_3)(H){-}C(=O){-}SCoA$ $\xrightarrow{\text{mutase}}$ $^{\ominus}OOC{-}CH_2{-}CH_2{-}C(=O){-}SCoA$

R-isomer of methylmalonyl-CoA

succinyl-CoA

Methylmalonyl-CoA racemase interconverts the *R*- and *S*-isomers of this methyl-branched CoA ester. The racemization proceeds with incorporation of solvent proton, but the chemical acidity of the C-2 hydrogen is sufficiently acidic that nonenzymatic exchange processes hinder mechanistic studies. The *aci*-thioester has been imputed to be a likely intermediate in the inversion of configuration at C-2.

$^{\ominus}OOC{-}C(CH_3)(H){-}C(=O){-}SCoA$ (Enz—B:) $\rightleftharpoons$ $^{\ominus}OOC{-}C(CH_3){=}C(O^{\ominus}){-}SCoA$ ($BH^{\oplus}$) $\rightleftharpoons$ $^{\ominus}OOC{-}C(H)(H_3C){-}C(=O){-}SCoA$

R-isomer

aci-thioester

S-isomer

19.A.1.d Mandelate racemase

The enzyme interconverting D- and L- mandelate has been induced in and purified from *Pseudomonas putida* (Hegeman et al., 1970; Kenyon and Hegeman, 1970). The protein is a tetramer of 280,000 mol wt; it has no detectable coenzymes, does require $Mg^{2\oplus}$ for catalysis, and shows a turnover number of 1,700 sec^{-1}. Studies by Hegeman, Kenyon, and colleagues indicate that racemization of α-[^{3}H]-L-mandelate produces D-mandelate containing ~80% of the tritium, with 20% exchange with solvent. This partial conservation of tritium in the product D-enantiomer suggests that a *single* active-site base may abstract the α-H and (as $BH^{\oplus}$) either exchange with solvent hydrogen (once in five times) or give back the substrate-derived proton in a stereochemically random manner (four in five times) to the bound intermediate. Fission of the C—H bond during racemization is rate-determining because 2-[^{2}H]-mandelates show a 5.4-fold deuterium isotope effect at V_{max}. Starting with D-mandelate and T_2O, incorporation parallels isomerization; however, tritium is equally distributed in the D- and L-isomers at early time here, indicating formation of an intermediate that partitions symmetrically. Kenyon et al. believe that a carbanion stabilized by the aryl system is a reasonable candidate, an explanation bolstered by the fact that electron-withdrawing ring substituents increase V_{max}. Such substitutents should increase the ability of the benzene ring to lower the energy of an adjacent carbanion by charge delocalization (resonance contribution).

If there is a single active-site base (partially sequestered from solvent) that donates the substrate-derived proton back to a stabilized carbanion, it is unclear how proton transfer to form the product enantiomer is achieved. Either the $BH^{\oplus}$ must be on a flexible arm, or the carbanion must flip over and present the other plane to the stationary $BH^{\oplus}$ (as with aconitase). The following is a possible mechanism involving flipping of a carbanion intermediate.

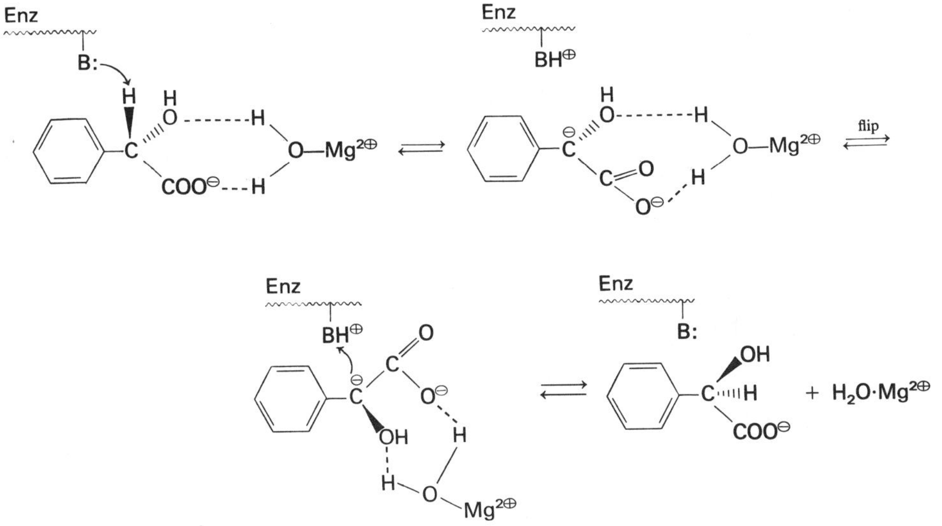

The carbanion intermediate probably has a planar geometry for maximal π-orbital overlap and stabilization with both the $COO^{\ominus}$ and the phenyl ring. The other alternative of a swinging-arm $BH^{\oplus}$ might require less reorientation and motion at the active site than the ring-flip mechanism. One clue to the nature of $BH^{\oplus}$ is that 80% internal transfer occurs. This rules out —$NH_3^{\oplus}$ as $BH^{\oplus}$ because the three hydrogens on —$NH_3^{\oplus}$ would be equivalent, and a maximum of 33% internal transfer should occur.

Note that this postulated one-base mechanism is distinct from the two-base mechanisms proposed for proline racemase and hydroxyproline epimerase. It does have analogy to a possible mechanism for ribulose-5-phosphate 4-epimerase (¶19.A.2).

In an effort to identify the nature of basic groups in the racemase active site, Fee, Hegeman, and Kenyon (1974) prepared an epoxide related to mandelate, α-phenylglycidate, as a candidate for an affinity label.

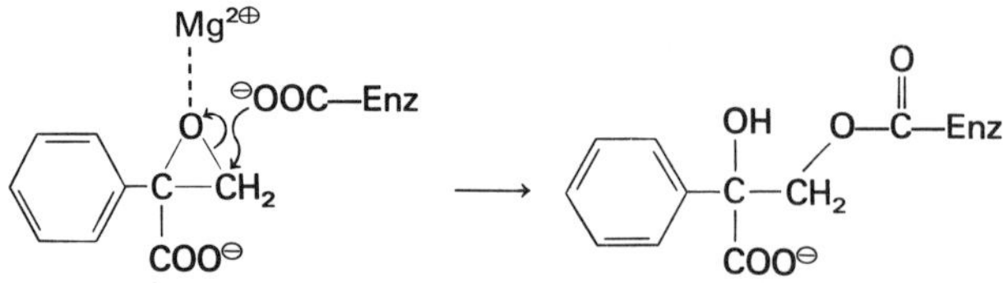

$Mg^{2\oplus}$; $^{\ominus}OOC$—Enz ; C—CH_2 ; $COO^{\ominus}$ $\longrightarrow$ OH ; O ; O—C—Enz ; C—CH_2 ; $COO^{\ominus}$

In the presence of enzyme and $Mg^{2\oplus}$ (which probably acts as Lewis acid for ring opening), the epoxide can be shown to bind reversibly and then inactivate the enzyme, as expected for inactivation proceeding from a Michaelis complex.

$$E + I \rightleftharpoons E \cdot I \longrightarrow E\text{—}I$$

Radioactive epoxide results in stoichiometric labeling of the enzyme. The nature of the linkage to the enzyme is not yet fully elucidated, although sensitivity to base and NH_2OH suggest that an oxygen ester may be the linkage, implying an enzyme $COO^{\ominus}$ as nucleophile. Thus, an aspartyl β-COOH or a glutamyl γ-COOH would provide a flexible, monoprotonic active-site $BH^{\oplus}$ in racemization.

19.A.1.e Ribulose-5-phosphate 3-epimerase

Ribulose-5-phosphate 3-epimerase purified from yeast catalyzes the reversible inversion of configuration at C-3 of the ketose substrate, providing entry into the xylulose series (Glaser, 1972).

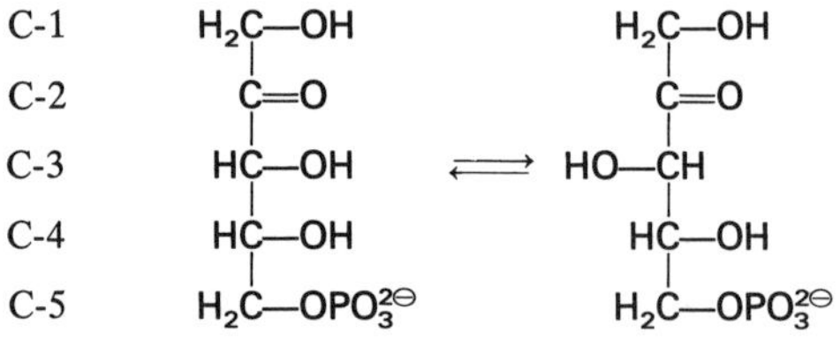

	D-ribulose-5-P		D-xylulose-5-P
C-1	H_2C—OH		H_2C—OH
C-2	C=O		C=O
C-3	HC—OH	$\rightleftharpoons$	HO—CH
C-4	HC—OH		HC—OH
C-5	H_2C—$OPO_3^{2\ominus}$		H_2C—$OPO_3^{2\ominus}$

One atom of solvent hydrogen is incorporated at C-3 during epimerization.

A reasonable mechanism here is abstraction of the C-3 proton by an enzymatic base as initial step. This proton abstraction is facilitated by the adjacent carbonyl group at C-2, which can stabilize the C-3 carbanion thus generated by formation of the enediolate anion.

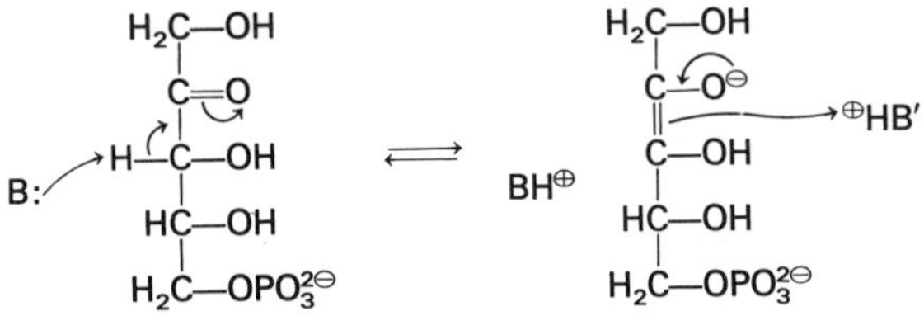

$H_2C{-}OH$, $C{=}O$, $B{:}\rightarrow H{-}C{-}OH$, $HC{-}OH$, $H_2C{-}OPO_3^{2\ominus}$ $\rightleftharpoons$ $BH^{\oplus}$ + $H_2C{-}OH$, $C{-}O^{\ominus}$, $C{-}OH$ $\rightarrow {}^{\oplus}HB'$, $HC{-}OH$, $H_2C{-}OPO_3^{2\ominus}$

D-ribulose-5-P　　　　enediolate anion

Addition of a solvent proton to the opposite face of the planar enediolate will generate xylulose-5-phosphate. This might be most easily accomplished by a second base (B′) protonated with a solvent hydrogen and properly situated at the face opposite to the base performing the initial abstraction.

19.A.2 Epimerizations and Racemizations Proceeding Without Solvent Proton Exchange

The two known examples of epimerases or racemases that do not use a nicotinamide coenzyme and yet do not exchange the migrating H with solvent are ribulose-5-phosphate 4-epimerase and lactate racemase.

The epimerase interconverts L-ribulose-5-P and D-xylulose-5-P by epimerization at C-4.

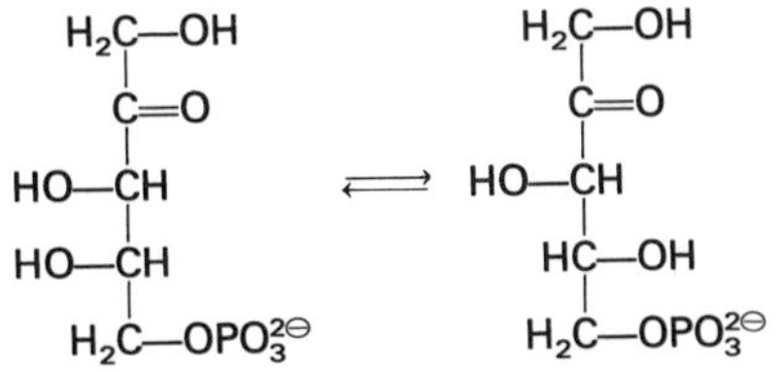

$H_2C{-}OH$, $C{=}O$, $HO{-}CH$, $HO{-}CH$, $H_2C{-}OPO_3^{2\ominus}$ $\rightleftharpoons$ $H_2C{-}OH$, $C{=}O$, $HO{-}CH$, $HC{-}OH$, $H_2C{-}OPO_3^{2\ominus}$

L-ribulose-5-P　　　　D-xylulose-5-P

The enzyme has been purified from *E. coli* (103,000 mol wt) and from *Aerobacter aerogenes* (114,000 mol wt). It has been conclusively shown to lack any known coenzymes or chromophoric prosthetic groups capable of mediating hydride transfer in a redox process, nor do added nicotinamides accelerate rates of epimerization (Glaser, 1972). Yet, 4-[^{3}H]-L-ribulose-5-P is isomerized to 4-[^{3}H]-D-xylulose-5-P with no loss of tritium to the solvent, in contrast to the tritium washout from C-3 of D-ribulose-5-P when it is epimerized by the ribulose-5-P 3-epimerase (¶19.A.1.e). The chemical problem facing these two epimerases also is different. The hydrogen at C-3 of ribulose-5-P is activated by the carbonyl at C-2; no such activation is transmitted to the hydrogen at C-4. Thus, a carbanion generated at C-4 would be much less stable than one generated at C-3 (stabilized as the enediolate). How the enzyme might increase the acidity of the C-4 proton for abstraction is not known. If a hydride transfer occurs in the enzymatic epimerization of C-4, no precedents are currently available.

We have noted examples of shielded proton transfer in catalysis by aconitase and mandelate racemase; those two enzymes and this epimerase (if it utilizes proton transfer) have to achieve antarafacial transfer of the sequestered proton. The protonated base must redeliver the C-4 proton to the opposite face of an intermediate carbanion (the data would seem to rule out a concerted pathway). In the aconitase reaction, a flip or rotation of the plane of the *cis*-aconitate intermediate was postulated. In the ribulose-5-P 4-epimerase reaction, Rose (1970, 1972) has suggested that it may be more likely that, rather than rotation of an intermediate carbanion to offer a different face to the protonated base, the mechanism involves a $BH^{\oplus}$ at the end of a flexible arm (e.g., the ε-$NH_3^{\oplus}$ of lysine or the γ-$COO^{\ominus}$ of glutamate). Such a flexible base could move relative to the bound carbanion.

The V_{max} for epimerization is 18.7 μmoles min^{-1} mg^{-1} at 37° C for the *E. coli* enzyme and is 12.5 μmoles min^{-1} mg^{-1} at 25° C for the *A. aerogenes* enzyme. These values correspond to turnover numbers of about 25 to 30 sec^{-1}. Rose has noted that rapid-reaction studies show that primary amino groups R—$NH_3^{\oplus}$ (e.g., that of lysine) dissociate at rates of about 10 sec^{-1}. Rates of dissociation of R—COOH are as high as 10^7 sec^{-1}. Although the ammonium ion might be appropriate kinetically and sterically, using the same argument as in the case of mandelate racemase (¶19.A.1.d), the fact that *complete transfer* of the C-4 hydrogen occurs suggests that $BH^{\oplus}$ is a *monoprotonic* base. The R—$NH_3^{\oplus}$ ought to yield only 33% direct hydrogen transfer in any catalytic cycle because abstraction of the C-4 proton by R—NH_2 generates three sterically indistinguishable protons in the R—$NH_3^{\oplus}$. Any one of the three should be transferred back with equal probability on such randomization. Thus, R—COOH is a more likely candidate. (Note that R—SH and the imidazolium ring of histidine residues are also appropriately monoprotonic in the $BH^{\ominus}$ form, and they, unshielded, would exchange with solvent at rates of about 10^2 to 10^3 sec^{-1}.)

Evidence for enzymatic racemization of lactic acid has been obtained in extracts of *Clostridium butylicium.* (In other bacteria, D- and L-lactate are interconverted by action of two NAD-dependent dehydrogenases, each reducing pyruvate to a single hydroxy-acid enantiomer.) The enzyme has recently been purified, and evidence has been accumulated that is interpreted in favor of a hydride shift (Shapiro and Dennis, 1965; Pepple and Dennis, 1976). During racemization, the C-2 hydrogen is transferred directly to the product enantiomer formed. A k_H/k_D of 2.14 implies that the C—H cleavage is somewhat rate-limiting. The V_{max} for racemization is identical starting with either D-lactate or

L-lactate. Dennis has shown that hydroxylamine inhibits racemization, suggesting a carbonyl intermediate. Borohydride reduction during racemase action purportedly produces lactaldehyde, presumably from a reduction of a pyruvylhemithioacetal intermediate (i.e., net reduction of both C-1 and C-2). Experiments with $B^3H_4^{\ominus}$ should be performed, and the amount of lactaldehyde generated was not quantitated. Dennis speculates that a reversible 1,2-hydride shift occurs, possibly by way of a thioester. It is not clear how the activation energy for initial formation of the activated lactyl thioester would be obtained. (Also, one would have expected that borohydride would reduce the lactaldehyde further to glycerol.)

Enz—SH + L-Lactate ⇌ H_3C—C(^{3}H)(O—H⋯:B)—C(=O)—S—Enz ⇌ H_3C—C(=O)—C($O^{\ominus}$)(^{3}H)—S—Enz (BH$^{\oplus}$) ⇌ H_3C—C(HO)(^{3}H)—C(=O)—S—Enz

L-lactyl enzyme | pyruvyl hemiothioacetal | D-lactyl enzyme

↓ trapping with NABH$_4$

H_3C—C(H)(OH)—C(H)=O

lactaldehyde

A lactate racemase has also been reported from *L. sake* and characterized as a species of 25,000 mol wt without obvious cofactors; yet it also carries out direct hydrogen transfer (Hiyama et al., 1968). It is possible that such a reversible intramolecular redox reaction with hydride shift operates for this enzyme with sterically random return of hydride from C-1 to C-2 to generate racemization.

19.A.3 Epimerases Proceeding with Redox Catalysis

We have previously discussed the NAD-dependent nucleotide-diphosphosugar epimerase, UDP-glucose epimerase, which epimerizes UDP-glucose at C-4 to UDP-galactose by way of a keto intermediate and enzyme-NADH. Glaser (1972) has summarized evidence for NAD$^{\oplus}$ requirements in the epimerizations of UDP-*N*-acetylglucosamine, UDP-glucuronic acid, and the C-2 position of the 3,6-dideoxy sugar CDP-paratose. An interesting addition to this contingent is dTDP-L-rhamnose synthetase, which converts dTDP-4-keto-6-deoxy-D-glucose and

NADPH to NADP$^{\oplus}$ and dTDP-L-rhamnose. We discussed enzymatic formation of the 4-keto-6-deoxy substrate in ¶17.B.3.a. The dTDP-L-rhamnose synthetase inverts configuration at both C-3 and C-5 during product formation. The hydrogens at both C-3 and C-5 are activated by the 4-keto group of the substrate, and equilibration of configuration via planar enolates is expected. In agreement, solvent hydrogen ends up at C-3 and C-5 of dTDP-L-rhamnose, while NADPH is used to reduce C-4 to the alcohol. Two polypeptide chains (Enz I and Enz II) comprise the enzyme activity, as is the case in the 3,6-dideoxyglucose-synthesizing system noted earlier (¶17.B.3.b). Enz II catalyzes epimerization first at C-3 and then at C-5, as indicated by allowing partial exchange to proceed in D_2O. Recovered molecules are monodeuterated at C-3 or dideuterated at C-3 and C-5, but essentially no molecules monodeuterated at C-5 are found.

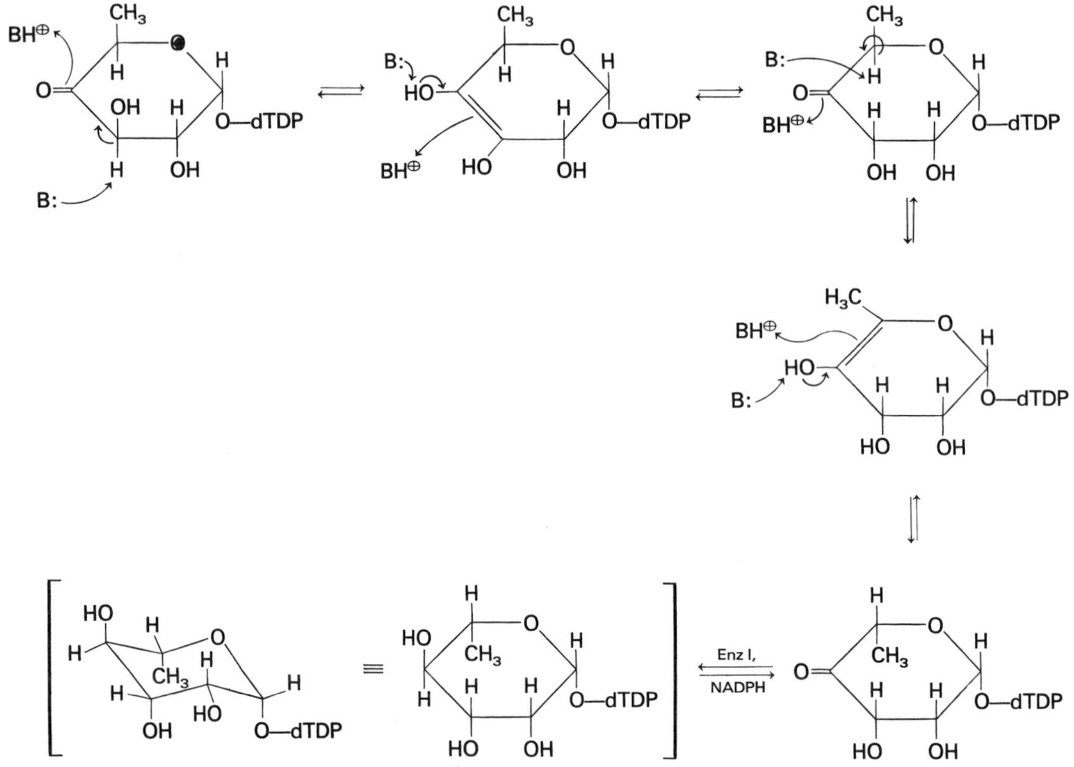

dTDL-L-rhamnose

In the final step shown, Enz I reduces the dTDP-4-keto-6-deoxy-L-mannose bound at the Enz II active site to the product dTDP-L-rhamnose.

Substrate deuterated at C-3 shows a k_H/k_D of 3.4 at V_{max}, whereas substrate deuterated at C-5 shows a k_H/k_D of 2.0 at V_{max}, indicating that deuterium substitution can slow down either epimerization step to the point where it represents a slow step in catalysis.

This concludes our discussion of pyridoxal-P–independent 1,1-hydrogen shift mechanisms. We now switch attention to 1,2-hydrogen shifts—migration, in a formal sense, of a hydrogen species between adjacent carbon atoms.

19.B 1,2–HYDROGEN SHIFTS: ALDOSE–KETOSE ISOMERASES

The enzymes discussed in this section carry out isomerizations of α-hydroxy aldehydes to β-keto alcohols. (One can think of this as an internal redox reaction between isomeric hydroxycarbonyl systems via a common enolate.)

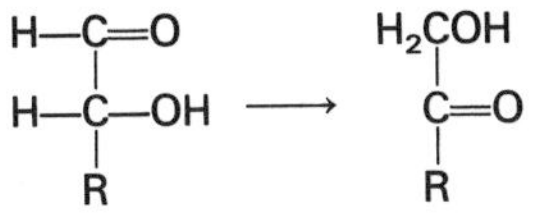

H—C=O | H—C—OH | R ⟶ H_2COH | C=O | R

α-OH-aldehyde (aldose) α-keto alcohol (ketose)

Common substrates range from the C_3 aldose glyceraldehyde-3-P to various aldopentoses and aldohexoses. In recent reviews, Rose (1975) and Rose and Hanson (1975) have noted that, in these tautomerizations, one of the prochiral hydrogens at C-1 of the ketose substrate is removed by the enzyme and transferred, at least partially, to C-2 of the aldose formed. The partial transfer is in keeping with the suggested formation of an enediol or enediolate intermediate and a $BH^{\oplus}$ during isomerization.

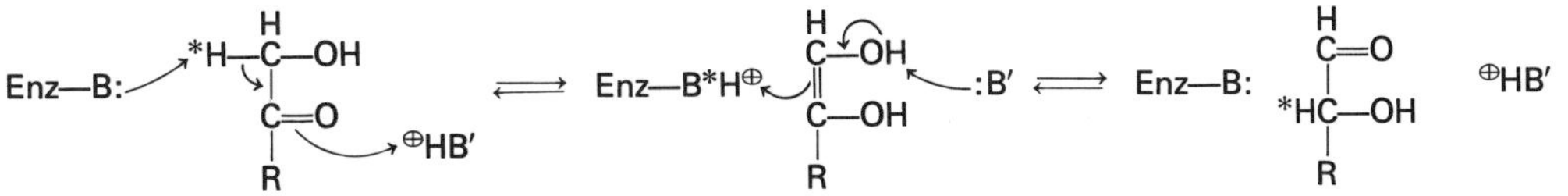

Enz—B: *H—C(H)—OH | C=O | R ${}^{\oplus}$HB′ ⇌ Enz—B*H$^{\oplus}$ HC—OH ‖ C—OH | R :B′ ⇌ Enz—B: H—C=O | *HC—OH | R ${}^{\oplus}$HB′

ketose *cis*-enediol aldose

The intramolecular hydrogen transfer implies that a *single base* removes hydrogen from C-1 and adds it back to C-2 of the enediol (or the opposite in the reverse direction). Knowledge of the configuration of the C-2 asymmetric center in the aldose substrate and of the prochiral position to which and from which *H is transferred at C-1 of the ketose has been accumulated for eight aldose–ketose isomerases. In all cases, the *cis* geometry is indicated for the enediolate, and a suprafacial transfer occurs along one side of the plane of the enediol intermediate (Rose, 1975). Rose and Hanson (1975) have noted that both suprafacial possibilities are represented among these isomerases. They have argued that the general use of a single base and a *cis*-enediol by these enzymes represents a "dual economy in the use of catalytic groups" and may contribute to catalytic efficiency. A *trans*-enediol would require two enzymatic base groups to polarize both the aldehyde of the aldose and the C-2 ketone of the ketose. Similarly, they note that, although a single base serves for suprafacial proton transfer along the enediol, antarafacial processes would require separate bases for C-1 and C-2.

We shall examine results for three specific enzymes: glucose-6-phosphate isomerase, mannose-6-phosphate isomerase, and triosephosphate isomerase. Also note that we discussed in Chapter 5 (on amino transfers) the formation of D-glucosamine-6-P from fructose-6-P and glutamine; this reaction also is an aldose-ketose isomerization.

19.B.1 Phosphoglucose isomerase

Phosphoglucose isomerase (glucose-6-P isomerase) catalyzes the first isomerization step in glucose fermentation and consequently is present in most organisms (Noltman, 1972). The enzyme interconverts glucose-6-P and fructose-6-P, with an equilibrium constant of 0.3 for the conversion as written.

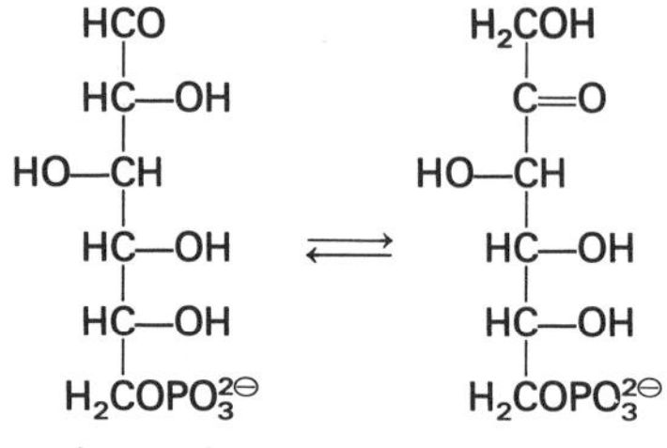

D-glucose-6-P		fructose-6-P
HCO		H_2COH
HC—OH		C=O
HO—CH		HO—CH
HC—OH	$\rightleftharpoons$	HC—OH
HC—OH		HC—OH
$H_2COPO_3^{2\ominus}$		$H_2COPO_3^{2\ominus}$

The enzyme from a number of sources appears dimeric (130,000 mol wt per dimer) and exhibits a turnover number of about 1,000 catalytic events per second. It has been established that the C-2 hydrogen of D-glucose-6-P is introduced into the pro*R*-position at C-1 of the fructose-6-P product. With deuterated substrates (1-[^{2}H]-fructose-6-P), a primary isotope effect is observed. In T_2O, the isomerization proceeds to yield product of low specific radioactivity, consistent with partial shielding of the itinerant hydrogen in the $EBH^{\oplus}$ form. It is estimated that proton transfer is 80% intramolecular in the aldose→ketose direction and is 50% intramolecular in the ketose→aldose direction with rabbit-muscle isomerase. The simplest explanation of the labeling data is suprafacial proton transfer mediated by a single base at the *re–re* face of a *cis*-enediolate intermediate.

Clearly, the preceding representation of phosphoglucose isomerase catalysis as an isomerization between the open-chain forms of the free aldehyde and free keto molecules is an unwarranted oversimplification. We have noted earlier that only trace amounts of acyclic aldose exist for glucose. With the glucose-6-P molecule, 38% of the population is α-anomer and 62% is β-anomer, which equilibrate at a nonenzymatic rate of about 0.06 sec^{-1} (physiological conditions) via free aldehyde.

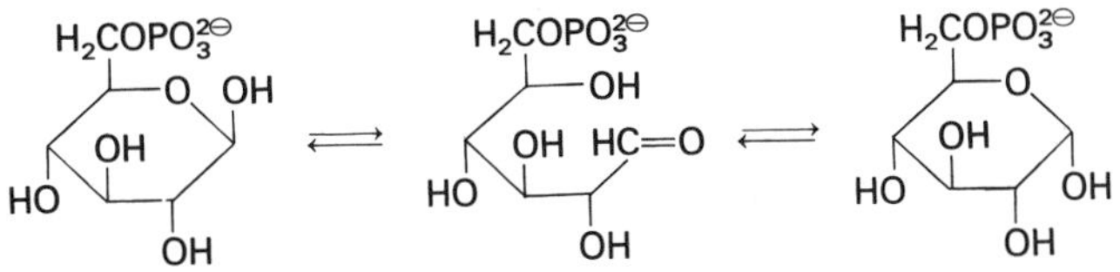

$H_2COPO_3^{2\ominus}$ ... β-anomer (62%) ⇌ free aldehyde (trace quantities) ⇌ α-anomer (38%)

Similarly, the ketose-6-P in aqueous solutions is as the cyclic ketal (but in a five-membered furanose ring rather than the six-membered pyranose ring), distributed 20% as α-anomer and 80% as β-anomer.

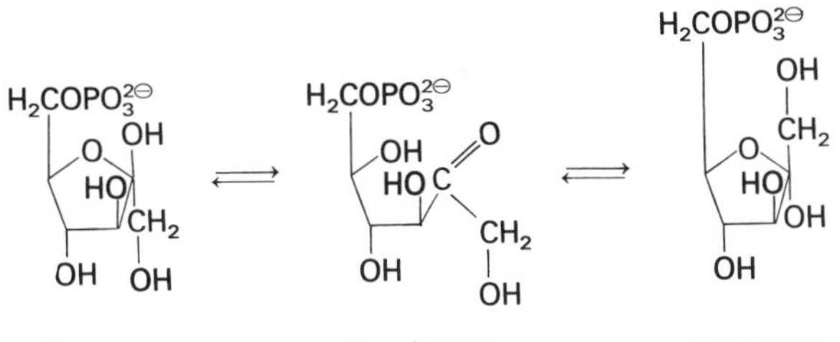

β-anomer (80%) ⇌ free ketone (trace quantities) ⇌ α-anomer (20%)

Elegant rapid-reaction kinetic experiments have indicated that phosphoglucose isomerase uses the α-anomers of the cyclic forms of glucose-6-P and fructose-6-P preferentially (Schray and Benkovic, 1976). Rose and colleagues have suggested that specific enzyme-bound conformers can explain this kinetic disposition. In the ketose→aldose direction, proton addition is at the *re* face of C-2 of the enediol. To obtain the α-anomer of the glucose-6-P hemiacetal, the C-5 hydroxyl oxygen should approach the C-1 carbonyl from the *si* face.

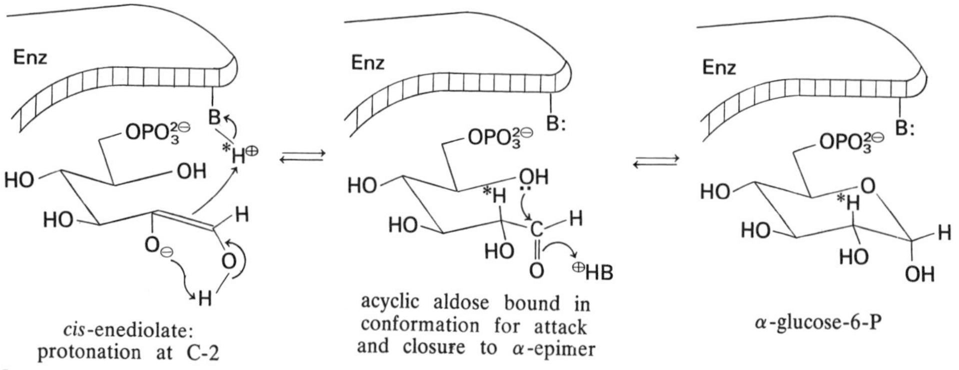

cis-enediolate: protonation at C-2

acyclic aldose bound in conformation for attack and closure to α-epimer

α-glucose-6-P

A similar scheme can be used to explain the preference for formation of α-fructose-6-P.

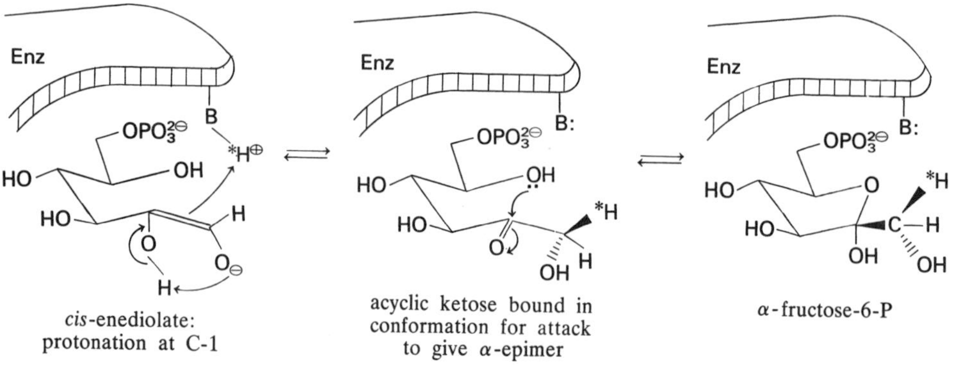

cis-enediolate: protonation at C-1

acyclic ketose bound in conformation for attack to give α-epimer

α-fructose-6-P

However, the enzyme also will utilize the β-anomers of glucose-6-P and fructose-6-P at about a tenfold slower rate. Because data exist against *trans*-enediols as intermediates in these conversions, it is suggested that the isomerase binds both anomers of each substrate (Rose and Hanson, 1975) and catalyzes ring opening to the bound acyclic form (nonenzymatic equilibration rates are 10^4-fold slower than V_{max}, arguing that the enzyme must catalyze ring opening). Rose suggests that then the enzyme permits rotation about the single bond between C-1 and C-2 of the acyclic sugar forms to produce, as the accumulating conformer, a common cisoid conformation of C-1 and C-2 oxygen substituents prior to proton abstraction. An additional stricture in this hypothesis is that after proton abstraction the *planar enediol cannot rotate* (otherwise some mannose-6-P, not an observed product, should form) (Rose, 1975). The following is one possible scheme.

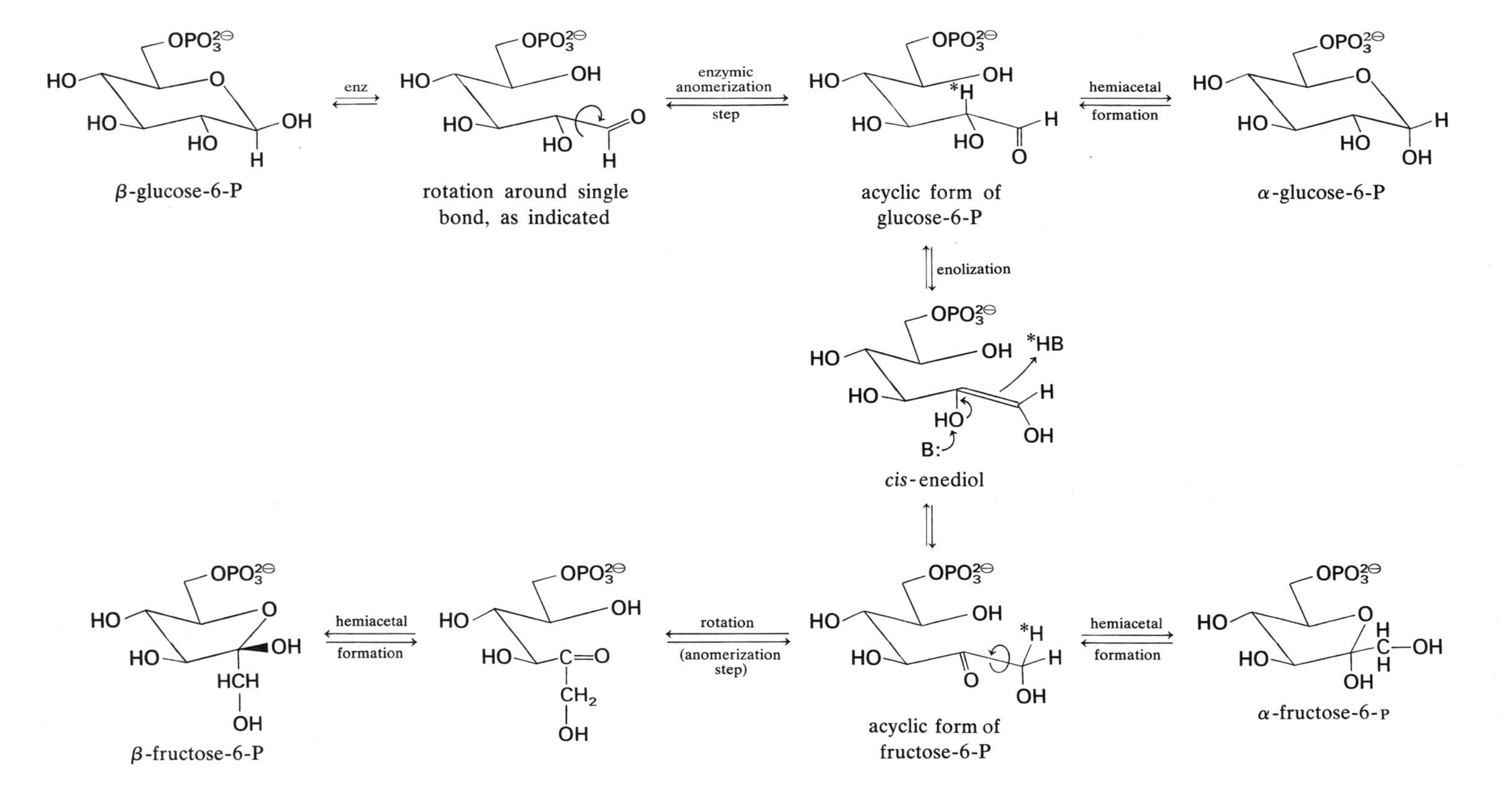

β-glucose-6-P
enz
rotation around single bond, as indicated
enzymic anomerization step
acyclic form of glucose-6-P
hemiacetal formation
α-glucose-6-P
enolization
cis-enediol
β-fructose-6-P
hemiacetal formation
rotation (anomerization step)
acyclic form of fructose-6-P
hemiacetal formation
α-fructose-6-P

Recently O'Connell and Rose (1973) have shown that the 1,2-anhydrohexitol-6-phosphates (epoxide-containing substrate analogues) are active-site–directed inactivators of the isomerase. The synthetic procedure yielded a mixture of four diastereomers, resulting from enzymatic condensation of D- or L-glycidaldehyde with dihydroxyacetone-P, followed by $NaBH_4$ reduction of the C-5 ketone group (yielding a mixture at C-4 and C-2). D-Glycidaldehyde gave a mixture of 2*R*,4*S* (L-gulitol) and 2*R*,4*R* (D-mannitol) anhydrohexitols. The L-glycidaldehyde yielded a 2*S*,4*R* (D-glucitol) and 2*S*,4*S* (L-iditol) epimeric mixture.

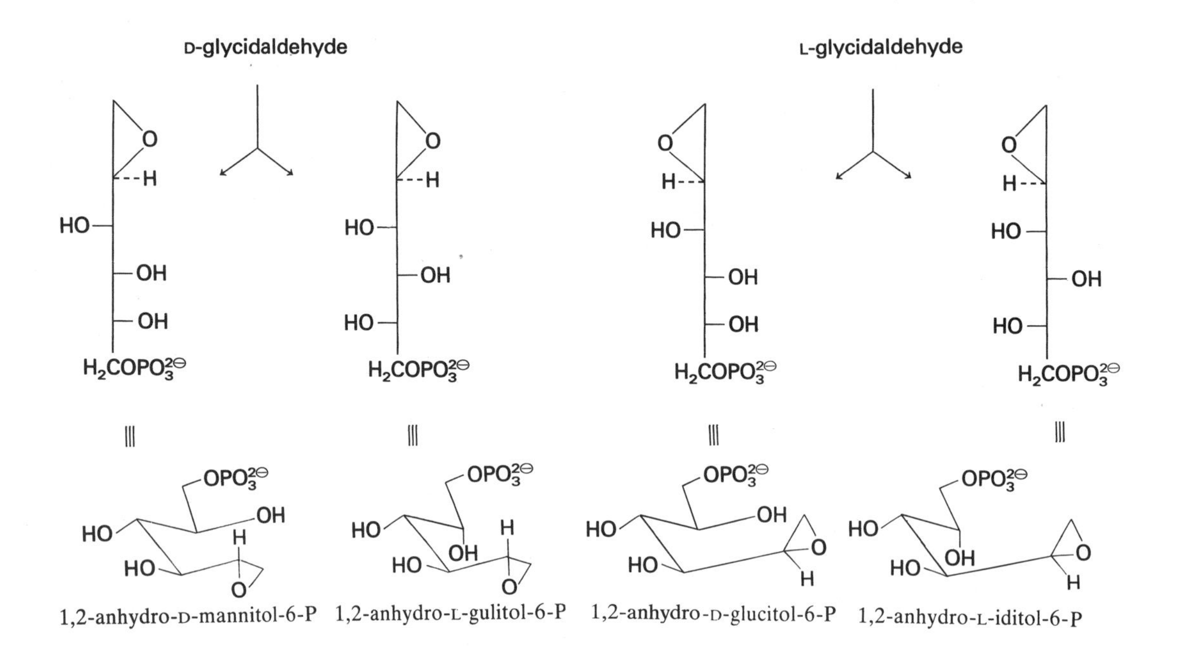

The 2*R* mixture (from D-glycidaldehyde) showed a K_i of 0.27 mM and a $T_{1/2}$ for irreversible inactivation of 28 min. The 2*S* mixture (from L-glycidaldehyde) showed a K_i of 0.47 mM but a $T_{1/2}$ of 23 hr. Each mixture yielded one label per subunit when 5-[^{3}H]–inactivating-epimers were used. Susceptibility to hydroxylamine in urea (to denature the enzyme) indicated an ester linkage, and isolation of a single tripeptide indicated a γ-glutamyl carboxyl of the enzyme as nucleophile. Rose suggested that the 2*R* epimers can achieve a favorable conformation for S_N2 backside attack by the glutamyl $COO^{\ominus}$ more often than can the 2*S* epimers. Presumably, the γ-$COO^{\ominus}$ is the single base responsible for proton abstraction and suprafacial transfer in the aldose–ketose isomerization carried out by glucose-6-P isomerase.

19.B.2 Phosphomannose isomerase

Mannose-6-phosphate (2*S*) is the C-2 epimer of glucose-6-P (2*R*), and it is isomerized to the same ketose (D-fructose-6-P) by a separate isomerase, phosphomannose isomerase.

D-glucose-6-P (2*R*) D-mannose-6-P (2*S*)

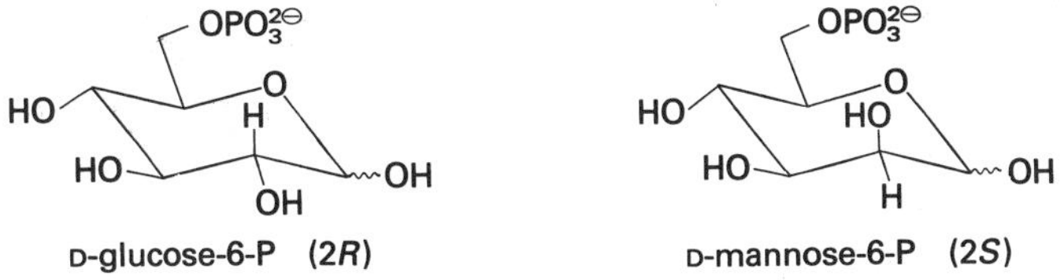

The physiological role of the enzyme is to convert ingested mannose (after phosphorylation by hexokinase) to the common ketose-6-P, which is then metabolized by the standard glycolytic pathway (Chapter 27). The enzyme has been purified from a number of sources and found to be a zinc enzyme, in contrast to the metal-free phosphoglucose isomerase. The zinc may chelate the oxygen atoms at C-1 and C-2. $Zn^{2\oplus}$ is a catalyst for the nonenzymatic aldose–ketose isomerization (Lobry de Brun–Alberda von Ekenstein reaction). The yeast enzyme has 45,500 mol wt with one zinc atom. This enzyme shows a molecular activity of about 600 sec^{-1} and shows a K_{eq} of 1.03 under physiological conditions. Much less intramolecular proton transfer occurs during epimerization of the 2*S*-aldohexose than was noted with the phosphoglucose isomerase; in only 5% to 10% of the isomerizations is the proton transferred without equilibration with solvent, suggesting less shielding of EB*H at the active site in the *cis*-enediol stage.

Because the mannose-6-P is epimeric at C-2 with glucose-6-P in the phosphomannose isomerase reaction, the intermediate *cis*-enediol is protonated suprafacially at the *si–si* face rather than at the *re–re* face (as is the case in the phosphoglucose isomerase reaction). In this event, in the ketose → aldose direction, the C-5 hydroxyl should approach the C-1 aldehyde group from the *re* face on hemiacetal formation before mannose-6-P release. Then the hemiacetal product released ought specifically to be the *β-anomer* of 6-phosphomannose (p. 592).

This prediction recently was confirmed by Rose (1975) and Schray. They took advantage of the fact that hexokinase rapidly phosphorylates both the α- and β-anomers of free mannose to α- and β-mannose-6-P, respectively. Thus, addition of α-mannose to excess hexokinase yields within five seconds only the α-mannose-6-P. Then the isomerase was allowed to react for five to ten seconds, and the reaction was quenched and analyzed for fructose-6-P content. Incubations were performed at 3° C and for such short intervals to minimize competing

putative conformer at mannose-P-isomerase active site

protonation at *si–si* face

β-mannose-6-P

rotation of *cis*-enediolate around bond between C-2 and C-3

protonation at *re–re* face

putative conformer at glucose-P-isomerase active site

α-glucose-6-P

nonenzymatic anomerization ($T_{1/2}$ at 20° C predicted to be 11.5 sec per anomerization event). In these experiments, freshly dissolved α-mannose produced no fructose-6-P in the two-enzyme couple. Conversely, the hexokinase-catalyzed β-mannose→β-mannose-6-P reaction was highly active with the isomerase, as predicted by the *cis*-enediol mechanism.

The apparent lack of reaction of the α-anomer also points out that the phosphomannose isomerase *does not catalyze ring opening* of the α-anomer to the acyclic form—unlike the phosphoglucose isomerase, which catalyzes ring opening of both α- and β-anomers of its substrates, glucose-6-P and fructose-6-P. Interestingly, Rose (1975), O'Connell, and Schray then showed that the phosphoglucose isomerase does hasten the anomerization of mannose-6-P as well. Thus, it binds either α- or β-mannose-6-P and catalyzes ring opening from each. However, no conversion to fructose-6-P is observed from the α-anomer. This corroborates the idea that *catalyzed ring opening of the anomeric hemiacetal forms of aldose substrates occurs prior to and independent of the proton-abstraction step from which isomerization then proceeds.* Indeed, because mannose is a 2*S* molecule and glucose a 2*R* molecule, the active-site base of phosphoglucose isomerase ought to be incorrectly situated (above instead of below the plane) to abstract the C-2 hydrogen (below the sugar-ring plane) of bound mannose-6-P molecules, and no 1,2-shift should be possible; this is in accord with the observations.

19.B.3 Triose-Phosphate Isomerase

Triose-P isomerase interconverts the three-carbon phosphorylated substrates glyceraldehyde-3-P and dihydroxyacetone phosphate, which are themselves generated by aldol cleavage of fructose-1,6-diphosphate (Chapter 23; Noltman, 1972).

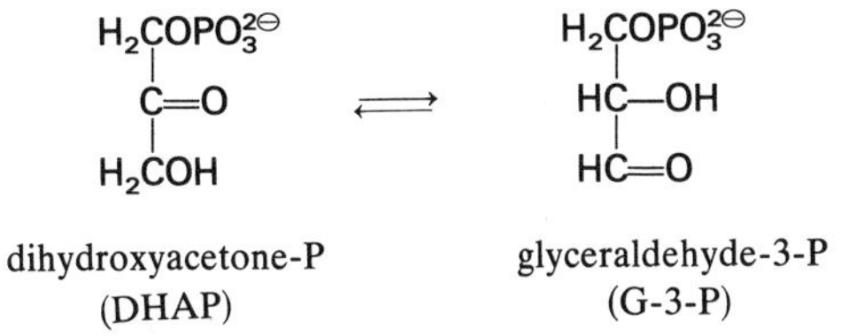

$$\begin{array}{c} H_2COPO_3^{2\ominus} \\ | \\ C{=}O \\ | \\ H_2COH \end{array} \rightleftharpoons \begin{array}{c} H_2COPO_3^{2\ominus} \\ | \\ HC{-}OH \\ | \\ HC{=}O \end{array}$$

dihydroxyacetone-P (DHAP) glyceraldehyde-3-P (G-3-P)

Subsequent enzymatic flux in glycolysis is through glyceraldehyde-3-P-dehydrogenase–mediated oxidation of the aldehyde to the acyl phosphate (Chapter 10). The isomerase serves then to draw dihydroxyacetone-P molecules through the glycolytic pathway by this initial isomerization to the usable aldehyde. The enzyme from several sources is a dimer of 50,000 mol wt without metals, isomerizing substrates with a V_{max} approaching 10^3 sec^{-1}. The equilibrium constant had been measured as 22 to 24 for DHAP/G-3-P with K_m values (in the case of the rabbit-muscle enzyme) of 0.32 mM for aldehyde and 0.62 mM for dihydroxyacetone-P. However, Pogson and colleagues have recently determined that, in neutral aqueous solutions, glyceraldehyde-3-P is 96.7% hydrated to the gem diol, and dihydroxyacetone-P is 44% hydrated (Trentham et al., 1964; Reynolds et al., 1971).

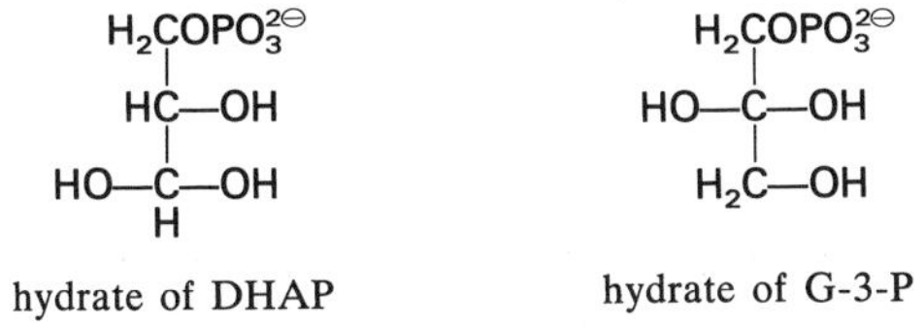

$$\begin{array}{c} H_2COPO_3^{2\ominus} \\ | \\ HC{-}OH \\ | \\ HO{-}C{-}OH \\ H \end{array} \qquad \begin{array}{c} H_2COPO_3^{2\ominus} \\ | \\ HO{-}C{-}OH \\ | \\ H_2C{-}OH \end{array}$$

hydrate of DHAP hydrate of G-3-P

Additionally, they determined that the enzyme binds only the unhydrated substrates, so the true K_m values are 11 μM for glyceraldehyde-3-P and 0.34 mM for dihydroxy acetone-P, leading to a corrected K_{eq} of 367 at 25° C.

As with other isomerases discussed in preceding subsections of ¶19.B, evidence has accrued for the *cis*-enediol intermediate; experimentally, the value of intramolecular proton transfer is low but real (between 3% and 6%). This washout, given a turnover number for the isomerization of 10^3 sec^{-1}, implies that $EBH^{\oplus}$ must exchange at least 15-fold faster than V_{max}, or at a rate of at least 1.5×10^4 sec^{-1}, which is considerably faster than rates estimated for dissociation of phenols, ammonium ions, or imidazolium ions in neutral solution (1.4, 25, and 1,600 sec^{-1}, respectively). A reasonable candidate for the $—BH^{\oplus}$ is a —COOH. If it has a pK_a of 6, then

$$K_{diss} = 10^{10}\,sec^{-1} \div 10^{6}\,sec^{-1} = 10^{4}$$

If an E—COOH had a pK_a of 5, K_{diss} would be 10^5. An enzymatic $COO^{\ominus}$ residue is thus a kinetically likely candidate for the active-site proton-abstracting base.

Knowles and colleagues have recently carried out a series of elegant and comprehensive experiments to probe the kinetics and energetics of triose-P-isomerase catalysis and provide the most detailed picture now available of the energy barriers in an enzymatic reaction (Herlihy et al., 1976; Maister et al., 1976; L. Fisher et al., 1976; Albery and Knowles, 1976, Knowles and Albery, 1977). To understand both the kinetic course of isomerization and the energetics of the component steps, these workers examined the fate of both the abstracted substrate hydrogens and the fate of solvent hydrogens incorporated into both substrate and product, and they analyzed these processes in both directions.

In the G-3-P→DHAP direction, the ketone product generated was trapped in situ by reduction to glycerol phosphate through inclusion of mixtures of NADH and purified α-glycerol-P dehydrogenase.

$$\underset{\text{G-3-P}}{\begin{matrix} H_2COPO_3^{2\ominus} \\ | \\ HC—OH \\ | \\ HC=O \end{matrix}} \xrightleftharpoons{\text{isomerase}} \underset{\text{DHAP}}{\begin{matrix} H_2COPO_3^{2\ominus} \\ | \\ C=O \\ | \\ H_2C—OH \end{matrix}} \xrightarrow[\text{dehydrogenase (trapping system)}]{\text{NADH,}} \underset{\text{glycerol-P}}{\begin{matrix} H_2COPO_3^{2\ominus} \\ | \\ HC—OH \\ | \\ H_2C—OH \end{matrix}}$$

The trap ensures that product inhibition does not occur *and, more importantly,* that DHAP molecules, once they have been *released* from the active site, are scavenged reductively; the net back reaction does not occur. One can take the same precautions when studying the isomerase in the opposite direction (DHAP→G-3-P) by using $NAD^{\oplus}$ and glyceraldehyde-3-P dehydrogenase as trap in arsenate buffer (the 1-arsenic ester of 3-phosphoglycerate is initial product; it hydrolyzes rapidly to 3-phosphoglycerate and $AsO_4^{2\ominus}$, displacing the equilibrium).

$$\begin{matrix} H_2COPO_3^{2\ominus} \\ | \\ C=O \\ | \\ H_2C—OH \end{matrix} \xrightleftharpoons{\text{isomerase}} \begin{matrix} H_2COPO_3^{2\ominus} \\ | \\ HC—OH \\ | \\ HC=O \end{matrix} \xrightarrow[\text{dehydrogenase, } AsO_4^{2\ominus}]{NAD^{\oplus},} \underset{\text{3-phosphoglycerate}}{\begin{matrix} H_2COPO_3^{2\ominus} \\ | \\ HC—OH \\ | \\ COO^{\ominus} \end{matrix}}$$

The kind of mechanistic information obtainable by following the path of solvent hydrogens into product or recovered substrate is twofold: (1) one can monitor isotopic discrimination in product formation; and (2) one can monitor the partitioning between exchange of solvent and substrate hydrogens in competition with conversion to isomerized product.

For example, incubation of DHAP, enzyme, and trapping system in ^3H_2O allows estimation of the specific radioactivity (sp rad) of 2-[^{3}H]-glyceraldehyde-3-phosphate compared to starting specific radioactivity of ^3H_2O.

$$\text{(sp rad of 2-[}^3\text{H]-G-3-P)/(sp rad of }^3\text{H}_2\text{O}) = 0.77$$

The solvent specific radioactivity is only 1.3-fold greater than that of product, a discrimination so slight that it probably does not reflect a primary kinetic isotope effect against ^{3}H (which might be about 8 to 15); it most likely is an equilibrium isotope effect, suggesting that the processes involved in solvent hydrogen incorporation are all in equilibrium. Because the *cis*-enediolate is the exchanging species, one expects enzyme species I, II, III, and IV in the following scheme all to be in rapid equilibrium.

Enz–C(=O)–O$^{\ominus}$ ··· $H_2COPO_3^{2\ominus}$–C=O–HC(H)–OH ⇌ I ⇌ III $\xrightarrow[\text{slow}]{k_{off}}$ 2-[^{1}H]-G-3-P

V: 1-[^{1}H]-DHAP

1*R*-[^{3}H]-DHAP

I ⇅ BH$^{\oplus}$ exchange step ⇅ II

VI ⇌ II ⇌ IV $\xrightarrow[\text{slow}]{k_{off}}$ 2-[^{3}H]-G-3-P

The enzyme species shown in this scheme are the following:

I = EBH$^{\oplus}$ · · · Enediolate

II = EB^{3}H$^{\oplus}$ · · · Enediolate

III = EB · · · 2-[^{1}H]-Glyceraldehyde-3-phosphate

IV = EB · · · 2-[^{3}H]-Glyceraldehyde-3-phosphate

V = EB · · · 1-[^{1}H]-Dihydroxyacetone phosphate

VI = EB · · · 1-[^{3}H]-Dihydroxyacetone phosphate

The cycle III⇄I⇄II⇄IV (including the $EBH^{\oplus}$⇄$EB^3H^{\oplus}$ equilibration step) can happen many, many times before the aldehyde product is released. One then expects that the slow step in the DHAP→G-3-P direction is *rate-determining product release*, a physical step. This example illustrates the use of information from isotopic discrimination during product formation.

If one also monitors whether the pool of DHAP molecules becomes radioactive during this ^3H_2O experiment, one gets information on how the enediolate intermediate *partitions* between going on to product or back to starting materials. Note that the coupling enzyme traps any G-3-P that is released from the active site, so no back reaction is possible, and introduction of ^3H into DHAP cannot be due to reversal of the overall isomerization sequence; it must stem from reversion of enzyme-bound enediolate. In fact, 1-[^3H]-DHAP (stereospecifically tritiated in the 1*R*-locus) is formed, and the specific radioactivity increases with time. These data are analogous to those found with enolase; the intermediate reverts back to substrate before going on to *released* product (not necessarily before giving *bound* product) (Figure 19-1). Analysis of specific radioactivity in dihydroxyacetone-P and in the trapped glyceraldehyde-3-P indicates the following partitioning of the enediolate:

$$\text{DHAP} \xleftarrow{1} \text{Enediolate} \xrightarrow{3} \text{G-3-P}$$

Suppose that one now performs analogous experiments in ^3H_2O, this time proceeding in the opposite direction (G-3-P→DHAP). The specific radioactivity of the 1-[^3H]-glycerol-P (from the trapping enzyme) to ^3H_2O is 0.13.

$$(\text{sp rad of 1-}[^3\text{H}]\text{-glycerol-P})/(\text{sp rad of }^3\text{H}_2\text{O}) = 0.13$$

This represents about an eightfold kinetic discrimination against tritium, large enough to be a primary kinetic isotope effect, not an equilibrium effect. In this direction, the exchanging species are not in equilibrium prior to DHAP (either [^1H]- or [^3H]-) release. (The exchanging species here are V⇄I⇄II⇄VI.) DHAP release is not rate-determining. Partitioning experiments allow assignment of all

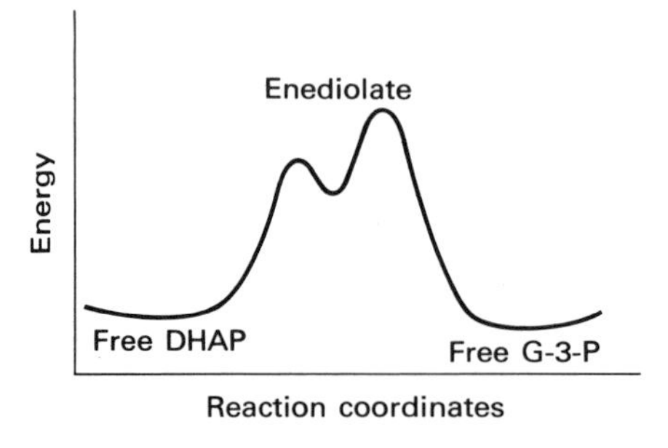

Figure 19-1
A hypothetical free-energy profile for triose-P isomerase catalysis indicating an enediolate intermediate. (Figure 19-2 shows a quantitative profile.)

the relative rates for breakdown of the enediolate; in the following diagram, we use the designations of enzyme species from the scheme given earlier.

$$
\begin{array}{ccccc}
\text{V} & \xleftarrow{9} & \text{I} & \xrightarrow{3} & \text{III} \\
 & & \Big\updownarrow 15{,}000\ \text{sec}^{-1} & & \\
\text{VI} & \xleftarrow{1} & \text{II} & \xrightarrow{3} & \text{IV}
\end{array}
$$

The experiments described thus far concentrated on the fate of solvent hydrogens that end up in substrate or product. One can also monitor the behavior of substrate hydrogens. For example, if 1R-[^{2}H]-DHAP (with deuterium in the transferable position) is incubated with isomerase and trapping enzyme, under initial-velocity conditions a V_{max} kinetic isotope effect of 2.9 exists at low (1.5%) amounts of conversion. One would guess that this implies a slow formation of the enediolate from DHAP. This step may be partially rate-determining, along with release of G-3-P (as discussed earlier). In the other direction, D-2-[^{2}H]- and D-2-[^{1}H]-G-3-P show the same V_{max} rates; there is no V_{max} kinetic isotope effect in the G-3-P→DHAP direction. This is consistent with the idea that, once the E—COO^{2}H · · · Enediolate has formed, it can exchange to E—COOH · · · Enediolate before slow conversion (by protonation at C-1) to DHAP.

With experiments such as those outlined, Knowles and colleagues constructed a Gibbs free-energy profile for the energetics of triose-P isomerase catalysis (Fig. 19-2). They chose as standard state concentrations of 40 μM for DHAP and G-3-P; these are the estimated concentrations in vivo. Transition state 4 is the highest barrier. The rate constant for conversion of E+G-3-P⇄E · G-3-P is

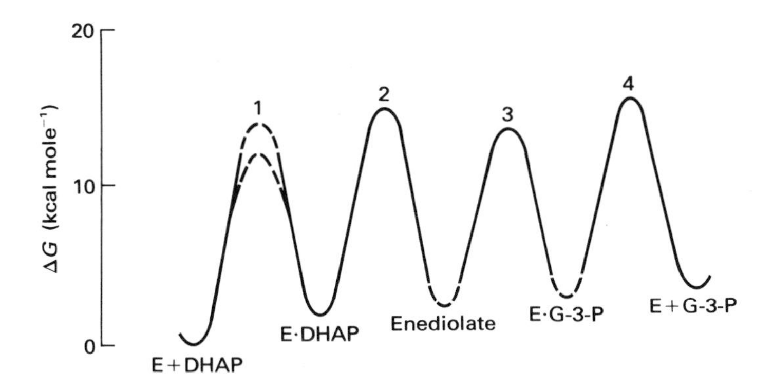

Figure 19-2
Free-energy profile for triose-P isomerase catalysis. (Reprinted with permission from Albery and Knowles, *Biochemistry* 16, 1977. Copyright © by the American Chemical Society.)

$3 \times 10^8\ \text{M}^{-1}\ \text{sec}^{-1}$, a value at the upper end of association rates for proteins and small ligands (Hammes and Schimmel, 1970) and close to or at the diffusion-controlled limit for reaction. This implies the interesting point (hinted earlier) that a physical step, not a chemical step, is mostly rate-determining and, because this physical step is essentially diffusion-limited, the enzyme cannot become any more efficient. An increase in the rate of any of the chemical steps could have no effect on turnover numbers in either direction. The enzyme "has reached the end of its evolutionary development" (Albery and Knowles, 1976). We have seen examples of this situation previously. NADH dissociation is the slow step for several dehydrogenases (Chapter 10), and it may be that some enzymes have evolved to maximal catalytic efficiency. One measure of catalytic efficiency is the ratio of k_{cat}/K_m for enzymes interacting with substrates. Recall from Chapters 2 and 3 that (at low substrate concentrations) this ratio is effectively the apparent second-order rate constant for E · S complex formation. A number of enzymes have k_{cat}/K_m ratios in the region of $10^8\ \text{M}^{-1}\ \text{sec}^{-1}$ (Chapter 2). Albery and Knowles (1976) point out that, if physiological concentrations of substrates are such that $[S] \leqslant K_m$, then these "enzymes may be under diffusion control." Physical adsorption and disorption, not the rate of chemical changes, control V_{max}.

In an effort to identify the catalytic base at the active site of triose-P isomerase, a number of affinity-labeling reagents have been utilized, either haloacetol phosphates or the epoxide glycidol-P.

$$\text{X}-\text{H}_2\text{C}-\text{C}(=\text{O})-\text{CH}_2-\text{OPO}_3^{2\ominus}$$

haloacetol-P
(X = Cl, Br)

$$\text{(epoxide, O)}-\text{CH}_2-\text{OPO}_3^{2\ominus}$$

glycidol-P

Hartman found that chloroacetol-P inactivates the enzyme and yields a single peptide with the inactivator fixed in ester linkage to a γ-carboxyl group of a glutamate (Hartman, 1971). The nucleophilicity of this $COO^{\ominus}$ moiety at the active site must be greatly enhanced, because no displacement of halide from haloacetol-P occurs with 1,000-fold higher concentration of free glutamate. This enhancement may facilitate catalytic efficiency of isomerization. Knowles and colleagues, using bromohydroxyacetone-P and chicken-muscle enzyme, isolated a peptide where the C-3 inactivator chain is in ether linkage to a tyrosine residue. The contradiction was resolved when they inactivated and immediately reduced the C-2 carbonyl of the acetol-P with BH_4. Now the inactivator indeed proved to be in isolated ester linkage to the γ-carboxyl of Glu^{165}. They suggested that initial addition is by the glutamate γ-carboxyl but, if the C-2 carbonyl is unreduced, migration to a neighboring phenolic hydroxyl of tyrosine can occur through S_N2 displacement of P_i by the phenolate moiety.

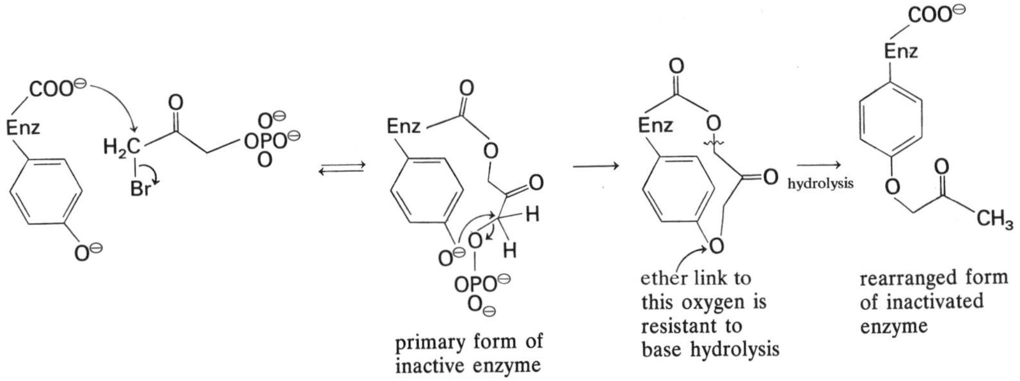

This glutamic acid (Glu[165]) is at the end of a pocket (probably the active site) as determined in the 2.5 Å X-ray map (Banner et al., 1975). The high-resolution map of the substrate–enzyme complex is in progress.

Finally, the experiments of O'Connell and Rose (1973) show that both the D-isomer and the L-isomer of glycidol-P cause covalent labeling (the L-isomer inactivates 10-fold faster) of the glutamyl residue. This Glu[165] γ-$COO^{\ominus}$ group is an attractive candidate for the active-site base.

This concludes our discussion of 1,2-proton shifts and will serve as basis for comparison to the 1,3-proton shifts discussed next.

19.C 1,3-PROTON SHIFTS: ALLYLIC ISOMERIZATIONS

There are five well-characterized enzymes that catalyze allylic isomerizations on their unsaturated substrates (Rose, 1970, 1972).

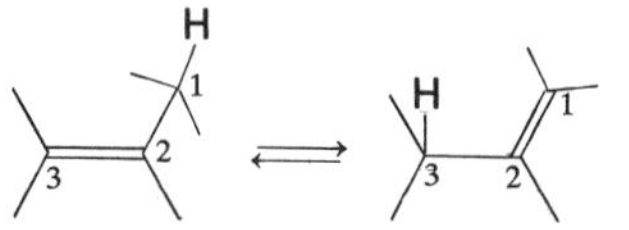

The pyridoxal-P–dependent transaminases provide a sixth and separate example—distinct in that the isomerization is azaallylic:

In their recent review of enzyme stereochemistry, Rose and Hanson (1975) summarize stereochemical and intramolecular-proton-transfer data for four of the allylic isomerases.

19.C.1 Δ^4-3-Ketosteroid Isomerase

The Δ^4-3-ketosteroid isomerase from *Pseudomonas testosteroni* is available in homogeneous form (Talalay and Benson, 1972); it utilizes a variety of Δ^4- and Δ^5-3-ketosteroids. Intramolecular transfer exceeds 95%. The following is a typical reaction. (In steroid structures, methyl groups attached to the rings are indicated simply by the carbon–carbon bonds.)

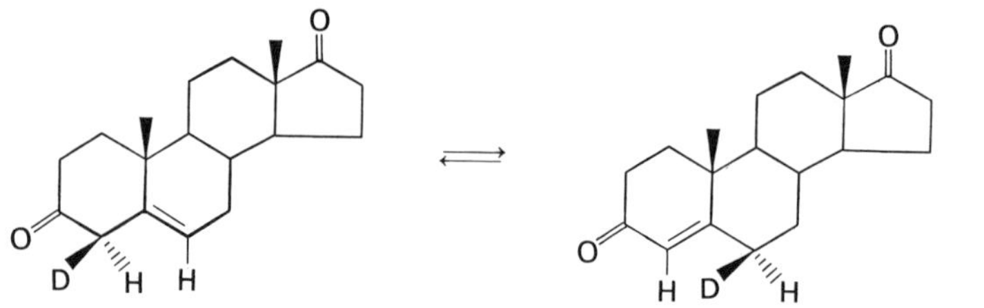

4-β-[^{2}H]-Δ^5-androstene-3,17-dione ⇌ 6-β-[^{2}H]-Δ^4-androstene-3,17-dione

19.C.2 Aconitate Isomerase

The aconitate isomerase from *Pseudomonas* interconverts the *cis* and *trans* isomers of the monoolefinic tricarboxylate aconitate. Klinman and Rose (1971) have carefully established that this is an allylic isomerization. There is 4% intramolecular 1,3-tritium transfer in D_2O from substrate tritiated in one transferable hydrogen locus.

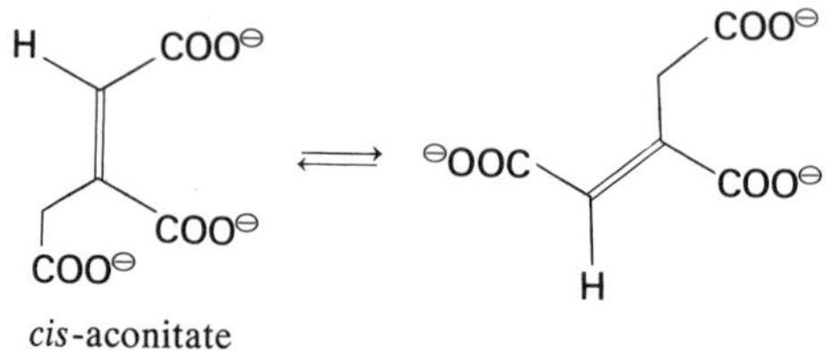

cis-aconitate ⇌

19.C.3 Vinylacetyl-CoA Isomerase

Vinylacetyl-CoA isomerase from ox liver converts β-methylvinylacetyl-CoA to β-methylcrotonyl-CoA. In ^3H_2O, no tritium is incorporated during isomerization again supporting a hypothesis of intramolecular proton transfer.

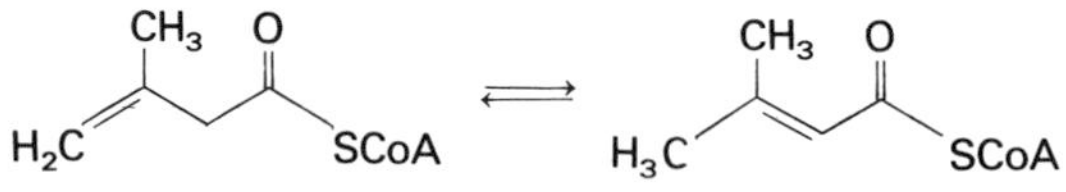

$H_2C{=}C(CH_3)CH_2C(O)SCoA \rightleftharpoons H_3C{-}C(CH_3){=}CHC(O)SCoA$

19.C.4 Mechanistic Studies

With all three of the enzymes just described, some intramolecular 1,3-proton transfer occurs. Further, the stereochemistry has been elucidated for each reaction, and the transfer is suprafacial in each instance. Rose (1970, 1972) has pointed out that the existence of the intramolecular hydrogen transfer is consistent with an allylic carbanion, but not with an allylic carbonium ion. Initial abstraction of the allylic hydrogen as a proton generates the stabilized carbanion, which can reprotonate with some internal transfer at C-1 to regenerate starting material, or at C-3 to produce labeled rearranged product.

B: *H $\underset{\longleftarrow}{\xrightarrow{-*H^{\oplus}}}$ B*H⊕ ⊖ $\rightleftarrows$ B: *H

The carbonium-ion alternative involves initial protonation of the double bond to yield the indicated carbocation.

*H H $\underset{\longleftarrow}{\xrightarrow{+H^{\oplus}}}$ H ⊕ *H B: H $\underset{\longleftarrow}{\xrightarrow{-*H^{\oplus}}}$ B*H⊕ H H

The carbocation species can lose back that proton just added to C-1 and produce starting material, or it can lose the labeled hydrogen at C-3 and yield allylically rearranged product. No internal 1,3-hydrogen transfer to rearranged product is possible with this mechanism. It is thus ruled out for the three allylic isomerases discussed thus far.

Given the allylic carbanion mechanism and the suprafacial, internal 1,3-proton transfer, the allylic transformation of the Δ^5-3-ketosteroid into conjugation is represented as follows (Talalay and Benson, 1972).

:B—Enz *H CH_3 O H H $\rightleftarrows$ ⊕H*B—Enz H_3C O H ⊖ H $\rightleftarrows$:B—Enz CH_3 O H *H H

Δ^5-3-keto-4-β-[*H]-steroid Δ^4-3-keto-6-β-[*H]-steroid

The ketosteroid isomerase is an impressively efficient catalyst, showing a V_{max} with the Δ^5-3,17-diketosteroid of 2.8×10^5 catalytic events per sec. Experiments designed to elucidate the single active-site base involved photoinactivation with specific loss of one histidine residue and chemical inactivation with diethylpyrocarbonate, which also modifies the same residue. It is likely that the monoprotonic base is the imidazole ring of this His residue. Given its normal pK_a value near 7.0, the maximal expected rate of proton transfer from imidazolium ion to solvent water is about 10^3 sec^{-1}, as we have noted earlier.

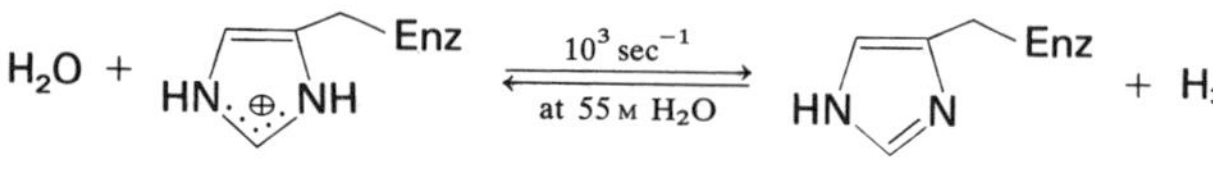

H_2O + HN⊕NH (imidazolium)–Enz ⇌ (10^3 sec^{-1}; at 55 M H_2O) HN–N (imidazole)–Enz + $H_3O^{\oplus}$

If this is indeed the catalytic base, then the complete internal hydrogen transfer is comprehensible. EB*H cannot dissociate fast enough to wash out the itinerant proton to solvent before it is transferred back to the allylic carbanion. The ratio $(2.8 \times 10^5)/(1 \times 10^3)$ suggests that dissociation would not occur more than once out of 280 turnovers. The rate-limiting value of 10^3 sec^{-1} is not relevant if the transferred proton cycles back and forth to regenerate the original acid or base. The rate of proton transfer within a tight reaction complex could in principle approach 10^{13} sec^{-1}. Rose and Hanson (1975) have termed this "intrinsic proton recycling," and they suggest that it is a selection favoring high catalytic capacity in enzymatic reactions.

Evidence supporting complete internal *proton* transfer (rather than a 1,3-internal *hydride* shift) in this allylic transformation is provided by studies with a competitive inhibitor, 19-nortestosterone, which lacks one of the substrate's angular methyl groups and is reduced at C-17. The molecule binds but is not isomerized. Binding produces a V_{max} shift of the bound steroid's spectrum (Fig. 19-3) consistent with enolate formation.

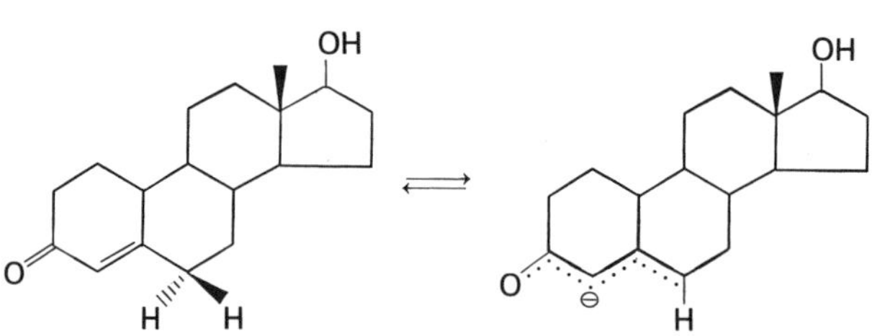

19-nortestosterone ⇌ enzyme-bound enolate

The enolate's existence is confirmed by ^3H incorporation from ^3H_2O into the recovered nortestosterone. This result supports enolization with the productive substrate.

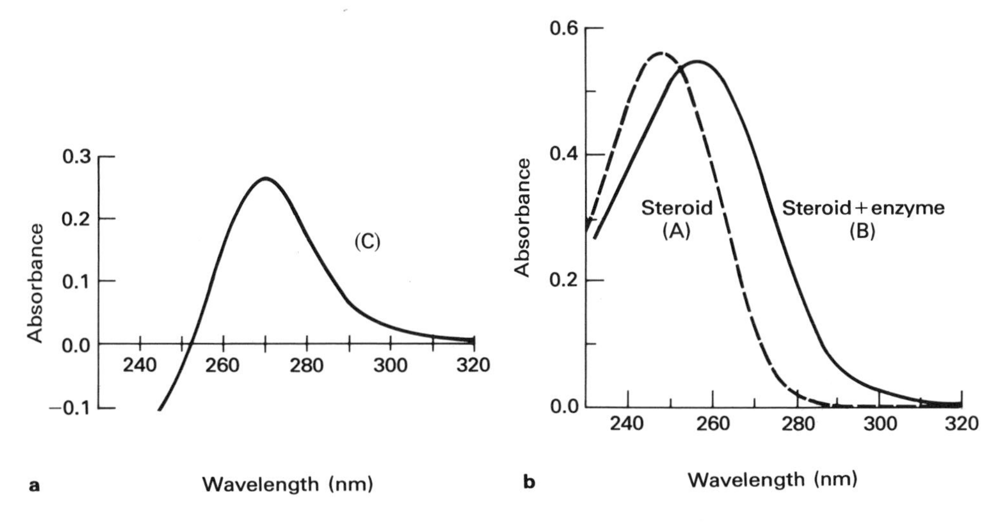

Figure 19-3
(**a**) Spectrum of bound enolate. (**b**) Ultraviolet absorption spectra showing the combination of 19-nortestosterone with crystalline Δ^5-3-ketosteroid isomerase from *P. testosteroni.* Curve A: 19-nortestosterone (33.0 μM) measured against a buffer blank. Curve B: a mixture of 19-nortestosterone (33.0 μM) and crystalline isomerase (18.3 μM) measured against a buffer blank containing isomerase (18.3 μM). Curve C: the calculated difference spectrum between B and A. All cuvettes contained 0.3 μmole of potassium phosphate, pH 7.0, and 0.01 ml of methanol in a final volume of 0.31 ml. (From Talalay and Benson, 1972, p. 608)

19.C.5 Isopentenyl-Pyrophosphate Isomerase

A fourth example of an enzymatic allylic isomerization is provided by isopentenyl-PP isomerase, which interconverts isopentenyl pyrophosphate (whose formation we discussed in Chapter 17) and its allylic isomer dimethylallyl pyrophosphate.

isopentenyl-PP $\rightleftharpoons$ dimethylallyl-PP

Subsequent biosynthesis of polyisoprenoid compounds involves the head-to-tail condensation of these two C_5 units, producing the C_{10} allylic product geranyl-PP (as discussed in more detail in Chapter 26).

$$\text{isopentenyl-PP} + \text{dimethylallyl-PP} \xrightarrow[\text{transferase}]{\text{prenyl}} \text{geranyl-PP} + PP_i$$

This C_{10} unit can add another isopentenyl-PP to generate the C_{15} allylic pyrophosphate farnesol-PP. Two molecules of farnesol-PP condense to the C_{30} acyclic polyene squalene, which is the biological precursor of steroids.

Cornforth and associates have shown that the stereochemistry of this allylic isomerization, unlike the three discussed previously, requires an antarafacial rearrangement: H removed from one face at C-1 and added to the opposite face at C-3 (Cornforth, 1973; Cornforth et al., 1972). Further, no intramolecular hydrogen transfer occurs. Both results mitigate against a single-base mechanism. Starting with dimethylallyl-PP, tritium is incorporated into the isopentenyl-PP product but not into the recovered starting material; similar results are obtained in the opposite direction. These results are consistent with the carbonium-ion hypothesis noted in ¶19.C.4, but also with a two-base carbanion mechanism with exchange restrictions (perhaps less likely).

19.C.6 β-Hydroxydecanoyl-Thioester Dehydrase

Our fifth example of an allylic isomerization is the reaction catalyzed by the bacterial enzyme β-hydroxydecanoyl-thioester dehydrase. In some bacteria, such as *E. coli*, the biosynthesis of long-chain olefinic fatty acids does not utilize O_2 in the desaturation step. This is in contrast to the eucaryotic acylthioester-desaturase systems that require O_2, cyctochrome-b_5 reductase (flavoprotein), and nonheme-iron components. (We discussed the animal stearyl-CoA desaturase in ¶15.A.5.)

The oxygen-independent bacterial route involves the dehydration of β-hydroxyacylthioesters preferentially at the C_{10} stage to β,γ-olefins that are not reduced by the normal complement of flavin-dependent acyl-CoA dehydrogenases in the cell (these use only α,β-enoylthioesters). Thus, the β,γ-monoolefin at the C_{10} level is elongated by three or four subsequent cycles of two-carbon addition in the normal fatty-acid synthesis process to give the C_{16} and C_{18} molecules that predominate as the end products released from fatty-acid synthetase complexes. The C_{16} molecules are $C_{16}\Delta^9$ (palmitoleate) and the C_{18} molecules are $C_{18}\Delta^{11}$ (oleate).

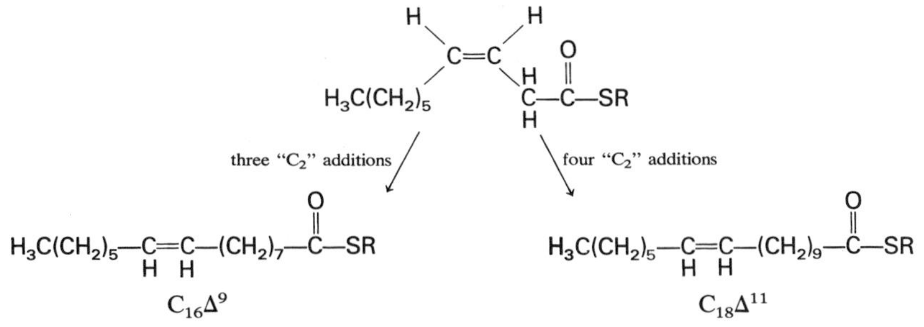

$$H_3C(CH_2)_5-CH=CH-CH_2-\overset{O}{\overset{\|}{C}}-SR \quad (cis)$$

three "C_2" additions ↙ ↘ four "C_2" additions

$$H_3C(CH_2)_5-\underset{H}{C}=\underset{H}{C}-(CH_2)_7-\overset{O}{\overset{\|}{C}}-SR \qquad H_3C(CH_2)_5-\underset{H}{C}=\underset{H}{C}-(CH_2)_9-\overset{O}{\overset{\|}{C}}-SR$$

$C_{16}\Delta^9$ $\qquad$ $C_{18}\Delta^{11}$

The β-hydroxydecanoyl-thioester dehydrase is the enzyme responsible for the physiological production of the nonreducible β,γ-enolates. Bloch (1969, 1972) and associates have purified the *E. coli* enzyme and characterized it as a dimer of 36,000 mol wt.

As a dehydrase, this enzyme catalyzes an elimination of the elements of water. In fact, it gives two dehydrated products: the α,β-decenoyl thioester and the β,γ-decenoyl thioester.

$$R-\underset{H}{\overset{H}{C}}-\underset{H}{\overset{OH}{C}}-\underset{H}{\overset{H}{C}}-\overset{O}{\overset{\|}{C}}-SR' \rightleftharpoons R-\underset{H}{\overset{H}{C}}-\underset{H}{C}=\overset{H}{C}-\overset{O}{\overset{\|}{C}}-SR' \text{ and } R-\underset{H}{C}=\underset{H}{C}-\underset{H}{\overset{H}{C}}-\overset{O}{\overset{\|}{C}}-SR'$$

$\gamma \ \beta \ \alpha$ $\qquad$ $\gamma \ \beta \ \alpha$ $\qquad$ $\gamma \ \beta \ \alpha$

β-OH-decanoyl thioester (70%) $\qquad$ *trans*-α,β-decenoyl thioester (27%) $\qquad$ *cis*-β,γ-decenoyl thioester (3%)

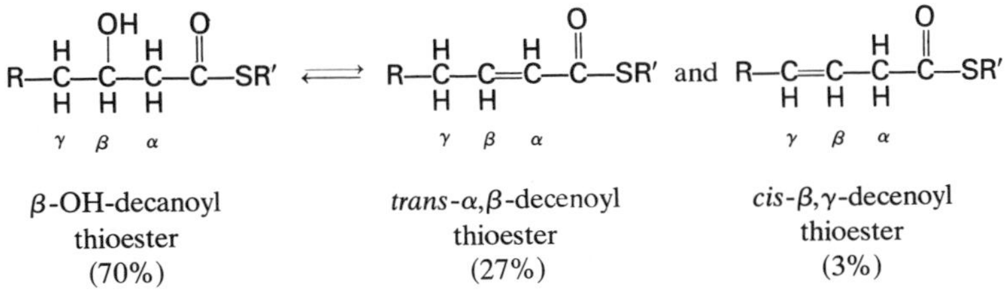

The proportions of the species at equilibrium are given above; note that the *cis*-β,γ-decenoyl thioester is the physiologically significant product, although it is the least abundant species at equilibrium.

In addition to catalyzing the elimination of water, however, this dehydrase also catalyzes an allylic rearrangement, a direct conversion of the α,β- to the β,γ-decenoyl thioester. In fact, it has been established that the β,γ product is not formed by direct dehydration of the β-OH substrate. It is formed *only* from allylic rearrangement of the initial dehydration product, the *trans*-α,β-decenoyl thioester. This is in keeping with the chemical fact that the α-H is activated for abstraction by its position adjacent to the thioester carbonyl. Its abstraction will yield the α,β-decenoyl thioester initially. The γ-H is chemically nonactivated. Dehydration directly to the β,γ-olefin would require a much higher activation energy.

Some evidence for this scheme was demonstrated by using α-dideutero-β-hydroxydecanoyl thioester as substrate. The β,γ-decenoyl thioester formed had only one deuterium, located at the α-C, indicating removal of one of the α-protons in forming the β,γ product.

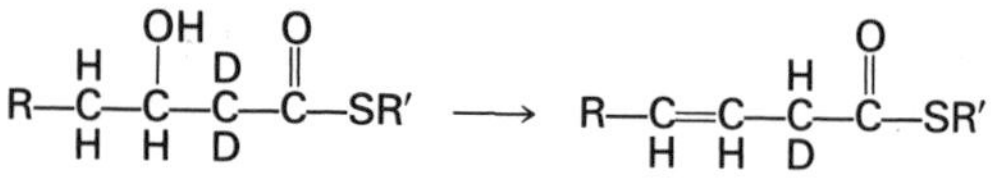

$$\text{R—}\overset{\text{H}}{\underset{\text{H}}{\text{C}}}\text{—}\overset{\text{OH}}{\underset{\text{H}}{\text{C}}}\text{—}\overset{\text{D}}{\underset{\text{D}}{\text{C}}}\text{—}\overset{\text{O}}{\overset{\|}{\text{C}}}\text{—SR}' \longrightarrow \text{R—}\underset{\text{H}}{\text{C}}\text{=}\underset{\text{H}}{\text{C}}\text{—}\overset{\text{H}}{\underset{\text{D}}{\text{C}}}\text{—}\overset{\text{O}}{\overset{\|}{\text{C}}}\text{—SR}'$$

α-dideutero-β-hydroxydecanoyl thioester — α-monodeutero-*cis*-β,γ-decenoyl thioester

Also, with α-dideutero-β-hydroxy substrate, there is a kinetic isotope effect in forming β,γ product, confirming obligate rate-determining removal of the α-H by the enzyme.

Lastly, kinetic analysis of product formation shows that α,β-product forms much faster than does β,γ- and so qualifies kinetically as an intermediate. Thus we arrive at the following scheme:

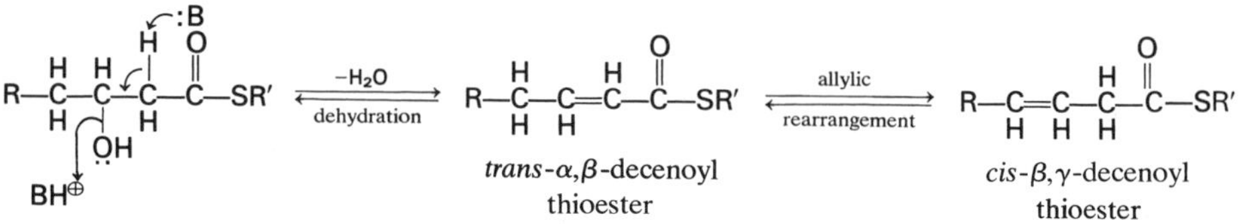

$$\text{R—}\overset{\text{H}}{\underset{\text{H}}{\text{C}}}\text{—}\overset{\text{H}}{\underset{\ddot{\text{O}}\text{H}}{\text{C}}}\text{—}\overset{\text{H}\,(\curvearrowleft :\text{B})}{\underset{\text{H}}{\text{C}}}\text{—}\overset{\text{O}}{\overset{\|}{\text{C}}}\text{—SR}' \quad (\text{BH}^{\oplus}) \underset{\text{dehydration}}{\overset{-\text{H}_2\text{O}}{\rightleftharpoons}} \text{R—}\overset{\text{H}}{\underset{\text{H}}{\text{C}}}\text{—}\underset{\text{H}}{\text{C}}\text{=}\overset{\text{H}}{\text{C}}\text{—}\overset{\text{O}}{\overset{\|}{\text{C}}}\text{—SR}' \underset{\text{rearrangement}}{\overset{\text{allylic}}{\rightleftharpoons}} \text{R—}\underset{\text{H}}{\text{C}}\text{=}\underset{\text{H}}{\text{C}}\text{—}\overset{\text{H}}{\underset{\text{H}}{\text{C}}}\text{—}\overset{\text{O}}{\overset{\|}{\text{C}}}\text{—SR}'$$

trans-α,β-decenoyl thioester — *cis*-β,γ-decenoyl thioester

Apparently, no large amount of internal transfer occurs in this 1,3-allylic rearrangement, and the stereochemistry has not been determined (it could be studied carefully with α,β-decenoyl-γ-[^{3}H] material). Whereas, as indicated above, use of α-D_2-β-OH-decanoyl thioester gives $k_H/k_D = 2.2$, the use of γ-D_2-β-OH-decanoyl thioester gives $k_H/k_D = 1.0$. Removal of α-H is slow; removal of γ-H is fast.

In the conversion of β-OH-decanoyl thioester to *cis*-β,γ-decenoyl thioester at early time points, addition of exogenous [^{14}C]-α,β-decenoyl thioester results in essentially no labeling of the β,γ-olefinic thioester produced. Therefore we conclude that the α,β-thioester formed as an intermediate in the dehydration–allylic-isomerization sequence remains enzyme bound in those turnovers where β,γ-olefinic product is generated.

19.C.7 Inactivation Studies with the Dehydrase

Bloch (1969, 1972) and colleagues tested a number of substances as inhibitors of β-hydroxydecanoyl-thioester dehydrase. They found that a β,γ-acetylenic thioester

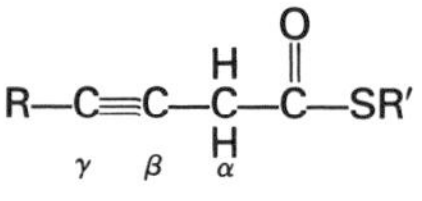

$$\text{R—}\underset{\gamma}{\text{C}}\equiv\underset{\beta}{\text{C}}\text{—}\overset{\text{H}}{\underset{\underset{\alpha}{\text{H}}}{\text{C}}}\text{—}\overset{\text{O}}{\overset{\|}{\text{C}}}\text{—SR}'$$

3-decynoyl thioester

is a potent irreversible inactivator leading to covalent attachment of the inactivator at an active-site histidine residue. The thioester moiety is required, and the positioning of the triple bond is critical: the 2-decynoyl (α,β-acetylenic) and the 4-decynoyl (γ,δ-acetylenic) thioesters are not inactivating agents.

A kinetic isotope effect of 2.6 in the rate of inactivation was observed with α-dideutero-3-decynoyl thioester, suggesting removal of the α-H. Thus, the acetylenic thioester functions as a substrate for the enzyme, apparently resembling the β,γ-decenoyl-thioester substrate sufficiently to undergo enzyme-catalyzed α-H abstraction. This would yield a β,γ-acetylenic-α-carbanion capable of rapid propargylic rearrangement to the allenic thioester.

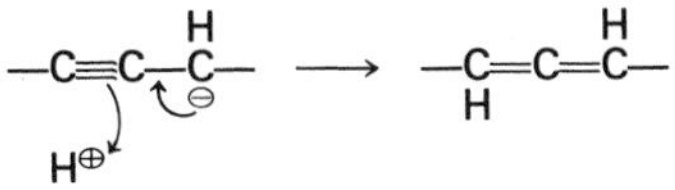

The observed reaction would be the following:

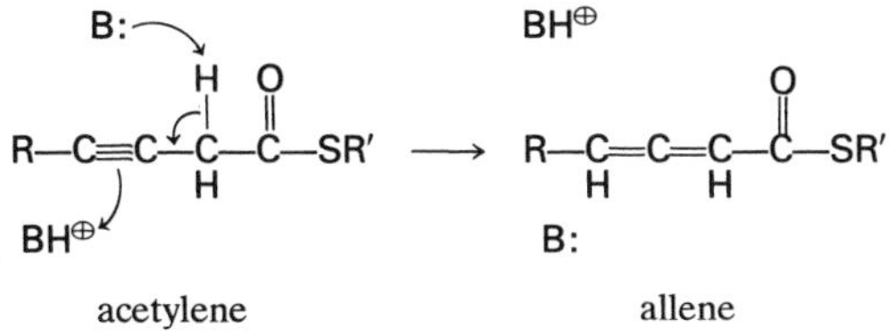

acetylene allene

The allenic thioester is the likely inactivating species, because conjugated allenes are reactive electrophiles, readily undergoing Michael-type addition by nucleophiles at the allene central carbon (Morisaki and Bloch, 1972).

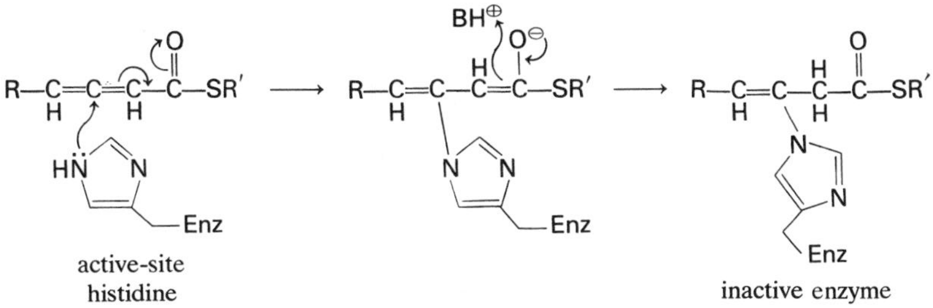

active-site histidine

inactive enzyme

In support of the idea that the dehydrase catalyzes isomerization of the acetylene to the allene, which then causes inactivation, Bloch and colleagues have found that synthetic 2,3-decadienyl thioester (the allene) inactivates at a rate 10 to 15 times faster than the acetylenic thioester, and that it hits the same active-site histidine.

Because there is restricted rotation in allenes (which leads to geometric isomers), the asymmetric allenes can be resolved. In this instance, only the dextrorotatory isomer (unknown absolute configuration) inactivates the enzyme (Morisaki and Bloch, 1972). This was the first known example of an acetylenic substrate analogue with the acetylene adjacent to a carbon where the enzyme would catalyze an α-carbanion that could rearrange to an inactivator. These have been termed *suicide substrates* because they are chemically unreactive until bound at the target-enzyme active site, where they undergo at least one step in the catalytic process until a latently reactive functional group is unmasked (Maycock and Abeles, 1976; C. Walsh, 1977). Specificity of inactivation of the desired target enzyme is ensured because only that enzyme can activate the latent "killer." The 3-decynoyl thioester has indeed been used to select bacteria that are lacking a functional dehydrase. These bacteria cannot make monounsaturated fatty acids and require olefinic supplements. These mutants have then provided a means of exogenously controlling the composition of unsaturated acyl residues in bacterial membranes, which in turn has permitted interesting conclusions about the role of fluidity (or its lack) in biological membrane function.

Subsequently, an acetylenic secosteroid analogue of the Δ^4-ketosteroid-isomerase substrates has been prepared and tested with that isomerase. It undergoes extremely rapid enzyme-catalyzed conversion to the allenic ketone. The enzyme has thereby generated a reactive electrophile that can then covalently modify the enzyme in an apparently active-site–directed process (Batzold and Robinson, 1975; Covey and Robinson, 1976). The identity of the enzyme nucleophile has not yet been established.

isomerization

:B—Enz

A (80%) and B (20%) (1)

inactivation and ketonization

X—Enz

alkylated enzyme (2)

All the acetylenic keto-secosteroid appears to have been isomerized prior to inactivation, a reflection of the high turnover number of the isomerase (8×10^4 sec^{-1} with normal substrate). An 8:2 mixture of the allenic diastereomers A and B is observed. Both A and B can then rebind to the active site and cause covalent inactivation. This sequence means that many reactive product molecules (the allenic ketone) would be generated in vivo before inactivation of this enzyme.

19.C.8 Azaallylic isomerases

The sixth and last enzymatic allylic isomerization we shall note here involves a nitrogen atom in place of C-2 and is thus an azaallylic system. This system thus contains an imine linkage, and the isomerization represents a 1,3-tautomeric shift of an imine to an alternative imine.

$$>C{=}\overset{H}{\underset{\oplus}{N}}{-}\overset{H}{C}< \;\rightleftharpoons\; >\overset{H}{C}{-}\overset{H}{\underset{\oplus}{N}}{=}C<$$

Because imines are in reversible equilibrium in aqueous medium with the amine and carbonyl compounds, the azaallylic isomerization results in reduction of the original carbonyl component to the amine and in oxidation of the amine to a carbonyl derivative. As we shall see in Chapter 24, this is precisely the general half-reaction catalyzed by pyridoxal-P–dependent transaminases, and we shall continue discussion of it there.

$$R^1R^2C{=}O + H_2N{-}CH(R^3)(R^4) \underset{-H_2O}{\rightleftharpoons} R^1R^2C{=}\overset{H}{\underset{\oplus}{N}}{-}CH(R^3)(R^4)$$

$$\Big\updownarrow \text{ azaallylic transformation}$$

$$R^1R^2CH{-}NH_2 + O{=}C(R^3)(R^4) \underset{+H_2O}{\rightleftharpoons} R^1R^2CH{-}\overset{H}{\underset{\oplus}{N}}{=}C(R^3)(R^4)$$

For now it serves to note that, with the pyridoxamine-pyruvate transaminase, Dunathan, Ayling, and Snell (1968) found 4% 1,3-intramolecular ^2H-transfer in H_2O and 50% ^1H transfer in D_2O. The stereochemistry demands a suprafacial transfer, and it is likely that the mechanism is analogous to those of the three allylic isomerases noted in ¶¶19.C.1–19.C.4. For example:

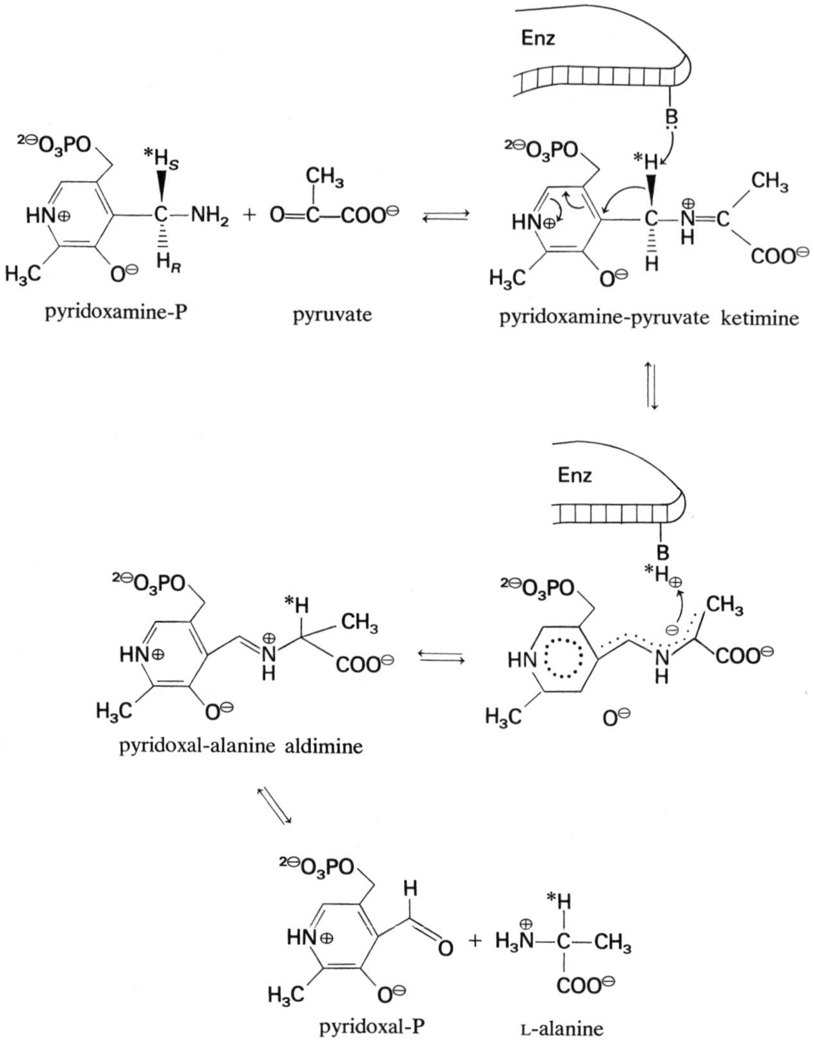

pyridoxamine-P pyruvate pyridoxamine-pyruvate ketimine

pyridoxal-alanine aldimine

pyridoxal-P L-alanine

The net conversion from the azaallylic isomerization is reductive transamination of pyruvate to L-alanine and concomitant oxidative transamination of pyridoxamine phosphate to pyridoxal phosphate. The azaallylic isomerization of pyridoxal-P–mediated reactions is accompanied by redox changes in the partners. Dunathan and Voet (1974) have recently summarized evidence that seven pyridoxal-P enzymes that can carry out transaminations of amino acids (either during normal catalytic half-reactions or, abortively, to yield inactive pyridoxamine-P–enzyme forms) all show the same absolute stereochemistry of protonation at the methine carbon of the pyridoxal-amino-acid aldimine in yielding the methylene group of the pyridoxamine–ketimine linkage (see also Hashimoto et al., 1973). In all cases, the proton is delivered from the *si* face of the enzyme-bound planar intermediate, ending up in the pro*S* position of the methylene group of pyridoxamine phosphate, as shown in the example just given. Perhaps one active-site geometry evolved first and has been maintained in all the vitamin-B_6–utilizing enzymes.

This concludes our analysis of 1,1-, 1,2-, and 1,3-hydrogen-shift reactions for now. We shall return to the coenzyme-B_{12}–dependent rearrangements at the end of Chapter 20. Now, however, we shall look briefly at enzymes that carry out apparent isomerization of phosphorylated sugars.

19.D PHOSPHOSUGAR MUTASES

The existence of enzymes catalyzing apparent intramolecular phosphoryl transfer between C-1 and C-6 of glucose or between C-2 and C-3 of glycerate was established about 40 years ago. Physiologically, the phosphoglucose mutases provide an enzymatic link between glucose-6-P (the phosphorylated form in enzymatic breakdown of glucose both anaerobically and aerobically) and glucose-1-P generated by the action of various glycan phosphorylases (Chapter 9). The phosphoglycerate mutases interconvert 3-phosphoglcerate (produced by action of phosphoglycerate kinase) and 2-phosphoglycerate (a substrate for enolase).

19.D.1 Phosphoglucomutase

Phosphoglucomutase from various sources appears to be a large single polypeptide chain of about 60,000 mol wt, with about 600 amino-acid residues (Ray and Peck, 1972). Depending on the source, the homogeneous enzyme is obtained with stoichiometric amounts of phosphate bound covalently in monoester linkage (phosphoenzyme, obtained from rabbit muscle and yeast) or as the dephosphoenzyme (from *E. coli*, *M. lysodekticus*, *B. cereus*). The resting form of phosphoglucomutase is a phosphoryl-enzyme derivative! Leloir and associates discovered that the yeast and rabbit-muscle enzymes require catalytic amounts of glucose-1,6-diphosphate to produce maximal turnover numbers, which approximate $10^3\ sec^{-1}$. The enzyme recognizes the α-anomer, but not the β-anomer, of glucose-1-P and glucose-6-P. Further, only the α-anomer of glucose-1,6-diphosphate is an activator.

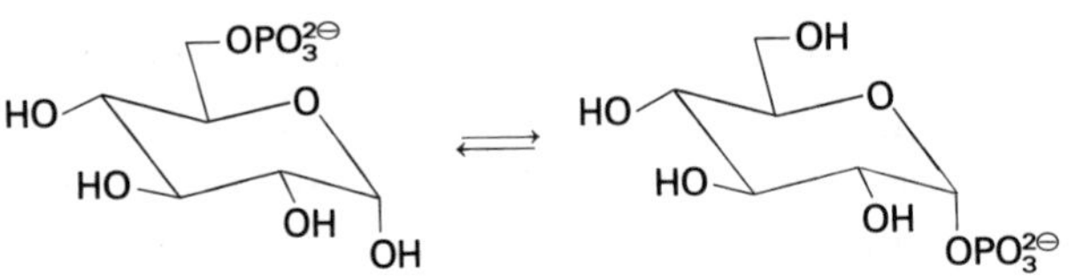

α-glucose-6-P $\rightleftharpoons$ α-glucose-1-P

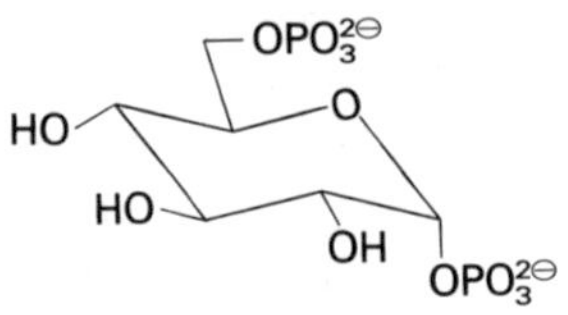

α-glucose-1,6-diphosphate (required in catalytic amounts for high turnover)

Isotope-exchange experiments corroborated that the hexose-diphosphate activator can participate in catalysis: exchange of both the glucosyl and phosphoryl moieties between substrates and the sugar-diphosphate activator was observed. The mechanism proposed for catalysis involves the recycling of enzyme between the phosphoenzyme form at the beginning and end of each catalytic cycle and dephosphoenzyme at an intermediate stage where phosphoryl transfer from E—$PO_3^{2\ominus}$ to either substrate yields bound glucose-1,6-diphosphate and dephosphoenzyme.

$$\underset{\text{resting enzyme}}{E\text{—}^{32}PO_3^{2\ominus}} + \alpha\text{-glucose-1-}PO_3^{2\ominus} \rightleftharpoons E\cdots\text{Glucose-1,6-}(PO_3^{2\ominus})_2$$

$$E\cdots\text{Glucose-1,6-}(PO_3^{2\ominus})_2 \rightleftharpoons E\text{—}PO_3^{2\ominus} + \alpha\text{-Glucose-6-}^{32}PO_3^{2\ominus}$$

This scheme is in accord with the demonstration that isolated E—$PO_3^{2\ominus}$ reacts with either sugar monophosphate in a phosphoryl transfer, but not with hexose diphosphate. The dephosphoenzyme is, in complementarity, phosphorylated by added hexose diphosphate, but not by either monophosphate.

An obvious question, raised by the stimulatory effect of exogenously added glucose-1,6-diphosphate, is whether the E $\cdots$ glucose-1,6-$(PO_3^{2\ominus})_2$ dissociates freely during catalysis. Ray and associates showed that this dissociation occurs very infrequently with the rabbit-muscle enzyme. Incubation of resting enzyme, E—$^{32}PO_3^{2\ominus}$, with saturating α-glucose-1-$PO_3^{2\ominus}$ and excess glucose-1,6-$(PO_3^{2\ominus})_2$ produced α-glucose-6-$^{32}PO_3^{2\ominus}$, which contained all the radioactivity. If the E $\cdots$ glucose-1,6-[^{32}P]-$(PO_3^{2\ominus})_2$ species had dissociated at some stage before back phosphoryl transfer to the enzyme, it would have been replaced by the excess unlabeled glucose-1,6-$(PO_3^{2\ominus})_2$ from solution, and radioactive hexose diphosphate should have accumulated. Essentially none was detected, confirming that the intermediate stays bound. On the rare occasions when it does dissociate, the dephosphoenzyme is formed and is *catalytically inactive.* The dephosphoenzyme can be reactivated by exogenous hexose-1,6-diphosphate, which explains the stimulatory effect of this species on catalytic activity. The dissociation of the bound glucose-1,6-$(PO_3^{2\ominus})_2$ (step 3 in the following scheme) is presumably an abortive side reaction that the catalyst would like to avoid (if it had any feelings at all). In support of that notion, the K_{diss} value for E · · · glucose-1,6-$(PO_3^{2\ominus})_2$ is about 10^{-8} M, whereas K_m values for glucose-1-$PO_3^{2\ominus}$ and glucose-6-$PO_3^{2\ominus}$ are in the range around 10^{-5} M; the intermediate is bound 1,000 times more tightly than is either substrate. Isotope exchanges at chemical equilibrium confirm the expected exchange rates. The [^{14}C]-glucose-1-P ⇄ glucose-6-P exchange is twofold faster than the glucose-1-$^{32}PO_3^{2\ominus}$ ⇄ glucose-6-$PO_3^{2\ominus}$ exchange. (The reader can confirm in the following scheme that phosphoryl transfer from glucose-1-P to glucose-6-P requires two catalytic cycles.)

α-glucose-1-$PO_3^{2\ominus}$ + Enz—Ser—O—P(=O)(O$^\ominus$)—O$^\ominus$ + α-glucose-6-$PO_3^{2\ominus}$

① ② ③ slow ④ ⑤

Enz—Ser—OH + glucose-1,6-$(PO_3^{2\ominus})_2$ (inactive)

The mutase will catalyze phosphoryl transfer to C-1 or C-6 of other hexose monophosphates, although at much reduced rates. With fructose-6-P the turnover number ($k_{cat} \equiv V_{max}/[E_T]$) is only 10^{-6} times that for glucose-6-P. With the alternative substrate, the phosphoryl-transfer steps are now slower than dissociation of the bound hexose diphosphates from the enzyme, and the fructose-1,6-diphosphate intermediate dissociates freely and completely in each catalytic cycle. This generates ping-pong BiBi initial-velocity kinetics (in contrast to the UniUni kinetics seen with good substrates) and an *absolute dependence on added glucose-1,6-diphosphate for this catalysis.* (Fructose-1,6-diphosphate transfers the phosphoryl group at C-1 back to the enzyme only very slowly.)

$$\text{E—PO}_3^{2\ominus} + \text{Fructose-6-PO}_3^{2\ominus} \rightleftharpoons \text{E}(\text{PO}_3^{2\ominus})\cdots\text{Fructose-6-PO}_3^{2\ominus} \rightleftharpoons \text{E}\cdots\text{Fructose-1,6-(PO}_3^{2\ominus})_2 \rightleftharpoons \text{E} + \text{Fructose-1,6-(PO}_3^{2\ominus})_2 \quad (1)$$

$$\text{E} + \text{Glucose-1,6-(PO}_3^{2\ominus})_2 \rightleftharpoons \text{E}\cdots\text{Glucose-1,6-(PO}_3^{2\ominus})_2 \rightleftharpoons \text{E}(\text{PO}_3^{2\ominus})\cdots\text{Glucose-6-PO}_3^{2\ominus} \rightleftharpoons \text{E—PO}_3^{2\ominus} + \text{Glucose-6-PO}_3^{2\ominus} \quad (2)$$

The net reaction (sum of reactions 1 and 2) is the following:

$$\text{Fructose-6-PO}_3^{2\ominus} + \text{Glucose-1,6-(PO}_3^{2\ominus})_2 \rightleftharpoons \text{Fructose-1,6-(PO}_3^{2\ominus})_2 + \text{Glucose-6-PO}_3^{2\ominus} \quad (1 + 2)$$

Finally, even with the glucose-1-P and glucose-6-P substrates, with the phosphoglucomutases from *Bacillus cereus* and *Micrococcus lysodekticus* the glucose-1,6-diphosphate dissociates so rapidly that no E—$PO_3^{2\ominus}$ accumulates, and there is an absolute requirement for glucose-1,6-diphosphate, which then functions as a rephosphorylating substrate.

The chemical nature of the phosphoenzyme has been elucidated. The phosphate is in ester linkage to an active-site serine residue and is characteristically acid-stable and base-labile, in analogy with other phosphoenzymes such as alkaline phosphatase. Step 4 in the catalytic scheme shown earlier poses an interesting problem in catalysis. In step 2, the phosphate group of the active-site serine is close enough to the bound glucose-1-P for nucleophilic attack of the C-6 hydroxyl group of the sugar to effect the phosphoryl transfer. But in step 4 the serine—OH (presumably rendered a good nucleophile by general base catalysis must now be close, not to C-6, but to C-1 of the bound hexose diphosphate, so that it will be the phosphate at C-1 that undergoes phosphoryl transfer in regeneration of phosphoenzyme and production of the product isomer, glucose-6-P. There must be motion at the active site between steps 2 and 4, because the C-6 OH and the anomeric C-1 OH are on opposite sides of the plane of the pyranose ring. Whether the intermediate flips or whether a protein conformational change can move the active-site serine sufficiently is unknown.

The native rabbit-muscle Enz—Ser—$OPO_3^{2\ominus}$ will also carry out phosphoryl transfer to H_2O—in effect a phosphatase reaction that puts the enzyme into a dephospho (inactivated) form. This hydrolytic reaction proceeds at 3×10^{-10} times the rate of phosphoryl transfer to the C-6 hydroxyl of the substrate α-glucose-1-phosphate.

$$\text{E—Ser—OPO}_3^{2\ominus} + H_2O \xrightarrow{1} \text{E—Ser—OH} + \text{HOPO}_3^{2\ominus}$$

$$\text{E—Ser—OPO}_3^{2\ominus} + \alpha\text{-glucose-1-}OPO_3^{2\ominus} \xrightarrow{3 \times 10^{10}} \text{E—Ser—OH} + \text{glucose-1,6-bis}(OPO_3^{2\ominus})$$

Because H_2O is considerably smaller than the hexose phosphate, the specificity could not be due to steric occlusion and was used as an early example stimulating the idea of "induced fit" in enzymatic catalysis. This (3×10^{10})-fold "substrate-induced rate effect" corresponds to a 15 kcal/mole (63 kJ/mole) free-energy difference for productive phosphoryl transfer between hydroxyls of similar basicity (H_2O, C-6 of glucose). The free-energy difference must devolve ultimately from the increased binding interaction energy available from the nonreacting phosphoglucosyl moiety in the E $\cdots$ α-glucose-1-P complex compared to the $^{2\ominus}O_3PO$—E $\cdots$ H_2O complex. A careful study by Ray and colleagues suggests that, when the thermodynamic explanation is analyzed mechanistically, "neither enthalpic destabilization nor entropic immobilization of the reactant groups, nor increased binding interactions of the phosphoglucosyl moiety in the transition state, *when taken individually*, can account for a major fraction of the substrate-induced rate effect" (Ray et al., 1976; see also Ray and Long, 1976). They then analyzed individual factors that could additively produce the 10^{10} factor of specificity.

> The approach employed involves making a series of stepwise changes to convert the enzyme-water system into the enzyme · glucose-1-P system and evaluating the effect of each change on the rate of the corresponding PO_3-transfer process. Scheme I shows a possible sequence of this type; it suggests that an effect on the rate of the catalytic PO_3 transfer might be produced by (a) binding a phosphate ester group, (b) using an alcohol as an acceptor instead of water, (c) joining the phosphate group and the acceptor via a chemical bridge so that the transfer becomes an intramolecular process, and (d) adding the $(CHOH)_3$ portion of the glucose ring.
>
> SCHEME I. A Series of Stepwise Changes by Means of Which the Water Reaction Can Be Related to the Glucose-1-P Reaction. (The letter above each arrow serves to relate the indicated change to an increased efficiency of PO_3 transfer produced by that change [as noted above].)

$$E_P + H_2O \xrightarrow{a} E_P \cdot POCH_3 + H_2O \xrightarrow{b} E_P \cdot POCH_3 + CH_3OH \xrightarrow{c} E_P \cdot \begin{matrix} POCH_2O \\ | \\ HOCH_2CH_2 \end{matrix} \xrightarrow{d} E_p \cdot \text{Glc-1-P}$$

(Ray et al., 1976, p. 4011)

Experiments indicated that step a is worth 10^3-fold, step b 10^2-fold, step c (2.5×10^2)-fold, and step d (2×10^2)-fold. The total of (5×10^9)-fold indeed approximates the observed substrate-induced rate effect of (3×10^{10})-fold. Whether the rate accelerations for enzymatic reactions are generally due to a composite of such small effects or to an overwhelming contribution from a single factor should be the focus of research in the near future.

19.D.2 Phosphoglycerate mutase

The rabbit-muscle phosphoglycerate mutase using the C_3 acids as substrate is a dimer of 55,000 mol wt that Z. B. Rose (1970) has isolated as fully loaded phosphoenzyme. The reaction is mechanistically similar to the hexose transformation, and it is stimulated by the appropriate diphosphate, 2,3-diphosphoglycerate.

$\begin{matrix} & H & \\ H_2C & -C- & COO^{\ominus} \\ \vert & \vert & \\ {}^{2\ominus}O_3PO & OH & \end{matrix}$	$\rightleftarrows$	$\begin{matrix} & H & \\ H_2C & -C- & COO^{\ominus} \\ \vert & \vert & \\ HO & OPO_3^{2\ominus} & \end{matrix}$	$\begin{matrix} & H & \\ H_2C & -C- & COO^{\ominus} \\ \vert & \vert & \\ {}^{2\ominus}O_3PO & OPO_3^{2\ominus} & \end{matrix}$
3-phosphoglycerate		2-phosphoglycerate	2,3-diphosphoglycerate

Estimates of the leakage of bound 2,3-diphosphoglycerate molecules from the active site suggest that free, inactive dephosphoenzyme may form once in every 100 turnovers. The isolated phosphoenzyme is unexpectedly acid-labile, but base-stable. This puzzle was resolved by Rose's demonstration that the $E—PO_3^{2\ominus}$ intermediate is a phosphohistidine residue, in distinction to the phosphoserine in resting phosphoglucomutase.

In contrast to the phosphoglycerate mutases from animal and yeast cells, the wheat-germ phosphoglycerate mutase is *not* dependent on added 2,3-diphosphoglycerate for maintenance of activity. The mechanism of this cofactor-independent mutase is distinct: it catalyzes intramolecular phosphoryl transfer between C-2 and C-3 of the glycerate skeleton rather than intermolecular transfer. The intramolecularity was demonstrated by Gatehouse and Knowles (1977), who found no exchange of labeled phosphoryl group from mass-spectrometric analysis of products from isomerization of a mixture of 3-phospho-D-2-[^{2}H]-glycerate and 3-[^{18}O]-phospho-D-glycerate (i.e., there was no 2-[^{18}O]-2-[^{2}H]-2-phospho-D-glycerate). From studies with phosphorylated substrate analogues, Breathnach and Knowles (1977) suggested that the wheat-germ enzyme does however involve a phosphoryl-enzyme intermediate, now in similarity to the animal enzyme, from which the phosphoryl portion can be transferred back to either the 2- or the 3-hydroxyl group of bound D-glycerate before it ever dissociates.

19.E ENZYMATIC *cis–trans* ISOMERIZATION

Seltzer (1972) has classified *cis–trans* isomerases in two categories: the first involves isomerization without double-bond migration in the product, and the second category shows positional migration accompanying the geometrical isomerization.

The enzymes so far discovered in the first category require sulfhydryl groups, on the enzyme or in a specific cofactor, for activity; they isomerize maleyl groups in substrates to fumaryl moieties in products. An isomerase isomerizing free maleate to fumarate (thermodynamically more stable) has been purified from *Pseudomonas* strains grown on maleate (Seltzer, 1972). The fumarate then enters the normal metabolism carried out by tricarboxylate-cycle enzymes.

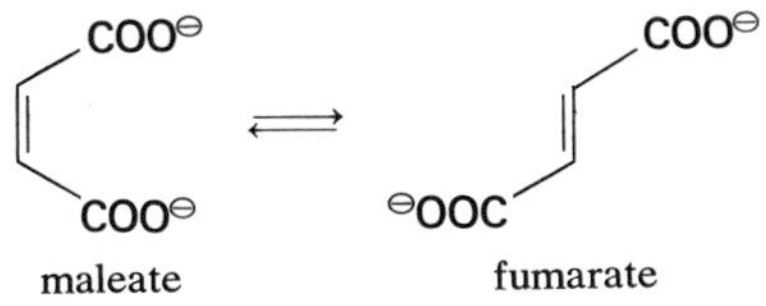

$COO^{\ominus}$ $COO^{\ominus}$ $\rightleftharpoons$ $COO^{\ominus}$ $^{\ominus}OOC$

maleate fumarate

The *P. fluorescens* enzyme has 74,000 mol wt, has a turnover of 300 sec^{-1}, and requires exogenous mercaptans to keep an active-site enzymatic thiol in reduced form.

The action of dioxygenases in aromatic ring openings can lead to acyclic maleyl derivatives as products. In the following expression R may be $COO^{\ominus}$ (gentisate→maleyl pyruvate) or $CH_2COO^{\ominus}$ (homogentisate→maleyl acetoacetate).

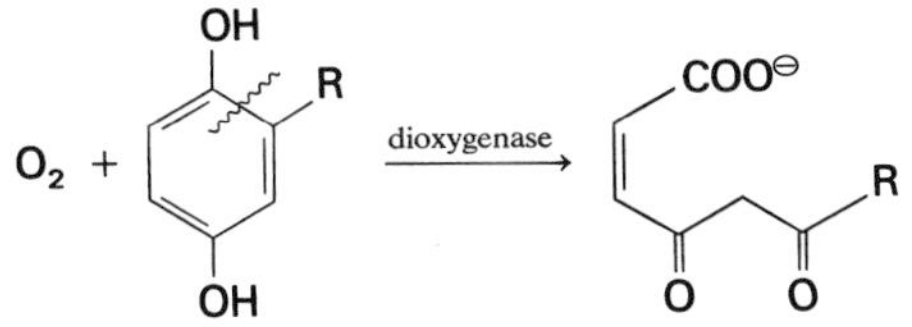

OH R O_2 + OH $\xrightarrow{\text{dioxygenase}}$ $COO^{\ominus}$ R O O

The *cis–trans* isomerases convert these maleyl keto acids to fumaryl derivatives, which are then substrates for hydrolytic cleavage to fumarate and acyl groups.

maleyl derivative ⇌ (GSH) fumaryl derivative → (hydrolases) fumarate + group

Glutathione (GSH) is a specific cofactor in each isomerase reaction. Isomerization of maleyl pyruvate has been studied in D_2O and found to proceed without incorporation of deuterium. The same result holds for a nonenzymatic isomerization of this substrate, catalyzed by glutathione (Seltzer, 1972). A mechanism for both the model and the enzymatic isomerizations involves (1) nucleophilic addition by the thiolate ion of the tripeptide glutathione, (2) rotation around the single bond to the more stable *trans* isomer, and (3) elimination of the added glutathione.

A particularly important *cis–trans* isomerization of this type occurs in animal-cell visual processes. The visual pigment consists of one isomer of vitamin-A aldehyde, specifically the 11-*cis* isomer of the retinal in imine linkage to the protein opsin, giving the complex rhodopsin, a purple (Fig. 19-4) holoprotein embedded as oriented arrays in internal membranes of the photoreceptor cells of the eye (Heller, 1972).

11-*cis*-retinal in imine linkage $\xrightarrow{\text{light}}$ all-*trans*-retinal in imine linkage

$\updownarrow$

free all-*trans*-retinal + $H_2\overset{\oplus}{N}$—Lys—Protein

Absorption of visual light by the conjugated polyene aldehyde produces a photoisomerization to the thermodynamically more stable all-*trans*-retinal bound to the protein. This photoprocess induces a complementary conformational change in the protein in the membrane milieu; this conformational change leads to increased permeability of the membrane to calcium ions, a consequent membrane depolarization, and subsequent propagation of an impulse to the optic nerve.

Although the initial *cis*-to-*trans* isomerization is photocatalyzed, the regeneration of the active form of rhodopsin (the conversion of all-*trans* to 11-*cis* isomer) is a dark reaction. It may occur while the *trans* aldehyde is still bound as the imine, or evidence has been advanced for isomerization after dissociation to the free *trans*-retinal. Hubbard (1956) has suggested reduction to retinol by retinol

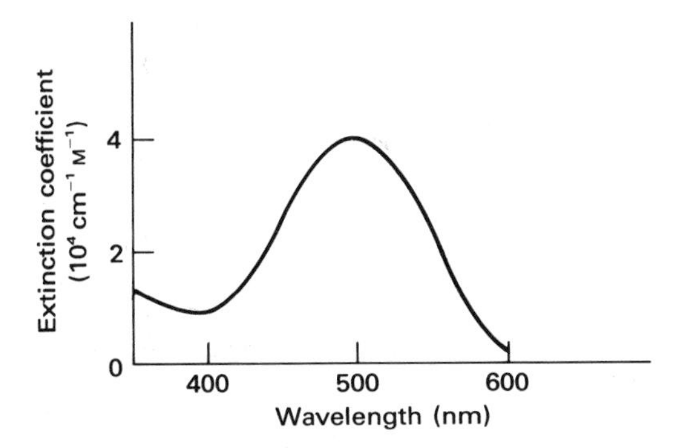

Figure 19-4
Absorption spectrum of rhodopsin. (From L. Stryer, *Biochemistry*, p. 800. W. H. Freeman and Company. Copyright © 1975.)

dehydrogenase, isomerization back to the 11-*cis* isomer of retinol by a poorly characterized retinol isomerase, and then reoxidation to the 11-*cis* retinal, which can recombine with opsin and regenerate the active visual pigment. As an alternative to this complex scheme, Futterman and Rollins (1973) have shown that bleached photoreceptor-membrane preparations will isomerize *trans*-retinal back to the bioactive 11-*cis* isomer, and they suggest that flavins may be involved. This suggestion stemmed from the observation that dihydroflavins at concentrations as low as 10^{-6} M could catalyze equilibration of *trans*-retinal with the 9-*cis* or 13-*cis* isomer. They note that *trans*-retinol (the alcohol) was *not* isomerized, and they suggest a nucleophilic role for the dihydroflavins. This could be an addition–elimination sequence as shown here, with free rotation around the single bond in the covalent adduct. This putative scheme suggests N-5 of the dihydroflavin as the attacking nucleophile in a vinylogous Michael addition. The reasons for lack of addition to C-11 in the model system are unclear.

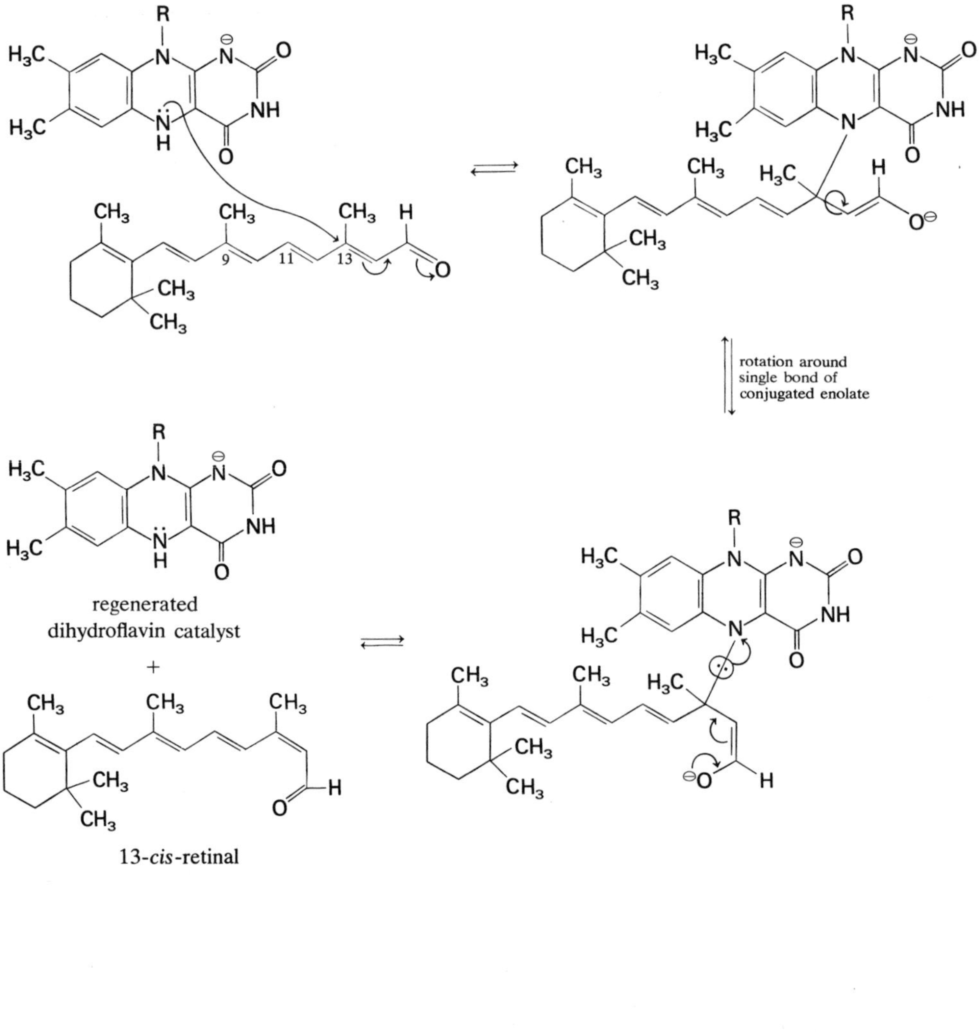

rotation around single bond of conjugated enolate

regenerated dihydroflavin catalyst

+

13-*cis*-retinal

In the second category of *cis–trans* isomerases (geometrical isomerization with bond migration) are poorly characterized enzymes responsible for *cis–trans* isomerization of unsaturated fatty acids such as the two shown here, moving the double bonds into conjugation in the isomerized products (Seltzer, 1972).

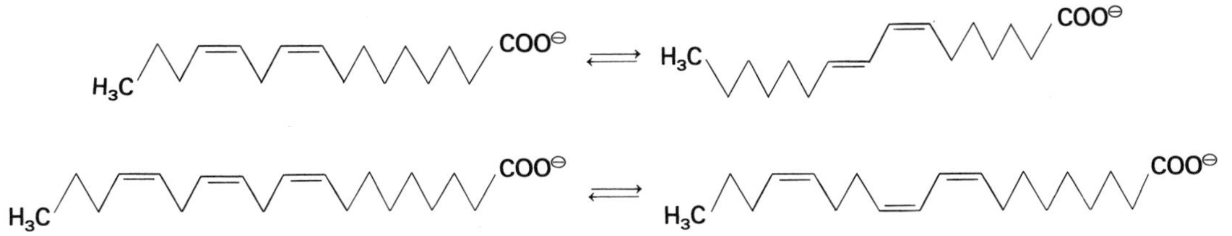

The enzyme activities responsible require reduced SH groups and divalent cations.

Also in this category one can list the aconitate isomerase noted earlier in the discussion of 1,3-hydrogen shifts (¶19.C.2). The *cis*- to *trans*-aconitate conversion proceeds with double-bond migration and suprafacial proton transfer as noted there. Also, the β-hydroxydecanoyl-thioester dehydrase (¶19.C.6) fits in this category because the 2-decenoyl-thioester substrate is *trans*, whereas the 3-decenoyl-thioester product is *cis*.

This concludes our discussion of enzyme-catalyzed isomerization reactions.

Chapter 20

Enzyme-Catalyzed Rearrangements

The enzymatic reactions discussed in this chapter involve rearrangements in the carbon skeleton of a substrate as it is converted to product. Rearrangement reactions are rather rare among those enzymatic systems where the catalysts have been purified. (There undoubtedly are several kinds of rearrangements in secondary plant metabolism involved in biosynthesis of alkaloids and terpenoid compounds, but the responsible enzymes have not been isolated, and these reactions are not discussed in this book; see Mayo, 1964, pp. 771–965). We shall focus on three distinct reaction types:

1. alkyl migrations in biosynthesis of branched-chain amino acids and of lanosterol;
2. Claisen rearrangements in chorismate metabolism;
3. 1,2-migrations requiring coenzyme B_{12}.

We might note at this point, however, that the conversion of *p*-hydroxyphenylpyruvate to homogentisate (discussed in ¶16.A.3 as an intramolecular example of one type of dioxygenase reaction) can also be categorized as a rearrangement of the carbon skeleton during catalysis.

$$O_2 + HO{-}C_6H_4{-}CH_2{-}C(=O){-}COO^{\ominus} \xrightarrow{\text{Enz—Fe}^{II}} HO{-}C_6H_3(OH){-}CH_2{-}COO^{\ominus}$$

homogentisate

20.A ALKYL MIGRATIONS

We shall comment here on two enzyme-mediated processes involving *intramolecular migrations* of alkyl (methyl or ethyl) groups. The first reaction has the characteristics of the acyloin rearrangement (Mayo, 1964, p. 965). The second involves purportedly simultaneous migrations of hydrogen and methyl groups in a carbonium-ion rearrangement scheme, as squalene epoxide forms lanosterol.

20.A.1 Acetohydroxy-Acid Isomeroreductase

In the biosynthesis of the branched-chain α-amino acids valine and isoleucine, microbial cells use identical enzymes for transformations of both valine and isoleucine precursors.

$$(H_3C)_2CH{-}CH(NH_3^{\oplus}){-}COO^{\ominus}$$

valine

$$H_3C{-}CH_2{-}CH(CH_3){-}CH(NH_3^{\oplus}){-}COO^{\ominus}$$

isoleucine

The branching at C-3 is introduced at the same stage for each by the acetohydroxy-acid isomeroreductase of the organism. In valine biosynthesis this enzyme converts acetolactate to α,β-dihydroxyisovalerate; in isoleucine biosynthesis it converts α-aceto-α-hydroxybutyrate to α,β-dihydroxy-β-methylvalerate.

$$H_3C{-}CO{-}C(OH)(CH_3){-}COO^{\ominus} + NADPH \rightleftharpoons H_3C{-}C(OH)(CH_3){-}CH(OH){-}COO^{\ominus} + NADP^{\oplus}$$

acetolactate → α,β-dihydroxyisovalerate

$$H_3C{-}CO{-}C(OH)(CH_2CH_3){-}COO^{\ominus} + NADPH \rightleftharpoons H_3C{-}C(OH)(CH_2CH_3){-}CH(OH){-}COO^{\ominus} + NADP^{\oplus}$$

α-aceto-α-hydroxybutyrate → α,β-dihydroxy-β-methylvalerate

As the name of the enzyme indicates, the reaction involves both a reduction (a keto group to an alcohol) and an isomerization. NADPH is the reducing agent, and with 4*S*-[^{3}H]-NADPH, tritium is introduced at C-2 of the product dihydroxy acid. The more interesting facet of these two conversions is the intramolecular alkyl-group migration from C-2 to C-3 at some stage of the reaction. Labeling studies have demonstrated the *intramolecular* nature of the methyl or ethyl migrations (Strassman et al., 1953, 1954; Adelberg, 1955).

The isomeroreductase has been purified to homogeneity from *Salmonella typhimurium* as a tetramer of 220,000 mol wt. It catalyzes the formation of dihydroxy acid with a pH optimum of 7.5; in the direction forming α-hydroxy β-keto acid, the pH optimum is 9.5. Turnover numbers are 1,100 min^{-1} with α-acetolactate as substrate and 4,700 min^{-1} with α-aceto-α-hydroxybutyrate (Arfin and Umbarger, 1969).

Given the structure of the products and the idea that NADPH is reducing an α-keto acid species (to incorporate tritium at C-2), a reasonable mechanistic scheme is an acyloin rearrangement (Mayo, 1964, p. 965), *a formal migration of* $R^{\ominus}$.

Thus, the enzyme could facilitate rearrangement of the bound α-hydroxy β-keto acid (substrate) to the β-hydroxy α-keto acid as bound intermediate. This in turn would be reduced at the α-carbonyl to yield the observed dihydroxy product.

E—B: H
O O
$H_3C—C—C—COO^{\ominus}$
R
rearrangement
EB: H
O O
$H_3C—C—C—COO^{\ominus}$
R
NADPH
HO OH
$H_3C—C—C—COO^{\ominus}$
R H

Some experimental evidence supporting this conjecture has been obtained with the *Salmonella* enzyme. Arfin and Umbarger (1969) prepared the α-keto acid (β-hydroxy-α-ketoisovalerate) and observed that it is a substrate for NADPH-dependent enzymatic reduction, the second step in the scheme just given. The pH optimum is 9.5. The fact that the ratio of isomeroreductase activity to this reductase activity remains constant at each purification step in the isolation of the enzyme is consistent with its existence as an intrinsic catalytic capacity of the isomeroreductase, rather than as an accompanying contaminating activity. It is also possible to demonstrate an isomerization step distinct from reduction. At pH 7.5, the enzyme isomerizes added β-hydroxy-α-ketoisovalerate back to acetolactate. This isomerization (the acyloin rearrangement) requires $Mg^{2\oplus}$ and NADPH. At this pH, reduction is minimal compared to acetolactate formation. Apparently NADPH is required for effective binding of the α-keto species and its subsequent isomerization. It has not been determined if ordered binding of NADPH and β-hydroxy-α-ketoisovalerate occurs. Thus, the β-hydroxy α-keto compound behaves as a chemically competent intermediate.

If this two-step scheme is in fact an accurate description of isomeroreductase action, then the β-hydroxy α-keto acid must remain tightly bound and nondissociable during dihydroxy-acid formation. This is inferred from the observation that

[^{14}C]-acetolactate is converted without dilution of specific radioactivity to [^{14}C]-α,β-dihydroxyisovalerate even in the presence of an excess of exogenous β-hydroxy-α-ketoisovalerate. There is no equilibration of any bound molecules with the pool of added intermediate in solution.

In subsequent metabolism, the two dihydroxy acids formed by this enzyme are next dehydrated via enol intermediates to the α-keto acids by a dihydroxy-acid dehydrase noted earlier (¶17.B.1.a).

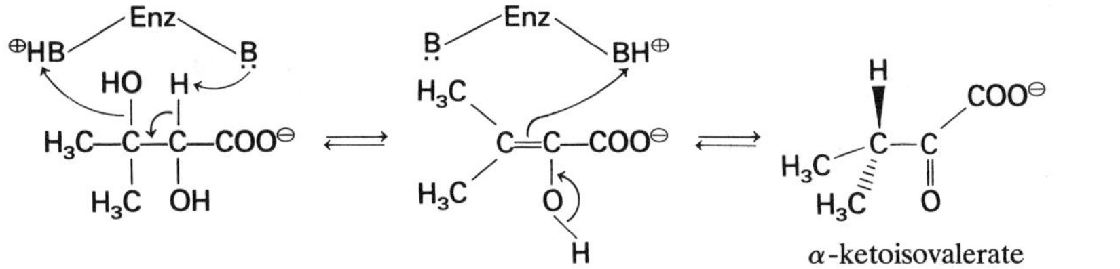

α-ketoisovalerate

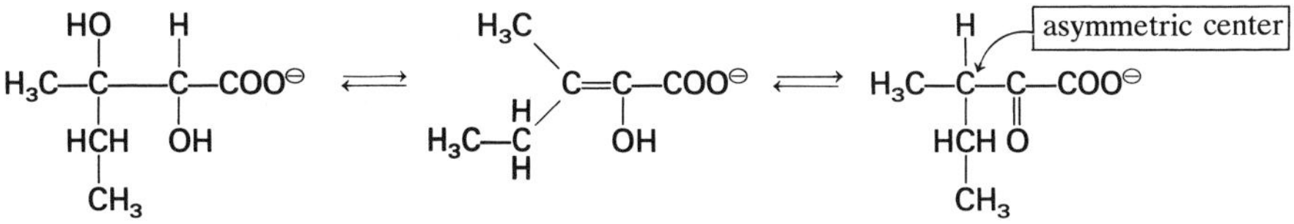

asymmetric center

α-keto-β-methylvalerate
(only one isomer generated)

Ketonization of the enol must be enzyme-catalyzed because C-3 in the α-keto-β-methylvalerate is asymmetric, and only one enantiomer of L-isoleucine is produced. The last biosynthetic step is a pyridoxal-P–mediated transamination (Section V) of the α-keto acids to α-amino acids. This must occur rapidly in vivo, before nonenzymatic enolization of the α-keto acids can equilibrate configuration at C-3.

20.A.2 Cyclization of Squalene Epoxide to Lanosterol

In Chapter 19 we noted that polyisoprenoids are synthesized from the two key C_5 isomers, Δ^3-isopentenyl pyrophosphate and dimethylallyl pyrophosphate (Δ^2-isopentenyl pyrophosphate). The tetracyclic sterols are formed from a C_{30} acyclic precursor, squalene. We shall discuss the formation of the C_{30} unit from two C_{15} (farnesyl-PP) molecules in Section V, where we discuss formation of carbon–carbon bonds during biosynthesis. In animals and fungi, the primary sterol product is lanosterol, which can be converted in several steps (involving oxidative demethylations, double-bond isomerizations, and saturations) to, among other molecules, cholesterol. Cholesterol is the major steroid found in animal-cell membranes. Cholesterol in turn is metabolizable to such molecules as steroid hormones and bile acids by monooxygenations and oxidative fragmentations, as noted in Chapter 15, by P_{450}-dependent monooxygenase systems.

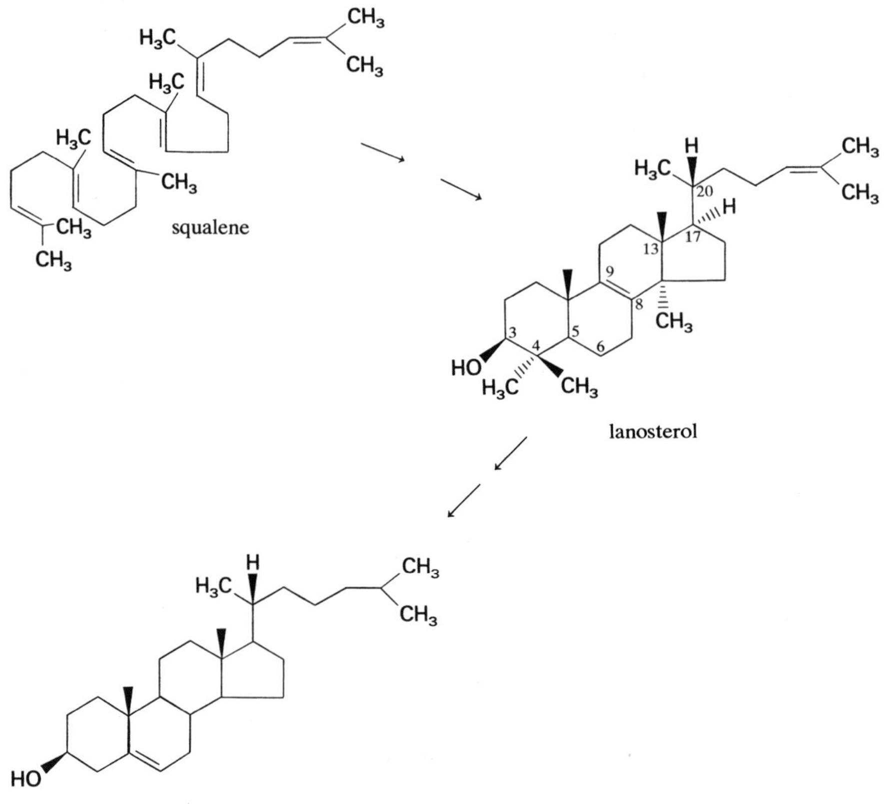

squalene

lanosterol

cholesterol

Early studies by Bloch and colleagues established that the C-3 β-hydroxyl of lanosterol derives from O_2, not H_2O—suggestive of monooxygenase involvement in conversion of squalene to lanosterol. This was a precursor to the discovery of the enzyme squalene epoxidase, a partially characterized system requiring NADPH and O_2, and probably containing flavin and nonheme iron (Tai and Bloch, 1972). The product from the epoxidase is the squalene β-2,3-epoxide. The conversion of the double bond to the electrophilic epoxide functionality is probably a variant of the monooxygenase mechanisms discussed in Section III, although it is unclear whether the flavin or the enzyme-bound iron is the species that combines with O_2 to form a reactive hydroxylating agent. Purification of the epoxidase has been difficult because it is a membrane protein (endoplasmic

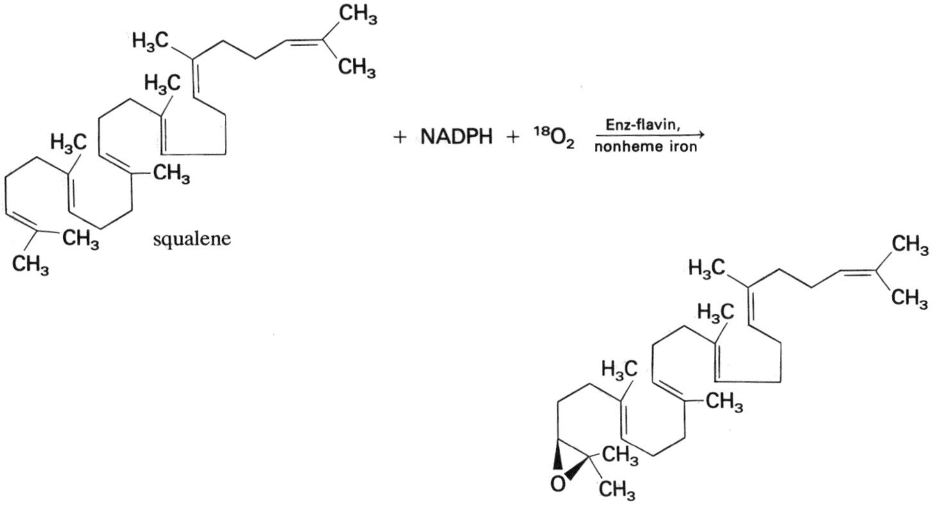

CH3 H3C H3C H3C CH3 CH3 CH3 squalene + NADPH + $^{18}O_2$ $\xrightarrow[\text{nonheme iron}]{\text{Enz-flavin,}}$

reticulum membranes of cells) and appears to have requirements for phospholipids as activators (Ono and Bloch, 1975). It is this squalene β-epoxide that then undergoes the putative electrophilic cyclization sequence to the tetracyclic product, lanosterol. However, rational cyclization mechanisms do not yield lanosterol as initial product. Rather, an initially cyclic protosterol is postulated to undergo two 1,2-hydrogen migrations (from C-13 to C-17, and from C-17 to C-20) and two methyl-group migrations (the C-8 and the C-14 methyls move).

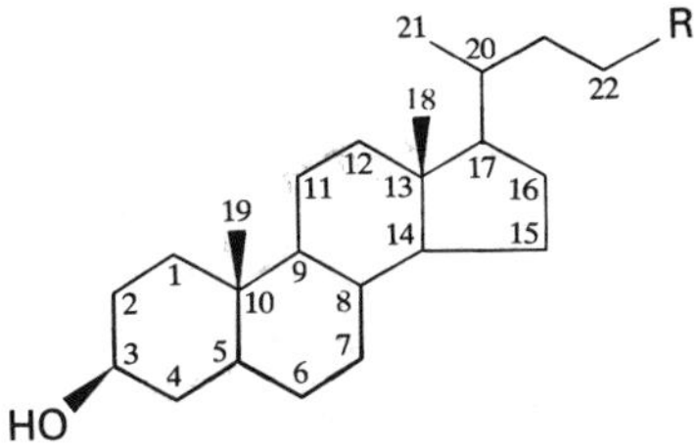

sterol numbering

Intensive isotopic studies have been carried out, which prove that the four migrations occur intramolecularly, and that the C-9 β-hydrogen is lost to the medium as a proton (Mulheirn and Ramm, 1972), but the membrane-bound oxidocyclase itself has been refractory to much purification so far, and reactions are carried out with crude membrane fragments.

The following is the postulated reaction mechanism.

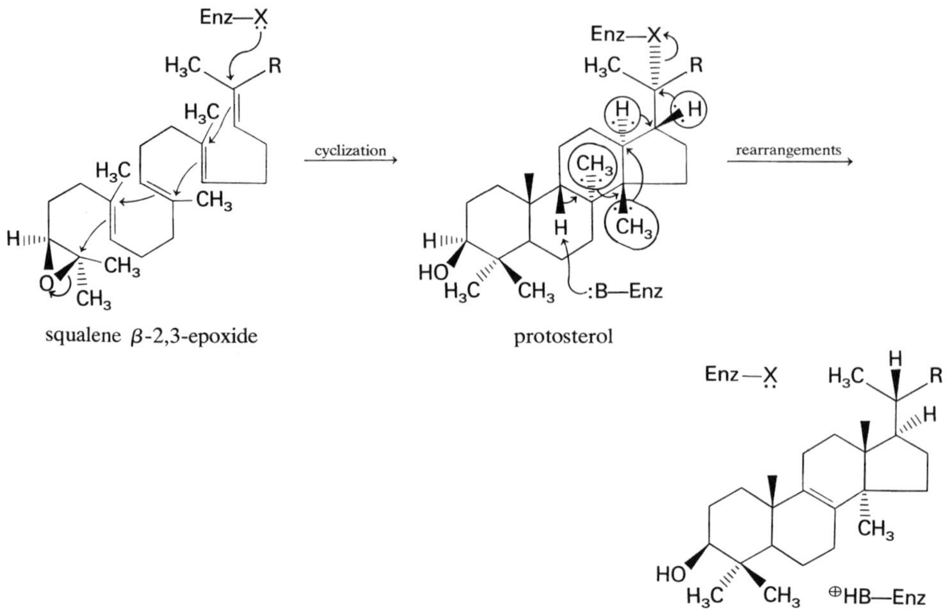

cyclization → rearrangements →

squalene β-2,3-epoxide

protosterol

lanosterol

It is as yet uncertain whether the cyclization proceeds with the squalene epoxide bound at the active site to generate a free protosterol carbonium ion from which the 1,2-shifts then occur, or whether cyclization is initiated as shown here by reversible attack of some enzymatic nucleophile at the terminus of the conjugated system. The protosterol carbonium ion or covalent enzyme derivative is then postulated to undergo the sequence of concerted 1,2-migrations shown, producing lanosterol *as the first stable product.* Abstraction of the 9-β-hydrogen by an enzymatic base will result in formation of the 8,9 olefinic linkage in lanosterol when the 8-α-methyl migrates to C-14. The 14-β-methyl in turn migrates to C-13, setting off two hydride migrations, 13-α-H to C-17, and 17-β-H to C-20. How concerted the four migratory events may be has not been determined. If X is the carbonium ion in the scheme above, this quenches the charge. If X is the enzyme nucleophile, it is displaced by the migrating hydride.

In higher plants and algae, the initial sterol from squalene-oxide cyclization is not lanosterol, but is the cyclopropane-containing cycloartenol. Basically the same electrophilic cyclization and rearrangement scheme as that just given is suggested, but in this instance the 9-β-H is postulated to migrate to C-8, leaving the electron-deficient center (the carbonium ion) at C-9 (Mayo, 1964, p. 1019). This specificity might be controlled in this enzyme by the lack of a basic group to abstract that 9-β-H as a proton; it may thus be left to undergo the 1,2-shift (now moving with its electron pair as a hydride ion).

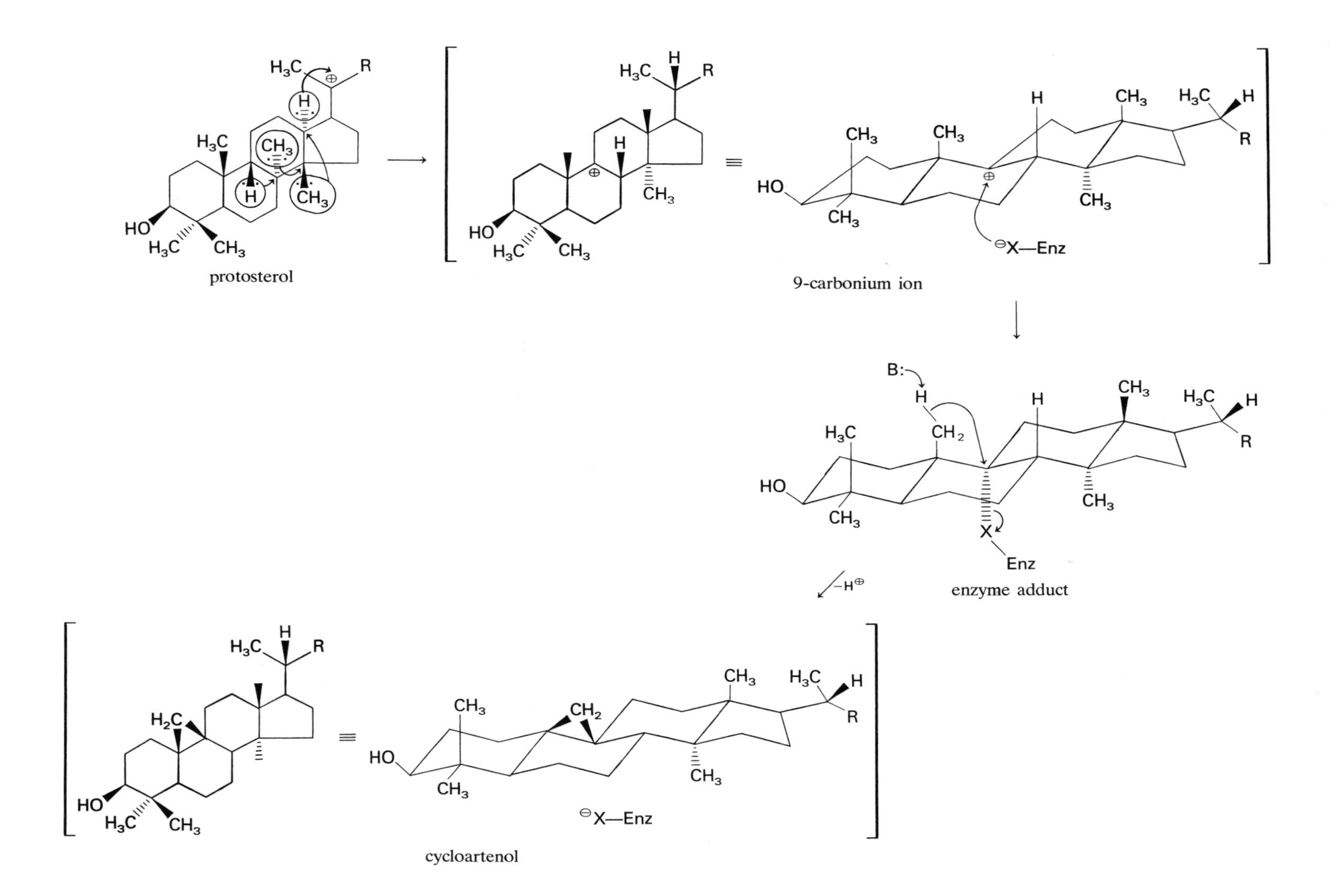
protosterol
9-carbonium ion
⊖X—Enz
B:
enzyme adduct
−H⊕
cycloartenol

One hypothesis (shown here) is that C-9 of the carbonium ion could be captured from below by an enzyme nucleophile. This adduct would have a geometry suitable for subsequent 1,3-*anti* elimination of (1) a proton from the C-19 angular methyl group and (2) the added nucleophile; the result would be cyclopropane formation to produce cycloartenol.

20.B ENZYME-CATALYZED REARRANGEMENTS OF CHORISMIC ACID

Microbes that synthesize aromatic amino acids such as phenylalanine, tyrosine, tryptophan, and aromatic vitamins—or their precursors such as *p*-aminobenzoate, folate, *p*-hydroxybenzoate, ubiquinone (coenzyme Q), 2,3-dihydroxybenzoate, and vitamin K—use a common biosynthetic pathway up to the key intermediate, chorismic acid. Chorismate is unusual for a molecule in biological systems, as a cyclohexa-1,5-diene carboxylic acid with an enolpyruvyl-ether linkage at C-3.

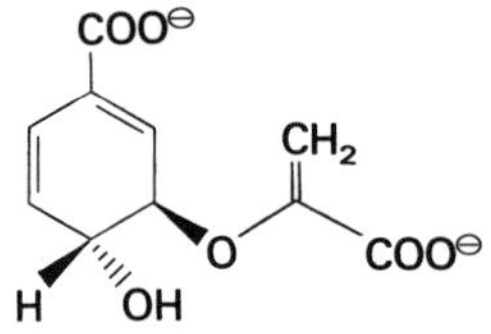

$COO^{\ominus}$ CH_2 O $COO^{\ominus}$ H OH

Chorismate then can undergo a variety of transformations, leading ultimately to the various aromatic molecules noted above (Gibson and Pittard, 1968). The chemical synthesis of chorismate has not yet been reported. It was initially isolated by Gibson (1968) from cultures of a strain of *Aerobacter aerogenes* with multiple genetic blocks in the enzymes that utilize chorismate. This strain then is an "overproducer" of chorismate and excretes it into the culture medium. From two liters of culture fluid, 500 to 800 mg of chorismate can be isolated. Structural determination was carried out on this material, and at present this preparation is the only way to obtain chorismate.

20.B.1 Chorismate Formation

Before considering the various rearrangements of chorismate to subsequent metabolites, we shall comment briefly on how the molecule is formed enzymatically and how the unusual enol ether linkage may be fashioned. In Chapter 17 (on enzyme-catalyzed eliminations of H_2O), we discussed the net *syn* elimination carried out by dehydroquinate dehydratase, producing dehydroshikimate. The α,β-unsaturated ketone of dehydroshikimate can be reduced with NADPH to the allylic alcohol in shikimate. The newly generated alcohol is then phosphorylated from ATP by shikimate kinase to generate shikimate-3-phosphate.

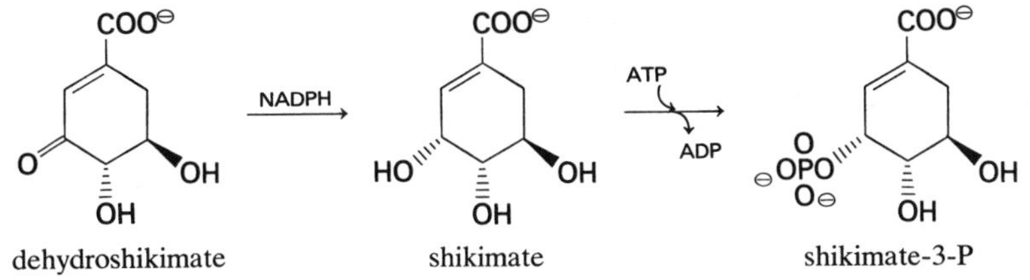

NADPH → ATP → ADP

dehydroshikimate　　shikimate　　shikimate-3-P

It is at this stage that the enolpyruvyl ether moiety is introduced by the enzyme 5-enolpyruvylshikimate-3-phosphate synthetase. The cosubstrate is phosphoenolpyruvate. Currently available data support a reversible addition-elimination mechanism in which the 5-hydroxyl group of shikimate-3-P acts as nucleophile in attacking C-2 of PEP, with protonation and generation of a methyl group at C-3 (Bondinelli et al., 1971). This initial adduct is now set up to eliminate inorganic phosphate by reabstraction of one of the methyl hydrogens as a proton. This elimination creates the olefinic linkage in the enolpyruvyl group of the product 5-enolpyruvylshikimate-3-phosphate.

shikimate-3-P + phosphoenolpyruvate

tetrahedral adduct

5-enolpyruvylshikimate-3-P + P_i

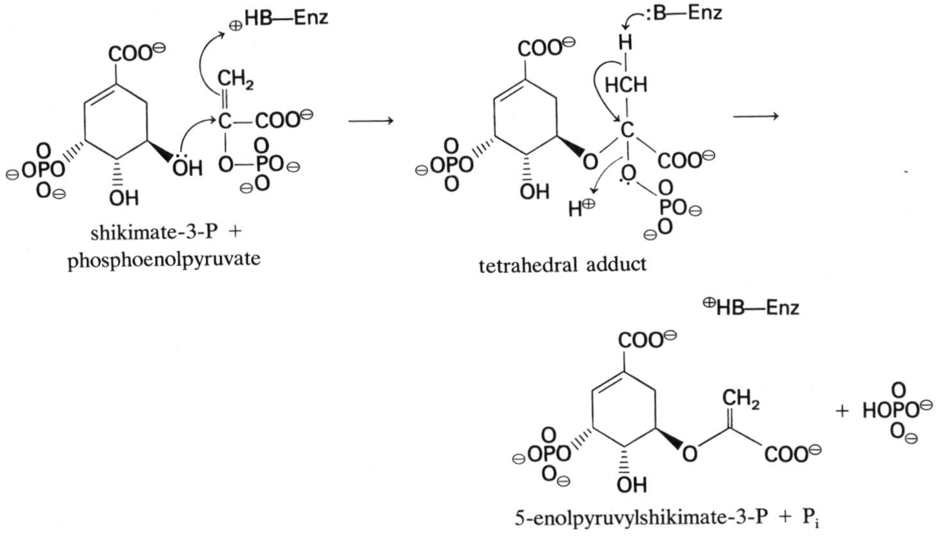

This mechanism is supported by labeling studies, including ^{18}O studies that establish C—O cleavage from C-2 of PEP (Gunetileke and Anwar, 1968). ($[^{18}O]$-PEP labeled in the $C—^{18}O—PO_3^{2\ominus}$ bridge oxygen yields $[^{18}O]$-P_i.) In passing, we might note that only one other enzymatic example of transfer of the intact enolpyruvyl group of PEP is known. That is in the formation of UDP-*N*-acetylenolpyruvylglucosamine (Gunetileke and Anwar, 1968). The enolpyruvyl moiety there is subsequently reduced to a lactyl ether that persists as a structural element in the muramic-acid component (Chapter 9) of bacterial cell walls.

We have noted previously (¶17.B.5) that chorismate synthetase converts 5-enolpyruvylshikimate-3-P to chorismate by a 1,4-elimination of the elements of phosphate (Welch et al., 1974).

5-enolpyruvylshikimate-3-P → chorismate + P_i

20.B.2 Rearrangements of Chorismate

Of the biosynthetic rearrangements of the chorismate skeleton that occur in procaryotes and in lower and higher plants, four enzymes have been at least partially characterized, involving primary transformations of chorismate into phenylpyruvate, *p*-hydroxyphenylpyruvate, anthranilate, and *p*-aminobenzoate (Fig. 20-1). A common precursor, prephenate, is involved in formation of both phenylpyruvate and *p*-hydroxyphenylpyruvate. These two keto acids are subsequently transaminated to phenylalanine and tyrosine, respectively. Anthranilate (*o*-aminobenzoate) is a precursor of tryptophan, whereas *p*-aminobenzoate is a component of the folate coenzymes discussed in Section V. We shall now consider what is known about each of these four catalyses.

20.B.3 Chorismate Mutase–Prephenate Dehydrase

The catalytic activity converting chorismate to phenylpyruvate targeted for phenylalanine synthesis has been purified to homogeneity from an *E. coli* K12 strain (Davidson et al., 1972). It is observed that a single protein has both chorismate mutase activity (chorismate→prephenate) and prephenate dehydrase activity (prephenate→phenylpyruvate). The single protein thus is a bifunctional catalytic complex. It exists as a dimer of 85,000 mol wt, with identical subunits. Because there are separate genes for each activity, it seems apparent that (for this protein and for others to be discussed in this chapter) multifunctional proteins

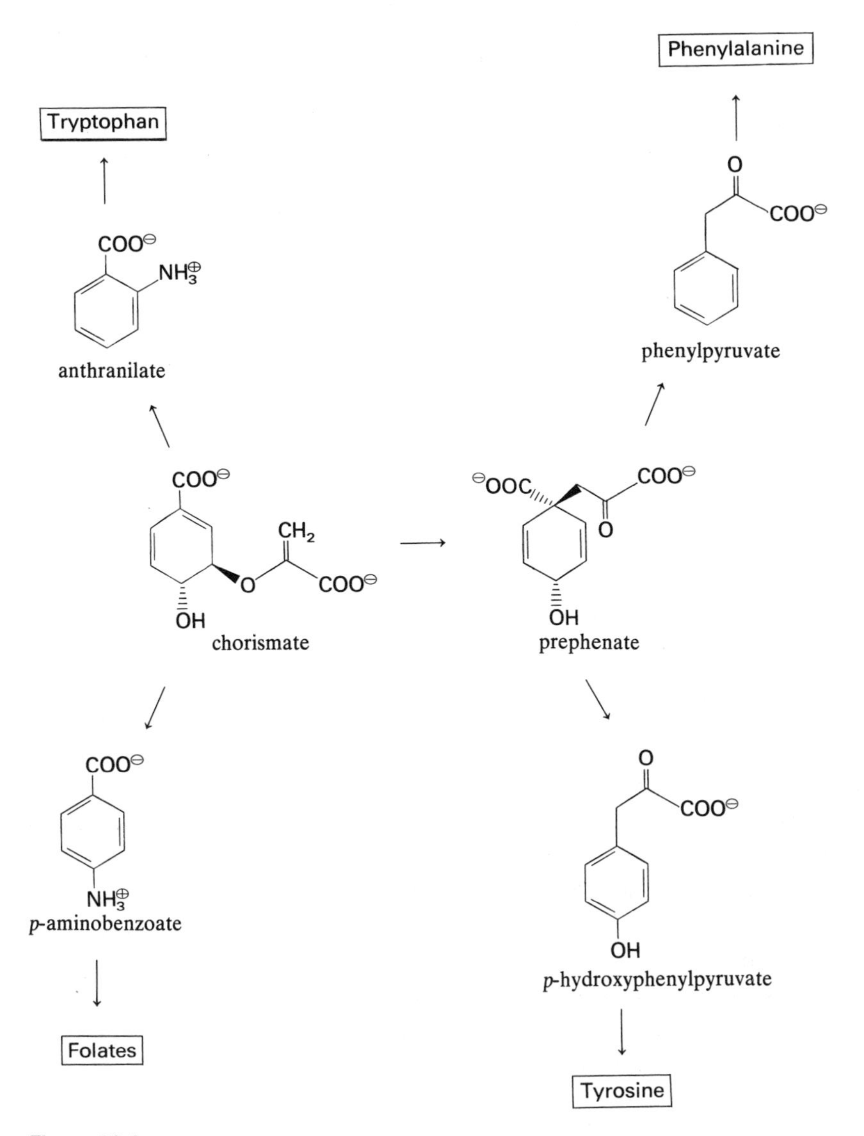

Figure 20-1
Biosynthetic rearrangements of the chorismate carbon skeleton.

arise from fusion of adjacent genetic loci in the bacterial DNA. The mutational loss of a stop signal at the end of the first gene would allow uninterrupted readthrough of RNA polymerase to the end of the second gene. The resulting messenger RNA would be used to make a protein with two separate catalytic domains. It may be that there is some evolutionary advantage from possible increased catalytic efficiency for chorismate→phenylpyruvate: the prephenate intermediate may never get free into solution.

The skeletal-rearrangement step is in the chorismate-to-prephenate conversion, and it can also occur nonenzymatically during the breakdown of the relatively unstable chorismate molecule.

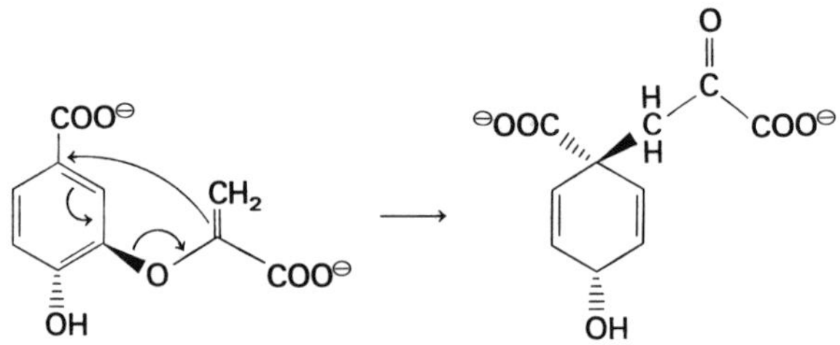

chorismate

prephenate

This rearrangement is a clear example of a Claisen rearrangement, a concerted intramolecular rearrangement process proceeding through a chairlike transition state. More generally, it can be described as a 3,3-sigmatropic shift. As categorized by the Woodward-Hoffman rules for such concerted reactions, the chorismate-to-prephenate rearrangement is a thermally allowed suprafacial one (Woodward and Hoffman, 1969). At pH 7.5 and 37° C, the nonenzymatic rearrangement proceeds at 2.6×10^{-5} sec^{-1} (Gill, 1968). The enzyme accelerates this process by a factor of about 2×10^{6}. The NMR of chorismate in solution indicates, as expected, that the predominant conformer has the substituents on the cyclohexadiene ring in quasiequatorial positions (Gill, 1968). This will not be the reactive conformation for the Claisen rearrangement. It is likely that the enzyme binds chorismate and achieves a bound rotamer that has the side chain oriented over C-1 and has the puckered-ring conformation with substituents quasiaxial, so that sigmatropic shift can occur readily.

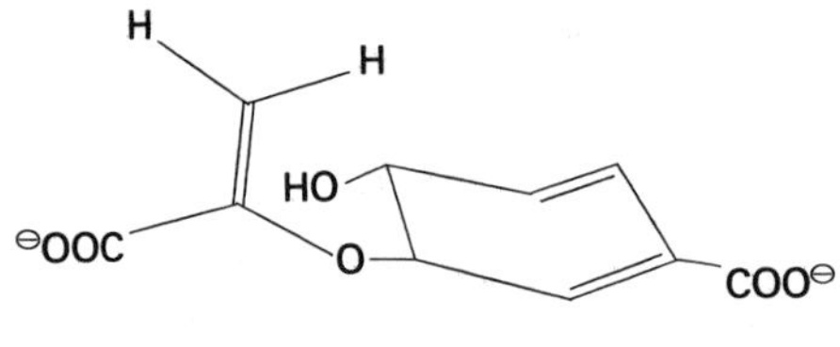

predominant conformer in solution

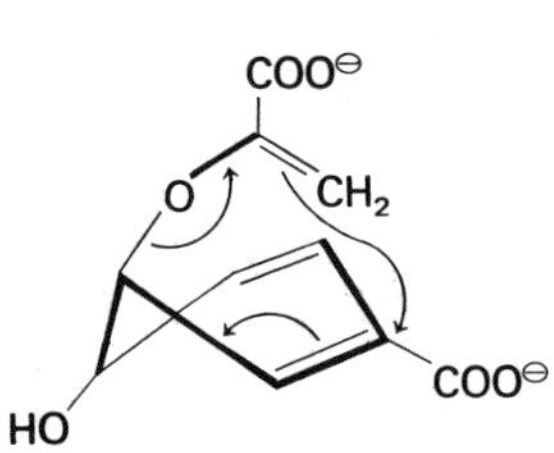

possible conformer bound at active site

It is likely that prephenate remains as a bound intermediate and suffers rapid enzymatic decarboxylation in the normal operation of the chorismate-mutase–prephenate-dehydrase complex, and in the analogous chorismate-mutase–prephenate-dehydrogenase complex (¶20.B.4). As we shall emphasize in discussing enzyme-catalyzed decarboxylations in Section V, the prime factor in catalysis of decarboxylation is stabilization of the incipient carbanion generated as the electrophilic CO_2 is produced. In the dehydratase reaction, this is achieved by elimination of the *para*-OH group, probably protonated to H_2O by an enzymatic base (general acid catalysis) as it leaves. The ensuing aromatization also is a driving force for decarboxylation.

Enz—BH$^{\oplus}$

prephenate

phenylpyruvate
+ CO_2 + H_2O

subsequent transamination

phenylalanine

20.B.4 Chorismate Mutase–Prephenate Dehydrogenase

Another bifunctional enzyme, chorismate mutase–prephenate dehydrogenase, has been purified from both *Aerobacter aerogenes* and *E. coli* as a dimer of 80,000 mol wt (Koch et al., 1971). It converts chorismate to *p*-hydroxyphenylpyruvate, which is one transamination step removed from tyrosine. (Animals cannot synthesize aromatic amino acids de novo, and dietary phenylalanine is hydroxylated to tyrosine by the monooxygenase phenylalanine hydroxylase, as noted earlier.) It has been suggested that this complex and the mutase-dehydrase complex may have evolved from a common ancestor. The formation of prephenate is undoubtedly similar in the two enzymes. However, decarboxylation of prephenate to *p*-hydroxyphenylpyruvate means that the *para*-hydrogen, not the *para*-hydroxyl, must be the leaving group $X^{\ominus}$. That is, the *para*-hydrogen must depart with its electrons as a formal hydride ion. This is consonant with the observation that the prephenate dehydrogenase activity requires $NAD^{\oplus}$ as cosubstrate (and hydride acceptor), reducing it to NADH.

Interestingly, in a *Bacillus subtilis* strain, the chorismate mutase activity is fused, not to either of the prephenate-utilizing enzymes, but to the very first

prephenate p-hydroxyphenylpyruvate tyrosine

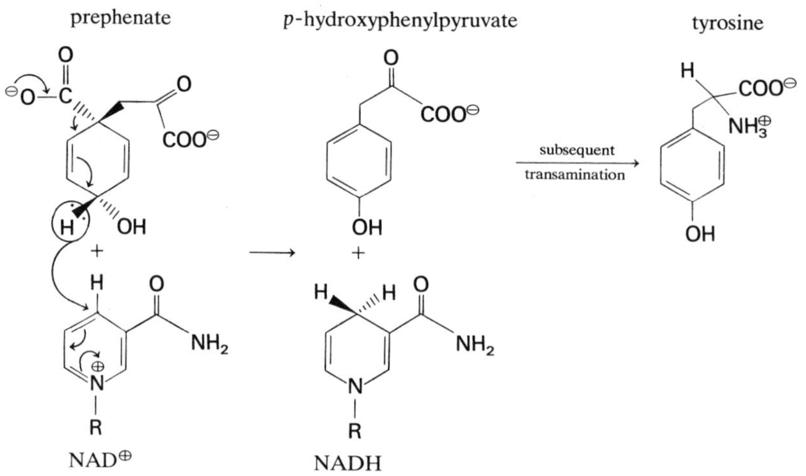

subsequent transamination

NAD⊕ NADH

enzyme in the aromatic biosynthetic pathway, 3-deoxy-D-arabinoheptulosonate-7-phosphate (DAHP) synthetase (Nakatsukasa and Nester, 1972; Altendorf et al., 1971).

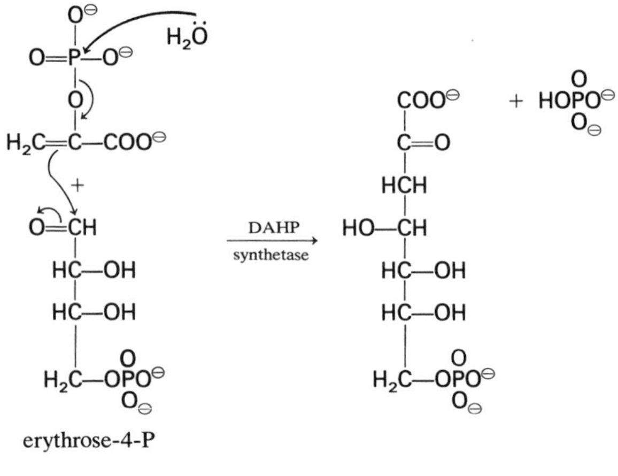

DAHP synthetase

erythrose-4-P

20.B.5 Anthranilate Synthetase and *p*-Aminobenzoate Synthetase

Each of the other two enzymes in Figure 20-1 catalyzes synthesis of an isomer (*ortho* or *para*) of aminobenzoate, using glutamine as the physiological amino donor. Although the anthranilate synthetase has been purified and studied (Zalkin, 1973), the *p*-aminobenzoate enzyme (or enzymes) is not well understood.

The anthranilate synthetase may be a separate activity, or it may be fused with other enzymes of the tryptophan biosynthetic pathway. The enzyme from *Serratia marcescens* is an example of the first type, whereas the enzyme from

Salmonella typhimurium is physically associated with the next enzyme of the tryptophan pathway, the PRPP-dependent phosphoribosyl transferase.

anthranilate + PRPP $\xrightarrow{\text{transferase}}$ 5-phosphoribosyl-1-anthranilate + PPi

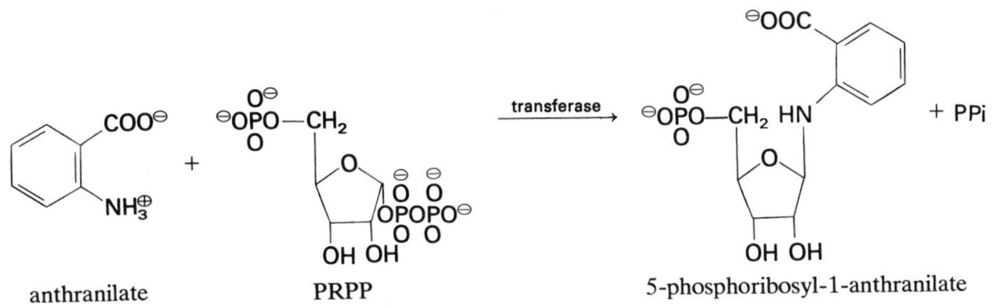

In common with a number of glutamine-utilizing enzymes discussed in Chapter 5, the anthranilate synthetase (of either type) is composed of two nonidentical polypeptide subunits, one with glutaminase activity and the other capable of synthesizing anthranilate from NH_3 in the absence of the complementary subunit. The subunit with glutaminase activity appears (as usual) to activate glutamine by forming a bound γ-glutamyl thioester and "nascent ammonia." The *Salmonella* enzyme is tetrameric when active, composed of two anthranilate-forming subunits and two glutamine-activating subunits. The latter subunit is the species fused with the phosphoribosyl transferase.

Although the role of glutamine in donating the amino group is relatively straightforward, the further details of anthranilate formation remain obscure, with no isolable intermediates yet reported. One speculation suggests attack of "nascent ammonia" to yield an amino intermediate that could then eliminate the keto form of pyruvate during aromatization.

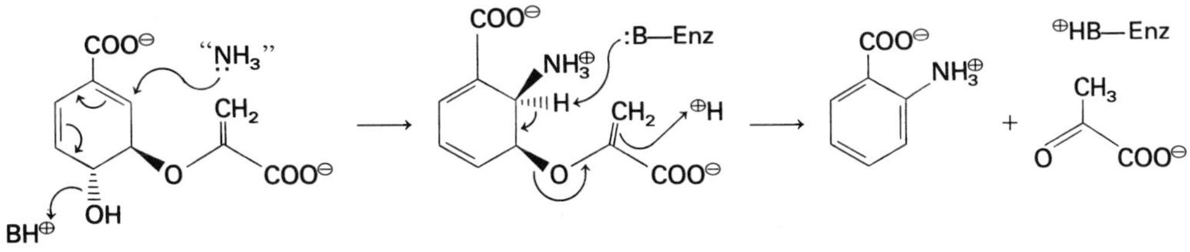

In systems that form *p*-aminobenzoate, it has been reported that an intermediate accumulates in the culture medium of an *A. aerogenes* mutant. The intermediate purportedly is an aminated cyclohexadiene carboxylate, but that structural assignment is (at best) tenuous (for discussion see Larsen et al., 1975; Larsen and Wieczorkowska, 1975).

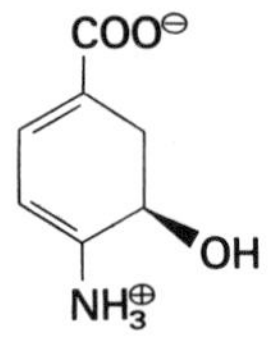

purported intermediate

One could also imagine a displacement reaction on chorismate to yield a 4-amino-3-enolpyruvylcyclohexadiene carboxylate, which might then eliminate pyruvate (Dardenne et al., 1975).

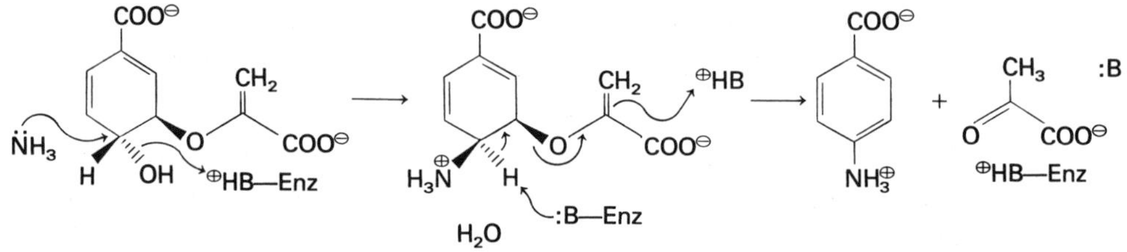

20.B.6 Other Metabolites from Chorismate

A variety of other metabolites have been demonstrated (by labeling studies in plants or bacteria) to derive from chorismate. Among these substances are *p*-hydroxybenzoate, itself a precursor of ubiquinone (coenzyme Q), and *trans*-2,3-dihydroxybenzoate, an intermediate in biosynthesis of naphthoquinones (vitamin K). In plants, *p*-aminophenylalanine and *p*-carboxy-*m*-aminophenylalanine are produced (Larsen et al., 1975; Larsen and Wieczorkowska, 1975).

The formation of *p*-hydroxybenzoate from chorismate is explicable. It even occurs nonenzymatically in chorismate samples. One can write a cyclic mechanism for pyruvate elimination, where aromaticity is developed in the transition state.

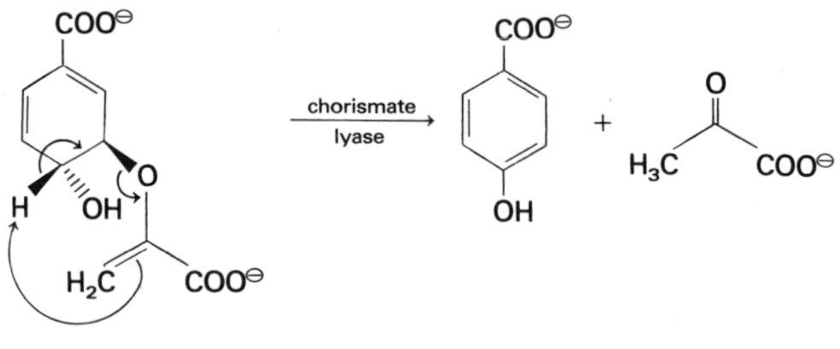

Whether any intramolecular transfer of hydrogen occurs in the reaction catalyzed by chorismate lyase has not been probed. The *p*-hydroxybenzoate is subsequently substituted by a C_{40} isoprenyl unit *ortho* to the OH, and this octaprenylhydroxybenzoate is eventually converted to coenzyme Q_8, an important mobile redox component in membrane electron-transport chains.

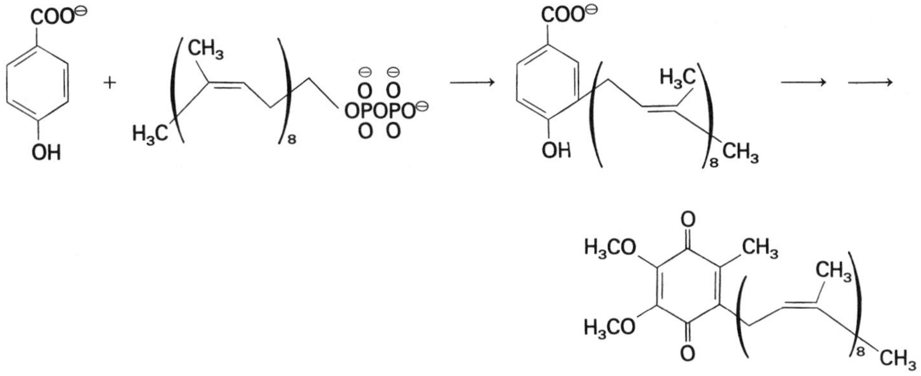

coenzyme Q
(ubiquinone)

Molecules of 2,3-dihydroxybenzoate most likely arise from prior conversion of chorismate to isochorismate (Larsen et al., 1975; Larsen and Wieczorkowska, 1975). The isochorismate could then yield either 2-hydroxybenzoate (also observed) or 2,3-dihydrodihydroxybenzoate. Both isochorismate and the dihydroxycyclohexadiene carboxylate have been observed in extracts of genetically blocked bacteria (Gibson and Pittard, 1968). Addition of NAD$^{\oplus}$ to these extracts results in formation of the aromatic derivative.

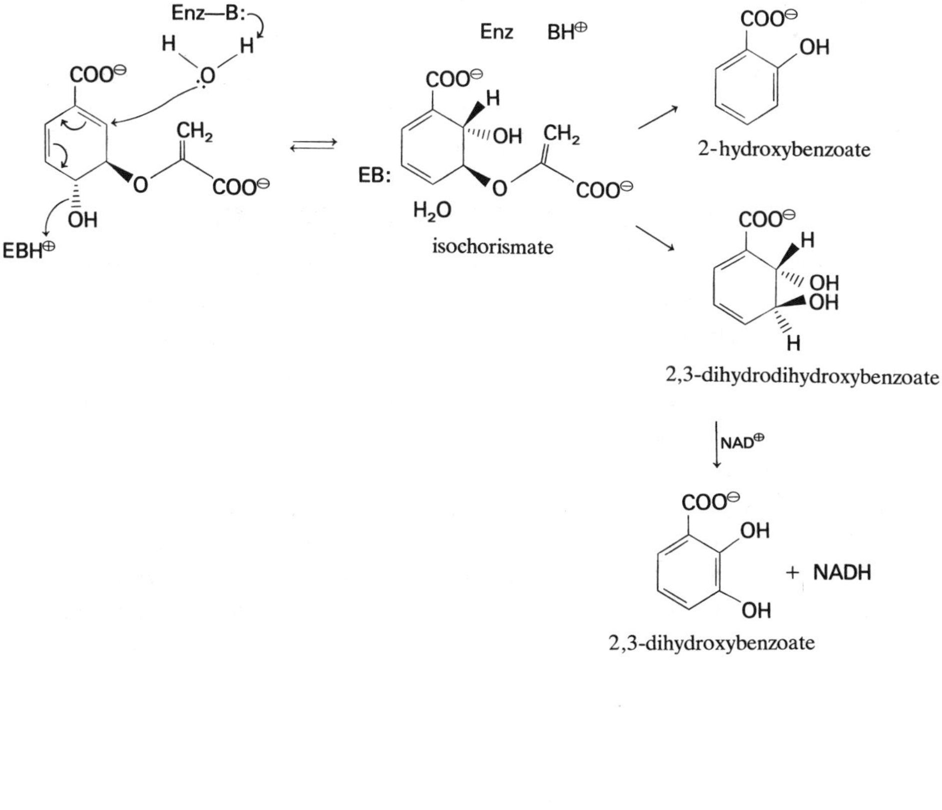

isochorismate

2-hydroxybenzoate

2,3-dihydrodihydroxybenzoate

2,3-dihydroxybenzoate

Molecules of *p*-aminophenylalanine could arise from the putative intermediate in *p*-aminobenzoate synthesis (the 4-amino-3-enolpyruvylcyclohexadiene carboxylate) by a concerted sigmatropic shift (Dardenne et al., 1975; Larsen et al., 1975; Larsen and Wieczorkowska, 1975), followed by a NAD-dependent decarboxylation of the type noted earlier for prephenate to yield *p*-aminophenylpyruvate.

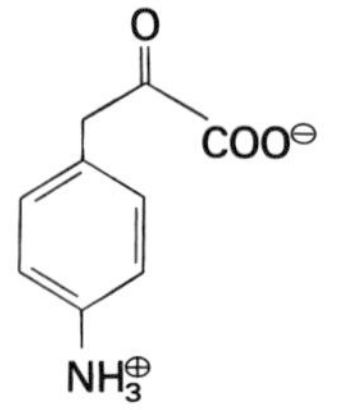

O
$COO^{\ominus}$
$NH_3^{\oplus}$

p-aminophenylpyruvate

Similarly, Pederson has suggested rearrangement of the intermediate in anthranilate biosynthesis to arrive at a precursor for the plant metabolite *p*-carboxy-*m*-aminophenylalanine (Dardenne et al., 1975).

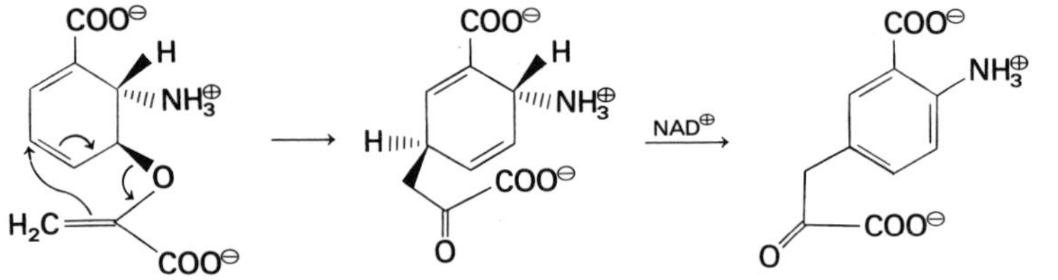

$COO^{\ominus}$ H $NH_3^{\oplus}$ O H_2C $COO^{\ominus}$ → H $COO^{\ominus}$ H $NH_3^{\oplus}$ $COO^{\ominus}$ O $\xrightarrow{NAD^{\oplus}}$ $COO^{\ominus}$ $NH_3^{\oplus}$ O $COO^{\ominus}$

The chemical versatility of chorismate for addition of nucleophiles and for isomeric Claisen rearrangements gives logic to its central placement in the biosynthetic metabolism of aromatic amino acids.

20.C REARRANGEMENTS DEPENDENT ON COENZYME B_{12}

Coenzyme B_{12} was initially detected in 1926 as a factor in raw liver effective in the treatment of pernicious anemia. Some twenty years were then required for its isolation as a pure compound and for its crystallization. Chemical studies indicated a complex structure, and the final structural determination (Fig. 20-2) resulted from X-ray analysis (Lenhard and Hodgkin, 1961), eventually leading to a Nobel prize for Dorothy Hodgkin.

The coenzyme contains a modified porphyrin ring system, with two of the four pyrrole rings fused directly (without an intervening methine bridge). The pyrrole-ring substituents are highly reduced. The modified porphyrin system binds an atom of cobalt in the +3 oxidation state; it is the only naturally occurring coenzyme with functional cobalt, and this is called a corrin ring system. The corrinoid coenzyme B_{12} has the Co^{III} bound so that the four pyrrole nitrogens are

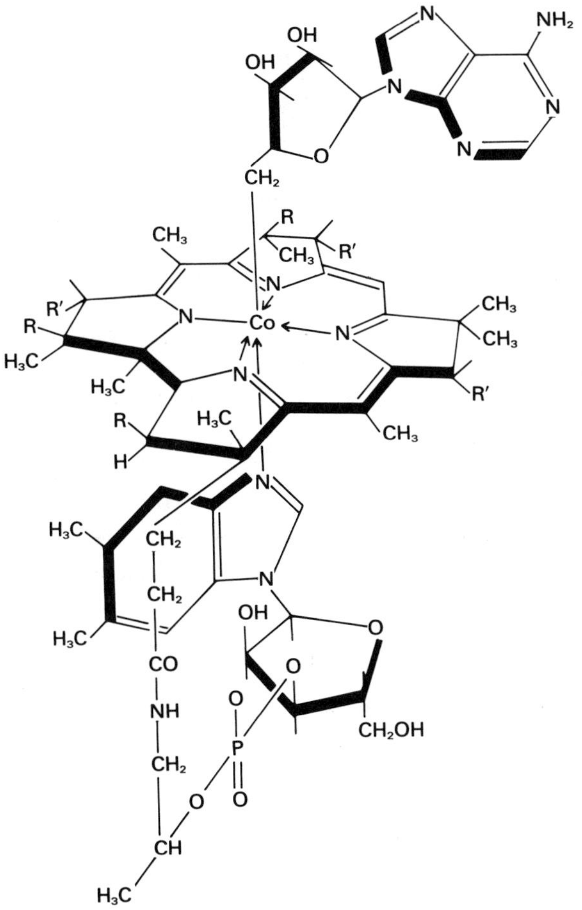

Figure 20-2
Structure of adenosylcobalamin. (From *Federation Proceedings* [1966] 24: 1623–1627)

equatorial ligands, whereas other groups occupy the axial ligands. The bottom axial ligand can vary depending on biological source, but the most frequently observed group is a benzimidazole nucleotide coordinated through one of the nitrogens of the imidazole ring. The 5,6-dimethylbenzimidazole ribotide (DMB) is shown in Figure 20-2; it is covalently linked to one of the pyrrole-ring substituents via a phosphodiester bond. The top axial ligand in the isolated vitamin B_{12} is cyanide ion, and this species is also known as cyanocobalamin. However, the cyanide is an adventitious ligand picked up during isolation and purified as a high-affinity derivative. As with many other vitamins, cyanocobalamin is not coenzymatically active, and the ingested (or de novo synthesized) vitamin must be enzymatically modified. Depending on its cellular function, different modifications are performed. Coenzyme B_{12} is involved in two distinct types of enzymatic catalysis: (1) rearrangements involving 1,2-shifts, and (2) methyl transfer in methionine biosynthesis and in biomethylation of trace metals (which we shall consider in Section V). Only the rearrangements are under discussion here.

For the rearrangements, the active form of the coenzyme B_{12} contains as top axial ligand a 5′-deoxyadenosyl moiety with a covalent carbon–cobalt bond between the $Co^{3\oplus}$ and the 5′-methylene carbon of the ribose group. The inactive vitamin is converted to active coenzyme in an unusual reaction involving ATP and FAD (Friedman, 1975). Although it is poorly characterized, it is likely that the enzyme responsible uses the flavin to reduce the corrin-ring cobalt from $Co^{3\oplus}$ to $Co^{\oplus}$.★ This extraordinarily potent nucleophile (Abeles and Dolphin, 1976) attacks C-5 of the ribose moiety of ATP, expelling tripolyphosphate as leaving group and forming the carbon–cobalt bond in an unusual adenosyl transfer (one of two known examples; see Chapter 25 for the other). This is the only known natural organometallic compound functional in metabolism in biological systems. Figure 20-3 shows the optical absorption spectra of vitamin B_{12} and of the DMB form of the B_{12} coenzyme. Figure 20-4 shows the absorption spectra of cyanocobalamin (cyanide as top axial ligand) that has Co^{III} in the macrocycle along with spectra of the corresponding Co^{II} and Co^{I} species, which may be intermediates in some of the enzymatic reactions discussed later in this chapter. Spectroscopic characteristics of various corrinoid derivatives are presented in Table 20-1.

vitamin B_{12} (Co^{III}) —$FADH_2$→ (Co^{I}) → coenzyme B_{12} (Co^{III})

★Recent studies have suggested that two flavoenzymes may serve sequentially as the physiological reducing system from NADH to the B_{12}-dependent methionine synthase (Chapter 25) in *E. coli.*

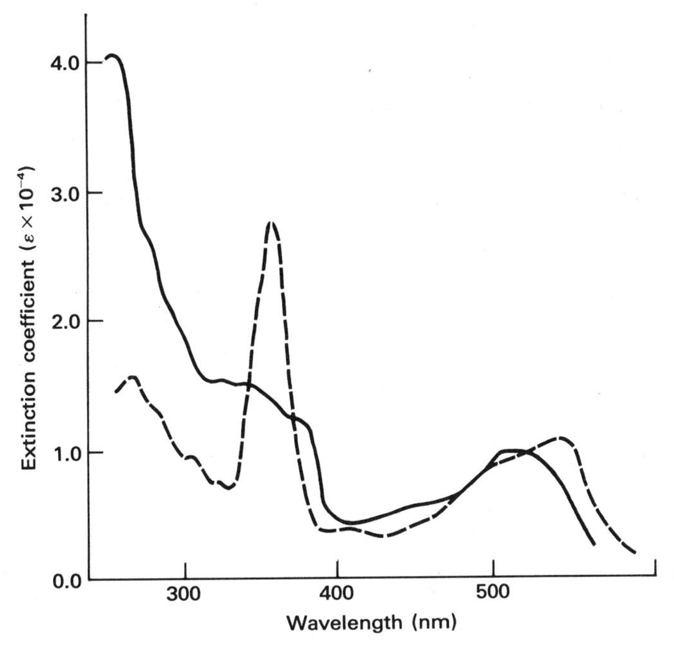

Figure 20-3
Optical absorption spectra of vitamin B_{12} with H_2O as top axial ligand (*dotted line*) and of coenzyme B_{12} with deoxyadenosyl as top axial ligand (*solid line*). (Reprinted with permission from Abeles and Dolphin, *Accounts of Chemical Research* [1976] 9: 115. Copyright by the American Chemical Society.)

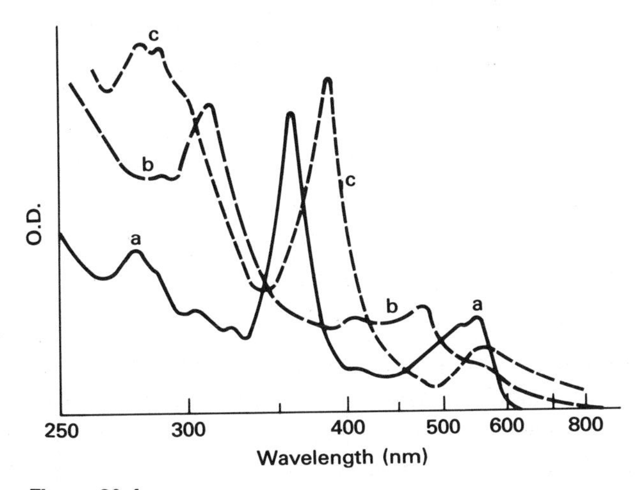

Figure 20-4
Absorption spectra of (**a**) cyanocobalamin, (**b**) cob(II)alamin, and (**c**) cob(I)alamin. (From B. Babior, ed., *Cobalamins*, p. 35. Copyright © 1975. Reprinted by permission of John Wiley & Sons, Inc.)

Table 20-1
Electronic absorption spectra of corrinoids

Corrinoid	Principal absorption band (nm) ($\varepsilon \times 10^{-2}$ in parentheses)						
Cyanocobalamin	—	278 (16.3)	305 (9.7)	322 (7.9)	361 (28.1)	518 (7.4)	550 (8.7)
Aquocobalamin	—	274 (20.6)	—	317 (6.1)	351 (26.5)	499 (8.1)	525 (8.6)
Hydroxocobalamin	—	278 (19.1)	—	325 (11.4)	358 (20.6)	516 (8.9)	535 (9.3)
Adenosylcobalamin	262.0 (35.1)	290 (18.2)	—	318 (13.0)	341 (12.8)	376 (11.0)	522 (8.0)
Methylcobalamin	266.0 (19.1)	280 (18.3)	289 (17.1)	315 (12.5)	340 (13.3)	373 (10.7)	519 (8.7)
Cob(II)alamin	—	288	—	311 (27.5)	—	402 (7.5)	473 (9.2)
Cob(I)alamin	280.5 (29.1)	288 (29.4)	386 (28.0)	455 (2.5)	545 (2.8)	680 (1.7)	800 (1.4)

SOURCE: B. Babior, ed., *Cobalamins* (New York: Wiley, 1975), p. 55.

X-ray analysis indicates that the Co—C bond is a long bond and somewhat labile. It is particularly labile to light-catalyzed homolysis, yielding as initial products the 5′-deoxyadenosyl radical and a Co^{II} form of the coenzyme.

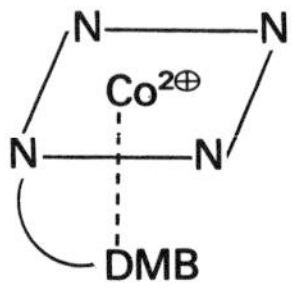

Various apoproteins that bind coenzyme B_{12} can stabilize it somewhat against photolysis, but the usual precaution is to work *in darkness* (as close to total darkness as possible) because the deleterious effect of light on a population of coenzyme molecules is cumulative. In general, coenzyme B_{12} stays tightly bound to apoenzymes during catalysis; it does not dissociate freely at the end of each catalytic cycle.

There are about a dozen enzymes known that carry out B_{12}-dependent rearrangements (Abeles and Dolphin, 1976; Babior, 1975*a*). For almost all of them, the generalized stoichiometry is the following:

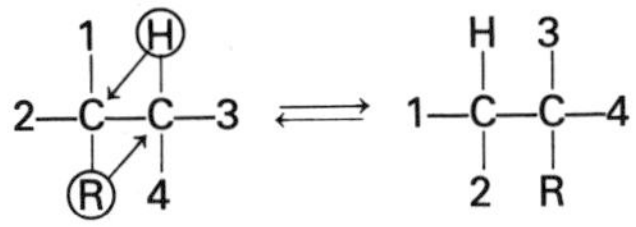

$$2\text{—}\overset{1}{\underset{(R)}{C}}\text{—}\overset{(H)}{\underset{4}{C}}\text{—}3 \rightleftharpoons 1\text{—}\overset{H}{\underset{2}{C}}\text{—}\overset{3}{\underset{R}{C}}\text{—}4$$

where R may be a carbon atom with substituents, an oxygen atom of an alcohol, or an amino group. We shall first discuss the category involving alcoholic substrates, because initial experiments unraveling the role of the coenzyme were elucidated in that system. We shall then generalize to the carbon-skeleton rearrangements and to the amino-group migrations.

20.C.1 Dehydration with Rearrangements: Internal Redox Reactions

The class of dehydrations with rearrangements includes three enzymes (Abeles and Dolphin, 1976; Babior, 1975*a*; Abeles, 1972):

1. propanediol dehydrase converts 1,2-propanediol to propionaldehyde;
2. glycerol dehydrase acts on glycerol to form β-hydroxypropionaldehyde (which rapidly loses another mole of H_2O nonenzymatically to form acrolein);
3. ethanolamine deaminase transforms ethanolamine into acetaldehyde and ammonia.

The stoichiometry of the reactions mediated by propanediol dehydrase and by glycerol dehydrase is similar to that noted in Chapter 17 for the dihydroxy-acid dehydrase in valine and isoleucine biosynthesis. The dihydroxy-acid dehydrase generates an enol, then ketonizes it to the acetol with solvent incorporation, but this mechanism is *not* used in the B_{12}-dependent dehydrases. There is no exchange with or incorporation of solvent hydrogens at any stage in the reaction.

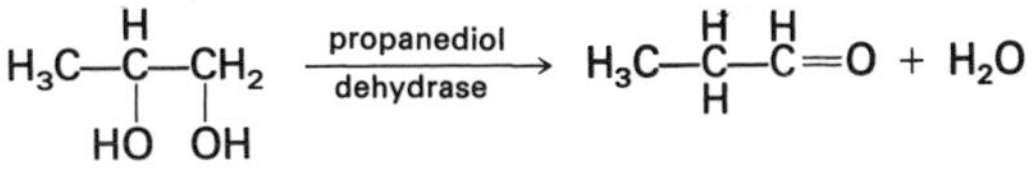

$$H_3C\text{—}\underset{HO}{\overset{H}{C}}\text{—}\underset{OH}{CH_2} \xrightarrow[\text{dehydrase}]{\text{propanediol}} H_3C\text{—}\underset{H}{\overset{H}{C}}\text{—}\overset{H}{C}{=}O + H_2O \quad (1)$$

$$\underset{HO}{H_2C}\text{—}\underset{OH}{\overset{H}{C}}\text{—}\underset{OH}{CH_2} \xrightarrow[\text{dehydrase}]{\text{glycerol}} H_3C\text{—}\underset{OH}{\overset{H}{C}}\text{—}\overset{H}{C}{=}O + H_2O \quad (2)$$

$$\downarrow \text{nonenzymatic}$$

$$H_2C{=}\overset{H}{C}\text{—}\underset{H}{C}{=}O + H_2O$$

$$\underset{HO}{H_2C}\text{—}\underset{NH_3^{\oplus}}{CH_2} \xrightarrow[\text{deaminase}]{\text{ethanolamine}} H_3C\text{—}\overset{H}{C}{=}O + NH_4^{\oplus} \quad (3)$$

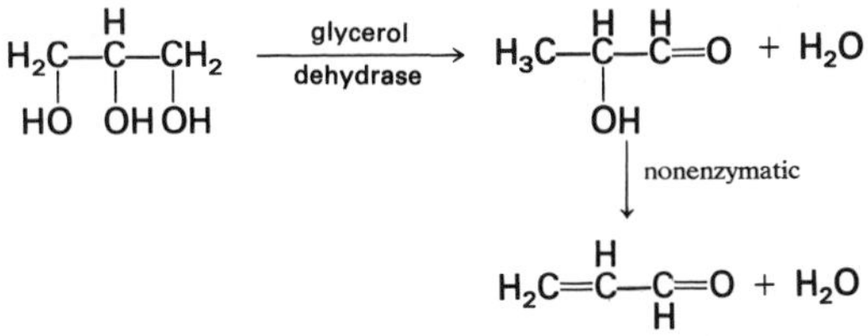

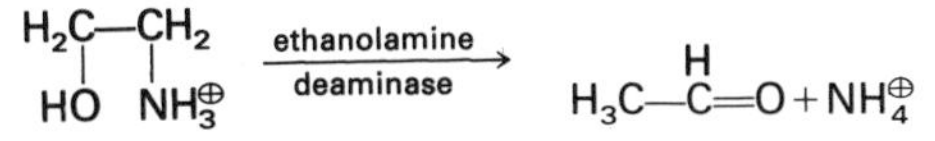

We shall consider the mechanistic information obtained by studies with the propanediol dehydrase because these elucidated the basic role of the coenzyme. Then we shall briefly discuss schemes for the other enzymes.

Propanediol dehydrase is induced in *Aerobacter aerogenes* and has been purified essentially to homogeneity as a complex of 240,000 mol wt, with two subunits having apparent molecular weights of 20,000 and 220,000. The larger subunit probably binds the coenzyme. Although binding of coenzyme B_{12} to the diol dehydrase is tight ($K_{diss} < 10^{-12}$ M) and functionally irreversible, the holoenzyme is photolyzable; thus it is most convenient to purify the light-stable apoprotein, then add back B_{12} coenzyme to reconstitute holoenzyme.

20.C.1.a Direct hydrogen transfer and stereospecificity of the 1,2-shifts

The stereochemistry of substrate utilization has been examined and, curiously, both the 2*R*- and the 2*S*-propanediol molecules react. (The V_{max} for dehydration of the 2*R*-isomer is twofold higher than that for dehydration of the 2*S*-isomer.) One of the initial experiments showing that dehydration proceeds by way of a formal 1,2-shift was carried out with substrates deuterated at C-1 (which then also becomes a chiral center). With 1*R*,2*R*-1-[^{2}H]-propanediol, the enzyme produces 2-[^{2}H]-propionaldehyde of 2*S*-configuration.

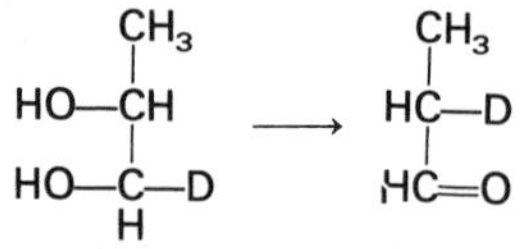

```
    CH3              CH3
     |                |
 HO—CH      ——→      HC—D
     |                |
 HO—C—D              HC=O
     H
   1R,2R      2S-[2H]-propionaldehyde
```

Formation of 2*S*-deuteropropionaldehyde shows that inversion occurred at C-2 with backside attack of deuterium from C-1. Further, because it was the deuterium (not the hydrogen) that migrated, the enzyme is catalyzing specific migration of the *R*-substituent at C-1. The other 1*R*-1-[^{2}H]-diastereomer (1*R*,2*S*) was similarly prepared by reduction of *S*-lactaldehyde with 4*R*-[^{2}H]-NADH and alcohol dehydrogenase. This 1*R*,2*S*-1-[^{2}H]-propanediol produces a 1-[^{2}H]-propionaldehyde after incubation with diol dehydrase. This time, migration of the *S*-substituent at C-1 (the hydrogen rather than the deuterium) has occurred.

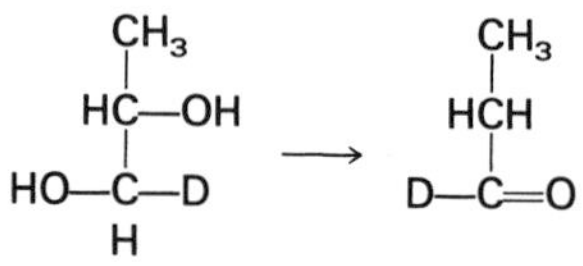

```
     CH3              CH3
      |                |
     HC—OH            HCH
      |     ——→        |
 HO—C—D             D—C=O
     H
   1R,2S      1-[2H]-propionaldehyde
```

Thus, for the two C-2 enantiomers of 1,2-propanediol, the enzyme displays *opposite* stereospecificities of migrating group at C-1.

It seemed quite likely that formation of propionaldehyde involves a concomitant 1,2-shift of an OH from C-2 to C-1 as a hydrogen species migrates from C-1 to C-2; this would yield as initial product a 1,1-gemdiol, the hydrated aldehyde. Arigoni and colleagues demonstrated elegantly and cleanly that subsequent dehydration to the unhydrated aldehyde is both enzyme-catalyzed and stereospecific (Retey et al., 1966). They prepared either 2*S*-1-[^{18}O]-1,2-propanediol or 2*R*-1-[^{18}O]-1,2-propanediol by chemical exchange of H_2O into either *S*- or *R*-lactaldehyde, followed by enzymatic reduction. After reaction with diol dehydrase, they assayed the propionaldehyde product for ^{18}O content, with the following results. (H_S migration is shown here as a direct transfer; it is not. See ¶20.C.1.b.)

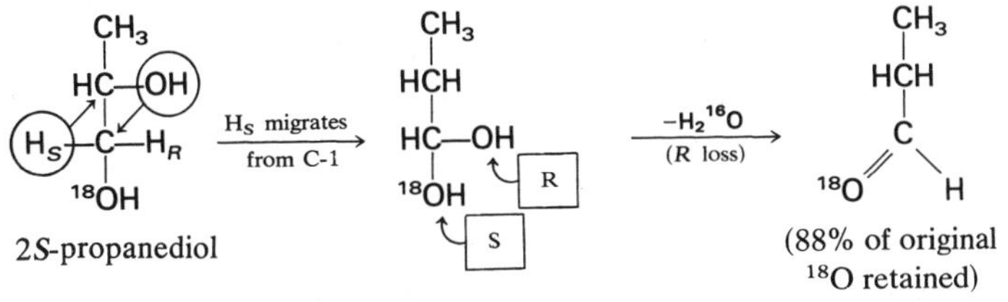

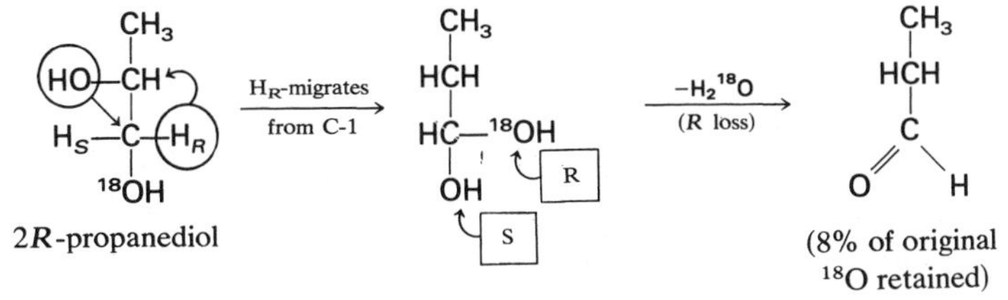

2*S*-propanediol — H_S migrates from C-1 → — $-H_2{}^{16}O$ (*R* loss) → (88% of original ^{18}O retained)

2*R*-propanediol — H_R-migrates from C-1 → — $-H_2{}^{18}O$ (*R* loss) → (8% of original ^{18}O retained)

Therefore, the diol dehydrase removes the pro*R*-hydroxyl group from the 1,1-gemdiol specifically, independent of which enantiomeric substrate is used (the gemdiol is the same from each). With 2*S*-propanediol, the OH migrating from C-2 to C-1 ends up pro*R* and is removed. With 2*R*-propanediol, the migrating OH ends up in the pro*S*-position at C-1 and is retained.

20.C.1.b Hydrogen transfer actually is a multistep process

As suggested by the intramolecular deuterium-transfer experiments, there is no equilibration with solvent hydrogen (no tritium incorporation from tritiated water during propionaldehyde formation). Studies on the rate of hydrogen transfer then were carried out with *R,S*-1-[$^2H_2$]-propanediol. The V_{max} data, compared with those for the diprotio substrate, revealed a deuterium kinetic isotope effect of 10 to 12 and indicated that some hydrogen-transfer step (more than one are involved!) is fully rate-determining.

Indirect evidence that *hydrogen transfer must be a multistep process* was obtained using two substrates in a single reaction mixture. In addition to accepting propanediol, the dehydrase also will utilize ethylene glycol, converting it to acetaldehyde. A substrate mixture of unlabeled ethylene glycol and 1-[^{3}H]-propanediol was used, and the product aldehydes were examined for tritium content. Both the propionaldehyde and the acetaldehyde were radioactive, showing that tritium transfer can be *intermolecular.*

$$\underset{\text{HO}\quad\text{OH}}{H_2C{-}CH_2} + \underset{\text{HO}\quad\text{OH}}{H_3C{-}\overset{\text{H}}{C}{-}\overset{\text{H}}{C}{-}{}^3H} \longrightarrow {}^3H{-}\underset{\text{H}}{\overset{\text{H}}{C}}{-}\overset{\text{H}}{C}{=}O + H_3C{-}\underset{{}^3H}{\overset{\text{H}}{C}}{-}\overset{\text{H}}{C}{=}O$$

These data strongly supported the hypothesis of an *intermediate* that contains the migrating substrate-derived hydrogen and also additional hydrogens to equilibrate with the substrate-derived species. This must be so if tritium is transferred *in* to the intermediate, but a proton is transferred *back out* to form unlabeled propionaldehyde. In a subsequent turnover, a proton is transferred in from a molecule of ethylene glycol, and a tritium atom is transferred back out to form [^{3}H]-acetaldehyde. (With only a single type of substrate present, hydrogen transfer *can* be intramolecular.)

The likely candidate for the hydrogen-bearing intermediate was the B_{12} coenzyme. Direct evidence for this role stemmed from experiments with 1-[^{3}H]-propanediol of high specific radioactivity, so that labeled coenzyme (present in catalytic amounts bound to the enzyme) could be detected. Incubations were quenched after short intervals (10 to 60 sec) and the intact coenzyme isolated in the dark. The specific radioactivity of the coenzyme increases with time (Fig. 20-5); the coenzyme accumulates tritium during turnover. To determine the locus

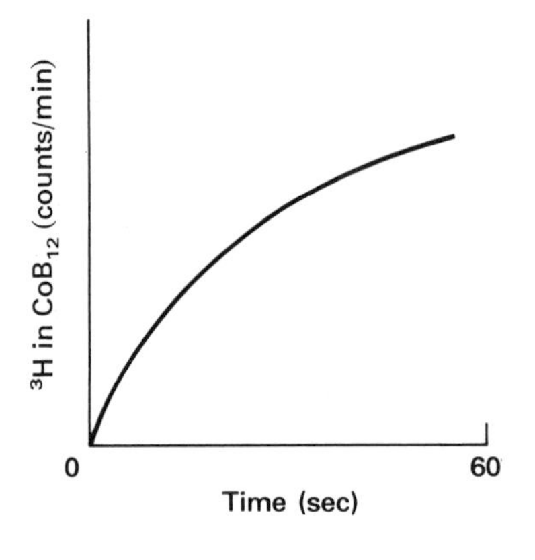

Figure 20-5
Incorporation of substrate-derived tritium into the C-5′ methylene group of coenzyme B_{12} during catalysis.

of the tritium in the coenzyme, it was photolyzed either anaerobically (yielding the 5′,8-cycloadenosine from the deoxyadenosyl axial ligand) or aerobically (producing aquocobalamin and adenosine 5′-aldehyde) (Abeles, 1972).

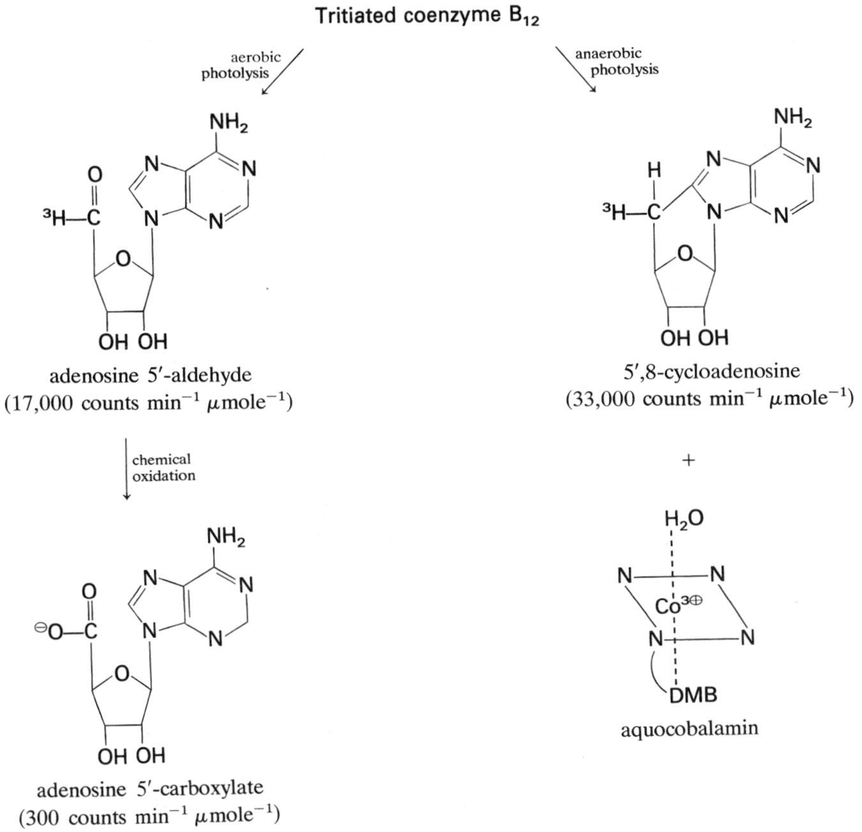

Tritiated coenzyme B_{12}

aerobic photolysis

anaerobic photolysis

adenosine 5′-aldehyde
(17,000 counts min^{-1} $\mu mole^{-1}$)

chemical oxidation

adenosine 5′-carboxylate
(300 counts min^{-1} $\mu mole^{-1}$)

5′,8-cycloadenosine
(33,000 counts min^{-1} $\mu mole^{-1}$)

+

aquocobalamin

As indicated, the aldehyde is radioactive (17,000 counts min^{-1} $\mu mole^{-1}$) and the aquocobalamin is nonradioactive. Tritium has been introduced into the deoxyadenosyl ligand during turnover. Subsequent chemical oxidation of the aldehyde to the acid, adenosine 5′-carboxylate, indicates that the acid is virtually unlabeled (300 counts min^{-1} $\mu mole^{-1}$). Thus, the radioactivity in adenosine 5′-aldehyde is all in the aldehyde hydrogen, and C-5′ is the specific locus to which substrate tritium was transferred. Interestingly, the 5′,8-cycloadenosine from anaerobic homolysis has a specific radioactivity of 33,000 counts min^{-1} $\mu mole^{-1}$, twice that of the adenosine 5′-aldehyde. The cyclic derivative has both of the coenzyme's C-5′ methylene hydrogens, whereas the aldehyde retains only one. The observed specific radioactivities suggest that *tritium is distributed equally (and randomly)* between both methylene hydrogens of the C-5′ position of the coenzyme.

20.C.1.c Lack of chirality indicates an intermediate symmetric with respect to transferable hydrogens

In a complementary experiment, the enzymatically generated [^{3}H]-coenzyme B_{12} was reincubated with apoenzyme and unlabeled propanediol. As expected, 2-tritiopropionaldehyde was produced, showing the chemical competence of the 5′-[^{3}H]-coenzyme B_{12} as an intermediate hydrogen carrier. Although there is an isotope selection against tritium, all of the tritium is eventually transferable out of the coënzyme, further indicating that tritium at both the H_R and H_S positions is transferable back to the developing propionaldehyde (Abeles, 1972).

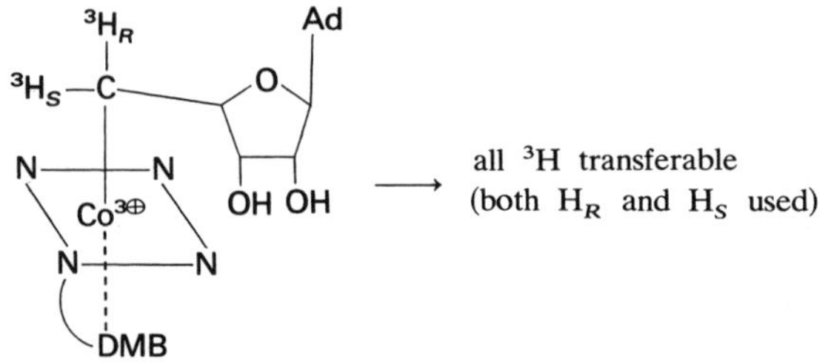

^3H_R, ^3H_S—C, Ad, O, N, N, $Co^{3\oplus}$, N, N, DMB, OH OH → all ^{3}H transferable (both H_R and H_S used)

Final proof that the chiral enzyme does not distinguish between H_R and H_S at C-5′ of the deoxyadenosyl axial ligand came with the chemical synthesis of the *R*,*S*-5′-[^{3}H]-coenzyme B_{12}, which is a mixture of the *R*- and *S*-tritio species.

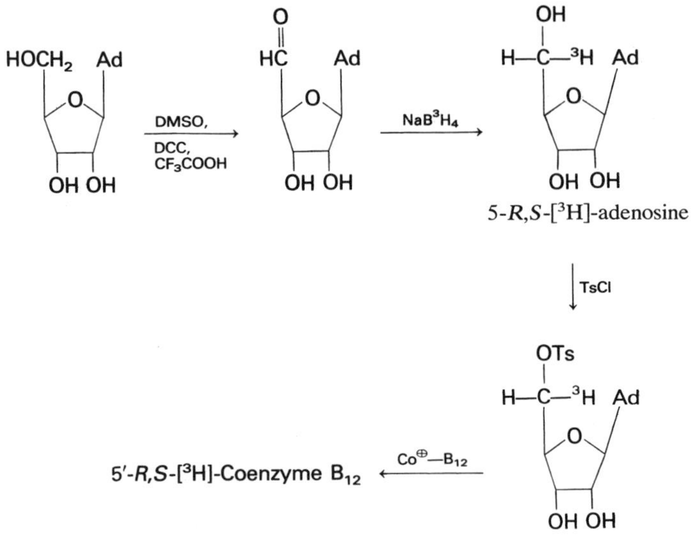

$HOCH_2$ Ad, O, OH OH —(DMSO, DCC, CF_3COOH)→ HC(=O) Ad, O, OH OH —(NaB^3H_4)→ H—C(OH)—^{3}H Ad, O, OH OH

5-*R*,*S*-[^{3}H]-adenosine

↓ TsCl

H—C(OTs)—^{3}H Ad, O, OH OH —($Co^{\oplus}$—B_{12})→ 5′-*R*,*S*-[^{3}H]-Coenzyme B_{12}

It was possible, though unlikely, that the sodium borotritide reduction proceeded with some degree of unwanted stereoselectivity. To avoid this uncertainty, the synthetic tritiated coenzyme B_{12} was photolyzed aerobically back to the 5′-[^{3}H]-aldehyde, which was then reduced with $NaBH_4$ (which would now give the opposite pattern of stereoselectivity) and the B_{12} coenzyme resynthesized. Both species of [^{3}H]-coenzyme B_{12} behaved identically. All the tritium was transferable to the propionaldehyde formed. Great effort was employed to establish this lack of chiral action at C-5′ of the coenzyme because this result was a priori unlikely, but once verified it had dramatic mechanistic significance.

Having thus unequivocally established that *the enzyme does not distinguish between the two prochiral hydrogens* at C-5′, Abeles and colleagues were faced with the need to explain how substrate hydrogen is equilibrated with these methylene hydrogens (¶20.C.1.b). They proposed that the two prochiral hydrogens and the substrate-derived hydrogen *become equivalent* during catalysis through fragmentation of the carbon–cobalt bond and formation of a C-5′ methyl group on addition of the C-1 hydrogen of propanediol. A methyl group is torsiosymmetric, unlike a prochiral methylene group. Free rotation of the methyl group would equilibrate the three hydrogens and cause the observed loss of stereochemical information.

$$\mathrm{S{-}^{*}H} + \mathrm{H_S{-}C(R)(Co^{3\oplus}){-}H_R} \qquad \left[\mathrm{R{-}CH_2{-}^{*}H} \underset{}{\overset{\text{fast}}{\rightleftharpoons}} \mathrm{R{-}CH_2{-}^{*}H} \overset{\text{fast}}{\rightleftharpoons} \mathrm{R{-}CH_2{-}^{*}H}\right] + \dot{\mathrm{S}} + \mathrm{Co^{2\oplus}}$$

Before we consider direct evidence for 5′-deoxyadenosine formation in each catalytic cycle, it is worth noting again that kinetic experiments were performed to show that coenzyme B_{12} is a kinetically competent hydrogen carrier, as measured by tritium transfer in from substrate and back out to product. During these experiments, a tritium kinetic isotope selection of 250 was observed in the half-reaction of 5′-[^{3}H]-coenzyme B_{12} going to product aldehyde. Correcting for the statistical factor of three hydrogens in the 5′-methyl group of the deoxy-adenosine species, a value of 125 is obtained for k_H/k_T. This is an unusually and inexplicably large kinetic isotope discrimination.

20.C.1.d Evidence for carbon–cobalt bond fragmentation during catalysis

Direct evidence that pseudosubstrates can cause breakage of the carbon–cobalt bond in the enzyme evolved initially from studies with glycolaldehyde (Wagner et al., 1966) and with 2-chloroacetaldehyde (Findlay et al., 1972).

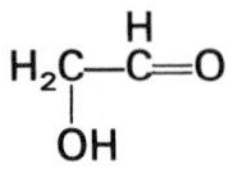

$$\underset{\text{glycolaldehyde}}{H_2C(OH){-}C(H){=}O} \qquad \underset{\text{chloroacetaldehyde}}{H_2C(Cl){-}C(H){=}O}$$

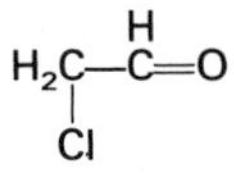

Inactivation ensues when glycolaldehyde is mixed with the CoB_{12}–diol-dehydrase complex; the coenzyme is converted to what appears from the characteristic UV–visible spectrum to be a transient $Co^{2\oplus}$ species (as expected for homolytic cleavage). The formation of 5′-deoxyadenosine occurs concomitant with inactivation. The same phenomena are observable with chloroacetaldehyde, although in this case the molecule is an alkylating agent, which can obscure the reasons for loss of catalytic activity (Findlay et al., 1972). Chloroacetaldehyde can undergo at least the hydrogen-abstraction step, because addition of 5′-[^{3}H]-CoB_{12} and enzyme results in tritium exchange into chloroacetaldehyde. Although these experiments proved that deoxyadenosine (with a 5′-methyl group) is detectable, it remained possible that it does not form as an intermediate in ordinary catalysis, but rather occurs only as a consequence of inactivation—that its formation signals an abortive fragmentation process.

Recently, studies by Carty, Abeles, and Babior on the related enzyme ethanolamine deaminase strengthened the case for reversible carbon–cobalt bond fission in every turnover (Carty et al., 1971; Babior et al., 1974). Ethanolamine (the physiological substrate) is converted to acetaldehyde and $NH_4^{\oplus}$.

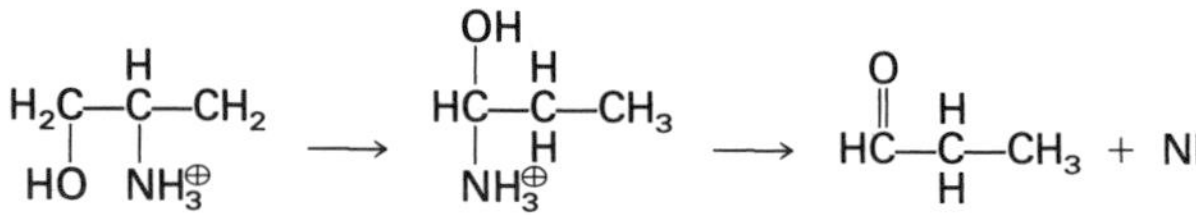

$$H_2C(HO){-}CH_2(NH_3^{\oplus}) \longrightarrow H_3C{-}C(H)(NH_2){-}OH \longrightarrow H_3C{-}C(H){=}O + NH_4^{\oplus}$$

It was observed that 2-aminopropanol is a poor substrate, yielding propionaldehyde, again with amino-group migration to C-1.

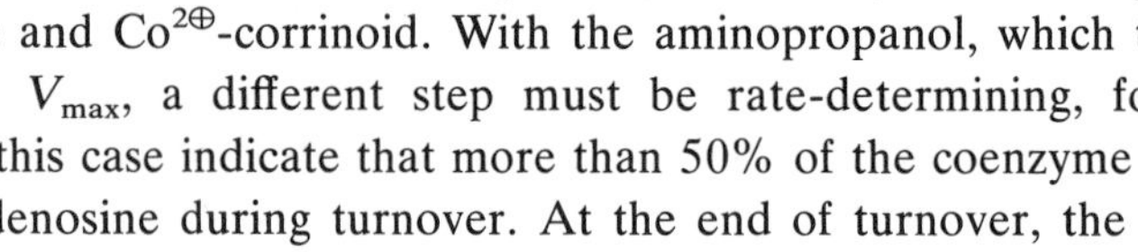

$$H_2C(HO){-}C(H)(NH_3^{\oplus}){-}CH_2 \longrightarrow HC(OH)(NH_3^{\oplus}){-}C(H)(H){-}CH_3 \longrightarrow HC(=O){-}C(H)(H){-}CH_3 + NH_4^{\oplus}$$

Rapid-quench experiments with ethanolamine during steady-state turnover indicated that 2% to 4% of the coenzyme exists as the cleaved species, 5-deoxyadenosine and $Co^{2\oplus}$-corrinoid. With the aminopropanol, which turns over at much lower V_{max}, a different step must be rate-determining, for quench experiments in this case indicate that more than 50% of the coenzyme is cleaved to free deoxyadenosine during turnover. At the end of turnover, the coenzyme has been reconverted to the form with the covalent cobalt–carbon bond, confirming that 5′-deoxyadenosine formation is reversible. This per se provides no information on what the oxidation state of the cobalt may be in such a cleaved intermediate. That knowledge would, in turn, indicate whether Co—C fission occurs homolytically (to radicals) or heterolytically.

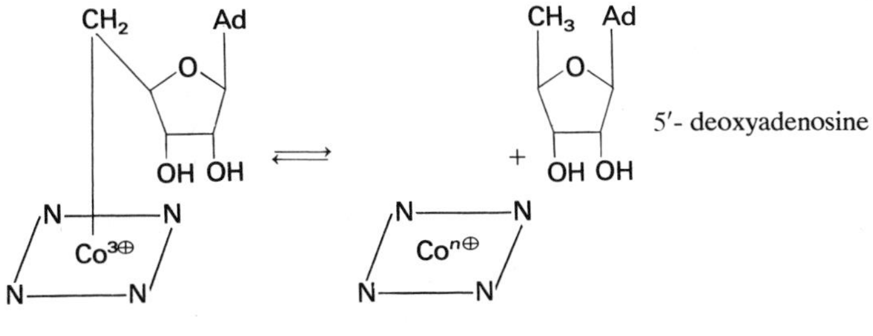

5′- deoxyadenosine

20.C.1.e Evidence for homolytic fission

This question of the oxidation state of cobalt during catalysis may also be related to whether the migrating hydrogen of the substrate is abstracted as a hydride ion (H:) or as a hydrogen atom (H·). (A proton abstraction seems unlikely.) If Co—C bond cleavage occurs heterolytically (to Co^{I} and $^{\oplus}CH_2R$), and then hydride transfer occurs from substrate to form the C-5′ methyl of 5′-deoxyadenosine, the $Co^{3\oplus}$ in the resting corrin ring should undergo two-electron reduction to Co^{I}, a powerful nucleophile and also a diamagnetic cation.

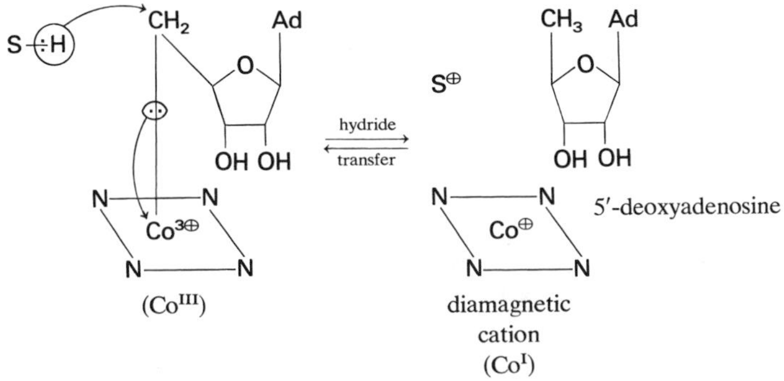

hydride transfer

5′-deoxyadenosine

(Co^{III})

diamagnetic cation (Co^{I})

On the other hand, if homolytic fission occurs (to Co^{II} and $\cdot CH_2R$), with subsequent H· transfer from substrate to form the 5′-deoxyadenosine, then a one-electron reduction to the *paramagnetic* Co^{II} ion would occur. Whereas a substrate carbonium ion would form in the hydride-transfer case, a substrate radical would form in the case of hydrogen-atom transfer (top of p. 654).

Provided that the substrate radical and the $Co^{2\oplus}$ corrinoid do not spin pair, the second mechanism should generate the characteristic eight-line hyperfine structure of the $Co^{2\oplus}$ radical during turnover. Quantitation of the radical depends

S–H CH_2 Ad O OH OH (Co^III^) N N N N $Co^{3\oplus}$

hydrogen-atom transfer

$S^{\cdot}$ N N N N $Co^{2\oplus}$ paramagnetic cation (Co^{II}) + CH_3 Ad O OH OH 5′-deoxyadenosine

on what the predominant form of the holoenzyme is during turnover or in rapid-mixing experiments. Indeed, recent ESR (electron spin resonance) experiments on ribonucleotide reductase, glycerol dehydrase, ethanolamine deaminase, and propanediol dehydrase all detect $Co^{2\oplus}$ corrin during turnover (Fig. 20-6; Abeles and Dolphin, 1976; Babior, 1975*a*). Recent ESR rapid kinetic studies

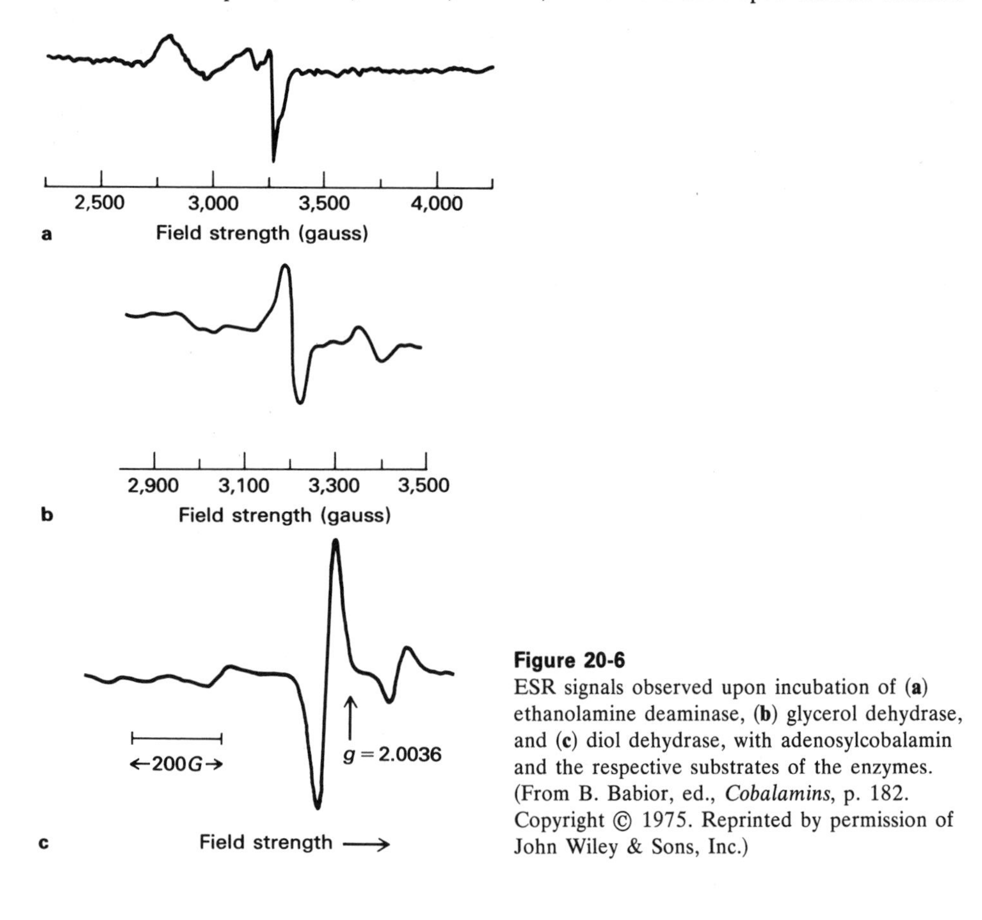

Figure 20-6
ESR signals observed upon incubation of (**a**) ethanolamine deaminase, (**b**) glycerol dehydrase, and (**c**) diol dehydrase, with adenosylcobalamin and the respective substrates of the enzymes. (From B. Babior, ed., *Cobalamins*, p. 182. Copyright © 1975. Reprinted by permission of John Wiley & Sons, Inc.)

with diol dehydrase established that homolytic cleavage of the C—Co bond is complete within 5 msec (the dead time for the instrument) as measured by $Co^{2\oplus}$ appearance (Findlay et al., 1973). Additionally, other radical signals exist that differ with different substrates and that probably represent complex mixtures of the three other possible radicals: substrate, product, and 5′-deoxyadenosyl radicals.

With this information, Abeles has postulated a scheme equivalent to that shown below for diol dehydrase catalysis, where SH = propanediol, and the H is specifically the transferable hydrogen atom at C-1.

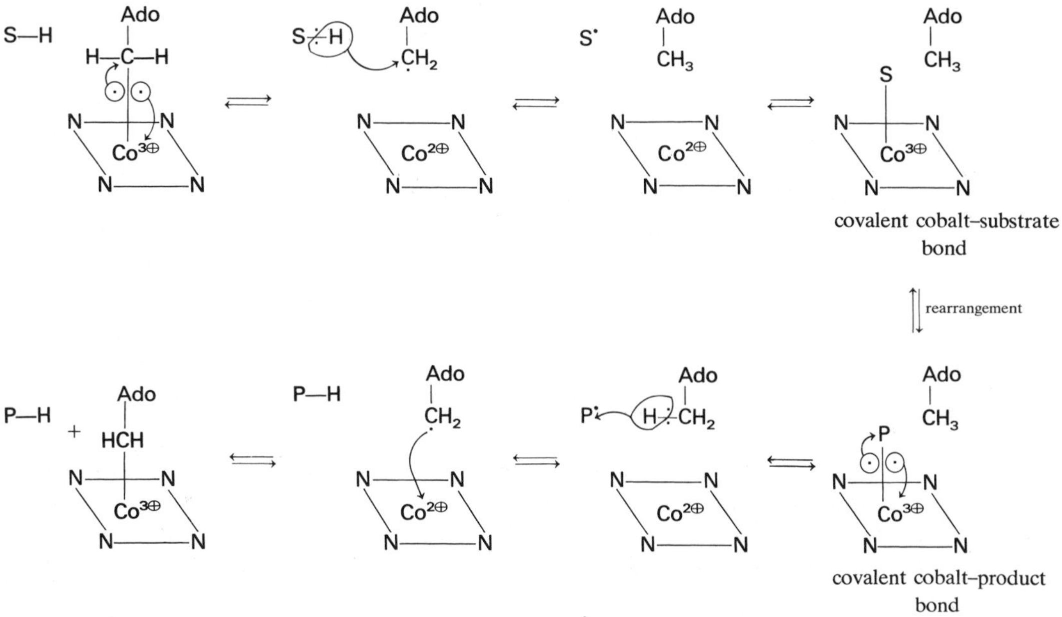

covalent cobalt–substrate bond

rearrangement

covalent cobalt–product bond

Coenzyme homolytic fission is followed by hydrogen-atom abstraction from substrate by the 5′-deoxyadenosyl radical. In the resultant 5′-deoxyadenosine, rapid free rotation of the 5′-methyl generates steric equivalence among the three hydrogens. Collapse of the substrate radical and $Co^{2\oplus}$ then generates an alkyl cobalt compound. The mysterious migration of the OH from C-2 to C-1 then occurs by an unknown process (labeled "rearrangement" in the scheme). If this product–cobalt species is obtained, then homolysis yields $Co^{2\oplus}$ and $P^{\bullet}$, from which a retracing of steps yields the product (with hydrogen transferred to C-2) and the $Co^{2\oplus}$–deoxyadenosyl radical pair, which collapses to reform coenzyme B_{12}.

If this mechanistic scheme is in essence correct for fission and reformation of the Co—5′-CH_2 bond and transfer of hydrogen from substrate to the 5′-deoxyadenosine intermediate and back to rearranged product, it is likely to apply not only to the three enzymes in this subcategory, but also to those carrying out CoB_{12}-dependent carbon-skeleton rearrangements and to those catalyzing amino-group migrations.

The rearrangement of the putative substrate–cobalt σ-complex to the product–cobalt σ-complex is (as noted earlier) unclear and could occur by any of a variety of mechanisms (Abeles and Dolphin, 1976; Babior, 1975*a*). For example, one possible mechanism would involve a cobalt π-complex intermediate:

H H
C—C—OH
HO H
Co^{III}
substrate
σ-complex

$\xrightarrow{-OH}$

H
C—$CH_2^{\oplus}$
HO
Co^{III}

$\rightleftarrows$

H
C—CH_2
HO
Co^{III}
π-complex

$\xrightarrow{H_2O}$

OH
H
C—CH_2
HO Co^{III}
product
σ-complex

Although there is some consensus for homolytic fission of the Co—C^5 bond of the B_{12} coenzyme during catalysis, at least two groups of researchers are unconvinced that the rearrangement of substrate to product skeleton need proceed via a covalent σ-bond between cobalt and substrate or product fragment (Krower et al., 1978; Retey et al., 1978). Retey and colleagues have also pointed out that there need be no common stereochemical outcome in the various B_{12}-mediated enzymatic rearrangements. Inversion is seen with propanediol dehydrase, retention with methylmalonyl-CoA mutase (discussed below; Sprecher et al., 1966; Retey and Zagalak, 1973), and racemization with chirally labeled 2-aminopropanal and ethanolamine deaminase (Retey et al., 1974). The particular apoprotein must control the geometry of bond formation in the rearranging intermediate; racemization mandates that free rotation around the C^{α}—C^{β} bond of the intermediate must be fast.

20.C.2 Carbon Skeletal Rearrangements

The second subgroup of rearrangements mediated by CoB_{12}-dependent mutases comprises carbon skeletal rearrangements. There are three characterized examples (Babior, 1975*a*; H. Barker, 1972):

1. glutamate mutase;
2. methylmalonyl-CoA mutase;
3. methyleneglutarate mutase.

Enzymes 1 and 3 apparently are confined to clostridia (anaerobic bacteria), whereas enzyme 2 is also present in animal cells and is to date the only identified B_{12}-dependent rearrangement enzyme in higher organisms.

Glutamate mutase was the first enzyme identified as a B_{12}-dependent entity (H. Barker, 1972). It has been purified to a state containing two components: a corrin coenzyme-binding subunit of 120,000 mol wt, and an air-sensitive sulfhydryl-containing polypeptide of 17,000 mol wt. The roles of these subunits in catalysis have not been determined. L-Glutamate is converted reversibly to *threo*-β-methylaspartate, which in turn is converted to mesaconate and ammonia by the carbon–nitrogen lyase we discussed in Chapter 18. Labeling studies have shown that the two-carbon fragment comprising C-1 and C-2 of glutamate migrates from C-3 in that molecule to C-4, while a C-4 hydrogen is transferred to C-3, yielding the methyl group.

5 4 3 2 1

$^{\ominus}OOC-CH_2-CH_2-CH(NH_3^{\oplus})-COO^{\ominus}$ $\underset{\text{mutase}}{\xrightleftharpoons{CoB_{12}\text{-dependent}}}$ $^{\ominus}OOC-CH(CH_3)-CH(NH_3^{\oplus})-COO^{\ominus}$ $\xrightleftharpoons{\text{lyase}}$ $^{\ominus}OOC(H)C=C(CH_3)COO^{\ominus}$ + $NH_4^{\oplus}$

formerly C-3

formerly C-4

L-glutamate

threo-β-methyl-L-aspartate

The α-methyleneglutarate mutase is a recently described activity induced by growth of *Clostridium barkeri* on nicotinate (Kung et al., 1970). It may produce β-methylitaconate by the general path outlined for the glutamate mutase.

$^{\ominus}OOC-CH_2-CH_2-C(=CH_2)-COO^{\ominus}$ $\rightleftharpoons$ $^{\ominus}OOC-CH(CH_3)-C(=CH_2)-COO^{\ominus}$

α-methyleneglutarate

β-methylitaconate

The third example, methylmalonyl-CoA mutase, is responsible for converting the branched-chain methylmalonyl-CoA to the more prosaic metabolite succinyl-CoA.

$H_2C-CH(COSCoA)-COO^{\ominus}$ $\rightleftharpoons$ $^{\ominus}OOC-CH_2-CH_2-COSCoA$

R-methylmalonyl-CoA

succinyl-CoA

In bacterial and animal cells, this enzyme is important in catabolism of valine and isoleucine, both of which funnel into the three-carbon metabolite propionyl-CoA, which in turn undergoes carboxylation to the *S*-isomer of methylmalonyl-CoA. *R*- and *S*-isomers of the branched-chain acyl-CoA are isomerized by the methylmalonyl-CoA isomerase mentioned in Chapter 19.

The intimate details of the three carbon-skeleton rearrangements are even less well understood than those of the CoB_{12}-dependent rearrangements discussed in ¶20.C.1. The overall processes can be described as transformations of what look to be unactivated methyl groups into methylene moieties and their concomitant insertion into the backbone carbon chain of the product. The 1,2-transfer of hydrogen is *intermolecular* to coenzyme B_{12} and back to product. Whether the migrating carbon fragment of substrate attaches to the coenzyme as well (e.g., by cobalt–carbon bond formation) had been unclear, but recent model studies by Dowd and colleagues have shed light on some relevant chemistry. They synthesized a derivative of vitamin B_{12} with the methyl carbon of methylitaconate attached to the cobalt. On standing in the dark, this substance underwent a slow (~200 hr) reaction. Three products derived from the original organic ligand (methylitaconate) were detected:

1. *β*-methylitaconate (25%);
2. butadiene-2,3-dicarboxylate (50%);
3. *α*-methyleneglutarate (25%).

$H_2C{=}C(COOR)CH(COOR)CH_2{-}Co \longrightarrow$ $H_2C{=}C(COOH)CH(CH_3)COOH$ (1) + $H_2C{=}C(COOH)C(COOH){=}CH_2$ (2) + $H_2C{=}C(COOH)CH_2CH_2COOH$ (3)

The *α*-methyleneglutarate is the product formed in the enzymatic rearrangement of methylitaconate. Thus, the likelihood of formation of organocorrin intermediates and their subsequent rearrangement seems quite high in the enzymatic process as well. (Dowd has speculated that a cyclopropyl isomer may possibly be involved in the rearrangement.) The ability of the corrinoid cobalt to exist in the +1, the +2, or the +3 oxidation state leaves several possibilities open for the electronic configuration of any rearranging species.

Scott and Kang (1977), Dowd and Shapiro (1976), and Flohr et al. (1976) have also mimicked the rearrangement of the methylmalonic-to-succinic skeleton rearrangement (methylmalonyl-CoA mutase). Thus, the indicated methylmalonyl-thioester–cobalt diester, on incubation in the dark for 24 hr, gave as the only rearranged product the methylsuccinate thioester (Scott and Kang, 1977). The presence of the —COSR grouping improves rearrangement yield and directs exclusive thioester migration (as seen enzymatically).

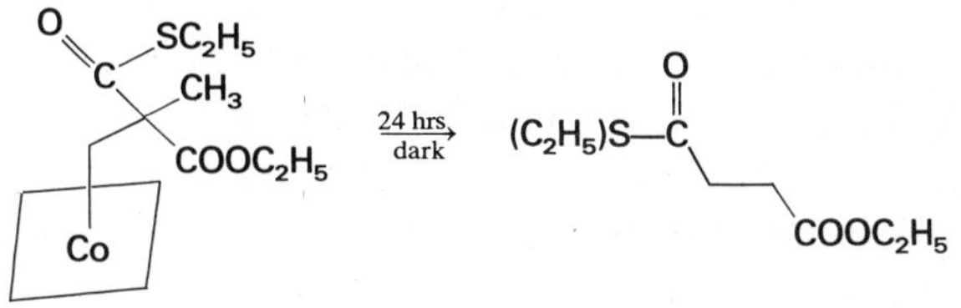

O, SC_2H_5, CH_3, $COOC_2H_5$, Co $\xrightarrow[\text{dark}]{\text{24 hrs}}$ $(C_2H_5)S—C(=O)$, $COOC_2H_5$

20.C.3 Amino-Group Migrations

There are three enzymes known, all from clostridia, that catalyze the migration of the ω-amino group of diamino acids (Babior, 1975*a*; T. Stadtman, 1973). These enzymes are important in fermentative metabolism when the anaerobic bacteria are grown on lysine as carbon source. The three distinct enzymes have been purified to homogeneity, are of similar size, and have all been shown to require pyridoxal phosphate as well as coenzyme B_{12}. In all three reactions there is a 1,2-shift of a hydrogen (from C-5 in the C_6 substrates, from C-4 in the C_5 substrates) that occurs without solvent hydrogen exchange.

The first enzyme is L-3,6-diaminohexanoate (β-lysine) mutase. Studies with ^{15}N have confirmed that the amino-group migration is intramolecular with this enzyme (Stadtman, 1973).

H_2C, $NH_3^{\oplus}$, $NH_3^{\oplus}$, $COO^{\ominus}$ $\longrightarrow$ H_3C, $NH_3^{\oplus}$, $NH_3^{\oplus}$, $COO^{\ominus}$

β-lysine 3,5-diaminohexanoate

The second enzyme is D-α-lysine mutase.

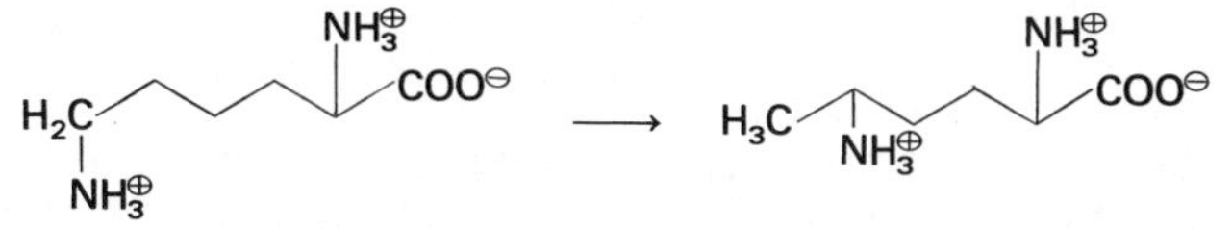

$NH_3^{\oplus}$, $COO^{\ominus}$, H_2C, $NH_3^{\oplus}$ $\longrightarrow$ $NH_3^{\oplus}$, $COO^{\ominus}$, H_3C, $NH_3^{\oplus}$

α-lysine 2,5-diaminohexanoate

The third enzyme is D-ornithine mutase.

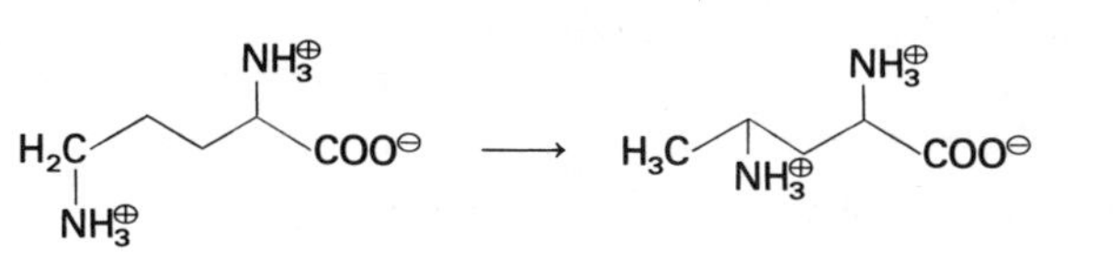

$NH_3^{\oplus}$, H_2C, $COO^{\ominus}$, $NH_3^{\oplus}$ $\longrightarrow$ $NH_3^{\oplus}$, H_3C, $NH_3^{\oplus}$, $COO^{\ominus}$

ornithine 2,4-diaminopentanoate

The role of the pyridoxal phosphate in these reactions is as yet unclear. It is likely that it engages in imine linkage at C-6 of the lysine substrates and at C-5 of the ornithine. This expectation is borne out by the fact that the apoenzyme (with B_{12} removed) of D-α-lysine mutase will incorporate solvent hydrogen slowly into C-6 of D-lysine on addition of pyridoxal phosphate (T. Stadtman, 1973). Thus, the function of the B_6 coenzyme (pyridoxal-P) may be to retain the migrating amino moiety at the active site (in a process not yet well formulated). Whether a substituted aziridine intermediate may form during the movement of the amino group between adjacent carbons is unclear. For example:

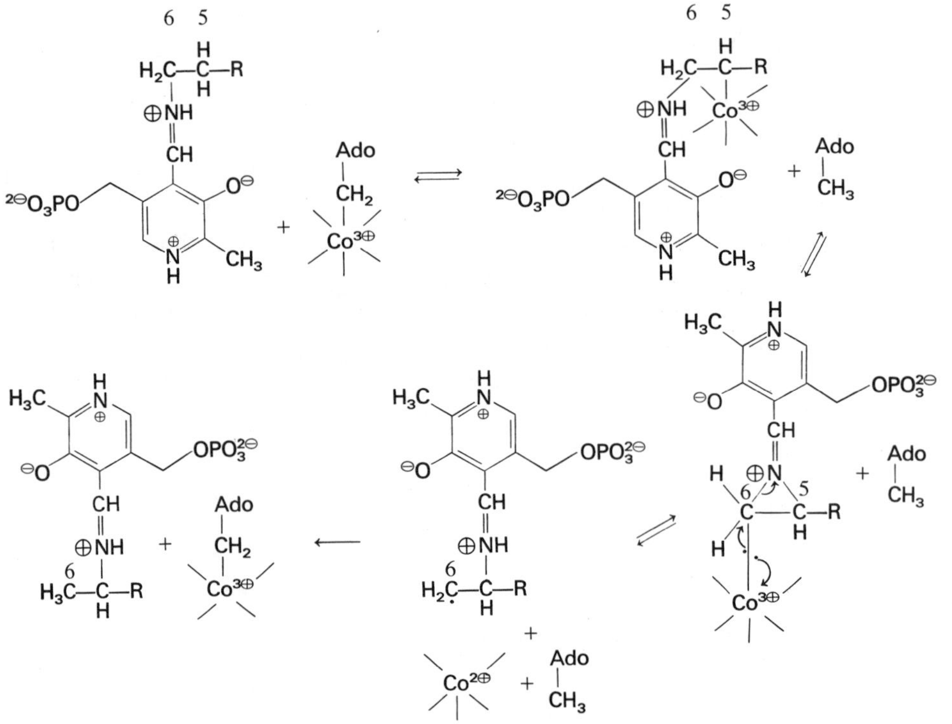

As mentioned, all three of these mutases are involved in fermentative pathways in the clostridia. The D-α-lysine and D-ornithine substrates arise from action of lysine racemase and ornithine racemase on the respective L-isomers.

On the other hand, L-β-lysine is formed from L-α-lysine by a migration of the amino group at C-2 to C-3, catalyzed by lysine 2,3-aminomutase.

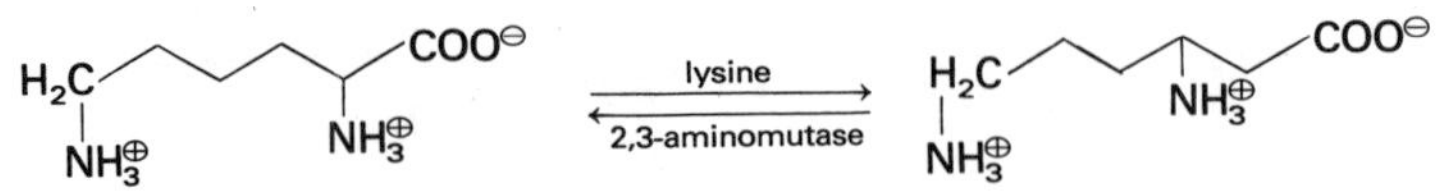

L-α-lysine ⇌ (lysine 2,3-aminomutase) L-β-lysine

This enzyme is extremely oxygen-labile, but it has been purified by Barker and associates from *Clostridium* SB4 as an $Fe^{2\oplus}$-requiring dimer of 285,000 mol wt (Chirpich et al., 1970). The homogeneous enzyme is yellow in color and has two moles of pyridoxal phosphate per mole of enzyme, in keeping with the pyridoxal requirement for the other mutases. However, there is *no corrinoid coenzyme* in the pure enzyme, nor is there any requirement for (or stimulation by) exogenous coenzyme B_{12}. *This rules out obligatory participation of coenzyme* B_{12} *in a reaction of this type.* Enzymatic incubations were then conducted in D_2O and T_2O to determine if solvent hydrogen is incorporated into product; essentially no exchange occurs during catalysis. Although this result could be in part related to isotopic discrimination against solvent hydrogen incorporation, it strongly indicates a shift of hydrogen from C-3 of α-lysine to C-2 of β-lysine, a point not yet directly tested. If this is the case, then the reaction may occur by carbanionic mechanisms with shielded proton transfers. The mechanism of the pyridoxal-P–dependent amino-group migration, like the CoB_{12}-dependent reactions discussed earlier, is obscure. Again, it may be that an aziridine intermediate is involved. The degree of similarity between the CoB_{12}-independent and the CoB_{12}-dependent lysine-aminomutase mechanisms will be of interest.

20.C.4 Ribonucleotide Reductase

The conversion of ribonucleotides to 2′-deoxyribonucleotides is of prime physiological importance in the metabolic connections between ribonucleic acids and deoxyribonucleic acids. Two distinct enzymatic mechanisms have evolved for the reduction of the 2′-hydroxyl of the ribose moiety of nucleotides—one CoB_{12}-dependent and the other CoB_{12}-independent. The CoB_{12}-dependent enzymes have been observed in lactobacilli, clostridia, and the facultative photosynthetic microbe *Euglena gracilis.* The CoB_{12}-independent ribonucleotide reductase is an $Fe^{2\oplus}$-containing protein found in microbes such as *E. coli* and in higher organisms (including mammals). We shall consider this latter enzyme first; then we shall compare the information available on the CoB_{12}-dependent reductases.

20.C.4.a CoB_{12}-independent reductase

The *E. coli* ribonucleotide reductase (CoB_{12}-independent) utilizes nucleoside *diphosphates* as substrates, converting them to 2′-deoxyribonucleoside diphosphates. In ^3H_2O, tritium is incorporated during the reduction at C-2 *with retention,* arguing against a simple S_N2 displacement process.

ribonucleotide
(CDP)

$\xrightarrow[^3H_2O]{\text{enzyme,}}$

2′-deoxyribonucleotide
(2′-dCDP)

Reduction implies the existence of a prosthetic group that can supply the necessary two electrons. Although small dithiols—such as dithiothreitol or 5,8-dithiooctanoate (lipoate)—serve in vitro, undergoing oxidation to disulfides during product formation, the direct reductant in vivo may be the redox-active protein thioredoxin (or, in thioredoxin-negative cells, an analogous glutaredoxin). This is a small protein (108 amino acids; 11,657 mol wt) that can shuttle between a reduced form with cysteines at positions 32 and 35 and the disulfide in the oxidized form. The reduced form of thioredoxin is, in turn, generated by the NADPH-dependent flavoenzyme thioredoxin reductase (¶11.B.3). The sequence of electron transfers from NADPH to oxidized thioredoxin is probably first via $FADH_2$ at the active site of the reductase. (Unlike the lipoamide dehydrogenase and the glutathione reductase discussed in Chapter 11, this reductase does not have a reducible active-site disulfide as additional electron sink.) In its reduced form, the dihydroflavin moiety of the reductase reduces oxidized thioredoxin.

$$\text{NADPH} + \text{E—FAD} \rightleftharpoons \text{NADP}^{\oplus} + \text{E—FADH}_2 \quad (1)$$

$$\text{E—FADH}_2 + \text{Thioredoxin(S—S)} \rightleftharpoons \text{E—FAD} + \text{Thioredoxin(HS SH)} \quad (2)$$

$$\text{Thioredoxin(HS SH)} + \text{Enz}{<}^{S}_{S} \rightleftharpoons \text{Thioredoxin(S—S)} + \text{Enz}{<}^{SH}_{SH} \quad (3)$$

It is not clear how the actual reduction of the 2′-hydroxyl of the substrate to the 2′-methylene of the product is effected. The *E. coli* ribonucleotide reductase has been purified; it consists of two unequal subunits termed B_1 (160,000 mol wt) and B_2 (78,000 mol wt) (Reichard, 1968; Thelander, 1973, 1974). Neither subunit alone shows any detectable catalytic activity. The large subunit (B_1) has the substrate-binding site and a dihydrothioredoxin-reducible disulfide unit. The small subunit (B_2) contains two atoms of nonheme high spin Fe^{III} and an organic free-radical signal recently identified as unpaired electron density at the β-carbon of a tyrosyl residue in the subunit (Sjoberg et al., 1977). Growth of *E. coli* on

β,β-dideuterotyrosine specifically collapsed the ESR hyperfine signal, proving that the free radical is centered at that benzylic carbon. The tyrosyl β-radical may be generated by interaction with one of the bound iron atoms, but its role in catalysis (if any) is as yet unclear. (It could be involved catalytically in subsequent one-electron chemistry in the substrate-reduction phase of catalysis.)

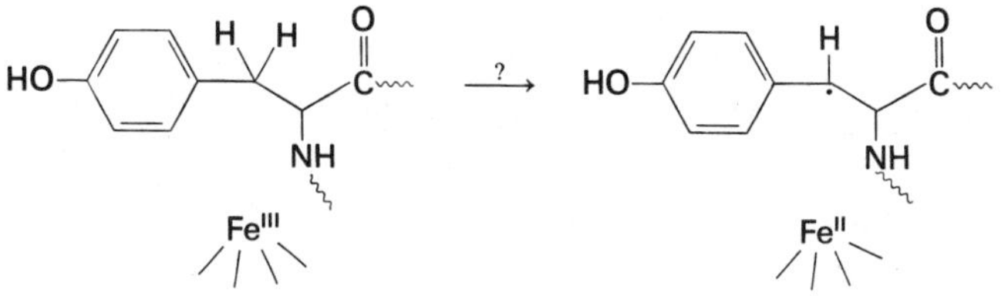

The nonheme iron does not appear to be an intermediate electron carrier (Atkin et al., 1973). Thus, the immediate reductant may be a complex of $B_1(SH)_2 \cdot B_2$ (reductase) and ribonucleoside diphosphate. The tritium incorporation from solvent probably derives from fast preequilibrium exchange of solvent hydrogens with the thiol protons. One could imagine acid catalysis of loss of the 2′-hydroxyl to a carbonium ion, which would be shielded from backside attack (to explain observed stereochemistry) and capturable by a hydride ion from the dithiol with overall retention, or the redox process could proceed via one-electron transfers and protein-thiol radicals (or protein-tyrosyl radicals).

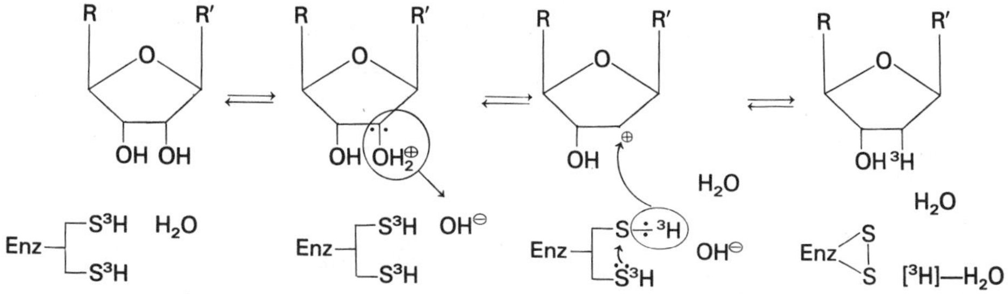

20.C.4.b CoB_{12}-dependent reductases

The CoB_{12}-dependent ribonucleotide reductase has been purified and studied from *Lactobacillus leichmanii* (Follman and Hogenkamp, 1969, 1970). In contrast to the CoB_{12}-independent *E. coli* enzyme, it utilizes ribonucleoside *triphosphates* for reduction. But the enzymes are similar in that, in ^3H_2O, tritium is incorporated into the 2′-deoxynucleoside triphosphate, with overall retention. This solvent exchange is *unique* for a CoB_{12}-dependent reaction and again probably is mediated via exchanges involving the dithiol required for catalysis. It is likely again that a protein similar to thioredoxin is the dithiol reductant in vivo of a disulfide in the *L. leichmanii* ribonucleotide reductase.

$$\text{Protein—(SH)}_2 + \text{XTP} \xrightarrow[\text{enzyme}]{\text{CoB}_{12},} \text{2'-dXTP} + \text{Protein—S}_2 + H_2O$$

The two-electron–reduced dithiol form of the enzyme is the active form that can then deliver reducing equivalents to the bound nucleoside triphosphate. The hydrogens of the 5′-methylene group of CoB_{12} will exchange with solvent protons in the presence of thiol and substrate, supporting the idea that the coenzyme does function in this instance also as an intermediate hydrogen carrier and that it may be the *immediate hydrogen donor* to C-2 of the ribose moiety of the substrate. Radical intermediates (suggested as $Co^{2\oplus}$, RS·, 5-deoxyadenosyl·) are detectable during catalysis (J. A. Hamilton et al., 1969, 1972). Studies with 3′-[^{18}O]-ATP as substrate show 3′-[^{18}O]-2′-dATP will form without isotope dilution (Follman and Hogenkamp, 1969, 1970), ruling out any migration of the 3′-hydroxyl analogous to propanediol dehydrase catalysis (and also ruling out 2′-hydroxyl departure through neighboring group participation by the 3′-hydroxyl followed by *random* reopening of the epoxide; the epoxide could form, but it would have to be opened with 100% regioselectivity). The fact that, in ^3H_2O, only the 2′-carbon of the deoxynucleotide product (and not the 3′-carbon) acquires tritium rules against several variations of mechanisms involving olefinic intermediates. One might *imagine* that catalysis occurs with homolytic fission of the coenzyme to yield the $Co^{2\oplus}$-corrin and the 5′-deoxyadenosyl radical, which then abstracts a hydrogen atom from the active-site dithiol, generating 5′-deoxyadenosine and the thiolate radical. The thiol radical and the corrinoid $Co^{2\oplus}$ species could collapse to form a Co—S bond.

The reduction at C-2 could then proceed by hydride transfer from the methyl of 5′-deoxyadenosine to the substrate in a controlled frontside attack on a C-2′ shielded carbonium ion. Whether this occurs is unknown, but it might conceivably occur as follows. Disulfide formation might generate a transient $Co^{\oplus}$ corrin, a powerful nucleophile that can attack the 5′-deoxyadenosine, re-forming the

$Co^{3\oplus}$–carbon bond of the intact coenzyme B_{12} with expulsion of the hydride, which goes to product.

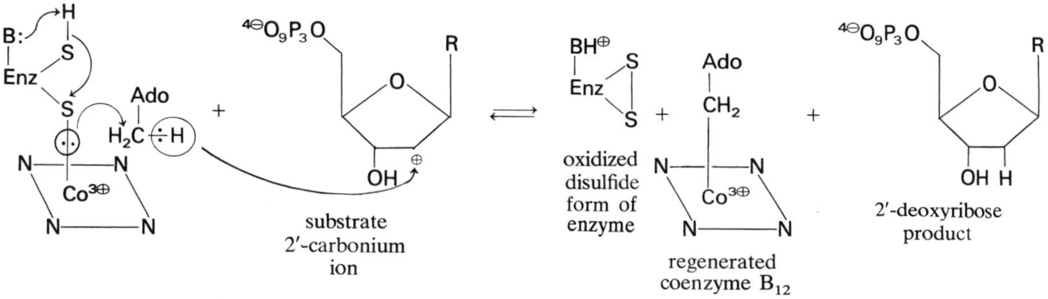

substrate 2′-carbonium ion

oxidized disulfide form of enzyme

regenerated coenzyme B_{12}

2′-deoxyribose product

This hypothetical scheme would suggest that tritium incorporation from solvent into the 5′-deoxyadenosine methyl group would occur by reversible formation of 5′-deoxyadenosine many times for each time that intermediate goes on to product. The two-step scheme proposes both one-electron and two-electron chemistry for the active-site dithiol/disulfide of ribonucleotide reductase.

This completes our discussion of CoB_{12}-dependent rearrangements. (B_{12} will be discussed as a methyl donor in Chapter 25.) It also concludes this chapter on rearrangements and Section IV of the book, dealing with enzymatic eliminations, isomerizations, and rearrangements. In the next Section we deal with reactions that make and break carbon–carbon bonds. Clearly, the carbon skeletal rearrangements discussed in this chapter fall under that definition. However, we have discussed them separately here as rearrangements, because this categorization seems more useful in probing their mechanisms.

Section V

ENZYMATIC REACTIONS THAT MAKE AND BREAK CARBON–CARBON BONDS

In this fifth section of the text, we focus our attention on a large and diverse group of enzyme-catalyzed reactions that have in common the reversible formation or cleavage of carbon–carbon bonds in biological systems. This group includes many of the interconversions involved both in biosynthesis of organic molecules and in their subsequent breakdown that are catalyzed by well-characterized enzymes.

As in the nonenzymatic models from organic chemistry, much of the enzyme chemistry involved in reactions altering carbon skeletons is carbanion chemistry, and catalysis is achieved by selective stabilization of carbanionic intermediates or transition states. We shall see that the substrates themselves can stabilize negative charge developed during catalysis, or that the enzymes will utilize coenzymes capable of delocalizing incipient electron density.

We shall begin with enzymatic decarboxylation of β-keto acids and examine the amine-catalyzed pathways used by the decarboxylases, then proceed to oxidative decarboxylations of β-hydroxyacids that appear to yield the β-keto acids through initial oxidation. A variety of α-keto acid decarboxylation reactions are enzyme-catalyzed. Unlike β-keto acid substrates, however, α-keto acids cannot stabilize carbanions developing during decarboxylation, and these enzymatic reactions all require thiamine pyrophosphate as coenzyme to form covalent adducts with the substrate and provide extended electron sinks.

Then we shall discuss enzymatic carboxylations (the reverse of decarboxylations), which are physiologically significant routes for CO_2 assimilation. These enzymes fall into three categories. One group uses ATP to activate the carbon dioxide and features the obligate use of biotin as a coenzyme that serves as an intermediate CO_2 carrier. The second category includes various enzymes that carboxylate phosphoenolpyruvate (PEP) without any requirement for ATP. The third class includes those enzymes that use neither ATP and biotin nor PEP as carboxylation substrate. The best-characterized example is ribulose-diphosphate carboxylase, the enzyme responsible for the initial step in photosynthetic carbon assimilation in green plants and photosynthetic bacteria.

Enzymatic aldol reactions are considered next, followed by analogous Claisen condensations, many of which add the stabilized C-2 carbanion of the acetyl moiety of acetyl-CoA to the electrophilic carbonyl carbon of a variety of keto-acid substrates. Although aldol and Claisen reactions comprise two of the major enzymatic routes to new carbon–carbon bonds, neither is the dominant reaction in the biological formation of isoprenoid compounds (such as sterols) and other terpenes, where carbon-alkylation reactions account for the introduction of $(C_5)_n$ units derived from dimethylallyl pyrophosphate and Δ^3-isopentenyl pyrophosphate. The prenylation mechanisms will be summarized.

In many enzymatic transformations of α-amino acids, the coenzyme form of vitamin B_6 (pyridoxal phosphate) is used as an electron sink to provide a low-energy carbanionic path for transformations at the α, β, and γ carbons of specific amino-acid substrates. A variant of pyridoxal-P–dependent enzymatic catalysis, which may be more primitive in an evolutionary sense, involves the use of pyruvyl groups covalently bound to the enzyme protein rather than pyridoxal-P.

The enzyme that interconverts serine and glycine, serine hydroxymethylase, requires not only pyridoxal-P but also an additional coenzyme, tetrahydrofolic acid, which serves as the acceptor of the C_1 fragment released as serine undergoes aldol cleavage to glycine. This one-carbon fragment is not free formaldehyde, but rather the methylene derivative of tetrahydrofolate. This tetrahydropteridine coenzyme is one of the two involved in one-carbon metabolism, and we shall see that it mediates redox changes in the C_1 fragment from the redox level of $—CH_3$ to $—CH_2OH$ to —CHO to formate. The derivatized tetrahydrofolate serves as the direct precursor of methyl groups in only a few enzymatic cases, perhaps the most interesting mechanistically being thymidylate synthetase. The preferred biological methyl donor is a methylated sulfonium cation, *S*-adenosylmethionine, which is involved in enzymatic formation of *C*-methyl, *N*-methyl, *O*-methyl, and *S*-methyl groups in product molecules. The last example of enzymatic methyl transfer we shall consider is the formation of the methyl group of methionine itself, which can occur by two routes—one of them involving the corrinoid coenzyme B_{12} in a different role from that in the CoB_{12}-dependent enzymatic rearrangements that were discussed in Section IV.

Chapter 21

Enzyme-Catalyzed Decarboxylations

As an introduction to this section focusing on making and breaking carbon–carbon bonds in enzymatic reactions, we shall take up two sides of a simple process: decarboxylases in this chapter and carboxylases in the next, making that division both on the basis of physiological functions (to remove or add CO_2) and on the basis of mechanism.

There are three common types of organic acids that undergo enzyme-catalyzed decarboxylations:

1. β-keto acids;
2. β-hydroxy acids;
3. α-keto acids.

We shall examine each category in turn, and then we shall note a fourth category of miscellaneous reactions.

21.A DECARBOXYLATION OF β-KETO ACIDS

Probably the prime electronic requirement to facilitate decarboxylation in catalysis is the ability of a substrate to stabilize the developing carbanionic transition state (lower the activation energy) as decarboxylation proceeds (Jencks, 1969, p. 116 ff).

$$R{-}\underset{H}{\overset{H}{C}}{-}\overset{O}{\overset{\|}{C}}{-}O^{\ominus} \longrightarrow \underset{\text{product carbanion}}{[R\overset{\ominus}{C}H_2 + CO_2]} \xrightarrow{H^{\oplus}} RCH_3 + CO_2$$

A β-keto-acid substrate has this structural requirement built in; the ready possibility for enolization allows the β-carbonyl group to act as an electron sink, an example of electrophilic assistance to catalysis.

$$R{-}\overset{O}{\overset{\|}{C}}{-}\underset{H}{\overset{H}{C}}{-}\overset{O}{\overset{\|}{C}}{-}O^{\ominus} \longrightarrow R{-}\overset{O^{\ominus}}{\overset{|}{C}}{=}CH_2 + CO_2 \xrightarrow[H^{\oplus}]{\text{ketonization}} R{-}\overset{O}{\overset{\|}{C}}{-}CH_3 + CO_2$$

β-keto acid — enolate — ketone

Protonation of the β-carbonyl group would generate a better electron sink

$$R{-}\overset{OH^{\oplus}}{\overset{\|}{C}}{-}\underset{H}{\overset{H}{C}}{-}COO^{\ominus}$$

and would give the free enol as product. However, the low basicity of the carbonyl group provides an energy barrier to this pathway under physiological conditions. (Such a protonated species does not exist in significant concentration.)

For nonenzymatic decarboxylation of acetoacetate, Westheimer has suggested a cyclic six-membered transition state when the free acid is the reactant, leading to the enol as initial product (Westheimer and Jones, 1941; Westheimer, 1963).

$$H_3C{-}\overset{O\quad H{-}O}{\overset{\|}{C}}{-}\underset{H}{\overset{H}{C}}{-}C{=}O \longrightarrow H_3C{-}\overset{OH}{\overset{|}{C}}{=}CH_2 + O{=}C{=}O$$

acetoacetic acid — enol

It has long been known that amines are effective catalysts of β-keto-acid decarboxylations. A reasonable mechanism involves formation of an imine or Schiff's base prior to carbon–carbon bond cleavage.

$$R{-}\underset{O}{\underset{\|}{C}}{-}\underset{H}{\overset{H}{C}}{-}COO^{\ominus} + R'{-}NH_2 \longrightarrow R{-}\underset{\underset{R'}{\underset{|}{N}}}{\underset{\|}{C}}{-}\underset{H}{\overset{H}{C}}{-}COO^{\ominus} + H_2O$$

β-keto acid — amine — imine

The catalytic advantage of the imine is that the nitrogen of the imine is much more basic than the carbonyl oxygen and so is easily protonated to generate the cationic imine.

$$R{-}C({=}N{-}R'){-}CH_2{-}COO^{\ominus} \rightleftharpoons R{-}C({=}\overset{\oplus}{N}H{-}R'){-}CH_2{-}COO^{\ominus} \qquad pK_a \approx 12$$

imine iminium cation

With its positive charge, the iminium cation is an exceptionally good sink for stabilizing the incipient carbanion during decarboxylation.

$$R{-}C({=}\overset{\oplus}{N}H{-}R'){-}CH_2{-}C({=}O){-}O^{\ominus} \longrightarrow R{-}C({-}\ddot{N}H{-}R'){=}CH_2 + CO_2 \longrightarrow R{-}C({=}\overset{\oplus}{N}H{-}R'){-}CH_3 + CO_2$$

iminium cation of substrate; eneamine (a low-energy carbanion); iminium cation of product

21.A.1 Acetoacetate Decarboxylase

The prototypic β-keto-acid decarboxylase is the acetoacetate decarboxylase purified from *Clostridium acetobutylicum* (Fig. 21-1) and studied by Westheimer and colleagues (Hamilton and Westheimer, 1959; Warren et al., 1966; Fridovich, 1972*b*). This organism and its relatives were in some demand during World War I when acetone was produced primarily by clostridial fermentations.

$$H_3C{-}C({=}O){-}CH_2{-}COO^{\ominus} \longrightarrow CO_2 + H_3C{-}C({=}O){-}CH_3$$

acetoacetate acetone

Westheimer suspected that the enzyme, by analogy to nonenzymatic amine-catalyzed decarboxylations, would use a cationic imine species as an electron sink to stabilize the charge density formed at C-2 in the transition state.

Several elegant and now classical experiments from Westheimer's laboratory supported this premise (Warren et al., 1966; Zerner et al., 1966). When acetoacetate ^{18}O-labeled in the β-keto group is incubated with decarboxylase in $H_2{}^{16}O$, the acetone formed has only ^{16}O (no ^{18}O above background).

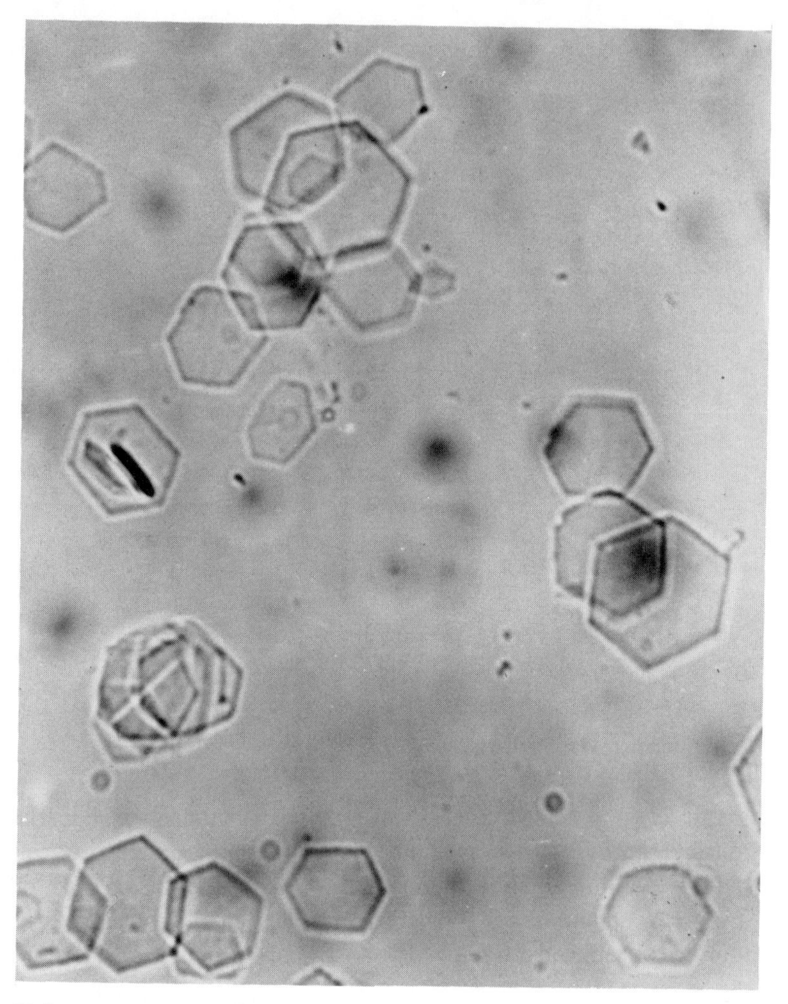

Figure 21-1
Photomicrograph of crystals of acetoacetate decarboxylase. (Reprinted with permission from Zerner et al., *Biochemistry*, 1966. Copyright by the American Chemical Society.)

$$H_3C-C(={}^{18}O)-CH_2-C(=O)-O^{\ominus} \xrightarrow[H_2{}^{16}O]{enz,} H_3C-C(={}^{16}O)-CH_3 + CO_2 + H_2O^{18} \quad \begin{pmatrix}\text{stoichiometric}\\ \text{amount}\end{pmatrix}$$

The converse experiment utilizes ^{18}O in the medium ($H_2{}^{18}O$) and ^{16}O-labeled β-keto acid. Again, the acetone product has the oxygen content of the surrounding medium—in this case, ^{18}O-enrichment.

$$H_3C-C(={}^{16}O)-CH_2-C(=O)-O^{\ominus} \xrightarrow[H_2{}^{18}O]{enz,} H_3C-C(={}^{18}O)-CH_3 + CO_2$$

These complementary experiments strongly imply the obligate formation of an imine intermediate at the β-position of the keto-acid substrate. Reversible imine formation would explain the oxygen-exchange data. When decarboxylation proceeds in D_2O, deuteroacetone is the detected ketone product. Incorporation of deuterium would be explained by generation of an eneamine of acetone as initial product, followed by ketonization and hydrolysis of the resultant imine.

If imine intermediates form between substrate and some enzymatic amino group, these should be trappable by reduction with sodium borohydride. Indeed, treatment of solutions of acetoacetate decarboxylase in the presence of acetoacetate with $NaBH_4$ leads to the time-dependent irreversible loss of enzymatic activity expected for such a derivatization of a hydrolytically unstable imine to a stable secondary amine. If the enzyme is treated with borohydride in the absence of substrate, no inactivation ensues, confirming that the loss of activity in the presence of substrate is mechanistically significant. When 3-[^{14}C]-acetoacetate is used, $NaBH_4$ reduction produces stoichiometric labeling of the now inactive enzyme. Total acid hydrolysis of the modified enzyme allows isolation of a single radioactive compound identified as ε-isopropyl-*N*-lysine. This secondary amine is also identified when the normal decarboxylation product (acetone) is admixed with enzyme and $NaBH_4$ at low ($\sim 10^{-5}$ M) concentrations of acetone.

Evidently, the imine that accumulates and then is reduced by $NaBH_4$ is the *product imine* (after decarboxylation) and *not the initial substrate imine,* which would still have the $COO^{\ominus}$ group affixed. (Alternatively, the substrate imine has the $COO^{\ominus}$ negative charge and so may provide electrostatic repulsion to $BH_4^{\ominus}$, a repulsion not open to the product imine. See Fridovich, 1972*b*.)

$$CH_3C(=O)CH_2COO^{\ominus} + H_2N\text{-Lys-Enz} \longrightarrow \longrightarrow \longrightarrow \text{“}H^{\ominus}\text{”} \to (CH_3)_2C{=}\overset{\oplus}{N}H\text{-Lys-Enz} + CO_2 \longrightarrow (CH_3)_2CH\text{-}NH\text{-Lys-Enz} \xrightarrow{\Delta,\ H^{\oplus}} (CH_3)_2CH\text{-}NH\text{-}(CH_2)_4\text{-}CH(\overset{\oplus}{N}H_3)COO^{\ominus}$$

ε-isopropyl-*N*-lysine

The amino group provided by enzyme is, not unexpectedly, the ε-amino of a specific lysine residue. A minimal mechanism thus can be indicated as follows (Tagaki and Westheimer, 1968; Tagaki et al., 1968; O'Leary, 1968).

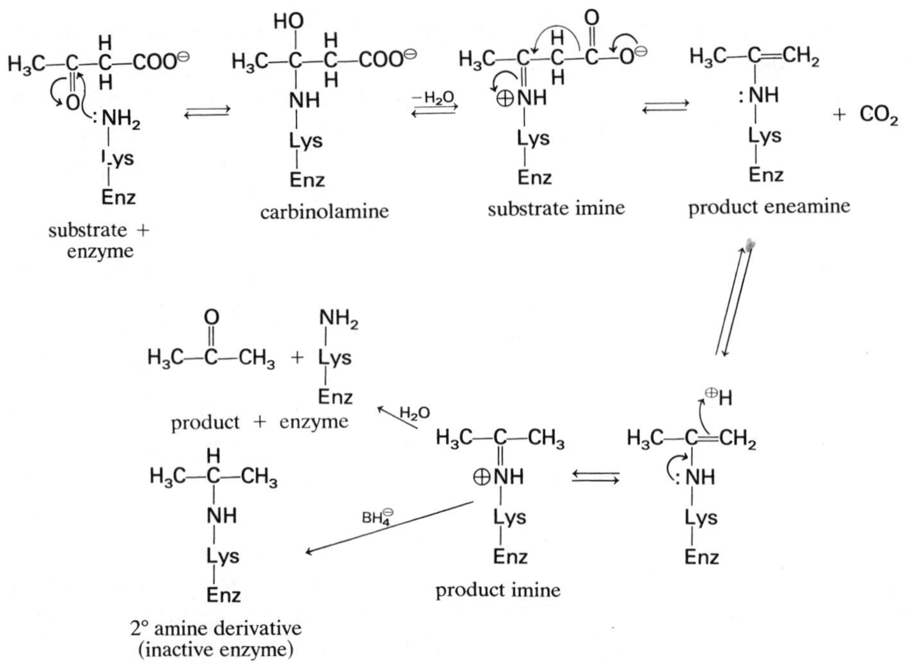

$H_3C—C(=O)—CH_2—COO^{\ominus}$ + :NH_2—Lys—Enz ⇌ $H_3C—C(OH)(NH—Lys—Enz)—CH_2—COO^{\ominus}$ ⇌ ($-H_2O$) $H_3C—C(=\oplus NH—Lys—Enz)—CH_2—C(=O)—O^{\ominus}$ ⇌ $H_3C—C(=CH_2)—$:NH—Lys—Enz + CO_2

substrate + enzyme

carbinolamine

substrate imine

product eneamine

$H_3C—C(=O)—CH_3$ + NH_2—Lys—Enz

product + enzyme

H_2O

$H_3C—C(=\oplus NH—Lys—Enz)—CH_3$ ⇌ $H_3C—C(=CH_2)—$:NH—Lys—Enz ($^{\oplus}H$)

product imine

$BH_4^{\ominus}$

$H_3C—CH(NH—Lys—Enz)—CH_3$

2° amine derivative
(inactive enzyme)

The primary sequence of the active-site peptide from which the $BH_4^{\ominus}$-reduced isopropyl-ε-N-lysine derives is the following (with the derivatized lysine indicated by an asterisk):

Gly—Leu—Ser—Ala—Tyr—Pro—*Lys—Lys—Leu
(with NH_2 on each Lys)

A study of the effect of pH on V_{max} and V_{max}/K_m suggested that groups with apparent pK_a values of 5.5 and 6.5 are important for catalysis. The unusual feature of the active-site peptide is a second lysyl group proximal to the first. One might expect two adjacent ε-NH_3 groups to provide mutual electrostatic destabilization. A potential solution would be an unusual depression of the pK_a of one of the ε-amino-group ionizations so that it would become a stronger acid, give up its proton more readily, and thus lose its cationic charge.

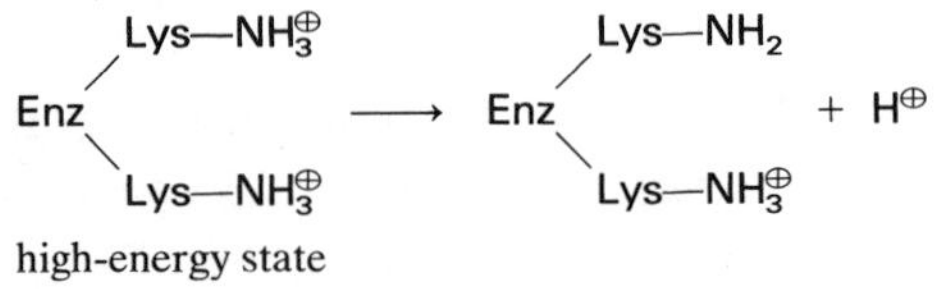

Enz(Lys—$NH_3^{\oplus}$)(Lys—$NH_3^{\oplus}$) ⟶ Enz(Lys—NH_2)(Lys—$NH_3^{\oplus}$) + $H^{\oplus}$

high-energy state

Studies with the acylating agent 2,4-dinitrophenyl propionate provided evidence suggesting that the pK_a of one of the lysyl residues is indeed drastically perturbed. Acylation and inactivation are measurable by release of the yellow 2,4-dinitrophenolate anion (Schmidt and Westheimer, 1971).

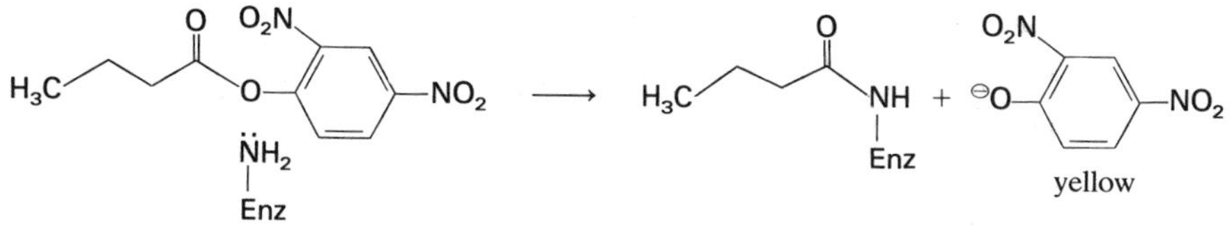

yellow

The pH rate for acylation shows a pK_a of 5.9 for the active-site amine nucleophile, presumably the active-site lysine. This would represent a 10^4-fold increase in acidity for a lysine ε-amino group at the active site of acetoacetate decarboxylase, compared to the acidity of the group free in solution. Given a pH optimum for decarboxylation of 5 to 6, it would be catalytically advantageous for the enzyme to have a high concentration of unprotonated ε-NH_2 as a good nucleophile to initiate attack on the β-carbonyl of bound acetoacetate to initiate imine formation.

A final interesting feature of acetoacetate decarboxylase is the reversible inhibition by acetopyruvate (2,4-dioxopentanoate) as α-keto analogue of the substrate.

acetopyruvate

acetoacetate

Acetopyruvate is not decarboxylated, but it shows reversible inhibition with three interesting features. (1) The K_i for acetopyruvate is 10^{-7} M, 10^4-fold lower than the K_m for acetoacetate (Tagaki and Westheimer, 1968; Tagaki et al., 1968; O'Leary, 1968). (2) Although inhibition is reversible, the $T_{1/2}$ for dissociation of the bound inhibitor is about 10 min at 30° C, an extraordinarily slow rate (too slow by orders of magnitude for most simple dissociations). (3) The enzyme binds one mole of acetopyruvate per active site with generation of a new chromophore ($\lambda_{max} \approx 325$ nm and $\varepsilon \approx 14{,}000$). The high extinction suggested generation of a highly conjugated chromophore. The chromophore was not reducible with borohydride; no covalent derivative could be formed, in contrast to the trapping of acetoacetate. The spectroscopic data and the nonreducibility suggest a conjugated eneamine arising from attack of the active-site ε-amino group (putatively at the C-2 carbonyl of acetopyruvate) and subsequent isomerization into eneamine in conjugation with the C-4 carbonyl. (Initial attack could be at the C-4 carbonyl; the isomeric eneamine would result.)

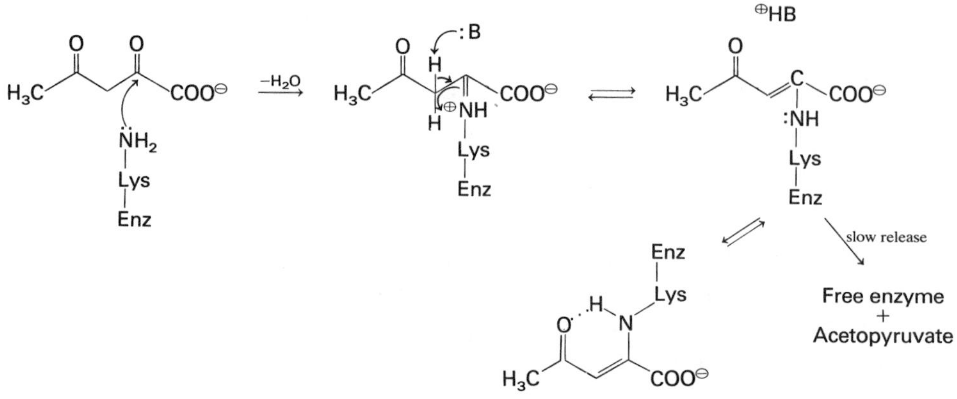

slow release

Free enzyme + Acetopyruvate

cis-eneamine
($\lambda_{max} \approx 325$ nm)

This formulation would also explain slow release of bound inhibitor. A similar eneamine has been postulated for inhibition of the flavoenzyme D-amino-acid oxidase (¶11.D.1) during oxidation of D-2-amino-4-pentynoate (Marcotte and Walsh, 1976).

The pioneering experiments on clostridial acetoacetate decarboxylase, establishing initial imine formation between a substrate carbonyl and an enzyme amine grouping, have been repeated many times since. They lead to the following general conclusion.

> A GENERAL COMMENT: When carbanions are generated during enzymatic catalysis, one general mechanism for lowering the free energy of activation and thereby accelerating the reaction involves the formation of imines at adjacent carbonyls to delocalize and stabilize the developing negative charge via the electron sink of the cationic imine.

Not only will this be true for other β-keto-acid decarboxylations; it is the quintessence of the strategy for enzymatic decarboxylations of α-keto acids. We noted (¶17.B.2.b) that the enzyme ketoarabinate dehydrase uses this device to activate an initially nonacidic hydrogen. We shall pursue this idea in later chapters when we examine aldol cleavages (and condensations) and pyridoxal-P–dependent catalyses.

21.A.2 A Second Class of Acetoacetate Decarboxylases

A second class of acetoacetate decarboxylases has been observed. In contrast to the data from the clostridial enzyme, enzyme isolated from some other microorganisms shows—

1. no $BH_4^{\ominus}$ inactivation of substrate;
2. no ^{18}O exchange out of the keto carbonyl of acetoacetate.

Each of these points mitigates against the formation of an imine between substrate and enzyme.

Instead, these enzymes require a divalent metal ion for catalysis. The metal ion in these decarboxylases presumably acts as a superacid catalyst, polarizing the oxygen and making it a better electron sink to spread the developing charge. (This is an example of a type of enzymatic electrophilic catalysis that is less common than the cationic route.)

coordinated enol
as initial product

A nonenzymatic model for such a metal-promoted catalysis is the $Mn^{2\oplus}$-catalyzed decarboxylation of dimethyl oxaloacetate to the chelated enolate as initial product (Steinberger and Westheimer, 1951).

21.B DECARBOXYLATION OF β-HYDROXY ACIDS

There are a number of examples of metabolically important decarboxylations of β-hydroxy acids. We shall look in some detail at the conversion of isocitrate to α-ketoglutarate. Then we shall take a brief look at the four other known β-hydroxy-acid decarboxylases.

21.B.1 Isocitrate Dehydrogenase

Isocitrate dehydrogenase, a component enzyme of the citrate (tricarboxylate) cycle, catalyzes the conversion of the monohydroxytricarboxylate isocitrate to α-ketoglutarate (Plaut, 1963). The dehydrogenase requires $Mn^{2\oplus}$ for catalysis, presumably to coordinate with bound substrate and act as superacid.

$$\underset{\text{isocitrate}}{H_2C(COO^{\ominus})-HC(COO^{\ominus})-HC(OH)-COO^{\ominus}} \xrightarrow[\text{dehydrogenase}]{Mn^{2\oplus},\ \text{isocitrate}} \underset{\alpha\text{-ketoglutarate}}{H_2C(COO^{\ominus})-HCH-C(=O)-COO^{\ominus}} + CO_2$$

A general remark might be made that decarboxylation of β-hydroxy acids should be considerably less facile than that of β-keto acids, because the β-OH moiety cannot function as a useful electron sink. Thus there is no obvious structural element in the substrate to facilitate decarboxylation. Therefore, a reasonable path for decarboxylation of a β-hydroxy acid might be its prior conversion to a β-keto acid, which does have a low-energy path for decarboxylation. Indeed, inspection of the overall formula for conversion of isocitrate to α-ketoglutarate shows that (in addition to the loss of CO_2) that product has a keto group where the substrate had a hydroxyl group. The reaction stoichiometry for isocitrate dehydrogenase is, in fact, incomplete as written. There is an absolute requirement for stoichiometric quantities of an oxidized nicotinamide coenzyme: $NADP^{\oplus}$ for the cytoplasmic isocitrate dehydrogenase, and $NAD^{\oplus}$ for a genetically distinct mitochondrial isozyme. The former, due to greater ease of isolation, has been the better-studied mechanistically (Plaut, 1963). It requires $Mg^{2\oplus}$ or $Mn^{2\oplus}$ for catalysis.

$$\underset{NADP^{\oplus}}{\text{NADP}^{\oplus}} + \underset{2R,3S\text{-isocitrate}}{H_2C(COO^{\ominus})-HC(COO^{\ominus})-{}^{*}H\text{-}C(OH)-COO^{\ominus}} \rightleftharpoons \underset{\text{NADPH}}{\text{NADPH } (^{*}H)} + \underset{\alpha\text{-ketoglutarate}}{H_2C(COO^{\ominus})-HCH-C(=O)-COO^{\ominus}} + CO_2$$

In keeping with other nicotinamide-mediated oxidations, it is likely that the alcohol functionality of isocitrate is oxidized to the keto group of oxalosuccinate concomitant with $NADP^{\oplus}$ reduction to NADPH by direct hydrogen transfer. Oxalosuccinate is both an α-keto acid and a β-keto acid (the central carboxylate).

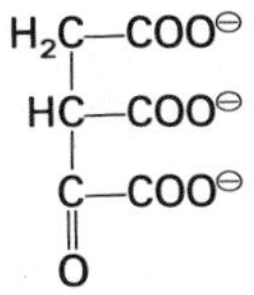

$$\begin{array}{l} H_2C—COO^{\ominus} \\ \quad | \\ HC—COO^{\ominus} \\ \quad | \\ \;\; C—COO^{\ominus} \\ \quad \| \\ \;\; O \end{array}$$

oxalosuccinate
(a putative intermediate)

As we shall note in ¶21.C, α-keto acids do not decarboxylate as readily as β-keto acids do, and it is the central $COO^{\ominus}$ that should be lost to produce the observed product, α-ketoglutarate. The bound $Mn^{2\oplus}$ at the enzyme active site could polarize the ketocarbonyl as follows.

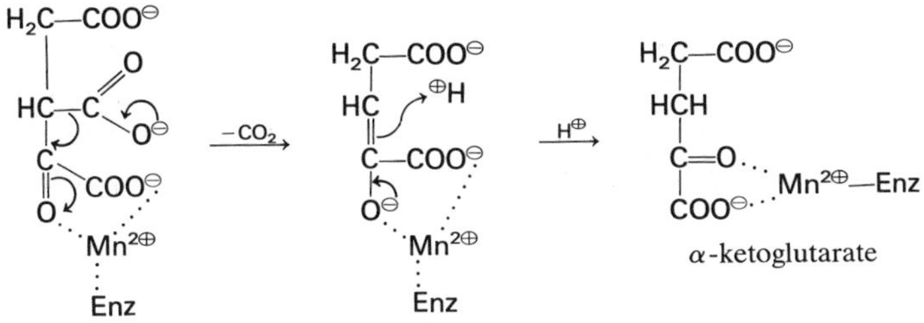

oxalosuccinate intermediate $\xrightarrow{-CO_2}$ bound enolate $\xrightarrow{H^{\oplus}}$ α-ketoglutarate

Supporting this scheme is the observation that synthetic oxalosuccinate is acted on by the $Mn^{2\oplus}$–enzyme complex, undergoing catalytic decarboxylation independent of the presence of nicotinamide coenzyme (Siebert et al., 1957; Moyle and Dixon, 1955). In the presence of NADPH, exogenous oxalosuccinate can be reduced by the enzyme back to isocitrate. However, although each of the expected partial reactions does occur, neither is kinetically competent. Further, oxalosuccinate cannot be trapped during normal catalysis by carbonyl reagents (e.g., borohydride). Also, when [^{14}C]-isocitrate is undergoing oxidative decarboxylation, addition of unlabeled exogenous oxalosuccinate to incubations does not dilute the specific radioactivity of the α-ketoglutarate product. If, as seems chemically reasonable, oxalosuccinate is a finite intermediate, it must remain tightly bound and nondissociable before decarboxylation occurs. The failure of added reagents to trap or reduce the keto group of bound oxalosuccinate could reflect its failure to accumulate significantly before decarboxylation, or it could reflect a steric block to the trapping reagents. (An additional possibility is that oxidation and hydride transfer are actually concerted with decarboxylation, and oxalosuccinate is not formed as a discrete intermediate.)

The stereochemistry of this enzymatic reaction has been scrutinized. The substrate is 2*R*,3*S*-isocitrate, and direct hydrogen transfer occurs from the pro*S* hydrogen at C-2 to the pro*R* position at C-4 of the nicotinamide coenzyme. In D_2O, one atom of deuterium is incorporated, forming 3S-[^{2}H]-2-ketoglutarate, a stereospecific reprotonation of the enzyme-bound enol with overall retention (Alworth, 1972).

21.B.2 Other β-Hydroxy-Acid Dehydrogenation/ Decarboxylation Sequences

There are four other enzymes known that catalyze β-hydroxy-acid decarboxylations (Rose and Hanson, 1975). We commented in Chapter 19 on enzymatic isomerizations involving ribulose-5-phosphate. This C_5 keto sugar is formed during aerobic metabolism of glucose-6-phosphate. First, the hexose phosphate undergoes NADP-dependent oxidation at C-1 to 6-phosphogluconate. The oxidative decarboxylation to ribulose-5-phosphate is then carried out by 6-phosphogluconate dehydrogenase, probably via a 3-keto intermediate that precedes CO_2 loss.

$$\text{NADP}^{\oplus} + \begin{array}{rl} 1 & \text{COO}^{\ominus} \\ 2 & \text{HC—OH} \\ 3 & \text{HO—CH} \\ & \text{HC—OH} \\ & \text{HC—OH} \\ & \text{H}_2\text{C—OPO}_3^{2\ominus} \end{array} \longrightarrow \begin{array}{rl} 2 & \text{H}_2\text{C—OH} \\ 3 & \text{C=O} \\ & \text{HC—OH} \\ & \text{HC—OH} \\ & \text{H}_2\text{C—OPO}_3^{2\ominus} \end{array} + \overset{1}{\text{CO}_2} + \text{NADPH}$$

6-phosphogluconate ribulose-5-phosphate

Another example of widespread biological distribution is the malate enzyme forming pyruvate from malate, with intermediate generation of oxaloacetate.

$$\text{NADP}^{\oplus} + \begin{array}{c} \text{COO}^{\ominus} \\ \text{HC—OH} \\ \text{HCH} \\ \text{COO}^{\ominus} \end{array} \underset{\text{Mn}^{2\oplus}}{\overset{\text{malate enzyme,}}{\rightleftarrows}} \begin{array}{c} \text{COO}^{\ominus} \\ \text{C=O} \\ \text{CH}_3 \end{array} + \text{CO}_2 + \text{NADPH}$$

The enzyme is believed to function physiologically in the decarboxylation direction, producing NADPH as reducing equivalent for other cellular reactions, among them fatty-acid biosynthesis. The malate oxidation–decarboxylation bears some analogy to the more straightforward oxidation of L-malate to oxaloacetate carried out by the L-malate dehydrogenase (Chapter 10) functioning in the citrate cycle.

$$\text{L-Malate} + \text{NAD}^{\oplus} \rightleftharpoons \text{Oxaloacetate} + \text{NADH}$$

When one examines malate enzymes from various biological sources, in fact, categories can be seen for these catalysts on two bases: (1) some use $NAD^{\oplus}$, others use $NADP^{\oplus}$; (2) some (e.g., pigeon-liver enzyme) will decarboxylate added oxaloacetate, others (e.g., the enzyme from lactate bacteria) will not, perhaps indicating a decarboxylation concerted with hydride transfer in the latter case.

The purified pigeon-liver malate enzyme has, all in all, five detectable catalytic activities (Hsu, 1970).

$$\text{L-Malate} + \text{NADP}^{\oplus} \xrightleftharpoons{Mn^{2\oplus}} CO_2 + \text{Pyruvate} + \text{NADPH} \qquad (1)$$

$$\text{Oxaloacetate} \xrightleftharpoons{Mn^{2\oplus}} CO_2 + \text{Pyruvate} \qquad (2)$$

$$\text{Pyruvate} + \text{NADPH} \xrightleftharpoons{Mn^{2\oplus}} \text{L-Lactate} + \text{NADP}^{\oplus} \qquad (3)$$

$$\text{Oxaloacetate} + \text{NADPH} \xrightleftharpoons{Mn^{2\oplus}} \text{L-Malate} + \text{NADP}^{\oplus} \qquad (4)$$

$$\text{L-Malate} + \text{NADP}^{\oplus} \xrightleftharpoons{Mn^{2\oplus}} \text{L-Lactate} + CO_2 + \text{NADP}^{\oplus} \qquad (5)$$

Reaction 1 is the physiological oxidative decarboxylation. Reaction 2 is decarboxylation of the presumed intermediate. Reactions 3, 4, and 5 are slow. A kinetic study indicates that there is an ordered mechanism (as with some other nicotinamide-requiring enzymes; see Chapter 10). The following scheme can account for all the partial reactions (Hsu, 1970).

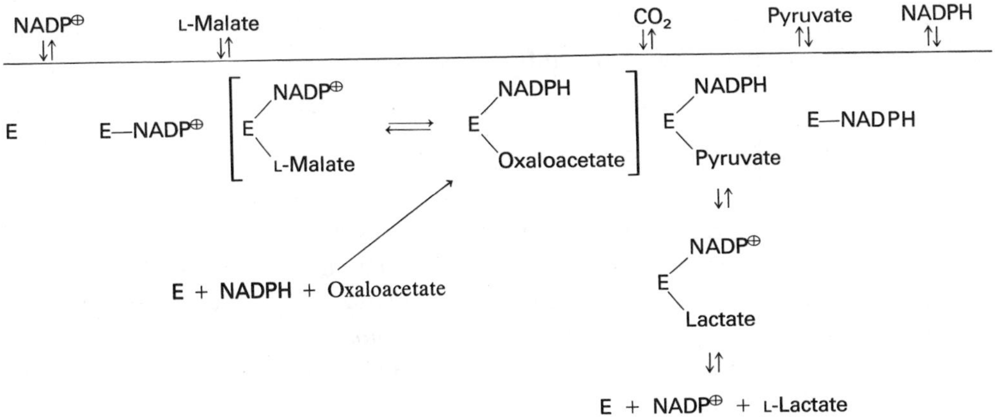

NADP⊕ ↓↑ | L-Malate ↓↑ | | CO₂ ↓↑ | Pyruvate ↑↓ | NADPH ↑↓

E — E—NADP⊕ — [E(NADP⊕)(L-Malate) ⇌ E(NADPH)(Oxaloacetate)] — E(NADPH)(Pyruvate) — E—NADPH

E + NADPH + Oxaloacetate → E(NADPH)(Oxaloacetate)

E(NADPH)(Pyruvate) ↓↑ E(NADP⊕)(Lactate) ↓↑ E + NADP⊕ + L-Lactate

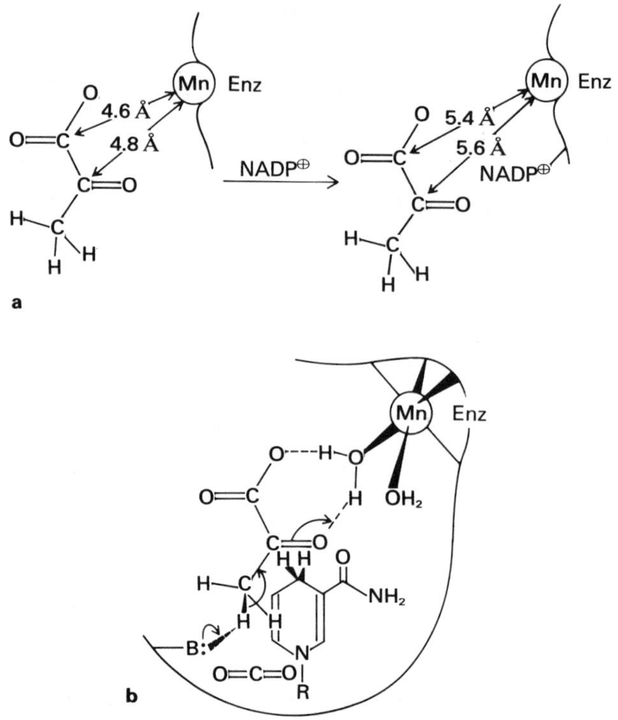

Figure 21-2
(**a**) Geometry of the $Mn^{2\oplus}$ · pyruvate interaction on malate enzyme: ternary enz · $Mn^{2\oplus}$ · pyruvate complex (*left*) and quaternary enz · $Mn^{2\oplus}$ · NADP · pyruvate complex (*right*). (**b**) Role of $Mn^{2\oplus}$ in the enolization of pyruvate catalyzed by malate enzyme. (From R. Y. Hsu et al., 1976, p. 6582)

Hsu and colleagues have recently mapped the distances between $Mn^{2\oplus}$ and malate at the active site of malate enzyme, and they have suggested a possible active-site geometry, based on NMR measurements (Fig. 21-2).

21.C DECARBOXYLATION OF α-KETO ACIDS

A variety of enzyme-catalyzed decarboxylations of α-keto acids are known (including some metabolically central decarboxylations of pyruvate). They can be divided into two categories: *nonoxidative decarboxylations* yielding aldehydes (or their condensation products), and *oxidative decarboxylations* that generate products at the acyl level of oxidation (a change in oxidation state from the α-ketone moiety in the substrate).

Among the examples of nonoxidative decarboxylation we shall consider in detail are the conversion of pyruvate to acetaldehyde

$$H_3C{-}\overset{\overset{\displaystyle O}{\|}}{C}{-}COO^{\ominus} \longrightarrow H_3C{-}\overset{\overset{\displaystyle O}{\|}}{C}H + CO_2$$

and the synthesis of acetolactate from pyruvate, followed by its decarboxylation to acetoin.

$$2\,H_3C{-}\overset{\overset{\displaystyle O}{\|}}{C}{-}COO^{\ominus} \longrightarrow H_3C{-}\overset{\overset{\displaystyle O}{\|}}{C}{-}\underset{\underset{\displaystyle OH}{|}}{\overset{\overset{\displaystyle CH_3}{|}}{C}}{-}COO^{\ominus} + CO_2$$

$$\downarrow$$

$$H_3C{-}\overset{\overset{\displaystyle O}{\|}}{C}{-}\underset{\underset{\displaystyle OH}{|}}{\overset{\overset{\displaystyle CH_3}{|}}{C}}H + CO_2$$

In the oxidative process, there clearly must be requirements for electron acceptors in catalysis. Among the examples we shall consider are the conversion of pyruvate to acetyl-CoA

$$H_3C{-}\overset{\overset{\displaystyle O}{\|}}{C}{-}COO^{\ominus} \xrightarrow[+\text{CoASH}]{-2\,e^{\ominus}} H_3C{-}\overset{\overset{\displaystyle O}{\|}}{C}{-}SCoA + CO_2$$

and the conversion of pyruvate to free acetate ion.

$$H_3C{-}\overset{\overset{\displaystyle O}{\|}}{C}{-}COO^{\ominus} \xrightarrow{-2\,e^{\ominus}} H_3C{-}\overset{\overset{\displaystyle O}{\|}}{C}{-}O^{\ominus} + CO_2$$

One problem in α-keto-acid decarboxylation mechanisms is similar to that experienced by the β-hydroxy acids. The α-keto acids lack a suitably placed electron sink to stabilize negative-charge development during decarboxylation. The α-carbonyl cannot be used. For this reason all of the four enzymatic decarboxylations just listed require an organic coenzyme that reacts with the α-keto-acid substrate to form a covalent adduct as a prerequisite for decarboxylation. We shall see that formation of such an adduct sets up an efficient electron sink in the form of a cationic imine in β-relation to the carboxylate moiety to be lost as CO_2, thus providing a low-energy path for the decarboxylation.

The coenzyme is thiamine pyrophosphate (the coenzyme form of vitamin B_1), whose formation by pyrophosphoryl transfer we discussed in Chapter 8. It consists of a pyrimidine nucleus and a thiazolium ring, connected by a methylene bridge.

C-2

thiamine-PP
(TPP)

A three-dimensional representation of TPP and its UV spectrum are shown in Figure 21-3. The nitrogen of the thiazolium ring is quarternized, and this is a key element in coenzyme function. The elucidation of how TPP functions was somewhat tortuous. For mechanistic reasons, a TPP carbanion seemed a likely candidate for an active species, and Breslow (1958) determined that, in fact, the C-2 hydrogen of the thiazolium ring is a very acidic proton. By nuclear magnetic resonance experiments in D_2O, one can watch the loss of the C-2 hydrogen signal

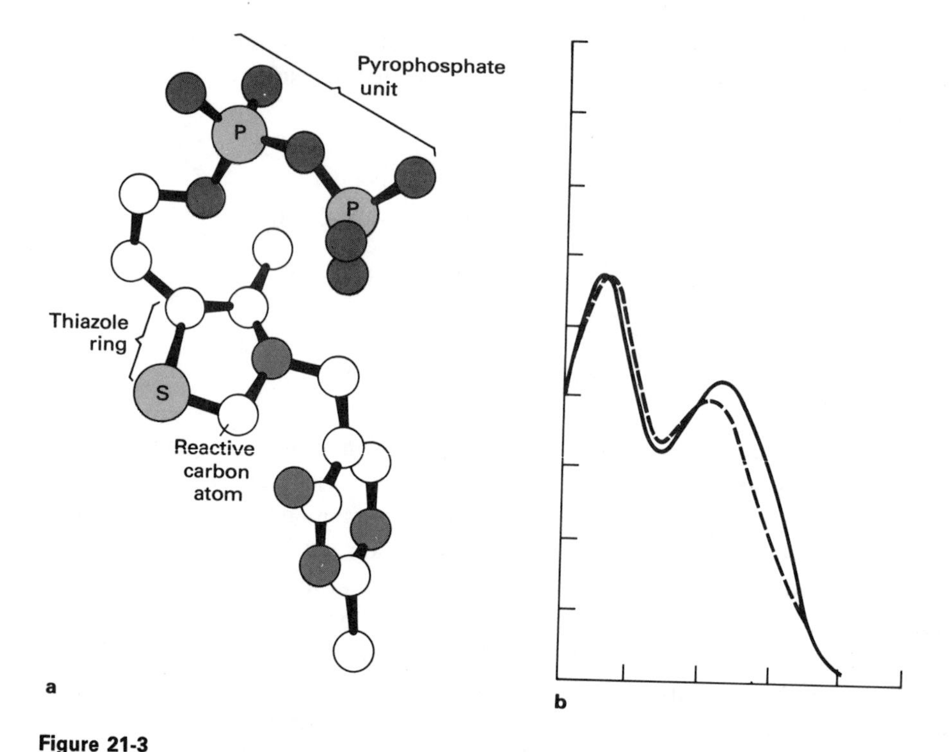

Figure 21-3
(**a**) Molecular model of thiamine pyrophosphate (TPP). (From L. Stryer, *Biochemistry*, p. 800. W. H. Freeman and Company. Copyright © 1975.) (**b**) Ultraviolet absorption spectrum of thiamine in 0.02 M phosphate buffer. (Adapted from D. Metzler, 1960)

as it exchanges and becomes C-2 deuterium, a process with $T_{1/2}$ of 2 min at pH 5 and physiological temperature (Breslow, 1961).

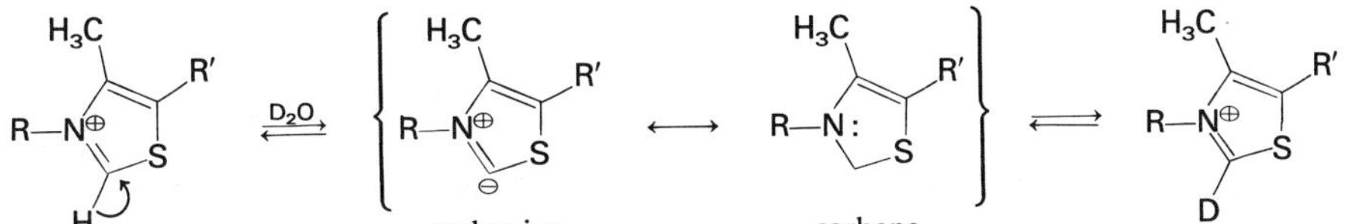

carbanion contributor

carbene contributor

The exchange must proceed through the C-2 carbanion, which appears unusually stable due to electrostatic interaction with the cationic nitrogen. Also, *d–p* orbital overlap of the negative charge with the adjacent sulfur atom has been suggested, but C-2 of oxazolium ions is actually more acidic and, in that heterocycle, oxygen is adjacent to the carbanion (and oxygen has no *d* orbitals).

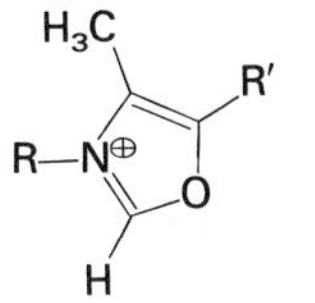

oxazolium ion

On the other hand, despite the increased acidity of oxazolium ions at C-2, they are not catalysts for such model reactions as the benzoin condensation to be discussed shortly. Lowe and Ingraham (1975, p. 71 ff) have suggested that the C-2 oxazolium ion is too stable to add to benzaldehyde (or other weak electrophiles). They suggest that "the anion must be stable enough to form but not so stable that it is unreactive."

Duclos and Haake (1974) have compared both oxazolium and imidazolium systems with thiazolium ions and noted that oxazolium ions are unsuited for coenzymatic function because they are too unstable to ring opening (to inactive forms) at neutral pH. Both thiazolium and imidazolium ions are thermodynamically stable at pH 7. But the imidazolium ring is very slow to generate the ylid (carbanion) required for nucleophilic attack on the electrophilic group of the substrate. Thus the thiazolium ring system is the only one of the three that is suitable on thermodynamic and kinetic grounds.

Before discussing how the C-2 thiazolium carbanion is utilized to assist enzymatic α-keto-acid decarboxylations, we shall examine a well-characterized model reaction, the benzoin condensation, which is a good mimic of enzyme-catalyzed formation of acetoin (a principal flavor component in butter) from pyruvate (see the second example of nonoxidative decarboxylation at the beginning of this section). In the model reaction, two molecules of benzaldehyde are condensed to benzoin with specific catalysis by cyanide ion.

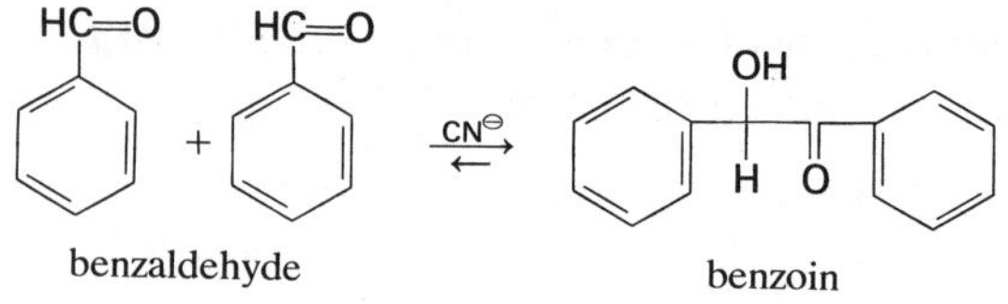

benzaldehyde benzoin

This reaction proceeds by nucleophilic attack of $CN^{\ominus}$ on the benzaldehyde electrophilic carbonyl to give a cyanohydrin with an acidic α-hydrogen (much more acidic than the original aldehyde hydrogen).

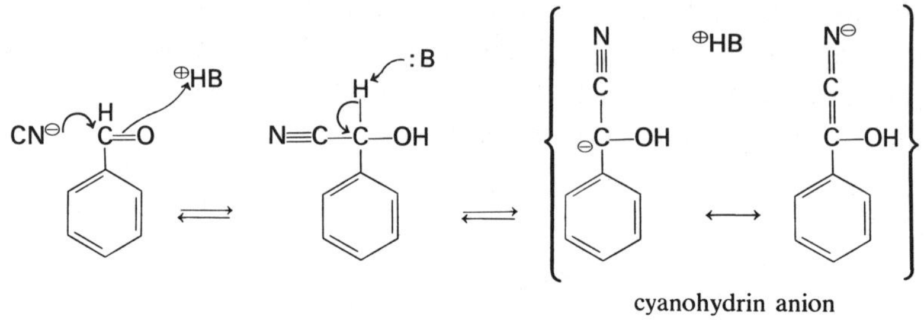

cyanohydrin anion

The carbanion of the cyanohydrin is a suitably reactive species for condensation with the second molecule of aldehyde, from which loss of the good leaving group $CN^{\ominus}$ then produces benzoin.

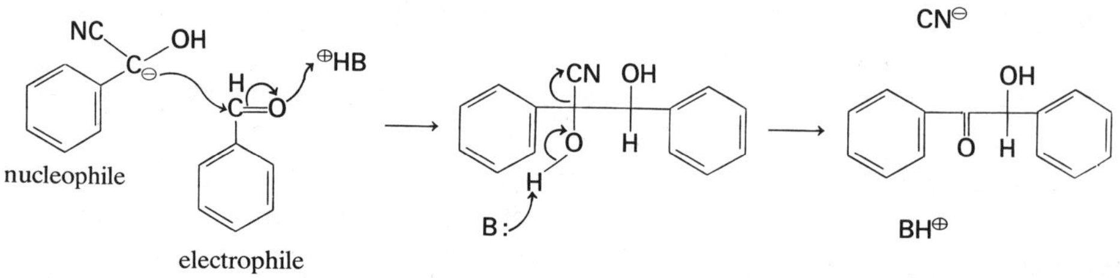

nucleophile

electrophile

So, the specific catalysis by $CN^{\ominus}$ is explicable by its addition to the carbonyl and stabilization of the carbanion used as the attacking species in the subsequent condensation.

This is precisely the function of thiamine-PP and, if anything, it may be a more effective electron sink (cationic imine) than is the cyanide ion. In the enzymatic reactions we discuss here involving TPP, two steps can be identified:

1. carbon–carbon bond cleavage in a substrate-TPP adduct;
2. *either* dissociation of the cleaved fragment from TPP, *or* its reaction as a nucleophile in a condensation with some electrophilic carbonyl-acceptor compound.

The TPP is bound noncovalently to the specific apoenzymes. It is bound so tightly as to be nondissociating in such enzymes as pyruvate decarboxylase, acetolactate synthetase, glyoxylate carboligase, and the pyruvate-dehydrogenase complex, but it must be added to preparations of *E. coli* pyruvate oxidase (Krampitz, 1957).

21.C.1 Nonoxidative Decarboxylations

Of the nonoxidative decarboxylations of α-keto acids, we shall discuss here the pyruvate decarboxylase and acetolactate synthetase mentioned earlier, as well as glyoxalate carboligase.

21.C.1.a Yeast pyruvate decarboxylase

In yeast, pyruvate is metabolized to acetaldehyde by pyruvate decarboxylase (175,000 mol wt). (The acetaldehyde is subsequently reduced to ethanol in fermenting yeast by NADH-dependent alcohol dehydrogenase molecules.) Catalysis proceeds by attack of the TPP C-2 carbanion on the α-keto group of a pyruvate molecule at the active site. Decarboxylation is facilitated in the adduct by delocalization of electrons into the iminium electron sink. The initial product is the stabilized anion of hydroxyethyl-TPP (HETPP), which then expels the initially added coenzyme carbanion.

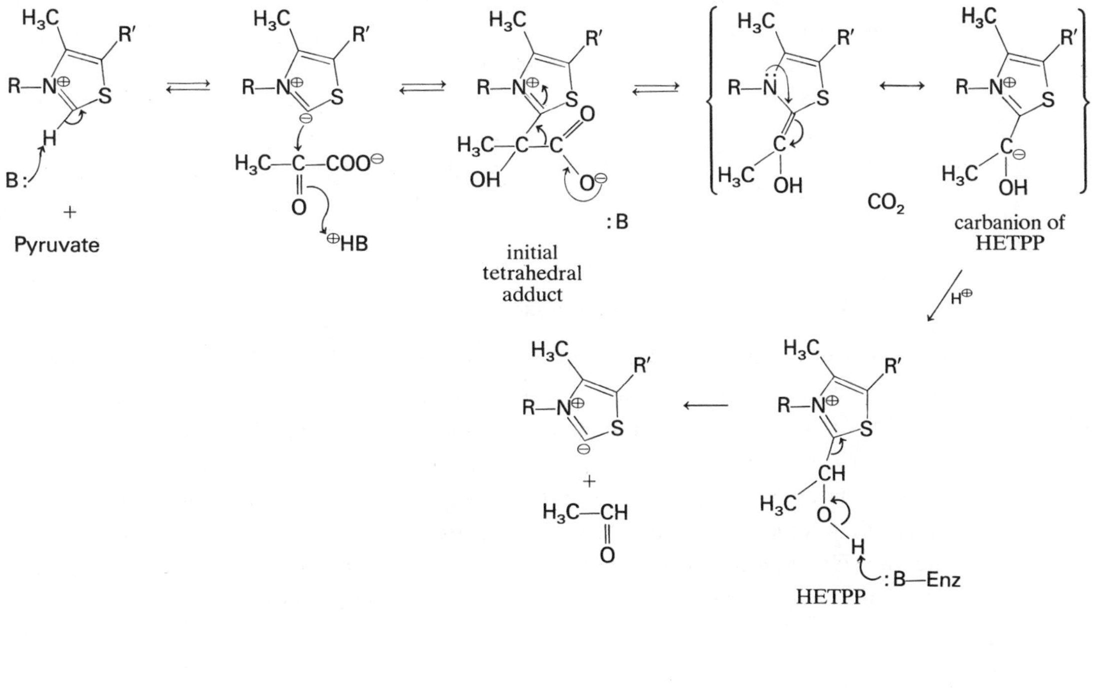

Pyruvate

initial tetrahedral adduct

CO_2

carbanion of HETPP

HETPP

Note that there is no change in the oxidation state of C-2 of pyruvate as it becomes C-1 of acetaldehyde. HETPP has been synthesized chemically, and it undergoes rapid enzyme-catalyzed conversion to acetaldehyde and TPP. In the absence of enzyme, HETPP is relatively resistant to nonenzymatic reversion to acetaldehyde and TPP anion. An analogue of the initial tetrahedral adduct has been prepared and studied for its rate of decarboxylation (Crosby et al., 1970).

analogue of initial adduct

transition state

The results of this study suggest a striking catalytic effect of desolvation. When rates of decarboxylation in ethanol were compared to rates in H_2O, the ethanol rate was 10^4-fold to 10^5-fold faster. In polar aprotic solvents, the rate was even faster. The acceleration may be due to a selective stabilization of the transition state relative to the starting material in polar aprotic media and in media with lower dielectric constant. The starting adduct is charged, but the charge is much more diffuse and dispersed in the transition state; the transition state is *less destabilized* than the ground state. Jencks (1975*a*) has commented that the active site of pyruvate decarboxylase appears to be hydrophobic, and the above acceleration may account for the bulk of the enzymatic acceleration to pyruvate decarboxylation. The adduct may be held at the hydrophobic active site by interaction energy provided by the pyrimidine and pyrophosphate groups (nonreacting moieties) with the enzyme.

21.C.1.b Acetolactate synthetase

We noted in Chapter 20 that acetolactate is a precursor of valine in bacterial cells and undergoes a skeletal rearrangement, catalyzed by the isomeroreductase, to α,β-dihydroxyvalerate. Acetolactate is synthesized in a TPP-dependent condensation of two molecules of pyruvate (Krampitz, 1975; Stormer and Umbarger, 1964; Glatzer et al., 1972). The initial steps up to HETPP formation proceed as noted for pyruvate decarboxylase. Then (rather than breakdown to acetaldehyde and the TPP anion) the hydroxyethyl adduct acts as nucleophile on the α-keto group of a second molecule of pyruvate, and only then is TPP expelled. Specificity may be imposed by the enzyme protein on control of the rate at which the TPP anion is expelled (i.e., before or after another molecule of substrate can diffuse in to the active site).

$$\text{TPP-C(CH}_3\text{)(OH)}^{\ominus} + \text{H}_3\text{C—CO—COO}^{\ominus} \longrightarrow \text{TPP—C(CH}_3\text{)(OH)—C(CH}_3\text{)(O}^{-}\text{)—COO}^{\ominus} \longrightarrow \text{TPP}^{\ominus} + \text{H}_3\text{C—C(=O)—C(CH}_3\text{)(OH)—COO}^{\ominus}$$

acetolactate

Note that acetolactate is a β-keto acid and, in addition to enzymatic processing by the isomeroreductase, it is readily decarboxylated by acetolactate decarboxylase to acetoin, one of the end products of bacterial metabolism.

21.C.1.c Glyoxalate carboligase

Glyoxalate carboligase is induced in various bacteria by growth on oxalate or glycolate (Gupta and Vennesland, 1964; Chung et al., 1971). It is mechanistically analogous to the acetolactate synthetase, converting two molecules of glyoxalate to one molecule of CO_2 and one of tartronate semialdehyde.

$$\text{HC(=O)—COO}^{\ominus} + \text{HC(=O)—COO}^{\ominus} \underset{}{\overset{\text{Enz—TPP}}{\rightleftharpoons}} \text{HC(=O)—CH(OH)—COO}^{\ominus} + CO_2$$

The aldehyde product can undergo subsequent NADH-dependent reduction to the more common metabolite, glycerate. Curiously, in addition to the TPP requirement, the homogeneous enzymes from *E. coli* and *Pseudomonas oxalyticus* contain tightly bound FAD. This requirement is not obvious because (unlike the succeeding reactions we shall discuss) the carboligase reaction involves no obvious redox change. Enzyme from which the FAD has been removed is inactive, but it is reactivated on addition of exogenous FAD. It seems quite likely that the bound FAD does not undergo redox change but rather may play a structural role in catalysis (Cromartie and Walsh, 1976). (Removal of FAD results in dissociation of active oligomeric enzyme.) On examination, acetolactate synthetases from some sources appear to be flavoproteins, but those purified from other organisms do not; these observations argue against the catalytic essentiality of flavin (Stormer and Umbarger, 1964; Glatzer et al., 1972).

21.C.2 Oxidative Decarboxylations of α-Keto Acids

We turn now to the decarboxylations of α-keto acids that involve oxidations. Here we shall consider the conversions of pyruvate to acetyl-CoA and to free acetate ion, and then we shall briefly mention some recently discovered enzymes that catalyze the back reactions (synthesis of α-keto acids).

21.C.2.a α-Keto-acid dehydrogenase complexes

The α-keto-acid dehydrogenase complexes come to mind immediately in any consideration of oxidative decarboxylations of α-keto acids by TPP-requiring enzymes. These are multienzyme complexes functioning as catalytic aggregates. One such complex, pyruvate dehydrogenase, converts pyruvate to acetyl-CoA and CO_2, a major route to "active" acetate.

$$H_3C{-}\overset{O}{\overset{\|}{C}}{-}COO^{\ominus} + CoASH \xrightarrow[\text{dehydrogenase}]{\text{pyruvate}} H_3C{-}\overset{O}{\overset{\|}{C}}{-}SCoA + CO_2$$

The other well-known complex, α-ketoglutarate dehydrogenase, functions in the tricarboxylic-acid (Krebs) cycle to produce succinyl-CoA from α-ketoglutarate.

$$^{\ominus}OOC{-}CH_2{-}CH_2{-}\overset{O}{\overset{\|}{C}}{-}COO^{\ominus} + CoASH \xrightarrow[\text{dehydrogenase}]{\alpha\text{-ketoglutarate}} {}^{\ominus}OOC{-}CH_2{-}CH_2{-}\overset{O}{\overset{\|}{C}}{-}SCoA + CO_2$$

As in other multienzyme aggregates we have discussed, one imagines that aggregation may affect both catalytic efficiency of the individual enzyme components and their regulation (Reed, 1966, 1974, and references therein). One could well imagine that channeling of metabolites occurs in such complexes and that the high microconcentration of catalysts prevents loss of intermediates by diffusion, breakdown, or diversion by other cellular enzymes. This could lead to enhanced catalytic efficiency; k_{cat}/K_m might not be limited in principle (¶2.C.3) by the upper limit of $10^{10}\ \text{M}^{-1}\ \text{sec}^{-1}$ that characterizes diffusion control in aqueous solutions. The stoichiometries shown above are incomplete. They conceal the fact that the oxidative decarboxylations are complex processes requiring a total of five coenzymes before catalytic formation of the activated acyl thioester is complete. Both the pyruvate dehydrogenase complexes and the α-ketoglutarate dehydrogenase complexes are complexes of three distinct enzyme components:

1. pyruvate decarboxylase (or α-ketoglutarate decarboxylase);
2. dihydrolipoamide transacetylase (or dihydrolipoamide transsuccinylase);
3. dihydrolipoamide dehydrogenase.

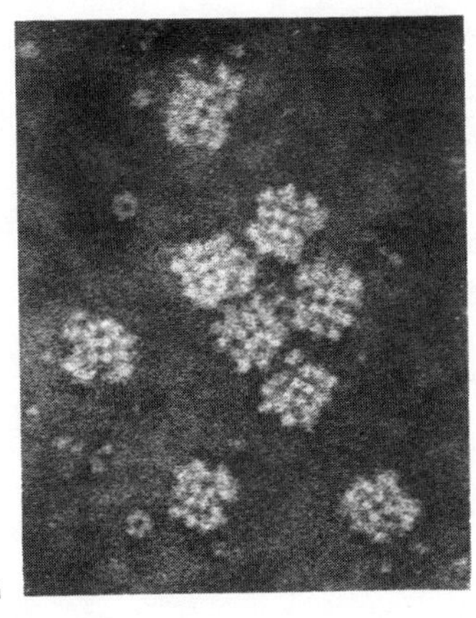

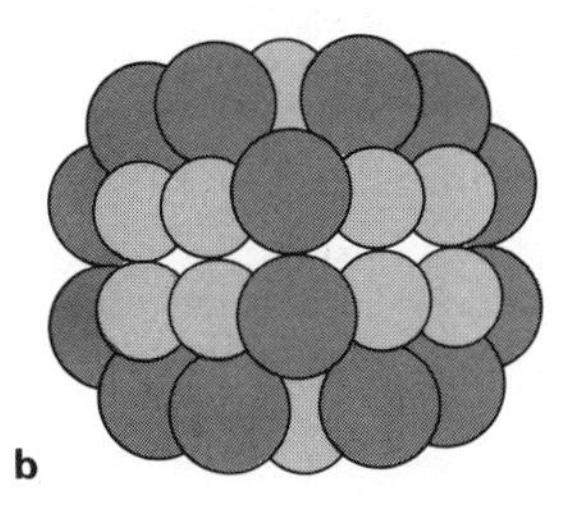

a b

Figure 21-4
Electron micrograph images and interpretative model. (**a**) *E. coli* pyruvate dehydrogenase complex. (**b**) Model of the *E. coli* pyruvate dehydrogenase complex. The 12 pyruvate dehydrogenase dimers are placed on the twofold positions (i.e., on the edges) of the transacetylase cube, and the 6 dihydrolipoamide dehydrogenase dimers are placed on the fourfold positions (i.e., in the faces). The electron micrograph was taken by Robert M. Oliver. The samples were negatively stained with sodium methyl phosphotungstate. Magnification is 300,000×. (Reprinted with permission from L. Reed, *Accounts of Chemical Research*, vol. 7, 1974, p. 42. Copyright by the American Chemical Society.)

The *E. coli* pyruvate dehydrogenase complex has 24 chains (12 dimers) of TPP-dependent decarboxylase of 92,000 mol wt, 24 chains of transacetylase of 65,000 mol wt, and 12 chains (6 dimers) of the flavoprotein dehydrogenase of 56,000 mol wt, for a total particle weight of 4,600,000. The corresponding α-ketoglutarate dehydrogenase complex has 12, 24, and 12 chains of these respective enzymes. Models of the multimeric structure have been constructed; Figure 21-4 indicates the shape and disposition of components. Although the catalytic mechanism for the bovine-kidney pyruvate dehydrogenase complex seems to be identical, the composition of the complex in that case is 80, 60, and 10 chains of the respective enzymes, where the first enzyme is an $\alpha_2\beta_2$ tetramer (20 tetramers per complex).

We shall comment explicitly on the conversion of pyruvate to acetyl-CoA, but the transformations undergone by α-ketoglutarate appear fully analogous mechanistically. Considering the individual enzymes in sequential action, we find that the pyruvate decarboxylase component is the TPP-utilizing enzyme, converting pyruvate to CO_2 and the activated C_2 fragment HETPP. This adduct is freely dissociable and identifiable with the separated decarboxylase component. Unlike the yeast pyruvate decarboxylase, this enzymes *does not* convert HETPP to acetaldehyde and TPP (Krampitz, 1957).

The second enzyme, dihydrolipoamide transacetylase, has the following stoichiometry.

$$\text{CoASH} + \text{HETPP} \rightleftharpoons \text{TPP (C-2–H)} + H_3C\text{—}\overset{O}{\overset{\|}{C}}\text{—SCoA} + \text{TPP}$$

HETPP

The TPP is regenerated, and the hydroxyethyl group has undergone oxidation and acyl transfer to the thiol group of CoASH to form the observed reaction product, acetyl-CoA. This is the key step in the multienzyme's catalysis. Knowledge of how this oxidative transfer may be effected is aided by the discovery that the transacetylase contains a covalently bound coenzyme that acts as the intermediate carrier of the two-carbon substrate moiety between thiamine pyrophosphate (TPP) and coenzyme A (CoA). This molecule is 6,8-dithiooctanoic acid, commonly known as dihydrolipoate.

$$\text{dihydrolipoate (HS, HS; COO}^{\ominus}\text{)} \underset{+2e^{\ominus}}{\overset{-2e^{\ominus}}{\rightleftharpoons}} \text{lipoate (S—S; COO}^{\ominus}\text{)} + 2H^{\oplus}$$

The dithiol grouping is susceptible to facile two-electron oxidation with intramolecular ring closure to the five-membered cyclic disulfide (oxidized lipoate). The coenzyme is bound to the transacetylase apoprotein via the valerate side chain in amide linkage to an ε-amino group of a lysine residue at the active site. Reed (1966, 1974), whose laboratory has concentrated on these enzymes, has suggested that this attachment generates a flexible side chain of length 14 Å, which may be important in the transfer through space of bound substrate from the TPP-binding site to the CoASH binding site, a swinging-arm mechanism (Fig. 21-5).

A reasonable suggestion for the role of the lipoamide coenzyme here would be initial addition of the HETPP carbanion to the disulfide group of oxidized lipoamide, generating a hemithioacetal. We have previously noted this kind of grouping in enzyme-catalyzed oxidations—e.g., with glyoxalase and glyceraldehyde-3-P dehydrogenase (Chapter 10).

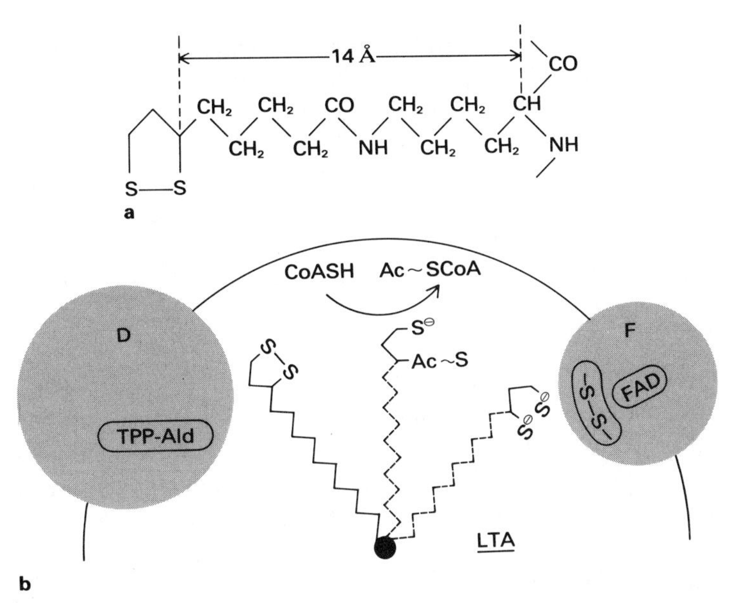

Figure 21-5
(**a**) Functional form of lipoic acid in the *E. coli* pyruvate and α-ketoglutarate dehydrogenase complexes. The carboxyl group of lipoic acid is bound in amide linkage to the ε-amino group of a lysyl residue, providing a flexible arm about 14 Å long for the reactive dithiolane ring. (**b**) A schematic representation of the possible rotation of a lipoyllysyl moiety between α-hydroxyethyl thiamine pyrophosphate (TPP-Ald) bound to pyruvate dehydrogenase (D), the site for acetyl transfer to CoA, and the reactive disulfide of the flavoprotein (F). the lipoyllysyl moiety is an integral part of dihydrolipoamide transacetylase (LTA). (Reprinted with permission from L. Reed, *Accounts of Chemical Research*, vol. 7, 1974, p. 43. Copyright by the American Chemical Society.)

Subsequent expulsion of the stable TPP carbanion would constitute the actual oxidation step, converting the sp^3 carbon of the hemithioacetal to the sp^2 carbon of the acetylthioester linkage of an acetyl lipoamide.

H_3C R′ R—N⊕ S H_3C—C—OH ⊖ S S BH⊕ Transacetylase component ⇌ H_3C R′ R—N⊕ S H_3C—C—O—H B—Enz S HS R″ B: hemithioacetal —oxidation⇌ H_3C R′ R—N⊕ S ⊖ + H_3C—C=O S HS R″ ⊕HB—Enz acetyl lipoamide

Now the C_2 fragment is at the product oxidation state and is a thioester, another case of enzymatic oxidation concerted with acyl activation. Simple acyl transfer from the thiol of lipoamide to the thiol of CoASH can follow as a transthiolation.

$$H_3C{-}\overset{O}{\overset{\|}{C}}{-}S\text{(HS, R'')} + {}^{\ominus}SCoA \rightleftharpoons H_3C{-}\overset{SCoA}{\overset{|}{C}}{-}S \text{(}O^{\ominus}\text{, HS, R'')} \rightleftharpoons \boxed{H_3C{-}\overset{O}{\overset{\|}{C}}{-}SCoA} + {}^{\ominus}S\text{(HS, R'')}$$

At this juncture, the overall conversion of substrate α-keto acid to product has been completed, but the lipoamide coenzyme has been left in the 2-$e^{\ominus}$–reduced state, as it must because it was the acceptor for the two electrons removed during substrate oxidation. In the dihydrolipoamide form, the transacetylase is not competent to carry out another turnover. It is the job of the third enzyme in the complex to reoxidize the dihydrolipoamide back to the active, disulfide form. Again, the swinging arm with 14 Å radius could move the dihydrolipoamide to the active site of the third enzyme. This simple thought would place the transacetylase at the center of a sphere of 28 Å diameter, with the active sites of the other two enzymes at some point within that volume.* The third enzyme, dihydrolipoamide dehydrogenase, is a flavoprotein dehydrogenase using FAD as the coenzyme (as noted briefly in Chapter 11). The dithiol form of the substrate reduces the bound FAD to $FADH_2$ and is oxidized to the disulfide. Note that the substrate moiety is a covalent portion of the transacetylase. There is a cystine disulfide at the active site of the dehydrogenase so that artificial reductants can put four electrons into the enzyme (two into $FADH_2$ and two into reduction of the disulfide), but in catalytic turnover only two electrons are put into the reduced enzyme. Both model studies (Loechler and Hollocher, 1975) and enzymatic experiments (Thorpe and Williams, 1976) suggest (as noted in Chapter 11) that this class of dithiol-oxidizing flavoenzymes effects electron transfer via attack of thiolate anion at C-4a of the oxidized FAD to produce 4a-covalent adducts as redox intermediates (see top of p. 695).

One last task remains before the pyruvate dehydrogenase complex has completed a catalytic cycle. As a flavoenzyme *dehydrogenase*, the $FADH_2$ in dihydrolipoamide dehydrogenase is not rapidly reoxidized by O_2 in air, and some other reoxidant must be used. In this case, that reoxidant is $NAD^{\oplus}$. Note that this

---

* Recent studies have attempted to map distances between the active sites of the subunits by energy-transfer measurements between fluorescent donor–acceptor pairs immobilized at specific points on adjacent subunit surfaces. These studies suggest that the active sites are in fact more than 28 Å apart and would, if valid, argue against a simple swinging-arm mechanism (Shepherd and Hammes, 1977).

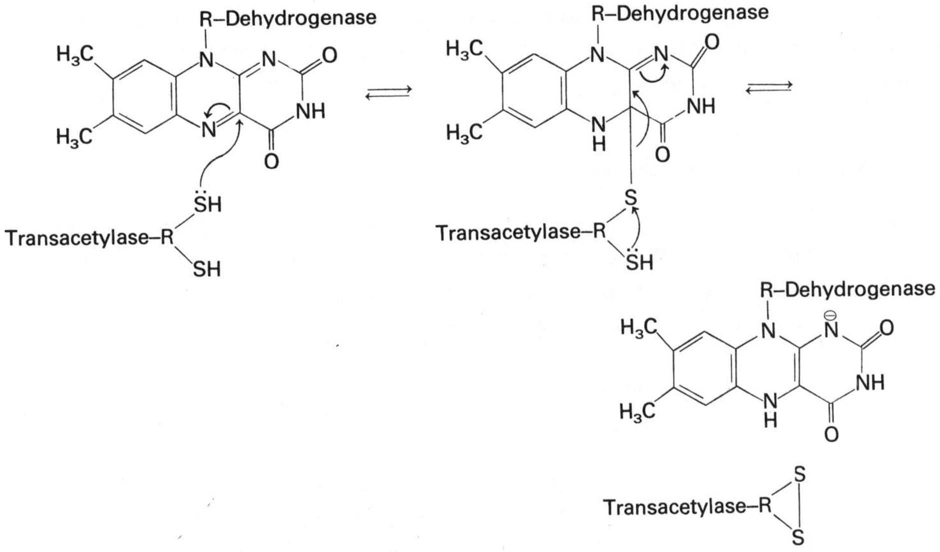

is an example of an enzyme that will catalyze exchange of the C-4 hydrogen (pro*S*) of NADH with solvent protons because equilibration occurs at the E—$FADH_2$ level (see Chapter 11).

$$E\text{—}FADH_2 + NAD^{\oplus} \rightleftharpoons E\text{—}FAD + NADH + H^{\oplus}$$

The overall stoichiometry of the pyruvate dehydrogenase complex is the following, with TPP, FAD, and lipoamide required catalytically.

$$H_3C\text{—}\overset{\displaystyle O}{\overset{\|}{C}}\text{—}COO^{\ominus} + NAD^{\oplus} + CoASH \longrightarrow H_3C\text{—}\overset{\displaystyle O}{\overset{\|}{C}}\text{—}SCoA + NADH + CO_2$$

One might expect that this production of the key biosynthetic metabolite acetyl-CoA would be under cellular regulation, and indeed the complex is subject to a number of reversible regulatory constraints, including covalent modification. The enzyme in eucaryotes is mitochondrial and (from sources as diverse as *Neurospora* and bovine kidney) undergoes phosphorylation (Reed, 1966, 1974). The enzyme from bacterial sources does not. One finds that the purified pyruvate dehydrogenase complex from bovine kidney has about 5 molecules of a kinase enzyme tightly bound, specifically to transacetylase chains (of which there are 60 in the complex). The site of phosphorylation is the pyruvate decarboxylase component, which is an $\alpha_2\beta_2$ tetramer, with the α-chain containing decarboxylase activity and the β-chain carrying out reductive acetylation of the transacetylase with HETPP. A unique serine on one of the α-chains is phosphorylated, leading to inactivation of the tetramer. When reactivation is desired, a phosphatase (also associated with the complex) removes the phosphate as $HOPO_3^{2\ominus}$ in a reaction stimulated by calcium ions.

21.C.2.b Pyruvate oxidase

An alternative oxidative decarboxylation of pyruvate occurs in certain strains of *E. coli.* The product from the action of this pyruvate oxidase is free acetate ion, not the activated acetyl-CoA, and the reaction is catalyzed by a single enzyme rather than a multienzyme complex.

$$H_3C{-}C({=}O){-}COO^{\ominus} \xrightarrow[\text{FAD}]{\text{Enz—TPP,}} H_3C{-}COO^{\ominus} + CO_2$$

The crystalline enzyme requires TPP, as expected, and also has stoichiometric amounts of FAD as the redox-active component. In contrast to the nonoxidative glyoxalate carboligase (¶21.C.1.c), the FAD bound to pyruvate oxidase undergoes demonstrable reduction. The physiological reoxidants of this membrane-bound flavoenzyme are electron-transfer components of the respiratory chain of the membrane. The role of the enzyme may be to funnel electrons into the membrane and thus to be used to energize membrane functions such as active transport of solutes. It is likely that the initial HETPP undergoes two-electron oxidation by the bound FAD to acetyl-TPP, which then undergoes hydrolysis.

$$\text{Enz—FAD} + \text{HETPP (thiazolium C-2: } H_3C{-}CH{-}OH) \rightleftharpoons \underset{\text{acetyl-TPP}}{\text{thiazolium C-2: } H_3C{-}C{=}O} + \text{Enz—FADH}_2 \xrightarrow{H_2O} \text{TPP (C-2—H)} + H_3C{-}COO^{\ominus}$$

Lienhard (1966) has prepared an acetyl thiamine and found that it is stable enough to be isolated. In weak base the acyl adduct exists as the hydrate, but it will produce acetate in H_2O.

$$\text{hydrate (C-2: } H_3C{-}C(OH){-}OH) \xleftarrow{H_2O,\ OH^{\ominus}} \text{acetyl thiazolium (C-2: } C({=}O){-}CH_3) \xrightarrow{H_2O,\ H^{\oplus}} \text{thiazolium (C-2—H)} + H_3C{-}COOH$$

21.C.2.c Ferredoxin-dependent α-keto-acid synthases

The pyruvate dehydrogenase and α-ketoglutarate dehydrogenase complexes are functionally irreversible, but enzymes that catalyze the back reaction have recently been described from fermentative bacteria and also from anaerobic photosynthetic bacteria (B. Buchanan, 1972). The driving force for CO_2 assimilation and α-keto-acid generation is provided by the low-potential (−400 mV) iron–sulfur protein, *reduced ferredoxin* (Chapter 13). Strictly speaking, these enzymes such as pyruvate synthase and α-ketoglutarate synthase are *carboxylases, requiring TPP.* To date, the instability of enzymatic preparations has precluded extensive analysis, but it may be that these synthases have endogenous iron–sulfur chromophores that may mediate electron transfer between ferredoxins and substrates.

$$\text{Ferredoxin}_{(red)} + \text{Acetyl-CoA} + CO_2 \rightleftharpoons \text{Ferredoxin}_{(ox)} + \text{CoASH} + \text{Pyruvate} \quad (1)$$

$$\text{Ferredoxin}_{(red)} + \text{Succinyl-CoA} + CO_2 \rightleftharpoons \text{Ferredoxin}_{(ox)} + \text{CoASH} + \alpha\text{-Ketoglutarate} \quad (2)$$

Each enzyme catalyzes an acyl-CoA–independent partial exchange between α-keto acid and CO_2, consistent with reversible formation of the hydroxyalkyl-TPP adducts. One can imagine a reaction scheme that is consistent with the steps we have noted in the preceding TPP-dependent enzymes—ending with attack, in the case of pyruvate synthase, of the HETPP anion on the electrophilic CO_2. (Presumably CO_2 and not $HCO_3^{\ominus}$ is the actual substrate.)

H_3C R′ R—N⊕ S ⊖ H_3C—C—SCoA ‖ O ⇌ H_3C R′ R—N⊕ S H_3C—C—SCoA O⊖ ⇌ H_3C R′ R—N⊕ C=O H_3C + CoASH

acetyl-TPP

$+CO_2$ 2 e⊖ from $Fd_{(red)}$

H_3C R′ R—N⊕ S ⊖ + H_3C—C=O COO⊖ ⇌ H_3C R′ R—N⊕ S H_3C—C—OH C=O ⊖O ⇌ H_3C R′ R—N⊕ S H_3C—C—OH ⊖ + O=C=O

pyruvate

21.C.3 α-Ketol Transfers

Before leaving the subject of TPP-mediated enzymatic catalysis, we must consider the other major type of reaction in which this coenzyme is required: ketol transfers.

The most common metabolic transfer of a ketol group occurs during the aerobic metabolism of glucose in the pentose-phosphate pathway, catalyzed by an enzyme aptly designated transketolase. We have previously (Chapter 19) discussed how ribulose-5-P is formed from glucose-6-P and how this ketose is isomerized in a 1,2-shift to the aldose ribose-5-P. These two phosphorylated pentoses are substrates for transketolase, generating a C_3 aldose (glyceraldehyde-3-P) and a C_7 ketose (sedoheptulose-7-P) as products.

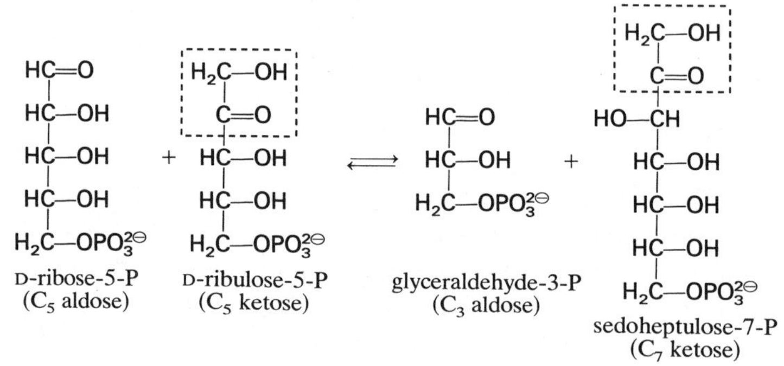

D-ribose-5-P (C_5 aldose) + D-ribulose-5-P (C_5 ketose) ⇌ glyceraldehyde-3-P (C_3 aldose) + sedoheptulose-7-P (C_7 ketose)

The conversion involves the shuttle of the C_2 ketol unit between the aldose acceptor molecules. We note explicitly here that TPP fulfills exactly the same function as it does in α-keto-acid decarboxylations, providing a good electron sink so that a stabilized carbanion can attack the aldose acceptor.

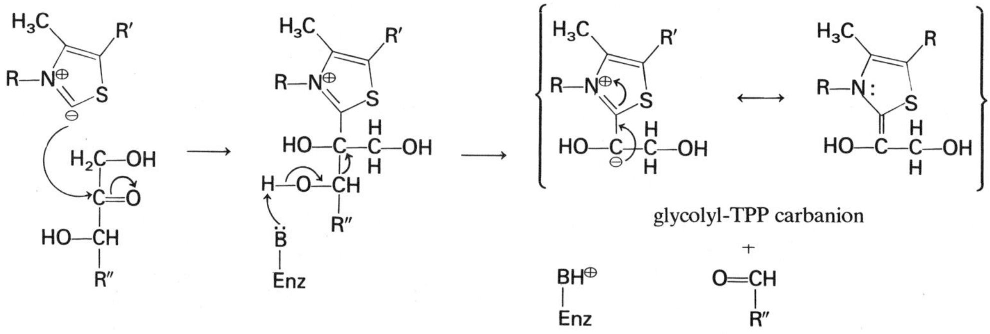

glycolyl-TPP carbanion

\+

O=CH–R″

In transketolase, this glycolyl-TPP carbanion is held at the enzyme site long enough for the aldose moiety (ribose-5-P) just formed to depart and the new aldose (glyceraldehyde-3-P) to enter and undergo the reverse of the above process. The glycolyl-TPP carbanion is the key intermediate, and it is generated by an aldol cleavage whose driving force is the fact that the glycolyl anion is stabilized by the electron sink of the coenzyme.

A second example of a ketol transfer is provided by the enzyme phosphoketolase (Krampitz, 1957), which is found in anaerobic bacteria where it catalyzes formation of the "high-energy" acetyl phosphate by an oxidation process without expenditure of ATP (e.g., compare with acetate kinase). Because anaerobes are inefficient organisms in terms of ability to generate ATP (no oxidative phosphorylation), this device is useful to them.

$$\underset{\text{ketose-6-P}}{H_2C(OH){-}C(=O){-}CH(OH){-}CH(OH){-}CH(OH){-}H_2C{-}OPO_3^{2\ominus}} + \underset{P_i}{HOPO_3^{2\ominus}} \xrightleftharpoons{\text{Enz—TPP}} \underset{\text{acetyl-P}}{H_3C{-}\overset{O}{\overset{\|}{C}}{-}OPO_3^{2\ominus}} + \underset{\text{erythrose-4-P}}{HC(=O){-}HC(OH){-}HC(OH){-}H_2C{-}OPO_3^{2\ominus}}$$

The reaction involves a phosphorylytic cleavage of fructose-6-P to acetyl-P and the aldose erythrose-4-P. Acetyl-P can then be funneled into acetyl-CoA by phosphotransacetylase. It is known that tritium appears in the methyl group of acetyl-P when incubations are performed in ^3H_2O. Also, glycolaldehyde is a substrate, reacting with P_i to give acetyl-P. This latter evidence suggests that glycolyl-TPP is a likely candidate in both reactions (from glycolaldehyde or from fructose-6-P) and suggests analogy to transketolase.

$$\text{TPP ylide } (H_3C, R', R{-}N^{\oplus}, S, C^{\ominus}) + O{=}C(CH(OH)R''){-}CH_2OH \rightleftharpoons \text{adduct: } HO{-}C(TPP)(CH(OH)R''){-}CH_2OH,\ \text{with } \ddot{B}{-}Enz \rightleftharpoons \left\{ R{-}\ddot{N}{:}\ \text{enamine } HO{-}C{=}C_2(TPP){-}CH_2OH \longleftrightarrow R{-}N^{\oplus}\ HO{-}\overset{\ominus}{C}(TPP){-}CH_2OH \right\} + BH^{\oplus}{-}Enz + O{=}CH{-}R''$$

Given the glycolyl-TPP, formation of acetyl-P requires reduction of the hydroxymethyl group to a methyl group and oxidation of the C-2 from an aldehyde to the acyl oxidation state. One can imagine this occurring by loss of water and then a prototropic shift to form acetyl-TPP, accounting also for solvent incorporation data. Phosphorolysis would produce the product acyl-P and regenerate the coenzyme.

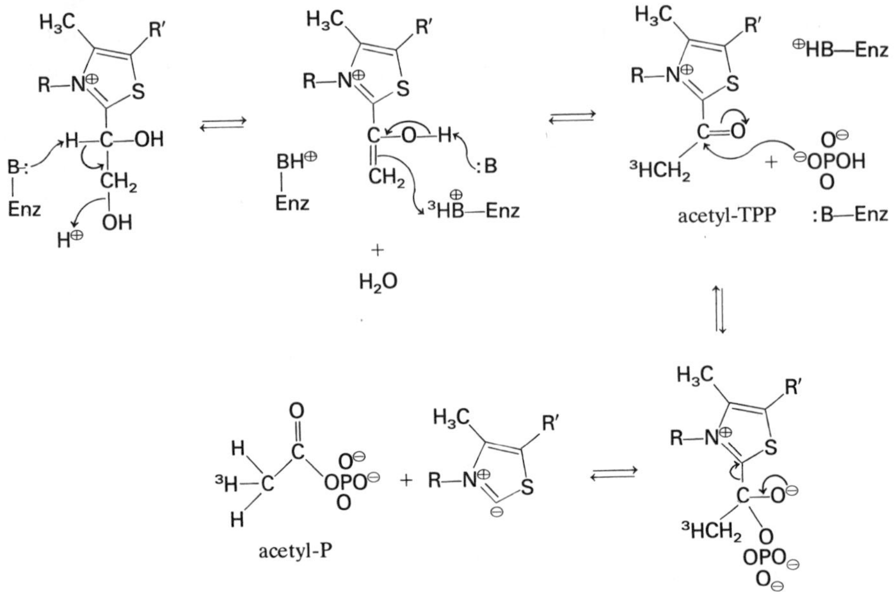

acetyl-TPP

acetyl-P

This concludes our discussion of the coenzymatic functioning of TPP.

There is evidence that some form of thiamine is important in nerve-cell metabolism in animals. In fact, thiamine deficiency has long been known to generate neurological disorders. The triphosphate of thiamine can be detected in nerve cells, as can the kinase enzyme responsible for its synthesis.

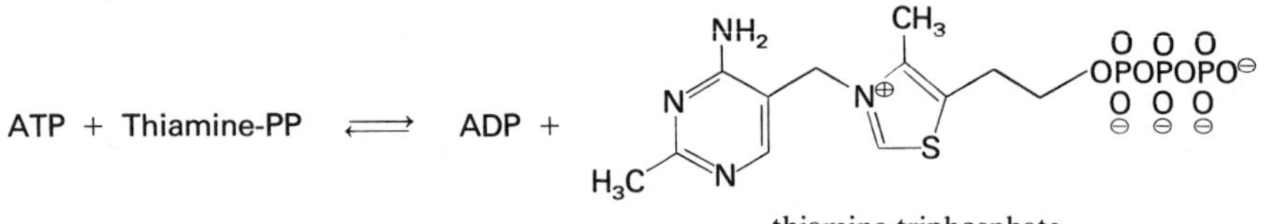

ATP + Thiamine-PP ⇌ ADP + thiamine triphosphate

thiamine triphosphate

21.C.4 Miscellaneous Decarboxylations

A number of other enzyme-catalyzed decarboxylations occur on molecules that are neither β-keto acids (nor their precursors) nor α-keto acids. In some cases, the driving force for CO_2 is obvious; in others it is not.

In Chapter 20, we discussed the decarboxylation of the dihydroaromatic compound prephenate, with elimination of either the *para* OH (prephenate dehydrase) or the *para* H (prephenate dehydrogenase). Aromatization can be the driving force.

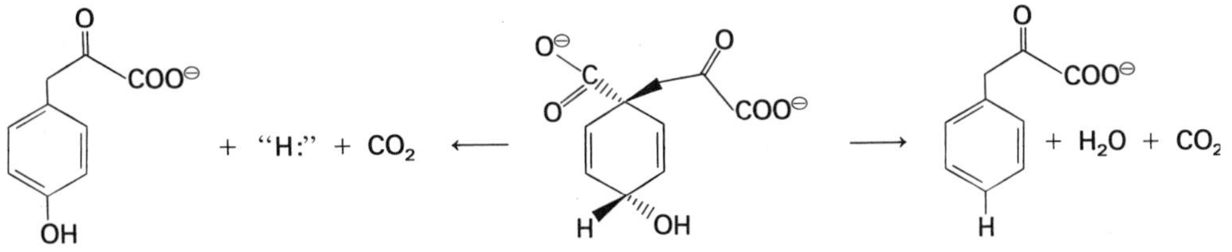

O, COO⊖, OH $+$ "H:" $+ CO_2 \longleftarrow$ O⊖, C, O, O, COO⊖, H, OH $\longrightarrow$ O, COO⊖, H $+ H_2O + CO_2$

Two other enzymes are orotidylate decarboxylase and α-picolinate decarboxylase, which catalyze loss of CO_2 from sp^2 centers.

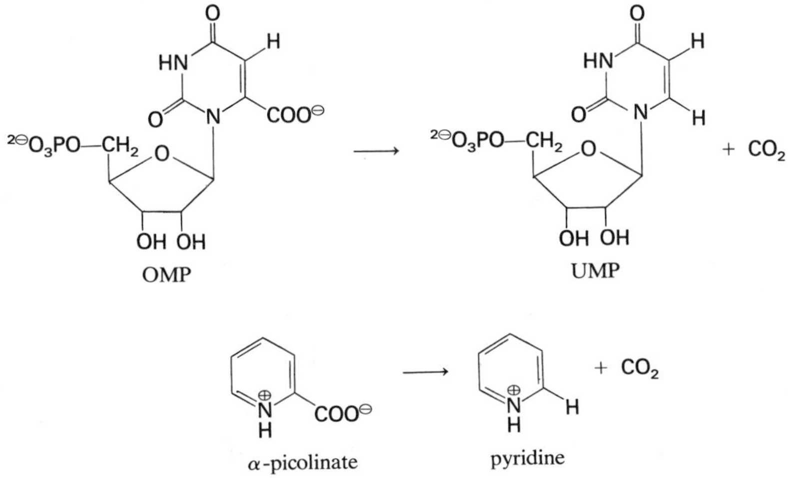

O, HN, H, O, N, COO⊖, ²⊖O₃PO—CH₂, O, OH OH

OMP

$\longrightarrow$ O, HN, H, O, N, H, ²⊖O₃PO—CH₂, O, OH OH $+ CO_2$

UMP

N⊕, H, COO⊖ $\longrightarrow$ N⊕, H, H $+ CO_2$

α-picolinate pyridine

Mechanistic studies on these enzymes are essentially nonexistent. In the OMP decarboxylase, one possibility is reversible addition of an enzyme nucleophile at C-5 to produce an sp^3 hybridization at both C-5 and C-6, from which CO_2 loss could be smooth (albeit with possible *syn* geometry).

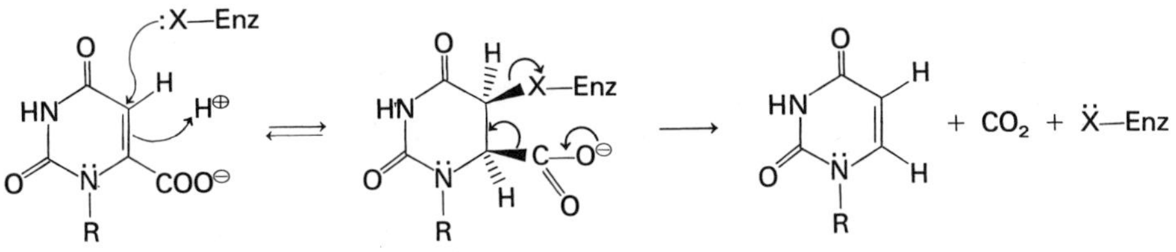

:X—Enz, O, H, HN, H⊕, O, N, COO⊖, R $\rightleftharpoons$ O, H, X—Enz, HN, C, O⊖, O, N, H, O, R $\longrightarrow$ O, H, HN, O, N, H, R $+ CO_2 +$ X—Enz

Or, CO_2 could be lost from the sp^2 centers directly to yield an sp^2 carbanion that could be stabilized by electrostatic interaction with an adjacent quaternary nitrogen center. For example,

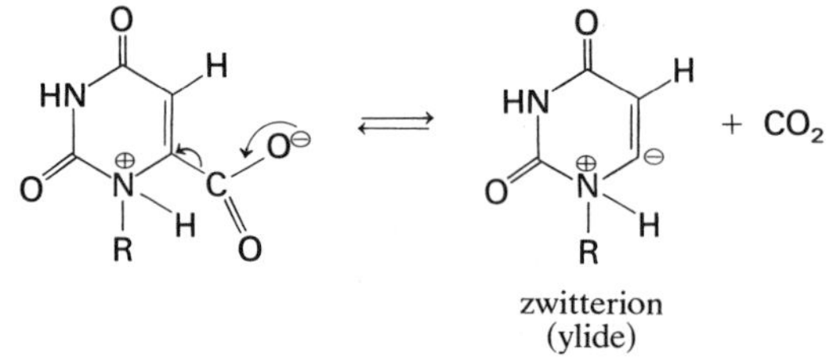

zwitterion
(ylide)

Such a zwitterion could also form in the picolinate decarboxylase sequence. These zwitterions, of course, have analogy to the electrostatically stabilized C-2 carbanion of TPP.

Two other decarboxylations worth brief mention are (1) the conversion of salicylate to resorcinol (a replacement of $COO^{\ominus}$ by OH) carried out by the flavin-dependent monooxygenase salicylate hydroxylase (Chapter 12),

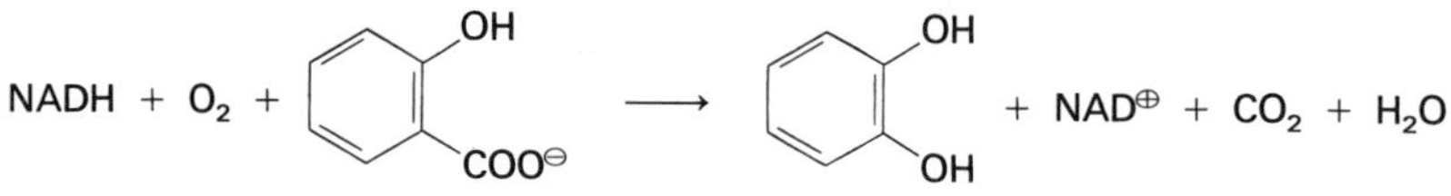

$$NADH + O_2 + \text{salicylate} \longrightarrow \text{resorcinol} + NAD^{\oplus} + CO_2 + H_2O$$

and (2) the decarboxylation of *p*-hydroxybenzoate (¶20.B.5) during coenzyme-Q biosynthesis.

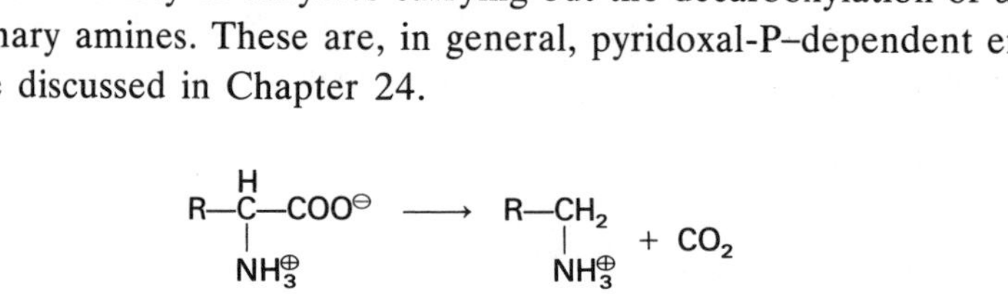

There are a variety of enzymes carrying out the decarboxylation of α-amino acids to primary amines. These are, in general, pyridoxal-P–dependent enzymes, and they are discussed in Chapter 24.

$$R{-}CH(NH_3^{\oplus}){-}COO^{\ominus} \longrightarrow R{-}CH_2{-}NH_3^{\oplus} + CO_2$$

Chapter 22

Enzymatic Carboxylations

Physiological *carboxylation* reactions are the reverse of the *decarboxylation* reactions we have just discussed. Enzymatic CO_2 fixation involves the addition of a good electrophile, CO_2, to a substrate carbanion. For effective catalysis, the molecules undergoing carboxylation ought to have structural features that can stabilize the carbanion so that the energy barrier to reaction is feasibly low. We shall note this feature in the biological molecules that react with CO_2.

A second problem to be solved by carboxylases is what species of CO_2 to utilize. Unhydrated CO_2 is a good electrophile at carbon, but is present at low concentrations in aqueous solutions at neutral pH. The bulk species of the equilibrium is the hydrate (bicarbonate, $HCO_3^{\ominus}$). On the other hand, although bicarbonate is plentiful in the milieu, it is nowhere near as good an electrophile as CO_2. The anionic $HCO_3^{\ominus}$ must be activated to a more electrophilic species, either by metal-ion coordination or via dehydration at the enzyme active site (or by both).

A kinetic method for determining the reactive species has been developed in Wood's laboratory by taking advantage of the fact that, when CO_2 is added to aqueous solutions, the chemical hydration takes several seconds to come to equilibrium (Cooper et al., 1968).

$$H_2O + CO_2 \rightleftharpoons H_2CO_3 \rightleftharpoons HCO_3^{\ominus} + H^{\oplus}$$

Enzymatic incubations (at concentrations of CO_2 near the K_m) with a carboxylase are performed with enough carboxylase so that a significant rate of product formation occurs during those several seconds. The kinetic course is monitored. If

CO_2 is the substrate, the initial rate will be faster than the rate after the hydration equilibrium has been established. If $HCO_3^{\ominus}$ is the enzymatic substrate, then the rate in the first few seconds will lag behind the subsequent rate (still in the initial-velocity period).

The incubations can be repeated using solid $HCO_3^{\ominus}$ to initiate the carboxylation sequence, again monitoring whether a kinetic burst or lag occurs. Finally, as a control to ensure that any nonlinear behavior seen in these experiments in the initial seconds is meaningful, one can initiate by either CO_2 or $HCO_3^{\ominus}$ addition to solutions containing *both* the carboxylase of interest and the enzyme carbonic anhydrase. Carbonic anhydrase accelerates the nonenzymatic equilibration of CO_2 and $HCO_3^{\ominus}$ dramatically with a turnover number of $\sim 10^6 \text{ sec}^{-1}$, so that hydration becomes instantaneous on the experimental time scale. The lag or burst should be abolished. We shall see that, although a given carboxylase will carboxylate only CO_2 or only $HCO_3^{\ominus}$, there are representatives of each class. For instance, the data of Figure 22-1a suggest that the enzyme PEP carboxytransphosphorylase uses CO_2, whereas those of Figure 22-1b indicate that pyruvate carboxylase uses $HCO_3^{\ominus}$.

In previous chapters, we have already noted two enzymes that act physiologically as carboxylases. One is carbamoyl-P synthetase, which uses ammonia, ATP, and bicarbonate to form the activated carbamoylating agent, carbamoyl phosphate (¶5.C.3.c). Pyruvate synthetase, noted in the last chapter as a TPP- and ferredoxin-requiring enzyme, also fixes CO_2 into an organic substrate (¶21.C.2.c). The enzyme could easily have been placed in this chapter rather than the last.

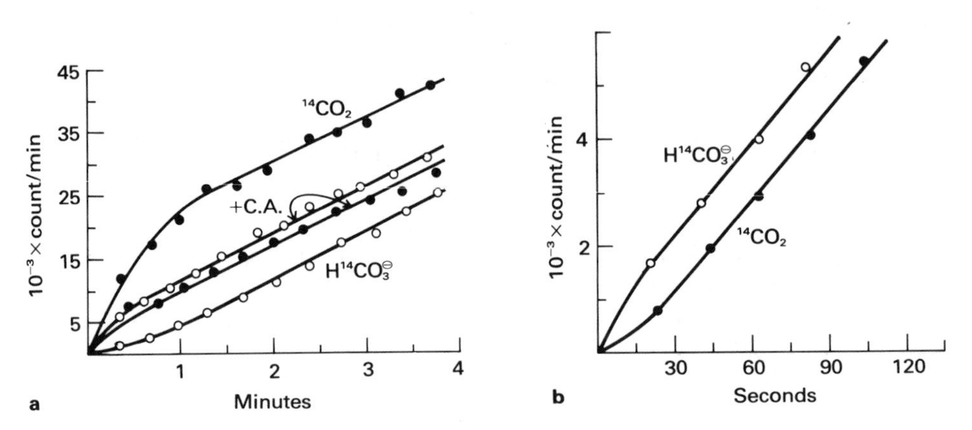

Figure 22-1
(**a**) Radiochemical assay of PEP carboxytransphosphorylase activity when the initial mixture includes either $H^{12}CO_3^{\ominus}$ plus $^{14}CO_2$ (•) or $H^{14}CO_3^{\ominus}$ plus $^{12}CO_2$ (∘). The curves labeled +C.A. represent the same samples with the addition of carbonic anhydrase. (**b**) Radiochemical assay of pyruvate carboxylase activity when the initial mixture includes either $H^{12}CO_3^{\ominus}$ plus $^{14}CO_2$ (•) or $H^{14}CO_3^{\ominus}$ plus $^{12}CO_2$ (∘). (From Cooper et al., 1968, p. 3862)

In this chapter, we shall divide enzymatic carboxylations into four categories. The first comprises enzymes that carboxylate the substrate phosphoenolpyruvate. The second contains the biotin-requiring carboxylases that use ATP as cosubstrate and that carboxylate either α-keto acids or acyl-CoA thioesters. The third category is represented by ribulose-1,5-diphosphate carboxylase, a key enzyme in photosynthetic plants and microorganisms. The fourth category contains a fascinating (albeit less well-characterized) carboxylation sequence, dependent on vitamin K, of certain glutamyl residues in specific proteins involved in blood coagulation or bone metabolism.

22.A ENZYMES THAT CARBOXYLATE PEP

We shall comment here on the three enzymes known to carboxylate PEP to oxaloacetate.

$$H_2C{=}C(OPO_3^{2\ominus})COO^{\ominus} + RX + \text{“}CO_2\text{”} \longrightarrow {}^{\ominus}OOC{-}C(=O){-}CH_2{-}COO^{\ominus} + RXPO_3^{2\ominus}$$

The phosphoryl acceptor RX may be water, a nucleoside diphosphate, or P_i. The first enzyme, PEP carboxylase, uses H_2O as phosphoryl acceptor. This enzyme is restricted to bacteria and plants, and it may be less sophisticated than PEP carboxykinase (yeast, plants, and higher animals), which uses a nucleoside diphosphate (XDP) as phosphoryl acceptor. The third enzyme, PEP carboxytransphosphorylase, is of restricted distribution—to date found only in bacteria that ferment propionic acid (*Propionobacterium*); it uses P_i as phosphoryl acceptor.

It has been argued that *the common active species in these three enzymatic reactions is the enol form of pyruvate, bound and stabilized at the active site of each enzyme* (Utter and Kollenbrander, 1972). Enolpyruvate has the requisite carbanion character at C-3 to attack the carbon dioxide or activated bicarbonate and produce oxaloacetate. Recall that enolpyruvate also has been postulated as an intermediate in pyruvate kinase action, on the basis of partial isotopic exchange during enolization (¶7.D.1).

22.A.1 PEP Carboxylase

With PEP carboxylase, the substrate is $HCO_3^{\ominus}$, the enzyme requires $Mg^{2\oplus}$, and the pH optimum is 8.3 on the basis of exchange results in ^3H_2O (Utter and Kollenbrander, 1972).

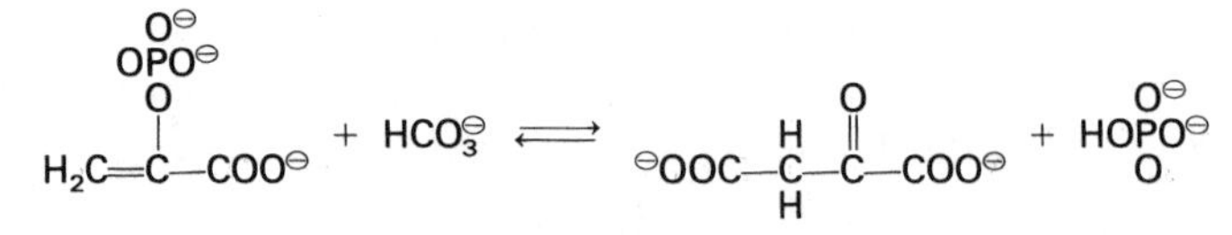

$$H_2C{=}C(OPO_3^{2-})\text{—}COO^- + HCO_3^- \rightleftharpoons {}^-OOC\text{—}CH_2\text{—}C(=O)\text{—}COO^- + HOPO_3^{2-}$$

The reaction is functionally unidirectional for oxaloacetate formation, driven by hydrolysis of the high group-transfer potential of the enol-phosphate linkage in PEP. With [^{18}O]-bicarbonate, Lane has observed that one ^{18}O atom ends up in P_i and two end up in the oxaloacetate. There are no detectable partial exchange reactions that would indicate formation of discrete, freely reversible intermediates. Although a concerted four-center mechanism (Fig. 22-2) has been written as a possibility, we can speculate on alternative proposals. One such would use the phosphoryl group of PEP to *activate* the bicarbonate for the eventual

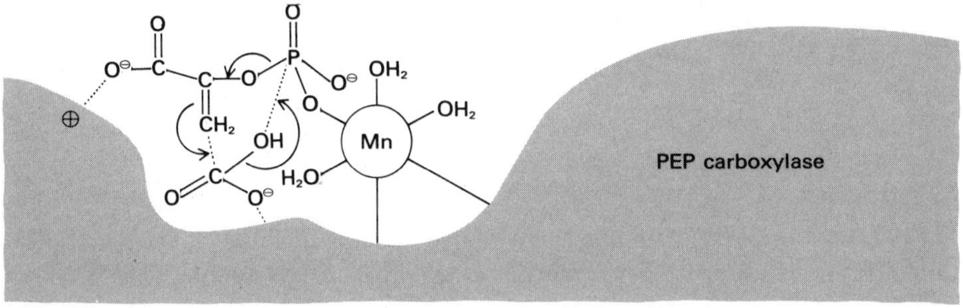

Figure 22-2
Postulated concerted mechanism for PEP-carboxylase–catalyzed reaction. (From Utter and Kollenbrander, 1972, p. 167)

dehydration that must ensue in oxaloacetate formation, by formation of the mixed anhydride of carbonic and phosphoric acid, carbonyl phosphate. This intermediate was postulated for carbamoyl-P synthetase (¶5.C.3.c) and is a reasonable species whenever bicarbonate is the molecule undergoing enzyme-catalyzed carboxylation (Meister, 1976; Sauers et al., 1976). A reasonable two-step sequence involves a bicarbonate oxygen as nucleophile toward phosphorus in step 1 to produce the mixed carbonic–phosphoric anhydride and to release the enolate anion of pyruvate.

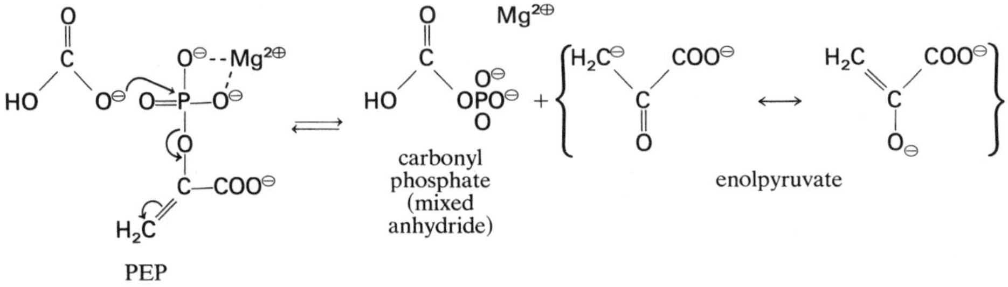

$$HO\text{—}C(=O)\text{—}O^- + \underset{\text{PEP}}{H_2C{=}C(COO^-)\text{—}O\text{—}PO_3^{2-}\cdots Mg^{2+}} \rightleftharpoons \underset{\substack{\text{carbonyl phosphate}\\ \text{(mixed anhydride)}}}{HO\text{—}C(=O)\text{—}OPO_3^{2-}} + \underset{\text{enolpyruvate}}{\left\{ {}^-H_2C\text{—}C(=O)\text{—}COO^- \longleftrightarrow H_2C{=}C(O^-)\text{—}COO^- \right\}} \quad (1)$$

These two species can now react in step 2, with the carbanionic carbon of enolpyruvate attacking the electrophilic carbon of carbonyl phosphate with expulsion of inorganic phosphate. The two-step scheme would account for observed ^{18}O-transfer data.

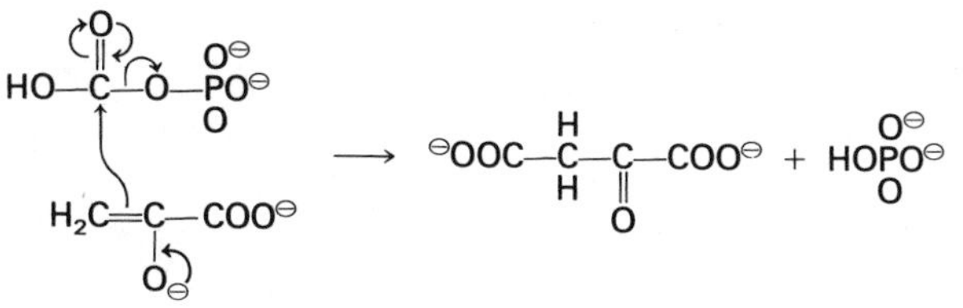

$$HO{-}C({=}O){-}O{-}PO_3^{2-} + H_2C{=}C(O^-){-}COO^- \longrightarrow {}^-OOC{-}CH_2{-}C({=}O){-}COO^- + HOPO_3^{2-} \quad (2)$$

22.A.2 PEP Carboxykinase

PEP carboxykinase differs from PEP carboxylase in that the carboxykinase catalyzes a demonstrably reversible reaction by virtue of using a nucleotide as additional cosubstrate (Utter and Kollenbrander, 1972). The mammalian enzyme uses either guanosine diphosphate (GDP) or inosine diphosphate (IDP); the yeast enzyme uses ADP.

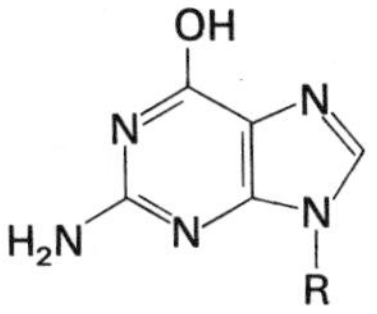

GDP 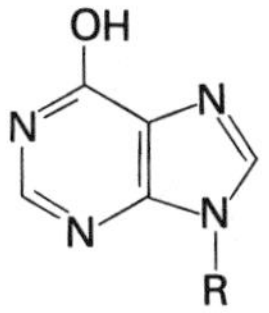 IDP 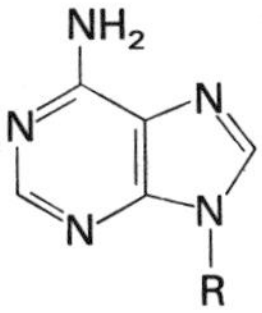 ADP

During carboxylation of PEP to oxaloacetate, the phosphoryl group is transferred to the nucleoside diphosphate and produces GTP, ITP, or ATP.

$$\text{PEP} + CO_2 + \text{XDP} \xrightleftharpoons{\text{Enz—Mg}^{2+}} \text{Oxaloacetate} + \text{XTP}$$

The reaction is reversible because, in the back direction, the high-energy enolphosphate bond can be formed at the expense of β,γ-phosphoric-anhydride bond breakage in the XTP and carbon–carbon bond cleavage in CO_2 release. As the stoichiometry indicates, unhydrated CO_2 is the substrate bound and then attacked in this instance. No partial exchange reactions have been detected in careful experiments performed with liver carboxykinase.

On the other hand, the yeast enzyme has been reported to display a kinase activity separable from carboxylation—i.e., a pyruvate kinase behavior (Cannata and Stoppani, 1963).

$$\text{PEP} + \text{ADP} \xrightleftharpoons{\text{kinase activity}} \text{ATP} + \text{Pyruvate}$$

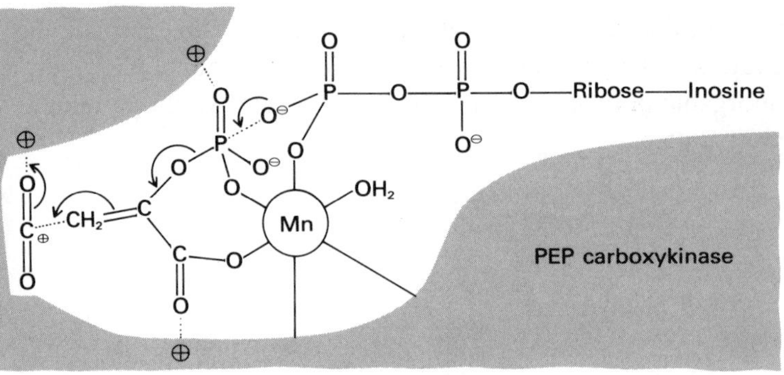

Figure 22-3
Postulated concerted mechanism for PEP-carboxykinase–catalyzed reaction. (From Utter and Kollenbrander, 1972, p. 167)

No phosphoenzyme is detectable, and no independent enolization of pyruvate (as is seen with authentic pyruvate kinase) has been reported. The yeast enzyme also displays an oxaloacetate decarboxylase activity independent of nucleotide (also a property of authentic pyruvate kinase) (Creighton and Rose, 1976).

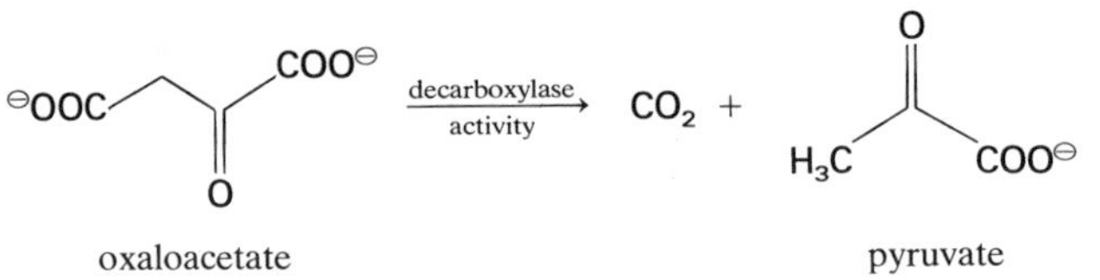

$$\text{oxaloacetate} \xrightarrow[\text{activity}]{\text{decarboxylase}} CO_2 + \text{pyruvate}$$

One thus can imagine that the enzyme from either source may proceed via a ternary complex mechanism during normal turnover, with pyruvate-kinase–like and oxaloacetate-decarboxylase–like steps in sequence. That is, PEP could phosphorylate ADP to yield ATP and enolpyruvate, which in turn then attacks bound CO_2. An alternative concerted scheme (Fig. 22-3) has also been postulated.

22.A.3 PEP Carboxytransphosphorylase

PEP carboxytransphosphorylase has been purified from *Propionobacterium shermanii* and has been shown to catalyze the functionally irreversible carboxylation of PEP with phosphoryl transfer to P_i as acceptor, generating inorganic pyrophosphate in addition to oxaloacetate (Utter and Kollenbrander, 1972). This stoichiometry is similar to that of the PEP carboxykinase reaction with P_i replacing XDP.

$$\text{PEP} + CO_2 + P_i \longrightarrow \text{Oxaloacetate} + PP_i$$

Furthermore, both enzymes use CO_2 rather than bicarbonate, and in both instances there is no ^{18}O incorporation in products when incubations are conducted in $H_2{}^{18}O$. In the absence of CO_2, a proton can act as "alternative electrophile" to form pyruvate and PP_i as products (Siu and Wood, 1962; Lochmuller et al., 1968; J. Davis et al., 1969). This reaction is analogous to the kinase activity of the yeast PEP carboxykinase.

$$\text{PEP} + H^{\oplus} + P_i \longrightarrow \text{Pyruvate} + PP_i$$

No partial isotope exchanges are demonstrable for either activity, suggesting either concerted reaction or retention of initial bound product until after the remaining substrates are bound to the active site and react. There are complex requirements for divalent cations such as $Mg^{2\oplus}$, $Mn^{2\oplus}$, $Cu^{2\oplus}$, or $Co^{2\oplus}$ during catalysis. One speculation has suggested attack by an oxygen of inorganic phosphate on the phosphorus of PEP to yield a transient pentacovalent enolpyruvate, which can then attack CO_2 or undergo protonation to pyruvate.

enolpyruvate

CO_2 → Oxaloacetate

$H^{\oplus}$ → Pyruvate

Rose and Wood undertook stereochemical experiments designed to determine whether the addition of CO_2 or of a proton occurs at the *si–si* or at the *re–re* face of the PEP substrate (Willard and Rose, 1973; O'Brien et al., 1973). We shall first consider the stereochemistry of oxaloacetate formation (CO_2 addition); then we shall look at that of pyruvate formation (the nonphysiological proton addition).

The carboxylation stereochemistry could be solved by using PEP samples stereospecifically labeled at C-3. We noted in our discussion of enolase catalysis (¶17.A.3) that it has been possible to prepare the 3-[^{3}H]-PEP where the tritium is *cis* to the phosphate group, (*Z*)-3-[^{3}H]-PEP (*Z* for *zusammen* = "together"), and also the 3-tritio-PEP where the tritium is *trans* to the phosphate group, (*E*)-3-[^{3}H]-PEP (*E* for *entgegen* = "opposite") (Alworth, 1972, pp. 146–148). Carboxylation of either stereospecific tritiated PEP would yield a 3-[^{3}H]-oxaloacetate molecule. However, C-3 of oxaloacetate is adjacent to the C-2 carbonyl group and thus is subject to nonenzymatic enolization that would lead to randomization of stereochemistry at C-3, as the planar enolate could be reprotonated equally well at either face. Thus, the oxaloacetate is reduced in situ to 3-[^{3}H]-malate with

NADH and malate dehydrogenase. Now one can determine at leisure whether the tritiated malate is 3*R* or 3*S*. We noted in Chapter 17 that fumarase carries out a *trans* elimination of water from 2*S*,3*R*-malate, removing only the hydrogen species at the 3*R* position as ^3H_2O. This fact allows determination of the PEP carboxylation stereochemistry.

Consider the two alternatives open to PEP carboxytransphosphorylase carboxylating (*Z*)-3-[^{3}H]-PEP molecules. (Note that the 2-*si*–3-*re* face corresponds to the 2-*re*–3-*re* face in the unlabeled molecule because of priority changes in going from protio- to tritio-PEP. See Alworth, 1972, pp. 146–148.)

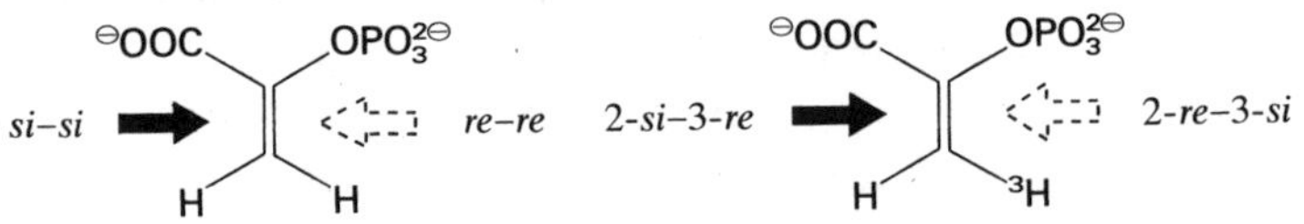

si–si ⊖OOC OPO$_3^{2\ominus}$ H H re–re
PEP

2-si–3-re ⊖OOC OPO$_3^{2\ominus}$ H ^{3}H 2-re–3-si
(*Z*)-3-[^{3}H]-PEP

Addition of CO_2 to the 2-*si*–3-*re* face (*si*–*si* of unlabeled PEP) will generate 3*R*-[^{3}H]-oxaloacetate and then 3*R*-[^{3}H]-malate, which will lose tritium to the medium when allowed to react to equilibrium with fumarase. Conversely, addition of CO_2 to the 2-*re*–3-*si* face (*re*–*re* of unlabeled PEP) will produce 3*S*-[^{3}H]-malate and result in retention of tritium in the fumarate product, as indicated in the following scheme.

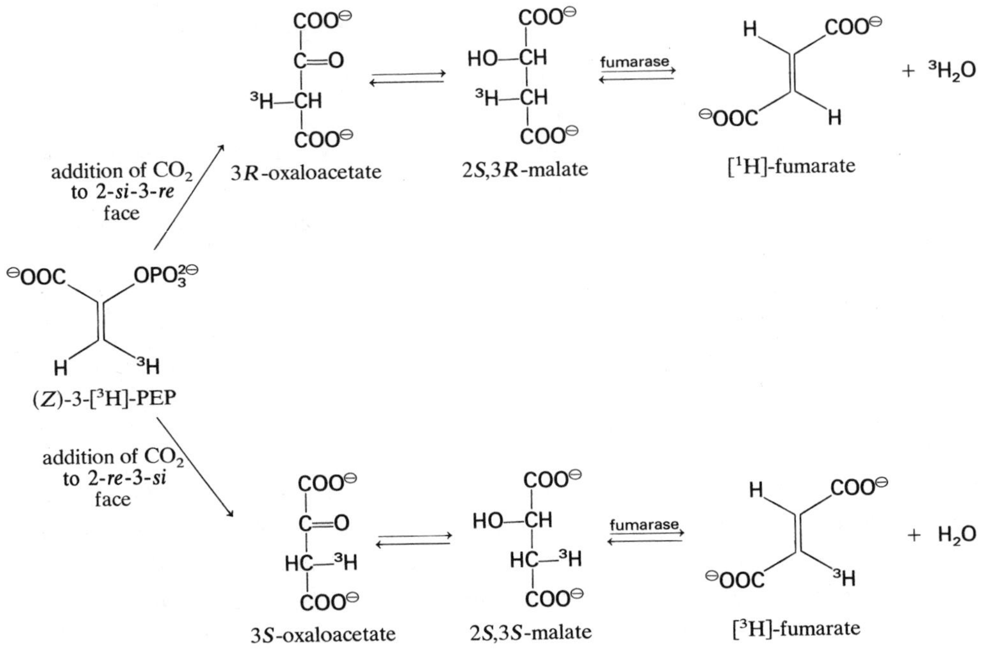

(*Z*)-3-[^{3}H]-PEP (⊖OOC, OPO$_3^{2\ominus}$, H, ^{3}H)

addition of CO_2 to 2-*si*-3-*re* face → 3*R*-oxaloacetate (COO⊖, C=O, ^{3}H—CH, COO⊖) ⇌ 2*S*,3*R*-malate (COO⊖, HO—CH, ^{3}H—CH, COO⊖) ⇌ (fumarase) [^{1}H]-fumarate (H, COO⊖, ⊖OOC, H) + ^3H_2O

addition of CO_2 to 2-*re*-3-*si* face → 3*S*-oxaloacetate (COO⊖, C=O, HC—^{3}H, COO⊖) ⇌ 2*S*,3*S*-malate (COO⊖, HO—CH, HC—^{3}H, COO⊖) ⇌ (fumarase) [^{3}H]-fumarate (H, COO⊖, ⊖OOC, ^{3}H) + H_2O

The (*Z*)-3-[^{3}H]-PEP yields 98% of its tritium as ^3H_2O after fumarase equilibration, establishing that, with protio-PEP, the carboxytransphosphorylase catalyzes addition of the electrophilic CO_2 to the *si–si* face. The complementary experiment with (*E*)-3-[^{3}H]-PEP yields the same conclusion, based on the retention of 98% of the radioactivity in fumarate after a similar incubation (Willard and Rose, 1973). By essentially identical experiments, it was possible to show that the other two enzymes, PEP carboxylase and PEP carboxykinase, also catalyze addition of CO_2 to the 2-*si*–3-*si* face of PEP, indicating that they all bind PEP such that the *re–re* face is inaccessible to incoming CO_2 (Rose et al., 1969). This stereochemical unanimity suggests a possible close common evolutionary origin for the active sites of these proteins.

22.A.4 Stereochemical Analysis of Chiral Methyl Groups

Rose next turned his attention to the carboxytransphosphorylase-dependent conversion of PEP to pyruvate in the absence of CO_2 (Willard and Rose, 1973). This problem is considerably more complex to solve because C-3 of pyruvate is a methyl group, not a methylene prochiral center as in oxaloacetate. Until very recently, it has been impossible to determine the stereochemistry of formation of a methyl group because there was no way of distinguishing the three hydrogens on it. Because the methyl group can freely rotate ($T_{1/2} \approx 10^{-12}$ sec), the three hydrogens are sterically equivalent. The methyl is said to be *torsiosymmetric.*

Ingenious recent experiments have resulted in preparations of chiral acetates and chiral pyruvates, samples where the methyl groups contain all three isotopes of hydrogen (^{1}H, ^{2}H, ^{3}H) and thus are chiral.

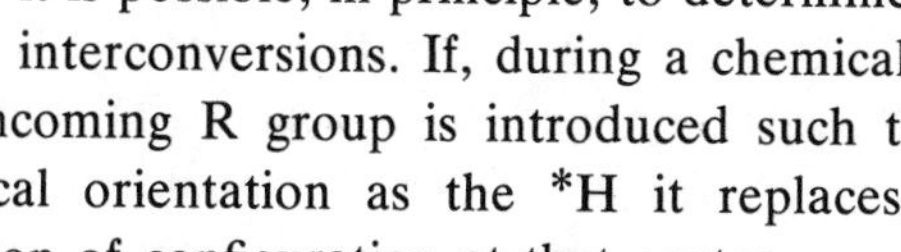

^{2}H
^{1}H···C—X
^{3}H

chiral *R*-isomer

^{2}H
X—C···^{1}H
^{3}H

chiral *S*-isomer

With these substrates it is possible, in principle, to determine the stereochemistry of methyl–methylene interconversions. If, during a chemical reaction at a chiral methyl group, the incoming R group is introduced such that it has the same relative stereochemical orientation as the *H it replaces, then the reaction proceeds with retention of configuration at that center.

$$R + *H-\overset{}{C}(H)(H)-X \rightleftharpoons R-C(H)(H)-X + *H$$

If the new carbon–carbon bond has the opposite stereochemical orientation, the transformation will involve inversion. To determine the actual result, one must—

1. know the starting chirality (and know the degree of chiral purity);

2. have a method of analysis for the chirality generated in the methylene of the product;

3. know that the removal of a hydrogen species at the chiral center proceeds with a kinetic isotope discrimination against deuterium and tritium.

We shall examine each of these conditions before discussing the actual carboxy-transphosphorylase data.

The groups of Cornforth and Eggerer and of Arigoni have succeeded in the preparation of chiral 2*R* and 2*S* ($^1H,^2H,^3H$)-acetic acids where the syntheses lead unambiguously to each isomer (Alworth, 1972, pp. 193–211, 234–240; Cornforth et al., 1969; Luthi et al., 1969; Retey, Luthi, and Arigoni, 1970). As we have noted several times, tritium is normally employed at tracer levels, whereas deuterium is present in all the molecules. These syntheses represented no exception, and only a small percentage (about 1 molecule in 10^9) contained tritium and was chiral; all the others were prochiral ($—CH_2D$). This problem is obviated, however, by the fact that, in all these experiments, *one assays for radioactivity*, measuring only those molecules that contain tritium, and thus selectively examining the behavior of the chiral molecules in a sea of unlabeled, achiral species. Every molecule containing tritium must also contain deuterium in these preparations. We shall now see how measurement of radioactivity can be used to define stereochemistry *with the explicit provision that the interconversion show a normal kinetic isotope effect.* The following explanation is essentially that offered by Luthi et al. (1969; Retey, Luthi, and Arigoni, 1970) and by Alworth (1972, pp. 193–211, 234–240).

Consider a conversion of the *R*- or *S*-acetate into a methylene group, where we specify that the incoming X replaces the removed hydrogen species with retention. Given no steric discrimination between H, D, and T, all six conformations (*a* through *f* in the following expressions) should be equally probable at the active site of an enzyme carrying out the replacement as indicated.

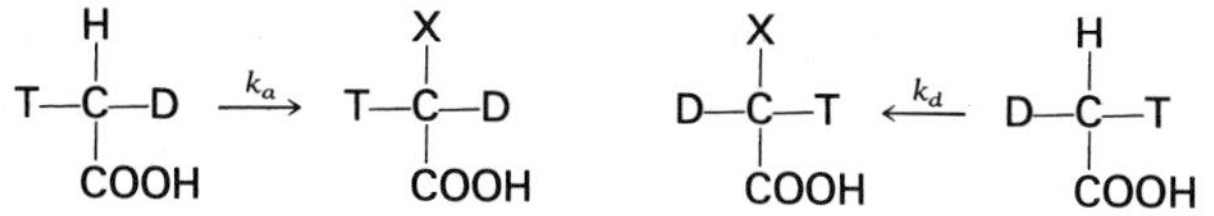

```
     H                    X                    X                     H
     |                    |                    |                     |
  T—C—D    --ka-->     T—C—D                D—C—T     <--kd--    D—C—T
     |                    |                    |                     |
    COOH                 COOH                 COOH                  COOH
R-substrate a         product 1            product 4           S-substrate d

     D                    X                    X                     D
     |                    |                    |                     |
  H—C—T    --kb-->     H—C—T                T—C—H     <--ke--    T—C—H
     |                    |                    |                     |
    COOH                 COOH                 COOH                  COOH
R-substrate b         product 2            product 5           S-substrate e

     T                    X                    X                     T
     |                    |                    |                     |
  D—C—H    --kc-->     D—C—H                H—C—D     <--kf--    H—C—D
     |                    |                    |                     |
    COOH                 COOH                 COOH                  COOH
R-substrate c         product 3            product 6           S-substrate f
```

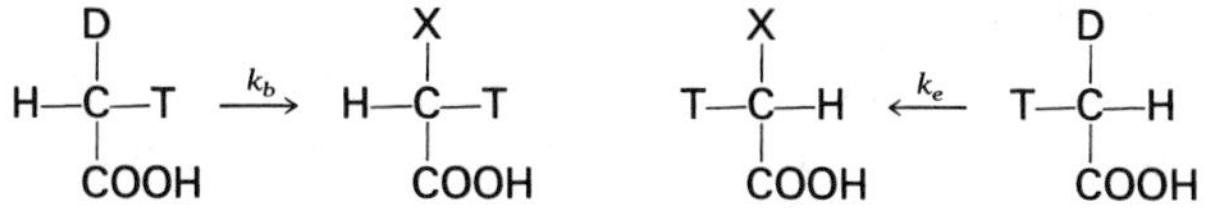

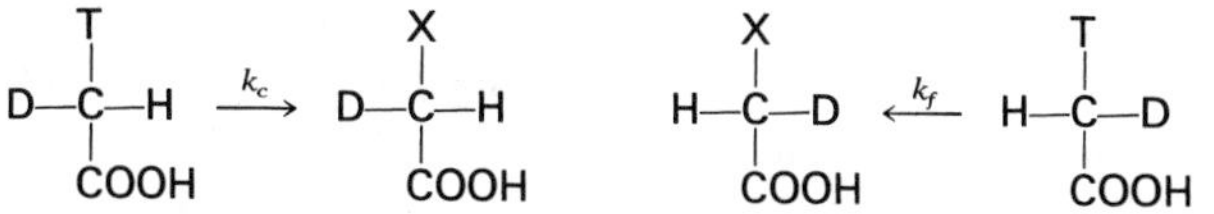

Now, if all six bound conformations react *at equal rates* to form product, the stereochemical course (retention) *will not be determinate* even from acetate samples of 100% chiral purity. (Reaction at equal rates means that inherent differences in the rates of C—H, C—D, and C—T bond breakage do not show up in V_{max}—that is, these bond-breakage elementary steps are not rate-determining.) This indeterminacy is easily demonstrated by considering the products from one isomer—*R*-acetate, for example. Product molecules 1, 2, and 3 will form at equal rates, but 3 will not be detectable because it bears no tritium, so only 1 and 2 are catalogued. Suppose X is an OR group. Then 1 and 2 have the forms shown here as 1A and 2A.

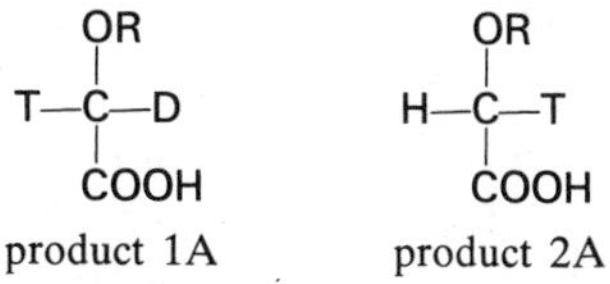

```
     OR                  OR
     |                   |
  T—C—D               H—C—T
     |                   |
    COOH                COOH
 product 1A          product 2A
```

The product 1A has tritium in the *S*-position, whereas 2A has tritium in the *R*-position. Thus, the 2*R*-[^{1}H,^{2}H,^{3}H]-acetate will generate products whose methylene groups have *equal amounts* of tritium at 2*R* and 2*S* positions. Exactly the same conclusion will derive from the 2*S*-chiral acetate, where products 4 and 5 are formed. As Alworth has pointed out, this result cannot then distinguish among inversion, retention, or a random, nonstereospecific process.

On the other hand, suppose that, in the above reaction of acetate, a kinetic isotope effect is manifested with $k_H > k_D > k_T$. Again we can monitor formation of tritium-containing products 1 and 2 from *R*-acetate. But note that formation of

1 involves C—H cleavage and formation of 2 involves C—D cleavage, so 2 must form more slowly than 1. Product 1 should accumulate preferentially under V_0 conditions. Note that the amount of enrichment is directly related to the intramolecular k_H/k_D for the reaction. For example, in formation of products 1A and 2A with an observed k_H/k_D of 4.0 in this transformation, use of *R*-acetate will produce an 80:20 mixture of 1A:2A: that is, 80% of the tritiated product molecules have tritium in the *S*-position of the methylene group. Now, if a method exists for evaluating tritium content at the *S*-position at the methylene C-2 of the products 1A and 2A, the stereochemistry of methyl→methylene conversion is solved. This analytical methodology is of the type we noted earlier in the fumarase reaction (Willard and Rose, 1973; Alworth, 1972, pp. 146–148).

In the first reported example of this technique, the separated chiral *R*- and *S*-acetates were converted via acetate kinase and phosphotransacetylase to *R*- and *S*-acetyl-CoA samples (Cornforth et al., 1969; Luthi et al., 1969; Retey, Luthi, and Arigoni, 1970). These samples then were incubated one at a time with glyoxalate and the enzyme malate synthase (an enzyme we shall discuss in more detail in the following chapter on enzymatic Claisen condensations). This synthase was used because of its availability and because of the known existence of an isotope effect ($k_H/k_T = 2.7$) with [^{3}H]-acetyl-CoA in the rate of *S*-malate formation (Eggerer and Klette, 1967).

$$\underset{\underset{\displaystyle O}{\|}}{HC}—COO^{\ominus} + \text{Acetyl-CoA} \rightleftharpoons \text{CoASH} + \begin{array}{c} COO^{\ominus} \\ | \\ H_S—C—H_R \quad \boxed{\text{C-3}} \\ | \\ H—C—OH \quad \boxed{\text{C-2}} \\ | \\ COO^{\ominus} \end{array}$$

2*S*-malate

The other reason for choosing this enzyme is that the *S*-malate can then be submitted to the action of fumarase, which removes the hydrogen species at the 3*R*-position of 2*S*-malate. Monitoring of the amount of ^3H_2O formed indicates what percentage of the tritium ends up in the *S*-position at C-3 of the malate. As a control, it was established that achiral [^{3}H]-acetate yields 50% of the tritium as ^3H_2O after equilibration by fumarase. Finally, the 2*S*-acetyl-CoA yields a malate that loses 76.5% of its tritium into water in the presence of fumarase. The malate from *R*-acetyl-CoA loses only 23.3% of its tritium. This result indicates that the major product from *S*-acetyl-CoA is 2*S*,3*R*-[^{3}H]-malate, whereas this is the minor product from *R*-acetyl-CoA. The reader should confirm that these results indicate that the glyoxylate fragment adds in a stereochemical sense opposite to that of the departing proton—*condensation proceeds with inversion at the methyl group.* (For detailed structural presentation of these results, see pp. 193–211 and 234–240 of the book by Alworth, 1972.)

Given the knowledge that malate synthetase shows a kinetic isotope effect and catalyzes condensation with inversion, the malate synthase–fumarase couple can be (and is) used as a convenient analytical system to determine the absolute configuration of acetate molecules of unknown chirality. Two cautions are worth voicing. Whenever possible, the experiments should be done with both isomers of chiral acetate and complementary results obtained. Note further that the preferential accumulation of one isomer of malate is related not only to the kinetic isotope effect* but also to the chiral purity of the acetate samples. Less than absolute chiral purity will diminish the real differentials between rates of formation of the major and minor isomers of tritiated malate.

At this point one can see that the methodology can also be applied for enzymatic methylene→methyl conversions, the reverse of the methyl→ methylene transformation. Given a methylene group stereospecifically labeled with two isotopes of hydrogen, if the enzyme cleaving the methylene group to a methyl moiety introduces the third hydrogen isotope in a stereospecific manner, the resultant product will have a chiral methyl group. It is worth noting explicitly that *in this direction (methylene → methyl), an isotope effect is not required for accumulation of a chiral product species.*

With this explanation of the principles and methodology of chiral methyl-group determination, we can return to the carboxytransphosphorylase reaction and the CO_2-independent production of pyruvate from PEP. Rose first used (*Z*)-[^{3}H]-PEP and P_i in D_2O with the transphosphorylase to generate molecules of pyruvate with all three hydrogen isotopes in the methyl group (Willard and Rose, 1973).

$$\mathrm{(Z)\text{-}[^3H]\text{-}PEP\ (^3H,\ OPO_3^{2\ominus};\ H,\ COO^{\ominus})} \xrightarrow{D_2O} \text{3-}[^1H,^2H,^3H]\text{-Pyruvate} \xrightarrow{H_2O_2} \text{2-}[^1H,^2H,^3H]\text{-Acetate} + CO_2$$

$$\text{2-}[^1H,^2H,^3H]\text{-Acetate} \xrightarrow{\text{acetate kinase}} \text{2-}[^1H,^2H,^3H]\text{-Acetyl-}PO_3^{2\ominus}$$

$$\text{2-}[^1H,^2H,^3H]\text{-Acetyl-CoA} \xleftarrow[\text{CoASH}]{\text{phosphotransacetylase,}} \text{2-}[^1H,^2H,^3H]\text{-Acetyl-}PO_3^{2\ominus}$$

$$\text{2-}[^1H,^2H,^3H]\text{-Acetyl-CoA} \xrightarrow{\text{glyoxalate, malate synthase}} \text{3-}[^2H,^3H]\text{-Malate}$$

The pyruvate sample was decarboxylated with H_2O_2 to acetate, which then was subjected sequentially to acetate kinase, phosphotransacetylase, malate synthase, and fumarase. Surprisingly, fumarase treatment liberated only 50% of the

---

* The relevant kinetic isotope effect is the one describing *intramolecular* competition between H, D, or T removal from a *given* chiral methyl group (Eggerer and Klette, 1967).

tritium of 3-[^{3}H]-malate into ^3H_2O. As a check, the experiment was repeated, starting this time with (*E*)-[^{3}H]-PEP. Again, 50% washout of tritium by fumarase indicates an equal population of 3*R*- and 3*S*-tritiomalates. Because the conversion by the transphosphorylase is a methylene-in-PEP to methyl-in-pyruvate conversion, no kinetic isotope effect is needed for selective enrichment of one chiral pyruvate species, and this cannot be the factor producing racemic mixtures of tritiated product. The mechanistic conclusion then was that, in the carboxytransphosphorylase-mediated conversion of PEP to pyruvate, *proton addition occurs randomly to either face of the enolpyruvate.* Because the addition of CO_2 to the enzyme-bound enolpyruvate is at least 98% to the *si* face (¶22.A.3), Rose concluded that, *in the absence of* CO_2, the enolpyruvate anion is *released* from the active site and protonated randomly in solution. Pyruvate formation occurs only by artifactual leakage of enolpyruvate anion into solution. Consistent with this idea, carboxytransphosphorylase will not detritiate added pyruvate, apparently not recognizing it as a substrate molecule for enolization. These results on carboxytransphosphorylase contrast with the ability of pyruvate kinase to carry out pyruvate enolization (Chapter 7). (In the kinase case, chiral pyruvate is indeed formed from (*Z*)- or (*E*)-[^{3}H]-PEP, by *si* addition at C-3.)

22.B ATP- AND BIOTIN-DEPENDENT CARBOXYLASES

There are six well-characterized carboxylases that have in common a requirement for ATP and bicarbonate as substrates and for the coenzyme biotin as obligate cofactor (Fig. 22-4; Moss and Lane, 1971; Alberts and Vagelos, 1972). The specific substrates are either α-keto acids (pyruvate), acyl-CoA thioesters (acetyl-CoA, propionyl-CoA, β-methylcrotonyl-CoA, geranyl-CoA), or urea. Acetyl-CoA and propionyl-CoA are carboxylated at the α-carbon, whereas β-methylcrotonyl-CoA and geranyl-CoA are carboxylated at the γ-carbon, which is activated by the α,β-olefinic linkage (a vinylogous enolate acting as nucleophile).

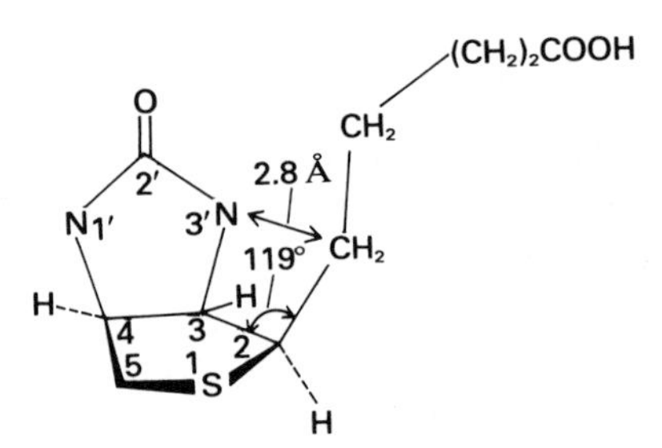

Figure 22-4
Spatial representation of the structure of *d*-biotin. (Reproduced from T. C. Bruice and S. Benkovic, *Bioorganic Mechanisms*, vol. 2. With permission of the publishers, Addison Wesley/W. A. Benjamin Inc., Advanced Book Program, Reading, Massachusetts, U.S.A.)

$$ATP + HCO_3^{\ominus} + \text{pyruvate} \longrightarrow \text{oxaloacetate} + ADP + P_i$$

$$ATP + HCO_3^{\ominus} + \text{acetyl-CoA} \longrightarrow \text{malonyl-CoA} + ADP + P_i$$

$$ATP + HCO_3^{\ominus} + \text{propionyl-CoA} \longrightarrow \text{methylmalonyl-CoA} + ADP + P_i$$

$$ATP + HCO_3^{\ominus} + \beta\text{-methylcrotonyl-CoA} \longrightarrow \beta\text{-methylglutaconyl-CoA} + ADP + P_i$$

$$ATP + HCO_3^{\ominus} + \text{geranyl-CoA} \longrightarrow (\text{carboxygeranyl-CoA}) + ADP + P_i$$

$$ATP + HCO_3^{\ominus} + \text{urea} \longrightarrow N\text{-carboxyurea} + ADP + P_i$$

A seventh carboxylase, transcarboxylase, does not use ATP and bicarbonate, but it shuttles CO_2 between keto-acid and acyl-CoA substrates.

The structure of biotin (Fig. 22-4) was determined by DuVigneaud's group; it was synthesized by Folker's laboratory at Merck. It has an imidizalone ring *cis*-fused to a tetrahydrothiophene ring, which in turn has a valerate side chain.

$$\text{biotin}\ (\ldots(CH_2)_4COO^{\ominus}) + H_2N{-}(CH_2)_4{-}CH(NH{-})C(=O){-}\text{Enz} \longrightarrow \text{biotinyl-lysyl-Enz}$$

In all the biotin-dependent carboxylases examined, the coenzyme is attached covalently at the active site, in amide linkage with the ε-amino group of a lysine residue. Formation of the holocarboxylases from specific apoenzymes is accomplished by a specific loading enzyme that requires ATP along with the biotin, cleaving the triphosphate to AMP and PP_i (Lane and Lynen, 1963). As expected from the product pattern, synthetic biotinyl-AMP will re place biotin and ATP with the loading enzyme. This linkage of coenzyme to enzyme is entirely analogous to the linkage of lipoamide to the lipoamide transacetylase and, in this instance also, it has been hypothesized that the flexible anchoring allows the coenyzme the mobility (at the end of this swinging arm of 14 Å radius) to move between component subunits of the carboxylases, all of which are multimeric.

One diagnostic test for biotin involvement in an enzymatic transformation is to test for inhibition of catalysis by addition of the protein *avidin*, purified from egg white (Green, 1966). The protein is named for its avidity in complexing with biotin, either free in solution or bound to carboxylases. Avidin is a glycoprotein tetramer of 68,000 mol wt, with one biotin-combining site per subunit (17,000 mol wt per subunit). The K_D for dissociation of biotin from an avidin-biotin complex has been estimated at a phenomenal 10^{-15} M, with an "on" rate (complex formation) of $7 \times 10\ \text{M}^{-1}\ \text{sec}^{-1}$ and a slow unimolecular "off" rate of $9 \times 10^{-8}\ \text{sec}^{-1}$ ($T_{1/2} \approx 8 \times 10^7$ sec, or about 2.5 years).

$$\text{Avidin} + \text{Biotin} \underset{k_{off}}{\overset{k_{on}}{\rightleftharpoons}} \text{Avidin} \cdots \text{Biotin} \qquad K_{eq} = 10^{15}$$

If one examines the ATP-dependent biotin carboxylases for their ability to carry out partial exchange reactions, two generally are exhibited. One is an exchange of [^{14}C]-ADP into ATP, dependent on the presence of bicarbonate.

$$\text{ATP} + [^{14}\text{C}]\text{-ADP} \xrightarrow{\text{enz—biotin, } HCO_3^{\ominus}} [^{14}\text{C}]\text{-ATP} + \text{ADP}$$

The other is an exchange between specific substrate and product. That is, [^{14}C]-pyruvate$\rightleftharpoons$oxaloacetate exchange with pyruvate carboxylase, or [^{14}C]-acetyl-CoA$\rightleftharpoons$malonyl-CoA exchange with acetyl-CoA carboxylases. For example,

$$^{\ominus}OOC{-}CH_2{-}C(=O){-}COO^{\ominus} + H_3{}^{14}C{-}C(=O){-}COO^{\ominus} \xrightarrow{\text{enz—biotin}} {}^{\ominus}OOC{-}{}^{14}CH_2{-}C(=O){-}COO^{\ominus} + H_3C{-}C(=O){-}COO^{\ominus}$$

The latter exchange reaction suggests the reversible formation of a "CO_2–enzyme" intermediate, which is indeed real and has been characterized, as we shall note. The first exchange reaction suggests the reversible formation of the mixed anhydride of carbonic and phosphoric acids—the carbonyl-phosphate species we have invoked earlier as a reasonable candidate for activation of bicarbonate for subsequent attack by a nucleophile (Meister, 1976; Sauers et al., 1976).

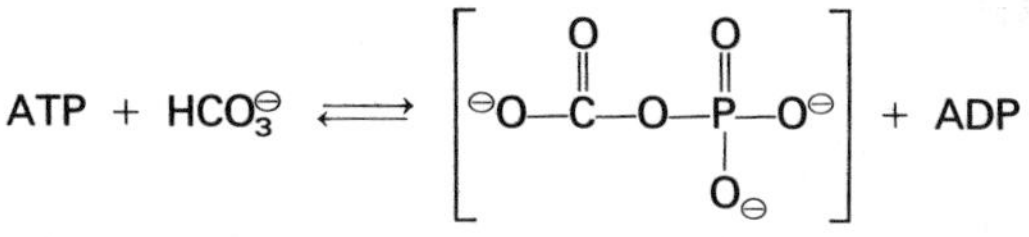

$$ATP + HCO_3^{\ominus} \rightleftharpoons \left[{}^{\ominus}O{-}\overset{O}{\overset{\|}{C}}{-}O{-}\overset{O}{\overset{\|}{\underset{O^{\ominus}}{\underset{|}{P}}}}{-}O^{\ominus} \right] + ADP$$

Again, ^{18}O data are consistent with such a species, one oxygen atom from bicarbonate ending up in the inorganic phosphate formed (Moss and Lane, 1971).

Direct evidence for the existence of a carboxylated-enzyme intermediate has been obtained for a number of the biotin-dependent carboxylases (Alberts and Vagelos, 1972), but the initial data were obtained by Lynen et al. (1959, 1961) with the β-methylcrotonyl-CoA carboxylase, crystallized from an *Achromobacter* bacterium. This enzyme carries out a useful model reaction, the carboxylation of free biotin. Incubation of ATP, $H^{14}CO_3^{\ominus}$, and free biotin with the enzyme produces ADP, P_i, and a ^{14}C-labeled biotin derivative, [^{14}C]-carboxybiotin. This product is quite labile; at pH 4.5 and 0° C, 97% of the radioactivity is released as $^{14}CO_2$ in 25 min. More stable in base, the compound releases 7% of its radioactivity after 30 min at 0° C in 33 mM KOH. Reaction of $^{14}CO_2$–biotin with diazomethane stabilizes the compound as its methyl ester, allowing determination that this ester is identical to synthetic N^1-methoxycarbonylbiotin methylester.

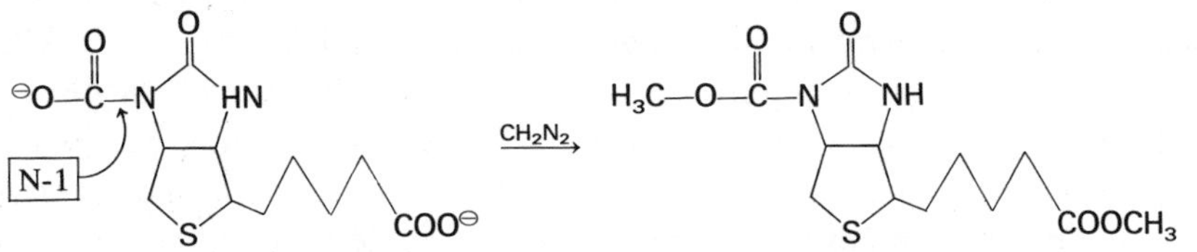

N-carboxybiotin → (CH_2N_2) → N^1-methoxycarbonylbiotin

The localization at the sterically less-hindered N-1 was proven by X-ray analysis of the bis(*p*-bromoanilide) derivative of the carboxybiotin (Bannemere et al., 1965).

If one returns to the normal reaction catalyzed by β-methylcrotonyl-CoA carboxylase, one can show that incubation of ATP and $H^{14}CO_3^{\ominus}$ with the holoenzyme will produce a $^{14}CO_2$–enzyme species that can again be stabilized with diazomethane and then proteolyzed to release the same methoxy derivative as that formed with free biotin, suggesting that N^1-carboxybiotinyl–enzyme is formed as a normal intermediate in this and other such carboxylase reactions (Fig. 22-5). The carboxybiotinyl-enzyme is, in fact, chemically competent to carboxylate β-methylcrotonyl-CoA or (on addition of ADP and P_i) to reform ATP.

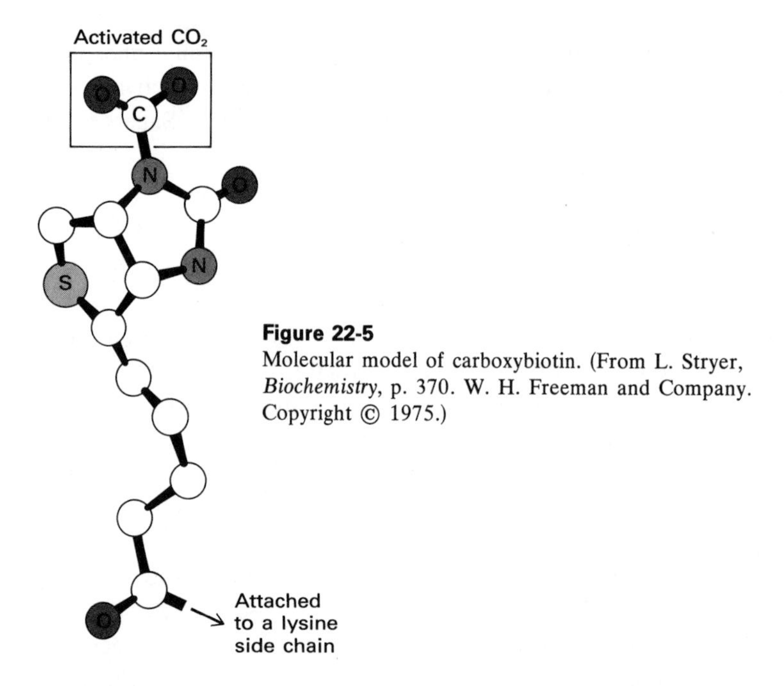

Figure 22-5
Molecular model of carboxybiotin. (From L. Stryer, *Biochemistry*, p. 370. W. H. Freeman and Company. Copyright © 1975.)

22.B.1 Acetyl-CoA Carboxylase

As we mentioned earlier, all the biotin-dependent carboxylases are multimeric species. Sheep-liver pyruvate carboxylase (Scrutton and Young, 1972; Utter et al., 1975) and pig-heart propionyl-CoA carboxylase (Alberts and Vagelos, 1972) are tetramers in the range of 700,000 mol wt, with each subunit possessing one biotinyl group and one active site. Acetyl-CoA carboxylase shows markedly different structural properties depending on the biological source (Alberts and Vagelos, 1972). Its physiological role, both in mammalian liver and in *E. coli*, is the initial step in biosynthesis of fatty acids (Chapter 27). The animal enzyme in its active form has a molecular weight of several million and, on electron-microscopic observation, is a filamentous polymer 70 to 100 Å wide by 5,000 Å in length (Fig. 22-6; Moss and Lane, 1971). The polymer can be induced to dissociate into essentially inactive monomers of 410,000 mol wt containing one biotinyl group (Fig. 22-7). This monomer is, in turn, composed of four chains of about 100,000 mol wt each, only one of which has the biotin. Citrate is known to activate the liver acetyl-CoA carboxylase by increasing V_{max} without altering K_m values for substrates (Moss and Lane, 1971). This is accomplished by inducing monomers to polymerize to the active form of the enzyme. The carboxybiotinyl–enzyme molecules, in the absence of acetyl-CoA, undergo depolymerization.

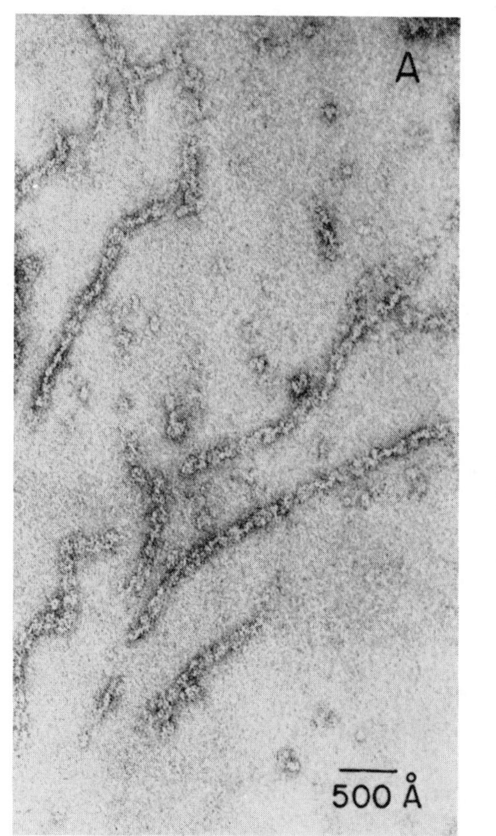

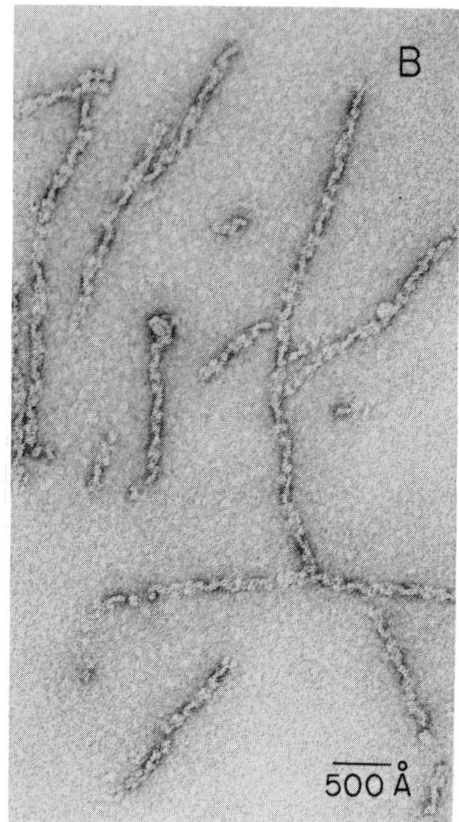

Figure 22-6
Filamentous forms of acetyl-CoA carboxylases from avian liver (**A**) and from bovine adipose tissue (**B**), in the presence of citrate. The preparations were stained with 4% aqueous uranyl acetate. (From A. K. Kleinschmidt et al., *Science*, vol. 166, p. 1276. Copyright © 1969 by the American Association for the Advancement of Science.)

Citrate apparently accelerates decarboxylation of the E—CO_2 species to CO_2 and free enzyme, which then repolymerizes (Alberts and Vagelos, 1972).

The situation is quite different with the *E. coli* enzyme, which has allowed considerable structural and mechanistic insight into acetyl-CoA carboxylase function. At the initial stages of purification, the enzyme dissociates into three components, each retaining activity for its component reaction in catalysis (Alberts and Vagelos, 1972). The three components are—

1. biotin-carboxylase component;
2. carboxyl-carrier protein;
3. carboxyl-transferase component.

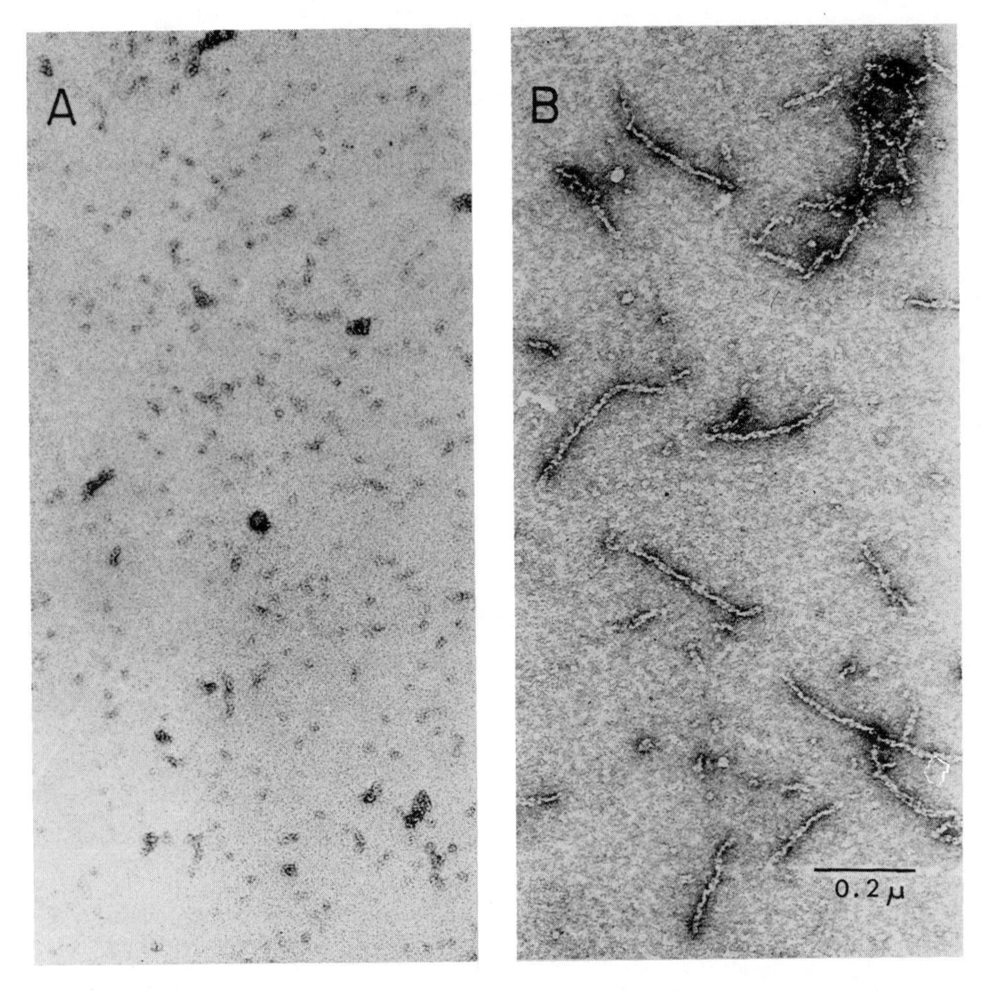

Figure 22-7
Protomer–polymer transitions of avian-liver acetyl-CoA carboxylase. The protomer (**A**) of 410,000 mol wt and $S_{20,w} = 13.1$ S is essentially inactive. On treatment with citrate and malonyl-CoA or with ATP, $Mg^{2\oplus}$, and $HCO_3^{\ominus}$, it forms the filamentous polymer (**B**) of 4 million to 10 million mol wt and 47 S to 59 S, which is catalytically active. (From Moss and Lane, 1971, p. 374)

Each protein has been purified to homogeneity (Guchhait, Polakis, Dimroth et al., 1974). Figure 22-8 shows crystals of component 1. The biotin-carboxylase component is a dimer of 100,000 mol wt that displays ADP→ATP exchange activity in the presence of bicarbonate. It will interact with the carboxyl-carrier protein, which contains the covalently linked biotin, to form $^{14}CO_2$-labeled biotinyl carrier protein (Polakis et al., 1974). Alternatively, the biotin-carboxylase dimer will

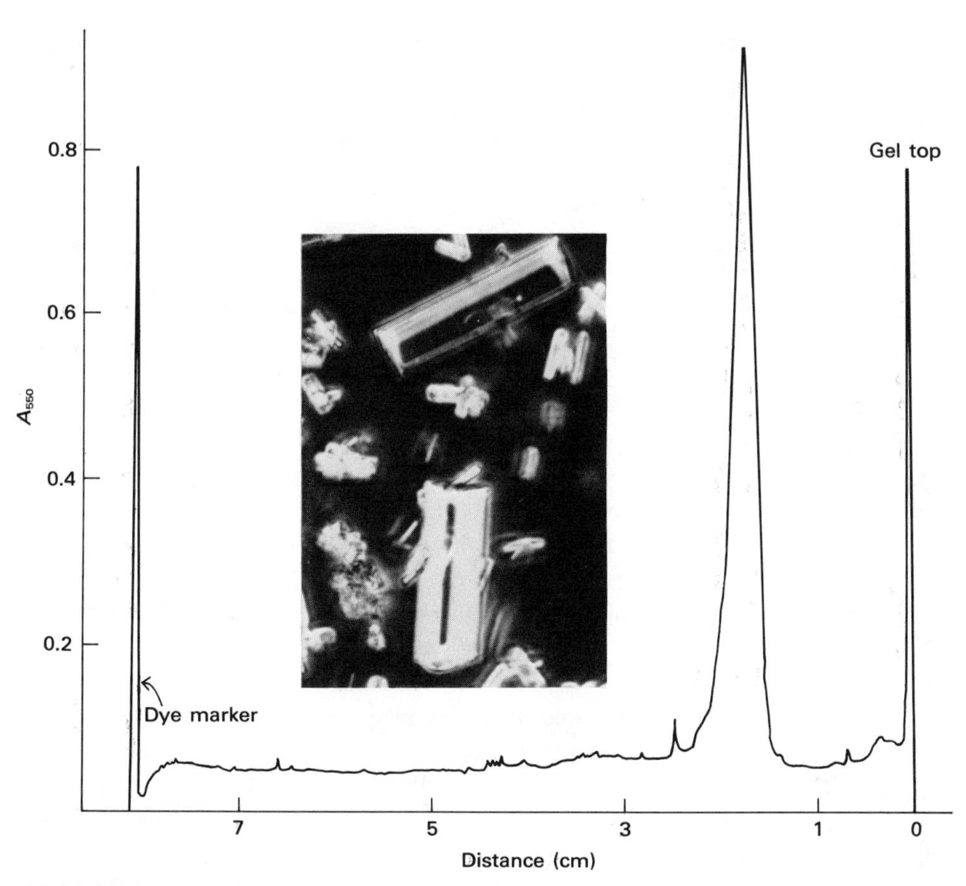

Figure 22-8
Polyacrylamide-gel electrophoresis pattern of purified biotin carboxylase. The inset shows crystalline biotin carboxylase prepared from a solution containing 4 mg of enzyme per ml. (From Guchhait, Polakis, Hollis, et al., 1974, p.6637)

carboxylate free biotin in a reaction analogous to that described earlier for β-methylcrotonyl-CoA carboxylase. The biotin-containing carboxyl-carrier protein was purified in part by an affinity step. A column of avidin bound to Sepharose (essentially a porous Sephadex) was percolated with the unpurified mixture of cellular proteins, and the biotin-containing protein was retained selectively on the column. Elution with six molar guanidine–HCl disrupted the avidin–biotin complex sufficiently to elute the carrier protein at a high degree of purity with ~22,000 mol wt (Alberts and Vagelos, 1972; Guchhait, Polakis, Dimroth et al., 1974). The third component, the carboxyl transferase, is an $\alpha_2\beta_2$ tetramer of 130,000 mol wt, with the α and β subunits of 35,000 and 30,000 mol wt, respectively. This protein interacts with the carboxyl-carrier protein to carboxylate acetyl-CoA (Polakis et al., 1974). These observations clearly establish separate active sites on different protein components for the half-reactions of the acetyl-CoA carboxylase reaction and establish that the biotinyl-carrier protein (biotinyl-CP) can interact with each specific subunit as a mobile CO_2 carrier (Polakis et al., 1974).

$$H_3C{-}\overset{O}{\overset{\|}{C}}{-}SCoA + \text{Carboxyl-transferase component} + {}^{\ominus}CO_2\text{–biotinyl-CP} \rightleftarrows$$

$$^{\ominus}OOC{-}\underset{H}{\overset{H}{C}}{-}\overset{O}{\overset{\|}{C}}{-}SCoA + \text{Carboxyl-transferase component} + \text{Biotinyl-CP}$$

malonyl-CoA

Experiments with the isolated biotin-carboxylase component of *E. coli* acetyl-CoA carboxylase have furthered identification of N^1-carboxybiotinyl–enzyme as the carboxyl carrier. Although the initial work of Lynen on its isolation was never in doubt, it had been pointed out by Bruice and coworkers that the nucleophilicity of the amide nitrogens of the ureido moiety in biotin should be low (Bruice and Hegarty, 1970; Hegarty and Bruice, 1970). This is true for imidazolone itself and might cast doubt on a mechanism invoking the nitrogen as nucleophile attacking an activated form of bicarbonate. Furthermore, model studies suggested that the ureido oxygen might be at least as attractive a nucleophile and that, in biotin catalysis, the initial intermediate (and the one active in catalysis) might be the *O*-carboxybiotin, which could have rearranged to the thermodynamically more stable *N*-carboxybiotin and then been trapped by diazomethane (i.e., the *N*-carboxy species is a side product not on the main reaction path).

$$\text{Biotin} + \text{ATP} + HCO_3^{\ominus} \longrightarrow O\text{-carboxybiotin} \longrightarrow N\text{-carboxybiotin}$$

O-carboxybiotin *N*-carboxybiotin

This possibility seems not to occur in acetyl-CoA carboxylase catalysis because chemically synthesized N^1-carboxybiotin reacts with the biotin-carboxylase component and ADP and P_i to support ATP formation (Guchhait, Polakis, Hollis, et al., 1974). Similarly, N^1-carboxybiotinol, enzyme, and [^{14}C]-acetyl-CoA yield [^{14}C]-malonyl-CoA. Still unresolved is the apparently insufficient nucleophilicity of the ureido nitrogen, but this may be less of a problem if the reaction is subject to general acid catalysis at the enzyme active site, as has been postulated to explain the lability of the *N*-carboxybiotin. Protonation of the carbonyl oxygen facilitates decarboxylation and expulsion of biotin as the isourea form.

Similarly, the isourea species might be important in the reverse of decarboxylation, acting as nucleophile in the carboxybiotin-forming step.

22.B.2 Transcarboxylase

An interesting variant among the biotin-containing carboxylases is the enzyme transcarboxylase, purified by H. G. Wood (1972) and colleagues from *Propionobacterium shermanii* (Wood and Zwolinski, 1976). This enzyme does not use ATP or $HCO_3^{\ominus}$ as substrates, but rather transfers a CO_2 moiety between the acyl-CoA and α-keto-acid substrates indicated.

In some analogy to the *E. coli* acetyl-CoA carboxylase, transcarboxylase is composed of three types of subunits. One large subunit is specific for the acyl-CoA substrates; one medium-size subunit is specific for the α-keto-acid substrates and requires cobalt or zinc as chelating metal; the third subunit again is a small protein with the covalently bound biotin coenzyme. This biotinyl carrier protein may carry the transferring CO_2 group between the keto-acid and the acyl-CoA subunits in the swinging-arm mechanism as proposed in Figure 22-9.

For example, the biotinyl group might swing to the active site of the acyl-CoA subunit and acquire a CO_2 unit from methylmalonyl-CoA bound there. The carboxybiotin so produced might then wander to the active site of the keto-acid subunit. There a molecule of pyruvate coordinated to a cobalt or zinc atom as the enolate could attack the carboxybiotin, effecting carboxyl transfer to yield oxaloacetate and the free biotinyl group still tethered by its side chain to the small carrier subunit. The active form of the enzyme is at least a hexamer of the

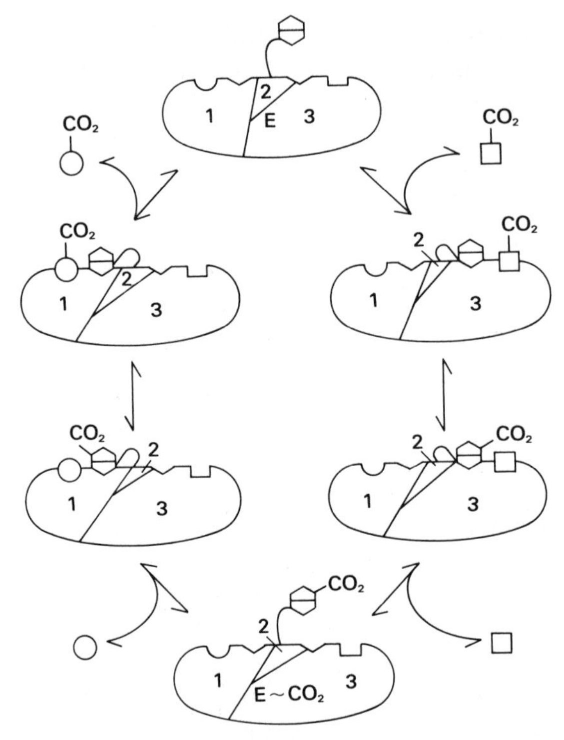

Figure 22-9
Pictorial model of the transcarboxylase reaction. The free circle is pyruvate, the carboxylated circle oxaloacetate; the free square is propionyl-CoA, the carboxylated square is methylmalonyl-CoA; the hexagon is biotin, the carboxylated hexagon carboxybiotin. E is one of possibly six reactive enzyme centers of transcarboxylase, with 1 representing the keto-acid subunit, 2 the biotinyl carrier subunit, and 3 the acyl-CoA subunit. The two substrate sites may be on different subunits—one associated with the 12S component and the other with the 6S biotin,Co,Zn subunit. (From H. G. Wood, 1972, p. 113)

basic structure elucidated above—that is, at least six copies of each kind of subunit in each multimeric enzyme species.

In a mechanistic sense, the transcarboxylase-mediated reaction is simply the sum of the two substrate-specific half-reactions carried out by the two ATP- and biotin-dependent carboxylases, pyruvate and propionyl-CoA carboxylases (Scrutton and Young, 1972; Utter et al., 1975). Homogeneous pig-liver propionyl-CoA carboxylase is metal-free, but pure samples of pyruvate carboxylase require manganous ion for catalysis, probably both for structural purposes in positioning reactants and to increase the stability of enzyme-bound enolate molecules.

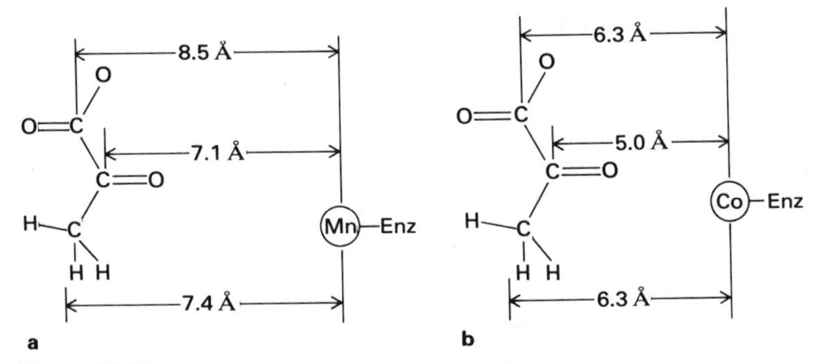

Figure 22-10
Estimated distances (by NMR measurements) from the bound metal to substrate carbons at the active sites of pyruvate carboxylase (**a**) and of transcarboxylase (**b**). (From Mildvan, 1974, p. 366. Reproduced with permission from *Annual Review of Biochemistry*, vol. 43. Copyright © 1974 by Annual Reviews Inc. All rights reserved.)

Mildvan (1974) and colleagues have been involved in NMR and ESR techniques to measure distances from paramagnetic metal cations to various atoms of bound substrate molecules at the active sites of transcarboxylase, pyruvate carboxylase, and other enzymes. Estimated distances from Co^{III} to pyruvate carbons are 5 to 6.3 Å, and they are 7.1 to 8.5 Å for the same distances from Mn^{II} in pyruvate carboxylase, as indicated in Figure 22-10. These distances are too great for direct (or first coordination sphere) chelation with the Mn^{II} and Co^{III}, and Mildvan (1974) has postulated that second-sphere complexes with an intervening water molecule (in the first sphere) between the metal and the chelating substrate are the general rule.

binding site for biotinyl group

22.B.3 Stereochemistry of Biotin-Dependent Carboxylase Action

Stereochemical studies have been carried out on propionyl-CoA carboxylase, pyruvate carboxylase, and transcarboxylase, where the results of each half-reaction of the third enzyme corroborate the data from the first two enzymes. In propionyl-CoA carboxylation by either the ATP-dependent enzyme or transcarboxylase, the pro*R* hydrogen at C-2 of propionyl-CoA is removed and replaced by the incoming CO_2 of carboxybiotin with retention (Rose et al., 1976). In both enzymes, the rate of release of tritium from 2*R*-[^{3}H]-propionyl-CoA is identical to the rate of methylmalonyl-CoA formation, with no preequilibrium washout or kinetic isotope selection. For pyruvate carboxylation, the stereochemistry can be determined by the chiral methyl-group methodology described in ¶22.A.4. Because pyruvate-to-oxaloacetate is a methyl-to-methylene conversion, a kinetic isotope selection at V_{max} is required for obtainment of stereochemical information. This selection exists for pyruvate carboxylase ($k_H/k_D = 3.1$)★ (Cheung and Walsh, 1976*b*) and for transcarboxylase ($k_H/k_D = 2.1$) (Cheung et al., 1975). The 3-[^{3}H]-oxaloacetates formed on carboxylation from 3*S*-[$^1H, ^2H, ^3H$]-pyruvate have excess molecules of 3*R*-[^{3}H]-oxaloactetate over 3*S*-[^{3}H]-oxaloacetate as assayed by malate dehydrogenase and fumarase action (most of the ^{3}H is released into water). These data then also imply retention of configuration in this half-reaction.

None of these stereochemical experiments distinguishes between carboxylation by concerted processes or that by carbanionic intermediates, but frontside attack must occur in either mechanistic instance. A concerted cyclic frontside displacement has been written for pyruvate carboxylase as one possibility, with the bound manganese at the active site acting as chelate (Scrutton and Young, 1972; Utter et al., 1975). In this putative mechanism for pyruvate carboxylase action, the transfer of CO_2 from *N*-carboxybiotin occurs as follows.

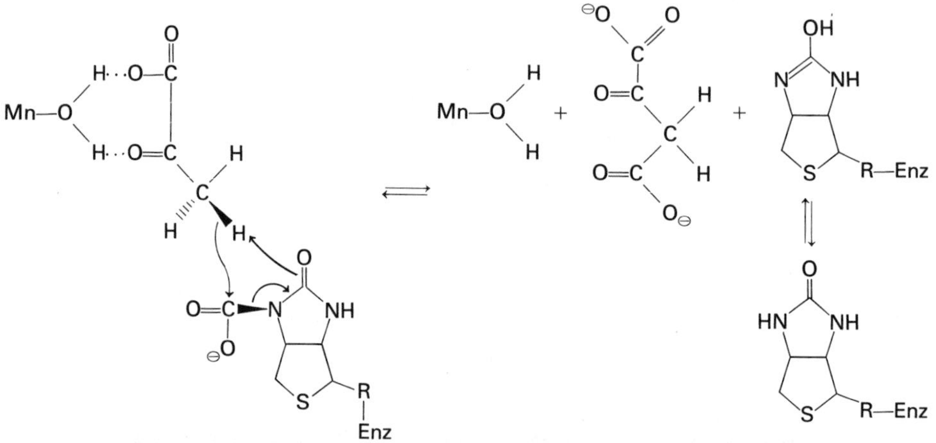

★This is an intramolecular isotope discrimination for removing H vs. D from a CH_2D group in a bound pyruvate molecule. With $CD_3COCOO^\ominus$, there is an isotope effect of 2.1 on V_{max}/K_m, but none on V_{max} (see ¶4.A.5; Cheung and Walsh, 1976*b*). The isotope effect on the elementary chemical step is not expressed at V_{max}; some other step is slow.

In the carboxylation step from ATP and $HCO_3^{\ominus}$, again the nitrogen of the isourea isomer might attack a carbonyl-P species (presumably also chelated to the metal?).

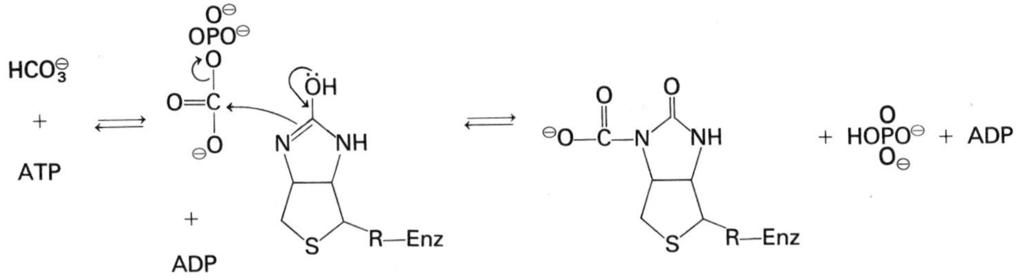

A specific test for a concerted six-center mechanism in transcarboxylase catalysis has recently been carried out by Rose et al. (1976). With 3-[^{3}H]-pyruvate and unlabeled propionyl-CoA, they observed that transcarboxylase transfers about 5% of the tritium atoms labilized during the reaction to propionyl-CoA.

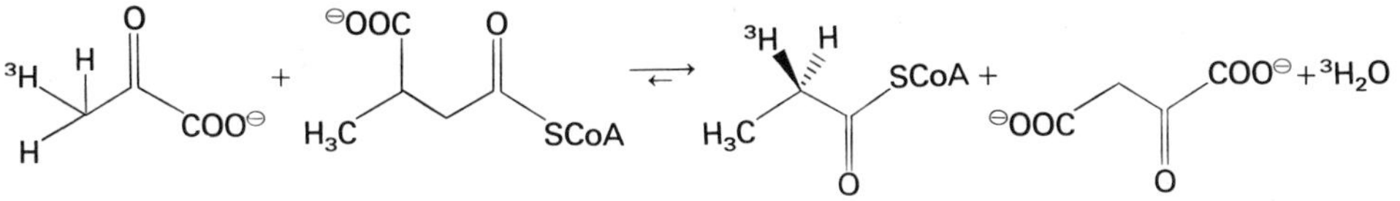

[^3H_2O]/[2*R*-[^{3}H]-propionyl-CoA] = 19/1

Although distinct basic groups, enz—B^{3}H$^{\oplus}$, can be imagined on each subunit, it is economical to suppose that biotin itself is the intermediate ^{3}H-carrier as the isoamide tautomer. Dissociation of this enol form to the enolate and $^3H^{\oplus}$ (step 3 in the following scheme) is only about an order of magnitude faster than transfer to C-2 of the propionyl-CoA anion, hence the partial transfer. Cleland (1977) reviews the following possible scheme for intermolecular ^{3}H transfer, based on the work of Rose and colleagues.

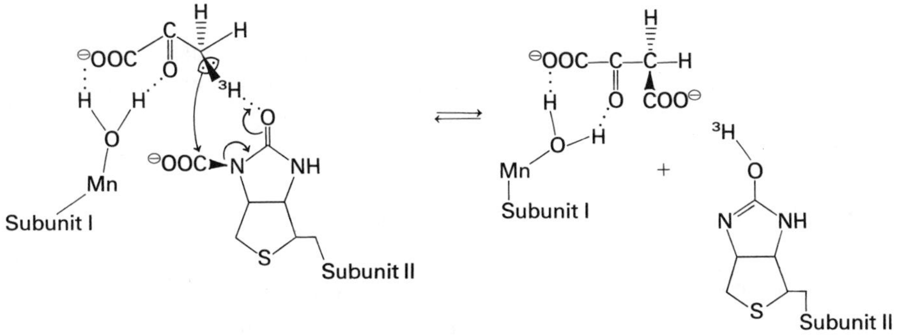

(1)

isoamide

Now the isoamide is rotated and moved from the active site of subunit I to the active site of subunit III (metal-free).

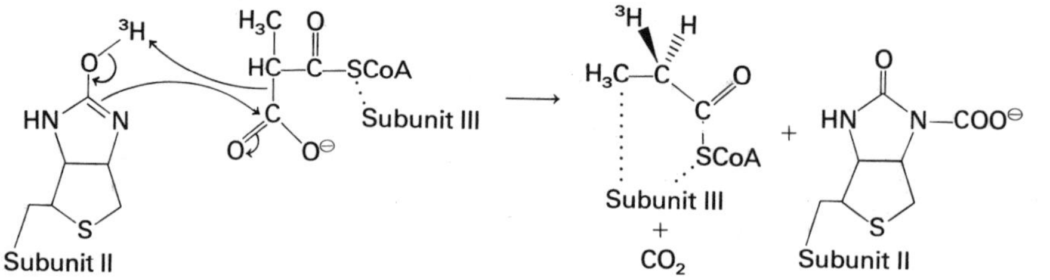

(2)

The exchange of the migrating $^3H^{\oplus}$ could occur in kinetic competition with reaction 2:

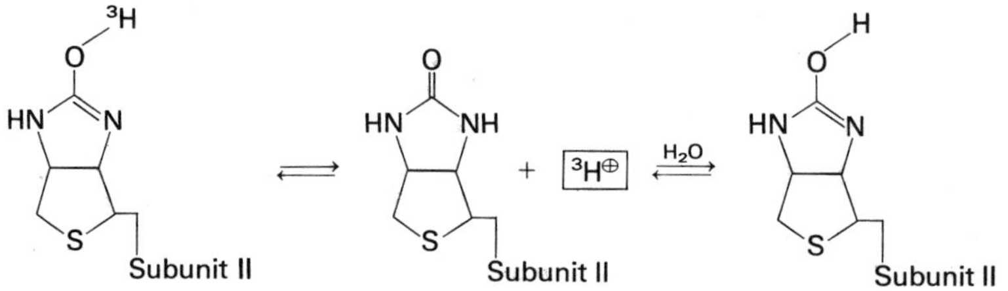

(3)

At this point then, there is not yet unambigous kinetic evidence for a discrete carbanionic substrate intermediate attacking carboxybiotin but, given the chemical stabilization of carbanions adjacent to keto and acylthioester groups, it is difficult to accept that they are not the attacking nucleophiles in these reactions.

22.C OTHER CARBOXYLASES

We turn now to the last two categories of enzymatic carboxylations: the ribulose-1,5-diphosphate carboxylase, and the vitamin-*K*-dependent carboxylation of certain glutamyl residues in proteins involved in blood coagulation.

22.C.1 Ribulose-1,5-Diphosphate Carboxylase

As a representative of those carboxylases that do not use ATP, do not have biotin cofactors, and do not carboxylate phosphoenolpyruvate, ribulose-1,5-diphosphate carboxylase is probably the best-characterized. It also is centrally important for sustenance of life on the planet. Ribulose diphosphate is carboxylated as the first step of carbon assimilation in photosynthetic carbon fixation in green plants and photosynthesizing bacteria. In green plants, the enzyme is localized exclusively in the chloroplasts and can account for up to 16% of the protein in spinach leaves, perhaps indicative of its central metabolic role. The spinach enzyme is one of several purified to homogeneity (Guchhait, Polakis, Dimroth, et al., 1974). It is an $\alpha_8\beta_8$ complex of 560,000 mol wt. The α-subunit has 56,000 mol wt, the smaller β-subunit has 14,000 mol wt. The spinach enzyme contains Cu^{II} as isolated; it also requires $Mg^{2\oplus}$ or $Mn^{2\oplus}$ for carboxylation. CO_2 (not bicarbonate) is the reactive species. Mildvan (1974) has suggested that the distance from Mn^{II} to bound CO_2 is 5.4 Å, again consistent with a second-sphere coordination distance.

The product from combination of the CO_2 with the pentulose is not a six-carbon sugar but rather two molecules of the C_3 acid D-3-phosphoglycerate (in a functionally irreversible process). Labeling studies indicate the distribution of the substrate carbons into product as indicated here.

C-1 $H_2C{-}OPO_3^{2\ominus}$
C-2 $C{=}O$
C-3 $HC{-}OH$
C-4 $HC{-}OH$
C-5 $H_2C{-}OPO_3^{2\ominus}$

$+ \; CO_2 \xrightarrow[Mg^{2\oplus}]{Enz{-}Cu^{2\oplus}}$

C-1 $H_2C{-}OPO_3^{2\ominus}$
C-2 $HC{-}OH$
$COO^{\ominus}$
+
C-3 $COO^{\ominus}$
C-4 $HC{-}OH$
C-5 $H_2C{-}OPO_3^{2\ominus}$

The CO_2 incorporated has been attached to C-2, suggesting that C-2 has added nucleophilically to the one-carbon electrophile (CO_2). Yet, C-2 of ribulose diphosphate is itself electrophilic as a ketone carbon. This reaction presents three puzzles to be explained (Siegel et al., 1972).

1. The C-2 of substrate apparently reverses its polarity prior to carbon–carbon bond formation with CO_2.
2. Additionally, C-2 of the substrate undergoes a two-electron reduction in forming product, whereas C-3 undergoes a net oxidation of four electrons.
3. The substrate later undergoes cleavage between C-2 and C-3. (We shall soon examine the evidence on which this statement is based.)

A number of experimental observations have been accumulated that are pertinent to outlining the mechanism of this catalysis (Siegel et al., 1972). No partial exchange reactions (e.g., $^{14}CO_2 \rightleftarrows$ 3-phosphoglycerate exchange) are detectable. One tritium atom is incorporated at C-2 of one of the 3-phosphoglycerate product molecules. That tritiated phosphoglycerate is also the one that incorporates the radioactivity from $^{14}CO_2$. The other 3-phosphoglycerate molecule (stemming from C-3, C-4, and C-5 of the ribulose substrate) is unlabeled. With 3-[^{3}H]-ribulose 1,5-diphosphate, there is a kinetic isotope selection ($k_H/k_T = 4$ to 6) at V_{max}, indicating that breakage of the bond between C-3 and ^{3}H is a slow step in catalysis.

Some twenty years ago, Calvin (1954) postulated a C_6 compound as the initial product from CO_2 addition. The compound was suggested to have the structure that follows and, although its intermediacy is not yet fully proven, it is the chemically likely precursor of the two phosphoglycerate molecules.

```
        H2C—OPO3(2-)               H2C—OPO3(2-)            H2C—OPO3(2-)
         |                          |                       |
  ⊖OOC—C—OH                  ⊖OOC—C—OH                N≡C—C—OH
         |                          |                       |
         C=O                       HC—OH                   HC—OH
         |                          |                       |
        HC—OH                      HC—OH                   HC—OH
         |                          |                       |
        H2C—OPO3(2-)               H2C—OPO3(2-)            H2C—OPO3(2-)
```

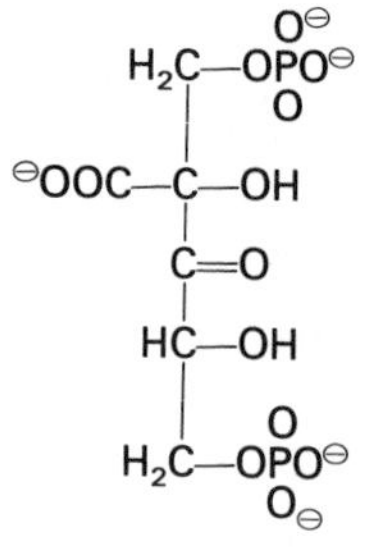

Calvin's intermediate 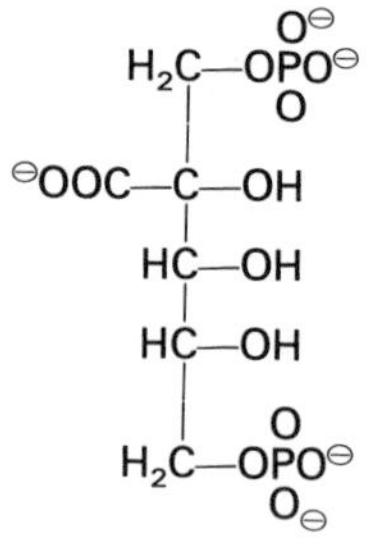 — carboxyribitol diphosphate 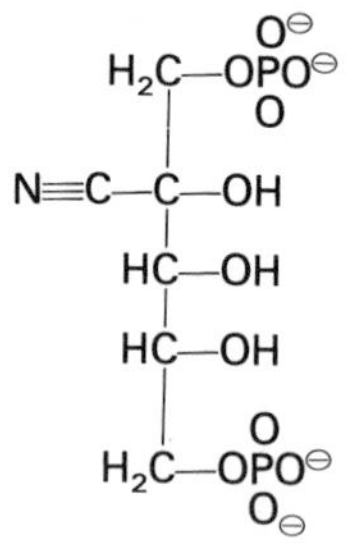 — cyanohydrin

The probable role of Calvin's intermediate is corroborated by the finding that the other two molecules shown, each structurally related to the putative intermediate, are good reversible inhibitors of the carboxylase. These molecules are carboxy-D-ribitol 1,5-diphosphate and the cyanohydrin that results from cyanide addition to ribulose diphosphate (Siegel et al., 1972).

We can now outline a reasonable mechanistic proposal. First, rate-determining abstraction of the acidic proton at C-3 of the substrate generates an anion that is a 2,3-enediolate (Calvin, 1954). This delocalized anion has the appropriate carbanionic character at C-2 for attack on carbon dioxide. One resonance form of that anion is the C-2 anion of 3-ketoribulose diphosphate, which has C-2 reduced by two electrons and C-3 oxidized by two electrons. Attack on CO_2 would, of course, produce Calvin's intermediate, the 2-carboxy-3-keto compound.

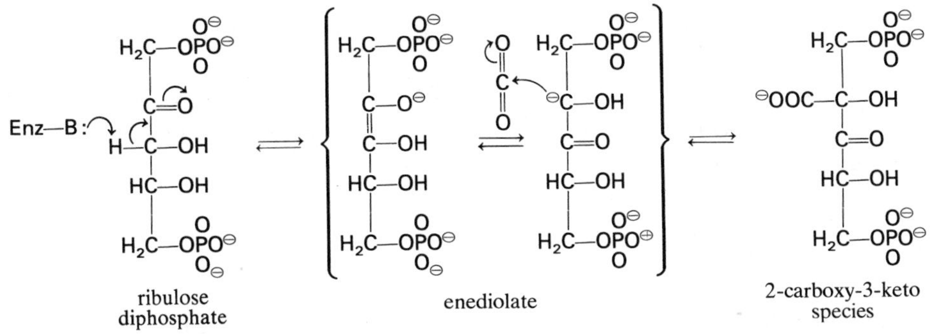

To proceed to products, attack by a water molecule (coordinated to one of the active-site metal ions?) at the 3-keto group is preliminary to an aldol cleavage that constructs the acid group of one phosphoglycerate molecule and, by expulsion, the C-2 carbanion of the other PGA molecule. Quenching of this carbanion with a solvent-derived proton accounts for the experimental results in ^3H_2O.

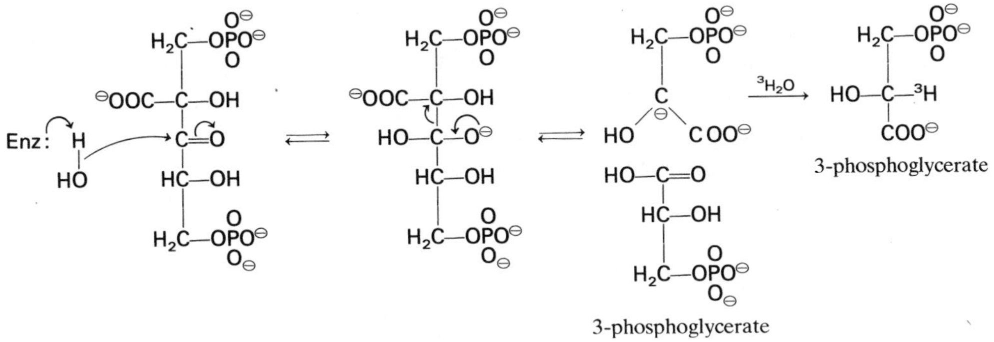

Let us now return to the experiments that demonstrated that the solvent hydrogen is incorporated at C-2 of the 3-phosphoglycerate molecule in which the C-1 carboxylate derives from CO_2 rather than in the 3-phosphoglycerate molecule whose carboxylate derives from the original C-3 of the substrate (Mullhofer and Rose, 1965). In the reaction scheme we have written, carboxylation occurs at C-2. This is chemically more reasonable than carboxylation at C-4, but an experiment to rule out carboxylation at C-4 is complicated by the symmetry of the two product molecules. With 2-[^{14}C]-ribulose diphosphate in D_2O, one can ask whether the products are from carboxylation at C-2 or from the less likely carboxylation at C-4. If the carboxylation occurs at C-2, all the [^{14}C]-3-PGA molecules will contain deuterium at C-2.

$$\begin{array}{ll} \boxed{\text{C-1}} & H_2C{-}OPO_3^{2-} \\ \boxed{\text{C-2}} & C{=}O \\ \boxed{\text{C-3}} & HC{-}OH \\ \boxed{\text{C-4}} & HC{-}OH \\ \boxed{\text{C-5}} & H_2C{-}OPO_3^{2-} \end{array} + {}^{14}CO_2 \xrightarrow{D_2O} \begin{array}{ll} \boxed{\text{C-1}} & H_2C{-}OPO_3^{2-} \\ \boxed{\text{C-2}} & D{-}C{-}OH \\ & {}^{14}COO^{\ominus} \end{array} + \begin{array}{ll} \boxed{\text{C-5}} & H_2C{-}OPO_3^{2-} \\ \boxed{\text{C-4}} & H{-}C{-}OH \\ \boxed{\text{C-3}} & COO^{\ominus} \end{array}$$

On the other hand, if the carboxylation occurs somehow at C-4, no [^{14}C]-3-PGA molecules will contain deuterium.

$$\text{Ribulose diphosphate} + {}^{14}CO_2 \xrightarrow{D_2O} \begin{array}{ll} \boxed{\text{C-1}} & H_2C{-}OPO_3^{2-} \\ \boxed{\text{C-2}} & D{-}C{-}OH \\ \boxed{\text{C-3}} & COO^{\ominus} \end{array} + \begin{array}{ll} \boxed{\text{C-5}} & H_2C{-}OPO_3^{2-} \\ \boxed{\text{C-4}} & H{-}C{-}OH \\ & {}^{14}COO^{\ominus} \end{array}$$

Mullhofer and Rose (1965) utilized glycolate oxidase to distinguish between these possibilities, taking advantage of the kinetic isotope effect that occurs during glycolate oxidation ($V_{max(H)}/V_{max(D)} = 3.5$). The mixture of [^{14}C]- and 2-[^{2}H]-3-PGA molecules resulting from the experiment with 2-[^{14}C]-ribulose and D_2O was converted chemically to the corresponding 1-[^{14}C]- and 2-[^{2}H]-glycolate molecules. The mixture was then incubated with NAD$^{\oplus}$ and glycolate oxidase.

$$\underset{\substack{\text{glycolate} \\ (\text{pro}S\text{ removal})}}{H{-}CH(OH){-}COO^{\ominus}} + NAD^{\oplus} \rightleftharpoons \underset{\text{glyoxalate}}{HC({=}O){-}COO^{\ominus}} + NADH$$

If the [^{14}C]-glycolate molecules also contain the deuterium, the rate of [^{14}C]-glyoxalate production will show the kinetic isotope effect, and the glyoxalate product will have a low specific radioactivity. This was in fact the observed result. Had the ^{14}C label been on molecules distinct from those containing deuterium, there would have been no lag in the rate of [^{14}C]-glyoxalate production (or in its specific activity). This latter result was observed when 4-[^{14}C]-ribulose diphosphate was incubated in D_2O with the carboxylase, confirming the validity of the assumptions underlying the experiment.

Recent studies with the apparently homogeneous ribulose-diphosphate carboxylase have uncovered a novel catalytic activity: a monooxygenase capacity

(Andrews et al., 1973). It had long been known that, in addition to 3-phosphoglycerate, 2-phosphoglycolate is a major radioactive product of photosynthetic $^{14}CO_2$ fixation. Incubation of ribulose 1,5-diphosphate in the presence of O_2 (but not CO_2) leads to one molecule of phosphoglycerate and one of phosphoglycolate.

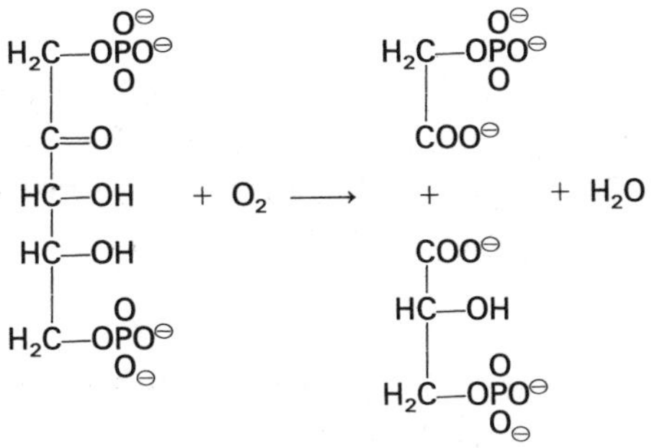

$$H_2C(OPO_3^{2-})\text{—}C{=}O\text{—}HC\text{—}OH\text{—}HC\text{—}OH\text{—}H_2C(OPO_3^{2-}) + O_2 \longrightarrow H_2C(OPO_3^{2-})\text{—}COO^{\ominus} + COO^{\ominus}\text{—}HC\text{—}OH\text{—}H_2C(OPO_3^{2-}) + H_2O$$

The V_{max} for oxygenation is 60% of that for carboxylation, and the pH optimum is more alkaline (9.3 rather than 7.8). With $^{18}O_2$, one atom of ^{18}O ends up in the phosphoglycolate carboxylate, the other in H_2O. Phosphoglycerate is unlabeled. A tentative proposal for attack of the enediolate on O_2 has been advanced. The resulting peroxide would be decomposed by hydroxide into observed products. How such a peroxide might form is unspecified, but other copper-containing monooxygenases may provide precedents as indicated in Chapter 14.

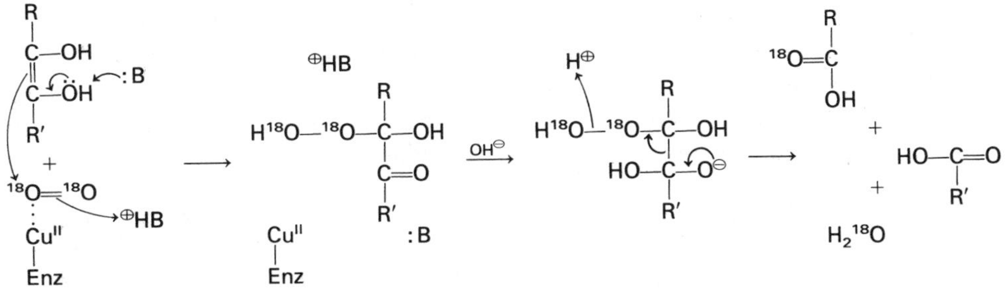

R, C—OH, C—ÖH, :B, R′ + $^{18}O{=}^{18}O$, ⊕HB, Cu^{II}, Enz ⟶ ⊕HB, $H^{18}O\text{—}^{18}O\text{—}C(R)(OH)\text{—}C{=}O(R')$, Cu^{II}, Enz, :B $\xrightarrow{OH^{\ominus}}$ $H^{\oplus}$, $H^{18}O\text{—}^{18}O\text{—}C(R)\text{—}OH$, HO—C—$O^{\ominus}$ (R′) ⟶ $^{18}O{=}C(R)\text{—}OH$ + HO—C=O (R′) + $H_2{}^{18}O$

For ribulose-diphosphate carboxylases from other sources, which may not be cuproenzymes (e.g., C. Walsh, 1978), the electron-rich enediolate could transfer one electron to O_2 at the active site to yield superoxide and the stable enediolate radical, which could then undergo rapid radical recombination to the hydroperoxide. No inhibition of oxygenase activity occurs on addition of superoxide dismutase, suggesting the lack of free $O_2^{\ominus\cdot}$. The mechanistic proposal just given has analogy to the nonenzymatic oxidation of α-diketones to two carboxylates by alkaline hydrogen peroxide, in which the initial hydroperoxy adduct decomposes by an addition and elimination of hydroxide (Gleason and Barker, 1971).

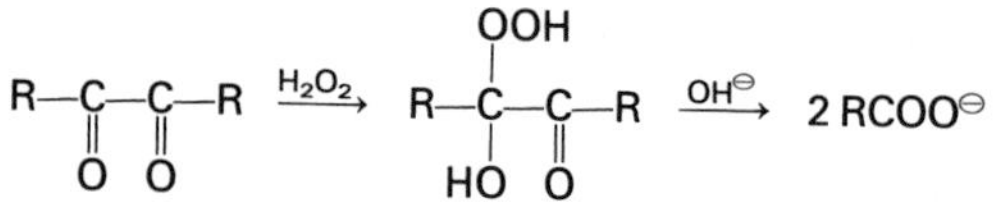

$$\mathrm{R{-}\underset{\displaystyle\|}{\overset{}{C}}{-}\underset{\displaystyle\|}{C}{-}R} \;(\text{C=O, C=O}) \xrightarrow{H_2O_2} \mathrm{R{-}C(OOH)(HO){-}C(=O){-}R} \xrightarrow{OH^{\ominus}} 2\,\mathrm{RCOO^{\ominus}}$$

The K_m for O_2 in the ribulose-diphosphate oxygenase reaction is greater than 0.25 mM, the normal concentration of dissolved O_2 in air-saturated buffer but, in 21% O_2 and 0.03% CO_2 found in air, the oxygenase activity may still compete well with action as a carboxylase in the leaf.

22.C.2 Vitamin-K–Dependent Carboxylation of Proteins Involved in Blood Coagulation

A fascinating story is now beginning to unfold involving enzymatic carboxylation of glutamyl residues in several of the glycoproteins that act to initiate blood clotting. It has been known for some years that blood coagulation is a cascade process, in which a dozen or so serum glycoproteins function catalytically and sequentially with the ultimate conversion of fibrinogen to fibrin, the major cross-linking protein in clots. The various protein components exist in normal plasma as inactive or precursor forms until the clotting sequence is initiated (Davie et al., 1975). About half of these components are endopeptidases with active-site serine residues (Chapter 3); their specificity is for limited proteolysis of the specific protein component functioning distally in the cascade. Figure 22-11 shows a scheme for the coagulation cascade.

A variety of nonprotein factors are required for normal coagulation, among them calcium ions and the naphthoquinone vitamin K_1.

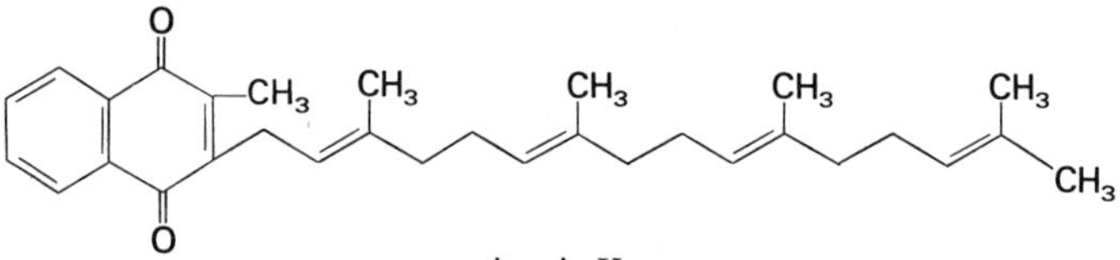

vitamin K_1

This function is the only well-defined requirement for vitamin K in animal systems. A number of compounds structurally related to vitamin K can function as anticoagulants, particularly those with coumarin ring systems. Two of note are dicumarol (a clinically effective anticoagulant) and warfarin (used predominantly as a rodenticide).

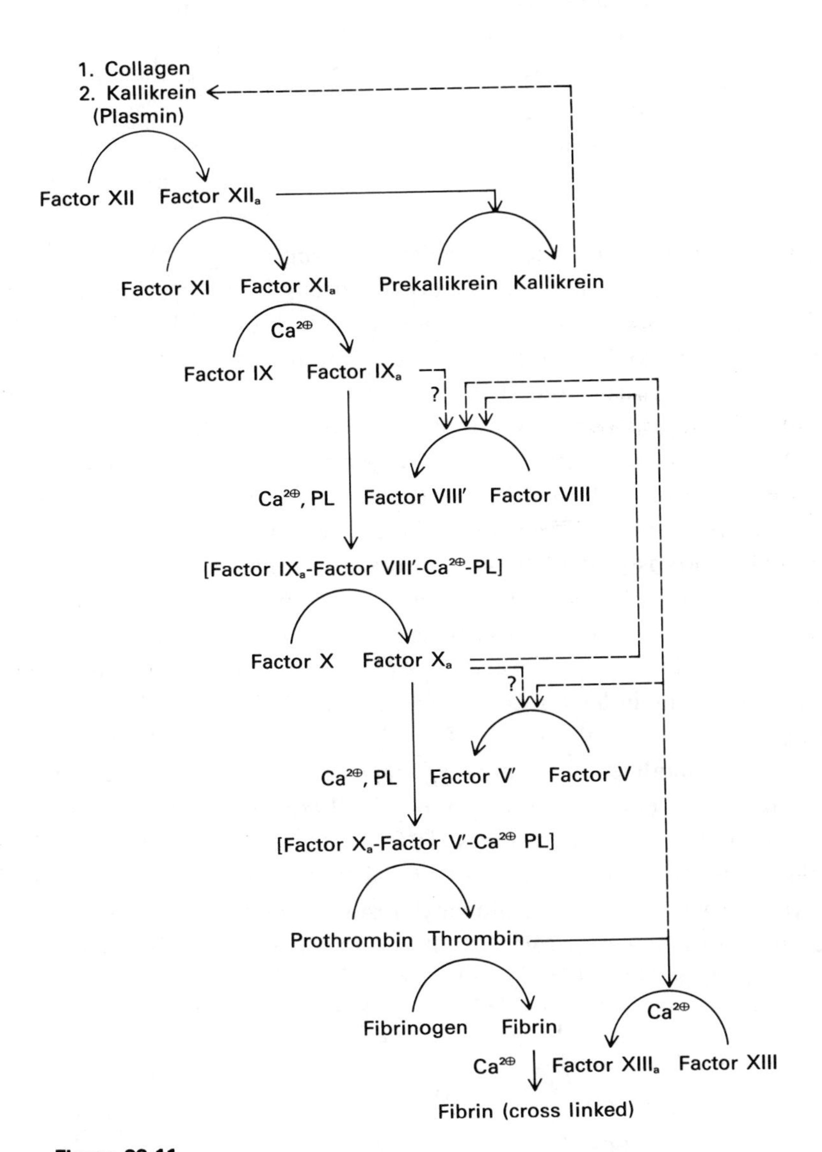

Figure 22-11
Tentative mechanism for the initiation of blood clotting in mammalian plasma in the intrinsic system. PL = phospholipid. (From Davie et al., 1975, p. 66)

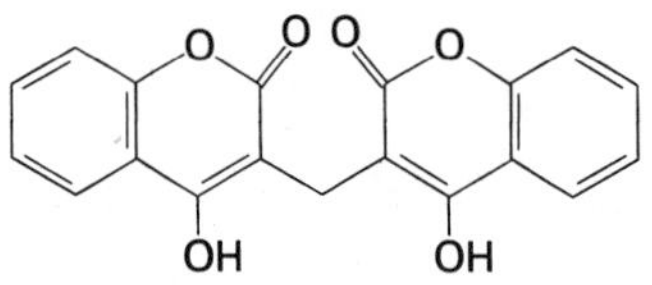

dicumarol

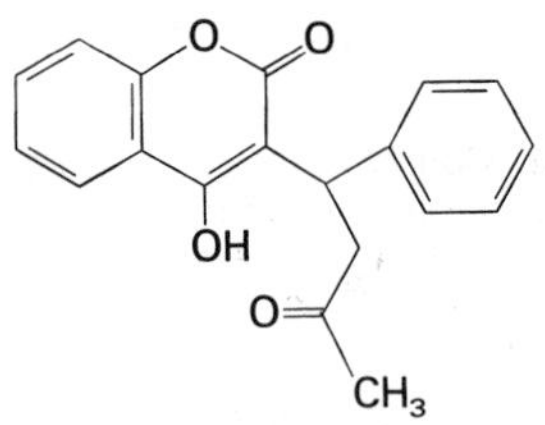

warfarin

The antidote in each instance is vitamin K, suggesting impairment of some vitamin-K–mediated process. The abnormal clotting is due to immature forms of serum factors VII, IX, and X, and prothrombin (Jackson et al., 1975). Prothrombin has been most extensively characterized. Normal prothrombin binds 10 to 12 calcium ions per molecule and, as this calcium complex, forms an aggregate with phospholipid surfaces, with factor V_a, and with activated factor X. In this aggregate, factor X (an active-site serine peptidase with trypsinlike specificity) initiates cleavage of prothrombin to the active protease thrombin (Davie et al., 1975; Jackson et al., 1975). Thrombin is the carboxy-terminal fragment of prothrombin and has proteolytic activity toward fibrinogen.

The immature form of prothrombin does not bind $Ca^{2\oplus}$ tightly, and it thus will not bind to phospholipid surfaces; the in vivo cleavage to thrombin probably is impaired for this reason. Structural studies on mature prothrombin and on immature prothrombin have recently revealed the molecular differences. Normal prothrombin has ten of the glutamyl residues in the first 42 residues from the amino terminus modified as γ-carboxyglutamyl residues (Fig. 22-12). The immature prothrombin does not (Stenflo et al., 1975; Magnusson et al., 1975). Incubation of liver-cell microsomes, $[^{14}C]$-$HCO_3^{\ominus}$, vitamin K_1, and immature prothrombin leads to incorporation of ^{14}C-label and derivatization of some of the glutamyl residues to γ-carboxyglutamyl residues. Figure 22-12 shows the sequence and positioning of γ-carboxyglutamyl residues in the amino-terminal fragment of normal prothrombin (Magnusson et al., 1975).

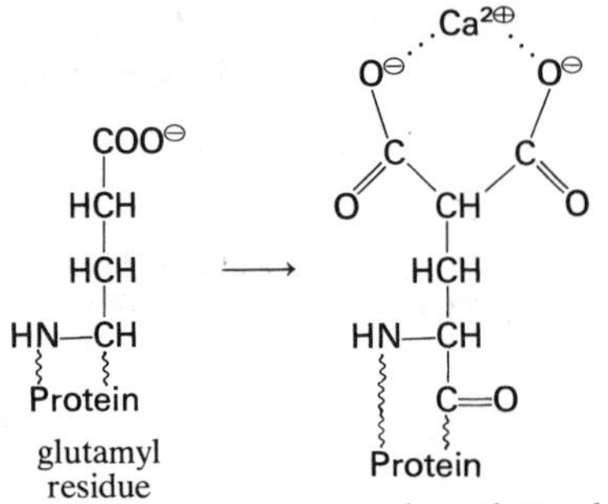

glutamyl residue

γ-carboxyglutamyl residue

```
                                   (7)                 (8)
Ala-Asn-Lys-Gly-Phe-Leu-NH · CH——CO-HN——CH · CO-Val-Arg-Lys-
                             |             |
                             CH2           CH2
                             |             |
                             CH            CH
                            /  \          /  \
                        HOOC    COOH  HOOC    COOH

            (15)              (17)                  (20)          (21)
-Gly-Asn-Leu-HN · CH · CO-Arg-HN · CH · CO-Cys-Leu-HN · CH——CO-HN——CH · CO-Pro-
                  |                |                    |            |
                  CH2              CH2                  CH2          CH2
                  |                |                    |            |
                  CH               CH                   CH           CH
                 /  \             /  \                 /  \         /  \
             HOOC    COOH     HOOC    COOH         HOOC    COOH HOOC    COOH

            (26)              (27)                  (30)                    (33)
-Cys-Ser-Arg-HN · CH——CO-NH——CH · CO-Ala-Phe-HN · CH · CO-Ala-Leu-HN · CH · CO-
                  |             |                   |                   |
                  CH2           CH2                 CH2                 CH2
                  |             |                   |                   |
                  CH            CH                  CH                  CH
                 /  \          /  \                /  \                /  \
             HOOC    COOH  HOOC    COOH        HOOC    COOH        HOOC    COOH

-Ser-Leu-Ser-Ala-Thr-Asp-Ala-Phe-Trp-
```

Figure 22-12
Amino-acid sequence of the calcium-binding, vitamin-K–dependent part of prothrombin (residues 1–42). Normal amino-acid residues in the usual three-letter code; γ-carboxyglutamate residues in positions 7, 8, 15, 17, 20, 21, 26, 27, 30, and 33. (From Magnusson et al., 1975, p. 134)

The γ-carboxyglutamyl residues are good chelators of $Ca^{2\oplus}$ ions and provide the specific $Ca^{2\oplus}$–prothrombin complex required for binding to phospholipid surfaces (possibly via interactions with the phosphate group of the phospholipid). The cleavage in vivo to thrombin splits off the γ-carboxyglutamyl-containing fragment; it does not appear in thrombin.

Given a knowledge of the structure of the carboxylated protein product, it is now possible to look for the enzyme carrying out this carboxylation to try to define the mechanism of CO_2 activation and incorporation and the role of vitamin K_1. The carboxylations are posttranslational events, occurring after the prothrombin has come off the ribosomes and while the protein is passing through the endoplasmic reticulum and Golgi membranes on its way into the serum. Glycosylation probably occurs at this physical stage as well.

Suttie and colleagues have solubilized the membrane enzyme activity with Triton X-100, a nonionic detergent, and have observed that ATP is not required nor does biotin appear to be involved (Esmon and Suttie, 1976). This would seem to rule out derivatization of the γ-carboxylate of glutamyl residues (e.g., to a thioester) as prelude for activating one of the γ-methylene hydrogens and generation of a γ-glutamyl carbanion. In the crude soluble system, two other facts are known. The dihydro form of vitamin K_1 (the hydroquinone) is the oxidation state that is active, and O_2 is required. The metabolic fate of O_2 (e.g., H_2O_2 or H_2O?) has not yet been ascertained. These two observations suggest an electron-transport scheme. Dihydro vitamin K may pass electrons to microsome-derived cytochromes (e.g., b_5) and then to O_2. The potential drop experienced by the electron could be used to provide energy for the carboxylation. No mechanistic route for such a process has been discussed. At this rudimentary stage, the possibility also exists that dihydro vitamin K_1 could act as an intermediate CO_2 carrier. For example,

B: O=C=O :OH CH₃ R OH → O COO⊖ CH₃ R OH B: COO⊖ H—C—H HCH HN—CH C=O Protein → γ-Carboxyglutamyl residue + Dihydro vitamin K

The requirement for O_2 is unclear in such a scheme; variants involving vitamin-K peroxides could be envisaged (C. Walsh, 1978). Given the dramatic physiological importance of vitamin-K–dependent carboxylations (certain bone-marrow proteins also have been found to have γ-carboxyglutamyl residues), much attention will be brought to bear on the molecular mechanisms of this type of CO_2 fixation in the near future.

Chapter 23

Enzyme-Catalyzed Aldol and Claisen Condensations

Prominent among the enzyme-catalyzed reactions that lead (in the biosynthetic direction) to carbon–carbon bond formation or (in the degradative direction) to carbon–carbon bond cleavage are aldol condensations and the related Claisen condensations. The ribulose-1,5-diphosphate carboxylase reaction discussed at the end of the preceding chapter involves an aldol cleavage, as does the TPP-requiring phosphoketolase catalysis discussed in Chapter 21.

Nonenzymatically, the aldol condensation involves the *condensation* (usually most effective under basic conditions) *of the α-carbon of an aldehyde or ketone to the carbonyl carbon of another aldehyde or ketone.* The hydrogen that is α to the carbonyl group in one molecule is abstracted in base to form the enolate, which is the nucleophile. One resonance contributor to the enolate structure has the negative charge at the α-carbon, emphasizing the carbanion character of the enolate. When this carbanion attacks the electron-deficient carbonyl carbon in the second molecule of aldehyde (or ketone), the new carbon–carbon bond is produced. The condensation product from two molecules of acetaldehyde is called aldol, which provides the generic name for the reaction. The equilibrium in simple cases is near unity, stressing the reversibility of the reaction and suggesting that cells can use this reaction for synthesis or for degradation.

$$2\,H_3C{-}\overset{O}{\overset{\|}{C}}H \rightleftharpoons H_3C{-}\overset{OH}{\underset{H}{C}}{-}\overset{H}{\underset{H}{C}}{-}\overset{O}{\overset{\|}{C}}H$$

aldol

The generalized aldol condensation has the following form;

$$R{-}\overset{O}{\overset{\|}{C}}{-}\overset{H}{\underset{H}{C}}{-}R' \;\; (B{:}) \rightleftharpoons BH^{\oplus} \left\{ R{-}\overset{O^{\ominus}}{\overset{|}{C}}{=}\underset{R'}{CH} \longleftrightarrow R{-}\overset{O}{\overset{\|}{C}}{-}\overset{\ominus}{\underset{R'}{CH}} \right\} + R''{-}\overset{O^{\delta\ominus}}{\overset{\|}{C}}H^{\delta\oplus} \rightleftharpoons R{-}\overset{O}{\overset{\|}{C}}{-}\overset{H}{\underset{R'}{C}}{-}\overset{OH}{\underset{H}{C}}{-}R''$$

enolate

new carbon–carbon bond

The related Claisen condensation can involve condensation between two ester molecules where one becomes acylated at the α-carbon, or it can involve acylation of the ester component by an aldehyde or ketone molecule. Again, the reactive species is the enolate of the ester, attacking the electrophilic carbonyl carbon of its condensation partner. In biological systems, the nucleophilic component is the α-anion of *acylthioesters* rather than acyl oxygen esters. As in the aldol condensation, the first step is the abstraction of acidic hydrogen to yield the stabilized carbanion.

$$R^1{-}\overset{H}{\underset{H}{C}}{-}\overset{O}{\overset{\|}{C}}{-}S{-}R^2 \;\; (B{:}) \longrightarrow \left\{ R^1{-}\overset{H}{C}{=}\overset{O^{\ominus}}{\overset{|}{C}}{-}S{-}R^2 \longleftrightarrow R^1{-}\overset{H}{\underset{\ominus}{C}}{-}\overset{O}{\overset{\|}{C}}{-}S{-}R^2 \right\} \;\; BH^{\oplus}$$

nucleophilic partner component

The nucleophilic partner then condenses with another ester or a ketone as the electrophilic component.

thioester enolate + thioester → new carbon–carbon bond + $R^4S^{\ominus}$

thioester enolate + ketone (neither R^3 nor R^4 good leaving groups) → new carbon–carbon bond

In a dozen or so characterized enzymatic *aldol reactions*, the nucleophilic components form stabilized enolates by two general mechanisms entirely analogous to the two mechanisms seen for acetoacetic decarboxylases.

1. In some cases, the substrate forms an imine linkage with an ε-amino group of a lysine at the active site as prelude to eneamine (low-energy carbanion) formation.
2. In the other cases, the enzyme active site contains a divalent cation that coordinates to the substrate's carbonyl oxygen and facilitates enolate formation by carbonyl polarization.

The nucleophilic components may be α-keto acids such as pyruvate,

$$\left\{ H_2C{=}C(O^{\ominus}){-}COO^{\ominus} \longleftrightarrow H_2\overset{\ominus}{C}{-}C({=}O){-}COO^{\ominus} \right\}$$

or ketones such as dihydroxyacetone phosphate,

$$\left\{ \begin{array}{c} H_2C{-}OPO_3^{2\ominus} \\ | \\ C{=}O \\ | \\ H\underset{\ominus}{C}{-}OH \end{array} \longleftrightarrow \begin{array}{c} H_2C{-}OPO_3^{2\ominus} \\ | \\ C{-}O^{\ominus} \\ \| \\ HC{-}OH \end{array} \right\}$$

or glycine (as an imine bound to pyridoxal phosphate).

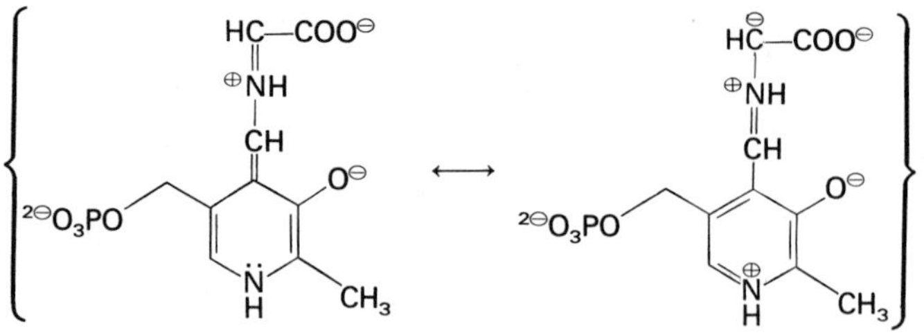

Usually, one can detect enzyme-catalyzed enolization occurring in the absence of the electrophilic cosubstrate. Rose and Hanson (1975) have noted that, in the six enzymatic aldol reactions whose stereochemistry has been determined, the electrophilic carbonyl component always approaches the same face of the enol or eneamine from which the proton was abstracted—that is, there is overall retention of configuration at the carbon that acts as carbanion. As yet, the mechanistic significance of this uniformity is unclear.

A variety of *Claisen condensations* are represented by the enzymes known as acyl-CoA ligases (Abiko, 1975)—for instance, where the α-carbanion of acetyl-CoA attacks a specific keto acid such as pyruvate (Higgins et al., 1972).

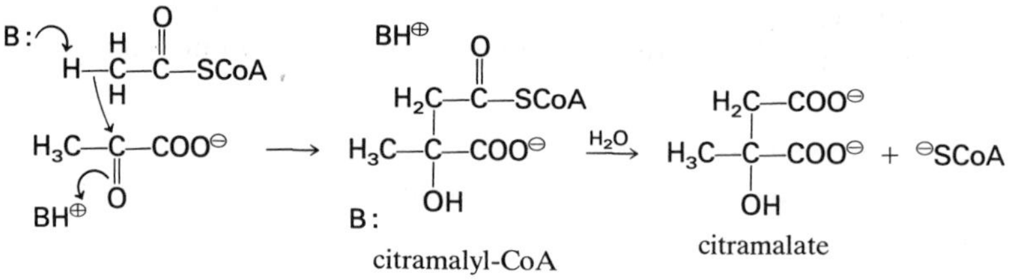

citramalyl-CoA

citramalate

The citramalyl-CoA may be released or it may be deacylated with transfer to water, producing citramalate and CoASH. Acetyl-CoA can also condense head-to-tail with itself, in the reaction catalyzed by thiolase, to yield acetoacetyl-CoA and CoASH (Gehring and Lynen, 1972).

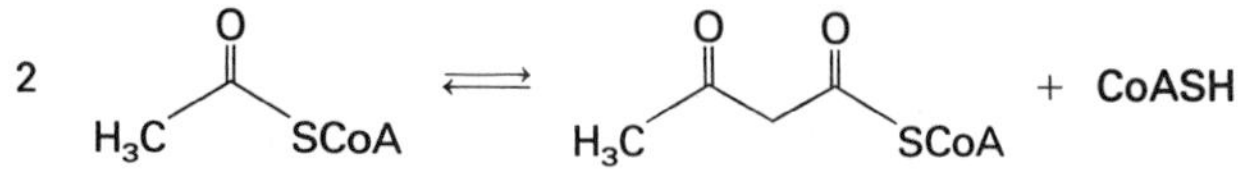

Other acylthioesters such as succinyl-CoA can be utilized, as we shall note later in this chapter. Unlike the related aldolase reactions, the exchange of α-hydrogen of the nucleophilic component (e.g., loss of ^{3}H from 2-[^{3}H]-acetyl-CoA) proceeds either not at all or quite slowly in the absence of cosubstrate or a cosubstrate analogue. This raises the possibility that a reaction may involve the carbanionic species only as a fleeting transition state and *not* as a discrete intermediate formed reversibly in the absence of the electrophilic cosubstrate. Another contrast of uncertain significance between the biological aldol and Claisen condensations is that the seven Claisen condensations studied proceed with *inversion* of configuration as the acetyl methyl (the nucleophilic carbon) goes to a methylene group during carbon–carbon bond formation (Rose and Hanson, 1975).

We shall first discuss aldol reactions and then turn to the Claisen variants.

23.A ALDOL REACTIONS

Susceptible substrates for enzymatic aldol cleavage (or condensation) have the following partial structure. (The X and Y groups are the electronegative atoms oxygen or nitrogen, to stabilize incipient negative charge density during the reaction.)

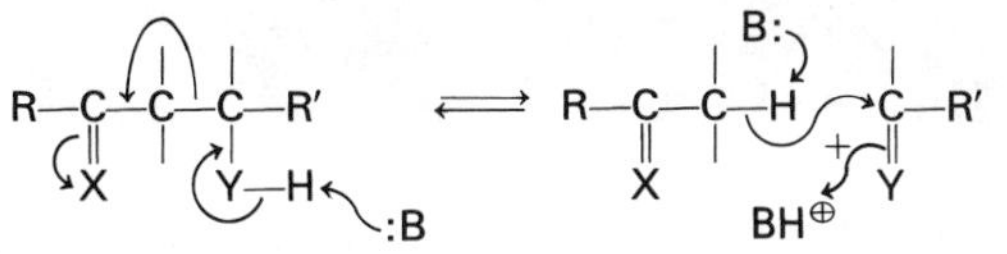

The arrows on the lefthand structure shows the pattern of electron flow during cleavage; the arrows on the righthand side show the pattern of electron flow during condensation (C—C bond formation). When X is oxygen, the enolate is the attacking species in the condensation direction and the initial product in the cleavage direction. When X is nitrogen, the eneamine plays the corresponding roles.

As noted earlier, there are two types of aldolases. Usually, cationic imines provide the requisite electron sink in animals and higher plants, whereas divalent cations act as electrophilic catalysts in procaryotes. We examine four aldolase enzymes in the following subsections.

23.A.1 Fructose-1,6-Diphosphate Aldolase

A well-studied example of the eucaryotic type of aldolase is rabbit-muscle fructose-1,6-diphosphate aldolase (Horecker et al., 1972; Morse and Horecker, 1968). It is an $\alpha_2\beta_2$ tetramer of 158,000 mol wt; it catalyzes the net cleavage reaction during periods of glucose breakdown and the net condensation reaction during periods of cellular gluconeogenesis. Figure 23-1 shows a photomicrograph of aldolase crystals.

Figure 23-1
Photomicrograph of crystals of aldolase. (Courtesy of David Eisenberg)

The hexulose diphosphate is converted to the three-carbon ketose dihydroxyacetone phosphate and the three-carbon aldose glyceraldehyde 3-phosphate.

$H_2C{-}OPO_3^{2\ominus}$
$C{=}O$
$HO{-}CH$
$HC{-}O{-}H$ (3,4-*trans*) :B—Enz
$HC{-}OH$
$H_2C{-}OPO_3^{2\ominus}$
fructose 1,6-diphosphate
(FDP)

$\rightleftharpoons$

$H_2C{-}OPO_3^{2\ominus}$
$C{=}O$
$HC{-}OH$
H
dihydroxyacetone-P (DHAP)
(produced by ketonization of
the initial enolate)

+

$O{=}CH$
$HC{-}OH$
$H_2C{-}OPO_3^{2\ominus}$
glyceraldehyde-3-P
(G-3-P)

The key aldol partial structure in fructose diphosphate is indicated by the dashed box.

The following paragraphs summarize some mechanistically useful observations on fructose-diphosphate aldolase.

23.A.1.a Enolization

When the enzyme and dihydroxyacetone phosphate are incubated in D_2O, one atom of solvent deuterium is incorporated at C-3.

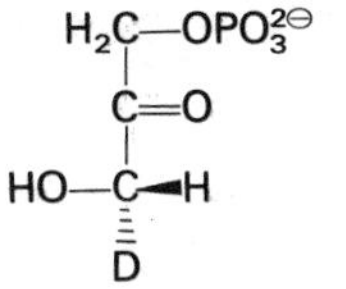

$H_2C—OPO_3^{2\ominus}$
C=O
HO—C—H
D
3*R*-[^{2}H]-DHAP

Even at very long incubation times, the second hydrogen at C-3 is not replaced by deuterium. This is predictable because C-3 of dihydroxyacetone-P is a prochiral center, and the enzyme acts with complete stereospecificity, removing only the pro*S* hydrogen and generating 3*S*-[^{2}H]-DHAP specifically (Alworth, 1972, p. 260). (One can compare this enzyme-catalyzed enolization result with the enolization of pyruvate in D_2O catalyzed by pyruvate kinase; see Chapter 7. In that instance, *all three hydrogens* of the pyruvate methyl are exchangeable with solvent deuterium because a methyl group is torsiosymmetric, and all three hydrogens are sterically equivalent.)

This chiral exchange process proves that the enzyme, in the absence of other substrates, can catalyze rapid enolization and stereospecific reprotonation of dihydroxyacetone-P with reversible addition and loss of $D^{\oplus}$ to and from only one face of the enolate double bond. For example,

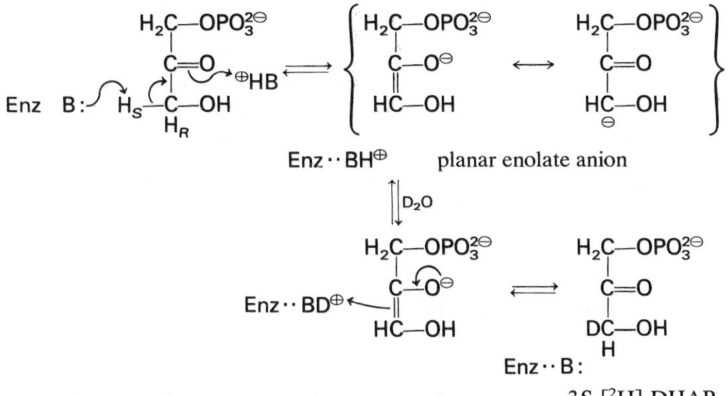

Enz B: H_S—C—OH (H_R), C=O, $H_2C—OPO_3^{2\ominus}$, $^{\oplus}$HB ⇌ {$H_2C—OPO_3^{2\ominus}$, C—$O^{\ominus}$, HC—OH ⟷ $H_2C—OPO_3^{2\ominus}$, C=O, HC$^{\ominus}$—OH}
Enz··$BH^{\oplus}$ planar enolate anion
⇅ D_2O
Enz··$BD^{\oplus}$ $H_2C—OPO_3^{2\ominus}$, C—$O^{\ominus}$, HC—OH ⇌ $H_2C—OPO_3^{2\ominus}$, C=O, DC—OH (H)
Enz··B:
3*S*-[^{2}H]-DHAP

23.A.1.b Partial exchange rates of [^{14}C]-triose phosphates

Measurements have been made of the rate of exchange of radioactive cleavage products into hexulose-diphosphate substrate (Rose et al., 1965). Radioactivity

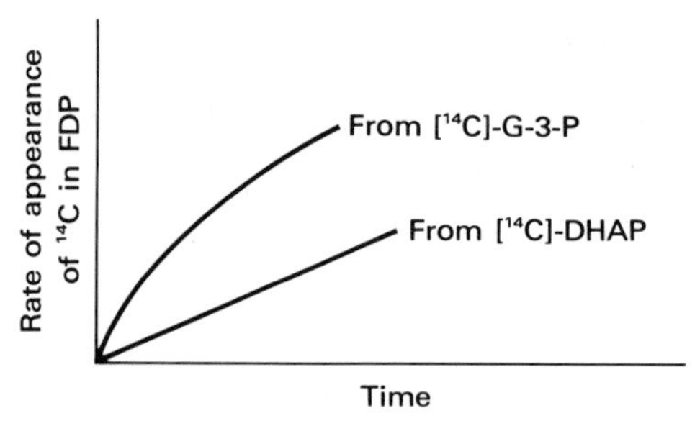

Figure 23-2
Different rates of appearance of label into fructose diphosphate from the C_3 fragments of glyceraldehyde-3-P and dihydroxyacetone-P.

from [^{14}C]-glyceraldehyde-3-P appears in fructose diphosphate at a rate faster than if [^{14}C]-DHAP is used (Fig. 23-2). The experimental conditions involve starting with nonradioactive FDP in each case with enzyme. As the front-direction reaction (cleavage) proceeds, addition of either labeled 3-carbon product allows measure of the *flux* of the back reaction (condensation). These data suggest that there is ordered product release and that the glyceraldehyde-3-P comes off (k_2) before the DHAP, which is released only slowly (k_3).

$$\text{Enz}\cdots\text{FDP} \underset{k_{-1}}{\overset{k_1}{\rightleftharpoons}} \text{Enz}\begin{matrix}\cdot\cdot\,\text{G-3-P}\\ \cdot\cdot\,\text{DHAP}\end{matrix} \underset{k_{-2}}{\overset{k_2}{\rightleftharpoons}} \begin{matrix}\text{Enz}\cdots\text{DHAP}\\ + \\ \text{G-3-P}\end{matrix} \underset{k_{-3}}{\overset{k_3}{\rightleftharpoons}} \text{Enz} + \text{DHAP}$$

Such a kinetic scheme would allow preferential kinetic equilibration of radioactive glyceraldehyde-3-P label into fructose diphosphate. Also, this ordered kinetic mechanism implies that, in the condensation direction, DHAP may bind first. It also is consistent with the solvent hydrogen exchange in the absence of other substrates noted in ¶23.A.1.a.

23.A.1.c Borohydride trapping of an imine intermediate

Incubation of [^{14}C]-DHAP and aldolase, followed by $NaBH_4$ addition, yields an inactive enzyme with one mole of radioactive label, an experiment that parallels the acetoacetate-decarboxylase experiments (Chapter 21). This trapped enzyme-substrate derivative was characterized, after acid hydrolysis and phosphatase action on the radioactive protein, as N^6-β-glyceryl lysine (Speck et al., 1963). The secondary amine linkage must form by reductive trapping of an imine between DHAP and an active-site ε-amino group of a lysine residue.

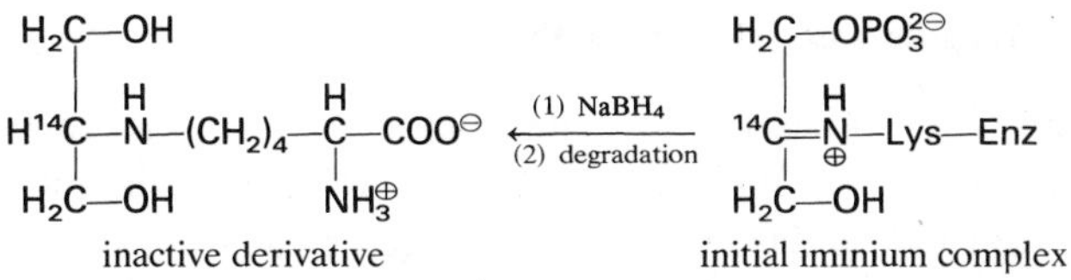

A similar inactivation occurs in the presence of $NaBH_4$ if fructose diphosphate is used, trapping the initial substrate imine in this case as well. But no such loss of enzyme activity occurs when either enzyme alone or enzyme and G-3-P are treated with borohydride. Thus the aldehyde carbonyl of G-3-P is not bound in such a way that it can form an imine linkage with the active-site lysine.

All this information (and more not explicitly discussed here, such as ^{18}O studies of the type used for acetoacetate decarboxylase) suggests obligate imine formation during aldolase catalysis to stabilize or generate carbanionic character at C-3 of the dihydroxyacetone-P moiety in the condensation direction.

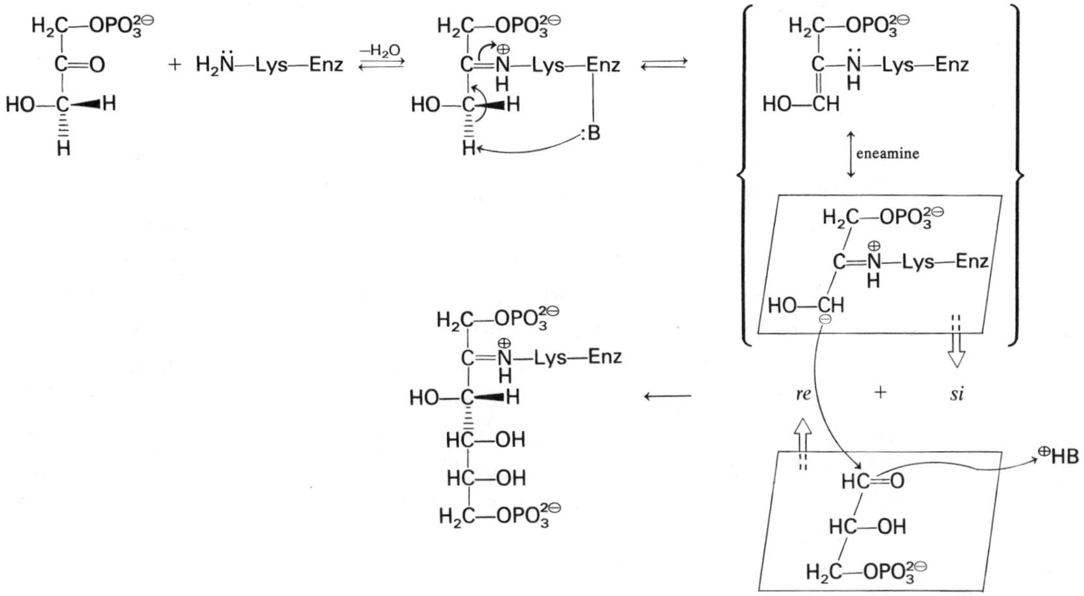

Note that the enzyme controls attack such that the bound DHAP eneamine attacks on only one face of the bound aldehyde, generating only one product isomer (with the OH groups *trans* on C-3 and C-4: it is 3*S*, 4*R*-FDP imine). The new carbon–carbon bond at C-3 to C-4 of the FDP has the same orientation as the pro*S* hydrogen initially abstracted from C-3 of the DHAP-imine complex: there is overall retention of configuration at C-3 (Alworth, 1972).

23.A.1.d Trapping carbanions with oxidizing agents

Studies by Riordan, Christen, and colleagues have provided further data interpreted as evidence for the finite existence of a DHAP carbanionic species (e.g., the eneamine) during aldolase catalysis (Christen and Riordan, 1968, 1969).

The reagent tetranitromethane, $C(NO_2)_4$, has been used as an electrophilic nitrating agent for aromatic residues reacting rather rapidly with the activated phenoxide ring of accessible tyrosyl residues to produce *ortho* nitration (Means and Feeny, 1971, p. 183).

nitroform anion (yellow)

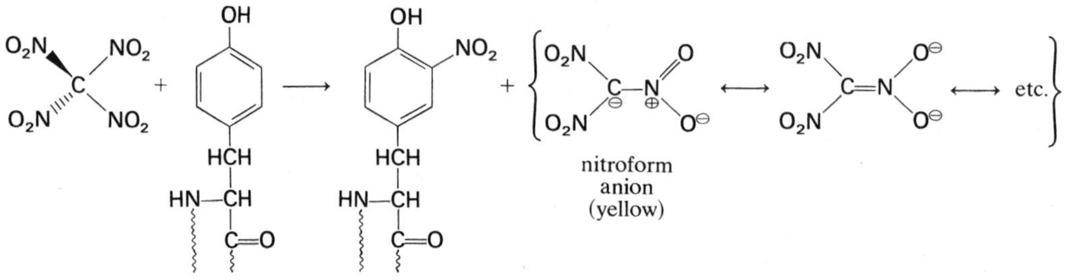

The reaction can be followed by the detection of the intensely yellow coproduct, the resonance-stabilized nitroform anion, which has a λ_{max} at 350 nm and an extinction coefficient of 14,400. The reagent also can oxidize the easily polarized sulfhydryl groups of cysteine (Means and Feeny, 1971, p. 183). Although aldolase from rabbit muscle is inactivated by such sulfhydryl-group oxidation on tetranitromethane addition, addition of DHAP or FDP protects against loss of enzyme activity. Yet, the rate of nitroform production then becomes catalytic rather than stoichiometric, suggesting that tetranitromethane is now reacting with and drawing off some reactive catalytic intermediate. Both fructose diphosphate and dihydroxyacetone phosphate promote enzyme-catalyzed generation of $C(NO_2)_3^{\ominus}$ (Fig. 23-3), but glyceraldehyde-3-phosphate does not. Subsequent experiments indicated that consumption of FDP or DHAP is occurring during this reaction. Model studies show that tetranitromethane reacts rapidly to produce nitroform with such carbanions as the enolate of acetone or with thiamine pyrophosphate at pH 8.0. For example,

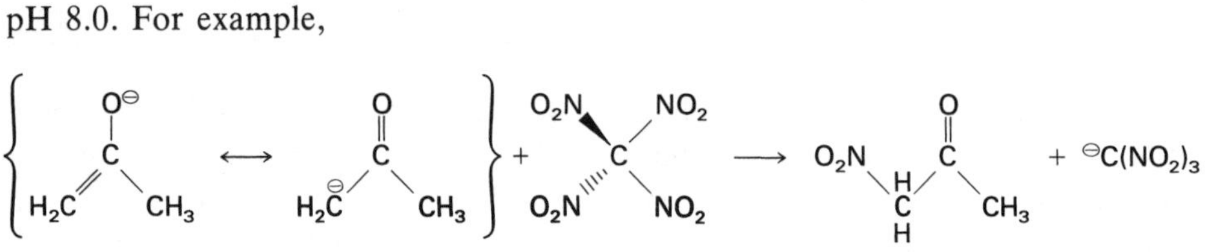

The following is a possible two-electron pathway for the TPP reaction. (One-electron pathways could be operating instead.)

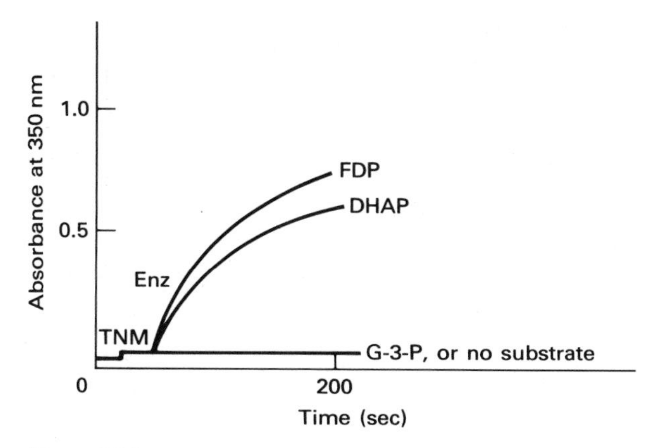

Figure 23-3
Aldolase-catalyzed production of nitroform anion in presence of fructose diphosphate or dihydroxyacetone-P. TNM = tetranitromethane; Enz = FDP aldolase.

$$\text{thiazolium ylide (R–N}^{\oplus}\text{, }H_3C\text{, CH}_2\text{CH}_2\text{OPOPO}^{\ominus}\text{)} + C(NO_2)_4 \longrightarrow {}^{\ominus}C(NO_2)_3 + \text{2-}NO_2\text{-thiazolium}$$

The reactive species thus was presumed to be the enzyme–DHAP complex in its eneamine form where C-3 possesses carbanionic character. The derivative of DHAP formed in enzyme-catalyed reaction with tetranitromethane was subsequently identified as hydroxypyruvaldehyde phosphate, an overall oxidation of the alcohol group of DHAP to the aldehyde (Healey and Christen, 1972). The tetranitromethane is correspondingly reduced by two electrons to one equivalent of nitrite ion along with the equivalent of nitroform anion. No reaction of DHAP with $C(NO_2)_4$ occurs in the absence of enzyme.

$$\text{Enz} + H_2C(OPO_3^{2\ominus})\text{–C=O–}H_2C\text{–OH} \rightleftharpoons H_2C(OPO_3^{2\ominus})\text{–C}{=}\overset{\oplus}{N}H\text{–Lys–Enz}\text{–}{}^{\ominus}CH\text{–OH} \xrightarrow{C(NO_2)_4} H_2C(OPO_3^{2\ominus})\text{–C=O–CHO} + {}^{\ominus}C(NO_2)_3 + NO_2^{\ominus} + \text{Enz}$$

hydroxypyruvaldehyde phosphate; nitrite

The oxidative step could involve radical intermediates generated through single-electron transfer steps, or a transient 3-nitrodihydroxyacetone-P that falls apart in an internal redox reaction. With the idea that electron-rich enzyme-carbanionic intermediates might be detectable by their ease of oxidation, Christen and Healey showed that ferricyanide and other redox agents such as 2,6-dichlorophenol indophenol and phenazine methosulfate would undergo reduction by the complex of aldolase and dihydroxyacetone phosphate (Healey and Christen, 1973; Birchmeier et al., 1973).

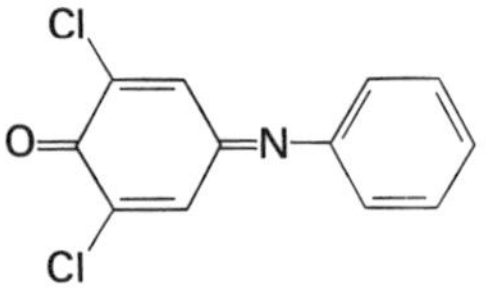

2,6-dichlorophenol indophenol

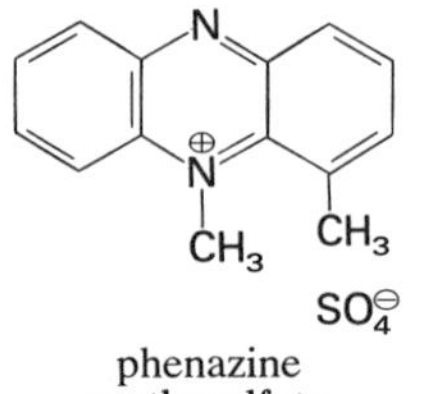

phenazine methosulfate

The enzyme-catalyzed reduction of redox indicators is not limited to the aldolase case, but also occurs with other enzymes thought to generate substrate-carbanionic intermediates (Healey and Christen, 1973; Birchmeier et al., 1973). One is yeast pyruvate decarboxylase in the presence of pyruvate, presumably reflecting an oxidative trapping of the anion of bound hydroxyethylthiamine pyrophosphate (Chapter 21). Another instance is the incubation of gluconate-6-P dehydrogenase in the presence of ribulose-5-P (¶21.B.2). The intermediate oxidatively trapped is presumably the 1,2-enediolate anion of ribulose-5-P. On these bases, electron acceptors might be useful as mechanistic probes for detection of enzyme-generated carbanionic intermediates that are accessible and of finite lifetime. (They obviously are not diagnostic of carbanions for enzymes whose normal role involves redox processes, because that capacity will engender rapid dye reduction independent of the nature of any transient species.)

23.A.2 KDPG Aldolase

A second aldolase that has been studied in some detail is the 2-keto-3-deoxy-6-phosphogluconate (KDPG) aldolase of *Pseudomonas* species (W. A. Wood, 1972). It is representative of a group of 2-keto-3-deoxycarboxylate aldolases found variously in microorganisms, plants, and animals. The substrate of KDPG aldolase is derived from 6-phosphogluconate by dehydrase action (Chapter 17). This aldolase produces pyruvate and glyceraldehyde-3-P (G-3-P) as cleavage products.

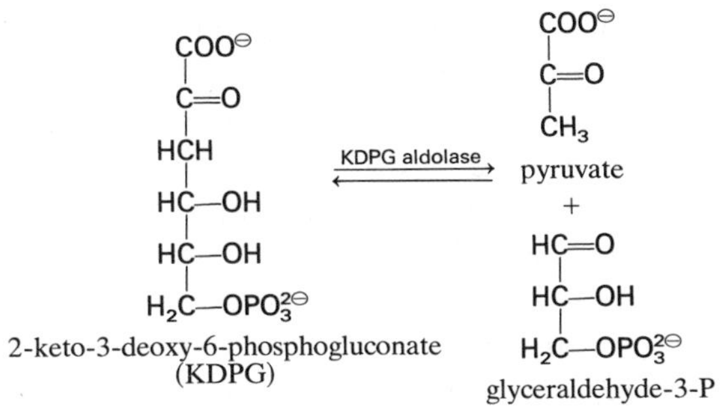

$$\begin{array}{c} COO^{\ominus} \\ | \\ C{=}O \\ | \\ HCH \\ | \\ HC{-}OH \\ | \\ HC{-}OH \\ | \\ H_2C{-}OPO_3^{2\ominus} \end{array} \underset{}{\overset{\text{KDPG aldolase}}{\rightleftharpoons}} \begin{array}{c} COO^{\ominus} \\ | \\ C{=}O \\ | \\ CH_3 \end{array} + \begin{array}{c} HC{=}O \\ | \\ HC{-}OH \\ | \\ H_2C{-}OPO_3^{2\ominus} \end{array}$$

2-keto-3-deoxy-6-phosphogluconate (KDPG) — pyruvate — glyceraldehyde-3-P

The enzyme from *P. saccharophilia* has a molecular weight of 72,000 and may be a functional trimer. The turnover number in the cleavage direction is 17,250 moles min^{-1} (mole enzyme)$^{-1}$ at 28°C. This aldolase functions quite analogously to the imine-utilizing FDP aldolase. Kinetic isotope studies indicate that carbanion formation is about twofold faster than the cleavage to products. Whereas FDP aldolase exchanges only the pro*S* hydrogen at prochiral C-3 of dihydroxyacetone phosphate, the KDPG aldolase will exchange all three hydrogens of the torsiosymmetric methyl group of pyruvate.

Meloche (1970, 1973) has carried out a number of interesting experiments on KDPG aldolase using the β-halo-α-keto acid bromopyruvate.

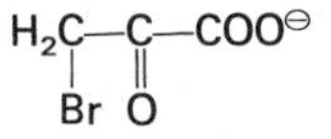

$$\begin{array}{l} H_2C{-}C{-}COO^{\ominus} \\ \;\;\,|\;\;\;\;\,\| \\ \;\;Br\;\;O \end{array}$$

This substrate analogue of pyruvate has a reactive α-haloketone moiety, qualifying it as an alkylating agent of the type we have discussed before (Chapter 2). Bromopyruvate does indeed inactivate the enzyme, and the rate of inactivation is slowed competitively by pyruvate but not by glyceraldehyde-3-P. Kinetic evidence also favors initial active-site binding prior to inactivation to a covalent adduct. The K_i for inactivation is 1 mM, and k_2 is 0.00115 sec^{-1} at pH 6 and 24° C.

$$E + Br{-}Pyruvate \underset{k_{-1}}{\overset{k_1}{\rightleftharpoons}} E \cdots Br{-}Pyruvate \xrightarrow{k_2} E{-}Pyruvyl + Br^{\ominus}$$

The nucleophile trapped by alkylation can be either an enzyme carboxyl group (at low salt conditions, where the enzyme is most active) or a thiolate anion (at high salt and a less active enzyme conformer). Although both groups are accessible nucleophiles, the carboxylate would appear to be the preferred candidate for the enzymatic base that facilitates enolization of pyruvate during catalysis.

Meloche suggested that bromopyruvate is initially bound in imine linkage to the active-site lysine residue and is then in position to undergo attack by the carboxylate anion of the enzyme. This view is substantiated by experiments with *R,S*-3-[^{3}H]-bromopyruvate (unlike pyruvate, bromopyruvate has a prochiral C-3), which showed that *the enzyme catalyzes detritiation before it undergoes alkylation and inactivation.* Detritiation proceeds at a V_{max} of 0.6 sec^{-1}, some 50-fold faster than inactivation, with a K_m of 1 mM, identical to the K_i for alkylation. Thus, the active-site carboxylate carries out its normal role as a general base abstracting a hydrogen for enolization with bromopyruvate, but once every 50 events behaves instead as a nucleophile toward carbon, and this behavior produces inactivation. The V_{max} for detritiation of pyruvate is 367 sec^{-1}, some 640-fold faster than detritiation of bromopyruvate.

Because C-3 of bromopyruvate is prochiral, one might expect that KDPG aldolase acts stereospecifically to labilize only one hydrogen during enolization. This could be proved in the following way. *R,S*-[^{3}H]-bromopyruvate is detritiated to release 50% of the radioactivity as ^3H_2O faster than the enzyme is inactivated. The chirality of the residual tritiated bromopyruvate is determined by hydroxide-catalyzed S_N2 displacement (with inversion) to 3-[^{3}H]-hydroxypyruvate, which is in turn decarboxylated with hydrogen peroxide to 2-[^{3}H]-glycolate (Meloche, 1970, 1973). The glycolate sample is oxidized with the O_2-dependent glycolate oxidase known to remove the pro*R* hydrogen at C-2 (Alworth, 1972, p. 246). Some 80% to 85% of the tritium in the glycolate sample is removed on oxidation to glyoxalate, indicating that the sample was predominantly 2*R*-[^{3}H]-glycolate. This is produced from 3*R*-[^{3}H]-hydroxypyruvate, which in turn arose from 3*S*-[^{3}H]-bromopyruvate. Therefore, the KDPG aldolase specifically removes the pro*S* hydrogen at C-3 of *R,S*-bromopyruvate to yield 3*S*-[^{3}H]-bromopyruvate.

Because detritiation is so much faster than the alkylation of the enzyme when *R,S*-3-[^{3}H]-bromopyruvate is used as inactivator, it is actually the 3*S*-[^{3}H] species that alkylates and then produces ^{3}H-labeled inactive enzyme. The ester linkage in the tritiated enzyme is susceptible to hydroxyaminolysis, releasing tritiated hydroxypyruvate, which on decarboxylation and reaction with glycolate oxidase proves to be 3*R*-[^{3}H]-hydroxypyruvate. As expected, the attack by the enzyme carboxylate on the 3*S*-[^{3}H]-bromopyruvate imine occurs with backside approach and resultant configurational inversion at C-3 in the inactive enzyme adduct.

Some indication of specificity in the ability of bromopyruvate to partition between enzymatic enolization (general base reaction with hydrogen) or alkylation (nucleophilic attack reaction with carbon) with KDPG aldolase is shown by the study of α-keto-β-deoxyhexonate aldolase from *E. coli* (metalloenzyme) and the 2-keto-4-hydroxyglutarate aldolase from liver (imine-forming) (Meloche and Mehler, 1973). Each uses pyruvate as condensation substrate, but neither of these enzymes will detritiate *R,S*-3-[^{3}H]-bromopyruvate *or* undergo inactivation.

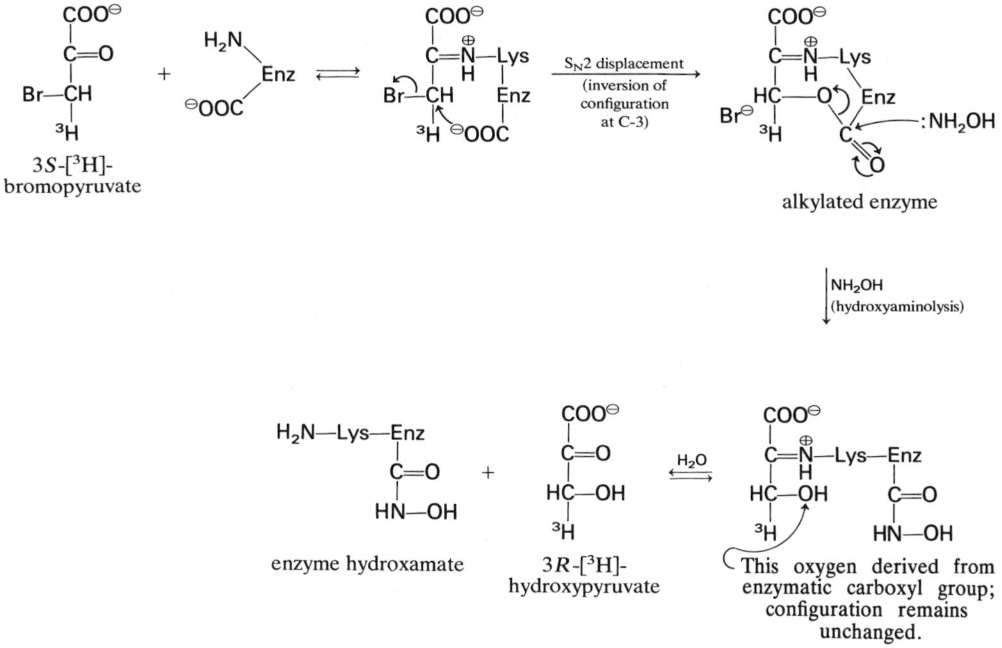

3S-[^{3}H]-
bromopyruvate

S_N2 displacement
(inversion of configuration at C-3)

alkylated enzyme

NH_2OH
(hydroxyaminolysis)

H_2O

enzyme hydroxamate

3R-[^{3}H]-
hydroxypyruvate

This oxygen derived from enzymatic carboxyl group; configuration remains unchanged.

23.A.3 Transaldolase

Transaldolase is of wide distribution, functioning during the aerobic (as opposed to fermentative or glycolytic) oxidation of glucose in the pentose-P pathway. The stoichiometry of the transformation catalyzed by this aldolase is the following.

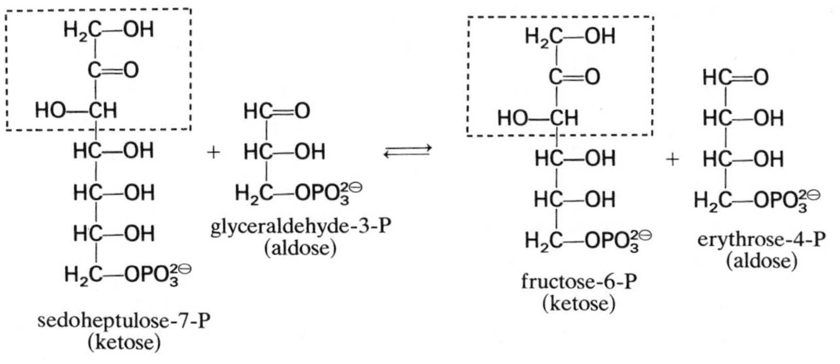

sedoheptulose-7-P
(ketose)

glyceraldehyde-3-P
(aldose)

fructose-6-P
(ketose)

erythrose-4-P
(aldose)

Erythrose 4-phosphate is a precursor for microbial biosynthesis of aromatic amino acids, and the sedoheptulose phosphate is a component of some bacterial outer membranes. The enzyme from the yeast *Candida utilis* has been purified as a dimer of 66,000 mol wt, with apparently only one active site per dimer (Tsolas and Horecker, 1972). This enzyme appears to operate by the cationic lysine-imine electron-sink mechanism, independent of the biological source examined.

Clearly, the enzyme catalyzes the shuttle of a three-carbon (dihydroxy-acetone) unit between the aldose acceptor molecules. One might contrast this reaction with that catalyzed by transketolase (¶21.C.3), in which a two-carbon ketol unit is transferred between aldose acceptors. In the case of the transketolase, the C_2 fragment is held on the enzyme by covalent attachment to the tightly bound coenzyme thiamine pyrophosphate. In the case of transaldolase, the C_3 fragment is retained by virtue of its imine linkage to an active-site lysine residue. The DHAP eneamine is held at the active site sufficiently long (without imine hydrolysis) for the glyceraldehyde-3-P product (from cleavage of a molecule of bound fructose diphosphate) to diffuse out and for the other aldose (the cosubstrate erythrose-4-P) to diffuse in, bind, and then react to give the C_7 ketose product—all before the DHAP moiety is released free into solution. *The major mechanistic difference* between transaldolase and FDP aldolase (¶23.A.1) is that release of DHAP from the enzyme · DHAP complex is much more rapid in the FDP-aldolase case.

Indeed, the essential similarity in the reactions catalyzed by transaldolase and FDP aldolase was underscored by controlled modification of the FDP aldolase through brief exposure to carboxypeptidase—exposure long enough for removal of 3 or 4 carboxyl-terminus residues (tyrosines) of the aldolase by action of the exopeptidase (Rose et al., 1965; Rose and O'Connell, 1969). The proteolyzed FDP aldolase is still active, but its conformation appears to be sufficiently altered to exhibit altered rates of component steps in catalysis.

The V_{max} for fructose-diphosphate cleavage is diminished 25-fold in the carboxypeptidase-treated enzyme, whereas the detritiation rate of $3S$-[^{3}H]-DHAP is diminished some 200-fold, a selective slowdown in enolization rate. Consistent with these data is the observation that, although $3S$-[^{2}H]-DHAP shows a V_{max} identical to protio-DHAP in the condensation with glyceraldehyde-3-P for native enzyme (i.e., enolization is a fast step in catalysis), for the treated enzyme there is a k_H/k_D of 7 for the condensation reaction. Analogously, although $3S$-[^{3}H]-DHAP is detritiated 3.4-fold faster than it is condensed by native enzyme, detritiation is some 3-fold slower than condensation in the proteolyzed enzyme.

In addition to the conclusion that proton abstraction from DHAP has now become rate-determining, it is worth noting that the *proteolyzed FDP aldolase has much increased capacity for transaldolation* with other aldehydes. The enzyme that has been treated with carboxypeptidase holds the DHAP eneamine for a longer interval before protonation and release than does the native enzyme.

23.A.4 δ-Aminolevulinate Dehydratase

The final example we shall note of an enzyme-catalyzed aldol condensation is that carried out by the enzyme δ-aminolevulinate dehydratase (Shemin, 1972). This enzyme carries out the condensation of two molecules of the C_5 substrate to a monopyrrole, porphobilinogen, which is the precursor of porphyrin (Chapter 15) and corrin (Chapter 20) ring systems in nature.

δ-aminolevulinates porphobilinogen

Aminolevulinate (ALA) itself is formed in a Claisen reaction, as we shall note in ¶23.B. The best-characterized enzyme is from the photosynthetic bacterium *Rhodopseudomonas spheroides* and appears to contain six subunits of 40,000 mol wt each, arranged as a dimer of trimers. As Shemin (1972) has indicated in a recent review, the enzyme catalyzes at least three steps:

1. an aldol condensation in which one ALA molecule acts as nucleophile and the second ALA molecule serves as electrophile;
2. a dehydration; and
3. cyclic imine formation to complete the pyrrole ring.

Not unexpectedly, incubation of [^{14}C]-aminolevulinate with enzyme and $NaBH_4$ produces an inactive, covalent, radioactive protein—suggesting an imine mechanism with an active-site lysine to generate nucleophilic character on the β-carbon of one bound ALA molecule (eneamine) prior to its attack at the keto-carbonyl γ-carbon of the second ALA molecule in the aldol condensation.

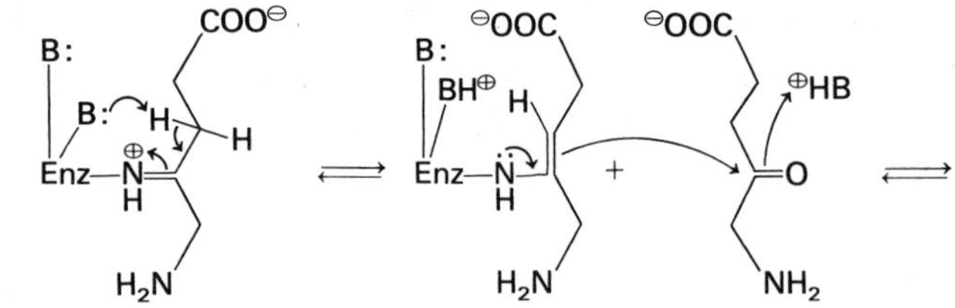

Enz · ALA imine

Enz · ALA eneamine

second ALA

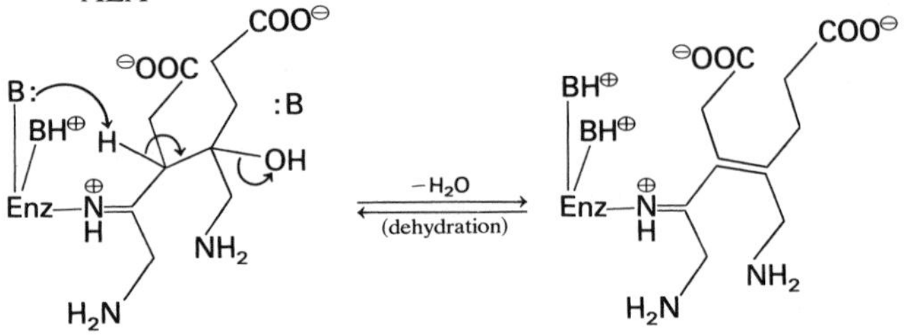

initial aldol product

$-H_2O$ (dehydration)

conjugated acyclic intermediate

The remaining β-hydrogen of the left half of the initial aldol product is still acidic, and facile elimination of water is a common fate for aldol condensations, yielding α,β-unsaturated carbonyl products. Dehydration leads to a conjugated acyclic intermediate that can rotate to become aligned for cyclization. Cyclization to a five-membered geminal diamine is chemically feasible and in this instance is favored because the cationic imine carbonyl undergoing attack is highly electron-deficient. Indeed, such transaldiminations proceed very rapidly in analogous model systems.

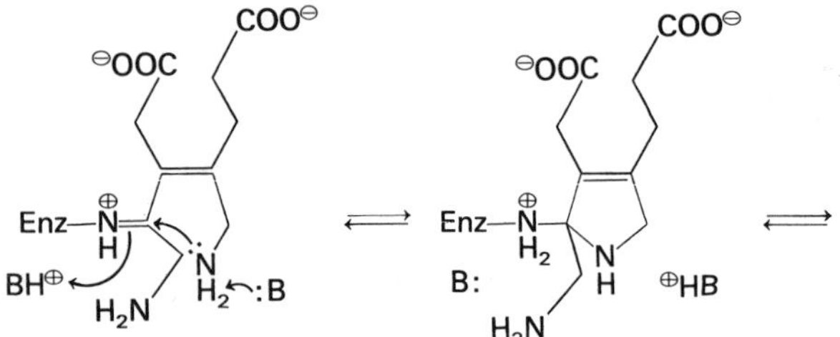

acyclic intermediate

geminal diamine

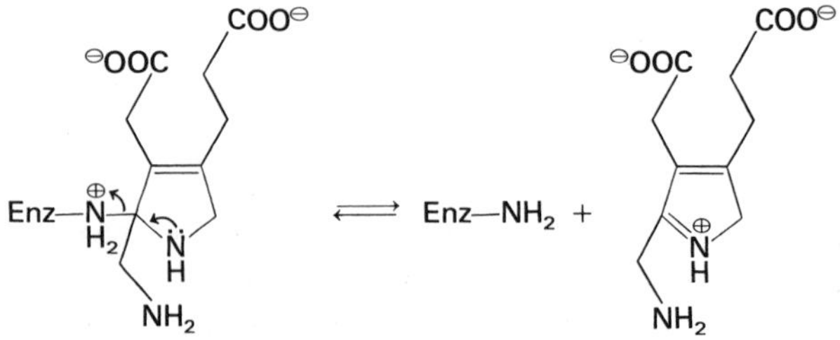

$Enz{-}NH_2$ +

pyrrole imine

The geminal diamine can fall apart back to the acyclic intermediate or alternatively, by elimination of the enzyme-lysine amino component, can form the imine isomer of the pyrrole product. Prototropic shifts then yield the pyrrole product.

$COO^{\ominus}$ $^{\ominus}OOC$:B—Enz H H N H NH_2 pyrrole imine ⇌ $COO^{\ominus}$ $^{\ominus}OOC$ $^{\oplus}HB$—Enz N H NH_2 pyrrole product

Presumably, a driving force in the overall equilibrium is the resonance energy gained in the formation of the pyrrole system. This represents a remarkably efficient and regioselective construction by a single enzyme of one of the key heterocycles found in biochemical metabolism.

23.B ENZYME–CATALYZED CLAISEN CONDENSATIONS

The enzymes catalyzing Claisen condensations or retrocondensations (cleavages) use chemistry very similar to the aldol chemistry, but they differ from the aldolases in that the nucleophilic component is the α-carbanion of an ester, which can attack a variety of carbonyl-containing cosubstrates. *There are few intermediates in metabolism that are oxygen esters* (acetylcholine and aminoacyl-tRNA species are notable exceptions). Rather, as we have indicated repeatedly in Section II of this text, activation of acyl groups may proceed via phosphate esters (acyl phosphates), but subsequent acyl transfer to thiols generally follows to produce *acylthioesters* as the general currency for acyl-transfer processes. Alternatively, we have noted oxidative reactions that produce acylthioesters directly by removal of two electrons from thiohemiacetals (masked aldehydes).

Although teleological arguments in favor of acylthioesters over acyl oxygen esters are at best unprovable, it is quite clear that the α-hydrogens of thioesters are more acidic than those of oxygen esters (Bruice and Benkovic, 1965–1966, vol. 1, p. 259 ff), and thus the α-carbanions of thioesters required in the enzymatic Claisen reactions are of lower energy than corresponding enolates of oxygen esters. Wessely and Lynen (1953) noted that the pK of the α-hydrogen of *S*-acetoacetyl-*N*-acetylthioethanolamine is 8.5, whereas that of ethyl acetoacetate is

10.5, a 100-fold difference in acidity. This distinction also is bolstered by rate data on reactions involving nonenzymatic Claisen condensations for each type of ester: the increased acidity of α-carbanions of thioesters leads to faster rates, a useful advantage for catalysis. The bases for acidity differences have been assigned to both resonance and inductive effects (Bruice and Benkovic, 1965–1966, vol. 1, p. 259 ff). Whereas resonance structures with an alkyl oxonium ion are thought to be fairly representative of an oxygen ester description, it is felt that corresponding sulfur cation structures are not significant contributors to thioester structure, in part due to the large bulk of the sulfur atom. The lack of importance of this resonance contributor permits the carbonyl of a thioester to display more of the properties of a simple ketone, including stabilized enolate formation.

$$\left\{ R-\overset{O}{\overset{\|}{C}}-O-R' \longleftrightarrow R-\overset{O^{\ominus}}{\overset{|}{C}}=\overset{\oplus}{O}-R' \right\} \text{ but } \left\{ R-\overset{O}{\overset{\|}{C}}-S-R' \not\longleftrightarrow R-\overset{O^{\ominus}}{\overset{|}{C}}=\overset{\oplus}{S}-R' \right\}$$

Inductively, on the other hand, sulfur and carbon have about the same electronegativity, which is less than that of oxygen, and thus enolates of thioesters should be less destabilized than those of oxygen esters—i.e., the negative charge density may be more evenly shared between carbon and sulfur. It may be that this energy difference is sufficient to put the energy barriers to thioester α-carbanion formation selectively within reach of the enzymes acting as catalysts.

We shall discuss malate synthase, three related Claisen reactions for making or cleaving citrate, and the thiolase-mediated formation and cleavage of acetoacetyl-CoA, and then we shall comment briefly on hydroxymethylglutaryl-CoA synthetase and also δ-aminolevulinate synthetase. Even this brief listing shows that Claisen reactions are pervasive among reactions that make and break the carbon–carbon bonds of many key compounds of central metabolism. We shall discuss another metabolically important example in Chapter 27 in our discussion of fatty-acid biosynthesis.

Note also that the biotin-dependent carboxylases acetyl-CoA carboxylase and propionyl-CoA carboxylase discussed in Chapter 22 can be construed as Claisen-condensation catalysts. They generate transition states where the α-carbon of these acylthioesters acts as nucleophile toward the electrophilic *N*-carboxybiotin.

23.B.1 Malate Synthase

In Chapter 22 we noted the existence of malate synthase as a common analytical system for evaluating the chiral purity of asymmetric samples of acetate, chiral in the methyl group (¶22.A.4). *In all the acetyl-CoA–utilizing Claisen enzymes, the condensation involves conversion of the acetyl methyl group into a methylene grouping,* and all thus far examined proceed with inversion (Rose and Hanson, 1975; Alworth, 1972, p. 260). Eggerer's research group has examined the mechanism by which acetyl-CoA and glyoxalate react to form malate and CoASH (Eggerer and Klette, 1967).

$$\mathrm{HC(=O)-COO^{\ominus}} + \mathrm{H_3C-C(=O)-SCoA} \rightleftharpoons \mathrm{CoASH} + \mathrm{{}^{\ominus}OOC-CH_2-CH(OH)-COO^{\ominus}}$$

2*S*-malate

Regarding the question of discrete α-carbanion formation, the enzyme will detritiate 2-[^{3}H]-acetyl-CoA in a $Mg^{2\oplus}$-dependent enolization, but only at 10^{-3} times the rate of malate formation. Yet, α-keto-acid analogues of glyoxalate stimulate exchange, with pyruvate raising the rate some 10^3-fold, such that enolization approaches the rate of condensation with glyoxalate (Eggerer and Klette, 1967). Pyruvate apparently acts solely as effector, because no citramalate or citramalyl-CoA (the expected condensation products between pyruvate and acetyl-CoA) is detectable. No evidence for covalent acetyl-enzyme intermediates is detected here, in contrast to such evidence for some of the other acyl-CoA ligases noted below.

Because condensation ought to yield *S*-malyl-CoA initially, the enzyme ought to be able to hydrolyze added malyl-CoA. This activity is observed, with *S*-malyl-CoA undergoing acyl transfer to water faster than does *R*-malyl-CoA. Chemical evidence suggests that malyl-CoA (and citryl-CoA) hydrolyze by neighboring-group participation intramolecularly via the cyclic substituted succinic anhydride, which is kinetically labile in aqueous media (Higuchi et al., 1967).

malyl-CoA ⇌ [tetrahedral intermediate] ⇌ CoAS$^{\ominus}$ + malic anhydride $\xrightarrow{H_2O}$ malate

Eggerer feels that this route may also occur on malate synthase. The absence of any detectable preequilibrium enolization prior to condensation with glyoxylate is consistent with the idea that enolization is rate-determining to some extent, an expectation validated by the stereochemical data on methyl⇄methylene conversion, which requires a kinetic isotope discrimination in order to be observed (Alworth, 1972, p. 205 ff).

23.B.2 Claisen Reactions Leading to or from Citrate

The tricarboxylate citrate, a central molecule in the metabolism of most organisms (e.g., the citrate cycle), is acted on by four different enzymes. One is aconitase (discussed in some detail in Chapter 17), catalyzing isomerization to isocitrate via *cis*-aconitate. The other three enzymes carry out Claisen reactions interconverting citrate, oxaloacetate, and acetate. The first two of these enzymes use acetyl-CoA, and the third uses free acetate. The first is ubiquitous; the second is found in mammalian cytoplasm; and the third is restricted to bacteria, and in many cases is of inducible status there. The three enzymes are—

1. citrate synthase ("condensing enzyme"),
2. ATP–citrate lyase ("citrate-cleavage enzyme"),
3. citrate lyase ("citritase").

We shall discuss each of these enzymes in some detail.

23.B.2.a Citrate synthase

Citrate synthase is the first enzyme of the citrate, or tricarboxylate, cycle (see Chapter 27). Its physiological role in bacteria or in the mitochondria of plants and animals is to catalyze citrate formation, and the equilibrium is driven in that direction thermodynamically by hydrolysis of the "high-energy" acylthioester bond at some stage during the reaction (Spector, 1972).

$$H_3C{-}\overset{\overset{\displaystyle O}{\|}}{C}{-}SCoA + {}^{\ominus}OOC{-}\underset{\underset{\displaystyle O}{\|}}{C}{-}\overset{H}{\underset{H}{C}}{-}COO^{\ominus} \underset{\longleftarrow}{\xrightarrow{k\,=\,8\times10^3}} HO{-}\overset{\overset{\displaystyle H_2C{-}COO^{\ominus}}{|}}{\underset{\underset{\displaystyle H_2C{-}COO^{\ominus}}{|}}{C}}{-}COO^{\ominus} + CoASH$$

No divalent cation requirements are known. The homogeneous *E. coli* enzyme is of 280,000 mol wt, whereas the enzyme from pig heart is of 100,000 mol wt. On the basis that malate synthase is a prototype for citrate synthase, one might expect citryl-CoA as a bound intermediate. Further, for the enzyme from most sources, radioactive acetyl-CoA produces citrate with only the pro*S* arm radioactive

(Srere, 1975). Thus, addition of the acetyl-CoA carbanion is always to the *si* face of the oxaloacetate carbonyl at the active site.

$$\text{Enz—B:} \; H_2{}^{14}C\text{—}C(=O)\text{—SCoA} + {}^{\ominus}OOC\text{—}C(=O)\text{—}CH_2\text{—}COO^{\ominus} \rightleftharpoons \underset{S\text{-citryl-CoA}}{H_2{}^{14}C(\text{—}C(=O)\text{—SCoA})\text{—}C(OH)(COO^{\ominus})\text{—}H_2C\text{—}COO^{\ominus}\cdots\text{Enz}} \underset{}{\overset{H_2O}{\rightleftharpoons}} H_2{}^{14}C(COO^{\ominus})\text{—}C(OH)(COO^{\ominus})\text{—}H_2C\text{—}COO^{\ominus} + \text{CoASH}$$

si

pro*S* arm

pro*R* arm

There are, however, a few ornery bacteria that add the acetyl-CoA α-carbanion to the opposite (*re*) face of the oxaloacetate keto group to yield label from the C_2 fragment in the pro*R* arm (Srere, 1975). The *re* citrate synthase is of limited biological distribution, but inversion occurs at the acetyl methyl in both the *re* and the *si* synthases. Two important testable points in catalysis are (1) is enzyme-catalyzed enolization of acetyl-CoA detectable, and (2) does citryl coenzyme A function as a finite intermediate?

For the first point, one can ask whether the enzyme will enolize acetyl-CoA in the absence of oxaloacetate (analogous to FDP-aldolase enolization with DHAP). Exhaustive search for this exchange met only with failure.

Eggerer (1965) eventually devised an experiment to look for the enolization (^3H_2O into acetyl-CoA) by using malate as a substrate analogue of oxaloacetate. Malate, though, has an OH in place of the carbonyl oxygen and cannot undergo aldol condensation. Malate is a chiral molecule, and Eggerer later used the separated 2*R* and 2*S* isomers. Table 23-1 shows the results.

$$\underset{\text{oxaloacetate}}{O{=}C(COO^{\ominus})\text{—}H_2C\text{—}COO^{\ominus}} \qquad \underset{S\text{-malate}}{HO\text{—}C(H)(COO^{\ominus})\text{—}H_2C\text{—}COO^{\ominus}}$$

Table 23-1
Effector role of *S*-malate in partial exchange reaction with citrate synthase

Additions	^3H into acetyl-CoA (counts/min)
None	200
R,S-Malate	2,530
R-Malate	210
S-Malate	6,440

The data in Table 23-1 imply that *S*-malate acts synergistically to generate the active-site conformation of the enzyme necessary for partial exchange and enolization to occur.

Eggerer also solved the second question—that of the discrete involvement of the citryl-CoA (Eggerer, 1963; Eggerer and Remberger, 1963*a*). To prepare the acylthioester derivitizing only one of the primary COOH groups, Eggerer blocked the tertiary COOH and carried out the indicated preparation. Synthetic citryl-CoA is a good substrate, proving to be a chemically competent intermediate both for aldol cleavage and thioester hydrolysis.

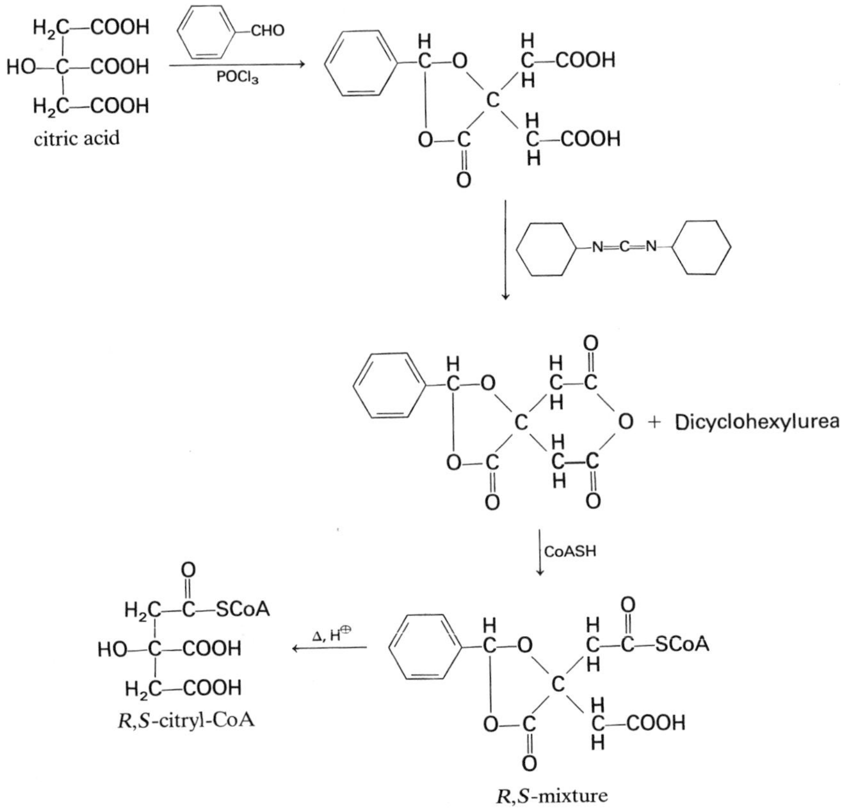

H_2C—COOH
HO—C—COOH
H_2C—COOH
citric acid

$POCl_3$

CoASH

+ Dicyclohexylurea

Δ, $H^{\oplus}$

H_2C—C(=O)—SCoA
HO—C—COOH
H_2C—COOH
R,S-citryl-CoA

R,S-mixture

Using limited quantities of *R,S*-citryl-CoA, Eggerer could show that citrate synthase uses up only 50%, implying chiral recognition of only one isomer, presumably *S*-citryl-CoA. (The enzyme also will hydrolyze *S*-malyl-CoA, but not *R*-malyl-CoA.)

What about the back reaction? Granted it is an unfavorable equilibrium, but how does the enzyme activate a citrate $COO^{\ominus}$ to make the thermodynamically activated citryl-CoA intermediate? Perhaps by intramolecular anhydride formation, with the enzyme generating a favorable rotamer distribution for anchimeric assistance by one primary carboxylate on the other, a possible entropic contribution to catalysis (Spector, 1972; Wunderwald and Eggerer, 1969).

H_2O

HSCoA

→ Citryl-CoA

cyclic anhydride
(low concentration)

In fact, in aqueous solutions of citric acid, there is a small but finite amount of the six-membered cyclic anhydride present in equilibrium with citrate (Higuchi et al., 1967).

23.B.2.b ATP–citrate lyase

The second enzyme acting on citrate, ATP–citrate lyase, also interconverts acetyl-CoA, oxaloacetate, and citrate (Spector, 1972; Srere, 1975). The enzyme is located in the cytoplasm of mammalian cells, and isotope studies have indicated that citrate cleavage provides the acetyl-CoA molecules that are used directly for fatty-acid synthesis (the first step is the acetyl-CoA carboxylase reaction) and for synthesis of the neurotransmitter acetylcholine in neural cells. When ATP–citrate lyase activity was initially discovered by Srere and Lipmann (1953), it was unclear why cells should possess both this activity and citrate synthase. The answer lies in intracellular compartmentation of the two enzymes—the citrate synthase is confined to mitochondria (as is the whole tricarboxylate cycle)—and in the fact that mitochrondrial membranes appear to be impermeant toward acetyl-CoA (caused by the anionic phosphate moieties in the CoA). On the other hand, citrate (as the magnesium chelate) appears to pass readily through the intracellular membrane barrier. Thus, intramitochondrial acetyl-CoA (produced in large part from oxidative decarboxylation of pyruvate; see Chapter 21) can be converted to extramitochondrial acetyl-CoA (for biosynthetic purposes) by condensation to citrate, passage through the membrane, and cleavage back to acetyl-CoA.

The ATP–citrate lyase from rat liver is multimeric, of 500,000 mol wt, with no detectable cofactors or bound prosthetic groups. Under physiological conditions, the enzymatic cleavage proceeds to an equilibrium favoring acetyl-CoA and oxaloacetate by about 10. Yet, the equilibrium for citrate synthase is about 10^4 in the other direction, favoring condensation, driven by acetyl-CoA hydrolysis. Thus, one expects some additional input of energy in the reaction catalyzed by ATP–citrate lyase and, in fact, ATP is a required cosubstrate, undergoing cleavage to ADP and P_i.

$$\text{Mg} \cdot \text{ATP} + \text{Citrate} + \text{CoASH} \rightleftharpoons \text{Mg} \cdot \text{ADP} + P_i + \text{Oxaloacetate} + \text{Acetyl-CoA}$$

Endergonic formation of acetyl-CoA is just about balanced by exergonic hydrolysis of ATP. At physiological concentrations, 10^{-2} to 10^{-3} M, the cleavage of one citrate into two product molecules provides some additional entropic driving force from left to right.

Because the substrate citrate contains a carboxylate (in the pro*S* arm) that is activated as a thioester in the acetyl-CoA produced, we would suspect that the ATP is used for the now-familiar acyl-activation pattern of intermediate formation of an acyl phosphate, *S*-citryl 1-phosphate, which could then be attacked by CoASH to yield *S*-citryl-CoA. In analogy to its postulated role in citrate synthase, the acyl-CoA would then undergo Claisen cleavage to the products. The intermediacy of such an acyl phosphate is supported by ^{18}O transfer from citrate to the inorganic phosphate formed in each turnover and by the observation that one isomer of *R,S*-citryl-1-P is utilized by the enzyme in place of ATP and citrate (Walsh and Spector, 1969). Similarly, *R,S*-citryl-CoA is cleaved by the enzyme to products until one of the isomers (presumably the *S*-isomer) is used up (Eggerer and Remberger, 1963*b*; Srere and Bhaduri, 1964).

$$S\text{-citryl-1-P} + \text{CoASH} \xrightarrow{\text{enz}} P_i + \text{Acetyl-CoA} + \text{Oxaloacetate}$$

However, the ATP–citrate lyase reaction clearly involves more than two intermediates. The enzyme shows a [^{14}C]-ADP$\rightleftharpoons$ATP exchange in the absence of other substrates, and a competent phosphoenzyme is isolable on incubation of enzyme with Mg · ATP (Inoue et al., 1967, 1968). Addition of 1,5-[^{14}C]-citrate to solutions of [^{32}P]-phosphoenzyme results in attachment of ^{14}C radioactivity and discharge of [^{32}P]-P_i, implicating a covalent citryl-enzyme intermediate that may have formed via an intermediate enzyme–citryl-P complex (Inoue et al., 1967, 1968).

$$\text{Mg} \cdot \text{ATP} + \text{Enz} \rightleftharpoons \text{Enz—X—}PO_3^{2\ominus} + \text{ADP}$$
$$\text{Enz—X—}PO_3^{2\ominus} + \text{Citrate} \rightleftharpoons \text{Enz—X—Citrate} + P_i$$

Addition of CoASH to the citryl enzyme produces acetyl-CoA and oxaloacetate, a transformation that could proceed via the enzyme–citryl-CoA complex. Thus, one can write a reaction scheme involving the following intermediates, with each step chemically reasonable (Walsh and Spector, 1969).

$$ATP^{4\ominus} + Enz{-}\ddot{X} \rightleftharpoons Enz{-}XPO_3^{2\ominus} + ADP^{3\ominus} \quad (1)$$

phosphoryl
enzyme
(covalent)

$$Enz{-}X{-}PO_3^{2\ominus} + \begin{matrix} COO^{\ominus} \\ | \\ CH_2 \\ | \\ HO{-}C{-}COO^{\ominus} \\ | \\ CH_2 \\ | \\ COO^{\ominus} \end{matrix} \rightleftharpoons Enz{-}X\!: \quad \begin{matrix} O \\ \| \\ C{-}OPO_3^{2\ominus} \\ | \\ CH_2 \\ | \\ HO{-}C{-}COO^{\ominus} \\ | \\ CH_2 \\ | \\ COO^{\ominus} \end{matrix} \quad (2)$$

enz · citryl-P
(noncovalent)

$$Enz{-}X\!: \; \begin{matrix} O \\ \| \\ C{-}OPO_3^{2\ominus} \\ | \\ CH_2 \\ | \\ HO{-}C{-}COO^{\ominus} \\ | \\ CH_2 \\ | \\ COO^{\ominus} \end{matrix} \rightleftharpoons \begin{matrix} O \\ \| \\ Enz{-}X{-}C \\ | \\ CH_2 \\ | \\ HO{-}C{-}COO^{\ominus} \\ | \\ CH_2 \\ | \\ COO^{\ominus} \end{matrix} + HOPO_3^{2\ominus} \quad (3)$$

citryl enzyme
(covalent)

$$\begin{matrix} O \\ \| \\ Enz{-}X{-}C \\ | \\ CH_2 \\ | \\ HO{-}C{-}COO^{\ominus} \\ | \\ CH_2 \\ | \\ COO^{\ominus} \end{matrix} + {}^{\ominus}S{-}CoA \rightleftharpoons Enz{-}X\!: \quad \begin{matrix} O \\ \| \\ C{-}SCoA \\ | \\ CH_2 \\ | \\ HO{-}C{-}COO^{\ominus} \\ | \\ CH_2 \\ | \\ COO^{\ominus} \end{matrix} \quad (4)$$

enz · citryl-CoA
(noncovalent)

$$\begin{matrix} Enz{-}X\!: \\ | \\ B\!: \end{matrix} \quad \begin{matrix} O \\ \| \\ C{-}SCoA \\ | \\ CH_2 \\ | \\ H{-}O{-}C{-}COO^{\ominus} \\ | \\ CH_2 \\ | \\ COO^{\ominus} \end{matrix} \xrightleftharpoons{\text{Claisen cleavage}} \begin{matrix} Enz{-}X\!: \\ | \\ BH^{\oplus} \end{matrix} + \begin{matrix} H_2\overset{\ominus}{C}{-}\overset{O}{\overset{\|}{C}}{-}SCoA \\ + \\ O{=}C{-}COO^{\ominus} \\ | \\ CH_2 \\ | \\ COO^{\ominus} \end{matrix} \xrightleftharpoons{H^{\oplus}} H_3C{-}\overset{O}{\overset{\|}{C}}{-}SCoA \quad (5)$$

A caveat can be raised about the citryl-P and citryl-CoA data, in that these species have not been isolated as intermediates starting from ATP, citrate, and enzyme. Although the rapid and stereospecific processing of added citryl-1-P and citryl-1-CoA is consonant with recognition of a normal intermediate by the lyase, both substances are reactive chemical acylating (citrylating) agents in their own right and could conceivably be acting to acylate the correct active-site nucleophile to yield citryl enzyme. In this regard, Inoue et al. (1969) observed enzymatic catalysis of a [^{3}H]-CoASH$\rightleftarrows$acetyl-CoA exchange, suggestive of an acetyl enzyme as an intermediate. This could arise from cleavage of a citryl enzyme, as we shall note below in discussing citrate lyase, and then the acetyl enzyme could be attacked by CoASH.

$$\text{Citryl enzyme} \rightleftarrows \text{Oxaloacetate} + \text{Acetyl enzyme} \xrightarrow{\text{CoASH}} \text{Acetyl-CoA} + \text{Enzyme}$$

The identity of the nucleophile component in the phosphoenzyme is in dispute. Srere has isolated a histidinyl *N*-phosphate after base hydrolysis of Enz—X—$PO_3^{2\ominus}$, whereas Inoue and colleagues converted the phosphoenzyme to a hydroxamate and, after Lossen rearrangement (via the isocyanate), reported isolation in uncertain yield of the amine derived from glutamate (Srere, 1975). It is possible that phosphoryl transfer from one nucleophile to the other occurs on workup, in analogy to the problem noted in ¶19.B.3 with triose-P isomerase, and it is unclear which residue is the catalytically significant one. In similar experiments, it has been claimed that a glutamate γ-carboxyl is the nucleophile at the enzyme active site that acts as nucleophile in formation of the covalent citryl enzyme (Srere, 1975).

At this juncture, one can say with some certainty that this may be one of the most complex series of transformations occurring at the active site of a single enzyme.

23.B.2.c Bacterial citrate lyase

Citrate cleavage by the bacterial citrate lyase occurs reversibly in the absence of other substrates (e.g., CoASH) to yield oxaloacetate and the unactivated acetate ion. It has the appearance of a simple aldol cleavage rather than a Claisen variant. Perhaps one expects a primitive, unsophisticated enzyme in comparison to the two just discussed. That would be an unwarranted expectation.

The enzyme is inducible by growth of bacteria on citrate. The *Klebsiella aerogenes* and the *Streptococcus diacetilis* enzymes have been purified to

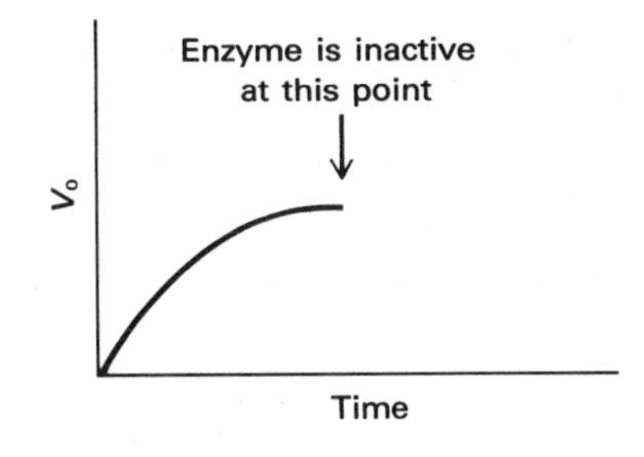

Figure 23-4
Autoinactivation of citrate lyase during catalytic turnover with citrate.

homogeneity, with ~550,000 mol wt (Srere, 1975), and they show requirements for a divalent cation (e.g., $Mg^{2\oplus}$, $Mn^{2\oplus}$), presumably as chelator with substrate oxygen functionalities. Curiously, this enzyme is observed to commit suicide during catalysis of the citrate aldol cleavage (Fig. 23-4); this was quite unexpected and rather unusual for a biological catalyst (Wheat and Ajl, 1955; Singh and Srere, 1971). On examination of inactive enzyme, both Srere and Eggerer observed that either acetyl-CoA or, equally, acetic anhydride (i.e., a good acetylating agent) can reactivate the enzyme (Buckel et al., 1969; Srere, 1972). With [^{14}C]-acetic anhydride, ^{14}C-labeled active enzyme was produced, implying that acetylation gives reactivation, suggesting that the active enzyme is acetylated *before and after* it completes a catalytic cycle. The [^{14}C]-acetyl enzyme was then incubated with [^{12}C]-citrate. The resulting enzyme in turn was treated with 1,5-[^{14}C]-citrate to yield the products in reaction 2.

$$[^{14}C]\text{-Acetyl-Enz} + [^{12}C]\text{-Citrate} \rightleftharpoons [^{14}C]\text{-Acetate} + [^{12}C]\text{-Oxaloacetate} + ([^{12}C]\text{-Acetyl-Enz})? \tag{1}$$

not detectable
because nonradioactive

$$[^{12}C]\text{-Acetyl-Enz} + 1,5\text{-}[^{14}C]\text{-Citrate} \rightleftharpoons [^{14}C]\text{-Enz} + [^{12}C]\text{-Acetate} + [^{14}C]\text{-Oxaloacetate} \tag{2}$$

These results of experiments 1 and 2 show that the acetyl group on the enzyme turns over in every catalytic cycle.

Eggerer has suggested the following scheme.

$$\text{Citrate} + [^{14}C]\text{-Acetyl-Enz} \rightleftharpoons [^{14}C]\text{-Acetate} + [^{12}C]\text{-Citryl-Enz}$$

$$\downarrow \text{Claisen cleavage}$$

$$[^{12}C]\text{-Acetyl-Enz} + \text{Oxaloacetate}$$

That is, the catalysis involves two discrete stages:

1. an acyl-exchange step; then
2. an aldol or Claisen cleavage (acyl lyase).

How could such a mechanism be effected? First one must know the nature of the acetyl-enzyme (and citryl-enzyme) intermediate. The acetyl enzyme is dischargeable with neutral NH_2OH, forming acetyl hydroxamate and inactive enzyme (Buckel et al., 1969; Srere, 1972); it is acid-stable, base-labile. All these observations are consistent with an acetylthioester as the resting acetyl enzyme. Thus, one pictures attack of a citrate carbonyl oxygen to produce a noncovalently-bound mixed acetic–citric anhydride and a free Enz—SH. Attack by the thiol (or thiolate anion) on the other carbonyl carbon of the anhydride would generate the acetate product and the *S*-citryl enzyme (Srere, 1975).

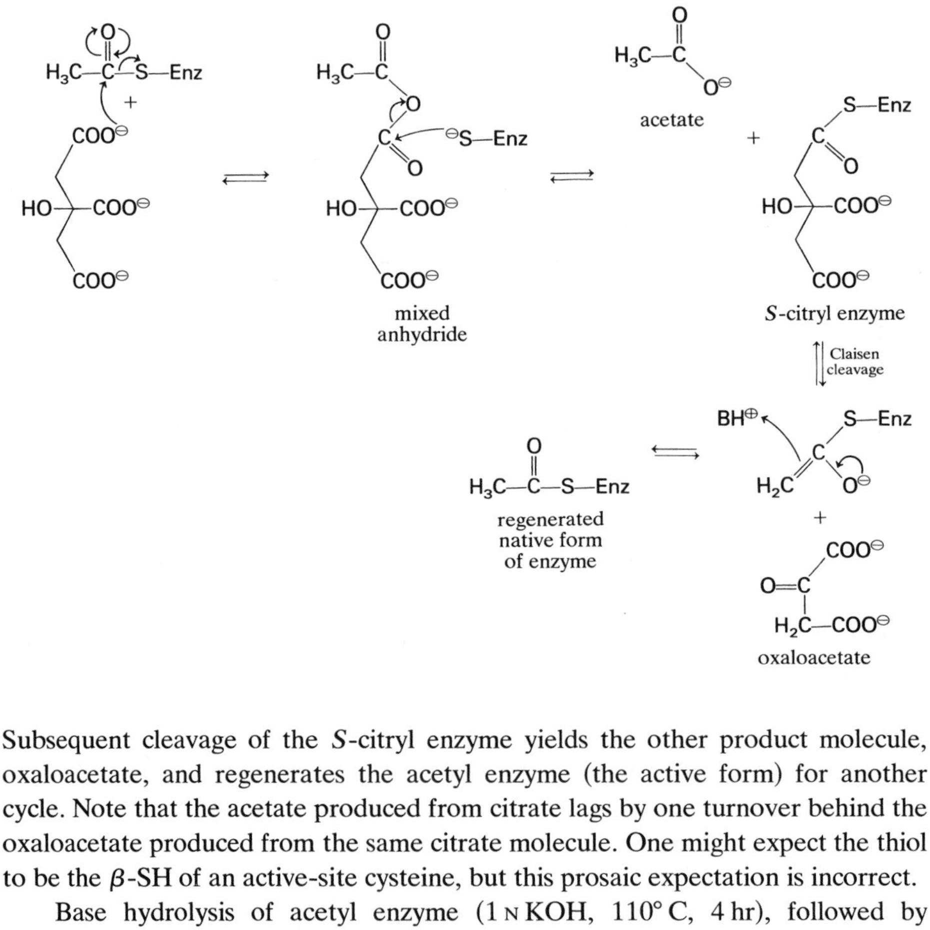

mixed anhydride

acetate

S-citryl enzyme

Claisen cleavage

regenerated native form of enzyme

oxaloacetate

Subsequent cleavage of the *S*-citryl enzyme yields the other product molecule, oxaloacetate, and regenerates the acetyl enzyme (the active form) for another cycle. Note that the acetate produced from citrate lags by one turnover behind the oxaloacetate produced from the same citrate molecule. One might expect the thiol to be the β-SH of an active-site cysteine, but this prosaic expectation is incorrect.

Base hydrolysis of acetyl enzyme (1 N KOH, 110° C, 4 hr), followed by alkaline phosphatase treatment, produces a compound that will support the growth of a strain of *Lactobacillus plantarum* with an absolute growth requirement for pantothenate (a microbiological assay).

The actual cofactor molecule in citrate lyase is not simply an ester of 4-phosphopantetheine with some active-site amino-acid residue, but coenzyme A covalently bound in the enzyme (Dimroth, 1975, 1976).

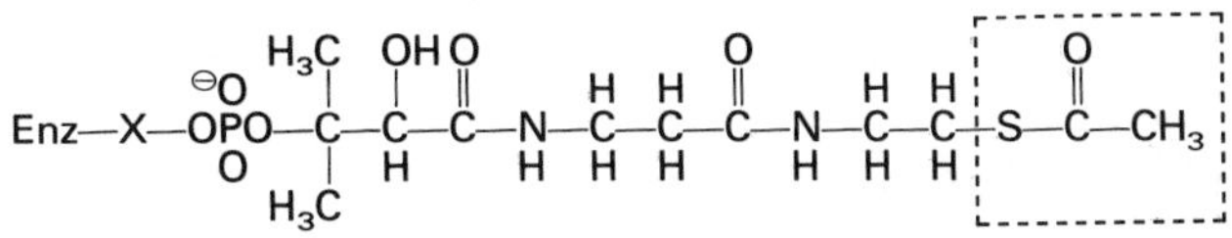

$$\text{Enz—X—}\overset{\ominus O}{\underset{O}{OPO}}\text{—}\overset{H_3C}{\underset{H_3C}{C}}\text{—}\overset{OH}{\underset{H}{C}}\text{—}\overset{O}{\overset{\|}{C}}\text{—}\underset{H}{N}\text{—}\overset{H}{\underset{H}{C}}\text{—}\overset{H}{\underset{H}{C}}\text{—}\overset{O}{\overset{\|}{C}}\text{—}\underset{H}{N}\text{—}\overset{H}{\underset{H}{C}}\text{—}\overset{H}{\underset{H}{C}}\text{—}\boxed{S\text{—}\overset{O}{\overset{\|}{C}}\text{—}CH_3}$$

acetyl citrate lyase

Other characterized examples of phosphopantetheine prosthetic groups covalently bound to enzyme are in gramicidin- and tyrocidin-biosynthesizing enzymes (Chapter 19; Lipmann et al., 1968) and in fatty-acid biosynthesis, where malonyl-CoA generated by acetyl-CoA carboxylase undergoes transthiolation from the CoA thiol to the thiol of a phosphopantetheine group bound to the acyl-carrier protein component of the fatty-acid-synthetase complex (Chapter 27; Prescott and Vagelos, 1971).

When native citrate lyase (550,000 mol wt) containing a [^{14}C]-acetyl group is dissociated in sodium dodecylsulfate, one radioactive subunit of 10,000 mol wt is seen, plus an unlabeled subunit of 32,000 mol wt, with these subunits appearing in the ratio of 1:4. The small subunit is an *acyl-carrier protein* analogous to (but distinct from) that used by the *K. aerogenes* fatty-acid-synthetase complex (Srere, 1975).

The postulated stoichiometry is 20 subunits total, with four functional pentameric units. Whether all bacterial citrate synthases function by so elaborate a covalent catalytic mechanism is unclear. For instance, not all of the bacterial enzymes examined in crude extracts catalyze their own suicides. This example of a resting form of the enzyme containing some fragment of a substrate molecule bound covalently at the beginning and end of each catalytic cycle is unusual, but not unique. Recall that phosphoglucomutase is isolated as an Enz—Ser—$OPO_3^{2\ominus}$ (and that adventitious loss of Enz · · · 1,6-diphosphoglucose leads to inactivation there also; ¶19.D.1). Also, the bacterial enzyme responsible for cleaving citramalate, citramalate lyase, exists as an *S*-acetyl enzyme in the resting form and presumably functions by a mechanism analogous to that of citrate lyase (Buckel and Bobi, 1976).

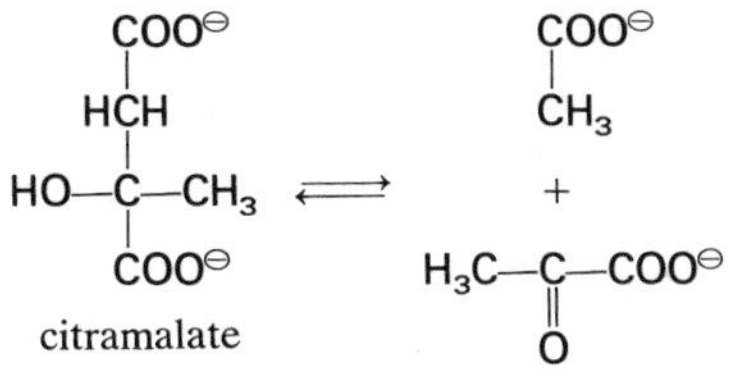

$$\underset{\text{citramalate}}{\begin{array}{c} COO^{\ominus} \\ | \\ HCH \\ | \\ HO\text{—}C\text{—}CH_3 \\ | \\ COO^{\ominus} \end{array}} \rightleftharpoons \begin{array}{c} COO^{\ominus} \\ | \\ CH_3 \\ + \\ H_3C\text{—}\underset{\underset{O}{\|}}{C}\text{—}COO^{\ominus} \end{array}$$

23.B.3 Isocitrate Lyase

An enzymatic reaction bearing some overall similarity to the cleavage of citrate by citrate lyase is a similar aldollike interconversion of isocitrate to glyoxalate and succinate (Spector, 1972).

2*R*,3*S*-isocitrate ⇌ glyoxalate + succinate

At physiological substrate concentrations (10^{-4}M), cleavage is favored because two products form from one substrate molecule. In plants and bacteria, the enzyme is involved in the so-called glyoxalate cycle, where carbons of the C_2 fragment glyoxalate are assimilated into tricarboxylate-cycle metabolism. Succinate is added to the *re* face of glyoxalate to yield 2*R*,3*S*-isocitrate. Cleavage in D_2O yields 2*S*-[$^2H_1$]-succinate, defining the process as an inversion (Spector, 1972).

In the reversible cleavages noted above between acetyl-CoA and such α-keto acids as oxaloacetate or glyoxalate, cleavage initially yields the α-carbanion of acetyl-CoA, and we have commented on the increased acidity of the α-hydrogens of acylthioesters, such that the α-carbanion is not a forbiddingly high-energy species. In contrast, the stoichiometry of the citrate-lyase and isocitrate-lyase cleavages appears to generate the α-carbanion of free acetate or free succinate, respectively. Or, examining the condensation direction, each enzyme must abstract the α-hydrogen of the acetate and succinate anions—hydrogens that are very nonacidic, with pK_a values estimated to be ~24 (vs. ~14 for acetylthioester α-hydrogens). We have just seen how citrate lyase solves this problem. It is not free citrate that is cleaved, but rather the *S*-citryl-enzyme intermediate, so the initial product is not the energetically unstable α-carbanion of acetate, but rather the stabilized α-carbanion of the acetylthioester enzyme species. This species protonates subsequently, before acetyl transfer to water occurs.

citryl-enzyme thioester ⇌ oxaloacetate + Enz—BH⊕ + {$H_2C{=}C(O^{\ominus})$—S—Enz ⟷ $H_2C^{\ominus}$—C(=O)—S—Enz}

It was then reasonable, with this precedent, to suppose that isocitrate lyase uses a similar scheme to increase the acidity of the 2*S*-hydrogen of succinate removed during catalysis. If the active form of native enzyme were a succinyl-thioester, the α-hydrogen would be much more acidic. However, experiments by Eggerer and colleagues appear to rule out the finite existence of a succinyl enzyme as the resting form of active enzyme (Dimroth et al., 1975). Hydroxylamine at up to 1 M concentrations does not inactivate homogeneous isocitrate lyase from *Pseudomonas indigofera*, whereas the acetyl enzyme of citrate lyase is completely deacylated (with inactivation) by this nucleophile. Further, when 5,6-[^{14}C]-isocitrate is incubated with isocitrate lyase and NH_2OH, no radioactive succinic monohydroxamate is detectable during or after cleavage of the substrate, so no activated succinate species (thioesters or anhydrides) are trappable. Also ruling out succinic-anhydride species is the fact that cleavage in $H_2{}^{18}O$ incorporates no ^{18}O into the succinate formed. So, at present one may state that it remains obscure how the enzyme can activate a pro*S* hydrogen of succinate to generate the requisite nucleophile for attack on glyoxalate. Indeed, a large k_H/k_T (of 6 to 15) suggests that proton abstraction is the rate-determining step (Spector, 1972).

23.B.4 Thiolase

The enzyme thiolase is more properly known as β-ketoacyl thiolase and is found in most organisms (Gehring and Lynen, 1972). In animal cells, thiolase is found both in mitochondria and in the cytoplasm, and it may have distinct physiological roles in each cell compartment. In bacteria and in *mitochondria* of eucaryotes, the enzyme functions as the last step in fatty-acid breakdown, an oxidative energy-generating sequence, a chain-shortening sequence removing two carbons at a time as acetyl-CoA (see also Chapter 27). The reaction involves thiolytic cleavage of β-ketoacyl-CoA substrates by free CoASH.

$$\text{CoASH} + \text{R}-\overset{\text{O}}{\overset{\|}{\text{C}}}-\text{CH}_2-\overset{\text{O}}{\overset{\|}{\text{C}}}-\text{SCoA} \rightleftharpoons \text{R}-\overset{\text{O}}{\overset{\|}{\text{C}}}-\text{SCoA} + \text{H}_3\text{C}-\overset{\text{O}}{\overset{\|}{\text{C}}}-\text{SCoA}$$

When acetoacetyl-CoA is the molecule cleaved, two molecules of acetyl-CoA are generated. Thus, in the back reaction, *acetyl-CoA can act as nucleophilic partner and electrophilic partner in a "head-to-tail" condensation* to form acetoacetyl-CoA. This equation clearly points out that acetyl-CoA is a *doubly activated biological molecule*, activated as an electrophile for acyl transfer and as a nucleophilic carbanion for its own acylation, *a versatility in chemical reactivity that accounts for its central role in metabolism.*

There is also thiolase activity in mammalian *cytoplasm* and, in this compartment, the role of the enzyme is to generate acetoacetyl-CoA as a *biosynthetic intermediate.* The bulk of the acetoacetyl-CoA so formed is drawn off through another Claisen condensation, condensation with yet another molecule of acetyl-CoA, catalyzed by hydroxymethylglutaryl-CoA synthase. The 3*S*-3-hydroxy-3-methylglutaryl-CoA product represents the first committed step in the biosynthesis of polyisoprenoid molecules such as terpenes, carotenoids, and sterols, as we shall note in Chapter 26.

$$H_3C{-}\overset{O}{\overset{\|}{C}}{-}CH_2{-}\overset{O}{\overset{\|}{C}}{-}SCoA + H_3C{-}\overset{O}{\overset{\|}{C}}{-}SCoA \xrightleftharpoons[\text{glutaryl-CoA synthase}]{\text{hydroxymethyl-}} H_3C{-}C(OH)(CH_2COO^{\ominus}){-}CH_2{-}\overset{O}{\overset{\|}{C}}{-}SCoA + CoASH$$

3*S*-hydroxymethylglutaryl-CoA

↓

polyisoprenoid biosynthesis

The thiolase from pig-heart mitochondria has been crystallized and is a tetramer of 170,000 mol wt, with acetoacetyl-CoA as the preferred β-ketoacyl-CoA substrate (Gehring and Lynen, 1972). There is a variety of evidence indicating that, in the condensation direction, a nucleophilic group on the enzyme forms a covalent linkage with one of the acetyl-CoA molecules. This acetyl enzyme is then attacked by the second molecule of acetyl-CoA, acting as nucleophilic α-carbanion. The covalent intermediate appears to be a thioester between a substrate acetyl group and the thiol of a cysteine of the enzyme. First, the enzyme is inactivated by [^{14}C]-iodoacetamide, forming (after hydrolysis) one mole of [^{14}C]-carboxymethylcysteine per mole of enzyme. The substrate protects against inactivation. The enzyme shows acetyl-transfer activity from one thiol nucleophile to another (from CoASH to *N*-acetylcysteamine) and will catalyze appropriate partial exchange reactions. Incubation of pure thiolase with [^{14}C]-acetyl-CoA allows isolation of an acetyl-enzyme intermediate with the proper reactivity characteristics. This radioactive acetyl enzyme is degradable to one radioactive peptide by protease treatment, and the peptide is identical to that generated with [^{14}C]-iodoacetamide inactivation. The stability characteristics of the acetyl enzyme also are consistent with its thioester nature (acid-stable, base-labile).

The following is a reasonable scheme.

$$\text{Enz—}\ddot{\text{S}}\text{H} + \text{CH}_3\text{C(=O)S—CoA} \rightleftharpoons \text{Enz—S—C(CH}_3\text{)(O}^{\ominus}\text{)—S—CoA} \rightleftharpoons \text{Enz—S—C(CH}_3\text{)=O} + {}^{\ominus}\text{SCoA} \qquad (1)$$

S-acetyl enzyme

$$\text{Enz—S—C(CH}_3\text{)=O} + \text{H}_3\text{C—C(=O)—SCoA} \rightleftharpoons \text{Enz—S—C(CH}_3\text{)(O}^{\ominus}\text{)—H}_2\text{C—C(=O)—SCoA} \rightleftharpoons \text{Enz—SH} + \text{CH}_3\text{C(=O)CH}_2\text{C(=O)SCoA} \qquad (2)$$

acetoacetyl-CoA

When longer-chain C_n β-ketoacyl-CoA molecules are substrates, the products of cleavage are the C_{n-2} acyl-CoA and C_2 acetyl-CoA.

$$\text{R—CH}_2\text{—C(=O)—CH}_2\text{—C(=O)—SCoA} + {}^{\ominus}\text{S—Enz} \rightleftharpoons \text{R—CH}_2\text{—C(O}^{\ominus}\text{)(S—Enz)—CH}_2\text{—C(=O)—SCoA} \longrightarrow$$

$$\text{R—CH}_2\text{—C(=O)—S—Enz (acyl enzyme)} \xrightarrow{\text{CoASH}} \text{R—CH}_2\text{—C(=O)—SCoA} + {}^{\ominus}\text{S—Enz}$$

$$\text{H}_2\text{C=C(O}^{\ominus}\text{)—SCoA} \underset{}{\overset{\text{H}^{\oplus}}{\rightleftharpoons}} \text{H}_3\text{C—C(=O)—SCoA} \qquad (2a)$$

In the cleavage direction, the enzyme's cysteinyl sulfur is the initial nucleophile. In the condensation direction, the acyl-enzyme intermediate is attacked by the acetyl-CoA α-carbanion. The kinetics of enolization vs. condensation have not been examined carefully, but it is clear that the enzyme can abstract a proton from C-2 of acetyl-CoA. The stereochemistry of the reaction is inversion at the methyl group of acetyl-CoA that is converted to the methylene group in acetoacetyl-CoA (Willadsen and Eggerer, 1975).

The capacity of the enzyme to carry out substrate-proton abstraction has recently been utilized in affinity inactivation of thiolase, in experiments patterned after the acetylenic-CoA→allenic-CoA isomerization carried out by the β-hydroxydecanoylthioester dehydrase discussed in Chapter 19. The acetylenic thioester 3-butynyl-CoA is a potent inactivator with an inactivation rate constant of 1.15 min^{-1} at 10 μM inactivator. The allenic analogue, 2,3-butadienyl-CoA, acts even faster, and it is likely that the enzyme abstracts a C-2 hydrogen from the butynyl-CoA, and the propargylic anion rearranges to the conjugated allenic species, which then undergoes alkylation by an active-site nucleophile (perhaps the active-site base) (Holland et al., 1973).

23.B.5 Hydroxymethylglutaryl-CoA Synthase

The mechanism of hydroxymethylglutaryl-CoA synthetase is probably similar in many respects to the thiolase reaction. The initial-velocity studies indicate a ping-pong kinetic mechanism, and the implicated covalent intermediate has been characterized as an *S*-acetyl enzyme (Miziorko et al., 1975). Yet, a postulated mechanism for this enzyme is somewhat distinct from the thiolase case. To obtain β-hydroxymethylglutaryl-CoA, it must be the acetyl-enzyme α-carbanion that acts as nucleophile on the C-4 carbonyl of acetoacetyl-CoA, *not* the enolate anion of acetoacetyl-CoA attacking the electrophilic carbonyl carbon of the acetyl enzyme. A second covalent enzyme derivative, involving a molecule of hydroxymethylglutaryl-CoA in thioester linkage to the same enzyme cysteinyl sulfur, is thereby generated (Miziorko and Lane, 1977).

This example completes our discussion of enzymatic aldol and Claisen reactions for this chapter. In the next chapter, we look at some examples of aldol and Claisen condensations that require pyridoxal phosphate—such as the reactions catalyzed by tryptophan synthetase and δ-aminolevulinate synthetase.

Chapter 24

Enzymatic Reactions Requiring Pyridoxal Phosphate

Vitamin B_6, known also as pyridoxine or pyridoxol, is not the biologically active molecule in enzymatic reactions using pyridoxal coenzymes. Rather, it is the aldehyde oxidation state that is required for coenzymatic function. However, the free aldehyde exists in large part as the intramolecular cyclic hemiacetal.

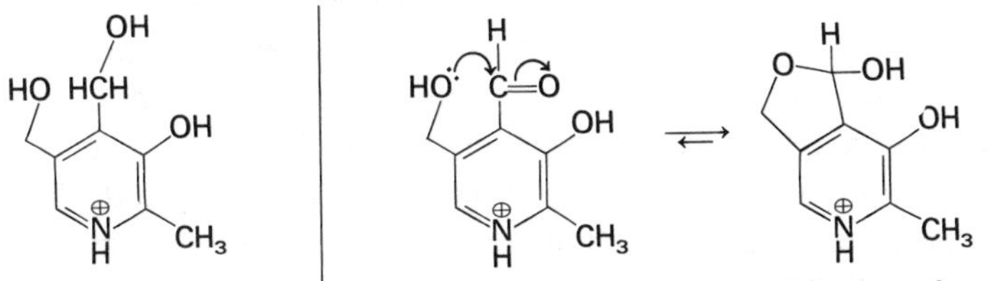

pyridoxine (pyridoxol)

pyridoxal (free aldehyde)

predominant form as cyclic hemiacetal

The actual coenzyme form, pyridoxal phosphate, has the hydroxymethyl tied up as a phosphate ester and thereby has the aldehyde group free.

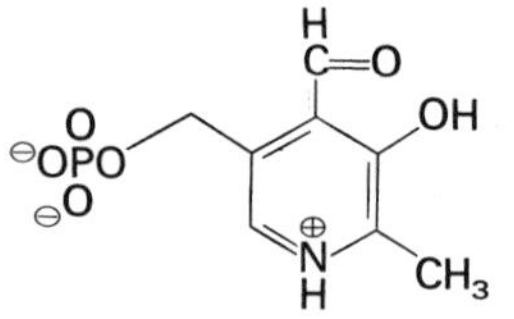

pyridoxal phosphate

Pyridoxal-P is an obligate coenzyme for the great majority of enzymes catalyzing some chemical change at the α-, β-, or γ-carbons of the common α-amino acids.

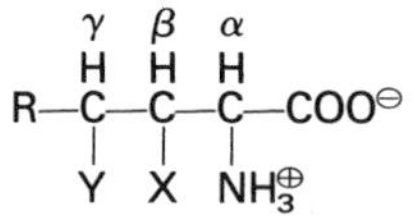

The reactions between amino acids and pyridoxal-P–containing enzymes (in general, the coenzyme does not dissociate freely, and holoenzymes exist as stable complexes) are similar to the β-keto-acid decarboxylations and the aldolase reactions discussed in preceding chapters in that *cationic imines* form and play key roles in lowering energy barriers to catalysis. In the decarboxylases and aldolases, the substrates provide the carbonyl component, and the enzyme provides the amine. In pyridoxal-P–dependent enzymes, *it is the substrate that provides the amine component,* and the coenzyme provides the carbonyl component.

Thus, the role of pyridoxal-P is *to act as electron sink to stabilize carbanionic intermediates that develop during enzymatic catalysis.* This particular electrophilic catalysis is very efficient because the cationic imine is in conjugation with the heteroaromatic pyridine ring to provide extensive charge delocalization. All pyridoxal-dependent enzymes thus function via (1) initial imine formation, (2) chemical changes via carbanionic intermediates, and (3) hydrolysis of a product imine. We shall note that more than one product imine can form, and the specificity of a given apoenzyme may be primarily in imposing a kinetic preference for proton transfers favoring formation of one imine over the other.

An additional feature of pyridoxal chemistry with amino acids is the *potential for reversed polarity at the β-carbon of the amino-acid substrate,* a feature exploited in β-elimination and replacement reactions. Although the simple eneamines (e.g., of the types of β-keto-acid decarboxylase and aldolase) have carbanionic character at the β-carbon (rendering that center nucleophilic as we have noted), eneamine formation with pyridoxal-P enzymes can lead to resonance structures that place *either cationic or anionic character at the β-carbon* (Snell and DiMari, 1970). Eneamines formed by abstraction of a substrate α-proton and then a β-proton have *β-carbanion character* (nucleophilic) only, analogous to simple eneamines.

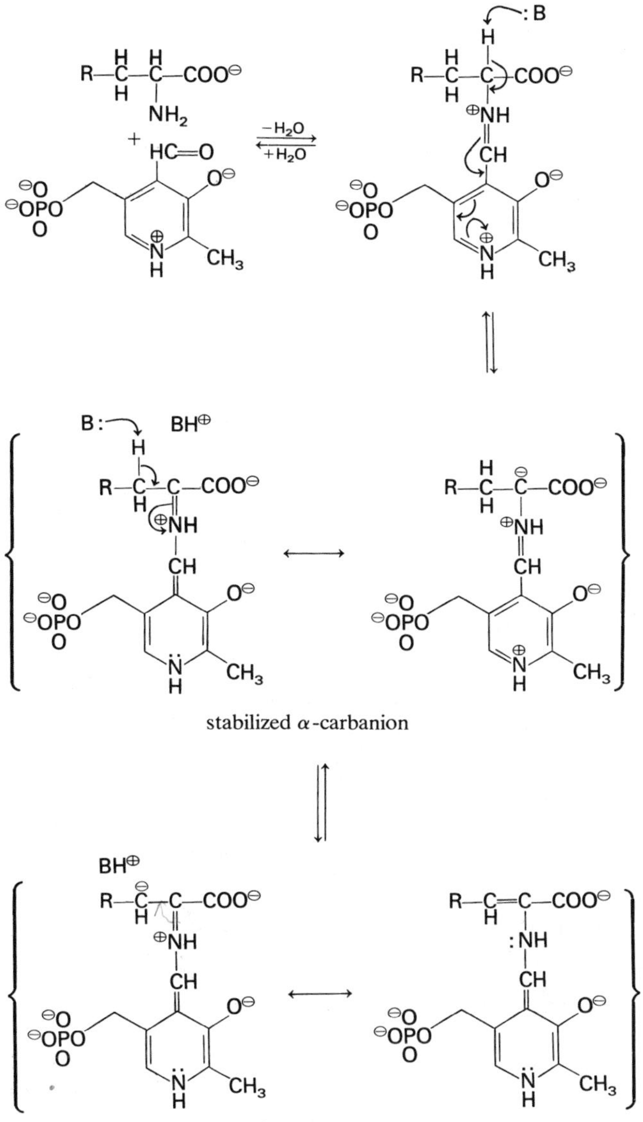

stabilized α-carbanion

eneamine

This eneamine must act as nucleophile, because the pyridine ring already has two electrons stored in it.

But, the eneamine formed by expulsion of a good leaving group at the β-carbon, using the electrons from the stabilized α-carbanion, has *electrophilic character* or charge deficiency at the β-carbon. We shall return to this reversed polarity and its important functional consequences later in the chapter.

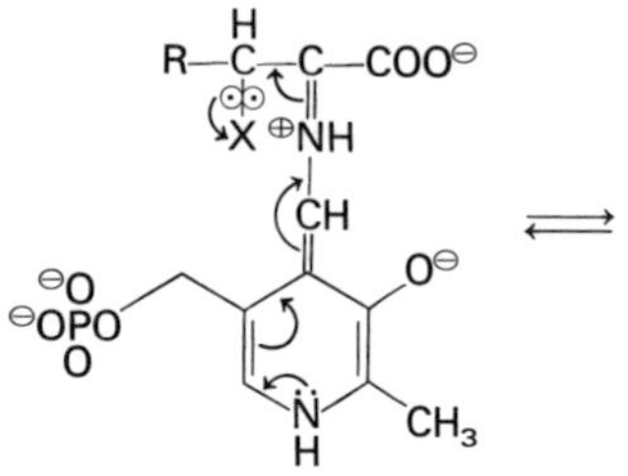

α-carbanion

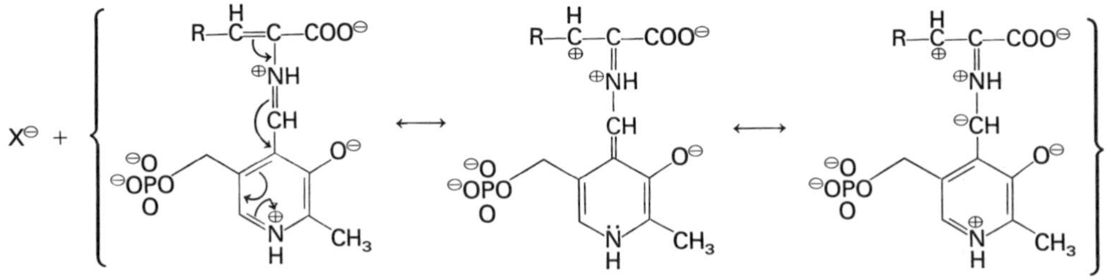

eneamine

We shall divide the pyridoxal-P–dependent enzymes into three major categories, according to whether they catalyze transformations at the α-, β, or γ-carbon of an amino-acid substrate, with subdivisions of each category:

1. α-carbon reactions
   a. transamination
   b. racemization
   c. decarboxylation
2. β-carbon reactions
   a. elimination
   b. replacement
3. γ-carbon reactions
   a. elimination
   b. replacement

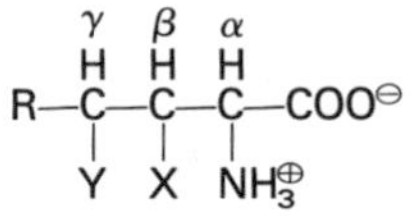

generalized amino acid

All three substituents at the α-carbon are subject to labilization in the three types of α-carbon reactions listed. Reactions at the β-carbon predominantly involve the elimination of the group X if it is a good leaving group and/or its replacement by a nucleophile X′. Similarly, reactions at the γ-carbon involve elimination of nucleophilic group Y and/or its replacement by Y′. The mechanistic constraints for reactions at the γ-carbon seem to have limited to a small group the number of enzymes known to perform such transformations, as we shall see. We turn first to a discussion of the α-carbon reactions.

24.A REACTIONS AT THE α-CARBON OF AMINO ACIDS

The α-carbon reactions are subdivided according to which of the α-C substituents is labilized. In transamination reactions, the amino group is labilized; in racemization reactions, the hydrogen is labilized; and, in decarboxylation reactions, the carboxyl group is labilized.

24.A.1 Transaminations

When amino acids and pyridoxal are mixed together nonenzymatically in aqueous solutions, one can detect a number of products, including the corresponding α-keto acids arising from transamination (Bruice and Benkovic, 1965–1966, vol. 2, p. 226 ff). Model studies have determined that the slow step in such transaminations is the prototropic shift in the aldimine⇄ketimine tautomerization, and that this step can be accelerated by general acid–base catalysis (i.e., imidazole, imidazolium buffers) to the point where this transamination is almost quantitative with suppression of other side reactions. This proton removal may also be the slow step in many enzymatic transformations, as indicated by substrate isotope effect studies. The stoichiometry of the enzymatic transaminations is the following:

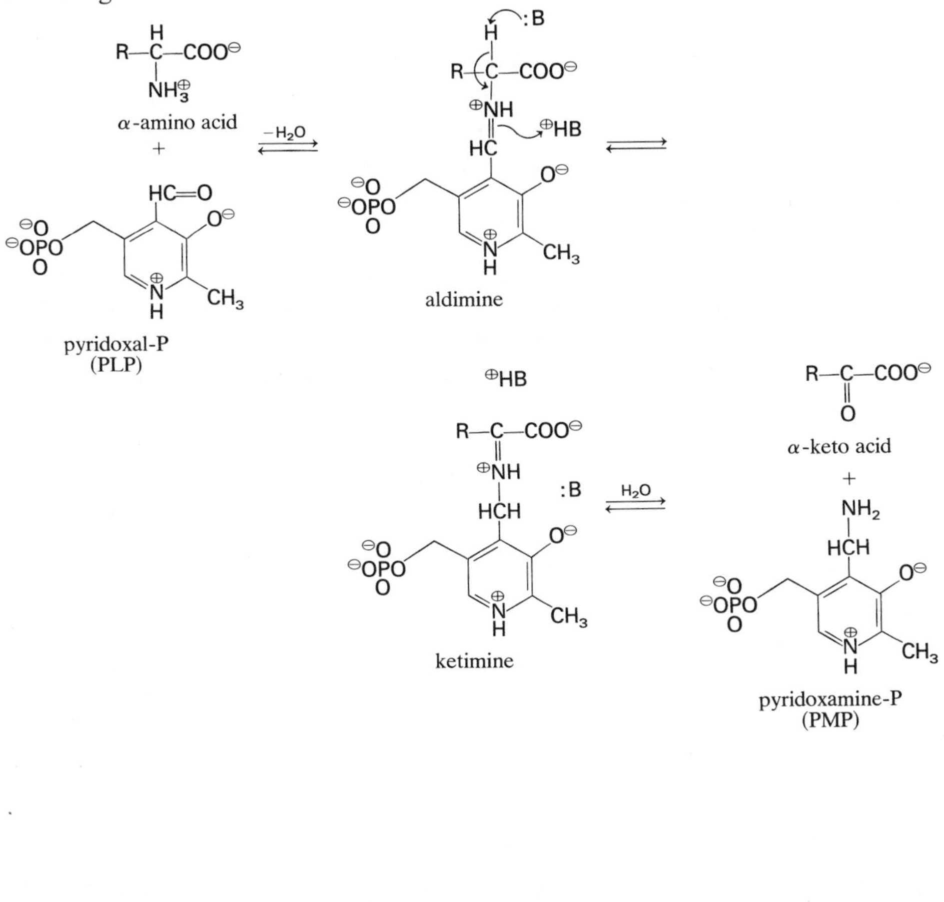

α-amino acid + pyridoxal-P (PLP) $\underset{}{\overset{-H_2O}{\rightleftarrows}}$ aldimine ⇄ ketimine $\overset{H_2O}{\rightleftarrows}$ α-keto acid + pyridoxamine-P (PMP)

24.A.1.a Aspartate transaminase

Although more than thirty discrete transaminases (aminotransferases) have been detected, the prototype that has been most extensively characterized with respect to structure and catalytic mechanism is aspartate–α-ketoglutarate transaminase, which catalyzes the following interconversion (Braunstein, 1973).

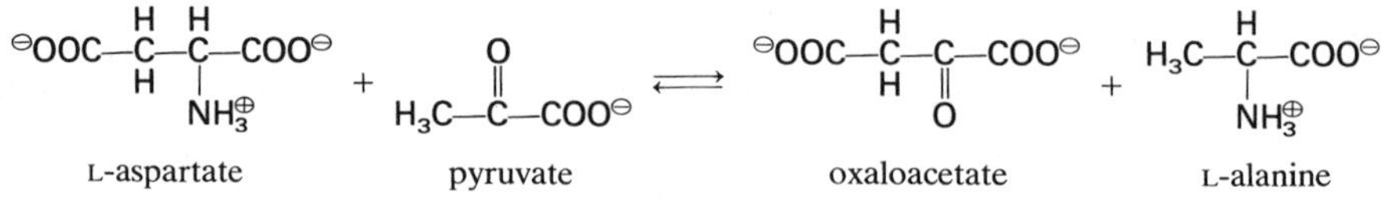

$$^{\ominus}OOC{-}CH_2{-}CH(NH_3^{\oplus}){-}COO^{\ominus} + H_3C{-}CO{-}COO^{\ominus} \rightleftharpoons {}^{\ominus}OOC{-}CH_2{-}CO{-}COO^{\ominus} + H_3C{-}CH(NH_3^{\oplus}){-}COO^{\ominus}$$

L-aspartate, pyruvate, oxaloacetate, L-alanine

Although aspartate is the preferred substrate, a variety of structurally related amino acids and α-keto acids serve as substrates. The enzyme has been purified from pig heart, where it exists as two genetically distinct isozymes—one cytoplasmic, the other mitochondrial. The cytoplasmic form is the one that has been most thoroughly examined. A dimer of 93,000 mol wt, it contains one tightly bound coenzyme molecule per subunit. Three subforms (α, β, and γ) have been described, the heterogeneity presumably a function in part of different sugar residues, because this enzyme does contain covalently bound carbohydrate. When pyridoxal phosphate (PLP) is free in solution at pH 7.0, it shows two λ_{max} values at 330 nm ($\varepsilon = 2{,}500$) and 388 nm ($\varepsilon = 4{,}900$), the latter accounting for its observed yellow color (Fig. 24-1a). When bound to aspartate transaminase, the λ_{max} is shifted—below pH 7.5 to 430 nm, or above pH 7.5 to 362 nm. A pK_a of 7.4 presumably represents deprotonation of some coenzyme species with concomitant shift of absorption to shorter wavelength (Fig. 24-1b). We shall return later to the various spectroscopic species.

In all of the PLP-dependent enzymes yet examined, the coenzyme is bound to the active site in imine linkage to the ε-amino group of a lysine residue, and this aldimine is the active form of resting enzyme, accounting for the species with $\lambda_{max} = 430$ nm (most likely the internally hydrogen-bonded species indicated).

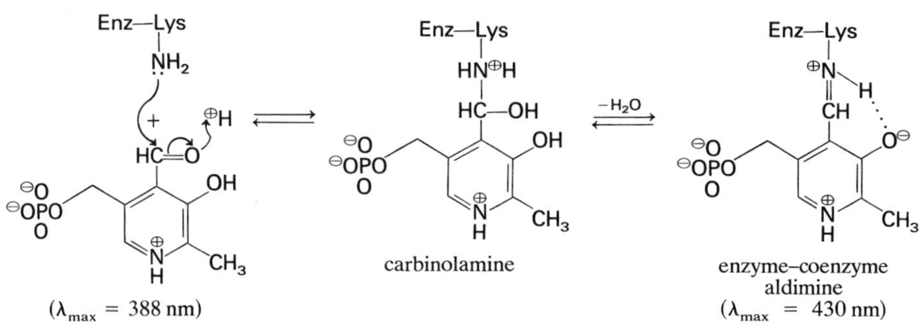

Enz—Lys—NH_2 + PLP (λ_{max} = 388 nm) ⇌ carbinolamine (Enz—Lys—$HN^{\oplus}H$—HC—OH) $\underset{}{\overset{-H_2O}{\rightleftharpoons}}$ enzyme–coenzyme aldimine (Enz—Lys—$^{\oplus}N$=CH, H···$O^{\ominus}$) (λ_{max} = 430 nm)

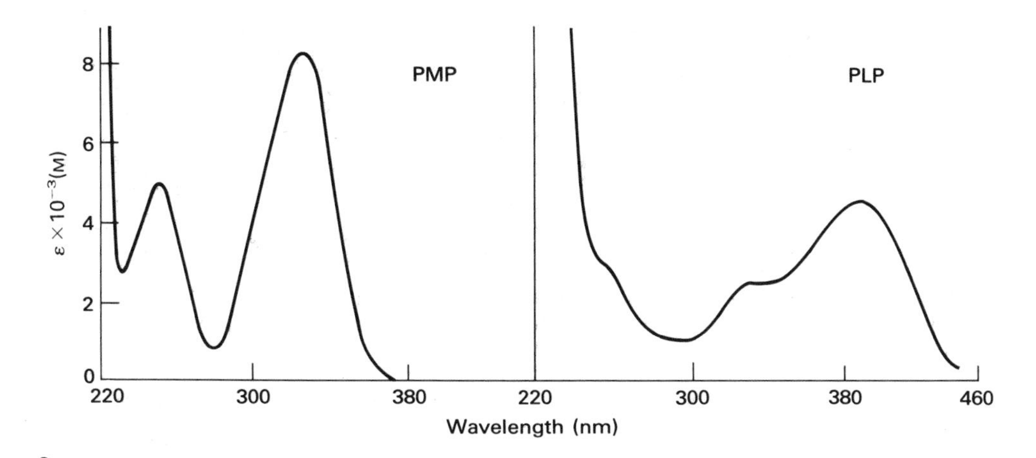

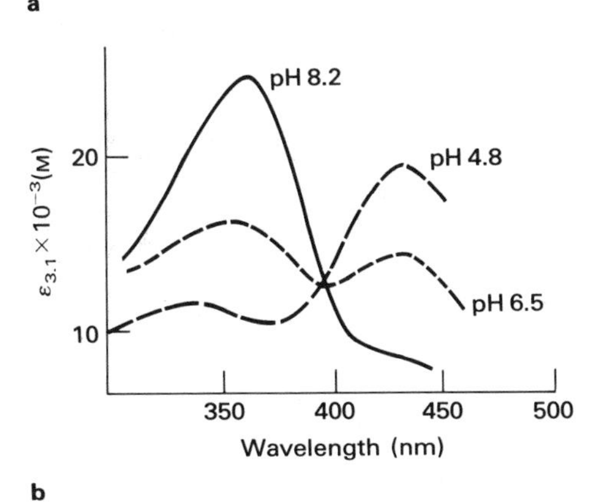

Figure 24-1
(**a**) Absorption spectra of PMP (pyridoxamine-P) and PLP (pyridoxal-P) in sodium phosphate buffer, pH 7.0 (Adapted from Braunstein, 1960, p. 131) (**b**) Absorption spectra of the PLP form of aspartate transaminase at various values of pH. (Adapted from Braunstein, 1973, p. 409)

Extensive evidence exists that all transaminases catalyze two discrete half-reactions; this evidence includes initial-velocity pingpong BiBi kinetic patterns, as well as isolation of the enzyme–PMP form at the end of the first half-reaction.

For the first half-reaction, aspartate enters into imine linkage with the coenzyme in a transaldimination, freeing the active-site lysine amino group (to act as a general base during subsequent catalysis?). Once this initial imine (aldimine) has formed, the acidity of the aspartate α-hydrogen has been increased because the carbanion, on proton removal, will be extensively stabilized by resonance. Thus, the abstraction of that proton is readily accomplished by an active-site base.

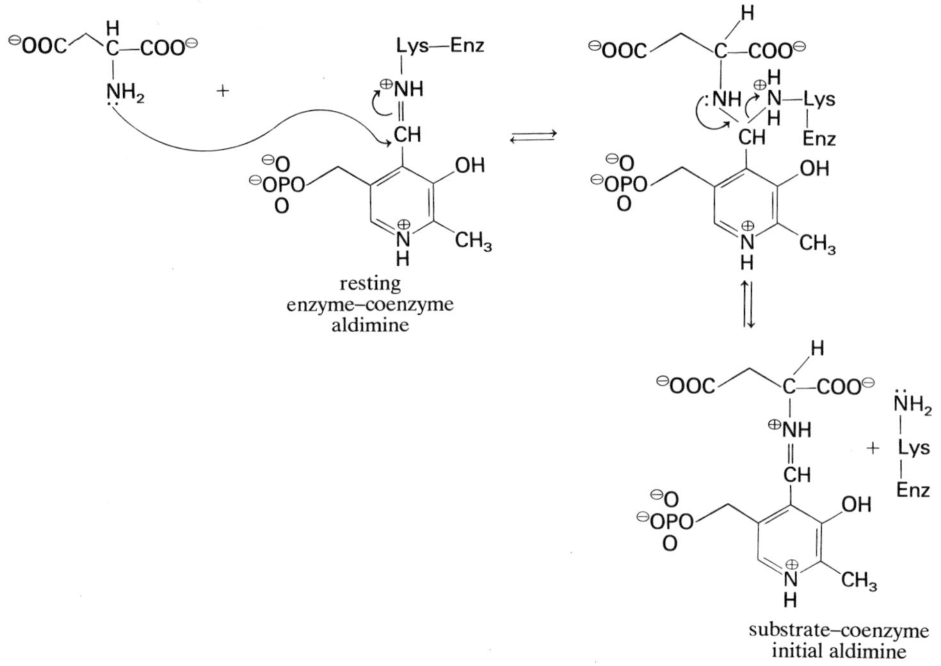

resting
enzyme–coenzyme
aldimine

substrate–coenzyme
initial aldimine

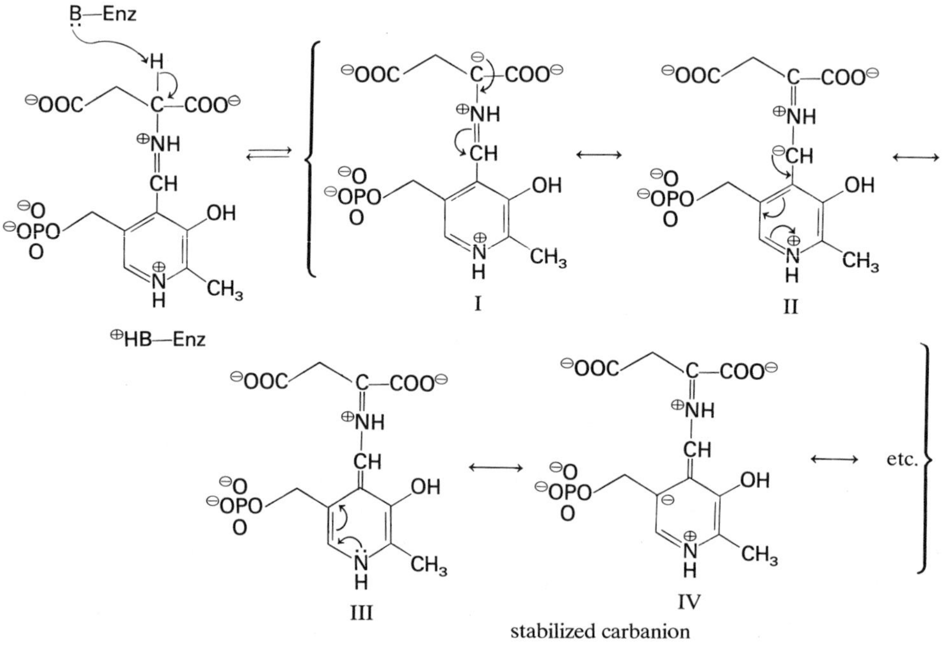

I

II

III

IV
stabilized carbanion

Note that structure III is a *para*-quinoid form.

The stabilized carbanion could be reprotonated at the substrate α-carbon to regenerate starting materials *or* (as structure II implies) protonation could occur at the carbon that was initially the aldehyde carbon of the coenzyme. This net prototropic shift would yield the ketimine product.

II ⇌ ketimine $\underset{H_2O}{\rightleftharpoons}$ oxaloacetate + PMP

If the enzyme promotes hydrolysis of the ketimine, the products will be oxaloacetate and the amine form of the coenzyme, pyridoxamine phosphate. Oxaloacetate is, in fact, the observed product from aspartate, representing a two-electron oxidation at the α-carbon. The coenzyme has undergone the corresponding necessary two-electron reduction from the aldehyde oxidation state to that of a primary amine in the PMP form. One might compare this half-reaction to that catalyzed by the flavoenzyme D-amino-acid oxidase, where the flavin coenzyme is reduced as the amino-acid substrate is oxidized (Chapter 11). As with the reduced flavoenzyme, the aspartate-transaminase–PMP complex is in the wrong oxidation state for reaction with another molecule of aspartate. The coenzyme must be reoxidized back to the aldehyde stage to complete a single catalytic cycle.

As the 2-$e^{\ominus}$–reduced PMP form, the coenzyme is however in the correct oxidation state for imine formation with the α-keto acid pyruvate, the second substrate of the enzyme. The second half-reaction then constitutes reversal of the steps of the first half-reaction, with pyridoxamine oxidized back to pyridoxal level and the pyruvate imine reduced by two electrons to alanine, adding a proton at the *si* face specifically to produce L-alanine. Expulsion of L-alanine from the product–coenzyme imine is initiated by attack of the active-site lysine's ε-amino group to yield free L-alanine and regenerate the resting-enzyme–coenzyme aldimine, ready for another catalytic cycle.

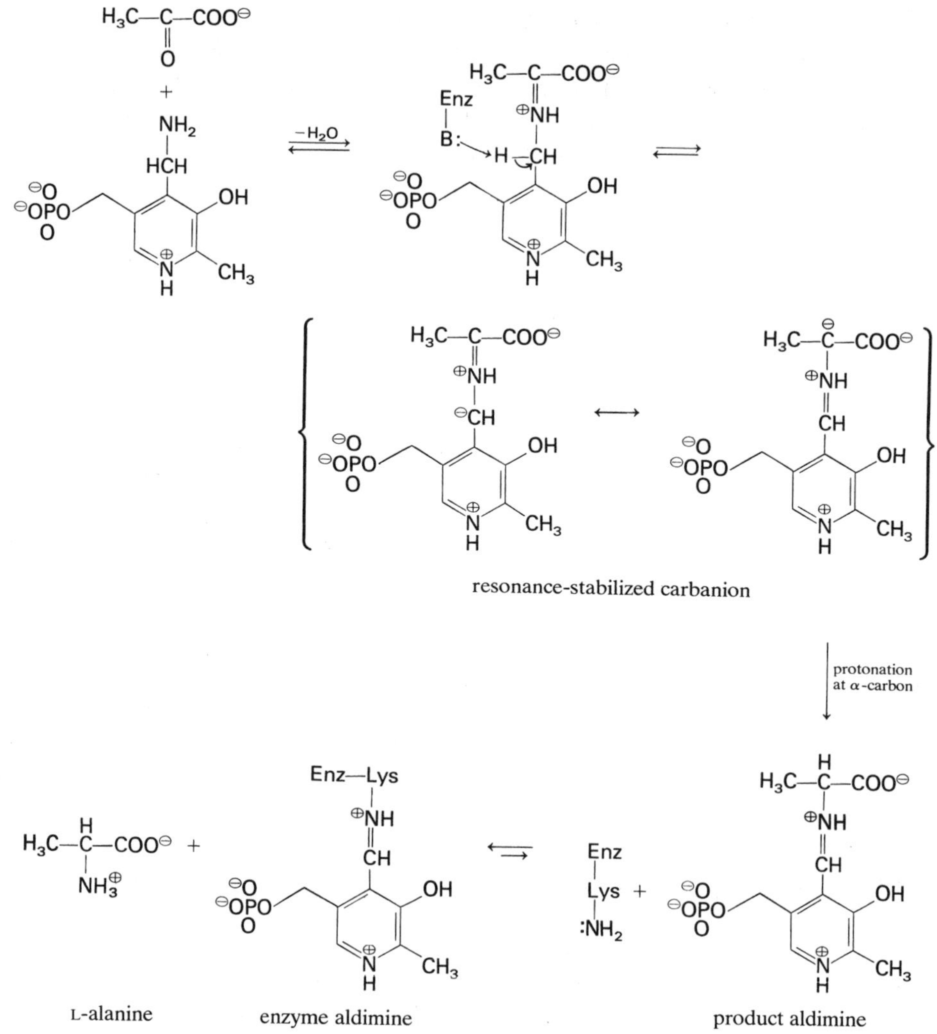

resonance-stabilized carbanion

protonation at α-carbon

L-alanine

enzyme aldimine

product aldimine

Note that this mechanistic formulation emphasizes that transaminases carry out redox catalysis. The two-step mechanism is substantiated by a wealth of data in addition to the kinetic pattern noted earlier. If the transaminase is incubated with L-aspartate alone, stoichiometric quantities of oxaloacetate are produced, and the enzyme–coenzyme complex now shows a λ_{max} at 332 nm with an ε of ~9,000, closely correlatable with the spectrum of authentic PMP, which absorbs maximally at 327 nm with $\varepsilon = 9{,}400$ (Fig. 24-1a). One can reconvert the enzyme–PMP back to starting enzyme by addition of pyruvate (Braunstein, 1973).

Direct evidence for the existence of the resonance-stabilized substrate–PLP α-carbanion has been obtained using a pseudosubstrate with the aspartate transaminase. Of the two β-hydroxy-L-aspartates, the *threo* diastereomer is a substrate undergoing transamination at an appreciable rate, but the *erythro* diastereomer does not undergo detectable catalytic turnover (Jenkins, 1961).

threo-β-hydroxy-L-aspartate

erythro-β-hydroxy-L-aspartate

Rather, addition of the *erythro* isomer to the enzyme–pyridoxal complex produces a slow absorbance change leading to a species absorbing maximally at 490 nm (Fig. 24-2). This long-wavelength absorbance is attributed to an extensively conjugated system and is assigned the quinoid structure shown as canonical form III of the stabilized carbanion earlier in this discussion. This quinoid form is the major contributor to the carbanion structure obtained on removal of the α-H of *erythro*-β-hydroxy-L-aspartate by the enzyme. This 490-nm form does not accumulate during transamination of the *threo* diastereomer of β-hydroxyaspartate, but rapid kinetic studies indicate its fleeting presence during normal catalysis also (Braunstein, 1973; Jenkins, 1961). With the *erythro* diastereomer of β-hydroxyaspartate, the hydroxyl group may provide steric hindrance to enzyme-promoted protonation at the coenzyme's methine carbon, thereby preventing completion of the transamination first half-reaction.

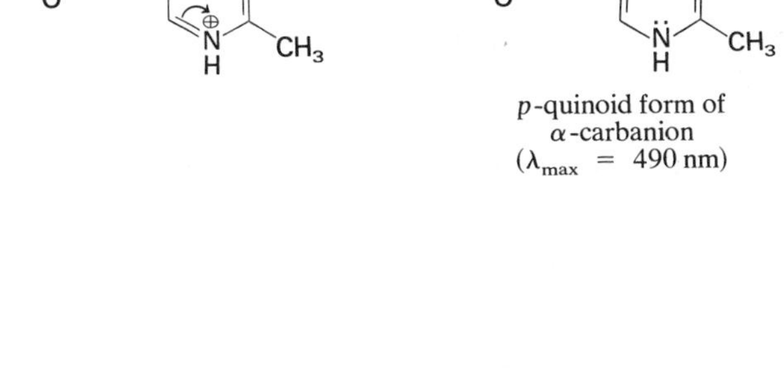

p-quinoid form of
α-carbanion
(λ_{max} = 490 nm)

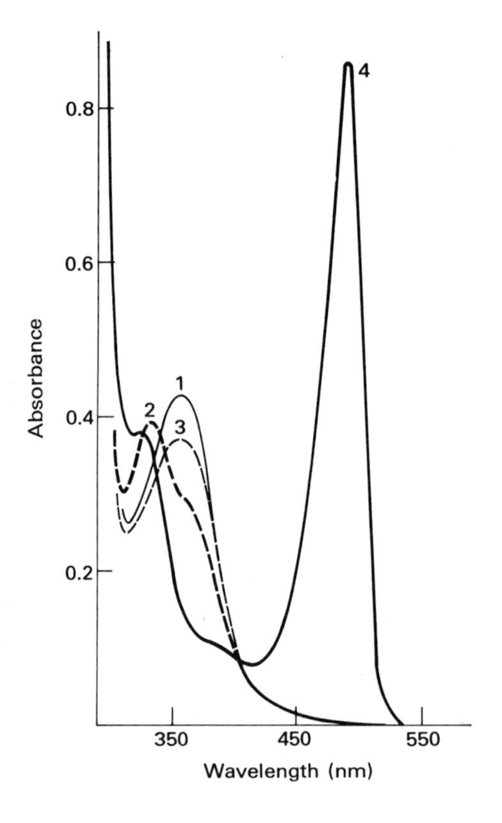

Figure 24-2
Spectra of the pig-heart aspartate transaminase in the presence of β-substituted aspartic acids in 0.1 M ethylene diaminetetraacetate buffer, pH 8.75, 25° C. Additions: (1) none; (2) 0.02 M *threo*-β-hydroxy-D,L-aspartate; (3) 0.02 M *threo*-β-hydroxy-D,L-aspartate + 0.004 M α-ketoglutarate; (4) 0.02 M *erythro*-hydroxy-D,L-aspartate. (From Jenkins, 1961, p. 1122)

With α-[^{2}H]-L-alanine as substrate and large quantities of enzyme–PLP, one can isolate the PMP produced and analyze it for deuterium content (Dunathan, 1971). The methylene carbon of the 2-e$^{\ominus}$–reduced coenzyme contains 0.04 atoms of deuterium, a small but significant amount of deuterium. This is consistent with utilization of a *single base* to abstract the α-hydrogen of the substrate and then protonate the carbanion of PMP. Partial transfer also implies that solvent competes when the ^{2}H is bound as Enz—B^{2}H$^{\oplus}$; 96% of the time Enz—B^{2}H$^{\oplus}$ dissociates and reprotonates with a solvent-derived hydrogen, and 4% of the time the transfer is direct. We noted in ¶19.C.8 that the transaminations are azaallylic isomerizations and that seven enzymes carrying out transaminations all show the same absolute stereochemistry of protonation at the methine carbon of the pyridoxal–amino-acid aldimine to yield the methylene group of the pyridoxamine-ketimine linkage: delivery of the proton from the *si* face of the planar intermediate carbanion so that it ends up in the pro*S* position in the pyridoxamine phosphate (Fig. 24-3).

PMP is not the only form of the coenzyme with maximal absorption in the 330-nm region. Subform γ of the cytoplasmic aspartate transaminase from pig

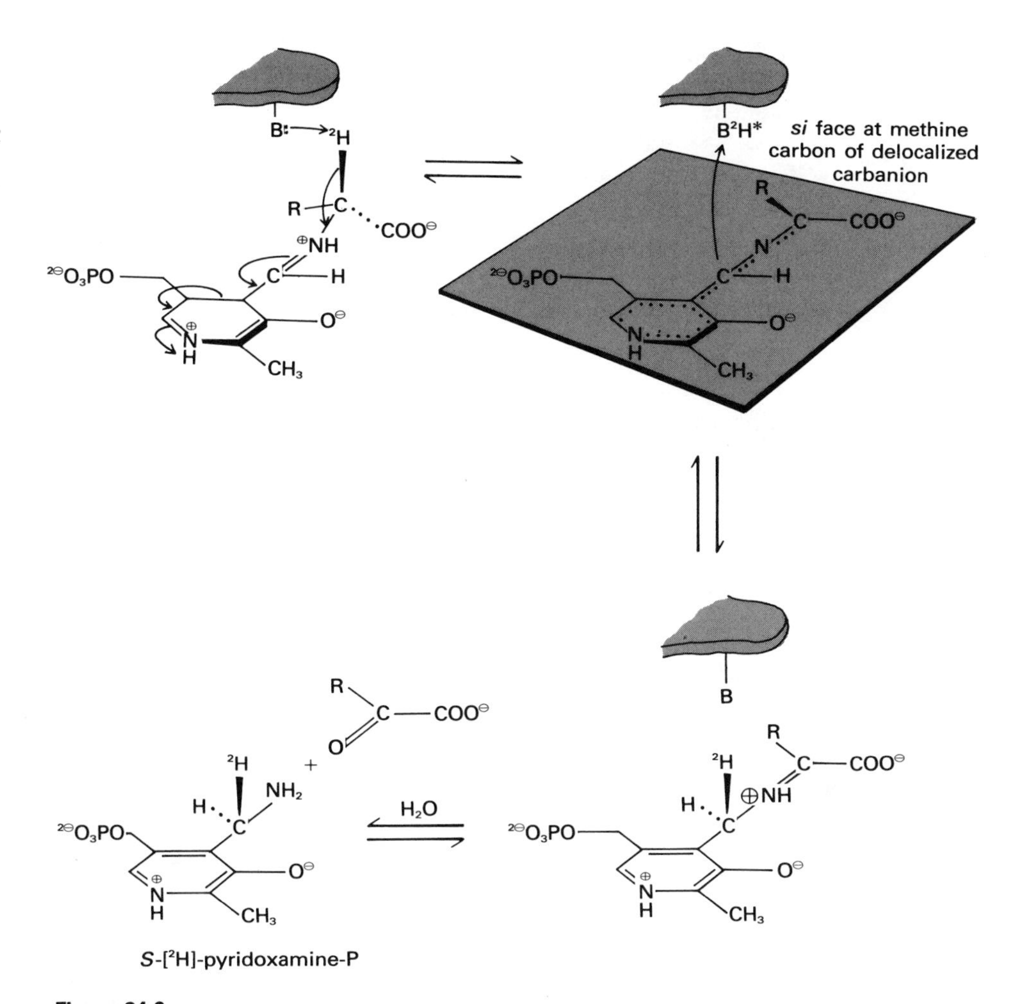

Figure 24-3
Removal of the α-deuterium from a 2*S*-2-[^{2}H]-amino-acid–PLP complex, and specific protonation at the trigonal *si* face of the delocalized carbanion to yield *S*-[^{2}H]-pyridoxamine-P.

heart has maximal absorption at 340 nm, but does not contain 2-e$^{\ominus}$–reduced coenzyme (Braunstein, 1973). A variety of other pyridoxal-P enzymes undergo transition from 430-nm peaks to peaks at 330 to 340 nm at high pH with loss of catalytic activity (Fig. 24-1b). Lowering the pH reverses the effect. On occasion, spectral changes have been correlated with formation of substituted aldimines (Anderson and Chang, 1965) by addition of nucleophilic side chains of amino acids at the active site to the aldimine carbon (where X is probably oxygen, nitrogen, or sulfur).

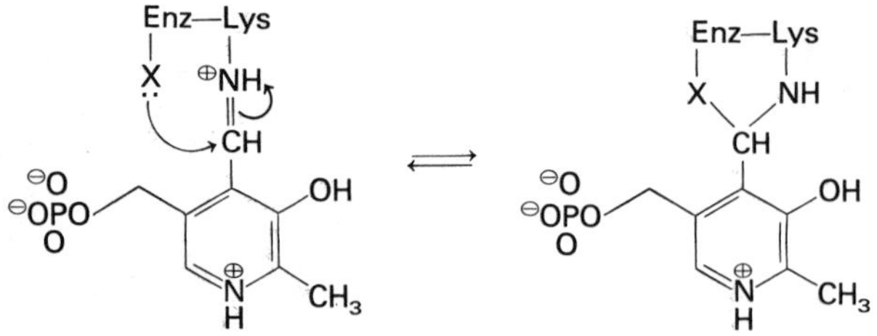

(λ_{max} ~ 430 nm)

inactive, substituted aldimine
(λ_{max} ~ 340 nm)

Indeed, free cysteine can be added to resolve pyridoxal-P holoenzymes into apoenzyme after dialysis of the substituted aldimine off the enzyme. This is a common preparative method for apoenzyme formation, although it is not universally successful.

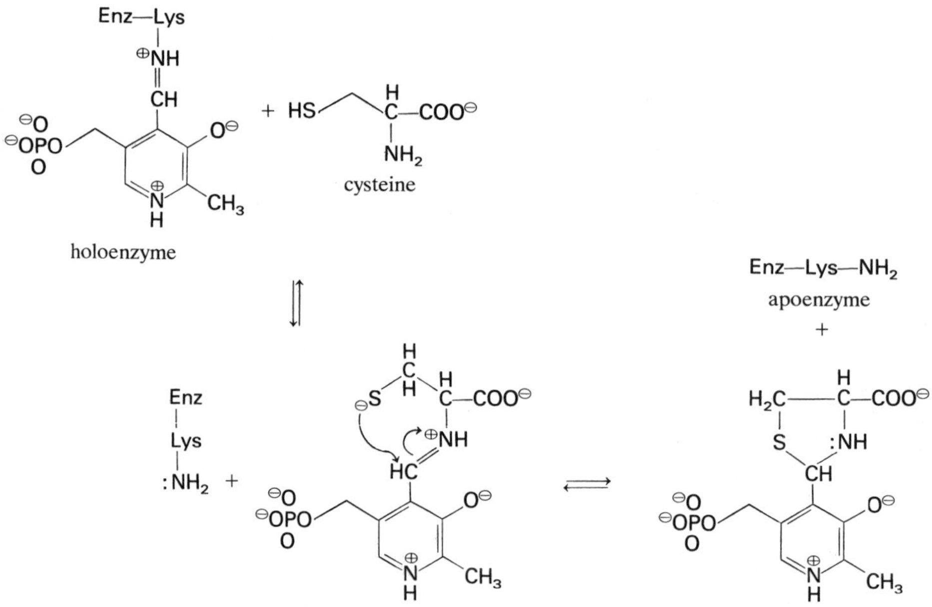

holoenzyme

cysteine

Enz—Lys—NH_2
apoenzyme
+

thiazolidine
derivative of PLP

Irreversible inactivation of aspatate transaminase by substrate analogues. Aspartate transaminase is irreversibly inactivated by L-cycloserine, a cyclic *O*-substituted hydroxamate, which is sterically cognate to an α-amino acid (Braunstein, 1973).

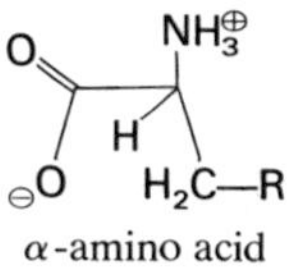

α-amino acid

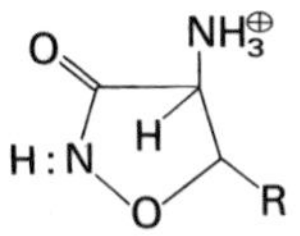

cycloserine (R = H) and derivatives

Only the pyridoxal form of the holoenzyme is susceptible to inactivation by cycloserine. Use of [^{14}C]-cycloserine produces radioactive acylated enzyme and PMP. This suggests binding at the active site and initial aldimine formation, followed by α-hydrogen abstraction and isomerization to the ketimine, with the coenzyme at the pyridoxamine stage. Attack by an active-site nucleophile at the carboxamate carbonyl would generate ring opening to an *O*-substituted hydroxylamine, now covalently linked to the active site through an acylation of the active-site nucleophile.

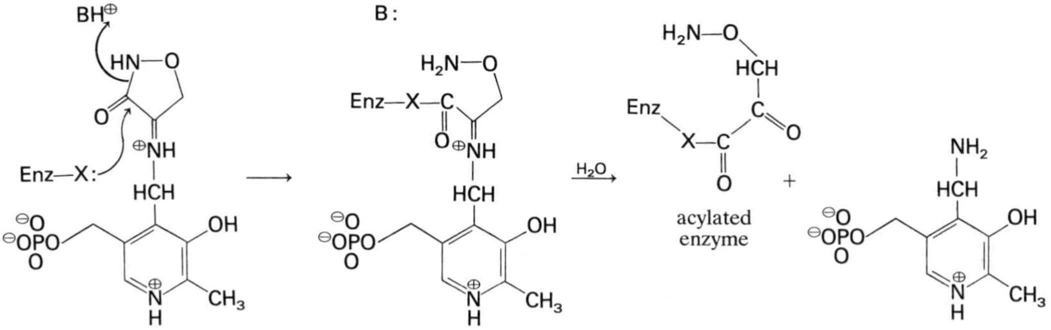

ketimine

acylated enzyme

PMP

Other pyridoxal-P–utilizing enzymes are also inactivated by cycloserine. One of these is alanine racemase of bacterial cell walls (¶24.A.2.a; Niehaus, 1967; Rando, 1975; Wang and Walsh, 1978), irreversibly modified by D-cycloserine.

A recent report has described a similar active-site–directed inactivation of aspartate transaminase by the olefinic amino acid 2-amino-3-butenoate (vinylglycine) (Rando, 1974*b*).

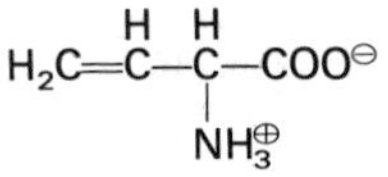

$$H_2C{=}CH{-}CH(NH_3^{\oplus}){-}COO^{\ominus}$$

vinylglycine

It is likely that this substrate undergoes transamination to the conjugated, β,γ-unsaturated ketimine, at which point it is activated for Michael addition by an active-site amino acid (the ε-amino group of the active-site lysine; Christen and Rando, 1978), again leading to covalent alkylation. A similar scheme appears to operate during inactivation of a *B. subtilis* D-specific D-aspartate transaminase, possibly involved in bacterial cell-wall synthesis, by the D-isomer of vinylglycine (Soper et al., 1977).

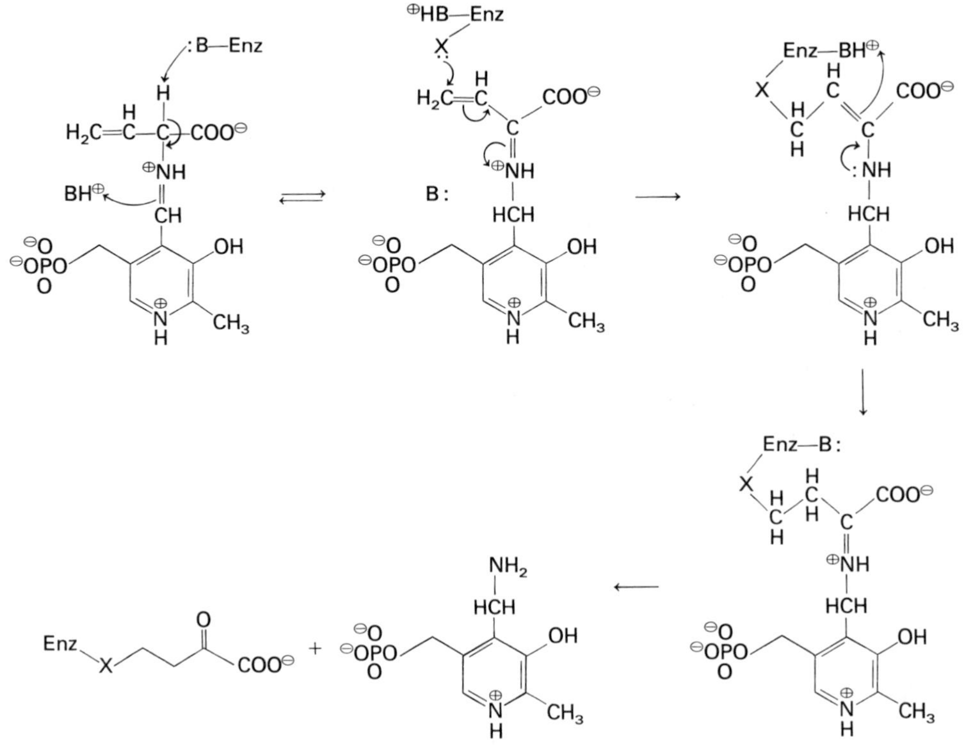

alkylated enzyme

PMP

24.A.1.b L-Alanine transaminase

L-Alanine transaminase, also purified from pig heart, oxidizes L-alanine to pyruvate while α-ketoglutarate is reduced to L-glutamate (Saier and Jenkins, 1967). In terms of transamination, this catalyst appears to function analogously to aspartate transaminase. It is mentioned here, then, only to constrast the susceptibilities of alanine transaminase and aspartate transaminase to the vinylglycine just noted and also to an acetylenic amino acid, 2-amino-4-pentynoate (propargylglycine).

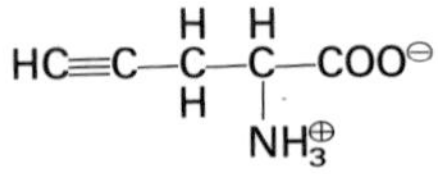

propargylglycine

The alanine transaminase is not inactivated by vinylglycine, which may be a slow substrate for this transaminase. If transamination does occur, then fast release of the conjugated keto-acid product could outcompete Michael attack in the alanine transaminase case. On the other hand, alanine transaminase is inactivated irreversibly by propargylglycine, whereas aspartate transminase is unaffected by this acetylenic amino acid* (Marcotte and Walsh, 1975). A clue to the susceptibility of the alanine transaminase is provided by the observation that addition of L-alanine to alanine transaminase in D_2O leads to rapid loss of the methyl signal as the enzyme catalyzes exchange of the methyl protons with D_2O (Babu and Johnston, 1974).

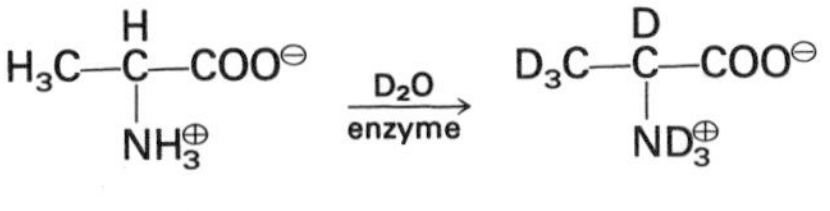

$$H_3C-\underset{NH_3^{\oplus}}{\overset{H}{C}}-COO^{\ominus} \xrightarrow[\text{enzyme}]{D_2O} D_3C-\underset{ND_3^{\oplus}}{\overset{D}{C}}-COO^{\ominus}$$

This exchange must proceed via an intermediate with carbanionic character at the β-carbon of alanine and might be explicable as follows.

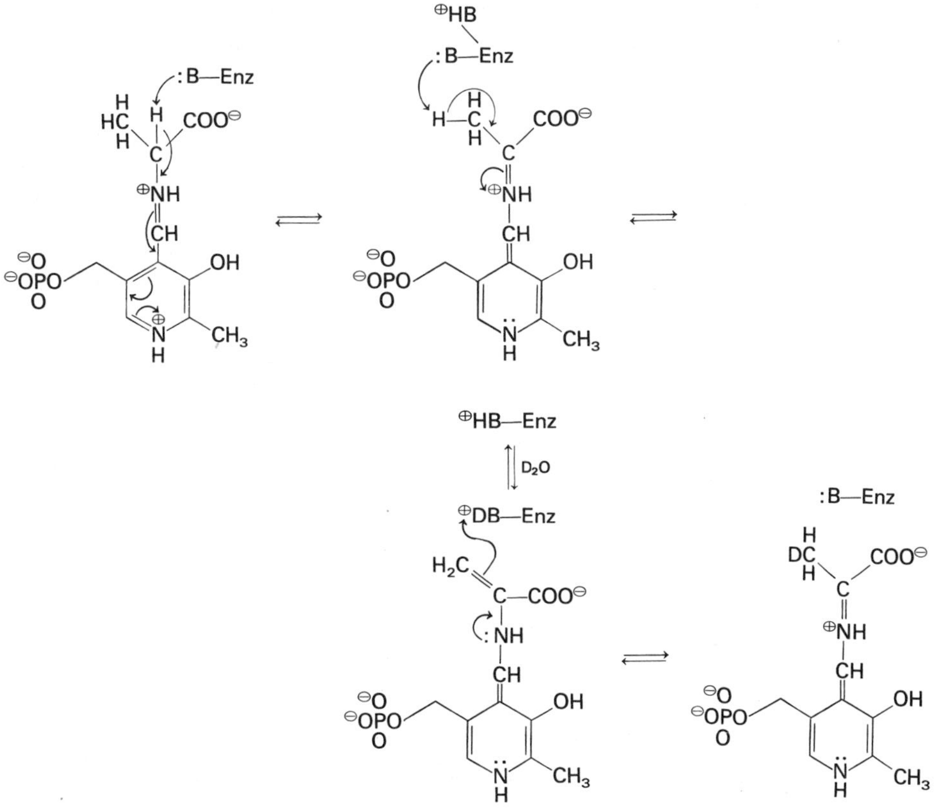

:B—Enz
⊕HB
:B—Enz
⊕HB—Enz
D_2O
⊕DB—Enz
:B—Enz

* When 3.2 M formate is added with L-propargylglycine to L-aspartate aminotransferase, active-site–directed inactivation does ensue (Tanase and Morino, 1976).

Because all three methyl hydrogens are sterically equivalent, the exchange with solvent deuterons, followed by conversion of the quinoid ketimine back to aldimine and then hydrolysis, will yield fully deuterated L-alanine. The key catalytic element is the ability of the enzyme to abstract a β-hydrogen from the quinoid form of the substrate carbanion. No such activity is seen with aspartate transaminase, nor is this catalytic capacity required for normal transamination by either enzyme.

However, the generation of a β-carbanionic species of a propargylglycine–coenzyme intermediate by alanine transaminase could explain why propargylglycine functions as a suicide substrate (Marcotte and Walsh, 1975). The initial α-carbanion is insulated from the acetylenic moiety by the β-methylene group, but a β-carbanion would be analogous to the inactivating reagents discussed earlier for thiolase (¶23.B.4) and β-hydroxydecanoylthioester dehydrase (¶19.C.7)—i.e., a propargylic anion capable of ready rearrangement to an allene. In this instance, the allene will be conjugated to the π-system of the coenzyme ketimine and susceptible to ready Michael attack and covalent modification of the enzyme.

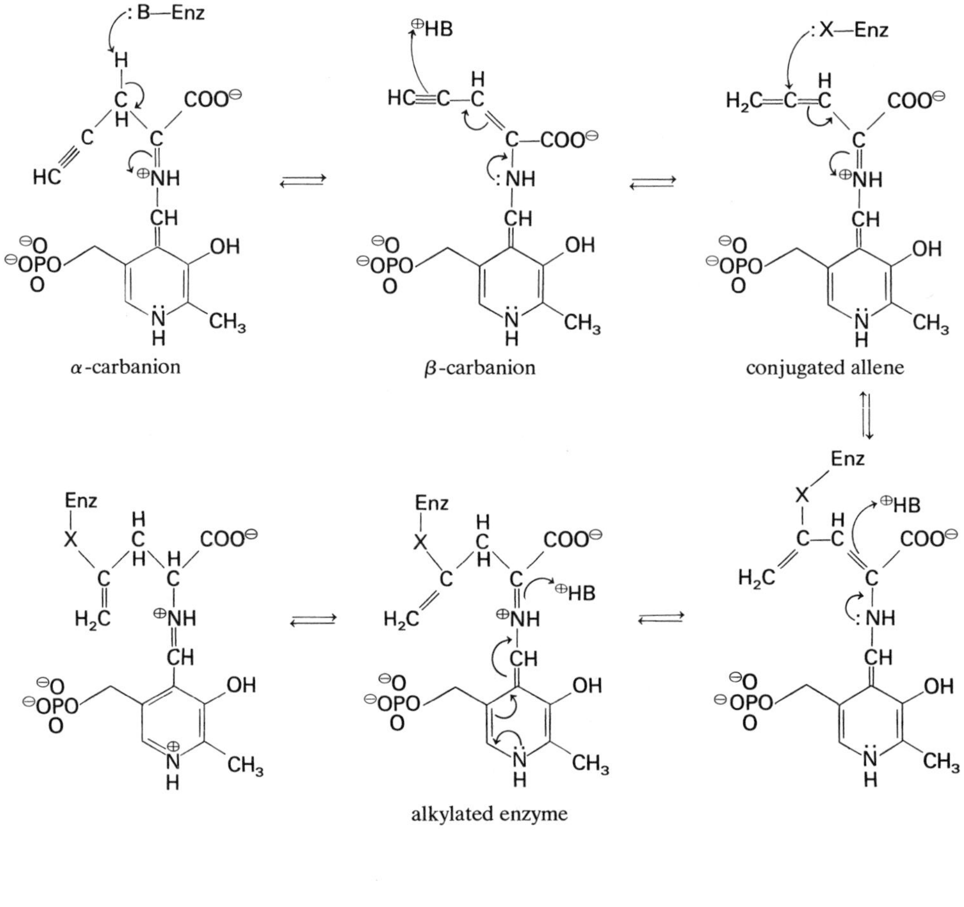

α-carbanion ⇌ β-carbanion ⇌ conjugated allene

⇅

alkylated enzyme

At this point, it may not be surprising to note that some of the other categories of pyridoxal-P enzymes we shall be discussing next carry out *adventitious transamination* reactions occasionally (Snell and DiMari, 1970; Braunstein, 1973). Because this rare abnormal event leaves the coenzyme in the reduced (PMP) form, the process is often detected as a slow inactivation of the enzyme during catalysis with the normal substrate. If transamination has been the culprit, addition of an α-keto acid reactivates the holoenzyme by converting the coenzyme back to its oxidized, active (PLP) form. These abortive transaminations occur slowly with such enzymes as arginine racemase, aspartate β-decarboxylase, serine hydroxymethylase, and kyneurinase.

The detection of catalysis of an abnormal side reaction raises the question of how a given holoenzyme imposes specificity on the pyridoxal chemistry, both in the bond to be labilized at the α-carbon and (as we shall note later) in the rate of protonation of the stabilized carbanionic intermediates at the α-carbon (so that amino acid is released) or at the pyridoxamine methylene carbon (so that α-keto acid is released). Dunathan (1971) has suggested that free rotation around the substrate C^{α}—N bond is suppressed (frozen out) in the initial substrate–coenzyme aldimine complex, and that a given apoenzyme will specifically orient one of three α-substituents (by ionic interactions, and by positioning of active-site bases, steric bulk of substrates, etc.) perpendicular to the plane of the aldimine π-system. Then this particular σ-bond, perpendicular to the plane of the imine system, will be cleaved in a facilitated manner with maximal gain in resonance energy and extended conjugation.

Thus, C—H bond breakage will occur from conformer 1 in transaminases and racemases (Fig. 24-4), whereas C—$COO^{\ominus}$ cleavage occurs from 2 in decarboxylases, and C^{α}—C^{β} cleavage occurs from 3 in the aldol cleavage catalyzed by serine hydroxymethylase (Schirch and Mason, 1962, 1963).

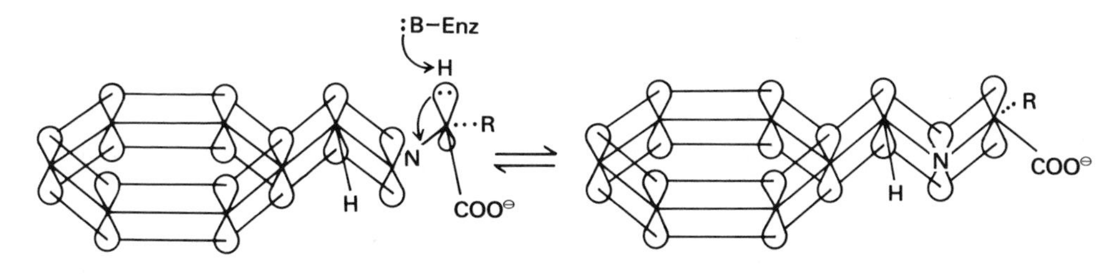

Figure 24-4
Orientation of one particluar α-substituent for bond cleavage, through orientation of that σ-bond perpendicular to the plane of the imine system. The example shown here is conformer 1, oriented for cleavage of the bond to the α-hydrogen.

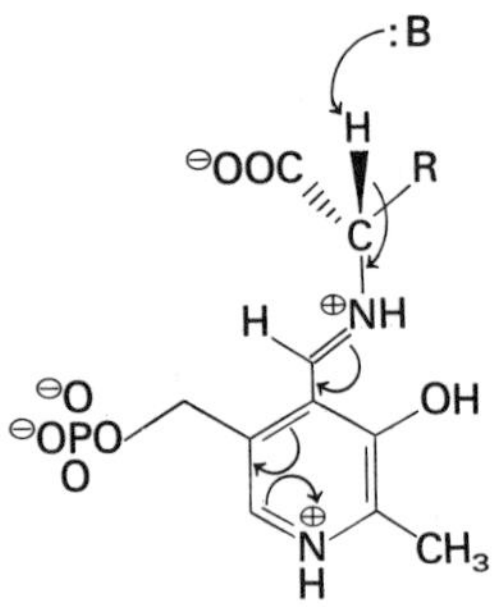

:B
H
$^{\ominus}$OOC R
C
$^{\oplus}$NH
H
$^{\ominus}$O
$^{\ominus}$OPO
O
OH
$\overset{\oplus}{N}$
H
CH_3
(1)

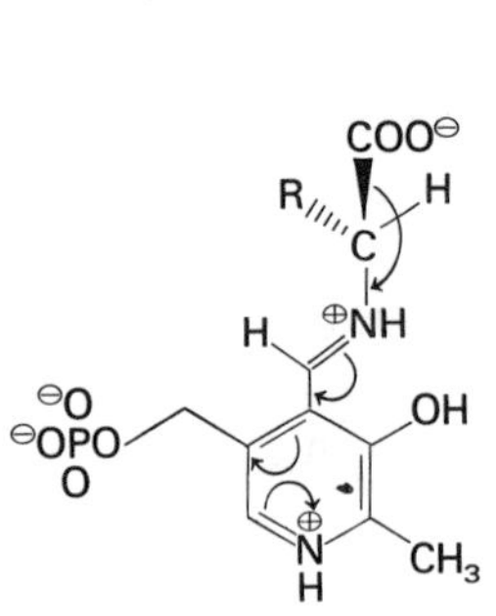

COO$^{\ominus}$
R H
C
$^{\oplus}$NH
H
$^{\ominus}$O
$^{\ominus}$OPO
O
OH
$\overset{\oplus}{N}$
H
CH_3
(2)

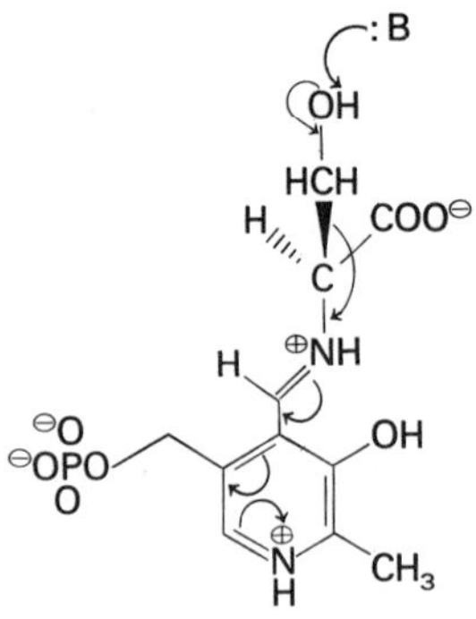

:B
OH
HCH
H COO$^{\ominus}$
C
$^{\oplus}$NH
H
$^{\ominus}$O
$^{\ominus}$OPO
O
OH
$\overset{\oplus}{N}$
H
CH_3
(3)

24.A.1.c γ-Aminobutyrate transaminase

An example of a pyridoxal-P–dependent transaminase that oxidizes amino groups distant from the carboxylate functionality is provided by γ-aminobutyrate (GABA) transaminase. The C_4 amino acid is produced by enzymatic decarboxylation of the α-carboxylate of glutamate (¶24.A.3.a).

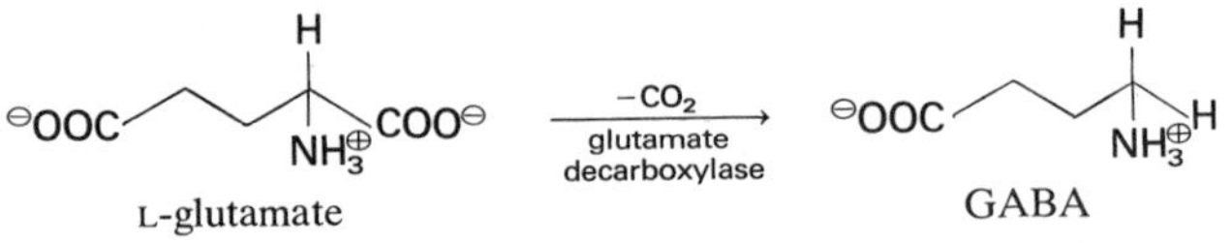

H
$^{\ominus}$OOC $NH_3^{\oplus}$ COO$^{\ominus}$
L-glutamate
$-CO_2$
glutamate decarboxylase
H
$^{\ominus}$OOC $NH_3^{\oplus}$ H
GABA

In animals, GABA appears to function as a (inhibitory) neurotransmitter substance in the central nervous system, and its concentration is regulated by both glutamate decarboxylase and GABA transaminase, which removes it to succinic semialdehyde while pyruvate is reductively aminated to L-alanine. The succinic semialdehyde is separately oxidized to the familiar metabolite succinate.

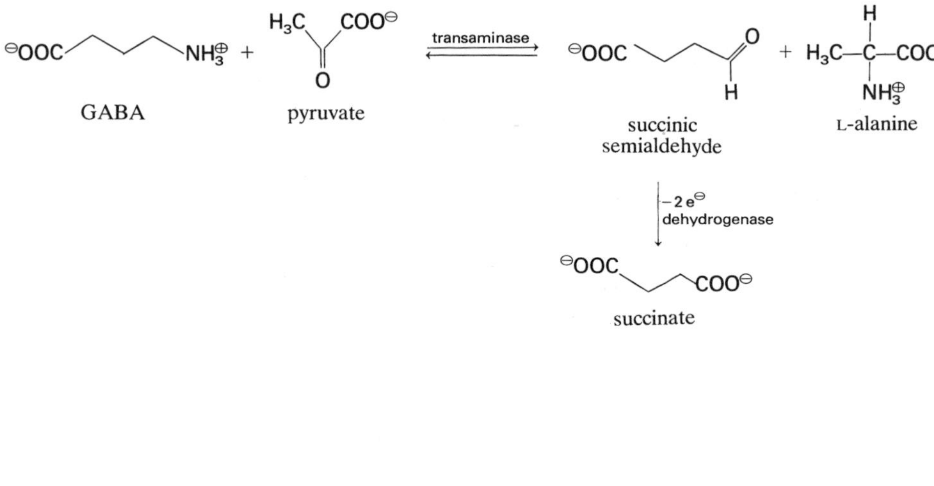

$^{\ominus}$OOC $NH_3^{\oplus}$ +
GABA
H_3C COO$^{\ominus}$
O
pyruvate
transaminase
$^{\ominus}$OOC O
H
succinic semialdehyde
+ H_3C—C—COO$^{\ominus}$
H
$NH_3^{\oplus}$
L-alanine

$-2\,e^{\ominus}$
dehydrogenase

$^{\ominus}$OOC COO$^{\ominus}$
succinate

A variety of mechanism-based inactivators of brain GABA transaminase have been reported, among them ethanolamine *O*-sulfate (Fowler and John, 1972), γ-vinyl-GABA and γ-acetylenic GABA (Lippert et al., 1977), and the natural product gabaculine (Rando and Bangerter, 1976, 1977; Rando, 1977).

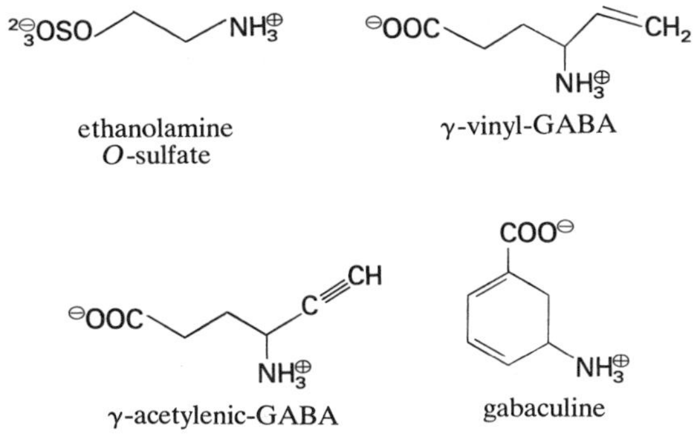

ethanolamine *O*-sulfate

γ-vinyl-GABA

γ-acetylenic-GABA

gabaculine

Ethanolamine *O*-sulfate probably inhibits after α,β-elimination of sulfate. The γ-vinyl- and γ-acetylenic-GABA can offer bound, electrophilic conjugated systems after enzymatic transamination; γ-acetylenic-GABA can also undergo propargylic rearrangement as just outlined for propargylglycine and alanine transaminase. Gabaculine is a rigid analogue of the extended conformation of GABA. It has a K_i for mouse-brain transaminase of 5.8×10^{-7} M, three orders of magnitude lower than the K_m for GABA (Rando and Bangerter, 1976). Irreversible inactivation ensues, apparently by aromatization of the transaminated gabaculine still in imine linkage to the PMP form of the coenzyme (Rando, 1977).

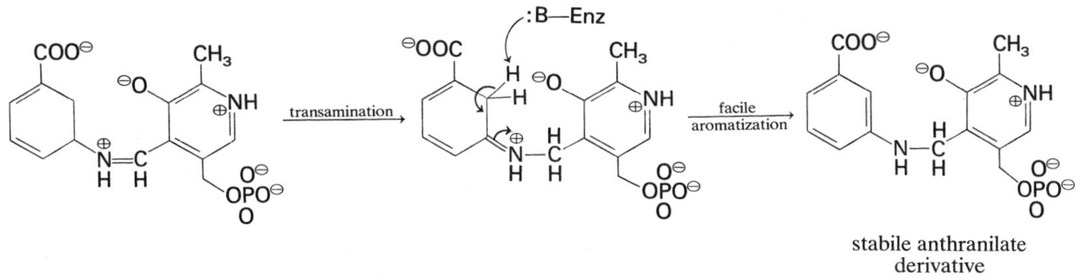

transamination

facile aromatization

stabile anthranilate derivative

This proton shift generates a *meta*-anthranilate skeleton, now in secondary amine linkage to the pyridoxamine-P at the active site. The amine linkage, unlike the normal substrate and product *imine* linkages, is *stable to hydrolysis*, and this mechanism is a unique variant for a suicide substrate.

24.A.2 Racemases

In Section IV we discussed enzyme-catalyzed racemizations as 1,1-proton shifts. We noted then that there are at least three known examples of enzymatic formation of a racemic mixture of amino acids by equilibration of configuration at the α-carbon that do not require pyridoxal-P. Two of these involve racemization of proline and epimerization of 4-hydroxyproline, the two common imino acids. The third is the ATP-dependent racemization of phenylalanine, an integral step in the synthesis of the cyclic decapeptide antibiotic gramicidin S, a synthesis that involves carboxyl activation and racemization (Chapter 19).

It is generally assumed that the enzyme activities responsible for racemizing other primary α-amino acids require pyridoxal-P. This expectation is confirmed for homogeneous enzymes racemizing alanine and arginine, but these represent the only two well-documented cases of purified racemases as pyridoxal-P–dependent holoenzymes (Snell and DiMari, 1970; Adams, 1976; Wang and Walsh, 1978). In fact, the catalytic details for pyridoxal-P–dependent racemases are not well understood, although a reasonable scheme is easily envisioned. Thus, for conversion of D-alanine into the racemic D,L-mixture, one imagines initial aldimine formation, with abstraction of the α-hydrogen to yield the stabilized carbanion postulated in transaminase action. Now racemization will occur if the enzyme can deliver a proton back to either face of the planar imine anionic intermediate, a sterically random return of hydrogen to the α-carbon.

D-alanine aldimine ⇌ planar stabilized carbanion ⇌ L-alanine aldimine

It is not yet clear how much transfer of the abstracted α-H occurs back to the opposite face, or whether one or two bases are involved at the active site.

24.A.2.a Alanine racemase

Alanine racemase has been purified to homogeneity from *Pseudomonas putida* (Adams 1976). This enzyme, on the basis of a molecular weight of 60,000, shows a high turnover number (about 4×10^3 racemizations sec^{-1}) but has a surprisingly high K_m of 30 mM for L-alanine. (The K_m for the D-isomer was not reported.) As expected, the equilibrium constant is unity when determined experimentally. This enzyme is probably the major supplier of the D-alanine used in cell walls of most bacteria. Indeed, to date, amino-acid racemases appear to be restricted to procaryotes. This distribution corresponds with the lack of any known metabolic utilization of D-amino acids in higher organisms, save their oxidation (and detoxification) by the flavoprotein D-amino-acid oxidases.

In view of its important role in bacterial physiology, the alanine racemase is a reasonable target for antibiotics. The partially purified *E. coli* enzyme is susceptible to inactivation by cycloserine as noted earlier (Niehaus, 1967; Rando, 1975). It also is blocked irreversibly (Wang and Walsh, 1978) by exposure to either the D- or L-isomers of β-fluoroalanine (Kahan et al., 1975), β-chloroalanine (Kaczorowski, Shaw, Laura, and Walsh, 1975), or trifluoroalanine (Silverman and Abeles, 1976).

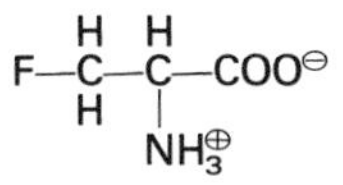

$$F-CH_2-CH(NH_3^{\oplus})-COO^{\ominus}$$

L-β-fluoroalanine

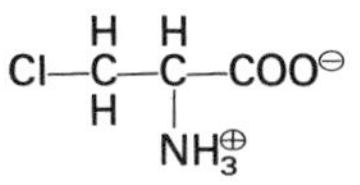

$$Cl-CH_2-CH(NH_3^{\oplus})-COO^{\ominus}$$

L-β-chloroalanine

$$F_3C-CH(NH_3^{\oplus})-COO^{\ominus}$$

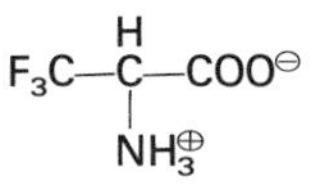

trifluoro-L-alanine

24.A.2.b Arginine racemase

The other well-characterized pyridoxal-P–dependent racemase is the arginine racemase from *Pseudomonas graveolans.* It is a tetramer of 170,000 mol wt with one coenzyme per subunit, exhibiting a turnover number of about $10^3\ sec^{-1}$ (Yorifugi et al., 1971). The K_m for D-arginine is reasonable (1 mM), and the coenzyme is tightly bound ($K_m = 4 \times 10^{-7}$ M); binding energy of interaction is provided by the imine link to the active-site lysine and by electrostatic interactions between coenzyme and enzyme. The pH optimum for racemization of arginine is 10.0, suggesting that the guanidinium group (pK ~11 in solution) may be deprotonated at the active site. Actually, lysine is racemized at a relative rate 100% that of arginine, and the 2,5-diaminopentanoate (ornithine) at 86% the arginine rate. During ornithine racemization, the enzyme undergoes slow inactivation due to an unwanted transamination of ornithine to the α-keto acid, which then cyclizes to Δ^1-pyrroline carboxylate, drawing the equilibrium (as written below) to the right and leaving holoenzyme in the inactive pyridoxamine form (Fig. 24-5).

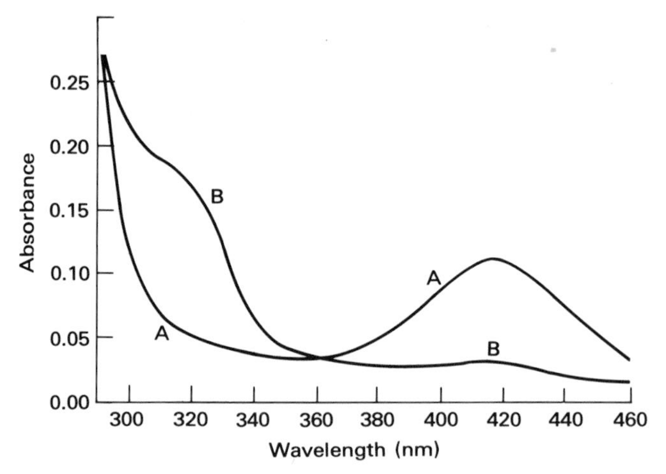

Figure 24-5
Spectral shifts on addition of L-ornithine to arginine racemase. Curve A was obtained with a 0.059% solution of holoenzyme in 0.01 M potassium phosphate buffer, pH 7.3. For curve B, the enzyme was incubated with 1.8×10^{-3} M L-ornithine in 0.02 M glycine–KCl–KOH buffer, pH 10.0, at 37° C for 80 min. (From Yorifugi et al., 1971, p. 5095)

$H_3\overset{\oplus}{N}$–(CH₂)₃–CH($NH_3^{\oplus}$)–$COO^{\ominus}$ + Enz—PLP ⇌ $H_2\ddot{N}$–(CH₂)₃–C(=O)–$COO^{\ominus}$ + Enz—PMP (inactive) ⇌ (cyclic carbinolamine: N–H ring with OH and $COO^{\ominus}$) $\underset{-H_2O}{\rightleftharpoons}$ Δ^1-pyrroline carboxylate

Addition of pyruvate or oxaloacetate reactivates the enzyme. The pH optimum for ornithine-pyruvate transaminations is pH 11, whereas the optimum for ornithine racemization is between pH 6.5 and pH 9. A calculation of the relative V_{max} values for racemization and transamination indicates that the spurious oxidation of ornithine occurs only 4×10^{-5} times as often as its racemization; this is the frequency of transaminative error.

24.A.3 Decarboxylations

At present, specific decarboxylases are known for at least half of the twenty common amino acids that remove the α-carboxyl group as CO_2 (Boeker and Snell, 1972). There is one characterized enzyme responsible for decarboxylating the β-carboxylate of aspartate to yield L-alanine and CO_2 (there is a distinct α-aspartate decarboxylase for synthesizing β-alanine) (Nishimura et al., 1962; Novogrodsky et al., 1963; Williamson, 1977). Additionally, there are enzymes that catalyze initial condensation of an amino acid with an electrophile and subsequent decarboxylation of the product. Almost all these enzyme-catalyzed decarboxylations require tightly-bound pyridoxal-P as coenzyme. The two exceptions are a *Lactobacillus* histidine decarboxylase (Riley and Snell, 1968; Lane and Snell, 1976; Lane et al., 1976) and *E. coli* *S*-adenosylmethionine decarboxylase (Wickner et al., 1970; Tabor and Tabor, 1976). These two enzymes use the same chemical mechanism but with an altered prosthetic group (in this instance, a covalently bound pyruvate) as electron sink (¶24.A.4).

The physiological role of amino-acid α-decarboxylases in bacteria is not fully elucidated, but inducible decarboxylases for glutamate, arginine, and lysine have been purified to homogeneity and their catalytic roles examined (Boeker and Snell, 1972). In mammals the decarboxylases are responsible for the production of a number of pharmacologically active amines. Glutamate decarboxylase produces γ-aminobutyrate, a neurotransmitter with inhibitory properties toward nerve firing. Histidine decarboxylase yields histamine, a molecule active in promoting gastric secretion as well as allergic and other hypersensitivity reactions; it is also active as a regulator of peripheral blood circulation. A major purified mammalian decarboxylase is dopa (dihydroxyphenylalanine) decarboxylase (Christenson et al., 1970; Boeker and Snell, 1972, p. 221 ff). It decarboxylates both 3,4-dihydroxyphenylalanine (produced from tyrosine hydroxylase action; see Chapter 12) to 3,4-dihydroxyphenylethylamine (dopamine), and also 5-hydroxytryptophan (from tryptophan hydroxylase action) to 5-hydroxytryptamine (serotonin). Dopamine is a precursor to epinephrine and norepinephrine; serotonin is neuroactive itself, especially in the central nervous system.

L-dopa —Enz—PLP→ dopamine + CO_2

5-hydroxytryptophan —Enz—PLP→ serotonin + CO_2

Decarboxylases have been reported for the nonphysiological, prochiral aminomalonate.

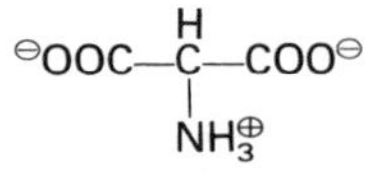

$$^{\ominus}OOC-\overset{H}{\underset{NH_3^{\oplus}}{C}}-COO^{\ominus}$$

aminomalonate

However, on purification, these enzymes turn out to be either aspartate β-decarboxylase (producing only S-[^{3}H]-glycine in ^3H_2O) or serine hydroxymethylase (producing both R- and S-[^{3}H]-glycine in ^3H_2O) (Palekar et al., 1970, 1973).

24.A.3.a Glutamate decarboxylase

The bacterial glutamate decarboxylase (Boeker and Snell, 1972), active in the pH region of 3 to 5, displays the 415–430-nm absorbance typical of a hydrogen-bonded aldimine at its pH optimum, but at pH 6 the absorbance is shifted to 340 nm, consistent with a substituted aldimine arising from nucleophilic attack by some active-site residue on the aldimine. The 340-nm form of the decarboxylase is inactive and is not reduced by NaB^3H_4, in contrast to the easily reduced and catalytically active aldimine.

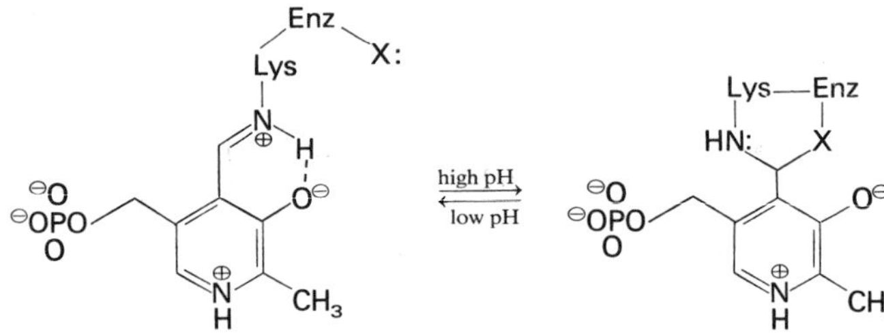

Enz X: Lys N H O$^{\ominus}$ $^{\ominus}$O $^{\ominus}$OPO O N H CH$_3$

high pH / low pH

Lys—Enz HN: X $^{\ominus}$O $^{\ominus}$OPO O O$^{\ominus}$ N H CH$_3$

active form (λ_{max} = 420 nm)

inactive form (λ_{max} = 340 nm)

There is internal hydrogen bonding of phenolate and iminium nitrogen in the active form.

In contrast, the β-aspartate decarboxylase of *Alcaligenes faecalis* shows a pH-independent absorption maximum at 360 nm in the native, active enzyme. This linkage, reducible by borohydride, appears to be the aldimine without internal hydrogen bonding to the phenoxide substituent on the pyridine ring (Bailey and Dempsy, 1967).

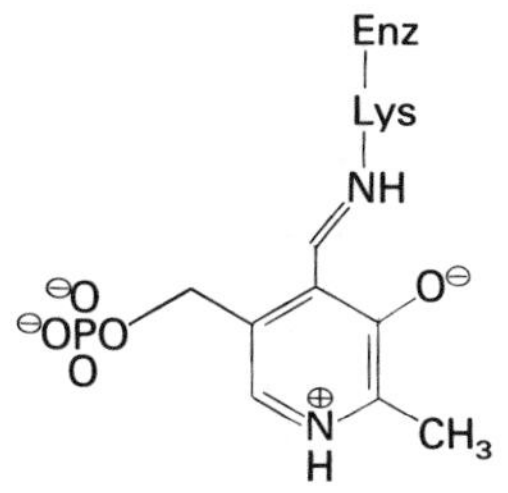

Enz
Lys
NH
$^{\ominus}$O
$^{\ominus}$OPO
O
O$^{\ominus}$
N$^{\oplus}$
H
CH_3

In our discussion of decarboxylations in Chapter 21, we noted that acceleration by enzymes could arise if the carbanionic species arising during the transition state could be stabilized. We noted how charge delocalization is used, especially in TPP-mediated decarboxylations of α-keto acids. In facilitating α-amino-acid decarboxylations, pyridoxal-P compares favorably with TPP as an electron sink.

Thus, one can write the following sequence.

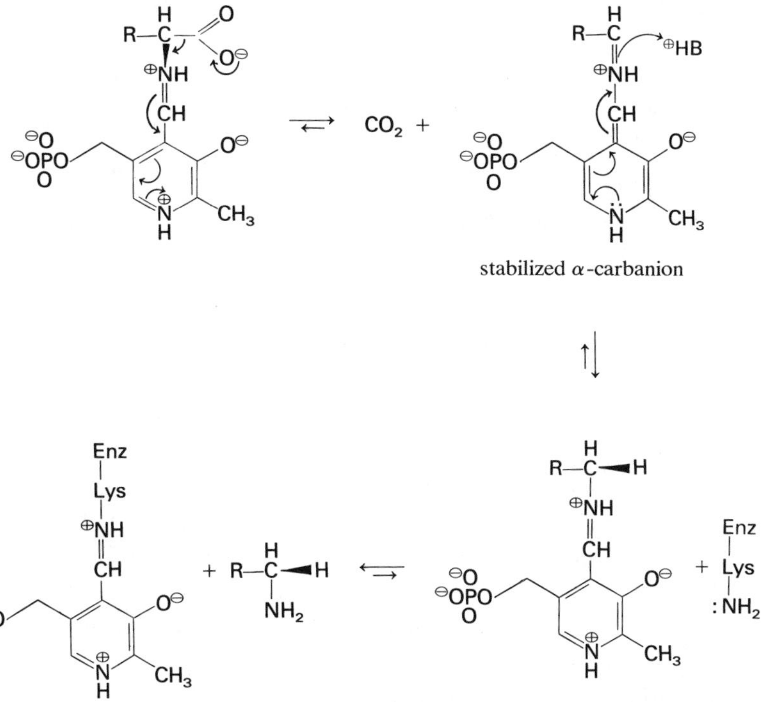

H
R—C
O
O$^{\ominus}$
$^{\oplus}$NH
CH
$^{\ominus}$O
$^{\ominus}$OPO
O
O$^{\ominus}$
N$^{\oplus}$
H
CH_3

$\rightleftharpoons$ CO_2 +

H
R—C
$^{\oplus}$HB
$^{\oplus}$NH
CH
$^{\ominus}$O
$^{\ominus}$OPO
O
O$^{\ominus}$
N
H
CH_3

stabilized α-carbanion

$\uparrow\downarrow$

Enz
Lys
$^{\oplus}$NH
CH
$^{\ominus}$O
$^{\ominus}$OPO
O
O$^{\ominus}$
N$^{\oplus}$
H
CH_3

+ R—C(H)(H)—NH_2 $\leftrightarrows$

H
R—C—H
$^{\oplus}$NH
CH
$^{\ominus}$O
$^{\ominus}$OPO
O
O$^{\ominus}$
N$^{\oplus}$
H
CH_3

+ Enz—Lys—:NH_2

The most distinguishing feature of this scheme is the implication that *the amino-acid α-hydrogen is not abstracted as a proton* in one of the initial steps of catalysis. This is in distinction to the mechanisms proposed for enzymatic transamination and racemization, as well as the mechanisms we shall discuss for most reactions at the β- and γ-carbons of amino acids (except serine hydroxymethylase). Experimental evidence confirms that the substrate α-H need not be removed for decarboxylation to the amine. Note that the primary amine product is prochiral. In D_2O, the product from decarboxylation of glutamate, tyrosine, or lysine has one atom of deuterium, yielding a chiral product. In the tyrosine decarboxylase reaction, it has been determined that the solvent deuteron is introduced in the same relative configuration as the departed carboxylate—i.e., with retention (Belleau and Burba, 1960). Similarly, α-[^{2}H]-amino acids yield α-[$^2H_1$]-amine without exchange with solvent H_2O. A final indication that α-H abstraction is nonessential comes from the observation that dopa decarboxylase uses α-methyl substrate analogues at 1% the normal V_{max} (Boeker and Snell, 1972). This reduction in rate by α-methyl substitution means that α-methyldopa can effectively block dopa decarboxylation in vivo. This methyl analogue (known commercially as Aldomet) is a clinically important agent for control of renal hypertension by virtue of this decarboxylase inhibition, although the major pharmacological effects may come from the enzymatically decarboxylated α-methyldopamine.

$$(HO)_2C_6H_3{-}CH_2{-}C(CH_3)(NH_3^{\oplus}){-}COO^{\ominus} \xrightarrow{\text{Enz}} (HO)_2C_6H_3{-}CH_2{-}CH(CH_3){-}NH_3^{\oplus} + CO_2$$

α-methyldopa α-methyldopamine

24.A.3.b α-Dialkylamino-acid transaminase

We noted earlier that enzymes using pyridoxal-P as coenzyme probably control the course of reaction by optimizing the geometry for labilization of the desired α-carbon substituent (R, H, or $COO^{\ominus}$). Bailey and colleagues cleverly showed that the orientation of the substrate does indeed determine whether the C^{α}—H or the C^{α}—$COO^{\ominus}$ bond is labilized (Bailey and Dempsy, 1967). They examined a bacterial α-dialkylamino-acid transaminase that uses α-aminoisobutyrate as substrate, decarboxylating it to the imine and releasing the aldehyde and enzyme–PMP. Pyruvate is the second substrate, undergoing transamination to L-alanine and regenerating the active pyridoxal-P form of the holoenzyme.

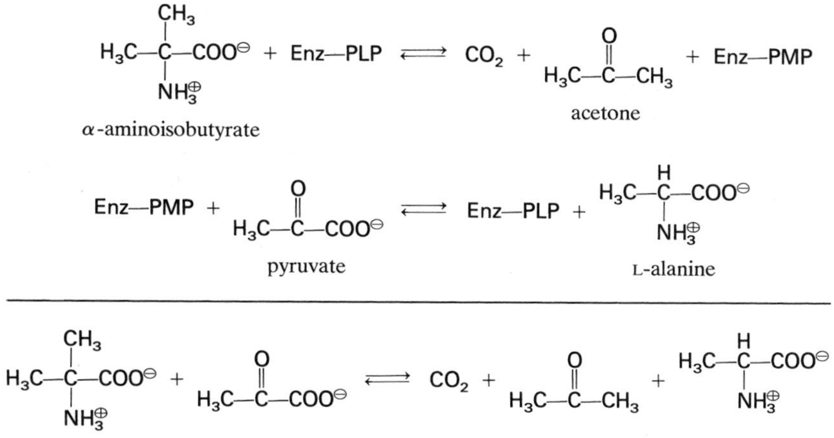

$$H_3C-C(CH_3)(NH_3^{\oplus})-COO^{\ominus} + \text{Enz—PLP} \rightleftharpoons CO_2 + H_3C-C(=O)-CH_3 + \text{Enz—PMP} \quad (1)$$

α-aminoisobutyrate acetone

$$\text{Enz—PMP} + H_3C-C(=O)-COO^{\ominus} \rightleftharpoons \text{Enz—PLP} + H_3C-CH(NH_3^{\oplus})-COO^{\ominus} \quad (2)$$

pyruvate L-alanine

$$H_3C-C(CH_3)(NH_3^{\oplus})-COO^{\ominus} + H_3C-C(=O)-COO^{\ominus} \rightleftharpoons CO_2 + H_3C-C(=O)-CH_3 + H_3C-CH(NH_3^{\oplus})-COO^{\ominus} \quad (1 + 2)$$

The enzyme will accept both D- and L-alanine as slow substrate; 1-[^{14}C]-D-alanine undergoes the decarboxylation–transamination sequence (C^{α}—$COO^{\ominus}$ breakage), whereas 1-[^{14}C]-L-alanine undergoes only normal transamination (C^{α}—H breakage).

1-[^{14}C]-D-Alanine + Pyruvate $\rightleftharpoons$ $^{14}CO_2$ + Acetaldehyde + L-Alanine
1-[^{14}C]-L-Alanine + Pyruvate $\rightleftharpoons$ [^{14}C]-Pyruvate + L-Alanine

24.A.3.c δ-Aminolevulinate synthetase

An additional variant on α-decarboxylase activity is shown by enzymes such as δ-aminolevulinate synthetase, which catalyzes a condensation prior to removal of CO_2 (Jordan and Shemin, 1972). We noted in Chapter 23 that δ-aminolevulinate is the direct acyclic precursor of pyrroles, which are incorporated into heme, corrin, and chlorophyll molecules. The synthetase may be an important control enzyme for heme biosynthesis.

The synthetase condenses glycine and succinyl-CoA to form δ-aminolevulinate, coenzyme A, and CO_2.

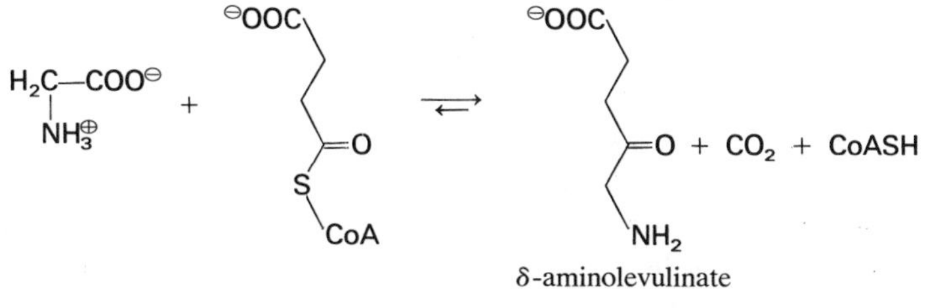

$$H_2C(NH_3^{\oplus})-COO^{\ominus} + {}^{\ominus}OOC-CH_2-CH_2-C(=O)-S-CoA \rightleftharpoons {}^{\ominus}OOC-CH_2-CH_2-C(=O)-CH_2-NH_2 + CO_2 + CoASH$$

δ-aminolevulinate

The reaction involves attack of the α-carbon of glycine on the carbonyl of the thioester of succinyl-CoA with displacement of the $CoAS^{\ominus}$ thiolate anion as leaving group. The role of the enzyme-bound pyridoxal-P is to convert the unreactive methylene carbon of glycine into a good nucleophile as the stabilized carbanionic adduct. The initial condensation adduct then undergoes decarboxylation before release, with removal of the original carboxylate of glycine.

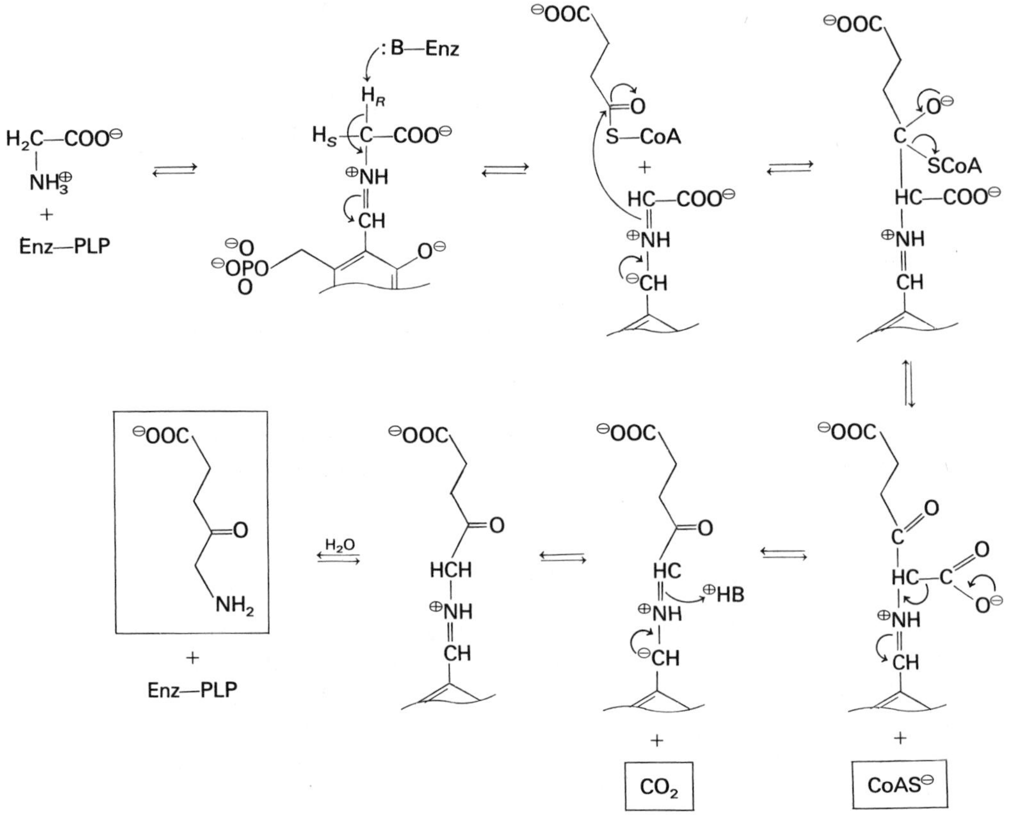

Stereochemical analysis has indicated that the pro*R* hydrogen of glycine is specifically labilized during this reaction (Akhtar and Jordan, 1969). Whether the new carbon–carbon bond is formed with retention or with inversion is not yet known.

24.A.4 Pyruvate-Containing Decarboxylases

Snell and colleagues have demonstrated that a histidine decarboxylase purified from a *Lactobacillus* strain does not contain pyridoxal phosphate, nor does addition of the coenzyme accelerate the observed V_{max} for decarboxylation (Riley and Snell, 1968; Lane and Snell, 1976; Lane et al., 1976).

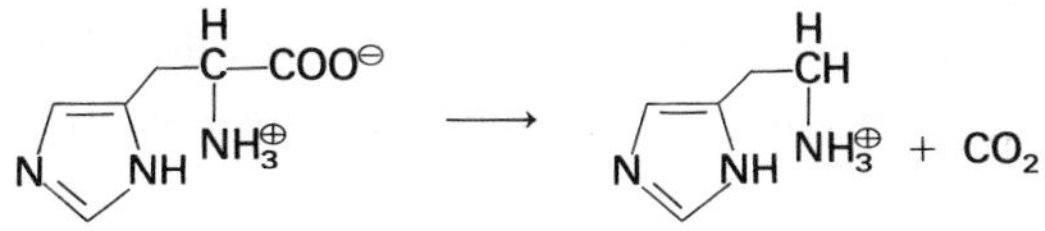

On the other hand, both sodium borohydride and phenylhydrazine completely inactivate the enzyme. Each of these reagents reacts with carbonyl groups—borohydride by reduction, and phenylhydrazine by conversion to phenylhydrazone.

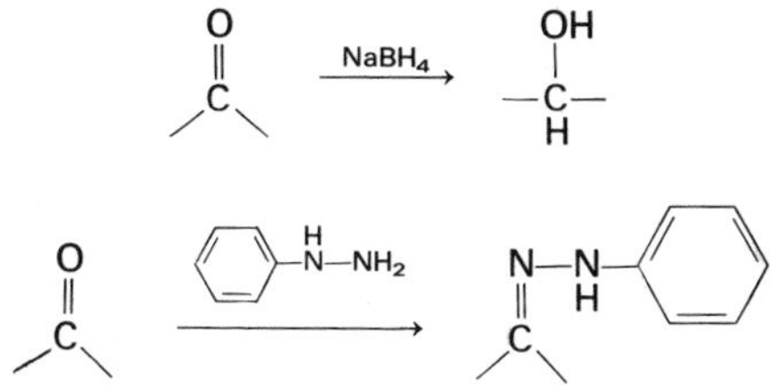

The presence of substrate histidine retards the rate of inactivation by either carbonyl reagent, suggesting active-site–directed attack. Using NaB^3H_4 leads to a tritiated inactive enzyme, which yields α-[^{3}H]-lactate on acid hydrolysis of the protein, implicating the reduction of a pyruvyl group. Correspondingly, [^{14}C]-phenylhydrazine produces a labeled protein that, after controlled proteolytic digestion, allows isolation of a phenylhydrazone of pyruvylphenylalanine.

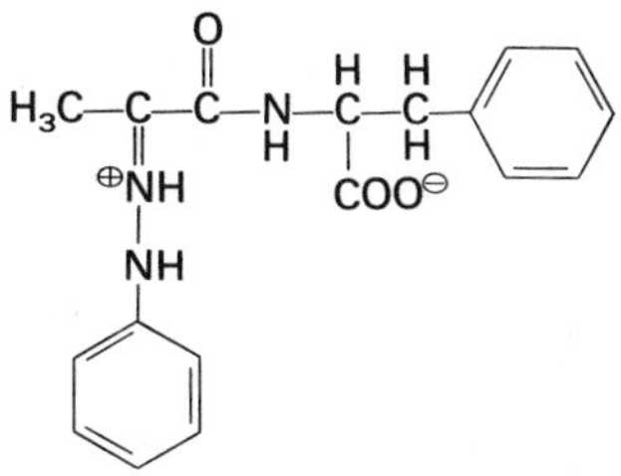

These data implicate the carbonyl group at the active site as a pyruvyl group in amide linkage to a phenylalanine residue.

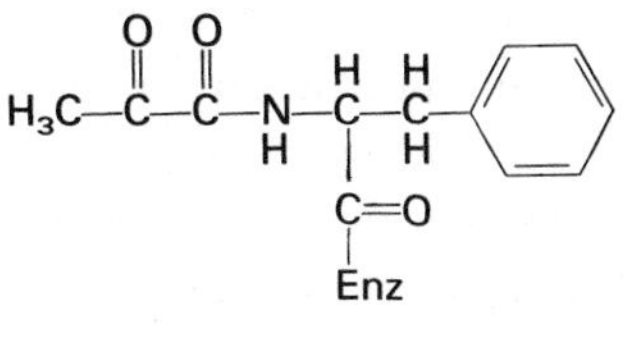

The functionality of the pyruvyl group as an electron sink was tested with [^{14}C]-histidine, followed by immediate borohydride addition to reduce any imine intermediate. Acid hydrolysis now yielded two radioactive products, the carboxyethyl derivatives of both substrate histidine and product histamine, indicating that both substrate and product imines had been present and trapped.

[^{14}C]-carboxyethylhistidine

[^{14}C]-carboxyethylhistamine

The pyruvyl group is at the amino terminus of five of the subunits of the histidine decarboxylase. Each subunit is about 30,000 mol wt, totaling to an $\alpha_5\beta_5$ decamer. Growth of the *Lactobacillus* on [^{14}C]-serine implicates this amino acid as precursor. Either the precursor to active enzyme has serine–phenylalanine as amino terminus of the α-subunits, or (less likely) a pyruvyl-tRNA^{Ser} is involved at the level of the initial peptide bond. Given a Ser–Phe–polypeptide, conversion to a pyruvyl-Phe–subunit involves an α,β-elimination of water, and this is probably catalyzed by a pyridoxal-P–requiring enzyme (e.g., a serine deaminase or a serine dehydrase).

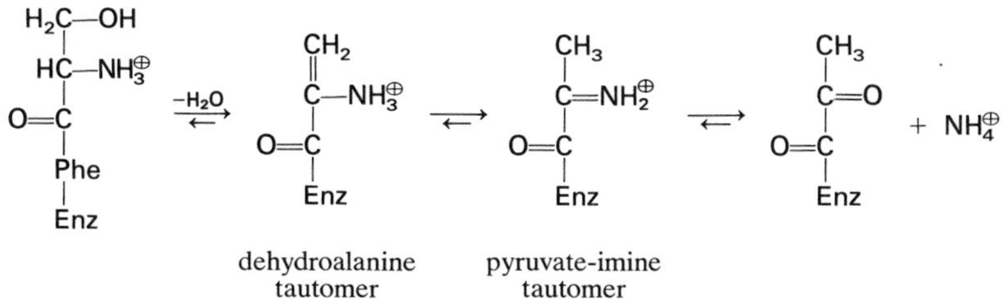

dehydroalanine tautomer

pyruvate-imine tautomer

A reasonable scheme can be written for the pyruvamide group as an electron sink to stabilize an anionic transition state in histidine decarboxylation.

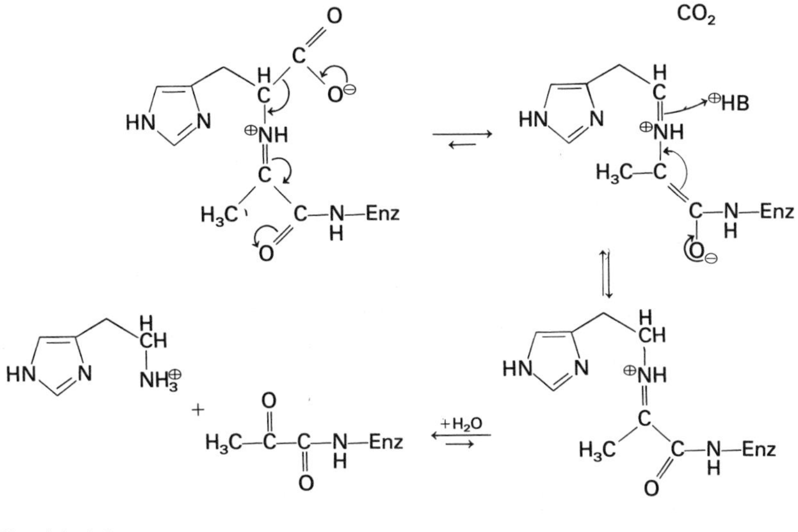

The histidine decarboxylase is not a unique example of a substitution of a pyruvyl group for the expected pyridoxal-P. The *S*-adenosylmethionine decarboxylase both from *E. coli* and from animal cells is another instance (Wickner et al., 1970; Tabor and Tabor, 1976).

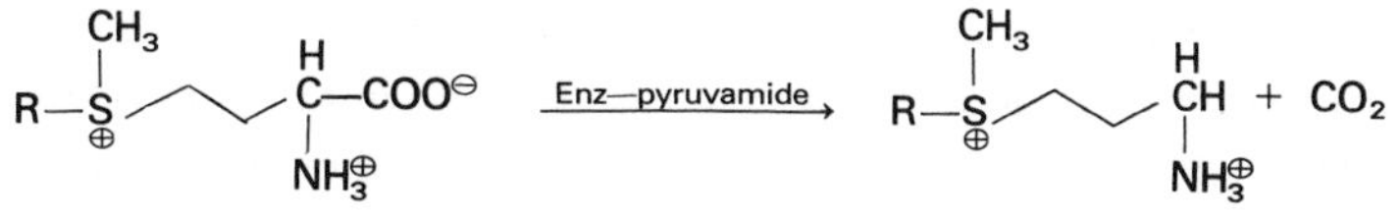

S-adenosylmethionine
(R = 5′-adenosyl)

S-adenosylhomocysteamine

The *S*-adenosylhomocysteamine undergoes subsequent expulsion of 5-thiomethyladenosine and propylamino-group transfer to amine acceptors (as we shall note in Chapter 25.) On the other hand, the mammalian histidine decarboxylase appears to use pyridoxal-P. It has been speculated that the pyruvamide moiety may represent a more primitive variety of electron sink than pyridoxal-P: it is an electrophilic catalyst much less formidable to biosynthesize, but undoubtedly less efficient at carbanion stabilizations (Riley and Snell, 1968; Lane and Snell, 1976; Lane et al., 1976; Wickner et al., 1970; Tabor and Tabor, 1976).

24.A.5 Cleavage of the C^{α}—C^{β} Bond

Cleavage of the C^{α}—C^{β} bond of an amino-acid substrate would complete the possibilities of labilization of substituents at the α-carbon.

$$R\text{—}C^{\beta}\text{—}\overset{\displaystyle H}{\underset{\displaystyle NH_3^{\oplus}}{C^{\alpha}}}\text{—}COO^{\ominus}$$

Such a reaction is catalyzed by the pyridoxal-P-dependent enzyme serine hydroxymethylase. This is an aldol cleavage of serine to glycine and formaldehyde. However, the physiological product is not free formaldehyde, but rather a derivative of formaldehyde, $N^{5,10}$-methylenetetrahydrofolate. Therefore, we shall defer discussion to Chapter 25, which deals with catalyses involving this pteridine cofactor.

24.B REACTIONS AT THE β-CARBON OF AMINO ACIDS

A number of pyridoxal-P-dependent enzymes catalyze transformations at the β-carbon of α-amino acids when the substrates possess substituents at the β-carbon that can function as good leaving groups (Kumagoi and Miles, 1971). This leaving group (X) can be $COO^{\ominus}$, OH or OR, SH or SR, a halogen such as Cl, or an aromatic nucleus such as a phenyl or indole ring (Table 24-1). The initial species generated on loss of X will be an α,β-dehydroamino-acid adduct with the coenzyme. This adduct can break down chemically, either to the free eneamino acid followed by tautomerization to *pyruvate imine*, or through a prior prototropic shift and then hydrolysis to *alanine.*

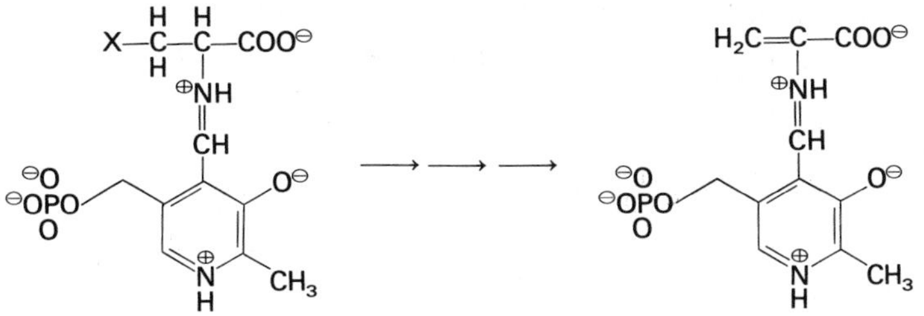

$\longrightarrow \longrightarrow \longrightarrow$

α,β-dehydroamino-acid adduct
(initial elimination product)

We shall see that the specific enzymes control which product is formed by controlling the rates of the prototropic shifts from the adduct.

We shall be selective in our coverage of reactions at the β-carbon, discussing aspartate β-decarboxylase, a β-elimination catalyzed by aspartate transaminase, a comparison of tryptophanase and tryptophan synthetase, and finally the enzyme kyneurinase.

Table 24-1
Substrates with good leaving groups at the β-carbon

Generalized substrate structure	X	Substrate
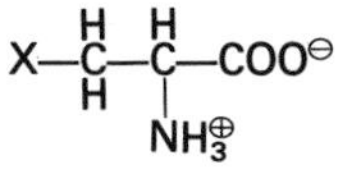 $X{-}CH_2{-}CH(NH_3^{\oplus}){-}COO^{\ominus}$	$COO^{\ominus}$	Aspartate
	OH	Serine
	SH	Cysteine
	Cl	β-Chloroalanine
	Indole	Tryptophan
	Phenol	Tyrosine

24.B.1 Aspartate β-Decarboxylase

We noted earlier that the pyridoxal-P–dependent aspartate β-decarboxylase differs from the other decarboxylases just discussed in that it alone catalyzes removal of a β-carboxyl group of a substrate. The enzyme from the bacterium *Alcaligenes faecalis* is a dodecamer (675,000 mol wt) of probably identical subunits (Nishimura et al., 1962; Novogrodsky et al., 1963).

The clear mechanistic difference between aspartate β-decarboxylase and the amino-acid α-decarboxylases is immediately demonstrated by the fact that the α-hydrogen of the aspartate (and not the α-carboxylate) is the first substituent labilized, and the stabilized carbanion formed has lost a proton rather than CO_2.

This carbanionic intermediate can serve as an electron sink for still another pair of electrons, those released on loss of the β-carboxylate as CO_2.

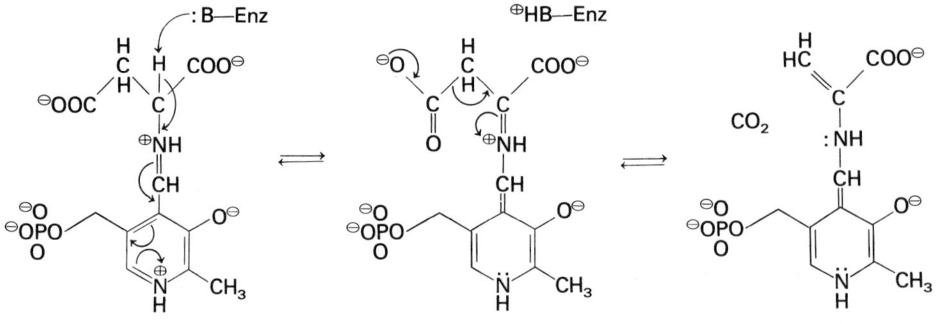

1

Structure 1 shows the versatility and capacity of the pyridoxal-P coenzyme to function as a stabilizing electron sink. It has two electron pairs stored temporarily, one on the pyridine-ring nitrogen, the other at the exocyclic amine. The β-carbon of the substrate moiety in 1 has carbanionic character and can protonate to form the methyl group of the eventual product alanine and produce adduct 2, which is the quinoid canonical form possessing electron density both at the substrate α-carbon and at the coenzyme methine carbon, as we have discussed in the transaminase catalyses.

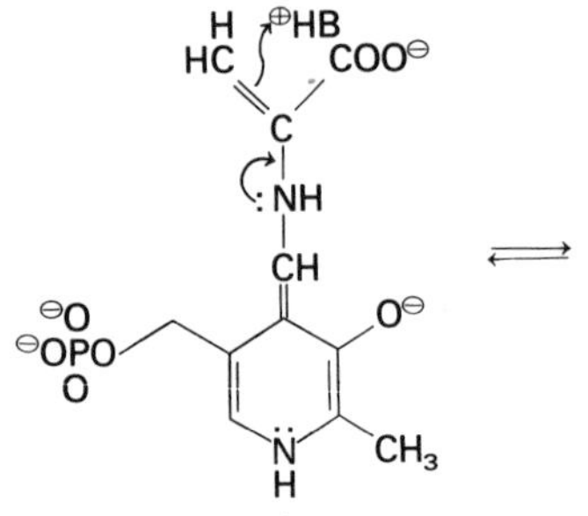

1

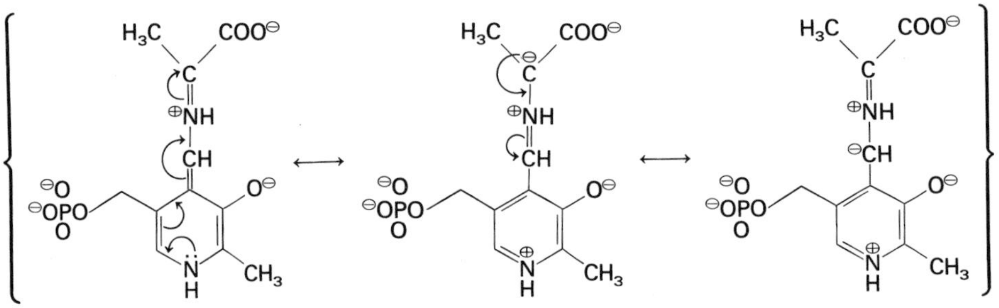

2

The normal product from aspartate β-decarboxylase action is L-alanine, so adduct 2 is protonated at the *si* face of the α-carbon and the aldimine is hydrolyzed.

$$\text{L-Aspartate} \longrightarrow \text{L-Alanine} + CO_2$$

However, this enzyme is incompletely specific in this protonation step and *occasionally* protonates at the methine carbon and allows ketimine hydrolysis to ensue, yielding pyruvate and enzyme–PMP—essentially a half-reaction for transamination.

$$\text{L-Aspartate} \rightleftharpoons \text{Pyruvate} + CO_2 + NH_4^{\oplus} + \text{Enz—PMP}$$

The pyruvate binds well enough so that the enzyme is not locked in inactive enzyme–PMP, but is in equilibrium with the pyridoxal-P form during steady-state catalysis, which runs at about 30% of the V_{max} calculated if all the active sites remained as PLP. Removal of pyruvate from the equilibrium (e.g., NADH and lactate dehydrogenase) does lead to accumulation of enzyme in the inactive pyridoxamine form.

The two courses of protonation open to the carbanionic intermediate (adduct 2) yield products in two different oxidation states at the α-carbon, and we shall see that this is a recurring option in enzymatic reactions at the β-carbon of amino acids.

As we intimated earlier, the aspartate β-decarboxylase will decarboxylate aminomalonate to glycine by α-carboxylate removal; in ^3H_2O, only S-2-[^{3}H]-glycine is produced (Palekar et al., 1970). The enzyme will also carry out elimination of the β-chloro group from chloroalanine.

$$H_2C(Cl)\text{—}CH(NH_3^{\oplus})\text{—}COO^{\ominus} \longrightarrow H_3C\text{—}C(=O)\text{—}COO^{\ominus} + Cl^{\ominus} + NH_4^{\oplus}$$

There is a *partitioning* between release of the eneamino-acid (dehydroalanine) intermediate and a covalent modification in reaction with a specific glutamate γ-carboxylate group at the active site (Relyea et al., 1974).

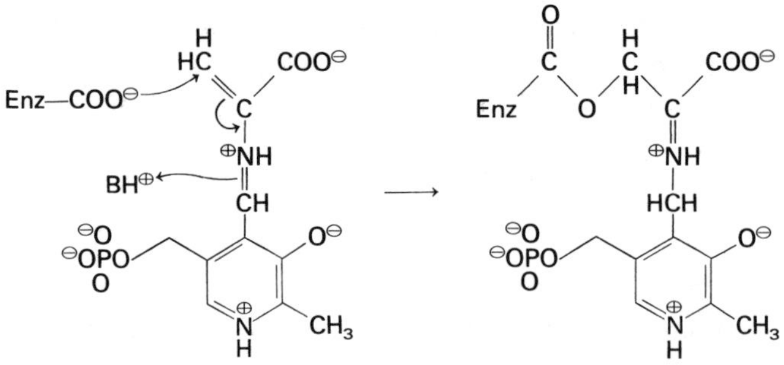

dehydroalanine–coenzyme intermediate

inactive enzyme

Note that this dehydroalanine-coenzyme intermediate differs from the one produced by β-decarboxylation of asparate (structure 1 above), which had two more electrons delocalized within it. This difference shows up in the fact that the *eneamine species from chloroalanine elimination has electrophilic (carbonium-ion) character at the β-carbon, whereas that from aspartate decarboxylation has only nucleophilic (carbanionic) character.* The dehydroalanine-coenzyme intermediate can undergo active-site alkylation; structure 1 cannot (at least, not by a nucleophile).

24.B.2 Reaction of Aspartate Transaminase with β-Chloroglutamate

In addition to its catalytic activity in transamination, the pig-heart aspartate aminotransferase will carry out α,β-elimination of HCl from both the *erythro* and the *threo* diastereomers of β-chloroglutamate (Manning et al., 1968; Antonini et al., 1970).

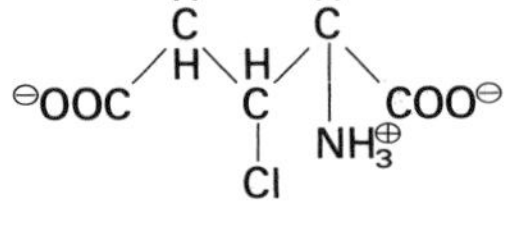

β-chloroglutamate

No transamination is observed with this synthetic substrate, suggesting that the β-chloro moiety is such a good leaving group that it eliminates much faster than protonation occurs at the coenzyme's methine carbon, effectively suppressing the normal transamination pathway kinetically. Because the reaction is catalytic, the enzyme must remain in the pyridoxal form at the end of each turnover, and therefore the eneamino acid must suffer hydrolysis in each turnover. The free eneamino acid is, of course, rapidly tautomerized to the imine and hydrolyzed, so that observed products are α-ketoglutarate and ammonia.

eneamino-acid–aldimine complex

(free eneamino acid) + Enz—PLP

Using [^{14}C]-N-ethylmaleimide, the fleeting eneamine can be trapped by virtue of its action as a nucleophile adding 1,4 to the maleimide (Manning et al., 1968; Antonini et al., 1970).

initial adduct

24.B.3 β-Eliminations versus β-Replacements

The conversion of β-chloroglutamate to α-ketoglutarate, $NH_4^{\oplus}$, and $Cl^{\ominus}$ is a β-elimination reaction; the product has experienced a 2-$e^{\ominus}$ oxidation at the α-carbon. This is a characteristic event in pyridoxal-mediated β-eliminations (Davis and Metzler, 1972). Thus, serine dehydrase converts L-serine to pyruvate, water, and ammonia. It is worth noting how these α-keto-acid products arise. The β-elimination enzymes hydrolyze the eneamine product to regenerate bound pyridoxal-P and a free product that, on paper, looks like an α,β-olefinic amino acid. The key point to remember is that, in aqueous solutions, the thermodynamically favored form of this compound is α-keto acid and ammonia, and this equilibrium form also is attained rapidly.

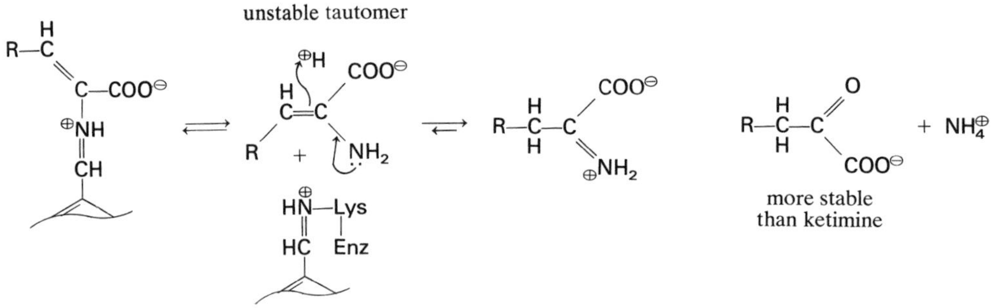

unstable tautomer

more stable than ketimine

Although the two-electron oxidation of amino-acid substrate to keto-acid product in β-elimination is reminiscent of a transaminase half-reaction, the difference is that the coenzyme remains as pyridoxal-P in β-eliminations; it has not suffered net 2-$e^{\ominus}$ reduction. *The coenzyme has not been the electron acceptor* (*as it is in the transaminase case*); *rather the two electrons have departed with the leaving group as* $X^{\ominus}$.

One way of looking at the difference between enzymatic β-eliminations and the related β-replacements is that the *β-replacement enzymes do not catalyze rapid hydrolysis of the eneamino–pyridoxal-P adduct.* Rather, this adduct is maintained long enough for the leaving group $X^{\ominus}$ to depart from the active site by diffusion, and for the replacement nucleophile $X'^{\ominus}$ to diffuse in, bind, and react as nucleophile on the β-carbon of the eneamine–coenzyme complex (made chemically possible by the electrophilic character of this particular eneamine at the β-carbon). As an efficient catalyst, the β-replacement enzyme avoids protonating the methine carbon (which must lead to pyridoxamine-P), but rather promotes protonation at the α-carbon of product to yield the product aldimine, which in turn will, on hydrolysis, regenerate native enzyme and β-substituted amino-acid product.

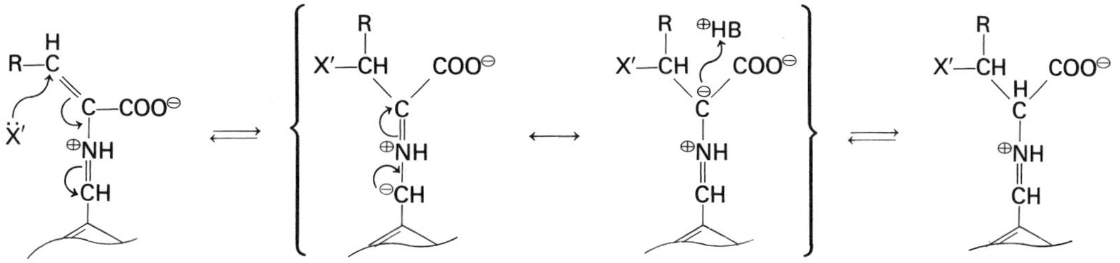

this resonance contributor protonated specifically

product aldimine

Thus, examination of the reaction stoichiometry permits immediate classification as a β-elimination or a β-replacement. *Replacement* of X by X′ yields an amino-acid product, not a keto acid. No net oxidation occurs at the α-carbon. (By this classification, the conversion of L-aspartate to L-alanine by aspartate β-decarboxylase is a replacement of $COO^{\ominus}$ by a proton.) The difference in release rates (and fates) of the common eneamine species in β-eliminations vs. β-replacements are reminiscent of the aldolase and transaldolase catalyses (Chapter 23).

24.B.3.a Tryptophanase versus tryptophan synthetase

As specific examples of related β-elimination and β-replacement enzymes, we now consider the functioning of tryptophanase and tryptophan synthetase from *E. coli.* Tryptophanase is a tetramer (220,000 mol wt) catalyzing the following reaction sequence.

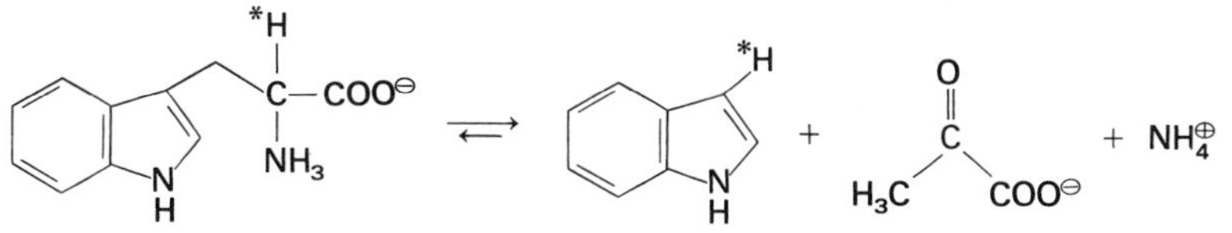

L-tryptophan ⇌ indole + pyruvate + $NH_4^{\oplus}$

L-tryptophan

indole

The stoichiometry clearly reflects the elimination process. A variety of other amino acids (such as serine, *S*-methylcysteine) with β-leaving groups can serve as substrates. Even alanine, which cannot eliminate, proceeds partway through catalysis as pseudosubstrate (Morino and Snell, 1967). In ^3H_2O, the enzyme incorporates tritium into the α-position (only) of L-alanine, and indeed the steady-state form of the alanine–holoenzyme complex is the quinoid intermediate, absorbing maximally at 500 nm. For the true substrates, labilization of α-hydrogen is fast, and the actual elimination of the β-leaving group is slow. Interestingly, recent studies on the origin of the hydrogen incorporated at the β-carbon of the indole product show that part of the time it derives from the hydrogen initially at C-2 of L-tryptophan as indicated above (Floss et al., 1976). Floss and colleagues thus arrived at the following conclusions.

1. A single base at the active site catalyzes the proton abstraction at C-2 *and* the reprotonation of the indole moiety.
2. This proton transfer must be *suprafacial*, and therefore the α,β-elimination of the elements of indole is characterized as an overall *syn* elimination sequence.

The scheme shown in Figure 24-6 is modified from their suggestion.

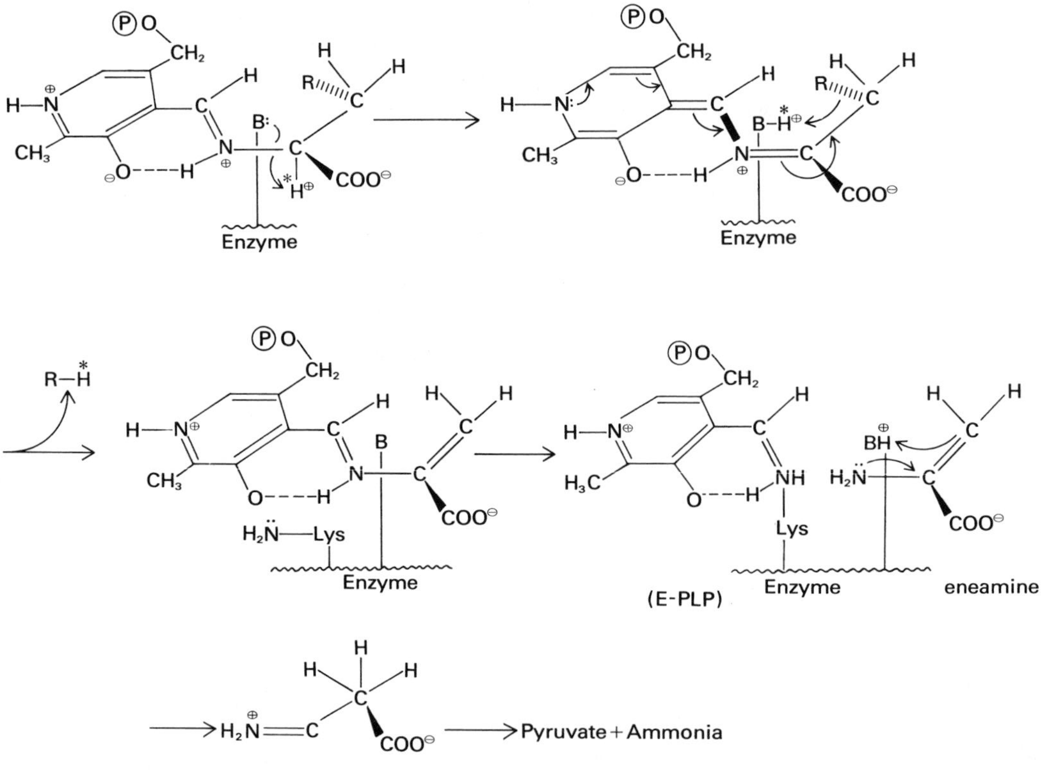

R—H*

Enzyme

Enzyme

H₂N̈—Lys

Enzyme

(E-PLP)

Lys

Enzyme

eneamine

→ Pyruvate + Ammonia

Figure 24-6
Syn stereochemistry of β-elimination by tryptophanase. (Reprinted with permission from E. Schleicher et al., *Journal of the American Chemical Society*, vol. 98, 1976, p. 1044. Copyright by the American Chemical Society.)

The other enzyme, the tryptophan synthetase, is an $\alpha_2\beta_2$ tetramer, with the α-subunit of 29,000 mol wt and the β-subunit of 44,000 mol wt (Davis and Metzler, 1972). One can determine that there are several reactions catalyzed by the complete tetrameric species. The physiological reaction to form tryptophan utilizes indole-3-glycerol phosphate as precursor; it involves a β-replacement of the hydroxyl group of serine by the indole ring. The other products are a molecule of water and one of glyceraldehyde-3-P.

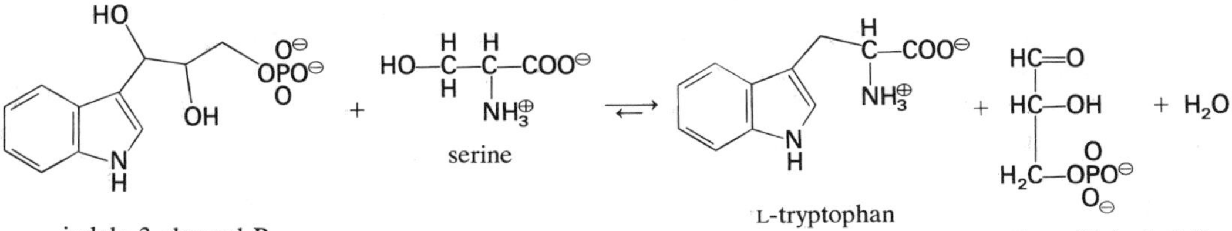

indole-3-glycerol-P + serine $\rightleftharpoons$ L-tryptophan + glyceraldehyde-3-P + H_2O

The following reactions also are catalyzed.

$$\text{Indole} + \text{L-Serine} \rightleftharpoons \text{L-Tryptophan} + H_2O \qquad (1)$$

$$\text{L-Serine} \rightleftharpoons \text{Pyruvate} + \text{Ammonia} \qquad (2)$$

$$\text{Indole-3-glycerol-P} \rightleftharpoons \text{Indole} + \text{Glyceraldehyde-3-P}$$

Reaction 1 is a β-replacement; it is catalyzed by the $\alpha_2\beta_2$ tetramer or by the isolated β_2 dimer. Reaction 2 is a β-elimination; it also is catalyzed by the tetramer or by the β_2 dimer. Reaction 3, however, is an aldol cleavage; it is catalyzed by the pyridoxal-free tetramer or by the isolated α-subunit.

These additional reactions represent two distinct chemical reaction types, and the enzyme has two distinct types of subunits that can be isolated free of each other (Kumagoi and Miles, 1971). The isolated β-subunits have the pyridoxal phosphate and carry out the β-replacements and β-eliminations. The isolated α-subunit does not have or require the coenzyme, and it possesses the aldolase activity. Thus, an overall picture of catalysis might involve sequential reactions on α- and β-subunits of the tetrameric holoenzyme.

The indole ring contains an eneamine moiety, and free indole shows nucleophilicity at the β-carbon. The enzyme α-subunit may facilitate tautomerization of the eneamine form of indole-3-glycerol-P to the iminium form, which would have the requisite electron sink to stabilize the carbanionic transition state arising during aldol cleavage. The free indole thus produced remains enzyme-bound to act as X′ in the subsequent β-replacement catalyzed by the β-subunit.

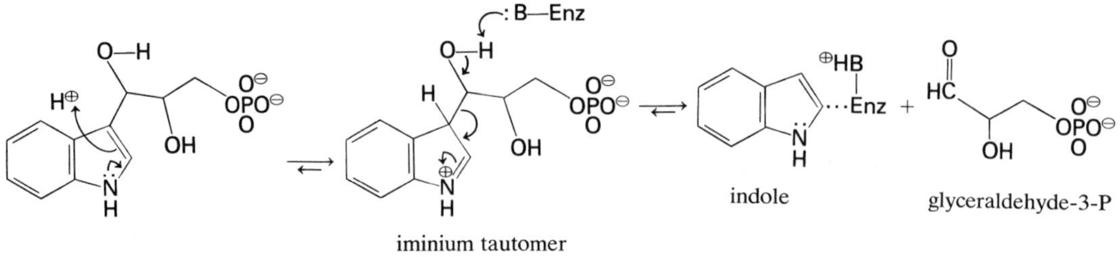

iminium tautomer

indole

glyceraldehyde-3-P

The serine is bound to pyridoxal-P at the active site of the β-subunits, where it is activated to aminoacrylate.

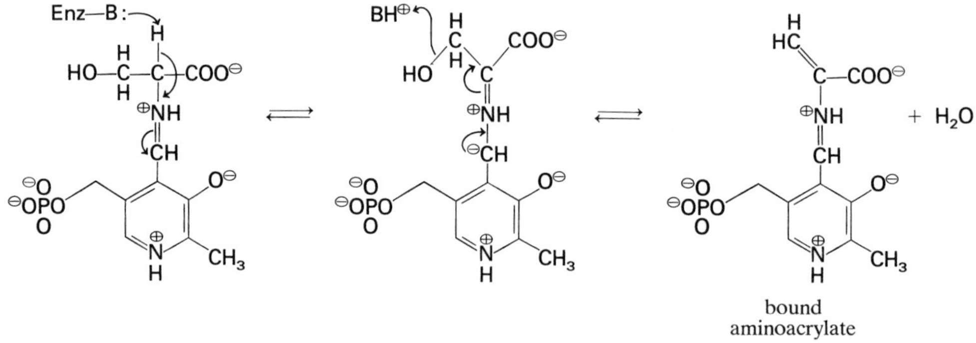

bound aminoacrylate

In the absence of an acceptor X′, the bound aminoacrylate can hydrolyze to pyruvate, $NH_4^{\oplus}$, and enzyme–PLP. However, in the physiological reaction, the indole bound at the α-subunit active site acts as X′ in the β-replacement reaction at the β-subunit.

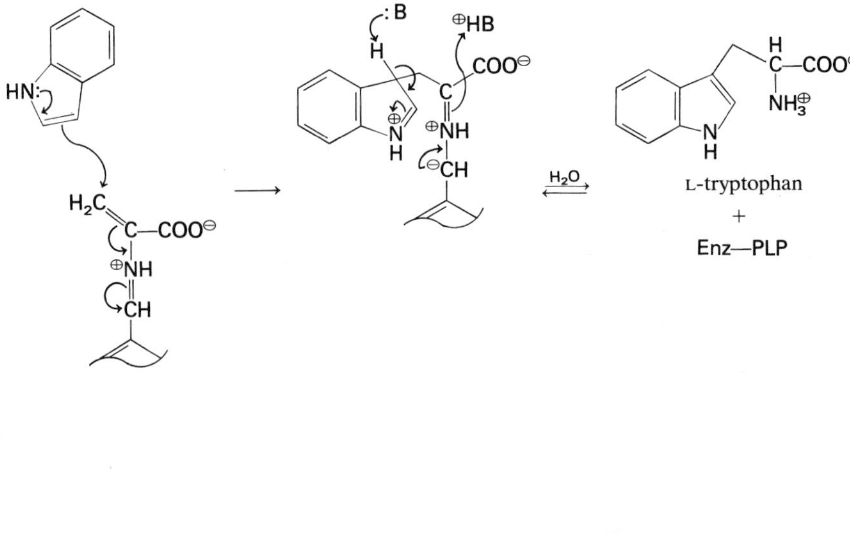

L-tryptophan

+

Enz—PLP

Tryptophan synthetase forms the new carbon–carbon bond in the L-tryptophan product with retention of configuration at C-3. The OH departs from the β-carbon of serine, and the incoming β-carbon of indole is added in the same relative orientation (Skye et al., 1974).

24.B.3.b Kyneurinase

An unusual variant of a β-replacement reaction is that catalyzed by hydroxykyneurinase (Moriguchi et al., 1973). This pyridoxal enzyme produces L-alanine and 3-hydroxyanthranilate from the substrate 3-hydroxykyneurinine, an intermediate in tryptophan catabolism.

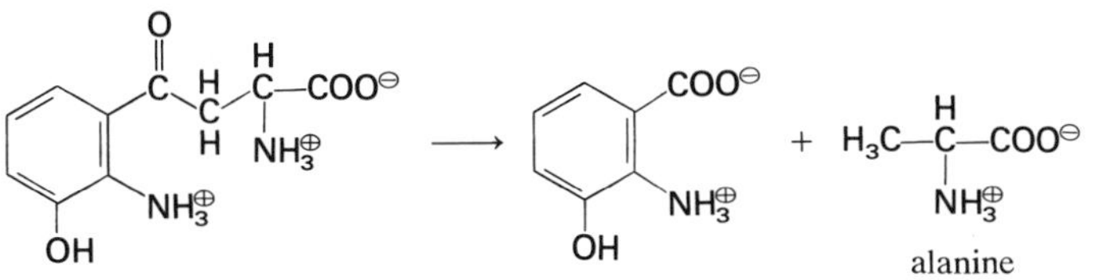

3-hydroxykyneurinine → 3-hydroxyanthranilate + alanine

Note that the β-keto group in the substrate becomes the carboxylate of the hydroxyanthranilate product. Thus, the process cannot be a simple β-replacement of a proton for the alanyl group, which would generate an aldehyde product.

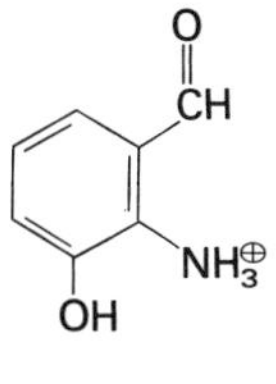

Little is known about this reaction, but we have precedents for speculating on how oxidation and fragmentation might occur. Although a molecule of water could attack a β-keto group directly, we have earlier (Chapter 10) seen that addition of enzyme nucleophiles to carbonyl groups is often used to generate covalent tetrahedral adducts that can then decompose with expulsion of a leaving group and simultaneous oxidation to an acyl-enzyme species, which could well be hydrolytically labile (e.g., glyceraldehyde-3-P dehydrogenase).

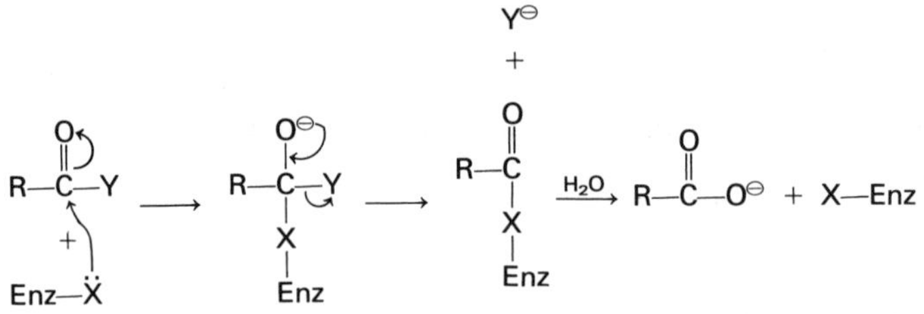

In this instance, Y is the alanyl moiety of the substrate, and one can write an adduct of pyridoxal-P and hydroxykyneurinine where such a β-carbanion would be stabilized. (Attempts to detect an acyl-enzyme intermediate have not been systematically reported.) For example,

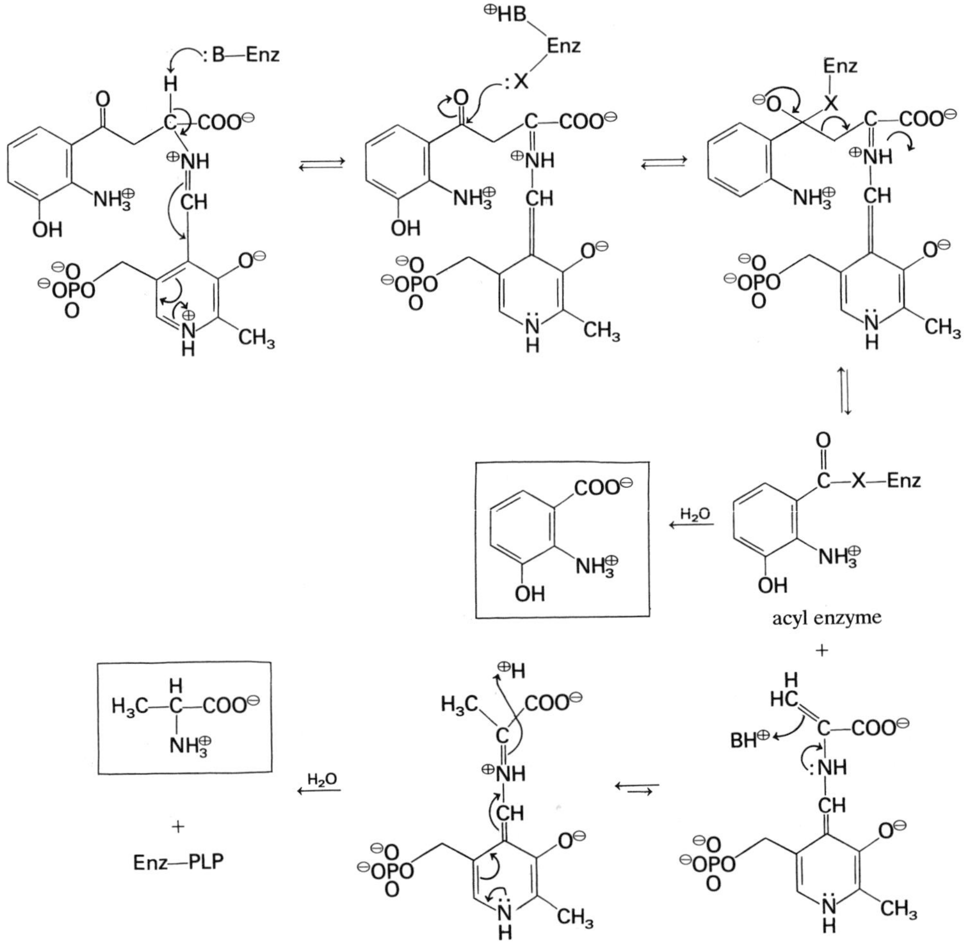

acyl enzyme

24.C REACTIONS AT THE γ-CARBON OF AMINO ACIDS

Although there are a wide variety of pyridoxal-P–dependent enzymes that carry out transformations at the β-carbon of susceptible substrates, there are only a very few enzymes able to effect chemical change at the γ-carbon. As in the β-carbon reactions, we can divide reactions at the γ-carbon into two types: elimination or replacement. Again, the stoichiometry of a γ-elimination is to produce an α-keto-acid product, whereas γ-replacement leads to an α-amino-acid product (Davis and Metzler, 1972).

As noted for most of the pyridoxal-P–mediated enzymatic reactions so far discussed, the enzymes operating at the substrate γ-carbon abstract the α-hydrogen as a proton at an early step. *The unique requirement of this class of reactions is the necessity for subsequent formation of a β-carbanion as well by labilization of a substrate β-hydrogen.* None of the other types of pyridoxal-P catalysis show this requirement. The electron density at the β-carbon is then used to provide anchimeric assistance to elimination of the γ-substituent Y. This event generates the fully conjugated, β,γ-unsaturated imine intermediate I, which is common to both γ-elimination and γ-replacement routes.

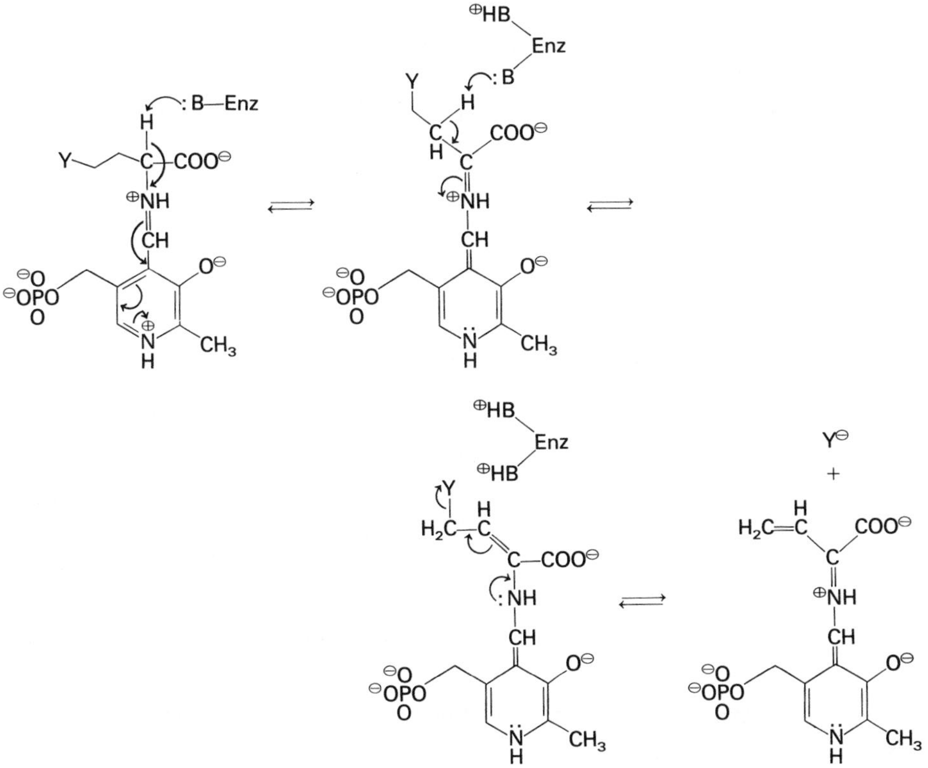

intermediate I

Product control depends on how the enzyme controls the relative rates of the protonation leading to γ-elimination and the attack by a nucleophile leading to γ-replacement, both reactions proceeding from the fully conjugated intermediate I. The γ-elimination involves protonation of the γ-carbon and imine hydrolysis.

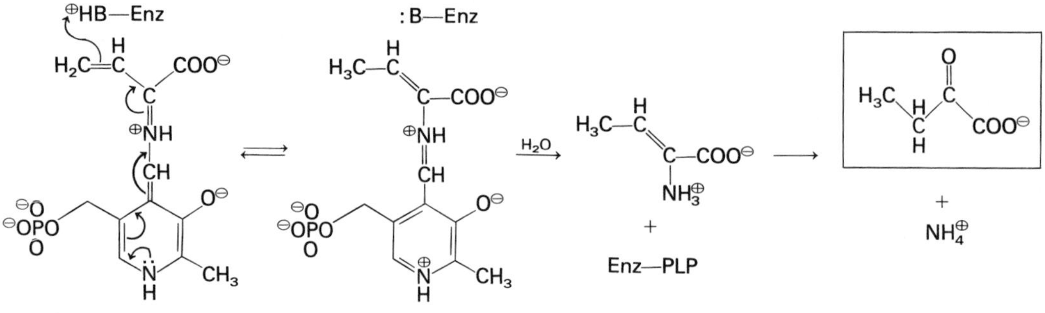

$^{\oplus}$HB—Enz
H
H_2C=C COO$^{\ominus}$
C
$^{\oplus}$NH
CH
$^{\ominus}$O
$^{\ominus}$OPO
O
O$^{\ominus}$
N
H
CH_3
intermediate I

:B—Enz
H
H_3C—C
C—COO$^{\ominus}$
$^{\oplus}$NH
CH
$^{\ominus}$O
$^{\ominus}$OPO
O
O$^{\ominus}$
$^{\oplus}$N
H
CH_3
intermediate II

H_2O

H
H_3C—C
C—COO$^{\ominus}$
$NH_3^{\oplus}$
+
Enz—PLP

O
H_3C H C
C COO$^{\ominus}$
H
+
$NH_4^{\oplus}$

The γ-replacement involves attack by a replacement nucleophile Y′ (at a γ-carbon that is electrophilic in intermediate I).

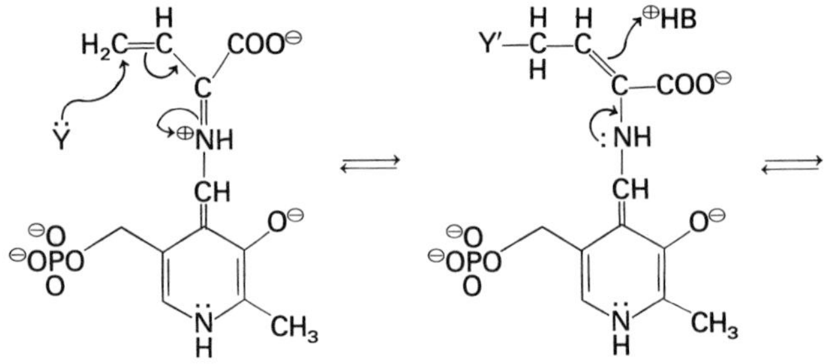

H
H_2C=C COO$^{\ominus}$
C
Y
$^{\oplus}$NH
CH
$^{\ominus}$O
$^{\ominus}$OPO
O
O$^{\ominus}$
N
H
CH_3
intermediate I

H H $^{\oplus}$HB
Y′—C—C
H
C—COO$^{\ominus}$
:NH
CH
$^{\ominus}$O
$^{\ominus}$OPO
O
O$^{\ominus}$
N
H
CH_3

$^{\oplus}$H
COO$^{\ominus}$
Y′ C
$^{\oplus}$NH
CH
$^{\ominus}$O
$^{\ominus}$OPO
O
O$^{\ominus}$
N
H
CH_3

H_2O

H
Y′ C—COO$^{\ominus}$
$NH_3^{\oplus}$
+
Enz—PLP

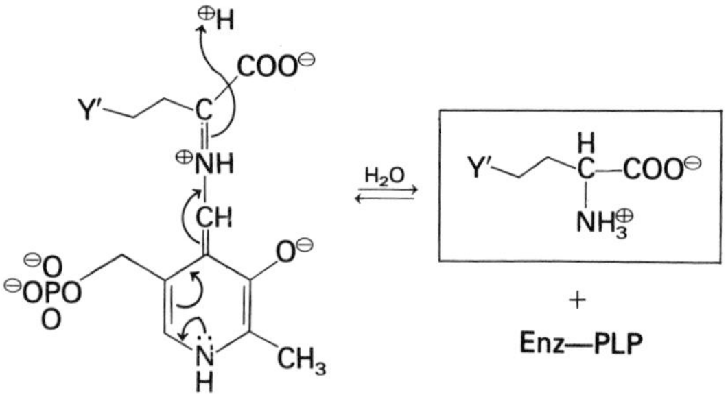

We shall examine two examples of this type of enzyme: γ-cystathionase and cystathionine γ-synthetase.

24.C.1 γ-Cystathionase

The enzyme γ-cystathionase (190,000 mol wt) is found in mammalian liver and in the mold *Neurospora crassa* (to date). The physiologically important reaction is breakdown of the C_7 sulfur-containing amino acid cystathionine to α-ketobutyrate and cysteine.

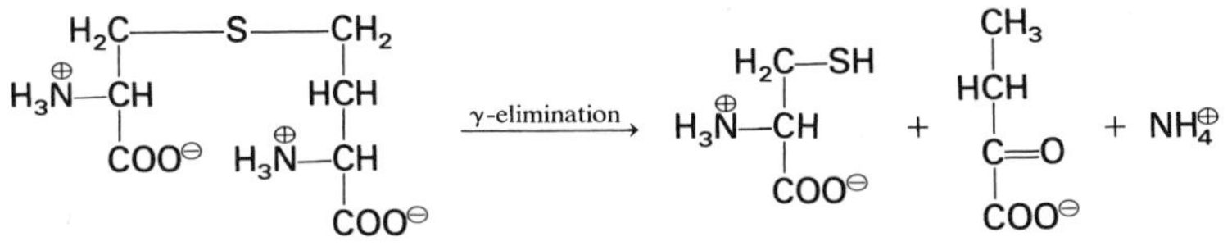

$$\text{cystathionine} \xrightarrow{\gamma\text{-elimination}} \text{cysteine} + \alpha\text{-ketobutyrate} + NH_4^{\oplus}$$

The enzyme also will carry out γ-elimination from homoserine

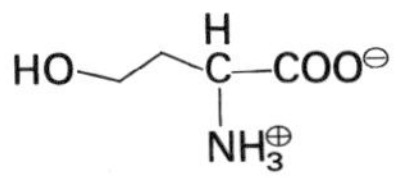

$$HO\text{—}CH_2\text{—}CH_2\text{—}CH(NH_3^{\oplus})\text{—}COO^{\ominus}$$

to yield α-ketobutyrate and H_2O (Davis and Metzler, 1972). The mechanism is likely to be the generalized mechanism for γ-elimination sketched out above. Note that cystathionine is composed of a C_3 arm and a C_4 arm. The γ-elimination yields a keto acid from the C_4 arm and an amino acid from the C_3 arm. The *Neurospora* enzyme (but not the rat-liver enzyme) will also show β-cystathionase activity, producing homocysteine and pyruvate from cystathionine.

Evidence has been obtained that the enzyme catalyzes the tautomerization of free eneamino acid to α-iminobutyrate before release from the active site. In D_2O, homoserine is converted to a 3-[^{2}H]-2-ketobutyrate that is exclusively the 3*S* stereoisomer (Krongelb et al., 1968). Tautomerization free in D_2O solution should lead to an equal 3*R*,3*S* mixture.

24.C.2 Cystathionine γ-Synthetase

Cystathionine γ-synthetase is a γ-replacement enzyme that functions in bacteria and plants in the biosynthetic pathway to methionine (animals do not synthesize methionine de novo; they require it in the diet). The homogeneous *Salmonella typhimurium* enzyme (a tetramer of 160,000 mol wt) uses *O*-succinylhomoserine as cosubstrate with cysteine (Kaplan and Flavin, 1966).

O-succinyl-L-homoserine + L-cysteine ⇌ cystathionine + succinate

On the other hand, the *Bacillus subtilis* enzyme uses *O*-acetylhomoserine as cosubstrate, and plant enzymes use both *O*-phosphorylhomoserine and *O*-malonylhomoserine (Datko et al., 1974). Acylation of the homoserine provides a better leaving group Y (an acyl anion, then, rather than $OH^{\ominus}$). The cystathionine undergoes β-elimination in a subsequent enzymatic step, and the homocysteine produced is then methylated on sulfur (see Chapter 25) to yield methionine. The turnover number for cystathionine formation is 2,500 moles min^{-1} (mole enzyme)$^{-1}$. In the absence of cysteine, the enzyme carries out γ-elimination at 500 min^{-1} from the *O*-succinyl-L-homoserine.

In support of the idea that the synthetase generates β-carbanionic species, we may note that, in D_2O, exchange of one of the two β-hydrogens of *O*-succinylhomoserine occurs. Exchange can occur independently of elimination because the enzyme will form 3-[$^2H_1$]-2-aminobutyrate in D_2O (Guggenheim and Flavin, 1969) to yield as 3*S*-[$^2H_1$]-amino acid.

We noted earlier (¶23.A.1.b) that the L-alanine transaminase catalyzes its own suicide with the acetylenic amino acid 2-amino-4-pentynoate (propargylglycine) by virtue of its unusual ability to abstract a C-3 hydrogen as a proton to yield an acetylenic carbanion that can readily rearrange to the conjugated allene, which reacts covalently by attack of an active-site nucleophile (presumably at C-4).

$$HC{\equiv}C{-}CH_2{-}CH(NH_3^{\oplus}){-}COO^{\ominus} \longrightarrow \longrightarrow \longrightarrow H_2C{=}C{=}CH{-}C({=}\overset{\oplus}{N}H{-}HCH{-}\text{enzyme}){-}COO^{\ominus}$$

Propargylglycine also is a suicide substrate both for γ-cystathionase (Abeles and Walsh, 1973) and for cystathionine γ-synthetase (Marcotte and Walsh, 1975), but not for other pyridoxal-P enzymes that generate α-carbanions (but not β-carbanions) during catalysis. In fact, L-propargylglycine is a natural product originally isolated from a mold (Scanell et al., 1971) and was first characterized as a compound with antibacterial activity whose effect is reversed by addition of methionine to the growth medium (Gershon et al., 1949). This old observation suggests that the γ-synthetase, involved in bacterial biosynthesis of methionine, may well be the target in vivo of propargylglycine.

An additional PLP-dependent enzyme carrying out a γ-elimination has been purified. Known as methionase, the enzyme cleaves methionine into methanethiol, α-ketobutyrate, and ammonium ion as a catabolic device.

$$H_3C{-}S{-}CH_2{-}CH_2{-}\underset{NH_3^{\oplus}}{\overset{H}{C}}{-}COO^{\ominus} \xrightarrow{\text{methionase}} H_3C{-}SH + H_3C{-}\underset{H}{\overset{H}{C}}{-}\underset{\underset{O}{\|}}{C}{-}COO^{\ominus} + NH_4^{\oplus}$$

This concludes our analysis of how various enzymes harness the chemical possibilities of the pyridoxal-P coenzyme to effect specific transformations.

Chapter 25

Enzymatic C_1-Group Transfers Requiring Tetrahydrofolate or *S*-Adenosylmethionine

In this chapter we shall focus on transfer of one-carbon fragments between nucleophilic acceptor molecules. We shall observe that these reactions, which form and cleave carbon–carbon bonds, are mediated largely by one of two possible carrier molecules: tetrahydrofolate, a pteridine coenzyme derivative (recall Chapter 12), or *S*-adenosylmethionine, a sulfonium cation acting as active methyl donor. We shall also examine the role of the organometallic methyl coenzyme B_{12} in the biosynthesis of the $S—CH_3$ of methionine and in methyl transfer to toxic elements in the environment. And we shall briefly note two novel coenzymes involved in bacterial production of methane from CO_2, an eight-electron reduction. The first example of this chapter bridges the pyridoxal-P–dependent enzymes of Chapter 24 and the alkyl-transfer enzymes of this chapter.

25.A SERINE HYDROXYMETHYLASE

The enzyme serine hydroxymethylase is an unusual pyridoxal-P-requiring enzyme (Schirch and Mason, 1962, 1963; Schirch and Diller, 1971; Schirch and Jenkins, 1964; Benkovic and Bullard, 1973). It catalyzes the reversible conversion of L-serine to glycine and a one-carbon fragment at the oxidation level of formaldehyde. The transformation is an aldol cleavage with expulsion of the stabilized glycine–coenzyme carbanion. One might predict, then, that the enzyme does not abstract the α-hydrogen of serine as a necessary step in catalysis, and this is confirmed experimentally. In the glycine-to-serine direction, it has been shown that the enzyme abstracts only the pro$S$ hydrogen of glycine, and the hydroxymethyl group is introduced with retention of configuration at the α-carbon (Jordan and Akhtar, 1970). For example,

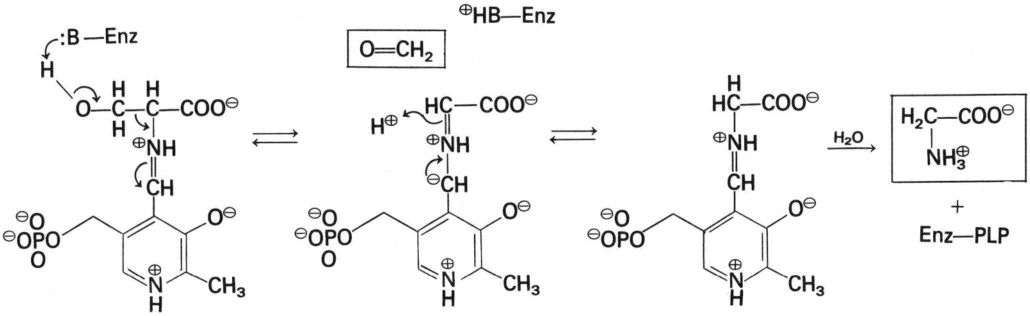

More interestingly, the cleavage of serine does not proceed at significant rates in the absence of an additional cofactor, tetrahydrofolate (tetrahydropteryl-glutamate) (Benkovic and Bullard, 1973; Bruice and Benkovic, 1965–1966, vol. 2, p. 350; Blakely, 1969).

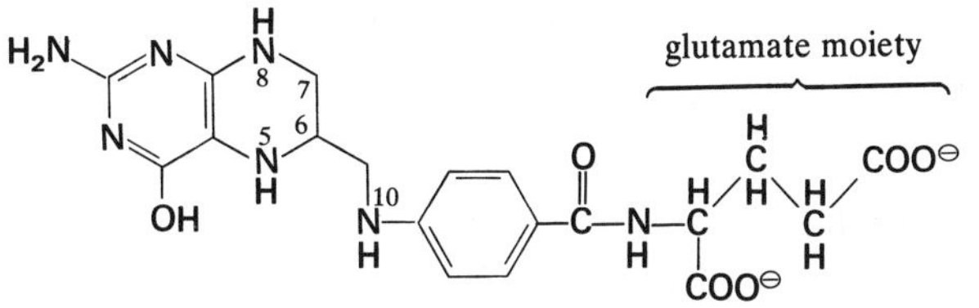

tetrahydrofolate (THF)

The structure shown is the monoglutamate form of tetrahydrofolate. The coenzyme forms containing up to seven glutamates (in γ-glutamyl-amide linkages) are not uncommon, and various enzymes show different preferences for polyglutamate forms. The substituted pteridine ring is a 5,6,7,8-tetrahydropteridine; it is linked to a *para*-aminobenzoyl moiety, which in turn is linked to the glutamate moiety.

We have encountered a pteridine coenzyme previously (¶12.D) as a redox cofactor for the phenylalanine, tyrosine, and tryptophan hydroxylases. In that system, Kaufman and colleagues could use a tetrahydropterin, 5,6,7,8-tetrahydrobiopterin, which is oxidized by two electrons to the 7,8-dihydrobiopterin during the hydroxylation of the amino acid. A separate enzyme, dihydropterin reductase, then takes the dihydro form back to the tetrahydro species. In the reactions of serine hydroxymethylase. however, the tetrahydrofolate coenzyme *does not undergo any redox change.* In fact, of the variety of reactions involved in C_1-fragment transfer, *the pteridine ring of the coenzyme undergoes oxidation only in the thymidylate-synthase type of one-carbon transfer* (discussed in ¶25.C).

All the C_1-fragment transfers involving tetrahydrofolate (THF) as carrier use a covalent bond between the fragment and either N-5 or N-10 or both, as we shall note. Although the C_1 fragment may be at any of three oxidation states (formate, formaldehyde, or methanol), in the serine hydroxymethylase reaction the C_1 unit is at the formaldehyde level. Addition of formaldehyde to THF in solution leads to rapid nonenzymatic formation of the cyclic adduct, the N^5,N^{10}-methylene-THF, with an equilibrium constant of 3×10^4 M at pH 7.2 in favor of the cyclic structure (Benkovic and Bullard, 1973).

carbinolamine   iminium ion N^5,N^{10}-methylene-THF

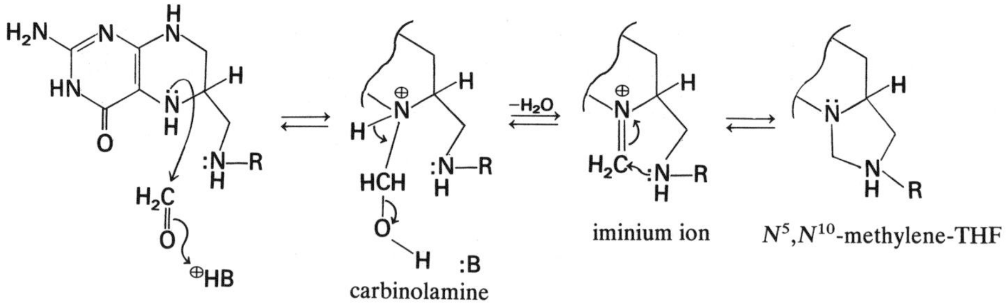

The nitrogen at position 5 is more basic (pK_a of conjugate acid = 4.8) than that at position 10 (pK_a of conjugate acid = −1.25), and one expects that N-5 is the initial nucleophile (Blakely, 1969). It has been suggested that the carbinolamine may dehydrate to the N-5 iminium cation prior to cyclization by attack of the proximate N-10 as shown here (Benkovic and Bullard, 1973).

Thus, the correct stoichiometry for the reaction catalyzed by serine hydroxymethylase is the following:

$$\text{L-Serine} + \text{THF} \rightleftharpoons \text{Glycine} + N^5,N^{10}\text{-methylene-THF}$$

How then might the folate cofactor function in this conversion? One could imagine a standard β-elimination reaction on serine bound to pyridoxal-P at the active site, followed by replacement with N-5 of THF as the incoming nucleophile and subsequent aldol cleavage to liberate the N^5,N^{10}-methylene-THF and glycine.

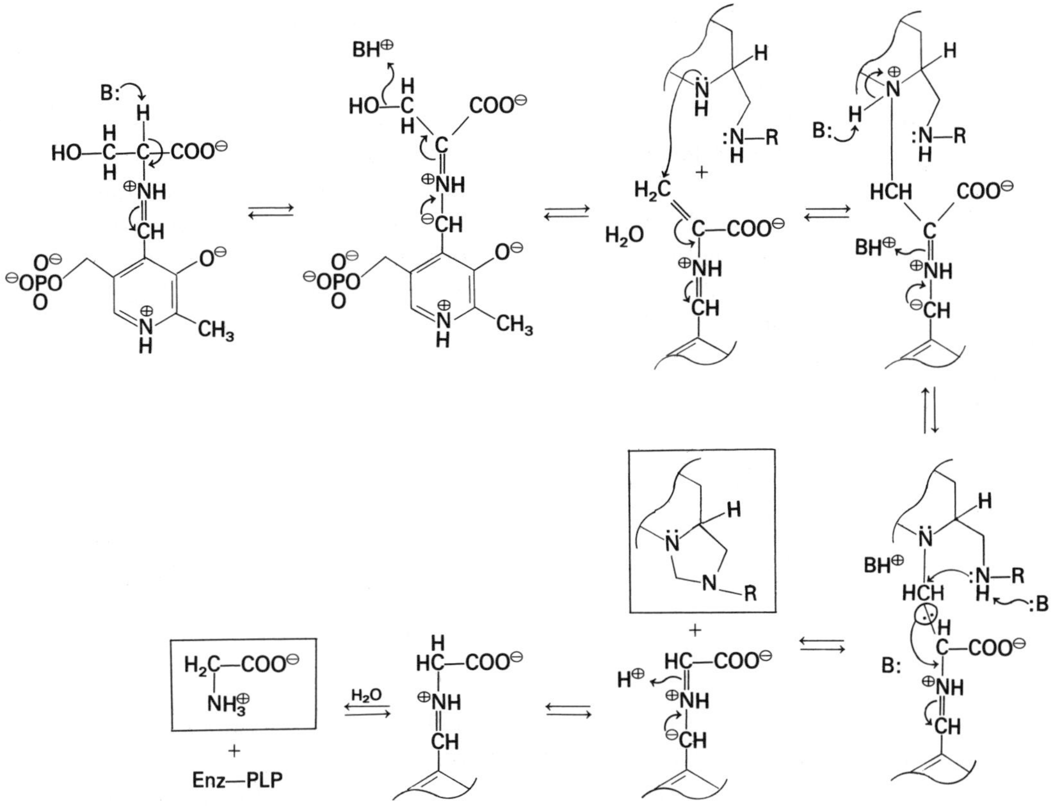

This sequence, however, would predict exchange of the serine α-hydrogen (or both glycine methylene hydrogens), which is not observed. Another alternative, consistent with no labilization of the serine α-H, is a direct S_N2 attack by one of the THF nitrogens (N-5 or N-10) on the serine β-carbon with expulsion of the β-hydroxyl. This, however, is unattractive in comparison to an aldol cleavage, where a glycine carbanion fragment can be stabilized by the pyridoxal-P.

It could be that the role of tetrahydrofolate is simply to act as a trap for formaldehyde at the active site after the aldol cleavage has occurred. The folate coenzyme then would serve to carry formaldehyde away from the active site. This could be beneficial to catalysis if the nascent formaldehyde is otherwise only slowly released, perhaps forming imine linkages with active-site lysine residues.

Recent experiments by Tatum et al. (1977) with serine chirally labeled with tritium at C-3 (3*R*-CHTOH or 3*S*-CHTOH) indicate that, as the chiral hydroxymethyl group is converted to the methylene (CHT) of N^5,N^{10}-methylene-THF, a partial loss of chirality occurs (a 24:76 split is observed rather than the expected

0:100 ratio for tritium release/retention for the 3*R*- and 3*S*-enantiomers). This observation suggests that nascent formaldehyde is produced as a free intermediate within the active site, and that on the average 24 of 100 molecules are free long enough to rotate and produce the other, indistinguishable face of the $H_2C{=}O$ carbonyl to the attacking nitrogen of bound THF.

Consistent with the idea that THF may act predominantly as a formaldehyde trap and not be mechanistically essential for the actual cleavage of serine to occur are observations that the homogeneous enzyme cleaves other β-hydroxy-α-amino acids in a process dependent on pyridoxal-P but independent of THF. Thus, both L-threonine and L-allothreonine (the *erythro* diastereomer) are cleaved to glycine and acetaldehyde (Schirch and Gross, 1968).

L-allothreonine ⇌ (Enz—PLP) $H_3C{-}CH{=}O$ + $H_2C(NH_3^{\oplus}){-}COO^{\ominus}$ ⇌ (Enz—PLP) L-threonine

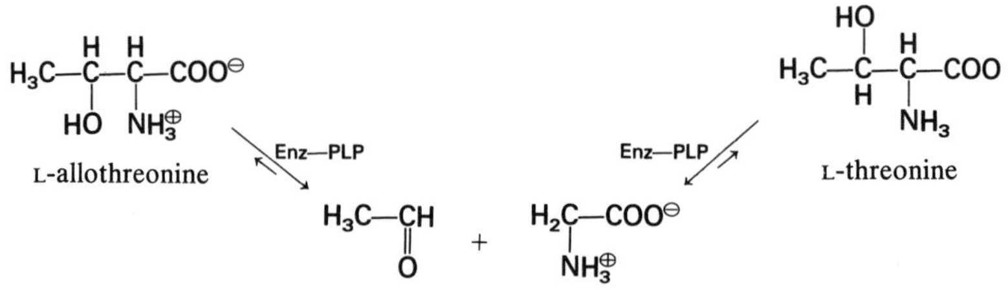

The *erythro* isomer is a better substrate both for cleavage and for condensation of acetaldehyde with glycine in the reverse direction (Palekar et al., 1973). Similarly, both *erythro*- and *threo*-L-phenylserines are cleaved to benzaldehyde and glycine (Schirch and Diller, 1971; Ulevitch and Kallen, 1977). The fact that both *erythro*- and *threo*-β-hydroxy acids are reversibly cleaved means that this enzyme shows an unusual lack of stereospecificity: in the condensation direction, the glycine carbanion equivalent must attack both the *si* and *re* faces of acetaldehyde and benzaldehyde.

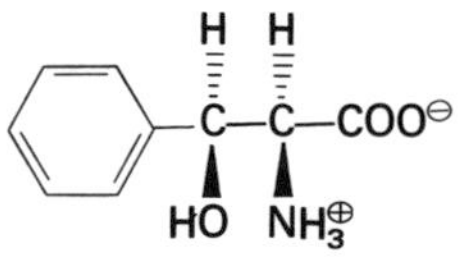

erythro-L-phenylserine

D-Alanine is slowly transaminated (at 10^{-3} times the rate of serine cleavage) with enzyme inactivation, suggesting that the enzyme can abstract an α-hydrogen when it occupies the space normally filled by the hydroxymethyl group of L-serine (Schirch and Gross, 1968). We noted in Chapter 24 that aminomalonate is decarboxylated in ^3H_2O to an equal mixture of R- and S-2-[^{3}H]-glycines, suggesting that it reacts as both a D- and an L-amino acid (Palekar et al., 1973).

THF may also alter active-site geometry on binding. The presence of THF accelerates the exchange of protons from the glycine–pyridoxal-P aldimine complex as though the pK_a of the glycine C—H bond were decreased by a factor of 100 (Schirch and Diller, 1971).

The slow transamination of D-alanine is also accelerated some hundredfold in rate by addition of saturating amounts of tetrahydrofolate. Again this is selective increase of abstraction as a proton of the H_R at C-2 of a bound amino acid. Maybe THF enables an active-site conformation facilitating close approach of the requisite active-site base to that locus. Neither dihydrofolate nor oxidized folate is effective. In this connection, a model for the preferred conformation of the tetrahydropyrazine ring of tetrahydrofolate has recently been proposed from NMR studies (Poe and Hoogsteen, 1978). A half-chair conformation is suggested, as shown here for the R-diastereomer.

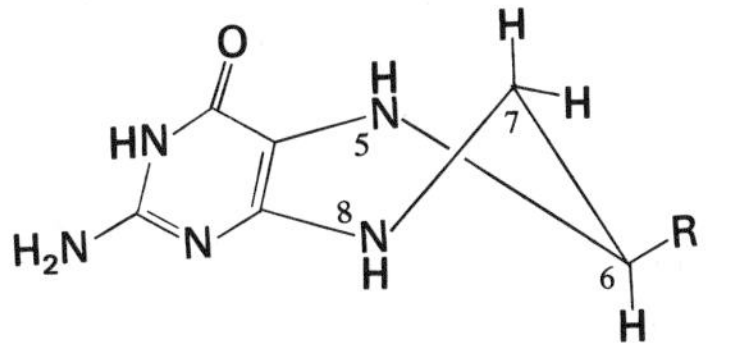

The dihydrofolate and fully oxidized folate would not have this shape.

With the evidence that serine hydroxymethylase removes the H_R but not the H_S proton at C-2 of glycine and D-alanine, Wang et al. (1978) exposed the homogeneous enzyme from lamb liver to both β-fluoro-D-alanine and β-fluoro-L-alanine; very slow rates of α,β-elimination of HF, concomitant with pyruvate production, ensued. On addition of THF, the HF-elimination rate from the D-fluoroalanine (but not from the L-fluoroalanine) increased 250-fold, in keeping with the proton-labilizing role noted above. Now turnover proceeded at $\sim 30\ \text{min}^{-1}$ and, after 60 turnovers per active site, the enzyme was irreversibly inactivated, presumably by alkylative capture of the aminoacrylate intermediate before release.

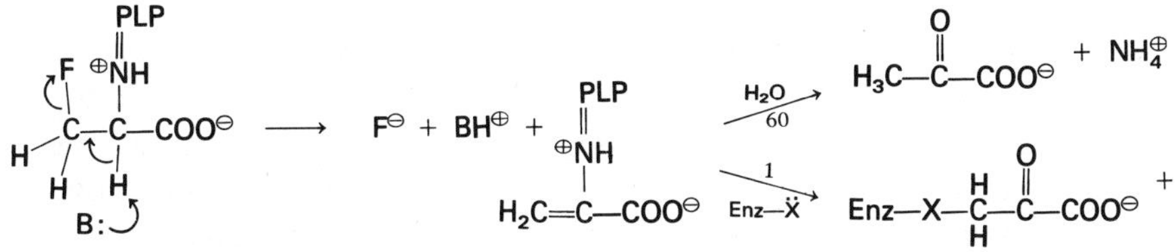

25.B SPECTRUM OF ONE–CARBON TRANSFERS INVOLVING TETRAHYDROFOLATE AS CARRIER

One-carbon fragments are reversibly transferred between THF and specific substrates at three oxidation states: the formate, formaldehyde, and methanol levels. We shall next look at examples of each of these levels.

25.B.1 Transfer of the C_1 Unit at the Formate Level

At the level of formate, enzyme-catalyzed formyl transfers occur to amino-group acceptor nucleophiles. In procaryotic protein synthesis, *N*-formylmethionine is the common amino-terminal residue in all proteins (Ochoa and Mazumder, 1973), and the amino group is enzymatically blocked by formylation at the aminoacyl-tRNA level.

N^{10}-Formyl-THF + Met-tRNA^{fMet} ⟶ *N*-formyl-Met-t-RNA^{fMet} + THF

This formylation ensures that the amino group of the methionyl residue *cannot initiate peptide-bond formation* with some other aminoacylated tRNA when bound to ribosomes; thus it must be the N-terminal residue of the growing protein chain. The formyl donor in formation of *N*-formylmethionine is specifically the N^{10}-formyl-THF.

Other enzyme-catalyzed *N*-formylations from N^{10}-formyl-THF occur during biosynthetic construction of the purine ring system. For example, the molecule formylglycinamide ribonucleotide (discussed in ¶5.C.3.a as a substrate for a glutamine-mediated amination) is biosynthesized by attack of the amino group of glycinamide ribonucleotide on the formyl group of N^{10}-formyl-THF.

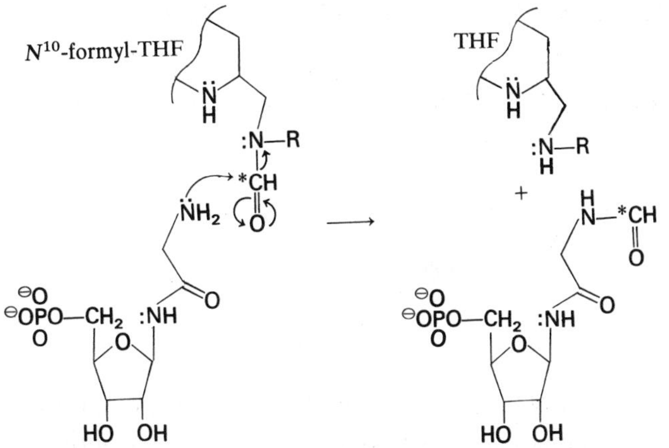

N^{10}-formyl-THF

:N—R

*CH

$\ddot{N}H_2$

O

$^{\ominus}O$

$^{\ominus}OPO—CH_2$:NH

O

O

HO OH

glycinamide ribonucleotide

→

THF

:N—R

H

+

H

N—*CH

O

$^{\ominus}O$

$^{\ominus}OPO—CH_2$:NH

O

O

HO OH

N-formylglycinamide ribonucleotide

A second *N*-formylation occurs later in purine-ring assembly, so that both C-2 and C-8 of the purine ring system originate as the transferable formyl group of N^{10}-formyl-THF.

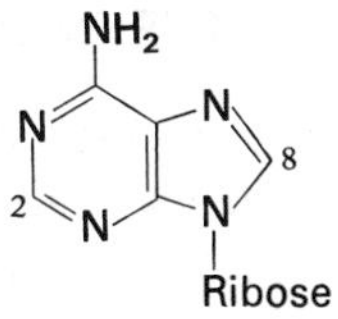

NH_2

N N

8

2 N N

Ribose

adenosine

25.B.2 Transfer of the C_1 Unit at the Formaldehyde Level

In addition to functioning as donor of a formaldehyde unit in serine formation, the N^5,N^{10}-methylene-THF is the reactive species in forming 5-hydroxymethylcytidine and in forming thymidylate, the latter involving a two-electron reduction of the transferred unit to a methyl group (Friedkin, 1973). We shall discuss the synthesis of thymidylate in detail in ¶25.C.

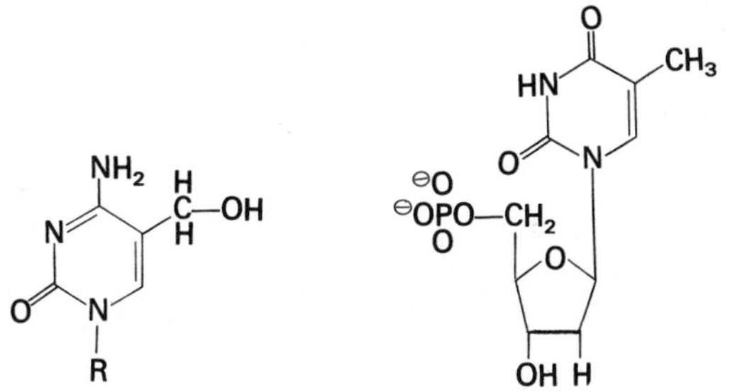

NH_2 H

C—OH

N H

O N

R

5-hydroxymethylcytidine

O

HN CH_3

O N

$^{\ominus}O$

$^{\ominus}OPO—CH_2$

O

O

OH H

thymidylate
(5-methyl-2′-deoxy-UMP)

25.B.3 Transfer of the C_1 Unit at the Methanol Level

Prominent among the reactions at the methanol level is a *methyl transfer* to homocysteine, forming the *S*-methyl group of methionine. The reactive species in this catalysis is the N^5-methyl-THF.

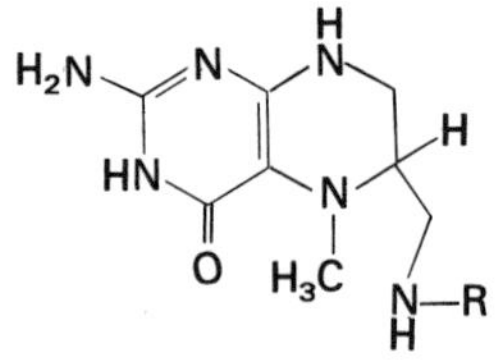

It has been claimed that N^5-methyl-THF is also the methyl donor for methylation of some neuroactive amines such as tryptamine and norepinephrine, as well as at least one nucleoside in a bacterial tRNA molecule (Rabinowitz, 1975). It is clear, though, that N^5-methyl-THF is not the immediate donor in most enzyme-catalyzed methylations, as we shall soon see.

25.B.4 Redox Interconversions of THF Derivatives

The various active forms of THF can undergo enzyme-catalyzed interconversion by a series of two-electron redox changes. The N^5-methyl-THF is oxidized by a flavoprotein dehydrogenase that transfers the two electrons to $NADP^{\oplus}$ (Rader and Huennekens, 1973). The initial oxidized product probably is the N-5 iminium compound, which is then rapidly attacked intramolecularly by N-10 to yield the N^5,N^{10}-methylene-THF. The Enz · $FADH_2$ so generated is reoxidized by $NADP^{\oplus}$. The methylamine portion of the substrate has been oxidized to an imine in the initial product, a typical flavoenzyme oxidative process.

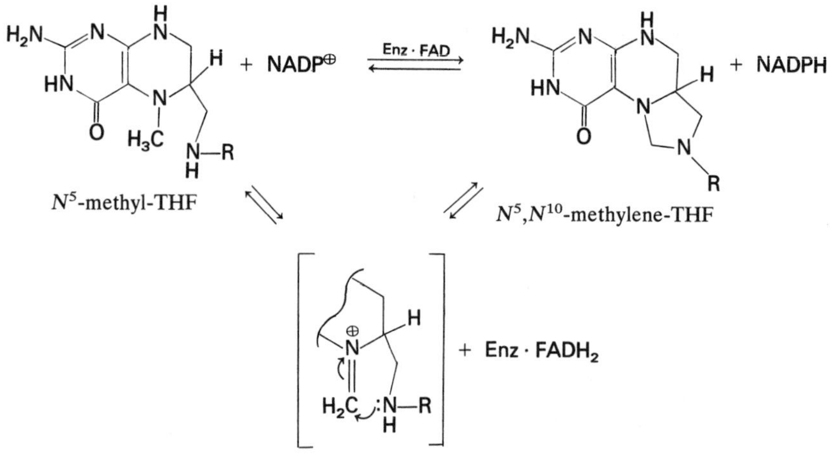

N^5-methyl-THF + $NADP^{\oplus}$ $\xrightleftharpoons{\text{Enz} \cdot \text{FAD}}$ N^5,N^{10}-methylene-THF + NADPH

[iminium intermediate] + Enz · $FADH_2$

The N^5,N^{10}-methylene-THF, in turn, can be oxidized by an NADP$^{\oplus}$-dependent dehydrogenase up to the formate oxidation state to yield as initial product a still-cyclized form, N^5,N^{10}-methenyl-THF (Rader and Huennekens, 1973). The oxidation could be initiated by use of the lone electron pair on N-5 or N-10 to assist in transfer of a hydride equivalent to NADP$^{\oplus}$. For example, N-5 initiation could give the following mechanism.

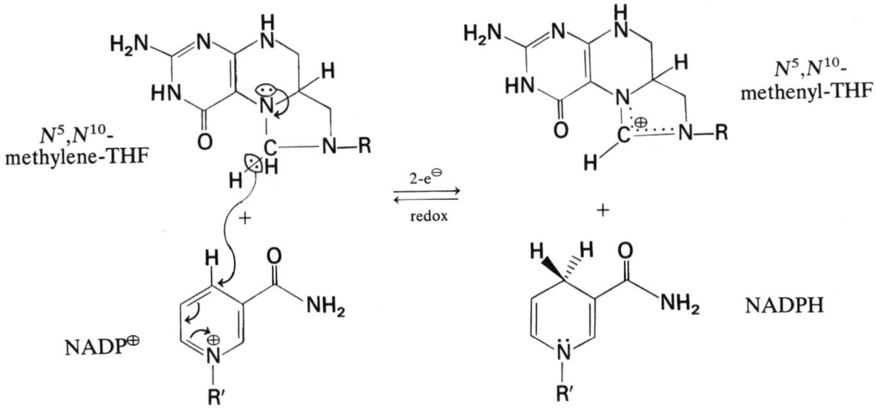

N^5,N^{10}-methylene-THF

NADP$^{\oplus}$

2-e$^{\ominus}$

redox

N^5,N^{10}-methenyl-THF

NADPH

The absolute chirality of prochiral-hydrogen removal from the N^5,N^{10}-methylene group has yet to be determined. The cyclic N^5,N^{10}-methenyl adduct is then convertible to either the N^5-formyl or the N^{10}-formyl species *with no change in oxidation level* by enzymes that cleave ATP to ADP and P_i in the process (Rader and Huennekens, 1973). Little mechanistic information is available on how these processes occur.

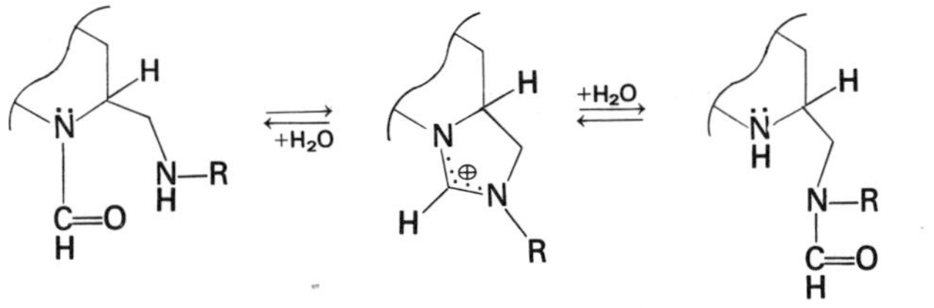

$+H_2O$

$+H_2O$

N^5-formyl-THF $\quad$ N^5,N^{10}-methenyl-THF $\quad$ N^{10}-formyl-THF

Another enzymatic entry to the N^{10}-formyl-THF specifically is provided by N^{10}-formyl-THF synthetase (Blakely, 1969).

$$HCOO^{\ominus} + ATP + THF \rightleftharpoons ADP + P_i + N^{10}\text{-formyl-THF}$$

The enzyme has been crystallized from *Clostridium cylindrosporum.* Although one might expect formation of the mixed anhydride of formate and phosphate as an intermediate, with a formyl group activated for subsequent loss of one of the initial carboxylate oxygens, no partial reactions have been detected. Thus, if such a covalent species has finite existence, it breaks down to products before ADP is released. There may be analogies here with glutamine synthetase, where detection of the putative intermediate γ-glutamyl-P has been so difficult (Chapter 5), or the postulated steps 1 and 2 could be concerted.

$$\mathrm{HC(=O)-O^{\ominus}} + \mathrm{ATP^{4\ominus}} \rightleftharpoons \mathrm{HC(=O)-OPO_3^{2\ominus}} + \mathrm{ADP^{3\ominus}}$$

formyl-P

$$\text{THF (:N–R, H)} + \mathrm{HC(=O)-OPO_3^{2\ominus}} \rightleftharpoons N^{10}\text{-Formyl-THF} + \mathrm{HOPO_3^{2\ominus}}$$

Recent studies indicate that the enzyme can synthesize ATP from ADP and carbamyl-P, used as an acyl-P analogue of the proposed formyl-P.

25.C THYMIDYLATE SYNTHETASE

Thymidylate is 2′-deoxy-5-methyl-UMP and is one of the four constituent heterocyclic bases of DNA. This 2′-deoxynucleotide was first isolated from thymus nucleic acid; hence the name. The enzymatic reaction of thymidylate synthesis is a key step in DNA biosynthesis and has been a target of research in cancer chemotherapy in attempts to selectively inhibit DNA synthesis in rapidly growing tumorous cells (Heidelberger, 1970; Sigman and Mooser, 1975). The enzyme catalyzes de novo formation of a methyl group and is one of three well-established methylations using a tetrahydrofolate species rather than *S*-adenosylmethionine as methyl donor.

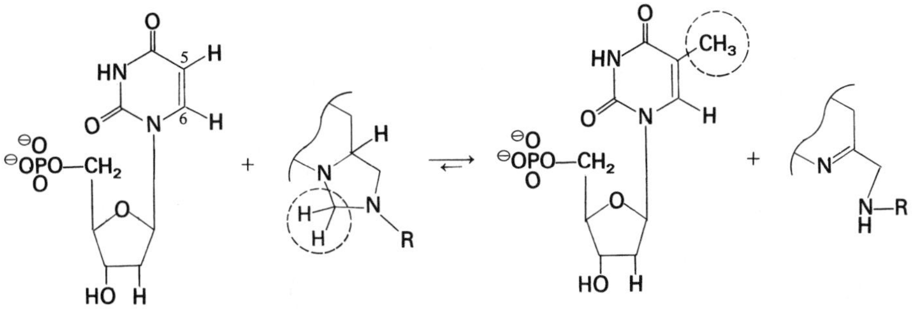

2′-deoxy-UMP (dUMP) $N^{5,10}$-methylene-THF thymidylate (TMP) 7,8-dihydrofolate

The stoichiometry indicates that the transferred one-carbon fragment has experienced a two-electron reduction (from formaldehyde to methanol level), while the carrier coenzyme has undergone a corresponding two-electron oxidation from the 5,6,7,8-*tetrahydro*pteridine to the 7,8-*dihydro*pteridine structure in the product (Friedkin, 1973). Figure 25-1 shows electronic spectra of folate, dihydrofolate, and tetrahydrofolate.

Thus, we might properly have discussed this reaction in Section II with other enzymatic oxidations. Because the folate is in the wrong oxidation state at the end of a turnover, it must be reduced for subsequent reutilization. This is accomplished in vivo by dihydrofolate reductase, which uses NADH as reductant (Benkovic and Bullard, 1973). With 4*R*-[^{3}H]-NADH, one can prepare radioactive tetrahydrofolate tritiated at C-6, which has been useful for mechanistic experiments (Friedkin, 1973). Note that the 7,8-dihydrofolate is a distinct tautomer from the paraquinoid dihydropterin species proposed in phenylalanine hydroxylation (Chapter 12).

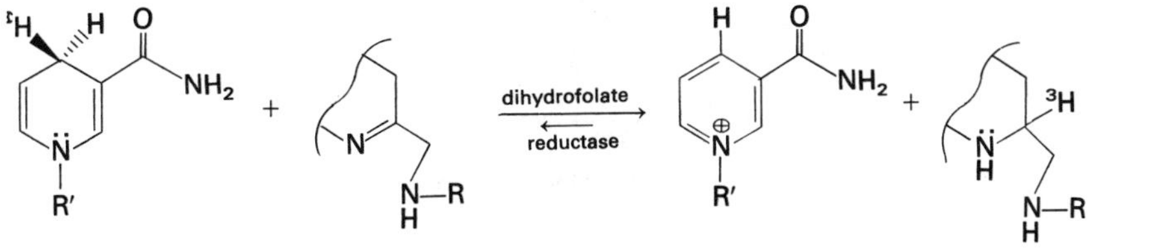

dihydrofolate reductase

4*R*-[^{3}H]-NADH 7,8-dihydrofolate NAD$^{\oplus}$ 6-[^{3}H]-tetrahydrofolate

Clearly, a functional dihydrofolate reductase must be coupled to a functional thymidylate synthetase, and the reductase also is a prime target of drugs for cancer chemotherapy. Two analogues of folate—aminopterin and amethopterin, also known as methotrexate—are competitive inhibitors of the coenzyme. They bind so tightly ($K_D < 10^{-9}$ M) that *they produce functionally irreversible inhibition* without any covalent modification of the reductase (Blakely, 1969). Both aminopterin and methotrexate are 2,4-diaminopteridines; methotrexate is also

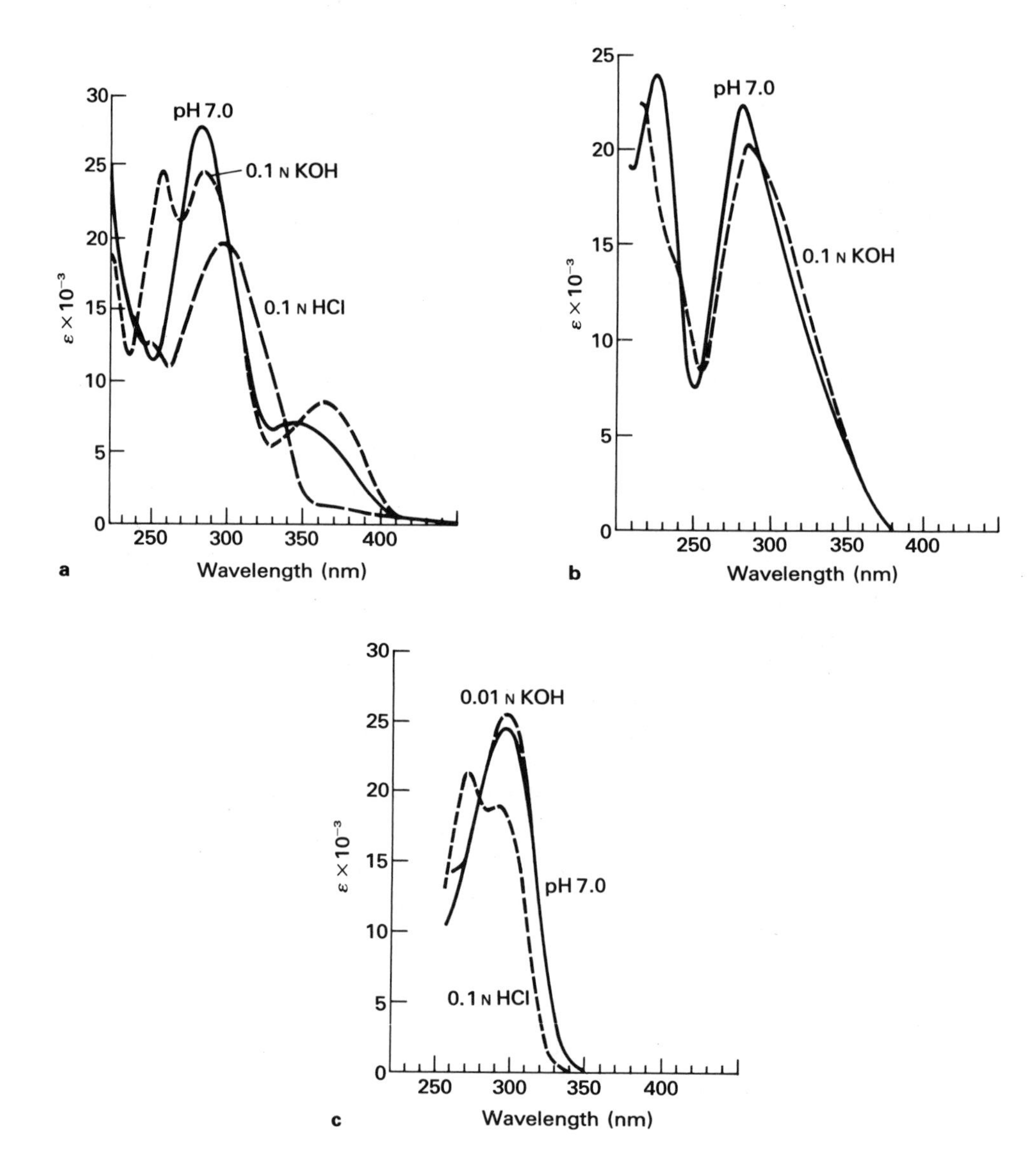

Figure 25-1
(**a**) Absorption spectrum of folate. The spectrum at neutral pH was determined in 0.1 M potassium phosphate buffer at pH 7.0 (From Rabinowitz, 1960, p. 195) (**b**) Absorption spectrum of dihydrofolate prepared by the reduction of folic acid with sodium hydrosulfite. The spectrum at neutral pH was determined in 0.1 M potassium phosphate buffer at pH 7.0 (From Rabinowitz, 1960, p. 197) (**c**) Absorption spectrum of tetrahydrofolate in the media indicated, made 1.0 M with respect to 2-mercaptoethanol. The spectrum at neutral pH was determined in 0.02 M tris(hydroxymethyl)aminomethane buffer, pH 7.0. (From Rabinowitz, 1960, p. 199)

methylated at N-10 to impair its functioning as a carrier. (In the following structure, R is CH_3 in methotrexate or H in aminopterin.) Methotrexate is clinically important in cancer chemotherapy.

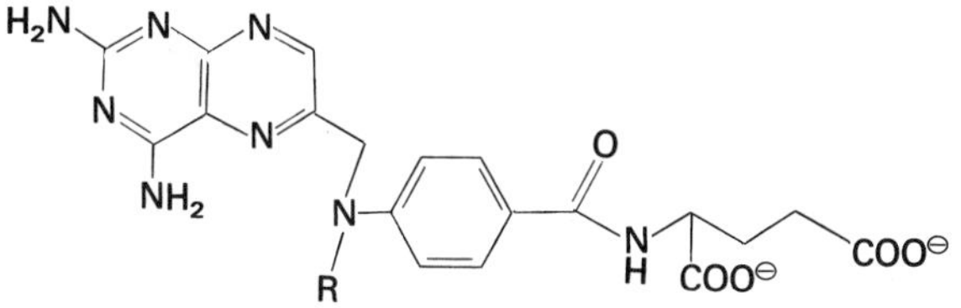

A preliminary publication on the X-ray structure of the 1:1 crystalline complex between dihydrofolate reductase and methotrexate has been presented (Matthews et al., 1977). Figure 25-2 shows a tracing of the polypeptide-chain backbone and the orientation of the methotrexate, bound presumably at the enzyme's active site. A detailed analysis of active-site interactions with the coenzyme analogue may aid in design of other clinically effective folate analogues.

Extensive investigations have been conducted on how thymidylate synthetase effects its catalysis (Friedkin, 1973, Sigman and Mooser, 1975). Use of the N^5,N^{10}-methylene-6-[^{3}H]-THF with enzyme and dUMP produces thymidylate containing tritium in the methyl group, with no loss in specific radioactivity. *This direct hydrogen transfer is consistent with a hydride-ion shift.* A kinetic analysis indicated a value for $(V/K)_H/(V/K)_T$ of 5.2 at 20% conversion of dUMP to TMP, indicating that this hydrogen transfer from C-6 is the rate-determining step (Friedkin, 1973). Control experiments with N^5,N^{10}-methylene-7-[^{3}H]-THF indicated, as expected, no transfer of that tritium to the thymidine formed.

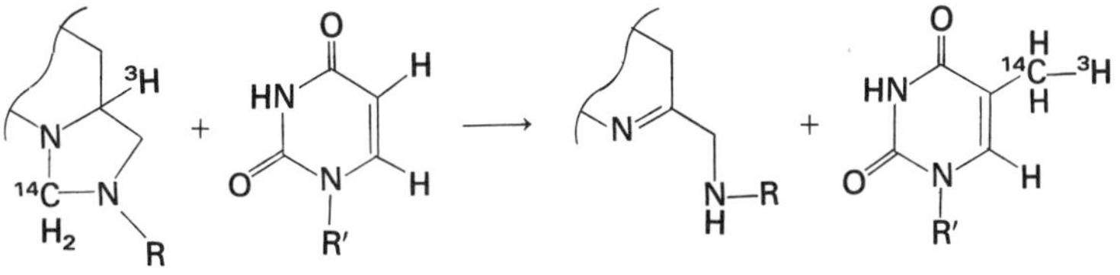

When 5-[^{3}H]-dUMP is utilized as substrate, the rate of tritium release into solvent is 80% to 85% the rate of dihydrofolate production. This differential is real and suggests a k_H/k_T secondary isotope effect consistent with a *rehybridization* (e.g., $sp^2 \rightarrow sp^3$) *at C-5* at some stage in catalysis (see Chapter 4).

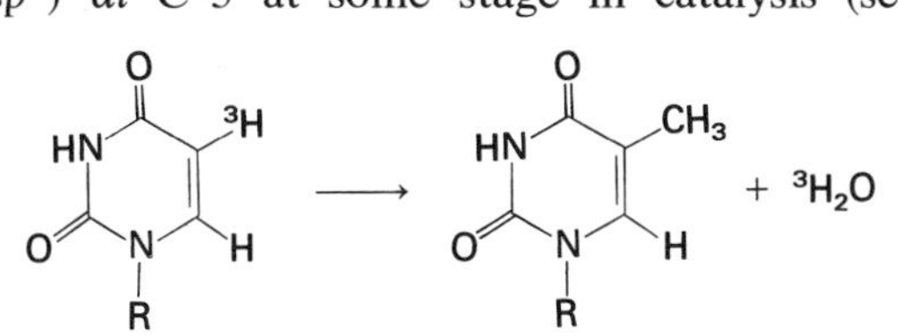

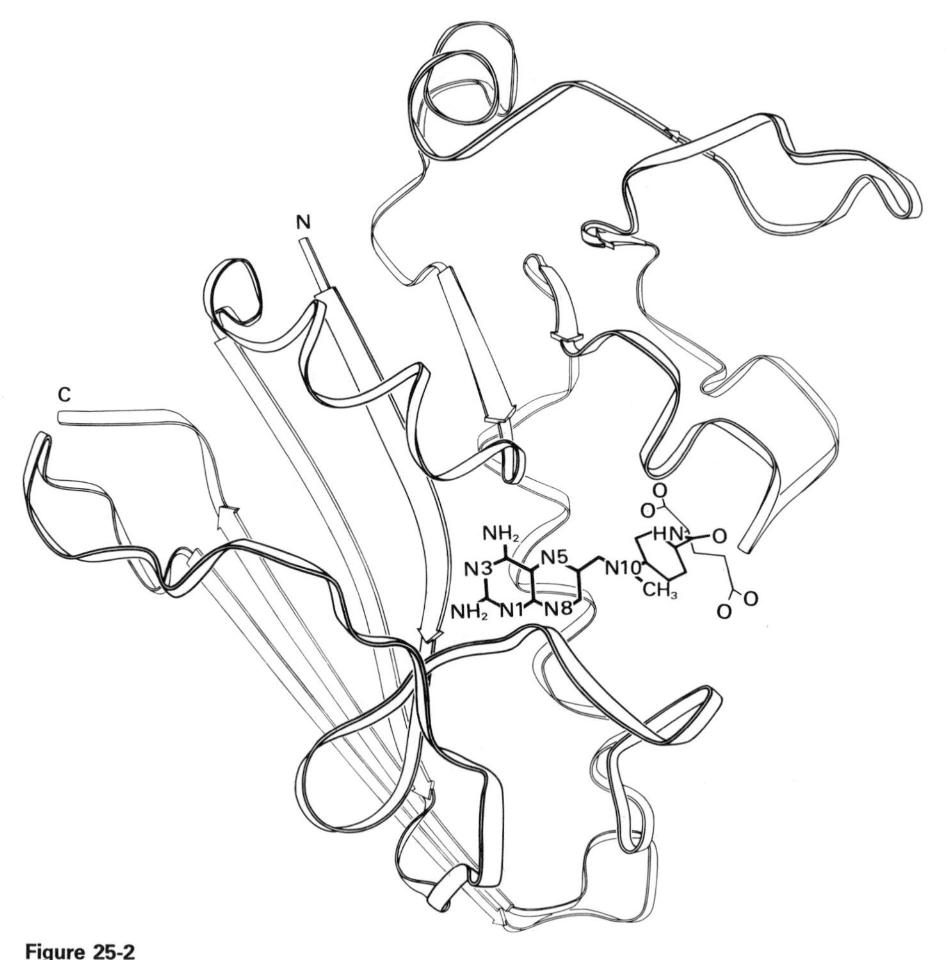

Figure 25-2
Tracing of polypeptide-chain backbone and orientation of methotrexate from preliminary X-ray structure of 1:1 complex between dihydrofolate reductase and methotrexate. (After Matthews et al., 1977)

A somewhat similar enzyme elaborated by bacteriophages uses N^5,N^{10}-methylene-THF and dUMP to make 5-hydroxymethyl-dUMP and THF, although it is a reaction with *no redox* character to it (Friedkin, 1973).

$$\text{dUMP (5-}^3\text{H)} + N^5,N^{10}\text{-Methylene-THF} \rightleftharpoons \text{5-}CH_2OH\text{-dUMP} + \text{THF} + {}^3H_2O$$

With this hydroxymethylase and 5-[^{3}H]-dUMP in the absence of the folate species, a partial exchange of the ^{3}H with H_2O occurs. This washout indicates that exchange (and enolate-anion formation at C-5) is an early step in catalysis here and, by inference, possibly in the thymidylate synthetase reaction as well.

As a model for how such enolization might occur, Santi and colleagues demonstrated a reasonable pathway for hydrogen exchange at C-5 of uracils, by prior addition of nucleophiles to C-6 (Santi and Brewer, 1968).

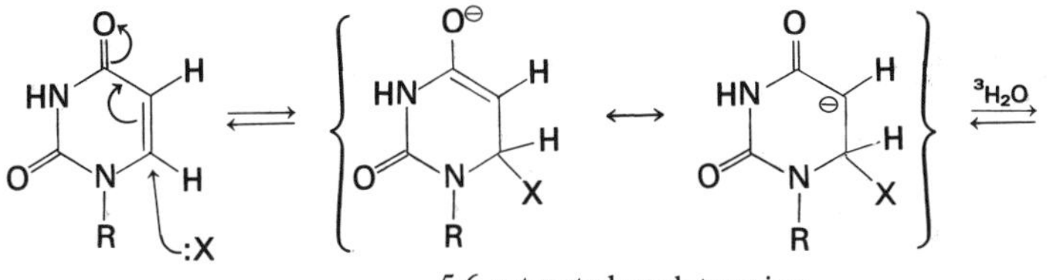

5,6-saturated enolate anion

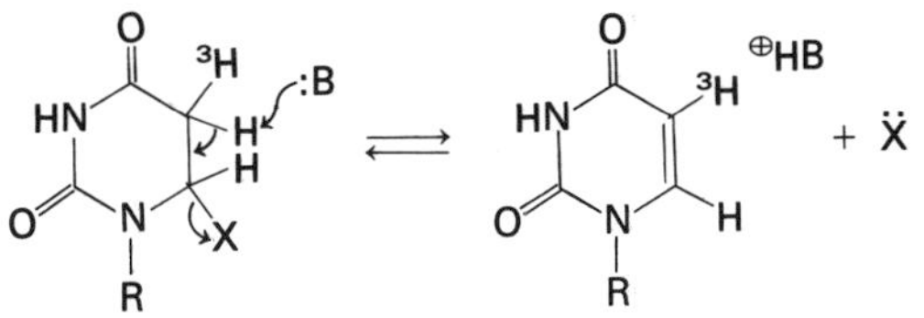

With this information available, one can suggest that the thymidylate synthetase catalysis proceeds in three discrete steps: (1) addition of an enzyme active-site nucleophile to C-6 of dUMP to form an adduct that is 5,6-saturated; (2) attack by the resulting C-5 enolate anion on methylene-THF to add an equivalent of formaldehyde to C-5; (3) the redox change, occurring via a 1,3-hydride shift (presumably suprafacial; Fig. 25-3).

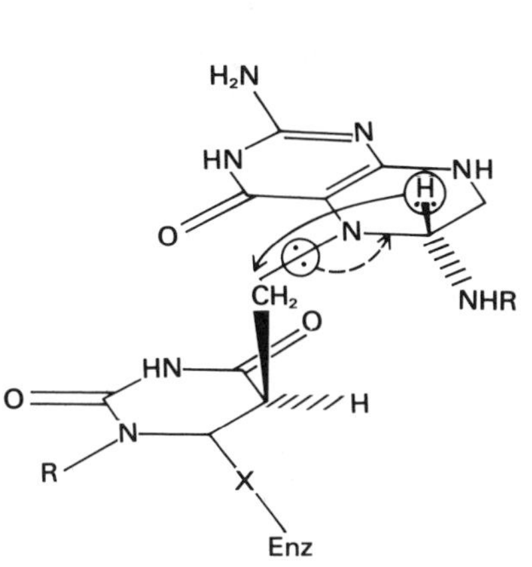

Figure 25-3
A hypothetical suprafacial 1,3-hydride-shift scheme, with arbitrary designation of relative stereochemistry in the intermediate.

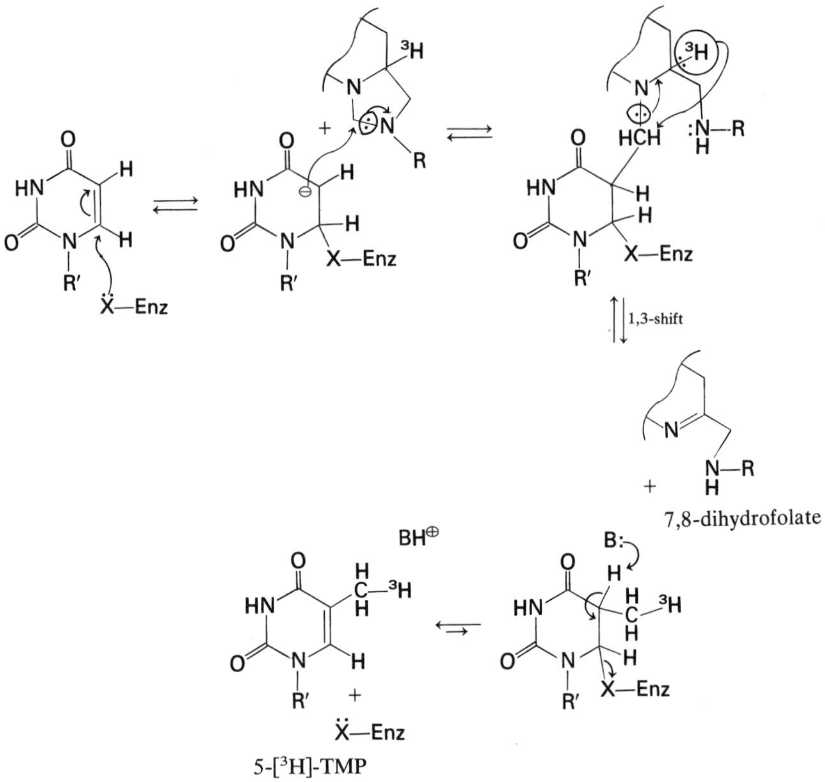

1,3-shift

7,8-dihydrofolate

5-[^{3}H]-TMP

Whether the C-5 anion of the enzyme–dUMP adduct attacks the 5,10-methylene or a 5-iminium cation is unclear. The driving force for the putative 1,3-hydride shift to yield the thymidylate–enzyme adduct might conceivably be the expulsion of the dihydro form of folate.

Experiments with 5-fluorodeoxyuridylate (F-dUMP), a powerful inhibitor of this enzyme and a clinically useful cancer chemotherapy agent, support the above mechanistic scheme (Heidelberger, 1970; Sigman and Mooser, 1975).

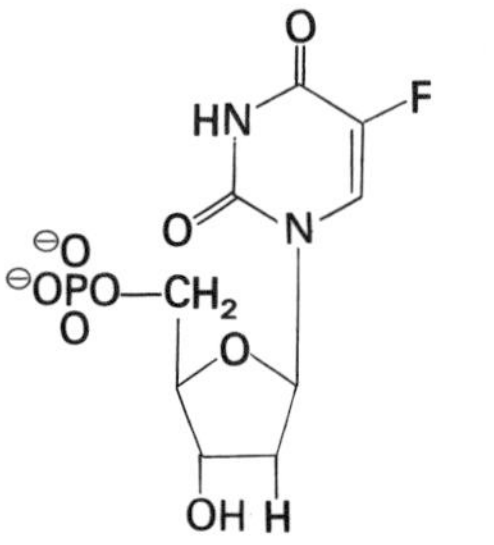

5-fluoro-2′-deoxyuridylate
(F-dUMP)

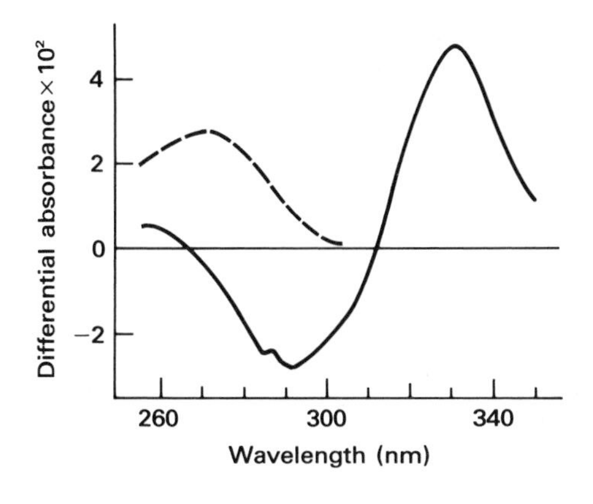

Figure 25-4
Difference spectra showing the loss of F-dUMP and methylene-THF absorbance and the appearance of a new peak at 330 nm. Reference and sample cuvettes contain equal amounts of enzyme. F-dUMP is added to the sample cuvette (*dashed line*), and then equal amounts of methylene-THF are added to both cuvettes (*solid line*). (Reprinted with permission from D. Santi et al., *Biochemistry*, vol. 13, 1974, p. 477. Copyright by the American Chemical Society.)

The fluoro analogue was originally believed to bind reversibly as a tight competitive inhibitor of deoxyuridylate. However, recent studies suggest that inhibition is due to formation of a ternary F-dUMP–N^5,N^{10}-methylene-THF–enzyme complex. Both Santi and colleagues and Heidelberger's group noted that the absorbance maximum at 269 nm of F-dUMP disappears on addition of stoichiometric quantities of enzyme (Fig. 25-4), substantiating the idea that an enzyme nucleophile adds across the 5,6-double bond (Santi and McHenry, 1972; Santi et al., 1974; Danenberg et al., 1974). With 6-[^{3}H]-F-dUMP, a tritiated enzyme complex forms that is not dissociated by chaotropic agents such as 6 M urea or 5 M guanidine hydrochloride, which should disrupt noncovalent complexes (Sigman and Mooser, 1975). Finally, the rate of inactivation caused by F-dUMP is potentiated by N^5,N^{10}-methylene-THF, resulting in binding one equivalent of radioactive F-dUMP per mole of enzyme. One can titrate as little as 3×10^{-4} nmoles of enzyme in this way.

It had been suggested that F-dUMP functions as a substrate *most of the way through the catalytic cycle*, but that reaction is arrested at the stage where $F^{\oplus}$ cannot be abstracted by the enzymatic base. For example,

:B—Enz
O F
HN CH$_3$ ⇸
H
O N X—Enz
R

However, it would appear that catalysis may actually stop at the previous step. Reaction with [^{3}H]-F-dUMP and [^{14}C]-N^5,N^{10}-methylene-THF results in irreversible binding of one mole of each isotopic species. The ultraviolet spectrum of the folate species indicates that that chromophore is no longer the cyclic 5,10-methylene species. It may have the following structure, analogous to the postulated intermediate in normal catalysis (see Sigman and Mooser, 1975).

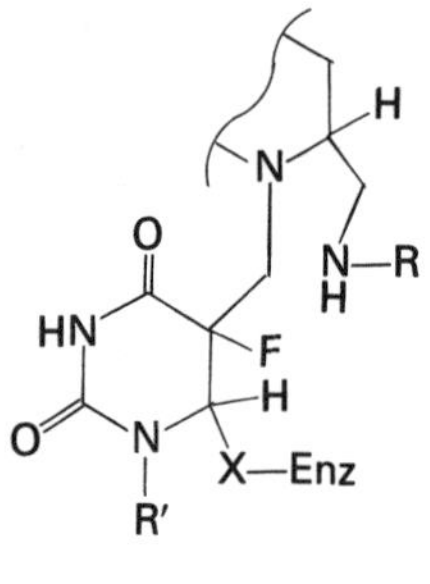

The half-time for breakdown (by reversal) of the ternary complex is 14 hr at 23° C in the presence of excess methylene-THF (Santi et al., 1974). Protease digestion of the radioactive enzyme complex yields a single peptide with both isotopes still present. The enzyme nucleophile that initially adds to the 5,6-double bond appears to be a cysteine sulfhydryl (Pogolotti et al., 1976). When the 6-[^{3}H]-F-dUMP–[^{14}C]-methylene-THF–enzyme complex dissociates and the ^{14}C rate is equated with the ^{1}H rate (i.e., the rate of dissociation of protio species; recall Raftery's experiments on lysozyme, discussed in Chapter 9), a k_H/k_T value of 1.23 is obtained—a secondary kinetic isotope effect, consistent with rehybridization at C-6 ($sp^3 \rightarrow sp^2$) occurring on release of the covalently bound 6-[^{3}H]-F-dUMP moiety from the enzyme (Santi et al., 1974).

25.D METHIONINE FORMATION

The other defined system in which the folate coenzyme serves as methyl donor is in the conversion of homocysteine to methionine, catalyzed by N^5-methyl-THF–homocysteine methyltransferase (Taylor and Weissbach, 1973). The stoichiometry is the following.

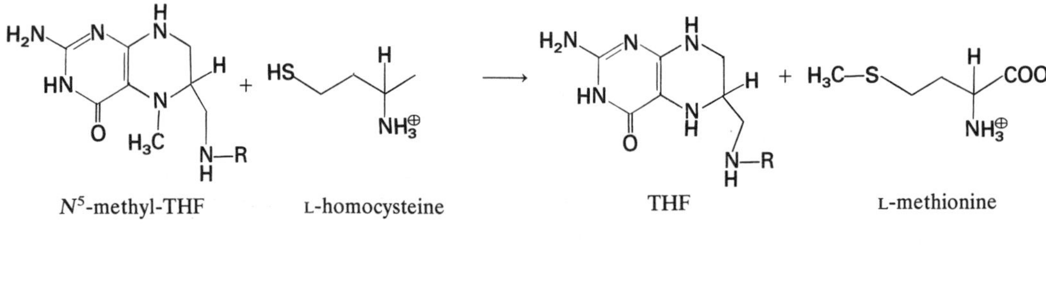

N^5-methyl-THF + L-homocysteine ⟶ THF + L-methionine

Unlike the thymidylate synthetase reaction, this reaction shows no net redox change during the methyl transfer from nitrogen to sulfur.

Two distinct homocysteine methyltransferases of identical overall stoichiometry have been detected in *E. coli*: one requires a form of coenzyme B_{12}; the other does not. The B_{12}-dependent enzyme has been analyzed by Taylor and Weissbach (1973; see also Poston and Stadtman, 1975). Clearly, the function of the corrin coenzyme here is distinct from its role in the 1,2-rearrangements discussed in Chapter 20.

The exact form of the corrin coenzyme in the methyltransferase holoenzyme is not known, In native, resting enzyme, the UV spectrum implicates a Co^{II} species, but no ESR signal is detectable. It has been speculated that it is a Co^{III} form with an OH as upper axial ligand and with the lower axial ligand (dimethylbenzimidazole) uncoordinated ("base off") (Taylor and Weissbach, 1973). Such a situation would undoubtedly modify both the electronic spectrum and the chemistry of the chromophore.

As isolated, the methyltransferase requires an exogenous reducing system for activity—e.g., dithiols or 1,5-dihydroflavins such as $FADH_2$. An additional requirement for turnover is the coenzyme *S*-adenosylmethionine, a molecule we shall discuss in ¶25.E.

On the other hand, methyl-CoB_{12} (Fig. 25-5) with a H_3C—Co bond, generated by chemical synthesis, and homocysteine will react with the enzyme (in the

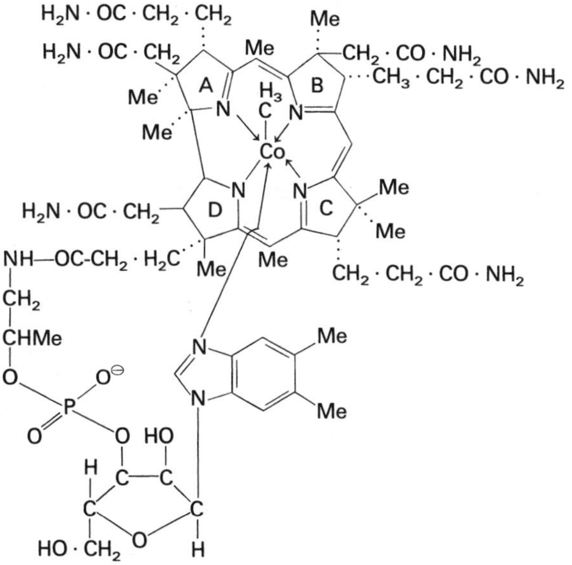

Figure 25-5
Structure of methyl-CoB_{12} (methyl-5,6-dimethylbenzimidazolyl cobalamide). (From Taylor and Weissbach, 1973, p. 142)

dark to avoid photolabilization of the carbon–cobalt bond) to produce methionine. This observation suggests that the thiol of homocysteine can attack the methyl carbon of the coenzyme adduct with expulsion of what should be a Co^I form of the coenzyme. Co^I is an extraordinarily good nucleophile (Hogenkamp, 1975). The species isolated is a Co^{III}–H_2O complex, which could have arisen from reaction of Co^I with H_2O.

How then might the enzyme catalyze formation of methyl-CoB_{12} as an intermediate? The $FADH_2$ might be used to convert the Co^{III} form of the coenzyme to the reactive Co^I, which could then attack the N^5-methyl-THF. Consonant with this idea is the finding that, on mixing N^5-[^{14}C]-methyl-6-[^3H]-THF with the holoenzyme and $FADH_2$, a complex is isolable, either by gel filtration on Sephadex or by acid precipitation. This complex contains the ^{14}C label, but not the tritium label of the folate ring system. Proof that this complex contains [^{14}C]-methyl-CoB_{12} was achieved by extraction of the coenzyme, purification and photolysis of the Co—$^{14}CH_3$ bond aerobically to produce labeled formaldehyde (Taylor and Weissbach, 1973).

Thus, one expects the following sequence.

$FADH_2$ + Enz··· Co^{III}(OH_2)(DBC) $\xrightarrow{\text{2-e}^\ominus \text{ reduction of cobalt}}$ Enz··· Co^I(DBC) + FAD (1)

N^5-methyl-THF + Enz··· Co^I(DBC) ⟶ Enz··· Co^{III}(CH_3)(DBC) (methyl-CoB_{12}) + THF (2)

Enz··· Co^{III}(CH_3)(DBC) + $^\ominus$S-homocysteine ($COO^\ominus$, $NH_3^\oplus$) ⟶ Co^I(DBC) (trapped as Co^{III}–OH acquocobalamin product) + H_3C–S-methionine ($COO^\ominus$, $NH_3^\oplus$) (3)

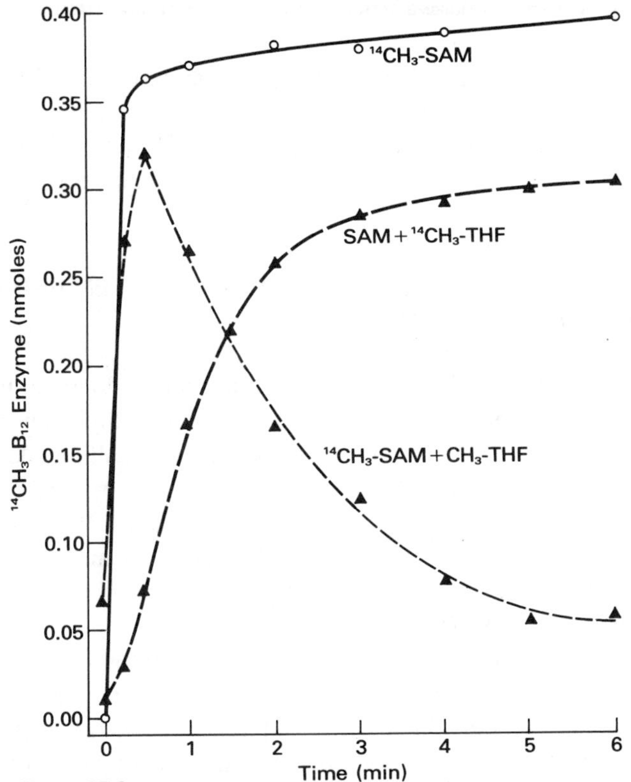

Figure 25-6
Time dependence of [^{14}C]-methyl-CoB_{12}–enzyme formation with [$^{14}CH_3$]-SAM and N^5-[^{14}C]-methyl-THF. All reaction mixtures (0.2 ml) contained 0.46 millimole of CoB_{12}–enzyme, 10 millimoles of SAM (either unlabeled, or [$^{14}CH_3$]-SAM with 19,000 counts per minute per millimole), and, where indicated, 10 millimoles of *dl*-N^5-methyl-THF (either unlabeled, or N^5-[^{14}C]-methyl-THF with 73,000 counts per minute per millimole). Incubations were under H_2 gas at 37° C for the times indicated, and they were initiated within 3 min after the injection at 0° C of reduced flavin. (From Taylor and Weissbach, 1973, p. 146)

However, the fitness of this speculation is somewhat imperiled by the facts that the enzyme shows a requirement for *S*-adenosylmethionine (SAM), and that [^{14}C]-methyl-*S*-adenosylmethionine yields both a radioactive methyl-CoB_{12}–enzyme complex and [^{14}C]-methionine in stoichiometric experiments. A rate study of utilization of methyl groups provided by SAM or N^5-methyl-THF shows the profile indicated in Figure 25-6. Taylor and Weissbach (1973) suggest that the *S*-adenosylmethionine methyl group is used to prime the inactive enzyme in the first turnover, and that all subsequent turnovers use the N^5-methyl-THF. The reasons behind the need for such proposed dichotomous behavior are unclear, if indeed Co^{III}-enzyme is first reduced to Co^{I}-enzyme by $FADH_2$ before any methyl transfers begin.

Unfortunately, the CoB_{12}-independent homocysteine methyltransferase has received essentially no mechanistic study, and it is unclear how methyl transfer from N-5 to the homocysteine sulfur is catalyzed, although direct nucleophilic displacement by the homocysteine thiolate anion is readily conceivable.

25.E METHYLATION OF TOXIC ELEMENTS IN THE ENVIRONMENT BY METHYL COENZYME B_{12}

In addition to its incompletely determined role as methyl donor to homocysteine, methyl-CoB_{12} appears to function in a second biomethylation category, the methylation of elements such as mercury, palladium, thallium, lead, platinum, gold, tin, arsenic, selenium, and chromium. Although no work has been done with purified methyl-transfer enzymes effecting such alkylations, Wood and colleagues have adduced evidence (much of it in model work) that methyl-CoB_{12} and not N^5-methyltetrahydrofolates or *S*-adenosylmethionine (discussed in ¶25.F) are involved (Ridley et al., 1977).

Methyl transfer from *S*-adenosylmethionine (and methyl-THF) occurs to C, O, N, and S atoms acting as nucleophiles: the methyl fragment is transferred as a $—CH_3^{\oplus}$ equivalent (certainly in a concerted reaction where a *free* methylcarbonium ion is *not* generated). On the other hand, biomethylation of $Hg^{2\oplus}$ salts to the potent neurotoxin methyl mercury has the aspect of transfer of a methylcarbanion ($—CH_3^{\ominus}$) equivalent. Ridley et al. (1977) note that methylcobalamin may well cleave heterolytically to yield electron-deficient or electron-rich methyl equivalents (and Co^{I} or Co^{III} oxidation states, respectively), depending on the electronic nature of the methyl acceptor. Further, they claim that heterolytic $H_3C—Co^{III}$ cleavage also occurs with such elements as $Sn^{3\oplus}$ to yield methyl tin derivatives. In the following scheme, the methyl anion, radical, and cation are not held to be free entities but are shown only to indicate the mechanistic equivalents involved.

$$\text{(a)}\quad CH_3\text{–}Co^{III}(BZ) \rightarrow Co^{III}(BZ) + \text{“}CH_3^{\ominus}\text{”} + Hg^{2\oplus} \longrightarrow H_3C\text{—}Hg$$

$$\text{(b)}\quad CH_3\text{–}Co^{III}(BZ) \rightarrow Co^{II}(BZ) + \text{“}CH_3^{\cdot}\text{”} + \dot{S}nX_3 \longrightarrow H_3C\text{—}Sn\text{—}X_3$$

$$\text{(c)}\quad CH_3\text{–}Co^{III}(BZ) \rightarrow Co^{I}(BZ) + \text{“}CH_3^{\oplus}\text{”} + {}^{\ominus}SR \longrightarrow H_3C\text{—}S\text{—}R$$

The redox states and redox potentials of the various elements that may undergo corrinoid-mediated methylation may condition whether mechanism a (the carbanion equivalent) or mechanism b (the radical equivalent) proceeds. Thus, Ridley et al. (1977) report that, in chemical studies, elements that react by the electrophilic heterolytic cleavage mechanism (a) have reduction potentials more positive than +0.8 V. These include $Pb^{4\oplus}$, $Tl^{2\oplus}$, $Se^{6\oplus}$, $Pd^{2\oplus}$, and $Hg^{2\oplus}$. Although reduced arsenic salts can attack *S*-adenosylmethionine (see ¶25.F), it is possible that they react with methyl coenzyme B_{12} by radical processes. The molecular toxicology of movement of toxic elements in the biosphere by enzymatically catalyzed processes obviously will be a subject of increased scrutiny.

25.F TRANSMETHYLATIONS INVOLVING *S*-ADENOSYLMETHIONINE

Du Vigneaud's group in the 1940s determined methionine to be the source of many methyl groups in natural metabolites, based on the feeding of $[^{14}CD_3]$-methionine to rats.

$D_3{}^{14}C$ S $COO^{\ominus}$ $NH_3^{\oplus}$

Cantoni, a decade later, found that the amino acid itself is not the immediate methyl donor, but is first activated by ATP to form the active methylating agent, *S*-adenosylmethionine (Fig. 25-7), in which the transferable methyl is attached now to a sulfonium cation (Shapiro and Schlerk, 1965; Cantoni, 1975).

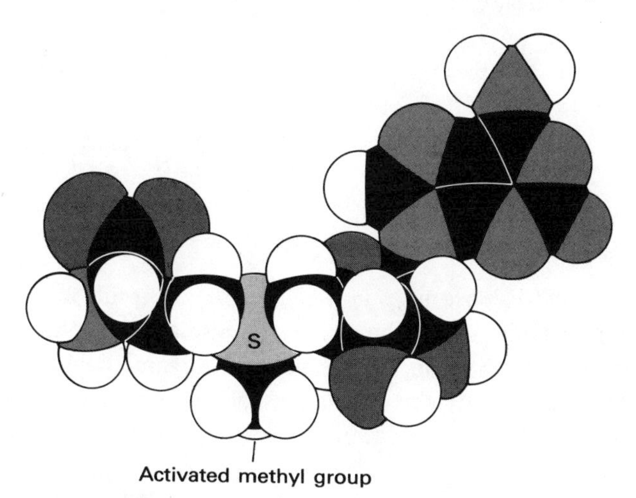

Figure 25-7
Space filling model of *S*-adenosylmethionine. (From L. Stryer, *Biochemistry*, p. 508. W. H. Freeman and Company. Copyright © 1975.)

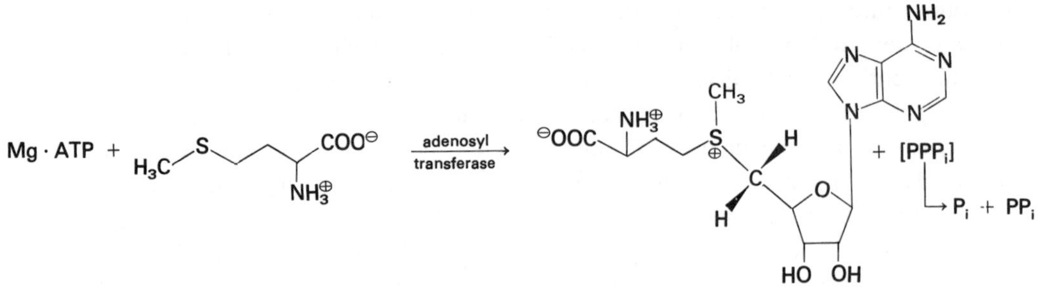

Mg · ATP + H_3C–S–…–$COO^{\ominus}$ ($NH_3^{\oplus}$) $\xrightarrow[\text{transferase}]{\text{adenosyl}}$ S-adenosylmethionine (SAM) + $[PPP_i]$ → $P_i + PP_i$

S-adenosylmethionine
(SAM)

The reaction catalyzed by methionine adenosyl transferase involves an alkylation at C-5′ of the ribose moiety of ATP with displacement of the triphosphate chain. This adenosyl transfer is an unusual enzymatic reaction; the nucleophilic sulfur atom of methionine has attacked at the ribose carbon rather than at the α-, β-, or γ-phosphorus atom (Mudd, 1973). The only other adenosyl transfer known occurs during CoB_{12} formation when the nucleophilic cobalt of the corrin ring (at the Co^{I} oxidation state?) attacks the same C-5′ of ribose to form the carbon–cobalt bond in the active form of the coenzyme.

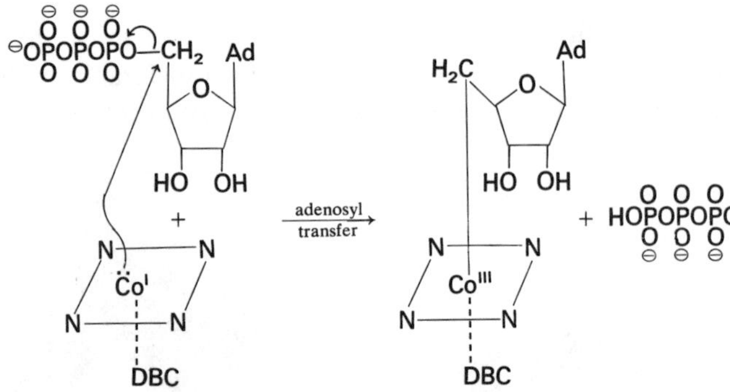

$\xrightarrow[\text{transfer}]{\text{adenosyl}}$

coenzyme B_{12}

The observed P_i and PP_i products in SAM formation result from cleavage of the initial triphosphate while still at the active site, displacing the equilibrium in favor of SAM formation. The enzyme shows potent inorganic triphosphatase activity.

The positive charge of the sulfonium ion destabilizes the SAM molecule thermodynamically ($\Delta G^{0\prime} = -13$ kcal/mole $= -54$ kJ/mole) and has converted the unreactive methylthiol group of methionine into *an active alkylating agent*, susceptible to attack by nucleophiles, yielding the methylated nucleophile and the alkyl sulfide (*S*-adenosylhomocysteine) as a good leaving group. The displacement stoichiometry for methyl transfer is the following.

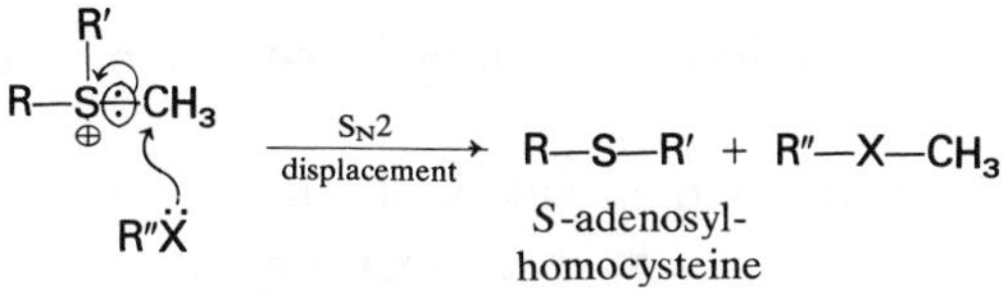

$$R{-}\overset{\oplus}{S}(R'){-}CH_3 + R''\ddot{X} \xrightarrow[\text{displacement}]{S_N2} R{-}S{-}R' + R''{-}X{-}CH_3$$

S-adenosyl-
homocysteine

It has been estimated that *S*-adenosylmethionine is some 10^3-fold more reactive with polarizable nucleophiles (oxygen, nitrogen, sulfur) than is N^5-methyl-THF (although quaternization at nitrogen could lower this differential), and this probably explains why SAM is the preferred biological methylating agent (Cantoni, 1975). Further indication that SAM is highly reactive stems from its facile conversion at 25° C, 10′, in 0.1 N NaOH to adenine and a pentosylmethionine.

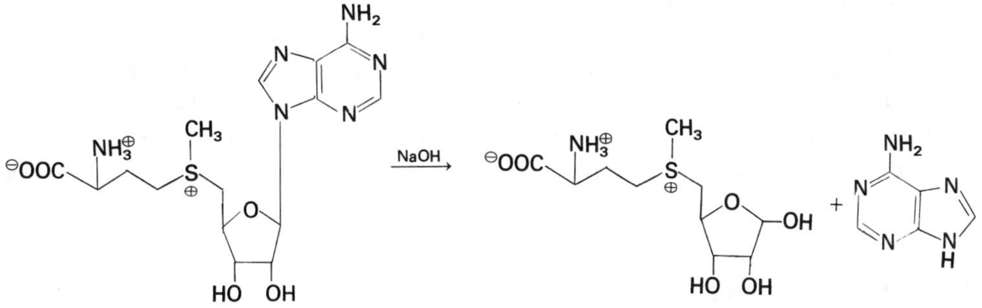

NaOH

The activating effect of the positive charge on the sulfonium ion has caused an unusual labilization of the glycosidic bond, five carbons away. (The only other labile glycosidic link in biological systems is in NAD$^{\oplus}$ and NADP$^{\oplus}$, where the positive charge is on the pyridine nitrogen.) One possible mechanism for the labilization of the glycosidic linkage in SAM is the following.

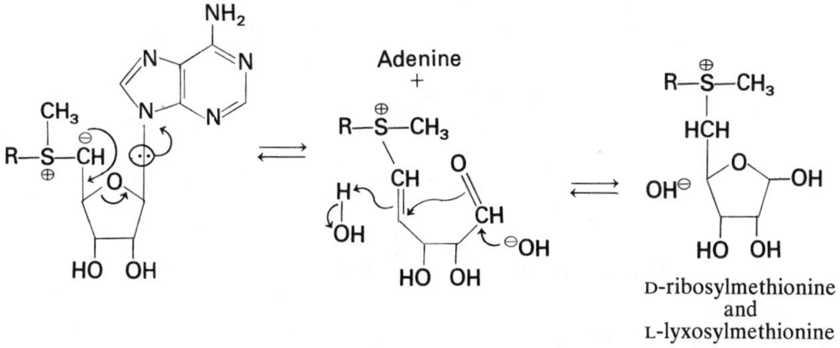

Adenine
+

D-ribosylmethionine
and
L-lyxosylmethionine

The adjacent S$^{\oplus}$ stabilizes such a carbanion (probably via d-p^{π} orbital overlap), and ylid species are, of course, important in the chemical reactivity of sulfonium and phosphonium compounds, *allowing for facile carbanion formation.* One sees loss of chirality at C-4 of the pentosyl product (i.e., both the D-ribosylmethionine and L-lyxosylmethionine products are formed), consistent with an unsaturated intermediate.

The biomolecules methylated enzymatically from SAM include all the major structural types. DNA and tRNA molecules have methylated bases (5-methyl-U, 5-methyl-C, 2-methyl-A, 7-methyl-G) at maturation (Kerr and Borek, 1973), as does the 5′-end of viral mRNA molecules (7-methyl-G). A variety of *N*-methylations of proteins are known, forming ε-*N*-methyl lysines, N^1- and N^3-methyl histidines, and *N*-methylarginine isomers. Some protein carboxylate methylations also have been reported, but the methylester products are often labile to hydrolysis (Cantoni, 1975). Most of the methylated proteins are structural or carrier proteins (histones, flagellar proteins, myosin, actin, rhodopsin) and, although cytochrome *c* in some organisms is methylated, no methylated amino acids have yet been detected in enzymes (Cantoni, 1975). A number of *O*-methylated polysaccharides (6-*O*-methylglucosyl groups) are known. Ethanolamine-containing phospholipids are converted to choline-containing phospholipids after three successive methylations. We shall consider representative methylations where sulfur, oxygen, nitrogen, and carbon, respectively, serve as nucleophiles.

25.F.1 Attack by Sulfur

Transsulfurations are relatively rare. One of possible physiological significance is formation of *S*-methylcysteine.

$$R-S^{\oplus}(CH_3)-CH_2CH_2-CH(NH_3^{\oplus})-COO^{\ominus} + HS-H_2C-CH(NH_3^{\oplus})-COO^{\ominus} \rightleftharpoons$$

SAM cysteine

$$H_3C-S-CH_2-CH(NH_3^{\oplus})-COO^{\ominus} + R-S-CH_2CH_2-CH(NH_3^{\oplus})-COO^{\ominus}$$

S-methylcysteine *S*-adenosylhomocysteine (SAH)

25.F.2 Attack by Oxygen

One characterized *O*-methyltransferase is the catechol *O*-methyltransferase, found in both the central and peripheral nervous systems. It shows broad specificity for metabolizing catecholamines such as ephinephrine (adrenalin).

HO, HO epinephrine + SAM (R–S⊕–CH₃ … COO⊖, NH₃⊕) ⟶ HO, H₃C—O metaphrine + SAM

Hegazi et al. (1976) measured V_{max} rates both with $[CH_3]$- and with $[CD_3]$-*S*-adenosylmethionine (on dihydroxyacetophenone as cosubstrate) and found a secondary deuterium isotope effect (Chapter 4) of $V_H/V_D = 0.86 \pm 0.04$, ruling against substantial rehybridization of the migrating methyl carbon, but consistent with a direct transfer in an S_N2 transition state.

25.F.3 Attack by Nitrogen

Methyl transfer to amine nitrogen is a very common process (choline formation, creatine formation) and is illustrated by the conversion of norepinephrine to epinephrine by the phenylethanolamine *N*-methyltransferase in the adrenal medulla.

SAM + norepinephrine ⟶ epinephrine + SAH

25.F.4 Methyl Transfer to Carbon

As suggested earlier, it is likely that all the alkylations discussed thus far occur by straightforward attack of sulfur, oxygen, or nitrogen lone electron pairs on the methyl group of *S*-adenosylmethionine. Alkylation at carbon clearly involves carbanion attack in some cases, but some reactions involving *C*-methylation at olefinic centers are mechanistically more obscure.

25.F.4.a Modification of bases in tRNA

The enzyme uracil-RNA methylase converts certain UMP residues in transfer RNA substrates to the 5-methyl-UMP derivatives. The enzyme also will convert cytidylate residues to 5-methylcytidylates.

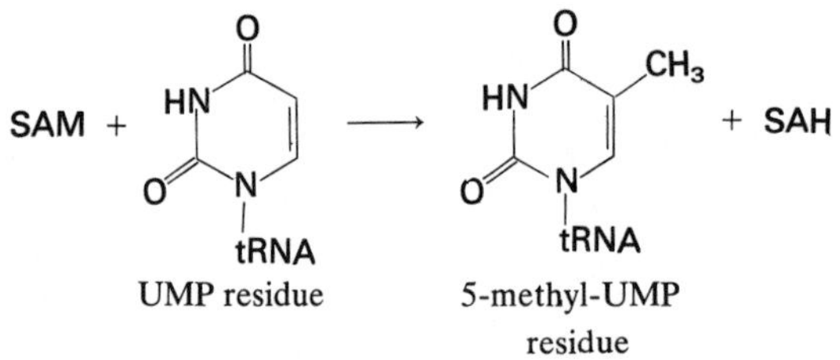

SAM + [UMP residue in tRNA] ⟶ [5-methyl-UMP residue in tRNA] + SAH

UMP residue

5-methyl-UMP residue

Labeling studies with [CD_3]-SAM prove that [CD_3]-5-methyluracil is the product; the methyl group is transferred intact (Kerr and Borek, 1973). One might compare this reaction with that catalyzed by thymidylate synthetase (¶25.C). In the tRNA methylase, one expects that a uracil C-5 enolate anion is the attacking nucleophile. That enolate may be formed by initial addition of an enzyme nucleophile at C-6 of the uracil ring.

Floss and colleagues have recently reported the preparation of [$^1H^2H^3H$]-SAM (only one enantiomer of the chiral methyl group was present) and have used it to probe the stereochemical outcome of a methyl transfer to carbon in fungal production of indolmycin (Mascaro et al., 1977). The chirality of the $C^1H^2H^3H$ group in the product was determined after its degradative extrusion as acetic acid and processing by malate synthase (as described in Chapters 22 and 23). The attacking nucleophile is presumed to be the enolate carbanion and proceeds with inversion at the transferring, chiral methyl center. Inversion is consistent with a direct S_N2 attack by the carbanion equivalent.

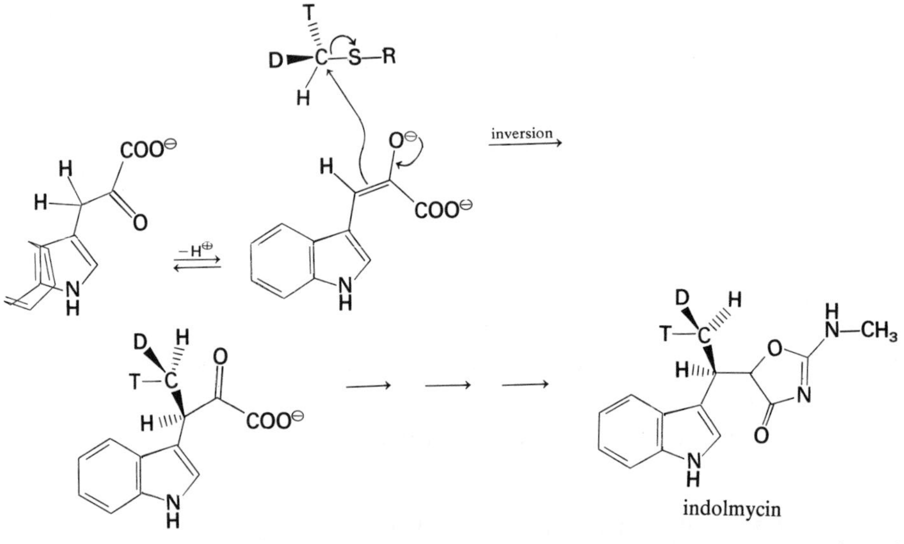

inversion ⟶

$-H^{\oplus}$

indolmycin

Chiral [$^1H,^2H,^3H$]-*S*-adenosylmethionine also has been used by Arigoni and colleagues to show that methyl transfer to carbon, nitrogen, or sulfur in other cases also goes with inversion (D. Arigoni, personal communication).

25.F.4.b Methyl transfer to an olefinic linkage, case 1

The tuberculosis-causing pathogen *Mycobacterium tuberculosis* synthesizes 10-methylstearate (tuberculostearate) from oleate, the cis-Δ^9-monounsaturated C_{18} fatty acid.

SAM + oleate (*cis*-Δ^9) → SAH + 10-methylstearate

Use of [CD_3]-SAM gives [CD_2H]-10-methylstearate, suggesting that the methyl group becomes a methylene group at some stage in the transformation (Law, 1971). Indeed, 10-methylenestearate is isolable from the organism. Also, 9,10-[$^2H_1$]-dideuterooleate yields 9-[$^2H_2$]-product, suggesting a 1,2-hydride (deuteride) shift from C-10 to C-9. Hydride shifts signal carbonium-ion intermediates, and one can hypothesize the following sequence: attack on the sulfonium cation, issuing from the π-electron cloud of the olefin, leads to an initial C-9 carbonium ion, which rearranges to the more stable tertiary carbonium ion at C-10 by the indicated hydride shift.

attack by π-e$^\ominus$ → SAH + carbonium ion; carbonium ion rearranges; elimination and olefin formation → 10-methylenestearate; reductive saturation? → 10-methylstearate

Abstraction of a methyl proton would quench the tertiary carbonium ion and generate the olefinic 10-methylenestearate as first product. A reductive saturation (perhaps a flavoprotein) might then proceed to convert this species to 10-methylstearate.

25.F.4.c Methyl transfer to an olefinic linkage, case 2

Olefinic linkages of fatty acyl groups can also be saturated by *S*-adenosylmethionine in a different mode, by conversion of olefinic double bond to a cyclopropane group. In contrast to 10-methylstearate formation, cyclopropanoid formation occurs on acyl groups at the phospholipid level. An oleyl phospholipid is converted to a dihydrosterculyl phospholipid as *E. coli* grows into the stationary phase, producing physical changes in membrane properties (Knivett and Cullen, 1965).

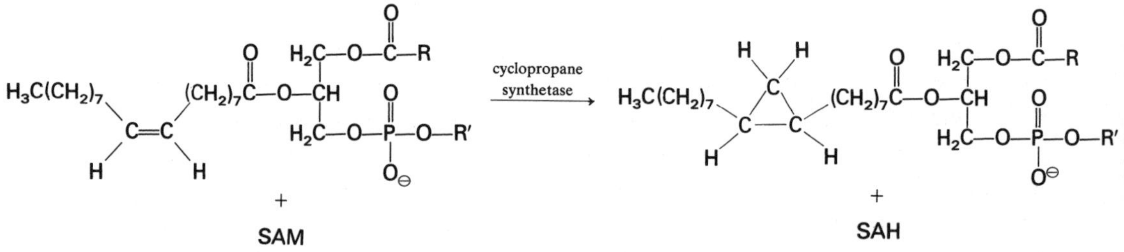

$H_3C(CH_2)_7$ CH=CH $(CH_2)_7$C(=O)—O—CH(CH_2—O—C(=O)—R)—CH_2—O—P(=O)(O$^\ominus$)—O—R′ + SAM $\xrightarrow{\text{cyclopropane synthetase}}$ $H_3C(CH_2)_7$ (cyclopropane, CH_2 bridge) $(CH_2)_7$C(=O)—O—CH(CH_2—O—C(=O)—R)—CH_2—O—P(=O)(O$^\ominus$)—O—R′ + SAH

No hydride shift occurs between C-9 and C-10, arguing by this slender criterion against the methylene fatty acid as intermediate (as was postulated for case 1). As might be expected, [CD_3]-SAM produces the [CD_2]-cyclopropane product (Law, 1971). Due to problems of enzyme instability, no more is known about the mechanism of catalysis (including the nature of any intermediate species).

Other cyclopropanoid molecules are known, but they need not derive the methylene bridge from SAM. One such example is cycloartenol (¶20.A.2), where the cyclopropylmethylene derives from an angular methyl group, in turn biosynthetically derived from an acetate methyl group. Cyclopropene fatty acids are found in seed oils, where they arise by dehydrogenation of the cyclopropane precursors (Hooper and Law, 1965).

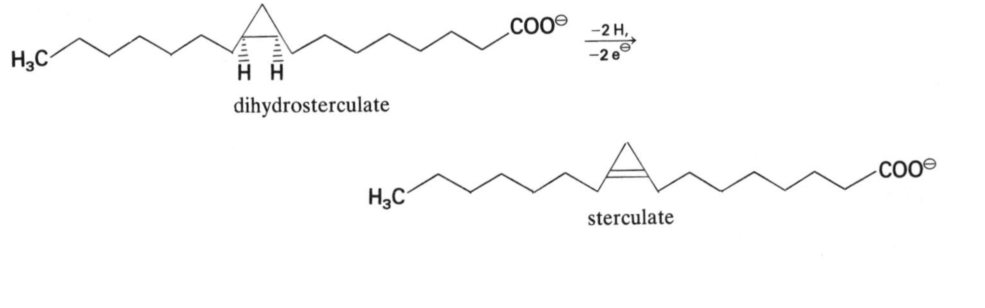

H_3C—(chain)—cyclopropane(H H)—(chain)—$COO^\ominus$ $\xrightarrow[-2e^\ominus]{-2H^+}$

dihydrosterculate

H_3C—(chain)—cyclopropene—(chain)—$COO^\ominus$

sterculate

Essentially nothing is known about the oxidative desaturation producing the highly strained cyclopropenes.

25.F.5 Transfer of the Other Alkyl Substituents of *S*-Adenosylmethionine

To this point, we have discussed the enzyme-catalyzed attack of various biological nucleophiles on the methyl carbon of *S*-adenosylmethionine. The methyl group in question is only one of the three alkyl substituents at the sulfonium cation and, in principle, nucleophiles could attack the other alkyl groups attached to sulfur.

$H_3C{-}\overset{\oplus}{S}{-}CH_2$ Ad
γ HCH
β HCH
α $H_3\overset{\oplus}{N}{-}CH$ HO OH
$COO^{\ominus}$

(1) RẌ attacks methyl
(2) RẌ attacks γ-methylene of methionine side chain
(3) RẌ attacks ribose C-5′

$H_3C{-}X{-}R$
(methyl transfer)

$R{-}X{-}CH_2{-}CH_2{-}CH(NH_3^{\oplus}){-}COO^{\ominus}$
(aminobutyryl transfer)

$R{-}X{-}CH_2$ Ad, HO OH
(adenosyl transfer)

Recent evidence has accrued for pathway 2, and we shall also examine a hint that pathway 3 may be operant.

25.F.5.a Pathway 2: *S*-adenosylmethionine decarboxylation and aminopropyl transfer

The attack of an amine nucleophile on the γ-methylene group occurs in the enzymatic synthesis of spermidine (Tabor and Tabor, 1976), but only after initial decarboxylation of *S*-adenosylmethionine to *S*-adenosylhomocysteamine and CO_2 by the enzyme SAM decarboxylase (¶24.A.3). As we noted in passing in Chapter 24, this decarboxylase does not contain pyridoxal-P as cofactor but rather appears to use a covalently bound pyruvyl group (or its functional equivalent at an internal locus in the polypeptide chain, a dehydroalanyl residue) as electron sink.

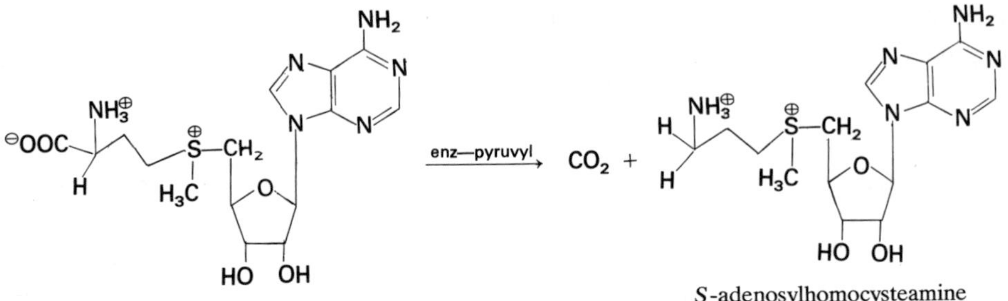

SAM $\xrightarrow{\text{enz—pyruvyl}}$ CO_2 + *S*-adenosylhomocysteamine

This decarboxylase activity is high in pancreatic cells and in prostatic tissues and seminal fluid. The *S*-adenosylhomocysteamine then serves as substrate for donation of the propylamine group to 1,4-diaminobutane (putrescine), catalyzed by an aminopropyl transferase (Bowman et al., 1973). The triamino C_7 product is spermidine; the coproduct is 5-thiomethyladenosine.

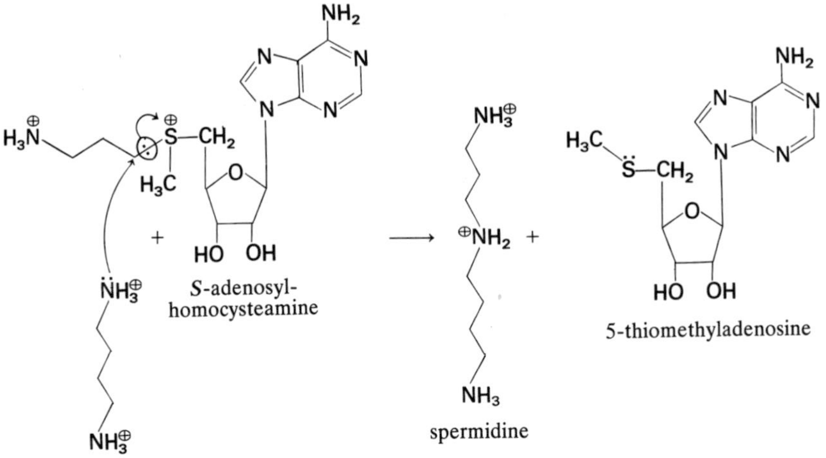

S-adenosyl-homocysteamine + putrescine ⟶ spermidine + 5-thiomethyladenosine

These two polyamines, putrescine and spermidine, are of physiological importance both for microorganisms and for animal cells (Tabor and Tabor, 1976). They are found in high concentrations in actively growing animal cells. Several suggestions have been made that, because these polycations bind tightly to the anionic nucleic acids (both DNA and RNA), these polyamines may control rates of nucleic-acid biosynthesis (Tabor and Tabor, 1976). The possibility has further been raised that their metabolism is intimately connected with increased rates of nucleic-acid metabolism in rapidly growing tumor cells. Tabor and Tabor (1976) have noted that putrescine and spermidine are found in bacteria, whereas spermidine and a higher polyamine (spermine) accumulate in eucaryotes. The major path for putrescine formation is probably via ornithine decarboxylase (Tabor and Tabor, 1976).

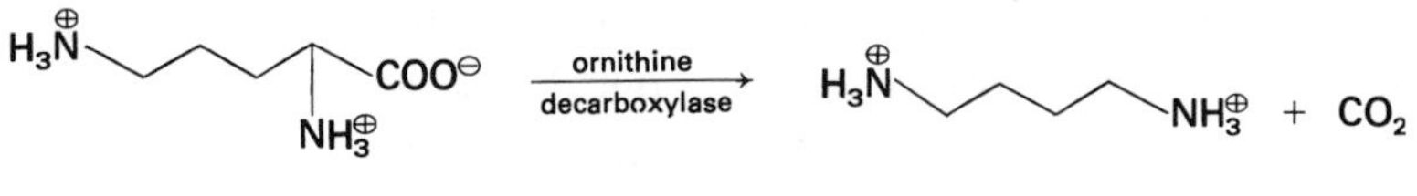

Spermine is formed by a second aminopropyl transfer from *S*-adenosylhomocysteamine, this time to spermidine as nucleophile. Of the three possible amine nucleophiles in spermidine, the primary amine of the C_4 chain is used specifically.

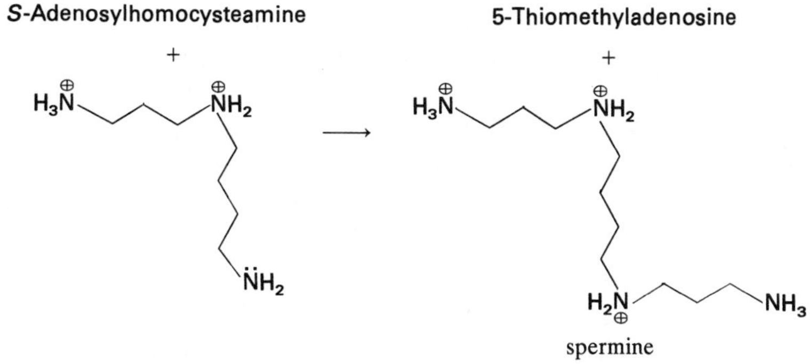

S-Adenosylhomocysteamine + spermidine → 5-Thiomethyladenosine + spermine

spermine

The names spermine and spermidine derive from the initial isolation of these compounds from semen. (Crystals as the phosphate salt were prepared from human semen in 1677 by the Dutch anatomist Leeuwenhoek.) The recent review by Tabor and Tabor (1976) can be consulted for a good summary of polyamine metabolism.

25.F.5.b Pathway 3: a possible adenosyl transfer from *S*-adenosylmethionine with pyruvate–formate lyase

Recent experiments by Knappe and colleagues suggest that SAM may be cleaved to L-methionine and 5′-deoxyadenosine (a net reduction at C′-5 of the ribose!) during the conversion of the enzyme pyruvate–formate lyase from an inactive to an active form (Knappe et al., 1974; Knappe and Schmitt, 1976).

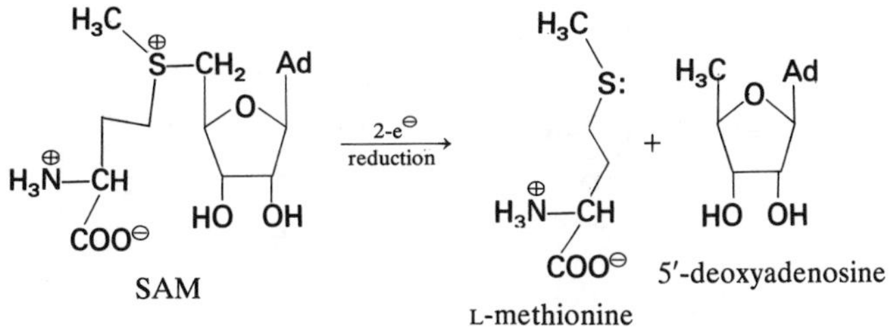

SAM —(2-e⁻ reduction)→ L-methionine + 5′-deoxyadenosine

Pyruvate–formate lyase has been of interest for some time, and some comments on the enzyme are required to put the above cleavage in perspective. The stoichiometry of the lyase reaction is the following.

$$H_3C{-}CO{-}COO^{\ominus} + CoASH \rightleftharpoons H_3C{-}CO{-}SCoA + HCOO^{\ominus}$$

pyruvate ... formate

The α-keto acid pyruvate is cleaved to acetyl-CoA and formate. This enzyme appears to be the bacterial anaerobic counterpart to the pyruvate-dehydrogenase complex (¶21.C.2.a), which cleaves pyruvate to acetyl-CoA and CO_2 during aerobic metabolism. When *E. coli* are switched from aerobic to anaerobic growth, the pyruvate–formate lyase activity appears within minutes and represents the major source of active acetyl groups during bacterial anaerobic metabolism. In marked mechanistic contrast to the pyruvate-dehydrogenase complex—and to the pyruvate-oxidase enzyme as well (¶21.C.2.b)—the pyruvate–formate lyase *does not use thiamine-PP as coenzyme.* Also in contrast is the fact that the C_1 fragment in this reversible lyase reaction is *formate, not carbon dioxide.*

The enzyme is designed for function in anaerobic environments and is extremely oxygen-labile, a fact that slowed mechanistic study until methods were developed for isolation as an inactive form that could be converted at will to an active form. (We shall discuss these methods shortly.) The homogeneous activated enzyme shows a turnover number of 1,100 sec^{-1} for cleavage and 380 sec^{-1} for condensation. Three lines of evidence support involvement of a covalent acetyl-*S*-enzyme intermediate: (1) pingpong initial-velocity kinetics; (2) a CoASH-independent partial exchange of [^{14}C]-HCOO$^{\ominus}$ label into pyruvate; (3) a radioactive enzyme that can be precipitated with trichloroacetic acid on incubation of 2-[^{14}C]-pyruvate (but not of 1-[^{14}C]-pyruvate) and identified as an acetyl-cysteinyl-enzyme form.

The mechanistic problem then reduces to a question of how formate is formed. An initial tetrahedral adduct as shown can yield an acetyl enzyme only by expulsion of formate anion, a wretchedly unlikely (high-energy) acyl anion as leaving group.

$$H_3C{-}CO{-}COO^{\ominus} + Enz{-}S^{\ominus} \rightleftharpoons H_3C{-}C(O^{\ominus})(S{-}Enz){-}C(=O){-}O^{\ominus} \overset{?}{\rightleftharpoons} H_3C{-}C(=O){-}S{-}Enz + [{}^{\ominus}C(=O){-}O^{\ominus}] \overset{H^{\oplus}}{\rightleftharpoons} HCOO^{\ominus}$$

Viewed in the condensation direction, formate would have to be converted to the acyl anion as nucleophile to attack the acetyl-enzyme intermediate. There is no obvious chemical precedent for such a postulate. The fate of the acetyl-*S*-enzyme would be straightforward acyl transfer to the thiolate anion of CoASH.

SAM is not required during catalytic turnover, but it is one of several components required to convert the isolable, inactive form of pyruvate–formate lyase into the catalytically active form. Pyruvate serves an effector role (it can be replaced by oxamate). Reduced flavodoxin (¶11.B.4) appears to be the electron donor ultimately used for reductive cleavage of SAM. An Fe^{II}-protein also is required for the lyase activation.

$$\text{Inactive pyruvate–formate lyase} \xrightarrow[\text{reduced flavodoxin}]{\text{SAM, pyruvate, Fe}^{II}\text{-protein,}} \text{Active pyruvate–formate lyase}$$

There appears to be stoichiometric cleavage of SAM into L-methionine and 5′-deoxyadenosine for each molecule of lyase activated. There is no detectable, stable adenosylation of either the lyase or the Fe^{II}-protein component, but a transient adenosylation of the iron protein is postulated. The production of 5′-deoxyadenosine is reminiscent of the CoB_{12}-dependent enzymes catalyzing 1,2-rearrangements (¶20.C.1), where the cobalt–5′-CH_2 bond is cleaved during catalysis. There is no cobalt in the lyase-activation system, but there is that Fe^{II} atom. Knappe and Schmitt (1976) have suggested that an Fe^{II}–CH_2–adenosyl intermediate may form during activation and then undergo reductive cleavage, with two electrons supplied by reduced flavodoxin (quite possibly by one-electron transfers?).

SAM —(Fe^{II} · protein · lyase complex)→ Methionine + Protein···Fe^{II}—CH$_2$ Ad (ribose ring with O, HO OH)

↓ FMNH$_2$ · Flavodoxin

Fe^{II} · Protein + H$_3$C Ad (ribose ring with O, HO OH) + FMN · Flavodoxin

We can hardly even guess at the purpose of the reductive cleavage of SAM. Perhaps it induces a conformational change in the Fe^{II} · protein · lyase complex to activate the lyase enzyme. Clearly, more work is required on both the activation and the mechanism of action of this interesting enzyme.

25.G METHANE FORMATION

Earlier in this chapter we have indicated the role of folate in mediating redox changes of one-carbon units attached to nitrogen at position 5 or 10. Certain anaerobic bacteria, such as those in ruminant stomachs, produce methane from CO_2 without the use of tetrahydrofolate. Methyl-CoB_{12} could be involved, but there is not much direct evidence for it. The net reaction is an 8-$e^{\ominus}$ reduction of carbon dioxide, with the oxygen atoms ending up as H_2O.

$$CO_2 + 8\,e^{\ominus} + 8\,H^{\oplus} \longrightarrow CH_4 + 2\,H_2O$$

Although the ultimate details of this complex multistep reductive process are unclear, Wolfe and colleagues have made great progress with extracts of pure methanogenic cultures. The methanogens use H_2 as the ultimate source of eight electrons, with a hydrogenase as the first enzyme.

$$H_2 \xrightarrow{\text{hydrogenase}} 2\,H^{\oplus} + 2\,e^{\ominus}$$

Subsequent electron flow appears to involve two novel cofactors unique so far to these bacteria: coenzyme M and factor F_{420} (McBride and Wolfe, 1971; Taylor and Wolfe, 1974; Tzeng et al., 1975). Coenzyme M is β-mercaptoethane sulfonate, and it functions as the carrier of the C_1 fragment. (For instance, extracts will not make methane from methyl-CoB_{12} without coenzyme M.) The intermolecular disulfide and methyl coenzyme M also were detected by Taylor and Wolfe (1974).

$$(S\text{-}CH_2CH_2\text{-}SO_3^{\ominus})_2 \xrightarrow{2\,e^{\ominus}} 2\,HS\text{-}CH_2CH_2\text{-}SO_3^{\ominus} \xrightarrow{CO_2} \longrightarrow \longrightarrow$$

coenzyme M

$$H_3C\text{—}S\text{-}CH_2CH_2\text{-}SO_3^{\ominus} \xrightarrow[\text{complex step}]{2\,e^{\ominus}} CH_4 + HS\text{-}CH_2CH_2\text{-}SO_3^{\ominus}$$

methyl coenzyme M

The disulfide undoubtedly is in redox equilibrium with the thiol form of the coenzyme. The methyl-CoM undergoes reduction to methane and free CoM by a complex methyl-reductase system whose mechanism has not yet been unraveled. Gunsalus et al. (1976) have suggested a possible scheme for formation of methyl-CoM from CO_2 and the free thiol form of the cofactor, although no direct evidence is yet available. A total of six electrons are required, as suggested in steps a, b, and c.

$$CO_2 + {}^{\ominus}S\text{—CoM} \rightleftharpoons {}^{\ominus}O_2C\text{—S—CoM} \xrightarrow[a]{2e^{\ominus}} [HC(O^{\ominus})(OH)\text{—S—CoM}] \xrightarrow{-H_2O} HC(=O)\text{—S—CoM (aldehyde)}$$

$$HC(=O)\text{—S—CoM} \xrightarrow[b]{2e^{\ominus}} H_2C(OH)\text{—S—CoM (alcohol)} \xrightarrow[c]{2e^{\ominus}} H_3C\text{—S—CoM (methyl-CoM)}$$

The proximal reductant in these three projected 2-e$^{\ominus}$ steps could be NADPH, which would deliver both the 2 e$^{\ominus}$ and the hydrogen that becomes affixed to the carbon center undergoing reduction.

The fluorescent factor F_{420} (it absorbs maximally at 420 nm) is an obligate intermediate for passage of electrons from H_2 to NADP$^{\oplus}$ to generate the NADPH. The reduction potential of H_2/2 H$^{\oplus}$ (the hydrogen electrode) is −0.42 V at pH 7; that of NADPH/NADP$^{\oplus}$ is −0.32 V. The reduction potential of factor F_{420} is −0.37 V, properly poised for such a role (Wolfe, personal communication). The structure of factor F_{420} has recently been proposed as an 8-hydroxy-7-demethyl-5-deaza-FMN derivative (Eirich, 1977; Eirich et al., 1978).

The chromophore is essentially identical to authentic, synthetic 8-hydroxy-5-deazariboflavin (C. Walsh et al., 1978). This is the only example of the natural occurrence of a 5-deazaflavin chromophore and, together with coenzyme M, gives some extra credence to the recent suggestions that methanogens represent an evolutionarily highly divergent form of life (Balch et al., 1977). One expects that oxidized F_{420} accepts electrons from reduced hydrogenase (Gunsalus et al., 1976) and, as the 1,5-dihydro-5-deaza species, then transfers a hydride equivalent to $NADP^{\oplus}$.

F_{420}

1,5-dihydro-F_{420} $NADP^{\oplus}$

Note that the 5-deaza substitution lowers the redox potential from that of a normal 8-hydroxyflavin species, which would be *reduced* by NADPH, to a level where it can instead reduce the oxidized nicotinamide. The 5-carba substitution alters the thermodynamics of the isoalloxazine system to suit this new purpose in the methanogens.

This concludes our discussion of one-carbon group-transfer processes, mediated by a variety of coenzymes that serve as covalent carriers.

Chapter 26

Enzyme-Catalyzed Alkylations Involving Prenyl-Group Transfer

In the preceding chapters on enzyme-catalyzed formation and cleavage of carbon–carbon bonds, we have discussed two major categories of reactions. One category represents the controlled generation of substrate enolates (or eneamines) as carbanion sources in carboxylation reactions, and also in aldol and Claisen condensations. We have focused on the structural elements used for stabilization of the carbanionic transition states and intermediates involved (use of biotin, thiamine-PP, or pyridoxal-P, for instance). A second category, covered in Chapter 25, is represented by transfer of one-carbon fragments via such carriers as tetrahydrofolate or *S*-adenosylmethionine. The methyl transfers represent one important example of biological alkylations where a single carbon atom is added on to some substrate molecule that has acted as a nucleophile; in alkyl transfers, the attacking nucleophilic atom becomes alkylated by the transferring electrophilic alkyl group.

In this chapter, we again focus on alkylation reactions, but now on alkylation processes responsible for the *assembly of carbon skeletons* of an enormous variety of biological molecules. In contrast to methyl transfers, which involve a one-carbon alkyl group, the fundamental alkyl transfer in this chapter is enzyme-catalyzed transfer of five-carbon isoprenoid units—*prenyl transfers.* Because carbon–carbon bond formation is at issue, the essential sequence will involve *prenylation of some carbon nucleophile,* with each alkylative condensation increasing the length of the carbon chain in the product by an isoprenyl unit or some multiple thereof.

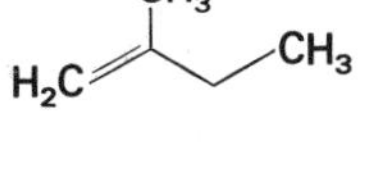

isoprene

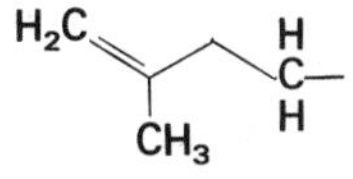

isoprenyl unit

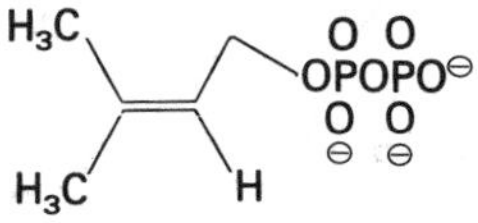

dimethylallyl-PP (Δ^2-isopentenyl-PP)

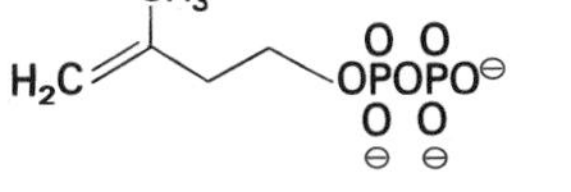

Δ^3-isopentenyl-PP

After some brief comments about the structural versatility of important metabolites derived from isoprenoid building blocks, we shall proceed in the following sequence. First, we shall examine the mode of synthesis of the C_5 isomeric pentenyl pyrophosphates, Δ^3-isopentenyl-PP and dimethylallyl-PP, *which are the biological isoprene units.* Second, we shall examine the nature of the basic chain-elongation step. Third, we shall examine reactions forming carbon–carbon bonds, involving both *head-to-tail* condensations of prenyl groups and *tail-to-tail* condensations of prenyl groups. Finally, we shall look at *cyclizations* to polycyclic products such as sterols.

26.A SOME EXAMPLES OF NATURAL PRODUCTS ENZYMATICALLY SYNTHESIZED FROM PRENYL GROUPS

An enormous variety of natural structures are enzymatically synthesized from prenyl groups, both in primary metabolism and especially in secondary plant metabolism (the terpenoid family of natural products; Simonsen, 1951). We cannot hope to give a catalogue of structural types and, in fact, shall limit this survey to compounds of primary metabolism only; even then, the particular choices are merely illustrations of a much wider range of known examples.

At the ten-carbon level (one prenyl transfer between two monomers) are such monoterpene compounds as geraniol (from geraniums) and the bicyclic ketone camphor (which obviously has undergone some rearrangement from the initial enzymatic C_{10} compound), whose metabolism we discussed in Chapter 15 as a substrate for the bacterial cytochrome-P_{450} monooxygenase system.

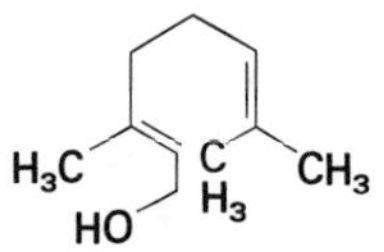

geraniol 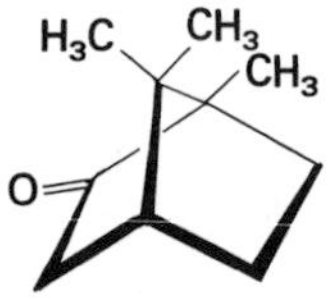 camphor

There are many C_{15} molecules—especially sesquiterpenes in plant secondary metabolism (Hendrickson, 1965)—that are derived biogenetically from three C_5 prenyl units. One specific C_{15} molecule of note is the juvenile hormone of certain insects (Fleming, 1965, p. 165 ff).

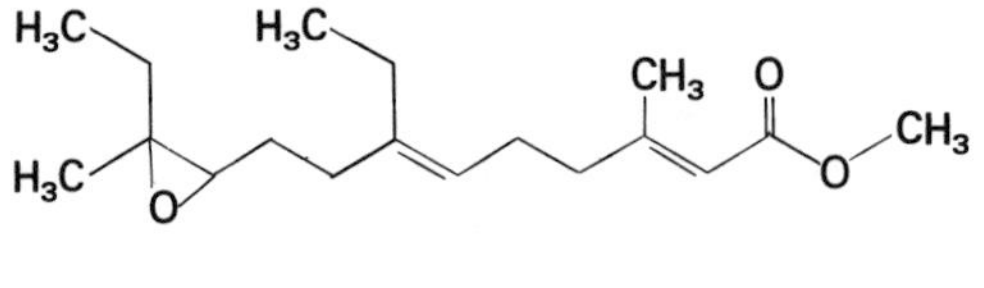

This methyl ester of farnesyl epoxide is a molecule that induces molting behavior in susceptible insects that undergo a series of successive larval and pupal developmental stages.

At the level of molecules derived from four isoprene units are the C_{20} compounds. One obviously important example is retinol, also known as vitamin A.

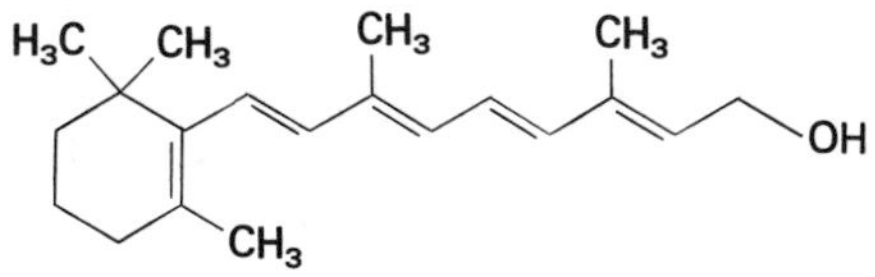

all-*trans*-retinol

This long-chain alcohol is a metabolic precursor of retinal (vitamin-A aldehyde), the actual light-active chromophore in vertebrate retina (¶19.E).

In mammalian tissues, retinol is usually derived from C_{40} polyisoprenoid β-carotene molecules ingested in the diet. β-Carotene is cleaved in the plasma by a poorly characterized dioxygenase (Chapter 16) to two molecules of retinal. The aldehyde product undergoes NADH-mediated reduction to retinol as a storage form.

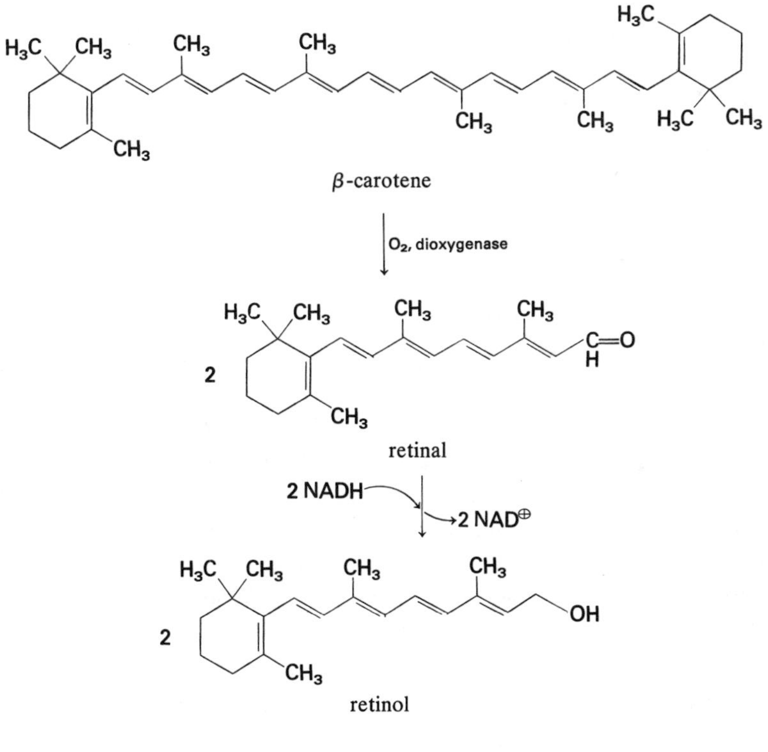

β-carotene

O_2, dioxygenase

2 retinal

2 NADH → 2 NAD$^{\oplus}$

2 retinol

The retinol then binds to a specific carrier protein in plasma (Kanac et al., 1968; Heller and Horowitz, 1974, 1975). By such specific and stoichiometric complexation to the retinol binding protein, the C_{20} alcohol is maintained in a stable form in plasma while the 1:1 complex circulates to the retina, and the retinol can be delivered to that target tissue. The β-carotene, a C_{40} carotenoid, is itself derived by tail-to-tail condensation of two identical C_{20} polyisoprenoids, geranylgeranyl pyrophosphate (¶26.E.3).

Perhaps the best-known class of polyisoprenoid molecules in nature is the steroids, tetracyclic structures derived biosynthetically from six "biological isoprene" units. The first major mammalian steroidal biosynthetic product is cholesterol, a 3-β-hydroxy C_{27} compound that has lost three carbon atoms from the expected C_{30} precursor.

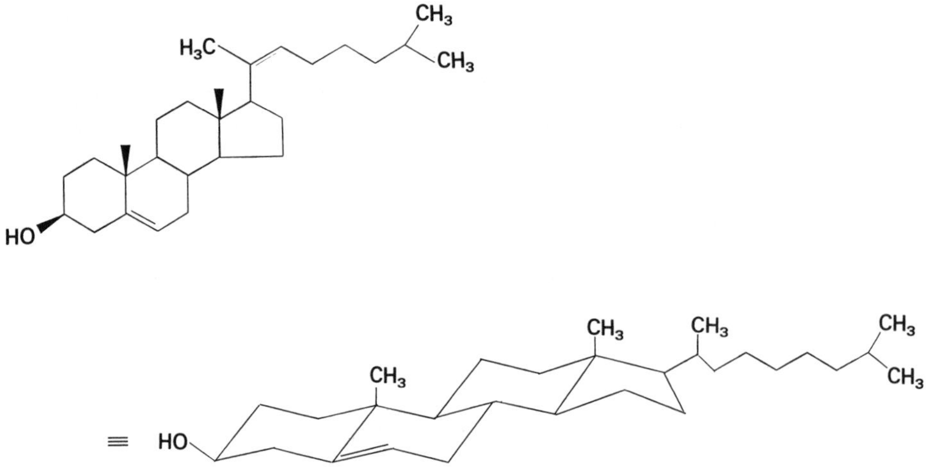

cholesterol

Cholesterol is a major component of many animal-cell membranes, its rigid structure tempering the fluidity of the hydrocarbon phase. It is also the key biosynthetic precursor itself to further steroidal metabolites such as bile acids and various adrenal (Chapter 15) and sex hormones.

Two other types of molecules derived at least in part from isoprenoid building blocks close out this brief introduction. One type is the ubiquitous benzoquinone (coenzyme Q) and naphthoquinone (vitamin K) derivatives, which have two structural features: the redox-active quinone groups and the isoprenyl chains ($n = 1$ to 12 are known).

oxidized naphthoquinone

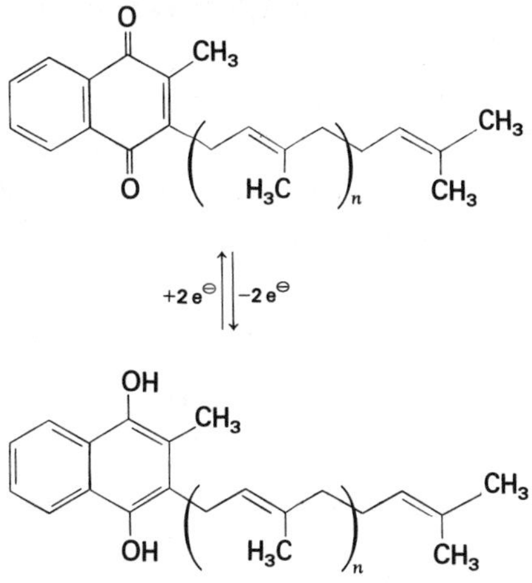

$+2e^{\ominus}$ / $-2e^{\ominus}$

benzoquinone

reduced naphthoquinone

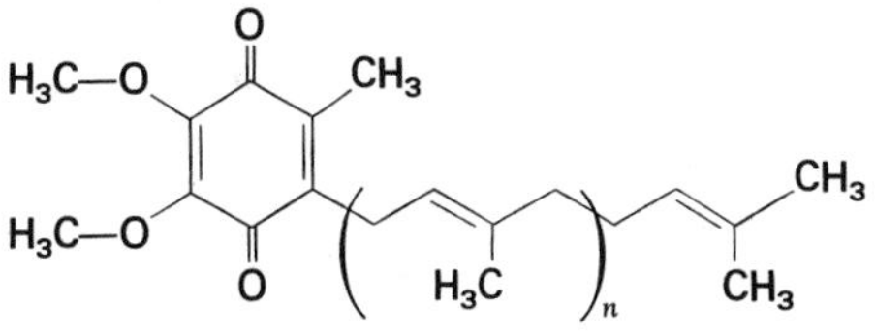

Because both 1-e$^{\ominus}$–reduced (semiquinone) and 2-e$^{\ominus}$–reduced (hydroquinone) forms are stable species, these molecules sit (like the flavin coenzymes) at the crossroads of two-electron and one-electron redox sequences in membrane electron-transport chains. The hydrophobic side chains anchor the molecules in the lipid phase of the membranes but, as small molecules, the quinones have the lateral mobility (in the plane of the membrane) to act as mobile redox shuttles between various membrane proteins such as flavoproteins and cytochromes.

A less widely distributed isoprenoid-based natural product is batrachotoxin, the active constituent in the arrow poison used by some Colombian Indians.

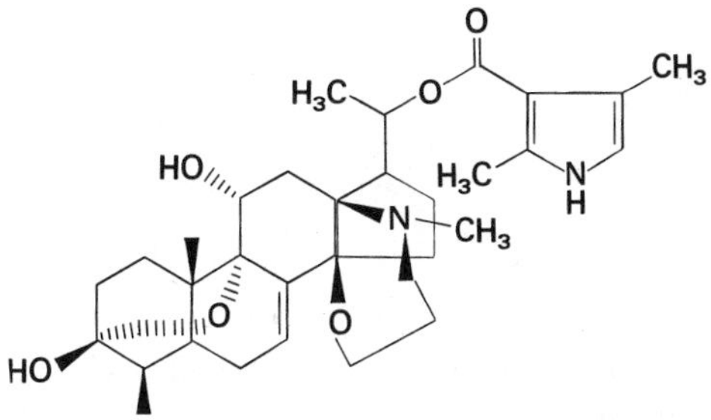

The compound is isolated from the skin of certain species of Colombian frogs and acts as a neurotoxin, binding to acetylcholine receptors (Chapter 4). On a weight basis, it is one of the most potent neurotoxins known.

26.B SYNTHESIS OF THE ISOMERIC ISOPRENYL MONOMERS: Δ^3-ISOPENTENYL-PP AND 3,3-DIMETHYLALLYL-PP

We noted in ¶19.C.5 that Δ^3-isopentenyl pyrophosphate and 3,3-dimethylallyl pyrophosphate are interconverted enzymatically by a 1,3-allylic isomerization process that is antarafacial and does not involve any detectably *intramolecular* 1,3-migration of hydrogen (Cornforth, 1973; Dugan and Porter, 1976).

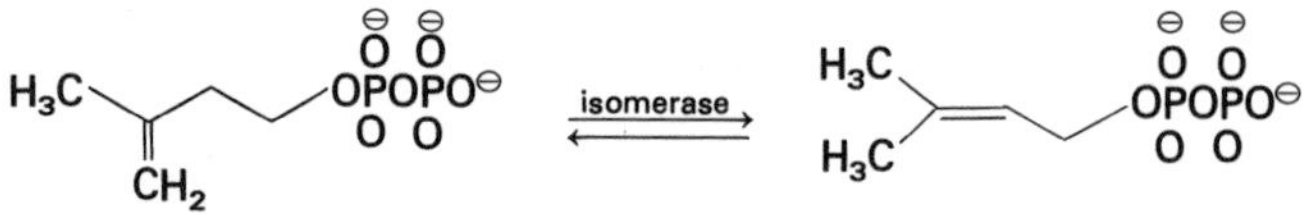

isomerase

Δ^3-isopentenyl-PP dimethylallyl-PP

Biosynthetically, the isopentenyl-PP is the isomer formed first in an apparently concerted reaction from 3-phosphomevalonate-5-PP, a decarboxylative sequence that we noted briefly in ¶17.B.5.

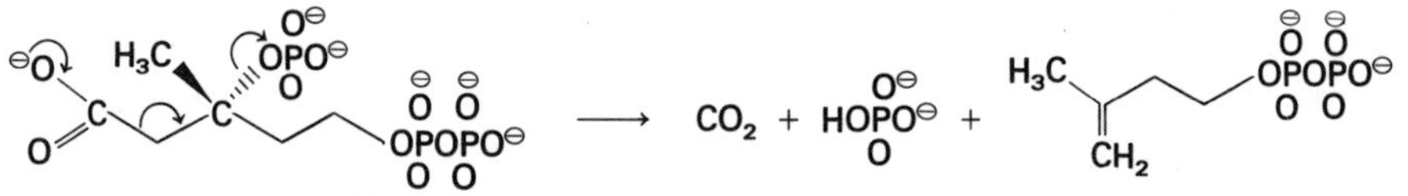

$\longrightarrow CO_2 + HOPO_3^{2-} +$ isopentenyl-PP

The mevalonate skeleton derives directly from β-hydroxymethylglutaryl-CoA, a molecule whose formation by Claisen condensation between acetyl-CoA and acetoacetyl-CoA we discussed in ¶23.B.5. That glutaryl derivative experiences a double reduction process involving two equivalents of NADH (as donor of four electrons) at the active site of a reductase enzyme (Popjack and Cornforth, 1966).

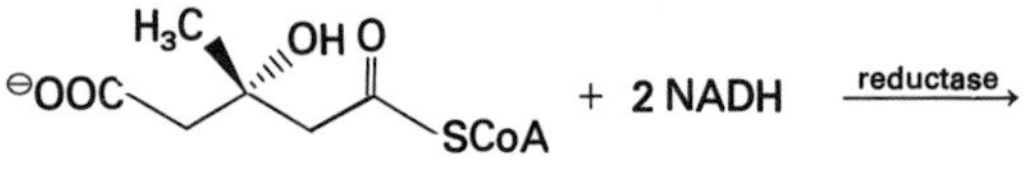

$+ 2\ NADH \xrightarrow{\text{reductase}}$

β-hydroxymethylglutaryl-CoA

$2\ NAD^{\oplus} + CoASH +$

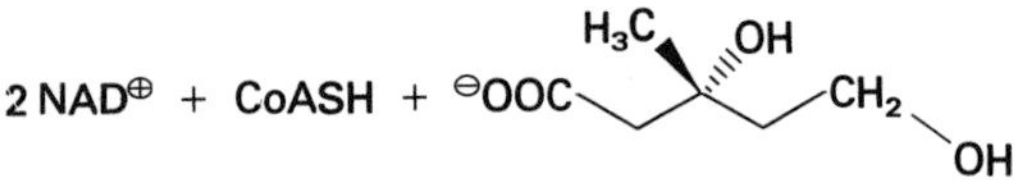

mevalonate

The acylthioester functionality of hydroxymethylglutaryl-CoA is thereby reduced to the primary alcohol group of mevalonate. It is likely that the four-electron reduction occurs in two discrete steps, producing the hemithioacetal (at the aldehyde oxidation state) as an *enzyme-bound* intermediate that does not dissociate before the second molecule of NADH can bind and deliver the second 2-$e^{\ominus}$–reducing equivalent.

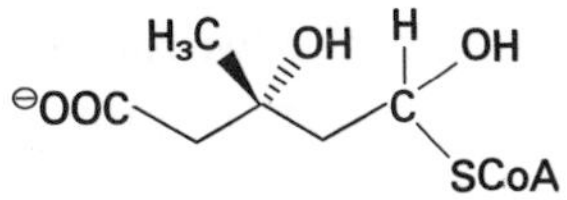

hemithioacetal

One can add the free aldehyde, mevaldate, to the resting enzyme along with NADH, and reduction to mevalonate will occur. Indeed, by using 4-[^{3}H]-NADH in this artificial second half-reaction, the Cornforth group achieved the enzymatic synthesis of chiral 5-[^{3}H]-mevalonate, specifically 5*R*-[^{3}H]-mevalonate (Alworth, 1972, p. 217).

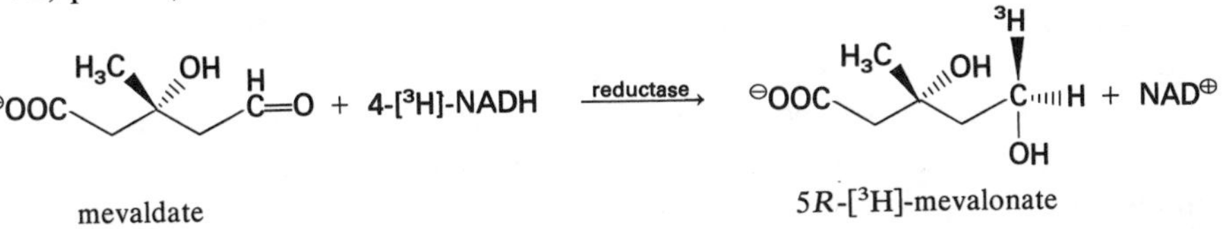

mevaldate + 4-[^{3}H]-NADH —reductase→ 5*R*-[^{3}H]-mevalonate + NAD$^{\oplus}$

Because C-5 of mevalonate becomes C-1 of the isopentenyl-PP and dimethylallyl-PP and this is the locus of carbon–carbon bond formation in the prenyl-transfer steps discussed next, this chiral molecule has been useful in determining the stereochemistry of those enzymatic alkylations, as we shall see.

Mevalonate, a C_6 molecule, then undergoes three successive phosphoryl transfers from ATP for the purpose of activating the two hydroxyl groups in the molecule. The primary hydroxyl at C-5 is sequentially converted to the 5-phosphate and then to the 5-pyrophosphate by kinases (Alworth, 1972, p. 222).

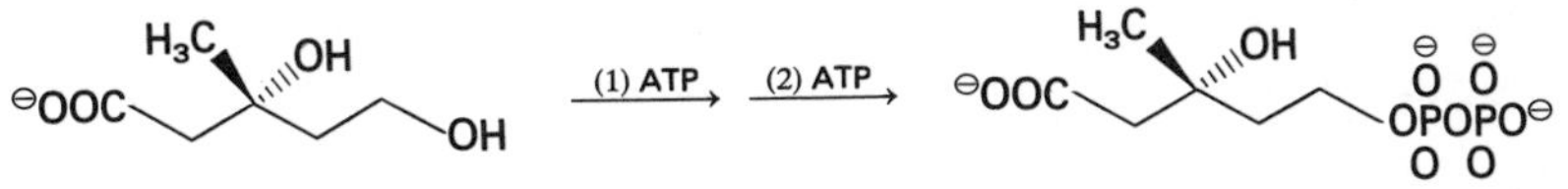

(1) ATP → (2) ATP →

This sequence has converted the primary alcohol into a group that can subsequently be displaced readily with *carbon–oxygen fission*; release of pyrophosphate as a leaving group takes with it the oxygen atom initially present as the C-5 alcoholic oxygen. This chemical activation is the *key preparation* for the carbon–carbon bond-forming steps that occur from these isoprenyl units.

When C-5 has been activated for subsequent reactions, the tertiary alcohol at C-3 is then phosphorylated to lower the activation energy for decarboxylation at C-1 and loss of the oxygen atom at C-3 in the decarboxylation step noted earlier.

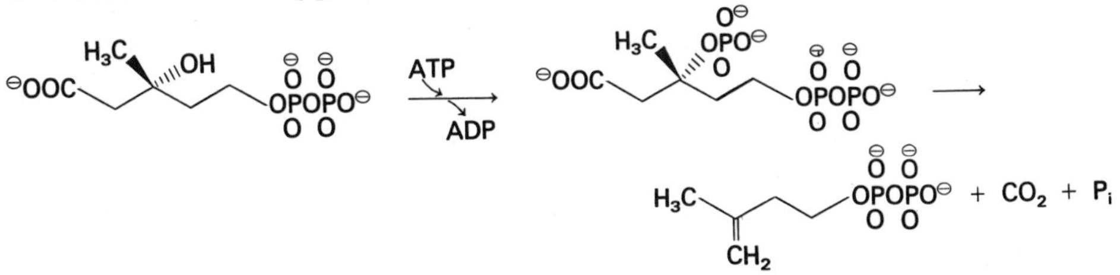

ATP → ADP

→ + CO_2 + P_i

The chemical logic of the activation of the two alcoholic groups in mevalonate by sequential phosphorylations is thus evident and provides an especially clearcut use of ATP for such group activation—the conversion of poor leaving groups to good ones.

26.C PRENYL TRANSFER: THE FUNDAMENTAL STEP IN CHAIN ELONGATION

Given the two isomeric pentenyl pyrophosphates, how are they converted enzymatically into the C_{10} to C_{110} acyclic isoprenoids found in nature as stable products and as precursors of sterols and other cyclic derivatives? The detailed mechanism of carbon–carbon bond formation in the fundamental chain-elongation step has been far from clear, but mechanistic experiments have been facilitated by the recent purification to crystalline homogeneity of the prenyl transferase from pig liver (Eberhardt and Rilling, 1975; B. Reed and Rilling, 1975).

The pig-liver prenyl transferase will carry out a condensation reaction between the Δ^3-isopentenyl-PP and the allylic partner, dimethylallyl-PP, to yield a single C_{10} product isomer, geranyl-PP.

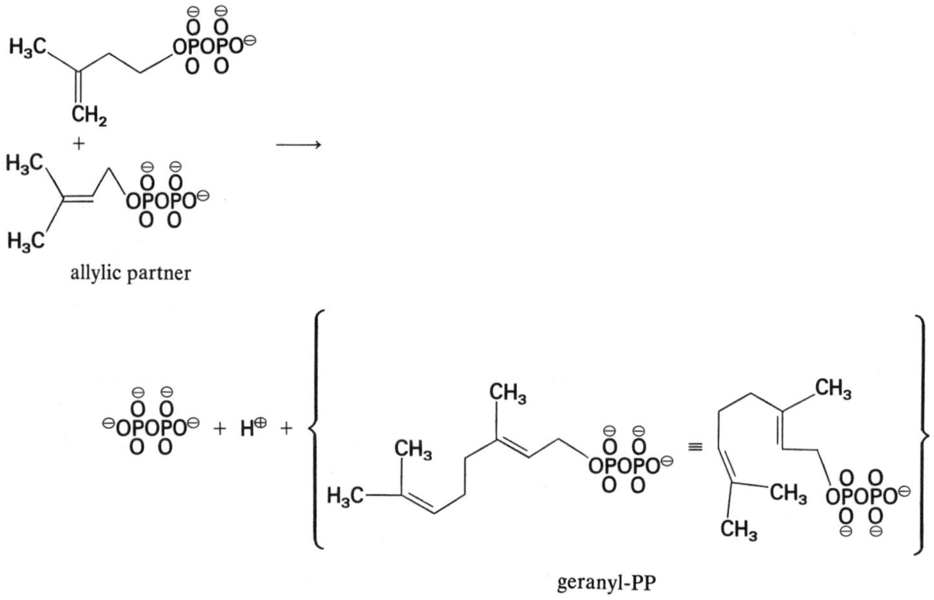

allylic partner

geranyl-PP

Specificity studies indicate a narrow specificity for the isopentenyl-PP substrate, but a broader range of specificities for the allylic-pyrophosphate partner. For instance, geranyl-PP is an excellent allylic substrate partner, and the enzyme will convert the C_{10} allylic pyrophosphate in the presence of Δ^3-isopentenyl-PP to a C_{15} product (three C_5 units), farnesyl-PP.

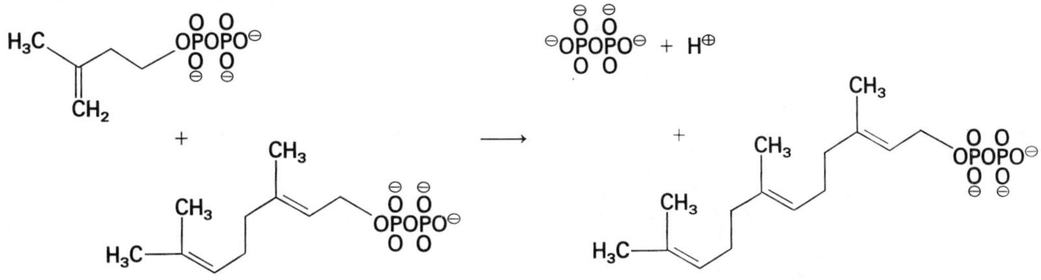

allylic partner

farnesyl-PP

Similarly, condensation of the C_{15} farnesyl-PP (as allylic partner) with another molecule of Δ^3-isopentenyl-PP will proceed enzymatically to yield the C_{20} pyrophosphate geranylgeranyl-PP.

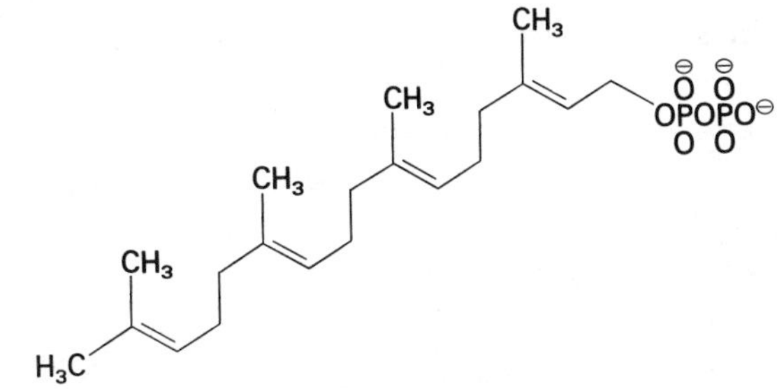

This C_{20} product is the largest one this enzyme will form. Other prenyl transferases exist in different organisms that will produce eight to ten linked monomers (C_{40} to C_{50} molecules; Allen et al., 1967) or even eighteen to twenty-two linked monomers (C_{90} to C_{110}; Beytia and Porter, 1976).

Thus, chain elongation proceeds by adding isopentenyl-PP monomers to an allylic-PP cosubstrate to produce the five-carbon homologue of the allylic-PP as product, along with a molecule of liberated inorganic pyrophosphate. The generalized stoichiometry is the following

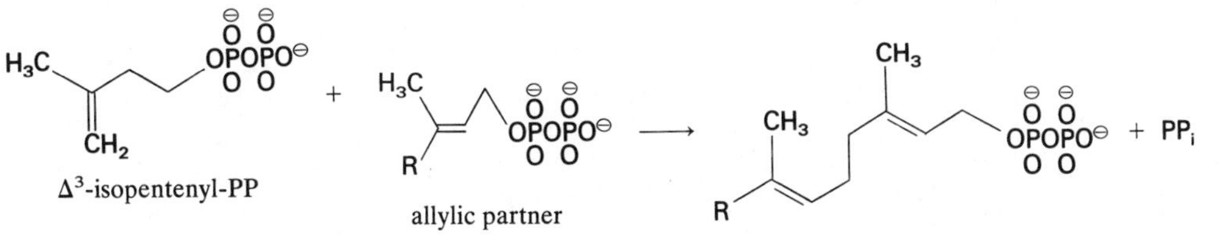

Δ^3-isopentenyl-PP

allylic partner

In each instance, the inorganic pyrophosphate released stems from the allylic partner in the alkylation sequence. *The new carbon–carbon bond is formed between* C-4 *of the isopentenyl substrate and* C-1 *of the allylic substrate, with cleavage of the bond between* C-1 *and oxygen as pyrophosphate* (*not a pyrophosphoryl group!*) *is eliminated.*

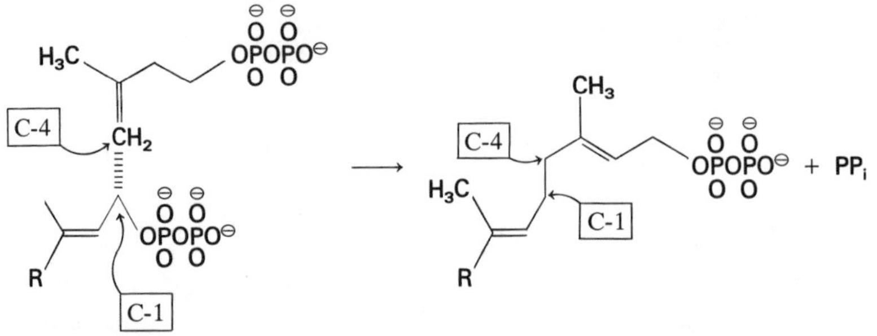

C-4 C-1 + PP_i

This represents a *head-to-tail joining in the* C—C *bond-forming step,* the tail being the C-1–pyrophosphate end, and the head being the 3,4-double bond of Δ^3-isopentenyl-PP. Two chemically reasonable schemes for the mechanism of this ubiquitous alkylative condensation at the active site of prenyl transferase have been advanced on several occasions and recently summarized by Poulter and Rilling (1976, 1978).

The first alternative is the heterolytic cleavage (i.e., a carbonium-ion formation) of the bond between C-1 and PP of the allylic substrate partner to initiate catalysis. This would lead to formation of a resonance-stabilized allylic carbonium ion, a well-known S_N1 chemical intermediate.

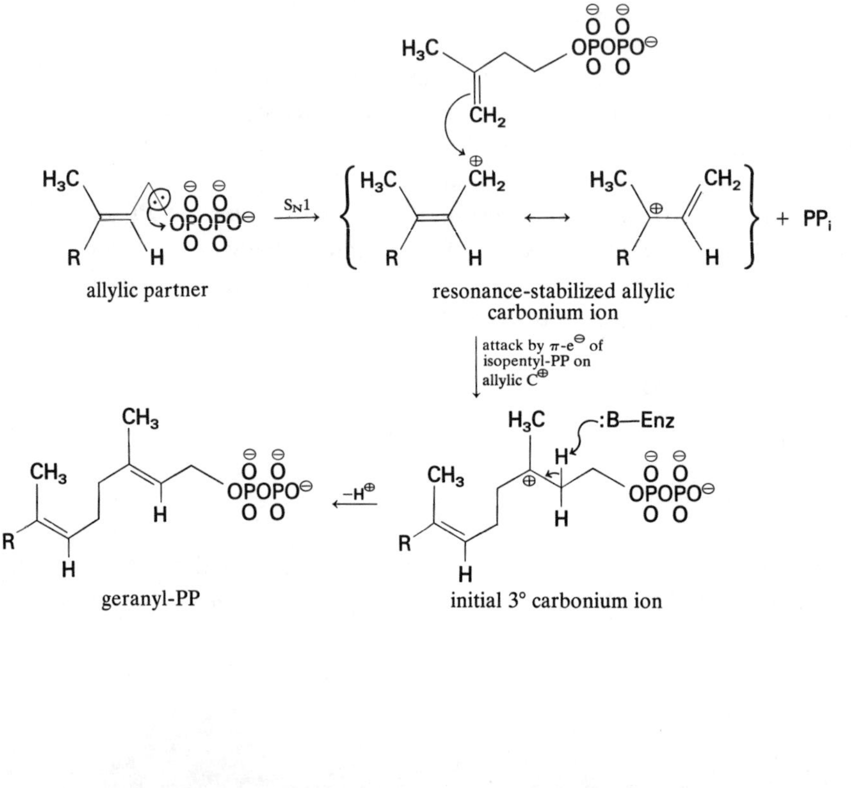

allylic partner

S_N1

resonance-stabilized allylic carbonium ion

+ PP_i

attack by π-$e^{\ominus}$ of isopentyl-PP on allylic $C^{\oplus}$

:B—Enz

$-H^{\oplus}$

geranyl-PP

initial 3° carbonium ion

The carbonium ion would be captured specifically at C-1 by attack of the π-electrons of the 3,4-double bond of the Δ^3-isopentenyl-PP cosubstrate. This generates a C_{10} tertiary carbonium-ion species that would lose a C-3 hydrogen as a proton, quenching the adjacent charge deficiency and producing the observed diolefinic product, geranyl-PP. Poulter and Rilling (1976) call this mechanism an *ionization–condensation–elimination* path.

The second alternative, on the other hand, would involve an enzyme nucleophile initiating condensation by attack on the Δ^3-isopentenyl-PP double bond at C-3, with synchronous attack of the C-4 end of that molecule at C-1 of the allylic substrate and concomitant cleavage between C-1 and O as the pyrophosphate leaving group is eliminated. This is a *displacement–elimination* mechanism—essentially a S_N2 substitution mechanism.

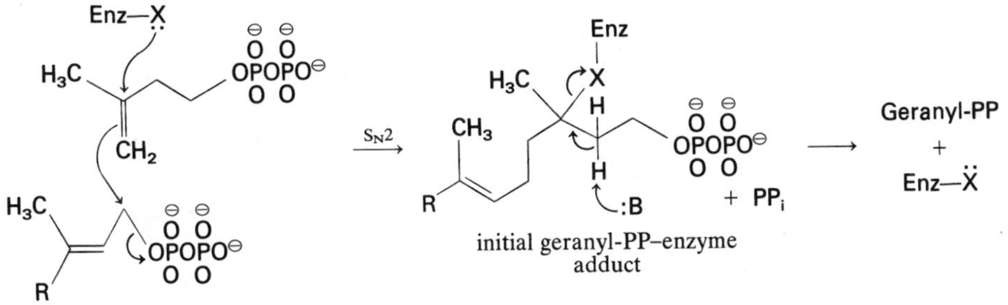

initial geranyl-PP–enzyme adduct

Recent evidence from the work of Poulter and Rilling (1976, 1978) leads them to favor the first alternative, involving the allylic carbonium ions as intermediates. First, the homogeneous prenyl transferase shows hydrolase activity toward the allylic-PP substrate in the absence of the Δ^3-isopentenyl-PP. The V_{max} for geranyl-PP hydrolysis (to geraniol and PP_i) is 1.7% of the V_{max} for alkylative condensation (to C_{15} farnesyl-PP when Δ^3-isopentenyl-PP is present). In $H_2{}^{18}O$, the geraniol product has one atom of ^{18}O, indicating the unusual pattern of cleavage between C-1 and PP that also characterizes the normal condensation mechanism. The stereochemistry at C-1 of the allylic-PP substrate is known to be an inversion of configuration in the alkylative condensation (Cornforth, 1968). Similar inversion occurs in the hydrolytic step (Poulter and Rilling, 1976), suggesting that the hydrolytic activity has mechanistic similarity to the normal condensation. The ^{18}O transfer data in hydrolysis seem to mandate *an allyl-cation intermediate that is held at the active site so that* H_2O *approaches from the backside only* (rather than equally from backside and frontside, as would occur when free in solution).

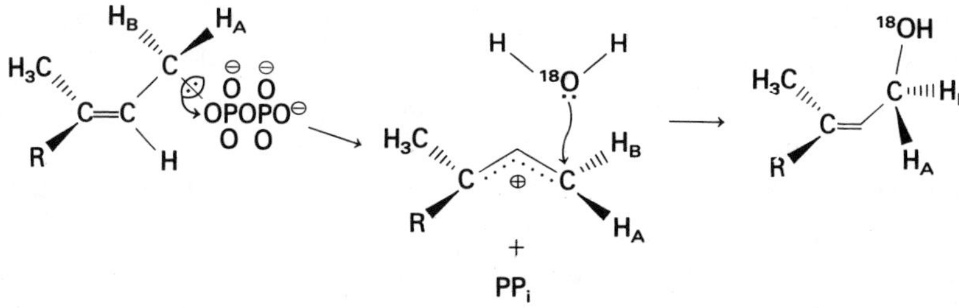

H_3C H_B H_A C C=C R H $\ominus$ $\ominus$ O O OPOPO$^\ominus$ O O → H_3C C ⊕ C H_B H_A R H ^{18}O H + PP_i → H_3C C= C ^{18}OH H_B H_A R

In a similar fashion during carbon–carbon bond formation, the π-e$^\ominus$ of the isopentenyl-PP could attack the allyl cation *from the backside.*

In a complementary study, Poulter et al. (1976) synthesized an allylic-PP substrate where the R group was not CH_3 but CF_3, a strongly electron-withdrawing substituent. They reasoned that the inductive effect should strongly retard ionization of the allylic-PP (S_N1), but should have only a slight effect on the rate of direct nucleophilic displacement (S_N2) at C-1. They observed that the *E*-isomer of trifluoromethyl-2-butenyl pyrophosphate does react enzymatically with isopentenyl-PP, but at a rate 1.5-million-fold slower than that of the normal dimethylallyl-PP substrate.

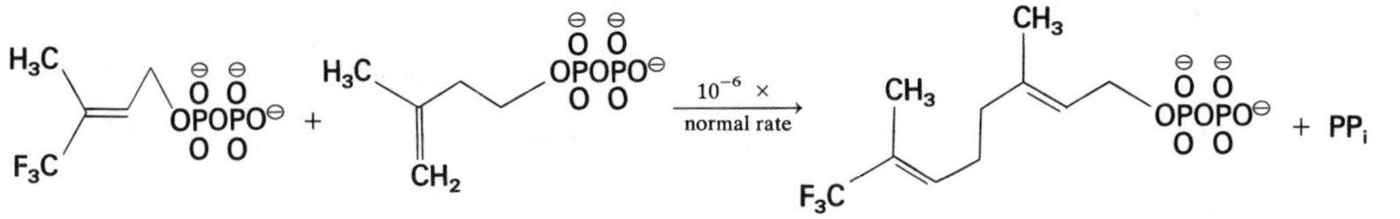

H_3C F_3C OPOPO$^\ominus$ + H_3C CH_2 OPOPO$^\ominus$ $\xrightarrow[\text{normal rate}]{10^{-6}\times}$ CH_3 CH_3 F_3C OPOPO$^\ominus$ + PP_i

Chemical model studies confirm that S_N1 solvolysis (carbonium-ion mechanism) of the trifluoromethylbutenyl-PP is severely retarded, whereas an S_N2 displacement is, in fact, slightly accelerated compared to dimethylallyl-PP. These data strongly suggest the involvement of allyl-cation intermediates during enzyme-catalyzed carbon–carbon bond formations in the prenyl-transfer steps. One potential criticism, that the trifluoromethyl analogue reacts so slowly as to vitiate meaningful measurement and comparison, has been rebutted by repetition of these experiments with the monofluoromethylbutenyl-PP; now the analogue reacts several hundredfold more slowly, consistent with inductive expectations, but at a readily measurable rate (Poulter and Rilling, 1978).

We noted earlier that the specificity of the particular pig-liver prenyl transferase that has been isolated and studied extends to formation of C_{20} geranylgeranyl compounds, but not to larger molecules. The C_{15} species, farnesyl-PP, is probably the physiologically significant product. Longer-chain acyclic isoprenoid alcohols are, however, found in microorganisms, plants, and animal cells. The long-chain alcohols that are lipid-carrier intermediates in microbial cell-wall polysaccharides are C_{35} to C_{55} isoprenoids (Gennis and Stominger, 1976). For instance, bactoprenol (undecaprenol) is a C_{55} alcohol.

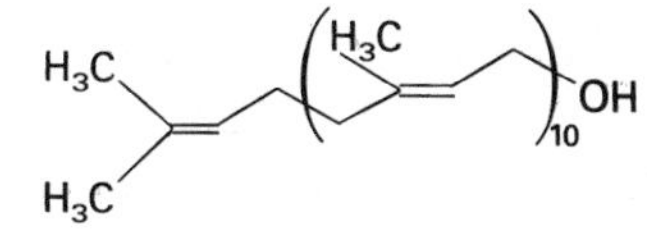

The synthesis of undecaprenol pyrophosphate and the lower homologues has been detected in cell-free extracts and partially purified enzymes from *Micrococcus lysodeikticus* (Gennis and Stominger, 1976). In nucleated cell types, the long-chain polyprenols are dolichols, ranging from C_{80} to C_{110}. Some of the olefinic links in the isoprenyl units are *cis*, some *trans*, and the α-isoprenyl unit is saturated (see Chapter 9). Although synthesis in vitro has been reported (Beytia and Porter, 1976), no mechanistic studies have been carried out. All the monomeric isopentenyl units condense sequentially head-to-tail with the lengthening allylic-PP chain.

26.D TAIL-TO-TAIL JOINING: SQUALENE SYNTHETASE

An alternate type of joining is also employed enzymatically, both at the level of C_{15} (farnesyl) and C_{20} (geranylgeranyl) polyprenyl pyrophosphates—a *tail-to-tail joining* (C-1 to C-1) that occurs by a distinct mechanism, more complex than the chain-elongation step. For instance, in sterol biosynthesis, two farnesyl-PP molecules are joined in a doubling step to yield (ultimately) the symmetric C_{30} hydrocarbon squalene and two molecules of inorganic pyrophosphate. In this carbon–carbon bond-formation sequence, both alcohol functionalities of the substrates have been eliminated at the two carbons reductively condensed. The requirement for a reductant is fulfilled by a molecule of NADPH. The stoichiometry of partially purified yeast squalene synthetase is the following (Quereshi et al., 1973).

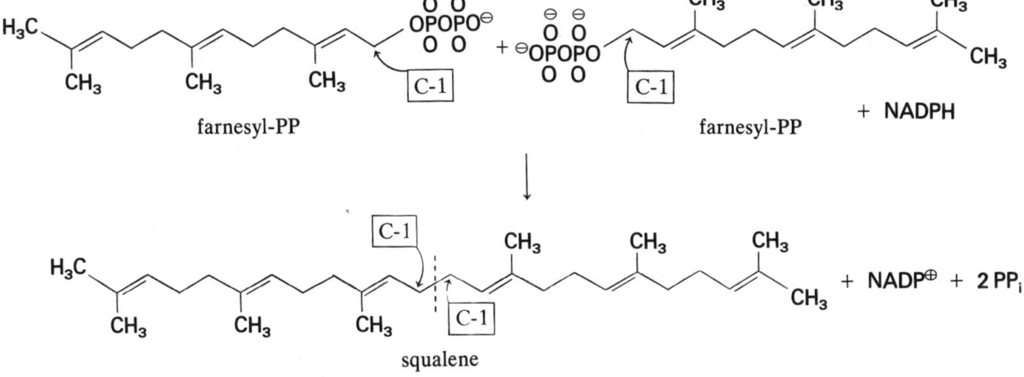

farnesyl-PP + farnesyl-PP + NADPH

↓

squalene + $NADP^{\oplus}$ + 2 PP_i

Analogously, in the biosynthesis of the colorful C_{40} carotenoid hydrocarbon pigments of bacteria and plants, two molecules of C_{20} geranylgeranyl-PP are joined tail-to-tail to yield the symmetric C_{40} hydrocarbon, lycopersene.

The stoichiometry of squalene synthesis suggests a complex process, and intermediates have, in fact, been detected and structures suggested. An intermediate in squalene biosynthesis was first isolated by Rilling (1966) and colleagues when farnesyl-PP was incubated with partially purified enzyme in the absence of NADPH. Structural elucidation of the accumulating intermediate, confirmed by eventual chemical synthesis, permitted identification of this species as a molecule called presqualene pyrophosphate, whose most salient feature is a three-membered cyclopropane ring (Epstein and Rilling, 1970). Although there was some initial speculation that the isolation of presqualene-PP was a side product from incomplete incubation mixtures (Cornforth, 1973; Beytia and Porter, 1976), subsequent studies have shown its transient formation in complete reaction mixtures (Muscio et al., 1974; Beytia and Porter, 1976) and its accumulation in the presence of inhibitors blocking subsequent reductive conversion of presqualene-PP to squalene (Corey and Volante, 1976). Thus, a reasonable mechanistic beginning is a two-step reaction for squalene synthetase as follows.

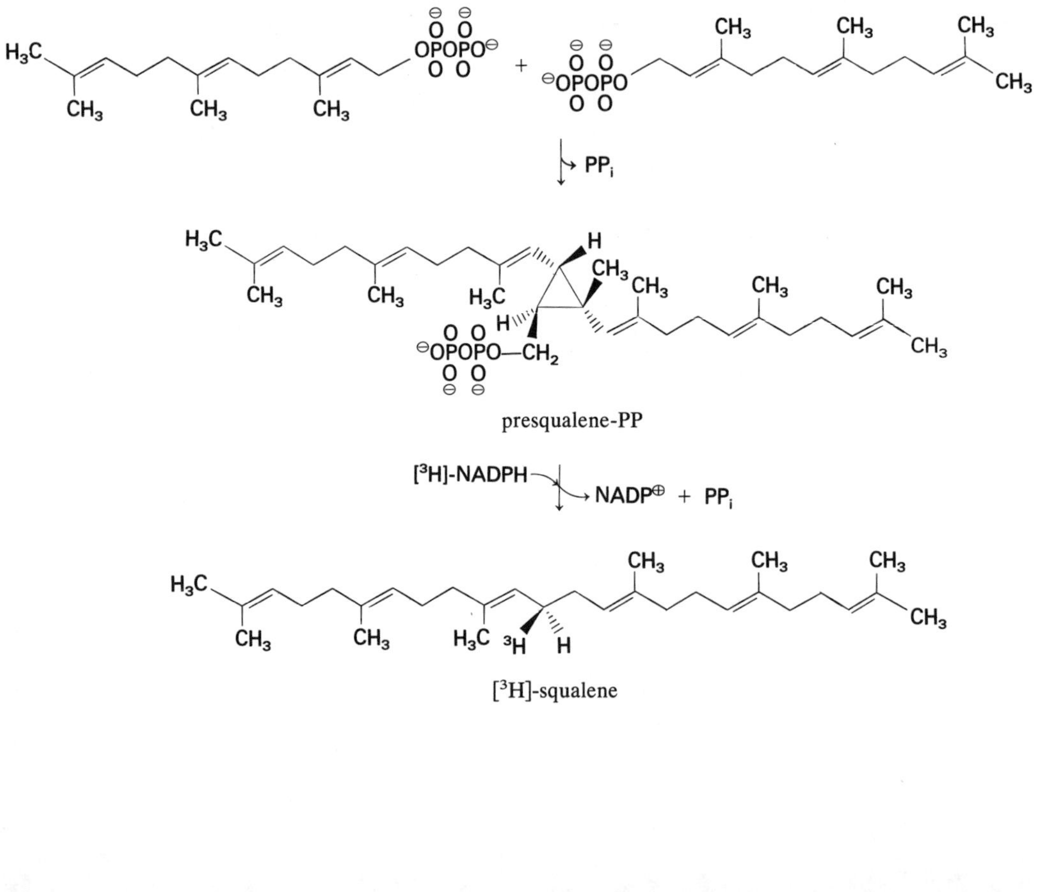

$\downarrow$ PP_i

presqualene-PP

$[^3H]$-NADPH $\rightarrow$ $NADP^{\oplus}$ + PP_i

$[^3H]$-squalene

Given the intermediacy of presqualene-PP, one can ask how the cyclopropane ring is formed enzymatically, and then how it is opened reductively with hydride addition from NADPH and elimination of the pyrophosphate group, again with cleavage between C-1 and PP. The mechanistic facts are few, and so the following comments represent what we hope are reasonable speculations (adapted from Beytia and Porter, 1976).

For presqualene-PP formation, kinetic studies with the yeast squalene synthetase indicate a pingpong kinetic mechanism (Beytia et al., 1973), suggesting that a farnesyl-enzyme covalent intermediate forms from one molecule of farnesyl-PP and a nucleophilic Enz—Ẍ group before the second substrate molecule is involved.

Attack of another enzymatic nucleophile (Enz—Ẍ′) on the second farnesyl-PP molecule could initiate the carbon–carbon bond-forming step, with attendant expulsion of the initially added enzyme nucleophile (Enz—Ẍ) and generation of the acyclic C_{30} pyrophosphoryl-enzyme intermediate. Finally, proton abstraction (C—H breakage) by an enzyme base could initiate cyclopropane ring formation, concomitant with expulsion of the added Enz—Ẍ′ to generate the presqualene-PP (see top of p. 882).

The breakdown of that cyclopropyl alcoholic pyrophosphate to squalene is also a complex transformation. Edmonds et al. (1971) have noted that (1) a new carbon–carbon bond must be formed between C-3 of the cyclopropane ring of presqualene-PP and the primary alcoholic carbon; (2) the (C-1)–PP group is broken with C—O cleavage and expulsion of inorganic pyrophosphate; (3) a hydride ion from NADPH is added to C-3 of the ring; and (4) a *trans* double bond is generated between C-1 and C-2 of the cyclopropyl ring.

Edmonds and colleagues proposed the speculative two-step transformation on p. 882 (bottom). First, presqualene-PP rearranges by ring enlargement and pyrophosphate migration to produce a (presumably transient) *cyclobutyl-PP tautomer.* Then stereospecific backside addition of hydride from NADPH could induce cyclobutyl-ring fragmentation, PP_i expulsion, and *trans* olefin formation.

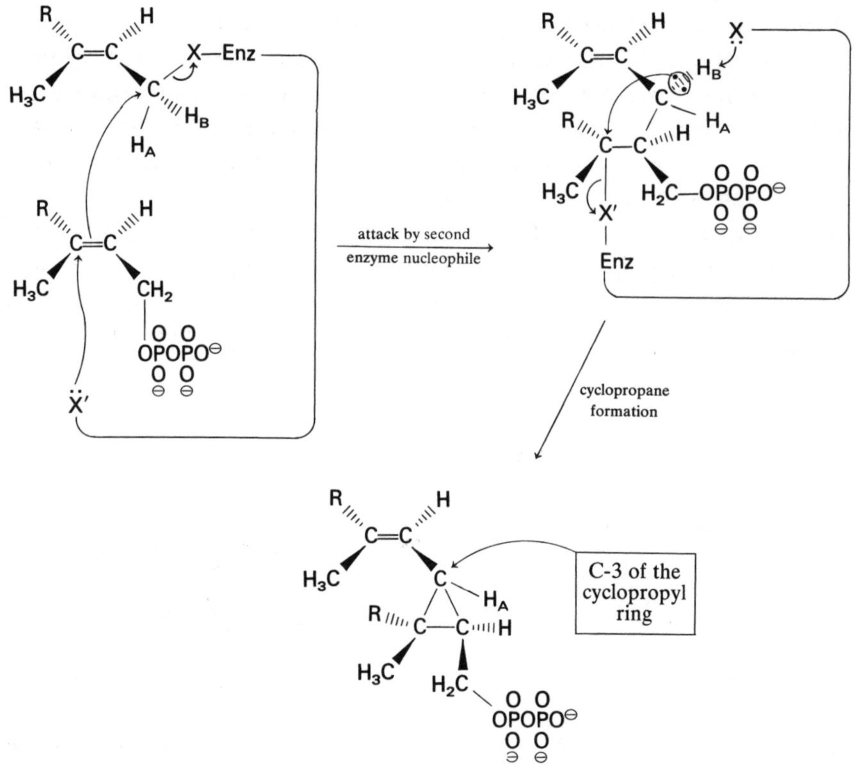

R H
C=C X—Enz
H_3C C
H_B
H_A
R H
C=C
H_3C CH_2
O O
OPOPO⊖
O O
⊖ ⊖
X′

attack by second enzyme nucleophile

R H
C=C X
H_B
H_3C C
R H_A
C—C H
H_3C H_2C—OPOPO⊖
O O
O O
⊖ ⊖
X′
Enz

cyclopropane formation

R H
C=C
H_3C C
H_A
R C—C H
H_3C H_2C
O O
OPOPO⊖
O O
⊖ ⊖

C-3 of the cyclopropyl ring

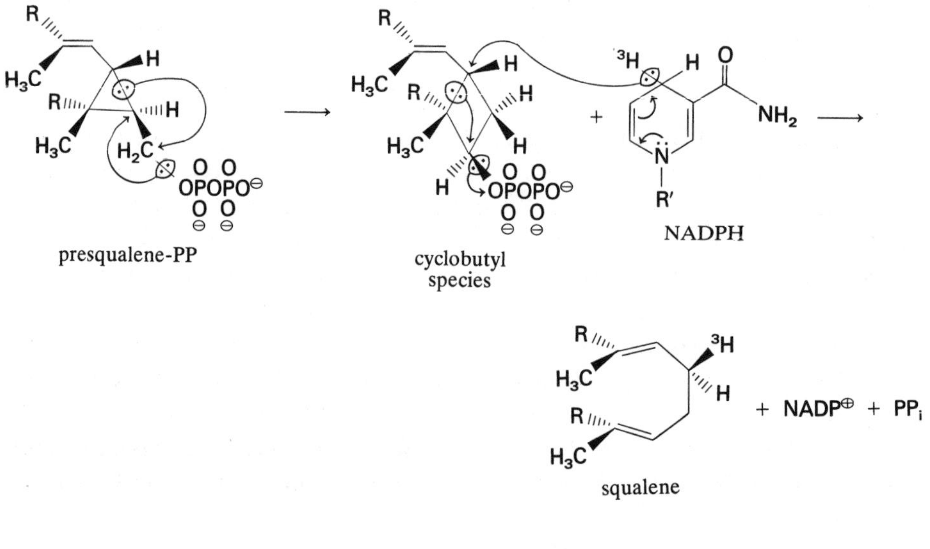

R
H_3C H
R H
H_3C H_2C
O O
OPOPO⊖
O O
⊖ ⊖

presqualene-PP

R
H_3C H
R H
H_3C H
H
O O
OPOPO⊖
O O
⊖ ⊖

cyclobutyl species

+

^3H H O
NH_2
N
R′

NADPH

R ^3H
H_3C H
R
H_3C

squalene

+ $NADP^{\oplus}$ + PP_i

Clearly, the tail-to-tail joining of two farnesyl-PP molecules to form the new carbon–carbon bond in squalene represents an extremely complex piece of enzyme chemistry, and much ingenuity will be required to decide if these speculations correctly describe that chemistry.

26.E CYCLIZATION OF SQUALENE TO LANOSTEROL AND OTHER STEROIDS

At this point, having considered the two ways (head-to-tail, then tail-to-tail) in which carbon–carbon bonds are formed enzymatically in the construction of squalene from the two C_5 isoprenyl-PP building blocks, we shall consider one further type of carbon–carbon bond formation in the metabolism of squalene—namely, the cyclization of this open-chain C_{30} hydrocarbon to the tetracyclic skeleton of the steroids. We noted this process briefly in Chapter 20 when we discussed the subsequent rearrangement of lanosterol, and it might be useful to review those comments along with this presentation.

26.E.1 Squalene→Lanosterol

Rather than drawing the structure of squalene in the fully extended conformation, it is useful to draw it in a conformation where the cyclization process can be more readily visualized. (In these structures, open-ended bonds represent methyl groups.)

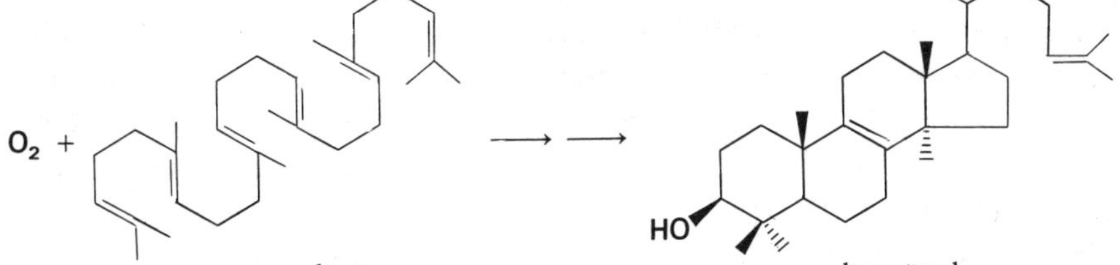

O_2 + squalene ⟶ ⟶ lanosterol

In animal cells, lanosterol is the first stable sterol product that can be isolated. It obviously has undergone both hydride shifts and methyl migrations from the expected initial cyclization product (¶20.A.2). Lanosterol also contains one oxygen atom present as the C-3β-alcohol group, and ^{18}O studies indicate molecular oxygen as the source.

This observation was explained by the demonstration of an enzymatic activity in liver extracts termed squalene epoxidase, an NADPH- and O_2-dependent system converting squalene to the β-2,3-epoxide. This squalene epoxide is then the actual substrate for subsequent enzymatic cyclization.

$$\text{Squalene} + \text{NADPH} + O_2 \longrightarrow \text{NADP}^{\oplus} + H_2O + \text{squalene-2,3-epoxide}$$

squalene-2,3-epoxide

The epoxidase appears to be an integral protein of the intracellular microsomal membranes in liver cells and has been difficult to purify and characterize (Tai and Bloch, 1972). The reaction stoichiometry is reminiscent of both flavin-linked and iron-dependent monooxygenase systems discussed in Section III. Preliminary data on solubilized enzyme preparations suggest flavoprotein involvement and not microsomal hemoproteins, but little else can be said about the mechanism at present.

A separate enzymatic activity, also membrane-bound in the microsomes, termed squalene oxidocyclase (Schechter et al., 1970, and references therein) does indeed cause cyclization of squalene-2,3-epoxide to sterol products, of which lanosterol is the major isolable product. The question of mechanism of stereospecific polyene cyclizations can be viewed as a *choice* between *stepwise* carbon–carbon bond-forming reactions, involving partially cyclized intermediates, and a *concerted* process in which all four rings are generated concurrently with synchronous formation of all the new carbon–carbon bonds. Studies with nonenzymatic cationic polyene cyclizations to steroidal structures are consistent with concerted "zipperlike" mechanisms, in that no intermediates are detectable kinetically (Johnson, 1976).

In the enzymatic process, the oxidocyclase probably has an amino-acid side chain at the active site that acts as a general acid catalyst to protonate the epoxide group of the substrate, but probably has few other nucleophiles that might capture the cyclizing carbonium species. Johnson (1976) has suggested that protonation of epoxide may generate "an incipient cationic center at C-2 that attacks the 6,7-olefinic bond initiating formation of the 2,7-σ-bond."

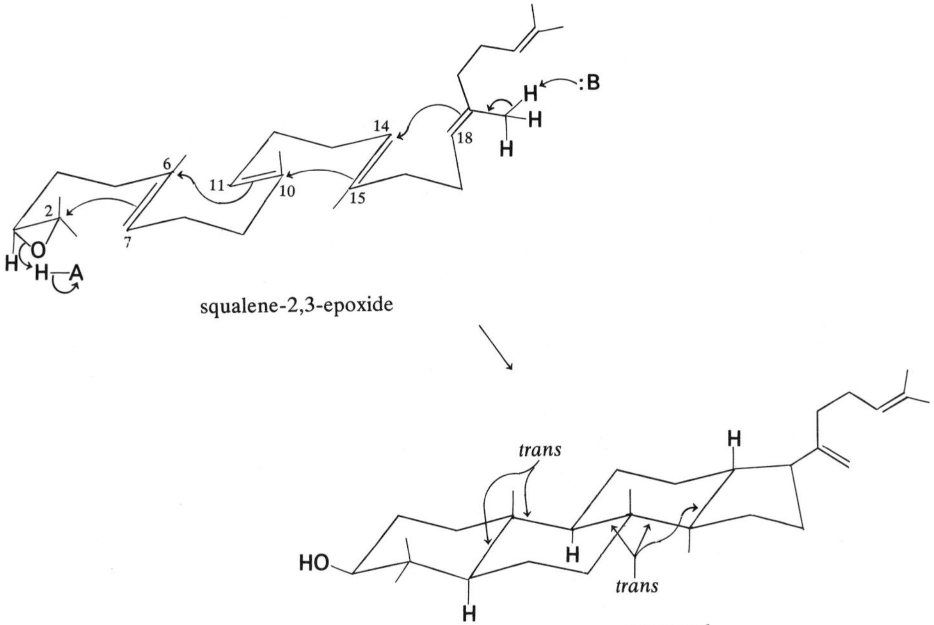

squalene-2,3-epoxide

protosterol

"Concomitantly, the cationic center developing at C-6 starts an electrophilic attack on the 10,11-olefinic bond generating the 6,11-σ-bond, etc. The addition of C-2 and C-11 to the 6,7-olefinic bond occurs in the *trans* manner, yielding the *trans*-fused ring system found in the product [protosterol]. Thus, the all-*trans* geometry of the olefinic bonds in squalene results in formation of a *trans,trans,-trans,trans,trans* fusion of the four rings in the product." (Johnson, 1976). The protosterol is thought to undergo enzyme-mediated rearrangements to lanosterol as outlined previously in ¶20.A.2.

26.E.2 Lanosterol→Cholesterol

The biosynthesis of cholesterol from lanosterol involves (1) conversion of a C_{30} steroid alcohol to a C_{27} steroid alcohol; (2) saturation of both the Δ^9-double bond and of the side-chain double bond; and (3) introduction of the Δ^5-double bond. Each demethylation must involve a carbon–carbon bond-cleavage step, and close examination has indicated that the two methyls at C-4 and the one at C-14 are lost as carbon dioxide, with the order of removal being C-14, C-4$^\alpha$, and then C-4$^\beta$ (Hamberg, Samuelsson, et al., 1974).

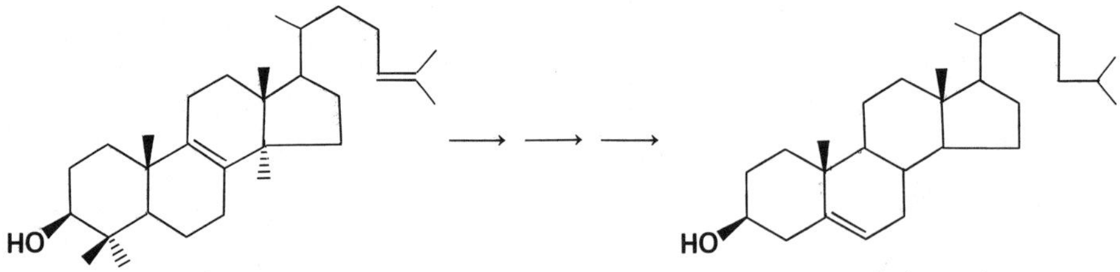

lanosterol → → → cholesterol

The oxidative demethylation sequence apparently occurs via three successive NADPH- and O_2-dependent hydroxylations catalyzed by a microsomal monooxygenase, probably not a heme enzyme (Gaylor and Mason, 1968; Hamberg, Samuelsson, et al., 1974, pp. 55–57; Gaylor, 1974). For example, loss of the 4^{α}-methyl might proceed as follows.

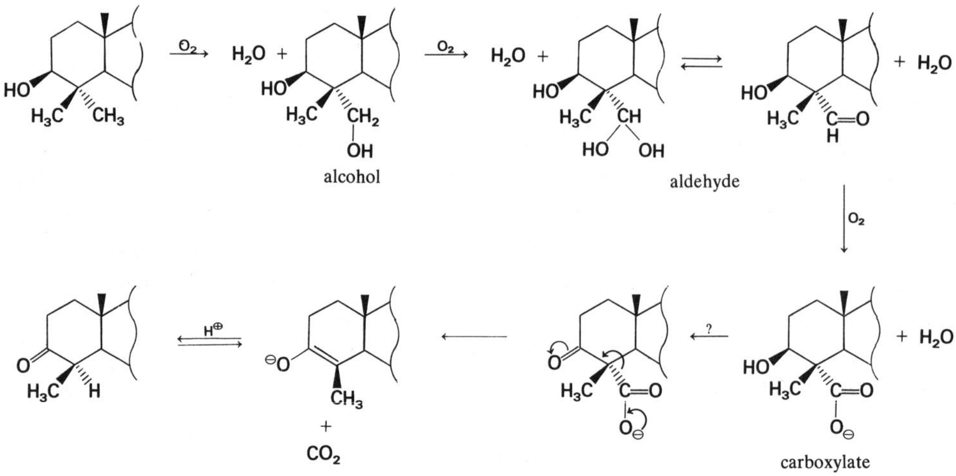

alcohol

aldehyde

carboxylate

The mechanism of decarboxylation of the 4-carboxy intermediates is not clearly understood, but it may proceed after transient oxidation of the 3-hydroxy group to a 3-keto group as indicated (and then rereduction back to the 3-ol). This strategy would yield a β-keto acid that would decarboxylate readily. For the removal of the CH_3 group at position 14, a Δ^7- or Δ^8-double bond is a structural requirement, suggesting that a reasonable scheme for this decarboxylation can occur with double-bond migration (Gaylor and Mason, 1968; Hamberg, Samuelsson, et al., 1974, pp. 55–57; Gaylor, 1974).

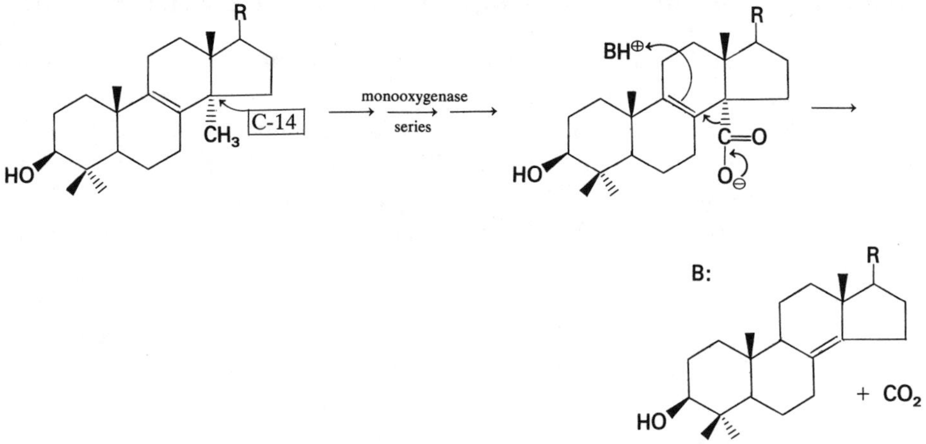

monooxygenase series

B: + CO_2

The saturation and desaturation steps are nicotinamide-linked and proceed with direct hydrogen transfer. Recent studies by Reddy et al. (1976, 1977) suggest that, in the subsequent processing of the Δ^7-ene to the $\Delta^{5,7}$-diene (which is in turn reductively saturated to the Δ^5-ene$\equiv$cholesterol), cytochrome b_5 may be involved. The overall stoichiometry is the following.

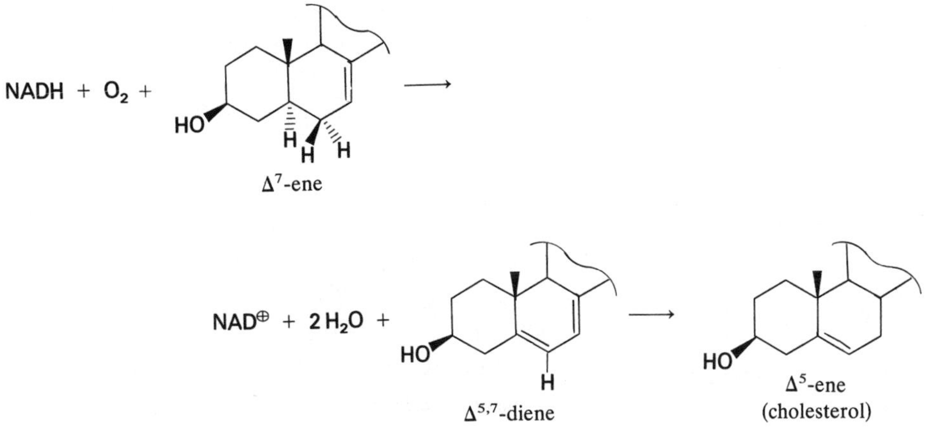

NADH + O_2 + Δ^7-ene $\longrightarrow$

$NAD^{\oplus}$ + $2H_2O$ + $\Delta^{5,7}$-diene $\longrightarrow$ Δ^5-ene (cholesterol)

This stoichiometry is reminiscent of the cytochrome-b_5–requiring stearyl-CoA desaturase complex (Chapter 15) also in its requirement for NADH and net 4-$e^{\ominus}$ reduction of O_2. The overall dehydrogenation is a net *syn* loss of 5^{α}- and 6^{α}-hydrogens, but it is unclear whether a 6-OH intermediate is formed and then dehydrated.

26.E.3 Cyclization of Geranylgeranyl-PP Without Prior Epoxide Formation

Although the sterols and many tricyclic terpenes of plants proceed by initial cyclization of a 2,3-epoxide of squalene or other acyclic isoprenoid pyrophosphates, cyclization can occur without the epoxy functionality present. For example, geranylgeranyl-PP is cyclized to copalyl-PP and that, in turn, is converted to kaurene, a precursor of the gibberellin plant hormones, in the absence of O_2. Protonation of the terminal double bond could assist cyclization (Coates et al., 1976).

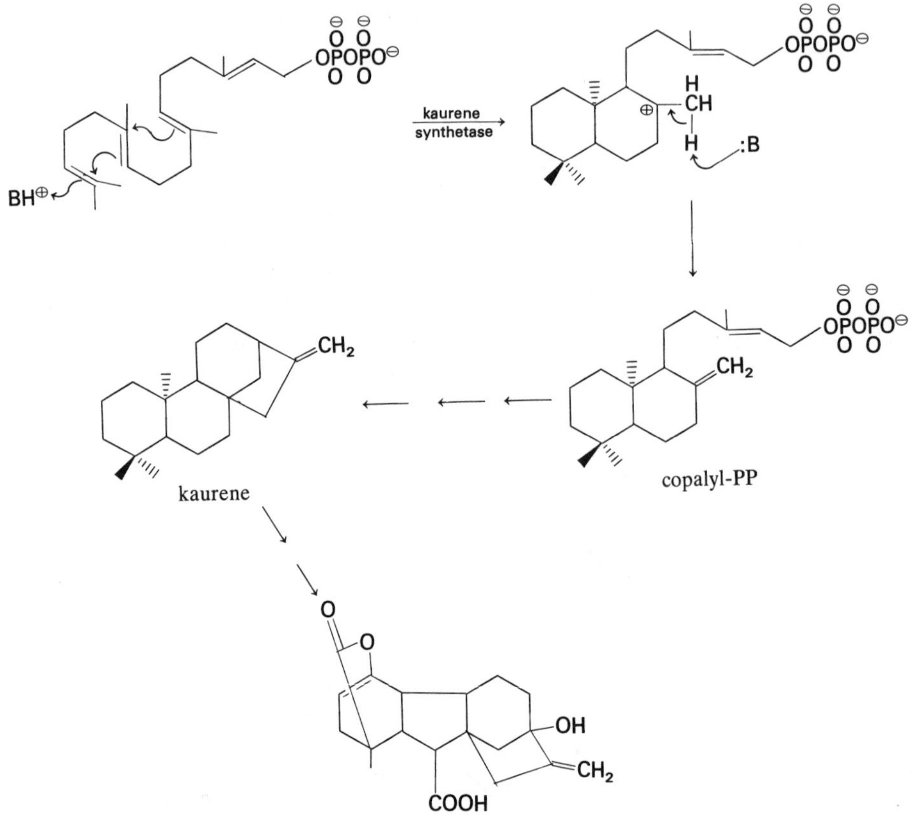

kaurene synthetase

copalyl-PP

kaurene

gibberellic acid

26.F PRENYLATION OF QUINONES IN BIOSYNTHESES OF VITAMIN K AND COENZYME Q

We noted at the outset of this chapter that the redox-active benzoquinones (coenzyme Q) and naphthoquinones (vitamin K) have isoprenyl side chains, where the number of carbons varies between 5 and 60 (i.e., 1 to 12 prenyl units). (In *E. coli*, CoQ_8 predominates with a 40-carbon isoprenyl side chain made up of 8 prenyl units.) Recent studies on biosynthesis of vitamin K in *E. coli* have suggested at what stage the prenylation of the quinone nucleus occurs and what the nucleophile attacking the long-chain allylic pyrophosphate might be.

The immediate precursor of the prenylated naphthoquinone is the hydroquinone carboxylate. A partially purified enzyme preparation carries out decarboxylation and prenyl transfer with C-1 of the original long-chain allylic alcohol-PP cosubstrate becoming bonded to the naphthohydroquinone carbon that has undergone decarboxylation (Shineberg and Young, 1976). A reasonable scheme might involve decarboxylation of a tautomer of the hydroquinone (perhaps selectively stabilized at the active site) that can delocalize the carbanion that must arise during loss of CO_2. The resulting enolate anion could serve as a carbanion nucleophile for displacement of pyrophosphate at C-1 of the polyprenyl-PP. The initial adduct can then lose the proton initially added from solvent and rearomatize to the hydroquinone form of vitamin K.

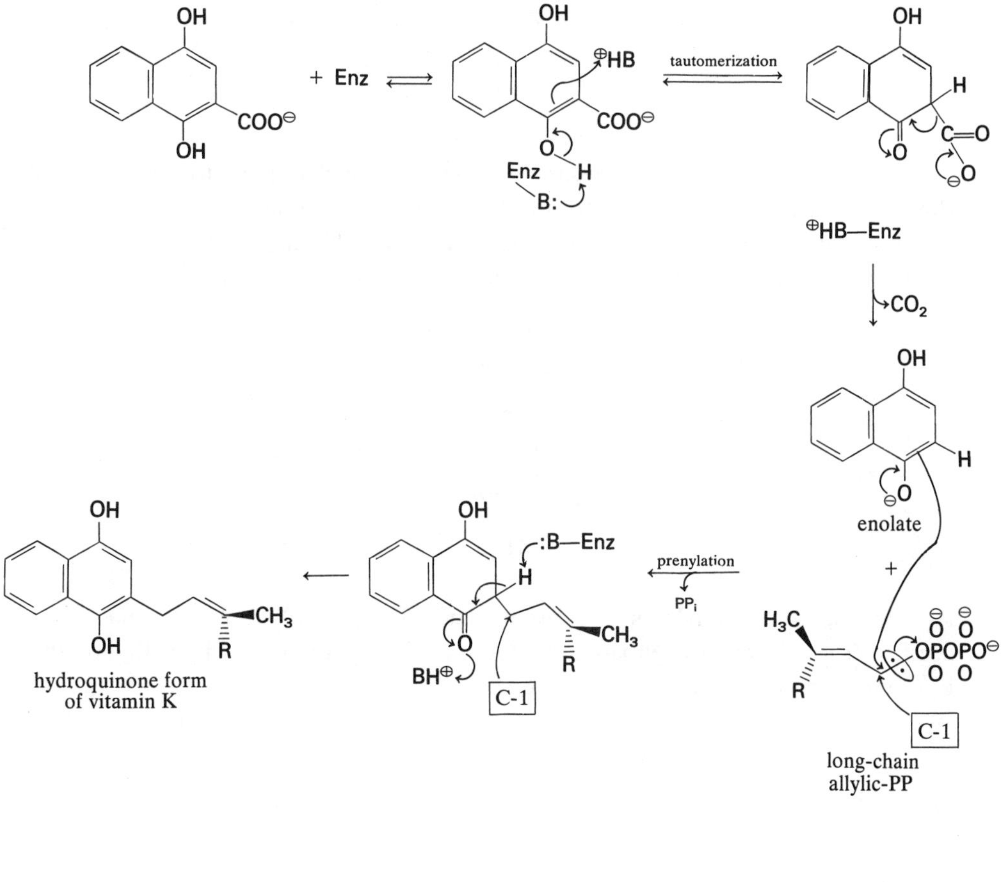

It is not known yet whether decarboxylation proceeds in the absence of the allylic-PP cosubstrate. A fundamentally similar mechanism probably exists for affixing the prenyl side chain during benzoquinone CoQ synthesis.

26.G MONOPRENYLATIONS

A variety of monoprenylations, generally of aromatic ring systems (McGrath, 1977), occur in secondary metabolism. An early step in biosynthesis of ergot alkaloids is the transfer of a dimethylallyl group to position 4 of tryptophan (Heinstein et al., 1971).

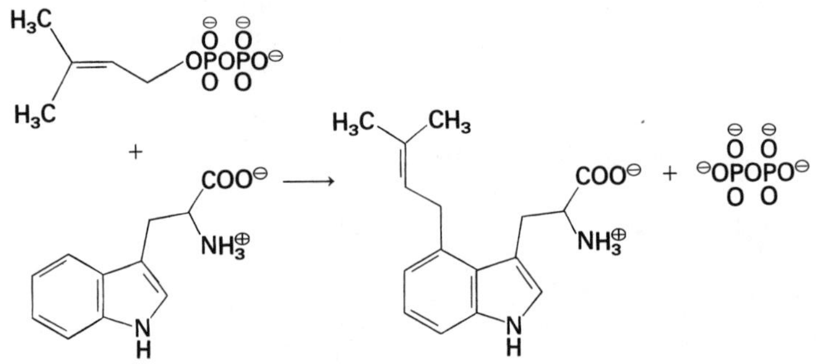

The aromatic ring locus presumably acts as nucleophile toward C-1 of dimethylallyl-PP, but the mechanism of such an attack is unclear.

Certain growth-promoting species, termed cytokinins, in plant tissues are dimethylallyl-substituted purines such as the adenosine shown here.

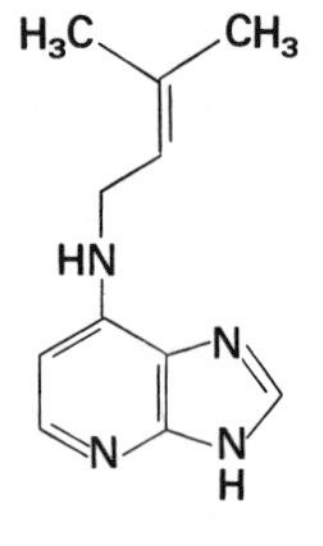

An enzyme that prenylates adenosine residues in tRNA molecules has been purified from *E. coli* (Rosenbaum and Gefter, 1972), but no activity acting on the free purine base has yet been detected.

Chapter 27

The Chemical Logic of Metabolic Pathways

In the preceding chapters, we have examined many enzymatic transformations by analyzing the chemical nature of the reactions, grouping several specific examples together, and then noting the distinctive and general patterns for that kind of chemical process in biological systems. In so doing, we have covered almost all of the *types* of reactions that occur in the primary metabolisms of microbes, plants, and animals.

In this last chapter, we shall switch focus and try to see how these few types of chemical mechanisms are used by organisms in *linked*, sequential reactions. We shall have a selective look at some important metabolic pathways to attempt to understand the underlying chemical logic of metabolism. We shall see that the routes organisms choose for degradation or synthesis of central metabolites are constrained both by the nature of the desired end product (there are economically few major primary metabolites) and by the chemical requirement that enzymes choose low-energy pathways for the small variety of chemical mechanisms available to them under physiological conditions.

In a single chapter, we could not hope to be comprehensive in sorting out patterns of metabolic diversity. Instead, we shall look at the following *three* topics to illustrate the chemical logic of metabolism, both for the individual enzymes and in the scope of the overall pathway. We shall also emphasize the combinatorial use in vivo of the types of reactions discussed in Chapters 3 through 26.

First, we shall focus on oxidative metabolism of the predominant cellular hexose, glucose, by the *glycolytic pathway*. Complete oxidation involves action of the *citrate cycle* as well. The sequential action of these two pathways involves oxidation of all six carbons of glucose to carbon dioxide. The overriding metabolic concern is to trap some of the energy released in these oxidations in a form (e.g., chemical energy) that can be used later by the organism. This concern controls and directs the chemical logic of these pathways.

The second set of metabolic sequences we shall examine includes the complementary breakdown and biosynthesis of fatty acids. We shall again focus on the redox steps, because these are either energy-yielding or energy-requiring, and on the steps involving formation or cleavage of carbon–carbon bonds. Additionally, we shall see what devices organisms use to segregate the degradative and biosynthetic pathways. We shall see how both thermodynamic and mechanistic considerations condition the chemistry employed.

The third and final major variety of metabolic transformations we shall address intertwines enzymology of aromatic-ring functionalization, degradation, and biosynthesis—focusing on phenylalanine as specific example. The enzymes that functionalize and degrade benzene rings are monooxygenases and dioxygenases; these topics illustrate the major patterns of oxygen metabolism. The microbial biosynthesis of phenylalanine, although quite intricate and complex, can be analyzed in terms of simple chemical strategems for construction of an aromatic ring with an exocyclic carbon side chain.

With these examples in mind, an interested reader can analyze the chemical logic of other metabolic pathways treated in passing or fragmentarily in earlier chapters. These include lanosterol biosynthesis (i.e., analyze the few ways in which carbon–carbon bonds are made in that C_{30} molecule, and how the mode of C—C bond formation changes from early in the pathway to later in the pathway); activation of D-galactose and its convergent incorporation into glycoproteins and isomerization to glucose at the level of UDP-galactose; and biosynthesis of purine and pyrimidine ring systems—to name just a few of the sequences of potential interest.

27.A OXIDATIVE METABOLISM OF GLUCOSE

In preceding chapters we have repeatedly discussed the enzymology of the C_6 aldohexose D-glucose. Its predominant role in animal, plant, and microbial cells is to serve as an energy source, although it can be rerouted into storage and/or structural polysaccharides in times of surfeit.

How can glucose serve chemically as an energy source? The most obvious way is to undergo a series of controlled oxidations with the released energy trapped in a chemically useful form. We have noted that ATP and acetyl-CoA represent two of the common chemical energy currencies in cells, the phosphoric anhydrides in ATP and the activated acyl group in acetyl-CoA being thermodynamically destabilized with respect to hydrolysis products. These molecules, therefore, can be used thermodynamically and mechanistically to do cellular work. A third useful form of chemical energy storage in cells is represented by the 2-$e^{\ominus}$–reduced form of the nicotinamide coenzymes NADH and NADPH, which are both mobile and kinetically stable in the cellular milieu. Yet they can be readily oxidized by the membraneous flavoprotein NADH (NADPH) dehydrogenases, in mitochondria or in prokaryotic cytoplasmic membranes, and can donate two electrons at low potential ($E^{0'} = -320$ mV) down respiratory chains to O_2 as ultimate electron acceptor ($O_2 \rightarrow H_2O$, $E^{0'} = +820$ mV). The potential drop of 1.14 V represents a large amount of energy that can be stored first in a transmembrane electrochemical potential and then used to make three ATP molecules (oxidative phosphorylation).

Thus, one would predict that glucose, as a reduced organic substrate, must be oxidized enzymatically. CO_2 and H_2O would represent completely efficient oxidation and would minimize problems of waste removal. During such oxidation, one should look for formation of molecules such as ATP, acetyl-CoA, or NADH as an index of energy capture and conversion to biochemically useful forms by the enzymatic processes. Indeed, the oxidative metabolism of glucose is described below in just these terms and is divided into two segments:

1. the multistep conversion of glucose to two molecules of acetyl-CoA and two molecules of CO_2—the *glycolytic* (literally, "sugar-cleaving") *pathway*; and then
2. the complete oxidation of both carbons of acetyl-CoA to CO_2 and H_2O in the *citrate cycle.*

We shall look at how much energy released during these oxidations is trapped and by what mechanisms. We shall see, in fact, that we have already discussed each enzyme in glycolysis and the citrate cycle in preceding chapters.

27.A.1 Glycolysis

During glycolytic breakdown—also known as glucose fermentation (no O_2 is involved as cosubstrate during these reactions)—glucose is converted first to pyruvate and then to acetyl-CoA. A net production of two molecules of ATP occurs during this sequence, as we shall note.

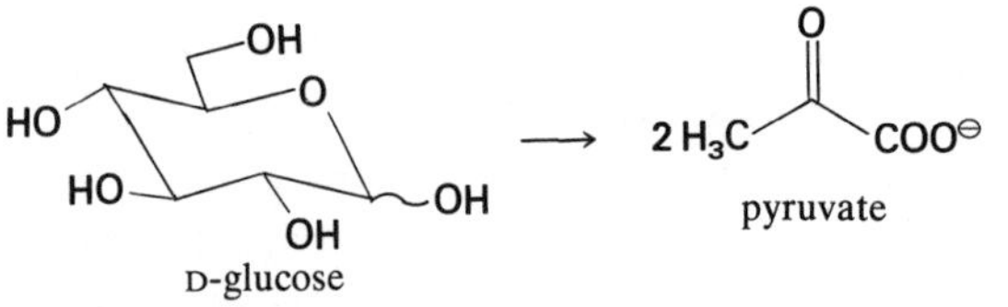

D-glucose ⟶ 2 H_3C–CO–$COO^{\ominus}$
pyruvate

This conversion involves fragmentation of the hexose into two C_3 products, so we shall look for a carbon–carbon cleavage reaction. It also involves reduction of some carbons and oxidation of others, so we shall look for redox processes. A polyhydroxy initial substrate produces two α-keto-acid products. Dehydrations are likely; we have noted that loss of OH often is facilitated enzymatically by prior conversion to phosphate esters. Enzymatic production of activated compounds of acyl phosphates or acylthioesters, we have seen, generally involves oxidation of unactivated intermediates.

The key determinant of carbon skeletal arrangements in the products is cleavage of the hexose to C_3 units, both of which can be processed eventually to pyruvate. The most likely mechanism in a hydroxy substrate would be an aldol cleavage, requiring a β-hydroxycarbonyl functionality. As an aldohexose, glucose *in the open-chain form* has a β-hydroxycarbonyl grouping, but aldol fragmentation of it would produce a C_4 and a C_2 unit. This is not the observed set of products in vivo.

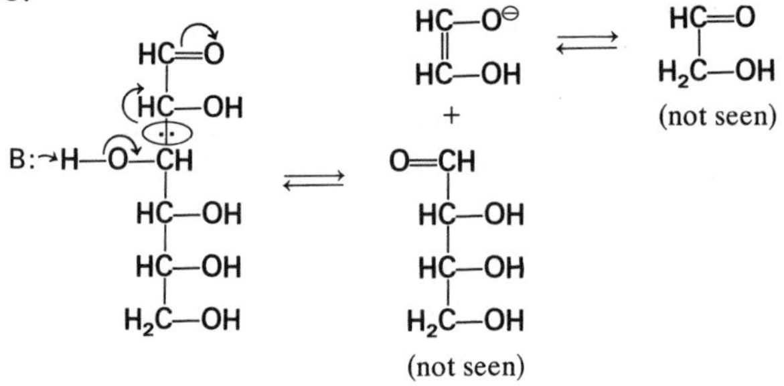

HC=O / HC–OH / B:→H–O–CH / HC–OH / HC–OH / H_2C–OH ⇌ HC–$O^{\ominus}$ / HC–OH + O=CH / HC–OH / HC–OH / H_2C–OH (not seen); HC–$O^{\ominus}$ / HC–OH ⇌ HC=O / H_2C–OH (not seen)

If the carbonyl group were not at C-1 but at C-2, the aldol cleavage *would* yield C_3 fragments, and this shift is the purpose of the first four enzymes of the glycolytic pathway, isomerization of D-glucose to D-fructose. The actual substrates are phosphorylated sugars. The first phosphoryl group traps glucose intracellularly as glucose-6-P; the second sets up fructose 1,6-diphosphate for cleavage to two C_3 fragments, each monophosphorylated and readily interconvertible. Two other features are noteworthy: (1) the phosphoryl groups may act as specificity handles, providing increased binding interaction energy with the enzymes; and (2) they will be progenitors of activated phosphoryl compounds in the latter half of glycolysis. We have discussed the first four enzymes in earlier chapters.

The first enzyme, hexokinase (¶7.D.1), catalyzes a nucleophilic attack on the γ-P of ATP by the C-6 hydroxyl of glucose, producing glucose-6-P.

D-glucose + $ATP^{4\ominus}$ $\xrightarrow[Mg^{2\oplus}]{\text{hexokinase}}$ D-glucose-6-P + $ADP^{3\ominus}$ (1)

The second enzyme, phosphoglucose isomerase (¶19.B.1), catalyzes an aldose–ketose isomerization (1,2-shift) with enzyme-catalyzed opening and closure of hemiacetals.

D-glucose-6-P ⇌ ⇌ ⇌ D-fructose-6-P (2)

The third enzyme, phosphofructokinase (¶7.D.1), catalyzes nucleophilic attack by the C-1 hydroxyl on the γ-P of ATP.

D-fructose-6-P + $ATP^{4\ominus}$ ⇌ D-fructose-1,6-diphosphate + $ADP^{3\ominus}$ (3)

The fourth enzyme, fructose-1,6-diphosphate aldolase (¶23.A.1), proceeds via cleavage to the DHAP eneamine as low-energy carbanion initial product; this is the C—C *fragmentation step* of glycolysis (see top of p. 896).

With the hexose diphosphate cleaved, subsequent metabolism proceeds only through glyceraldehyde-3-P, whose aldehyde functionality will be oxidized up to the acid grouping found in pyruvate. The fifth enzyme of the glycolytic pathway is another 1,2-isomerase converting dihydroxyacetone-P (DHAP) back to the utilizable glyceraldehyde-3-P, a beautifully economical device to use all the carbons of glucose (set up by the initial isomerization of the glucose skeleton to the fructose

$$\text{D-fructose-1,6-diphosphate} \rightleftharpoons \begin{array}{c} H_2C{-}OPO_3^{2\ominus} \\ | \\ C{=}O \\ | \\ HO{-}CH \\ | \\ HC{-}O{-}H \;\; :B \\ | \\ HC{-}OH \\ | \\ H_2C{-}OPO_3^{2\ominus} \end{array} \rightleftharpoons \begin{array}{c} H_2C{-}OPO_3^{2\ominus} \\ | \\ C{=}O \\ | \\ H_2C{-}OH \\ \text{dihydroxyacetone-P} \\ + \\ HC{=}O \\ | \\ HC{-}OH \\ | \\ H_2C{-}OPO_3^{2\ominus} \\ \text{glyceraldehyde-3-P} \end{array} \tag{4}$$

skeleton in step 2). This fifth enzyme is triose-P isomerase (¶19.B.3), catalyzing a 1,2-hydrogen shift.

$$\begin{array}{c} H_2C{-}OPO_3^{2\ominus} \\ | \\ C{=}O \\ | \\ H_2C{-}OH \\ \text{DHAP} \end{array} \rightleftharpoons \begin{array}{c} H_2C{-}OPO_3^{2\ominus} \\ | \\ HC{-}OH \\ | \\ HC{=}O \\ \text{glyceraldehyde-3-P} \end{array} \tag{5}$$

To this point, two ATP molecules have been consumed in the two phosphoryl transfers of steps 1 and 3. Two of the remaining steps in glycolysis resynthesize ATP. Because both C_3 pieces are metabolized, $2 \times 2 = 4$ ATP molecules are produced, a net gain of 2 ATP molecules per molecule of glucose cleaved to pyruvate. The first *energy-yielding step* is the oxidation catalyzed by glyceraldehyde-3-P dehydrogenase (¶10.C). Recall that the bound aldehyde substrate is attacked by Cys^{149} to give the thiohemiacetal, which is oxidized directly to the activated 3-phosphoglycerylthioester and then phosphorylized, with preservation of the acyl-group activation, to the observed *acyl-phosphate* product, 1,3-diphosphoglycerate. This is the energy-capture step. An additional energy dividend is the stoichiometric formation of NADH, which is a reduced electron source for oxidative phosphorylation in later steps of the cycle.

$$\begin{array}{c} H_2C{-}OPO_3^{2\ominus} \\ | \\ HC{-}OH \\ | \\ HC{=}O \\ \text{glyceraldehyde-3-P} \end{array} + HOPO(O)O^{\ominus}_{\,O^{\ominus}} + NAD^{\oplus} \rightleftharpoons NADH + \begin{array}{c} H_2C{-}OPO_3^{2\ominus} \\ | \\ HC{-}OH \\ | \\ C{=}O \\ | \\ OPO_3^{2\ominus} \\ \text{1,3-diphosphoglycerate} \end{array} \tag{6}$$

The seventh enzyme, 3-phosphoglycerate kinase, simply harvests the phosphoryl group from the doubly activated acyl phosphate (activated acyl and phosphoryl groups as a mixed acyl-phosphoric anhydride), transferring $-PO_3^{2\ominus}$ from the kinetically labile 1,3-diphosphoglycerate ($T_{1/2} = 30$ min under physiological conditions) to the kinetically stable and hence more useful ATP. An oxygen anion of ADP is the attacking nucleophile.

$$\begin{array}{c} H_2C{-}OPO_3^{2\ominus} \\ | \\ HC{-}OH \\ | \\ C{=}O \\ | \\ OPO_3^{2\ominus} \end{array} + ADP^{3\ominus} \rightleftharpoons \begin{array}{c} H_2C{-}OPO_3^{2\ominus} \\ | \\ HC{-}OH \\ | \\ C{=}O \\ | \\ O^{\ominus} \end{array} + ATP^{4\ominus} \tag{7}$$

1,3-diphosphoglycerate 3-phosphoglycerate

The remaining strategy is to convert 3-phosphoglycerate (3-PGA) to pyruvate via a chemically activated intermediate that can be harvested enzymatically. Phosphoenolpyruvate (PEP) springs to mind as just such a pyruvate precursor, but PEP has the phosphoryl substituent at C-2, not C-3.

$$\begin{array}{c} OPO_3^{2\ominus} \\ | \\ H_2C{=}C{-}COO^{\ominus} \end{array}$$

PEP

Thus, the eighth enzyme is phosphoglycerate mutase (¶19.D.2), which moves the phosphoryl group from unactivatable C-3 to activatable C-2. The reaction involves intermediate formation of 2,3-diphosphoglycerate in some cases, but is direct intramolecular 1,2-transfer in others.

$$\begin{array}{c} H_2C{-}OPO_3^{2\ominus} \\ | \\ HC{-}OH \\ | \\ COO^{\ominus} \end{array} \xrightleftharpoons[\text{of 2,3-diPGA}]{\text{catalytic amounts}} \begin{array}{c} H_2C{-}OH \\ | \\ HC{-}OPO_3^{2\ominus} \\ | \\ COO^{\ominus} \end{array} \tag{8}$$

3-PGA 2-PGA

The 2-phosphoglycerate (2-PGA) is then processed by enolase (¶17.A.3); an unactivated phosphate ester is dehydrated to an enol, trapped in that form as a phosphoryl derivative so it cannot ketonize to the more thermodynamically stable α-keto-acid tautomer. Thus, a "high-energy" enol phosphate is generated in an internal redox step by the facile dehydration. This is the second *energy-yielding step* in glycolysis, and the mode of energy capture can be contrasted with that in step 6. The reaction proceeds by a carbanion intermediate.

$$B{:}\curvearrowright \begin{array}{c} H_2C{-}OH \\ | \\ H{-}C{-}OPO_3^{2\ominus} \\ | \\ COO^{\ominus} \end{array} \rightleftharpoons H_2O + \begin{array}{c} CH_2 \\ \| \\ C{-}OPO_3^{2\ominus} \\ | \\ COO^{\ominus} \end{array} \tag{9}$$

2-PGA PEP

Again, the energy-capture step is followed by an enzymatic step, pyruvate kinase (¶7.D.1), where the activated phosphoryl group in PEP is transferred (downhill thermodynamically; see Table 7-1) to ADP to form ATP. The oxygen anion of ADP acts as nucleophile during the phosphoryl transfer.

$$\underset{\text{PEP}}{\begin{matrix} CH_2 \\ \| \\ C{-}OPO_3^{2\ominus} \\ | \\ COO^{\ominus} \end{matrix}} + ADP^{3\ominus} \rightleftharpoons \begin{matrix} CH_2 \\ \| \\ C{-}O^{\ominus} \\ | \\ COO^{\ominus} \\ + \\ ATP^{4\ominus} \end{matrix} \rightleftharpoons \underset{\text{pyruvate}}{\begin{matrix} CH_3 \\ | \\ C{=}O \\ | \\ COO^{\ominus} \end{matrix}} \quad (10)$$

The chemical logic of the ten enzymatic steps in conversion of glucose to two molecules of pyruvate has been charted, and the energy yield thus far is 14 kcal (59 kJ) per mole of glucose, a meager return, but the major oxidative steps remain to come.

Depending on the metabolic state of the particular cell in question, *three fates* can await pyruvate, involving three obvious chemical steps for processing of an α-keto acid: nonoxidative decarboxylation, reduction to lactate, or oxidative decarboxylation.

In fermenting yeast, pyruvate is nonoxidatively decarboxylated via the TPP-dependent pyruvate decarboxylase (¶21.C.1.a) to acetaldehyde and CO_2. The acetaldehyde is then reducible to ethanol (¶10.C.2), consuming the NADH produced in step 6 by glyceraldehyde-3-P dehydrogenase—an important balancing maneuver to maintain a reasonable concentration of $NAD^{\oplus}$ for other dehydrogenases (in fermenting yeast, the culture is anaerobic, and NADH cannot be converted back to $NAD^{\oplus}$ via a respiratory chain using O_2 as electron acceptor).

$$\underset{\text{pyruvate}}{H_3C{-}C({=}O){-}COO^{\ominus}} \xrightarrow{\text{Enz—TPP}} CO_2 + \underset{\text{acetaldehyde}}{H_3C{-}C({=}O){-}H} \xrightarrow[\text{NADH} \rightarrow \text{NAD}^{\oplus}]{\text{alcohol dehydrogenase}} \underset{\text{ethanol}}{H_3C{-}CH_2{-}OH} \quad (11)$$

During local anaerobiosis in actively metabolizing tissues (e.g., leg-muscle cells during a mile race), an alternative way to control NADH levels is to reduce pyruvate to lactate via lactate dehydrogenase (¶10.C). The lactate can be reoxidized to pyruvate when the cells become fully aerobic again, and it can then be used for acetyl-CoA formation.

$$\underset{\text{pyruvate}}{H_3C{-}C({=}O){-}COO^{\ominus}} + NADH \rightleftharpoons \underset{\text{L-lactate}}{H_3C{-}CH(OH){-}COO^{\ominus}} + NAD^{\oplus} \quad (12)$$

A third enzyme-catalyzed reaction of α-keto acids is TPP-dependent *oxidative* decarboxylation, where C-2 is oxidized to the acyl level in the product. This key step is carried out by the oxidative three-enzyme complex pyruvate dehydrogenase (¶21.C.2), composed of a TPP-dependent decarboxylase, a lipoamide-bearing transacetylase, and a dihydrolipoyl reductase flavoprotein.

$$H_3C{-}C(=O){-}COO^{\ominus} + CoASH + NAD^{\oplus} \xrightarrow[\text{lipoamide}]{\text{TPP, FAD,}} H_3C{-}C(=O){-}SCoA + CO_2 + NADH \quad (13)$$

pyruvate → acetyl-CoA

This multistep process generates the activated C-2 acyl unit of acetyl-CoA. With the carboxyl group of acetate so derivatized, the further oxidation of pyruvate-generated carbon can go all the way to CO_2 in the citrate cycle, an efficient ambient-temperature "combustion" of pyruvate. Thus, step 13 normally follows step 10 in aerobic metabolism of glucose, leading into the citrate cycle.

The thirteen reactions just discussed are recognizable as group transfers, isomerization or elimination steps, C—C fragmentations (such as decarboxylations or aldol cleavages), or redox catalyses. The redox steps are the energy-yielding steps, just as they are in the citrate cycle to which we now turn our attention.

27.A.2 Citrate Cycle

The citrate cycle (tricarboxylate cycle, or Krebs cycle) is the common final oxidation pathway for any metabolic fuel source that can be presented as an acetyl group or as some C_4 dicarboxylic acid. Thus, the metabolism of sugars, fatty acids (¶27.B), and amino acids leads into the citrate cycle (Srere, 1975; Stryer, 1975, p. 307 ff). Although the idea of a cycle is useful in many ways, the key chemical principle here is that, for each turn of the cycle, the two carbons of acetyl-CoA are oxidized to CO_2.

$$H_3C{-}C(=O){-}SCoA \longrightarrow HSCoA + 2\,CO_2$$

This transformation involves a six-electron oxidation of the methyl group and a two-electron oxidation of the carbonyl carbon—eight electrons removed in all. Thus, all three carbon atoms of pyruvate (and so all six carbons of glucose) are completely oxidized to carbon dioxide, and the various decarboxylation steps represent C—C bond fragmentations that are terminal elements in the processing

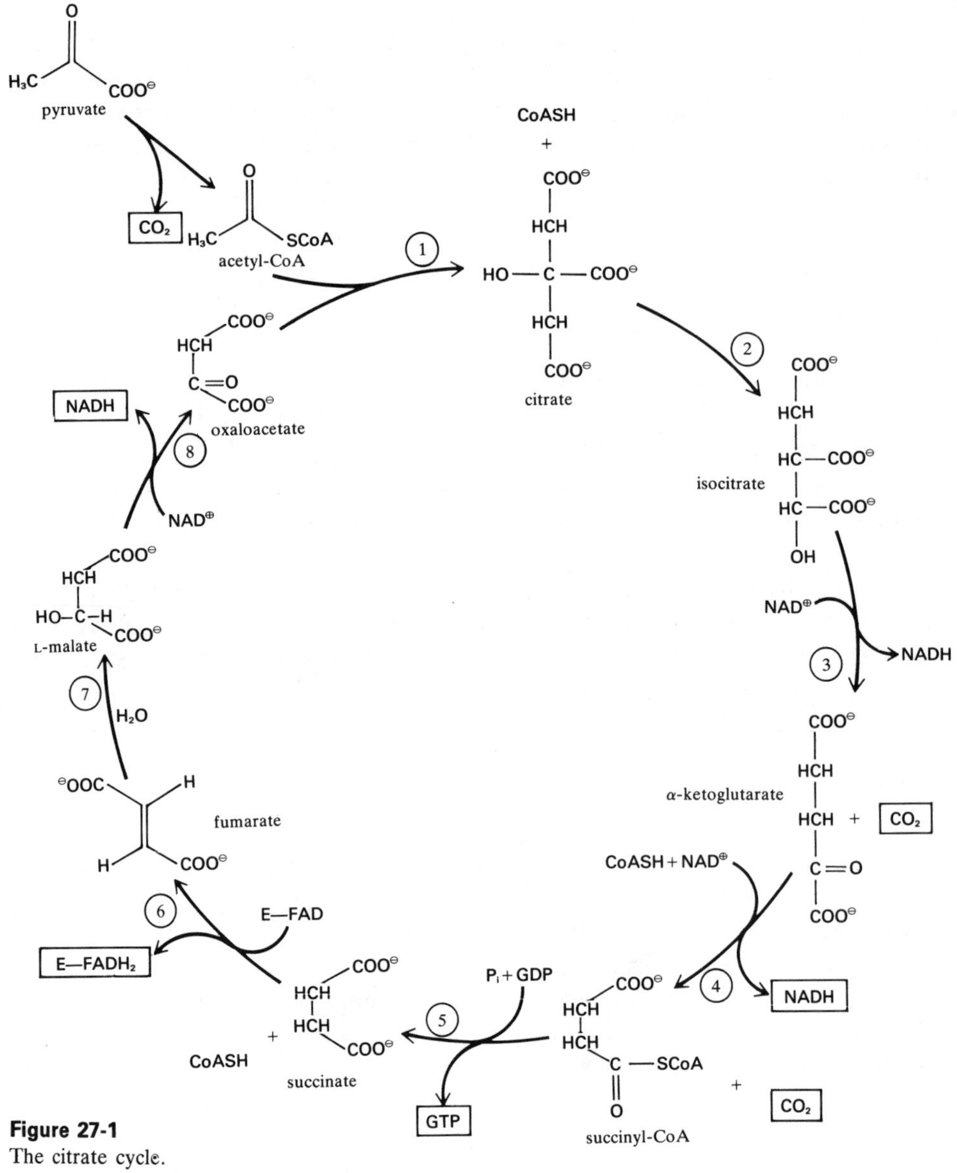

Figure 27-1
The citrate cycle.

of reduced organic molecules. Ultimately, as we shall note, the eight electrons removed are used to produce twelve ATP molecules.

The operating mechanism of the citrate cycle (Fig. 27-1) is to condense the acetyl unit of the entering acetyl-CoA in a Claisen condensation with the C_4 electrophilic acceptor oxaloacetate to yield the C_6 tricarboxylate citrate (step 1) by action of citrate synthase (¶23.B.2.a). This C_6 species is the vehicle for oxidation. The tertiary hydroxyl group of citrate is not properly placed in the carbon skeleton for the key oxidative decarboxylations to follow, and so it is isomerized via the dehydration–rehydration sequence of aconitase action (¶17.A.1) in step 2. Now the hydroxyl group of isocitrate is β to the tertiary carboxylate, and an oxidative decarboxylation via a poised β-keto-acid intermediate is readily envisaged; this happens in step 3 by action of isocitrate dehydrogenase (¶21.B.1). The actual oxidative step generates NADH from $NAD^{\oplus}$ and yields bound oxalosuccinate, which loses CO_2 to yield α-ketoglutarate. Thus, one CO_2 has been produced by decarboxylation of a β-keto acid. Step 4 is the α-keto-acid variant of oxidative decarboxylation; α-ketoglutarate dehydrogenase (¶21.C.2.a) is functionally analogous *in toto* to the pyruvate-dehydrogenase complex. This production of the second CO_2 of the cycle is an efficient energy trap; a molecule of NADH is formed, and the product succinyl-CoA has an activated acyl group as well. This chemical energy is harvested in step 5, a conversion of succinyl-CoA to enzyme-bound succinyl-P (Bridger, 1974) and then reversible phosphoryl transfer from that acyl phosphate to GDP, producing a nucleoside triphosphate GTP, an energy-currency exchange. The enzyme of step 5, succinyl-CoA synthetase, appears to be a one-enzyme analogue of the phosphotransacetylase–acetate-kinase couple discussed in ¶7.D.3.

With both molecules of CO_2 produced, the remaining enzymes of the cycle are concerned simply with conversion of the C_4 succinate to the C_4 oxaloacetate so that another turn of the oxidative wheel can occur.

$^{\ominus}OOC$–HCH–HCH–$COO^{\ominus}$	$^{\ominus}OOC$–HCH–C=O–$COO^{\ominus}$
succinate	oxaloacetate

This must involve four-electron oxidation of the succinate methylene to the keto functionality of oxaloacetate. As one might predict, two 2-$e^{\ominus}$ steps are involved. Step 6 involves flavoprotein dehydrogenative catalysis by succinate dehydrogenase (¶11.C.2) to fumarate. The two electrons in the dihydroflavin coenzyme are eventually passed into the respiratory chain (¶27.A.3) and yield two ATP molecules via oxidative phosphorylation. Fumarate is hydrated to malate by fumarase (¶17.A.2) in step 7. The malate is then dehydrogenated with attendant NADH production in step 8 by NADH dehydrogenase, completing one turn of the cycle.

From a chemical point of view, steps 6 through 8 make up an elegant three-enzyme–controlled functionalization sequence to convert a methylene into a carbonyl group, and the sequence *flavoprotein dehydrogenase, hydratase, and nicotinamide dehydrogenase* is the common enzymatic solution to this chemical problem (as we shall note in discussing fatty acid metabolism in ¶27.B) and represents a *complementary* role for flavin and nicotinamides in soluble oxidative catalysis.

$$H_2O + \text{HCH} \rightleftharpoons \text{>C=O} + 4H^{\oplus} + 4e^{\ominus}$$

Note the source of the two CO_2 molecules produced in the cycle (boxed in Fig. 27-1). They derive in sequence from the tertiary $COO^{\ominus}$ of citrate and the primary carboxylate of the pro*R* arm of citrate, respectively, as inspection will show. In turn, these two $COO^{\ominus}$ groups derive from the two oxaloacetate carboxylate groups, because step 1 introduces the acetyl group of acetyl-CoA exclusively into the pro*S* arm of citrate. This means that, although the *net working* of any turn of the cycle is conversion of an acetate fragment to two molecules of CO_2 and eight electrons, the actual carbon atoms removed as CO_2 *do not derive* from the acetyl fragment just introduced. Nor, correspondingly, do all eight electrons get removed from the pro*R* carboxymethyl arm of citrate. One can trace the path of the acetyl group through subsequent cycles, both on paper and in the laboratory, but this exercise is complicated by the fact that succinate has a plane of symmetry, as does fumarate, and the origin of the various carbon atoms is scrambled there even to the view of the chiral enzyme reagents.

Although the stereochemistry of the various enzymes of the cycle has been elegantly established, it is not discussed here; the information is available both in discussions of the individual enzymes in preceding chapters and in the book by Alworth (1972).

Inspection of the types of catalysis represented by the enzymes of the citrate cycle reveals *four redox steps* (two electrons each) for removal of the requisite eight substrate electrons; this categorization *emphasizes the redox nature of the cycle* and underlines the chemical logic of it. Note that the four redox enzymes represent four distinct variants: β-keto-acid decarboxylation, α-keto-acid decarboxylation, flavoenzyme desaturase, and nicotinamide-dependent alcohol-group oxidation. One C—C bond is formed in the initial Claisen condensation; two are broken by the successive decarboxylation steps.

Our last comment about the citrate cycle deals with the energy yield. Three of the redox steps produce NADH (or NADPH); one produces enzyme-bound $FADH_2$. Additionally, a molecule of GTP is gleaned directly. We shall note in the following paragraphs that the respiratory chain can make three ATP molecules per NADH-derived electron pair, but only two ATP molecules from $FADH_2 \cdot$ succinate dehydrogenase. Thus the yield is eleven ATP plus one GTP, or a net of twelve nucleoside triphosphate molecules per acetyl-CoA oxidation in a turn of the citrate cycle. Given two molecules of acetyl-CoA (from two pyruvates) per initial glucose molecule, then 24 ATP molecules (or equivalents) worth of energy are harvested in the citrate cycle. Adding the net of two ATP from glycolysis and four NADH from separate action of glyceraldehyde-P dehydrogenase and the pyruvate-dehydrogenase complex, a total of 38 ATP molecules derive from the combined action of glycolytic enzymes and citrate-cycle enzymes, taking into account passage of all the substrate-derived electrons. At a value of −7.3 kcal/mole (−31 kJ/mole) for $\Delta G^{0'}$ of ATP hydrolysis under physiological conditions, we obtain $38 \times (-7.3) = -277$ kcal $= -1{,}160$ kJ per mole of glucose. The total combustion of a mole of glucose in a calorimeter yields −686 kcal/mole (−2,870 kJ/mole). So, the organisms using this sequence of coupled enzyme action are $277/686 = 40\%$ efficient in harvesting the available energy released on isothermal oxidation of a reduced organic molecule by removal of twenty electrons in the logical nineteen-step chemical sequence indicated.

27.A.3 Oxidative Phosphorylation

Probing the murky depths of oxidative phosphorylation is beyond the scope of this discussion (see Stryer, 1975, p. 331 ff; Racker, 1976). However, some brief comments are germane to the preceding discussion, because the *catalytic* action of the enzymes of glycolysis and the citrate cycle must involve regeneration of the initial oxidized forms of flavin and nicotinamide cofactors before processing of another molecule of glucose can occur. The *respiratory chain* of the mitochondrial membrane in eukaryotes or the cytoplasmic membrane of prokaryotes *is the electron acceptor* from NADH and $FADH_2$ in *aerobic metabolism.* (In anaerobes, other solutions are found: either there are alternative electron acceptors—e.g., sulfide or fumarate as terminal respiratory-chain acceptor—or membrane electron-transport chains may not be involved in regeneration of oxidized coenzymes.) Also, although we have discussed oxidative metabolism in aerobic cells where oxygen is readily available, note that none of the twenty enzymes mentioned thus far in the chapter interacts with O_2 directly. Among other things, *this decreases the toxicity of oxygen metabolism.* Cytochrome oxidase, the terminal electron-chain component, reduces O_2 smoothly (if obscurely) by four electrons to water. No amount of 1-$e^{\ominus}$–reduced superoxide ion or 2-$e^{\ominus}$–reduced hydrogen peroxide (both toxic metabolites) is produced.

As noted briefly in Chapter 15, the membraneous components of electron-transport chains are quinones, flavoproteins, metalloflavoproteins, heme proteins, and the copper heme protein cytochrome oxidase. Electrons are fed in from NADH at −0.32 V to the metalloflavoenzyme NADH dehydrogenase or from the reduced flavoenzymes of butyryl-CoA dehydrogenase ($E^{0\prime} = -0.01$ V) and succinate dehydrogenase ($E^{0\prime} = 0.03$ V), passing down to cytochrome *a* ($E^{0\prime} = +0.29$ V), which stores up four electrons and then transfers them all out to O_2, reducing it to H_2O.

Fall of the substrate-derived electrons (at low potential) through this potential drop of (up to) 1.0 to 1.1 V releases electrical energy that (if it can be trapped and converted to chemical energy) can be used by the cell as a dividend (indeed, the major dividend) from oxidative metabolism. The molecular mechanisms of energy storage and transduction during electron flow are still uncertain, but some form of the chemiosmotic hypothesis of Mitchell (1966) seems very probable (Harold, 1972). In essence, that hypothesis states that electron flow down such respiratory chains leads to anisotropic acid–base chemistry by the membrane proteins, such that protons are pumped vectorially to one face of the mitochondrial (or bacterial) membrane. The energy-generating cellular membrane systems are relatively impermeable to protons, and so this oriented proton pumping induces both a charge and a mass separation, yielding an electrochemical potential of ~200 mV across the membrane. This potential is an electrical form of stored work. By mechanisms as yet incompletely resolved, this field gradient across the membrane can be utilized to drive either active transport of metabolites or synthesis of ATP from ADP and P_i. Many studies have indicated that about three ATP molecules are formed per electron pair from NADH and about two ATP molecules per electron pair from reduced flavins. For a drop of 1.1 V (NADH electrons→O_2), $\Delta G^{0\prime} = -nF\,\Delta E^{0\prime} = -52.1$ kcal/mole = −218 kJ/mole of energy released. Three ATP molecules synthesized represent −22 kcal/mole (−92 kJ/mole) of energy captured and ultimately stored as phosphate-anhydride bond energy, a 40% yield. Recall that, of the 38 molecules of ATP obtainable by combined glycolytic and citrate-cycle oxidation of glucose to CO_2, 34 derive ultimately from capture of electrical energy released during NADH and $FADH_2$ reoxidation by respiratory electron-transport chains. This transfer of electrons to O_2 is the major energy-harvesting sequence in cells that operate in aerobic environments.

27.B CATABOLIC AND ANABOLIC METABOLISM OF FATTY ACIDS

As an adjunct to our discussion of the chemistry involved in oxidative metabolism of glucose, we shall examine the chemistry used by enzymes either for degradation of fatty acids for energy generation or, complementarily, for the biosynthesis of fatty acyl chains for storage of reduced carbon units and for production of the hydrophobic diacyl phospholipids that are the hydrophobic backbone of cellular membrane structures. The predominant fatty acids in most organisms are C_{16} and C_{18} saturated (palmitate and stearate) species, and the corresponding monounsaturated C_{16} *cis*-Δ^9 (palmitoleate) and C_{18} *cis*-Δ^9 (oleate) species.

27.B.1 Enzymatic Degradation of Fatty Acids

In view of our discussion of the efficient oxidative processing of acetyl-CoA by the citrate cycle, it is not surprising to find that enzymatic breakdown of fatty acids yields acetyl-CoA units to feed into that cycle. For example,

$$\underset{\text{palmitate}}{H_3C(CH_2)_{14}C(=O)O^{\ominus}} + 8\ \text{CoASH} \longrightarrow 8\ \underset{\text{acetyl-CoA}}{H_3C{-}C(=O){-}SCoA}$$

Thus, the enzymatic strategy is to break down long-chain fatty acids two carbons at a time—for instance, converting a C_{16} fatty acyl-CoA to a C_{14} product (an $n-2$ unit) and an acetyl-CoA unit. The chain-shortening process then continues for another six cycles.

$$H_3C(CH_2)_{12}\underset{\beta}{CH_2}\underset{\alpha}{CH_2}C(=O)SCoA + \text{CoASH} \longrightarrow H_3C(CH_2)_{12}\underset{\beta}{C}(=O)SCoA + \underset{\alpha}{H_3C}C(=O)SCoA$$

The chemical problem to be solved by the degrading enzymes is clear. The β-methylene group of the substrate acyl-CoA must be functionalized to yield the acyl group of the $n-2$ product acyl-CoA. Two points are apparent: (1) this process will involve a four-electron oxidation at the β-carbon, so *redox cofactors* must be involved; and (2) a *carbon–carbon bond cleavage* between the β- and α-carbons of the substrate acyl-CoA must occur. Inspection of the products suggests that, (on paper) in the back direction, C—C bond formation would proceed by a Claisen condensation with the α-anion of acetyl-CoA acting as nucleophilic carbanion to attack the carbonyl carbon of the C_{14} acyl-CoA. Thus, the physiological *cleavage* reaction would most likely be a *retro-Claisen*, from a β-keto C_{16} acyl-CoA. For example,

R—C(=O)—SCoA + { $H_2C{=}C(O^{\ominus})$—SCoA ⟷ $H_2C^{\ominus}$—C(=O)—SCoA } ⇌ R—C($O^{\ominus}$)(SCoA)—CH_2—C(=O)—SCoA ⇌ R—C(=O)—CH_2—C(=O)—SCoA + $CoAS^{\ominus}$

β-keto acyl-CoA

With this view, the chemical problem devolves to oxidation of the β-methylene group of the substrate fatty acid to a β-keto group and derivitization of the carboxylate as a thioester to *stabilize* the α-carbanion that will arise at the C—C cleavage step. The formation of the acylthioester occurs first in a classical acyl-activation pattern (Chapter 8) of ATP fragmentation to PP_i and the acyl-AMP derivative, which is then attacked by the nucleophilic thiolate anion of $CoAS^{\ominus}$. In addition to its acyl-activation function, the large CoA moiety probably provides specificity by the large amount of binding interaction energy available on combination of acyl-CoA with the enzymes.

$$H_3C(CH_2)_{12}CH_2CH_2\text{—C(=O)—}O^{\ominus} + \text{ATP} \rightleftharpoons H_3C(CH_2)_{12}CH_2CH_2\text{—C(=O)—O—AMP} + PP_i \xrightarrow{CoAS^{\ominus}} H_3C(CH_2)_{12}CH_2CH_2\text{—C(=O)—SCoA} + \text{AMP} \quad (1)$$

In the citrate cycle, we noted that one of the methylene groups of succinate is converted to the keto group of oxaloacetate in a three-enzyme sequence involving two-electron removal by a flavoenzyme dehydrogenase, hydration of the resulting olefin to the alcohol, and then nicotinamide-dependent dehydrogenase action to the keto group. This type of elegant controlled-oxidation sequence is also used in each fatty-acyl β-oxidation cycle.

$$H_3C(CH_2)_{12}CH_2CH_2C(=O)SCoA + E\text{—}FAD \xrightarrow[\text{dehydrogenase}]{\text{acyl-CoA}} E\text{—}FADH_2 + H_3C(CH_2)_{12}CH{=}CHC(=O)SCoA \qquad (2)$$

trans-α,β-enoyl-CoA

$$H_3C(CH_2)_{12}CH{=}CHC(=O)SCoA + H_2O \xrightarrow[\text{hydrase}]{\text{enoyl-CoA}} H_3C(CH_2)_{12}CH(OH)CH_2C(=O)SCoA \qquad (3)$$

L-β-hydroxyacyl-CoA

$$H_3C(CH_2)_{12}\text{—}CH(OH)\text{—}CH_2\text{—}C(=O)\text{—}SCoA + NAD^{\oplus} \xrightarrow[\text{dehydrogenase}]{\beta\text{-OH-acyl-CoA}} NADH + H_3C(CH_2)_{12}\text{—}C(=O)\text{—}CH_2\text{—}C(=O)\text{—}SCoA \qquad (4)$$

β-ketoacyl-CoA

Presumably, the flavoenzyme-mediated oxidative desaturation to the *trans*-olefinic acyl-CoA is initiated by abstraction of the acidic α-H (acidic because the cell has made the acylthioester first) as a proton and the β-H as a hydride equivalent. The hydration by enoyl-CoA hydrase (crotonase, ¶17.A.4) is *trans* and to a single face of the planar olefin to give the L-isomer of the β-hydroxyacyl-CoA.

The β-ketoacyl-CoA molecule is a substrate for β-ketoacyl thiolase (¶23.B.4), where a nucleophilic cysteinyl sulfur at the enzyme active site attacks the β-carbonyl first, and that tetrahedral adduct expels the enolate of the C_2 product fragment acetyl-CoA while collapsing to the $n-2$ acyl enzyme. Transthiolation will then liberate the Enz—SH and the $n-2$ acyl-CoA. Thiolytic (rather than hydrolytic) cleavage of the β-ketoacyl-CoA has the virtue of generating the n-2 activated acyl-CoA, ready for another cycle, rather than the unactivated acyl carboxylate. Thus, *only one ATP is expended* to form the initial C_{16} acyl-CoA.

$$H_3C(CH_2)_{12}\text{—}C(=O)\text{—}CH_2\text{—}C(=O)\text{—}SCoA + Enz\text{—}S^{\ominus} \longrightarrow H_3C(CH_2)_{12}\text{—}C(O^{\ominus})(S\text{—}Enz)\text{—}CH_2\text{—}C(=O)\text{—}SCoA \longrightarrow$$

$$H_3C(CH_2)_{12}\text{—}C(=O)\text{—}S\text{—}Enz + \left\{ H_2C{=}C(O^{\ominus})SCoA \longleftrightarrow H_2\overset{\ominus}{C}\text{—}C(=O)SCoA \right\} \qquad (5a)$$

$$\downarrow H^{\oplus}$$

$$\boxed{H_3C\text{—}C(=O)\text{—}SCoA}$$

$$H_3C(CH_2)_{12}\text{—}\overset{O}{\overset{\|}{C}}\text{—}S\text{—Enz} \;\;(\leftarrow CoAS^{\ominus}) \rightleftharpoons \text{Enz—}S^{\ominus} + H_3C(CH_2)_{12}\text{—}\overset{O}{\overset{\|}{C}}\text{—SCoA} \qquad (5b)$$

All seven other acyl-CoA species (e.g., $n-2$, $n-4$, $n-6, \ldots$), in seven turns of the cycle to yield eight acetyl-CoA products, are generated by acyl transfer of enzyme thioester to cosubstrate $CoAS^{\ominus}$ without additional requirement for energy input.

The chemical logic of enzymatic β-oxidation of fatty acids is apparent; what is the energy yield? For the five-step sequence just outlined, the NADH-generated in step 4 yields three ATP via oxidative phosphorylation, and the $FADH_2 \cdot$ acyl-CoA dehydrogenase ($E^{0'} = -0.01$ V) yields two ATP by that route. One ATP was expended in step 1, so a net of four ATP are produced in the first turn of $C_{16} \rightarrow C_{14}$ acyl-CoA + acetyl-CoA. We noted earlier that processing of acetyl-CoA by the citrate cycle yields twelve ATP, for a total of sixteen ATP, or $16 \times (-7.3) = -116.8$ kcal/mole $= -489$ kJ/mole for total oxidative processing of the —$CH_2COO^{\ominus}$ in the starting palmitate. Each subsequent β-oxidation yields five ATP in acetyl-CoA production plus twelve ATP from the citrate cycle for $17 \times (-7.3) = -124.2$ kcal/mole $= -520$ kJ/mole. The total energy captured by complete oxidation of palmitate would be $(7 \times 124.2) + 116.8 = 986$ kcal/mole $=$ 4,130 kJ/mole, an impressively large number.

We shall not discuss oxidation of unsaturated fatty acids here, except to note that they suffer similar breakdown by C_2 units until the locus of the double bond is reached as a β,γ-olefinic acyl-CoA. This nonreducible olefin is isomerized enzymatically to the thermodynamically more stable conjugated α,β-enoyl-CoA, which is then processed by steps 3 through 5 above.

Odd-carbon fatty acids (e.g., C_{17}, C_{19}) are relatively rare but, when encountered, are processed by β-oxidation. A C_{17} heptadecanoyl-CoA would undergo seven cycles to yield seven acetyl-CoA and a residual propionyl-CoA. To bring the C_3 propionate carbons into the mainstream of metabolic chemistry, organisms choose to elongate to a C_4 dicarboxylate derivative by carboxylation. The problem with this strategy is that the methyl group is β (not α) to the —C(=O)SCoA in propionyl-CoA and so is unactivated; the hydrogens are not sufficiently acidic (the resulting carbanionic transition state too unstable) to be readily removed. So carboxylation by the biotin-dependent propionyl-CoA carboxylase (¶22.B) is constrained chemically to occur at the α-carbon, because that is the locus where an attacking carbanion equivalent can be formed. The product is *not a linear* C_4 diacyl compound, but a *branched-chain* methylmalonyl-CoA, specifically the S-isomer.

$$HCO_3^{\ominus} + ATP + H_3C{-}CH_2{-}C(=O){-}SCoA \xrightleftharpoons{\text{Enz—biotin}} ADP + P_i + HC(COO^{\ominus})(CH_3){-}C(=O){-}SCoA$$

Thus limited by the low-activation-energy chemical mechanism, the enzymatic reaction has generated a compound not yet readily processable. It is, on the other hand, isomeric to succinyl-CoA, a citrate-cycle metabolite. The enzyme-catalyzed isomerization to succinyl-CoA requires two steps. First, the *S*-enantiomer of methylmalonyl-CoA is isomerized to the *R*-isomer (¶19.A.1.c), and then the actual carbon skeletal rearrangement follows, the only known CoB_{12}-dependent rearrangement in mammalian metabolism, to generate the linear succinyl-CoA (¶20.C.2).

$$\underset{S\text{-methylmalonyl-CoA}}{HC(COO^{\ominus})(CH_3){-}C(=O){-}SCoA} \xrightleftharpoons{\text{methylmalonyl-CoA recemase}} \underset{R\text{-methylmalonyl-CoA}}{HC(CH_3)(COO^{\ominus}){-}C(=O){-}SCoA}$$

$$\underset{R\text{-methylmalonyl-CoA}}{HC(CH_3)(COO^{\ominus}){-}C(=O){-}SCoA} \xrightleftharpoons{\text{mutase—}CoB_{12}} \underset{\text{succinyl-CoA}}{CoAS{-}C(=O){-}CH_2{-}CH_2{-}COO^{\ominus}}$$

The chemical logic in conversion of propionyl-CoA to succinyl-CoA is to carry out carboxylation as constrained by the chemical requirements for carbanion formation and then use a carbon skeletal rearrangement to obtain a useful metabolite.

27.B.2 Enzyme-Catalyzed Biosynthesis of Fatty Acids

In the dynamics of cellular metabolism, fatty acids are biosynthesized for membrane biogenesis, or energy storage, as well as being degraded. A comparison of biosynthesis with β-oxidation reveals a complementary chemical logic attended by two specificity devices. One device is to use distinct intracellular loci for degradation and biosynthesis. Fatty acids are degraded for energy in mitochondria, the powerhouse organelles of the cell. Fatty acids are biosynthesized on cytoplasmic membranes known as the endoplasmic reticulum, a distinct subcellular compartmentalization. The second device involves use of NADPH in biosynthesis and $NAD^{\oplus}$ (not $NADP^{\oplus}$) in degradation. Also, the β-hydroxyacyl-CoA intermediate in β-oxidation is the L-isomer at the alcoholic carbon; in biosynthesis, the D-isomers of β-hydroxyacylthioesters are formed. Acyl-CoA species are the thioesters used in β-oxidation; acylthioesters where the thiol group comes from a low-molecular-weight protein are used in biosynthesis. These distinctions are a use of stereospecificity and cosubstrate specificity to differentiate the anabolic and catabolic processes.

The major chemical problems in fatty acid synthesis are (1) how to form carbon–carbon bonds in a Claisen condensation *with the equilibrium displaced significantly* in favor of C—C *bond synthesis,* and (2) how to convert β-keto groups to methylene groups (a problem already solved twice in the reverse direction in this chapter). A third problem is an organizational one: because C_{16} and C_{18} fatty acids are the principal products of long-chain fatty-acid biosynthesis, there must be an efficient processing of the intermediate-length chains (e.g., C_6, C_8, C_{10}, C_{12}, C_{14}) along this route.

The first problem is solved by using a malonate derivative as the source of the C_2 fragment added in each condensative step.

$$^{\ominus}OOC-CH_2-\overset{O}{\overset{\|}{C}}-X$$

During the enzymatic C—C bond condensation, the malonyl fragment is also decarboxylated, driving the equilibrium irreversibly toward the biosynthetic condensation.

$$^{\ominus}O-\overset{O}{\overset{\|}{C}}-CH_2-\overset{O}{\overset{\|}{C}}-X + R-\overset{O}{\overset{\|}{C}}-Y \xrightarrow[\text{step}]{\text{irreversible}} CO_2 + R-\underset{O^{\ominus}}{\underset{|}{\overset{H_2C-\overset{O}{\overset{\|}{C}}-X}{\overset{|}{C}}}}-Y \rightleftarrows Y^{\ominus} + R-\underset{O}{\underset{\|}{\overset{H_2C-\overset{O}{\overset{\|}{C}}-X}{\overset{|}{C}}}}$$

added C_2 unit

β-ketoacyl derivative

Let's look at the first two enzymes involved: acetyl-CoA carboxylase and β-ketoacyl-ACP synthase (ACP = acyl-carrier protein). Acetyl-CoA carboxylase is a typical biotin-dependent carboxylase; it was discussed at length in ¶22.B.1. The product of interest is malonyl-CoA, an activated C_2 donor.

$$\text{Enz—Biotin} + \text{ATP}^{4\ominus} + \text{HCO}_3^{\ominus} \rightleftharpoons \text{Enz—Biotin—CO}_2 + \text{ADP}^{3\ominus} + \text{P}_i \quad (1)$$

$$\text{H}_3\text{C—C(=O)—SCoA} + \text{Enz—Biotin—CO}_2 \xrightleftharpoons[\text{carboxylase}]{\text{acetyl-CoA}} \text{Enz—Biotin} + \text{H}_2\text{C(COO}^{\ominus}\text{)—C(=O)—SCoA} \quad (2)$$

malonyl-CoA

The fundamental chain-elongation step in fatty-acid biosynthesis is carried out by the next enzyme, β-ketoacyl-ACP synthase. (Compare this process with the isoprenyl elongations discussed in Chapter 26.) The enzyme uses acylthioesters of varying chain lengths (C_2 through C_{14}) as electrophilic partners and a malonyl-thioester repetitively as the nucleophilic (attacking) partner. However, the synthase does not use CoA-thioesters. The thiol group to which the acyl and malonyl groups are attached is provided by the SH of a 4′-phosphopantetheinyl group, itself covalently bonded to a low-molecular-weight protein called acyl-carrier protein (ACP).

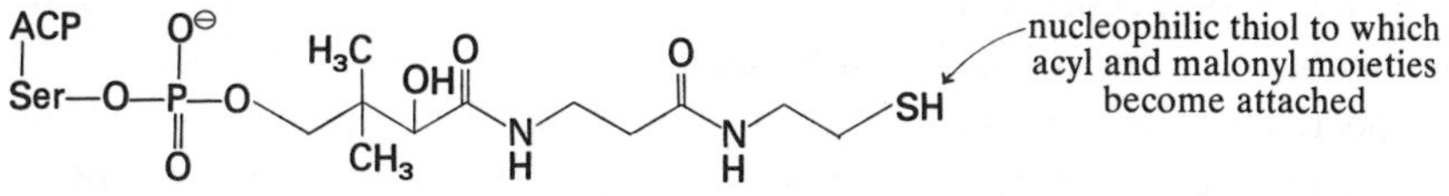

ACP
Ser—O—P(=O)(O$^\ominus$)—O—CH$_2$—C(CH$_3$)$_2$—CH(OH)—C(=O)—NH—CH$_2$CH$_2$—C(=O)—NH—CH$_2$CH$_2$—SH ← nucleophilic thiol to which acyl and malonyl moieties become attached

4′-phosphopantetheinyl group in ACP

The *E. coli* ACP is a small protein of 77 amino-acid residues (10,000 mol wt) (Volpe and Vagelos, 1973; Prescott and Vagelos, 1971). A specific serine residue is covalently bonded to the phosphopantetheine group as shown in Fig. 27-2.

NH_2–Ser(1)—Thr–Ile—Glu–Glu–Arg–Val–Lys–Lys–Ile—Ile—Gly–Glu–
Gln—Leu–Gly–Val—Lys–Gln–Glu–Glu–Val–Thr–Asp–Asn–Ala–Ser–
(P)—Pantetheine—SH (attached to Ser 36)
Phe—Val–Glu–Asp–Leu–Gly–Ala–Asp–Ser(36)–Leu–Asp–Thr–Val–Glu–
Leu—Val–Met–Ala—Leu–Glu–Glu–Glu–Phe–Asp–Thr–Glu–Ile—Pro–
Asp–Glu–Glu–Ala—Glu–Lys–Ile—Thr–Thr–Val—Gln–Ala–Ala–Ile–
Asp—Tyr–Ile—Asn–Gly–His–Gln–Ala(77)–COOH

Figure 27-2
The complete amino-acid sequence of ACP from *E. coli.* (Reproduced, with permission, from J. Volpe and P. R. Vagelos, *Annual Review of Biochemistry*, vol. 42, p. 37.

Recall that phosphopantetheine is one-half of the coenzyme-A molecule. Thus, before the first condensation step of fatty-acid biosynthesis, a molecule of acetyl-CoA undergoes prior acyl transfer to ACP-SH to yield acetyl-*S*-ACP and CoASH. This reaction is catalyzed by an ACP transacetylase.

$$H_3C\text{-}C(=O)\text{-}SCoA + HS\text{—}ACP \rightleftharpoons \underset{\text{acetyl-ACP}}{H_3C\text{-}C(=O)\text{-}S\text{—}ACP} + CoASH \qquad (3)$$

Similarly, the malonyl-CoA from step 2 must be converted to malonyl-ACP.

$$^{\ominus}OOC\text{-}CH_2\text{-}C(=O)\text{-}SCoA + HS\text{—}ACP \rightleftharpoons {}^{\ominus}OOC\text{-}CH_2\text{-}C(=O)\text{-}S\text{—}ACP + CoASH \qquad (4)$$

In fact, the ACP molecule forms a central core for the other enzymes of fatty-acid synthesis to cluster around in a multienzyme complex (Fig. 27-3), and the pantetheinyl group represents a flexible, swinging arm of 20 Å radius (in analogy to the swinging-arm proposals for lipoamide and biotinamide; Chapters 21 and 22, respectively). Because only *one* molecule of ACP is involved, although both electrophilic and nucleophilic components in the Claisen condensation must initially be as ACP derivatives, an ordered mechanism must pertain, and a second thiol group must be available. The second thiol group is an active-site cysteine on the β-ketoacyl-ACP synthase, and the condensation occurs in the following two-step (at least) sequence. The electrophilic acyl partner (in this case, the acetyl group) undergoes attack by the synthase's cysteinyl thiolate and is transferred to yield an acetyl-*S*-enzyme intermediate and free up the pantotheinyl thiolate of ACP.

$$H_3C\text{-}C(=O)\text{-}S\text{—}ACP + {}^{\ominus}S\text{—}Cys\text{—}Enz \rightleftharpoons H_3C\text{-}C(=O)\text{-}S\text{—}Enz + {}^{\ominus}S\text{—}ACP$$

This pantetheinyl thiolate then reacts with malonyl-CoA in step 4 to produce malonyl-ACP. *The nucleophilic partner is on the ACP; the electrophile is on the synthase's cysteine residue.* Condensation now occurs *with transfer of the electrophile* (the acetyl or, more generally, the acyl group) *to the nucleophile.* The C_4 β-ketoacyl product *is attached to the ACP pantetheinyl group, not to the synthase active site.* This attachment result is not random but is controlled by the nature of the acyl-group transfer chemistry.

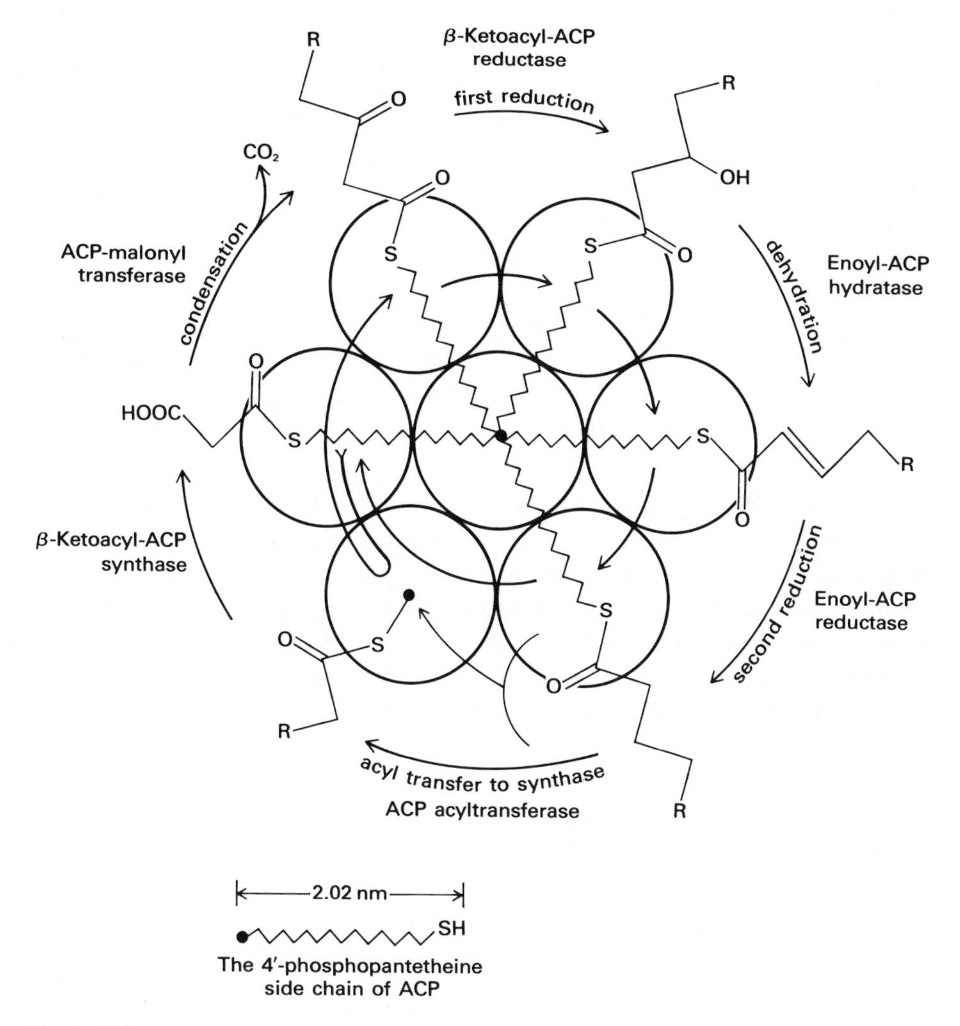

Figure 27-3

A schematic representation of the fatty-acid–synthetase complex. The central protein molecule is ACP. Its long phosphopantetheine side chain apparently serves as a swinging arm to carry acyl groups from one enzyme molecule to the next, to accomplish the six steps needed for addition of each C_2 unit. (From A. L. Lehninger, *Biochemistry*, Worth Publishers, New York, 1975, p. 664)

There is some debate about whether the malonyl-ACP is first decarboxylated to an enolate prior to attack, or whether decarboxylation and C—C bond formation are concerted (D'Agnolo et al., 1975). The synthase will slowly decarboxylate malonyl-*S*-ACP in the absence of electrophile, but we shall arbitrarily show the mechanism as concerted here. (The pantetheinyl group is abbreviated here as Pan.)

$$H_3C\text{-}C(=O)\text{-}S\text{—}Cys\text{—}Enz \;+\; {}^{\ominus}O\text{-}C(=O)\text{-}CH_2\text{-}C(=O)\text{-}S\text{—}Pan\text{—}ACP \xrightarrow[\text{irreversible}]{\text{functionally}} CO_2 + H_3C\text{-}C(O^{\ominus})(S\text{—}Cys\text{—}Enz)\text{-}CH_2\text{-}C(=O)\text{-}S\text{—}Pan\text{—}ACP$$

$$\rightleftharpoons H_3C\text{-}C(=O)\text{-}CH_2\text{-}C(=O)\text{-}S\text{—}Pan\text{—}ACP + {}^{\ominus}S\text{—}Cys\text{—}Enz \qquad (6)$$

The β-ketoacyl-*S*-ACP product (here acetoacetyl-*S*-ACP) so generated is now a prime candidate for the reverse of the β-oxidation steps: reduction of ketone to alcohol using NADPH specifically, dehydration of the D-β-hydroxyacyl-CoA to the *trans*-olefinic acyl-CoA, and then saturation. It is reported that NADPH is the donor in the saturation of the olefin to the butyryl-CoA, rather than using a flavoprotein reductase. If so, one hydrogen will derive from NADPH and the other from solvent in the saturated butyryl-*S*-ACP, but this has not been reported.

$$H_3C\text{-}C(=O)\text{-}CH_2\text{-}C(=O)\text{-}S\text{—}ACP + NADPH \rightleftharpoons H_3C\text{-}CH(OH)\text{-}CH_2\text{-}C(=O)\text{-}S\text{—}ACP + NADP^{\oplus} \qquad (7)$$

$$H_3C\text{-}CH(OH)\text{-}CH_2\text{-}C(=O)\text{-}S\text{—}ACP \;(\text{H}\ \text{:B}) \rightleftharpoons H_3C\text{-}CH{=}CH\text{-}C(=O)\text{-}S\text{—}ACP + H_2O \qquad (8)$$

trans-α,β-butenyl-*S*-ACP

$$H_3C\text{-}CH{=}CH\text{-}C(=O)\text{-}S\text{—}ACP + NADPH + H^{\oplus} \rightleftharpoons H_3C\text{-}CH_2\text{-}CH_2\text{-}C(=O)\text{-}S\text{—}ACP + NADP^{\oplus} \qquad (9)$$

Given the efficient production of butyryl-*S*-ACP in this multienzyme fatty-acid synthetase complex as suggested by Table 27-1, it appears that the butyryl-*S*-ACP is efficiently processed for another cycle rather than being lost to the medium by acyl transfer back to CoA to yield butyryl-CoA. That is, the butyryl acyl moiety is transferred now to the cysteinyl sulfur at the active site of the β-ketoacyl-ACP synthase and serves as C_4 electrophile. The ACP-thiol group, thus

Table 27-1
Activity of β-ketoacyl-ACP synthetase with various intermediates in fatty-acid synthesis in *E. coli*

Intermediate	K_m (μM)	V_{max} (μmole product/min/mg)
Acetyl-ACP	0.52	2.8
Decanoyl-ACP	0.33	2.8
Dodecanoyl-ACP	0.27	0.97
Tetradecanoyl-ACP	0.28	0.31
Hexadecanoyl-ACP	—	No activity
cis-3-Decenoyl-ACP	0.71	1.9
cis-5-Dodecenoyl-ACP	0.20	1.7
cis-9-Hexadecenoyl-ACP	0.37	0.37
cis-11-Octadecenoyl-ACP	—	No activity

SOURCE: From Prescott and Vagelos (1971), p. 298.

liberated, can be reacylated with another malonyl group to give back the malonyl-*S*-ACP state. Condensation with a malonyl-*S*-ACP yields a C_6 β-ketoacyl-*S*-ACP, which is then saturated in the above three steps (7–9) to the *n*-hexanoyl-*S*-ACP.

This efficient iterative sequence occurs until a C_{16} acylthioester of ACP is produced (this is a palmityl-*S*-ACP, or hexadecanoyl-*S*-ACP). The β-ketoacyl-*S*-ACP synthase will accept a C_{14} acyl chain to make a tetradecanoyl-*S*-enzyme intermediate, but its active-site cysteinyl does not rapidly attack a C_{16} acyl chain (Table 27-1). This discrimination (probably on the basis of size) controls the specificity of products formed in fatty-acid biosynthesis. The major product is C_{16} palmitate, with some C_{18} stearate as well. A discrete multienzyme-complex entity in fatty-acid synthesis that rapidly elongates bound intermediates ensures that the general cellular concentration of intermediates of four to fourteen carbons in length is kept low. This is economical in that such intermediates appear to serve no important physiological role in cells.

We shall not discuss biosynthetic desaturation to monoolefinic acids here, but we note that the bacterial systems introduce a β,γ-olefinic link at the C_{10} level. This double bond persists in subsequent elongations to be a Δ^9 in the C_{16} and a Δ^{11} in the C_{18} fatty acids (See ¶19.C.6). The desaturation of stearyl-CoA to olelyl-CoA in animal cells occurs in endoplasmic reticulum membranes and is a complex O_2-requiring process involving the flavoenzyme cytochrome-b_5 reductase, the hemoprotein cytochrome b_5, and the nonheme-iron–containing enzyme stearyl-CoA desaturase (¶15.A.5).

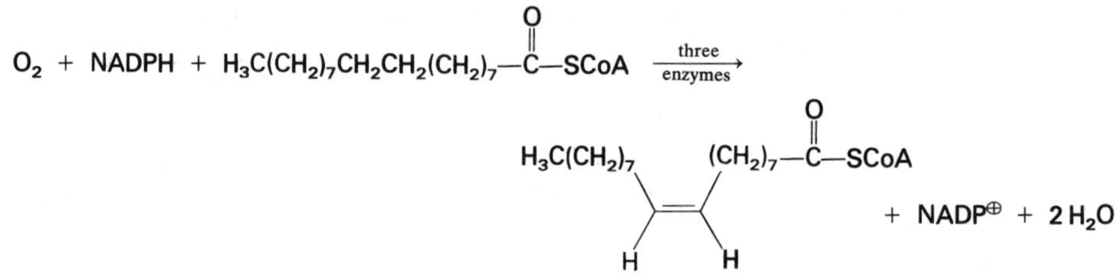

$$O_2 + NADPH + H_3C(CH_2)_7CH_2CH_2(CH_2)_7\text{—}\overset{O}{\overset{\|}{C}}\text{—}SCoA \xrightarrow[\text{enzymes}]{\text{three}}$$

$$\underset{cis\text{-}\Delta^9\text{-octadecenoyl-CoA}}{\overset{H_3C(CH_2)_7 \qquad (CH_2)_7\text{—}\overset{O}{\overset{\|}{C}}\text{—}SCoA}{\underset{H \qquad\quad H}{\text{\textbackslash C=C/}}}} + NADP^{\oplus} + 2\,H_2O$$

Molecular oxygen is reduced by 4 $e^{\ominus}$ during this complex desaturation. One might contrast this O_2-dependent desaturation to an olefin with the simpler O_2-independent mechanisms just discussed during breakdown and synthesis of saturated fatty acids. It is chemically more difficult to introduce an unactivated double bond (e.g., at C-9 or C-11) than it is to introduce one α,β to a carbonyl group.

27.C A LOOK AT THE PATTERNS OF AROMATIC RING METABOLISM: FUNCTIONALIZATION, DEGRADATION, AND BIOSYNTHESIS OF AROMATIC RINGS

In the primary metabolism of bacteria, plants, and animals, there are a number of aromatic compounds containing benzenoid ring systems—among them the common amino acids phenylalanine, tyrosine, and tryptophan. In this last section of the chapter, we shall look at the basic strategy used by animal cells to functionalize or degrade the aromatic ring of phenylalanine. This example points out some major roles of O_2 in animal metabolism. Animals cannot synthesize the benzene nucleus; phenylalanine must be obtained in the diet.

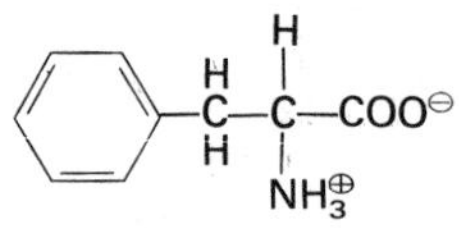

$$C_6H_5\text{—}CH_2\text{—}CH(NH_3^{\oplus})\text{—}COO^{\ominus}$$

L-phenylalanine

We shall then look at the pattern of phenylalanine synthesis in those bacteria that make it.

27.C.1 Phenylalanine→Epinephrine

In the adrenal glands of animals, phenylalanine is processed enzymatically to norepinephrine and epinephrine (noradranalin and adrenalin), substances that act as chemical neurotransmitters. The most obvious features in these conversions are the three hydroxyl groups introduced, two on the ring to yield a catecholic structure, and one at the benzylic position.

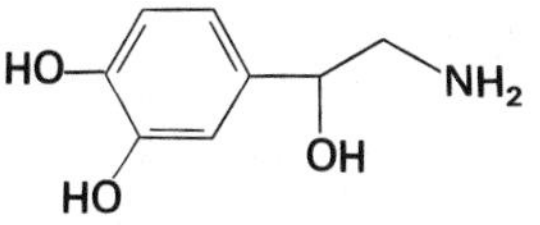

HO, HO, OH, NH_2
norepınephrine

HO, OH, H, N, CH_3
ephinephrine

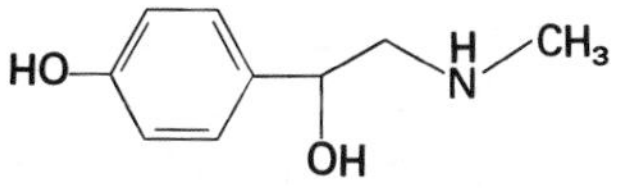

Enzymatic hydroxylations almost invariably proceed using some activated, bound form of molecular oxygen, as we discussed in detail in Chapters 13 through 15—a monooxygenation process. Actually, the *ortho*-dihydroxy groups of the catecholamine products could, in principle, have arisen from action of a dioxygenase followed by a reductase, but such a pathway to catechols is rare, occurring only occasionally in bacteria and never in animals. In addition to the three reactions involving molecular oxygen, a decarboxylation of the amino acid to the amine is required, a flag signaling a pyridoxal-P–dependent decarboxylase. Epinephrine has the nitrogen methylated; that amine group must have acted as nucleophile in a transmethylase reaction with *S*-adenosylmethionine (SAM) as cosubstrate. The first hydroxylation is catalyzed by phenylalanine hydroxylase.

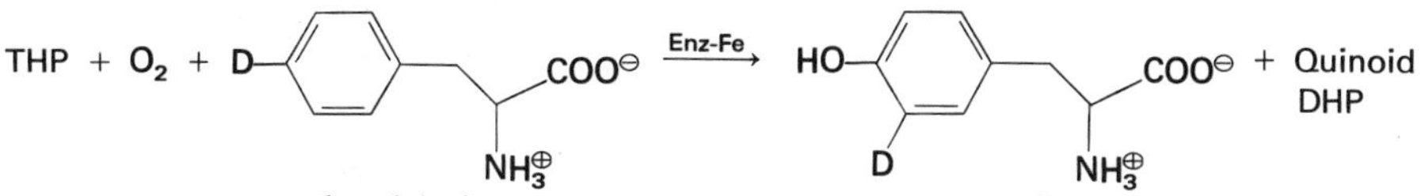

THP + O_2 + D–(phenylalanine, $COO^{\ominus}$, $NH_3^{\oplus}$) $\xrightarrow{\text{Enz-Fe}}$ HO–(tyrosine, D, $COO^{\ominus}$, $NH_3^{\oplus}$) + Quinoid DHP (1)

where THP = tetrahydropterin, and DHP = dihydropterin. The second hydroxylation is catalyzed by tyrosine hydroxylase.

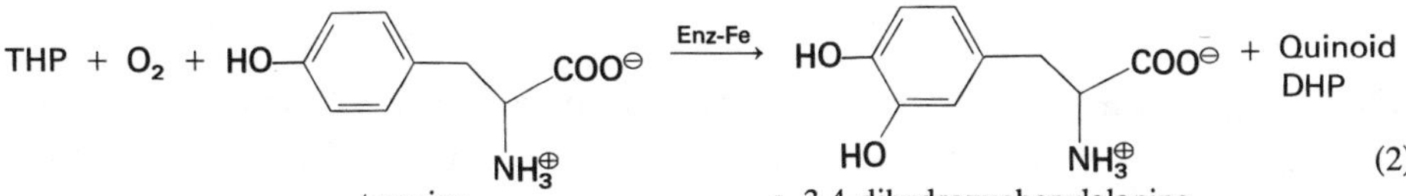

THP + O_2 + HO–(tyrosine, $COO^{\ominus}$, $NH_3^{\oplus}$) $\xrightarrow{\text{Enz-Fe}}$ HO–(HO, $COO^{\ominus}$, $NH_3^{\oplus}$; L-3,4-dihydroxyphenylalanine (L-dopa)) + Quinoid DHP (2)

Both reactions involve a possible pterin hydroperoxide (Chapter 12) as active hydroxylating species for each enzyme, perhaps yielding the arene oxide as initially functionalized ring product. The putative first arene oxide would open with an N.I.H. shift; the second (a hydroxy arene oxide) would not show the hydride shift in L-dopa formation.

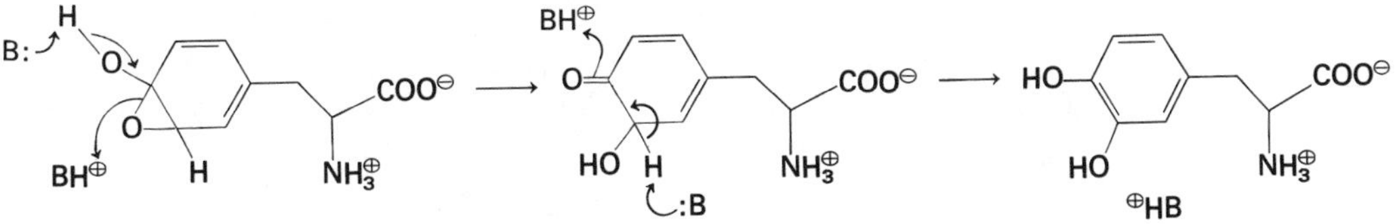

B:, H, O, O, $BH^{\oplus}$, H, $COO^{\ominus}$, $NH_3^{\oplus}$ ⟶ $BH^{\oplus}$, O, HO, H, :B, $COO^{\ominus}$, $NH_3^{\oplus}$ ⟶ HO, HO, $^{\oplus}HB$, $COO^{\ominus}$, $NH_3^{\oplus}$

Hydroxylation is the most common (if not the only) mechanism organisms use to functionalize benzene rings in substrates, either to generate neuroactive compounds such as these catecholamines or 5-hydroxytryptamine, or to detoxify aromatic xenobiotic compounds. It is likely that both types of aromatic-ring hydroxylases of animal cells (the pterin–Fe enzymes of Chapter 12 and the hemoprotein P_{450} monooxygenases of Chapter 15) work by construction of arene oxides as first products. In the case of the xenobiotic aromatics, those arene oxides may be proximal carcinogens.

After L-dopa is formed, it is decarboxylated in a PLP-dependent reaction to yield dopamine, which is substrate for the second variety of hydroxylase in this sequence, the copper-containing dopamine β-hydroxylase, which introduces a hydroxyl group at the benzylic position with retention of configuration.

L-dopa —(decarboxylase · PLP)→ dopamine + CO_2 (3)

dopamine —(β-hydroxylase-Cu)→ norephinephrine (4)

We commented in ¶14.C.1 on the mechanism of dopamine β-hydroxylase; here we simply note again that this is the only well-characterized copper-dependent monooxygenase yet studied. It is a key enzyme in mammalian metabolism, because the hydroxylated product is the neurotransmitter, norepinephrine.

The final reaction is indeed a methyl transfer to nitrogen from S-adenosylmethionine (SAM).

norepinephrine + SAM → epinephrine + SAH (5)

Other potential nucleophiles in norepinephrine are the three hydroxyl groups; various catechol-*O*-methyl transferases exist for methylation of one or both of the catecholic oxygens.

27.C.2 Phenylalanine Degradation: Phenylalanine→Acetoacetate+Fumarate

Having considered the pattern of functionalization of phenylalanine for biosynthetic purposes, we can now ask what is the logic for degradation of the benzene ring as part of the dynamic turnover of cellular constituents in metabolism? We commented in Chapter 16 that the key enzymatic step in catabolism of benzene (and indole and pyridine) rings is dioxygenase-mediated fission. This may be the sole mechanism for aromatic-ring breakdown, and it emphasizes yet another essential chemical role of O_2 in aerobic organisms.

The final breakdown products of L-phenylalanine in animals are acetoacetate and fumarate. These account for eight of the nine carbons of phenylalanine; the ninth is lost as CO_2.

L-phenylalanine ⟶ acetoacetate + fumarate + CO_2

Degradation of phenylalanine actually proceeds via processing of tyrosine, which is produced by action of phenylalanine hydroxylase in step 1 of the functionalization sequence discussed in ¶27.C.1.

$$\text{THP} + O_2 + \text{L-phenylalanine} \xrightarrow{\text{Enz-Fe}} \text{L-tyrosine} + \text{Quinoid DHP} \quad (1)$$

Then there are actually two consecutive dioxygenase steps involved in catabolism, representing both known varieties (Chapter 16): one of the type requiring α-keto-acid, and the other of the ring-fission variety. The α-keto-acid–requiring dioxygenase is the only known intramolecular example of that category; it is the *p*-hydroxyphenylpyruvate hydroxylase (misnamed because it is, in fact, a dioxygenase).

Tyrosine can be readily converted to *p*-hydroxyphenylpyruvate by a pyridoxal-P–dependent transaminase—for instance, with pyruvate as cosubstrate.

HO–C6H4–CH2–CH(NH3⊕)–COO⊖ + CH3COCOO⊖ ⇌(Enz—PLP) H3C–CH(NH3⊕)–COO⊖ + HO–C6H4–CH2–CO–COO⊖ (2)

L-tyrosine   pyruvate   L-alanine   *p*-hydroxyphenylpyruvate

The processing of *p*-hydroxyphenylpyruvate to homogentisate was discussed in detail in ¶16.A.3 and ¶16.A.4; it involves decarboxylation (the CO_2-production step in catabolism), side-chain migration, and presumed intermediate formation of arene oxide as hydroxylating agent.

$^{18}O_2$ + HO–C6H4–CH2–CO–COO⊖ →(Enz-Fe) homogentisate (HO, ^{18}OH, CH2–C(^{18}O)O⊖) + CO_2 (3)

p-hydroxyphenylpyruvate   homogentisate

Homogentisate is a benzenoid compound properly functionalized for dioxygenative fission between the OH and acetate side chains to yield ring-opened maleylacetoacetate, possibly by way of a cyclic endoperoxide species.

homogentisate + $^{18}O_2$ →(Enz—Fe) [cyclic endoperoxide (^{18}O–^{18}O), HO, OH, COO⊖] → enol intermediate (H⊕, HÖ, ^{18}O, C(^{18}O)OH, COO⊖) ⇌ maleylacetoacetate (4)

homogentisate   maleylacetoacetate

Cleavage as indicated can account for introduction of one atom of the O_2 on each side of the cleaved C—C bond as a ketone and a carboxylate, respectively. Once the resonance energy of the benzene ring has thus been broken, the hydroxyl group that had been a phenolic hydroxyl in substrate is an enol in the initial product and will exist predominantly as the more stable keto tautomer, maleylacetoacetate.

Because a *ring* was cleaved, the ring-opened product is necessarily a *cis* (maleyl) derivative rather than a *trans* (fumaryl) derivative. Yet, although fumarate is a common metabolite (e.g., in the citrate cycle), the *cis*-olefinic dicarboxylate maleate is not.

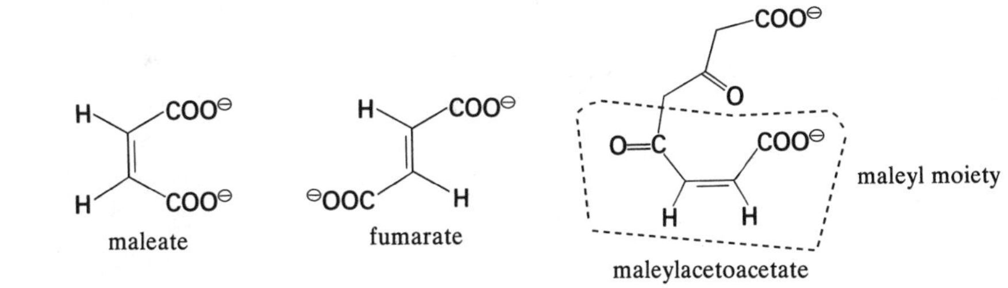

maleate

fumarate

maleyl moiety

maleylacetoacetate

Thus, to maximize the usability of the product, cells at this juncture perform a *cis–trans* isomerization on the ring-opened maleylacetoacetate, converting it to fumarylacetoacetate. As noted in ¶19.E, this is probably effected by a nucleophilic addition–elimination sequence to generate an intermediate with a single bond free to rotate to the *more stable trans*-conformation. One could argue that the cell makes use of thermodynamics in deciding whether to use maleate or fumarate.

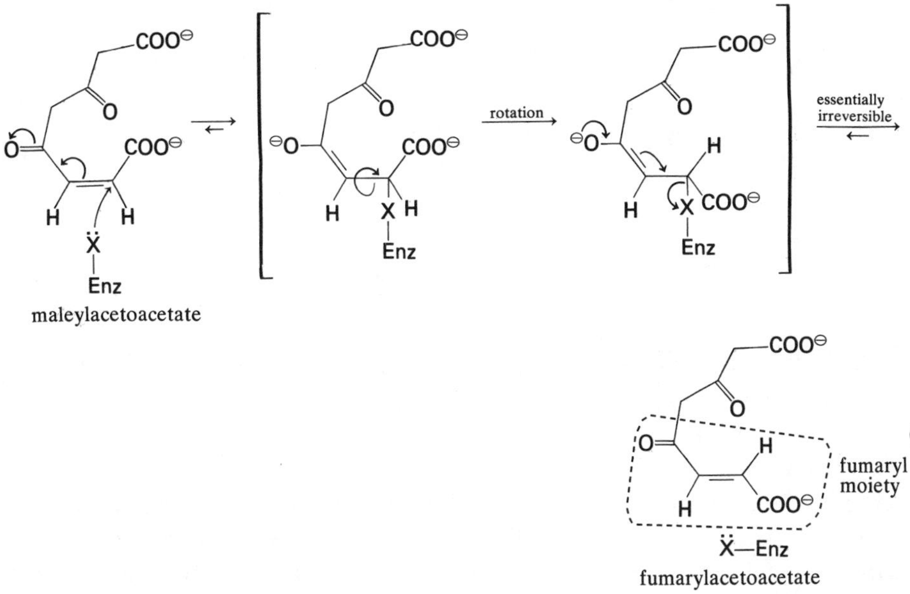

maleylacetoacetate

rotation

essentially irreversible

(5)

fumaryl moiety

fumarylacetoacetate

With the generation of fumarylacetoacetate, the cell is only one step away from the common metabolites, acetoacetate and fumarate. Inspection of the fumarylacetoacetate skeleton suggests that, if H_2O could add nucleophilically to the keto group of the fumaryl moiety in a simple formation of a carbonyl tetrahedral adduct, then that adduct could not only revert to starting carbonyl, but it could also (in a low-energy path) *expel the stable enolate anion of acetoacetate.*

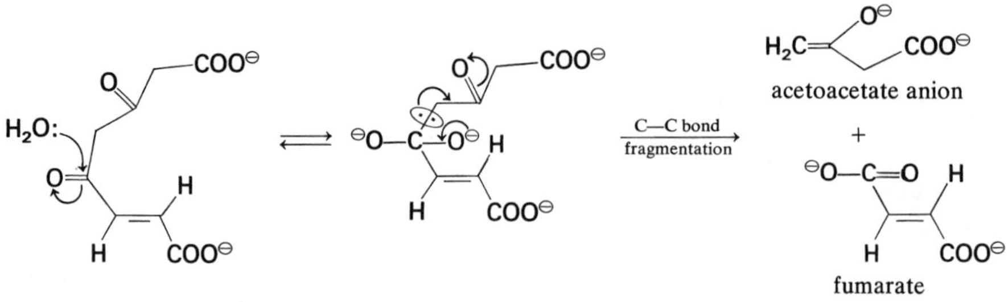

C—C bond fragmentation

acetoacetate anion

fumarate

(6)

This process is essentially an *aldol cleavage* from a β-dicarbonyl compound. The tetrahedral adduct in equilibrium with the dicarbonyl is competent for aldol cleavage. Because water could add to either carbonyl, cleavage could have occurred the other way around to yield malonate and the olefinic methyl ketone. The enzyme imposes specificity such that only the desired aldol cleavage occurs (only the desired tetrahedral adduct is formed, by controlling to which carbonyl the H_2O can productively add).

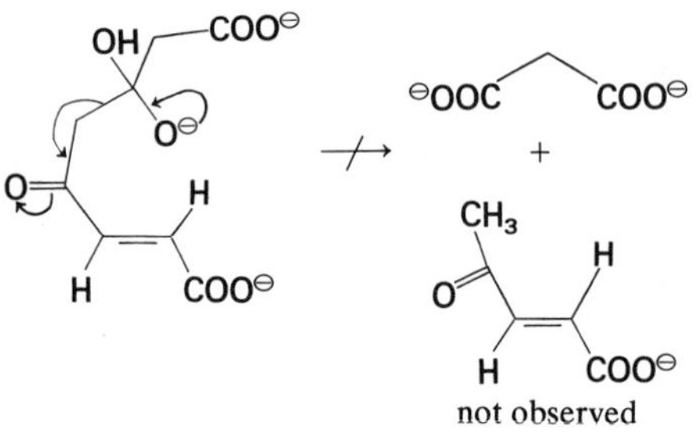

not observed

Thus, this degradation pathway from phenylalanine has some diverse and interesting chemistry: one monooxygenation and two dioxygenations, each of the latter attended by C—C bond breakage (a decarboxylation and fission of an aromatic ring). The *cis–trans* isomerization and aldol cleavage are imposed, not so much for chemical requirements, but for economy—to yield readily usable products. A key point in this pathway is the homogentisate formation; the rearrangement in its formation was not capricious. If the introduction of the second hydroxyl group on the benzene ring were *not* accompanied by side-chain migration, the subsequent dioxygenative fission at any site on the ring *could not produce the final acyclic products.*

27.C.3 The Strategy of Aromatic Ring Construction in Microbial Biosynthesis of Aromatic Amino Acids

We have noted how animal cells, which cannot synthesize benzene rings, functionalize and/or degrade them. It may be instructive now to look at the strategy used by microorganisms that can elaborate aromatic rings. The major pathway (termed the shikimate pathway, after one of the intermediates) leads to phenylalanine (and hence to tyrosine) and tryptophan and thereby to aromatic constituents in secondary metabolism (Haslam, 1974).

A priori, one expects that a benzene ring will derive from a more saturated six-membered cyclohexane, cyclohexene, or cyclohexadiene precursor. One obvious way of introducing the three double bonds in benzene is by loss of the elements of water (or P_i) from adjacent carbons. For example,

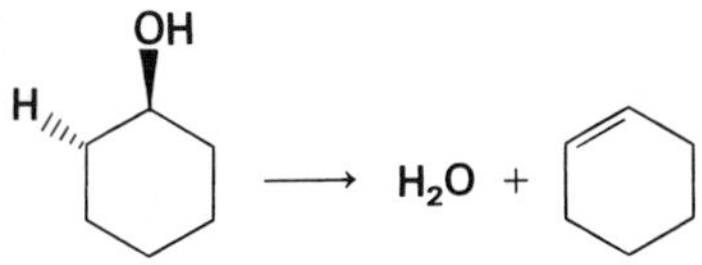

OH
H
⟶ H_2O +

Thus, a polyhydroxylated cyclohexane would be a reasonable cyclic precursor. If we look at the arsenal of common C—C bond-forming mechanisms available to enzymes, an *intramolecular aldol* seems most likely, where a carbanion site would condense with a carbonyl group six carbons away in the acyclic molecule. For example,

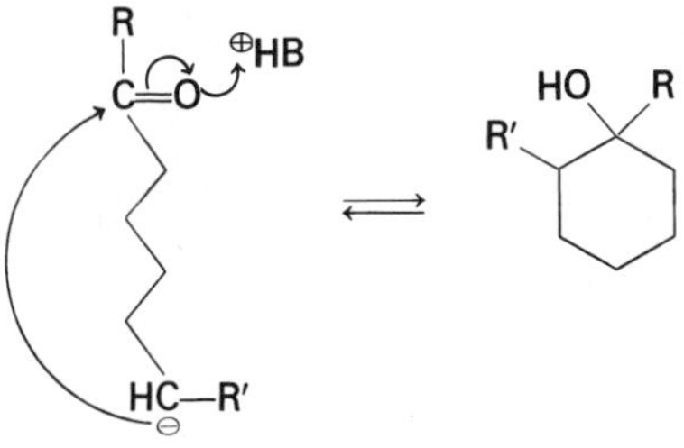

R
C=O ⊕HB
HC—R′
⊖
⇌
HO R
R′

The most likely polyhydroxy acyclic precursor with a carbonyl group is a sugar. A minimum of six carbons is needed to make benzene, but a benzene ring *once formed* is difficult to functionalize with a carbon side chain under physiological reaction conditions. Aromatic rings resist electrophilic addition and nucleophilic substitution (under mild conditions). Thus, an acyclic compound with seven or more carbons is a likely precursor (we shall return to this point soon).

The actual starting material employed is a phosphorylated C_7 sugar acid, 2-keto-3-deoxyarabinoheptulosonate-7-P, which is itself formed from PEP and erythrose-4-P by an aldol condensation.

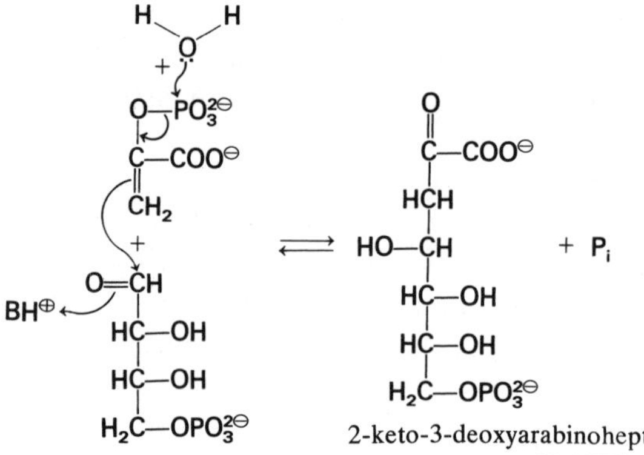

2-keto-3-deoxyarabinoheptulosonate-7-P
(DAHP)

The C-2 carbonyl of the phosphorylated C_7 sugar acid (DAHP) is a suitable electrophile for intramolecular attack. To form a cyclohexane ring, C-7 (the phosphorylated alcoholic carbon) must act as carbanion nucleophile. The chemical problem is: how does an enzyme generate a carbanion equivalent at C-7? The solution is ingenious and is suggested by the stoichiometry of the reaction, which is catalyzed by dehydroquinate synthase.

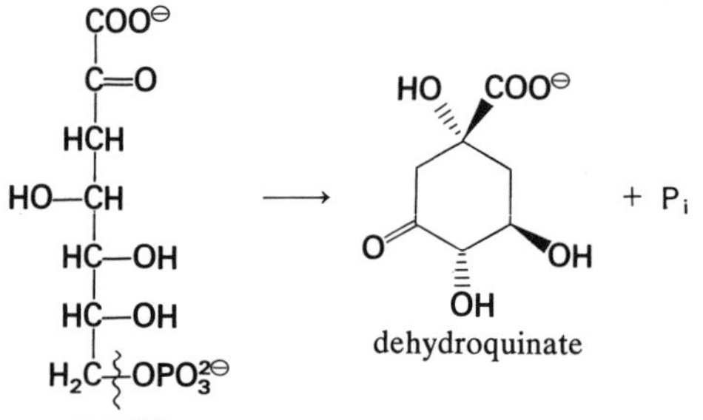

DAHP → dehydroquinate + P_i

The cyclic product does not contain the phosphoryl group. It has been lost as P_i, meaning that the C^7–O bond has been cleaved as it would be in an elimination of $H^{\oplus}$ at C-6 and $OPO_3^{2\ominus}$ at C-7. Such an elimination would yield an olefin that is an *enol* because C-6 has an OH substituent. As such, it has the requisite carbanionic character at C-7 to initiate ring closure, and the hypothetical cyclized product would indeed be dehydroquinate.

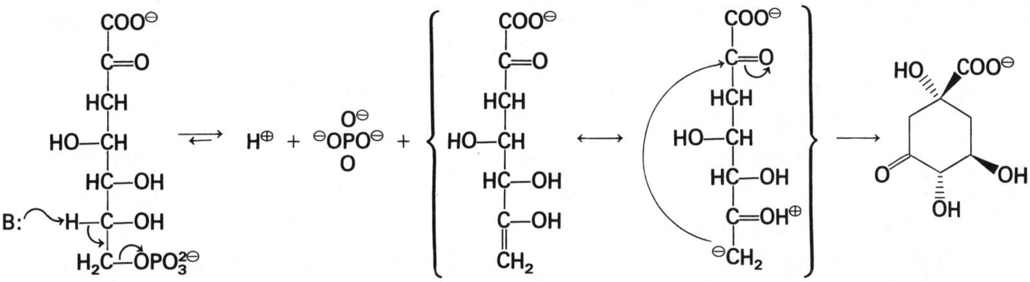

However, there is one problem with the chemical logic of this scheme. We emphasized in Chapter 17 that enzymes act as catalysts for dehydrations by taking advantage of (or by *increasing*) the acidity of carbon-bound hydrogens to initiate eliminations by their abstraction as *protons*. This is possible only if the substrate carbanionic transition-state species so generated can be stabilized; otherwise, the free energy of activation for C—H bond breakage is too high. The hydrogen bonded to C-6 of DAHP is *not particularly acidic* (although the net elimination of phosphoric acid would be reasonable). An obvious chemical device to increase acidity would be to generate a carbonyl group at the adjacent carbon—e.g., oxidation of the C-5 alcoholic group to a C-5 ketone. Then the C-6 hydrogen can be abstracted to yield an enolate anion. The biological solution does, in fact, use this chemical logic. $NAD^{\oplus}$ is a required cofactor for dehydroquinate synthase for reversible oxidation (and eventual rereduction) of C-5. The 5,6-enediolate anion can expel the C-7 phosphate and thereby yield (in low-energy steps) the required enolate anion for cyclization.

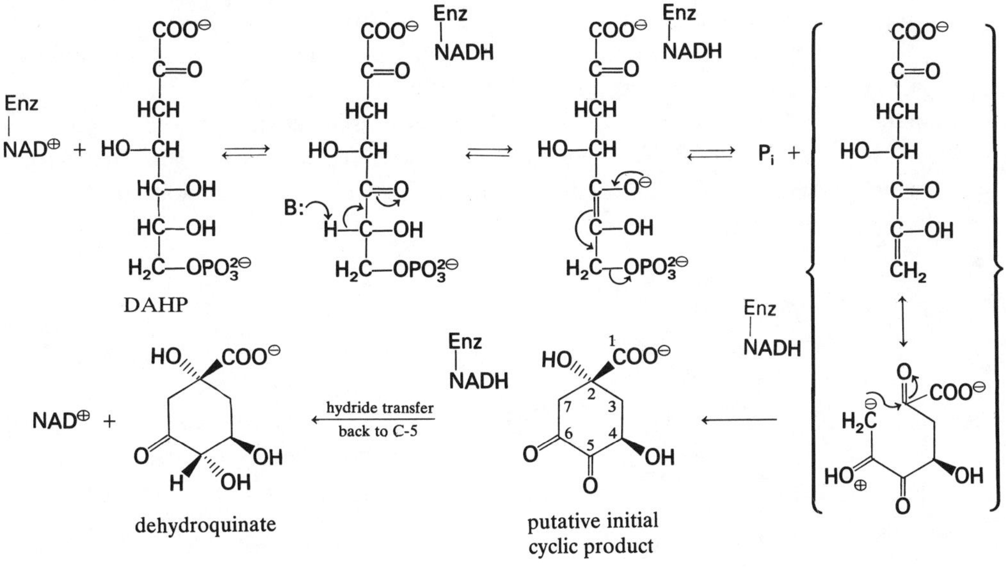

(1)

DAHP

hydride transfer back to C-5

dehydroquinate

putative initial cyclic product

The putative initial α-dicarbonyl cyclized product must undergo both regiospecific (i.e., C-5 not C-6) and stereospecific reduction to the C-5 α-alcohol of dehydroquinate. The logic of this enzyme-catalyzed cyclization sequence is comprehensible in chemical terms.

One additional comment is that the deoxyketoarabinoheptulosonate (DAHP) substrate probably exists predominantly as the cyclic hemiacetal, and the enzyme may or may not catalyze ring opening at its active site. Clearly, it must be the *acyclic* form that is undergoing cyclization to the cyclohexane product; in the hemiacetal forms, the carbonyl is masked and not susceptible to carbanion attack.

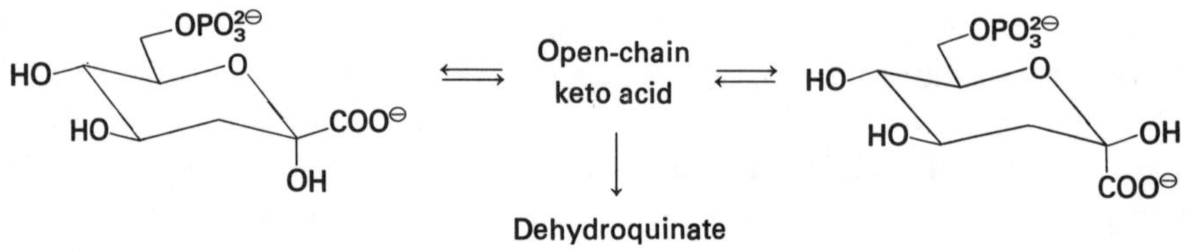

$OPO_3^{2\ominus}$ HO HO O $COO^\ominus$ OH ⇌ Open-chain keto acid ⇌ $OPO_3^{2\ominus}$ HO HO O OH $COO^\ominus$

↓

Dehydroquinate

With the highly oxygenated cyclohexane ring of dehydroquinate constructed, the organisms must now introduce three double bonds to aromatize the ring and must also remove all four of the oxygen substituents to arrive at the benzene ring of phenylalanine. Also, dehydroquinate is a C_7 compound, whereas phenylalanine and its direct precursor, phenylpyruvate, are C_9 compounds, so the carbon skeleton must be extended.

The first two double bonds are introduced by successive 1,2-elimination of $H^\oplus + OH^\ominus$ and 1,4-elimination of $H^\oplus + OPO_3^{2\ominus}$ to yield a cyclohexadiene chorismate as a dihydroaromatic precursor. We have discussed the particular reactions in Chapters 17 and 20. The sequential stoichiometry is indicated and shows that, prior to the second dehydration, the C-5 hydroxyl is functionalized as an enolpyruvyl group, as discussed in ¶20.B.1.

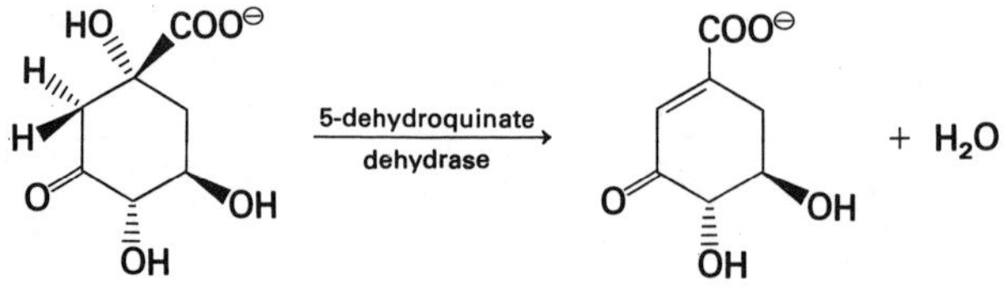

dehydroquinate (HO, $COO^\ominus$, H, H, O, OH, OH) $\xrightarrow[\text{dehydrase}]{\text{5-dehydroquinate}}$ dehydroshikimate ($COO^\ominus$, O, OH, OH) + H_2O (2)

$$\text{dehydroshikimate} + NADH + ATP \xrightarrow[\text{2. shikimate kinase}]{\text{1. dehydroshikimate reductase}} \text{shikimate-5-P} + ADP + NAD^{\oplus} \quad (3)$$

$$\text{shikimate-5-P} + PEP \xrightarrow[\text{synthetase}]{\text{3-enolpyruvylshikimate-5-P}} P_i + \text{3-enolpyruvylshikimate-5-P} \quad (4)$$

$$\text{3-enolpyruvylshikimate-5-P} \xrightarrow[\text{synthetase}]{\text{chorismate}} P_i + \text{chorismate} \quad (5)$$

Chorismate is the key intermediate in this biosynthetic pathway. As we noted in ¶20.B.2, it undergoes the only enzyme-catalyzed Claisen rearrangement in primary metabolism. The enolpyruvyl group connected by an oxygen atom to C-3 in chorismate forms a new carbon–carbon bond during the rearrangement, properly introducing the pyruvyl side chain of phenylpyruvate. The initial product, however, is still a dihydroaromatic known as prephenate.

$$\text{chorismate} \xrightarrow[\text{(rearrangement)}]{\text{chorismate mutase}} \text{prephenate} \quad (6)$$

Aromatization and the consequent gain in resonance energy of the product provide the driving force for decarboxylation and concomitant loss of either the *para*-OH (undoubtedly assisted by general acid catalysis) or the *para*-H as a *hydride ion* to $NAD^{\oplus}$.

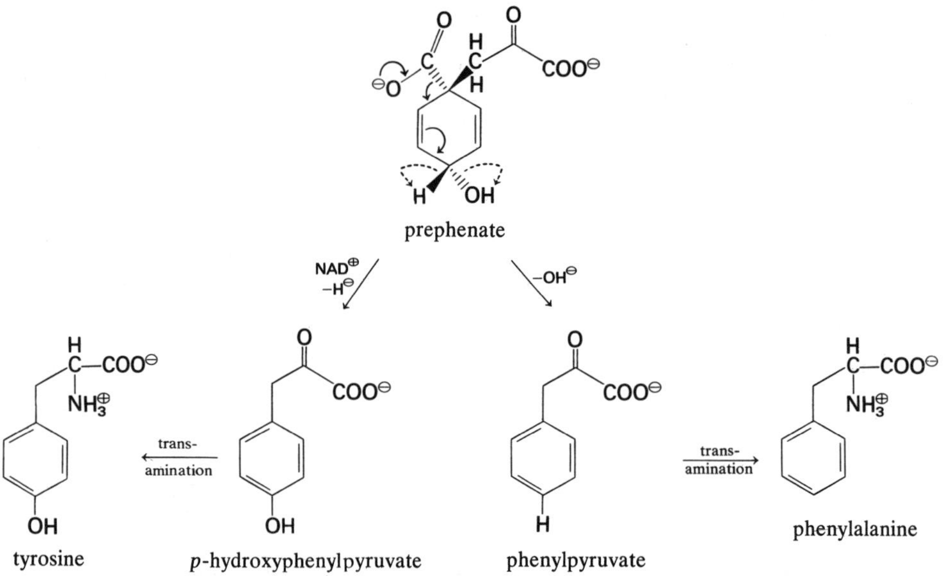

prephenate

NAD$^{\oplus}$ $-H^{\ominus}$ $-OH^{\ominus}$ (7)

tyrosine   ← trans-amination   *p*-hydroxyphenylpyruvate   phenylpyruvate   trans-amination →   phenylalanine

We noted that functionalization of an aromatic ring to form new carbon–carbon bonds in biological systems is difficult, and that might be why a C_7 (rather than a C_6) acyclic precursor was used, to provide an exocyclic carbon from which the alanyl side chains of phenylalanine and tyrosine could be elaborated. In fact, the exocyclic $COO^{\ominus}$ is not a building block for the side chain, but is used for the third and final dehydration in the pathway. Instead, the C_3 side chain is efficiently inserted intact by the Claisen concerted rearrangement, an unusual solution for side-chain emplacement at the dihydroaromatic stage. Of the four oxygens in the initial cyclic intermediate dehydroquinate, three were predictably removed by dehydration sequences; the fourth ended up as the keto group of phenylpyruvate.

Chorismate can also be routed, in glutamine-dependent amination (¶5.C and ¶20.B.4), to anthranilate (plus pyruvate). The *ortho*-aminobenzoate is the direct precursor of the indole ring of tryptophan.

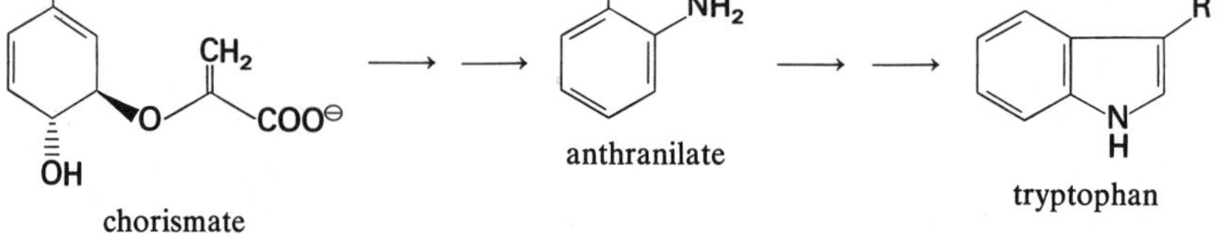

chorismate → → anthranilate → → tryptophan

With these ideas as backing, one could go on and dissect the formation of other aromatic and heteroaromatic ring systems that are important in biochemical molecules. Many of these heterocycles are the active portions of coenzyme molecules, so an understanding of their formation and degradation is important. These coenzymes include riboflavin, folate, thiamine, biotin, pyridoxal, and nicotinamides, as well as purines and pyrimidines and the pyrrole rings of porphyrins and corrins. In some instances, the pathways are understood in some detail and the chemical logic is clear. In other instances, there are large gaps in our knowledge, and understanding of the constraints of biochemical and chemical logic may be useful for future research. At this point, the reader may wish to evaluate the patterns of other metabolic pathways as a check of his or her predictive or analytic abilities.

One ultimate goal in understanding the logic of metabolic pathways is to predict how organisms will process any organic or inorganic molecule presented, either as a normal constituent or as an abnormal or xenobiotic component of metabolism. Having explored the few general classes of chemical reactions used by enzymatic catalysts, we are in a position to gain this predictive ability, which is useful not only for understanding what intermediates form and what their reactivity might be, but also for design and testing of compounds that might selectively block one specific mechanistic step in a pathway for antibiotic or therapeutic purposes.

References

GENERAL BIOCHEMISTRY REFERENCES

Lehninger, A. 1975. *Biochemistry*, 2d ed. New York: Worth.

Mahler, H., and E. Cordes. 1971. *Biological Chemistry*, 2d ed. New York: Harper & Row.

Metzler, D. 1977. *Biochemistry: The Chemical Reactions of Living Cells*. New York: Academic Press.

Segel, I. H. 1975. *Biochemical Calculations*, 2d ed. New York: Wiley. [How to do simple calculations encountered in biochemistry.]

Stryer, L. 1975. *Biochemistry*. San Francisco: W. H. Freeman and Company.

GENERAL REFERENCES ON ENZYMES

Advances in Enzymology. New York: Wiley. [Current editor is A. Meister; 46 volumes have been published to date.]

Annual Review of Biochemistry. Palo Alto, Calif.: Annual Reviews. [Yearly review volume; generally has several chapters on various aspects of enzymology.]

Barman, T. 1974. *Enzyme Handbook*. 2 vols. & suppl. New York: Springer-Verlag. [A tabulation of various kinetic constants and binding parameters for enzymes.]

Bender, M. 1971. *Homogeneous Catalysis from Protons to Proteins*. New York: Wiley.

Blackburn, S. 1976. *Enzyme Structure and Function*. New York: Marcel Dekker. [Eleven chapters on enzymes whose X-ray structure have been determined.]

Boyer, P. D., ed. 1970–1976. *The Enzymes*, 3d ed. New York: Academic Press. [The comprehensive multivolume treatise in the field of enzymology. The third edition is now complete in 13 volumes. Volumes 1 and 2 are of general scope and are available bound together in a paperback volume; the other volumes deal with specific enzymes. Each enzyme is covered by a separate reviewer, and the amount of mechanistic information provided varies widely from chapter to chapter.]

Dayhoff, M. O., ed. 1968–1972. *Atlas of Protein Sequence and Structure*, vols. 3–5. Silver Spring, Md.: National Biomedical Research Foundation. [A compilation of primary sequence data and three-dimensional structures of proteins and enzymes whose crystal structures had been solved at the time of publication.]

Dickerson, R. E., and I. Geis. 1969. *The Structure and Action of Proteins*. New York: Benjamin.

Dixon, M., and E. Webb. 1964. *The Enzymes*, 2d ed. New York: Academic Press. [A somewhat dated compendium.]

Fersht, A. R. 1977. *Enzyme Structure and Mechanism*. Reading, England, and San Francisco: W. H. Freeman and Company.

Gray, C. 1971. *Enzyme-Catalyzed Reactions*. New York: Van Nostrand–Rheinhold.

Horizons in Biochemistry and Biophysics. Reading, Mass.: Addison-Wesley. [Four volumes published to date in this continuing series.]

Jencks, W. P. 1969. *Catalysis in Chemistry and Enzymology*. New York: McGraw-Hill.

Lowe, J., and L. Ingraham. 1974. *An Introduction to Biochemical Reaction Mechanisms*. Englewood Cliffs, N.J.: Prentice-Hall.

Methods in Enzymology. New York: Academic Press. [More than 52 volumes to date. The source book for methods of purification, assay, and structural characterization of substrates and enzymes, and also for methods in protein, lipid, and carbohydrate chemistry relevant to enzymology.]

Zeffren, E., and P. Hall. 1973. *The Study of Enzyme Mechanisms*. New York: Wiley.

GENERAL REFERENCES ON BIOORGANIC CHEMISTRY

Bruice, T., and S. Benkovic. 1965. *Bioorganic Chemistry*. 2 vols. New York: Benjamin.

Kaiser, T., and F. Kezdey, eds. *Progress in Bioorganic Chemistry*. New York: Wiley. [Four volumes have appeared to date in this continuing series.]

GENERAL REFERENCES ON ENZYME KINETICS

Gutfreund, H. 1972. *Enzymes: Physical Principles*. New York: Wiley.

Segel, I. 1975. *Enzyme Kinetics*. New York: Wiley.

Westley, J. 1969. *Enzymic Catalysis*. New York: Harper & Row.

GENERAL REFERENCE ON ENZYME STEREOCHEMISTRY

Alworth, W. 1972. *Stereochemistry and Its Application in Biochemistry*. New York: Wiley.

REFERENCES CITED IN TEXT

COMMON ABBREVIATIONS USED: *ABB* = *Archives of Biochemistry and Biophysics; BBA* = *Biochimica et Biophysica Acta; BBRC* = *Biochemical and Biophysical Research Communications; JACS* = *Journal of the American Chemical Society; JBC* = *Journal of Biological Chemistry; PNAS* = *Proceedings of the National Academy of Science of the U.S.A.*

Abbott, M., and S. Udenfriend. 1974. In *Molecular Mechanisms of Oxygen Activation*, ed. O. Hayaishi (New York: Academic Press), p. 168.

Abbott, S., S. Jones, S. Weinman, and J. Knowles. 1978. *JACS* 100:2558.

Abeles, R. 1972. In *The Enzymes*, 3d ed., ed. P. Boyer (New York: Academic Press), vol. 5, p. 481.

Abeles, R., and D. Dolphin. 1976. *Accts. Chem. Res.* 9:114.

Abeles, R., B. Hutton, and F. Westheimer. 1957. *JACS* 79:712.

Abeles, R., and C. Walsh. 1973. *JACS* 95:6124.

Abiko, Y. 1975. In *Metabolic Pathways*, 3d ed., ed. D. Greenberg (New York: Academic Press), vol. 7, p. 1.

Adair, W., R. Gaugler, and O. Gabriel. 1971. *Fed. Proc. Amer. Soc. Exp. Biol.* 30:376.

Adams, E. 1970. In *The Enzymes*, 3d ed., ed. P. Boyer (New York: Academic Press), vol. 6, p. 479.

———. 1976. In *Advances in Enzymology and Related Areas of Molecular Biology*, vol. 44, ed. A. Meister (New York: Wiley), p. 69.

Adelberg, E. 1955. *JBC* 216:431.

Adman, E., L. Sieker, and L. Jensen. 1973. *JBC* 248:3987.

Ainslie, R., J. Shill, and K. Neet. 1972. *JBC* 247:7088.

Akhtar, M., and P. Jordan. 1969. *Tetrahedron Letters* 11:875.

Albert, T., G. Petsko, and D. Tsernoglu. 1976. *Nature* 263:297.

Alberts, A., and P. R. Vagelos. 1972. In *The Enzymes*, 3d ed., ed. P. Boyer (New York: Academic Press), vol. 6, p. 37.

Alberty, R., W. Miller, and H. Fisher. 1957. *JACS* 79:3973.

Albery, J., and J. Knowles. 1976. *Biochemistry* 15:5627, 5631.

Allen, C., W. Alworth, A. Macrae, and K. Bloch. 1967. *JBC* 242:1895.

Allison, W. 1976. *Accts. Chem. Res.* 9:293.

Alston, T., L. Mela, and H. Bright. 1977. *PNAS* 74:3767.

Altendorf, K., B. Gilch, and F. Lingens. 1971. *Fed. Eur. Biochem. Soc. Letters* 16:95.

Alworth, W. L. 1972. *Stereochemistry and Its Application in Biochemistry.* New York: Wiley.

Anderson, J., and H. Chang. 1965. *ABB* 110:346.

Andrews, T., G. Lorimer, and N. Tolbert. 1973. *Biochemistry* 12:11, 18.

Anfinsen, C. 1962. In *Basic Problems of Neoplastic Disease* (New York: Columbia Univ. Press), p. 112.

Angelides, K., and A. Fink. 1976. *Biochemistry* 15:5287.

Anson, M. L. 1937. *J. Gen. Physiol.* 20:663.

Anthony, R. S., and L. B. Spector. 1971. *JBC* 246:6129.

———. 1972. *JBC* 247:2170.

Antonini, E., M. Brunori, P. Fasella, R. Khomutov, J. Manning, and E. Severin. 1970. *Biochemistry* 9:1211.

Arfin, S., and H. Umbarger. 1969. *JBC* 244:1118.

Arias, L., and W. Jakoby, eds. 1976. *Glutathione: Metabolism and Function.* New York: Raven Press.

Atkin, C., L. Thelander, P. Reichard, and G. Lang. 1973. *JBC* 248:7464.

Babior, B. 1975*a*. *Accts. Chem. Res.* 8:376.

———, ed. 1975*b*. *Cobalamins.* New York: Wiley.

Babior, B., T. Carty, and R. Abeles. 1974. *JBC* 249:1689.

Babu, U., and R. Johnston. 1974. *BBRC* 58:460.

Bachovchin, W., and J. Roberts. 1978. *PNAS* (in press).

Bailar, J. 1958. *J. Inorg. Nucl. Chem.* 8:165.

Bailey, G., and W. Dempsy. 1967. *Biochemistry* 6:1526.

Baker, B. R. 1967. *Design of Active-Site-Directed Irreversible Enzyme Inhibitors.* New York: Robert E. Krieger.

Balch, W., L. Magrum, G. Fox, R. Wolfe, and C. Woese. 1977. *J. Mol. Evol.* 9:305.

Baldwin, A., and P. Berg. 1960. *JBC* 241:839.

Balls, A. K., C. E. McDonald, and A. S. Bracher. 1958. In *Proc. Int. Symp. Enzyme Chem., Tokyo, Kyoto* (Tokyo: Maruzen), p. 245.

Bannemere, C., J. Hamilton, L. Steinrauf, and J. Knappe. 1965. *Biochemistry* 4:240.

Banner, D., A. Bloomer, G. Petsko, D. Phillips, C. Pogson, and R. Offord. 1975. *Nature* 255:609.

Barker, H. 1972. In *The Enzymes*, 3d ed., ed. P. Boyer (New York: Academic Press), vol. 6, p. 509.

Barker, R. 1971. *Organic Chemistry of Biological Compounds.* Englewood Cliffs, N.J.: Prentice-Hall.

Bartlett, P. 1965. *Non-Classical Carbonium Ions.* New York: Benjamin.

Batzold, F., and C. Robinson. 1975. *JACS* 97:2576.

Beckwith, J., and D. Zipser, eds. 1970. *The Lac Operon.* Long Island, N.Y.: Cold Spring Harbor Press.

Bell, R. M., and D. E. Koshland. 1971. *Science* 172:1253.

Belleau, B., and J. Burba. 1960. *JACS* 82:5751, 5752.

Bender, M. 1971. *Homogeneous Catalysis from Protons to Proteins.* New York: Wiley.

Bender, M., and L. Brubacher. 1966. *JACS* 88:5880.

Bender, M., and G. Hamilton. 1962. *JACS* 84:2570.

Bender, M., and F. Kezdy. 1965. *Ann. Rev. Biochem.* 34:49.

Bender, M., F. Kezdy, and F. Wedler. 1967. *J. Chem. Educ.* 44:84.

Bender, M., E. Pollock, and M. Neveu. 1962. *JACS* 84:595.

Benkovic, S., and W. Bullard. 1973. In *Progress in Bioorganic Chemistry*, ed. E. Kaiser and F. Kezdy (New York: Wiley), vol. 2, p. 133.
Benkovic, S., and K. Schray. 1976. In *Advances in Enzymology and Related Areas of Molecular Biology*, vol. 44, ed. A. Meister (New York: Wiley), p. 139.
Bentley, R. 1969–1970. *Molecular Asymmetry in Biology*. 2 vols. New York: Academic Press.
Beytia, E., and J. Porter. 1976. *Ann. Rev. Biochem.* 45:113.
Beytia, E., A. Quereshi, and J. Porter. 1973. *JBC* 248:1856.
Birchmeier, W., P. Zaoralek, and P. Christen. 1973. *Biochemistry* 12:2874.
Birktoft, J. J., and D. M. Blow. 1972. *JMB* 68:187.
Blackburn, S. 1976. *Enzyme Structure and Function*. New York: Marcel Dekker.
Blake, C., G. Mair, A. North, D. Phillips, and V. Sarma. 1967. *Proc. Roy. Soc. Ser. B* 167:365, 378.
Blakely, R. 1969. *Biochemistry of Folic Acid and Related Pteridines*. New York: Wiley. Bloch, K. 1969. *Accts. Chem. Res.* 2:191.
———. 1972. In *The Enzymes*, 3d ed., ed. P. Boyer (New York: Academic Press), vol. 5, p. 441.
Bloch, W., and M. Schlessinger. 1974. *JBC* 249:1760.
Blow, D. M. 1969. *Biochem. J.* 112:261.
———. 1973. In *The Enzymes*, 3d ed., ed. P. Boyer (New York: Academic Press), vol. 3, p. 189.
Blow, D. M., J. J. Birktoft, and B. S. Hartley. 1969. *Nature* 221:337.
Bloxham, D. P., and H. A. Lardy. 1973. In *The Enzymes*, 3d ed., ed. P. Boyer (New York: Academic Press), vol. 8, p. 240.
Boeker, E., and E. Snell. 1972. In *The Enzymes*, 3d ed., ed. P. Boyer (New York: Academic Press), vol. 6, p. 217.
Bondinelli, W., S. Vnek, P. Knowles, M. Sprecher, and D. Sprinson. 1971. *JBC* 246:6191.
Boos, W. 1974. *Ann. Rev. Biochem.* 43:123.
Bowman, W., H. Tabor, and C. Tabor. 1973. *JBC* 248:2480.
Boyer, P., ed. 1970–1976. *The Enzymes*, 3d ed. 13 vols. New York: Academic Press.
Bramlett, R., and H. Peck. 1975. *JBC* 250:2979.
Branden, S., H. Jornvall, B. Eklund, and B. Furugren. 1975. In *The Enzymes*, 3d ed., ed. P. Boyer (New York: Academic Press), vol. 11, p. 104.
Braunstein, A. 1960. In *The Enzymes*, 2d ed., vol. 2, ed. P. Boyer, H. Lardy, and K. Myrback (New York: Academic Press), p. 131.
———. 1973. In *The Enzymes*, 3d ed., ed. P. Boyer (New York: Academic Press), vol. 9, p. 379.
Bray, R. C. 1975. In *The Enzymes*, 3d ed., ed. P. Boyer (New York: Academic Press), vol. 12, p. 300.
Breathnach, R., and J. Knowles. 1977. *Biochemistry* 16:3054.
Breslow, R. 1958. *JACS* 80:3719.
———. 1961. In *The Mechanism of Action of Water Soluble Vitamins*, ed. A. de Reuck and M. O'Connor (Boston: Little, Brown), p. 65.
Breslow, R., and D. Wernick. 1977. *PNAS* 74:1303.
Bridger, W. A. 1974. In *The Enzymes*, 3d ed., ed. P. Boyer (New York: Academic Press), vol. 10, p. 581.
Briggs, G. E., and J. B. S. Haldane. 1925. *Biochem. J.* 19:338.
Bright, H. 1964. *JBC* 739:2307.
———. 1967. *Biochemistry* 6:1191.
Bright, H., R. Lundin, and L. Ingraham. 1964. *Biochemistry* 3:1224.
Britton, L., and A. Markovetz. 1977. *JBC* 252:8561.
Brown, J., and B. S. Hartley. 1966. *Biochem. J.* 101:214.
Brown, L., and J. Drury. 1965. *J. Chem. Phys.* 43:1688.
Bruice, T. C. 1970. In *The Enzymes*, 3d ed., ed. P. Boyer (New York: Academic Press), vol. 2, p. 217.
———. 1976. *Ann. Rev. Biochem.* 45:331.
Bruice, T. C., and S. J. Benkovic. 1965–1966. *Bioorganic Mechanisms*. 2 vols. New York: Benjamin.
Bruice, T. C., and P. Bruice. 1976. *Accts. Chem. Res.* 9:378.
Bruice, T. C., and A. F. Hegarty. 1970. *PNAS* 65:805.

Buchanan, B. 1972. In *The Enzymes*, 3d ed., ed. P. Boyer (New York: Academic Press), vol. 6, p. 193.
Buchanan, J. 1973. In *Advances in Enzymology*, vol. 39, ed. A. Meister (New York: Wiley), p. 91.
Buckel, W., and A. Bobi. 1976. *Eur. J. Biochem.* 64:255.
Buckel, W., V. Buschmeier, and H. Eggerer. 1969. *Z. Physiol. Chem.* 350:1367.
Bruce, G., E. Paniago, and D. Margerum. 1975. *Chem. Comm.*, p. 261.
Burma, D., and B. Horecker. 1958. *JBC* 231:1053.
Bush, K., V. Shiner, and H. R. Mahler. 1973. *Biochemistry* 12:4802.
Butler, J., W. Alworth, and M. Nugent. 1974. *JACS* 96:1617.
Butler, L. G. 1971. In *The Enzymes*, 3d ed., ed. P. Boyer (New York: Academic Press), vol. 4, p. 529.
Cahn, R. S., C. K. Ingold, and V. Prelog. 1956. *Experientia* 12:81.
———. 1966. *Angew. Chem. Int. Ed.* 5:385.
Calvin, M. 1954. *Fed. Proc.* 13:697.
Canfield, R. 1963. *JBC* 238:2698.
———. 1965. *JBC* 240:1997.
Cannata, J., and A. Stoppani. 1963. *JBC* 238:1208.
Cannizzaro, S. 1912. *J. Chem. Soc.* 111:1677.
Cantoni, G. 1975. *Ann. Rev. Biochem.* 44:435.
Gardinale, G., and R. Abeles. 1968. *Biochemistry* 7:3970.
Carty, T., B. Babior, and R. Abeles. 1971. *JBC* 246:6313.
Cashell, M. 1975. *Ann. Rev. Microbiol.* 29:301.
Cashell, M., and S. Gallant. 1969. *Nature* 221:838.
Caughey, W. S., W. J. Wallace, T. A. Volpe, and S. Yashikawa. 1976. In *The Enzymes*, 3d ed., ed. P. Boyer (New York: Academic Press), vol. 13, p. 299.
Chamberlin, M. 1974. In *The Enzymes*, 3d ed., ed. P. Boyer (New York: Academic Press), vol. 10, p. 333.
Chambon, P. 1974. In *The Enzymes*, 3d ed., ed. P. Boyer (New York: Academic Press), vol. 10, p. 261.
Chan, T., and T. C. Bruice. 1977. *JACS* 99:2387.
Chang, C. K., and D. Dolphin. 1975. *JACS* 97:5948.
Cheung, Y., C. Fung, and C. Walsh. 1975. *Biochemistry* 14:2981.
Cheung, Y., and C. Walsh. 1976*a*. *Biochemistry* 15:2432.
———.1976*b*. *Biochemistry* 15:3907.
Childs, A., D. Davies, A. Green, and J. Rutland. 1955. *Brit. J. Pharmacol.* 10:462.
Chirpich, T., V. Zappia, R. Costilow, and H. Barker. 1970. *JBC* 245:1778.
Christen, P., and J. Riordan. 1968. *Biochemistry* 7:1531.
———. 1969. *Biochemistry* 8:2381.
Christenson, J., W. Dairman, and S. Udenfriend. 1970. *ABB* 141:356.
Chuang, H., D. Patek, and L. Hellerman. 1974. *JBC* 249:238.
Chung, S., R. Tan, and Y. Suzuki. 1971. *Biochemistry* 10:1205.
Cleland, W. 1970. In *The Enzymes*, 3d ed., ed. P. Boyer (New York: Academic Press), vol. 2, p. 1.
———. 1975. *Accts. Chem. Res.* 8:145.
———. 1977. In *Advances in Enzymology and Related Areas of Molecular Biology*, vol. 45, ed. A. Meister (New York: Wiley), p. 273.
Cleland, W., M. O'Leary, and D. Northrop, eds. 1976. *Sixth Steenbock Symposium: Isotope Effects on Enzyme-Catalyzed Reactions*. Baltimore, Md.: Univ. Park Press.
Clement, G. 1973. In *Progress in Bioorganic Chemistry*, ed. T. Kaiser and F. Kezdy (New York: Wiley), vol. 2, p. 178.
Coates, R., R. Conradi, D. Ley, A. Akeson, J. Garads, S. Lee, and C. West. 1976. *JACS* 98:4659.
Cogoli, A., and G. Semenza. 1975. *JBC* 250:7802.
Cohn, M., and T. Hughes. 1972. *JBC* 237:176.
Cohn, M., J. Pearson, E. L. O'Connell, and I. A. Rose. 1970. *JACS* 92:4095.

Coleman, J. E. 1971. In *Progress in Bioorganic Chemistry*, ed. T. Kaiser and F. Kezdy (New York: Wiley), vol. 1, p. 159.
Coleman, J. E., and B. L. Vallee. 1960. *JBC* 235:390.
Collier, R. 1967. *JMB* 25:83.
Colowick, S. 1973. In *The Enzymes*, 3d ed., ed. P. Boyer (New York: Academic Press), vol. 9, p. 1.
Conway, A., and D. E. Koshland. 1968. *Biochemistry* 7:4011.
Cook, G., and R. Stoddart. 1973. *Surface Carbohydrates of the Eukaryotic Cell*. New York: Academic Press.
Cooper, R., and H. L. Kornberg. 1974. In *The Enzymes*, 3d ed., ed. P. Boyer (New York: Academic Press), vol. 10, p. 631.
Cooper, T., T. Tchen, H. Wood, and C. Benedict. 1968. *JBC* 243:3857.
Corey, E. J., and R. Volante. 1976. *JACS* 98:1291.
Cormier, M. J. 1975. *Ann. Rev. Biochem.* 44:255.
Cormier, M. J., D. M. Hercules, and J. Lee, eds. 1973. *Chemiluminescence and Bioluminescence*. New York: Plenum Press.
Cornforth, J. 1968. *Angew. Chem. Int. Ed.* 7:903.
———. 1973. *Chem. Soc. Rev.* 2:1.
Cornforth, J., K. Clifford, R. Mallaby, and G. Phillips. 1972. *Proc. Roy. Soc. Ser. B* 182:277.
Cornforth, J., R. H. Cornforth, C. Donninger, G. Popjack, G. Ryback, and G. J. Schroepfer, 1966. *Proc. Roy. Soc. Ser. B* 163:436.
Cornforth, J., J. Redmond, H. Eggerer, W. Buckel, and G. Gutschow. 1969. *Nature* 221:1212.
Covey, D., and C. Robinson. 1976. *JACS* 98:5038.
Covitz, T., and F. H. Westheimer. 1963. *JACS* 85:1773.
Cram, D. 1965. *Fundamentals of Carbanion Chemistry*. New York: Academic Press.
Cramer, S., K. Hodgson, W. Gillum, and L. Mortenson. 1978. *JACS* 100:3398.
Creighton, D., and I. A. Rose. 1976. *JBC* 251:61, 69.
Cromartie, T., and C. Walsh. 1975. *Biochemistry* 14:3482.
———. 1976. *JBC* 251:329.
Crosby, J., R. Sine, and G. Lienhard. 1970. *JACS* 92:2891.
D'Agnolo, G., I. Rosenfeld, and P. R. Vagelos. 1975. *JBC* 250:5283, 5289.
Dahl, J., and L. Hokin. 1974. *Ann. Rev. Biochem.* 43:327.
Dahlquist, F., T. Rand-Meir, and M. Raftery. 1969. *Biochemistry* 8:4214.
Dalziel, K. 1975. In *The Enzymes*, 3d ed., ed. P. Boyer (New York: Academic Press), vol. 11, p. 1.
Danenberg, P., R. Langenbach, and C. Heidelberger. 1974. *Biochemistry* 13:926.
Dang, T. Y., Y. Cheung, and C. Walsh. 1976. *Biochem. Biophys. Res. Commun.* 72:960.
Dardenne, G., P. Larsen, and E. Wieczorkowska. 1975. *BBA* 381:416.
Das, S., and R. Fujimura. 1977. *JBC* 252:8700, 8708.
Datko, A., J. Giovanelli, and S. Mudd. 1974. *JBC* 249:1139.
Davidson, B., E. Blackburn, and T Dopheide. 1972. *JBC* 247:4441, 4447.
Davie, E., K. Fujikawa, M. Legaz, and H. Kato. 1975. In *Proteases and Biological Control*, ed. E. Reich, D. Rifkin, and E. Shaw (Long Island, N.Y.: Cold Spring Harbor Press), p. 65.
Davis, J., J. Willard, and H. Wood. 1969. *Biochemistry* 8:3127, 3137, 3145.
Davis, L., and D. Metzler. 1972. In *The Enzymes*, 3d ed., ed. P. Boyer (New York: Academic Press), vol. 7, p. 33.
Degani, C., and P. Boyer. 1973. *JBC* 248:8222.
Dela Fuente, G. 1970. *Eur. J. Biochem.* 16:240.
DeLange, R., and E. L. Smith. 1971. In *The Enzymes*, 3d ed., ed. P. Boyer (New York: Academic Press), vol. 3, p. 81.
DeLisi, C., and D. M. Crothers. 1973. *Biopolymers* 12:1689.
Delpierre, G. R., and J. Fruton. 1966. *PNAS* 56:1817.
Dicfalusy, U., P. Falardeau, and S. Hammarstrom. 1977. *Fed. Eur. Biochem. Soc. Letters* 84:271.

Dickerson, R. E., and R. Timkovich. 1975. In *The Enzymes*, 3d ed., ed. P. Boyer (New York: Academic Press), vol. 11, p. 397.
Dimroth, P. 1975. *Fed. Eur. Biochem. Soc. Letters* 51:100.
———. 1976. *Eur. J. Biochem.* 64:269.
Dimroth, P., W. Dittmar, G. Walther, and H. Eggerer. 1973. *Eur. J. Biochem.* 37:295.
Dimroth, P., K. Mayer, and H. Eggerer. 1975. *Eur. J. Biochem.* 51:267.
Dinovo, E., and P. Boyer. 1971. *JBC* 246:4580.
Dixon, M., and E. C. Webb. 1964. *Enzymes*, 2d ed. New York: Academic Press.
Dolphin, D., A. Forman, D. Borg, J. Fajer, and R. Felton. 1971. *PNAS* 68:614.
Dowd, P., and M. Shapiro. 1976. *JACS* 98:3752.
Drenth, J., J. N. Jansonius, R. Koekoek, H. M. Swen, and B. G. Wolthers. 1968. *Nature* 218:929.
———. 1970. *Phil. Trans. Roy. Soc. Ser. B* 257:231.
Duclos, J., and P. Haake. 1974. *Biochemistry* 13:5358.
Dugan, R., and J. E. Porter. 1976. In *The Enzymes of Cell Membranes*, ed. A. Martinosi (New York: Plenum Press), vol. 2, p. 161.
Dunathan, H. 1971. In *Advances in Enzymology*, vol. 35, ed. F. Nord (New York: Wiley), p. 79.
Dunathan, H., J. Ayling, and E. Snell. 1968. *Biochemistry* 7:4537.
Dunathan, H., and J. Voet. 1974. *PNAS* 71:3888.
Eady, R., and J. Postgate. 1974. *Nature* 249:840.
Eagon, R., J. Paxton, and F. Kuehl. 1976. *JBC* 251:7329.
Eberhardt, N., and H. Rilling. 1975. *JBC* 250:863.
Ebner, K. 1973. In *The Enzymes*, 3d ed., ed. P. Boyer (New York: Academic Press), vol. 9, p. 363.
Edlund, B., L. Rask, P. Olsson, O. Walinder, O. Zetterquist, and L. Engstrom. 1969. *Eur. J. Biochem.* 9:451.
Edmonds, J., G. Popjack, S. Wong, and V. Williams. 1971. *JBC* 246:6254.
Eggerer, H. 1963. *Ann. Chem.* 666:192.
———. 1965. *Biochem. Z.* 343:111.
Eggerer, H., and A. Klette. 1967. *Eur. J. Biochem.* 1:447.
Eggerer, H., and U. Remberger. 1963*a*. *Biochem. Z.* 337:202.
———. 1973*b*. *Biochem. Z.* 339:62.
Eichhorn, G. 1971. *Inorganic Biochemistry*. New York: Elsevier.
Eigen, M. 1964. *Agnew. Chem. Int. Ed.* 3:1.
Eil, C., and I. Wool. 1971. *BBRC* 43:1001.
Eirich, D. 1977. Ph.D. dissertation, Univ. of Illinois.
Eirich, D., G. Vogels, and R. Wolfe. 1978. *Biochemistry* 17:4583.
Eldred, E., and P. Schimmel. 1972. *JBC* 247:2961.
Enoch, H., and R. Lester. 1975. *JBC* 250:6693.
Entsch, B., D. Ballou, M. Hussain, and V. Massey. 1976. *JBC* 251:7367.
Entsch, B., D. Ballou, and V. Massey. 1976. *JBC* 251:2556.
Epstein, W., and H. Rilling. 1970. *JBC* 245:4597.
Ernster, L., J. Capdevila, G. Dallner, J. W. DePierre, S. Jakobsson, and S. Orrenius. 1976. In *The Structural Basis of Membrane Function*, ed. Y. Hatefi and D. Stiggall (New York: Academic Press), p. 389.
Esmon, C., and J. Suttie. 1976. *JBC* 251:6238.
Essigman, J., R. Croy, A. Nadzan, W. Busby, V. Reinhold, G. Büchi, and G. Wogan. 1977. *PNAS* 74:1870.
Ettinger, M. 1974. *Biochemistry* 13:1242.
Ettinger, M., and D. Kosman. 1974. *Biochemistry* 13:1247.
Faeder, E., P. Davis, and L. Siegel. 1974. *JBC* 249:1599.
Fasella, P. 1967. *Ann. Rev. Biochem.* 36:185.
Fastrez, J., and A. R. Fersht. 1973. *Biochemistry* 12:2025.
Fee, J., G. Hegeman, and G. Kenyon. 1974. *Biochemistry* 13:2533.
Fee, J., E. Shapiro, and T. Moss. 1976. *JBC* 251:6157.

Feigelson, P., and F. Brady. 1974. In *Molecular Mechanisms of Oxygen Activation*, ed. O. Hayaishi (New York: Academic Press), p. 87.
Feldman, F., and L. Butler. 1969. *BBRC* 36:119.
Fersht, A. R. 1974. *Proc. Royal Soc. (London) Ser. B* 187:397.
———. 1977. *Enzyme Structure and Mechanism.* Reading, England, and San Francisco: W. H. Freeman and Company.
Fersht, A. R., D. M. Blow, and J. Fastrez. 1973. *Biochemistry* 12:2035.
Fersht, A. R., and M. Kaethner. 1976. *Biochemistry* 15:3342.
Fielden, E. M., P. B. Roberts, R. C. Bray, D. Lowen, G. Mautner, G. Rotilio, and L. Calabrese. 1974. *Biochem. J.* 139:49.
Fieser, L., and M. Fieser. 1961. *Advanced Organic Chemistry*. New York: Reinhold.
Findlay, T., and E. Adams. 1970. *JBC* 245:5248.
Findlay, T., J. Valinsky, A. Mildvan, and R. Abeles. 1973. *JBC* 248:1285.
Findlay, T., J. Valinsky, K. Sato, and R. Abeles. 1972. *JBC* 247:4197.
Fink, A. 1977. *Accts. Chem. Res.* 10:223.
Fisher, H., E. Conn, B. Vennesland, and F. Westheimer. 1953. *JBC* 202:687.
Fisher, J., R. Spencer, and C. Walsh. 1976. *Biochemistry* 15:1054.
Fisher, J., and C. Walsh. 1974. *JACS* 96:4345.
Fisher, L., J. Albery, and J. Knowles. 1976. *Biochemistry* 15:5621.
Flashner, M., and V. Massey. 1974. In *Molecular Mechanisms of Oxygen Activation*, ed. O. Hayaishi (New York: Academic Press), p. 245.
Fleming, A. 1922. *Proc. Roy. Soc. (London) Ser. B* 93:306.
Fleming, I. 1965. *Selected Organic Syntheses*. New York: Wiley.
Flohr, H., W. Parnhost, and J. Retey. 1976. *Angew. Chem.* 88:613.
Floss, H., E. Schleicher, and R. Potts. 1976. *JBC* 251:5478.
Follman, M., and H. Hogenkamp. 1969. *Biochemistry* 8:4372.
———. 1970. *JACS* 92:671.
Fowler, L., and R. John. 1972. *Biochem. J.* 130:569.
Freer, S. J., J. Kraut, J. D. Robertus, H. T. Wright, and Ng. H. Xuong. 1970. *Biochemistry* 9:1997.
Fridovich, I. 1972*a*. In *Horizons in Biochemistry and Biophysics*, vol. 1, ed. E. Quaglierello (Reading, Mass.: Addison-Wesley), p. 1.
———. 1972*b*. In *The Enzymes*. 3d ed., ed. P. Boyer (New York: Academic Press), vol. 6, p. 255.
———. 1975. *Ann. Rev. Biochem.* 44:147.
Friedkin, M. 1973. In *Advances in Enzymology*, vol. 38, ed. A. Meister (New York: Wiley), p. 235.
Friedman, H. 1975. In *Cobalamins*, ed. B. Babior (New York: Wiley), p. 75.
Frode, H. C., and I. B. Wilson. 1971. In *The Enzymes*, 3d ed., ed. P. Boyer (New York: Academic Press), vol. 5, p. 87.
Fruton, J. 1971. In *The Enzymes*, 3d ed., ed. P. Boyer (New York: Academic Press), vol. 3, p. 179.
———. 1974. *Accts. Chem. Res.* 7:241.
Futterman, S., and M. Rollins. 1973. *JBC* 248:7773.
Gatehouse, J., and J. Knowles. 1977. *Biochemistry* 16:3045.
Gawron, O., and T. Fondy. 1954. *JACS* 81:6333.
Gawron, O., A. Glaid, T. Fondy, and M. Bechtold. 1962. *JACS* 84:3877.
Gawron, O., M. Kennedy, and M. Rainer. 1974. *Biochem. J.* 143:717.
Gaylor, J. 1974. In *Lipids*, vol. 4, ed. T. Goodwin (Baltimore, Md.: Univ. Park Press, International Review of Science: Biochemistry Section), p. 1.
Gaylor, J., and H. Mason. 1968. *JBC* 243:4966.
Gefter, M. 1975. *Ann. Rev. Biochem.* 44:45.
Gehring, U., and F. Lynen. 1972. In *The Enzymes*, 3d ed., ed. P. Boyer (New York: Academic Press), vol. 7, p. 391.
Gennis, R., and L. Hager. 1976. In *The Enzymes of Cell Membranes*, ed. A. Martinosi (New York: Plenum Press), vol. 2, p. 493.

Gennis, R., and J. Strominger, 1976. In *The Enzymes of Cell Membranes*, ed. A. Martinosi (New York: Plenum Press), vol. 2, p. 327.
George, P. 1956. In *Currents in Biochemical Research*, vol. 2, ed. D. Green (New York: Wiley), p. 358.
Gershon, H., J. Merk, and K. Dittmer. 1949. *JACS* 71:3573.
Gertler, A., K. A. Walsh, and H. Neurath. 1974. *Biochemistry* 13:1302.
Ghisla, S., and V. Massey. 1978. In *International Symposium on Oxidizing Enzymes*, ed. R. Ondarza and T. Singer (New York: Elsevier), p. 55.
Ghosh, S., and S. Roseman. 1965. *JBC* 240:1531.
Gibian, M., and R. Galway. 1977. In *Bioorganic Chemistry*, ed. E. Van Tamelen (New York: Academic Press), vol. 1, p. 117.
Gibson, F. 1968. *Biochem. Prep.* 12:94.
Gibson, F., and J. Pittard. 1968. *Bacteriol. Rev.* 32:465.
Gill, G. 1968. *Quart. Revs. Chem. Soc.* 22:338.
Giordano, R., R. Bereman, D. Kosman, and M. Ettinger, 1974. *JACS* 96:1023.
Givot, I., T. Smith, and R. Abeles. 1969. *JBC* 244:6341.
Glaser, L. 1972. In *The Enzymes*, 3d ed., ed. P. Boyer (New York: Academic Press), vol. 6, p. 355.
Glaser, L., and L. Ward. 1970. *BBA* 198:613.
Glaser, L., and H. Zarkowsky. 1971. In *The Enzymes*, 3d ed., ed. P. Boyer (New York: Academic Press), vol. 5, p. 465.
Glasstone, S. 1962. *Textbook of Physical Chemistry*. London: Macmillan.
Glatzer, L., E. Eakin, and R. Wagner. 1972. *J. Bacteriol.* 112:453.
Gleason, W., and R. Barker. 1971. *Can. J. Chem.* 49:1425.
Glusker, J. 1968. *J. Mol. Biol.* 38:149.
———. 1971. In *The Enzymes*, 3d ed., ed. P. Boyer (New York: Academic Press), vol. 5, p. 413.
Goff, C. 1974. *JBC* 249:6181.
Goldstein, M., T. Jon, and T. Garvey. 1968. *Biochemistry* 7:2724.
Goody, R., and F. Eckstein. 1971. *JACS* 93:6252.
Goody, R., F. Eckstein, and R. Schirmer. 1972. *BBA* 276:155.
Gould, E. 1959. *Mechanism and Structure in Organic Chemistry*. New York: Holt, Rinehart and Winston.
Graves, D., and J. Warn. 1972. In *The Enzymes*, 3d ed., ed. P. Boyer (New York: Academic Press), vol. 7, p. 435.
Gray, C. J. 1971. *Enzyme-Catalysed Reactions*. New York: Van Nostrand–Rheinhold.
Green, N. 1966. *Biochem. J.* 101:774.
Griffin, M., and P. Trudgill. 1972. *Biochem. J.* 129:595.
Groves, J., G. McClusky, R. White, and M. Coon. 1978. *BBRC* 81:154.
Groves, J., and M. Van der Puy. 1976. *JACS* 98:5290.
Guchhait, R., S. Polakis, P. Dimroth, E. Stoll, J. Moss, and M. D. Lane. 1974. *JBC* 249:6633.
Guchhait, R., S. Polakis, D. Hollis, C. Fenselan, and M. D. Lane. 1974. *JBC* 249:6646.
Guggenheim, S., and M. Flavin. 1969. *JBC* 244:6217.
Guidotti, G. 1976. In *Trends in Biochemical Sciences* (New York: Elsevier), vol. 1, p. 11.
Gunetileke, K., and R. Anwar. 1968. *JBC* 243:5770.
Gunsalus, I., J. Mecks, J. Lipscomb, P. Debrunner, and E. Munck. 1974. In *Molecular Mechanisms of Oxygen Activation*, ed. O. Hayaishi (New York: Academic Press), p. 561.
Gunsalus, R., D. Eirich, J. Romesser, W. Balch, S. Shapiro, and R. Wolfe. 1976. In *Microbial Production and Utilization of Gases (H_2, CH_4, CO) Symposium, Gottingen, Goltze* (Gottingen, Germany: Verlage).
Gupta, N. K., and B. Vennesland. 1964. *JBC* 239:3787.
Gupta, R., C. Fung, and A. Mildvan. 1976. *JBC* 251:2421.
Gupta, R., T. Kasai, and D. Schlessinger. 1977. *JBC* 252:8945.
Guroff, G., D. Jerina, J. Rensen, S. Udenfriend, and B. Witkop. 1967. *Science* 157:1524.

Gutfreund, H. 1972. *Enzymes: Physical Principles.* New York: Wiley.
Gutte, B., and R. Merrifield. 1969. *JACS* 91:501.
Habig, W., M. Pabst, and W. Jakoby. 1974. *JBC* 249:7130.
Hagihara, B., N. Sato, and T. Yamanaka. 1975. In *The Enzymes,* 3d ed., ed. P. Boyer (New York: Academic Press), vol. 11, p. 397.
Hall, S., A. Doweyko, and F. Jordan. 1976. *JACS* 98:7460.
Hamberg, M., and B. Samuelsson. 1973. *PNAS* 70:899.
Hamberg, M., B. Samuelsson, I. Bjorkhem, and H. Danielsson. 1974. In *Molecular Mechanisms of Oxygen Activation,* ed. O. Hayaishi (New York: Academic Press), p. 30.
Hamberg, M., J. Svensson, and B. Samuelsson. 1975. *PNAS* 72:2994.
Hamberg, M., J. Svensson, T. Wakabayashi, and B. Samuelsson. 1974. *PNAS* 71:345.
Hamilton, G. 1969. In *Advances in Enzymology,* vol. 35, ed. A. Meister (New York: Wiley), p. 55.
———. 1971. In *Progress in Bioorganic Chemistry,* ed. T. Kaiser and F. Kezdy (New York: Wiley), vol. 1, p. 83.
———. 1974. In *Molecular Mechanisms of Oxygen Activation,* ed. O. Hayaishi (New York: Academic Press), p. 405.
Hamilton, G., G. Dyrkacz, and R. Libby. 1976. In *Iron and Copper Proteins,* ed. K. Yasunobu, H. Mower, and O. Hayaishi (New York: Plenum Press), p. 489.
Hamilton, G., and F. Westheimer. 1959. *JACS* 81:6332.
Hamilton, J. A., R. Blakley, F. Looney, and M. Winfield. 1969. *BBA* 177:374.
Hamilton, J. A., Y. Tamao, R. Blakley, and R. Coffman. 1972. *Biochemistry* 11:4696.
Hammarstrom, S., and P. Falardeau. 1977. *PNAS* 74:3691.
Hammes, G., and P. Schimmel. 1970. In *The Enzymes,* 3d ed., ed. P. Boyer (New York: Academic Press), vol. 2, p. 67.
Hansen, J., E. Dinovo, and P. Boyer. 1969. *JBC* 244:6270.
Hanson, K., and E. Havir. 1970. *ABB* 141:1.
Hanson, K., and I. Rose. 1963. *PNAS* 50:981.
———. 1975. *Accts. Chem. Res.* 8:1.
Harold, F. 1972. *Bacteriol. Rev.* 36:172.
———. 1974. In *Current Topics in Membranes and Transport,* vol. 5, ed. A. Kelenzeller (New York: Academic Press), p. 2.
Harris, J. I., B. P. Meriwether, and J. H. Park. 1963. *Nature* 198:154.
Harris, J. I., and M. Waters. 1976. In *The Enzymes,* 3d ed., ed. P. Boyer (New York: Academic Press), vol. 13, p. 1.
Harris, M., D. Usher, H. Albrecht, G. Jones, and J. Moffat. 1969. *PNAS* 63:246.
Hartley, B. S. 1964. *Nature* 201:1284.
Hartley, B. S., and B. A. Kilby. 1954. *Biochem. J.* 56:288.
Hartman, F. C. 1971. *Biochemistry* 10:146.
Hartman, S. C. 1971. In *The Enzymes,* 3d ed., ed. P. Boyer (New York: Academic Press), vol. 4, p. 79.
Hartsuck, J. A., and W. N. Lipscomb. 1970. In *The Enzymes,* 3d ed., ed P. Boyer (New York: Academic Press), vol. 1, p. 71.
———. 1973. *PNAS* 70:3793.
Hashimoto, H., H. Gunther, and H. Simon. 1973. *Fed. Eur. Biochem. Soc. Letters* 33:81.
Haslam, A. 1974. *The Shikimate Pathway.* Toronto: Butterworths.
Hass, G. M., and J. Byrne. 1960. *JACS* 82:947.
Hass, G. M., and H. Neurath. 1970. *Biochemistry* 10:3535, 3541.
Hassid, W., and E. Neufeld. 1962. In *The Enzymes,* 2d ed., vol. 6, ed. P. Boyer (New York: Academic Press), p. 278.
Hastings, J., and C. Blany. 1975. *JBC* 250:7288.
———. 1976. *Biochemistry* 14:4719.
Hastings, J., C. Balny, C. LePeuch, and P. Douzou. 1973. *PNAS* 70:3468.

Hastings, J., A. Eberhard, T. Baldwin, M. Nicoli, T. Cline, and K. Nealson. 1973. In *Chemiluminescence and Bioluminescence*, ed. M. J. Cormier, D. M. Hercules, and J. Lee (New York: Plenum Press), p. 369.

Hatefi, Y., and D. L. Stigall. 1976. In *The Enzymes*, 3d ed., ed. P. Boyer (New York: Academic Press), vol. 13, p. 176.

Haugland, R. P., and L. Stryer. 1967. In *Conformation of Biopolymers*, ed. G. N. Rachmandran (New York: Academic Press), vol. 1, p. 321.

Havir̆, E., and K. Hanson. 1973. In *The Enzymes*, 3d ed., ed. P. Boyer (New York: Academic Press), vol. 7, p. 75.

Hayaishi, O., ed. 1974. *Molecular Mechanisms of Oxygen Activation*. New York: Academic Press.

———. 1976. In *Trends in Biochemical Sciences* (Amsterdam: Elsevier), vol. 1, p. 10.

Hayaishi, O., Y. Bai, and T. Hata. 1975. *JBC* 250:5221.

Hayaishi, O., M. Nozaki, and M. Abbott. 1975. In *The Enzymes*, 3d ed., ed. P. Boyer (New York: Academic Press), vol. 12, p. 120.

Hayaishi, O., and W. B. Sutton. 1957. *JACS* 79:4809.

Healey, M., and P. Christen. 1972. *JACS* 94:7911.

———. 1973. *Biochemistry* 12:35.

Hecht, S., and C. Chinault. 1976. *PNAS* 73:405.

Hegarty, A. F., and T. C. Bruice. 1970. *JACS* 92:6561, 6568.

Hegazi, M., R. Borchardt, and R. Schowen. 1976. *JACS* 98:3048.

Hegeman, G., E. Rosenberg, and G. Kenyon. 1970. *Biochemistry* 9:4029.

Heidelberger, C. 1970. *Cancer Res.* 30:1549.

———. 1975. *Ann. Rev. Biochem.* 44:79.

Heinstein, P., S. Lee, and H. Floss. 1971. *BBRC* 44:1244.

Heller, J. 1972. In *The Enzymes*, 3d ed., ed. P. Boyer (New York: Academic Press), vol. 6, p. 573.

Heller, J., and J. Horowitz. 1974. *JBC* 249:5933, 6308, 6317, 7181.

———. 1975. *JBC* 250:3021.

Hendrickson, J. 1965. *The Molecules of Nature*. New York: Benjamin.

Henkin, J., and R. Abeles. 1976. *Biochemistry* 15:3472.

Herlihy, J., S. Maister, J. Albery, and J. Knowles. 1976. *Biochemistry* 15:5601.

Hers, H. 1976. *Ann. Rev. Biochem.* 45:167.

Hersh, L., and M. Jorns. 1975. *JBC* 250:8728.

Higgins, M., J. Kornblatt, and H. Rudney. 1972. In *The Enzymes*, 3d ed., ed. P. Boyer (New York: Academic Press), vol. 7, p. 407.

Higuchi, T., G. Flynn, and A. Shah. 1967. *JACS* 89:616.

Hill, A. V. 1910. *J. Physiol.* (London) 40:IV.

Hill, R., and K. Brew. 1975. In *Advances in Enzymology and Related Areas of Molecular Biology*, vol. 44, ed. A. Meister (New York: Wiley), p. 411.

Hill, R., and J. Tiepel. 1971. In *The Enzymes*, 3d ed., ed. P. Boyer (New York: Academic Press), vol. 5, p. 539.

Hill, R., S. Yan, and S. Arfin. 1973. *JACS* 95:7857.

Hirschmann, R., R. Nutt, D. Veber, A. Vitali, S. Varga, T. Jacob, F. Holly, and R. Denkewalter. 1969. *JACS* 91:507.

Hiyama, T., S. Fukui, and K. Kitahara. 1968. *J. Biochem.* 64:99.

Hoffmann, T. 1974. *Adv. Chem. Ser.*, no. 136 (New York: American Chemical Society).

Hogenkamp, H. 1975. In *Cobalamins*, ed. B. Babior (New York: Wiley), p. 21.

Holland, P., M. Clark, and D. Bloxham. 1973. *Biochemistry* 12:3309.

Hollenberg, P., T. Rand-Meir, and L. Hager. 1974. *JBC* 249:5816.

Holm, R. 1975. *Endeavour* 34:38.

———. 1977. *Accts. Chem. Res.* 10:427.

Holme, E., G. Linstedt, S. Linstedt, and M. Tofft. 1968. *Fed. Eur. Biochem. Soc. Letters* 2:29.

Holzer, H., and W. Duntze. 1971. *Ann. Rev. Biochem.* 40:345.

Honjo, T., Y. Nishizuka, O. Hayaishi, and I. Kato. 1968. *JBC* 243:3553.
Hooper, N., and J. Law. 1965. *BBRC* 18:426.
Horecker, B., O. Tsolas, and C. Lai. 1972. In *The Enzymes*, 3d ed., ed. P. Boyer (New York: Academic Press), vol. 7, p. 213.
House, H. 1972. *Modern Synthetic Reactions*, 2d ed. New York: Benjamin.
Howell, L., and V. Massey. 1971. In *Flavins and Flavoproteins*, ed. H. Kamin (Baltimore, Md.: Univ. Park Press), vol. 3, p. 363.
Howell, L., T. Spector, and V. Massey. 1972. *JBC* 247:4340.
Hsu, I.-N., L. T. J. Delbaere, and M. N. G. James. 1977. *Nature* 266:140.
Hsu, R. 1970. *JBC* 245:6672.
Hsu, R., A. Mildvan, G. Chang, and C. Fung. 1976. *JBC* 251:6582.
Hubbard, R. 1956. *J. Gen. Physiol.* 39:935.
Hunkapiller, M. W., S. H. Smallcombe, D. R. Whitaker, and J. H. Richards. 1973. *Biochemistry* 12:4732.
Hurwitz, J., and B. Horecker. 1956. *JBC* 223:993.
Inoue, H., F. Suzuki, H. Tanioka, and Y. Takeda. 1967. *BBRC* 26:602.
———. 1968. *J. Biochem.* (Tokyo) 63:89.
Inoue, H., T. Tsunemi, F. Suzuki, and Y. Takeda. 1969. *J. Biochem.* (Tokyo) 65:889.
Inoue, M., S. Horiuchi, and Y. Morino. 1977. *Eur. J. Biochem.* 78:609.
Inouye, K., and J. Fruton. 1967. *Biochemistry* 6:1705.
International Union of Biochemistry. 1964. *Enzyme Nomenclature.* New York: American Elsevier.
Jackson, C., C. Esmon, and W. Owen. 1975. In *Proteases and Biological Control*, ed. E. Reich, D. Rifkin, and E. Shaw (Long Island, N.Y.: Cold Spring Harbor Press), p. 95.
Jacobson, G., and G. Stark. 1973. In *The Enzymes*, 3d ed., ed. P. Boyer (New York: Academic Press), vol. 9, p. 226.
Jansen, E. F., M. F. Nutting, and A. K. Balls. 1949. *JBC* 179:201.
———. 1950. *JBC* 185:209.
Jansen, E. F., M. F. Nutting, R. Jang, and A. K. Balls. 1949. *JBC* 179:189.
Jeffrey, A., K. Jennette, S. Blobstein, I. Weinstein, F. Beland, R. Harvey, H. Kasai, I. Miura, and K. Nakanishi. 1977. *JACS* 98:5714.
Jeffrey, A., H. Yeh, D. Jerina, T. Patel, J. Davey, and D. Gibson. 1975. *Biochemistry* 14:575.
Jencks, W. P. 1962. In *The Enzymes*, 2d ed., vol. 6, ed. P. Boyer (New York: Academic Press), p. 384.
———. 1969. *Catalysis in Chemistry and Enzymology.* New York: McGraw-Hill.
———. 1974. In *The Enzymes*, 3d ed., ed. P. Boyer (New York: Academic Press), vol. 9, p. 483.
———. 1975*a*. In *Advances in Enzymology and Related Areas of Molecular Biology*, vol. 43, ed. A. Meister (New York: Wiley), p. 219.
Jencks, W. P., and M. I. Page. 1972. *Proc. 8th Fed. Eur. Biochem. Soc. Mtg., Amsterdam* 29:45.
Jenkins, W. T. 1961. *JBC* 236:1121.
Jensen, L. 1974. *Ann. Rev. Biochem.* 43:461.
Jerina, D. 1973. *Chem. Technol.* 4:120.
Jerina, D., et al. 1977. In *Origins of Human Cancer*, ed. H. Hiatt, J. Watson, and J. Winsten (Long Island, N.Y.: Cold Spring Harbor Press), p. 639.
Johnson, W. S. 1976. *Bioorganic Chem.* 5:51.
Jordan, P., and M. Akhtar. 1970. *Biochem. J.* 116:277.
Jordan, P., and D. Shemin. 1972. In *The Enzymes*, 3d ed., ed. P. Boyer (New York: Academic Press), vol. 7, p. 339.
Jorns, M., and L. Hersh. 1974. *JACS* 96:4012.
Josse, J., and S. Wong. 1971. In *The Enzymes*, 3d ed., ed. P. Boyer (New York: Academic Press), vol. 4, p. 499.
Kaczorowski, G., L. Shaw, M. Fuentes, and C. Walsh. 1975. *JBC* 250:2855.
Kaczorowski, G., L. Shaw, R. Laura, and C. Walsh. 1975. *JBC* 250:8921.

Kahan, F., et al. 1975. In *Interscience Conference on Antimicrobial Agents and Chemotherapy, Washington, D.C., September 1975* (New York: American Chemical Society), meeting abstracts 100–103.

Kaiser, E. T., and B. L. Kaiser. 1972. *Accts. Chem. Res.* 5:219.

Kanac, M., A. Raz, and DeW. Goodman. 1968. *J. Clin. Invest.* 47:2025.

Kang, U., L. Nolan, and P. Frey. 1975. *JBC* 250:7099.

Kaplan, M., and M. Flavin. 1966. *JBC* 241:4463.

Kaplan, N., and M. Ciotti. 1956. *JBC* 221:823.

Katigiri, M., H. Maluo, S. Yamamoto, O. Hayaishi, T. Kitao, and S. Oal. 1965. *JBC* 240:3414.

Kaufman, S. 1976. In *Iron and Copper Proteins*, ed. K. Yasunobo, H. Mower, and O. Hayaishi (New York: Plenum Press), p. 91.

Kaufman, S., and D. B. Fisher. 1974. In *Molecular Mechanisms of Oxygen Activation*, ed. O. Hayaishi (New York: Academic Press), p. 285.

Kayne, F. J. 1973. In *The Enzymes*, 3d ed., ed. P. Boyer (New York: Academic Press), vol. 8, p. 353.

Kearny, E. B., and W. C. Kenney. 1974. In *Horizons in Biochemistry and Biophysics*, vol. 1 (Reading, Mass.: Addison-Wesley), p. 62.

Keele, B., J. McCord, and I. Fridovich. 1970. *JBC* 245:6176.

Kelly-Falcoz, F., H. Greenberg, and B. Horecker. 1965. *JBC* 240:2966.

Kemal, C., and T. Bruice. 1976. *PNAS* 73:995.

Kenney, W., D. L. Edmondson, and R. L. Seng. 1976. *JBC* 251:5386.

Kenyon, G., and G. Hegeman. 1970. *Biochemistry* 9:4036.

Kerr, S., and E. Borek. 1973. In *The Enzymes*, 3d ed., ed. P. Boyer (New York: Academic Press), vol. 9, p. 167.

Kido, J., K. Soda, T. Suzuki, and L. Asada. 1976. *JBC* 251:6994.

Kirsch, J. 1976. In *Sixth Steenbock Symposium: Isotope Effects on Enzyme-Catalyzed Reactions*, ed. W. Cleland, M. O'Leary, and D. Northrup (Baltimore, Md.: Univ. Park Press), p. 100.

Kirshner, N. 1957. *JBC* 226:821.

Kleinschmidt, A. K., J. Moss, and M. D. Lane. 1969. *Science* 166:1276.

Klinman, J. 1978. In *Advances in Enzymology and Related Areas of Molecular Biology*, vol. 46, ed. A. Meister (New York: Wiley), p. 415.

Klinman, J., and I. A. Rose. 1971. *Biochemistry* 10:2259.

Klotz, I. 1967. *Energy Changes in Biochemical Reactions*. New York: Academic Press.

Knappe, J., K. Biederbich, and W. Brummer. 1962. *Angew. Chem.* 74:433.

Knappe, J., H. Blaschkavski, P. Grobner, and T. Schmitt. 1974. *Eur. J. Biochem.* 50:253.

Knappe, J., and T. Schmitt. 1976. *BBRC* 71:1110.

Knappe, J., B. Wenger, and U. Wiegand. 1963. *Biochem. Z.* 337:232.

Knivett, V., and J. Cullen. 1965. *Biochem. J.* 96:771.

Koch, G., D. Shaw, and F. Gibson. 1971. *BBA* 229:795, 805.

Kohn, L. 1977. *Ann. Rept. Med. Chem.* 12:211.

Kohn, L., and H. Kaback. 1973. *JBC* 248:7012.

Komai, H., V. Massey, and G. Palmer. 1969. *JBC* 244:1692.

Kornberg, A. 1974. *DNA Synthesis*. San Francisco: W. H. Freeman and Company.

Kornfeld, R., and S. Kornfeld. 1973. In *The Enzymes*, 3d ed., ed. P. Boyer (New York: Academic Press), vol. 9, p. 217.

Koshland, D. E. 1970. In *The Enzymes*, 3d ed., ed. P. Boyer (New York: Academic Press), vol. 1, p. 341.

Kosower, E. M. 1956. *JACS* 78:3497.

———. 1975. In *Free Radicals in Molecular Biology and Pathology*, vol. 2, ed. W. A. Pryon (New York: Academic Press), chapt. 10.

Koster, J., and C. Veeger. 1968. *BBA* 151:11.

Krampitz, L. 1967. *Ann. Rev. Biochem.* 38:213.

Krongelb, M., T. Smith, and R. Abeles. 1968. *BBA* 167:473.
Krower, J., R. Schultz, and B. Babior. 1978. *JBC* 253:1041.
Kumagoi, H., and E. Miles. 1971. *BBRC* 44:1271.
Kung, H., S. Cederbaum, L. Tsai, and T. Stadtman. 1970. *PNAS* 65:978.
Kunitz, M., and J. H. Northrop. 1933. *Science* 78:558.
———. 1935. *J. Gen. Physiol.* 18:433.
Kurz, L., and C. Frieden. 1975. *JACS* 97:677.
Lane, M. D., and F. Lynen. 1963. *PNAS* 49:379.
Lane, R., J. Manning, and E. Snell. 1976. *Biochemistry* 15:4180.
Lane, R., and E. Snell. 1976. *Biochemistry* 15:4175.
Langan, T. A. 1968. *Science* 162:579.
Larsen, P., D. Ondokera, and H. Floss. 1975. *BBA* 381:397.
Larsen, P., and E. Wieczorkowska. 1975. *BBA* 381:409.
Laursen, R., and F. Westheimer. 1966. *JACS* 88:3426.
Law, J. 1971. *Accts. Chem. Res.* 4:199.
Legler, G. 1966. *Z. Physiol. Chem.* 345:197.
———. 1976. *Z. Physiol. Chem.* 348:1359.
———. 1970. *Z. Physiol. Chem.* 351:25.
Lehman, I. R. 1971. In *The Enzymes*, 3d ed., ed. P. Boyer (New York: Academic Press), vol. 4, p. 251.
———. 1974. In *The Enzymes*, 3d ed., ed. P. Boyer (New York: Academic Press), vol. 10, p. 237.
Lehninger, A. 1975. *Biochemistry*, 2d ed. New York: Worth.
Lenhart, P., and D. Hodgkin. 1961. *Nature* 192:937.
Levine, D., T. Reid, and I. Wilson. 1969. *Biochemistry* 8:2374.
Levitski, A., and D. E. Koshland. 1974. In *The Enzymes*, 3d ed., ed. P. Boyer (New York: Academic Press), vol. 10, p. 539.
Levitski, A., W. Stallcup, and D. E. Koshland. 1971. *Biochemistry* 10:3371.
Levy, H., F. Loewus, and B. Vennesland. 1957. *JACS* 79:2949.
Lienhard, G. 1966. *JACS* 80:5642.
———. 1973. *Science* 180:149.
Lipmann, F. 1971. *Science* 173:875.
Lipmann, F., W. Gevers, H. Kleinkauf, and R. Roskoski. 1968. In *Advances in Enzymology*, vol. 31, ed. A. Meister (New York: Wiley), p. 1.
Lippert, B., B. Metcalf, M. Jung, and P. Casara. 1977. *Eur. J. Biochem.* 74:441.
Lipscomb, W. N., G. N. Reeke, J. A. Hartsuck, F. A. Quiocho, and P. H. Bethge. 1970. *Phil. Trans. Roy. Soc. Ser. B* 257:177.
Lochmuller, H., H. Wood, and J. Davis. 1968. *JBC* 241:5678.
Loechler, L. E., and T. C. Hollocher. 1975. *JACS* 97:3235.
Loewus, F., F. Westheimer, and B. Vennesland. 1953. *JACS* 75:5018.
Loge, E. T., and M. J. Coon. 1973. In *Iron-Sulfur Proteins*, ed. W. Lovenberg (New York: Academic Press), vol. 1, p. 173.
Lovenberg, W., ed. 1973–1975. *Iron-Sulfur Proteins*. 3 vols. New York: Academic Press.
Lowe, T., and L. Ingraham. 1975. *An Introduction to Biochemical Reaction Mechanisms*. Englewood Cliffs, N.J.: Prentice-Hall.
Lowe, J., L. Ingraham, J. Alspach, and R. Rasmussen. 1976. *Biochem. Biophys. Res. Comm.* 73:465.
Lu, A., D. Ryan, D. Jerina, J. Daly, and W. Levin. 1975. *JBC* 250:8283.
Luthi, J., J. Retey, and D. Arigoni. 1969. *Nature* 221:1213.
Lynen, F., J. Knappe, E. Lorch, G. Jutting, and E. Ringelmann. 1959. *Angew. Chem.* 71:481.
Lynen, F., J. Knappe, E. Lorch, G. Jutting, E. Ringelmann, and J. P. Lachance. 1961. *Biochem. Z.* 335:123.
McBride, B., and R. Wolfe. 1971. *Biochemistry* 10:2317, 4312.

McCapra, F. 1976. *Accts. Chem. Res.* 9:201.
McCord, J., and I. Fridovich. 1969. *JBC* 244:6049.
McDonough, M., and W. Wood. 1961. *JBC* 236:1220.
McGrath, R. 1977. In *Bioorganic Chemistry*, ed. E. Van Tamelen (New York: Academic Press), vol. 2, p. 231.
MacLennan, D. 1975. *JBC* 249:974, 980.
Magnusson, S., T. Peterson, L. Sottrup-Jensen, and H. Claeys. 1975. In *Proteases and Biological Control*, ed. E. Reich, D. Rifkin, and E. Shaw (Long Island, N.Y.: Cold Spring Harbor Press), p. 123.
Mahler, H., and E. Cordes. 1966. *Biological Chemistry*. New York: Harper & Row.
Main, L., G. Kasperek, and T. Bruice. 1972. *Biochemistry* 11:3991.
Maister, S., C. Pett, J. Albery, and J. Knowles. 1976. *Biochemistry* 15:5607.
Makinen, M., K. Yamamura, and E. Kaiser. 1976. *PNAS* 73:3882.
Malkin, R. 1973. In *Inorganic Biochemistry*, ed. G. Eichorn (New York: Elsevier), vol. 2, p. 689.
Malmstrom, B., L. Andreasson, and B. Reinhammar. 1975. In *The Enzymes*, 3d ed., ed. P. Boyer (New York: Academic Press), vol. 12, p. 507.
Manning, J., R. Khomutov, and P. Fasella. 1968. *Eur. J. Biochem.* 5:199.
Manning, J., S. Moore, W. Rowe, and A. Meister. 1969. *Biochemistry* 8:2681.
Maradufu, A., G. Cree, and A. Perlin. 1971. *Can. J. Chem.* 49:3429.
Marayuma, H., R. Easterday, H. Chang, and M. D. Lane. 1966. *JBC* 241:2405.
Marcotte, P., T. Soper, J. Manning, and C. Walsh. 1977. *JBC* 252:1571.
Marcotte, P., and C. Walsh. 1975. *BBRC* 62:677.
———. 1976. *Biochemistry* 15:3070.
Margerum, D., K. Chellappa, F. Bossu, and G. Burce. 1975. *JACS* 97:6894.
Mark, D., and C. Richardson. 1976. *PNAS* 73:780.
Mascaro, J., R. Horhammer, S. Eisenstein, L. Sellers, K. Mascaro, and H. Floss. 1977. *JACS* 99:273.
Mason, H. S. 1965. *Ann. Rev. Biochem.* 34:595.
Massey, V. 1973. In *Iron-Sulfur Proteins*, ed. W. Lovenberg (New York: Academic Press), vol. 1, p. 301.
Massey, V., and D. Edmondson. 1970. *JBC* 245:6595.
Massey, V., and S. Ghisla. 1974. *Ann. N.Y. Acad. Sci.* 227:446.
———. 1975. *Proc. 10th Fed. Eur. Biochem. Soc. Mtg.*, p. 145.
Massey, V., and P. Hemmerich. 1975. In *The Enzymes*, 3d ed., ed. P. Boyer (New York: Academic Press), vol. 12, p. 191.
Massey, V., H. Komai, G. Palmer, and G. Elion. 1970. *JBC* 245:2387.
Massey, V., F. Muller, R. Feldberg, M. Schuman, P. Sullivan, L. Howell, S. Mayhew, R. Matthews, and G. Foust. 1969. *JBC* 244:3999.
Massey, V., S. Strickland, S. Mayhew, L. Howell, P. Engel, R. Matthews, M. Schuman, and P. A. Sullivan. 1969. *BBRC* 36:891.
Matthews, B. W., P. Sigler, R. Henderson, and D. M. Blow. 1967. *Nature* 214:652.
Matthews, D., R. Alden, J. Birktoft, S. Freer, and J. Kraut. 1977. *JBC* 252:8875.
Matthews, D., R. Alden, J. Bolin, S. Freer, R. Hamlin, N. Xuong, J. Kraut, M. Poe, M. Williams, and K. Hoogsteen. 1977. *Science* 197:452.
Maycock, A., and R. Abeles. 1976. *Accts. Chem. Res.* 9:313.
Maycock, A., R. Abeles, J. Salach, and T. Singer. 1976. *Biochemistry* 15:114.
Maycock, A., R. Suva, and R. Abeles. 1975. *JACS* 97:5613.
Mayhew, S. G., and M. Ludwig. 1975. In *The Enzymes*, 3d ed., ed. P. Boyer (New York: Academic Press), vol. 12, p. 72.
Mayo, P. de, ed. 1964. *Molecular Rearrangements*, vol. 2. New York: Wiley.
Mazumder, R., T. Sasakawa, Y. Kaziro, and S. Ochoa. 1962. *JBC* 237:3065.
Means, G., and R. Feeny. 1971. *Chemical Modification of Proteins*. San Francisco: Holden-Day.
Meister, A. 1973. *Science* 180:33.
———. 1974. In *The Enzymes*, 3d ed., ed. P. Boyer (New York: Academic Press), vol. 10, p. 699.

Meister, A. 1976. Abstract, 1976 Biochemistry Meetings, San Francisco.
Melander, L. 1960. *Isotope Effects on Reaction Rates.* New York: Ronald Press.
Meloche, H. P. 1970. *Biochemistry* 9:5050.
———. 1973. *JBC* 248:6945.
Meloche, H. P., and L. Mehler. 1973. *JBC* 248:6333.
Meloche, H. P., and W. A. Wood. 1964. *JBC* 239:3505, 3517.
Meloun, B., I. Kluh, V. Kostka, L. Moravek, Z. Prusik, J. Vanecek, B. Keil, and F. Sorm. 1966. *BBA* 130:543.
Menezes, L., and J. Puckles. 1976. *Eur. J. Biochem.* 16:240.
Meselson, M., R. Yuan, and J. Heywood. 1972. *Ann. Rev. Biochem.* 41:447.
Messner, B., H. Eggerer, J. Cornforth, and R. Mallaby. 1975. *Eur. J. Biochem.* 53:255.
Metzler, D. 1960. In *The Enzymes*, 2d ed., vol. 2, ed. P. Boyer, H. Lardy, and K. Myrback (New York: Academic Press), p. 306.
Michaelis, L., and M. L. Menten. 1913. *Biochem. Z.* 49:333.
Middlefort, C., and I. Rose. 1976. *JBC* 251:5881.
Mieyal, J., and R. Abeles. 1972. In *The Enzymes*, 3d ed., ed. P. Boyer (New York: Academic Press), vol. 7, p. 515.
Mildvan, A. 1970. In *The Enzymes*, 3d ed., ed. P. Boyer (New York: Academic Press), vol. 2, p. 446.
———. 1974. *Ann. Rev. Biochem.* 43:357, 382.
———. 1977. *Accts. Chem. Res.* 10:246.
Mildvan, A., D. Sloan, C. Fung, R. Gupta, and E. Melamud. 1976. *JBC* 251:2431.
Miller, D., and F. Westheimer. 1966. *JACS* 88:1507.
Milner, Y., and H. G. Wood. 1972. *PNAS* 69:2463.
Milstein, S., and L. A. Cohen. 1970. *PNAS* 67:1143.
Misra, P., and I. Fridovich. 1972. *JBC* 247:3170.
Mitchell, P. 1966. *Biological Revs.* 41:445.
Miyamoto, T., N. Ogino, S. Yamamoto, and O. Hayaishi. 1976. *JBC* 251:2629.
Miziorko, H., K. Clinkenbeard, W. D. Reed, and M. D. Lane. 1975. *JBC* 250:5768.
Miziorko, H., and M. D. Lane. 1977. *JBC* 252:1414.
Modrich, P., and D. Zabel. 1976. *JBC* 251:5866.
Moffet, F., and W. Bridger. 1970. *JBC* 245:2758.
Mondovi, B., G. Rotilio, M. Costa, and A. Finazzi-Agro. 1971. In *Methods in Enzymology*, vol. 7B, ed. S. Colowick and N. Kaplan (New York: Academic Press), p. 735.
Monod, J., J. Wyman, and J. P. Changeux. 1965. *J. Mol. Biol.* 12:88.
Moriguchi, M., T. Yamamoto, and K. Soda. 1973. *Biochemistry* 12:2969.
Morino, Y., and E. Snell. 1967. *JBC* 242:2800.
Morisaki, M., and K. Bloch. 1972. *Biochemistry* 11:309.
Morns, D., and R. Fillingame. 1974. *Ann. Rev. Biochem.* 43:303.
Morrison, J. F., and E. Heyde. 1972. *Ann. Rev. Biochem.* 41:29.
Morse, D., and B. Horecker. 1968. In *Advances in Enzymology*, vol. 31, ed. A. Meister (New York: Wiley), p. 125.
Mortensen, L. 1978. In *International Symposium on Oxidizing Enzymes*, ed. R. Ondarza and T. Singer (New York: Elsevier), p. 119.
Moss, J., and M. Lane. 1971. In *Advances in Enzymology*, vol. 35, ed. A. Meister (New York: Wiley), p. 321.
Moss, J., V. Magnaniello, and M. Vaughan. 1977. *PNAS* 73:4424.
Moss, J., and M. Vaughan. 1977. *JBC* 252:2455.
Moyle, J., and M. Dixon. 1955. *BBA* 16:434.
Mudd, S. 1973. In *The Enzymes*, 3d ed., ed. P. Boyer (New York: Academic Press), vol. 8, p. 121.
Mulheirn, L., and P. Ramm. 1972. *Chem. Soc. Revs.* 1:259.
Mullhofer, G., and I. A. Rose. 1965. *JBC* 240:1341.

Muscio, F., J. Carlson, L. Kuehl, and H. Rilling. 1974. *JBC* 249:3746.
Muth, W., and R. Costillow. 1974. *JBC* 249:7457.
Nakatsukasa, W., and E. Nester. 1972. *JBC* 247:5972.
Nakazawa, T., K. Hori, and O. Hayaishi. 1972. *JBC* 247:3439.
Nathans, D., and H. Smith. 1975. *Ann. Rev. Biochem.* 44:273.
Neal, R. 1970. *JBC* 245:2599.
Neims, A., H. DeLuca, and L. Hellerman. 1966. *Biochemistry* 5:203.
Neumann, R., R. Hevey, and R. Abeles. 1975. *JBC* 250:6362.
Neurath, H. 1975. In *Proteases and Biological Control*, ed. E. Reich, D. Rifkin, and E. Shaw (Long Island, N.Y.: Cold Spring Harbor Press), p. 51.
Newmark, A. K., and J. R. Knowles. 1975. *JACS* 97:3557.
Nicholls, P., and W. Elliot. 1974. In *Iron in Biochemistry and Medicine*, ed. A. Jacobs and M. Worwood (New York: Academic Press), p. 1.
Niehaus, F. C. 1967. In *Antibiotics*, ed. P. Shaw and D. Gottlieb (New York: Springer-Verlag), vol. 1, p. 40.
Nikkaido, H. 1973. In *Bacterial Membranes and Walls*, ed. L. Lieve (New York: Marcel Dekker), p. 131.
Nishimura, J., J. Manning, and A. Meister. 1962. *Biochemistry* 1:442.
Noda, L. 1973. In *The Enzymes*, 3d ed., ed. P. Boyer (New York: Academic Press), vol. 8, p. 279.
Noltman, E. 1972. In *The Enzymes*, 3d ed., ed. P. Boyer (New York: Academic Press), vol. 6, p. 272.
Nordlie, R. 1971. In *The Enzymes*, 3d ed., ed. P. Boyer (New York: Academic Press), vol. 4, p. 543.
Northrop, D. B. 1975. *Biochemistry* 14:2644.
Novogrodsky, A., J. Nishimura, and A. Meister. 1963. *JBC* 238:1903.
Nowak, T., and A. Mildvan. 1972. *Biochemistry* 11:2813.
Nozaki, M. 1974. In *Molecular Mechanisms of Oxygen Activation*, ed. O. Hayaishi (New York: Academic Press), p. 135.
O'Brien, W., R. Singleton, and H. Wood. 1973. *Biochemistry* 12:5247.
Ochoa, S., and R. Mazumder. 1973. In *The Enzymes*, 3d ed., ed. P. Boyer (New York: Academic Press), vol. 10, p. 1.
O'Connell, E., and I. Rose. 1973. *JBC* 248:2225.
Ohtushka, H., N. Rudie, and J. Wampler. 1976. *Biochemistry* 15:1001.
O'Leary, M. 1968. *Biochemistry* 7:913.
O'Leary, M., and M. D. Kluetz. 1972. *JACS* 94:3585.
Omura, T., and R. Sato. 1964. *JBC* 239:2370, 2379.
Ong, E. B., E. Shaw, and G. Schaellman. 1964. *JACS* 86:1271.
———. 1965. *JBC* 240:694.
Ono, T., and K. Bloch. 1975. *JBC* 250:1571.
Orf, H. W., and D. Dolphin. 1974. *PNAS* 71:2646.
Orme-Johnson, W. H. 1973. *Ann. Rev. Biochem.* 42:159.
Orr, G., J. Simon, S. Jones, G. Chin, and J. Knowles. 1978. *PNAS* 75:2230.
Orrenius, S., and L. Ernster. 1974. In *Molecular Mechanisms of Oxygen Activation*, ed. O. Hayaishi (New York: Academic Press), p. 215.
Overath, P., G. Kellerman, and F. Lymen. 1962. *Biochem. Z.* 335:500.
Oxender, D. 1972. *Ann. Rev. Biochem.* 41:777.
Palekar, A., S. Tate, and A. Meister. 1970. *Biochemistry* 9:2310.
———. 1973. *JBC* 248:1158.
Palmer, G. 1975. In *The Enzymes*, 3d ed., ed. P. Boyer (New York: Academic Press), vol. 12, p. 1.
Palmer, G., and V. Massey. 1968. In *Biological Oxidations*, ed. T. Singer (New York: Wiley), p. 263.
Parks, R. E., Jr., and R. P. Agarwal. 1973. In *The Enzymes*, 3d ed., ed. P. Boyer (New York: Academic Press), vol. 8, p. 307.
Pastan, I., G. Johnson, and W. Anderson. 1975. *Ann. Rev. Biochem.* 44:491.
Patel, D. J. 1969. *Nature* 221:1239.

Pauling, L. 1948. *Amer. Scientist* 36:58.
Paulson, L., and D. Ziegler. 1976. *Fed. Proc.* 35:1653.
Peisach, J., P. Aisen, and W. Blumberg, eds. 1966. *The Biochemistry of Copper.* New York: Academic Press.
Pepple, J., and D. Dennis. 1976. *BBA* 429:1036.
Perutz, M. 1970. *Nature* 228:726.
Peterkofsky, A. 1962. *JBC* 237:726.
Pinkus, L., and A. Meister. 1972. *JBC* 247:6119.
Plaut, G. 1963. In *The Enzymes*, 2d ed., vol. 7, ed. P. Boyer, H. Lardy, and K. Myrback (New York: Academic Press), p. 112.
Plowman, K. M. 1972. *Enzyme Kinetics.* New York: McGraw-Hill.
Poe, M., and K. Hoogsteen. 1978. *JRC* 253:543.
Pogolotti, A., K. Ivanetich, H. Sommer, and D. Santi. 1976. *BBRC* 70:972.
Polakis, S., R. Guchhait, E. Zwergel, M. D. Lane, and T. G. Cooper. 1974. *JBC* 249:6657.
Pontremoli, S., D. Prandinc, A. Bonsignore, and B. Horecker. 1961. *PNAS* 47:1942.
Popjack, G., and J. Cornforth. 1966. *Biochem. J.* 101:553.
Porter, D., and H. Bright. 1975. In *The Enzymes*, 3d ed., ed. P. Boyer (New York: Academic Press), vol. 12, p. 421.
Porter, D., J. Voet, and H. Bright. 1973. *JBC* 248:4400.
———. 1977. *JBC* 252:4464.
Portsmouth, D., A. Stoolmiller, and R. Abeles. 1967. *JBC* 242:2751.
Poston, J., and T. Stadtman. 1975. In *Cobalamins*, ed. B. Babior (New York: Wiley), p. 111.
Poulter, C. D., and H. Rilling. 1976. *Biochemistry* 15:1079.
———. 1978. *Accts. Chem. Res.* 11:307.
Poulter, C. D., D. M. Satterwhite, and H. C. Rilling. 1976. *JACS* 98:3376.
Powers, S., and A. Meister. 1976. *PNAS* 73:3020.
Poynton, R., and G. Schatz. 1975. *JBC* 250:752, 762.
Prescott, J., and P. R. Vagelos. 1971. In *Advances in Enzymology*, vol. 36, ed. A. Meister (New York: Wiley), p. 269.
Pressman, B. 1976. *Ann. Rev. Biochem.* 45:501.
Prough, R., and D. Ziegler. 1977. *ABB* 180:363.
Prusiner, S., and E. Stadtman, eds. 1973. *Enzymes of Glutamine Metabolism.* New York: Academic Press.
Quaroni, A., E. Gershon, and G. Semenza. 1974. *JBC* 249:6424.
Que, L., J. Lipscomb, E. Munck, and J. Wood. 1977. *BBA* 485:60.
Quereshi, A., E. Beytia, and J. Porter. 1973. *JBC* 248:1848.
Rabanc, J., D. Klung-Roth, and J. Lilie. 1973. *J. Phys. Chem.* 77:1169.
Rabinowitz, J. 1960. In *The Enzymes*, 2d ed., vol. 2, ed. P. Boyer, H. Lardy, and K. Myrback (New York: Academic Press), p. 195.
———. 1975. *PNAS* 72:528.
Rabinowitz, M., and F. Lipmann. 1960. *JBC* 235:1043.
Racker, E. 1976. *A New Look at Mechanisms in Bioenergetics.* New York: Academic Press.
Racker, E., and I. Krimsky. 1953. *Nature* 169:1043.
Rader, J., and F. Huennekens. 1973. In *The Enzymes*, 3d ed., ed. P. Boyer (New York: Academic Press), vol. 9, p. 197.
Rando, R. 1974*a*. *Science* 185:320.
———. 1974*b*. *Biochemistry* 13:3859.
———. 1975. *Accts. Chem. Res.* 8:281.
———. 1977. *Biochemistry* 16:4604.
Rando, R., and F. Bangerter. 1976. *JACS* 98:6762.
———. 1977. *JACS* 99:5141.
Rapkine, L. K. 1938. *Biochem. J.* 32:1729.

Ratner, S. 1973*a*. In *Advances in Enzymology*, vol. 39, ed. A. Meister (New York: Wiley), p. 3.
———. 1973*b*. In *The Enzymes*, 3d ed., ed. P. Boyer (New York: Academic Press), vol. 7, p. 168.
Ray, W., and J. Long. 1976. *Biochemistry* 15:3993.
Ray, W., J. Long, and J. Owens. 1976. *Biochemistry* 15:4006.
Ray, W., and E. Peck. 1972. In *The Enzymes*, 3d ed., ed. P. Boyer (New York: Academic Press), vol. 6, p. 408.
Rebeck, J., and F. Gavina. 1975. *JACS* 97:1591, 3221.
Reddy, V., and E. Caspi. 1976. *Eur. J. Biochem.* 69:577.
Reddy, V., D. Kupfer, and E. Caspi. 1977. *JBC* 252:2797.
Reed, B., and H. Rilling. 1975. *Biochemistry* 14:50.
Reed, L. 1966. In *Comprehensive Biochemistry*, ed. M. Florkin and E. Stolz (Amsterdam: Elsevier), vol. 14, p. 104.
———. 1974. *Accts. Chem. Res.* 7:40.
Reed, L. J., and D. J. Cox. 1970. In *The Enzymes*; 3d ed., ed. P. Boyer (New York: Academic Press), vol. 1, p. 213.
Reichard, P. 1968. *Biosynthesis of Deoxyribose.* New York: Wiley.
Reid, T., and I. B. Wilson. 1971. In *The Enzymes*, 3d ed., ed. P. Boyer (New York: Academic Press), vol. 4, p. 373.
Relyea, N., S. Tate, and A. Meister. 1974. *JBC* 249:1519.
Retey, J., J. Luthy, and D. Arigoni. 1966. *Nature* 226:519.
Retey, J., J. Seibl, D. Arigoni, J. W. Cornforth, G. Ryback, W. P. Zeylemiller, and C. Veeger. 1970. *Eur. J. Biochem.* 14:232.
Retey, J., E. Smith, and B. Zagalak. 1978. *Eur. J. Biochem.* 83:437.
Retey, J., C. Sucking, D. Arigoni, and B. Babior. 1974. *JBC* 249:6359.
Retey, J., A. Umani-Ronchi, and D. Arigoni. 1966. *Experientia* 22:502.
Retey, J., and B. Zagalak. 1973. *Angew. Chem. Int. Ed.* 12:671.
Reuben, J. 1971. *PNAS* 68:563.
Reynolds, S., D. Yates, and C. Pogson. 1971. *Biochem. J.* 122:285.
Riceberg, L., M. Simon, H. Van Vunakis, and R. Abeles. 1973. *Biochem. Pharmacol.* 24:119.
Richards, F. M., and H. W. Wyckoff. 1971. In *The Enzymes*, 3d ed., ed. P. Boyer (New York: Academic Press), vol. 4, p. 647.
Richards, J. H. 1970. In *The Enzymes*, 3d ed., ed. P. Boyer (New York: Academic Press), vol. 2, p. 321.
Richardson, J., K. Thomas, B. Rubin, and D. Richardson. 1976. *PNAS* 72:1351.
Ridley, W., L. Dizikes, and J. Wood. 1977. *Science* 197:329.
Riepe, M., and J. H. Wang. 1968. *JBC* 243:2779.
Riley, W., and E. Snell. 1968. *Biochemistry* 7:3520.
Rilling, H. 1966. *JBC* 241:3233.
Riordan, J. F., M. Sokolovsky, and B. L. Vallee. 1967. *Biochemistry* 6:3609.
Robbins, P., and A. Wright. 1971. In *Microbial Toxins*, ed. G. Weinbaum, S. Kadis, and S. Ajl (New York: Academic Press), vol. 4, p. 351.
Rogers, M., and P. Strittmatter. 1973. *JBC* 248:793, 800.
———. 1974. *JBC* 249:895, 5565.
Ronzio, R., W. Rowe, and A. Meister. 1969. *Biochemistry* 8:1066, 2674.
Rose, I. A. 1960. *JBC* 235:1170.
———. 1970. In *The Enzymes*, 3d ed., ed. P. Boyer (New York: Academic Press), vol. 2, p. 281.
———. 1972. In *Critical Reviews in Biochemistry* (Cleveland: Chemical Rubber), vol. 1, p. 33.
———. 1975. In *Advances in Enzymology and Related Areas of Molecular Biology*, vol. 43, ed. A. Meister (New York: Wiley), p. 491.
Rose, I. A., and E. O'Connell. 1967. *JBC* 242:1870.
———. 1969. *JBC* 244:126.
———. 1973. *JBC* 248:2214.

Rose, I. A., E. O'Connell, and A. Mehler. 1965. *JBC* 240:1758.
Rose, I. A., E. O'Connell, P. Noce, M. Utter, H. Wood, J. Willard, T. Cooper, and M. Benziman. 1969. *JBC* 244:6130.
Rose, I. A., E. O'Connell, and F. Solomon. 1976. *JBC* 251:902.
Rose, Z. B. 1970. *ABB* 140:508.
Rosen, O. M., and J. Erlichman. 1975. *JBC* 250:7788.
Rosenbaum, N., and M. Gefter. 1972. *JBC* 247:5675.
Rosenberg, T. L. 1975. In *Advances in Enzymology and Related Areas of Molecular Biology*, vol. 43, ed. A. Meister (New York: Wiley), p. 103.
Rossman, M., A. Liljas, C. Branden, and L. Banaszak. 1975. In *The Enzymes*, 3d ed., ed. P. Boyer (New York: Academic Press), vol. 11, p. 62.
Rosso, G., K. Takashima, and E. Adams. 1969. *BBRC* 34:134.
Roth, G., N. Stanford, and P. Majerus. 1975. *PNAS* 72:3073.
Rubenstein, P., and J. Strominger. 1974. *JBC* 249:3776, 3782, 3789.
Rubin, R., and P. Modrich. 1977. *JBC* 252:7265.
Rudnick, G., and R. Abeles. 1975. *Biochemistry* 14:4515.
Ruzicka, F., and H. Beinert. 1978. *JBC* 253:2514.
Rydstrom, J., J. Hoek, and L. Ernster. 1976. In *The Enzymes*, 3d ed., ed. P. Boyer (New York: Academic Press), vol. 13, p. 51.
Saier, M., and W. Jenkins. 1967. *JBC* 242:91.
Samuelsson, B., and R. Paoletti, eds. 1975. *Prostaglandins and Thromboxanes.* 2 vols. New York: Raven Press.
Santi, D., and C. Brewer. 1968. *JACS* 90:6236.
Santi, D., and C. McHenry. 1972. *PNAS* 69:1855.
Santi, D., C. McHenry, and H. Sommer. 1974. *Biochemistry* 13:471.
Sarma, R., V. Ross, and N. Kaplan. 1968. *Biochemistry* 7:3052.
Sauers, C., W. Jencks, and S. Groh. 1976. *JACS* 97:5547.
Saunders, W., and A. Cockerell. 1972. *Mechanisms of Elimination Reactions.* New York: Wiley.
Scanell, J., D. Preuss, T. Demny, F. Weiss, T. Williams, and A. Stempel. 1971. *J. Antibiotics* 4:239.
Schaffer, N. K., S. C. May, and W. H. Summerson. 1953. *JBC* 202:67.
Schechter, I., and A. Berger. 1967. *BBRC* 27:157.
———. 1968. *BBRC* 32:898.
———. 1970. *Phil. Trans. Roy. Soc. Ser. B* 257:249.
Schechter, I., F. Sweat, and K. Bloch. 1970. *BBA* 220:463.
Schimerlik, M., J. E. Rife, and W. W. Cleland. 1975. *Biochemistry* 14:5347.
Schirch, L., and A. Diller. 1971. *JBC* 246:3961.
Schirch, L., and T. Gross. 1968. *JBC* 243:5651.
Schirch, L., and W. Jenkins. 1964. *JBC* 239:3801.
Schirch, L., and M. Mason. 1962. *JBC* 237:2578.
———. 1963. *JBC* 238:1032.
Schirmer, R., E. Pai, and G. Schultz. 1978. In *International Symposium on Oxidizing Enzymes*, ed. R. Ondarza and T. Singer (New York: Elsevier), p. 17.
Schleicher, E., K. Mascaro, R. Potts, D. Mann, and H. Floss. 1976. *JACS* 98:1043.
Schlimme, E., G. Schafer, F. Eckstein, and R. Goody. 1973. *Eur. J. Biochem.* 40:485.
Schmidt, D., and F. Westheimer. 1971. *Biochemistry* 10:1249.
Schonbaum, G. R., and B. Chance. 1976. In *The Enzymes*, 3d ed., ed. P. Boyer (New York: Academic Press), vol. 13, p. 363.
Schonbrunn, A., R. Abeles, C. Walsh, H. Ogata, S. Ghisla, and V. Massey. 1976. *Biochemistry* 15:1798.
Schray, K., and S. Benkovic. 1973. In *The Enzymes*, 3d ed., ed. P. Boyer (New York: Academic Press), vol. 8, p. 201.

Schwartz, J. 1963. *PNAS* 49:871.
Schwartz, J., and F. Lipmann. 1961. *PNAS* 47:1996.
Scott, A. I., and K. Kang. 1977. *JACS* 99:1997.
Scrutton, M., and M. Young. 1972. In *The Enzymes*, 3d ed., ed. P. Boyer (New York: Academic Press), vol. 6, p. 1.
Secemski, I., S. Lehrer, and G. Lienhard. 1972. *JBC* 247:4740.
Secemski, I., and G. Lienhard. 1971. *JACS* 93:3549.
Segal, H. L., and P. D. Boyer. 1953. *JBC* 204:265.
Segel, I. H. 1975*a*. *Biochemical Calculations*, 2d ed. New York: Wiley.
———. 1975*b*. *Enzyme Kinetics*. New York: Wiley.
Seltzer, S. 1972. In *The Enzymes*, 3d ed., ed. P. Boyer (New York: Academic Press), vol. 6, p. 381.
Seltzer, S., G. Hamilton, and F. Westheimer. 1969. *JACS* 81:4018.
Seto, B., and T. Stadtman. 1976. *JBC* 251:2435.
Shannon, P., R. Presswood, R. Spencer, J. Becvar, S. Tu, J. Hastings, and C. Walsh. 1978. In *International Symposium on Oxidizing Enzymes*, ed. R. Ondarza and T. Singer (New York: Elsevier), p. 69.
Shapiro, S., and D. Dennis. 1965. *Biochemistry* 4:2283.
Shapiro, S., and F. Schlenk, eds. 1965. *Transmethylation and Methionine Biosynthesis*. Chicago: Univ. of Chicago Press.
Shaw, E. 1970. In *The Enzymes*, 3d ed., ed. P. Boyer (New York: Academic Press), vol. 1, p. 91.
Shaw, E., and J. Ruscica. 1971. *ABB* 145:484.
Shemin, D. 1972. In *The Enzymes*, 3d ed., ed. P. Boyer (New York: Academic Press), vol. 7, p. 323.
Shepherd, G., and G. Hammes. 1977. *Biochemistry* 16:5234.
Shineberg, B., and I. Young. 1976. *Biochemistry* 15:2754.
Shinkai, S., T. Kunitake, and T. Bruice. 1974. *JACS* 96:7140.
Shinomura, O., and F. Johnson. 1978. *PNAS* 72:1546.
Siebert, G., M. Cariostes, and G. Plaut. 1957. *JBC* 226:977.
Siebert, G., J. Dubuc, R. Warner, and G. Plaut. 1957. *JBC* 226:965.
Siegel, L., and P. Davis. 1974. *JBC* 249:1587.
Siegel, L., P. Davis, and H. Kamin. 1974. *JBC* 249:1572.
Siegel, M., M. Wishnick, and M. D. Lane. 1972. In *The Enzymes*, 3d ed., ed. P. Boyer (New York: Academic Press), vol. 6, p. 169.
Sigman, D., and G. Mooser. 1975. *Ann. Rev. Biochem.* 44:895.
Silverman, R., and R. Abeles. 1976. *Biochemistry* 15:4718.
Simon, H., and A. Kraus. 1976. In *Isotopes in Organic Chemistry*, vol. 2, ed. E. Buncel and C. Lee (New York: Elsevier), p. 153.
Simoni, R., and P. Postma, 1975. *Ann. Rev. Biochem.* 44:523.
Simonsen, J., ed. 1951. *The Terpenes*. 2 vols. New York: Cambridge Univ. Press.
Simpson, R. T., J. F. Riorden, and B. L. Vallee. 1963. *Biochemistry* 2:615.
Singer, T., and D. Edmondson. 1974. *Fed. Eur. Biochem. Soc. Letters* 42:1.
Singer, T., M. Gutman, and V. Massey. 1973. In *Iron-Sulfur Proteins*, ed. W. Lovenberg (New York: Academic Press), vol. 1, p. 225.
Singh, M., and P. Srere. 1971. *JBC* 246:3847.
Sinnott, M., and I. Souchard. 1973. *Biochem. J.* 133:89.
Siu, P., and H. Wood. 1962. *JBC* 237:3044.
Sjoberg, B., P. Reichard, A. Graslünd, and A. Ehrenberg. 1977. *JBC* 252:536.
Skye, G., R. Potts, and H. Floss. 1974. *JACS* 96:1593.
Sloan, D., L. Loeb, A. Mildvan, and R. Feldman. 1975. *JBC* 250:8913.
Smith, L., L. Mohr, and M. Raftery. 1973. *JACS* 95:7497.
Snell, E., and S. DiMari. 1970. In *The Enzymes*, 3d ed., ed. P. Boyer (New York: Academic Press), vol. 2, p. 335.

Soll, D., and P. Schimmel. 1974. In *The Enzymes*, 3d ed., ed. P. Boyer (New York: Academic Press), vol. 10, p. 489.

Sparrow, L., P. Ito, T. Sundaram, D. Zach, E. Nyns, and E. Snell. 1969. *JBC* 244:2590.

Speck, J., P. Rowley, and B. Horecker. 1963. *JACS* 85:1012.

Spector, L. B. 1972. In *The Enzymes*, 3d ed., ed. P. Boyer (New York: Academic Press), vol. 7, p. 357.

———. 1973. *Bioorg. Chem.* 2:311.

Spencer, R., J. Fisher, and C. Walsh. 1976. *Biochemistry* 15:1043.

Speyer, J., and R. Dickman. 1956. *JBC* 220:193.

Spiro, T., and P. Saltman. 1974. In *Iron in Biochemistry and Medicine*, ed. A. Jacobs and M. Worwood (New York: Academic Press), p. 1.

Sprecher, M., M. Clark, and D. Sprinson. 1966. *JBC* 241:872.

Sprinzl, M., and F. Cramer. 1975. *PNAS* 72:3049.

Srere, P. 1972. In *Current Topics in Cellular Regulation*, ed. B. Horecker and E. Stadtman (New York: Academic Press), vol. 5, p. 229.

———. 1975. In *Advances in Enzymology and Related Areas of Molecular Biology*, vol. 43, ed. A. Meister (New York: Wiley), p. 57.

Srere, P., and A. Bhaduri. 1964. *JBC* 239:714.

Srere, P., B. Bottger, and G. Brooks. 1972. *PNAS* 69:1201.

Srere, P., and F. Lipmann. 1953. *JACS* 75:4874.

Stadtman, E. 1973. In *The Enzymes*, 3d ed., ed. P. Boyer (New York: Academic Press), vol. 8, p. 1.

Stadtman, E., and A. Ginsburg. 1974. In *The Enzymes*, 3d ed., ed. P. Boyer (New York: Academic Press), vol. 10, p. 755.

Stadtman, T. 1973. In *The Enzymes*, 3d ed., ed. P. Boyer (New York: Academic Press), vol. 6, p. 539.

———. 1974. *Science* 183:915.

Stalmans, W., and H. Hers. 1973. In *The Enzymes*, 3d ed., ed. P. Boyer (New York: Academic Press), vol. 9, p. 310.

Stanbury, J., J. Wyngaarden, and D. Frederickson, eds. 1972. *The Metabolic Basis of Inherited Disease.* New York: McGraw-Hill.

Steenkamp, D., W. Kenney, and T. Singer. 1978. *JBC* 253:2812.

Stefani, A., M. Janett, and G. Semenza. 1975. *JBC* 250:7810.

Steinberger, R., and F. Westheimer. 1951. *JACS* 73:429.

Stenflo, J., P. Fernlund, and P. Roepstorff. 1975. In *Proteases and Biological Control*, ed. E. Reich, D. Rifkin, and E. Shaw (Long Island, N.Y.: Cold Spring Harbor Press), p. 111.

Stoolmiller, A., and R. Abeles. 1966. *JBC* 241:5764.

Storm, D. R., and D. E. Koshland. 1972. *JACS* 94:5805, 5815.

Stormer, F., and H. Umbarger. 1964. *BBRC* 17:587.

Strassman, M., A. Thomas, and S. Weinhouse. 1953. *JACS* 75:5135.

———. 1955. *JACS* 77:1261.

Streitweiser, A., R. H. Jagow, R. C. Fahey, and S. Suzuki. 1958. *JACS* 80:2326.

Strittmatter, P. 1974. *PNAS* 71:4565.

Stroud, R. M., M. Kneger, R. E. Koeppe, A. A. Kossiakoff, and J. L. Chambers. 1975. In *Proteases and Biological Control*, ed. E. Reich, D. Rifkin, and E. Shaw (Long Island, N.Y.: Cold Spring Harbor Press), p. 13.

Stryer, L. 1975. *Biochemistry.* San Francisco: W. H. Freeman and Company.

Stubbe, J. 1975. *Biochemistry* 14:3489.

Suelter, C. 1970. *Science* 168:789.

Sumner, J. B. 1926. *JBC* 69:435.

Sun, M., and P. Song. 1973. *Biochemistry* 12:4663.

Swain, C. G., E. C. Stevers, J. F. Reuwer, and L. J. Schaad. 1958. *JACS* 80:5885.

Switzer, R. 1974. In *The Enzymes* 3d ed., ed. P. Boyer (New York: Academic Press), vol. 10, p. 607.

Sy, J., and F. Lipmann, 1973. *PNAS* 70:306.
Tabor, C., and H. Tabor. 1976. *Ann. Rev. Biochem.* 45:285.
Tagaki, W., P. Guthrie, and F. Westheimer. 1968. *Biochemistry* 7:905.
Tagaki, W., and F. Westheimer. 1968. *Biochemistry* 7:891, 895, 901.
Tai, H., and K. Bloch. 1972. *JBC* 247:3767.
Takahashi, M., and T. Hofmann. 1972. *Biochem. J.* 127:35P.
———. 1974. *BBRC* 57:39.
Takeda, H., and O. Hayaishi. 1966. *JBC* 241:2733.
Talalay, P., and A. Benson. 1972. In *The Enzymes*, 3d ed., ed. P. Boyer (New York: Academic Press), vol. 6, p. 591.
Tanaka, H., N. Esaki, and K. Soda. 1977. *Biochemistry* 16:100.
Tanase, S., and Y. Morino. 1976. *BBRC* 68:1301.
Tang, J. 1971. *JBC* 246:4510.
———. 1976. In *Trends in Biochemical Sciences* (Amsterdam: Elsevier), vol. 1, p. 205.
Tate, S. S., and A. Meister. 1974. *JBC* 249:7593.
———. 1975. *JBC* 250:4619.
———. 1976. *Ann. Rev. Biochem.* 45:559.
Tatum, C., P. Benkovic, S. Benkovic, R. Potts, E. Schleicher, and H. Floss. 1977. *Biochemistry* 16:1093.
Taylor, C., and R. Wolfe. 1974. *JBC* 249:4879.
Taylor, R., and H. Weissbach. 1973. In *The Enzymes*, 3d ed., ed. P. Boyer (New York: Academic Press), vol. 9, p. 121.
Taylor, W., C. Taylor, and M. Taylor. 1974. *J. Bacteriol.* 119:98.
Tchen, T. T., and H. Van Milligan. 1960. *JACS* 82:4115.
Tetas, M., and J. Lowenstein. 1963. *Biochemistry* 2:350.
Thelander, L. 1973. *JBC* 248:4591.
———. 1974. *JBC* 249:4848.
Theorell, H., and B. Chance. 1951. *Acta. Chem. Scand.* 5:1127.
Thomas, E., J. McKelvy, and N. Sharon. 1969. *Nature* 222:485.
Thorpe, C., and C. Williams. 1976. *JBC* 251:7726.
Tiepel, J., G. Hass, and R. Hill. 1968. *JBC* 243:5684.
Tober, C. L., P. Nicholls, and J. Brodie. 1970. *ABB* 138:506.
Todhunter, J., and D. Purich. 1975. *JBC* 250:3505.
Trentham, D., C. McMurray, and C. Pogson. 1964. *Biochem. J.* 114:19.
Trotta, P., L. Estis, A. Meister, and R. Haschemeyer. 1974. *JBC* 249:482.
Tsolas, O., and B. Horecker. 1972. In *The Enzymes*, 3d ed., ed. P. Boyer (New York: Academic Press), vol. 7, p. 259.
Tzeng, S., R. Wolfe, and M. Bryant. 1975. *J. Bacteriol.* 121:184.
Ulevitch, R., and R. Kallen. 1977. *Biochemistry* 16:5342, 5350, 5355.
Ulrich, V., and W. Duppel. 1975. In *The Enzymes*, 3d ed., ed. P. Boyer (New York: Academic Press), vol. 12, p. 253.
Ulrich, V., I. Roots, A. Hildebrandt, R. Estabrook, and A. Conney, eds. 1977. *Microsomes and Drug Oxidations*. Oxford: Pergamon Press.
Umezawa, K., K. Takai, S. Tsuji, Y. Kurashima, and O. Hayaishi. 1974. *PNAS* 71:4598.
Usher, D. 1969. *PNAS* 62:661.
Usher, D., D. Richardson, and F. Eckstein. 1972. *Nature* 228:663.
Utter, M., R. Barden, and B. Taylor. 1975. In *Advances in Enzymology and Related Areas of Molecular Biology*, vol. 42, ed. A. Meister (New York: Wiley), p. 1.
Utter, M., and H. Kollenbrander. 1972. In *The Enzymes*, 3d ed., ed. P. Boyer (New York: Academic Press), vol. 6, p. 117.
Vallee, B. L., J. F. Riordan, and J. E. Coleman. 1963. *PNAS* 49:109.
Vanaman, T., K. Brew, and R. Hill. 1970. *JBC* 245:4583.

Van Der Werf, P., and A. Meister. 1975. In *Advances in Enzymology and Related Areas of Molecular Biology*, vol. 43, ed. A. Meister (New York: Wiley), p. 519.
Vane, J. 1971. *Nature* 231:232.
Van Eikeren, P., and D. Chipman. 1972. *JACS* 94:4788.
Vaneste, W. H., and A. Zuberbühler. 1974. In *Molecular Mechanisms of Oxygen Activation*, ed. O. Hayaishi (New York: Academic Press), p. 371.
VanLier, J., and M. Rousseau. 1976. *Fed. Eur. Biochem. Soc. Letters* 70:23.
VillaFranca, J., and A. Mildvan. 1971. *JBC* 246:772, 5791.
———. 1972. *JBC* 247:3454.
Volpe, J., and P. R. Vagelos. 1973. *Ann. Rev. Biochem.* 42:21.
Waechter, C., and W. Lennarz. 1973. In *The Enzymes*, 3d ed., ed. P. Boyer (New York: Academic Press), vol. 9, p. 95.
Wagner, O., H. Lee, P. Frey, and R. Abeles. 1966. *JBC* 241:1751.
Wallenfels, K., and R. Weil. 1972. In *The Enzymes*, 3d ed., ed. P. Boyer (New York: Academic Press), vol. 7, p. 575.
Walsh, C. 1977. In *Horizons in Biochemistry and Biophysics*, ed. E. Quaglieriello (Reading Mass.: Addison-Wesley), vol. 3, p. 36.
———. 1978. *Ann. Rev. Biochem.* 46:881.
Walsh, C., R. Abeles, and H. Kaback. 1972. *JBC* 247:7456.
Walsh, C., E. Krodel, V. Massey, and R. Abeles. 1973. *JBC* 248:1946.
Walsh, C., O. Lockridge, V. Massey, and R. Abeles. 1973. *JBC* 248:7049.
Walsh, C., A. Schonbrunn, and R. Abeles. 1971. *JBC* 246:6855.
Walsh, C., and L. Spector. 1969. *JBC* 244:4366.
Walsh, C., R. Spencer, J. Fisher, F. Jacobson, R. Brown, and W. Ashton. 1978. In *Flavins and Flavoproteins*, ed. K. Yagi and T. Yamano (Tokyo: Univ. of Tokyo Press). (In press)
Walsh, D. A., and E. G. Krebs. 1973. In *The Enzymes*, 3d ed., ed. P. Boyer (New York: Academic Press), vol. 8, p. 555.
Walsh, K. 1975. In *Proteases and Biological Control*, ed. E. Reich, D. Rifkin, and E. Shaw (Long Island, N.Y.: Cold Spring Harbor Press), p. 51.
Wang, E., and C. Walsh. 1978. *Biochemistry* 17:1313.
Wang, E., C. Walsh, and R. Kallen. (Manuscript in preparation).
Warren, S., B. Zerner, and F. Westheimer. 1966. *Biochemistry* 5:817.
Watson, J. D. 1975. *Molecular Biology of the Gene*, 3d ed. Menlo Park, Calif.: Benjamin.
Weinstein, I., A. Jeffrey, K. Jennette, S. Blobstein, R. Harvey, C. Harris, H. Autrup, H. Kasai, and K. Nakanishi. 1976. *Science* 193:593.
Weisblat, D., and B. Babior. 1971. *JBC* 246:6064.
Welch, G., K. Cole, and F. Gaertner. 1974. *ABB* 165:508.
Wentworth, R., and R. Wolfenden. 1974. *Biochemistry* 13:4715.
Wessely, L., and F. Lynen. 1953. *Fed. Proc.* 12:685.
Westheimer, F. 1963. *Proc. Chem. Soc.*, p. 253.
———. 1968. *Accts. Chem. Res.* 1:70.
Westheimer, F., and W. Jones. 1941. *JACS* 63:3283.
Westley, J. 1969. *Enzymatic Catalysis.* New York: Harper & Row.
Wheat, R., and S. Ajl. 1955. *JBC* 217:909.
Whitaker, J. R., and J. Perez-Villasenor. 1968. *ABB* 124:70.
White-Stevens, R., and H. Kamin. 1972. *JBC* 247:2358.
Wickner, R., C. Tabor, and H. Tabor. 1970. *JBC* 245:2132.
Willadsen, P., and H. Eggerer. 1975. *Eur. J. Biochem.* 54:253.
Willard, J., and I. A. Rose. 1973. *Biochemistry* 12:5241.
Williams, C. H. 1976. In *The Enzymes*, 3d ed., ed. P. Boyer (New York: Academic Press), vol. 13, pp. 90–173.
Williams, R. F., S. Shinkai, and T. C. Bruice. 1975. *PNAS* 72:1763.

Williams, R. J. P. 1974. In *Iron in Biochemistry and Medicine*, ed. A. Jacobs and M. Worwood (New York: Academic Press), p. 184.
Williamson, J. 1977. Ph.D. dissertation, Massachusetts Institute of Technology.
Wilson, I. B., and S. Ginsburg. 1955. *BBA* 18:168.
Wimmer, M., and I. Rose. 1978. *Ann. Rev. Biochem.* 46:1031.
Winstein, S., C. Lindegren, H. Marshall, and L. Ingraham. 1953. *JACS* 75:147.
Wold, F. 1971. In *The Enzymes*, 3d ed., ed. P. Boyer (New York: Academic Press), vol. 5, p. 499.
Wolfenden, R. 1972. *Accts. Chem. Res.* 5:10.
Wolin, M., F. Simpson, and W. Wood. 1957. *BBA* 29:635.
Wong, L. J., and P. A. Frey. 1974. *Biochemistry* 13:3889.
Wood, B., and L. Ingraham. 1965. *Nature* 205:291.
Wood, H. G. 1972. In *The Enzymes*, 3d ed., ed. P. Boyer (New York: Academic Press), vol. 6, p. 83.
———. 1977. In *Advances in Enzymology and Related Areas of Molecular Biology*, vol. 45, ed. A. Meister (New York: Wiley), p. 118.
Wood, H. G., and G. Zwolinski. 1976. In *Critical Reviews in Biochemistry* (Cleveland: Chemical Rubber), vol. 4, p. 47.
Wood, W. A. 1971. In *The Enzymes*, 3d ed., ed. P. Boyer (New York: Academic Press), vol. 5, p. 573.
———. 1972. In *The Enzymes*, 3d ed., ed. P. Boyer (New York: Academic Press), vol. 7, p. 281.
Woodward, R. B., and R. Hoffman. 1969. *Angew. Chem. Int. Ed.* 8:781.
Wu, W., and V. Williams. 1968. *JBC* 243:5644.
Wunderwald, P., and H. Eggerer. 1969. *Eur. J. Biochem.* 11:97.
Wyers, F., A. Sentenac, and P. Fromageot. 1976. *Eur. J. Biochem.* 69:377, 385.
Yamamoto, S., and K. Bloch. 1970. *JBC* 245:1670.
Yamazaki, I. 1974. In *Molecular Mechanisms of Oxygen Activation*, ed. O. Hayaishi (New York: Academic Press), p. 535.
Yang, S., D. McCourt, P. Roller, and H. Gelboin. 1976. *PNAS* 73:2594.
Yim, J. 1975. Ph.D. dissertation, Biology Dept., Massachusetts Institute of Technology.
Yokoe, Y., and T. C. Bruice. 1975. *JACS* 97:450.
Yorifugi, T., H. Misono, and K. Soda. 1971. *JBC* 246:5093.
Yorifugi, T., K. Ogata, and K. Soda. 1971. *JBC* 246:5085.
Yost, F., and I. Fridovich. 1973. *JBC* 248:4905.
Yount, R. 1975. In *Advances in Enzymology and Related Areas of Molecular Biology*, vol. 43, ed. A. Meister (New York: Wiley), p. 1.
Yu, C., and I. Gunsalus. 1974. *JBC* 249:94, 102.
Zalkin, H. 1973. In *Advances in Enzymology*, vol. 38, ed. A. Meister (New York: Wiley), p. 1.
Zatman, L., N. Kaplan, and S. Colowick. 1953. *JBC* 200:197.
Zatman, L., N. Kaplan, S. Colowick, and M. Ciotti. 1954. *JBC* 209:467.
Zeffren, E., and P. Hall. 1973. *The Study of Enzyme Mechanisms.* New York: Wiley.
Zerner, B., S. M. Cotts, F. Lederer, H. H. Waters, and F. H. Westheimer. 1966. *Biochemistry* 5:813.
Zillig, W., H. Fujiki, and R. Mailhammer. 1975. *J. Biochem.* 77:7.

INDEX